W9-AZF-967

Introduction to the Structure of the EARTH

SECOND EDITION

Introduction to the Structure of the EARTH

EDGAR W. SPENCER

Professor of Geology
Washington and Lee University

McGRAW-HILL BOOK COMPANY

New York St. Louis San Francisco
Auckland Bogotá Düsseldorf Johannesburg London
Madrid Mexico Montreal New Delhi Panama Paris
São Paulo Singapore Sydney Tokyo Toronto

Introduction to the Structure of the EARTH

5 6 7 8 9 0 DODO 8 3 2 1 0

This book was set in Palatino by Monotype Composition Company, Inc.
The editors were Robert H. Summersgill and James W. Bradley;
the designer was Nicholas Krenitsky;
the production supervisor was Robert C. Pedersen.
The drawings were done by Elizabeth Spencer and Dennis Slifer.
R. R. Donnelley & Sons Company was printer and binder.

Library of Congress Cataloging in Publication Data

Spencer, Edgar Winston.
Introduction to the structure of the earth.

Includes indexes.
1. Geology, Structural. I. Title.
QE601.S73 1977 551.1 76-27282
ISBN 0-07-060197-6

This book is dedicated to the memory of two outstanding teachers

Arie Poldervaart

Walter Bucher

Contents

	Preface	ix

PART ONE PRINCIPLES

Chapter 1	Introduction	3
Chapter 2	Nontectonic Structural Features	7
Chapter 3	Concepts of Stress and Strain	21
Chapter 4	Experimental Study of Rock Deformation	45
Chapter 5	Brittle Failure of Rocks	75
Chapter 6	Impact Features	101
Chapter 7	Faults—General Considerations	107
Chapter 8	Low-Angle Thrusting and Gravity Gliding	123
Chapter 9	High-Angle Faults—Normal Faults and Upthrusts	145
Chapter 10	Strike-Slip Faults	159
Chapter 11	Geometry of Folded Rocks	177
Chapter 12	Mesoscopic Features Commonly Associated with Folded Rocks	207
Chapter 13	Folding in Principle and Experiment	229
Chapter 14	Large-Scale Folds and Fold Systems	247
Chapter 15	Diapirs and Salt Domes	267

PART TWO TECTONICS

Chapter 16 Major Structural Elements of the Lithosphere 287
Chapter 17 Global Tectonics 303
Chapter 18 Continental Cratons 335
Chapter 19 Atlantic-Type Continental Margins 357
Chapter 20 Divergent Junctions 371
Chapter 21 Island Arcs 397
Chapter 22 The Appalachian Orogen 427
Chapter 23 The Alpine-Himalayan Orogen 475
Chapter 24 The Cordilleran Orogen 517

APPENDIXES

Appendix A Stress Theory 567
Appendix B Structural Maps and Cross Sections 573
Appendix C Orthographic Projection 585
Appendix D Stereographic Projection 601
Appendix E Structural Notation 613

Indexes 615

Name Index
Subject Index

Preface

The objective of this book is to provide broad, up-to-date coverage of the principles of structural geology and tectonics suitable for use in an introductory course in these subjects. The book is organized into three parts. Part One deals with the traditional subject matter of structural geology, the description of structural features, and the theoretical and mechanical explanations of their origin. Part Two is devoted to a discussion of tectonics and regional structural geology, especially as it has come to be viewed in terms of plate-tectonic theory within the last decade. A discussion of the regional structural geology of the Appalachians, the Alpine-Himalayan system, and the Cordilleran system is included in order to broaden the student's conception of the relation of structural principles to regional and tectonic problems and to illustrate the relation of structural geology to other branches of geology and geophysics. The third part, the appendixes, provides a brief survey of the methods, particularly the geometric projections, used by structural geologists, an introduction to map and cross-section preparation, and stress analysis.

It has been my pleasure and good fortune to have been associated as a student, colleague, and friend with a number of outstanding geologists, and I am grateful for the help and encouragement these friends have given me. I am especially indebted to my former teachers Walter Bucher, Arie Poldervaart, Marshall Kay, and Marcellus Stow, who first stimulated my interest in the problems of structural geology. I also wish to express my appreciation to F. Aldaya, S. Warren Carey, J. M. Fontboté, Manuel Julivert, H. P. Laubscher, Arnold Lillie, Alberto Marcos, and Rudolf Trümpy for introducing me to the geology of other parts of the world.

I am indebted to my students, who helped me turn a critical eye on the first edition of this book, and to the reviewers, especially Woody Hickcox and Kevin Coppersmith, who helped me improve earlier drafts of the manuscript for this edition. I wish to acknowledge with thanks the many reprints, photographs, and line drawings which have been made available to me for use in the preparation of this book, and the help of Elizabeth Brewbaker, Dorothy Tolley, Peggy Riethmiller, Virginia Gillen, and Anne Barger who helped me prepare the manuscript. A special note of gratitude is due Dennis Slifer, Edward Backus, and my wife, Elizabeth, for preparation of many of the new illustrations in this edition.

Edgar W. Spencer

Introduction to the Structure of the EARTH

ONE

PRINCIPLES

1

Introduction

Knowledge of *structural geology* is central to our understanding of the earth either as a whole or in its myriad parts. The scope of the field is vast, ranging from the broadest framework of the earth's interior and the major crustal elements to the fine detail of rock fabric. It includes description of the geometry and spatial relationships of rock bodies on the one hand and the processes by which these relationships come into existence on the other.

Structural geology deals with the techniques used to obtain structural data in the field and to produce representations of that data suitable for analysis and interpretation. Development of an understanding of the physical processes or principles which govern the development of structural features is also of primary importance. Many of the underlying principles have been derived primarily from field observations, but increasingly, structural geologists are turning to methods of experimental physics and mathematical analysis for a more complete explanation of the observational data. Structural geology is a basic ingredient of regional geology, and description of regional structural relations is an important part of the field. In a sense, regional structure constitutes much of the basic data of tectonics.

Many of the most exciting developments in science during the last decade have occurred in the field of tectonics. The emergence of seafloor spreading and plate tectonics as an explanation for continental drift has captured the imagination of most earth scientists and has provided the most powerful unifying concept yet developed in viewing the origin and evolution of large-scale structural features. The purpose of this book is to examine each of these major aspects of the field of structural geology.*

Structural geology as a discipline is closely related to many other fields of earth science. The study of petrofabrics links structure to mineralogy and petrology. The form and internal fabric of undeformed sedimentary rocks showing features developed during

* See the Glossary at the end of the chapter.

deposition and consolidation of sediment is a concern of both sedimentologists and structural geologists. Many geophysical methods are used primarily to delineate subsurface structure, and other geophysical studies are used to reveal large crustal structures and point to their origin. And finally, structure and stratigraphy are intimately related in regional mapping and in development of the geologic evolution of the earth. Knowledge of earth structure is also vital in the applied fields of geology, such as ground-water hydrology and engineering, petroleum, and mining geology, where it is of critical importance to learn the subsurface configuration of rock bodies as well as to know their surface distribution. Many petroleum traps are structural features, and the geologist must know the structural setting of stratigraphic traps such as pinch-outs, unconformities, and reefs in order to locate and develop production intelligently. The evaluation of shape, size, and depth of an ore body or an economically important rock body determine the feasibility of mining it. Knowledge of the existence of a fault which displaces the ore or coal may well mean the difference between financial success and failure. Many of the failures of large engineering structures can be directly attributed to imperfect evaluation of the structure of the foundation rock. Dams have been built on active faults, tunnels have failed for lack of adequate knowledge of subsurface structure, and many large buildings and highways have been damaged because their foundation rock contained structural flaws.

Methods of Structural Geology

Methods as varied as the subject are used to obtain structural knowledge. Geologic mapping is by far the main source of structural data. Maps show the areal distribution of rock units, and the attitude of contacts between units and the form of the units at depth can be made from them. Maps in igneous and metamorphic terranes may show the geographic distribution of planar and linear structures in the rock. The larger scale structure of the rock body can be interpreted from this pattern.

Subsurface structural methods include all the means by which the configuration of a rock body below the ground surface is obtained. These methods include the construction of cross sections and other projections at depth based on the geometry of the rock at the surface or actual control points obtained by drilling, geophysical methods, or in mine shafts. Among the most powerful of these tools are structure contour and isopach or thickness maps.

Geophysics is of great significance not only as a means of determining depth to stratigraphic horizons and their attitude in space but as our best source of information about the crust as a structural unit, and about the deeper structure and composition of the interior of the earth. A very large part of what we know about the structure of ocean basins is derived from geophysical methods. Geophysical methods help us locate anomalous and thus potentially interesting areas for detailed study.

Much structural knowledge is based on field observations of rock fabric and the relations of the various elements of the rock structure to one another. Such studies include geometrical analysis of joints, rock cleavages, foliations in metamorphic rocks, flow lines in igneous intrusions, bedding surfaces, fold axes and axial surfaces, or the relations of any of these planar and linear elements to one another. Such analyses have proved fruitful in metamorphic rocks and in regions subjected to multiple deformations where conventional methods of geologic mapping have failed to resolve the more complex geometries of the rock bodies. Some techniques of analysis have been carried over from petrofabric studies, employing statistical methods and symmetry concepts.

Petrofabric studies have been useful in establishing the changes of fabric when rocks are deformed or metamorphosed, such as the alignment of crystallographic axes in stress fields and the effects of recrystallization during and after deformation. Petrofabrics has proved particularly significant in connection with experimental methods of studying rock deformation.

Experimental studies have been highly successful in demonstrating the behavior of rocks and minerals when deformed under various environmental conditions (especially varying temperature and pressure conditions). The deformed specimens are analyzed by petrofabric methods, and the geometric relationship between fabric and the known applied stresses is established. This has provided an important basis for dynamic structural geology. Other experiments employing scale models and synthetic materials have been used to simulate large-scale structural features ranging from small-scale folds to mountain systems.

Mathematical theories of stress, strain, and elasticity have been applied to try to gain a better understanding of the manner of rock yield and failure. The large number of variables involved in natural rock deformation make rigid theoretical analysis difficult, but interesting conclusions have been reached through theory concerning the mechanisms of folding and faulting. These methods are particularly valuable in combination with experiments and comparative field studies.

Synthesis implies the bringing together of information and integrating it into a coherent picture. As such, synthesis is important as a method of structural analysis. This bringing together of the details of regional geology, subsurface configuration, lithology, and age is an important part of the reconstruction or synthesis of structural evolution which has led to so many major discoveries both for the exploration geologist and for the theorist. The concepts of geosynclines, continental drift, sea-floor spreading, plate tectonics, and orogeny are products of this approach.

GLOSSARY*

Diastrophism: The process or processes by which the crust of the earth is deformed, producing continents and ocean basins, plateaus and mountains, flexures and folds of strata, and faults. Also the results of these processes.

Dynamic structural geology: The study of the relationship between structural features (strains) and the stress conditions under which they form.

Kinematic structural geology: The study of the rock fabric of deformed rock based primarily on the movement pattern necessary to produce the observed strain and independent of the stresses responsible.

Petrofabrics: The study of rock fabric, involving analysis of the component parts of a rock—the fragments, crystal grains, and their sizes, shapes, arrangement, orientations in space, relations to one another, internal structure, and the movements and process which played a part in the formation of the fabric.

Structural feature: A feature produced in rock by movements after deposition, and commonly after consolidation, of the rock.

Structural geology: The study of the structural features of rocks, the geographical distribution of the features, and their causes.

Tectonics: The study of the broader structural features of the earth and their causes.

REFERENCES

Badgley, P. C., 1965, Structural and tectonic principles: New York, Harper & Row.

Bailey, E. B., 1935, Tectonic essays: Oxford, Clarendon.

Beloussov, V. V., 1962, Basic problems in geotectonics (J. C. Maxwell, ed. English edit.): New York, McGraw-Hill.

* Refer to Am. Geol. Inst. © 1972, Glossary of geology and related sciences.

Billings, M. P., 1972, Structural geology, 3d ed.: Englewood Cliffs, N.J., Prentice-Hall.

Bucher, W. H., 1933, The deformation of the earth's crust: Princeton, N.J., Princeton Univ.

Cloos, Hans, 1936, Einfuhrung in die geologie: Berlin, Gebruder Borntraeger.

Daly, R. A., 1940, Strength and structure of the earth: Englewood Cliffs, N.J., Prentice-Hall.

DeSaussure, Horace, 1796, Voyages dans Les Alpes, v. IV: Neuchâtel.

De Sitter, L. U., 1964, Structural geology, 2d ed.: New York, McGraw-Hill.

Eardley, A. J., 1951, Structural geology of North America, 2d ed.: New York, Harper & Row.

Fairburn, H. W., 1942, Structural petrology of deformed rocks: Cambridge, Mass., Addison-Wesley.

Geikie, James, 1940, Structural and field geology, 5th ed.: Edinburgh, Oliver & Boyd.

Goguel, Jean, 1965, Tectonics, 2d ed.: San Francisco, Freeman.

Griggs, D. T., and **Handin, John**, eds., 1960, Rock deformation: Geol. Soc. America Mem. 79.

Hills, E. S., 1953, Outlines of structural geology, 3d ed.: London, Methuen.

——— 1963, Elements of structural geology: New York, Wiley.

Irvine, T. N., ed., 1965, The world rift system: Geol. Survey Canada Paper 66-14, Dept. Mines and Technical Surveys, Ottawa.

Jeffreys, H., 1970, The earth, 5th ed.: London, Cambridge.

Kay, G. M., 1951, North American geosynclines: Geol. Soc. America Mem. 48.

Lotze, Franz, 1956, Geotektonisches symposium en ehren von Hans Stille: Stuttgart, Kommissions—Verlag von Ferdinand Enke.

Nevin, C. M., 1942, Principles of structural geology, 3d ed.: New York, Wiley.

Poldervaart, Arie, ed., 1953, The crust of the earth: Geol. Soc. America Spec. Paper 62.

Ramsay, J. G., 1967, Folding and fracturing of rocks: New York, McGraw-Hill.

Sander, B., 1930, Gefugekunde der Gesteine: Vienna, Springer.

——— 1948, Einfuhrung in die Gefugekunde der Geologischen Korper, pt. I: Vienna, Springer.

Schmidt, Walter, 1932, Tektonik und Verformungslehre: Berlin, Borntraeger.

Sonder, R. A., 1956, Mechanik der Erde: Stuttgart, E. Schweizerbartsche Verlags.

Stočes, Bohuslave, and **White, C. H.**, 1935, Structural geology: London, Macmillan.

Turner, F. J., and **Weiss, L. E.**, 1963, Structural analysis of metamorphic tectonites: New York, McGraw-Hill.

Umbgrove, J. H. F., 1947, The pulse of the earth, 2d ed.: The Hague, Nijhoff.

Whitten, E. H. Timothy, 1966, Structural geology of folded rocks: Chicago, Rand McNally.

Willis, Bailey, 1923, Geologic structures: New York, McGraw-Hill.

——— and **Willis, Robin**, 1934, Geologic structures, 3d ed.: New York, McGraw-Hill.

2

Nontectonic Structural Features

Distinguishing between deformed and undeformed rocks is of basic importance in structural geology. This distinction is not always obvious. All rocks possess a primary external geometrical form, and the constituents of the rock have an initial size, packing, or spatial arrangement, referred to as *texture* or *fabric,* which arise as a result of the special conditions of sedimentation or crystallization. The initial fabric may have distinctive features or forms called *primary structural features* formed during deposition of sediment or crystallization of melt.

Both the external form and internal fabric undergo changes as sediment becomes lithified. These changes include compaction, cementation, sometimes crystallization, dehydration, and desiccation. All these processes take place in the earth's gravity field, and the rock is subject to the stresses which arise as a result of gravity. The initial form and fabric may be distorted or change volume—in a very real sense the initial rock is strained or deformed. But the important distinction for us is one between changes that occur as a normal result of lithification on the one hand and strains imposed on the rock as a result of abnormal applied stresses on the other. Features produced as a result of strains imposed on sediment or rock during or after consolidation by external stress fields are called *secondary structural features.*

In summary, different fabrics and structural features in rock bodies arise under different conditions:

1. Primary structural features formed during sedimentation or crystallization.

2. Features formed in the rock during lithification and subject to no applied forces other than gravity.

3. Secondary structural features formed during lithification as a result of externally applied forces in addition to gravity.

4. Secondary structural features formed after lithification and as a result of external forces (both gravity and other forces).

It is often very difficult to make these basic distinctions. The last three classes involve strain of the initial form and fabric, although that strain arises under different circumstances.

NONDIASTROPHIC STRAINS

The term *diastrophism* carries a genetic connotation. Thus it is implied that nondiastrophic strains are formed by forces other than those associated with uplift and deformation of mountains or continental drift. The remaining significant forces are gravity and such forces as may arise from internal changes in the material. Many types of structural features form in this way. For example:

1. Features due to differential compaction of sediment.
2. Folds developed as a result of compaction over irregular topography.
3. Features formed as water-saturated sediment is dewatered.
4. Folds and faults associated with slumps, flowage phenomena, and sliding of subaqueous masses.
5. Fractures formed as a result of volume changes caused by drying of sediment, dehydration, crystallization, and thermal effects.

PRIMARY SEDIMENTARY STRUCTURES

Field geologists encounter a great variety of textures, fabrics, and structures in rocks, some of which form during deposition or during emplacement and crystallization of igneous rocks and others which result from postdepositional or postcrystallization deformation. Often it is difficult to distinguish these later, secondary structural features from the earlier, primary features. This is particularly true when a distinction is to be drawn between syndepositional and postdepositional deformation in soft sediment. Secondary structural features may form at any time after deposition, including the stage during which the sediment is unconsolidated.

Study of primary sedimentary structural features is important because it is through our understanding of the primary form and attitude of strata that we are able to perceive that they are deformed and are able to measure the amount of that deformation. In addition, we rely on primary features to determine which side of a sedimentary layer was originally the top.

Stratification

No feature of sedimentary rocks is more universal than stratification. This feature arises through slight variations in any of the physical or chemical characteristics of sediment that occur essentially normal to the bedding. Stratification arises as a result of differences between layers in (1) chemical composition, (2) grain size, (3) degree of sorting, (4) degree and type of packing, (5) cohesion, (6) cementation, and (7) permeability or porosity.

The conditions under which sediment is deposited impart distinctive characteristics to the layering. This is reflected in the nature of the layers, in the arrangement of the material within the strata, in features on the bedding planes, and sometimes in the shape of the sedimentary unit. Some stratification arises directly from sedimentation, as in varves, layering in deep-sea carbonate and siliceous oozes, evaporites in playa lakes and behind fringing reefs, and chemical precipitates in general. Layering is formed indirectly when the water is agitated so that sediment is reworked and previously deposited sediment is broken up and redeposited, as in graded bedding and flat pebble conglomerates. This is most likely to occur in shallow water. Still other layers form under combined conditions of direct and indirect stratification.

Direct Stratification

A few examples will illustrate the types of stratification. The term *varve* is applied to regular layering or lamination due to annual changes in sedimentation. Glacial varved clay and silt is an excellent example. This sediment forms in glacial lakes in which a fine dark layer is deposited during the winter when the surface of the lake is frozen and the source of sediment is cut off by ice. A thick, coarser, and lighter colored layer is deposited in spring and summer when the sediment size and quantity increases. Varves also form in lakes in temperate regions, where plankton make up a large part of the sediment [see Bradley (1929) on the Green River shales]. Plankton production reaches a peak in late spring, while evaporation with increased calcium carbonate deposition reaches a peak in late summer.

The Solenhofen limestone is a classic example of *still-water deposition*. This fine-grained, uniform limestone formed in a lagoon between reefs which acted as sediment traps to filter out coarser sediment. The water must have been unagitated, because the layering is even and fossils of such delicate features as feathers are preserved.

The deep ocean basins are largely covered by finely stratified sediment composed of skeletal remains of diatoms, foraminifera, and other plankton. Red clays presumably derived from clay and meteoritic dust also constitute widespread deposits in the deep seas. Volcanic ash is locally important, and closer to continents, the finer clays transported far out from streams and glaciers are important. All these deposits assume blanket-like form as the particles of sediment settle on the existing submarine topography.

Exotic Blocks

Exotic blocks are large fragments of rock which occur in deposits formed under conditions that seem incongruous with the deposition of the blocks. For example, large blocks of reefs are found in black shales in the Delaware basin, west Texas. These masses must have been derived from the reef which encircles the basin, and they slid or slumped from the reefs into the deeper part of the basin where shales were being deposited. Even today, large blocks moved by streams from the mountains along the California coast slide into the sea or are transported in turbidity currents into much deeper water than that in which they might normally be expected. Still other exotic blocks are thought to be transported from land on ice rafts (icebergs) and dropped into various sedimentary environments.

Graded Bedding

Sediments are "graded" when sorting of sizes and vertical separation within the sequence have occurred. Graded bedding can be produced simply by placing a poorly sorted sediment into a long cylinder, shaking it until all the sediment is in suspension, and then setting the cylinder down. Initially all sizes are distributed at random through the cylinder, and the only forces acting on the mixture are the momentum of the agitated water and gravity. The largest and most dense fraction settles fastest. Thus the lowest layer is composed mainly of the larger and more dense fragments mixed with a small percentage of finer materials that were close to the bottom when settling started. At a later time all but the smallest fraction will have settled out and the bedding will have a distinct graded character. Kuenen (1952), Kuenen and Migliorini (1950), and others have demonstrated the formation of *graded bedding* in clastic sediment that has slumped, formed a turbidity current, and settled out in a sedimentation tank. Among the notable instances of graded bedding in modern sediments are turbidity current deposits in the North Atlantic. Many of the cores containing coarse clastic sediments taken in the deep sea exhibit graded bedding.

Deposition from turbidity currents is one of the most important processes of sedimentation on the continental rises and in the deep ocean basins, particularly within basins surrounded by oversteepened submarine slopes.

Ideally graded deposits show a gradual decrease in grain size from bottom to top of a layer, thus indicating diminishing current strength; but unfortunately for those wishing to use graded bedding to distinguish the top of beds, conditions sometimes exist during sedimentation which give rise to reversed grading. Some of the abnormalities may be related to the rate and character of sediment supplied to the turbidity current and the tendency for lighter particles to be removed. Also, eddies may be effective in keeping smaller fractions in suspension.

Reworked Sediments

Beds composed of broken shell fragments, coquina, and edgewise and flat pebble conglomerates indicate reworking of sediment. Edgewise and flat pebble conglomerates are formed when thin-bedded sediments are broken up in agitated waters. The conglomerate is likely to be edgewise if settling is rapid and loose sediment in which platy fragments can become embedded is present. The fragments of flat pebble conglomerates settle approximately in the original plane of bedding. Edgewise conglomerates are often composed of angular fragments and may be confused with breccia if their relation to bedding is not observed.

DETERMINATION OF BED TOPS*

The following brief outline summarizes the characteristics of the most common types of primary structures of both organic and inorganic origin used in determining the top of beds.

* Students should refer to Shrock (1948) and Middleton (1965) for more complete details and to Pettijohn and Potter (1964) for some of the best photographs of primary sedimentary features.

Ripple marks (Fig. 2-1)

Asymmetrical (current) ripples: The top and bottom are usually indeterminant unless the internal structure of the ripple has been buried by fine sediment laid in horizontal beds. The coarser grains accumulate in the trough of ripples formed in water, but the reverse is usually true of wind-formed ripples.

Symmetrical ripples: Good for top determination. These ripples have pointed crests and rounded troughs resulting from oscillatory water motion.

Interference ripples: Usually the top is indeterminant. These are complex ripples formed by interference wave motion resulting from combined oscillation and current action. Many complex forms result, some of which have sharp crests that may be used for top determination.

Antidune ripples: Good for top determina-

FIGURE 2-1 Ripple marks. (*a*) Current ripples formed by currents moving from right to left; (*b*) oscillation ripples; (*c*) current ripples showing internal cross lamination and coarse particles collected in troughs (currents moving left to right); (*d*) current ripples showing a thin shale layer deposited with greatest thickness in troughs of ripples; (*e*) antidune ripples. (*From Shrock, 1948.*)

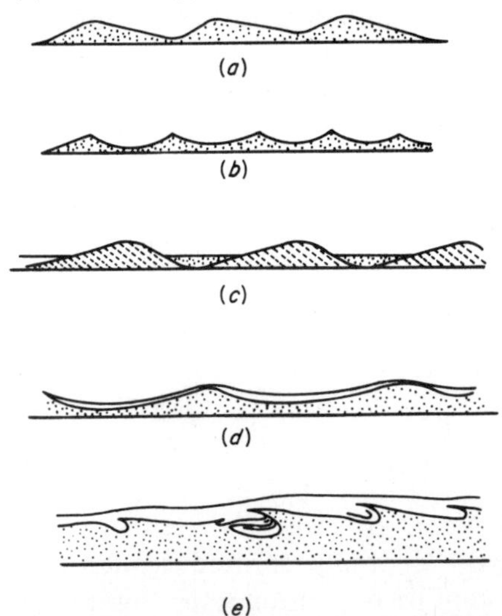

tion. These ripples form when the water velocity becomes so great that the sediment below the surface moves. These ripples move upstream by a process of erosion on the downstream face and deposition on the upstream face, resulting at some stages in the formation of sharp crests (Fig. 2-1*e*).

Linguoid ripples: The top is indeterminant. These are modified ripples which have a tongue-shaped form and may resemble mud flows.

Metaripples: Good for top determination. These are large ripples usually formed on a coarse base material, showing long wavelengths, and later altered as fine sediment is deposited to give the large ripples their asymmetry.

Channels (Fig. 2-2)

Tidal channels: Good for top determination. These are drainage channels usually dendritic in plan and of U shaped cross section developed on tidal flats as the tide goes out.

Rill marks: Good for top determination. These are small dendrite-shaped channel systems formed along beaches as the water flows seaward. They are modified where the flow is intercepted by rocks or shells; so, some rill systems join seaward and others split, forming distributaries seaward.

Lobate rill marks: Good for top determination. These are lobate-shaped, strongly asymmetric depressions aligned in the direction of flow of strong currents.

Grooves: Good for top determination. Under this heading we may include such features as grooves formed by floating or suspended objects which have projections that are dragged across a soft sediment base, leaving a groove.

Cut and fill structures: Good for top determination. These are usually short channels of lens-shaped cross section which are later filled. The manner of filling varies, but commonly the fill is the same as the material into which the channel was cut. The fill tends to possess structures somewhat similar to deltaic deposits. The layering is inclined into the cut channel, and generally these inclined beds have a slightly curved form, concave upward. Often previously filled channels are recut and one cut and fill feature is imposed on earlier ones. The truncated surfaces, since they are always on top, provide a good key to which way is up.

Other Inorganic Markings

Mud cracks: These features develop most readily in fine sediment subjected to alternate drying and wetting. The thin layers of mud tend to be broken into a complex system of polygonal-shaped pieces. The pieces curl upward as they dry. Usually the cracks are filled, and the top is easily recognized by the convex surfaces of the cracked layers. If the layer of cracked mud has been subjected to agitation sufficient to turn the plates over, great care should be taken in judging the top.

Raindrop impressions: Raindrops leave imprints in soft sediment in the form of small craters with rimmed edges. Depending on the consistency of the mud, it may rise slightly in the center of the crater.

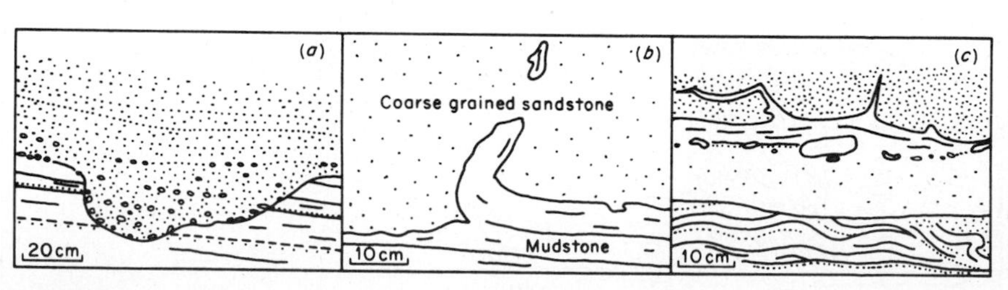

FIGURE 2-2 Sketches of sedimentary structures. (*a*) Scour channel at base of laminated pebbly sandstone; (*b*) overfold of mudstone at base of massive coarse-grained sandstone; (*c*) layer of pebbles along bedding plane. (*From Crowell and others, 1966.*)

Rocks falling into sediment: If the sediment is soft, the layers are likely to be strongly bent immediately under a falling rock to conform to its shape. In some instances the uppermost layers may even be penetrated. Essentially flat layers may form over the top of the rock as filling goes on, but if soft sediment is deposited over the rock, compaction will eventually cause the overlying layers to be gently arched.

In addition to the features described above, it is sometimes possible to identify the top of a bed by finding collapsed caves, cave deposits, buried soil and erosional features on unconformities, pot holes, etc.

Organic Markings

Among the multitude of animals which live in shallow water, many leave tracks or trails or bore into the sediment, leaving marks which are useful for the purpose of determining tops of beds. Usually the tracks and trails consist of furrows and ridges left in a pattern which is typical of the type of animal. In such cases it is necessary to know the pattern left by the particular animal. The top can then be determined from the relative distribution of ridges and furrows in the pattern. In simple cases the trail is a single, rather straight groove, but more complicated patterns are more typical. Footprints in soft sediment are more readily identified.

Many invertebrates burrow in sediment. One worm tube (*Scolithus*) is nearly always found perpendicular to bedding in sandstone or quartzite. It is a useful guide when the bedding is not apparent. Most borings are useful to determine the top of beds only if the starting point of the boring can be seen, but many worm borings are so complicated they are of little use. The burrows may have a small ridge left around the entrance to the hole where material removed from the hole has been built up.

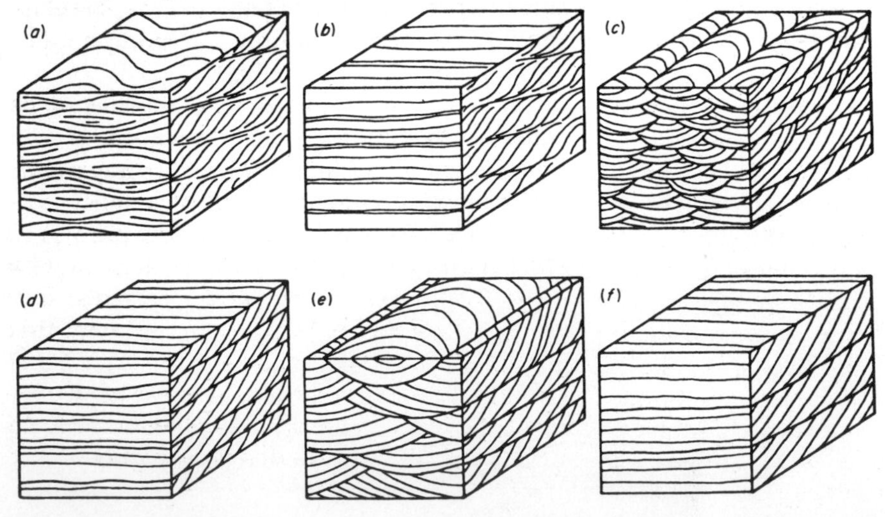

FIGURE 2-3 Common types of cross stratification considered to be generated by migration of asymmetrical ripple marks. (*a*) Small-scale cross-stratified sets showing pinch and swell structure and gradational contacts resulting from small-scale linguoid ripples; (*b*) cross-strata in small-scale planar sets with gradational contacts resulting from small-scale straight ripples; (*c*) small-scale trough cross-stratified sets resulting from small-scale linguoid ripples; (*d*) small-scale planar cross-stratified sets resulting from migration of small-scale straight ripples; (*e*) large-scale trough cross-stratified sets resulting from large-scale lunate and possibly linguoid ripples; (*f*) large-scale planar cross-stratified sets resulting from straight large-scale ripples. (*From Allen, 1963.*)

DIVERGENT STRUCTURES BETWEEN BEDDING PLANES

Of the various sedimentary structures found between bedding planes, two stand out as particularly significant: (1) *cross laminations* such as those arising on deltas, dune deposits, and scour and fill structures, and (2) *divergent structures* due to movement of materials within beds after initial deposition and before or during the processes of consolidation, compaction, and cementation. The first of these two categories is important as a means of identifying tops of beds, and the second because these structures can be easily confused with secondary deformation—particularly when they occur on a large scale.

Cross Laminations

Cross laminations of different origins have distinctive geometries, which usually makes differentiation possible. *Deltaic cross-bedding* is composed of three sets of bedding. The more thickly inclined beds, foreset beds, thin out in the direction of current movement and become tangential to the underlying bedding plane as bottomset beds. The topset beds are usually composed of sediment in the process of being moved toward the delta front, and consequently the topset beds are often eroded away, leaving truncated foreset beds. The truncated surface is the top. Such deltaic bedding is common in lakes, streams, and on coasts. The top edge in cross-bedded strata is the truncated edge, but it is important to know that the intersection of the foreset beds with the lower beds can be quite abrupt. This may lead to formation of cross-bedding which is apparently truncated at both top and bottom. Such cross-bedding forms under conditions of rapid sedimentation and is called *torrential cross-bedding* or *angular cross-bedding*.

Cut and fill structures with cross-laminated beds were referred to earlier. A great range of geometric shapes is possible in exposures of cross-laminated beds. This range arises both from the original form of the deposit and from the angle of the exposed section through it. Complication in form is also caused by changes in the direction of the currents. This is easy to visualize in the case of a braided stream in which the path of water is constantly changing in such a way that old deposits are being cut and removed as new ones are deposited. Allen (1963) believes that many common types of cross stratification are generated by migration of asymmetrical ripple marks (Fig. 2-3).

SOFT-SEDIMENT STRUCTURAL FEATURES

Compaction

Many of the structural features now found in hard, consolidated rocks exposed in folded mountain belts are thought to have formed when the sediment was unconsolidated or semiconsolidated. Major thrust faults, fold systems,* and even slaty cleavages have been attributed to conditions arising in unconsolidated rock.

The process of compaction sets in almost as soon as a sediment is deposited, and it continues as the deposit is buried under progressively greater thicknesses of overburden. *Differential compaction* is responsible for much of the thickening and thinning in sedimentary units composed of lenses of sandstone in shale, for warping of layers of sediment over enclosed consolidated rocks, and for the fishtail appearance of split coal seams; on a larger scale it plays a decisive role in the development of domes over buried topographic features.

The most marked differences in behavior are found between materials composed of colloidal-size particles and the larger fragmental materials such as sand or silt. Since most sediments are deposited in water, the process of compaction takes place in a two-phase system consisting of water and solid fragments. It is in the way water is held in the sediment that sand and clay differ so greatly.

Water in a sediment consisting of large fragments is held in pore spaces between the solid fragments, so porosity depends mainly on the

* See Chaps. 8 and 11.

shape, sorting, and packing of the sediment. If the sediment consists of irregularly shaped particles, a loosely packed and porous sediment may result. The platy fragments may form a boxwork fabric, or small cavities may be formed in loosely packed sediment, yielding very high porosities. Open cavities do not last long as compaction begins. As weight builds up, the loose cavities collapse and grains tend to shift in such a way as to reduce porosity. This process takes place if it is possible for the water to be expelled, but if the water cannot move out of the sediment, little compaction occurs, because water is not highly compressible. Most fragmental rocks are permeable, and the water can therefore be forced upward as compaction occurs. The porosity of a sand may be reduced from 30 to 40 percent to 20 to 30 percent as it becomes tightly compacted. The difference in the pore space due to compaction of coarse sediment, 10 to 20 percent, Fig. 2-4, is in marked contrast with clays.

Water is held in clays both between the individual particles or aggregates of particles and within the crystal lattice as absorbed water. *Porosity* depends on the clay mineral involved, the degree of dispersion of the clay particles, and the degree and arrangement of aggregates of particles. Some clays have porosity of as much as 60 to 80 percent. When compaction of clay takes place, water is first forced out of the spaces between aggregates. A film of water surrounds each clay colloid, and as the thickness of this film is reduced, the rate of water loss is retarded. Figure 2-5 shows the observed changes of porosity of clay-rich sediments from the Gulf Coast, Venezuela, and Oklahoma with depth in drill holes. Compare this with the porosity of sand at various depths, Fig. 2-6.

FIGURE 2-4 Variation of bulk density and porosity as a function of confining pressure (depth of burial) for dry sand and for sand with fluids in the pore spaces. (*From Maxwell, 1960.*)

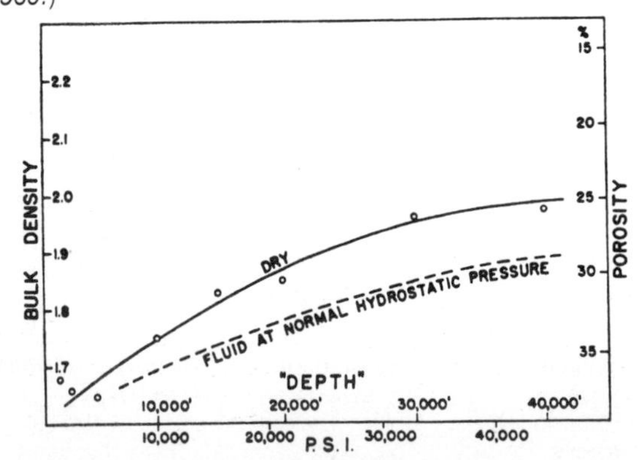

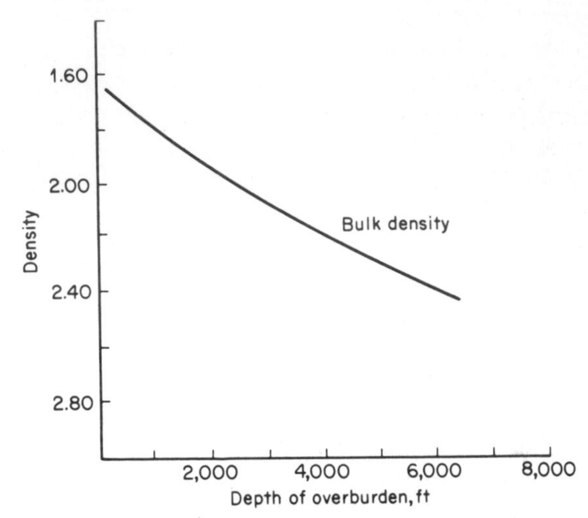

FIGURE 2-5 Relation of shale density to depth of overburden in Venezuelan wells. (*From Hedberg, 1936.*)

These differences in behavior account for many of the structures resulting from differential compaction. Sandstone lenses, concretions, or blocks of consolidated rocks found enclosed within clay compact less than the clay. Thus, as compaction occurs, the bedding in the clay is deflected over the sand or rock, producing a domical structure. Similar features form in tuffaceous deposits which compact more than buried objects such as lava flows.

Many structures in coal form as a result of shrinkage and compaction which accompanies the alteration of peat to coal. Between 10 and 30 m of peat is estimated to produce a 1-m coal seam. Since peat forms in swamps, coal deposits often contain sandy stream-channel deposits or deposits laid around the margin of the swamp. Conditions of long-term slow subsidence favor preservation of peat. Slight interruptions in subsidence accompanied by deposition of sand give rise to the splitting of coal seams, Fig. 2-7. Differential compaction which accompanies the transformation of peat to coal creates *fishtailing*, where a single thick coal seam splits into a large number of thinner seams laterally.

The famous Ten Yard coal seam of England changes from a single seam 10 m thick into 12 thinner seams interbedded with sandstone and shale, making a section 150 m thick in a distance of 8 km. Presumably the thickness of peat required to make the 10-m seam was nearly equal to the thickness of the peat, sand, and clay at the edge of the swamp when the peat was being deposited.

Supratenuous Folds

The name *supratenuous fold* is applied to folds formed as a result of compaction. Where topography is buried, sedimentation generally occurs on the flanks of the hills before the hills are submerged. Highs in the topography hold less sediment than the valleys until the relief disappears. A thinning of beds toward high topographic features is produced. If compaction goes on at a rate comparable to the rate of deposition, the beds in the valleys show a tilt

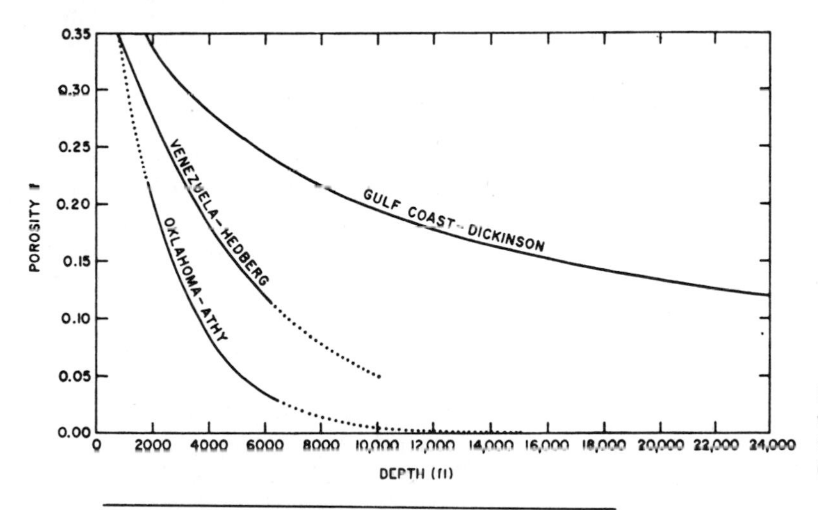

FIGURE 2-6 Variations in porosity with depth as observed in wells in the Gulf Coast, Venezuela, and Oklahoma. (*From Rubey and Hubbert, 1959.*)

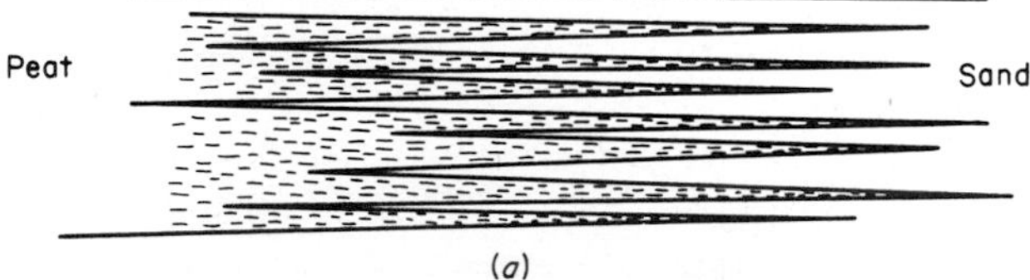

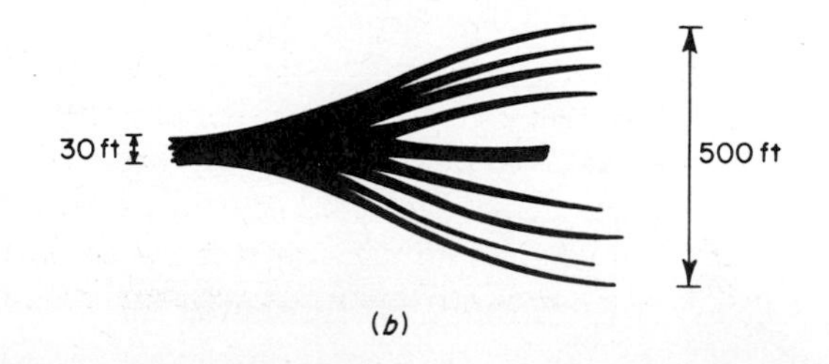

FIGURE 2-7 Cross sections of splitting of coal seams. (*a*) Interfingering of sand and peat as initially deposited near swamp margin; (*b*) fishtail resulting from differential compaction of a deposit like that shown above.

away from the hills by virtue of the variations in thickness even before the hills are buried.

Assume that a hill is buried until the surface of sediment is nearly level. The amount of compaction is a function of the type of sediment and the thickness of the section being compacted. Because this thickness is greater over low areas in the old topography, more compaction occurs there than over hills. This produces a fold in the units over the buried hill. If the hill is symmetrical, the compaction folds over it are also symmetrical and the line that bisects the hill also extends through the crest of the overlying folds. If the hill is asymmetrical, the compaction folds are also asymmetrical, but the line connecting the crest of the compaction folds does not coincide with the line that bisects the angle between the two sides of the hill. Instead, the crest line of the compaction folds is nearer the vertical. The top of each succeeding bed is moved in the direction of the lower slope of the buried topography as the series rises (Nevin and Sherrill, 1929).

Pore-Pressure Effects

Shales and clay are not generally very permeable; thus it is difficult to remove water from a thick deposit of clay. Water tends to be forced into sands and out along the water-sediment contact. Water can become confined either within the clay sediment or in pore spaces of sands from which the water is prevented from escaping by overlying impermeable layers. When this happens, pressures are exerted on the water by the weight of the overlying sediment-water mixture. High pressures can be built up in the water in pore spaces of the saturated sands. Pore pressure can reduce the internal friction of sand, inducing landslides if the sand is near the surface, and it may also be significant in large-scale thrust faulting. High pore pressure is of critical importance in the formation of many *soft-sediment structural features*. Sandstone dikes, called clastic dikes, found in coal and other sediments result from the injection of saturated sand under pressure into fractures. The presence of layers of high *pore pressure* would doubtless contribute to basin-directed slump and sliding of stratified sequences.

Convolute Bedding, Slump, and Load Casts

The *hydroplastic* character of clay and mud deposits, pore-pressure effects, the pressure of overlying deposits, and the initial dip of the stratification are important factors in the development of postdepositional deformation in unconsolidated sediments.

Convolute bedding (Fig. 2-8) is a name applied by Kuenen (1953) to peculiarly deformed laminations which have the following characteristics:

1. Crumpling of the laminations increases in intensity upward in a bed and then gradually dies out again so that the upper laminae of the same bed are again flat.
2. Laminae are not ruptured. The lamination planes can be traced across several undulations showing rounded distortions.

FIGURE 2-8 Diagram showing some soft-rock deformation types. (*a*) Load cast, developed by plastic deformation of a bed under a later load; (*b*) slump structures due to horizontal movement of beds after deposition; (*c*) convolute bedding developed during deposition of the bed without horizontal movement. (*From Kuenen, 1953.*)

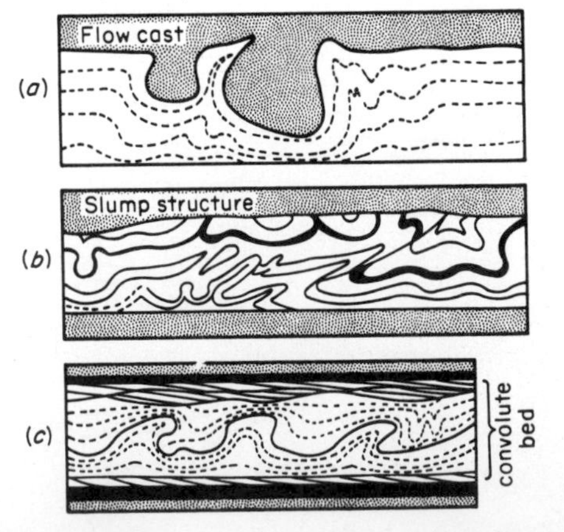

3. Only one bed is affected at a time, but this disturbance can be traced laterally for great distances in all directions within that bed.
4. The affected bed shows no external irregularities in thickness.
5. The degree of distortion is of almost equal intensity in two planes, one parallel, the other at right angles to the original slope.

Undisturbed laminae may occur below convoluted laminae in the same bed, and the undulations may be smoothed off before the deposition of the bed comes to an end. Hence the deformation takes place during sedimentation. Kuenen concludes that convolute beds are formed in turbidity current deposits, and this process cannot have occupied more than a few hours, probably much less.

Not all students of convolute bedding agree entirely with this conclusion. Among the ideas offered to explain its origin are (1) gravity-induced plastic flow, (2) injection of sediment from below, (3) plastic deformation of ripple marks coupled with loading pressure, (4) sliding of overlying sediment deforming a bed by shear, (5) quicksand movement, (6) deformation of cohesive sediment in response to shearing effects set up by high-velocity currents, (7) intrastratal laminar flow, and (8) localized thixotropic flow coupled with loading (Davies, 1966). It seems probable that not all convolute bedding originates in the same way.

Load casts or *flow casts* are bulges, bag-shaped, or bulbous protuberances of a coarse clastic sediment down into an underlying layer such as clay which has behaved as a hydro-

FIGURE 2-9 Photograph of flute cast. The pen provides a guide to the scale. (*From Marcos, 1973.*)

plastic material. These are good top-bottom indicators which are formed by unequal compaction and settling of unconsolidated sediment often caused by initial irregularities in the sediment thickness. Load casts are formed strictly by compaction and differ from another type of sole mark, *flute casts,* which are oblong, subconical, strongly aligned bulges found on the bottom side of sandstones or siltstones where a scoop-shaped or lobate depression, formed by scouring or current action, is filled with sediment, Fig. 2-9.

A distinction between load casts and convolute bedding is found in that convolute bedding occurs within the plastic layer itself, while load casts develop between two beds of different mechanical properties.

Kuenen (1953) describes the characteristics of *slump structures* as follows:

> Occurrence between undisturbed beds, rupture of the beds at the higher end (pull-apart structures, generally not exposed or not obvious), piling up or thickening of beds at the lower end, beheading of anticlines, the dying out downward of the deformation, less seldom also upwards, balled up structures of sandy or calcareous beds surrounded by more plastic material (clay or lime mud), sliding planes, faults, general irregularity of structures often showing similar intensity in sections parallel and at right angles to the slope, internal rupture of bedding planes, brecciated masses, abrupt changes in thickness of beds, the partaking of more than one bed in the movements, irregular surface sometimes smoothed off by a graded bed or by infilling with laminated deposits, the duplication of beds on top of each other.

REFERENCES

Allen, J. R. L., 1963a, Asymmetrical ripple marks and the origin of water laid cosets of cross strata: Liverpool Manchester Geol. Jour., v. 3, p. 187–236.

——— 1963b, The classification of cross-stratified units, with notes on their origin: Sedimentology, v. 2, p. 93–114.

Brace, W. F., 1967, Review of Coulomb-Navier fracture criterion, *in* Rock mechanics seminar, v. 1.

Bradley, W. H., 1929, The varves and climate of the Green River formation: U.S. Geol. Survey Prof. Paper 158, p. 87–110.

Brush, L. M., Jr., 1965, Sediment sorting in alluvial channels: Soc. Econ. Paleontologists and Mineralogists Spec Pub. no. 12, p. 25–34.

Cloos, Ernst, 1973, Microtectonics along the western edge of the Blue Ridge, Maryland and Virginia: Johns Hopkins University Studies in Geology, no. 20.

Cook, P. J., 1971, Illamurta diapiric complex and its position on an important central Australian structural zone: Amer. Assoc. Petroleum Geologists Bull., v. 55, no. 1, p. 64–79.

Crowell, J. C., and others, 1966, Deep-water sedimentary structures Pliocene Pico formation Santa Paula Creek, Ventura Basin, California: California Div. Mines and Geol. Spec. Rept. 89.

Davies, R., 1966, Concentration of mica by water flotation: Geol. Soc. America Bull., v. 77, p. 661–662.

Frazier, D. E., and **Osanik, A.**, 1961, Point-bar deposits, Old River Locksite, Louisiana: Gulf Coast Assoc. Geol. Soc. Trans., v. 11, p. 121–138.

Hedberg, H. D., 1936, Gravitational compaction of clays and shales: Am. Jour. Sci., v. 31, p. 241–287.

Heezen, B. C., Ericson, D. B., and **Ewing, Maurice**, 1952, Turbidity currents and sediments in North America: Am. Assoc. Petroleum Geologists Bull., v. 36, p. 489.

Kuenen, P. H., 1952, Classification and origin of submarine canyons: K. Nederlandse Akad. Wetersch. Proc., ser. b, v. 55, p. 464–473.

——— 1953, Significant features of graded bedding: Am. Assoc. Petroleum Geologists Bull., v. 37, p. 1044–1066.

——— and **Migliorini, C. I.**, 1950, Turbidity currents as a cause of graded bedding: Jour. Geology, v. 58, p. 91–127.

Lowry, W. D., and **Cooper, B. N.**, 1970, Penecontemporaneous downdip slump structures in Middle Ordovician limestone, Harrisonburg, Va.: Amer. Assoc. Petroleum Geologists Bull. v. 54/10 (pt. I), p. 1938ff.

McKee, E. D., and **Goldberg, M.**, 1969, Experiments on formation of contorted structures in mud: Geol. Soc. America Bull., v. 80, no. 2, p. 231–244.

Marcos, Alberto, 1973, Las series del Paleozoico inferior y la estructura Herciniana del Occidente de Asturias (NW de Espana): Trabajos de Geologia, no. 6.

Maxwell, J. C., 1960, Experiments on compaction and cementation of sand: Geol. Soc. America Mem. 79, p. 105–132.

Middleton, G. V., ed., 1965, Primary sedimentary structures and their hydrodynamic interpretation: Soc. Econ. Paleontologists and Mineralogists Spec. Pub. no. 12.

Moore, D. G., and **Scruton, P. C.**, 1957, Minor internal structures of some recent unconsolidated sediments (Gulf of Mexico): Am. Assoc. Petroleum Geologists Bull., v. 41, p. 2723–2751.

Nevin, C. M., and **Sherrill, R. E.**, 1929, Studies in differential compacting: Am. Assoc. Petroleum Geologists Bull., v. 13, p. 1–22.

Pettijohn, F. L., and **Potter, P. E.**, 1964, Atlas and glossary of primary sedimentary structures: New York, Springer-Verlag.

Powers, Sidney, 1922, Reflected buried hills and their importance in petroleum geology: Econ. Geology, v. 17, p. 233–259.

——— 1931, Structural Geology of north-eastern Oklahoma: Jour. Geology, v. 39, p. 117–132.

Raistrick, Arthur, and **Marshall, C. E.**, 1939, The nature and origin of coal and coal seams: London, English Universities Press.

Rubey, W. W., and **Hubbert, M. K.**, 1959, Role of fluid pressure in mechanics of overthrust faulting: Geol. Soc. America Bull., v. 70, p. 167–206.

Sanders, J. E., 1965, Primary sedimentary structures formed by turbidity currents and related resedimentation mechanisms, *in* Primary sedimentary structures and their hydrodynamic interpretation: Soc. Econ. Paleontologists and Mineralogists Spec. Pub. no. 12, p. 192–219.

Shrock, R. R., 1948, Sequence in layered rocks: New York, McGraw-Hill.

Von Herzen, R. P., **Hoskins, H.**, and **Van Andel, T. H.**, 1972, Geophysical studies in the Angola diapir field: Geol. Soc. America Bull., v. 83, no. 7, p. 1901ff.

3

Concepts of Stress and Strain

Much of structural geology deals with the description and genesis of strains of bodies of rock. Strain may be evidenced by changes in the internal fabric of the rock or by distortion of the primary shape of the rock mass, and it usually involves translational and/or rotational movement of the mass. *Dynamic structural geology* is concerned with the identification of the mechanisms by which the strains are impressed on the rock and ultimately with the explanation of these mechanisms in terms of the stresses acting within the earth's lithosphere.

NOTION OF STRESS

Concepts of stress are perhaps less familiar to most of us than the idea of force. A rock is in a state of stress when a force is applied to it. Forces are of two general types: those like gravity which act throughout the earth's crust, called *body forces*, and those that act on surfaces and cause neighboring parts of the medium to act on one another, called *surface forces*. Force is a vector quantity, Fig. 3-1, defined in terms of two values, magnitude and direction. So, unlike scalar quantities, a vector must be described in terms of some frame of reference, some coordinate system. Thus a force may be thought of as acting at some specific point. The forces within the earth act over surfaces or through volumes of material, and they generate stresses and strains which are three-dimensional entities called *tensors*. These are much more complicated than vectors and require either six or nine quantities for their description. To describe completely the stress in a body, the stress at each point in the body must be known, but the problem is greatly simplified if a homogeneous state of stress exists, such that the stress has the same magnitude and direction at every point in the body. This assumption is reasonable in some geological circumstances, especially if a small volume is considered, but this is not always the case in the formation of structural features.

The difference between the notion of stress and that of force may be illustrated by considering the conditions at a point somewhere in the interior of the earth where the only force acting is that due to gravity. The force due to gravity at this point is directed toward the center of the earth and has a net resultant magnitude that is a function of the depth. The stress condition, being a three-dimensional entity, must be considered in terms of a volume of rock—most conveniently in terms of a small cube of rock.

FIGURE 3-1 Vector representation of forces acting at a point. Top, two equal and opposite-directed forces act at point P, which is static and under compression. Middle, three forces act at point P, which is static. The resultant of forces F_1 and F_2 is equal and oppositely directed to F_3. Point P is in tension. Bottom, F_1 and F_2 act on point P with effects identical to that of the resultant.

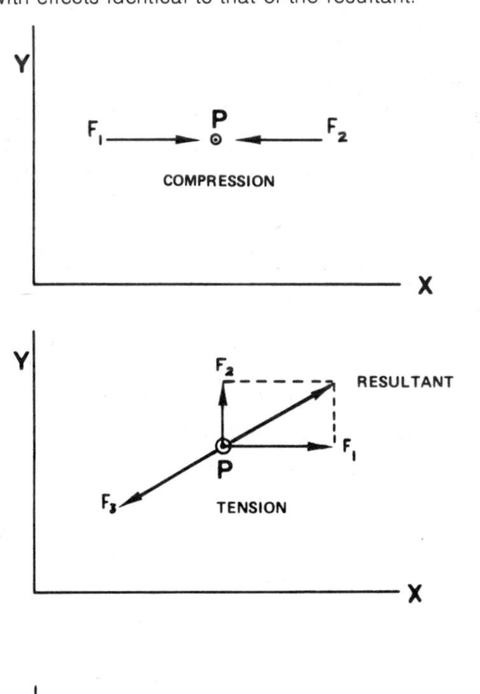

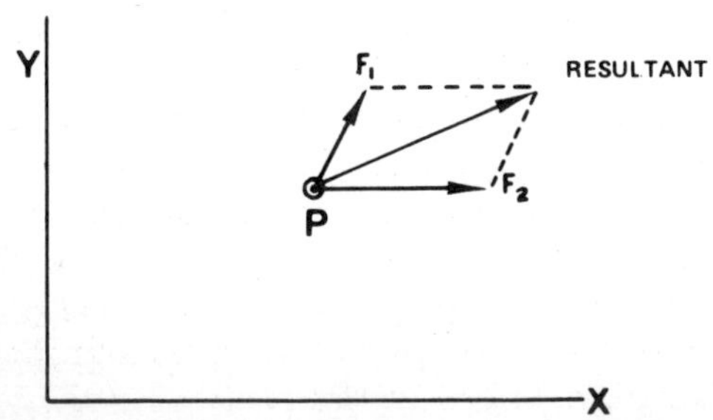

The concept of the *state of stress* at a point within a body connotes the normal and shearing stresses on planes of all possible orientations through that point. But, we generally are concerned with stresses on planes of particular orientations, and the stresses are described in terms of an orthogonal coordinate system.

In order to analyze the stresses at a point, the conditions on each surface of a small cube at the point are determined. This cube is chosen so small that uniform conditions exist across each of its faces. The force F of interaction between a face of the cube and the surrounding rock is applied uniformly across the face of the cube, and the force per unit area can be resolved into two components—one normal to the surface designated σ, called the *normal stress*, and a second designated τ which acts along the surface and is a *shearing stress*. The relative value of these two depends on the angular relationship between the force and the surface.*

The shearing stress is in turn resolved into two components oriented parallel to the axes of the coordinate system chosen, Fig. 3-2. Thus, we can describe the stresses on any surface in terms of one normal component and two shearing components, each of which acts parallel to the coordinate axes. To consider the total stress acting at some point within a body, the stress component on each surface of a

* *Notation conventions*

It is necessary to adopt a system of notation to designate the direction in which the stress components act. These notations are:

1. The stress components which act normal to the sides of the cube are designated σ_x, σ_y, and σ_z to indicate the coordinate axis which they parallel.
2. The shearing-stress components have two subscripts. The first subscript indicates the direction (coordinate axis) normal to the plane in which the shear lies. The second subscript indicates which coordinate axis is parallel to the stress component. (For example, τ_{xy} is a shearing stress acting in the surface normal to x and in the direction of the y axis.)
3. The normals to the planes of the cube point outward and are given + or − values, depending on the system used. Positive normal stress components may be tensile and negative normal stress components compressive. Others reverse this designation.

point-size cube is described. For convenience the cube is oriented so that its edges lie along the coordinate system. The force per unit area on each side of the cube can be resolved into three components (parallel to the coordinate axes) regardless of the orientation. A total of 18 stress components act on the 6 sides of the cube as shown, Fig. 3-3. However, if the cube is in a state of equilibrium—that is, if the cube is not moving either by translation or rotation —then several of the stress components are equal. The components of stress that would make the cube rotate must be zero or counterbalanced if no rotation occurs. Consider rotation about the z axis, for example. Shearing stresses τ_{xy} and τ_{yx} tend to cause such rotation. They must both be zero or equal since they have opposite senses of rotation. Similarly, $\tau_{yz} = \tau_{zy}$ and $\tau_{xz} = \tau_{zx}$. Thus only six of the nine stress components are independent, and state of stress at a point can be expressed in terms of these six components.

The Concept of Principal Stresses

Those who have worked with vectors are familiar with the combination of several vec-

FIGURE 3-2 Resolution of a force F into normal (sigma) and shear (tau) components. The shear is resolved into components parallel to the coordinate axes.

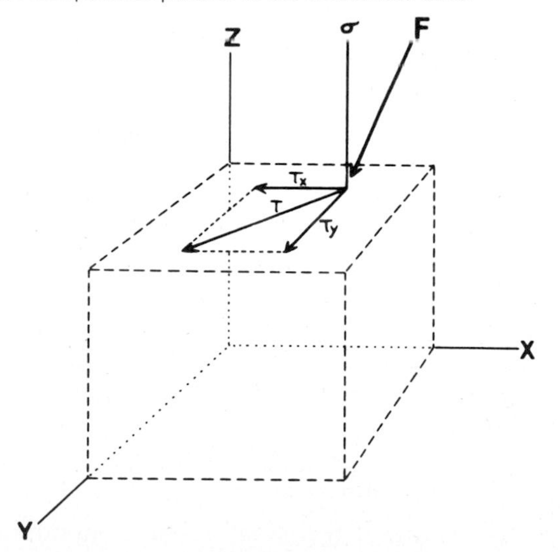

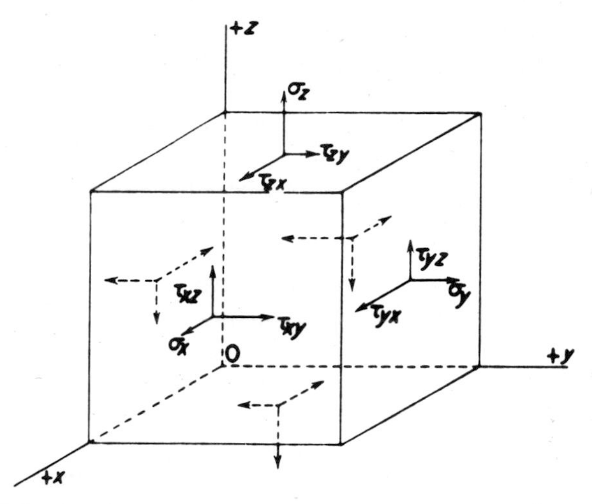

FIGURE 3-3 Stresses on a cubical element in equilibrium. (*From Varnes, 1962.*)

tor quantities to find a resultant vector. In dealing with force vectors; several different forces acting at a given point from different directions have exactly the same effect on the body at that point as a single resultant force found by adding the vectors.

The concept of *principal stresses* is a somewhat analogous way of examining stress conditions. We saw in the preceding section how a force/area acting in a given direction and with a given magnitude could be resolved into shearing and normal stresses on the faces of a cube of material oriented with the sides of the cube parallel to the coordinate axes of any orthogonal coordinate system we chose. Any number of forces acting on the cube from various directions can be resolved in the same way, and then all those components acting in any direction may be added to produce a set of resultant stress components on each face of the cube.

It is possible (as shown in Appendix A) to determine the value and direction of the shearing and normal stresses on any plane oriented in any direction we choose within the reference cube, given the six stress components. Certainly one of the most significant results obtained from the analysis of the way shearing and normal stresses vary with the orientation of the plane is the discovery that three mutually perpendicular planes exist in which shearing stresses vanish, and the stress condition on those planes can be described in terms of the normal stresses to each of those planes. Those three normal stresses are known as the *principal stresses,* and the directions in which they act are called the *principal stress directions.* The three principal stresses may be equal, as in a state of *hydrostatic stress;* one may be greater than the other two; or all three may be unequal, and they are then known as the *greatest, intermediate,* and *least* principal stresses. [The derivation of the principal stresses in the case of plane (two-dimensional) stress is given in Appendix A.] The concept of principal stresses greatly simplifies the problem we may face in trying to visualize stress conditions. No matter how complicated the applied stresses are, they can be resolved into three mutually perpendicular normal stresses, and if values are known for the applied stresses, the values and directions of the principal stresses can be calculated. Of course, the principal stress directions may be known in experimental work or postulated in theoretical work. For example, a specimen of rock may be placed in a press and squeezed. The maximum principal stress direction is then the axis of the press. Also, if the principal stresses are known, the values for the shearing and normal stresses on any other plane through the material may be determined as derived in Appendix A or by use of the Mohr circle described in the following section.

Stresses within the Earth

Stresses exist throughout the earth. These arise as a result of a wide range of natural phenomena. The most pervasive of these are the stresses due to the earth's gravity field, and these affect the development of many secondary structural features. Gravity produces stresses that are similar to the hydrostatic pressures in a liquid, as described later in this chapter. Other stress fields are superimposed on these gravitative stresses as a result of changes within the interior of the earth. The exact nature of these internal changes is still debated, but changes in the volume of materials in the mantle due to both phase changes and changes of state are often suggested. Isostatic movements due to removal of surface load are also expected. Special importance is attached to the idea that the materials within the mantle are moving as a result of internal temperature differences, giving rise to convection. Whatever the internal causes, the evidence which indicates that the continents have been and are moving is very strong. Obviously great force is required to move a continent, and very important stress fields must be associated with the lateral movements of the continents. These stresses would be significant in the creation of secondary structural features along the leading edge of the continental plates where compressive stresses would be expected to dominate and along the lateral and trailing edges where shear and extension would be expected.

Thus we may look to the interior of the earth for the ultimate source of the stresses in the lithosphere, but these primary sources of stress become altered in direction and magnitude as they are transmitted. Stresses vary from point to point within the lithosphere. Sometimes they may be uniform over considerable volumes of material, but in other places the variation is rapid, as across a fold after the initial displacements begin. Our ultimate objectives in studying stresses in the earth must be both to understand the local stress conditions which give rise to the small- and intermediate-scale features we see in outcrop and to understand the larger regional and global stress fields which are associated with the orogenic belts as a whole.

Mohr's Representation of Stress

In 1882, Otto Mohr, a German engineer, demonstrated one of the most useful methods of representing a state of plane stress. He made use of a circle plotted on a two-axis coordinate system in which the abscissa is the normal stress σ and the ordinate is the shearing stress τ, Fig. 3-4. The magnitude of the normal and shear stresses acting on any plane making an angle α with the direction of the least stress can be read directly from this plot, provided the magnitude of the maximum and minimum principal stresses at the point are known. The magnitude of the normal and shearing stresses on an arbitrary plane can be expressed in terms of the following three values as shown in Appendix A:

$$\frac{\sigma_1 + \sigma_3}{2} \qquad \frac{\sigma_1 - \sigma_3}{2} \qquad 2\alpha$$

If the point on the abscissa described by the first of these terms is taken as the origin of a radius vector which makes an angle 2α with the positive direction of the abscissa, and if the second of the above terms is taken as the length of the radius vector, then the coordinates of the end of the vector satisfy the equations for the values of the normal and shearing stresses on the plane which makes an angle α with the axis of least stress for every value of α. As α is changed, the radius vector describes a circle which is the locus of all the values for normal and shearing stress for all orientations of the reference plane.

The Mohr circle is easily read. When the reference plane is normal to the maximum principal stress direction (the angle α is equal to zero), the normal stress on the plane is the maximum principal stress, and the shearing stresses vanish. When the reference plane is perpendicular to the least principal stress ($\alpha = \pm 90°$), the normal stress on the plane is equal to the least principal stress, and the shearing stresses vanish; but for all other orientations both normal and shearing stresses exist. The magnitude of the maximum shearing stress is $\tau = (\sigma_1 - \sigma_3)/2$. This condition is obtained at two orientations of the reference plane when $2\alpha = 90°$ and $270°$ ($\alpha = 45°$ and $135°$). The two shear stresses thus produced are of equal magnitude but of opposite sense.

To find the principal stresses when σ and τ components on several planes are known, the points corresponding to these values are plotted, and a Mohr circle is constructed which will pass through them. Then σ_1 and σ_3 can be read from the graph.

Superimposed Stresses Using the Mohr Circle

Stress is not a vector quantity; thus two stresses which may be superimposed at a point cannot be added in the manner of vectors. It is, however, possible to add stress components acting on the same plane; so, we can determine the effect of superimposing two stress states by use of the Mohr circle in

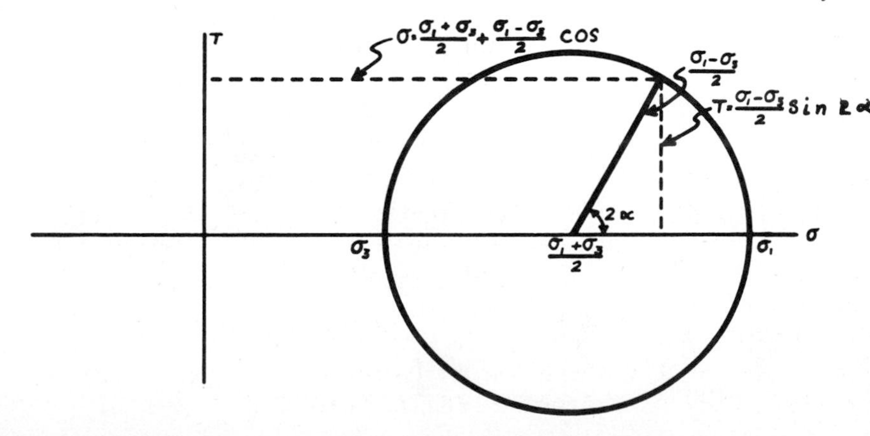

FIGURE 3-4 Mohr's stress circle depicts the variation of shearing and normal stresses on a plane oriented at an angle α to the least principal stress direction.

which stress components on any given plane can be determined as shown in Fig. 3-5. The top line shows the two stress states to be combined, but because the second stress is oriented in a different direction from the first they cannot be added directly. To solve this problem, the second stress condition is resolved into components on a square oriented parallel to the coordinate axes of the first using the Mohr circle, and the stress components on this square can be added to those on the first square.

Concept of Deviatoric Stresses

A *hydrostatic condition* is one in which the confining pressure increases uniformly with depth. In such a situation the three principal stresses are equal at any given point, $\sigma_1 = \sigma_2 = \sigma_3$. The pressure is equal to the weight of the overlying column of material. Such conditions are approximated in the earth when the only stresses are those due to gravity. The term *lithostatic pressure* is applied to such pressure effects.

In a truly hydrostatic condition the mean stress at a point is

$$\sigma_{\text{mean}} = \frac{\sigma_1 + \sigma_2 + \sigma_3}{3} = \frac{3\sigma_1}{3} = \sigma_1$$

When the stress conditions are nonhydrostatic but include a hydrostatic component, the mean stress is

$$\sigma_{\text{mean}} = \frac{\sigma_1 + \sigma_2 + \sigma_3}{3}$$

FIGURE 3-5 Schematic representation showing how two stress states can be added by conversion of both to the same coordinate system using the Mohr circle. (*After Brace, 1968.*)

TWO STRESS STATES

BY USE OF MOHR CIRCLE

TWO STRESS STATES SUPERIMPOSED RESULT

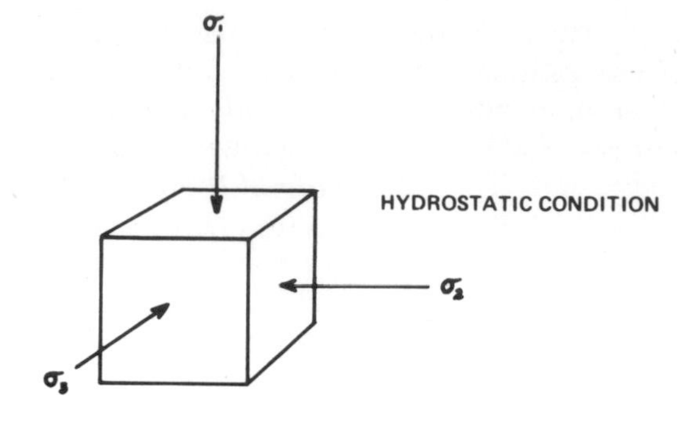

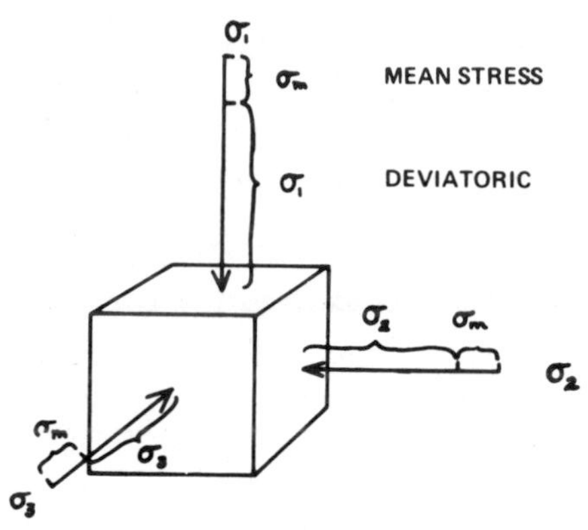

FIGURE 3-6 A hydrostatic stress condition in which $\sigma_1 = \sigma_2 = \sigma_3 = \sigma_{\text{mean}}$ (mean stress), top. A stress condition in which the principal stresses are shown divided into mean stress and deviatoric stresses.

The significance of this observation lies in the fact that uniform all-sided (confining) pressure alone does not appear to cause plastic deformation in crystalline materials such as rock unless the rock is highly porous. Once the porosity is reduced, little additional deformation can be attributed to confining pressure. (Note that confining pressure influences the behavior of rock but does not produce the differential pressures needed to cause deformation.) For these reasons, it is often useful to subtract the "hydrostatic" part of the stress from the total stress state, Fig. 3-6. The remainder is called the *deviatoric stress*.

Because "hydrostatic" does not contain shear components, the effects of hydrostatic or confining stress can be removed simply by subtracting that part from the normal stress components. When the stress state is not purely hydrostatic, the mean stress is taken as the hydrostatic component.

CONCEPTS OF STRAIN*

Materials respond to applied forces in three ways—change of position, change of orientation, and change of size or shape—and these taken together constitute what is meant by the total deformation. In its simplest form, change of position involves *pure translation* such that the internal structure of the material remains unchanged but the material as a unit is moved. Changes in orientation involve rotation of the material. The third type of strain refers to internal changes in the material which may involve *dilation,* change in volume and size, or distortion, change in shape.

Strains occur in the earth through a great range of scales, and the methods used to determine strain or total deformation must be appropriate to the scale of the deformation. Thus the efforts of the field geologist to accurately portray the shape of a folded and faulted sedimentary section that was originally horizontal may provide the basis for measurement of a large-scale strain, just as the analysis of distorted grains in a thin section does for finding the strain at a point.

Most of the mathematical theory of strain deals with *homogeneous strain,* in which all lines that were straight or parallel before strain remain straight or parallel after strain, Fig. 3-7. *Nonhomogeneous strain,* in which original straight lines become curved and values for the amount and type of strain vary, is difficult to deal with mathematically. Most strains in rock can be considered homogeneous if a sufficiently small volume of the material is examined. The strain of a large body of rock may be analyzed by considering the strain of many small parts of it and the way strain varies from point to point. Thus much of strain theory deals with the strain of small volumes of rock. The strain observed in a rock is a finite strain and is treated as the end product of a progressive deformation involving many small incremental changes called *infinitesimal strains.*

* See Bombolakis, *in* Riecker (1968), Jaeger (1969), or Ramsay (1967) for discussions of infinitesimal strain theory.

The various types of structural features commonly observed in strained rocks may conveniently be classed as follows:

Large-scale

Folds: Deflections of or changes of shape of an originally continuous marker (ideally a planar marker) such as a bed.

Faults: Displacements of rock masses across planes or zones (most evidently seen when involving displacement of a marker bed).

Massive solid-flowage features: Examples are found in the movements of salt and clay, sometimes forming features which intrude other rocks in piercement structures called diapirs.

Mesoscopic

Fractures: Planes in rocks across which the rock has lost cohesion, often due to brittle failure of the rock.

Cleavage and foliation: Closely spaced subparallel planes due to closely spaced fractures or to the alignment of mineral fabric.

FIGURE 3-7 Examples of homogeneous strain, in which straight lines remain straight, and nonhomogeneous strain, in which lines are distorted.

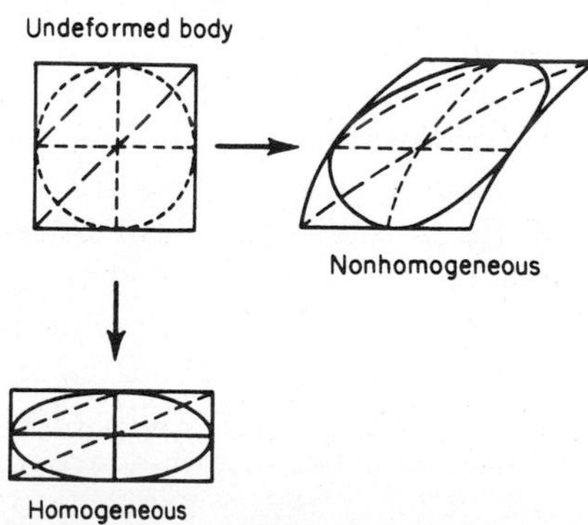

Rods: Pencil-shaped features (usually composed of quartz).
Mullions: Elongated oval (log-shaped) features found in some folded rock masses.
Boudins: Sausage-shaped features due to elongation of some layers more than the enclosing layers.
Mesoscopic folds: In a great variety of types.
Mesoscopic faults.

Microscopic

Flattened or elongated minerals: Especially evident in clays. (Some flattening and elongations in recrystallized rocks are due to syndeformational recrystallization.)
Slip within crystals.
Grain rotations.
Twinning of crystals (especially evident in calcite and dolomite).
Deformed oolites.
Deformed fossils.
Rotated crystals (notably rolled garnets and pyrite).

Deformation of rocks usually proceeds with changes in shape of the rock accomplished through changes in the fabric of the component parts of the rock mass. Many processes can be identified as contributing to these fabric changes. Notable among them are flattening and elongation of minerals which are ductile during the deformation; shattering or granulation of brittle components; rotation and rearrangement of grains in clastic rocks; microfracturing and microfaulting of grains (often important in the flattening process when the materials are relatively brittle); slips and twinning (which may develop within the crystal structure and are especially important in calcite); and recrystallization (which may proceed during deformation and may impart a new fabric as a result of pressure solution effects). All these processes result in creation of diagnostic fabrics.

The deformation of geological objects provides some of the best strain indicators. Deformation of oolites and spherulites has been especially successful because they are initially spherical (see Cloos, 1971). Fossils are valuable because their original shape is known, and deformed crystals are also useful (e.g., pyrite).

Translations

Translations can be determined only if the original position of the materials is known. Usually this is not known; but reconstructions may make it possible to estimate the translational deformation. As an example of extreme translational deformation, consider the reconstructions of predrift continental positions. Both amounts and rates of translation are being calculated on the basis of paleomagnetic studies.

At a smaller scale it is possible to estimate the amount of translation which is involved in the emplacement of some salt structures or other diapiric features in which the material has risen from a layer of known position. Similarly, the separations of boudins, Fig. 12-14, can be measured and afford a basis for estimating the translational component of the total deformation.

Translational movements are involved in all faults. Prefault reconstructions are frequently possible where the fault cuts a marker, and even in some instances of large-scale translation, reconstructions are attempted using bed thicknesses or facies relationships as illustrated, Fig. 23-9.

Rotational Deformation

Rigid rotations are involved in many types of deformation. Many faults involve some degree of rotation of the rock masses across the fault plane, resulting in a scissor-like movement across the fault. Such rotations can be determined from the rotation of marker beds cut by the fault. Essentially rigid rotations also make up a large component of the total deformation in many folds, especially folds that are angular, Fig. 11-10, or folds that are flattened, Fig. 11-15, so the limbs are rotated around the hinge.

At a smaller scale, crystals are valuable indicators of rotations; for example, pyrite, Fig. 3-8, which commonly occurs as cubes in black shales and limestones, may rotate as a rigid body relative to the enclosing rock, indicating the action of a shearing couple which is affecting the surrounding rock as well. Many such rigid rotations are accompanied by distor-

FIGURE 3-8 Pressure fringe on pyrite showing effects of a counterclockwise rotation of the pyrite relative to the enclosing rock.

FIGURE 3-9 Snowball garnet showing internal deformation caused by a counterclockwise rotation of the garnet, ×22. (*Courtesy Alan Spry.*)

tions. Rotations may be indicated by the presence of rolled garnets and by the reorientation of primary sedimentary features such as cross-bedding, Fig. 3-9.

Dilation

Because the original size of most geological objects varies through a considerable range, it is rarely possible to determine dilational changes. Even when the three principal stresses are unequal, a portion of the stress system acts as though it is hydrostatic; so dilational changes may be much more common than is generally recognized—especially when unconsolidated sediment is deformed. But for large strains in metamorphic and consolidated sedimentary rocks it is likely that the rock behaves as a nearly incompressible material.

Distortions

Distortions of the overall shape of deformed rock bodies result from a variety of types of displacement. These are conveniently classed as follows (for two-dimensional strains), Fig. 3-10:

1. Simple extension in one direction—uniextensional strain.
2. Extension in two directions at right angles—biextensional strain. (Pure shear is a special case in which no change in area occurs.)
3. Simple shear.
4. Superimposed shears and extensions.

Longitudinal Strains—Extension and Compression

In simple uniextensional strain, the amount of the strain e is the ratio of the change in length, $L_1 - L_0$, to the original length L_0:

$$e = \frac{L_1 - L_0}{L_0} = \frac{\Delta L}{L_0}$$

Generally when an object is extended in one direction it is shortened (negative extension) in a direction at right angles to this. Thus, transverse as well as longitudinal strain is produced. The ratio of these two strains, e_T/e_L, gives a value known as *Poisson's ratio* and is one of the elastic constants for the material.

Longitudinal strain may be produced by either compression or extension, and in both cases the forces are opposed by others acting along the same line. Biextensional strain may be produced by either compressing the material in one direction or extending it in a direction at right angles to the compression direction, or by some combination of these two.

Simple Shear

Simple shearing occurs when the forces acting do not act along the same line but cause parts

FIGURE 3-10 Top, uniextensional strain (the amount is equal to the ratio of the change in length to the original length); middle, biextensional strain. Poisson's ratio is the ratio of the transverse to the longitudinal strain; bottom. shear strain is measured by the tangent of the angle ϕ.

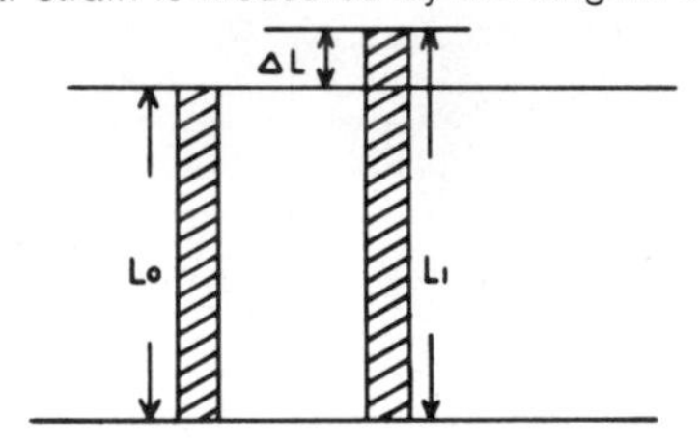

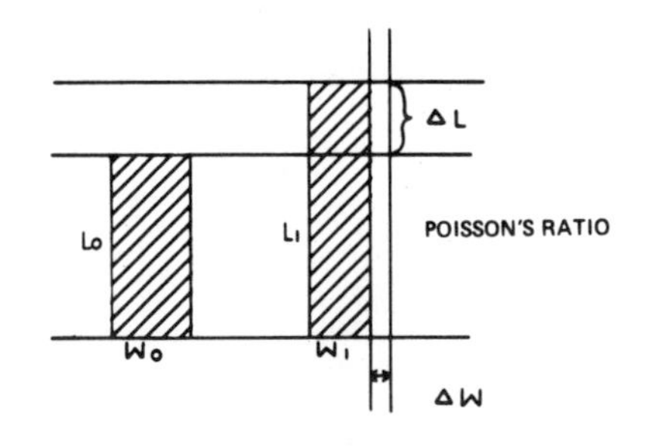

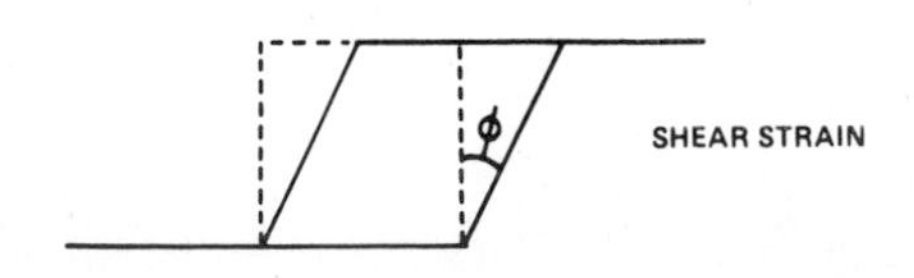

of the body to rotate relative to other parts Fig. 3-11. The angular rotation, which is a measure of the shear strain, can be determined if the change in angle between two originally perpendicular lines can be determined, Fig. 3-12.

Rotated pyrite and garnet crystals provide good examples of simple shear, because in both cases evidence of a rotational couple is clear. Pyrite, originally square in section, is often slightly distorted into a rhombic shape as well as rotated as shown by the pressure shadows. The deformed brachiopod, Fig. 3-13, on the other hand, appears sheared, but this strain could as easily have been caused by a longitudinal strain along the axis indicated.

The Strain Ellipse and Ellipsoid

The strain ellipse provides one of the most powerful tools for description of the homogeneous strain at a point. It can be shown mathematically that points originally on a

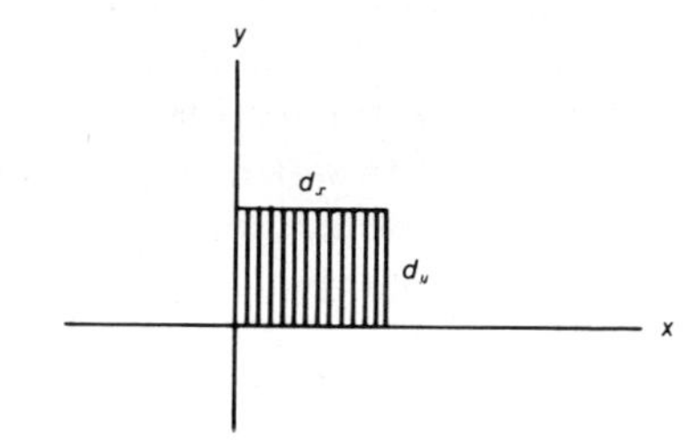

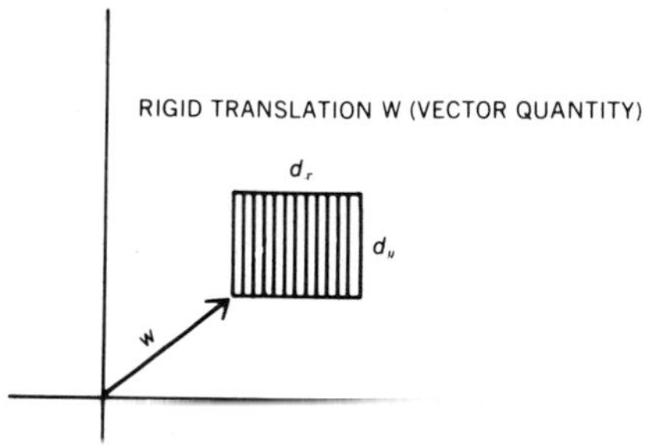

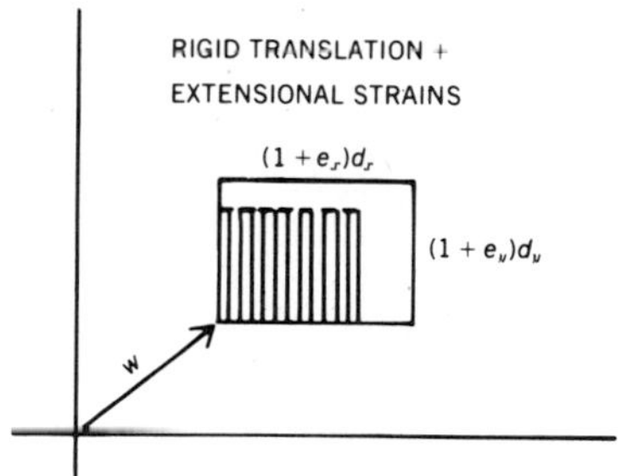

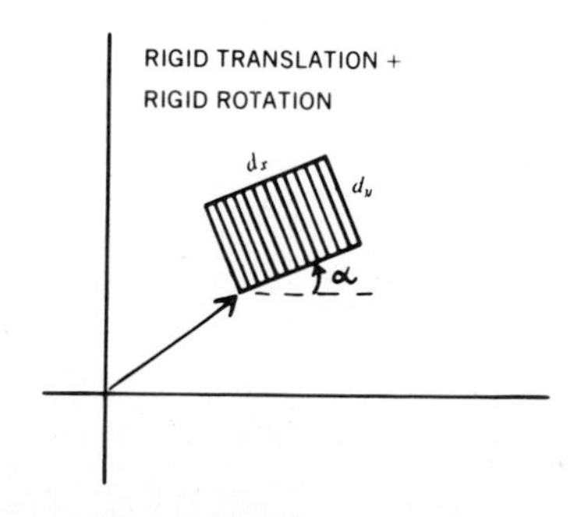

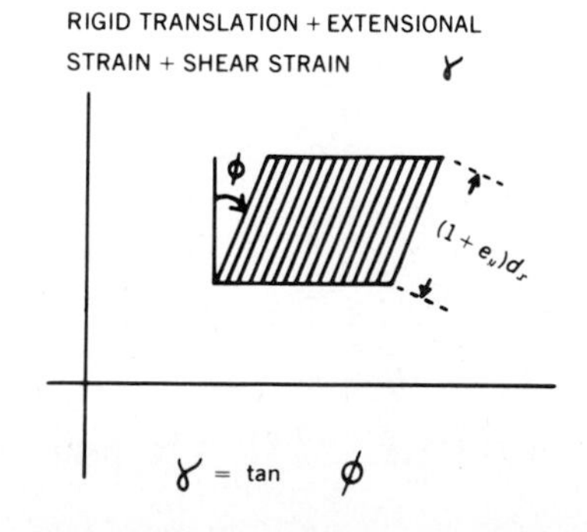

FIGURE 3-11 Schematic representation of translational and rotational movements and extensional and shear distortions.

circle before strain are transformed to points on an ellipse (for plane strain) in which the major and minor axes correspond to the maximum and minimum longitudinal strains and are called the *principal axes of strain*. In the three-dimensional case, a sphere becomes an ellipsoid in which there are three principal axes. These are parallel to and bear a reciprocal relationship to the axes of the stress ellipsoid.

The external form of a strain ellipse can be derived by pure shear and by dextral or sinistral simple shear, Fig. 3-12. Although the external shape of the strain ellipses deformed by pure and by simple shear are the same, the kinematics are significantly different. In simple shear the strained rhomb is shown with an external couple, but it is clear that movements within the ellipse have taken place by a process of slippage along closely spaced laminae. Note the offset of one diagonal marker and the lack of offset of the other diagonal marker. Compare this with similar markers in pure shear. The outer edges of the original square were subjected to a couple, but the outer boundaries of the square constrained internal rotations. By analogy simple shear is like laminar flow.

The strain ellipsoid is commonly used as a representation of the total (largely nonelastic) change of shape in a body after deformation, ideally illustrated by the deformation of spherical oolites, Fig. 3-14, to ellipsoids or deformation of pebbles or fossils the original shape of which is known.* The strain ellipsoid cannot be used to determine the exact stress history of the deformed rock, the manner of stress application, its duration, or its direction because a number of different types of stress

* See Elliott (1970) or Ramsay (1967) for discussion of measuring finite strain from oolites.

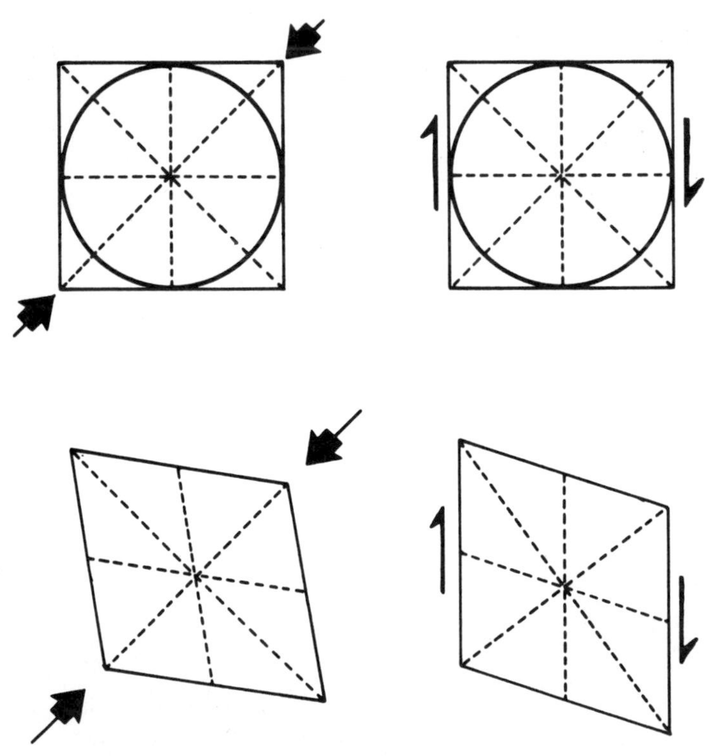

FIGURE 3-12 Modification of a reference figure deformed by pure shear, left, and simple shear, right.

application may produce the same strain form. The strain ellipsoid is valuable as a guide to visualizing the total strain and the relationships of various fabric elements of deformed rock.

From the foregoing, it should be clear that significant problems are involved in trying to deduce the stress system responsible for a particular strain. Unique solutions to this problem are rarely obtained. However, it is frequently possible to make a reasonable judgment of the original shape and position of a deformed body of rock by making use of elongated oolites, deformed fossils, rolled garnets, and deformed primary structures. The distinction drawn between pure and simple shear provides a useful guide to the manner of deformation. Some success has been achieved through petrofabric studies, in which it has been shown that the plane in

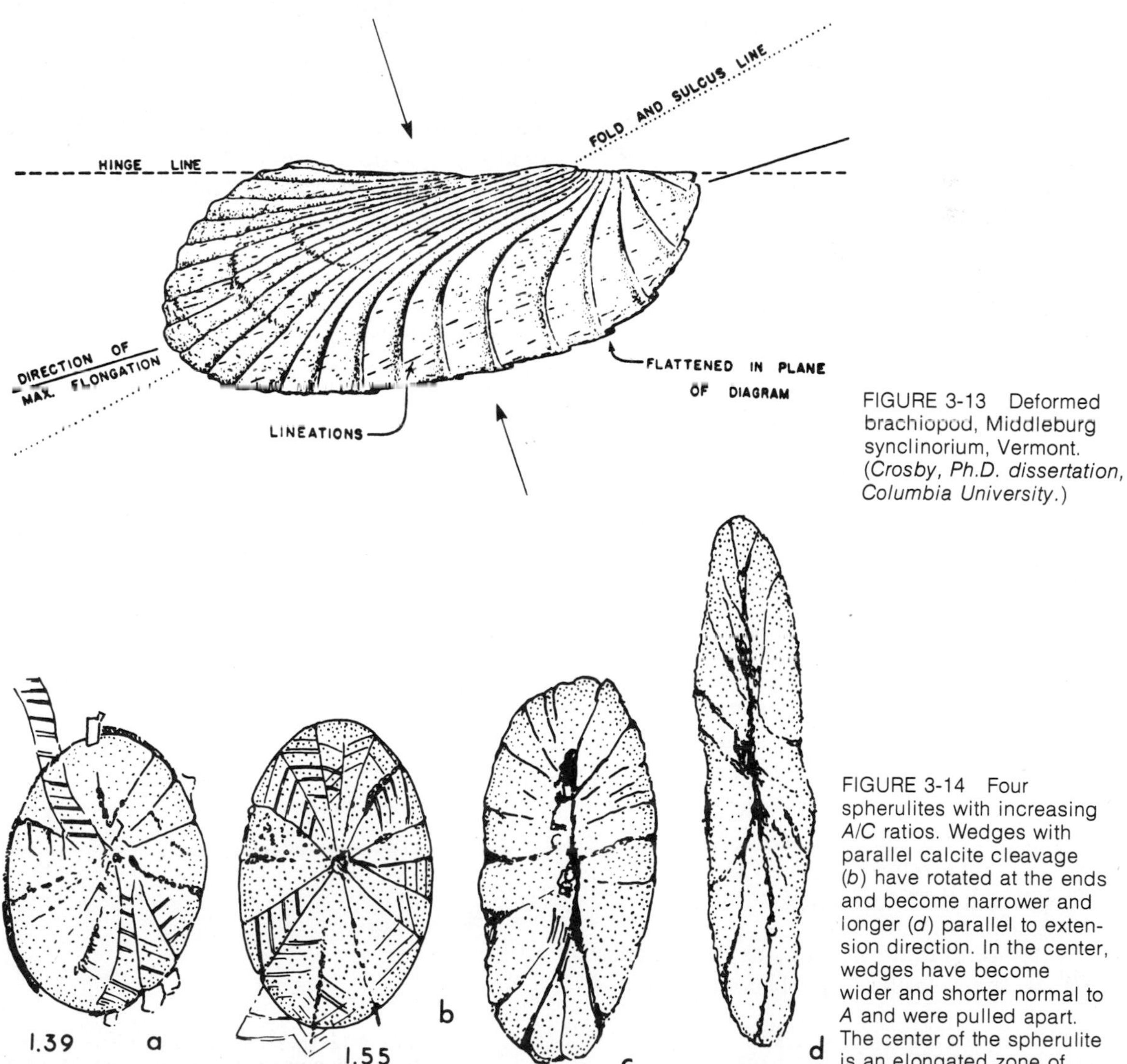

FIGURE 3-13 Deformed brachiopod, Middleburg synclinorium, Vermont. (*Crosby, Ph.D. dissertation, Columbia University.*)

FIGURE 3-14 Four spherulites with increasing *A/C* ratios. Wedges with parallel calcite cleavage (*b*) have rotated at the ends and become narrower and longer (*d*) parallel to extension direction. In the center, wedges have become wider and shorter normal to *A* and were pulled apart. The center of the spherulite is an elongated zone of debris. (*From Cloos, 1971.*)

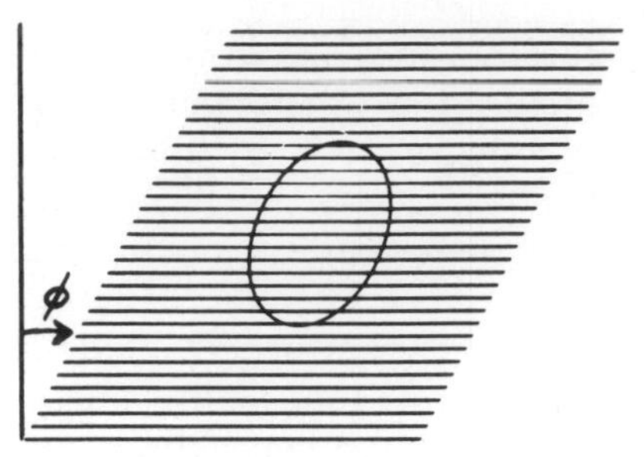

FIGURE 3-15 Simple shear produced by lateral movements on slip surfaces. Movements are like those produced in a stack of cards.

which slip occurs in crystals has a preferred orientation relative to the principal stress direction. Also, a positive relationship exists between the formation of certain types of fractures and the principal stress directions.

The strain ellipse provides a very useful device for the study of both the progressive changes in shape which accompany developing strain and the effects of superimposing one strain on another.* The progressive changes in form produced in simple shear deformation may be simulated by drawing a circle on the edge of a stack of cards which is then subjected to a simple shear, Fig. 3-15. As the cards are moved, the circle changes into an ellipse which becomes progressively flattened as shearing proceeds. Note that there are two mutually perpendicular diagonals in the unstrained circle which are still perpendicular in the final strained figure and that these are the principal strain axes. The lengths of the principal strain axes change systematically as the amount of shear increases. This makes it possible to determine the angle of shear by measuring the principal strains.

STRESS-STRAIN RELATIONSHIPS FOR IDEAL MATERIALS

The behavior of materials may be described in terms of the relationships between applied stresses and resulting strain and the way strain varies with time—both under carefully specified conditions of temperature, pressure, strain rate, and duration of load. Most of the early ideas grew out of studies in physics and engineering study of strength of materials. Later, as experimental work at higher temperatures and pressures became possible, the field was extended to include rocks and effects of long-term experiments. The evaluation of the time factor is of paramount importance in geologic considerations. From these studies a number of "ideal" models of material behavior have been established. Elastic, viscous, and plastic behavior are the simplest of these models; elasticoviscous, firmoviscous, and plasticoviscous models represent materials which combine characteristics of the first three.

Elastic Behavior

Robert Hooke described elastic behavior: strain in an elastic body is reversible (the strain disappears when the stress is removed) and the amount of strain is directly proportional to the load, Fig. 3-16. We may envision the process in terms of the work done in loading a spring. If no energy is dissipated during the loading process, then that energy may be

FIGURE 3-16 Typical form of stress-strain curve for deformation of most rocks at room temperature and atmospheric pressure. Behavior is elastic to the yield point and failure is by fracturing, as in all brittle materials.

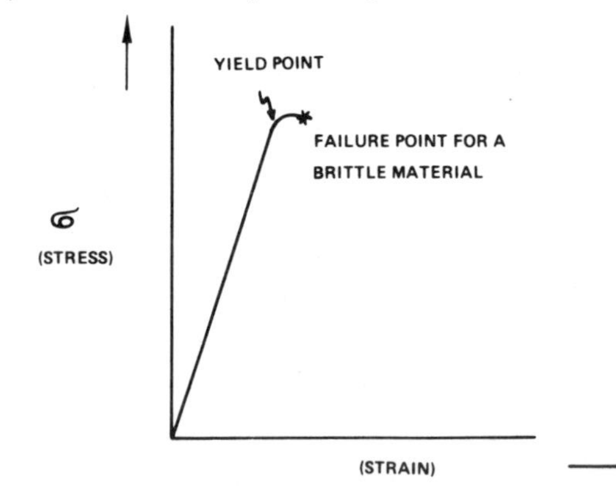

* Very interesting analysis of simple shear effects can be obtained by use of card decks as described especially by Ragan (1973) and O'Driscoll (1962, 1964).

thought of as having been transformed into elastic strain energy. This energy is then capable of doing work when the spring is released. An elastic material exhibits this reversible strain behavior, provided the load does not exceed a certain value (dependent on the material involved), called the *yield point*. Most rocks exhibit elastic behavior if deformed for relatively short periods and if strain does not exceed the elastic limit. If loads are applied above the yield point, strain is no longer proportional to stress—much larger strains occur with small increases in load, and at least part of the strain is a permanent deformation.

The elastic properties of a material may be described in terms of elastic moduli—Young's modulus, the shear modulus, and the bulk modulus—which are material constants.

Elastic moduli for several rocks are given in Clark (1966). These moduli are constant for isotropic materials, but rocks are rarely isotropic. The values, therefore, vary, depending on the fabric and structure of the rock and the direction along which it is deformed. Changes in values may result from variations in temperature and confining pressure also.

Stress-Strain Plots and Rock Strength

The concept of rock strength is easily understood in terms of the stress-strain plot. Two somewhat different concepts of strength are used. *Ultimate strength* is the maximum differential stress (deviatoric stress) that a given sample of rock can sustain without failure under specified conditions of deformation. In other words, ultimate strength is the highest stress obtained on a stress-strain curve for a material deformed until it fails. *Fundamental strength* is the greatest stress a material can sustain without continuous deformation (creep) over an essentially unlimited period of time; it cannot be determined from a stress-strain curve.

Viscous Behavior

Sir Isaac Newton formulated the concept of fluid viscosity. Mechanically it can be represented by the behavior of a fluid layer between two rigid plates. A tangential force is applied to the upper plate, causing the plate to move. Because the fluid adheres to both plates, the slip must occur in the fluid. The fluid is found to exert an internal frictional resistance, viscosity, to the movement; therefore, a force (a shearing stress) must be continually applied to keep the plate moving. For such fluids the strain is proportional to the stress (which is a simple shear) and to the time it acts, and inversely proportional to the viscosity η:

$$\epsilon = \frac{\sigma t}{\eta}$$

Movements within the fluid are represented by laminar flow or parallel shearing of the fluid on many fine laminae. Newton's law of viscosity states that the shearing stress applied to the plate is equal to the coefficient of viscosity (dynamic viscosity) multiplied by the shearing strain rate across the shear zone. The term *kinematic viscosity* is the ratio of the dynamic viscosity to the density of the fluid.

Newtonian fluids are those which exhibit the properties described by Newton. The coefficient of viscosity (dynamic viscosity) is independent of the velocity gradient or rate of application of the shearing stress, and the viscosity is constant for all rates of strain in true newtonian fluids. The viscosity of many fluids is found to be a function of the rate of application of the shearing stress. These are called *nonnewtonian* fluids. Most highly viscous materials are *nonnewtonian*. Variations in viscosity with strain rate are usually attributed to structural changes in the material during flow. Viscosity is particularly sensitive to temperature conditions for both types of fluids. A mechanical model for viscous flow is the dashpot.

Some materials exhibit unusual combinations of properties. Bentonite muds, for example, will flow under very low stress if they are first subjected to a sudden high stress. Otherwise they behave as solids. This type of behavior is called *thixotropy*.

Osborne Reynolds investigated the conditions under which different flow situations are dynamically similar. He showed that velocity of flow, V, diameter of a tube through which flow occurs, D, density of the fluid, ρ, and coefficient of viscosity, η, are related as follows:

$$\frac{VD}{\eta}\rho = \text{a constant (Reynolds' number)}$$

When the Reynolds' number reaches a certain value dependent on the fluid, laminar flow begins to break up into turbulent flow, but application of this idea to rocks suggests that turbulent flow will not occur even for the largest structural features.

Combinations of Elastic and Viscous Behavior

Characteristics of "ideal" elastic and viscous materials may be combined to produce other relatively simple behavioral models. The easiest way to envision these is in terms of their mechanical models, Fig. 3-17.

A Maxwell elasticoviscous substance (many waxes and pitches display this type of behavior) behaves as an elastic material for instantaneously applied stress, but if the stress is applied and held constant, the substance

FIGURE 3-17 Characteristics of several ideal bodies.

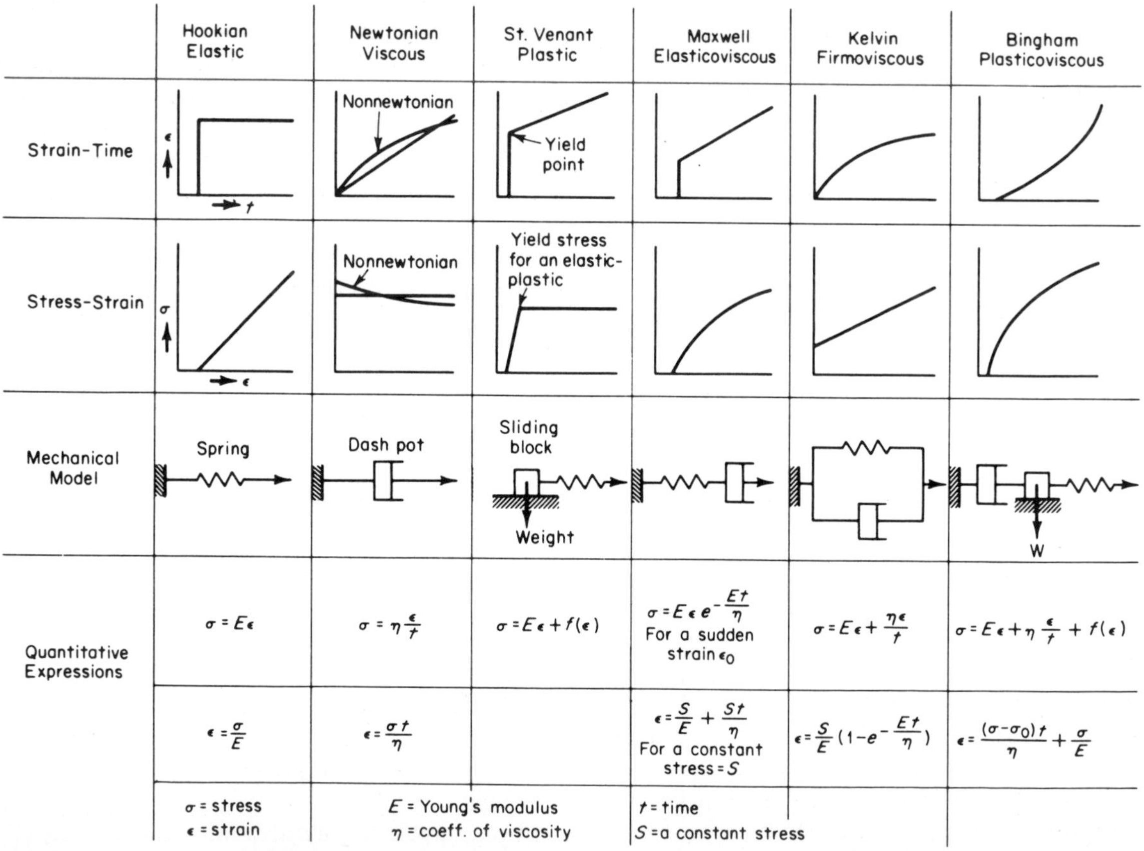

	Hookian Elastic	Newtonian Viscous	St. Venant Plastic	Maxwell Elasticoviscous	Kelvin Firmoviscous	Bingham Plasticoviscous
Strain-Time	ϵ vs. t	Nonnewtonian	Yield point			
Stress-Strain	σ vs. ϵ	Nonnewtonian	Yield stress for an elastic-plastic			
Mechanical Model	Spring	Dash pot	Sliding block; Weight			W
Quantitative Expressions	$\sigma = E\epsilon$	$\sigma = \eta\frac{\epsilon}{t}$	$\sigma = E\epsilon + f(\epsilon)$	$\sigma = E\epsilon e^{-\frac{Et}{\eta}}$ For a sudden strain ϵ_0	$\sigma = E\epsilon + \frac{\eta\epsilon}{t}$	$\sigma = E\epsilon + \eta\frac{\epsilon}{t} + f(\epsilon)$
	$\epsilon = \frac{\sigma}{E}$	$\epsilon = \frac{\sigma t}{\eta}$		$\epsilon = \frac{S}{E} + \frac{St}{\eta}$ For a constant stress $= S$	$\epsilon = \frac{S}{E}(1 - e^{-\frac{Et}{\eta}})$	$\epsilon = \frac{(\sigma - \sigma_0)t}{\eta} + \frac{\sigma}{E}$

σ = stress; ϵ = strain; E = Young's modulus; η = coeff. of viscosity; t = time; S = a constant stress

will yield continuously in the manner of a viscous substance after the initial elastic strain. The Maxwell body is represented by a spring and a dashpot connected in series. When stress is removed only the elastic component is recoverable. Any differential stress will produce continuous strain as in viscous substances. Stress relaxation is one characteristic of such materials. If the material is strained a certain amount by an applied stress, the amount of stress required to sustain the strain will decrease through time. The relaxation time is the time required for the stress to reduce to $1/e$ (e is the base of natural logarithms, 2.72) of its original value if the strain is held constant. Elastic substances theoretically will hold stresses permanently. The relaxation time is given by η/E, where E is Young's modulus and η is the coefficient of viscosity.

The Kelvin (viscoelastic or firmoviscous) substances are essentially solids, but they differ from elastic materials in that the strains are not instantaneous when stress is applied. Strain is taken up exponentially, and the time required for the strain to reach $1/e$ of its final value is called the *retardation time*. Strain also dissipates exponentially when the applied stress is released. The dashpot connected in parallel with the spring in the mechanical model of a Kelvin body acts to dampen the action of the spring.

Plastic Behavior

An ideal plastic body does not yield until a particular load, the yield stress, has been applied. Most materials that approach being plastic exhibit elastic characteristics (called *elastic-plastic*) below the yield point. Beyond the yield point a plastic material strains continuously; it flows, but it differs from viscous fluids which do not have a fundamental strength. Once plastic flow begins, it continues ideally without rupture as long as the yield stress is applied. Strain beyond the yield point is permanent. Some materials have a well-defined yield point at which an increase in deformation occurs without an increase in stress. More typically for rocks, the stress-strain curve continues to rise beyond the yield point but at a lower slope than is found in the elastic range. Such materials are said to show *strain hardening*.

Plastic deformation in metals and other crystalline solids is often accompanied by the development of physical discontinuities along which occur abrupt displacements called *slip lines*. They sometimes appear as orthogonal sets oriented at 30 to 45° to the maximum principal stress direction. Continuous internal changes occur during plastic deformation. Work is expended in producing new slip surfaces, in causing disorder in crystal lattices, etc., and this energy cannot be recovered as in elastic materials.

A generalized plastic material, also called *plasticoviscous* or *Bingham body*, may be represented by a sliding block, a spring, and a dashpot connected in series. The initial behavior is elastic, but as the level of stress is increased, the yield stress is reached and the material flows like a liquid. After release of applied stress, the elastic component is recovered but strain by flow remains permanent.

FIGURE 3-18 Strain-time curve for a series combination of a spring, dashpot, and Kelvin body. The top curve may be obtained by adding the lower three curves: (1) for elastic behavior, (2) for viscous behavior, and (3) for firmoviscous behavior. (*From Ramsay, 1967.*)

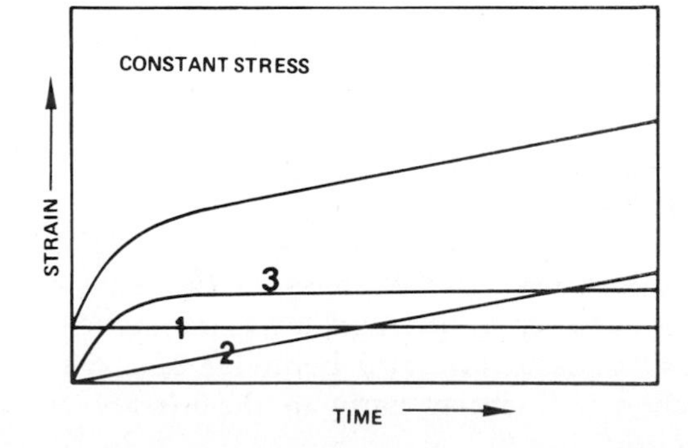

Rock Behavior and Composite Models

A number of attempts have been made to characterize the behavior of rocks in terms of some composite of ideal behavioral models. All agree that one such component is the ideal elastic material. It becomes more difficult to clearly identify other components. Ramsay (1967) suggests that the stress-strain curves of most rock experiments can be approximated most closely by a composite of the ideal elastic, viscous, and Kelvin elasticoviscous models, Fig. 3-18.

A more complex model is attributed to Andrade (1914) by Carey (1953). The components of the total strain in this model are: (1) a purely elastic component conforming closely to Hooke's law; (2) a time-dependent nonelastic nonrecoverable component (referred to as plasticity); (3) an elastic and viscous afterworking and "transient creep," which diminish with time, with or without some "work hardening"; and (4) a viscous component by which the strain increases at a slow but steady rate so that this part of the strain is directly proportional to the duration of the loading.

Total strain = elastic component
+ plastic component + transient creep
+ viscous strain

GLOSSARY FOR CONCEPTS OF STRESS AND STRAIN

I. General Terms and Stress

Affine deformation: Deformation that takes place in such a way that lines and planes that existed before deformation are transformed into new lines and planes, but these are not curved in the process.

Body force: A force acting on a material and proportional to the mass of the substance, e.g., gravity, centrifugal force, magnetic force.

Brittle behavior: The behavior of a material when it fractures early in the history of the stress-strain curve (total strain before fracturing is less than 5 percent). (Handin, 1966.)

Competent: A term applied to a bed or group of beds which, during folding, is able to lift not only its own weight but that of the overlying beds without appreciable internal flowage. (Willis.)

Confining pressure: An equal, all-sided pressure (i.e., geostatic) resulting from the load of overlying rocks.

Creep: The slow, permanent, and continuous deformation of a material under a constant load over a long period of time. (Nádai, 1950.)

Creep recovery: A process that occurs when a viscoelastic material is strained for a long period of time, after which the stress is removed; part of the creep strain will disappear or be recovered.

Deviatoric stress: That part of the stress system which deviates from the hydrostatic system and which causes distortion.

Ductile materials: Materials which can undergo very extensive plastic deformation before failing by rupture. (Turner and Verhoogen.) Handin (1966) uses total strain before fracture to approximate brittle-ductile behavior. A strain of 5 to 8 percent is considered moderately ductile, and over 10 percent is ductile.

Field: That area or space in which a given geophysical effect (stress, gravity, magnetism) occurs and is measurable.

Force: That which tends to put a stationary body in motion or to change the direction or speed of motion of a moving body.

Geostatic pressure: Syn. of **lithostatic pressure**.

Homogeneous strain: A state of strain in which the resultant form and orientation of the body is similar to the original form and orientation because the strain is the same at all points.

Ideal material: A hypothetical substance exhibiting distinctive physical properties as elasticity, plasticity, and viscosity.

Incompetent: A term applied to a bed that is

relatively weak and which thus cannot transmit pressure for any distance. (Structural Committee.*)

Impressed force: Syn. of **surface force.**

Isotropic substance: A material which has uniform mechanical properties in all directions in space. (Ant.: **anisotropic.**)

Kinematics: The study of strain based on the movement pattern without reference to force.

Lithostatic pressure: The vertical pressure at a point in the earth's crust equal to the pressure that would be exerted by a column of overlying rock or soil. (Syn.: **geostatic pressure.**)

Mohr circle: A graphic representation of the state of stress at a particular point at a particular time.

Normal stress: That part or component of a stress which acts perpendicular to a given surface. (Syn.: **normal traction.**)

Normal traction: Syn. of **normal stress.**

Plane stress condition: A state of stress in which two of the principal stresses are always parallel to a given plane and are constant in the normal direction.

Photoelasticity: The property of a transparent, isotropic solid by which it becomes doubly refracting under tensile or compressive stress.

Pressure gradient: The rate of variation of pressure in a given direction in space at a fixed time.

Principal stress directions: Three mutually perpendicular stress directions for which shear stresses are zero. Includes maximum, intermediate, and least stress directions.

Resultant force: The vector sum of several forces acting at a given point.

Rheid: A body of rock showing flow structure. (Carey, 1953.*)

Rheid folding: Folding accompanied by slippage along planes at an angle to the bedding or to earlier developed foliation. (Weiss, 1959.*)

* Refer to Am. Geol. Inst. © 1972, Glossary of geology and related sciences.

Rheidity: The capacity of material to flow within the earth. (Scheidegger, 1958.*)

Rheology: The science treating of the deformation and flow of matter. More specifically, it deals with the flow and permanent deformation of materials.

Shear stress: That part or component of a stress which acts tangentially to a given surface.

Stress: (1) Force per unit area, found by dividing the total force by the area over which the force is applied. (2) The intensity at a point in a body of the internal forces or components of force which act on a given plane through the point. As used in product specifications, stress is calculated on the basis of the original dimensions of the cross section of the specimen. (After Am. Soc. Testing Materials.)

Stress ellipsoid: A geometric representation of the state of stress at a point that is defined by the three mutually perpendicular principal stresses and their intensities.

Stress field: The state of stresses, either homogeneous or varying from point to point through time, in a given domain.

Stress trajectories: Lines on an arbitrary plane through a body showing the directions of the maximum and least principal stresses.

Surface force: A force which acts across an external surface of a body as a result of action and reaction between the body and another body with which it is in contact. (Syn.: **surface traction, external force, impressed force.**) (Turner and Weiss, 1963.)

Surface traction: Syn. of **surface force.**

Tectonite: (1) Any rock that owes its fabric to the summation of indirectly interrelated componental movements of the fabric elements, provided these movements take place in such a way that the spatial continuity of the rock is not impaired. (Knopf, 1938.) (2) A deformed rock the fabric of which is due to the systematic movement of the individual components under a common external force.

Tensile stress: A normal stress that tends to

cause separation across the plane on which it acts. (Ant.: **compressive stress.**)
Uniextensional strain: Distortional strain involving extension along one axis.
Vector: A graphic representation (eq., of force) showing magnitude and direction.
Yield point: A stress at which a body of a given material stressed in tension begins to stretch permanently.

II. Strain

Biextensional strain: Distortional strain involving extension along two axes.
Cold working: Syn. of **strain hardening.**
Dilation: A strain in which the change is purely one of volume.
Distortional strain: A strain which involves a change in shape.
Dynamics: The study of strain in terms of the stress fields which caused it.
Elastic strain: Strain which is instantaneously reversible, as when a body returns completely to its original unstrained state when the deforming stress is released.
Homogeneous strain: A state of strain in which the resultant form and orientation of the body is similar to the original form and orientation because the strain is the same at all points.
Infinitesimal strain: Material deformation involving elastic behavior in which small displacements and small strains of closely spaced elements in very small volumes occurs.
Plastic strain: The permanent strain or deformation developed in a material stressed beyond its elastic limit.
Principal directions of strain: The three mutually perpendicular axes of the strain ellipsoid which correspond to the directions of greatest, intermediate, and least strain. The axes are generally designated *A*, *B*, and *C* when *A* is the axis of maximum elongation and *A* greater than *B* greater than *C*.
Pure extension or compression: The phenomenon in which all points are displaced parallel to one coordinate axis. (Nádai, 1950.)
Pure rotation: Rotation of a rigid body.
Pure shear: A combination of pure extension in one coordinate direction and pure compression in another coordinate direction. (Nádai, 1950.)
Pure strain: Any general strain that lacks a rotational component.
Pure translation: A shift in position without rotation.
Simple shear: Shear that takes place when all points are displaced parallel to another coordinate axis. (Nádai, 1950.)
Strain: Any change occurring in the dimensions or shape of a body when forces are applied to it.
Strain ellipsoid: (1) In elastic theory, a sphere under homogeneous strain is transformed into an ellipsoid with this property; the ratio of the length of a line, which has given direction in the strained state, to the length of the corresponding line in the unstrained state, is proportional to the center radius vector of the surface drawn in the given direction. (Love.) (2) The ellipsoid whose half-axes are the principal strains. (Structural Committee.*)
Strain hardening: (1) A force-induced change in molecular structure of a crystalline material caused by bending or distortion and resulting in an increased resistance to further deformation. (2) The behavior of a material whereby each additional increment of strain requires an additional increment of differential stress. (Syn.: **work hardening, cold working.**)
Work hardening: Syn. of **strain hardening.**

III. Ideal Rheologic Substances

Bingham body: Syn. of **plasticoviscous substance.**
Elastic substance: A substance which behaves exactly as described by Hooke's law. Within the elastic limit, strain in the material is directly proportional to the applied stress. (Syn.: **Hookian solid.**)
Elasticoviscous solid: Syn. of **elasticoviscous substance.**

Elasticoviscous substance: In the ideal case, a substance which when stressed undergoes an instantaneous elastic strain that is followed by viscous behavior if the stress is continued. (Syn.: **Maxwell liquid, elasticoviscous solid.**)
Firmoviscous substance: A substance which has characteristics of viscous and elastic substances. Unlike elasticoviscous bodies in which elastic response to stress is instantaneous, firmoviscous substances require some time to react to the applied stress. (Syn.: **Kelvin body.**)
Hookian solid: Syn. of **elastic substance.**
Kelvin body: Syn. of **firmoviscous substance.**
Maxwell liquid: Syn. of **elasticoviscous substance.**
Newtonian fluid: Syn. of **pure viscous substance.**
Plastic substance: The ideal plastic material is one which deforms continuously when it is stressed above a certain critical yield stress.
Plasticoviscous substance: A material exhibiting a combination of the properties of plastic and viscous materials; a dense suspension of solid particles in a viscous fluid such that a stress must reach a certain finite or yield value before viscous flow can start. (Syn.: **Bingham body, viscoplastic material.**)
Pure viscous substance: A fluid characterized by purely viscous behavior such that the fluid deforms continuously and in a linear manner under any applied force. Most real liquids are not quite true newtonian fluids because they do not behave in a purely linear manner. (Syn: **newtonian fluid.**)
St. Venant body: Syn. of **plastic substance.**
Viscoplastic material: Syn. of plasticoviscous substance.

IV. Elastic Moduli and Effects

Bulk modulus: The bulk modulus B is the ratio of the increase in hydrostatic pressure P to the corresponding fractional decrease in volume $(-\Delta V/V_0)$:

$$B = \frac{P}{-\Delta V/V_0}$$

(Syn.: **volume elasticity, incompressibility modulus.**)
Compressibility: The reciprocal of the bulk modulus. It is defined as the fractional change in volume per unit increase in confining pressure.
Elastic aftereffect: A recovery effect in some materials, which undergo instantaneous elastic strain plus creep after being loaded for a time then recover the elastic strain immediately on unloading and slowly recover some of the creep deformation. (Syn.: **elastic afterworking.**)
Elastic afterworking: Syn. of **elastic aftereffect.**
Elastic limit: The stress at which a deviation from the straight-line relation on a stress-strain plot occurs. (Hooke's law does not apply strictly beyond this limit.)
Elastic modulus: The ratio of stress to the corresponding strain. It includes Young's modulus, the shear modulus, and the bulk modulus.
Hooke's law: A statement of elastic deformation that the strain is linearly proportional to the applied stress.
Incompressibility modulus: Syn. of **bulk modulus.**
Modulus of rigidity: Syn. of **shear modulus.**
Poisson's ratio: The ratio of the transverse contraction per unit dimension of a bar of uniform cross section to its elongation per unit length, when subjected to a tensile stress:

$$\sigma = -\frac{\Delta W/W_0}{\Delta L/L_0}$$

where σ = Poisson's ratio
W = transverse dimensions
L = length
The value is less than 0.5.

Relaxation: The phenomenon in which, if a spring is held in a stretched position for a sufficiently long time at high temperature, the force exerted by the spring will gradually start to diminish and may disappear entirely. (Nádai, 1950.)

Relaxation time: The time required for a substance to return to its normal state after release of stress.
Shear modulus: The ratio of a shearing stress to the corresponding shearing strain. (Syn.: **modulus of rigidity, torsion modulus.**)
Torsion modulus: Syn. of **shear modulus.**
Volume elasticity: Syn. of **bulk modulus.**
Young's modulus: The ratio of a tensile stress to the corresponding tensile strain. If a tensile force F applied across a cross-sectional area A produces a change in length ΔL in a bar of original length L, then Young's modulus is given by

$$Y = \frac{F/A}{\Delta L/L}$$

V. Strength

Cohesion: The maximum load a material can take impulsively (meaning that the load is applied so fast that no nonelastic deformation occurs) without rupture. (Carey.)
Fundamental strength: The stress a material is able to withstand, regardless of time, under any given set of conditions (temperature, pressure, solutions, etc.) without deforming continuously.
Practical strength: The maximum strength which may be sustained by a material within a definite time limit and within a specified sensitivity of measurement. (Carey.)
Strength of a material: (described qualitatively as resistance to failure): The force per unit area necessary to cause rupture at normal temperature and pressure conditions over a short period of time (crushing strength).
Ultimate strength: The highest stress a material attains on the stress-strain curve.

VI. Flow

Flow: Any deformation, not instantly recoverable, that occurs without permanent loss of cohesion. (Handin and Hager, 1957.)
Gliding flow: The type of solid flow which takes place by the combined mechanisms of translation and twin gliding.
Pseudoviscous flow: Load recrystallization. The type of solid flow which takes place under a strain and a stress too low to produce gliding flow, producing instead intergranular movement and dimensional orientation for the most part.
Reynolds' number: A number quantity used as an index to characterize the type of flow in a hydraulic structure in which resistance to motion is dependent upon the viscosity of the liquid in conjunction with the resisting force of inertia; i.e., it is the ratio of inertia forces to viscous forces.
Slip: Macroscopically discontinuous flow characterized by displacement on subparallel surfaces that pervade a rock. (Donath, 1963.)
Viscosity: A measure of the resistance of a fluid to flow. It may be thought of as the internal friction of a fluid. The coefficient of viscosity of a fluid can be measured by use of Stokes' law or with various types of viscosimeters.
Viscosity (nonnewtonian): Behavior in which the relationship of the shear stress to the rate of shear is nonlinear; i.e., flow of a substance in which viscosity is not constant.

REFERENCES

Andrade, E. N. C., 1914, Royal Soc. London Proc., v. A90; also 1911, v. A84.
Brace, W. F., 1960, Analysis of a large two dimensional strain in deformed rocks: Internat. Geol. Cong., 21st, Norden. v. 18, p. 261–269.
——— 1961, Mohr construction in the analysis of large geologic strain: Geol. Soc. America Bull., v. 72, p. 1059–1079.
——— 1968, Review of Coulomb-Navier fracture criterion, *in* Rock mechanics seminar, Riecker, R. E., ed., Terrestrial Sciences Laboratory Bedford, Mass.
Carey, S. W., 1953, The Rheid concept in geotectonics: Geol. Soc. Australia Jour., v. 1, p. 67–117.
Clark, S. P., Jr., ed., 1966, Handbook of physical constants: Geol. Soc. America Mem. 97.

Cloos, Ernst, 1971, Microtectonics: Johns Hopkins University Studies in Geology, no. 20.

Dixon, J. M., 1974, A new method of determining finite strain in models of geological structures: Tectonophysics, v. 24, p. 99–114.

Elliott, D., 1970, Determination of finite strain and initial shape from deformed elliptical objects: Geol. Soc. America Bull., v. 81, no. 8, p. 2221–2236.

——— 1972, Deformation paths in structural geology: Geol. Soc. America Bull., v. 83, no. 9, p. 2621FF.

Friedman, M., 1972, Residual elastic strain in rocks: Tectonophysics, v. 15, p. 297–330.

Ghosh, S. K., 1975, Distortion of planar structures around rigid spherical bodies: Tectonophysics, v. 28, p. 185–208.

Handin, J. W., and **Hager, R. V., Jr.,** 1957, Experimental deformation of sedimentary rocks under confining pressure—tests at room temperature on dry samples: Am. Assoc. Petroleum Geologists Bull., v. 41, p. 1–50.

Jaeger, J. E., 1936, Elasticity, fracture, and flow: London, Methuen.

——— 1969, Elasticity, fracture and flow with engineering and geological applications: London, Methuen; New York, Wiley.

Love, A. E. H., 1944, A treatise on the mathematical theory of elasticity: New York, Dover.

Mohr, Otto, 1914, Abhandlugen aus dem Gebiete der technischen Mechanik, 2d ed.: Berlin, Ernst.

Nádai, A., 1931, Plasticity: New York, McGraw-Hill.

——— 1950, Theory of flow and fracture of solids, 2d ed.: New York, McGraw-Hill.

O'Driscoll, E. S., 1962, Experimental patterns in superposed similar folding: Jour. Alberta Soc. Petrol. Geol., v. 10, p. 145–167.

——— 1964, Cross fold deformation by simple shear: Econ. Geology, v. 59, p. 1061–1093.

Oppel, G., 1961, Photoelastic strain gages: Experimental Mechanics, v. 1, p. 65–73.

Ragan, D. M., 1973, Structural geology: An introduction to geometrical techniques, 2d ed.: New York, Wiley.

Ramberg, H., 1975, Particle paths, displacement and progressive strain applicable to rocks: Tectonophysics, v. 28, p. 1–37.

Ramsay, J. G., 1967, Folding and fracturing of rocks: New York, McGraw-Hill.

Reiner, M., 1943, Ten lectures on theoretical rheology: Rubin Massada, Jerusalem.

Riecker, R. E., 1968, NSF advanced science seminar in rock mechanics for college teachers of structural geology, June 26–July 28, 1967: Rock mechanics seminar, v. 1.

Roberts A., 1964, Progress in the application of photoelastic techniques to rock mechanics: Rock mechanics symp., 6th, Rolla, Mo., p. 606–648.

Tan, B. K., 1973, Determination of strain ellipses from deformed ammonoids: Tectonophysics, v. 16, p. 89–101.

Timoshenko, S. P., and **Gere, J. M.,** 1961, Theory of elastic stability, 2d ed.: New York, McGraw-Hill.

Varnes, D. J., 1962, Analysis of plastic deformation according to Von Mises' theory, with application to the South Silverton area, San Juan Co., Colo.: U.S. Geol. Survey Prof. Paper 378-B, p. B1–B49.

Voight, Barry, 1967, On photoelastic techniques *in situ* stress and strain movement, and the field geologist: Jour. Geology, v. 75, no. 1, p. 46–58.

Whiten, E. H. Timothy, 1966, Structural geology of folded rocks: Chicago, Rand McNally.

4

Experimental Study of Rock Deformation

Great progress has been made in experimental and theoretical structural geology in recent years. Already the results of these studies are being applied in the field, but this phase of structural geology is in its early stages. Experiments can be run under carefully controlled conditions, and simplifying assumptions can be made in theoretical analysis, but the range of conditions and the large number of variables in natural situations make the application of theory and experiment difficult and often problematical.

Much of this chapter is devoted to showing how the behavior of selected rocks has been studied, to outlining the way in which the variables encountered in nature influence the behavior of these rocks, and to pointing out some of the textural and structural features which may be expected to be associated with certain conditions.

ANALYSIS OF ROCK DEFORMATION

The large number of variables involved poses a significant problem in the theoretical analysis of rock deformation. Variations in rock type, thickness, original structural configuration, and physical conditions under which deformation occurs lead to a high order of possible combinations. The scientific approach to this problem is to break it into its simplest component parts and evaluate each one. Thus we proceed from deformation of single crystals to small statistically homogeneous specimens of crystalline aggregates. These are deformed in tension, compression, torsion, under a variety of temperature and confining pressure conditions, in the presence of fluids, under various amounts of directed stress, and for differing amounts of time. Ultimately, we hope to derive general conclusions which can be applied to explain field observations.

Unfortunately the nature of the conditions favoring deformation of solid rocks, particularly the high confining pressures required and the long times involved, make experimental deformation of large bodies of rock impossible. Almost all experiments have been run on specimens no more than 2.5 cm in diameter and 5 cm long, but these are sufficient to reveal the change in fabric which results under specific conditions of deformation. Similar fabrics may be found in nature, and the character of the deformation inferred by analogy.

Mechanisms of deformation, faulting, fracture, flow, etc., can be inferred from the observed rock fabric, and these mechanisms are interpreted in terms of the stress field that acts within the rock body. Recent experimental and theoretical studies have put this method of analysis, known as *dynamics*, on a much firmer footing. Other subsidiary approaches are used in the problem of analyzing rock deformation. The kinematic approach proceeds from the observed fabric and determination of the geometry and symmetry of the fabric to a plan of movement which took place during deformation. In addition, stress analysis, as it has been developed in engineering materials studies of metals and to a lesser degree of ceramics and polycrystalline aggregates, is used. The stress analysis in a flexed elastic metal bar or plate is an example, but this is extended to include treatment of plastic and pseudoviscous materials under various types of lateral restraint and with layers enclosed in media of different kinds. Still another approach is through the use of model studies in which features of naturally deformed rock can be reproduced in scale models which have the significant value of helping us visualize what may happen, even though the materials used in the scale models are not rocks and their analogy to nature is uncertain.

In summary, the state of geologic analysis of large-scale rock deformation involves a combination of studies of overall structural form, of detailed examination of rock fabric which is interpreted in terms of our experience with the fabric of experimentally deformed rocks, of theoretical stress analysis of idealized configurations and materials, and of the behavior of simplified models.

A Review of Environmental Variables in the Earth

The environment in which rock deformation proceeds varies with depth because confining pressure, pore pressure, and temperature all increase with depth. Other conditions vary in a more irregular way, depending on geological conditions.

The principal stress on a body of rock being deformed is actually composed of three components:

Total max. stress	=	confining pressure	+	pore pressure	+	differential stress
σ_1	=	P_c	+	P_p	+	S_1

Confining pressure depends on the density of the overlying rock. It is least in unconsolidated sediments and greatest under consolidated sediments and crystalline rocks. The pressure gradient due to confining pressure is in the range of 200 to 300 bars/km (1 to 1.4 psi/ft). Pore pressure in porous and permeable materials has a gradient of 100 bars/km (0.5 psi/ft), and temperature increases on the average at a rate of 25 to 30°C/km in the crust and upper mantle. The rate at which rock becomes strained, called *strain rate*, is largely a function of the applied stress and the way it is applied and is expressed as the amount of strain per unit time. Strain is a ratio so it has no units. Strain rates in the earth lie in the following ranges:

Meteorite impact	10^3 to 10^6 sec^{-1}
Fracture (cataclasis)	10^{-1} to 10^3 sec^{-1}
Laboratory test	10^{-1} to 10^{-8} sec^{-1}

Intergranular glide	10^{-12} to 10^{-18} sec^{-1}
Pleistocene crustal rebound	10^{-14} sec^{-1}

Methods of Experimental Deformation

Frank Adams, one of the early students of experimental deformation, used equipment that consisted essentially of a nickel-steel cylinder with a 2-cm-diameter bore in the center and thicker walls at the ends. Specimens were cut to fit the bore or were embedded in some other material, pistons were inserted, and a load was applied to the pistons by a press. The sides of the cylinder acted to confine the specimen laterally, and the thickness of the walls of the cylinder was varied in successive experiments to change confining pressures. This arrangement was versatile, allowed variations of temperature, confining pressure, directed stress, and presence of fluids, and the results were good. The main disadvantages came about through boundary conditions arising from the contact of the specimen with the walls of the enclosing cylinder, and neither temperature nor confining pressure could be measured with precision.

These problems have been largely overcome in modern equipment such as that illustrated in breakaway view, Fig. 4-1. The sample, still small (1 to 2 cm in diameter; 2 to 4 cm long), is surrounded by a fluid in which confining pressure is controlled. The specimen is usually subjected to both confining pressure and an axial load, called *triaxial tests.* The specimen does not touch the walls of the "bomb," and temperature is controlled and measured. The equipment shown is for studies in axial compression; other types are used to apply torsion and tension.

Tests may also be run with the specimen encased in a rubber or soft metal jacket which isolates the specimen from the fluid in which confining pressure (P_c) is built up. Fluids may be injected into the specimen filling the pore spaces, and pore pressure (P_p) may be measured and controlled independently.

The axial load on the test specimen is transmitted through a hydraulic jack. Among the most frequent types of experiments are:

1. An axial differential stress is built up rapidly on the specimen and held constant as the specimen deforms. Strain is then measured as a function of time—a *creep test.*

2. The differential stress is applied in such a way that the rate of strain is constant, and changes in the applied stress are then plotted against shortening or elongation—a *stress-strain test.*

3. Constant differential stress is applied and the rate of strain is measured. The results are plotted as differential stress versus strain rate—a *strain-rate test.*

The usual procedure in designing an experiment is to control all variables except one and observe the effects of changing that variable. Variables involved in the deformation of rocks in the earth fall into two well-defined groups—those related to the environmental conditions under which deformation takes place and those related to the rock:

A. Environment of deformation

1. Confining pressure (P_c)
2. Pore pressure (P_p)
3. Amount of directed stress(S)
4. Manner of application of stress (compression, tension, torsion)
5. Chemically active solutions
6. Strain rate
7. Time
8. Temperature

B. Nature of the rock

1. Chemical and mineralogical compositions
2. Grain size, shape, orientation

3. Homogeneity
4. Anisotropy—effects of cleavages, bedding, etc.
5. Porosity and permeability
6. Intergranular cohesion and cementing

Of these, the time factor, although it may be the most important, is the most difficult to treat experimentally, for experiments are of short durations while diastrophism involves periods of perhaps millions of years.

DEFORMATION MECHANISMS IN SINGLE CRYSTALS

Most rocks are crystalline aggregates. The deformation of an aggregate depends on grain sizes, boundary shapes, and cement, as well as on the character of the individual mineral components. Yet, important insights to the behavior of the aggregate may be obtained by studying the characteristic deformation mechanisms of the individual crystals.

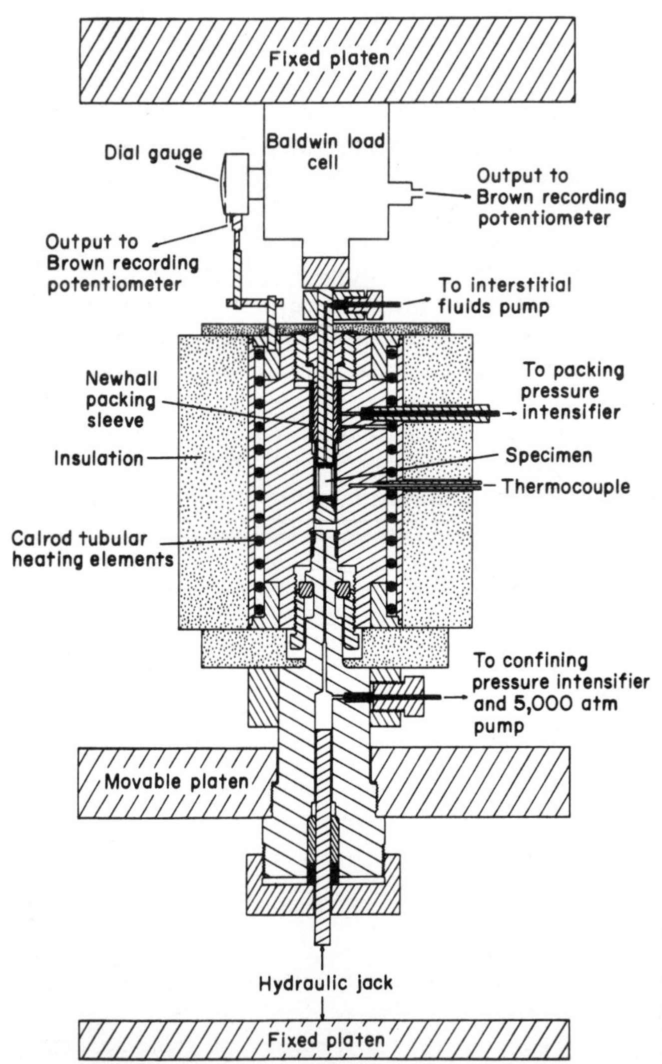

FIGURE 4-1 Schematic drawing of one of the types of presses used in experimental rock deformation. Temperature, confining pressure, and directed stress can all be varied with this press. It is designed for use in compression experiments. (*From Griggs, Turner, and Heard, 1960.*)

Deformation of crystals depends on the atomic configuration of the crystal structure, the strength of bonds holding atoms together, the presence of intracrystalline defects, and the conditions under which the deformation proceeds control the mechanism of deformation. Prominent among the mechanisms of deformation are the following:

1. *Fracture:* Loss of coherence across a plane within the crystal.

2. *Granulation:* Crushing of the crystal under conditions which do not allow openings to form.

3. *Slip:* Displacements produced by intracrystalline slip. Sometimes the slip is restricted to a plane or a set of parallel planes. In other instances a distinct band (kink) or lamella is formed within which rotation relative to material outside the band occurs. (Syn.: *slip band, deformation band, deformation lamella.*)
a. *Translation gliding:* A form of intracrystalline slip in which displacement along some lattice direction occurs without loss of cohesion.
b. *Twin gliding:* A slip mechanism by which the crystal is mechanically twinned.
c. *Kinking (kink bands):* A type of deformation band in which the lattice is changed or deflected by gliding or slip along slip planes. Most kink bands in crystals are characterized by abrupt bends at the boundary of the bands.

4. *Partings:* The breaking of a crystal along a plane of weakness caused by deformation such as a twin plane.

Translation Gliding

The atomic structure of most minerals is such that the density of atoms is greater in certain planes through the crystal than in others. If a crystal like that in Fig. 4-2 is compressed or

FIGURE 4-2 Translational slips formed in a single crystal of metal deformed by extension and without any external confinement.

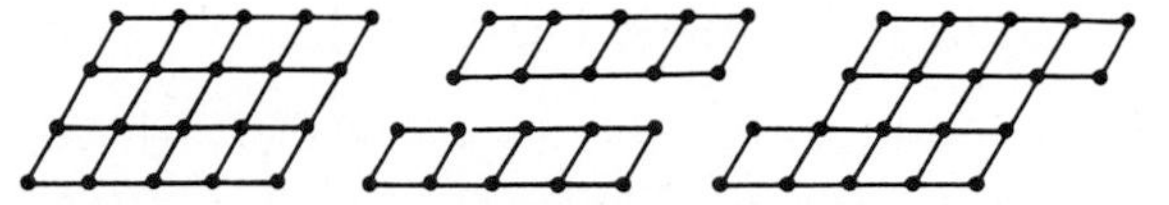

FIGURE 4-3 Schematic representation of a translational slip within a close-packed crystal. Centers of atoms are shown before, during, and after the slip.

extended, the crystal tends to deform by slip along planes of high atomic density. When the sides of the crystal are unrestricted, the slip planes go completely across the crystal, displacing the outer boundary of the crystal, Fig. 4-2. Slip of this type, called *translational gliding,* occurs when the energy supplied through shearing stress along a potential slip plane is great enough to lift the atoms in one plane out of their position in the spaces they occupy in the next lower sheet. This happens when the shear stress along the potential slip plane reaches some critical value, usually dependent on temperature and strain rate. Translation may result in movements of the sheets amounting to one or more interatomic distances, Fig. 4-3. A large number of closely spaced parallel slips is likely to occur.

Several potential planes of translation gliding occur in some crystals; for example, the planes containing high densities of atoms in a cubic crystal are illustrated, Fig. 4-4. The density of atoms is not uniform in all these planes, and for this reason the tendency for a slip to occur in all directions is not equal. The direction of movement within a plane and the relative sense of movement may be restricted by crystal structure. If a given crystal contains a number of planes of potential slip, slip is most likely to occur on the planes which are oriented in such a way that the shearing stress components on them are highest.

Bend gliding, Fig. 4-5, is a type of translation gliding in which the gliding planes undergo elastic deformation before the slip occurs. This occurs when the axis of loading is either parallel or perpendicular to the glide surface.

Effects of Crystal Defects

Translation gliding can be studied by deforming large rafts of similar-size bubbles in which bubbles simulate atoms and surface tension approximates atomic bonds. Gliding often takes place by movements in planes that contain defects which arise from the presence of a partially complete plane of atoms in the structure, producing a linear defect along the edge of the extra plane of atoms. Point defects occur where an atom is missing or where extra atoms or atoms of a different size occur, Fig. 4-6.

Gliding may consist of a simple shift of a whole plane of atoms over those below, but it is much more common in the bubble-raft experiments for a dislocation to "run" along the slip plane so that only a few of the bonds need to be broken at a time. Sheets of marbles may be used to visualize movements in gliding. Marbles in the upper sheet initially located in spaces between marbles in the lower sheet must be forced to rise up and over the top of a stable position. Bubble rafts differ in that the bubble is not rigid; shapes can be deformed and the movements can take place as a progressive process. Dislocations may occur in several ways as illustrated, Fig. 4-7.

Twin Gliding

A second important gliding mechanism is that accomplished through twinning by which the crystal structure of the mineral becomes essentially reversed on opposite sides of the twinning plane. The significant difference between translation gliding and twinning is found in the sense and amount of movement and in the rotation of crystallographic structure. All the layers of atoms between glide planes are displaced by the same amount in simple translation gliding, but each layer of atoms above the twin plane is displaced more than the one below in twin gliding, Fig. 4-8. In most minerals which show twin gliding, the crystal is divided by twin gliding planes into a number of parallel lamellae.

Inferring Stress Direction from Glide Systems

Gliding by translation or twinning occurs when the shearing stress directed parallel to

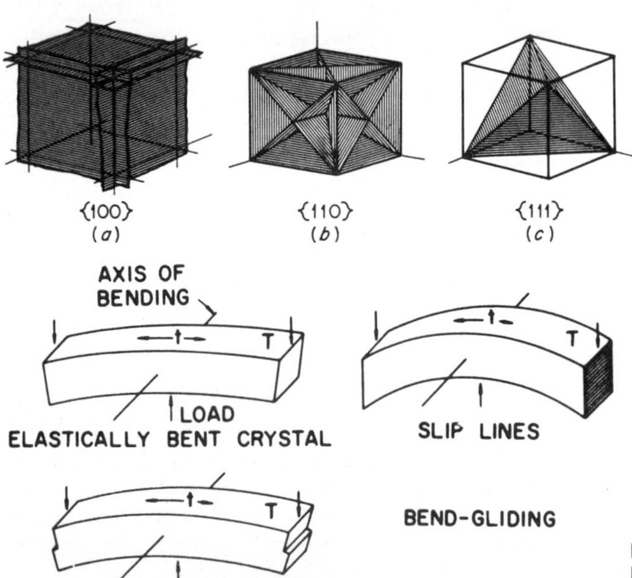

FIGURE 4-4 Planes of possible slip in a cubic crystal. (*a*) Three (100) planes; (*b*) six (110) planes; and (c) four (111) planes. Planes with higher indices are not shown.

FIGURE 4-5 Model of bend gliding. The crystal bends in a way analogous to that of a flexed deck of cards. (*From Friedman, 1964.*)

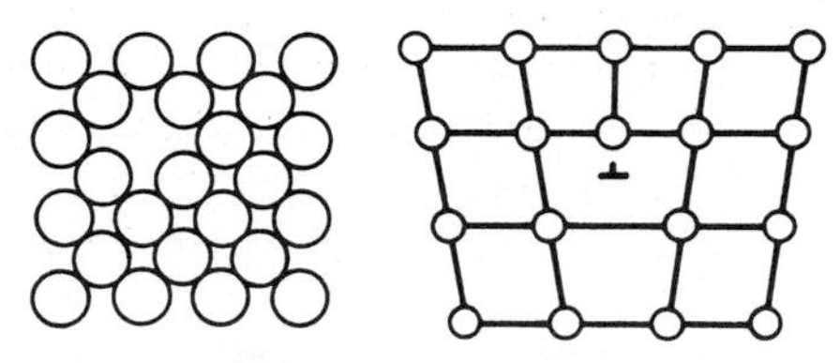

FIGURE 4-6 Examples of crystal defects. A line defect due to an extra row of atoms (right), and a point defect due to a missing atom (left).

the glide plane and in the right direction within that plane exceeds some critical value. This is most likely to happen when the component of shearing stress on the plane is a maximum. Friedman (1964) outlines the stress conditions in experimentally deformed calcite which are most favorable to gliding as follows, Fig. 4-9:

1. The intermediate stress direction lies parallel to the potential glide plane and perpendicular to the gliding direction.

2. The maximum stress direction is inclined 45° to the potential glide plane and is oriented in such a way as to produce the correct sense of shear on the glide plane.

3. The minimum stress direction is oriented at 45° to the glide plane and in the plane containing the maximum stress direction.

The idea that mineral and rock fabric can be directly related to stress conditions accompanying the formation of the fabric has been the basis for much work in petrofabrics. The study of gliding in crystalline aggregates has shown promise of being a fruitful approach to the problem of determining stress conditions. In studying an aggregate, it is necessary to analyze the fabric statistically. This is generally done by use of the stereographic projection to plot poles to twin planes and other fabric elements.

Although many crystals exhibit both translation gliding and twinning, only a few common rock-forming minerals display the properties. Of these, calcite and dolomite have been studied most thoroughly, but gliding systems for a large number of other crystals are known (Clark, 1966).

Kinking and Kink Bands*

The boundaries of most experimentally and naturally deformed crystals are restrained by adjacent crystals or by the specimen jacket. Under this type of restraint the gliding surfaces often show abrupt changes in orientation called *kinks* which usually appear as bands, called *kink bands*, which are bounded by surfaces which cut across the active glide planes. During deformation the material within the band is rotated relative to that outside the band, which may remain relatively undeformed. Rotation within the band is initially about an axis that lies in the glide plane and normal to the glide direction. As deformation proceeds, the material outside the band is rotated (externally) in the opposite direction from the internal sense of rotation along the glide surfaces, Fig. 4-10. The local strain from these rotations is:

$$S = \cot \alpha + \cot \beta$$

* The behavior of single crystals of biotite and crystalline aggregates containing large percentages of mica has been reviewed by Borg and Handin (1966).

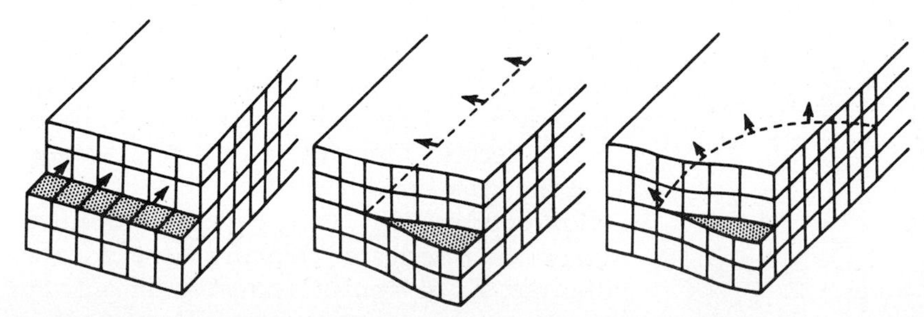

FIGURE 4-7 Slip within crystals may be accomplished by the propagation of a dislocation through the crystal. Three types of shear dislocations are illustrated, an edge dislocation (left), screw dislocations (center), and loop with edge and shear components (right).

where S is the shear strain and α and β are the angles between the gliding plane and the boundary measured within the band and outside the band, respectively (Turner, 1962; Borg and Handin, 1966).

Kinking is found in many types of crystals, but it is particularly well displayed in single crystals of mica and in rocks with a high percentage of micaceous minerals such as slate, phyllite, schist, and gneiss, Fig. 4-11.

Two types of ideas have been advanced to explain the development of kinks. According to the migration development hypothesis (Paterson and Weiss, 1961), kinking starts with gliding parallel to the foliation. A narrow kink band forms and the folia within the band are rotated. Additional shortening may then be accommodated by lateral migration of the band boundaries with no change in angle between glide surfaces in and outside the band or by further rotation of the foliation within the band.

FIGURE 4-8 Schematic drawing of twin gliding in crystals. (*a*) Atomic planes above the twin plane (K_1) have been sheared through a constant angle to a mirror reflection of the crystal structure below K_1. (*b*) Twin-gliding elements and conventions for angles viewed with the aid of a strain ellipsoid. (*From Carter and Raleigh, 1969.*)

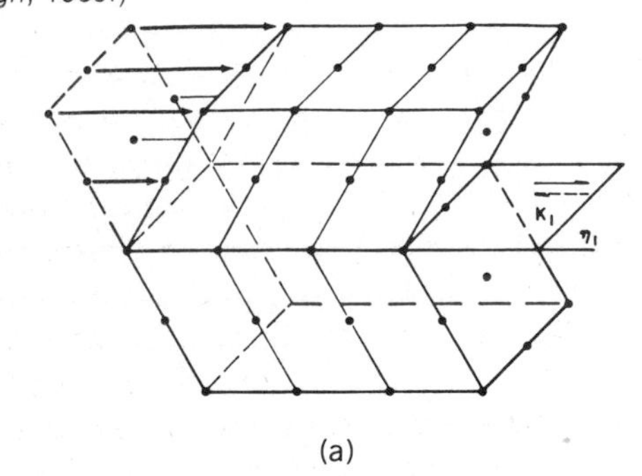

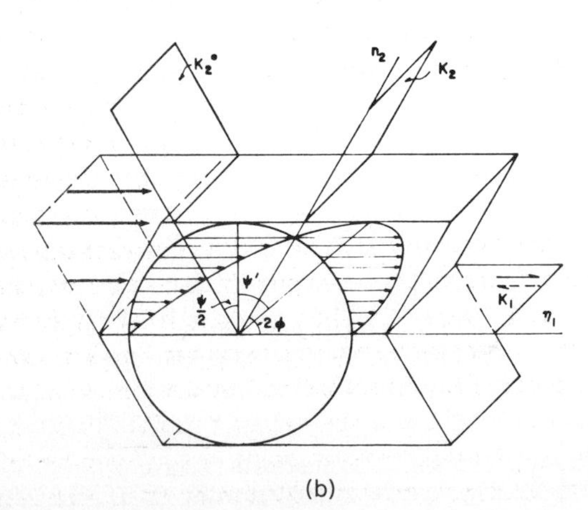

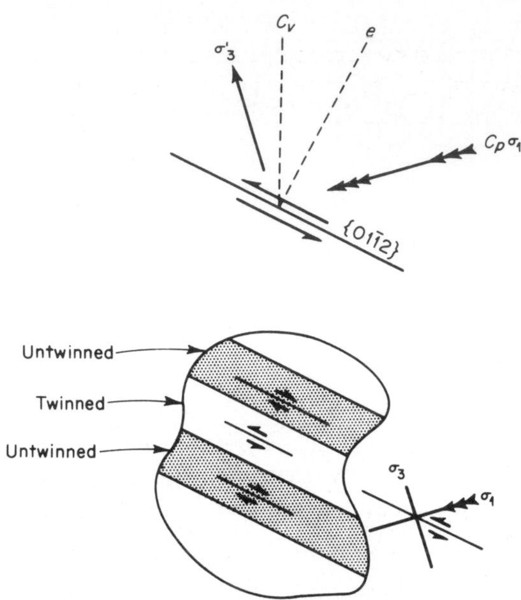

FIGURE 4-9 Relations between principal stress directions and twin gliding, showing maximum compressive stress σ_1' and maximum extension σ_3' axes oriented to produce glide on *e*. Section is normal to *e* plane and contains glide direction and optic axis, C_v. A section through a twinned calcite crystal is shown below. (*After Friedman, 1963.*)

An alternate theory of kinking is that the foliation is rotated within *fixed* kink-band boundaries (Donath, 1963) and that the kink boundary does not migrate as the kinking proceeds. Initial deformation is slip on the foliation which leads to instability that causes kinking. As deformation continues, foliation within the band is progressively rotated until a limiting orientation is reached, at which stage additional rotation is impossible. Then either a second kink forms or a fault forms along the kink boundaries.

Kink-band width varied with confining pressure in a series of Donath's experiments with slate. Wide bands (5 mm+) occur only at low pressures (200 bars or less), and band width becomes progressively narrower as confining pressure is increased. Thus band width may be a useful field indicator of depth of deformation, causing kinking.

Deformation Mechanisms of Calcite*

Calcite exhibits both twin and translational gliding, slip lamellae, partings, and kink bands, but the dominant mechanism of deformation is twin gliding, Fig. 4-12, on the plane designated as $e = \{01\bar{1}2\}$. This occurs whenever the orientation of the crystal is favorable for movement on this plane. At low temperatures, calcite is much stronger when the crystal is unfavorably oriented for this gliding to occur; at higher temperatures the difference disappears; so we should expect calcite-rich rock to flow much more readily at high temperatures.

Deformation Mechanisms of Quartz

The behavior of quartz is significant because it is one of the most common rock-forming minerals. Quartz lacks the closely spaced planar arrangements of atoms which make translation and twin gliding so common in other crystals, but quartz is deformed in other ways, Figs. 4-13 to 4-15. If quartz is deformed so that the crystal structure is bent, it exhibits undulatory extinction when viewed in polarized light. The margins of quartz grains are granulated in some specimens, quartz is commonly fractured in strongly deformed rocks, and deformation lamellae are visible in some deformed quartz. Deformation lamellae occur with three orientations in quartz deformed experimentally at high confining pressures (20 kbars) and high temperatures (500°C). The most common orientation of deformation lamellae is parallel to the *a* axes, a weaker orientation occurs at angles of 20 to 60° to *a*, and a third is subparallel to *c*. Lamellae are often accompanied by development of kinks in the slip planes. Slip on basal planes is most common among those grains, oriented in such a way that the base is a plane of high shearing stress.

* Refer to Turner, Griggs, and Heard (1954); Bell (1941); Friedman (1964); and Carter and Raleigh (1969).

FIGURE 4-10 Schematic drawings of four types of kink bands in crystals. (*a*) Kink bands in crystals loaded parallel to strong planar anisotropy. (*b*) Intersecting conjugate kinks loaded parallel to a strong planar anisotropy. (*c*) Symmetrical kink in crystal whose slip plane (T_1) is in an orientation of high shearing stress. (*d*) Asymmetrical kink in crystal whose slip plane is in an orientation of high shearing stress. (*From Carter and Raleigh, 1969.*)

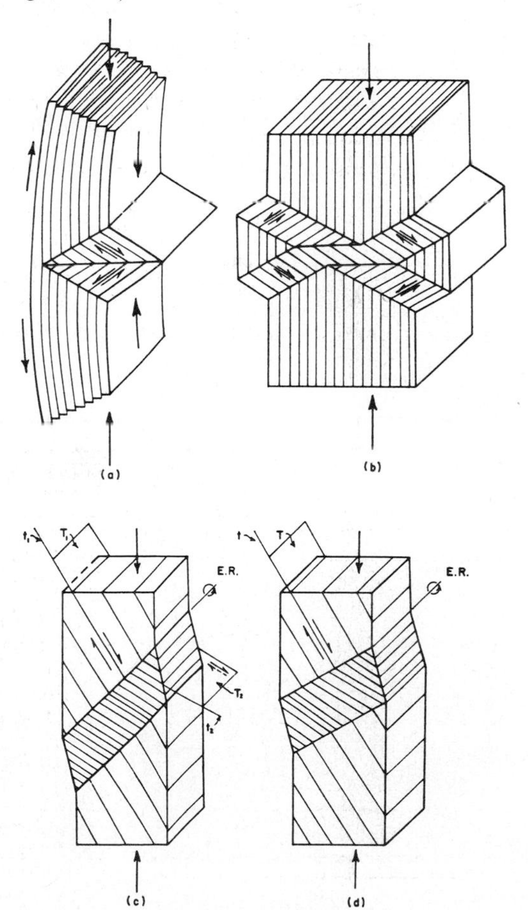

Deformation Mechanisms in Other Crystals*

Studies of several other common rock-forming minerals may be summarized as follows:

Dolomite: Deforms by translation gliding on the basal plane parallel to an *a* axis at low temperature and pressure. Favorably oriented crystals deform by basal slip at 5 kbars confining pressure, and at 300 to 400°C, twin gliding becomes important. Principle stresses can be determined from twinned dolomite as with calcite. One important discovery is that dolomite is much stronger and more brittle than limestone under comparable conditions (Handin and Fairbairn, 1955).

Olivine: Deforms by slip but on different planes, depending on the temperature and strain rate (Raleigh, 1968).

Pyroxene: Diopside deforms by slip and by twinning at moderate to high temperatures and 5-kbar confining pressure. It may also be used to determine stress orientations (Raleigh and Talbot, 1967).

* See Carter (1969).

Orthopyroxene: Undergoes translation gliding but does not mechanically twin.

Plagioclase: Can be twinned mechanically, but because twins are also growth features in plagioclase it cannot be used for stress analysis in tectonites. Kink bands have also been observed in deformed plagioclase.

Mica: Translation gliding and development of kink bands in biotite have already been described.

DEFORMATION OF CRYSTALLINE AGGREGATES*

One of the principal differences between single-crystal and aggregate deformation is the role played by the grain or individual crystal boundaries. Grain boundaries in most crystalline rocks form a mosaic pattern sep-

* Refer to Griggs and Handin (1960) and Heard, Borg, and Carter (1972) for a comprehensive discussion and a list of references.

FIGURE 4-11 Kinks in naturally deformed crystals (biotite at left and plagioclase at right). The scale bar is 0.1 mm long. (*From Carter and Raleigh, 1969.*)

arating domains in which the atomic configuration is different. Even in monomineralic rocks the crystallographic axes are likely to be oriented in different directions—some favorable for one or another of the glide systems, others not. Grain boundaries in clastic rocks are even more prominent unless the rock is tightly cemented, especially with silica. Although boundary effects are important, greater emphasis is placed on the bulk behavior, the stress-strain relations, of rocks under various conditions of temperature and confining pressure. These data are summarized for a number of common rocks in a plot of temperature versus compressive strength, Fig. 4-16.

The use of experimental results in interpretation of microstructures in the field is discussed by Tobin and Donath (1971) and Donath, Faill, and Tobin (1971).

TEMPERATURE AND CONFINING PRESSURE EFFECTS

Deformation of Marble and Limestone*

The deformation of aggregates of calcite crystals has been studied extensively with specimens taken from the Yule marble and the Solenhofen limestone. The results are most conveniently shown in graphs, Figs. 4-17 to 4-19.

Increasing confining pressure on limestone has the following effects both for extension and compression:

1. It raises the differential stress sustained before rupture.
2. It raises the differential stress required to produce a given amount of strain. (*Note:* This effect is

* Griggs and Miller (1951); Griggs and Handin (1960); Heard (1960).

FIGURE 4-12 Section of a specimen of Yule marble deformed by extension at 600°C and 3 kbars confining pressure. Note how the specimen necked near the center. An enlarged picture of this necked portion (right) shows how the calcite crystals have lost their original equidimensional shape and become drawn out as a result of twinning. Compare with the less deformed ends of the specimen. (*From Griggs, Paterson, and Heard, 1960.*)

true for temperatures less than 400°C and confining pressures less than 3,000 atm. At higher temperatures and pressures, less stress is required to produce a given amount of strain.)
3. It increases the amount of strain which takes place before rupture. In most experiments rupture takes place before the specimen is strained more than 10 percent, but at very high confining pressures, rupture may not occur.

Increasing temperature on limestone has three main effects both in extension and compression tests:

1. It increases the amount of strain at a given differential stress.
2. It increases the amount of strain before rupture.
3. It decreases the maximum amount of differential stress the specimen will stand before rupture.

The qualitative effects of increasing temperature and confining pressure are the same for extension and compression. Limestone's behavior changes from that of a brittle material to that of a ductile material* as temperature and confining pressures are increased. When the transition from brittle to ductile behavior is shown as a function of confining pressure and temperature, Fig. 4-20, the difference in behavior of the limestone deformed in extension and in compression becomes clear. Much higher confining pressures are required to obtain ductile behavior of materials

* The terms *competent* and *incompetent* are deeply ingrained in the literature of structural geology. Competence was used by Willis (1923, p. 149) as follows:

> In order that any stratum shall be competent it should possess certain inherent characteristics in a degree superior to that in which they are possessed by other strata. These qualities are (*a*) strength to resist shear; (*b*) capacity to heal fractures; (*c*) inflexibility. On the other hand, the conditions which favor incompetence of strata in folding are (*a*) lack of coherent strength; (*b*) lack of cementing quality; (*c*) flexibility. . . . Weak beds are often incompetent to lift any appreciable part of the confining pressures. They perforce move passively. Strong beds, on the other hand, are competent to move the passive beds and carry up the weight of overlying strata.

The terms competent and incompetent are still in use, but they are gradually being replaced by more precisely defined terms such as *ductility,* which is defined as the total percent deformation of a rock under given conditions before fracture.

FIGURE 4-13 Photomicrograph (bright-field illumination) of a set of several parallel northeast-trending kink bands. Bands containing abundant fractures at high angles to their boundaries are relatively less deformed. Lamellae are faintly visible in the clear, more highly deformed bands. Scale line beneath photo represents 0.1 mm. (*From Christie, Griggs, and Carter, 1964.*)

deformed by extension. The difference in behavior of limestone deformed by extension and compression can also be seen by comparing the stress-strain curves, Figs. 4-18 and 4-20. Lower differential stresses are required to produce any given amount of strain in extension. The ductility of a number of common rocks is easily compared in Fig. 4-21.

Experimental Deformation of Sandstone

Sandstone differs from marble and limestone not only in that most sandstone is composed of quartz, which is not as prone to gliding mechanisms as calcite, but also in the physical differences of individual grain boundaries.

The results of experiments by Borg and others (1960) with dry, disaggregated sand from the St. Peter sandstone are illustrated, Figs. 4-22 and 4-23. The sand is not cemented, so grain-boundary movements take place much more readily than in marble, etc. Fracturing is the dominant process of deformation, Fig. 4-24, and the fractures are oriented at small angles to the principal stress direction. Most fractures are formed normal to the axis of extension when the specimens are extended and are presumably tension fractures. It is remarkable that the fracturing shows a preferred orientation related to the applied stress system, because in sand the grains are in contact with other grains only at point contacts, producing a complex system of stresses which one might expect to produce a more random fracture pattern.

Grains show a tendency to be rotated in most extension tests if the grains are slightly elongated. Their long axis tends to become rotated until it is parallel to the direction of greatest extension. Quartz grains also tend to become oriented so that the optic axes assume a preferred alignment relative to the applied stress.

The deformation of sand is often characterized as cataclastic flow which consists of:

1. Fracturing by shear and tension
2. Rotation of grains
3. Eventual fragmentation of the grains

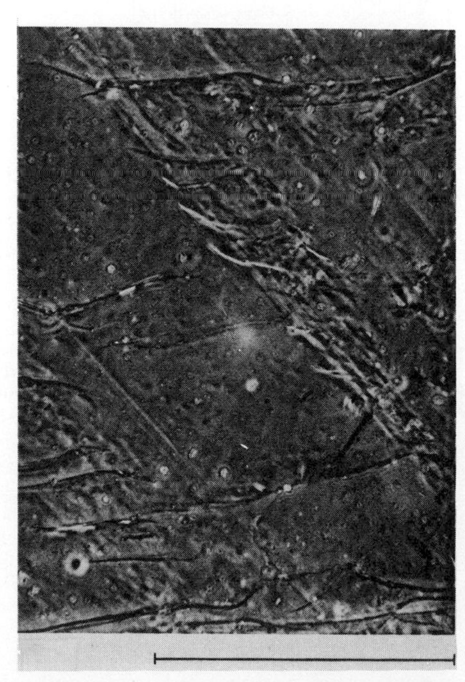

FIGURE 4-14 Slip bands seen on the polished surface (top) and deformation lamellae (bottom) of a quartz crystal compressed (normal to *r*) top to bottom, at 500°C, 20 kbars confining pressure, shortening about 1 percent. East-west cracks formed on unloading. (*From Christie, Griggs, and Carter, 1964.*)

Fracturing is extremely important in dry tests of sandstone, but grain breakage becomes progressively less important as pore pressure is increased because high pore pressure reduces internal friction.

Friedman (1963) examined deformed calcite-cemented sandstone and found that calcite deforms primarily by twin gliding and that quartz, feldspar, and garnet grains deform primarily by fracturing. Both the fractures and the twin lamellae are directly related to orientations of the principal stresses, Fig. 4-25.

Deformation of Igneous and Metamorphic Rocks*

A comparison of the behavior of crystalline rocks deformed under conditions simulating shallow depths of 3.5 km with others deformed under conditions of 18-km depth has been made by Borg and Handin (1966). Most rocks are brittle and largely elastic before failing by shear fracture at shallow depths. Only

* See Carter and Avé Lallemant (1970) for discussion of dunite and peridotite.

FIGURE 4-15 Quartz crystals compressed perpendicular to (01$\bar{1}$1). Compression direction *P* and orientation of the crystals are shown beneath the photographs. The large horizontal cracks are extension fractures produced on unloading the samples. Width of the specimens is approximately 3 mm. (*a*) Deformed at 400°C and 23 kbars confining pressure; sample was shortened 10 percent. Broad northwest trending bands are zones of undulatory extinction produced by bending the structure. (*b*) Deformed at 750°C and 22 kbars confining pressure; sample was shortened 15 percent. Deformation bands oriented subparallel to the *c* axis pervade the sample. (*From Carter, Christie, and Griggs, 1964.*)

schist faults under these conditions without loss of cohesion, and schist is also the weakest of these rocks, exhibiting an ultimate compressive strength of 2 kbars, compared with 3.3 kbars for granite and 8.2 kbars for granodiorite. Fine-grained rocks are stronger than their coarse-grained equivalents, and in general, rocks composed of quartz and feldspar are stronger than more mafic rocks.

The high temperature and pressure conditions at great depth result in a variety of modes of deformation. Nonmicaceous anisotropic rocks (amphibolite and peridotite) are brittle, and foliated rocks containing 10 percent or less mica are brittle or transitional. Of the richly micaceous rocks, schist is ductile regardless of its orientations, but slate is ductile only when the cleavage is inclined at high angles to the maximum principal compression.

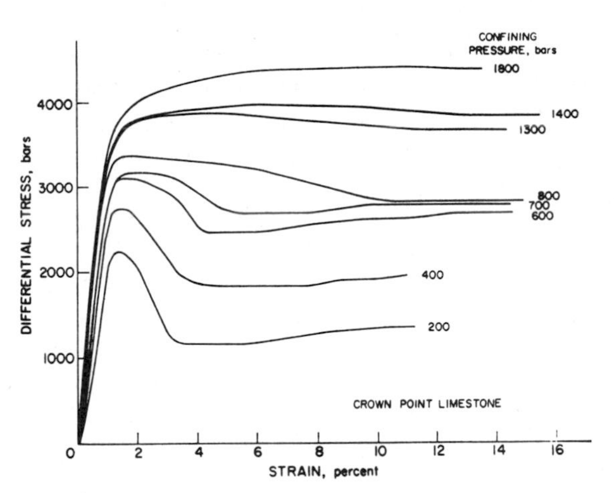

FIGURE 4-17 Effect of increased confining pressure on sustained differential stress (strength) of Crown Point limestone. Curves of differential stress versus longitudinal strain show the increase in strength caused by increased pressure and a change in character as the deformation changes from brittle at low confining pressure to ductile at high pressure. (*From Donath, 1970.*)

FIGURE 4-16 These graphs show the way compressive strength of various rocks and minerals varies as temperature changes. (*From Griggs and Handin, 1960.*)

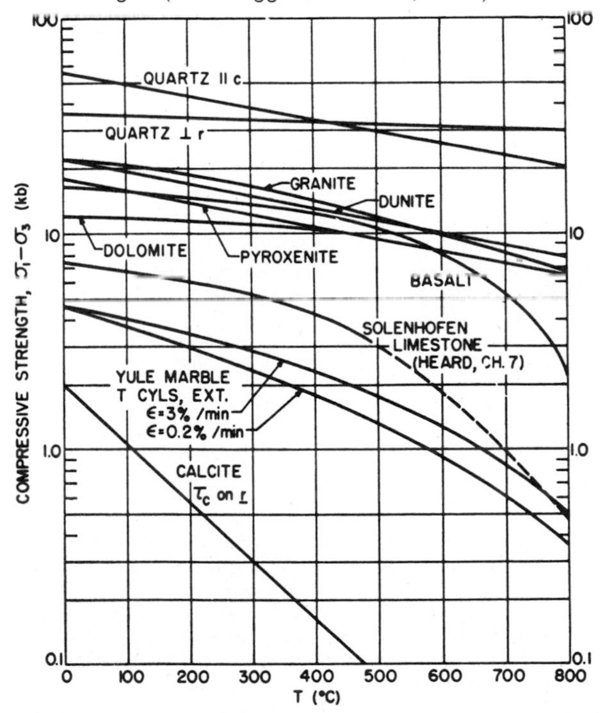

Shortening in the brittle rocks (generally shallow conditions) is due to offset along faults which are sharp breaks, often granulated, and generally cut through grains. The transitional rocks deform by shear or development of a broad zone of faults in which shear is distributed, and some intracrystalline flow and kinking of mica and pyroxene occurs. Ductile behavior (typical of deeper conditions) is characterized by intragranular flow in kink bands due to bend gliding in mica plus intergranular shear along *s* surfaces of properly oriented micaceous grains. Faulting occurs but only after large permanent strains.

Effects of Pore Pressure

Water is trapped in intergranular pore spaces in sediments during deposition and is absorbed in colloids such as clay. Some water is forced out during compaction and cementation, but much of it remains, and fluids may

TABLE 4-1
BEHAVIOR OF ROCKS AND STRENGTH WHEN COMPRESSED AT 5 kbars CONFINING PRESSURE, 500°C

State	*Rock type*	*Strength, kbars*
Brittle	Diabase	5.5
	Diorite	7.1
	Granite	8.3
	Rhyolite	10.5
	Amphibolite (normal to *s*)	11.4
	Biotite gneiss (normal to *s*)	11.3
	Granite gneiss (normal to *s*)	11.5
	Migmatite (normal to *s*)	9.9
	Peridotite (normal to *s*)	4.0
Transitional	Gabbro	8.2
	Pyroxenite	6.4
	Fordham gneiss (parallel to *s*)	8.8
	Migmatite (parallel to *s*)	6.7
Ductile	Fordham gneiss (normal to *s*)	4.3
	Slate (parallel to *s*)	6.3
	Schist (normal to *s*)	6.1

Source: After Borg and Handin (1966).

be present in pore spaces or as inclusions in igneous and metamorphic rocks as well. Fluids promote recrystallization and affect the mode of deformation of the rock in which they are trapped. The pressure in the pore fluids, called *pore pressure,* may become great even close to the confining pressure when impervious beds prevent the free movement of the pore fluids out of a formation. Pore pressure is important in fracturing and faulting and in the formation of certain sedimentary structures.

Experimental studies of pore-pressure effects are performed by placing specimens in impermeable jackets so that pressure can be built up in the pore spaces, P_p, independent of the confining pressure P_c surrounding the specimen. When this is done the pore pressure reduces the effects of the confining pressure so that

$$\text{Effective confining pressure} = P_c - P_p$$

Pore-pressure effects depend on the rock having sufficient permeability to ensure pervasion and uniform distribution of pore pressure. These conditions are fulfilled for sandstone and porous limestone but not for dolomite, shale, siltstone, limestone, or most igneous or metamorphic rocks.

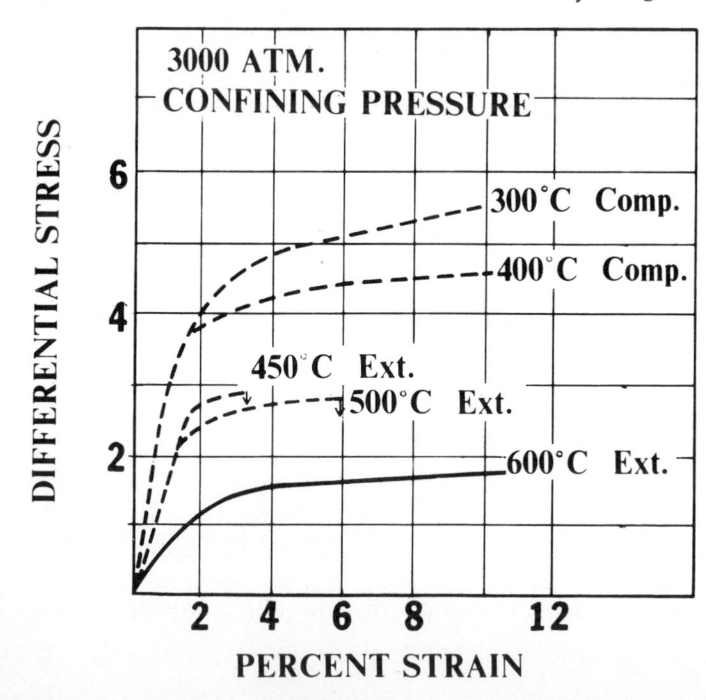

FIGURE 4-18 Temperature effects on strength are shown in this series of experiments with the Solenhofen limestone. All tests were run at 3,000 atm confining pressure. Strength is reduced by increased temperature for both extensional and compressional tests. (*After Heard, 1960.*)

The strength and ductility of water-saturated sediment can be predicted from experimental results which show that both ultimate strength and ductility are reduced by increased pore pressure in sandstone and porous limestone, Figs. 4-26 and 4-27. Yield strength also decreases as the pore pressure increases, and yield (points labeled *a–g* in Fig. 4-28) is followed by reduced strength referred to as *strain softening*,* a process by which the material becomes softened or loosened at the yield stress.

Fracture, Porosity, and Pore Pressure

Variations in pore pressure have a marked influence on the mechanics of deformation of sandstone,* affecting their strength, ductility, and porosity. As pore pressure increases, grain failure by fracture decreases, the rock becomes weak and brittle, and as zero effective confining pressure is approached, cohesive resistance of the rock is overcome, and deformation is governed entirely by intergranular movements as fractional resistance decreases.

At low effective confining pressure (0.25 kbar or less), frictional resistance to shear is low, and the specimen shears without grain breakage except in a narrow zone. The grains are essentially cushioned by the pore pressure. Much of the movement is intergranular, and the specimen may actually show a volume increase with deformation—it becomes dilatant.

At high effective confining pressure (1 to 2 kbars), grains are broken throughout the specimen. The grains are pressed together tightly so they cannot readily slip past one another. Stress concentrations at grain boundaries are great, resulting in extensive grain fracturing accompanied by large reductions in porosity as fractured grains fill pore spaces as deformation proceeds. Eventually the pore spaces are filled, additional grain fracturing is difficult, and a substantial increase in differential stress is required to deform the aggregate further.

* Strain hardening is common in metals; after an initial yielding, the metal becomes "hardened," and on reapplication of load the yield stress is found to be higher.

FIGURE 4-19 Effects of varying temperature on the stress-strain curves of dolomite deformed in compression and extension at 5 kbars confining pressure. (*From Griggs, Turner, and Heard, 1960.*)

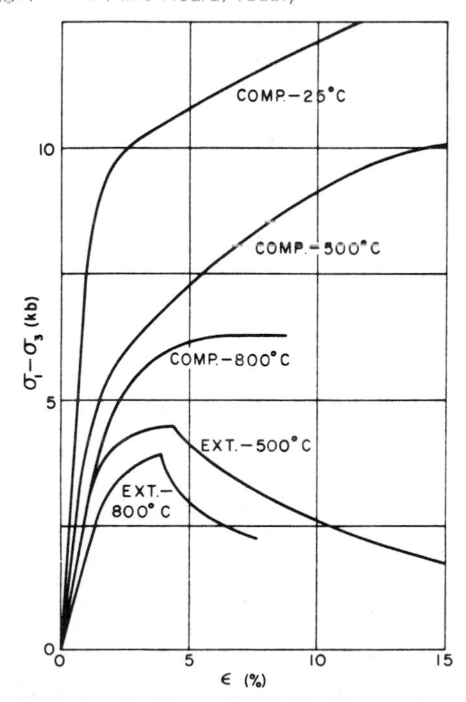

Effects of Fluids in Rock Deformation

Pore-pressure effects are but one of the consequences of solutions in rock masses subjected to deformation. Chemical reactions, recrystallization effects, and phase transformations may also lower the strength of rock. For example, the strength of gypsum drops to $\frac{1}{10}$ its value at lower temperatures as it reaches a temperature of 100°C and undergoes the transition to anhydrite:

Gypsum (at 100°C) $\rightleftharpoons$ anhydrite + water

* Handin (1969).

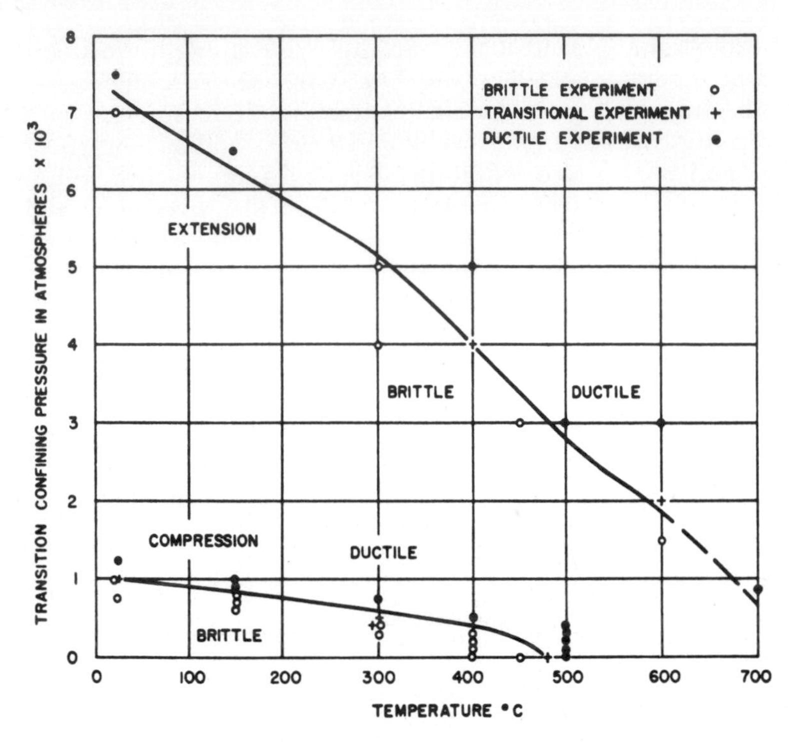

FIGURE 4-20 Temperature and confining pressures at which a transition from brittle failure to ductile deformation occurs in limestone deformed in compression (lower curve) and in extension (upper curve). (*From Heard, 1960.*)

FIGURE 4-21 Ductility (percent strain before fracture or faulting) of several common rocks as a function of confining pressure. The ductility is seen to increase with increasing confining pressure, but to differing degrees. (Data for several rocks are from Handin and Hager, 1957.) (*From Donath, 1970.*)

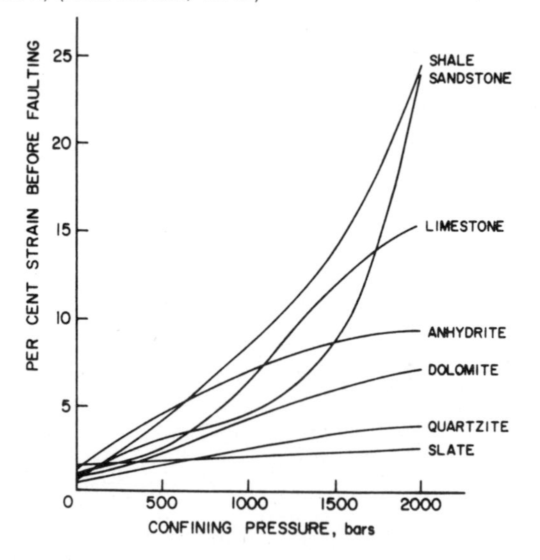

The water causes pore pressure to develop, and intergranular cohesion is apparently reduced (Heard and Rubey, 1966). Similar behavior is exhibited by serpentinites (Raleigh and Paterson, 1965) at the dehydration temperature of 300 to 500°C, and even the strength of quartz and other silicates is reduced by the presence of water (Griggs and Balacic, 1965).

ROLE OF RECRYSTALLIZATION

Both undeformed and deformed fossils* are found in metamorphic rocks. Little-deformed fossils in strongly foliated metamorphic rocks prove that the actual strain is much less than might otherwise be thought and proves that replacement of the fossils took place without mechanical flowage. Recrystallization also

* Examples of fossils in metamorphic rocks were reviewed by Bucher (1953).

occurs in rocks that are being deformed, and a considerable body of experimental information is now available which bears on these effects and processes.

Two types of deformation-related recrystallization are recognized. One type, called *syntectonic recrystallization,* takes place during deformation; the second type, called *annealing recrystallization,* occurs in materials that are first strained, then heated, and maintained under constant confining pressure while recrystallization takes place.

Whether a material will recrystallize or not depends on the internal strain energy, the temperature, and the presence of fluids.

A highly preferred mineral orientation develops in syntectonic recrystallization of mica, quartz, and calcite. Experiments with marble (Griggs) show that the *c* crystallographic axis of recrystallized grains is parallel to the maximum principal compressive stress even though unrecrystallized crystals of the host have other orientations; recrystallization starts at about 300°C and reaches a maximum at 400 to 600°C depending on strain rate. Recrystallization occurs both within crystals and be-

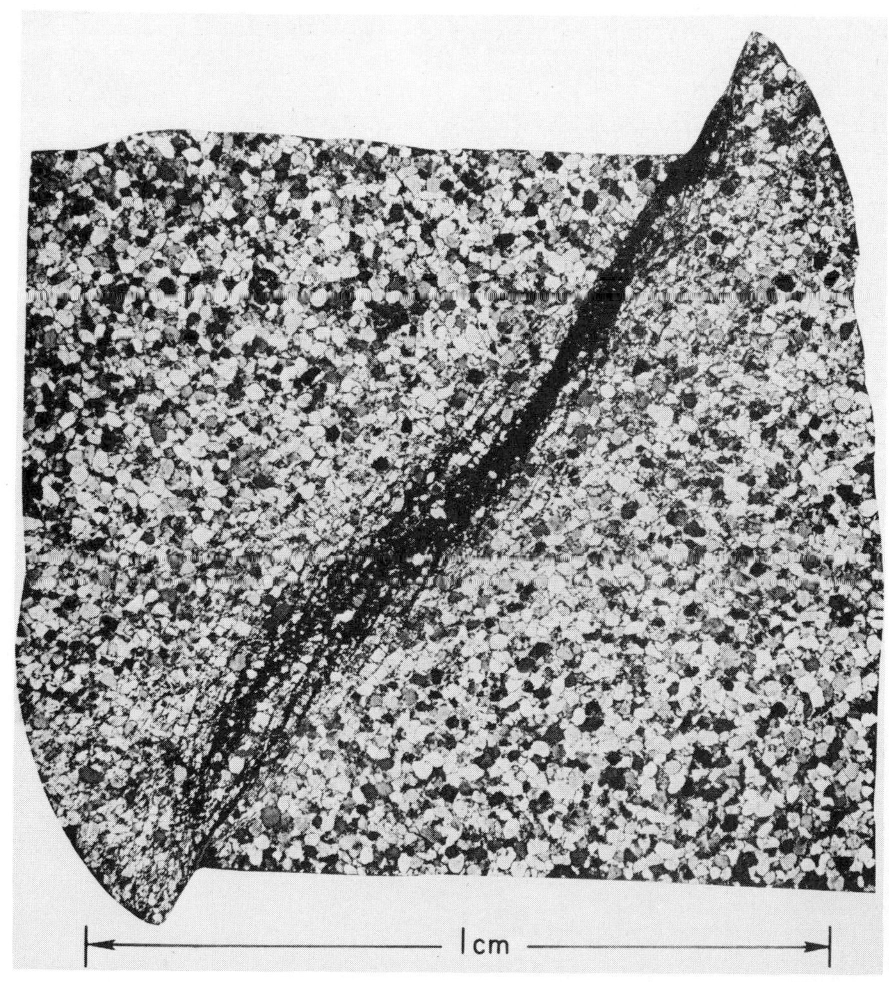

FIGURE 4-22 St. Peter sand specimen shortened 40 percent at 5,000 bars confining pressure, 1,000 bars interstitial water pressure, 500°C. Note lack of grain breakage except in gouge zone along thrust fault. (*From Griggs and Handin, 1960.*)

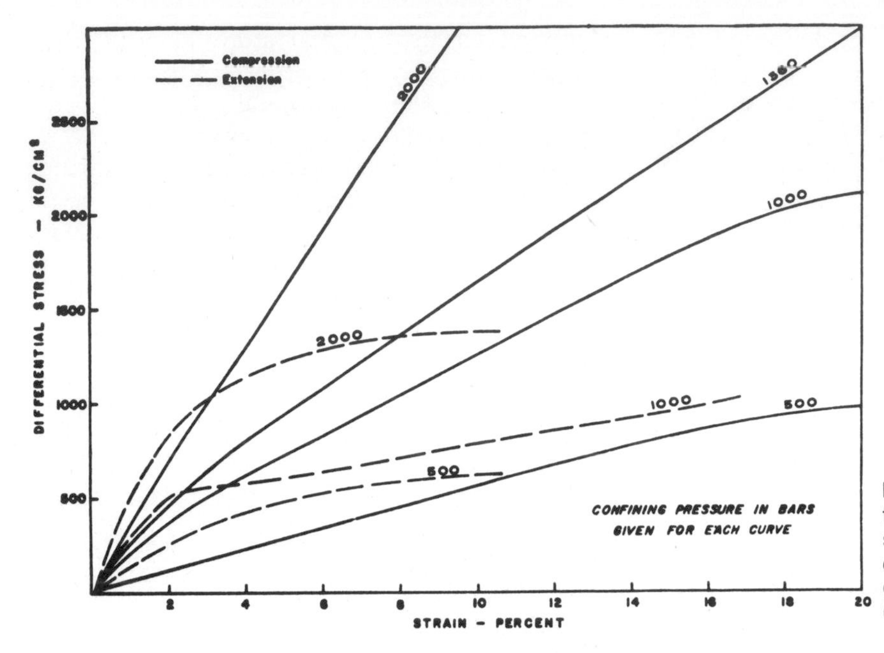

FIGURE 4-23 Stress-strain curves for dry, disaggregated St. Peter sand (250 to 300μ) deformed in compression and extension under different confining pressures. (*From Borg and others, 1960.*)

FIGURE 4-24 Cataclastic texture formed in a quartzite. (*From Higgins, 1971.*)

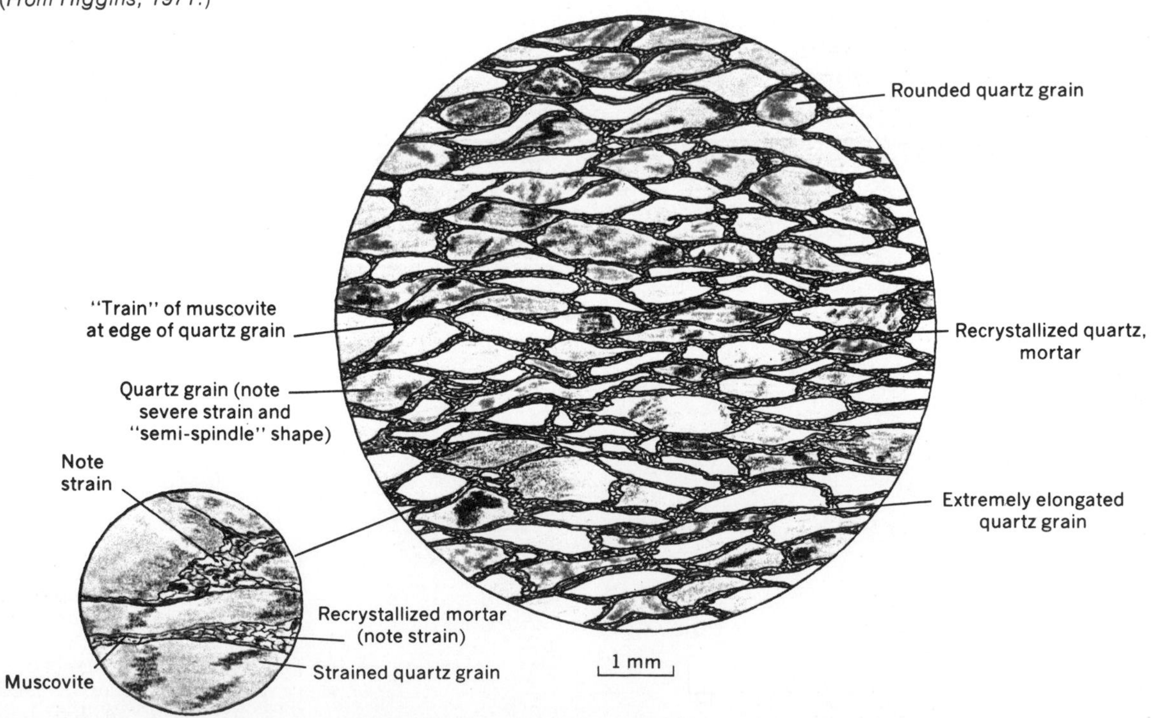

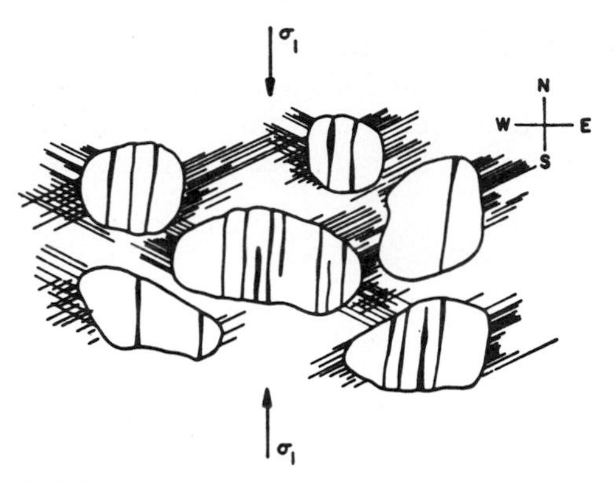

FIGURE 4-25 Sketch of fractured detrital grains and twin lamellae in a calcite crystal. Twin lamellae are best developed on either side of fractured grains and tend to die out into the interstices. Few lamellae are developed above or below grains. (*After Friedman, 1963.*)

tween grains along grain boundaries. Intergranular recrystallization is most pronounced in the portions of the specimens which are greatly elongated.

Syntectonic recrystallization of quartz is somewhat similar to that of calcite. The first indications of changes in boundaries of quartz grains occur at 1000°C (serrate edges appear on the grain boundaries). New grains crystallize near boundaries of deformation bands and grain boundaries, and at a later stage in recrystallization, original grains are completely replaced by aggregates of new grains which show strong preferred orientations. The *c* axes for deformed α quartz form a small circle on a stereographic plot around σ_1. The *c* axes for β quartz became oriented approximately parallel to σ_1. Larger grain sizes appear to result when recrystallization takes place in the presence of water. The resulting textures of some of the syntectonically recrystallized quartzites are remarkably similar to quartzites found in nature (Carter, Christie, and Griggs, 1964).

The critical temperature for annealing recrystallization of Yule marble is 500°C. The recrystallized specimens have textures like those of granoblastic or porphyroblastic metamorphic rocks, and the new crystals lack preferred orientations (Griggs, Paterson, Heard, and Turner, 1960). Annealing recrystallization has also been experimentally produced in quartz compressed 11 percent and annealed at 1500°C, 22 kbars confining pressure. New grains have a strong preferred orientation along zones of basal deformation lamellae and faults.

TIME FACTOR IN ROCK DEFORMATIONS

Evaluation of the effects of time in rock deformation is of critical importance. Short-

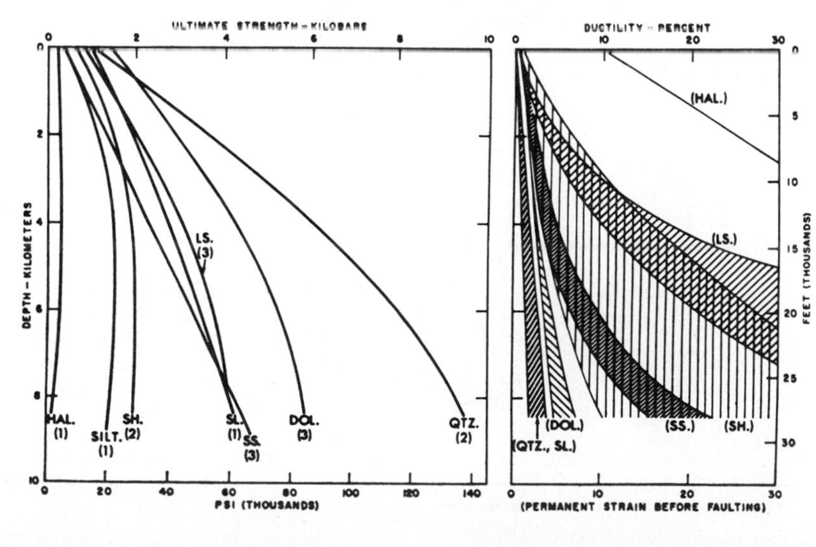

FIGURE 4-26 Ultimate compressive strengths and ductilities of dry rocks as functions of depth. Effects of confining (overburden) pressure and temperature (30°C/km) are included. (*From Handin and others, 1963.*)

duration experiments help clarify rock fabric and the mechanisms by which it is created, but many geological processes involve stresses applied over long periods of time. Most experiments last no more than a few days, but rocks being slowly lowered in a geosyncline or a subduction zone are subjected to gradually increasing confining pressure, and geological evidence indicates that long periods of time, measured in millions of years, are involved in orogeny. Two types of experiments are performed to evaluate time effects:

1. *Creep tests* in which environmental conditions and differential stress are maintained constant. A strain-time plot is made from the data, Fig. 4-29.

2. *Constant strain-rate tests* in which environmental variables are held constant and the rock is deformed so that it strains at a constant rate. Plots usually show stress difference versus strain.

Creep rate (the ratio of strain to time during steady deformation) for metals is an important factor in determining a metal's useful life. Creep effects in rocks are similar, and they are illustrated in Fig. 4-29. The material undergoes an elastic strain initially, followed by a period of adjustment along grain boundaries and within crystals, which constitutes a plastic strain. This is followed by a long period of slowly but steadily increasing strain and finally by a third stage of accelerated strain rate which continues until rupture occurs. Thus three stages of creep are recognized:

1. Instantaneous elastic deformation and transient creep—initial stages of deformation.

2. Steady-state creep—occurs when strain rate becomes steady.

3. Accelerating creep—final stage before rupture as strain rate increases rapidly.

Creep rate is highly sensitive to temperature, and much higher creep rates are observed at high temperatures.

When the load stress is removed from a rock that has been stressed long enough to show creep, a partial recovery takes place—the elastic strain and part of the creep strain, the transient creep, stage 1, is recovered (Fig. 4-30).

Studies of creep in Solenhofen limestone (Robertson, 1960) indicate that creep rate is decreased by increasing hydrostatic pressure. Highly fractured specimens display creep behavior that is similar to that of unfractured specimens. Fracturing is one of a number of mechanisms of creep which include gliding, recrystallization, and adjustments along grain boundaries. The pile-up of dislocations shown by intragranular slips is thought to occur in

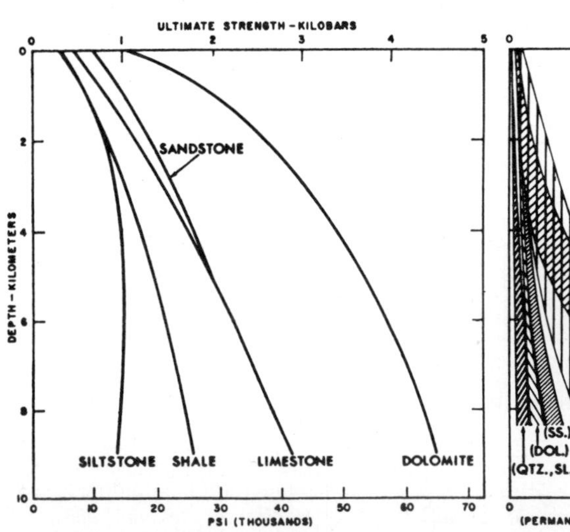

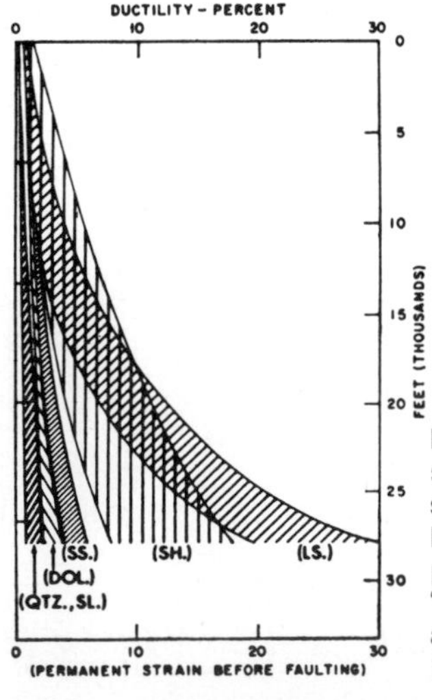

FIGURE 4-27 Ultimate compressive strengths and ductilities of water-saturated rocks as functions of depth. Effects of confining (overburden) pressure, temperature (30°C/km), and "normal" formation (pore) pressure are included. (*From Handin and others, 1963.*)

transient creep and gradually inhibits slip; recrystallization, intragranular slip, and diffusion are important in steady-state creep, and formation of voids and cataclastic flow appear in the late stages—accelerating creep. The relative importance of these is not clearly understood, although it is clear that fracture phenomena do become important in the last stage of accelerating creep.

Various mathematical expressions have been applied to creep.* Griggs (1939) suggests that the equation for creep has the form

$$\epsilon(t) = a + b \log t + ct$$

where $\epsilon(t)$ = creep strain
a = elastic strain
$b \log t$ = transient creep
t = time
a, b, and c = constants

The term *ct, pseudoviscous flow,* is important in long-period creep, and ultimately dominates the total strain.

* See other expressions in Heard (1963). Scholz (1968) emphasizes cracking as the most important creep mechanism in brittle rocks.

FIGURE 4-28 Pore-pressure effects on the stress-strain curves of Indiana limestone deformed in compression at confining pressures of 10,000 psig. (*From Robinson, 1959.*)

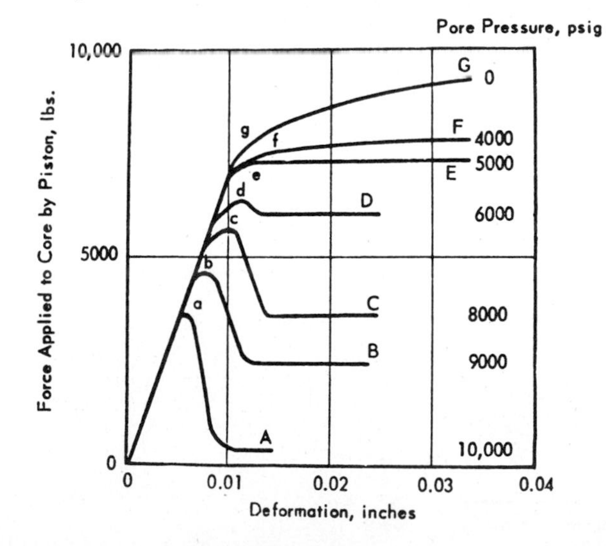

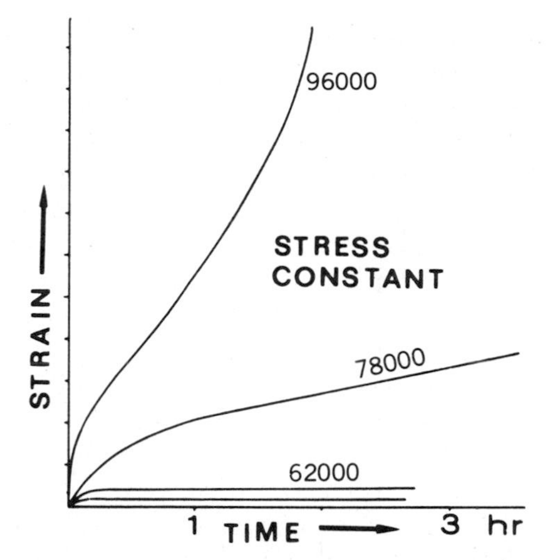

FIGURE 4-29 Strain-time (creep) curve for deformation of limestone. Loads (expressed in pounds per square inch) were applied and held constant and strain was measured. The lowest curve (47,000 lb/in.^{2}, or 3,290 kg/cm^2) showed little strain after initial elastic deformation. Continuous deformation is apparent at loads of 78,000 lb (35,400 kg) and strain accelerates with time for higher loads. (*From Griggs, 1939*)

STRAIN-RATE EFFECTS

Rates of strain in natural processes have a wide range of values, extending from the sudden impact of meteorites to relatively slow values associated with isostatic rebound. Movements of crustal plates may be as high as 10 to 12 cm/year; so values in the range of a fraction to 10 cm/year seem reasonable rates to look for in crustal shortening effects in orogenic belts. We might intuitively expect that strain rate as an environmental factor in rock deformation will affect and help determine which of the mechanisms of rock deformation occur just as temperature and confining pressure do. A considerable body of experimental data has now been obtained which demonstrates these effects.

Plots for constant strain-rate tests showing stress differences ($\sigma_1 - \sigma_3$) against time can be divided into four parts:

1. Elastic deformation.
2. Transient region—strain hardening effects are prominent here.

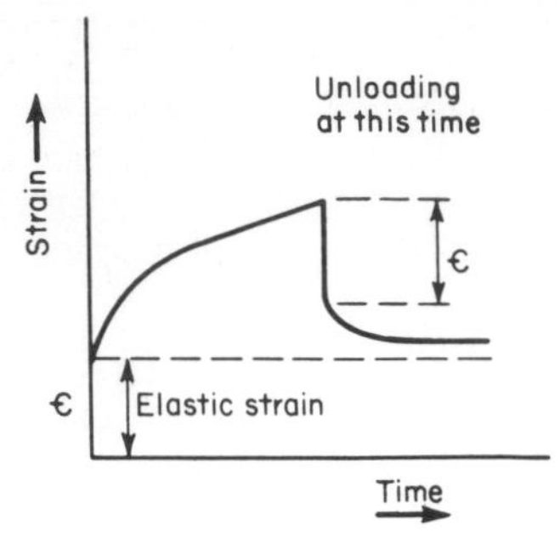

FIGURE 4-30 Strain-time curve with elastic and creep deformation showing afterworking effects. Note that elastic strain is removed almost instantaneously, but most of the plastic component is never recovered.

3. Steady-state flow at constant stress.

4. Decreasing stress until failure occurs.

These divisions are analogous to those of the creep-test curves. Reduction of strain rate in each case (limestone, dolomite, quartzite, and granite) causes ductility to increase. This is accompanied by a marked decrease in strength as well. At high strain rates quartzite deforms by cataclasis, with shears developing around and through grains, but at lower strain rates little internal fracturing of grains is seen —instead rotation of grains and intragranular slip is important.

THE RHEID CONCEPT

The great importance of the time factor in the behavior of rock being deformed is pointed out by Carey (1954), who has developed Andrade's model* as it applies to the earth. In geotectonic phenomena the duration of loading is generally very long—so long, in fact, that the transient creep and viscous strain (steady creep), both of which are time-dependent, become the dominant processes of deformation. For stort times the transient creep may be larger than the viscous strain because of the large coefficient of viscosity of rocks, but for longer times the viscous strain becomes the main component, Fig. 4-31.

* Total strain = elastic component + plastic component + transient creep + viscous strain.

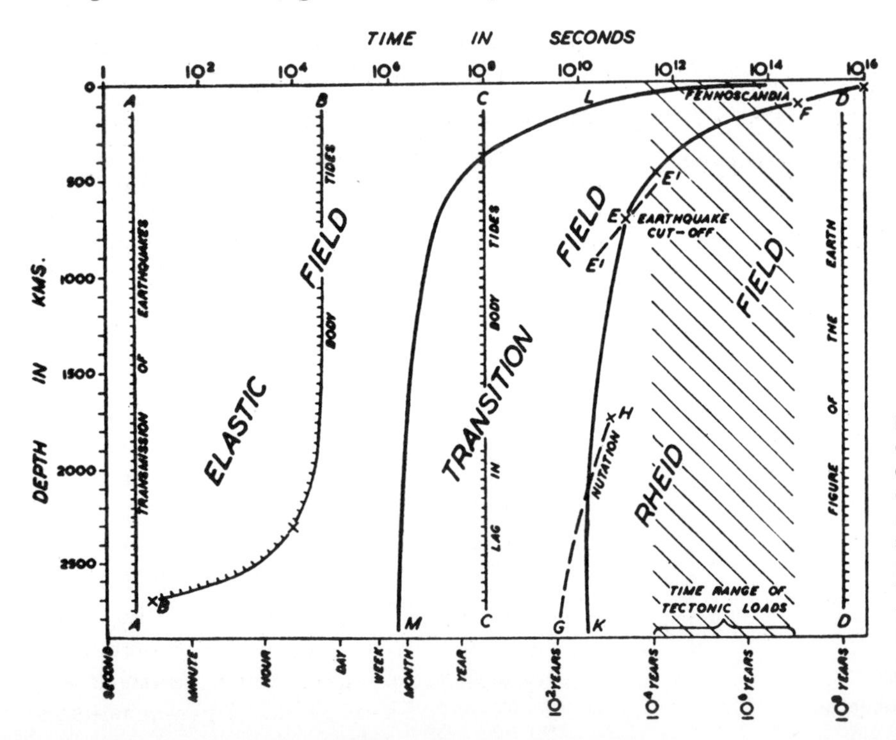

FIGURE 4-31 Rheid behavior of the mantle as a function of depth and time. Upper parts of the lithosphere and mantle are elastic even for very slow deformation, and the mantle is elastic throughout for seismic waves, but most of the mantle is a rheid for loads applied over the long time periods involved in tectonic deformations. (*From Carey, 1954.*)

The viscosity η of rocks is very poorly known, but experimental evidence suggests that it contains components of boundary layer flow (similar to newtonian viscosity) and movements such as intralattice slip, which are stress-dependent. Thus it is called *pseudo-viscosity,* and the magnitude of η is found to vary with the stress applied, according to the expression $\eta = e^{-A\sigma}$, where A = a constant, and with temperature according to $\eta = e^{B/kT}$, where B = activation energy, k = Boltzmann's constant, and T = absolute temperature.

Rheidity of a material is defined by the amount of time required for the viscous strain component to become 1,000 times as great as the elastic strain component. This is equal to $\eta/\mu \times 10^3$ sec. (The term η/μ is Maxwell's relaxation time.) This time is arbitrarily adopted as the threshold of wholly fluid behavior. "A *Rheid* is a substance whose temperature is below the melting point, and whose deformation by viscous flow during the time of the experiment is at least three orders of magnitude greater than the elastic deformation under the given conditions" (Carey, 1954). The approximate rheidities of some common earth materials are:

Ice	2 weeks
Rock salt	10^6 to 10^8 sec (10 years)
Gypsum	1 year
Serpentine	10,000 years

SUMMARY*

Experimental deformation of crystalline aggregates reveals the following effects (see the stress-strain curves, Figs. 4-17 to 4-19):

1. With temperature and strain rate constant:

Increasing effective confining pressure. In general the rock behavior undergoes a transformation from brittle to ductile behavior as effective confining pressure is increased.

* Handin, 1968.

a. Increases the ultimate strength (high point on the stress-strain curve) for low temperatures (400°C) and confining pressure (P_c = 3,000 bars)
b. Raises the *yield stress*
c. Increases *strain-hardening* effects
d. Increases *ductility*

2. With effective confining pressure constant:

Increasing temperature (at constant strain rate), or decreasing strain rate (at constant temperature), or increasing temperature and decreasing strain rate.

a. Reduces ultimate strength
b. Reduces yield stress
c. Eliminates strain hardening
d. Increases ductility

Three major classes of mechanisms are recognized for rock flowage:

1. Cataclastic flow
2. Intragranular slip
3. Recrystallization

Cataclastic flow consists of mechanical fracture crushing, faulting and granulation of mineral grains of the rock, and large-scale intergranular movements. This mechanism occurs most often under conditions which favor and in materials (e.g., quartz) which exhibit *brittle behavior.* Frictional phenomena are important. Conditions favorable to cataclastic flow are:

1. Low confining pressure (shallow depths)
2. High pore pressure
3. Low temperature
4. High strain rates
5. High differential stress

Intergranular slip is characterized by translation gliding and twinning phenomena as characterized by calcite. Some materials are much more susceptible to slip mechanisms than others as already noted. These phenomena are more common at greater depth and are favored by:

1. High confining pressure
2. Lower pore pressures
3. Higher temperatures
4. Lower strain rates
5. Lower differential stress

Recrystallization flow takes place largely as a result of molecular rearrangements due to solution and reprecipitation, local melting, and solid diffusion. These processes are favored by:

1. Presence of chemically active solutions
2. High temperatures
3. Low strain rates
4. Low differential stresses

Many rocks undergo transitions in the mechanism of deformation as the environmental conditions under which deformation occurs change. For many materials, the mechanisms change from cataclastic flow to intragranular slip as confining pressures or temperatures are increased, or as pore pressures or strain rates are decreased. These changes in mechanism of deformation coincide with changes in material behavior from brittle to ductile behavior.

REFERENCES

Adams, F. D., and **Nicholson, J. T.**, 1901, An experimental investigation into the flow of marble: Royal Soc. (London) Philos. Trans., v. 195, p. 363–401.

Andrade, E. N. C., 1914, Royal Soc. London Proc., v. A90.

Bailey, S. W., and others, 1958, Plastic deformation of quartz in nature: Geol. Soc. America Bull., v. 69, p. 1443–1466.

Bell, J. F., 1941, Morphology of mechanical twinning in crystals: Am. Mineralogist, v. 26, p. 247–261.

Borg, Iris, and **Handin, John**, 1966, Experimental deformation of crystalline rocks: Tectonophysics, v. 3, no. 4, Spec. Issue: New York, Elsevier.

——— and **Maxwell, J. C.**, 1956, Interpretation of fabrics of experimentally deformed sands: Am. Jour. Sci., v. 254, no. 2.

——— and **Turner, F. J.**, 1953, Deformation of Yule marble: Pt. VI, Identity and significance of deformation lamellae and parting in calcite grains: Geol. Soc. America Bull., v. 64, p. 1343–1352.

——— and others, 1960, Experimental deformation of St. Peter sand: A study of cataclastic flow, *in* Griggs, D. T., and Handin, John, eds., Rock deformation—A symposium: Geol. Soc. America Mem. 79, p. 133–191.

Bucher, W. H., 1953, Fossils in metamorphic rocks: Geol. Soc. America Bull., v. 64, p. 275–300.

——— 1956, Role of gravity in orogenesis: Geol. Soc. America Bull., v. 67, p. 1295–1318.

Byerlee, J. D., 1968, Brittle-ductile transition in rocks: Jour. Geophys. Res., v. 73, no. 14, p. 4741ff.

Carey, S. W., 1954, The rheid concept in geotectonics: Geol. Soc. Australia Jour. v. 1, p. 67–117.

——— 1962, Folding: Alberta Soc. Petroleum Geologists Jour., v. 10.

Carter, N. L., 1969, Principal stress directions from plastic flow in crystals: Geol. Soc. America Bull., v. 80, p. 1231–1264.

——— and **Avé Lallement, H. G.**, 1970, High temperature flow of dunite and peridotite: Geol. Soc. America Bull., v. 81, no. 8, p. 2181–2202.

——— and **Raleigh, C. B.**, 1969, Principal stress directions from plastic flow in crystals: Geol. Soc. America Bull., v. 80, p. 1231–1264.

———, **Christie, J. M.**, and **Griggs, D. T.**, 1964, Experimental deformation and recrystallization of quartz: Jour. Geology, v. 72, p. 687–733.

——— and others, 1961, Experimentally produced deformation lamellae and other structures in quartz sand: Jour. Geophys. Research, v. 66, p. 2518–2519.

Christie, J. M., Griggs, D. T., and **Carter, N. L.**, 1964, Experimental evidence of basal slip in quartz: Jour. Geology, v. 72, p. 734–756.

——— and **Raleigh, C. B.**, 1959, The origin of deformation lamellae in quartz: Am. Jour. Sci., v. 257, no. 6.

Clark, S. P., Jr., ed., 1966, Handbook of physical constants: Geol. Soc. America Mem. 97.

Crampton, C. B., 1958, Muscovite, biotite and quartz fabric reorientation: Jour. Geology, v. 66, no. 1.

Dieterich, J. H., 1972, Time-dependent friction in rocks: Jour. Geophys. Res., v. 77, no. 20, p. 3690ff.

Donath, F. A., 1963, Fundamental problems in dynamic structural geology, *in* Donnelly, T. W., ed., The earth sciences: Problems and progress in current research: Chicago, Univ. of Chicago, p. 83–103.

——— 1970, Some information squeezed out of rock: American Scientist, V. 58, no. 1, p. 54–72.

——— and **Fruth, L. S., Jr.,** 1971, Dependence of strain-rate effects on deformation mechanism and rock type: Jour. Geology, v. 79, no. 3, p. 347–371.

———, **Faill, R. T.,** and **Tobin, D. G.,** 1971, Deformational mode fields in experimentally deformed rock: Geol. Soc. America Bull., v. 82, no. 6, p. 1441ff.

Friedman, Melvin, 1963, Petrofabric analysis of experimentally deformed calcite-cemented sandstones: Jour. Geology, v. 71, p. 12–37.

——— 1964, Petrofabric technique for the determination of principal stress directions in rocks, *in* Judd, W. R., State of stress in the earth's crust: New York, Elsevier.

——— and **Stearns, D. W.,** 1971, Relations between stresses inferred from calcite twin lamellae and macrofractures, Teton anticline, Montana: Geol. Soc. America Bull., v. 82, no. 11, p. 3151ff.

Goranoon, R. W., 1940, Flow in stressed solids—an interpretation: Geol. Soc. America Bull., v. 51, p. 1023–1034.

Griggs, D. T., 1936, Deformation of rocks under high confining pressures: Jour. Geology, v. 44, p. 541–577.

——— 1939, Creep of rocks: Jour. Geology, v. 47, p. 255.

——— 1940, Experimental flow of rocks under conditions favoring recrystallization: Geol. Soc. America Bull., v. 51, p. 1001–1022.

——— and **Balacic, J. D.,** 1965, Quartz: anomalous weakness of synthetic crystals: Science, v. 147.

——— and **Bell, J. A.,** 1938, Experiments bearing on the orientation of quartz in deformed rocks: Geol. Soc. America Bull., v. 49, p. 1723–1746.

——— and **Handin, John,** eds., 1960, Rock deformation—A symposium: Geol. Soc. America Mem. 79.

——— and **Miller, W. B.,** 1951, Compression and extension experiments on dry Yule marble at 10,000 atmospheres confining pressure, room temperature, pt. 1, *in* Deformation of Yule marble (Colo.): Geol. Soc. America Bull., v. 62, p. 853–862.

———, **Paterson M. S., Heard, H. C.,** and **Turner, F. J.,** 1960, Annealing recrystallization in calcite crystals and aggregates, *in* Griggs, D. T., and Handin, John, eds., Rock deformation—A symposium: Geol. Soc. America Mem. 79, p. 21–39.

———, **Turner, F. J.,** and **Heard, H. C.,** 1960, Deformation of rocks at 500°C, *in* Griggs, D. T., and Handin, John, eds., 1960, Rock deformation—A symposium: Geol. Soc. America Mem. 79, p. 39–104.

——— and others, 1951, Effects at 150°C, Pt. 4, *in* Deformation of Yule marble (Colo.): Geol. Soc. America Bull., v. 62, p. 1385–1405.

——— 1954, Deformation of rocks at 500°C, 5000 atmospheres pressure: Geol. Soc. America Bull., v. 65, p. 1258.

Groshong, R. H., Jr., 1972, Strain calculated from twinning in calcite: Geol. Soc. America Bull., v. 83, no. 7, p. 2025ff.

Gzovsky, M. V., 1959, The use of scale models in tectonophysics: Internat. Geology Rev., v. 1, no. 4, p. 31–47.

Hahn, S. J., Ree, Taikyue, and **Eyring, H.,** 1967, Mechanism for the plastic deformation of Yule marble: Geol. Soc. America Bull., v. 78, p. 773–782.

Handin, John, 1969, On the Coulomb-Mohr failure criterion: Jour. Geophys. Res., v. 74, no. 22, p. 5343–5348.

——— and **Fairbairn, H. W.,** 1955, Experimental deformation of Hasmark dolomite: Bull. Geol. Soc. America, v. 66.

——— and **Hager, R. V., Jr.,** 1958, Experimental deformation of sedimentary rocks under confining pressure: Tests at high temperature: Am. Assoc. Petroleum Geologists Bull., v. 42, p. 2892–2934.

——— and others, 1963, Experimental deformation of sedimentary rocks under confining pressure: Pore pressure tests: Am. Assoc. Petroleum Geologists Bull., v. 47, p. 717–755.

Hansen, Edward, and **Borg, Iris,** 1962, The dynamic significance of deformation lamellae in quartz of a calcite-cemented sandstone: Am. Jour. Sci., v. 260, no. 5.

Heard, H. C., 1960, Transition from brittle fracture to ductile flow in Solenhofen limestones as a function of temperature, confining pressure, and interstitial fluid pressure, *in* Griggs, D. T., and Handin, John, eds., Rock deformation—A symposium: Geol. Soc. America Mem. 79.

——— 1963, Effect of large changes in strain rate in

the experimental deformation of Yule marble: Jour. Geology, v. 71, p. 162–195.

——— and **Raleigh, C. B.,** 1972, Steady-state flow in marble at 500° to 800°C.: Geol. Soc America Bull., v. 83, no. 4, p. 935ff.

——— and **Rubey, W. W.,** 1966, Tectonic implications of gypsum dehydration: Bull. Geol. Soc. America, v. 77.

———, **Borg, I. Y.,** and **Carter, N. L.,** eds., 1972, Flow and fracture of rocks: Am. Geophys. Union, Geophys. Monogr. no. 16.

Higgins, M. W., 1971, Cataclastic rocks: Geol. Survey Prof. Paper 687.

Higgs, D. V., and **Handin, John,** 1959, Experimental deformation of dolomite single crystals: Geol. Soc. America Bull., v. 70, no. 3, p. 245–277.

Hobbs, B. E., 1968, Recrystallization of single crystals of quartz: Tectonophysics, v. 6, no. 5, p. 353–401.

Hörz, Friedrich, 1960, Static and dynamic origin of kink bands in micas: Jour. Geophys. Res., v. 75, no. 5, p. 965–977.

Hubbert, M. K., 1937, Scale models and geologic structures: Geol. Soc. America Bull., v. 48, p. 1459.

——— 1951, Mechanical basis for certain familiar geologic structures: Geol. Soc. America Bull., v. 62, no. 4, p. 255–372.

Kamb, W. B., 1959, Theory of preferred crystal orientation developed by crystallization under stress: Jour. Geology, v. 67.

——— 1961, The thermodynamic theory of nonhydrostatically stressed solids: Jour. Geophys. Res., v. 66, p. 259–271.

Knoff, E. B., and **Ingerson, E.,** 1938, Structural petrology: Geol Soc. America Mem. 6.

Lomnitz, C., 1956, Creep measurement in igneous rocks: Jour. Geology, v. 64, p. 473–479.

Mogi, Kiyoo, 1972, Fracture and flow of rocks, *in* Ritsema, A. R., The upper mantle: Tectonophysics, v. 13, p. 541–568.

Mügge, O., 1898, Über translationen und Verwandte Erscheinungen in Krystallen, Neues Jahrb. Mineralogie, Geologie u. Paläontologie Abh. B, v. 1, p. 71–75.

Nádai, A., 1950, Theory of flow and fracture of solids, v. 1, 2d ed.: New York, McGraw-Hill.

Orowan, E., 1942, A type of plastic deformation new in metals: Nature, v. 149, p. 643–644.

——— 1952, Creep in metallic and non-metallic materials: Proc. 1st Natl. Cong. Appl. Mech., ASME, p. 453–472.

Paterson, M. S., 1958, The experimental deformation and faulting in Wombeyan marble: Geol. Soc. America Bull., v. 69, p. 465–476.

——— and **Weiss, L. E.,** 1961, Symmetry concepts in the structural analysis of deformed rocks: Geol. Soc. America Bull., v. 72, p. 841–882.

Raleigh, C. B., 1968, Mechanisms of plastic deformation of olivine: J. Geophys. Res., v. 73.

——— and **Paterson, M. S.,** 1965, Experimental deformation of serpentinite and its tectonic implications: Jour. Geophys. Res., v. 70.

——— and **Talbot, J. L.,** 1967, Mechanical twinning in naturally and experimentally deformed diopside: Am. Jour. Sci., v. 265, no. 2.

Robertson, E. C., 1965, Experimental study of the strength of rocks: Geol. Soc. America Bull., v. 66, p. 1275–1314.

——— 1960, Creep of Solenhofen limestone under moderate hydrostatic pressure, *in* Griggs, D. T., and Handin, John, eds., Rock deformation—A symposium: Geol. Soc. America Mem. 79.

Robinson, L. H., Jr., 1959, The effect of pore and confining pressure on the failure process in sedimentary rock: Colorado School Mines Quart., v. 54, no. 3, p. 177–199.

Rutter, E. H., 1972, The influence of interstitial water on the rheological behaviour of calcite rocks: Tectonophysics, v. 14, p. 13–33.

——— and **Schmid, S. M.,** 1975, Experimental study of unconfined flow of Solenhofen limestone at 500° to 600°C: Geol. Soc. America Bull., v. 86, p. 145–152.

Schock, R. N., Heard, H. C., and **Stephens, D. R.,** 1973, Stress-strain behavior of a granodiorite and two graywackes on compression to 20 kilobars: Jour. Geophys. Res., v. 78, no. 26, p. 5922ff.

Scholz, C. H., 1968, Mechanism of creep in brittle rock: Jour. Geophys. Res., v. 73, no. 10, p. 3295–3302.

Seifert, K. E., 1972, Strain rate vs. displacement rate: Geol. Soc. America Bull., v. 83, no. 6, p. 1853ff.

Spang, J. H., and **Van Der Lee, Joyceanne,** 1975, Numerical dynamic analysis of quartz deformation lamellae and calcite and dolomite twin lamellae: Geol. Soc. America Bull., v. 86, p. 1266–1272.

Tobin, D. G., and **Donath, F. A.**, 1971, Microscopic criteria for defining deformational modes in rock: Geol. Soc. America Bull., v. 82, no. 6, p. 1463ff.

Tullis, T. E., and **Wood, D. S.**, 1975, Correlation of finite strain from both reduction bodies and preferred orientation of mica in slate from Wales: Geol. Soc. America Bull., v. 86, p. 632–638.

Turner, F. J., 1962, Rotation of the crystal lattice in kink bands, deformation bands, and twin lamellae of strained crystals: Nat'l. Acad. Sci. Proc., v. 48, no. 6, p. 955–963.

——— and **Verhoogen, John**, 1960, Igneous and metamorphic petrology, 2d ed.: New York, McGraw-Hill.

——— **Griggs, D. T.**, and **Heard, H. C.**, 1954, Experimental deformation of calcite crystals: Geol. Soc. America Bull., v. 65, p. 883–934.

Van Vlack, L. H., 1964, Elements of materials science: Cambridge, Mass., Addison-Wesley.

Verhoogen, Jean, 1951, The chemical potential of a stressed solid: Am. Geophys. Union Trans., v. 32, p. 41–43.

Willis, Bailey, 1923, Geologic structures: New York, McGraw-Hill.

5

Brittle Failure of Rocks

ROCK FAILURE IN THEORY AND EXPERIMENT*

One of the most striking observations from experimental work is the consistency of the relationship between the attitude of zones of fracture and the applied stress direction for deformation of materials while they are brittle. When such materials are compressed, either they fail by development of longitudinal fractures oriented parallel to the principal stress direction or they deform by shearing along planes inclined at angles of 45° or less to the principal stress direction. When such materials are extended, they fail either by development of tension fractures normal to the direction of extension or by shearing with fractures oriented so that the direction of extension bisects obtuse angles between fractures, Figs. 5-1 and 5-2. Fractures with these types of angular relationships are commonly found in field studies.

The question of defining when a rock has failed in an experiment is not difficult when the rock is brittle and the material is deformed by extension. Failure is marked by the sudden formation of fractures, sudden loss of cohesion, and loss of resistance to differential stress. The time of failure is not clear when deformation occurs at high temperatures and high confining pressures and the behavior is ductile, or in brittle materials when they fail by shear. The fractures formed in experiments with marble and limestone ranging from brittle to ductile behavior are shown in Figs. 5-2 and 5-3. Griggs and Handin (1960) suggest that all rocks probably will be found to exhibit similar regimes of behavior (although the conditions may be quite varied). In general, materials exhibit brittle behavior at low temperatures and confining pressure. At higher temperatures and pressures, materials are able to sustain large permanent deformation. Continuous rehealing without loss of cohesion or

* The term *fracture* is restricted in its use here to apply to surfaces across which rock has lost cohesion, without displacement. Thus it is used synonymously with the term *joint*.

release of strain energy is considered to be an important factor in the continuous deformation of most rocks at high temperature and high pressure. In view of the transitional nature of brittle-ductile behavior, we should expect no clear definition of rock failure. Griggs and Handin (1960, p. 348) propose a threefold classification of fracture and flow:

1. An extension fracture is separation of a body across a surface normal to the direction of least principal stress. There is no offset parallel to the fracture surface. The correlation between extension fractures and the principal stresses follows from the criterion of no offset and involves no additional hypotheses. Separation is parallel to a plane of vanishing shear stress.

(The terms "extension fracture" and "tension fracture" are sometimes used as synonyms. The term "tension" implies that a particular state of stress existed at the time of fracture. Extension is a more general term and should be used unless stress conditions at time of fracture are known.)

The term "fracture" is used here advisedly since the phenomenon described involves total loss of cohesion, separation into two parts, release of stored elastic strain energy, and loss of resistance to differential stress. These characteristics are all associated with "fracture" (or "rupture") in the ordinary sense, so that the definition should be unambiguous.

2. A fault is a localized offset parallel to a more or less plane surface of nonvanishing shear stress. The surface may be inclined at from 45° to a few degrees to the direction of maximum principal (compressive) stress in homogeneous materials. As examples will show, there may or may not be total loss of cohesion, actual separation, release of stored elastic energy, or loss of resistance to differential stress. In other words, the phenomenon need not involve "fracture" in the ordinary sense. To be unequivocal the writers prefer "fault," and their usage of the term implies neither more nor less than that of the field geologist.

The name "shear fracture" is used by many metallurgists in reference to fractures parallel to planes of maximum shearing stress, hence only to planes inclined at 45° to the extreme principal stresses. This is a special case of faulting, as is the "shear fracture" of engineering which implies complete loss of cohesion.

3. Uniform flow denotes macroscopically homogeneous deformation. There are three principal flow mechanisms: (*a*) cataclasis, involving crushing, granulation, and intergranular adjustments for which friction is important, (*b*) intragranular gliding, involving translation (slip) and twinning for which friction is relatively unimportant, and (*c*) recrystallization by local melting or solid diffusion, or through the agency of solution. The writers wish to emphasize that these mechanisms are not always distinguishable from one another or from faulting on the basis of the stress-strain relations measured during a short-time triaxial test.

Significantly, it is possible for a material to fault without exhibiting loss of cohesion or fracture in the usual sense of the word. Thus

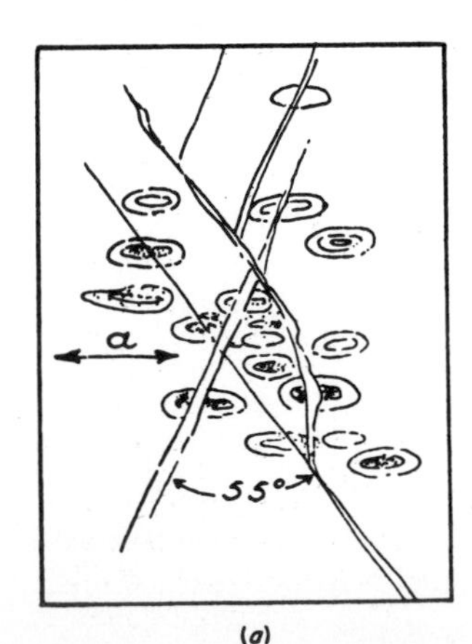

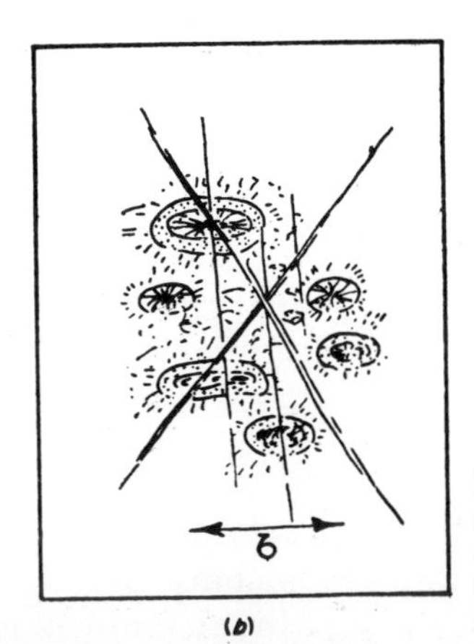

FIGURE 5-1 Deformed oolites and related shear and tension fractures from South Mountain area, Maryland. Both the fracture pattern and the flattening of the oolites indicate shortening from top to bottom in this rock. (*From Cloos, 1947.*)

in dealing with the phenomena of rock failure we must cope with a spectrum of behavior that varies with material type, conditions of confining pressure, temperature, strain rate, and type of stress application. This spectrum is divided here into five cases each for compression and extension, Fig. 5-2. The arbitrary definition of each case is based on percentage of strain (shortening or lengthening of specimen), and the boundaries of these various types of response are not rigid. They are shown overlapping. In general the spectrum represents the transition from brittle to ductile behavior of material; any material, therefore, which is brittle, case 1, under normal conditions may be expected to show behavior typical of higher cases as temperature or confining pressure is increased.

Brittle material deformed by extension deforms first by elastic strain followed by an abrupt failure by fractures oriented normal to the direction of extension. Under compression a brittle material deforms first by elastic strain followed by failure along fractures oriented parallel to the load axis. A ductile material extended undergoes elastic strain followed by plastic deformation accompanied by formation of a neck (narrowing of the specimen) and finally by failure in which a cup and cone structure is formed. Fibrous structure is often found in the break. Under compression, ductile materials undergo plastic deformation, or flow, after elastic strain. (See Fig. 5-3.)

Mohr Representation

The Mohr-circle method of representing stress is a valuable tool in visualizing the various criteria which have been used to study the theory of failure. Conditions of pure tension,

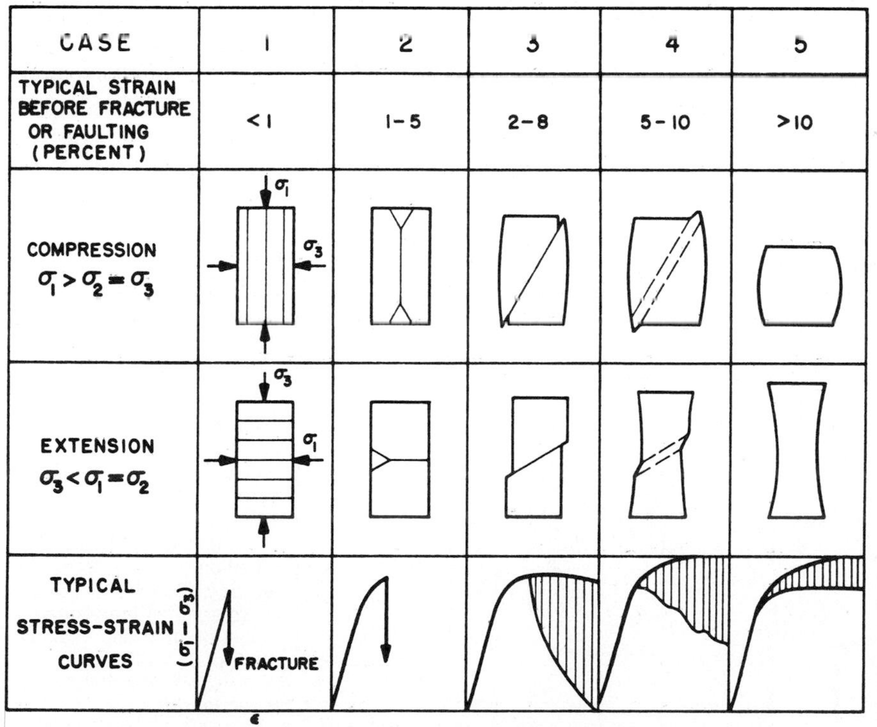

FIGURE 5-2 Schematic representation of the spectrum from brittle fracture to ductile flow, with typical strains before fracture and stress-strain curves for uniaxial compression and extension. The ruled portions of the stress-strain curves indicate the variation within each case and the overlap between cases 3, 4, and 5. (*From Griggs and Handin, 1960.*)

shear, and compression are shown in Fig. 5-4 on a Mohr diagram. Each of these is defined by the maximum and minimum values of the principal stresses. These stresses are, respectively, a negative value and zero in tension; equal negative and positive values for shear; zero and positive values in compression.

The stress conditions which a material can sustain without failure can be readily depicted on the Mohr plot.* Consider a material subjected to some particular stress conditions. The position of Mohr circle, Fig. 5-5, indicates the type of stress condition, and the size of the circle shows the difference between the maximum and least principal stress. Now if the diameter of the circle is progressively increased (indicating an increase in stress difference between the maximum and least principal stresses), at some particular value of stress difference the material fails.†

We may define the *strength* of the material as the maximum *stress difference* it will sustain without failure under some particular stress conditions. Strength depends on the length of time the stress is applied and on the physical conditions (e.g., temperature, confining pressure, strain rate, and manner of stress application) which prevail during deformation. Materials are generally much stronger under compression than tension—often by a factor of 10.

Now consider the representation of failure on a Mohr diagram. If a material has a tensile strength of given value, then the field of values of σ_3 less than that value falls in a realm of failure for that material, Fig. 5-5*a*.

If the stress conditions are uniaxial tension, then $\sigma_2 = \sigma_1 = 0$, and the condition of failure is represented by a point on the circle, Fig. 5-5*b*.

A second condition of failure might be expressed in terms of the maximum shearing stress a material can withstand, $\tau =$ some value A. This may be represented as shown, Fig. 5-5*c*, where the region of instability is shaded. The lines separating the stable from the unstable regions are termed a *Mohr envelope*. Note that the points of tangency on the envelope which denote failure correspond to planes 45° on either side of σ_3.

Now consider a material which is subjected to various stress conditions in a series of experiments and deformed under each condition

* See Brace, *in* Riecker (1968).

† Brittle failure means that the material fractures and the material separates into pieces which have undergone little permanent deformation (Brace).

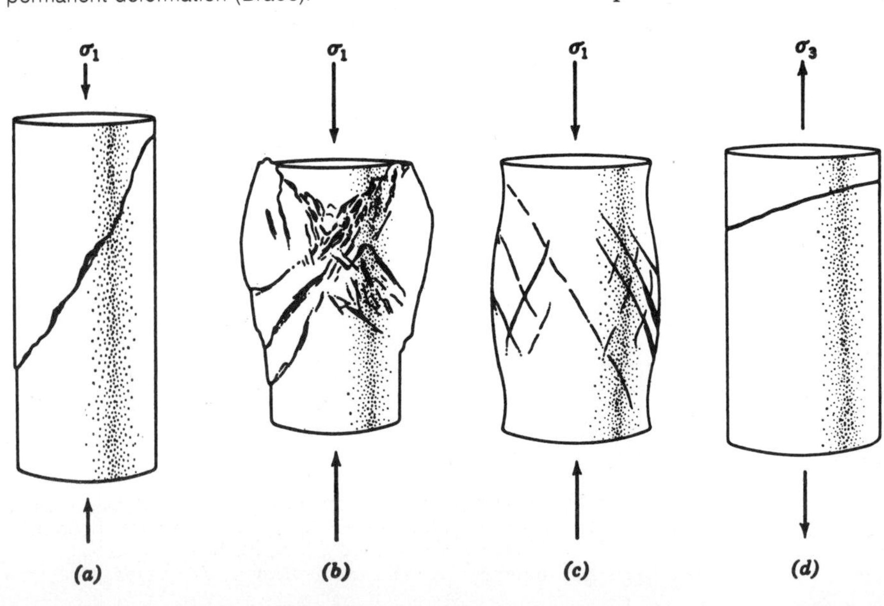

FIGURE 5-3 Shear fractures and faults characteristic of failure of dry limestone and marble. (*a*) Marble, brittle failure; 25°C, 35 bars, 1 percent strain. (*b*) Marble, transitional failure; 25°C, 280 bars, 20 percent strain. (c) Solenhofen limestone, almost ductile failure; 25°C, 1,000 bars, 11.2 percent strain. (*d*) Solenhofen limestone, ductile behavior followed by rupture; 150°C, 6,500 bars, 9.1 percent strain. (*After Heard, 1960.*)

until it fails. A Mohr circle can then be drawn for each experiment corresponding to the conditions and stress differences at failure. The result is a set of Mohr circles, Fig. 5-6. The line drawn tangent to the set of circles defines a Mohr envelope characteristic of the material. For most materials this has a parabolic shape. The Mohr circle for pure tension is small in diameter. The Mohr circle for uniaxial test is larger in size, and as the confining pressure increases in a triaxial test the circle becomes even larger, causing the parabola to open even more.

Criteria of Failure

A number of theories of failure have been formulated which involve the establishment of certain criteria by which the state of failure can be expressed mathematically. Among the more successful criteria are:

1. *Coulomb's criterion* (also called Coulomb-Navier criterion, a special case of Mohr's criterion). Failure is predicted when the shear stress along the planes of potential slip reaches a certain value.

2. *Tresca's and Von Mise's* (for plain strain) *maximum shear-stress criterion.* Failure by flow or by rupture occurs when the maximum shear stress reaches a constant value, which is characteristic of the material.

3. *Griffith's criterion.* Failure is related to the propagation of cracks originating from small flaws, cracks, or foreign matter.

Varnes (1962, p. B-17) points out the difficulty in applying these criteria to geological situations:

It should be emphasized that the laws governing fracture and flow of solids are still imperfectly understood. Internal flaws and duration of stress are but two of many factors that influence fracture, in addition to the overall relations of principal stress. Temperature and chemical action of fluids have potent effects on the strength and on the mechanical type of failure of many geologic materials.

Coulomb's Criterion of Failure

Normal and shearing stresses on a plane vary with the angle of inclination of the plane relative to the principal stress directions as shown in Fig. A-6. Shearing stresses reach a maximum on planes inclined at 45° to the principal stress direction. Experiments consistently indicate that when materials fail in shear, the shear plane lies at something less than 45° to the principal stress direction. This difference was attributed by Navier to internal friction

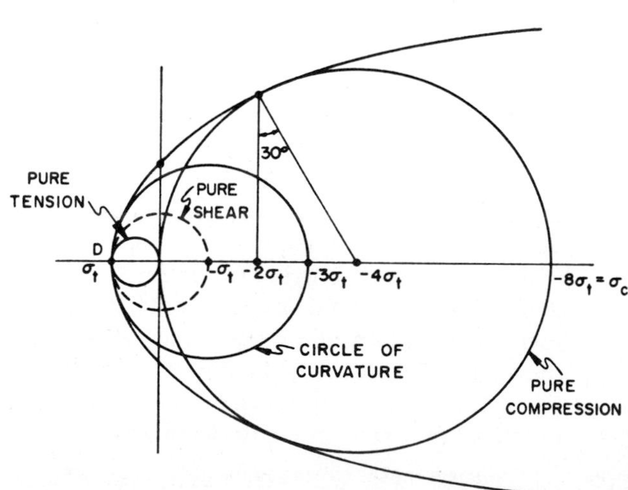

FIGURE 5-4 Stress circles for condition of pure tension, pure shear, and pure compression, showing the ratio between σ_c (compressive strength) and σ_t (tensile strength) for Griffith's hypothesis. (*From Odé, 1960.*)

and cohesive properties of the materials.* One method of evaluating the internal friction of sand or soil involves use of a specially designed container, Fig. 5-7. The amount of load N can be varied, and the amount of shearing stress necessary to start movement for each increment of load is determined. The ratio of shearing stress to normal stress is found to be a constant equal to tan ϕ, Fig. 5-8; ϕ is called the *angle of internal friction*. It is about 30 to 35° for sand.

For unconsolidated materials,

$$\frac{\tau}{N} = \frac{\tau}{\sigma} = \tan \phi$$

The French physicist Coulomb recognized the difference between cohesive materials such as rocks and noncohesive materials such as sand and formulated an equation to represent tan ϕ for cohesive materials as follows:

$$\tan \phi = \frac{\tau - \tau_0}{\sigma}$$

where τ_0 is the cohesive shear stress typical of the material. This equation can be rewritten as $\tau = \tau_0 + \sigma \tan \phi$, and it can be plotted on a diagram of τ against σ, lines of failure for the material with a specified τ_0 and ϕ, Fig. 5-8. Shear fractures are oriented so that the acute angle of intersection is bisected by the principal stress. The two conjugate* shears intersect in the intermediate stress direction.

A plot of Mohr circles showing successive increases in the amount of compression illustrates results of a triaxial test, Fig. 5-6. In these tests the confining pressure is built up to a certain value; then the directed pressure is increased until the maximum shearing stress is obtained and the material fractures. The confining pressure is then increased and the next test run. The tangent to these circles, called the *envelope*, approximates fracture orientations obtained in soil mechanics tests with loose granular material and in some ductile materials. The Coulomb criterion of failure which defines a pair of straight lines, Fig. 5-8, corresponding to fracture orientations yields a good agreement with experimental results in rock deformation for compressive tests, but the Coulomb criterion is not satisfactory for extension tests in which the observed angle is much smaller than the predicted angle.

* Brace (*in* Riecker, 1968, p. 82) points out the difficulty in explaining this cohesive property and prefers not to follow this usage.

* *Conjugate* means related to the same stress state.

FIGURE 5-5 Condition of failure. (*a*) The field of tensile failure is shaded. (*b*) A Mohr circle drawn for conditions of incipient tensile failure. (*c*) The field of failure due to shear. (*A*) indicates the shear strength of the material.

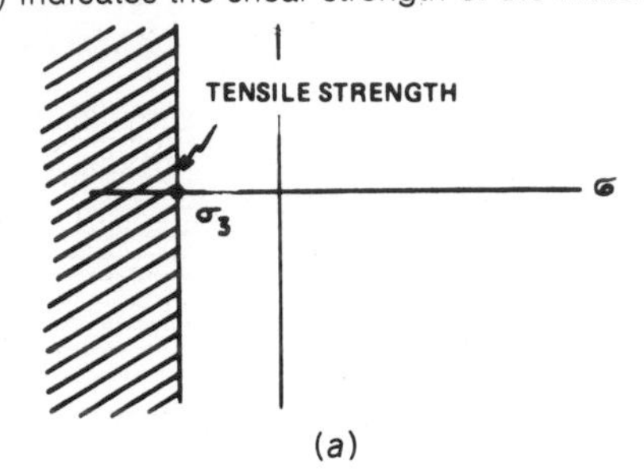

(*a*)

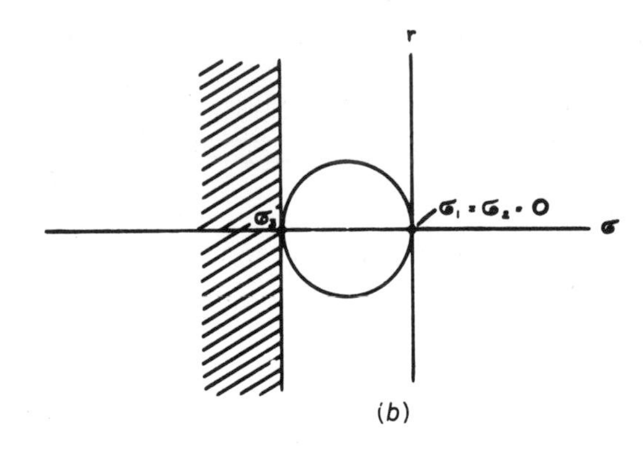

(*b*)

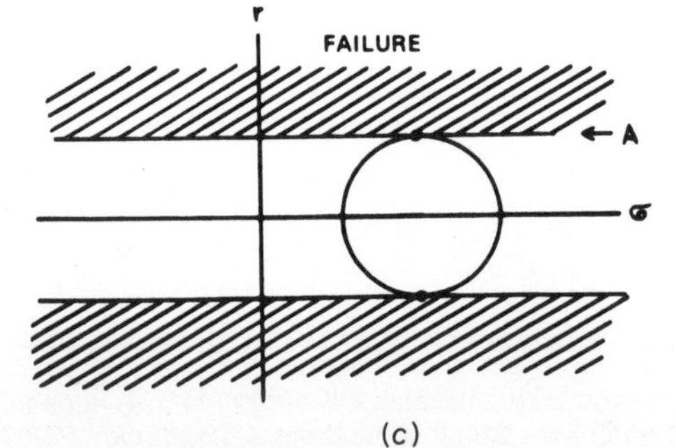

(*c*)

Mohr found that the angle ϕ does not remain constant for all pressures. It is not wholly a property of the material; it is also a function of the pressure. Thus Mohr envelopes are not necessarily straight lines.

Griffith's Theory of Brittle Fracture

Griffith (1920, 1924) formulated the theory that fractures in brittle materials (he used glass) are propagated from small, randomly oriented cracks in the material. The theory was first proposed for tensile fracture and later extended to failure in compression. These cracks have been thought of as highly eccentric ellipsoids in the mathematical development of the theory. Griffith analyzed the stresses on cracks of different orientations relative to the principal stresses to determine which cracks would be stable and which would be propagated. He found that the stress is increased on one edge of a favorably oriented flaw, causing the fracture to be propagated normal to the tensile stress. The flaws might be pore spaces, grain boundaries, inclusions, etc.

Odé (1960) summaries conclusions from the theory:

1. In pure tension the crack requiring the smallest amount of principal stress for instability is normal to the principal stress direction.
2. In compression the crack leading to instability is oriented at 30° to the direction of principal stress.
3. The envelope of all limiting stress circles for rupture can be computed and has the shape of a parabola (Fig. 5-4). This is significant because the envelopes derived from tests on specimens of many rocks under hydrostatic pressure are nearly parabolic.

There are a number of drawbacks to the Griffith theory, notably that the theory assumes elastic behavior of the material up to the moment of rupture—a condition which is not fulfilled in cases of sand, clay, and other materials under high confining pressure and temperatures.

Brittle Fracturing in Experiments

Experiments with brittle materials have repeatedly shown the development of fractures oriented parallel to the maximum stress direction (direction of the largest compressive load), called *axial fractures*, Fig. 5-9. They are oriented in the plane of the maximum and intermediate stress directions. If compression continues beyond the initial formation of these fractures, zones of fractures develop about 45° to the compressive stress, Fig. 5-10, and this orientation is the one most commonly reported in experiments and from field studies.

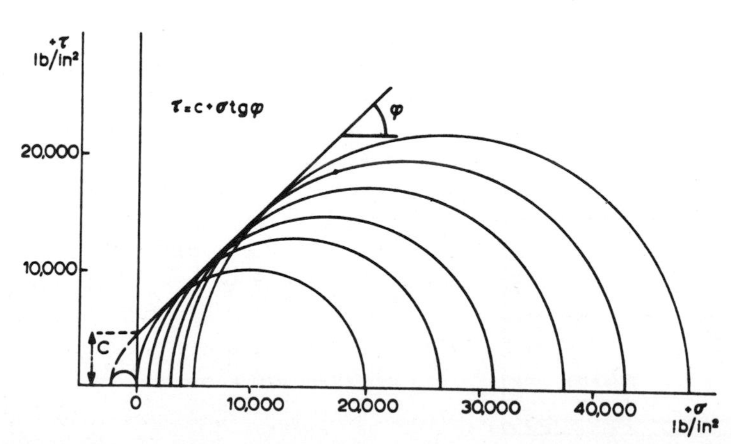

FIGURE 5-6 Mohr circles with envelope for Pennant sandstone. (*After Price, 1958.*)

One of the notable experiments of Gramberg (1965) involved compression of a cylinder of a single crystal of salt in which the cleavage planes were oriented at about 36° to the direction of maximum compression. Despite this, a crack which initially developed along a cleavage cut across the cleavages to become oriented parallel to the load axis. Thus this orientation is a highly preferred fracture orientation. If deformation is carried beyond the stage of initial formation of fractures, the rock becomes increasingly subject to cataclastic deformation. The specimen's shape changes by shortening and bulging, the volume increases, and an oblique shearing off occurs.

A representation of progressive stages in brittle deformation is shown in Fig. 5-10 on a Mohr diagram for compressive loading. The diagram is divided into zones (Gramberg, 1965):

A. Unfractured material. This is the zone of purely elastic behavior; there are no structural changes; and boundary I marks initial development of fractures.

B. The zone in which cataclastic fracturing is prominent. The test piece is now inhomogeneous, and lateral expansion occurs.

C. Cataclastic fracturing continues accompanied by development of shearing zones and sliding movements; "cataclastic plastic flow" results in collapse by shear.

D. Collapse by shearing off.

E. Zone of large cataclastic plastic flow. Envelope II is that predicted according to Mohr-Coulomb criteria; envelope III is a branch for large cataclastic plastic flow.

OCCURRENCE OF FRACTURES

Fractures occur in most igneous, metamorphic, and sedimentary rock bodies, and they are commonly found in unconsolidated sediment. Fractures usually occur as sets of subparallel trend and sometimes have even spacing that break the rock up into blocks; two or more sets are common. Fractures are important as reservoirs for ground water, oil, and gas; they are used in quarrying operations to remove large building stones; they often present structural problems in dam and foundation engineering; and they can provide valuable clues to deformational history.

Fractures are found in undeformed, unconsolidated sediment, in slightly arched coal, in all plutons, and in most extrusive bodies; highly complicated patterns are found in orogenic belts and in the Precambrian crystalline complexes. The widespread occurrence of fractures and their presence in all rock types suggests a multiplicity of causal conditions. The origin of fractures must be considered in the light of the particular geological circumstances in which they are found. The fractures which formed in experiments described earlier were the result of stresses applied externally to the specimen. These were imposed

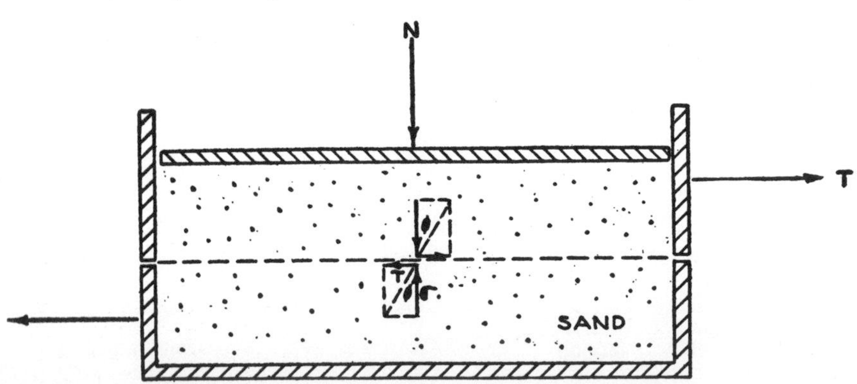

FIGURE 5-7 Test box for measuring τ/σ, the ratio at which slippage occurs. *(From Hubbert, 1951.)*

on the material, but many fractures must originate in the absence of external stresses. Fractures in undeformed lava flows and in dried lake muds prove this. It is likely that most fractures in dikes, sills, and plutons form as a result of internal stresses set up as a result of cooling and contraction. Fracture sets are usually well developed in flat-lying sedimentary rocks, both consolidated and unconsolidated, that show no sign of strong deformation. These and the beautifully developed cubic fractures so typical of bituminous coal, even where it is flat-lying, prompt us to look for explanations of these fractures in processes such as consolidation, compaction, and drying—the diagenetic processes. Such fractures may be intrinsic in the materials in which they occur.

Characteristic relationships between the direction of the deforming stresses and the attitude of the fractures under conditions of compression and extension have been found in experiment, confirmed in theory, and substantiated by numerous field observations. However, fracture patterns can be complex and can originate in various ways. Some are imposed on the rock by an external stress field of local character such as that associated with a fold or fault; some cannot be related to external stresses, but must arise as a result of internal strains; some are part of regional patterns the origin of which is usually open to debate; some result from superimposed deformation; some are intrinsic in the type of material and the response of that material to stress; and some arise from boundary conditions due to local irregularities.

In order to make the most intelligent interpretation of fractures, careful attention should be given to:

1. The pattern of the fracture system and the relative age of the different sets.
2. The nature of features on the fracture surfaces.
3. The geometrical relations of the fracture pattern to the structure of the rock mass in which the fractures occur.
4. The variations in pattern on various parts of the larger feature on which they occur.

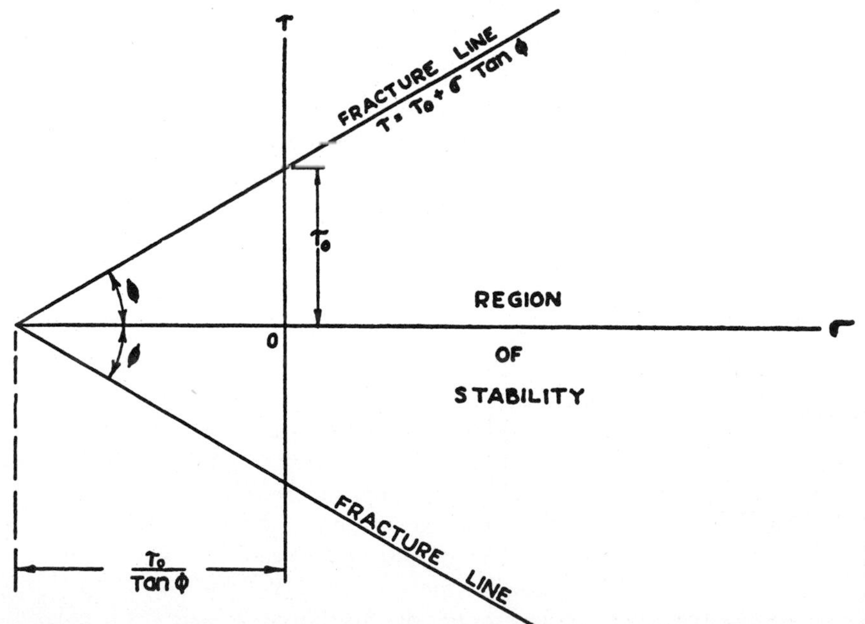

FIGURE 5-8 Position of the fracture lines on $\sigma\tau$ diagram for cohesive materials. (*After Hubbert, 1951.*)

5. The relationship of the fractures to the internal fabric of the rock.

6. The physical properties of the rock and other features of the rock which might indicate the conditions under which fracturing took place.

In addition, important conclusions may be reached from examining the spacing, regularity, number of directions, fillings, and relationship of fractures in one layer to those above and below.

Features on Fracture Surfaces*

The main fracture plane often contains ray-like ridges which diverge from a central axis to form a larger plumose structure similar to fracture surfaces produced experimentally by axial loading, Fig. 5-9. Many fractures of this character may form as a result of vertical loading where they are normal to bedding—particularly those in unfolded strata. The rays are the traces of cross fractures oriented in *en échelon* fashion on the main fracture surface.

* See Woodworth (1896), Hodgson (1961), and Gramberg (1966).

FIGURE 5-9 Brittle failure resulting in fractures oriented in the sigma 1 – sigma 2 plane. Note the feather-like markings on the fracture surfaces. (*From Gramberg, 1965.*)

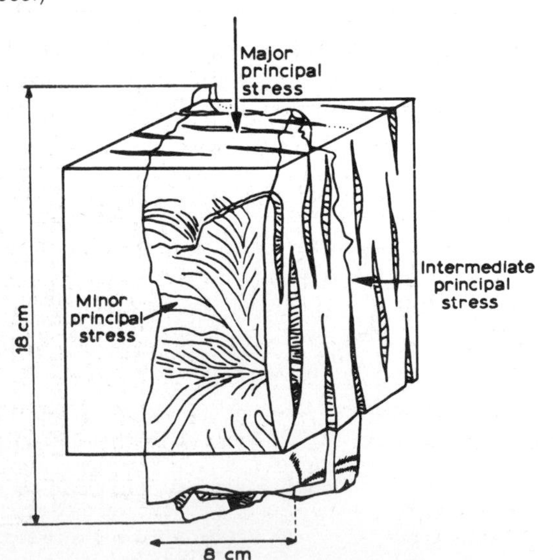

Slickensides may appear where slight displacement occurs along fractures. Such striations appear only when the surfaces of the fracture are forced together during slippage; they should not be expected on fractures formed in tension except in cases where the rock mass has been subjected to a second deformation.

Gramberg (1966) concludes that most fractures which do not have slickensided surfaces or show offsets are of tensional origin. A great many fractures show no offset, but fracture sets are frequently too irregular to allow proof of a consistent offset direction.

DETERMINATION OF THE AGE OF FRACTURES

Dating fractures with precision is very difficult. Even the obvious statement that the fractures in a sedimentary rock are formed after the rock was deposited is made complicated by the observation that fractures in sediment and in sedimentary rocks may be inherited from the underlying "basement" rock, as in the Grand Canyon (Hodgson, 1965). This is borne out by the similarity of fractures in Precambrian crystalline rocks in the Canadian Shield and in unconsolidated sediment overlying them. The fracture pattern formed under these circumstances may be related to a stress system which acted on the basement before the overlying cover was formed. The phenomena of inherited fractures seems verified, although the mechanism is poorly understood.

Fracture sets in layers of different lithology in a simply deformed stratified sequence often have different orientations. This stands in strong contrast to the idea of inherited fracture patterns. It is not valid to assume that fractures of different orientation in such a sequence were formed at different times or under different stress fields.

Open fractures are excellent avenues for the movement of fluids, and veins of calcite, quartz, and epidote are among the most common vein fillings found in fractures. Dikes and other types of fracture fillings provide the most useful means of dating fractures, or at least making a minimum age determination for the injected fracture set. The age of pegmatite dikes and veins can be dated by radiometric methods, and other dikes can be dated by the age of the youngest units intruded, by unconformities that cut the dike, and by other crosscutting relationships. A classic study of differentiation of dikes of different ages was made in Scotland by Richey (1939). Extensive dike swarms and related intrusions of different trends are found there. These dike swarms are thought to be related to different regional stress systems that affected the area at different times. The fracture sets may be dated relative to one another by the crosscutting relationships of one set to another. If each set of dikes can then be recognized by its particular petrography, the dikes of each age can be recognized in areas where cross-cutting is not found. Printz (1964) made effective use of this technique in a study of Precambrian dikes in the Beartooth Mountains of Montana and Wyoming.

A more tenuous argument about the charac-

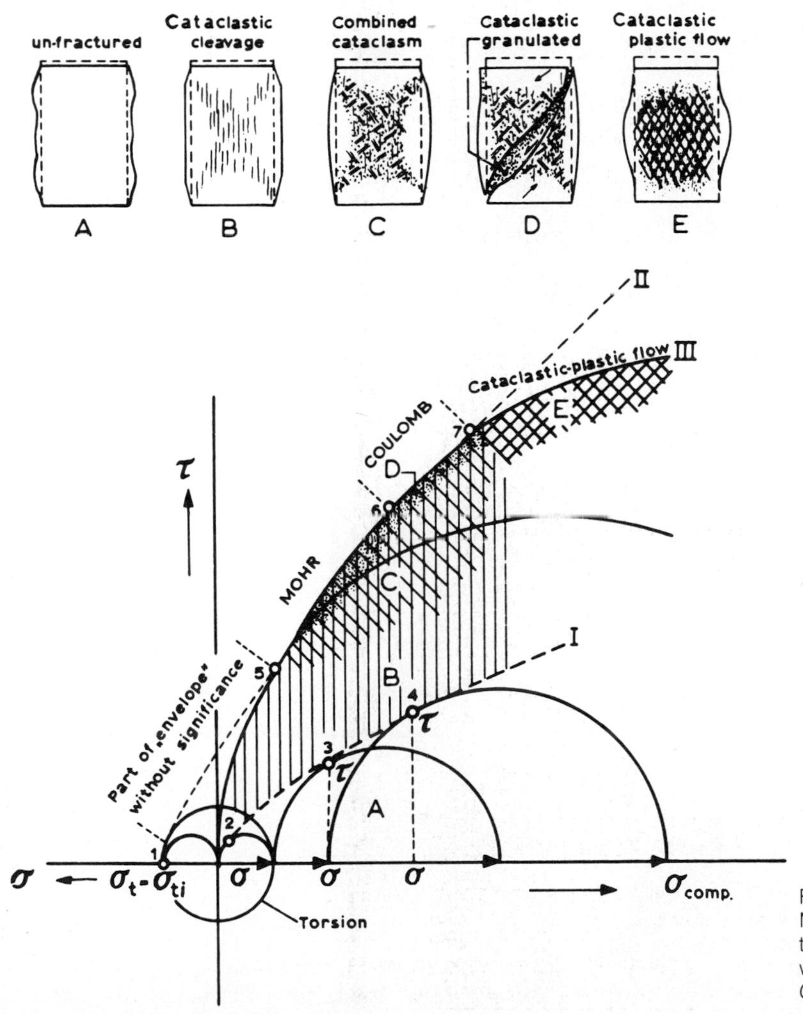

FIGURE 5-10 Representation by Mohr circles. Rock material passes through various structural stages when compressively loaded. (*From Gramberg, 1965.*)

ter of fractures during intrusion can be made on the basis of which fractures are injected. Intrusion is easier along fractures that are open at the time of intrusion. The opening of one or more sets suggests that the area might have been under tension along lines normal to the open sets.

DETERMINING THE FRACTURE PATTERN

Determination of the fracture pattern is the initial problem in most fracture studies. The pattern is known when it is possible to describe the number and orientation of fracture sets of parallel or subparallel trend at a point, and to describe the way the geometry of the fracture sets varies from place to place over the area or structural feature being studied. Various techniques are used to established the geographic distribution of the fracture sets. These usually involve determination of the number and orientation of fracture sets within a small area, called a *station*, and a comparison among the various stations. If the local geology is known, stations may be intentionally located on fold axes, fold limbs, or in other positions selected to establish a relationship between the fractures and the larger structural feature on which they occur. When the struc-

FIGURE 5-11 Aerial photograph of fractures in the Colorado Plateau. (*U.S. Geol. Survey.*)

ture is unknown, stations may be set up on intersection points on an arbitrarily established grid system. Regional patterns are often measured from aerial photographs on which lineaments can be seen, Fig. 5-11.

Various methods of sampling are used in efforts to determine fracture patterns. If only one set of fractures is present the task is simple. Usually, however, several sets are present, and slight variations occur in strike and dip of fractures belonging to a set. Thus a statistical approach is used (Pincus, 1951). The strike and dip of a large sample (usually 100) of fractures is measured and plotted on a point diagram (see Appendix D). The diagram is then contoured, and the centers of concentrations are used as the attitudes of the fracture sets.

One of the most difficult problems in the interpretation of the contoured point diagram is to determine which of the concentrations are significant. Sometimes only a few centers of concentration appear, but more often there are several low-value maxima. When the number of isolated concentrations is great, some criterion has to be used to separate those which are to be considered significant. The Poisson exponential binomial limit may be used to give an approximation of the probability that various levels of concentration will occur on a point diagram (Pincus, 1951; Spencer, 1959). A perfectly uniform distribution of 100 measurements would consist of one point in each 1 percent counting area. The Poisson exponential binomial limit gives the probability of finding more than each of the following number of points in any 1 percent area as follows:

Points	Probability
0	1.00
1	0.63
2	0.26
3	0.08
4	0.02
5	0.005
6	0.0006

The changes of having a 4 percent concentration in a random distribution are therefore approximately 1 in 50. The angle between frac-

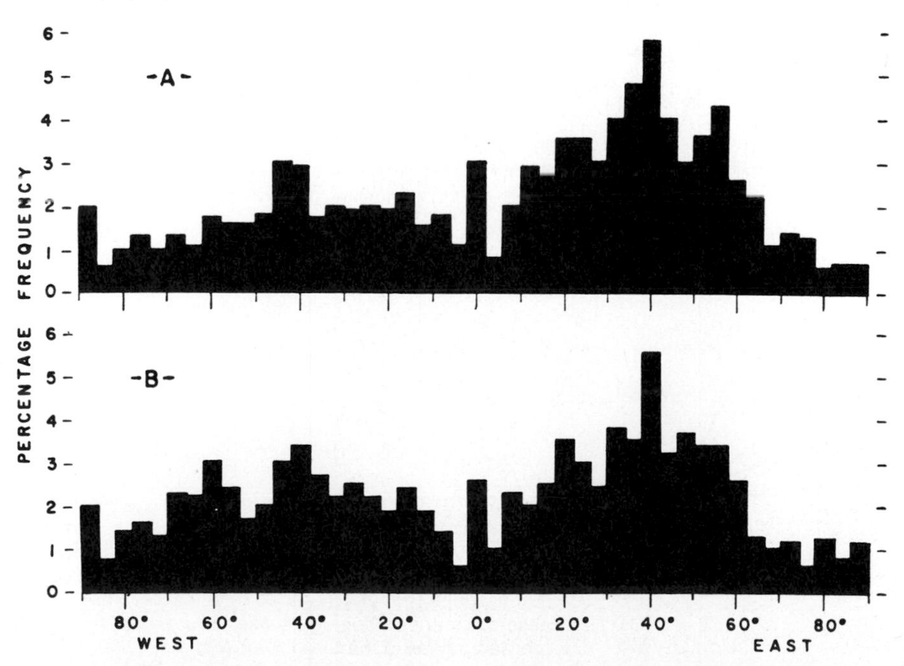

FIGURE 5-12 Histograms of lineaments identified as fractures on aerial photographs.

ture maxima is readily determined from the stereogram. If the fractures are all vertical or if a study is being made of fracture lineaments from aerial photographs, the data may be represented in a histogram or a rose diagram, Figs. 5-12 and 5-13.

Several sets of fractures frequently appear together, and sometimes, especially in areas of multiple deformation, the pattern is extremely complex. It is most helpful in later analysis if some means of differentiating sets in the field can be found. Among useful features are fraction fillings, features on fracture surfaces, and spacing of fractures. In the absence of suitable features to differentiate the various sets, a large random sample must be collected.

Fracture Spacing

Little attention has been given in field studies to fracture spacing; however, Stearns (1968) shows that this is a potentially valuable tool in fracture analysis. Most rocks behave brittly at shallow depths, but as confining pressure is increased in experiments, ductility contrast becomes more apparent.

Stearns (1968) studied the spacing of fractures in rocks of different types on an anticline that was buried under approximately 10,000 ft (roughly 3,000 m) of overburden at the time of deformation—enough to insure that effects of confining pressure on rock of different types should affect their relative ductility. The *fracture density* (intensity of fracturing of some authors) was measured by use of a fracture spacing index (FSI)—the average number of fractures per 100 feet of distance normal to the fracture plane (e.g., index of 100 means one fracture per foot). A good agreement of field measurements with laboratory predictions was found.

The rock types in order of increasing ductility were:

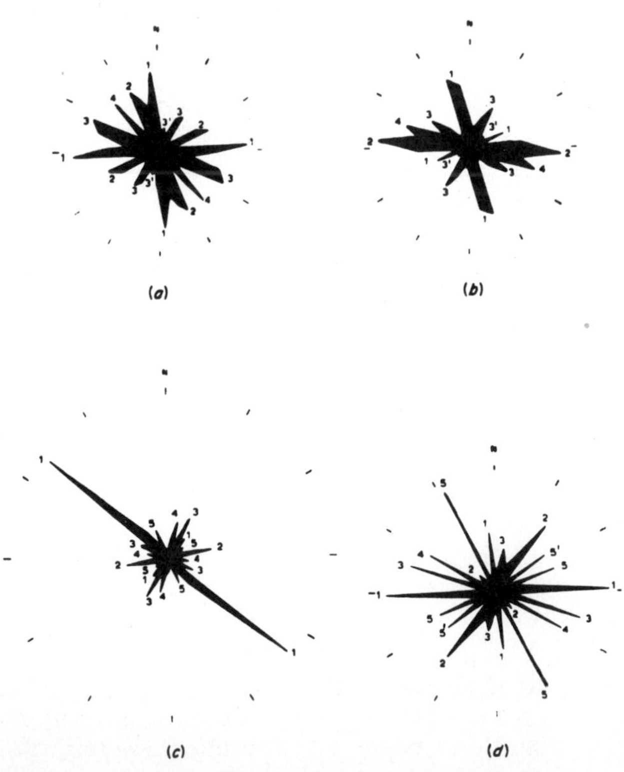

FIGURE 5-13 Rose diagrams showing the directions of lineations, revealed by aerial photographs, southern Saskatchewan. These directions are similar to those of the fracture patterns in the underlying bedrock. Each direction might be accompanied by a usually less-pronounced conjugate one, approximately perpendicular to the first. (*After Mollard, 1959.*)

High FSI	Quartzite	Brittle
	Dolomite	↓
	Silica-cemented sandstone	
	Calcite-cemented sandstone	
Low FSI	Limestone	Ductile

Fracture Patterns Related to Folds*

Well-defined fracture patterns can often be related to folds on which they occur. A number of mechanically different folding mechanisms are known (see Chap. 11). The fractures discussed here are those found on folds formed by a type of buckling mechanism in which individual layers in a sequence flex, and bedding plane slippage occurs as folding proceeds to allow development of concentric folds.

Folds of this type are seen in very simple development in shelf ice like that on the Ross Sea. The layer of ice is bounded by essentially free surfaces top and bottom, and beautifully developed doubly plunging folds are found in the ice. The most frequently encountered fractures on these ice folds are depicted schematically in Fig. 5-14. All fractures shown are vertical or steep and are labeled E for extensional and S for shear, with subscripts indicating fracture sets. Thus S_1 and E_1 belong to a single set. These fracture orientations are similar to those found in experimentally deformed rocks which bear the relationship shown in Fig. 5-2 to the principal stresses (σ_1 parallel to E_1; σ_2 is vertical; σ_3 is parallel to the fold axis). E_1 has the orientation of extension fractures described from experiments, and the two orientations of S_1 are conjugate shears bisected by σ_1. Note also that the fold axis is normal to the maximum principal stress, as we would expect it to be. Thus fractures E_1 and S_1 (set 1) may be related to the regional stress pattern responsible for the flexural folding. These fracture orientations are also frequently found on flexural folds in sedimentary rock sequences.

Another fracture direction (E_2) occurs parallel to the fold axis in strike and is particularly well developed along the crest. These fractures are often open in ice folds, and they form as the layer buckles. The top of the buckled layer is brought into tension as buckling occurs, and this tension is relieved by development of extension fractures along the fold crest. The fracture originates as a result of localized stresses related to the folding rather than to regional stress pattern.

Stearns (1968) describes a fracture study on the Teton anticlines located on the eastern side of the Sawtooth Range, Montana, in which four distinct sets of fractures are found. Each set consists of an extension and two conjugate shear fractures, Fig. 5-15. The relationship between the extension and conjugate shears is consistent, and these are thought to bear a consistent relationship to the principal stress orientations at the time of their formation. A simple way to describe these relation-

* Refer to Stearns (1968) for more details. Regional study of fractures on broad uplifts in the Colorado Plateau has been made by Kelley and Clinton (1960).

FIGURE 5-14 Sets of vertical extension fractures (E) and vertical shear fractures (S) formed in response to the pressure indicated by the arrows.

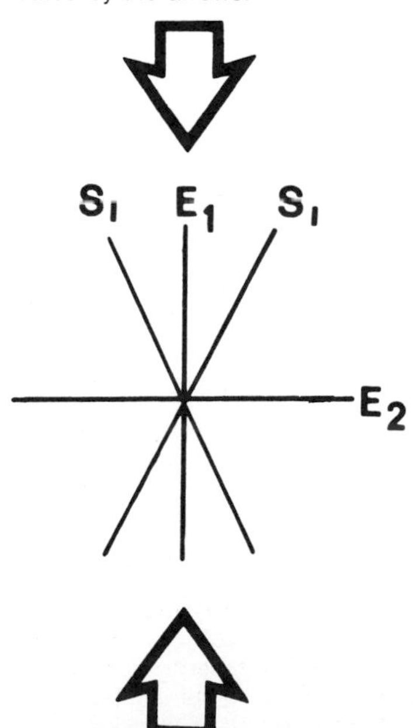

ships is to note that the three fracture orientations of each set intersect in a common line that corresponds to the orientation of the intermediate stress direction σ_2.

Set 1(Fig. 5-15). This has the orientation described for the folded ice. The fractures are orientated so that σ_1 is normal to the fold axis and σ_2 is vertical. These may be early formed fractures reflecting regional stresses. E_1 is perpendicular and normal to the fold axis as in experiments with brittle materials.

Set 2. These fractures are oriented as might be predicted to account for tensional stresses over the flexed layer. The shears are oriented as we might expect if σ_1 is parallel to the fold axis, but remember that shear fractures develop in cases of extension as shown in Fig. 5-2, case 3, without any applied compression.

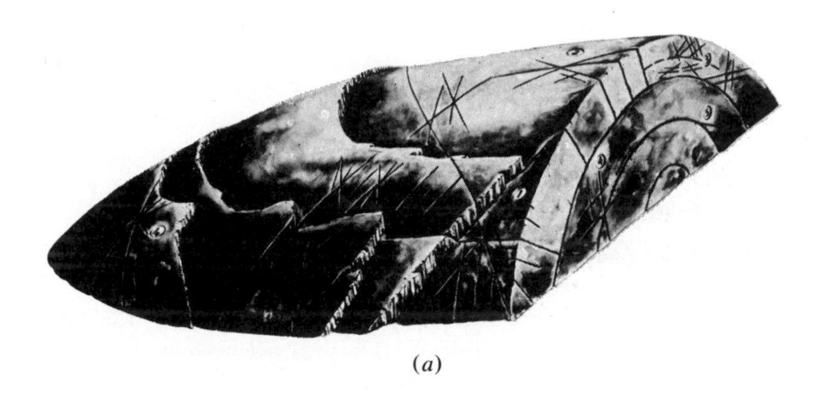

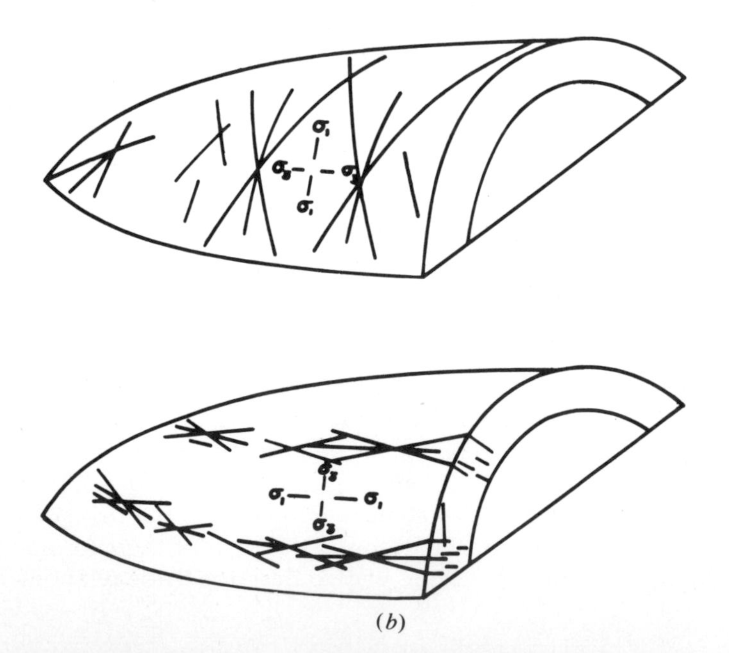

FIGURE 5-15 Schematic sketch of a folded sedimentary sequence showing the various sets of fractures which may be expected to form. Below, orientation of principle stresses responsible for some of the fracture sets. (*After Stearns, 1968.*)

E_1 is normal to the bedding and parallel to the fold axis in trend as expected for extension of the layer normal to the fold axis.

Set 3. The third set is oriented so that σ_1 appears vertical and σ_2 parallel to the fold axis. The shears are steep dipping relative to bedding. Stearns found this set only locally developed and the fractures short. The extension fracture that would be with this set is identical in orientation to that of set 2. The shears of set 3 are oriented relative to principal stresses as in normal faulting (Fig. 5-15). These may form in response to extension down dip as in set 2 but with a different orientation for σ_2.

Set 4. The shears of this set are low-dipping surfaces oriented as thrust faults and sometimes showing fault effects. The extension orientation for this set would coincide with bedding. This set, as in set 1, implies compression within layers across the fold axis. Thus elements of this set appear to reflect the regional stresses causing the folding as does set 1, but with relief (σ_3) upward instead of laterally as in set 1. Set 1 and E_2 are most frequently encountered. Sets 1 and 4 appear most closely related to regional stresses that cause the buckling. Sets 2 and 3 and possibly 4 form only after the fold is growing. Set 2, in particular, is related to stretching of beds normal to the fold axis.

Fractures in Plutons

Intrusions often follow preexisting fractures, but other fractures are formed by the forceful injection of magma. Fractures within a pluton appear to arise primarily from cooling and contraction of the magma, from fracturing along compositional layering or the planar flow structure, and as a result of denudation of the pluton and consequent development of sheeting.

Primary Fracture Systems*

Many fractures in intrusions are related to flow structures, Figs. 5-16 and 5-17. Balk

* See Hutchinson (1950) and Balk (1937).

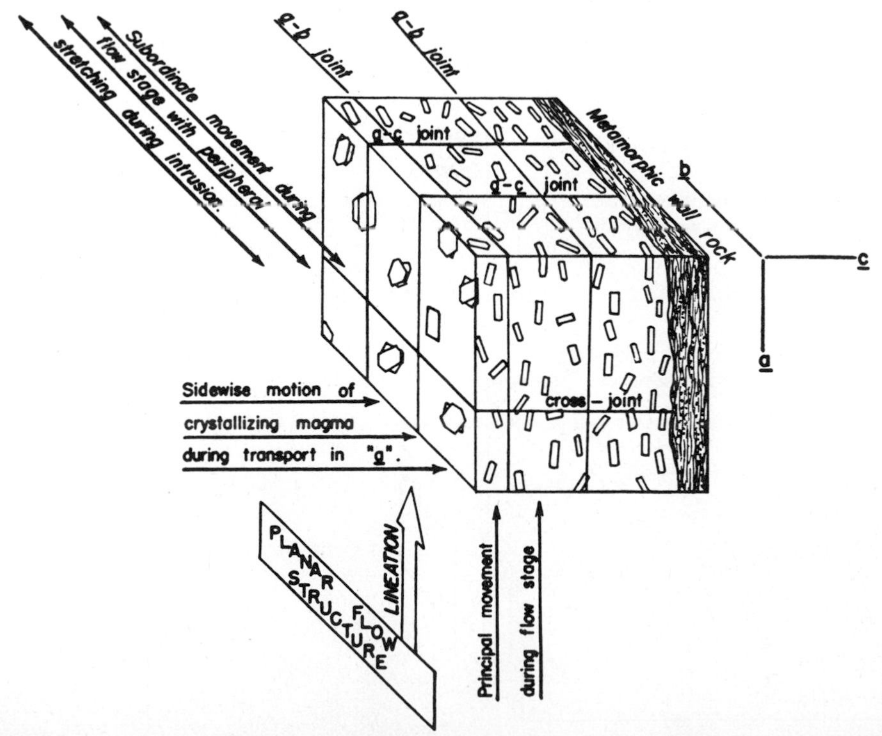

FIGURE 5-16 Relationship of primary fractures to primary flow structures. (*From Hutchinson, 1956.*)

named the fractures cross, longitudinal, or diagonal, based on their relationship to flow lines. The cross fractures are nearly perpendicular to flow lines; they are commonly filled with pegmatite, other hydrothermal deposits, and dikes. They are among the first fractures to form in the cooling magma and sometimes extend well beyond the intrusion's margins. Longitudinal fractures follow the physical weakness imparted to the igneous mass by flow structures and may be due to contraction on cooling, but they are not as commonly found as the cross fractures. Diagonal fractures strike across the flow lines at angles of of about 45° and usually occur as two intersecting sets most commonly interpreted as resulting from shear of the intrusion. Flat-lying fractures, as indicated by flat-lying dikes and vein filling, indicate that a consolidating igneous mass tends to break up into slabs separated by fractures of low dip. In some instances these joints parallel flat-lying flow layers, but in others they do not. Certainly the volume decrease of a cooling igneous intrusion would partially explain this fracture orientation, since the heat would tend to be lost most readily upward.

Sheeting

Fractures, called *sheeting,* in intrusions are often found subparallel to the topography and apparently unrelated to flow or other primary structures. Sheeting is sometimes referred to as a sort of large-scale exfoliation. Jahns (1943) has described sheeting in New England granites where sheets conform to the shape of the top surface of the granite near the surface, as on a hill, but they tend to flatten and become more nearly horizontal at depth. The spacing of sheets also increases with depth. Sometimes a single joint can be traced laterally into another, and others die out laterally or end abruptly against steep fractures. This last relation is important because it is taken to indicate that the sheeting formed later than the steep fractures. The sheeting cuts across flow lines, roof pendants, xenoliths, and late-stage pegmatites. At Quincy, Massachusetts, sheeting has been reported at depths of 97.5 m (320 ft). Sheeting in modern quarries is often subject to buckling as is seen at the Mount Airy quarry, North Carolina; similar broad, thin slabs break or "pop" up in New England quarries, showing that the rock mass is under compressive strain.

Most geologists are inclined to view the origin of sheeting as being due to dilation upon the relief of a primary confining pressure—to which the rock has become adjusted —through removal of superincumbent load.*

Fractures in Dikes, Sills, and Flows

Columnar fractures (joints) of varied polygonal cross section are well-known features of

* Jahns (1943) and Johnson (1970).

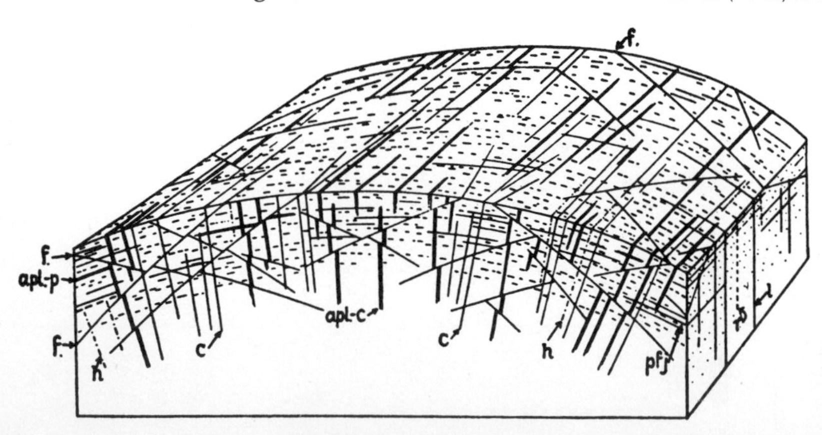

FIGURE 5-17 Primary structure elements and directions of parting in the Strehlen massif. Note sheeting fractures parallel to the surface, cross fractures that cut across the flow structures, and low-angle shears. (*Redrawn from H. Cloos by Balk, 1937.*)

tabular, shallow intrusions and of extrusions. Theoretically these fractures would form perfect hexagons if the cooling were uniform and the cooling rock homogeneous, and fractures in many bodies such as the Devil's Tower, Wyoming, and the Giants Causeway, Ireland, come close to outlining hexagonal patterns in sections, Fig. 5-18. The columns form approximately normal to the cooling face (and isotherms); therefore, in a flat-lying sill they are vertical. If the body varies in thickness the fractures fan out. A secondary fracture usually associated with the columnar joints lies in the plane normal to the long axis of the column and is commonly referred to as *sheeting*. These sheets are typically saucer-shaped, with the concave side facing upward.

FIGURE 5-18 Possible origin of curviplanar jointing. (*a*), (*b*), and (*c*) One hypothetical cross section at three different time intervals, spaced 1 msec apart, depicting the upward propagation of master columnar joints originating below, in the lower columnar zone. These have their leading edge convex-upward. (*d*) An interim period, duration unknown; stress is concentrated along the edges of the triangular prisms (here represented by triangles). (*e*) and (*f*) A hypothetical cross section spaced about 1 msec apart, showing curviplanar fractures beginning at edges of prisms and being propagated with their leading edge concave-upward and developing at right angles to the stress lines of (*d*). (*g*) The right-angle relationship sought, but not found, in the field between master columnar joints (planar joints) and curviplanar joints. (*Note:* "Edge" is used in this description in two different senses.) (*From Hill, 1965.*)

Although the simple pattern described above is common, it is by no means universal, and even in bodies which display well-developed columnar joints, the joints may be traced into fractures of much more complex character. Such is the case at the Palisades sill in New Jersey, where concentric cylinder-shaped joints, radiating planar fractures, and curviplanar fractures can all be seen near the top surface of the sill. All these are approximately normal to the sill surface. Similar cylindrical columnar fractures have been described in much thinner Precambrian dikes in the Beartooth Mountains (Prinz and Bentley, 1964). These dikes have three dominant joint sets: one parallel to the dike wall, the cylindrical set normal to the wall, and a set of radial fractures.

Hill (1965) describes probable modes of development of curviplanar and concentric fractures in a dolerite sill in Tasmania. He favors the idea that these patterns formed on the

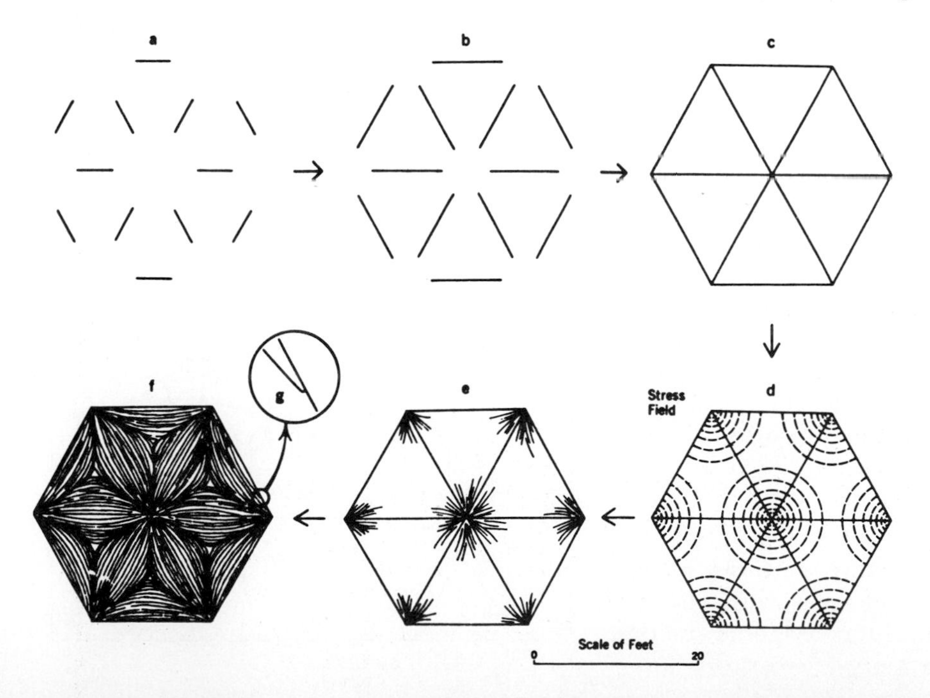

explosive release of stresses accumulated in the central part of the sill during cooling. In a cooling slab the surfaces are in tension as a result of contraction on cooling, and the center is brought under compression (Preston, 1926). As the cooling proceeds, fractures are extended from the margins into the center where the rate of advance becomes very rapid, virtually explosive. Figure 5-18 schematically illustrates the possible origin of the curviplanar fractures.

REGIONAL FRACTURE PATTERNS

A fracture pattern of definite geometry can usually be determined at any sufficiently large rock outcrop. Sometimes the pattern can be related to the structural feature (e.g., a fold, fault, diapir, or impact crater) of which the rock is a part in such a way that conclusions regarding the nature of the stress field responsible for the fracturing can be drawn. Interest in determination of regional patterns is partially derived from the hope that similar conclusions can be drawn on a regional basis where the pattern is consistent over a region or where regional components can be recognized. Where regional patterns do exist, it is important that they be identified if an attempt is to be made to relate other fractures to particular structural features. Regional patterns have been identified in many places, but explanations of their origin have not been widely accepted. Some of the more ambitious efforts have sought to establish patterns of worldwide scope. Several attempts have been made to relate most of the major faults and submarine lineaments to global shear patterns; others have concluded that most regional patterns are extensional in origin.

Regional Fractures in Foreland Regions

Analyses of regional fracture patterns have been made in the foreland areas west of the Appalachian and Ouachita orogenic belts (Parker, 1942; Nickelsen and Hough, 1967; Melton, 1929; and Friedman, 1964) and in the Canadian Rockies (Babcock, 1973). Asymmetric folds pass into broad open folds and arches before the beds flatten out and become essentially horizontal in each of these cases.

Numerous studies have been made of the relationship of fractures to folds in these foreland belts, and locally, fracture patterns can be explained in terms of the relationships of fractures to folds previously described. Especially common are near vertical fractures parallel to fold axes, at right angles to fold axes, and vertical conjugate shears, Fig. 5-19. All these can be explained in terms of shortening across the orogenic belt. However, both Parker and Nickelsen and Hough conclude that patterns in the Appalachians do not retain the above relationship to the folds everywhere. Instead, some of the sets gradually swing in trend so that their relationships to fold axes undergo gradual but continuous changes as they are traced from the southern part of the region into the Catskill Mountains at the northern end. In addition, fractures in coal and some

FIGURE 5-19 Regional fracture patterns in the Great Plains and in the Ouachita Mountain fold and thrust belts in Oklahoma. (*From Friedman, 1964.*)

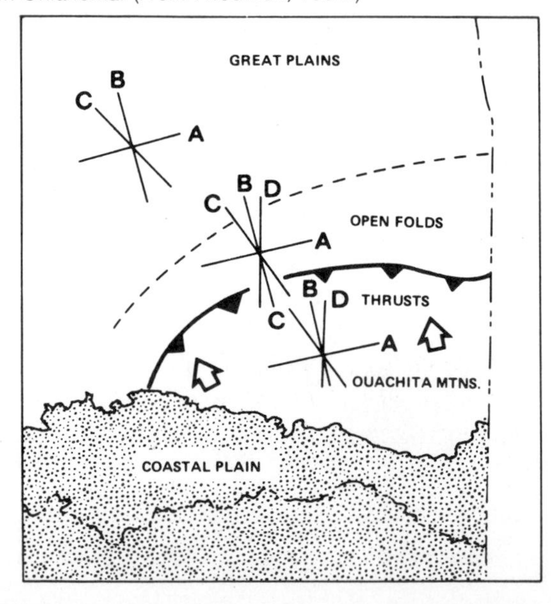

mudstones appear totally unrelated to regional stresses which caused folding, and they differ from patterns in other rocks with which they are interbedded.

Fractures in Basement Rocks*

Basement rocks in the Grand Canyon show three prominent fracture sets (Hodgson, 1961b). These fractures contain Precambrian pegmatite dikes. Faults with the same trend became activated during the Laramide deformation, and the same trends have become established in the overlying sedimentary cover. Many fracture patterns in sedimentary layers appear to have been inherited from the basement, and these zones of basement weakness have often had a controlling influence on the development of structural features in the overlying cover when they have been reactivated.

Although fracture directions in the sedimentary cover are the same as fracture sets in the basement, the Precambrian rocks often have a much more complex pattern consisting of other sets as well.† The fracture patterns are also governed by lithology and often show variation from layer to layer when compositions of layers are different.

Basement Fracture Zones

Crystalline basement rocks are almost universally broken by complex fracture systems. Sometimes fracturing is evident down to the individual feldspar crystal, but we are more aware of the larger fractures seen in outcrops and the fracture zones. Some of these zones are faults or weathered dikes, but in others, called *shatter zones,* faulting is uncertain. They are zones from a meter or so to some tens of meters wide, in which the rock is intensively fractured. These zones show up on aerial photographs as strong lineaments and are likely to be of great tectonic importance. The fracture zones typically have the same orientation as a prominent set of fractures in the rock outside the zones. These fracture zones have often localized basement yield during subsequent deformations after their initial formation, determining the position of faults and uplifts. These zones would also be the natural lines of weakness for lateral crustal movements along faults.

The origin of these basement fractures and zones of weakness is not known. They most certainly formed after the cooling of the magmas and after the high-grade metamorphism which affected the rocks in which they are developed. Mollard (1959) has pointed out that these strong lineaments also appear in flat-lying Paleozoic sedimentary rocks, and in Pleistocene deposits in the Canadian Shield. The surfaces on which they appear are as varied as glacial lake beds, alluvial flood plains, muskeg, marshes, and glacial scoured bedrock. Thus the trends are being impressed on unconsolidated materials at present.

Among the ideas which have been put forward to explain the origin of these fractures are:

1. Earth tides (these appear too weak at the present time to cause fractures initially, but they might be effective in propagating basement fractures upward into the cover and tides may have been stronger in the early Precambrian).
2. Oscillatory response (perhaps at resonance) to a nonoscillatory force such as earthquakes.
3. Crustal compression at depth.
4. Isostatic adjustments causing extension.
5. Expansion of the earth's interior causing extension in the crust.

Unfortunately, no criteria have yet been found by which we can conclusively evaluate these various possibilities.

* Gay (1973) summarizes and discusses a number of studies of basement trends.

† The complexity of basement fracture patterns is brought out in studies by Hoppin and Palmquist (1965), Wise (1964), Mollard (1959), Spencer (1959), and Spencer and Kozak (1964) in Precambrian gneissic rocks.

Gramberg (1966) advanced the idea that most vertical fractures in the crust, without shearing movement in the fracture plane, are caused by tensile stresses. Two types of external loading are recognized:

1. Direct tensile loading in which the crack forms normal to the tensile load; a single crack is formed leaving an open gap with rough walls.
2. Indirect or induced tensile loading in which the fracture plane extends in the direction of the major load axis and normal to the minor load.

These tension fractures are generated by some external process, such as convection flow in the mantle or expansion of the earth. Gramberg recognizes three zones:

1. A deep zone in which the rock is in a ductile state incapable of brittle fracture.
2. A zone in which rock is in a brittle state and deep enough that vertical lithostatic pressure is sufficient to cause indirect tensile fracturing.
3. A near-surface zone in which rock is brittle but too shallow for the lithological pressure to induce fracturing.

According to Gramberg's model, most regional fracture patterns, particularly those in basement rocks, form as a result of ductile flow in zone 1 which causes destressing in horizontal directions within zones 2 and 1. Two vertical fracture systems may form at right angles to one another if both the least and the intermediate stress directions are horizontal, and multiple vertical fracture directions may arise from several periods of destressing.

GLOSSARY FOR ROCK FAILURE —FRACTURE

Axial fracture: A type of fracture in which the fracture planes fan out at small to large angles on each side of the axial planes of folds.

Biextensional strain: A distortional strain involving simple extension along two axes.

Brittle: Said of a rock that fractures at less than 3 to 5 percent deformation or strain.

Cataclasis: Rock deformation accomplished by fracture and rotation of mineral grains or aggregates without chemical reconstitution.

Cataclastic flow: Flow involving intergranular movement, i.e., mechanical displacement of particles relative to each other.

Columnar jointing: Parallel, prismatic columns in basaltic flows and other intrusives and extrusives, formed as a result of contraction during cooling.

Compressive strength: The maximum compressive stress that can be applied to material, under given conditions, before failure occurs.

Coulomb's criterion of failure: A criterion of brittle shear failure based upon the concept that shear failure will occur along a surface when the shear stress acting in that plane is sufficiently large to overcome the cohesive strength of the material plus the frictional resistance to movement.

Cross fracture: A fracture plane oriented perpendicular to the major lineations and structural trends of a region.

Curviplanar fractures: Fractures whose surface is oriented from the curving of a plane about one or more axes.

Diagonal fracture: A fracture plane whose strike is oblique to that of the strike of the general structural trend of a region.

Ductile: Said of a rock that is able to sustain, under a given set of conditions, 5 to 10 percent deformation before fracturing or faulting.

***En échelon* (gash) fractures**: Small-scale tension fractures occurring at an angle to a fault arranged in an overlapping or staggered pattern and that tend to remain open.

Extension fracture: A fracture that develops perpendicular to the direction of greatest stress and parallel to the direction of compression.

Failure: Fracture or rupture of a rock or other

material that has been stressed beyond its ultimate strength.

Fracture (rupture): A general term for any break in a rock, whether or not it causes displacement, due to mechanical failure by stress. It includes cracks, joints, and faults.

Longitudinal fracture: A fracture plane which is oriented parallel to the general structural trend of a region.

Mohr envelope: An envelope of a series of Mohr circles; the locus of points whose coordinates represent the stresses causing failure.

Nonrotational strain: Strain at a point, in which the orientation of the axes of strain remain unchanged.

Plastic flow: Permanent deformation of the shape or volume of a substance, without rupture, and which, once begun, is continuous without increase in stress.

Pressure-release jointing: Exfoliation that occurs in once deeply buried rock that erosion has brought nearer the surface, thus releasing its confining pressure.

Primary flat joint: An approximately horizontal joint plane in igneous rocks.

Rotational strain: Strain in which the orientation of the axes of strain is changed.

Shatter zone: An area of randomly fissured or cracked rock that may be filled by mineral deposits, forming a network pattern of veins.

Shear: A strain resulting from stresses that cause or tend to cause contiguous parts of a body to slide relative to each other in a direction parallel to their plane of contact.

Shear fracture: A fracture caused by compressive stress.

Sheeting structure: The type of fracture or jointing formed by pressure-release jointing or exfoliation.

Tensile strength: The maximum applied tensile stress that a body can withstand before failure occurs.

Tension: A state of stress in which tensile stresses predominate; stress that tends to pull a body apart.

Uniaxial compression: A system of forces or stresses that tends to shorten a substance along a single axis.

Uniextensional strain: A distortional strain involving simple extension along a single axis.

Uniform flow: Flow in which there is neither convergence nor divergence.

REFERENCES

Babcock, E. A., 1973, Regional jointing in southern Alberta: Can. Jour. Earth Sci., v. 10.

Balk, Robert, 1937, Structural behavior of igneous rocks: Geol. Soc. America Mem. 5.

Bombolakis, E. G., 1968, Photoelastic study of initial stages of brittle fracture in compression: Tectonophysics, v. 6, no. 6, p. 461–473.

Bonham, L. C., 1957, Structural petrology of the Pico anticline: Jour. Sed. Petrology, v. 27.

Brace, W. F., 1960, An extension of the Griffith theory of fracture to rocks: Jour. Geophys. Res., v. 65, no. 10, p. 3477–3480.

——— 1968, Review of Coulomb-Navier fracture criterion, *in* Riecker, R. E., ed., Rock mechanics seminar: Bedford, Mass., Terrestrial Sciences Laboratory.

Bucher, W. H., 1920–21, Mechanical interpretation of joints: Jour. Geology, v. 28, p. 207.

Casella, C. J., 1976, Evolution of the lunar fracture network: Geol. Soc. America Bull., v. 87, no. 2, p. 226–234.

Chapman, C. A., and **Rioux, R. L.**, 1958, Statistical study of topography, sheeting and jointing in granite, Acadia National Park, Maine: Am. Jour. Sci., v. 256.

Chiliggar, G., and **Richards, C. A.**, 1954, Use of gash fractures in determining direction and relative amount of movement along faults: Compass, v. 31.

Cleary, J. M., 1958, Hydraulic fracture theory, pts. I and II: Illinois Geol. Survey Circs. 251 and 252.

——— 1959, Hydraulic fracture theory, pt. III—Elastic properties of sandstone: Illinois Geol. Survey Circ. 281.

Cloos, Ernst, 1947, Oolite deformation in the South Mountain fold, Maryland: Geol. Soc. America Bull., v. 58, p. 843–918.

——— 1955, Experimental analysis of fracture patterns: Geol. Soc. America Bull., v. 66, p. 241–256.
——— 1968, Experimental analysis of Gulf Coast fracture patterns: Am. Assoc. Petroleum Geologists Bull., v. 52, no. 3, p. 420–444.
Donath, F. A., 1961, Experimental study of shear failure in anistropic rocks: Geol. Soc. America Bull., v. 72, no. 6, p. 985–989.
——— 1963, Strength variation and deformational behavior in anistropic rock: Internat. Conf. State of Stress in Earth's Crust Proc., Santa Monica, Rand Corp.
Dunn, D. E., LaFountain, L. J., and **Jackson, R. E.,** 1973, Porosity dependence and mechanism of brittle fracture in sandstones: Jour. Geophys. Res., v. 78, no. 14, p. 2403ff.
Fairhurst, C., ed., 1963, Rock mechanics: New York, Pergamon.
Firman, R. J., 1960, The relationship between joints and fault patterns in the Eskdale granite (Cumberland) and the adjacent Borrowdale volcanic series: Geol. Soc. London Quart. Jour. v. CXVI, p. 317–347.
Friedman, Melvin, 1964, Petrofabric techniques for the determination of principal stress directions in rocks, *in* State of stress in the earth's crust: New York, Elsevier, 1964.
——— and **Logan, J. M.,** 1970, Influence of residual elastic strain on the orientation of experimental fractures in three quartzose sandstones: Jour. Geophys. Res., v. 75, no. 2, p. 387–405.
Garson, M. S., and **Krs, Miroslav,** 1976, Geophysical and geological evidence of the relationship of Red Sea transverse tectonics to ancient fractures: Geol. Soc. America Bull., v. 87, no. 2, p. 169–181.
Gay, N. C., and **Weiss, L. E.,** 1974, The relationship between principal stress directions and the geometry of kinks in foliated rocks: Tectonics, v. 21, p. 287–300.
Gay, S. P., Jr., 1973, Pervasive orthogonal fracturing in the earth's continental crust: Tech. Pub. no. 2, American Stereo Map Co., Salt Lake City.
Gramberg, J., 1965, Axial cleavage fracturing, a significant process in mining and geology: Eng. Geol., v. 1, p. 31–72.
——— 1966, A theory on the occurrence of various types of vertical and sub-vertical joints in the earth's crust: Internat. Soc. Rock Mechanics, Cong. Proc., 1st. p. 443–450.
Griffith, A. A., 1920, The phenomena of rupture and flow in solids: Royal Soc. (London) Philos. Trans., ser. A, v. 221, p. 163–198.
——— 1924, The theory of rupture: Proc. Internat. Cong. Appl. Mechanics. 1st, Delft, p. 55–63.
Griggs, D. T., and **Handin, John,** 1960, Observations on fracture and a hypothesis of earthquakes, *in* Griggs, D. T., and Handin, John, eds., Rock deformation—A symposium: Geol. Soc. America Mem. 79, p. 347–373.
Harris, J. F., 1960, Relation of deformation fractures in sedimentary rocks to regional and local structure: Am. Assoc. Petroleum Geologists Bull., v. 44, p. 1853–1873.
Heard, H. C., 1960, Transition from brittle to ductile flow in Solenhofen limestone as a function of temperature, confining pressure, and interstitial fluid pressure, in Rock Deformation, Griggs, D. T., and Handin, John, eds., Geol. Soc. America Mem. 79.
Hill, P. A., 1965, Curviplanar (radial, boe-tie, festoon) and concentric jointing in Jurassic dolerite, Mersey Bluff, Tasmania: Jour. Geology, v. 73, p. 255–271.
Hodgson, R. A., 1961a, Classification of structures of joint surfaces: Am. Jour. Sci., v. 259.
——— 1961b, Regional study of jointing in Comb Ridge—Navajo Mountain area, Arizona and Utah: Am. Assoc. Petroleum Geologists Bull., v. 45, p. 1–39.
——— 1965, Genetic and geometric relations between structures, *in* Basement and overlying sedimentary rocks, with examples from Colorado Plateau and Wyoming: Am. Assoc. Petroleum Geologists Bull., v. 49, p. 935.
———, **Gay, S. P., Jr.,** and **Benjamins, J. Y.,** eds., 1976, Proc. 1st Intern. conf. on the new basement tectonics: Salt Lake City, Utah Geol. Assoc. Pub. no. 5.
Hoppin, R. A., and **Palmquist, J. C.,** 1965, Basement influence on later deformation: The problem, techniques of investigation, and examples from Bighorn Mountains, Wyoming: Am. Assoc. Petroleum Geologists Bull., v. 49, p. 993–1004.
Hubbert, M. K., 1951, Mechanical basis for certain familiar geologic structures: Geol. Soc. America Bull., v. 62, no. 4, p. 255–372.
Hutchinson, R. M., 1956, Structure and petrology of the Enchanted Rock batholith, Llano and Gillespie Counties, Texas: Geol. Soc. America Bull., v. 67, p. 763–806.

Jaeger, J. C., 1960, Shear failure of anisotropic rocks: Geol. Mag., v. 97, p. 65–72.

Jahns, R. H., 1943, Sheet structure in granite: Its origin and use as a measure of glacial erosion in New England: Jour. Geology, v. 51, p. 71–98.

Kelley, V. C., and **Clinton, N. J.**, 1955, Fracture system and tectonic elements of the Colorado Plateau: Albuquerque, Univ. New Mexico Pub. in Geology No. 6.

——— 1960, Fracture systems and tectonic elements of the Colorado Plateau: Albuquerque, Univ. New Mexico Pub. in Geology No. 6.

Kutina, Jan, 1975, Tectonics development and metallogeny of Madagascar with reference to the fracture pattern of the Indian Ocean: Geol. Soc. America Bull., v. 86, p. 582–592.

Lachenbush, A. H., 1961, Depth and spacing of tension cracks: Jour. Geophys. Res., v. 66.

Lovering, T. S., 1928, The fracturing of incompetent beds: Jour. Geology, v. 36, p. 709–717.

McKinstry, H. E., 1953, Shears of the second order: Am. Jour. Sci., v. 251, no. 6, p. 401–414.

Melton, F. A., 1929, A reconnaissance of the joint systems in the Ouachita Mountains and Central Plains of Oklahoma: Jour. Geology, v. 37, p. 729–746.

Mollard, J. D., 1959, Aerial mosaics reveal fracture patterns on surface materials in southern Saskatchewan and Manitoba: Oil in Canada, v. 9, no. 40, p. 26–50.

Muehlberger, W. R., 1961, Conjugate joint sets of small dihedral angle: Jour. Geology, v. 69, no. 2 p. 211–219.

Nickelsen, R. P., and **Hough, V. N. D.**, 1967, Jointing in the Appalachian Plateau of Pennsylvania: Geol. Soc. America Bull., v. 78, p. 609–630.

Odé, Helmer, 1960, Faulting as a velocity discontinuity in plastic deformation, *in* Griggs, D. T., and Handin, John, eds., Rock deformation—A symposium: Geol. Soc. America Mem. 79, p. 293–321.

Orowan, E., 1949, Fracture and strength in solids: Repts. Prog. Physics, v. 12, p. 185–232.

——— 1960, Mechanism of seismic faulting, *in* Griggs, D. T., and Handin, John, eds., Rock deformation—A symposium: Geol. Soc. America Mem. 79, p. 323–346.

Parker, J. M., 1942, Regional jointing systematic in slightly deformed sedimentary rocks: Geol. Soc. America Bull., v. 53.

Pincus, H. J., 1951, Statistical methods applied to the study of rock fractures: Geol. Soc. America Bull., v. 62, p. 403–410.

Preston, F. W., 1926, The spalling of bricks: Am. Ceramic Soc. Jour., v. 9, p. 654–658.

Price, N. J., 1958, A study of rock properties in conditions of triaxial stress, *in* Walton, W. H., ed., Mechanical properties of non-metallic brittle materials: London, Butterworth, p. 106–122.

——— 1966, Fault and joint development in brittle and semi-brittle rock: New York, Pergamon.

Prinz, Martin, and **Bentley, R. D.**, 1964, Cylindrical columnar jointing in dolerite dikes, Beartooth Mountains, Montana–Wyoming: Geol. Soc. America Bull., v. 75, p. 1165–1168.

Richey, J. E., 1939, The dykes of Scotland: Geol. Soc. Edinburgh Trans., v. 13, no. 4, p. 393.

Roberts, J. C., 1961, Feather-fracture, and the mechanics of rock-jointing: Am. Jour. Sci., v. 259, no. 7.

Savage, J. C., and **Mohanty, B. B.**, 1969, Does creep cause fracture in brittle rocks?: Jour. Geophys. Res., v. 74, no. 17, p. 4329–4332.

Secor, D. T., Jr., 1965, Role of fluid pressure in jointing: Am. Jour. Sci., v. 263.

Spencer, E. W., 1959, Fracture patterns in the Beartooth Mountains, Montana and Wyoming: Geol. Soc. America Bull., v. 70, p. 467–508.

——— and **Kozak, S. J.**, 1976, Determination of regional fracture patterns in Precambrian rocks—A comparison of techniques, *in* Hodgson, R. A., Gay, S. P., Jr., and Benjamins, J. Y., eds., Proc. 1st intern. conf. on the new basement tectonics: Salt Lake City, Utah Geol. Assoc. Pub. 5.

Spry, A., 1962, The origin of columnar jointing, particularly in basalt flows: Geol. Soc. Australia Jour., v. 8, p. 191–216.

Stearns, D. W., 1968, Certain aspects of fractures in naturally deformed rocks, *in* Riecker, R. E., ed., Rock mechanics seminar: Bedford, Mass., Air Force Cambridge Research Laboratories.

Turner, F. J., and **Weiss, L. E.**, 1963, Structural analysis of metamorphic tectonites: New York, McGraw-Hill.

Varnes, D. J., 1962, Analysis of plastic deformation according to Von Mises' theory, with application to the South Silverton area, San Juan Co., Colo.: U. S. Geol. Survey Prof. Paper 378-B, p. B1–B49.

Ver Steeg, K., 1942, Jointing in the central coal beds of Ohio: Econ. Geology, v. 37.

Vinogradov, S. D., 1960, On the distribution of the

number of fractures in dependence of the energy liberated by the destruction of rocks: Izvestiya, Geofiz. Ser., Trans. 1959, no. 12, p. 1292–1293.

Vogt, P.R., 1974, Volcano spacing, fractures, and thickness of the lithosphere: Earth Planet. Sci. Letters, v. 21, p. 235–252.

Wallace, R. E., 1951, Geometry of shearing stress and relation for faulting: Jour. Geology, v. 59, no. 2, p. 118–130.

Wilson, C. W., Jr., 1934, A study of jointing in the Five Springs Creek areas, east of Kane, Wyoming. Jour. Geology, v. 42, p. 489–552.

Wise, D. U., 1964, Microjointing in basement, Middle Rocky Mountains of Montana and Wyoming: Geol. Soc. America Bull., v. 75, p. 287–306.

Woodworth, J. B., 1896, Fracture system of joints: Boston, Soc. Natural History Boston, v. 27.

6

Impact Features

IMPACT CRATERS AND STRUCTURES*

A number of oval or nearly circular structural features usually associated with localized brecciation and strong disruption of rocks in and immediately around a central depression have long been known on earth. Such features were often interpreted as cryptovolcanic structures, thought to form as a result of a gaseous volcanic explosion although ash and lavas were generally absent or present in small amounts within the central depression. The central depressions are often filled with water which hinders investigation, but the localized nature of the intensely deformed rock and the absence of similar or associated features in the surrounding area show clearly that the cause, whatever its origin, is localized. Many of these features have a distinct crater-like form, and meteorite impact rather than volcanic explosion is now the most widely accepted view

* *Cryptovolcanic features:* Circular features formed by near-surface rapid or explosive release of steam, gas, or lava as a result of volcanic activity.

Cryptoexplosive features: Circular features resulting from near-surface explosions of any origin.

Meteorite-impact features: Circular features due to high-speed impact of meteorites.

Astroblemes: Old eroded remanents of meteor-impact features.

Selected references on impact features:

Theory:
Gault (1974)
Gault and others (1968)
Short (1966)—a review

Earth craters:
Meteor Crater, Arizona—Bjork (1962) and Shoemaker (1960, 1963).
Gross Bluff, Australia—Milton and others (1972)
Brent, Colorado—Dence (1968)
Sudbury—Dietz (1964)
Bushveld, South Africa—Rhodes (1975)
West Hawk Lake, Canada—Short (1970)

Lunar features:
Oberbeck (1975)
Chao (1974)
Head (1972, 1975)

Explosion (*nuclear*):
Cratering symposium, *in* Jour. Geophy. Res., v. 66, no. 10, (1961).

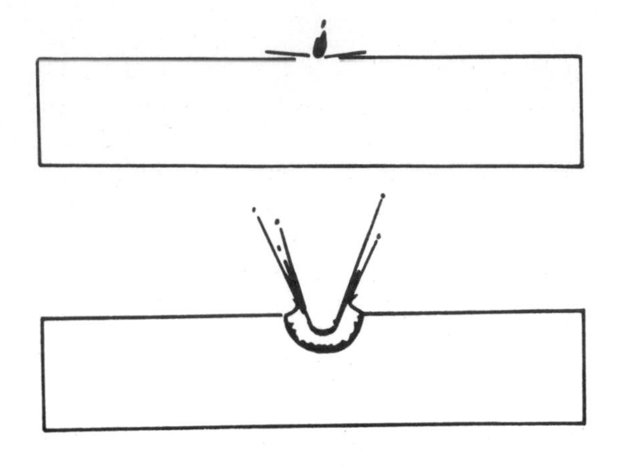

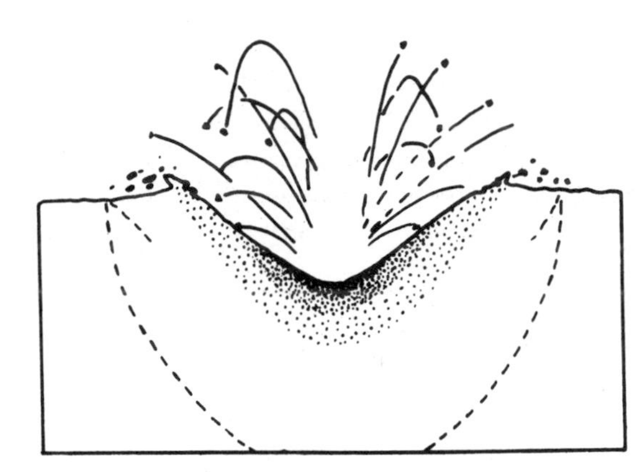

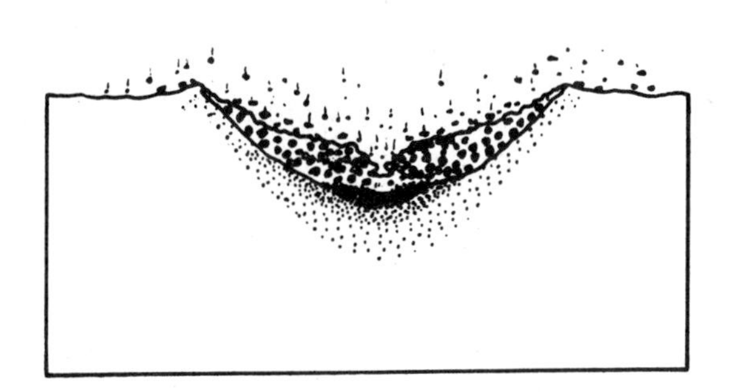

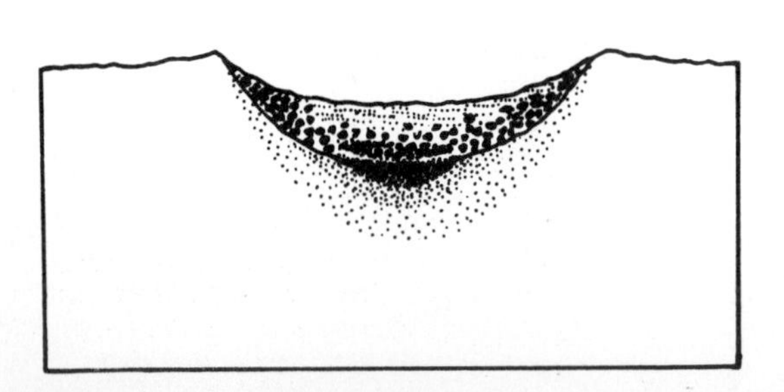

FIGURE 6-1 Schematic diagram of formation of Brent crater, a simple terrestrial impact crater, soon after the meteorite impacts (top sketch). A relatively small hemispherical volume of melted rock is produced, and ejection of some of it proceeds (second sketch). Shock-wave pressure decays enough outside this boundary so that lesser degrees of shock damage are produced in the rocks outside this region. However, melt rock is driven downward ahead of the meteoroid as the crater grows to its maximum size (third sketch). Fallback and collapse of material at the crater sides produce the final section. This contains melt rocks at the base of the transient crater and progressively less shock-damaged material with depth. Several reversals in shock damage occur above the melt rocks. [*After Dence, 1968. Reprinted from* Shock Metamorphism of Natural Materials *(1968), Mono Book Corporation, Baltimore, Maryland 21207.*]

of their origin. Investigation of meteorite impacts has been greatly accelerated as a part of lunar studies, and these investigations of lunar features are now shedding light on their earthly counterparts; craters formed by underground nuclear explosions are also yielding valuable information (Nordyke, 1961).

Some of these craters on earth are large, such as the Ries depression in Germany which measures 26 km in diameter, the Vredefort feature in South Africa which is over 40 km in diameter, and the Clearwater Lake feature in Quebec which is about 33 km in diameter. But most are about the size of Meteor Crater, Arizona (1.3 km), and Wells Creek, Tennessee (4.6 km), or even smaller, and none approach the size of the larger lunar craters, some of which are over 1,000 km in diameter.

METEOR CRATER, ARIZONA*

The meteor crater of Arizona is certainly one of the best known of the cryptoexplosion features on earth. It is an isolated depression surrounded by a rim of outwardly dipping strata located in an area of flat-lying sandstones and limestones which are generally undeformed. It is remarkably similar in shape to some of the craters formed by underground nuclear explosions, although they obviously originated by quite differ processes. The similarity

* See Shoemaker (1960), Short (1966), and Bjork (1962).

points up the explosive character of both phenomena and explains why sheets of ejected material surround large meteorite-impact features. The impact sets up a strong shock wave, both in the meteorite and the rock, which is reflected into the meteorite, shattering it into small fragments while the shock wave in the enclosing rock brecciates and in some cases melts it. Part of the breccia is ejected as a curtain of debris blown into the air and later settling back around the rim, but more of the breccia remains in the crater forming a layer of shattered rock on the crater floor.

Arizona's Meteor Crater, 1.3 km wide and nearly 300 m deep, is floored with brecciated rock and surrounded to a distance of several kilometers by soil which contains small fragments of iron estimated to be sufficient to account for a meteor of about 12,000 to 15,000 tons. The crater rim, which stands about 75 m above the surrounding plain, is composed of unsorted angular debris, ranging in size from splinters to blocks 30 m across which rest on upturned edges of the bedrock strata. Layers can be seen in the debris and be identified as having originated from the bedrock layers, but the debris layers are in reversed sequence from the corresponding bedrock layers, the highest of which are turned up steeply and locally overturned as though peeled back by the explosion. The upturned strata are broken by a number of small vertical faults which form a radiating pattern The regional joint pattern has influenced the shape of the crater and is responsible for the more or less rectilinear shape of the crater.

Characteristic Structural Features of Impact Craters

The form and types of features associated with impact craters show considerable variation related to both the size and velocity of the

* See Short (1966).

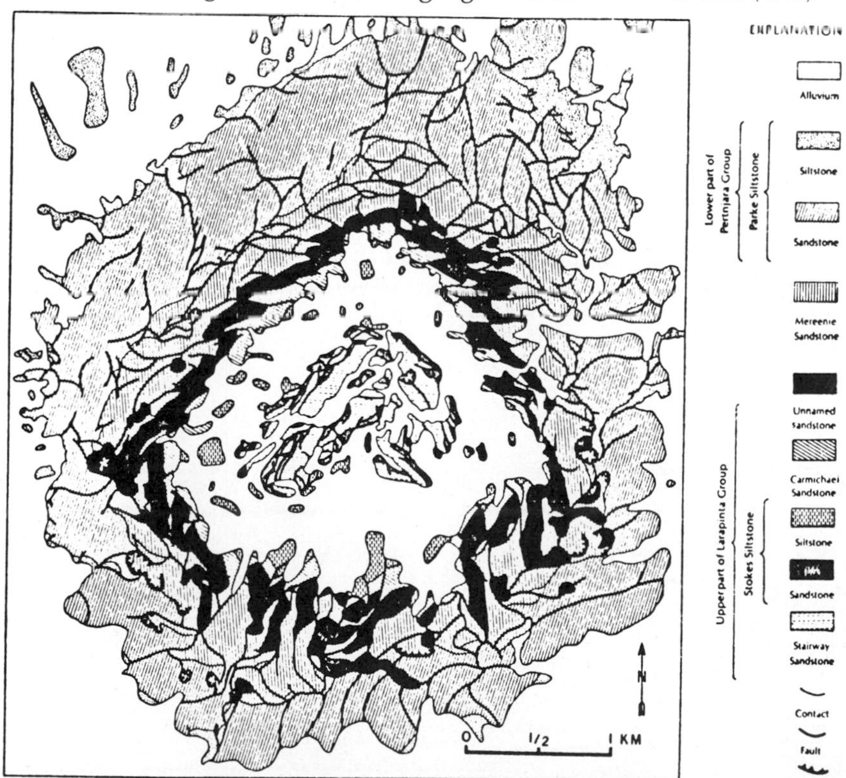

FIGURE 6-2 Geologic map of the central uplift area of the Gosses Bluff impact structure. Note the complex fault patterns around the center of the feature. (*From Milton and others, 1972. Copyright Am. Assoc. Adv. Sci.*)

meteorite and to the character (rock type, thickness, primary structure, and form) of the material that is hit. Small meteorites lose their cosmic velocities and approach earth's surface with free-fall velocities of 0.1 to 0.2 km/sec and do not form craters, but large meteorites may hit with velocities of 5 to 60 km/sec, and if they have masses greater than 10 tons they will create craters.

Breccias which result from the shock of impact resemble other types of cataclastic breccias, but they may contain lumps of glass and a variety of extremely high-pressure mineralogic (e.g., coesite and stishovite polymorphs of silica) and structural changes—referred to as *impact metamorphism*. These changes range from fracturing and damage to quartz and biotite to structural disordering of feldspars and amphibole expressed in their optical properties and finally to melting of rock. Kink bands or closely spaced planar features such as cleavages may be produced. Shatter cones, conical fracture surfaces with divergent striations along the length of the cone, are frequently present and apparently form in the compressive phase of the event so that the axes of the cones are normal to the shock front.

Fault and fracture patterns around impact craters range from highly complex localized patterns to the more characteristic combination of concentric and radial patterns such as those mapped at Grosses Bluff, Fig. 6-2, and at Wells Creek, Fig. 6-3.

REFERENCES

Bjork, R., 1962, Analysis of the formation of Meteor Crater, Arizona: Jour. Geophys. Res., v. 66, p. 3379–3387.

Bombolakis, E. G., 1973, Study of the brittle fracture process under uniaxial compression: Tectonophysics, v. 18, p. 231–248.

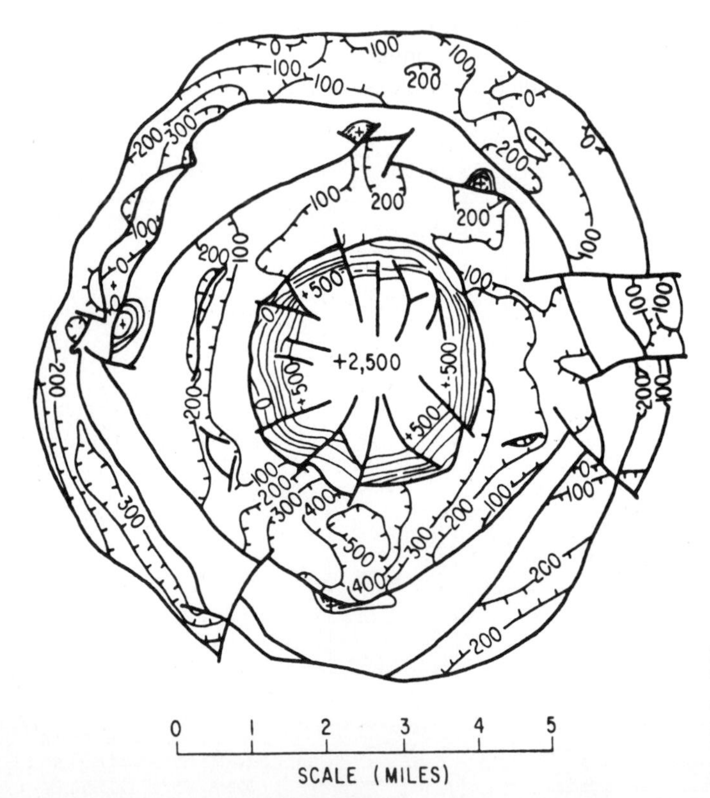

FIGURE 6-3 Structure of the Wells Creek, Tennessee, impact feature. Note the radial and concentric faults. [*From Stearns and others, 1968. Reprinted from* Shock Metamorphism of Natural Materials *(1968), Mono Book Corporation, Baltimore, Maryland 21207.*]

Chao, E. C. T., 1974, Impact cratering models and their application to lunar studies—A geologist's view: Proc. Fifth Lunar Science Conference, Supplement 5, v. 1, p. 35–52.

Dence, M. R., 1968, Shock zoning at Canadian craters: Petrography and structural implications, *in* Shock metamorphism of natural materials: Baltimore, Mono Book Corp.

Dietz, R. S., 1964, Sudbury structure as an astrobleme: Jour. Geology, v. 72, p. 412–434.

French, B. M., and **Short, N. M.,** 1968, Shock metamorphism of natural materials: Baltimore, Mono Book Corp.

Gault, D. E., 1974, Impact cratering, *in* A primer in lunar geology, Greeley, R., and Schultz, P., eds.: Ames Research Center, NASA.

———, **Quaide, W. L.,** and **Oberbeck, R.,** 1968, Impact cratering mechanics and structures: Shock metamorphism of natural materials, p. 87–99.

Head, J. W., 1972, Small scale analogs of the Cayley formation and Descartes Mountains in impact-associated deposits: Apollo 16 Preliminary Science Report, NASA Spec. Publ. SP-315.

——— 1975, Processes of lunar crater degradation: Changes in style with geologic time: Moon, v. 12.

——— 1976, Lunar volcanism in space and time: Rev. Geophys. Space Physics, v. 14.

Milton, D. J., and others, 1972, Grosses Bluff impact structure, Australia: Science, v. 175, no. 4027, p. 1199–1207.

Oberbeck, V. R., 1975, The role of ballistic erosion and sedimentation in lunar stratigraphy: Reviews of Geophysics and Space Physics, v. 13, no. 2, p. 337–362.

Rhodes, R. C., 1975, New evidence for impact origin of the Bushveld complex, South Africa: Geology, v. 3, no. 10, p. 549–554.

Shoemaker, E. 1960, Penetration mechanics of high velocity meteorites, illustrated by Meteor Crater, Arizona: Internat. Geol. Cong., 21st, Norden, 1960, section 18, pt. 18, p. 418–434.

——— 1963, Impact mechanics at Meteor Crater, Arizona, *in* Middlehurst, B. M., and Kuiper, G. P., eds., The solar system, v. 4, The moon, meteorites, and comets, pp. 301–336: Chicago, University of Chicago Press.

Short, N. M., 1966, Shock process in geology: Jour. Geol. Education, v. 14.

——— 1970, Anatomy of a meteorite impact crater: West Hawk Lake, Manitoba, Canada: Geol. Soc. America Bull., v. 81, p. 609–648.

Stearns, R. G., and others, 1968, The Wells Creek structure, Tennessee, *in* Shock metamorphism of natural materials: Baltimore, Mono Book Corp.

7

Faults—General Considerations

A *fault* is a plane or zone in solid rock parallel to which displacement has occurred. Faults occur in many structural situations, varying from minor faults associated with folds to major fault zones which form borders of mountain ranges and others which are major zones of weakness in the lithosphere. Fault zones can be traced to depths of thousands of meters, and their presence to depths as great as 700 km is inferred from the depth of deep-focus earthquakes.

A fault may be smooth, a finely striated or slickensided surface, but more commonly it is a zone ranging from a few to over a hundred meters in width in which the rock has been broken, brecciated, and sometimes ground into a powdery material, called *gouge,* or crushed and recrystallized into a flinty material called *mylonite.* Many fault zones are composed of a number of subparallel faults, *synthetic faults.* The bedding may be systematically displaced as in a step fault pattern, or the sections of rock caught between faults may be folded and rotated, forming a chaotic structural pattern.

Fault nomenclature is complex because the number of variables is great. The fault orientation, the position of the faulted beds relative to the fault, and the amount, direction, and type of movement between the blocks all vary. The movement may be such that adjacent points are displaced by translation, rotation, or both.

FAULT MOVEMENT—DISPLACEMENT, SLIP, AND SEPARATION

The amount and direction of movement between two rock masses separated by a fault is an important aspect of fault problems. Movement may be of critical importance when, for example, the fault involves a severed ore body, a displaced oil-bearing bed, a dam foundation, or a tunnel.

Determination of movement on faults requires thoughtful consideration of what is being measured, even when the fault has essentially planar shape and the movement is translational.

The amount of movement on a planar fault is usually expressed as *displacement* (syn.: *net slip*), which is the distance* between any two originally adjacent points on a fault plane after movement occurs. While this concept is simple its application is difficult or sometimes impossible because relatively few natural features appear as points. Most such points are formed by the intersection of a linear feature, such as those listed below with a fault plane (after Crowell, 1959):

1. Lines formed by intersection of two planes such as the line of intersection of two dikes, a dike with a bed, two veins, etc.

2. Lines formed by the trace of one plane on another:
a. Trace of a bed below or above an unconformity against the unconformity.
b. Any older structure terminating against an unconformity or older fault (faults, dikes, sills, veins, sheets, fold axial surfaces, etc.)

3. Linear geological features:
a. Buried river channels, shoestring sand, attenuated sand lines
b. Volcanic necks, ore shoots, etc.
c. Recent physiographic features along recent faults

4. Stratigraphic lines:
a. Pinch-out line
b. Lines formed by facies changes
c. Shoreline, basin marginal features

5. Constructed lines:
a. Isopach lines
b. Lithofacie lines
c. Axial and crestal lines

Displacement is but one of a class of terms, called *slip*, based on the actual relative movement between the fault blocks. Among these terms are:

Net slip: Actual displacement.

Dip slip: The component of net slip measured down the dip of the fault. (*Note:* Net slip may be equal to the dip slip.)

* Direction of the slip vector must also be known.

Strike slip: The component of net slip measured parallel to the strike of the fault.

Oblique slip: A slip which is oblique to the dip of the fault.

When slips cannot be determined, the best way to describe movements is in terms of *separations*—the distance between any two parts of an index plane disrupted by a fault and measured in some specified direction. Separations may be observed from a geologic map or from subsurface data. Separations are commonly measured:

1. Along the strike of the fault—strike separation

2. Down the dip of the fault—dip separation

3. In a vertical line (e.g., a well)—vertical separation

4. Perpendicular to beds—stratigraphic separation (e.g., the thickness of the beds omitted between units adjacent on opposite sides of the fault)

A variety of separations and components are illustrated, Fig. 7-1:

A = dip separation

B = strike separation

C = vertical separation

D = horizontal separation (in dip plane)

E = vertical component of dip separation

F = horizontal component of dip separation

G = stratigraphic or perpendicular separation

H = normal horizontal separation

Only the dip separation A and the strike separation B are used to classify a fault. Some possible net slips are shown by 1 to 5. Notice that 1 has a relative upward component of the *hanging wall** and that 5 has a right-lateral component of actual relative displacement.

* The block above the fault.

It is essential to keep the distinction between the concepts of slip and separation clearly in mind. The separation of marker horizons along the fault or down its dip are often known. The actual movements which produced a given separation are rarely known because many different movement patterns can produce the same separations. Usually the data available to solve a fault problem are derived from a geologic map or from a well. These readily yield information about separations and rarely about slip.

FAULT CLASSIFICATION

Several systems of classification are in use. Often terms from several classifications are used to give a more complete description of the fault. The most general and most commonly used terms are based on the apparent relative movement of the fault blocks as follows:

Normal fault: The hanging wall (block above an inclined fault) is down relative to the footwall (block below the fault).

Reverse fault: The hanging wall is up relative to the footwall.

Thrust fault: A low-angle reverse fault.

Strike-slip fault: Lateral movement of the blocks parallel to the strike of the fault.

Classification of the Type of Movement

Generally faults are thought of as involving *translational movement* of the blocks—movement such that all points on a block move in the same direction and undergo identical displacement. When the blocks undergo *rotational movement,* the application of many terms in the system of classification may lead to confusion. For example, if blocks separated by an inclined fault rotate with a scissor-like motion, the fault would be classified as a normal fault on one side of the pivot point and a reverse fault on the other side. This problem also arises if the fault plane is curved through the vertical.

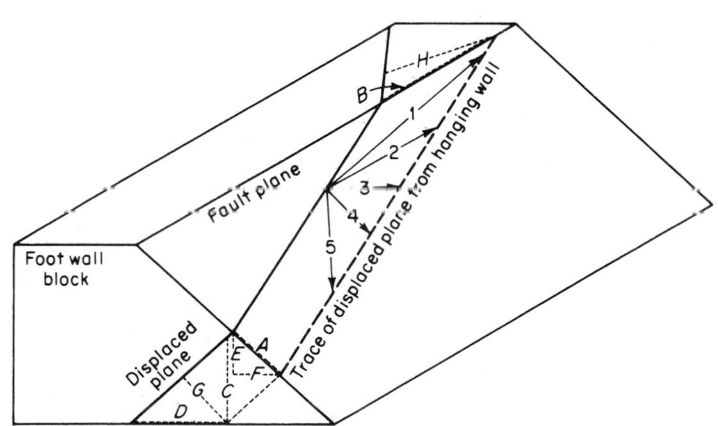

The strike-separation is *B*, the dip-separation is *A*. Other separations are:
C Vertical
D Horizontal
E Vertical component of dip
F Horizontal component of dip
G Stratigraphic
H Normal horizontal
Only the dip separation *A* and the strike separation *B* are used to classify a fault.
Some possible net-slips are shown by 1 through 5. Notice that (1) has a relative upward component of the hanging wall and that (5) has a lateral component of actual relative displacement.

FIGURE 7-1 The footwall block of a fault showing the trace of a marker plane (displaced plane) on the fault. The various types of separations are illustrated, and possible slips are shown by numbered lines. (*From M. L. Hill, 1947.*)

Classification Based on Slip

Terms based on slip or actual movement during faulting are useful when slip can be determined. Such faults are classified as follows:

Dip-slip fault: Movement is parallel to the dip of the fault. Terms such as *normal*, *reverse*, or *thrust* are added to specify the sense of movement.

Oblique-slip fault: Movement is oblique to the dip of the fault. Again normal, reverse, thrust, or other terms are needed to specify the sense of movement.

Strike-slip fault: Movement is parallel to the strike of the fault. The sense of movement across the fault is indicated by use of *left-lateral* and *right-lateral* to indicate the direction of offset across the fault as observed by a viewer facing the fault.

Classification Based on Separations and Slips

Actual slips can only rarely be determined, but it is generally possible to determine the apparent relative movements (separations) between fault blocks. Hill (1959) proposes a classification based on separation and, if possible, slip as well. When slip cannot be determined the fault is classified on the basis of separation alone as follows:

Apparent relative movement	Fault type	Actual movement	Fault type
Separation	Slip		
Dip separation	Normal Reverse Thrust	Dip slip	Normal slip Reverse slip Thrust slip
Strike separation	Right-lateral Left-lateral	Strike slip	Right-lateral slip Left-lateral slip
Combined dip and strike separations	Name after principal separation (e.g., normal left-lateral)	Oblique slip	Named after principal slip or combined (e.g., normal left-lateral slip)

Classification Based on the Relation of Fault to Bedding Orientation

The following terms are used to designate the relationship between the orientation of a fault and the strata cut by it:

Bedding fault: A fault that is parallel to bedding.

Dip fault: A fault that strikes approximately in the direction of the dip of the bedding.

Oblique fault: A fault that strikes obliquely across the strike of the beds.

Longitudinal fault: A fault that strikes parallel with the related structural features (e.g., parallel to a fold axis).

Transverse or tear fault: A strike-slip fault (often with rotational movement) that strikes transverse to the strike of the related structural features.

FAULTS AND STRESS ORIENTATIONS

"There can be no pressure or tension perpendicular to the surface, and no shearing force parallel to it, in its immediate vicinity." Thus Anderson (1951) argues that the normal to the ground surface tends to be one of the three principal stress directions. He assumes this is true for depths at which most faulting occurs. With this assumption three general cases of stress distribution are most likely (Fig. 7-2):

1. Maximum and intermediate principal stress are both horizontal—thrust faults.
2. Maximum stress is vertical—normal faults.
3. Maximum and least principal stress direction are horizontal—wrench faults (strike slip).

Mohr analysis of conditions favorable for development of normal and reverse faults is illustrated in Fig. 7-3. In reverse faulting, the least principal stress is vertical and remains constant, while the horizontal stress increases until the Mohr circle becomes tangent with the line of fracture. This occurs when

$$2\alpha = 90° + \phi$$

$$\alpha = 45° + \frac{\phi}{2}$$

For normal faulting the least principal stress is horizontal, and the maximum stress direction is vertical. For normal faults formed as a result of horizontal tension, the least stress becomes smaller, while the maximum stress remains constant. The Mohr circle increases in size as the magnitude of the least stress drops, until the Mohr circle again becomes tangent to the fracture line.

Fractures as a Guide to Fault Orientation

Fracture orientations in the vicinity of faults often provide a valuable clue to the stresses responsible for the fault if the fractures formed at the time of the faulting. The relationship between conjugate shear fractures and the principal stresses are depicted in Fig. 7-4. This relationship is well founded on the basis of experiment, theory, and field studies. Faults

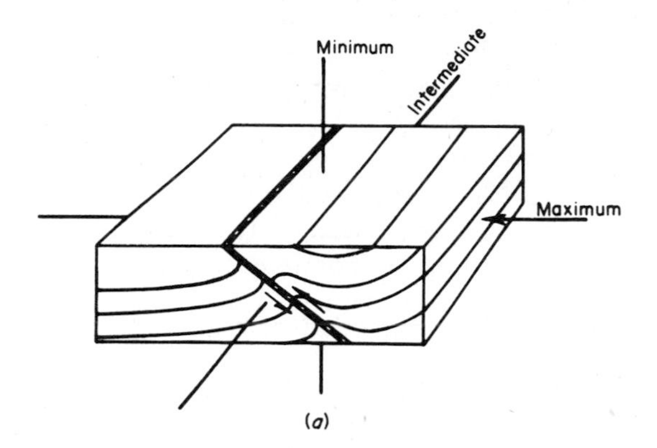

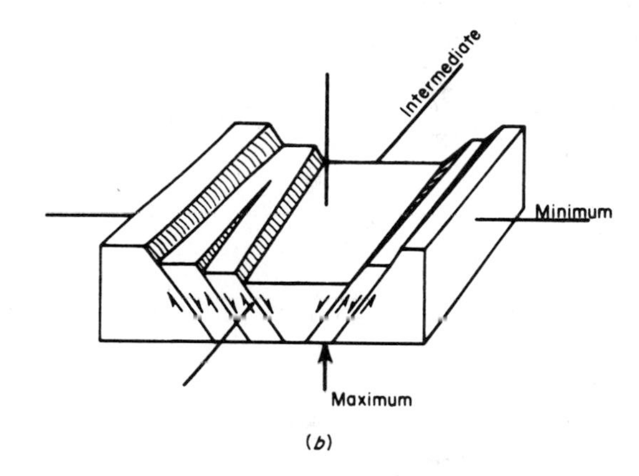

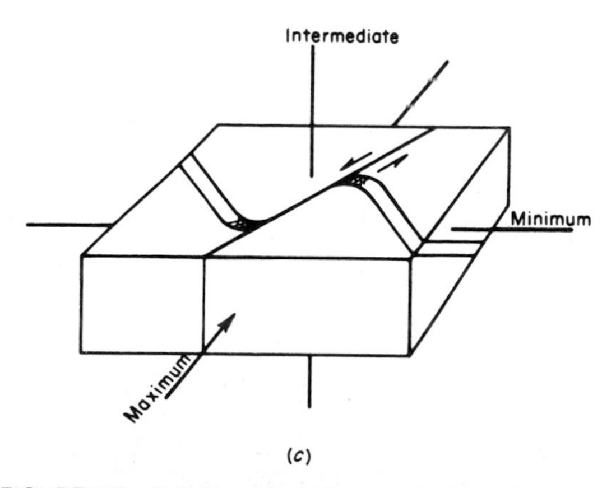

FIGURE 7-2 Relationship between principal stresses and common fault orientations, for thrust, normal, and strike-slip faults. (*After Anderson, 1951.*)

are essentially shear fractures along which displacement has been substantial. Thus shear fractures in the vicinity of a fault should include one orientation that coincides with that of the fault, and in some cases both fracture sets may be represented by faults, Fig. 7-4. Fractures may be rotated with the fault zone as a result of drag or postfaulting movements, and this must be considered when comparing fractures with fault orientations. This possibility may be checked in sedimentary rocks by looking for evidence of drag in the layering and by comparing fracture orientations with dip of the beds. Conditions at the time of fracturing can be reconstructed by rotation of beds and fractures back to the original regional dip.

FIGURE 7-3 Mohr-circle representation of the conditions leading to normal and reverse faulting. The minimum stress decreases in normal faulting; the maximum stress increases in reverse faulting. (*From Hubbert, 1951.*)

NORMAL FAULTING

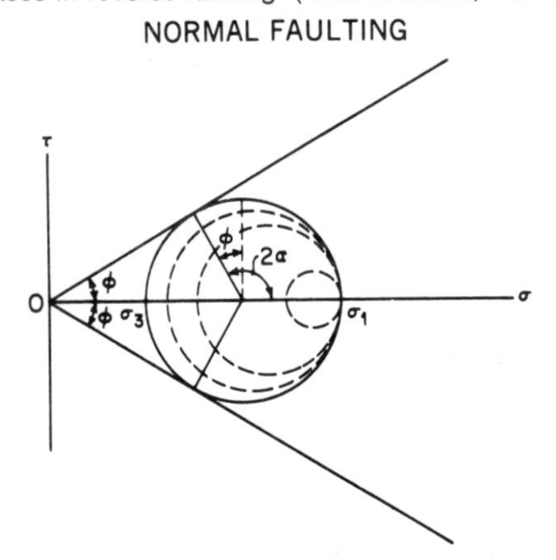

REVERSE FAULTING

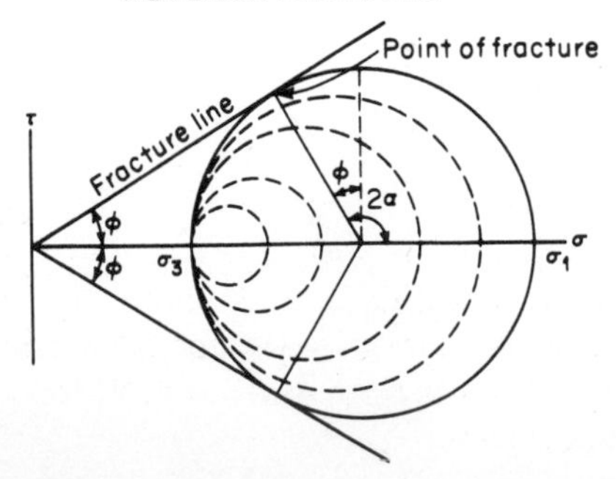

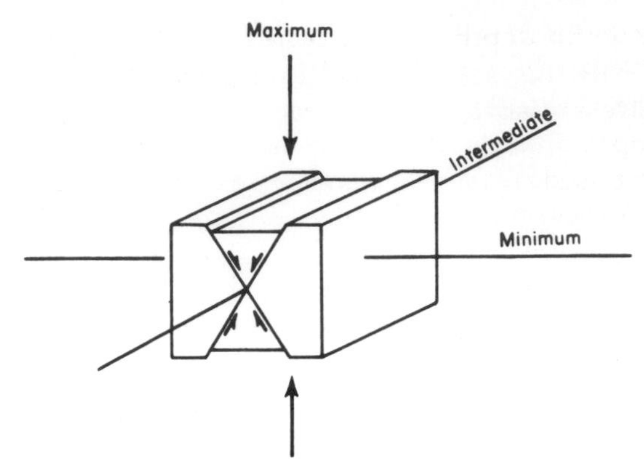

FIGURE 7-4 Relationship between principal stress axes and conjugate shear set of fractures or faults.

Stress Distribution and Faulting

Because theory and experimental studies show a clear relationship between applied stresses and fracture orientations, we should expect to find similar relationships between applied stresses and potential faults. Two approaches have been used successfully in such studies:

1. In the first method, a hypothetical system of stresses of known value is applied to an imaginary block of homogeneous material. The orientations of principal stresses at points within the block are calculated in terms of specified boundary conditions. Then the Mohr criterion of failure is used to determine the orientation of potential fault orientations.

2. In the second method, a system of stresses and potentially related faults are inferred from a known or given set of displacements along the borders of a block of material.

Hafner* analyzes a number of stress systems caused by various types of boundary forces which might be expected within the earth. His conclusions regarding the stress fields are highly significant because they indicate the pattern of potential fault surfaces most likely to be associated with each stress condition.

* Derivations of the stress systems are given in Hafner (1951). See also Chapple (1968) and Jaeger (1969).

The technique used in this analysis involves the assumption of a standard state on which supplementary boundary conditions which approximate plausible tectonic situations are imposed. A plane stress configuration is used. All diagrams are drawn in the *xy* plane; so, lines on the drawings represent planes perpendicular to the page. All diagrams are of arbitrary length and depth. The standard state, Fig. 7-5, includes a set of stresses due to gravity. Four boundary conditions for the standard state are:

1. The top surface of the block is the ground surface. It supports no shear, and only 1 atm of pressure acts on it from above. Since the top surface is a free-air boundary, the supplementary stresses must give zero stress here.

2. Lithostatic pressure increases with depth (*y* axis). Therefore

$\sigma_x = -\rho g y$ and $\sigma_y = -\rho g y$

where σ_x = horizontal stress component at depth y
ρ = density
g = gravitational acceleration

The value of σ_x is negative because compressive stresses are here considered negative.

3. At the bottom of the block, stresses directed upward just equal the weight of the block above.

4. No shear stresses act on the boundaries.

Three sets of supplementary conditions studied by Hafner are described below.

Case I. The supplementary stress is a uniform horizontal pressure applied along one end of the block, Fig. 7-6.

The top diagram depicts shearing and normal stresses acting on the boundaries of the block. These include lithostatic pressure that is a function of depth, indicated by ay, where y is depth and a equals density times the gravitational constant; a vertically directed pressure σ_y shown as constant across the base of the block; a horizontally directed stress cx which is defined in this case as being constant from the top to the bottom of the block (c is a constant expressed as a percentage or multiple of a). Shearing stresses must act on the faces of the block to fulfill the requirement for equilibrium (nonrotation) for the block. The trajectories of maximum and minimum principal pressure are calculated on the basis of stress theory (Hafner, 1951). The lateral pressure gradient of the superposed horizontal stress is expressed in terms of a fraction of the vertical pressure gradient (c/a). The lower diagram shows areas of stability, $c = a$, and potential fault planes. The positions of potential fault planes are determined by drawing them so that they cross each maximum principal pressure contour at an angle θ of about 32°, selected as the angle of fracture, and as predicted by the Mohr-Coulomb criterion of failure.

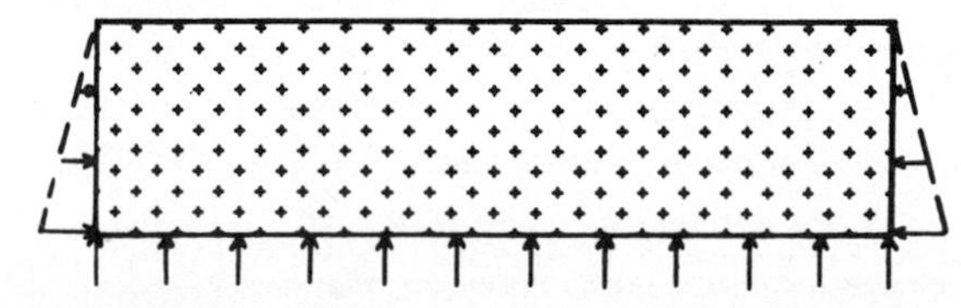

FIGURE 7-5 The standard state used in analysis of superimposed stress systems.

Case II. Horizontal compression with exponential lateral attenuation yields potential faults as illustrated in Fig. 7-7. Increase of lithostatic pressure with depth is taken into consideration but is not depicted in this figure. The zone of potential faulting is narrow and nearly vertical, but the potential fault system in the shallow parts of the block is similar to that in the first case. Horizontal thrust faults as well as high-angle reverse faults are possible in this system.

Case III. The case of sinusoidal vertical and shearing forces acting on the bottom of a block, Fig. 7-8, is pertinent to two general geological conditions, one involving vertical uplift and the other involving such frictional drag as is presumed to exist in convection current hypotheses.

Potential faults for any postulated system of boundary stresses can be found by use of Hafner's methods. The greatest difficulties arise in selecting boundary conditions which are geologically significant. We do not have methods of determining actual stress conditions for past times. Also, the earth's crust is rarely homogeneous, so actual crustal conditions are much more complicated than those postulated in this analysis. Yet this approach is valuable in several respects:

1. When the fault orientations in an area are known, this analysis may help clarify stress fields that were responsible for their formation.
2. When fault orientations in an area are not known or are partially known, it helps us postulate the position of a fault at depth.
3. It makes us aware that faults may have orientations other than the simple plane configurations, and it demonstrates other potential orientations.
4. The conditions assumed in case III come closest to fulfillment when larger relatively homogeneous masses (e.g., the entire crust) are considered, so it may be of particular value in studying large-scale earth structures.
5. It is apparent that the effects of boundary conditions and stress distributions can be quite complex and not readily predicted without mathematical analysis.

One striking conclusion from these diagrams is that the potential fault planes are curved surfaces. These studies indicate potential variations of fault orientations at depth,

FIGURE 7-6 Stress system consisting of superimposed horizontal pressure with constant lateral and vertical gradient. Stress trajectories (top) and orientations of maximum shear (potential faults) trajectories (below) are shown. (*From Hafner, 1951.*)

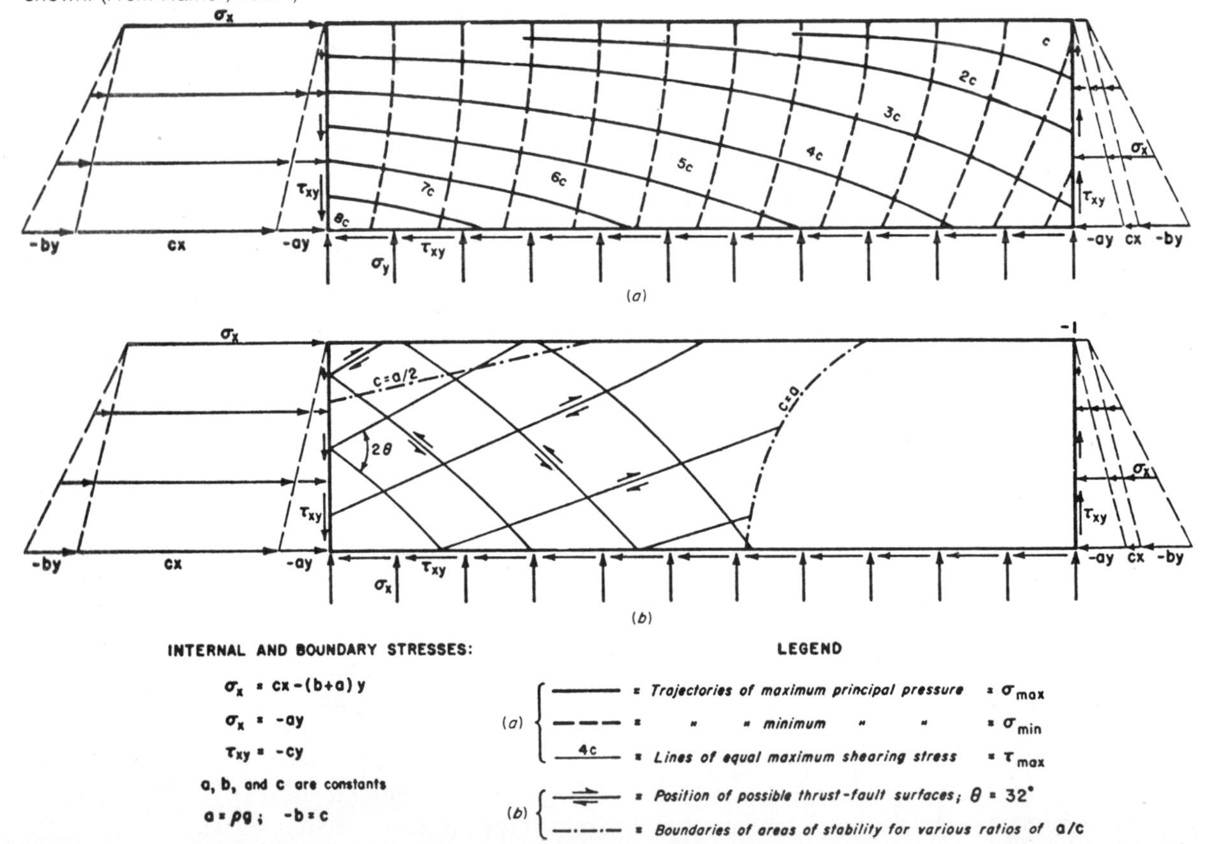

and at least imply that stress conditions may be inferred from fault patterns. We will compare these diagrams with actual field examples in the following chapters.

Prediction of Fault Orientation from Displacement Fields

Sanford (1959) demonstrates how fault orientations can be predicted from the distribution of vertical displacements along the base of a homogeneous elastic layer. This method can be applied more directly to field situations than Hafner's analysis because vertical displacements of a rigid basement (e.g., an igneous-metamorphic crystalline complex under a sedimentary veneer) can often be determined from field and subsurface observations, while the magnitude and distribution of stress along the base of the layer as required in Hafner's analysis cannot be determined.

The essential elements of Sanford's method are:

1. The vertical displacements which take place along the base of an elastic layer are specified or determined from field observations.

2. The stress trajectories that would be associated with the specified displacements are calculated for grid points distributed within the layer. Assumptions in this method are (*a*) that displacements and strains are small, and (*b*) that the material undergoing deformation is perfectly elastic, homogeneous, and isotropic. Application of the theory is greatly simplified if the lower boundary of a layer undergoes displacements in the *x* and *y* directions only and if these displacements are identical for all cross sections along the *z* axis. The problem is essentially reduced to a two-dimensional analysis, and it is applicable if a single cross section of a structure can be selected as approximating the structure for a long distance.

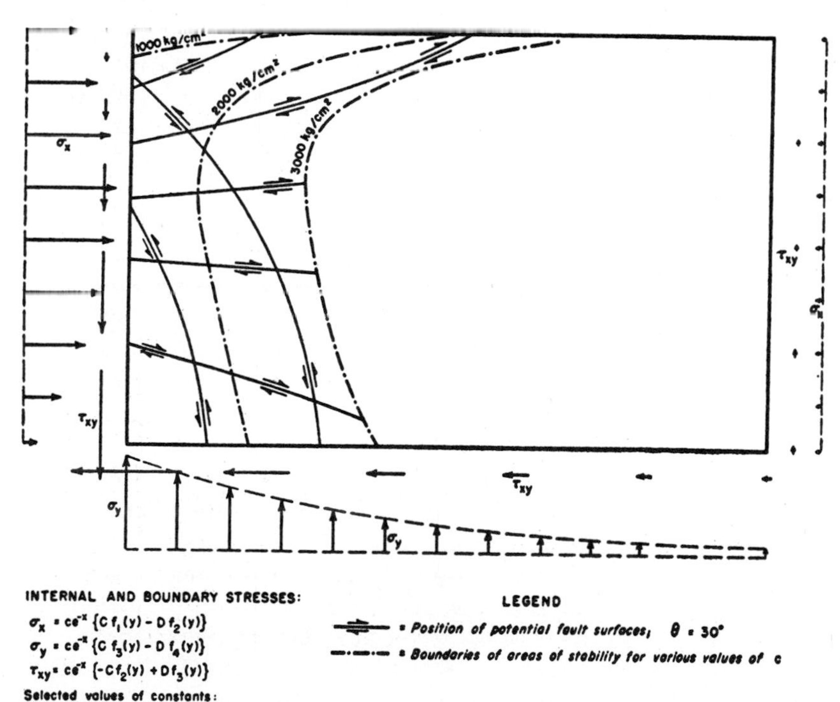

FIGURE 7-7 Shear trajectories (potential faults) which are postulated to result from a stress system consisting of superimposed horizontal stress which decreases exponentially from left to right. (*From Hafner, 1951.*)

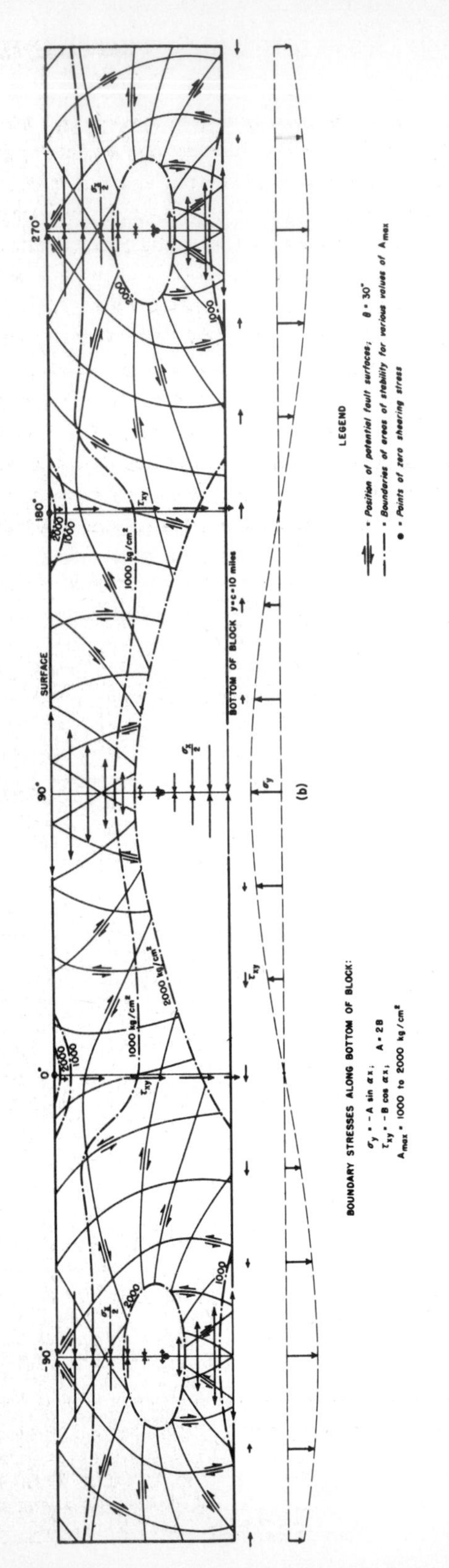

FIGURE 7-8 Potential fault orientations postulated to result from a stress system consisting of variable vertical and shearing stress along bottom of block. (*From Hafner, 1951.*)

3. The displacements at points within the deformed layer are determined and shown by means of displacement vectors.

4. Faulting presumably begins when the magnitude of the stresses and displacements are at the level required to cause failure at some point in the layer. The Mohr fracture criterion is used to determine this value. The point requiring the least displacement to originate fracture is taken as the point of fracture. Its orientation is determined from the stress trajectories.

Two examples of stress trajectories determined by this method are illustrated in Figs. 7-9 and 7-10. This method has been used to analyze the fault pattern of the Williams Range in Colorado by Howard (1966).

MAXIMUM DEPTH OF FAULTING

Earthquakes are associated with faulting because seismic waves appear to be caused by shear offsets propagated at elastic wave velocity. According to a widely accepted view, elastic strain is built up across a fault plane until it is released by a slip and elastic rebound. The deepest earthquakes have focal depths near 700 km, often considered the limiting depth for faulting. Experiments to determine rock behavior at high confining pressure indicate that brittle failure cannot be the cause of earthquakes at such great depths.

Confining pressure at 700 km is approximately 250,000 bars and temperatures are high; therefore the shear strength of rocks at that depth is probably low—on the order of 10 to 100 bars. The fracture theory of generating earthquakes at 700 km would appear to require a shear strength for the rock at that depth about 10,000 times greater than the strength predicted by experimental studies. All rocks except quartzite exhibit uniform flow or "faults" without a sudden release of elastic energy at 5,000 bars and 500°C. Thus even much shallower earthquakes probably cannot be generated by ordinary fracturing. Orowan (1960) estimates that the limiting depth for earthquakes caused by shear failure is 5 to 10 km. Thus the development of faults through brittle-shear failure is limited to the upper part of the earth's crust.

Several alternative hypotheses have been advanced to explain generation of earthquakes at moderate and great depths. Griggs (1954) suggests that deep-focus earthquakes are generated by shear melting. According to this hypothesis, melting is initiated at a flaw in the mantle (e.g., fluid-filled cavities or weak, tubular mineral grains) and as a result of shear stresses. The flaw is deformed, enlarged, and propagated with seismic velocity. Thus a rapid shear displacement occurs which sets up seismic waves.

Orowan (1960) proposed that deep earthquakes are due to instability of plastic deformation (creep). He argues that if creep produces structural changes which accelerate creep, then the deformation is gradually concentrated into thin layers in which very high rates of flow are developed. If the process proceeds fast enough, shear melting may occur. The idea of shear melting provides an explanation for the coincidence in time and space of earthquake and volcanic belts around the Pacific.

GLOSSARY FOR FAULT NOMENCLATURE

I. General Terms

Antithetic faults: Faults that dip in the opposite direction from that in which the associated sediments dip. (Cloos, H., 1936.*) (Ant.: **synthetic faults**.)

En échelon: Parallel structural features that are offset as are the edges of shingles on a roof when viewed from the side. (Structural Committee.*)

Fault: A fracture or fracture zone along which there has been displacement of the two sides

* Definitions based on the American Geological Institute's Glossary of geology and related sciences.

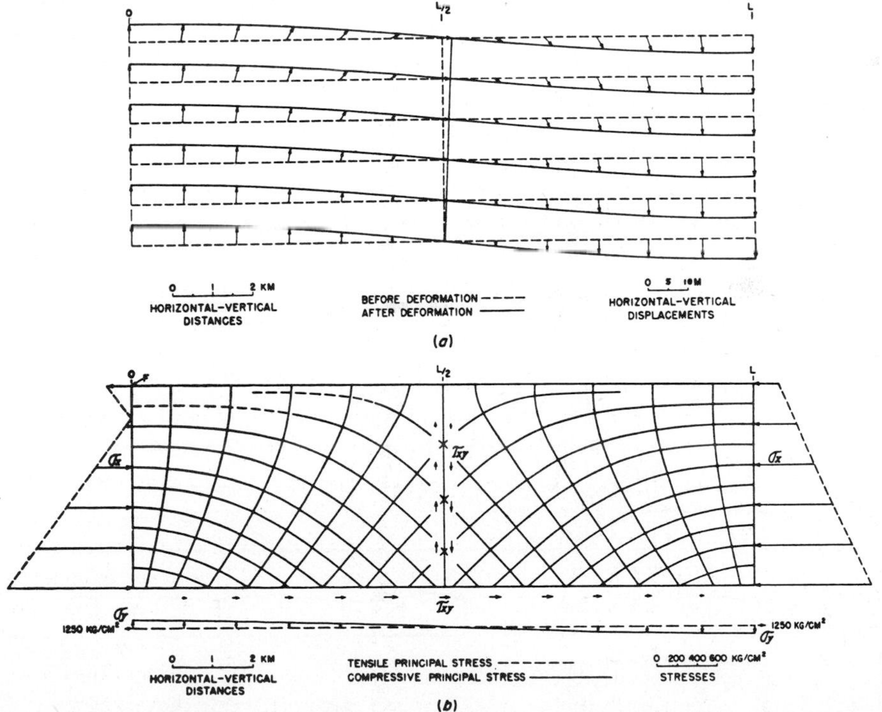

FIGURE 7-9 Displacement field (*a*) and stress distribution (*b*) calculated for an elastic layer 5 km thick and 15.7 km long which undergoes half a wavelength of sinusoidal vertical displacement. (*From Sanford, 1959.*)

relative to one another parallel to the fracture. (Reid, 1913.*) Reid's definition is adequate to explain faulting in brittle materials, but displacements in plastic materials are not always accompanied by loss of cohesion. A more general definition of a fault would be: A zone across which there has been displacement of the two sides relative to one another parallel to the zone.

Fault block: A mass bounded on at least two opposite sides by faults; it may be elevated or depressed relative to the adjoining region, or it may be elevated relative to the region on one side and depressed relative to that on the other. (Reid and others, 1913.*)

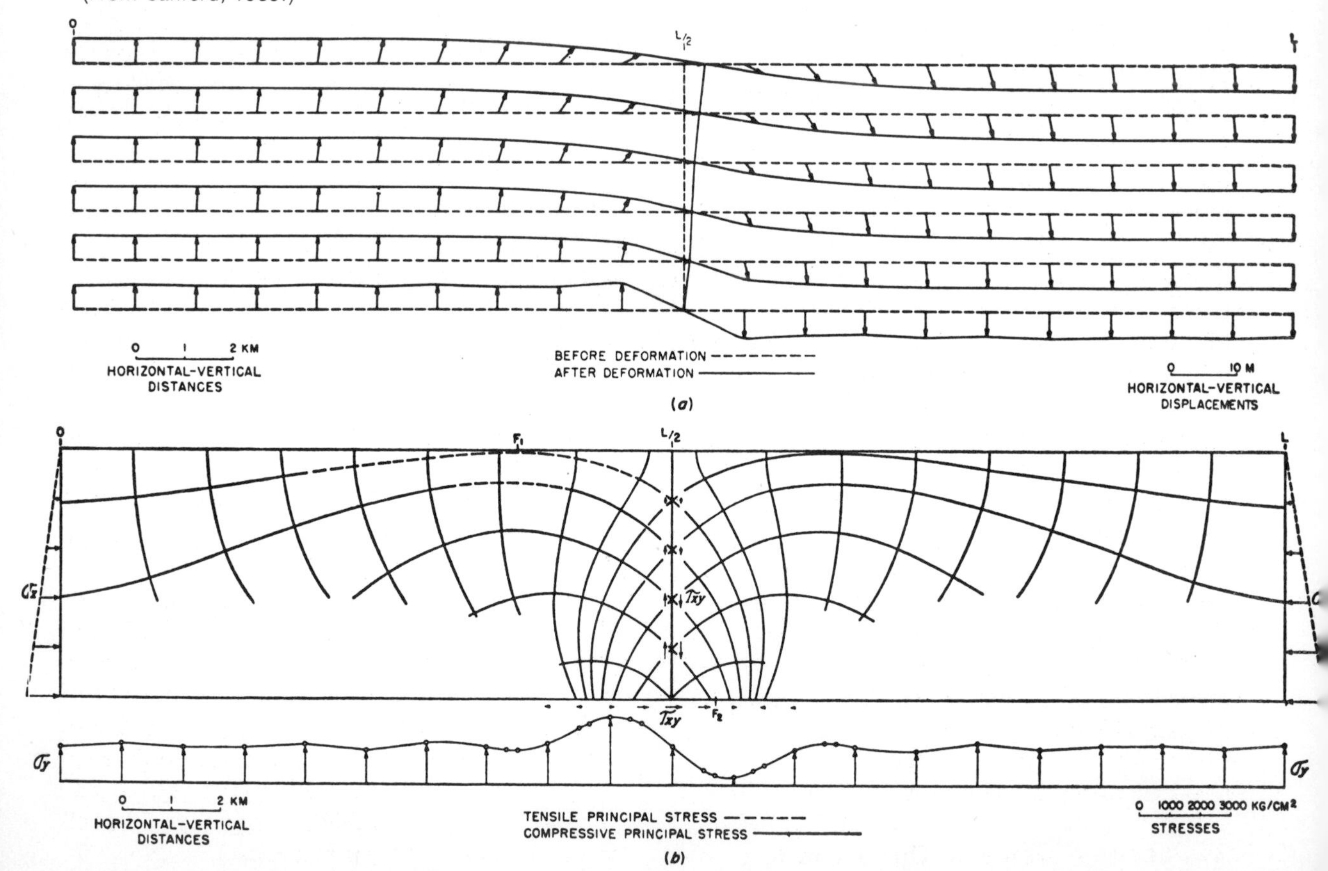

FIGURE 7-10 Displacement field (*a*) and stress distribution (*b*) calculated for an elastic layer the lower boundary of which undergoes a step-like displacement. (*From Sanford, 1959.*)

Fault breccia: The assembly of broken fragments frequently found along faults. (Lindgren, 1933.*) (Syn.: **crush breccia, fault gouge.**)

Fault fold: A fold accompanied by steep faults that are parallel to the fold and are contemporaneous with the folding. (Hills, J. M., 1953.*)

Fault plane: A fault surface.

Fault scarp: "The cliff formed by a fault. Most fault scarps have been modified by erosion since the faulting." (U.S. Geol. Survey Bull. 611.*)

Fault trace: The intersection of a fault surface with the surface of the earth or with any artificial surface of reference. (Lindgren, 1933.*) (Syn.: **fault outcrop, fault line.**)

Fault zone: A fault, instead of being a single clean fracture, may be a zone hundreds or thousands of feet wide; the fault zone consists of numerous interlacing small faults or a confused zone of gouge, breccia, or mylonite. (Billings, 1954.*)

Fenster: An exposure of the rock beneath a thrust sheet or recumbent fold produced where erosion has locally truncated the overlying rock units. The exposure is completely surrounded by units on the thrust sheet in a perfectly developed fenster.

Footwall: The mass of rock beneath a fault plane, vein, lode, or bed of ore. (Structural Committee.*)

Graben: An elongate fault block, usually a depression, bounded on two or more sides by normal or vertical faults and formed by the downthrow of the central block relative to the adjacent blocks.

Hade: The angle of inclination of a vein, fault, or lode measured from the vertical.

Hanging wall: The mass of rock above a fault plane, vein, lode, or bed of ore. (Structural Committee.*)

Horst: An elongate fault block, usually high, bounded on two or more sides by normal faults and formed by the upthrow of the central block relative to the adjacent blocks.

Klippe: An erosional remnant of a thrust sheet, nappe, or recumbent fold.

Megashear: A strike-slip fault whose horizontal displacement significantly exceeds the thickness of the crust. (Carey, 1958.*)

Mylonite: A fine-grained, laminated rock formed by extreme microbrecciation and milling of rocks during movement on fault surfaces. Metamorphism is dominantly cataclastic, with little or no growth of new crystals.

Parallel faults: A group of faults having essentially the same dip and strike. (Power.*)

Peripheral faults: Faults along the periphery of a geologically elevated or depressed region. (Reid, 1913.*)

Radial faults: A group of faults that, on a map, radiate from a common center. (Billings, 1954.*)

Rotational movement: This refers to movement on faults. Blocks rotate relative to one another about an axis perpendicular to the fault. Some straight lines on opposite sides of the zone and outside the dislocated zone, parallel before the displacement, are no longer parallel after it. (Reid, 1913.*)

Shear zone: A zone in which shearing has occurred on a large scale, so that the rock is crushed and brecciated. (La Forge.*)

Slickensides: Polished and striated (scratched) surface that results from friction along a fault plane. (Billings, 1954.*)

Step faults: "A series of parallel faults, which all incline in the same direction, gives rise to a gigantic staircase; hence these are called step faults. Each step is a fault block and its top may be horizontal or tilted." (Scott, 1922.*)

Synthetic faults: Subsidiary faults parallel to the master fault. (Cloos, H., 1936.*)

Translational movement: Movement along a fault such that all straight lines on opposite sides of the fault and outside the dislocated zone that were parallel before faulting are parallel after faulting. (Reid, 1913.*)

II. Terms Based on Actual Relative Movements

Dextral strike slip: Syn. of **right strike slip**.

Dip slip: The component of the slip parallel to the fault dip, or its projection on a line in the fault surface perpendicular to the fault strike. (Lindgren, 1933.*)

Displacement: Shift on opposite sides of fault measured outside the zone of dislocation. Whereas the net shift along the fault might be 1,000 ft, because of drag the net slip along the fault might be much less. (Reid, 1913.*)

Heave: In faulting, the horizontal component of the dip separation; that is, the apparent horizontal component of displacement of a disrupted index plane on a vertical cross section, the strike of which is perpendicular to the strike of the fault. (Structual Committee.*) (*Note:* The term is frequently applied to the horizontal component of the net slip. In such cases the meaning is ambiguous.)

Horizontal dip slip: The horizontal component of the dip slip (Gill, 1941.*)
Horizontal slip: In faulting, the horizontal component of the net slip. (Gill, 1941.*)
Left-lateral slip: Syn. of **left strike slip.**
Net slip: The total slip along a fault; the distance measured on the fault surface between two formerly adjacent points situated on opposite walls of the fault. It is the shortest distance measured in the fault plane between the two formerly adjacent points. (Reid, 1913.*) (Syn.: **displacement.**)
Oblique slip: A fault in which the net slip lies between the direction of dip and the direction of strike. (Lindgren, 1933.*)
Perpendicular slip: The component of the net slip measured perpendicular to the trace on the fault of the disrupted index plane (bed, dike, vein, etc.) in the fault plane. (Reid, 1913.*)
Right-lateral slip: Syn. of **right strike slip.**
Shift: The maximum relative displacement of points on opposite sides of the fault and far enough from it to be outside the dislocated zone. (Lindgren, 1915.*)
Sinistral strike slip: Syn. of **left strike slip.**
Slip: The relative displacement of formerly adjacent points on opposite sides of the fault, measured in the fault surface. (Lindgren, 1915.*)
Strike slip: The component of the slip parallel to the fault strike, or the projection of the net slip on a horizontal line in the fault surface. (Lindgren, 1915.*)
Throw: (1) The amount of vertical displacement occasioned by a fault. (Page, David, 1859.*) (2) More generally, the vertical component of the net slip.
Trace slip: Component of net slip parallel to the trace of an index plane (vein, bedding, etc.) on plane of the fault. (Reid, 1913.*)
Trace-slip fault: A fault on which the net slip is parallel to the trace of a bed (or some other index plane) on the fault. (Reid, 1913.*)
Vertical slip: Syn. of **throw.**

III. Terms Based on Apparent Relative Movements

Dextral: Syn. of **right strike separation.**
Dip separation: In faulting, the distance between two parts of a disrupted index plane (bed, dike, vein, etc.) measured in the fault plane parallel to its dip. (Billings, 1954.*)
Gap: In faulting, the *horizontal separation, q.v.,* can be measured parallel to the strike of the fault. Gap is the component of this separation that is measured parallel to the strike of the disrupted index plane (bed, dike, vein, etc.). **Overlap** is defined in the same way; however, gap is used when it is possible to walk at right angles to the strike of the disrupted index plane and miss it completely. Overlap is used when, under similar conditions, one would cross the index plane twice in certain places. (Reid, 1913.*)
Horizontal separation: Separation measured in any indicated horizontal direction.
Left-handed: Syn. of **left strike separation.**
Left-lateral: Syn. of **left strike separation.**
Normal horizontal separation: Syn. of **offset.**
Offset: Normal horizontal separation. Determined from outcrop of the index plane at the surface of the ground. In faulting, *horizontal separation, q.v.,* can be measured parallel to the strike of the fault. Offset is the component of this horizontal separation that is measured perpendicular to the strike of the disrupted index plane (bed, dike, vein, etc.). (Reid, 1913.*)
Overlap: See *gap* above.
Right-handed: Syn. of **right strike separation.**
Right-lateral: Syn. of **right strike separation.**
Separation: Indicates the distance between any two parts of an index plane (bed, vein, etc.) disrupted by a fault measured in some specified direction. Horizontal separation is separation measured in any indicated horizontal direction; vertical separation is measured along a vertical line; stratigraphic separation is measured perpendicular to the bedding planes. (Billings, 1954.*)
Sinistral: Syn. of **left strike separation.**

Stratigraphic separation: The stratigraphic throw. The stratigraphic thickness that separates two beds brought into contact at a fault. (Billings, 1954.*)
Strike separation: In faulting, the distance on a map between the two parts of an index plane (bed, vein, dike, etc.) where they are in contact with the fault and measured parallel to the fault. (Reid, 1913.*)

FAULT CLASSIFICATIONS

I. General Terms

Gravity: Syn. of **normal**.
Gravity-glide fault: A fault produced by the sliding of rock masses or strata downslope from an uplifted area. Nappes, recumbent folds, or low-angle overthrust faults may be associated with gravity gliding.
Normal: A fault at which the hanging wall has been depressed, relative to the footwall. (Lindgren, 1933.*) (Syn.: **gravity**.)
Overthrust: (1) A thrust fault with low dip and large net slip, generally measured in miles. (2) A thrust fault in which the hanging wall was the active element; contrasted with underthrust, but it is usually impossible to tell which was actively moved. (3) The process of thrusting the hanging wall (relatively) over the footwall.
Reverse: A fault along which the hanging wall has been raised, relative to the footwall. (Lindgren, 1933.*)
Thrust: A reverse fault that is characterized by a low angle of inclination with reference to a horizontal plane.
Underthrust: A thrust fault in which the footwall was the active element. In most instances, it is impossible to tell which element was active. (Billings, 1954.*)
Upthrust: A high-angle fault in which the relatively upthrown side was the active element.

II. Terms Based on Net Slip

Dip-slip fault: A fault in which the net slip is practically in the line of the fault dip. (Lindgren, 1933.*) (This could include normal, reverse, and thrust faults.)
Oblique-slip fault: A fault in which the net slip lies between the direction of dip and the direction of strike. (Lindgren, 1933.*) (This could include normal, reverse, and thrust faults.)
Strike-slip fault: A fault in which the net slip is practically in the direction of the fault strike. (Lindgren, 1933.*) (Syn.: **wrench, transcurrent**.)
Transcurrent: Syn. of **strike-slip fault**.
Wrench: Syn. of **strike-slip fault**.

III. Terms Based on Relation of Fault to Adjacent Strata

Bedding fault: A fault that is parallel to the bedding. (Reid, 1913.*)
Dip fault: A fault that strikes approximately perpendicular to the strike of the bedding or cleavage. (Billings, 1954.*)
Longitudinal fault: A fault whose strike is parallel with the general structure. (Lindgren, 1933.*)
Oblique fault: A fault whose strike is oblique to the strike of the strata. (Lindgren, 1933.*)
Strike fault: A fault whose strike is parallel to the strike of the strata. (Lindgren, 1933.*)
Tear fault: A strike-slip fault that tends transverse to the strike of the deformed rocks. (Hills, J. M., 1953.*). (Syn.: **transverse fault**.)
Transverse fault: Syn. of **tear fault**.

REFERENCES

Abendanon, E. C., 1914, Die Grossfalten der Erdrende: Leiden.
Anderson, E. M., 1951, The dynamics of faulting, 2d ed.: Edinburgh, Oliver & Boyd.
Billings, M. P., 1972, Structural geology, 3d ed.: Englewood Cliffs, N.J., Prentice-Hall.
Boyer, R. E., and **Muehlberger, W. R.**, 1960, Separation versus slip: Am Assoc. Petroleum Geologists Bull., v. 44, no. 12, p. 1938–1939 (Geological Notes).
Brace, W. F., 1972, Laboratory studies of stick-slip, and their application to earthquakes: Tectonophysics, v. 14, no. 3–4, p. 189–200.

Byerlee, J. D., 1970, The mechanics of stick-slip: Tectonophysics, v. 9, p. 475–486.
——— and **Brace, W. F.**, 1972, Fault stability and pore pressure: Seis. Soc. America Bull., v. 62, no. 2, p. 657–660.
Carey, S. W., 1958, The tectonic approach to continental drift, *in* Continental drift—A symposium: Geol. Dept. Univ. of Tasmania, p. 177–355.
Chapple, Wm. M., 1968, A mathematical theory of finite-amplitude rock-folding: Geol. Soc. America Bull., v. 79, p. 47–68.
Chinnery, M. A., 1961, The deformation of the ground around surface faults: Seis. Soc. America Bull., v. 51, p. 355–372.
Clark, S. K., 1943, Classification of faults: Am. Assoc. Petroleum Geologists Bull., v. 27, p. 1245–1265.
Crowell, J. C., 1959, Problems of fault nomenclature: Am. Assoc. Petroleum Geologists Bull., v. 43, p. 2653–2675.
Engelder, J. L., 1974, Cataclasis and the generation of fault gouge: Geol. Soc. America Bull., v. 85, no. 10, p. 1515.
Hafner, W., 1951, Stress distributions and faulting: Geol. Soc. America Bull., v. 62, no. 4, p. 373–398.
Gill, J. E., 1941, Fault nomenclature: Royal Soc. Canada Trans., Ser. 3, v. 35, sec. 4, p. 71–85.
——— 1971, Continued confusion on the classification of faults: Geol. Soc. America Bull., no. 5, v. 82, p. 1389–1392.
Griggs, D. T., 1954, High pressure phenomena with applications to geophysics, *in* Ridenour, L. N., ed., Modern physics for the engineer: New York, McGraw-Hill, p. 272–305.
Hill, M. L., 1947, Classification of faults: Am Assoc. Petroleum Geologists Bull., v. 31, p. 1669–1673.
——— 1959, Dual classification of faults: Am. Assoc. Petroleum Geologists Bull., v. 43, p. 217–222.
Howard, J. H., 1968, The role of displacements in analytical structural geology: Geol. Soc. America Bull., v. 79, no. 12, p. 1846–1852.
Hubbert, M. K., 1937, Theory of scale models as applied to the study of geologic structures: Geol. Soc. America Bull., v. 48, p. 1459–1519.
——— 1951, Mechanical basis for certain familiar geologic structures: Geol. Soc. America Bull., v. 62, p. 355–372.
——— 1972, Structural geology, New York, Hafner.
Jaeger, J. E., 1969, Elasticity, fracture and flow with engineering and geological applications, 2d ed.: London, Methuen; New York, Wiley.
Kelley, V. C., 1960, Slips and separations: Geol. Soc. America Bull., v. 71, p. 1545–1546.
King, Chi-Yu, 1972, A shallow-faulting model: Seis. Soc. America Bull., v. 62, no. 2, p. 551–559.
Odé, Helmer, 1960, Faulting as a velocity discontinuity in plastic deformation: Geol. Soc. America Mem. 79, p. 293–323.
Orowan, E., 1960, Mechanism of seismic faulting, *in* Griggs, D. T., and Handin, John, eds., Rock deformation—A symposium: Geol. Soc. America Mem. 79, p. 323–346.
Proctor, McKeown, 1972, Relation of known faults to surface ruptures, 1971 San Fernando earthquake: Geol. Soc. America Bull., v. 83, no. 6, p. 1601–1618.
Reid, H. F., and others, 1913, Report on nomenclature of faults: Geol. Soc. America Bull., v. 24, p. 163–186.
Rickard, M. J., 1970, Fault classification: Discussion: Geol. Soc. America Bull., v. 83, no. 8, p. 2545–2546.
Sanford, A. R., 1959, Analytical and experimental study of simple geologic structures: Geol. Soc. America Bull., v. 70, p. 19–52.
Threet, R. L., 1973, Classification of translational fault slip: Geol. Soc. America Bull., v. 84, no. 5, p. 1825–1827.
——— 1974, Down-structure method of viewing geologic maps to obtain sense of fault separation: Geol. Soc. America Bull., v. 84, no. 12, p. 4001–4004.
Walsh, J. B., 1969, Dip angle of faults as calculated from surface deformation: Jour. Geophys. Res., v. 74, no. 8, p. 2070–2080.

8

Low-Angle Thrusting and Gravity Gliding

The term *thrust fault* is applied to faults of low average dip in which the hanging wall has moved up relative to the footwall. When the hanging-wall block is the active member, the name *overthrust* is applied, and when it can be shown that the footwall moved, the name *underthrust* is used, but it may be difficult to make this distinction.

This definition provides a basis for recognition of thrusts, but we need to bear in mind that a thrust may be horizontal in one place and become nearly vertical in another. Horizontal transport of huge masses of rock for distances of many kilometers has taken place over many thrusts, but in most cases it still has not been determined if the movement is caused by horizontal compression generated by some type of local crustal shortening or whether the lateral movement arises from downhill, gravity induced creep, sliding, or spreading of the elevated rock masses.

THE CHARACTERISTICS OF THRUST FAULT ZONES

The actual zone of shearing along a thrust may consist of mylonite (flinty recrystallized material), gouge (powdery crushed rock), breccia, or in rare instances, one or more, smooth slickensided rock surfaces. Where the leading edge breaks through to the ground surface and the overthrust plate moves over the ground, soil and fans of sediment spread in front of the advancing sheet may become mixed with the fault breccia as it has in the Muddy Mountains of Nevada (Longwell, 1949).

The character of the fault zone depends especially on the lithologies involved, depth, temperature, strain rate, and the pore pressure in the rock adjacent to the fault zone. Experimental studies indicate that highly ductile rocks, high temperatures, high confining pressures, or low strain rates would favor creep phenomena, while brittle materials, low temperatures, shallow depths, and high strain rates would favor development of loose breccia in the zone. Important differences are also found where the fault is parallel to bedding as compared with zones that cut across lithologies.

Where thrusts have passed through thick sections of shale, the movement may be largely concentrated along bedding and cleavage planes. Alternatively, the fault appears as a thick zone of closely spaced shears forming a shaly cleavage in the fault; often the cleaved rock is laced with small thrusts. Thick sections of thinly bedded rocks are often characterized by highly complex folding rather than by brecciation or movement on cleavages.

Thrust faults may consist of breccia-free zones where the rock on either side has been highly ductile or where friction along the fault has been very low. If the thrust is deep, lithostatic pressure may be sufficient to promote ductile behavior and may inhibit the development of breccia, which is accompanied by an increase in volume. Friction may be reduced by the presence of zones of high pore pressure such as might be expected where a claystone caps a porous sandstone containing water under pressure.

FIGURE 8-1 Trace of two major thrust faults in central Virginia. The Staunton-Pulaski fault carries Cambrian units over Silurian and Devonian rocks. The fault along the Blue Ridge carries Precambrian and lower Cambrian units over Ordovician units. Note the partial window, Goose Creek, the spliting of the Staunton-Pulaski, and the irregular trace of both thrusts. R, L, and S mark the locations of Roanoke, Lexington, and Staunton, Virginia.

Structural Patterns Along Thrust Faults—Orientation of the Fault Surface

Complications in fault zones caused by folding and subsidiary faults which accompany a major fault, as well as effects of erosion, often make it difficult to find good exposures of faults or to determine the dip of thrust faults. A low-angle thrust may be recognized by the deflection of the fault trace in the topography where the relief is high enough. Where the thrust has a low dip and the relief is sufficient, erosion may produce isolated remnants of the overthrust sheet called *klippes*. Generally these may be recognized by the presence of older rocks faulted onto younger units. Erosion may also lead to the formation of windows through the thrust surface, exposing the underlying rock. Klippes, windows, and irregular fault traces which follow contour lines in places, Fig. 8-1, are excellent criteria for the recognition of thrusts.

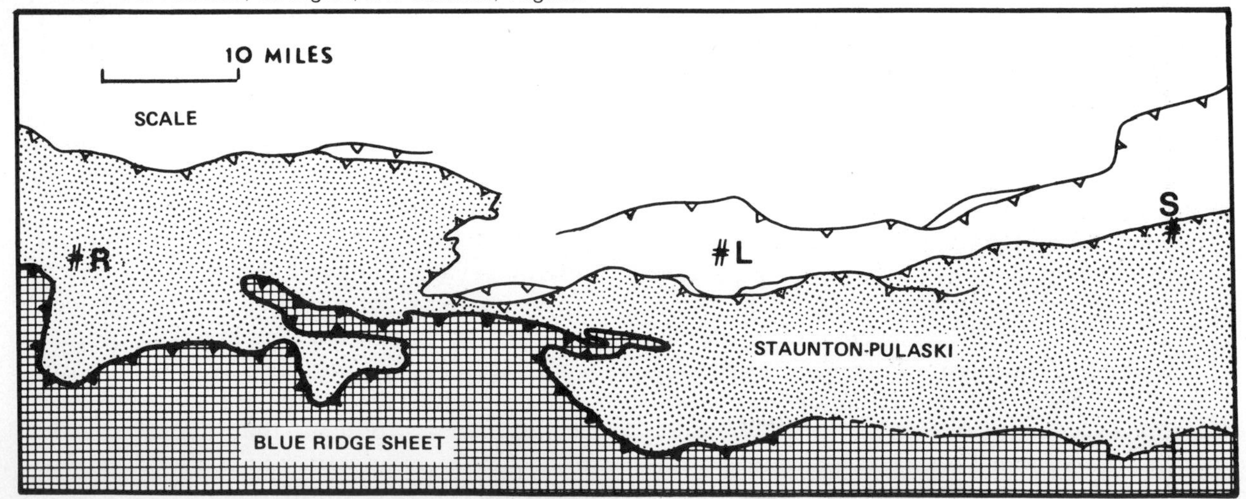

Certain characteristic patterns are associated with each major type of fault. Multiple subparallel faults, imbricate thrusts, characterize many thrust fault zones, and frequently a single fault may be followed into numerous branches, some of which may rejoin the master fault, whereas others reach the surface, die out in folds, along bedding, or at transverse faults. Branching is evident both on maps, and in cross sections and is responsible for formation of fault slices, Figs. 8-2 and 24-5.

Tear Faults

Small vertical faults oriented transverse to a major thrust occur where one part of the thrust sheet moves farther than adjacent parts. An excellent example of this is found at the northeastern corner of the Beartooth Mountains, Montana, Fig. 8-3, where tears formed in the sedimentary rocks along this mountain front as the sedimentary veneer was rotated to a vertical position during uplift and portions of the mountain front were thrust laterally. Some portions of the front are thrust nearly 1.6 km (a mile) farther than other parts across the tears. The orientation of the tears here is governed by fractures in the Precambrian rocks. Tears are usually oriented normal to a thrust or in positions which may be explained in terms of shear failure (two conjugate sets at ±30° to the direction of transport of the thrust) of the overthrust plate.

FIGURE 8-2 Cross section across the Rocky Mountain front in the Ghost River area of Alberta. Vertical and horizontal scales are the same. (*Redrawn from Fitzgerald, 1962.*)

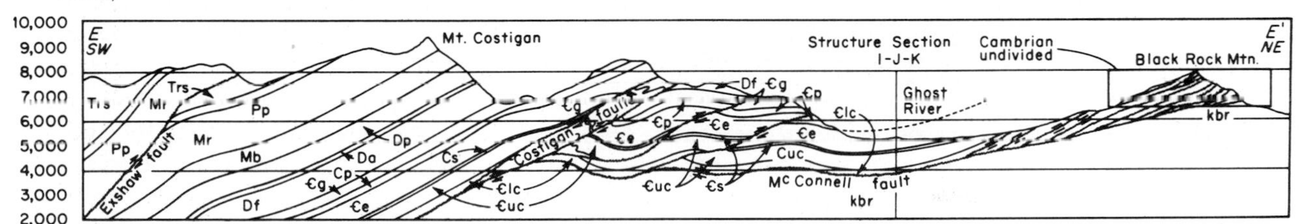

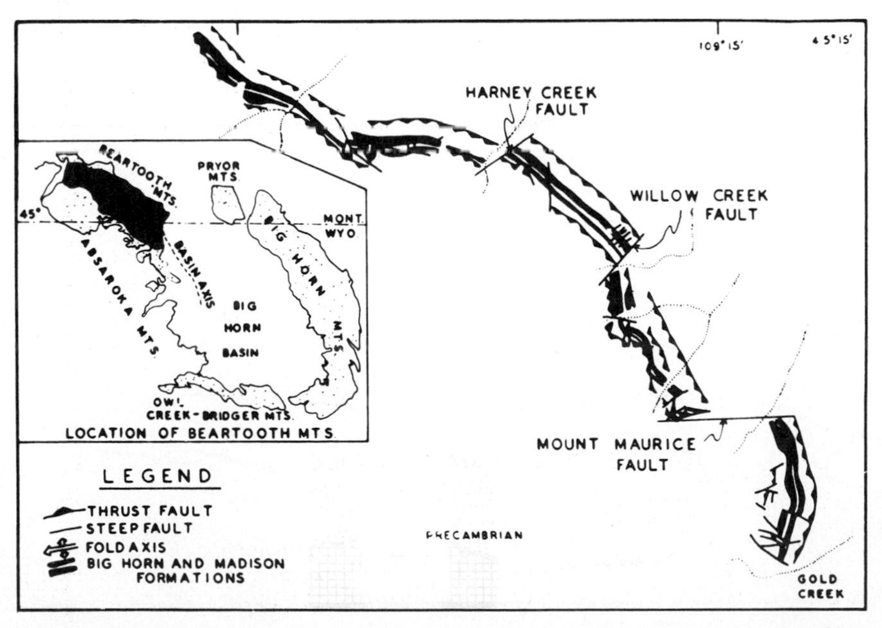

FIGURE 8-3 Thrust faults and structure at the northeast corner of the Beartooth Mountains, Montana. Note the vertical tear faults (Mount Maurice; Willow Creek, and Harney faults). (*From Casella, 1964.*)

Imbricate Faults

Thrusting is described as imbricate when a number of subparallel thrusts grow out of the same "sole" thrust, Figs. 8-2 and 8-4. Slice after slice of the same bed may be stacked on top of one another in such zones, and they commonly exhibit curved profiles.

Drag

Folding usually accompanies thrusting, and some folding can be attributed to drag along the fault zone, especially at the leading edge, but it is often difficult to establish whether the folding in beds over a thrust is a drag effect or if the thrust grew out of the fold. One common interpretation is that folding proceeds from open to asymmetrical to overturned folding, with a fault forming on the overturned limb and drawn out into a thrust fault. Many thrusts such as those that cut both the forelimb (the overturned limb) and the back limb in Turner Valley oil field (Fig. 8-4) do not fit this picture. Many geologists prefer the view that the fault forms first, often as a shear across brittle strata, following bedding in ductile (incompetent) beds, and that the fold forms as a result of drag as movement proceeds along the fault.

Back Thrusts and Underthrusts

Some thrust faults are oriented so that they dip in the same direction as the regional direction of tectonic transport. For example, most thrusts in the Appalachian region dip southeast, and most evidence indicates that the regional tectonic transport was from southeast to northwest; yet some thrusts dip to the northwest.* In these cases the block beneath the fault may be the active element, and the movement is essentially an underthrust and is called a *back thrust*. Most thrusts are interpreted as overthrust, in which the upper plate moved up and over the lower block, an interpretation favored on mechanical grounds, because pressure is less above. The weight of the hanging-wall block that must be raised over an underthrust becomes increasingly greater as movement proceeds. Therefore, we may anticipate that most such faults will not extend very deep or will not involve much lateral movement.

Back thrusting is now recognized in the Appalachians, where many back thrusts are formed on the western limbs of arches and folds generated where part of the sedimentary

* Gwinn (1964) and Bick (1973).

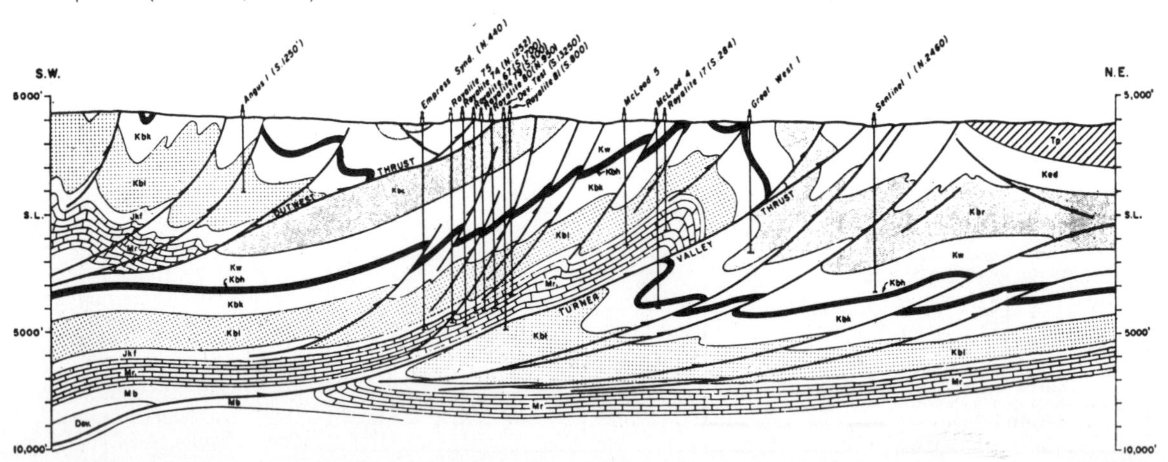

FIGURE 8-4 Cross section across the central sector of the Turner Valley. The Turner Valley fault is a forelimb thrust, but note the numerous faults on the back limb of this asymmetric anticline. Also note the two small thrusts that dip east. (*From Fox, 1959.*)

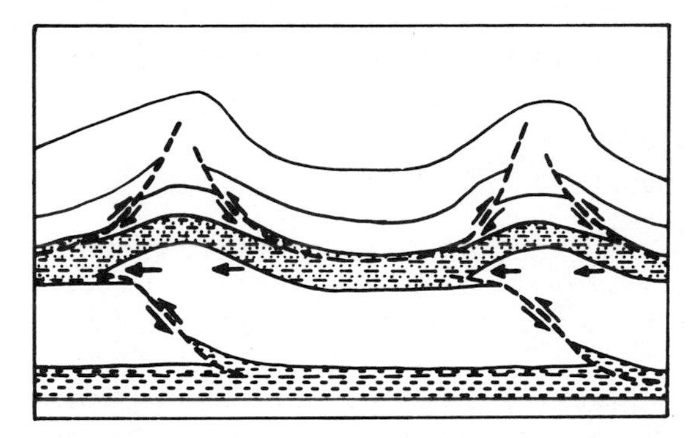

FIGURE 8-5 Schematic illustration showing a method by which folds flanked by thrusts directed toward the fold crest may originate as a result of *décollement* and thickening of the section over places where the thrust rises through competent beds. (*After Gwinn, 1964.*)

veneer slips laterally over some bed (called a surface of *décollement,* or literally, ungluing), as in the Summit oil field in Pennsylvania, Figs. 8-6 and 8-7. A larger scale interpretation of this process, Fig. 8-5, shows a major *décollement* zone at depth in which a bedding thrust occurs. Where this thrust breaks up out of the *décollement* across a competent section, the thrusting causes a localized thickening which results in formation of an arch or fold in the upper layers. Shortening in these layers is accommodated by folding and localized thrusting on both limbs of the fold, with formation of overthrusts on one limb and back thrusts on the other.

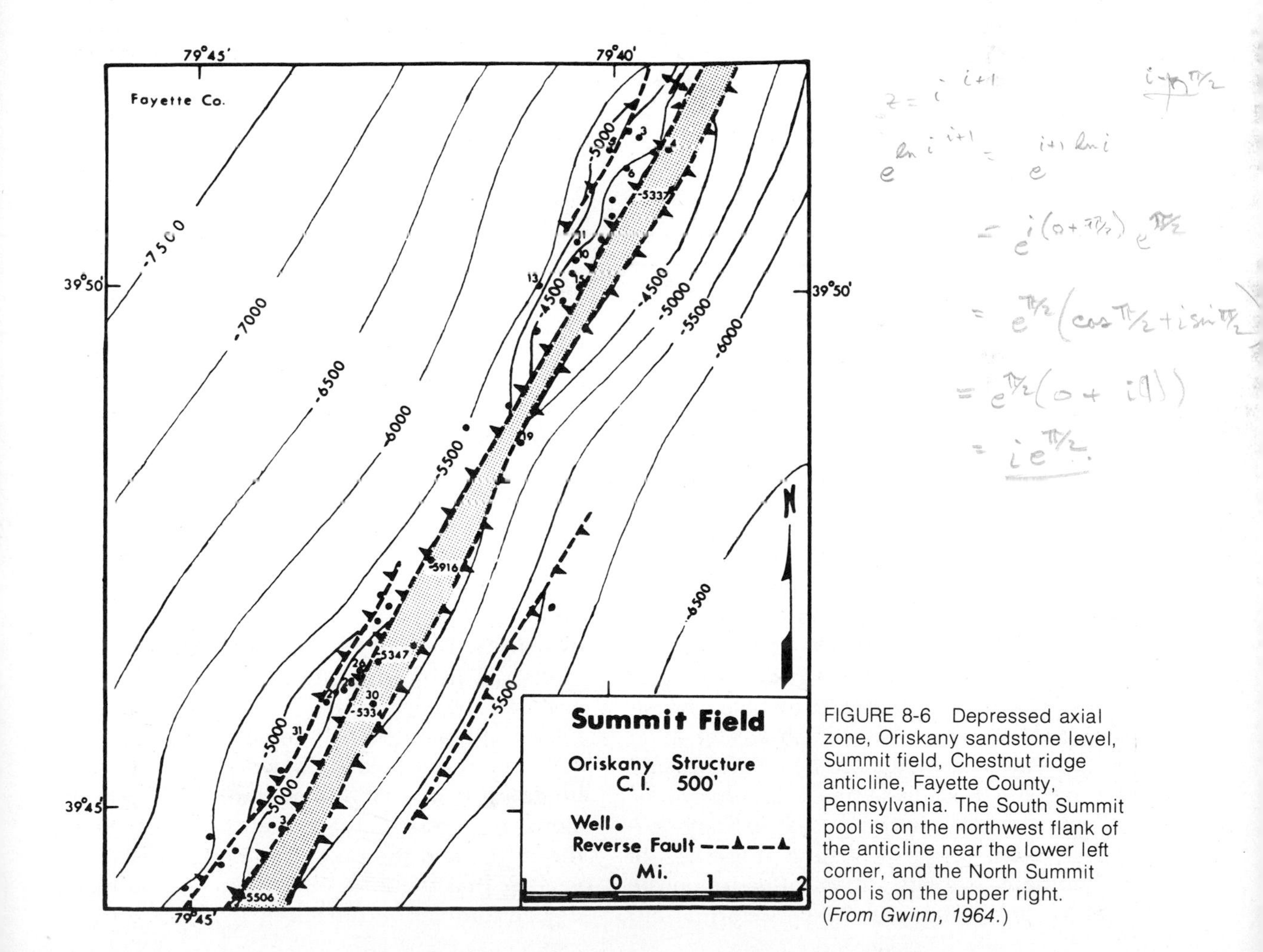

FIGURE 8-6 Depressed axial zone, Oriskany sandstone level, Summit field, Chestnut ridge anticline, Fayette County, Pennsylvania. The South Summit pool is on the northwest flank of the anticline near the lower left corner, and the North Summit pool is on the upper right. (*From Gwinn, 1964.*)

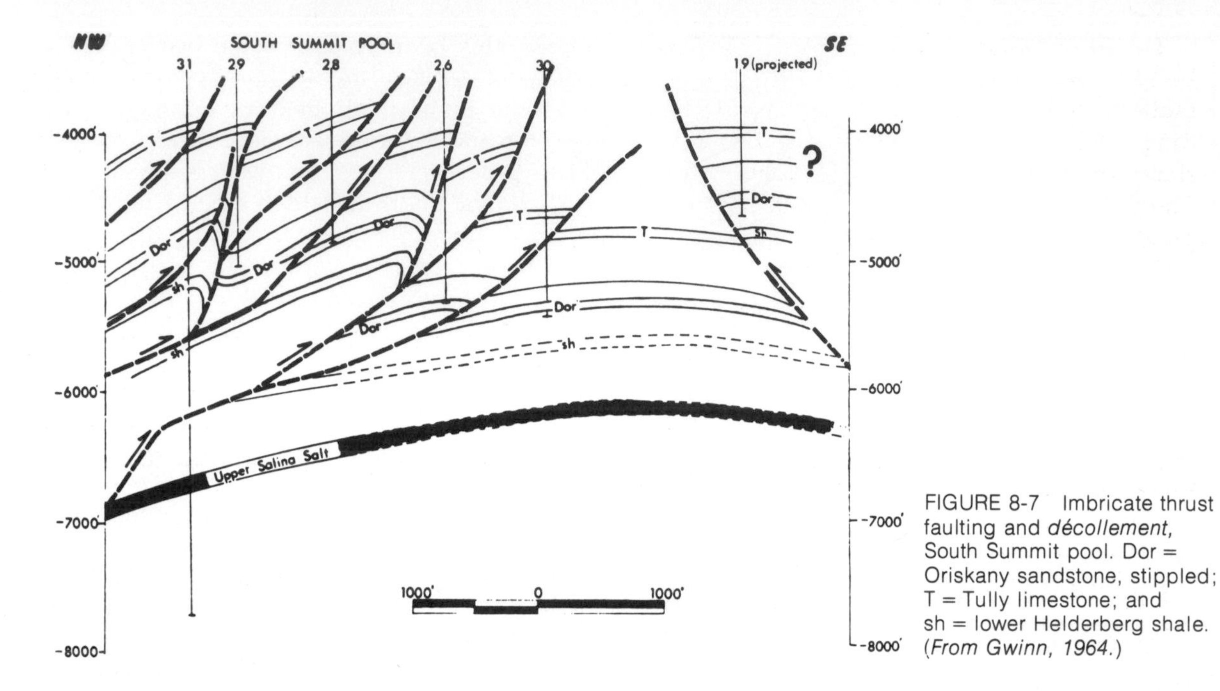

FIGURE 8-7 Imbricate thrust faulting and *décollement*, South Summit pool. Dor = Oriskany sandstone, stippled; T = Tully limestone; and sh = lower Helderberg shale. (*From Gwinn, 1964.*)

Stratigraphic Controls on Thrusting

Thrust faults are known to involve rocks of all ages and all common rock types. Some major thrusts in the European Alps and in the Appalachians bring metamorphic and igneous complexes into fault contact with sedimentary rocks; many of the thrusts in western North America and in the Appalachians, Ouachitas, and Marathon Mountains carry consolidated sedimentary rocks onto other sedimentary rock units; and some thrusts involve semiconsolidated and even unconsolidated units in the Gulf Coastal Plain and elsewhere. Obviously, then, the mechanics and probable causes of thrusting differ considerably from place to place. Thrusts which involve crystalline complexes which have much greater shear strength than sediments are most likely quite different from those in sedimentary units.

Where thrusts involve sedimentary sequences, variation in physical properties of the lithologies exerts a strong influence on the thrusting and total pattern of deformation. This is well known in the Appalachian Mountains, where the northwestward general transport of the sedimentary veneer has been accomplished in part by lateral movements on bedding faults. This is clearly shown in the Cumberland overthrust, a huge plate, 200 km long, which has moved nearly 10 km northwest along a thrust fault that is nearly flatlying under most of the block. The structure of this plate is known through surface mapping and by drilling in several oil and gas fields (Miller and Fuller, 1954). The leading edge of the thrust block is the westernmost major fault in the region. The block is bounded on the west by the Pine Mountain overthrust and on the north and south by tear faults, both of which become part of northeast-southwest-trending faults in the strongly folded and faulted belt to the east, Fig. 8-8. The overthrust block is warped to form a broad anticline and syncline, and erosion along the crest of the anticline has exposed the thrust and the units under the thrust, revealing that the Pine Mountain fault is made up of a complex of thin slices.

The Pine Mountain thrust is known to be a bedding fault under much of the Cumberland fault block, where the thrust is parallel to bedding under the syncline and is located in shales of Devonian and Mississippian age. The units beneath the fault are Silurian, so no great stratigraphic separation is apparent in wells, but the character of the fault is clear in the anticline, where wells penetrate Cambrian units above Silurian units, Fig. 8-9. Much of the lateral movement of this thrust block was accomplished by movements parallel to subhorizontal bedding. The beds in which shearing occurred are mainly shale beds of Cambrian (Rome formation) and Devonian age. The main thrust broke across the intervening limestones and dolostones at a steep angle along what are now called *ramp rises* as movement planes shifted from the lower to the higher shale. The edges of this fault block are neatly defined by the tear faults and the whole block moved northwest as a unit, suffering relatively little deformation outside the fault zones.

Structural evolution of the plate is described as follows by Harris and Zietz (1962):

> Major folding that apparently involved the basement initiated development of the Cumberland overthrust block. . . . The Pine Mountain fault later developed as a bedding-plane thrust in the incompetent Rome Formation. . . . As the fault began to grow it was deflected upward by the south limb of the primary fold from the Rome . . . across several thousand feet of beds to the Chattanooga Shale. Rocks stripped from the primary fold and moved toward the northwest formed the Powell Valley anticline by duplication of beds. Frictional drag on the fault caused the north limb of the nearly rootless Powell Valley anticline to fold into the Cumberland Mountain monocline. During the final phase of the orogeny the thrust plates locked in the eastern part of the area; this initiated arching of the Pine Mountain fault upward about 5,000 feet from its original position. In the western part of the area the plates did not lock, and movement continued, skewing the axis of the Powell Valley anticline northward about 1.5 miles. Displacement in the eastern part of the area is about 4 miles and in the western part about 5.6 miles.
>
> Analysis of the structure suggests that the initial dip of the Pine Mountain fault, which in the main is a bedding-plane thrust, was predetermined by the attitude of the country rock in which the fault developed. Size and shape of major folds in the moving plate were influenced by the initial fold, the strati-

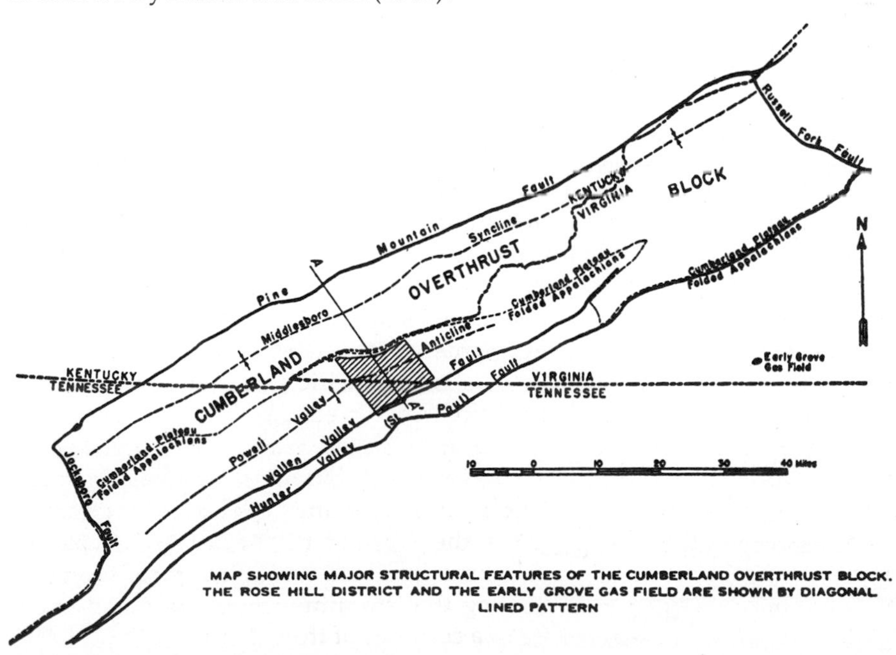

FIGURE 8-8 The Cumberland overthrust block, Pine Mountain thrust, and associated Jacksboro and Russell Fork tear faults. This is one of the classic examples of "thin-skinned" deformation in the Appalachian region. Compare with the cross section in the next figure. (*From Miller and Fuller, 1954.*)

graphic position of the fault, the amount of displacement and frictional resistance, root structures in the stationary block, and folding in the later stage of deformation.

MODE OF OCCURRENCE OF LOW-ANGLE THRUSTS

The general features of low-angle thrusts are summarized by King (1960) as follows:

1. Low-angle thrust faults do not originate from low-dipping shear planes which cut indiscriminately through heterogeneous rocks. They follow zones of weakness in the incompetent strata and shift abruptly from one zone to the next along diagonal shears in the intervening more competent strata.

2. Many low-angle thrust faults originate early in the orogenic cycle when the strata are little deformed. The thrust sheets move as broad plates and are not much folded, except where they are warped over the flats and pitches of the fault surface beneath. The overridden rocks are more folded, but probably mainly as a result of frictional drag of the thrust sheet above.

3. If a low-angle thrust fault develops in deformed rocks of varying competence, its initial rupture is not a smoothly dipping shear plane. Rupture is along bodies of incompetent rock or along zones of weakness already existing.

4. Thrusting of older rocks over younger is characteristic of most of the exposed thrust faults, but many thrust faults along which younger rocks override older have been discovered. Along the Pine Mountain fault, older rocks are thrust over younger along the pitches, younger rocks over older along the flats. Both relations are thus normal and expectable along low-angle thrust faults but occur in different parts of the same fault.

5. Low-angle thrust faults have finite breadth and end both rearward and forward. Rearward ends of low-angle thrust faults are seldom available for examination, but Nolan's suggestion is plausible that the faults die out rearward by decrease in contrasts between the deformation of the overriding and overridden blocks. Forward, a low-angle thrust fault ascends on a succession of pitches, and the highest pitch will bring it to the surface. These forward edges have long since been carried away by erosion in the Appalachian and Rocky Mountain regions, but they seem to be preserved along many of the thrust faults of the southern Great Basin.

6. Low-angle thrust faults likewise have finite length. Thrust sheets have limits of strength, depending on the materials which compose them, beyond which they cannot move as a single mass. Low-angle thrust faults must end laterally, either along transverse (tear) faults with strike-slip displacement or by loss of displacement along the thrust itself.

Décollement of Foreland Belts

The thrusts found in most of the foreland regions of orogenic belts exhibit many of the characteristics described by King in the preceding section. The greatest amount of lateral movement of the large overthrusts in the sedimentary cover appear consistently to occur in incompetent shales or in evaporite sequences

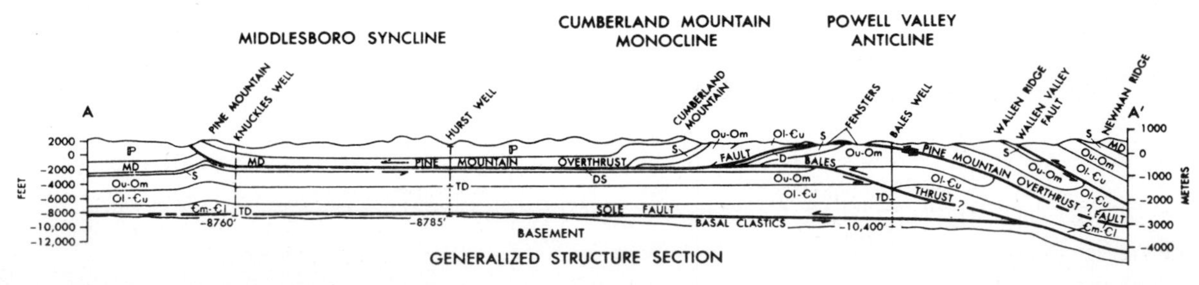

FIGURE 8-9 The Cumberland overthrust block and generalized structure section showing the stratigraphic positions of the Pine Mountain overthrust fault beneath the block. The heavy line marks the boundary between the Cumberland Plateau and the Valley and Ridge province. Note the windows in the Pine Mountain fault. (*From Miller, 1974.*)

as in the Jura Mountains, Fig. 23-6, and this lateral movement is largely parallel or subparallel to bedding. This movement of the cover, often many kilometers long, over such a layer requires that the cover becomes detached and is referred to as a *décollement*.

The level of the *décollement* is known, in many instances, to rise in the stratigraphic section. In some instances the rise is thought to occur where the thrust cuts steeply across a competent section, Fig. 8-5, such as quartzite or carbonates forming a *ramp-like rise*. In another interpretation the rises occur over high-angle faults, Fig. 8-10 (Jacobeen and Kanes, 1974).

FIGURE 8-10 Stages of development in formation of Broadtop and Whip Cove thrusts in the Valley and Ridge of the Appalachians. (*From Jacobeen and Kanes, 1974.*)

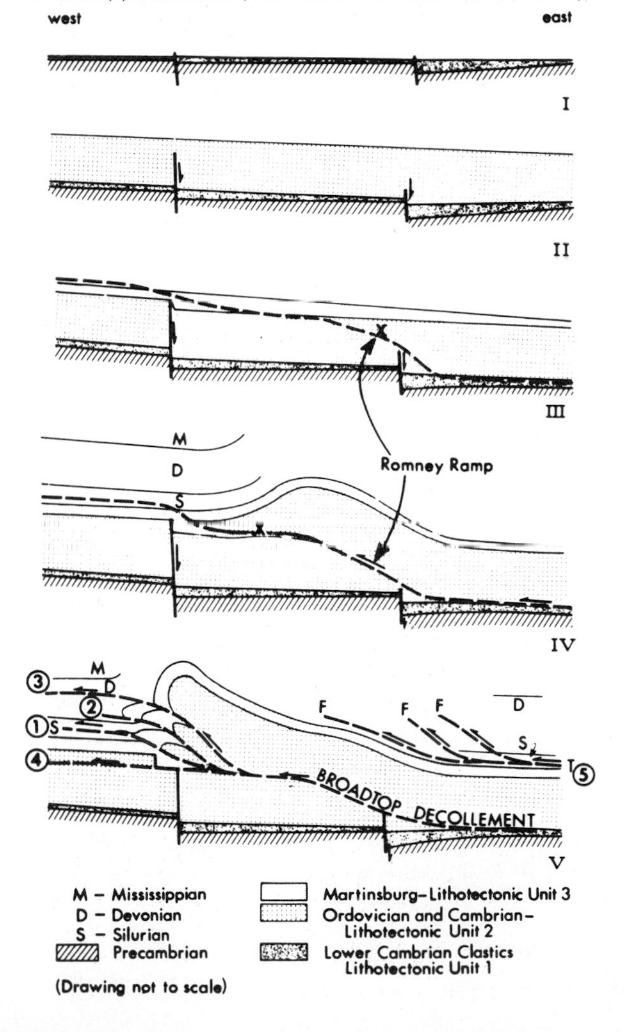

Gravity Gliding

Small-scale gravity-induced downslope movement of surficial materials is a familiar process we associate most readily with steep slopes, but it is well known also where conditions produce sliding surfaces with low frictional resistance. It is easy to envision similar processes acting on a very much larger scale, and some thrusts are best explained as the sliding surfaces over which large masses of material have shifted laterally under the influence of gravity. It is important to recognize that gravitative effects may be of at least two very distinctly different types. One is the sliding of a mass over a distinct zone of movement, the thrust fault in this case. The other is the lateral spreading of an uplifted mass of material as a result of gravitative effects. It is the first of these that is generally referred to as *gravity gliding*, or *sliding*.

Characteristics of Gravity Gliding

It is sometimes possible to identify a particular thrust block as a gravity-emplaced feature, but generally it is very difficult to determine if a particular mass slid into place or was forced into position by either crustal shortening or lateral spreading. De Sitter (1954) pointed out the following criteria which are useful in making the distinction between gravity-glide and other types of thrusts:

1. A slope is required to induce gravity gliding. The slope must have been in existence at the time of sliding, even though it may have subsequently been reduced or removed.

2. The rear end of the thrust should exhibit pull-apart features, Fig. 8-11, in cases of gravity gliding. The thrust often curves up to the surface and is expressed as a complex of normal faults.

3. When masses move downslope, different parts of the mass may move independently of one another. The structure may become chaotic and geometrically obscure. A gliding mass often lacks lateral continuity because the downslope movement is prompted by removal of support from the toe slope, and this tends to occur in stages rather than uniformly. When a mass is being moved by lateral compression, the mass tends to move more uniformly, the mass must in fact transmit the stress, and a more coherent and consistent internal and external geometry results.

4. Large masses of material are more likely to become inverted in gravity gliding than in compressional thrusting.

The Heart Mountain Thrust

For many years structural geologists were perplexed by the structural situation at Heart Mountain, Wyoming, where a block nearly 8 km across composed of flat-lying Paleozoic limestones rests on Eocene strata. This must be some type of thrust, but Heart Mountain is situated more than 19.3 km (12 miles) from the nearest large outcrops of Paleozoic units, and evidence of large-scale folding or forceful thrusting there, along the edge of the Absaroka Mountains, is lacking. Furthermore, the ground slope is very low from Heart Mountain, which is the eastern limit of thrusting to the Absaroka Mountains where Paleozoic rocks are in place (Fig. 8-12). Pierce (1957, 1963) describes nearly 50 fault blocks spread out over an area that is about 50 km wide and 100 km long. By tracing the fault carefully around each of these blocks, he is able to show that the fault surface cuts stratigraphically upward in a central area and that it is a bedding fault near the edge of the Absaroka Mountains (Fig. 8-12). Pierce suggests that the Heart Mountain thrust originated as a detachment thrust or bedding fault and that movements occurred as a result of gravitational sliding over a very low slope. Others have suggested the potential importance of lubricating effects from fluids or even gases associated with the volvanic activity in the Absaroka Mountains. In any case the absence of other structural features, the disconnected character of the blocks, and the disposition of the blocks all seem to exclude forceful movement resulting from lateral compression as a cause.

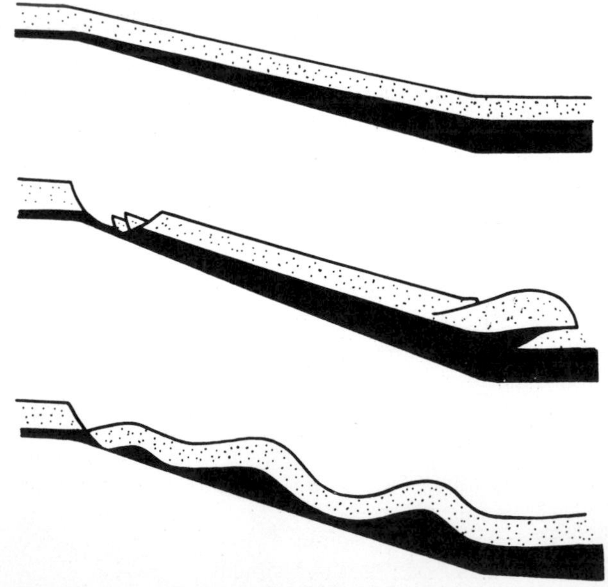

FIGURE 8-11 Sequential development of typical features associated with gravity gliding. Top shows rock sequence before slide. (Middle unit is assumed to be mobile shale or salt.) Middle diagram shows formation of pull-apart grabens updip and imbricate thrusts downdip. Lower diagram shows single pull-away with reverse drag updip and folding downdip. (*From Kehle, 1970.*)

Gravity Gliding and Gravity Spreading

Steep slopes in incompetent or highly ductile materials of any type are unstable. The material, whether it is unconsolidated sediment, water-saturated sediment, serpentine, or crystalline rock at high temperatures and pressures, will yield and spread laterally in response to gravity. This type of spreading may be accomplished in part by internal movements within the material, or it may be largely accommodated by lateral movements on a few discrete planes which we would call *thrusts* or *glide surfaces*. Gravity spreading and gravity gliding may be closely associated; they both derive by body forces due to gravity, but they are not identical processes. Gravity gliding takes place downslope along a distinct sloping surface. Spreading may occur laterally and even up adjacent sloping surfaces.

The olistostromes* in the Appennines provide a good example of a movement embodying both sliding and spreading. Some of these are simply submarine slumps in which masses break loose from steep slopes, become broken up, and slide downslope to come to rest within other sediment, but others appear to be slump and slide masses that form and move ahead of advancing thrust as illustrated, Fig. 8-13. In this example the movement is shown as proceeding downslope, but internal rearrangements due to plastic flowage, rotation of blocks, small-scale faults, as well as sliding play an important role in the overall deformation and movement pattern. The deformation toward the leading edge of the moving mass is caused primarily by the weight of the material back upslope from it.

The idea of gravity spreading has been used by Kozary (1968) to explain the emplacement of serpentine and ultramafic rocks along thrust in Cuba. The ultramafics occur along thrust zones and are found mixed with sediments. Kozary (1968, p. 2313) envisions their emplacement as follows:

* An *olistostrome* is a sedimentary deposit made up of a chaotic mixture of materials of different sizes that accumulated as a plastic or semifluid body by submarine sliding or slumping.

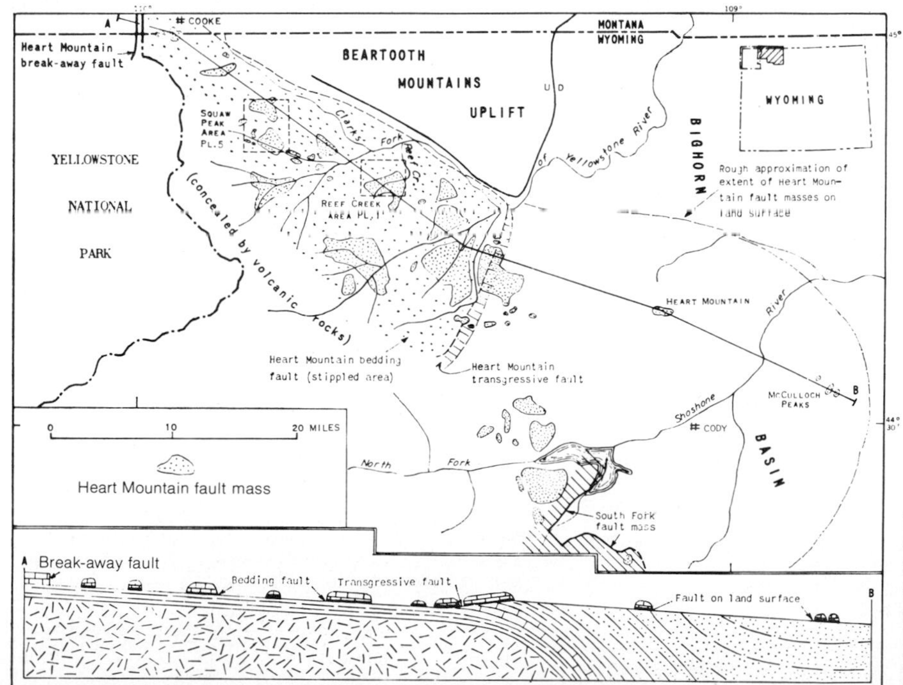

FIGURE 8-12 Distribution of Heart Mountain fault masses and cross section showing four types of faults constituting the Heart Mountain detachment fault, northwestern Wyoming. (*From Pierce, 1963.*)

The lower, undissected thrust plates, however, became more mobile because of the successively more complete hydration of the serpentine along the soles and the confining weight of the higher plates. During sliding, some of the plates were thrown into recumbent folds only 1 or 2 km in diameter. The serpentine within the folds, behaving nearly as a fluid, pierced the cores of the isoclinally folded plates, tore and separated the strata into thin sliver, and strung them out in relative order. The movement in the final state of the very mobile serpentine can be characterized as a "flow" thrusting similar to the types demonstrated in scale model experiments by Bucher (1956).

The resulting units are now exposed as illustrated, Fig. 8-14.

FIGURE 8-13 Schematic representation of gravitative tectonics in a transverse section across the Apennines about 50 km south of Florence. The three basins of deposition and the gravitational sliding and folding along the Upper Triassic are shown. (*After Elter and Trevisan, 1973.*)

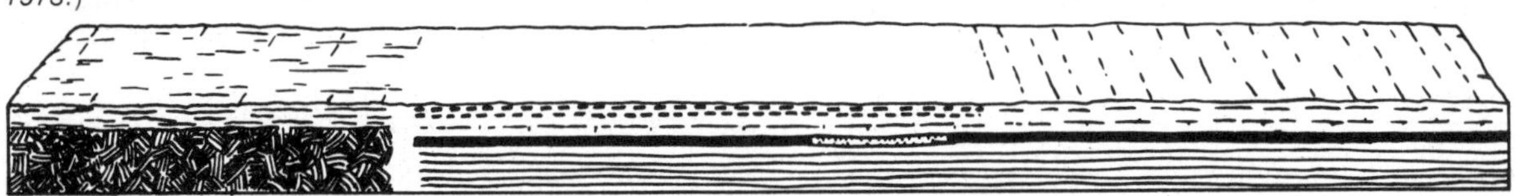

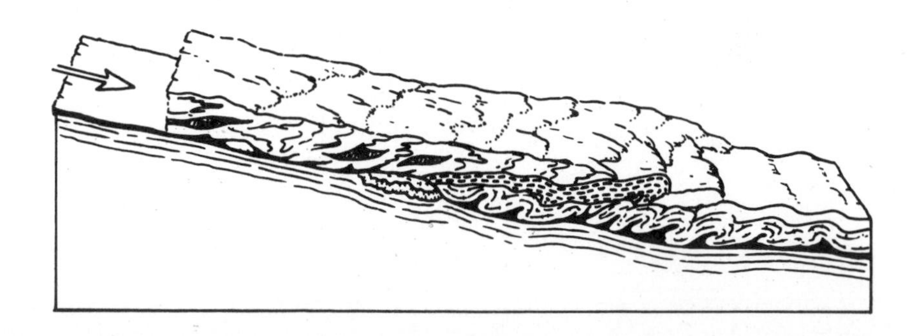

MECHANICAL CONSIDERATIONS IN THRUSTING

Basically, three mechanisms have been advanced to explain lateral transport of rock masses over surfaces of low dip: (1) gravity sliding of the thrust sheet down an inclined surface, (2) lateral compression resulting in movement of the thrust plate horizontally or up the inclined fault plane, and (3) lateral spreading of an uplifted rock mass. Lateral compression may be generated by a vise-like compression between more or less rigid plates or by the gravity-driven lateral spreading of an elevated pseudoviscous mass. The force in gravity gliding and gravity spreading is a body force, while the force in lateral compression is a directed push. If the thrust sheet may be thought of as a coherent mass that is strong, relatively brittle, and behaves as a unit or block, then the gravity-glide and lateral-compression models are both capable of simple representation by the inclined-plane problem of physics, Fig. 8-15. If accurate determinations can be made regarding the size of the thrust sheet and the coefficient of friction of the material, it is possible to reach conclusions regarding:

1. The angle of dip of a fault necessary for gravity sliding to occur.
2. The force necessary to push a thrust plate of given size horizontally or up an inclined fault.
3. The strength required for materials in the plate in order for the thrust plate to move as a plate and not fail by crushing.

The force required to move a block by a push from the rear is that necessary to overcome the frictional resistance between the thrust plate and the underlying rock:

FIGURE 8-14 Reconstructed section to illustrate the emplacement of "flow" thrust involving ultramafic rocks (wavy lines) and other rocks of various ages. (*From Kozary, 1968.*)

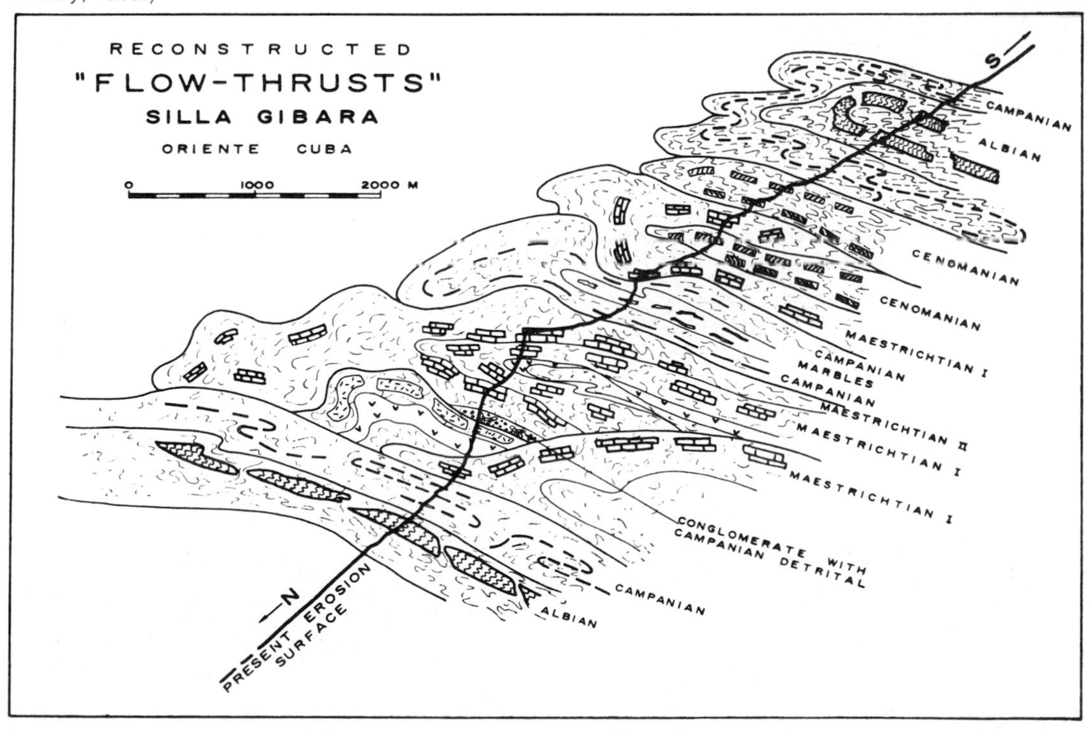

$$F = We = abc\rho e$$

where F = force required to move block
a, b, c = lengths of sides of block
W = weight of block (product of density ρ and volume abc)
e = coefficient of frictional resistance

The dimensions of many thrust plates can be closely approximated and the density of the rock easily determined. Unfortunately, actual values for the coefficient of frictional resistance along a thrust are difficult to determine. Smoluchowski (1909) calculated the force necessary to push a thrust plate along a horizontal surface, assuming the resistance equals that of iron on iron. He found that the strength of the rock in a plate 200 km long would have to be nearly seven times the crushing strength of granite at the ground surface. Thrust faulting along horizontal or up steeply inclined planes would seem unlikely unless the assumptions made are in error. One possible erroneous assumption has been pointed out by Price (1973). He draws attention to the probable mistake of using a model in which the entire thrust mass moves as a unit. Stresses are required in this model which are sufficient to overcome the entire frictional resistance of the mass. He proposes a more realistic model in which the movements occur on discrete slip surfaces within a coherent mass "... the net slip can be viewed as the cumulative effect of innumerable incremental displacements, each of which is initiated as a local shear failure that propagates as a dislocation on a scale and at a velocity that is small in comparison with the total area of the fault surface."

Smoluchowski concluded that thrusts must occur down inclined planes or that the coefficient of sliding friction must be much less than that assumed. Hubbert and Rubey (1959) reexamined the question of the maximum possible length for a thrust block, using Mohr's stress analysis methods and known crushing strengths for rock. They reach the conclusion that the maximum possible length for a thrust block pushed on a *horizontal* surface would be less than 30 km.

If the movement of the block is by means of gravity sliding, the strength of the material in the thrust plate ceases to be of critical importance, but the coefficient of friction continues to present a problem. The forces acting on a block on an inclined plane, Fig. 8-15 (where ρ = density, g = gravity force, and z_1 = height of a cross-sectional segment of the block of unit area) include:

1. The weight of the block acting along a vertical line and equal to $\rho g z_1$.

2. A component of the above force acting down the inclined plane with a magnitude that is a function of the slope of the plane and is equal to $\tau = \rho g z_1 (\sin \theta)$.

3. A component of the weight acting down and normal to the plane. This force is counterbalanced by the force the plane exerts on the block in the opposite direction. This counterbalanced force is equal to $\sigma = \rho g z_1 (\cos \theta)$.

4. Frictional resistance of the plane-block contact zone.

If the angle ϕ is the angle of inclination of the plane at which the block begins to slide, then $\tan \phi = \tau/\sigma$ (Fig. 8-15). For sliding, θ must equal ϕ. A similar condition applies to the angle of internal friction of sand. The angle is on the order of 30°. Thus for a mass of rock to shear off downslope under the force of gravity a slope on the order of 30° would be required. It seems highly improbable that any major low-angle thrust fault developed on a slope of that magnitude, and in the case of some of the best-documented gravity faults, it is improbable that the angle of inclination exceeded a few degrees.

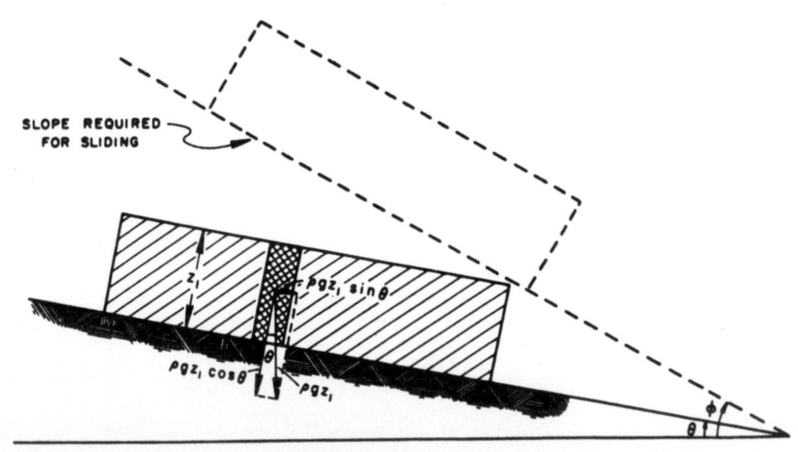

FIGURE 8-15 Gravitational sliding of a subaerial block, top. Normal and shear stresses on base of block inclined at angle θ are shown; angle ϕ is required for sliding. (*From Hubbert and Rubey, 1959.*)

Pore-Pressure Effects

Frictional drag between the thrust sheet and the underlying rock must be greatly reduced by some means in order to satisfy the observational data. Many major thrusts are bedding-plane thrusts situated in rocks composed of clay, shale, salt, or similar ductile materials, and various means of lubricating the fault surface have been sought. The most popular one in recent years is the buoyancy effect of high pressures in fluids held within pores of the rock, as described by Hubbert and Rubey (1959). Abnormally high pore pressures have been encountered in some deep wells in areas underlain by unconsolidated sediment.

High pore pressures are known in many of the world's deep sedimentary basins and geosynclines, such as the Gulf Coast of Texas and Louisiana and East Pakistan; and such abnormal pressures are found in regions that are now tectonically active, such as California, Trinidad, Burma, and Pakistan. The most likely mechanism for creation of the high pore pressures is found in compaction of sedimentary rocks which contain interbedded clays that have very low permeability and produce a self-sealing mechanism as compaction starts. The pore pressures in some cases come close to equaling the pressure due to the weight of the overlying water-saturated rock (Fig. 8-16). The water is in essence supporting a large part of the weight of the overlying rock in such a zone. This condition would be very significant for the models just considered in that the shear stress needed to slide the block depends only on that part of the pressure that is supported by the rock. As the pore pressure goes up, the shear stress required to move the block is decreased. Thus, frictional drag along a potential fault is greatly reduced in the first model, and the angle of slope required for gravitational sliding in the second is greatly reduced.

The interrelationship between plate width, angle of slope, and ratio of overburden weight to pore pressure is summarized in Fig. 8-17. Note especially that very large plates can be moved down slopes of only one or two degrees as the pore pressure exceeds 0.8 of the overburden weight.

Mechanics of Lateral Spreading

The potential importance of gravity-driven spreading was pointed out by Bucher (1956, 1962) and is illustrated by use of a group of model experiments described later (Chap. 13, Fig. 13-3). The uplifted wax in this experi-

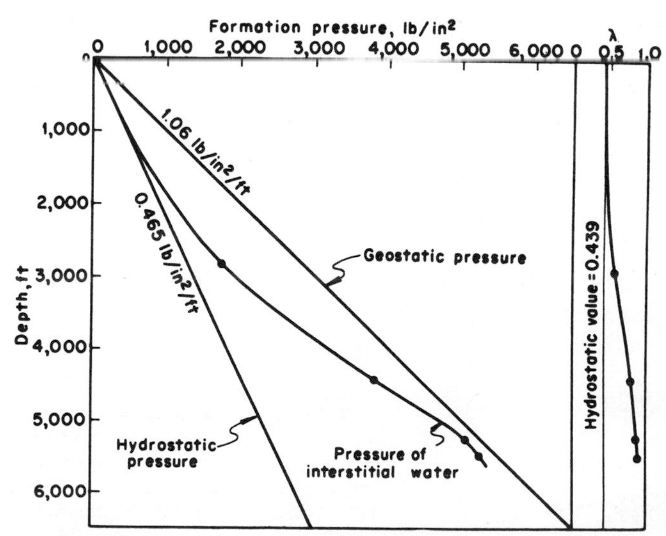

FIGURE 8-16 Variation of pressure, and of corresponding values of λ, with depth in Khaur field, Pakistan. (*From Keep and Ward, 1934.*)

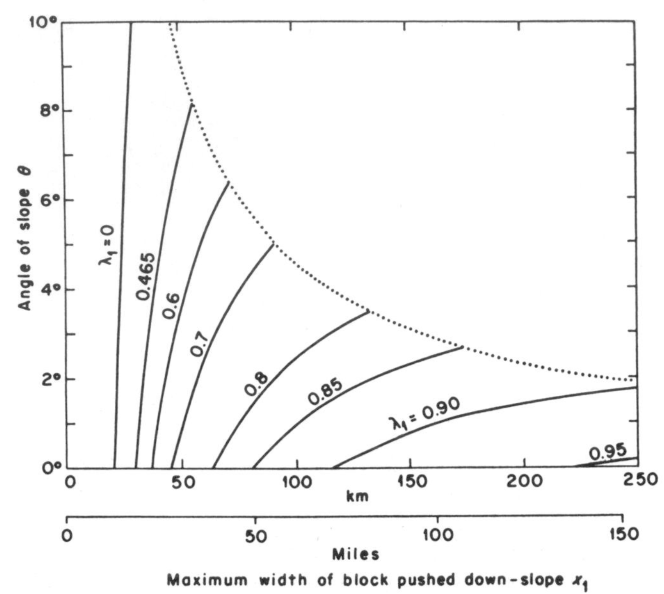

FIGURE 8-17 Width of plate, x_1, that can be pushed downslope, for various values of slope angle θ and fluid-pressure–overburden ratio λ_1. Thickness of plate constant at 6 km. Dotted line drawn at difference in elevation of 8 km between front and rear edge of plate. (*From Rubey and Hubbert, 1959.*)

ment spread laterally, folding and faulting the layers adjacent to it and producing folds and thrust that resemble those of many foreland belts, Fig. 8-18. More recently this mechanism has been employed by Price (1971) and Price and Mountjoy (1970) to explain characteristics of the structures in the Canadian Rockies, and Elliott (1976) presents a mechanical analysis of the process (see Fig. 8-19).

This model accounts for thrusts as a product of the lateral spreading of an uplifted area which is driven by the difference in surface elevation of the uplifted area and the flanking region. Elliott (1976) directs primary attention to the question of the shearing stress on the basal surface of movement and demonstrates that the regional average basal shear stress along a thrust which is deeper than 5 km is given by the relation $\tau = \rho g H \alpha$, where ρ is the density of the rock in the thrust sheet, H is the thickness of the sheet, and α is the slope of the ground surface. This relation applies to movements at great depths where the rock may be expected to yield by creep or a pseudoviscous flow process in a manner analogous to the basal flow of a glacier. Incidentally, this equa-

FIGURE 8-18 A model made with stiching wax, illustrating effects of gravity spreading. (*From Bucher, 1956.*)

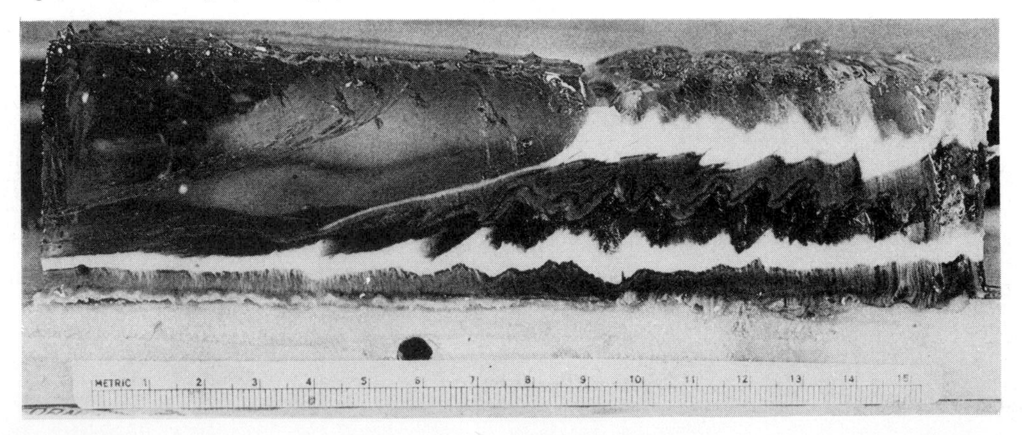

tion has long been used to calculate the shear at the base of ice. Locally the shear stresses may be magnified above the average regional value as a result of local stress concentrations on the leading edge of a growing thrust.

FIGURE 8-19 (*a*) Cross section through central Canadian Rockies (*after Price and Mountjoy, 1970, fig. 2-1*). The mid- and upper Cambrian formations (stippled) originally were an eastward-tapering wedge of shelf carbonates with a portion of shaly facies in the extreme west. Only major faults are shown. Undeformed basement is black. There is no vertical exaggeration. (*b*) Variation with distance of surface slope $\bar{\alpha}$. (*c*) Thickness *H* from surface to the sole *décollement*. Each value of $\bar{\alpha}$ and *H* is averaged over a series of overlapping 60-km lengths. (*d*) The regional basal shear stress $\bar{\tau}$ is equivalent to the down-surface slope stress $\rho g H \bar{\alpha}$. (*e*) Cross section shows surface topography, all major listric normal faults, and magnitude and sense of basal shear stress (arrows). Note coincidence of listric normal faults and reversed sense of shear stress. Footwall of basal *décollement* is stippled. (*From Elliott, 1976.*)

Elliott's analysis shows that it is the surface slope that determines the magnitude and sense of the shearing stress at the base and not the dip of the base. Therefore, the overthrust plate may move up the dip of the thrust plane.

This model has been successfully applied to the thrust in the Canadian Rockies as shown in Fig. 8-19. The graphs below the cross section show the average surface slope, the thickness of the overthrust plate, and the variation in shearing stress at the base. One of the most interesting aspects of this is that the area of reversed shearing stresses coincides with an area in which curved normal faults show some regional movements have taken place directed toward the internal parts of the belt. This could not have been the case when the main externally directed thrusting took place in the

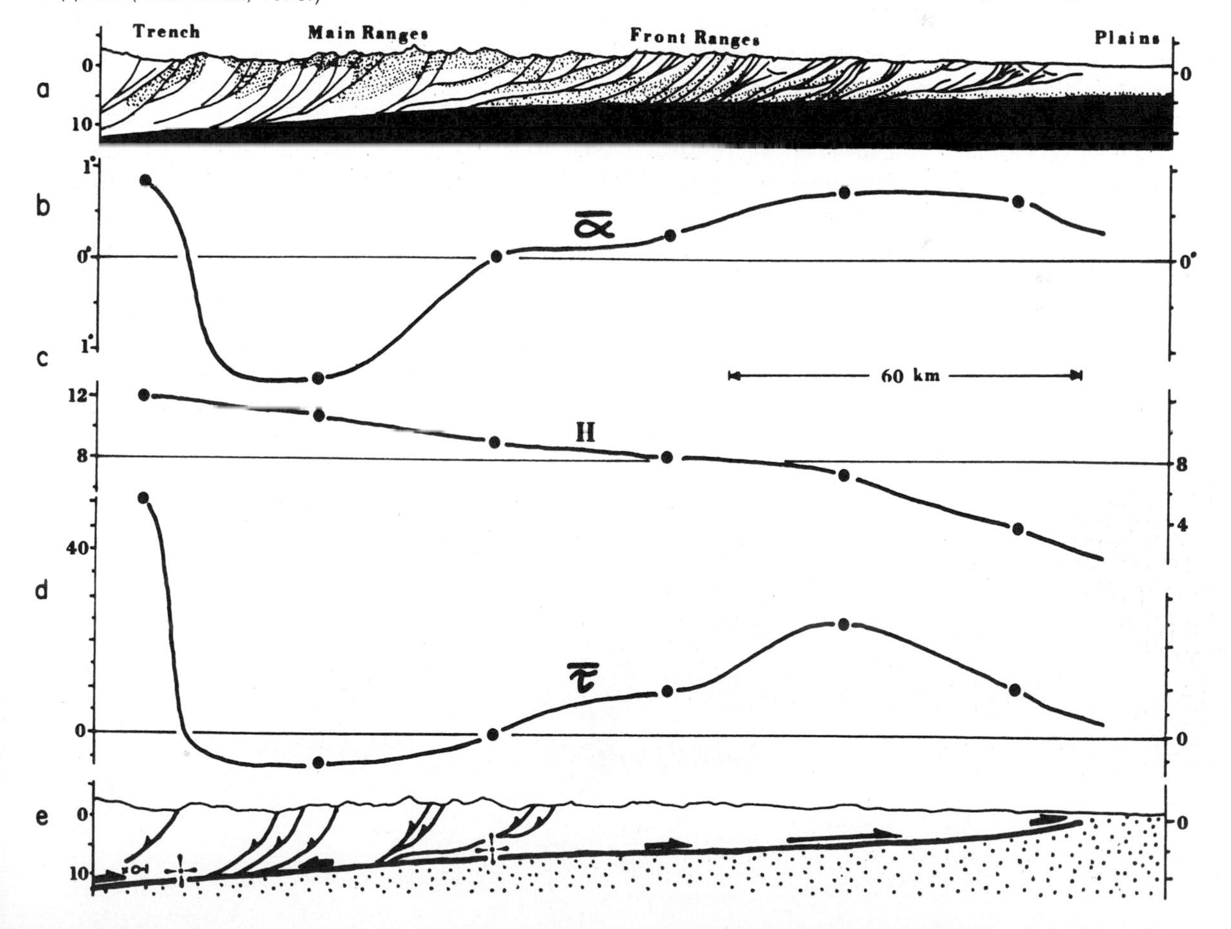

late Cretaceous when a greater surface slope of several degrees is needed to account for the eastward directed shearing. This slope could have been reduced by either regional tilting or erosion.

This model is very attractive as an explanation of thrusts in many orogenic belts. It provides a ready solution to the perplexing problem of how a thrust sheet with relatively slight strength can transmit stresses over great distances as is required by the lateral compression model, and it also avoids the problem of having thrust sheets move downslope and is compatible with the observation that most of them are not inclined down the dip of the thrust at the present time.

THRUSTING—TECTONIC SETTING

Thrusting is currently associated with two main tectonic settings, (1) orogenic belts—the folded mountain belts, and (2) subduction zones of modern plate tectonics.

Low-angle thrusts are characteristic features of the Caledonian-Appalachian-Ouachita-Marathon fold belt,* the Alpine-Himalayan belt, and the Cordilleran region among others. Within these belts the thrusts are perhaps best known where they occur in the thick sedimentary sections in foreland fold belts, in a marginal position relative to the central core of the mountains where metamorphic rocks dominate. The movement of the thrust masses is away from the core region. Thrusting also involves the crystalline rocks which may be carried laterally tens of kilometers over low-angle thrusts. Probably many more thrusts will be identified in the crystalline complexes as detailed mapping proceeds.

According to plate theory, the thrusts may be formed as a result of vise-like lateral compression created where continental plates collide. The Himalayan model is cited as an example of such a collision where one plate (the Asian) has overriden another (the Indian). Collisions of this sort are postulated for other folded–thrust-faulted orogenic belts of greater age.

Thrusting in the subduction zones is of a very different character. The oceanic plates are postulated as the mobile units, so the phenomenon is largely one of underthrusting. Because most areas postulated as modern subduction zones are located along continental margins and beneath deep-sea trenches, we have no opportunity to examine them. Older subduction zones have been postulated in places where ultrabasic rocks, especially those associated with trench sediments, are exposed on continents or on islands. The dip of most of these zones is steep or indeterminant, and the mode of emplacement of the ultrabasic rock is often questioned (see discussion in Chap. 17).

Gravity gliding occurs in a more or less pure form in some instances. Seismic reflection data obtained across the continental slope and

* See the regional discussions in Chaps. 22, 23, and 24.

FIGURE 8-20 Structure of the Kubor-Bismarck cordillera and the Papuan foreland folded belt, composite section based on the *décollement* principle of foreland folding. (*After Smith, 1965.*)

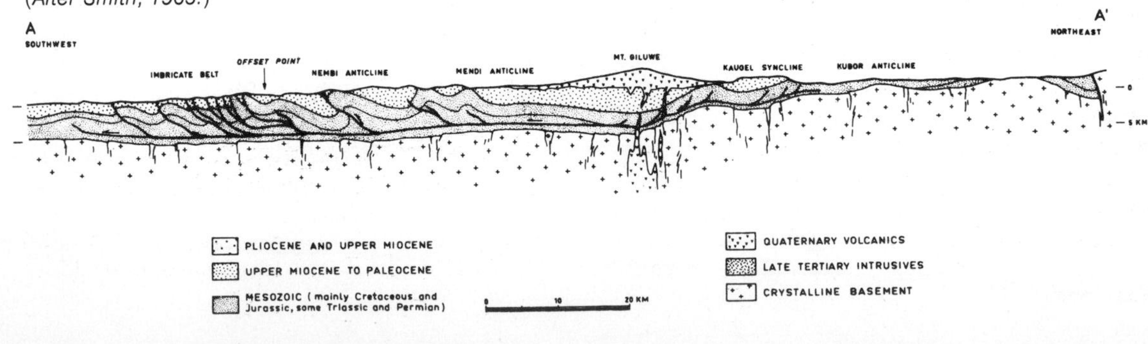

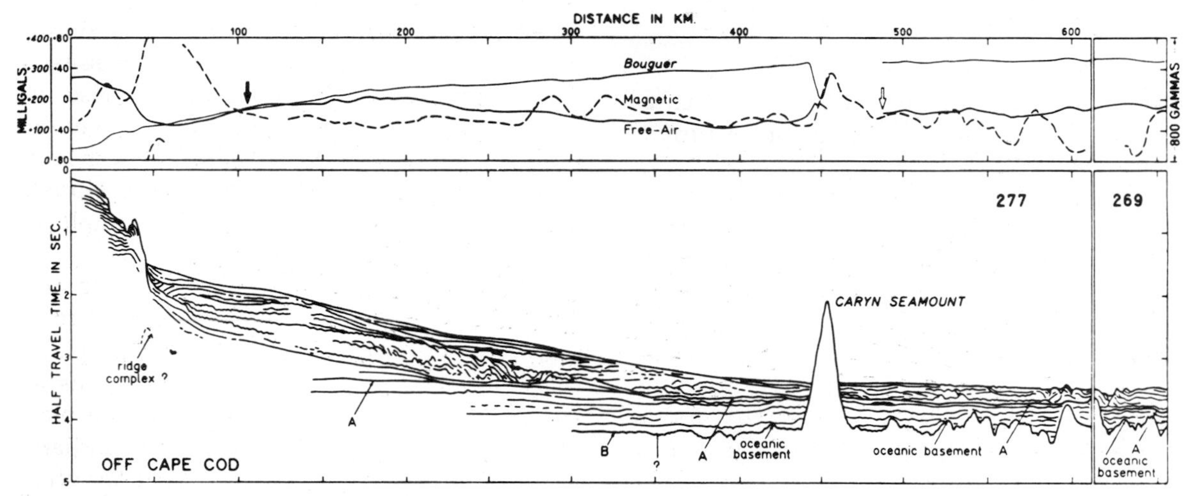

FIGURE 8-21 Continuous seismic profile and graphical data across the continental margin off Cape Cod. Caryn Seamount is an extinct volcano. Note the gravity and magnetic anomalies associated with it. Also note the disrupted bedding probably due to submarine slumping. (*From Emery and others, 1970.*)

rise of eastern North America, Fig. 8-21, show large discontinuous bodies of rock buried in the vast sedimentary prisms off the coast. The layering is sufficiently distinct to show clearly that the bedding is discontinuous and that some masses are internally deformed. These appear to be good examples of sediment probably originally deposited near the edge of the shelf which slid down the slope.

We should expect to find syndepositional basin-directed movements in many ancient sedimentary basins. Small-scale movements of this type have been described,* but large-scale movements of this type which might be confused with compressional thrusting are hard to prove.

Gravity gliding is widely accepted as a component of the lateral movement in most orogenic belts, but gravity spreading may be of far greater importance. Obviously slopes do exist as soon as an orogenic belt is uplifted, and gravitative forces would accompany other forces as a cause of the outwardly directed thrusting. The existence of the thrusts thus leaves unresolved the basic cause of the uplift, which some postulate to be primarily vertical movement while others prefer lateral shortening.

* See Cooper (1961).

REFERENCES

Barber, A. M., 1965, The history of the Moine thrust zone, Lochcarron and Lochalsh, Scotland: *Proc. Geol. Ass. London*, v. 76, p. 215–243.

Belyakov, L. V., 1969, The effect of pore pressure on the mechanism of large overthrusts: Geotectonics, no. 4 (1968), p. 214–221.

Bick, K. F., 1973, Complexities of overthrust faults in central Virginia: Am. Jour. Sci., Cooper vol., p. 343–352.

Birch, Francis, and others, 1961, Role of fluid pressure in overthrusting: Geol. Soc. America Bull., v. 72.

Bucher, W. H., 1956, Role of gravity in orogenesis: Geol. Soc. America Bull., v. 67, p. 1295–1318.

——— 1962, An experiment on the role of gravity in orogenic folding: Geol. Rundsch., v. 52, p. 804–810.

Carlisle, Donald, 1965, Sliding friction and overthrust faulting: Jour. Geology, v. 73, p. 271–292.

Casella, C. J., 1964, Geologic evolution of the Beartooth Mountains, Montana and Wyoming, pt. 4, Relationship between Precambrian and Laramide structures in the Line Creek area: Geol. Soc. America Bull., v. 75, p. 969–986.

Chapman, R. E., 1974, Clay diapirism and overthrust faulting: Geol. Soc. America Bull., v. 85, p. 1597–1602.

Cooper, B. N., 1961, Grand Appalachian field excursion: Blacksburg, Va. Poly. Institute.
Davis, G. H., 1975, Gravity-induced folding off a gneiss dome complex, Rincon Mountains, Arizona: Geol. Soc. America Bull., v. 86, p. 979–990.
De Sitter, L. U., 1954, Gravitational gliding tectonics —An essay in comparative structural geology: Am. Jour. Sci., v. 252, p. 321–344.
Eardley, A. J., 1969, Willard thrust and the Cache uplift: Geol. Soc. America Bull., v. 80, no. 4, p. 669–680.
Elliott, D., 1975, The energy balance and deformation mechanisms of thrust sheets: Proc. Roy. Soc., Ser. A. (In press.)
——— 1976, The motion of thrust sheets: Jour. Geophys. Res., v. 81, no. 5, p. 949–963.
Elter, P., and **Trevisan, L.**, 1973, Olistostromes in the tectonic evolution of the Northern Apennines, *in* DeJong, K. A., and Scholten, R., eds., Gravity and tectonics: New York, Wiley.
Emery, K. O., and others, 1970, Continental rise off eastern North America: Amer. Assoc. Petroleum Geologists Bull., v. 54, no. 1, p. 44–108.
Fitzgerald, E. L., 1962, Structure of the McConnell thrust sheet in the Ghost River area, Alberta: Alberta Soc. Petroleum Geologists Jour., v. 10, p. 553–574.
Forristall, G. Z., 1972, Stress distributions and overthrust faulting: Geol. Soc. America Bull., v. 83, no. 10, p. 3073ff.
Fox, P. P., 1959, Geology of Furnas Dam, Minas Gerais, Brazil: Geol. Soc. America Bull., v. 70, p. 1605.
Gwinn, V. E., 1964, Thin-skinned tectonics in the plateau and northwestern Valley and Ridge provinces of the central Appalachians: Geol. Soc. America Bull., v. 75, p. 863–900.
Harris, L. D., and **Zietz, Isidor**, 1962, Development of Cumberland overthrust block in vicinity of Chestnut Ridge fenster in SW Virginia: Am. Assoc. Petroleum Geologists Bull., v. 46, p. 2148–2160.
Hayes, C. W., 1891, The overthrust faults of the southern Appalachians: Geol. Soc. America Bull., v. 2, p. 141–154.
Howard, J. H., 1966, Structural development of the Williams Range thrust, Colorado: Geol. Soc. America Bull., v. 77, p. 1247–1264.
Hsü, K. Jinghwa, 1969, Role of cohesive strength in the mechanics of overthrust faulting and of landsliding: Geol. Soc. America Bull., v. 80, no. 6, p. 927–952.
Hubbert, M. K., and **Rubey, W. W.**, 1959, Role of fluid pressure in mechanics of overthrust faulting. I. Mechanics of fluid-filled porous solids and its application to overthrust faulting: Geol. Soc. America Bull., v. 70, no. 2, p. 115–166.
Jacobeen, Frank, Jr., and **Kanes, W. H.**, 1974, Structures of broadtop synclinorium and its implications for Appalachian structural style: Amer. Assoc. Petroleum Geologists Bull. v. 58, no. 3, p. 362–375.
Jenkins, David A. L., 1974, Detachment tectonics in western Papua, New Guinea: Geol. Soc. America Bull., v. 85, p. 533–548.
Keep, C. E., and **Ward, H. L.**, 1934, Drilling against high rock pressures with particular reference to operations conducted in the Khaur field, Punjab: Inst. Petrol. Technologists Jour. (London), v. 20, p. 990–1013.
Kehle, R. O., 1970, Analysis of gravity sliding and orogenic translation: Geol. Soc. America Bull., v. 81, no. 6, p. 1641–1664.
King, P. B., 1960, The anatomy and habitat of low angle thrust faults: Am. Jour. Sci., v. 258-A, p. 115–125.
——— 1964, Geology of the Central Great Smoky Mountains, Tenn.: U.S. Geol. Survey Prof. Paper 349-C, p. 1–148.
Kozary, M. T., 1968, Ultramafic rocks in thrust zones of northwestern Oriente province, Cuba: Amer. Assoc. Peroleum Geologists Bull., v. 52, no. 12, p. 2298–2317.
Laubscher, H. P., and others, 1960, Role of fluid pressure in mechanics of overthrust faulting: Geol. Soc. America Bull., v. 71, p. 611–628.
Longwell, C. R., 1949, Structure of the northern Muddy Mountain area, Nevada: Geol. Soc. America Bull., v. 60, p. 923–968.
——— 1951, Thrust faulting—What does it mean?: N.Y. Acad. Sci. Trans., ser. 2, v. 14, p. 2–5.
Lovering, T. S., 1932, Field evidence to distinguish overthrusting from underthrusting: Jour. Geology, v. 40, p. 651–663.
Malahoff, Alexander, 1970, Some possible mechanisms for gravity and thrust faults under oceanic trenches: Jour. Geophys. Res., v. 75, no. 11, p. 1992–2001.
Maxwell, J. C., 1959, Turbidity, tectonic and gravity transport, northern Appennine Mountains, Italy: Am. Assoc. Petroleum Geologists Bull., v. 43, p. 2701–2719.
Miller, R. L., and **Fuller, J. D.**, 1954, Geology and

oil resources of the Rose Hill district—The fenster area of the Cumberland overthrust block—Lee County, Virginia: Va. Geol. Survey Bull., v. 71.

Moore, Walter, 1961, Role of fluid pressure in overthrusting: Geol. Soc. America Bull., v. 72, notes.

Pierce, W. G., 1957, Heart Mountain and South Fork detachment thrust of Wyoming: Am. Assoc. Petroleum Geologists Bull., v. 41, p. 591–626.

——— 1963, Reef Creek detachment fault, northwestern Wyoming: Geol. Soc. America Bull., v. 74, p. 1225–1236.

Platt, L. B., 1962, Fluid pressure in thrust faulting—A corollary: Am. Jour. Sci., v. 260, no. 2.

Price, R. A., 1971, Gravitational sliding and the foreland thrust and fold belt of the North American Cordillera, discussion: Geol. Soc. America Bull., v. 82, p. 1133–1138.

——— 1973, Large-scale gravitational flow of supracrustal rocks, Southern Canadian Rockies, *in* DeJong, K. A., and Scholten, R., eds., Gravity and tectonics: New York, Wiley.

——— and **Mountjoy, E. W.,** 1970, Geologic structure of the Canadian Rocky Mountains between Bow and Athabasca Rivers—Progress report: Geol. Assoc. Can. Spec. Pap. 6, p. 7–25.

Raleigh, C. B., 1963, Effect of the toe in the mechanics of overthrust faulting: Geol. Soc. America Bull., v. 74, no. 7.

Rubey, W. W., and **Hubbert, M. K.,** 1959, Role of fluid pressure in mechanics of overthrust faulting: Geol. Soc. America Bull., v. 70, p. 167–206.

Seager, W. R., 1970, Low-angle gravity glide structures in the northern Virgin Mountains, Nevada and Arizona: Geol. Soc. America Bull., v. 81, no. 5, p. 1517–1538.

Shouldice, J. R., 1963, Gravity slide faulting on Bowes dome, Bearpaw Mountain area, Montana: Am. Assoc. Petroleum Geologists Bull., v. 47, p. 1943–1951.

Small, W. M., 1959, Thrust faults and ruptured folds in Roumanian oil fields: Am. Assoc. Petroleum Geologists Bull., v. 43, p. 455–472.

Smith, J. G., 1965, Orogenesis in western Papua and New Guinea: Tectonophysics, v. 2, p. 1–27.

Smoluchowski, M., 1909, Folding of the earth's surface in formation of mountain chains: Acad. Sci. Craiovie Bull., v. 6, p. 3–20.

Voight, Barry, 1974, Architecture and mechanics of the Heart Mountain and South Fork rockslides: Rock mechanics, Penn. State Univ., p. 26–36.

Wilson, H. H., 1969, Late Cretaceous eugeosynclinal sedimentation, gravity tectonics, and ophiolite emplacement in Oman Mountains, southeast Arabia: Am. Assoc. Petroleum Geologists Bull., v. 53, no. 3.

Woodward, L. A., 1972, Shears of 2nd order caused by flexure folding: Amer. Assoc. Petroleum Geologists Bull., v. 56, no. 3, p. 559–561.

9

High-Angle Faults—Normal Faults and Upthrusts

A normal fault is a fault in which the hanging wall appears to have moved downward relative to the footwall. This type of movement implies at least local extension of the rock masses involved and an absence of the types of compressive stresses often associated with thrusting and strike-slip faulting.

Character of Normal Fault Zones

Because normal faults generally form under conditions of extension, the fault zones associated with them are quite different from thrust or strike-slip zones. Normal faults usually have steep dips of 45 to 90°, the shear zone tends to be narrow, rock in the zone is not usually intensively brecciated, and mylonites are rare, although a thin zone of gouge or breccia may be present. The zone may be made up of a number of closely spaced faults parallel to the main plane of separation. Drag and slickensides are often present, but intensive small-scale folding, cataclastic deformation, and other high-pressure effects are usually lacking.

Structural Patterns along Normal Faults

The steep dips associated with normal faults cause the outcrop traces to be relatively straight lines on geologic maps. Usually, normal faults occur in sets that are subparallel and cause a step-like pattern in cross section. The steps die out laterally, and the subsidiary faults may merge to form a single fault on which the displacement gradually decreases downward.

Grabens and horsts are characteristic features produced by normal faulting due to lateral extension at right angles to the faults. The strata associated with grabens and horsts are usually undeformed, at least by the stresses that cause the normal faulting. The margins of many grabens and horsts are not straight but are made of straight-line fault segments, some of which trend as much as 30° off the strike of the major faults, as in the Rhine graben, in East Africa, and along the Red Sea, Fig. 9-1.

Antithetic Faults

A set of fractures is usually oriented parallel to any major normal fault. In addition, a set of fractures or faults with parallel strike often dips into the major fault. These faults are normal, have the orientation expected for a set of conjugate shears, Fig. 9-2, and are called *antithetic faults.* A third set of vertical fractures perpendicular to the fault may be present.

Drag

The distortion of layering where cut by normal faults may be caused by drag of the layering along the fault as a result of frictional forces set up in the fault zone, Fig. 9-3, but frequently the "drag" on the downthrown side of the fault appears to be in the wrong direction. It is called *reverse drag* or *downbending,* but it is not a true drag process at all. Such bending is encountered in subsurface studies in the Gulf Coastal Plain, and it is seen exposed in the Colorado Plateau. Downbending along some of the faults in the Grand Canyon has proceeded far enough to produce bedding

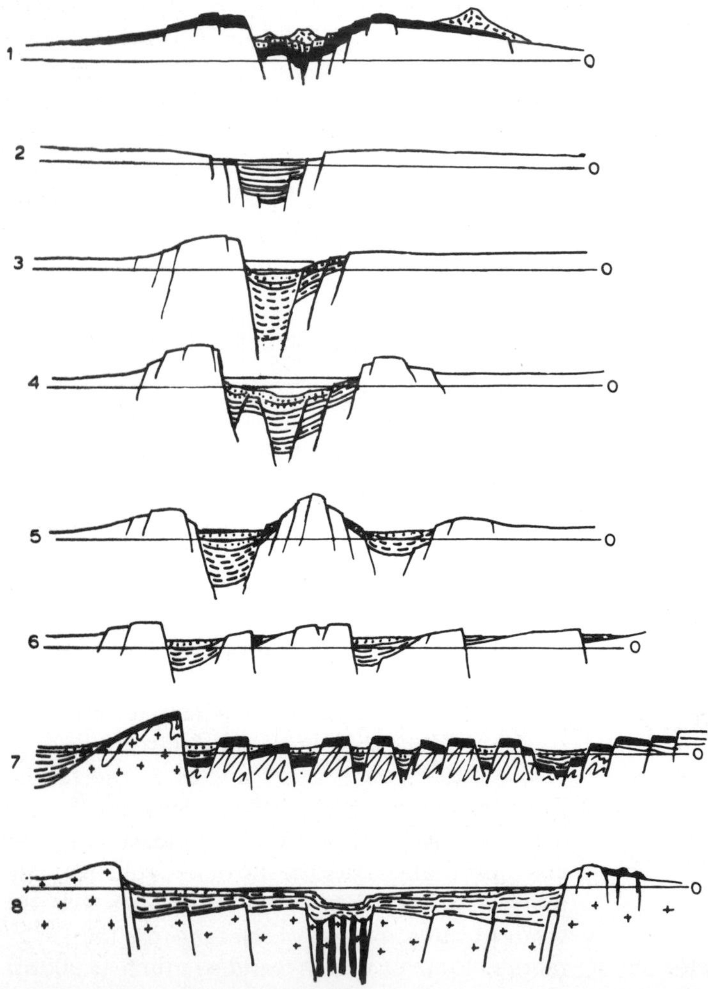

FIGURE 9-1 The characteristic structural types of rift zones (the vertical scale of the cross sections is exaggerated). 1 = arch-volcanic epiplatform rift zone (Kenyan rift zone according to B. H. Baker); 2–5 = crevice-like epiplatform rift zones: 2 = without marginal uplifts, 3 = with one marginal uplift, 4 = with two marginal uplifts, 5 = with internal uplift; 6 = zone of one-side tilted blocks; 7 = rift-like epiorogenic belt; 8 = intercontinental rift zone (rift zone of Red Sea according to C. L. Drake and R. W. Girdler). (*From Milanovsky, 1972.*)

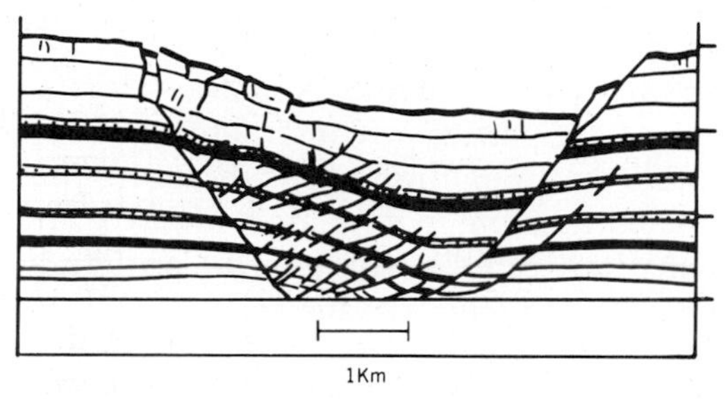

FIGURE 9-2 Graben formed in clay experimentally by lateral extension of the block by about 1 km. Note the step faults on both sides of the graben and the antithetic faults on the left. (*After Cloos, 1968.*)

dips of 26° toward the fault on the downthrown side (Hamblin, 1965). In places the downbending can be seen to pass into antithetic fault systems, and where relief is great enough the downbending is found to be related to normal faults which decrease in dip with depth. A similar relation is found in the Gulf Coast subsurface. This flattening of the dip at depth takes place as the two blocks are pulled apart. This in turn induces the material on the downthrown block to move laterally toward the fault plane to fill the gap. This movement may take the form of downbending, slumping, or the development of subsidiary fault systems (Fig. 9-3).

MECHANICS OF NORMAL FAULTING

Stress conditions favorable for creation of normal faults are obtained when the maximum principal stress is vertical and the least and intermediate principal stresses are horizontal, Fig. 7-2. It is important to remember that the localized principal stress orientations do not bear a unique relationship to the actual active forces causing deformation, but they may be identical for various types of force application. Favorable conditions exist when the active forces are vertically directed or when tensile forces act, pulling rock masses apart. Thus normal faults may be found over domes or where rocks are extended by upward folding or arching, and they may form where the crust is extended as a result of two blocks moving apart horizontally.

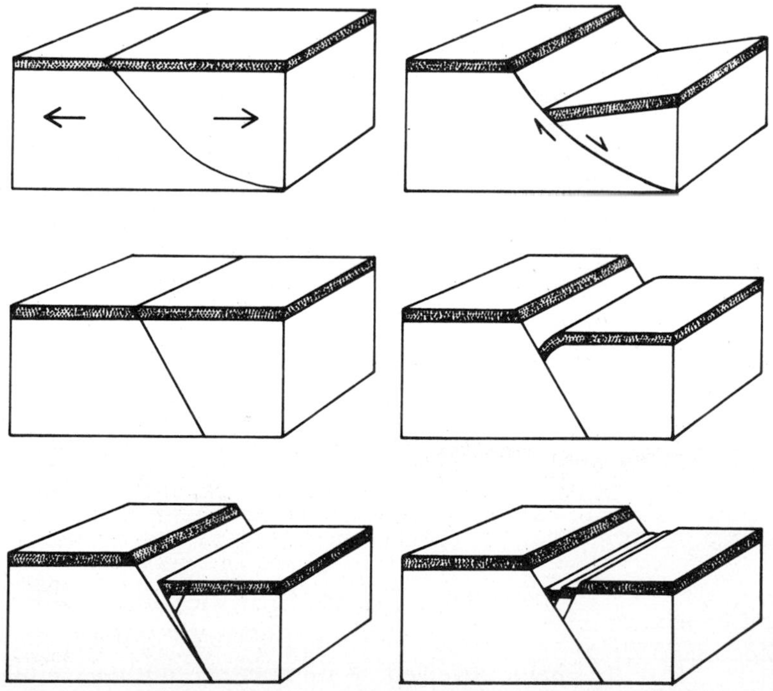

FIGURE 9-3 Features associated with normal fault zones. Top, rotation of one block on a curved fault. Middle, reverse drag. Bottom, a small graben formed along the fault zone.

Normal faults are sometimes described as "tension" faults, but even though the extension may actually be caused by tensile stress, the rock failure is by shearing. This explains both the conjugate relationship between major normal and the subsidiary antithetic faults and the dip of the faults.

Failure as exhibited on the Mohr diagram, Fig. 7-3, occurs when the Mohr circle (e.g., the combination of values of the principle stresses needed to cause failure) becomes tangent to the lines of failure for the particular material. This may occur as a result of the vertically directed σ_1 becoming increasingly larger (for uplift) or by σ_3, the least stress, becoming increasingly smaller as tensile stresses build up.

TECTONIC SETTINGS FOR NORMAL FAULTS

Many normal faults occur in the absence of orogenic deformation—under nondiastrophic conditions. They are commonly found over salt domes, in subsiding basins, and where gravity gliding takes place.

FIGURE 9-4 Schematic cross section of a graben showing amount of horizontal extension that accompanies the faulting, top. The bottom three sketches show how grabens may form as a result of uplift as well as lateral extension.

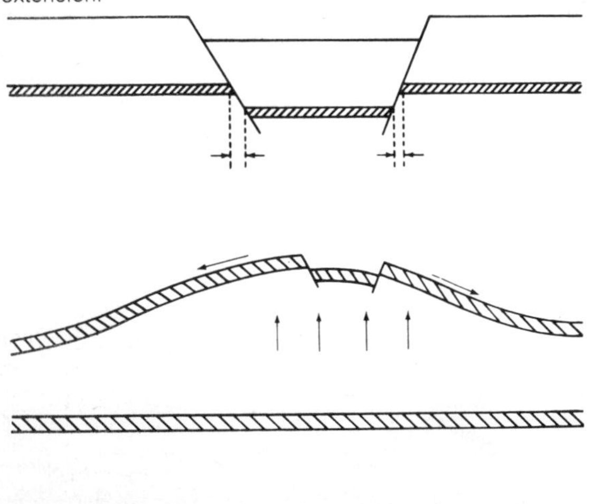

Over Salt Domes

Fault patterns over salt domes in the American Gulf Coast are particularly instructive because they occur in an environment where the processes causing them to form are still active and because they can be studied on structures known in detail from subsurface data.

The salt domes have moved up through a thick pile of unconsolidated and semiconsolidated sedimentary rocks. Evidence of horizontal compression is lacking; and folds or thrust faults related to horizontal compression are absent. The faults over the domes must be related to the vertically directed movement of the oval-topped salt plugs. This movement produces a dome in the overlying layered rock. As the doming continues, the layer is stretched and eventually fails by fracturing and faulting. When a homogeneous brittle material such as glass is deformed over a hemisphere, two systems of fractures usually occur—a radial set and a concentric set. Salt deformation, however, differs in that the maaterial is layered, often inhomogeneous, and the surface of the salt plug is not hemispherical; yet, both of these sets of fractures and faults are sometimes encountered, especially the radial sets (Fig. 9-5). The most persistent fault pattern over salt domes is that of a compound graben. The main graben splits at one or both ends, and step faults occur along many of the major faults. A less prominent set of normal faults perpendicular in trend to the main graben frequently intersects the main faults.

Thus it is evident in the case of salt domes that normal faults occur in such a way as to allow extension in the faulted layers over the dome. The radial pattern, the main graben, the step faults, and the normal faults perpendicular to the main faults all allow this extension.

Growth or Contemporaneous Faults

Subsurface studies in the Gulf Coastal region have revealed a number of normal faults which have continued to undergo movement over a period of time during deposition of the faulted units. These faults dip toward the center of the basins around which they occur, Fig. 9-6. Typically, the displaced units are thicker on the downthrown side, proving that the faults were active during deposition. Reverse drag is found on the downthrown side of some of these faults. Reversal in dip of the units due to sag or slumpage is possible only if the two blocks separated by the fault are tending to pull apart.

Some of the contemporaneous faults are related to differential compaction of the unconsolidated sediment. They occur on the seaward side of large, deeply buried masses of shale that are characterized by low bulk density and high fluid pressure. The faults form in sand beds which overlie these shale masses. Typically, the faults curve and converge at depth, Fig. 9-7, along the edge of the shale masses. Isolated curved normal faults form in this manner but usually a more complex pattern is formed (Bruce, 1973).

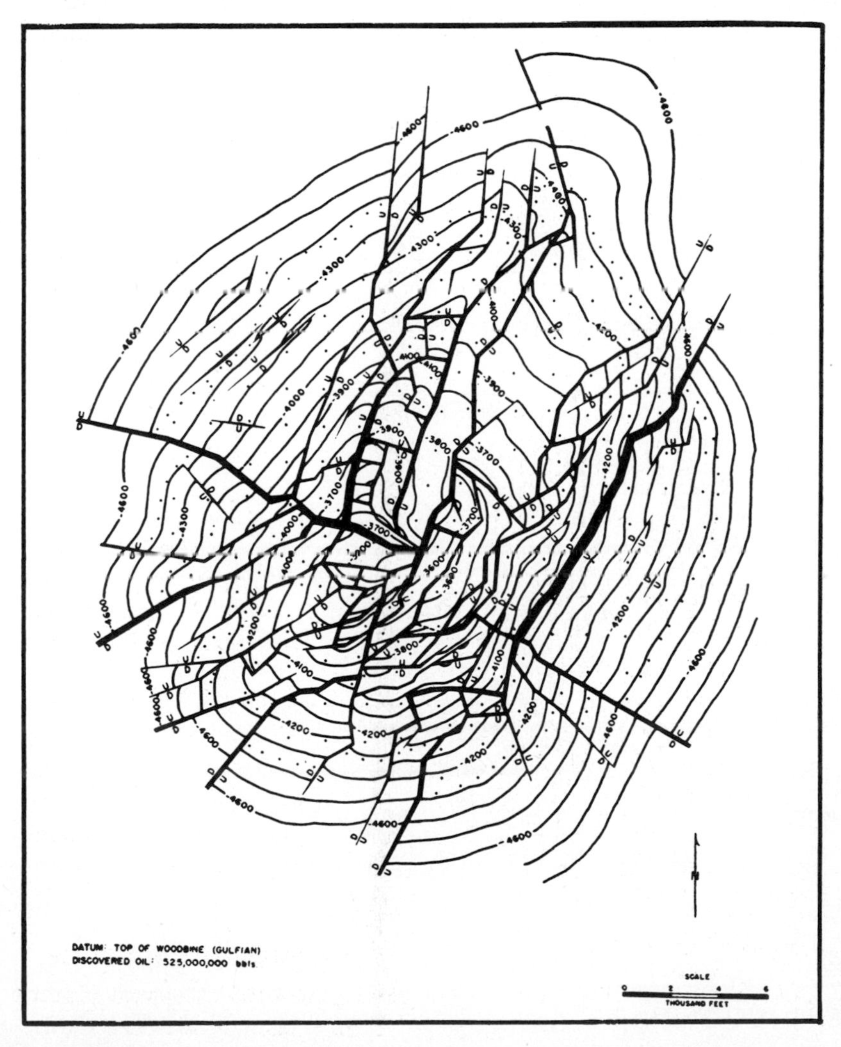

FIGURE 9-5 Fault pattern over the Hawkins field salt dome, Wood County, Texas. (*From Wendlandt, 1951.*)

Gravity Gliding

Normal faults may be expected where masses sliding downslope pull away at the upper edge of the slope, Fig. 8-11. Such high-angle faults may curve down and pass into the sliding surface (usually a bedding plane), and grabens are a characteristic feature of these zones.

TECTONIC SETTINGS FOR NORMAL FAULTS

Faults on Domes

Normal fault patterns on structural domes resemble those formed over salt domes. Good examples of normal faults around a dome are found along the margins of the Adirondack

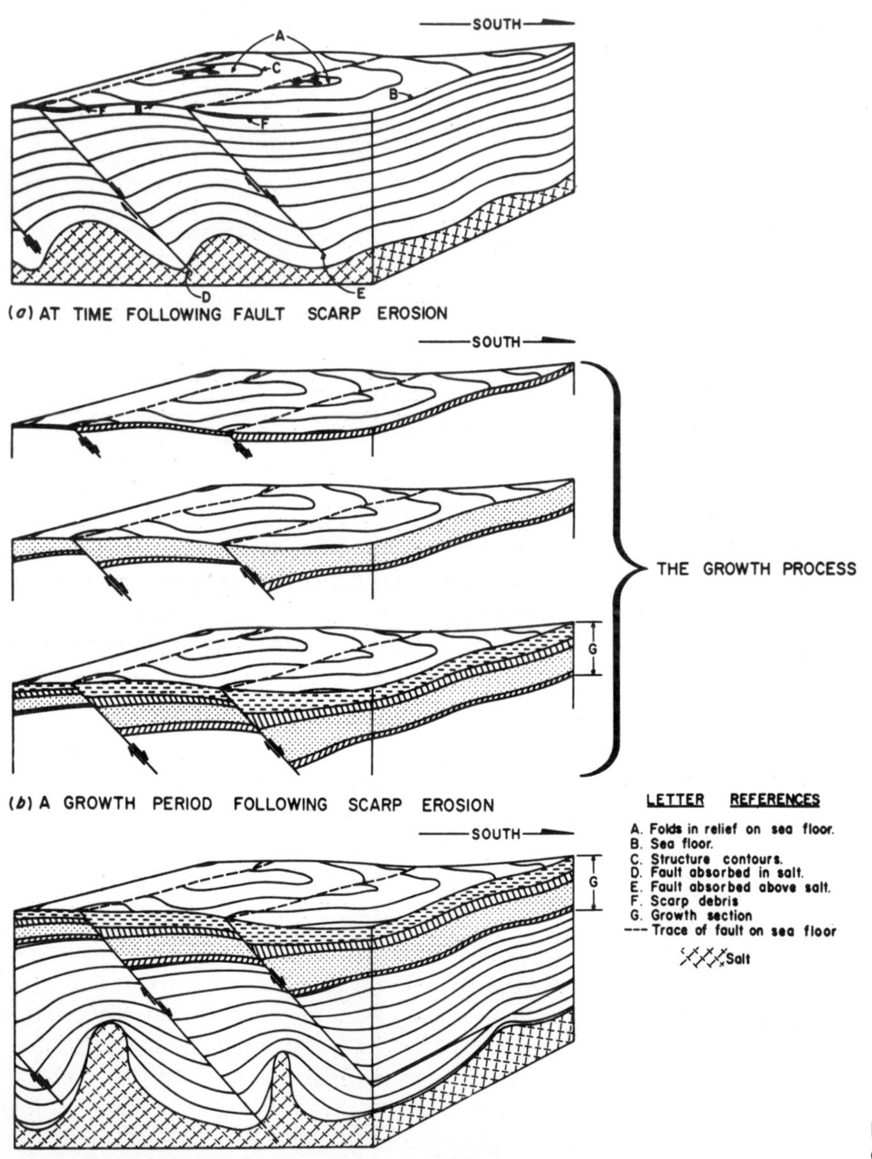

FIGURE 9-6 Development of growth-fault structures. (*From Ocamb, 1961.*)

Dome in New York. This dome has persisted since Paleozoic time and stands a thousand meters or more above sea level today. The normal faults form a complex pattern with radial components, grabens, and horsts.

Faults on Folds

Normal faults on folds occur in longitudinal, transverse, and oblique positions relative to the fold axis.

The Elk basin oil field of Wyoming is a good example of transverse faulting, and one longitudinal fault is also present, Fig. 9-8. The fold is a doubly plunging anticline. Extension of the rocks is mainly in two directions, one over the fold crest and the other along the axis. The orientation of the faults makes this explanation of their origin seem plausible, for they are oriented so as to allow extension in these two directions. Drilling on this structure has shown that the faults become steeper with depth.

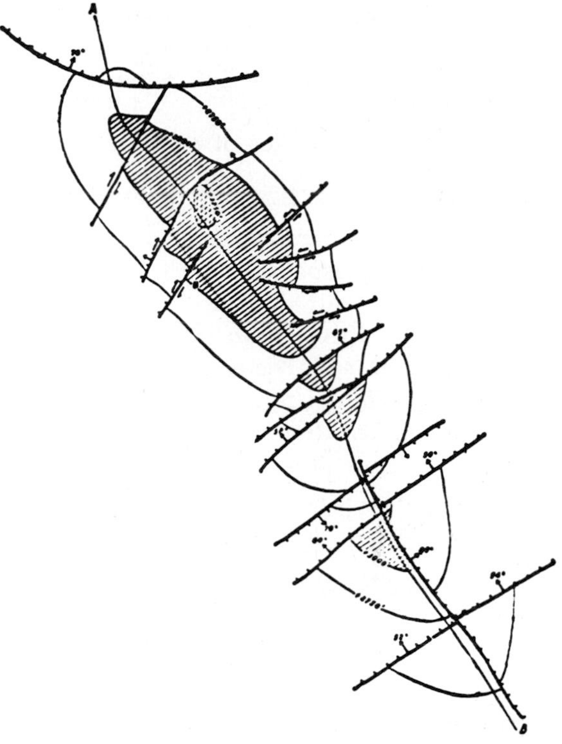

FIGURE 9-8 Faults in the Elk basin anticline (*By Korn, after Bartram, 1929.*)

Normal Faults Related to Thrust Faults

Normal faults may form along the leading edge of thrust sheet where the plate becomes arched as a result of drag. Also the leading edge of a thrust sheet may move unevenly, with resulting development of a curved thrust trace. The leading edge thus tends to become extended in a direction parallel to the trace, and normal faults may then form perpendicular to the thrust trace.

FIGURE 9-7 Seismic illustration of combination differential compaction and bedding-plane fault system. (AQUAPULSE—Courtesy Western Geophysical Company.) (*From Bruce, 1973.*)

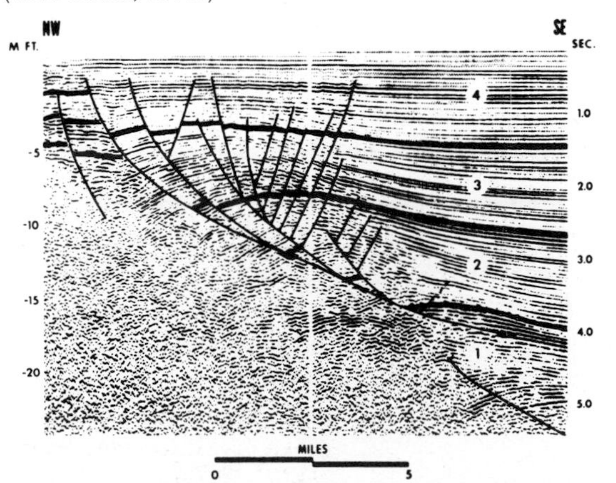

Divergent Plate Junctions (see Chap. 20)

Many large-scale crustal features are now thought to form along divergent junctions between lithospheric plates. The presence of grabens and normal faults is an important criterion for recognition of these junctions. Among modern examples are the grabens in Iceland, which are oriented parallel to the mid-Atlantic ridge; the East African rift valleys and the Red Sea, which are thought to mark a newly developed split between Africa and Asia; and the Gulf of California, where spreading appears to be in an active stage. Many graben-like valleys have been documented by sonic profiles of the oceanic ridge system, and step-like breaks on the flanks of the ridges may well be step faults.

Older sedimentary basins produced by normal faulting, especially the Triassic basins in the eastern United States, are commonly related to the rifting apart of North America from Africa. Indeed, the age of these grabens is a key bit of evidence used to date the break up of this supercontinent.

Block Faulting in Orogenic Belts

Normal faults form a prominent part of the structure of the Cordilleran belt of the western United States. Block faulting is a dominant structural style in the Basin and Range area of Utah, Nevada, California, and surrounding areas. These blocks are clearly part of the diastrophic effects in this area, but their relation to the overall pattern of orogeny there is not so clear (see Chap. 24).

Block Faulting in Continental Cratonic Areas

Many grabens, horsts, and normal faults have been mapped in the so-called stable cratons of the continents. They presumably represent local extensions of these masses, but generally their relationship to the larger tectonic framework is only vaguely known. Examples are found in the Rough Creek zone of southern Illinois, the Kentucky fault zone, the Ottawa grabens, and the normal faults in northeastern Oklahoma.

HIGH-ANGLE REVERSE FAULTS —UPTHRUSTS

A reverse fault is one in which the hanging wall moves up relative to the footwall. In the case of the faults considered here, the movement is essentially vertical, although the dip of the fault varies from being very steep at depth to being low near the surface. Such faults have been called *upthrust*.

Character of the Fault Zone

The surface expression of most upthrusts is very similar to that of low-angle thrusts; indeed, the faults are low-angle features at the surface, and gouge, breccia, shearing, and sometimes overriden surficial deposits are present in the fault zone. Where high-angle reverse faults are exposed, they, too, show effects of shearing, although the zones are not as wide nor is the intensity of shearing as great as that found in the strike-slip zones.

Structural Patterns along Upthrusts

Upthrusts are most often associated with block-like uplifts, and they may form one or more sides of a block. The blocks may be tilted blocks hinged along one edge and bound by upthrust or vertical faults on the other sides, or in some cases upthrusts may appear on opposite sides or corners of the block. An upthrust at one place along the block boundary may pass laterally into a vertical fault and then into a steep monoclinal flexure which may then die out. Structural features of this type are seen in the Rocky Mountain foreland and are especially well developed in Wyoming and Montana.

Dip of Upthrusts

The leading edges of some upthrusts are clearly low-angle thrusts. The northeastern corner of the Beartooth Range,* for example, is bounded by thrusts which are broken and offset along tear faults, Fig. 8-3. Some slight changes in dip are evident even here, and the fault steepens in dip toward the mountain front.

* Foose, Wise, and Garbarini (1961).

Evidence that these thrusts turn into much steeper faults at depth is postulated from several other types of arguments:

1. Observed gravity values across the faults are best interpreted by models in which the fault becomes steeper at depth, Fig. 9-9.

2. Some blocks have upthrust on opposite sides which would seem to be geometrically and mechanically impossible with low-angle thrusting.

3. Deep wells drilled adjacent to some of the upthrust mountain fronts have penetrated near-vertical sedimentary layers, Fig. 9-10, apparently rotated into that position by the vertical movements nearby.

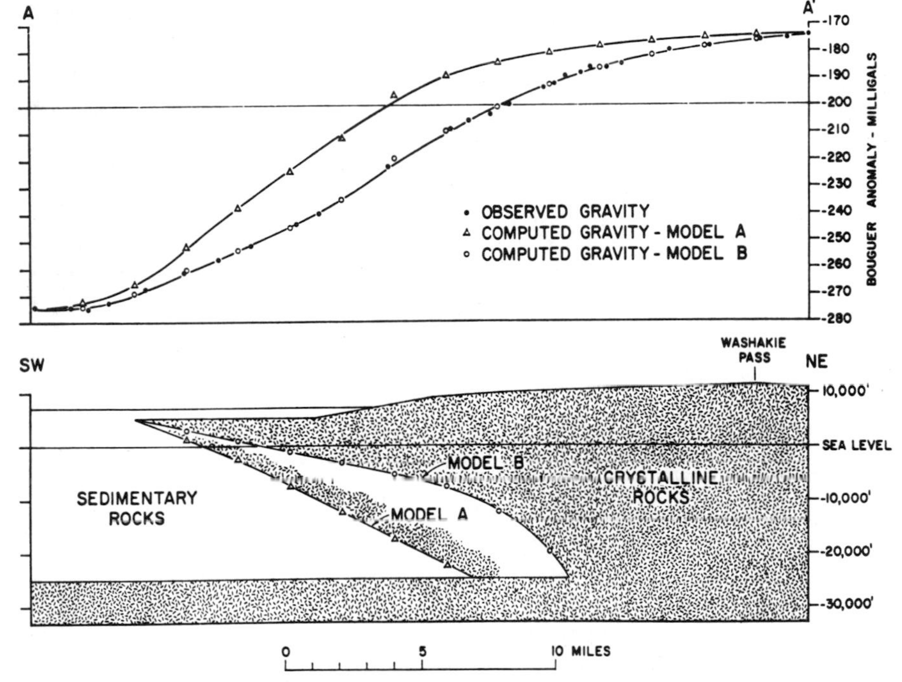

FIGURE 9-9 Gravity profiles on the southwest flank, Wind River Mountains, and interpretive structure models. Curves of gravity values calculated on the basis of two models (below) are shown above. (*From Berg and Romberg, 1966.*)

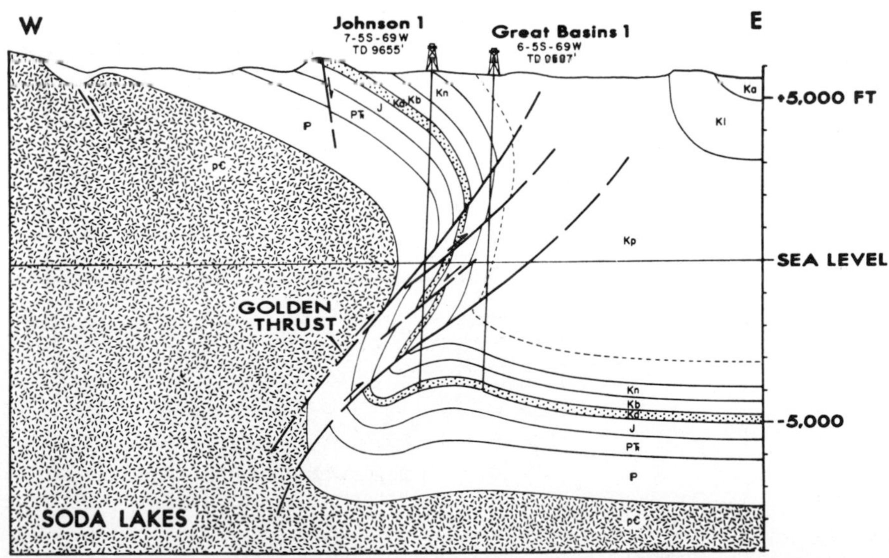

FIGURE 9-10 The Golden thrust in Jefferson County, Colorado. The faults curve at depth allowing a near vertical uplift of basement. (*From Berg, 1962.*)

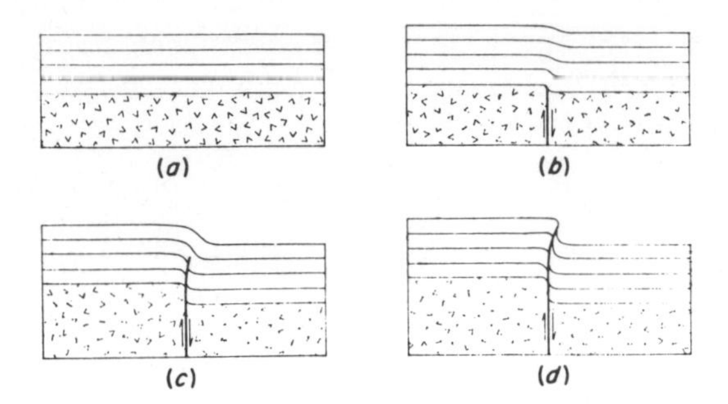

FIGURE 9-11 Postulated sequence in deformation of sedimentary rock beds overlying basement fault blocks. (*a*) Before faulting in basement. (*b*) Initial stage of faulting in basement. Sedimentary beds adjust by bending without faulting. (*c*) Intermediate stage in basement faulting. Lower sedimentary beds have exceeded critical degree of bending and are faulted; upper sedimentary beds are bent but not faulted. (*d*) Advanced stage of basement faulting. Faulting extends through sedimentary beds to surface. (*From Prucha, Graham, and Nickelsen, 1965.*)

Note also that features such as klippes and windows that occur along typical low-angle thrusts are rare or absent along upthrusts.

Evidence of the vertical relief between the uplifted blocks and adjacent basins is known in the Middle Rocky Mountains, where the tops of the block uplifts are Precambrian crystalline rocks and the depths of the basement in basins are known from seismic prospecting methods, Fig. 9-11. The amount of structural relief on this surface is as much as 12 km (e.g., between the Wind River Mountains and the Green River basin, Wyoming).

Synthetic and Antithetic Faults

The major upthrust may be accompanied by several subsidiary faults of similar character.

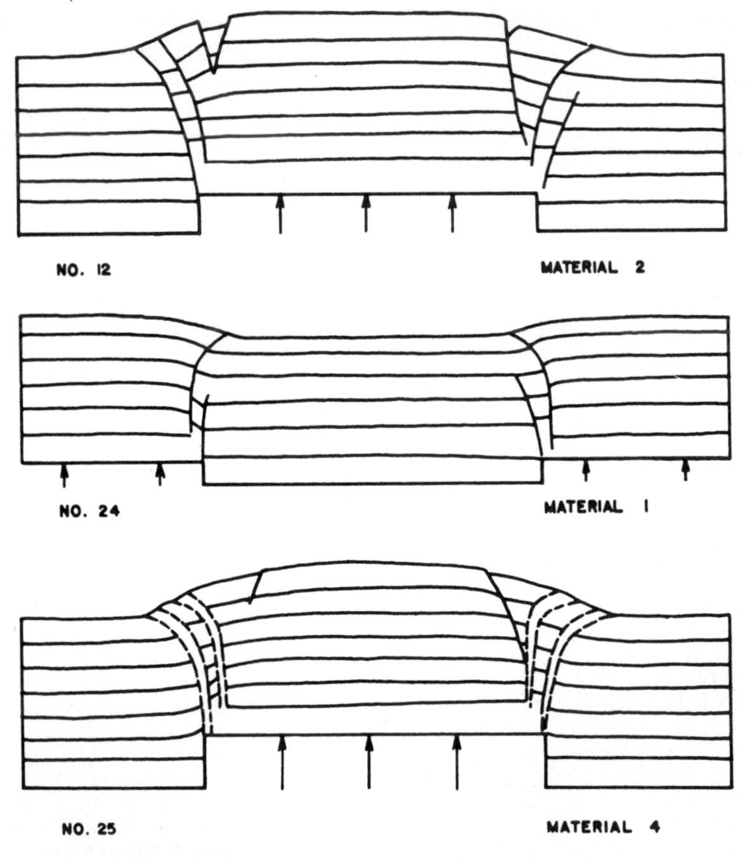

FIGURE 9-12 Model experiments in which a section of the underlying basement is moved vertically. (*From Sanford, 1959.*)

Model experiments, Fig. 9-12, suggest that these may also be arranged in *en échelon* patterns, with dips toward the uplifted side.

A system of antithetic fractures and faults is described by Lowell (1970) from experiments with upthrusts, Fig. 9-13. These form initially as one of a set of conjugate shears but become flattened as uplift proceeds.

Extensional Features

As the uplifted block rises, a high, relatively unconfined surface is formed along the margin. This surface tends to bend and expand, and the resulting extension leads to the development of grabens and step faults oriented parallel with the block margin, as seen along the Wind River Range border, Fig. 9-14. This extension may produce tensile fracturing of the margin of the block and open fractures initially formed by shear failure when uplift began. The lateral spreading of uplifted crystalline rocks may be a result of the opening of large numbers of small fractures induced by release of confining pressure and by stresses due to gravitative spreading.

MECHANICAL CONSIDERATIONS IN UPTHRUSTING

A sound mechanical basis for upthrusting is found in the theoretical work of Hafner and Sanford and in the experimental work of Sanford (1959) and Lowell (1970). Hafner's model

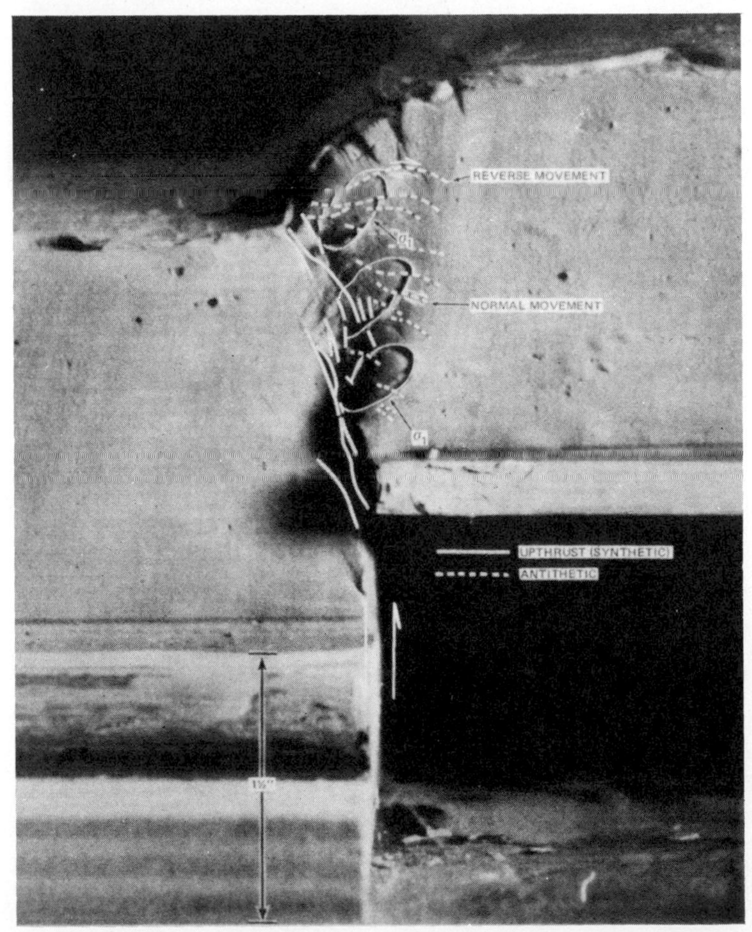

FIGURE 9-13 Clay-model experiment in which a basement block is uplifted, resulting in upthrusting and faults antithetic to upthrusting. Antithetic faults are commonly sigmoidal, range from horizontal to low dip, and show very small displacement of both normal and reverse sense. Note rotation of maximum principal compressive stress (σ_1) and hence change in orientation of antithetic faults along upthrust zone from lower part to upper part of clay. Ellipses are deformed circles. (*From Lowell, 1970.*)

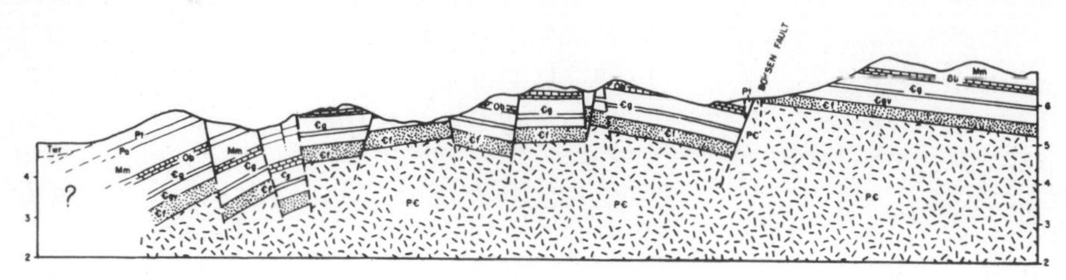

FIGURE 9-14 Structure section in Wind River Canyon area, Wyoming. (*From Wise, 1963.*)

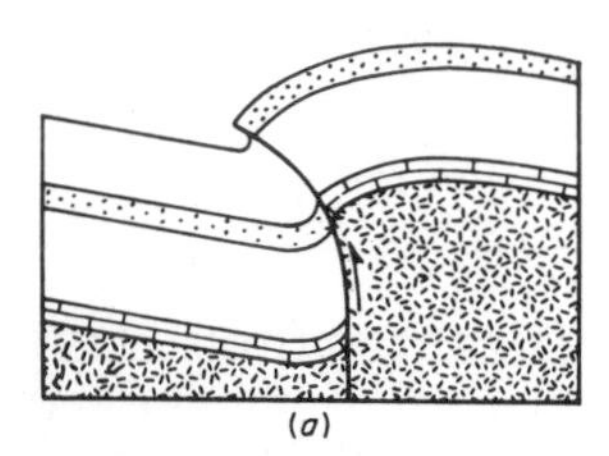

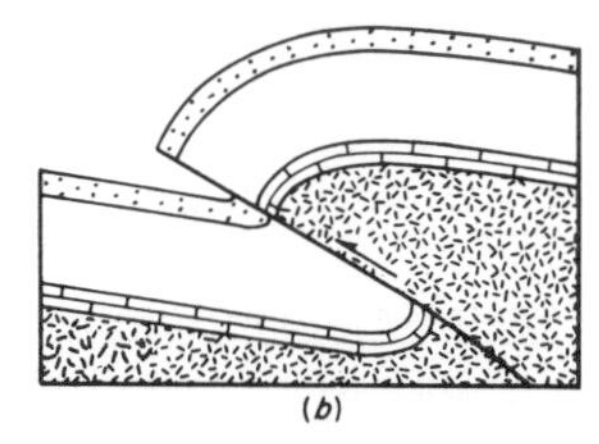

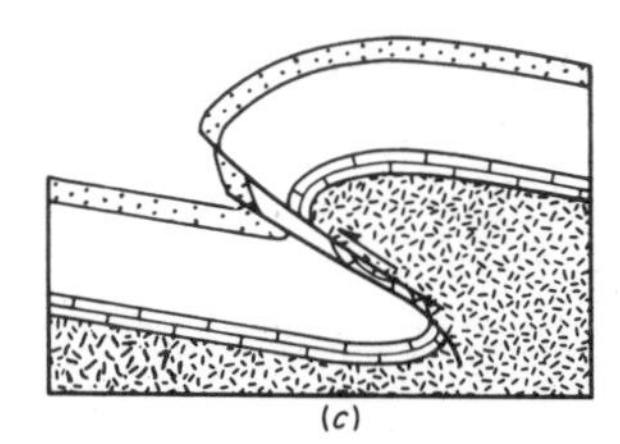

FIGURE 9-15 Diagrams illustrating hypotheses of mountain flank deformation commonly found in the Middle and Southern Rocky Mountains: (*a*) = block uplift, (*b*) = thrust uplift, (c) = fold-thrust uplift. (*Redrawn from Berg, 1962.*)

for vertical pressure of sinusoidal form applied to the bottom of a block led to development of potential failure surfaces that are similar in form to those postulated for upthrust, Figs. 7-7 and 7-8. And Sanford's model, Fig. 9-12, for vertical block-like uplifts produced not only the curved master fault but the arched block margins and associated extensional features. Howard (1966) used this method to develop a model of the Williams Range thrust of Colorado.

Hafner's analysis does not require the rigid basement used by Sanford and Lowell in their experiments, so it seems that upthrusting is not confined to areas where basement blocks bounded by faults are moved up; but upthrust may be a more general characteristic of vertical tectonics.

TECTONIC SETTING

The best documented upthrusts of North America are located in the Rocky Mountain region, Fig. 9-15. They are developed on the cratonic side of the mobile belt where the Paleozoic sedimentary veneer on the Precambrian basement is relatively thin. The upthrusts occur east of the more deeply subsiding geosynclinal trough in which Laramide orogenic deformation took the form of thrusting and folding directed toward the east. Thus the upthrusts are located in the zone which was subjected to the eastward tectonic transport of the orogenic belt as well as to vertically directed forces apparently acting from the mantle.

REFERENCES

Bartram, J. G., 1929, Elk Basin oil and gas field, Park County, Wyoming, and Carbon County, Montana, *in* Structure of typical American oil fields—A symposium, v. II: Am. Assoc. Petroleum Geologists, Tulsa, Oklahoma., p. 577–588.

Berg, R. R., 1962, Montain flank thrusting in Rocky Mountain foreland, Wyoming and Colorado: Am. Assoc. Petroleum Geologists Bull., v. 46, p. 2019–2032.

——— and **Romberg, F. E.**, 1966, Gravity profile across the Wind River Mtns., Wyoming: Geol. Soc. America Bull., v. 77, p. 647–656.

Bruce, Clemont H., 1973, Pressured shale and related sediment deformation: Mechanism for development of regional contemporaneous faults: Am. Assoc. Petroleum Geologists Bull., v. 57, p. 878.

Carl, J. D., 1970, Block faulting and development of drainage, northern Madison Mountains, Montana: Geol. Soc. America Bull., v. 81, no. 8, p. 2287–2298.

Carver, R. E., 1968, Differential compaction as a cause of regional contemporaneous faults: Am. Assoc. Petroleum Geologists Bull., v. 52, no. 3.

Casella, C. J., 1964, Geologic evolution of the Beartooth Mountains, Montana and Wyoming, pt. 4,

Relationship between Precambrian and Laramide structures in the Line Creek area: Geol. Soc. America Bull., v. 75, p. 969–986.

Cloos, Ernst, 1968, Experimental analysis of Gulf Coast fracture patterns: Am. Assoc. Petroleum Geologists Bull., v. 52, no. 3.

Donath, F. A., and **Juo, J. T.,** 1962, Seismic-refraction study of block-faulting, south central Oregon: Geol. Soc America Bull., v. 73, p. 429–434.

Drake, C. L., and **Girdler, R. W.,** 1964, A geophysical study of the Red Sea: Geophys. Jour. Royal Astronomical Soc., v. 8, no. 5, p. 473–495.

Foose, R. M., Wise, D. U., and **Garbarini, G. S.,** 1961, Structural geology of the Beartooth Mountains, Montana and Wyoming: Bull. Geol. Soc. America, v. 72.

Fuller, R. E., and **Waters, A. C.,** 1929, The nature and origin of the horst and graben structure of southern Oregon: Jour. Geology, v. 37, p. 204–238.

Gregory, J. W., 1920, The African Rift Valleys: Geog. Jour., v. 56.

Hager, D. S., and **Burnett, C. M.,** 1960, Mexia-Talco fault line in Hopkins and Delta Counties, Texas: Am. Assoc. Petroleum Geologists Bull., v. 44, p. 310–350.

Hamblin, W. K., 1965, Origin of "reverse drag" on the downthrown side of normal faults: Geol. Soc. America Bull., v. 76, p. 1145–1164.

Hardin, F. R., and **Hardin, G. C., Jr.,** 1961, Contemporaneous normal faults of the Gulf Coast and their relation to flexures: Am. Assoc. Petroleum Geologists Bull., v. 45, p. 238–248.

Heezen, B. C., and **Ewing, M.,** 1961, The mid-oceanic ridge and its extension through the Arctic basin, *in* Geology of the Arctic: Toronto, Univ. of Toronto, p. 622–642.

Howard, J. H., 1966, Structural development of the Williams Range thrust, Colorado: Geol. Soc. America Bull., v. 77, p. 1247–1264.

Hughes, D. J., 1960, Faulting associated with deep-seated salt domes in the northeast portion of the Mississippi salt basin: Gulf Coast Assoc. of Geol. Soc. Trans., v. 10, p. 155–173.

Irvine, T. N., ed., 1965, The world rift system; Geol. Survey of Canada Paper 66-14, Dept. of Mines and Technical Surveys, Ottawa, Canada.

Laughton, A. S., 1966, The Gulf of Aden, in relation to the Red Sea and the Afar depression of Ethiopia, *in* The world rift system: Geol. Survey of Canada Paper 66-14, p. 78–95.

Lensen, G. J., 1958, A method of graben and horst formation: Jour. Geology, v. 66, p. 579–587.

Lowell, James D., 1970, Antithetic faults in upthrusting: Am. Assoc. Petroleum Geologists Bull., v. 54/10, pt. I, p. 1946ff.

Milanovsky, E. E., 1972, Continental rift zones; their arrangement and development, *in,* Girdler, R. W., ed., East African rifts: Tectonophysics, v. 15.

Ocamb, R. D., 1961, Growth faults of south Louisiana: Gulf Coast Assoc. of Geol. Socs. Trans., v. XI, p. 139–175.

Prucha, J. J., and others, 1965, Basement-controlled deformation in Wyoming province of Rocky Mountains foreland: Am. Assoc. Petroleum Geologists Bull., v. 49, p. 966–992.

Sanford, A. R., 1959, Analytical and experimental study of simple geologic structures: Geol. Soc. America Bull., v. 70, p. 19–52.

Savage, J. C., and **Hastie, L. M.,** 1966, Surface deformation associated with dip-slip faulting: Jour. Geophys. Res., v. 71, no. 20, p. 4897–4904.

Shand, S. J., 1936, Rift valley impressions: Geol. Mag. (Great Britain), v. 123.

Shelton, J. W., 1968, Role of contemporaneous faulting during basinal subsidence: Am. Assoc. Petroleum Geologists Bull., v. 52, no. 3.

Thompson, G. H., 1959, Gravity measurements between Hazen and Auston, Nevada—A study of basin-range structure: Jour. Geophys. Res., v. 64, p. 217–229.

Tucker, D. R., 1967, Faults of south and central Texas, Transactions of the Gulf Coast Assoc. of Geol. Societies, v. XVII, p. 144–147.

——— 1968, Lower Cretaceous geology, northwestern Karnes County, Texas: Am. Assoc. Petroleum Geologists Bull., v. 52, no. 5, p. 820–851.

Wellman, H. W., 1955, New Zealand Quaternary tectonics: Geol. Rundschau, v. 43, p. 248–257.

Wendlandt, E. A., 1951, Hawkins field, Wood County, Texas: Austin, Univ. of Texas Pub. No. 5116, p. 153–158.

Willis, Bailey, 1936a, Rift-valley types: Pan Am. Geologist, v. 63, p. 304.

——— 1936b, East African plateaus and rift valleys: Carnegie Inst. Washington Pub. 470.

Wise, D. U., 1963, Keystone faulting and gravity sliding driven by basement uplift of Owl Creek Mountains, Wyo.: Am. Assoc. Petroleum Geologists Bull., v. 47, p. 586–598.

10

Strike-Slip Faults

Faults which have a large component of strike separation include some of the largest of all known faults. They have been called *rifts*, but this term is not good because it is also associated with graben structures such as the East African rift system. The term *transverse* applies where the fault cuts local structures, but these faults often parallel regional trends of deformed belts. More recently the term *wrench fault* has come into widespread use.* Anderson (1951) recognized wrench faults as one of the three general fault classes formed as a result of failure followed by faulting movement along one of a conjugate set of fractures, Fig. 7-2. It seems probable that some of these faults are not the result of pure shear. For this reason, and because the larger faults of this type almost certainly cut through all or a large part of the thickness of the lithosphere, Carey (1958) proposed the term *megashear* for them.

A special variety of strike-slip faults—the *transform fault*† (a strike-slip fault characteristic of mid-oceanic ridges and along which the ridges are offset)—is now an important conceptual part of sea-floor-spreading theory. Although the displacement is strike-slip, the sense of movement is the reverse of what is apparent from the offset of the ridges. Transforms may separate segments of a ridge which are growing at different rates, as described by J. T. Wilson (1965), or separate adjacent lines of spreading.

CHARACTERISTICS OF STRIKE-SLIP FAULT ZONES

Displacements along some of the larger strike-slip zones are measured in hundreds of kilometers. Failure appears to have taken place as a shear phenomena (both pure and simple shear have been postulated), and many geologists now view these faults as products of lateral compression (e.g., the fault is one of the conjugate shears). The zones usually have straight

* An excellent summary of wrench tectonics is available in Wilcox, Harding, and Seely (1973).

† See Chap. 17.

traces even where they cut across high and irregular topographic relief. Thus, although the true dip is not apparent, it is probable that the zones have steep dips.

Exposures within fault zones are rare, but cuts into a few major strike-slip faults reveal their internal details. Rock within the San Andreas zone is brecciated and mylonitized. Crushing, shearing, and mylonitization of rock in a belt up to 1.5 km wide is present in the Great Glen fault of Scotland and along the Alpine fault of New Zealand. The country rock has been converted into a cataclastic schist and numerous near vertical faults are present. Pegmatites are developed in some parts of the crushed rock.

Evidence of Displacements

Establishment of strike-slip displacement on faults is most conclusive when linear features are cut and displaced by the fault. Modern movements have taken place on the San Andreas and on the Alpine fault of New Zealand, and these movements show as lateral displacements of such physiographic features as recent alluvial fans, stream terrace deposits, glacial moraines, and coastal terraces. Streams reflect the offsets, and in some places the faults offset highways, railroads, fences, and other cultural features. Older movements are estimated from offsets of distinctive plutons (as along the Great Glen fault), structural features, strand lines, and sedimentary facies.

Evidence of Motion on the San Andreas Fault

A considerable literature has been assembled concerning the amount of displacement on the San Andreas fault, Figs. 10-1 and 10-2. The fault lies in a region of Cenozoic crustal deformation where stratigraphic and structural relationships are complex. Sedimentary basins adjacent to and between faults have formed while the fault system has been active, and accumulations of sediment, more than 15 km in places, have been deposited as the sea invaded and retreated from the region. The usual methods of estimating displacements by determination of separation of marker beds of known attitude is difficult to apply here because the marker beds have been deformed after the initial faulting. Crowell (1962) concludes that the best evidence supports the idea of a right-lateral movement of about 200 km on the San Andreas since early Miocene. A separation of 350 to 500 km since the Cretaceous is possible, although the evidence is "of a different order of acceptability." Accepting the first figures, the *average* rate of movement is 1 m per century.

Although the idea that most of the faults in this region are major strike-slip faults is widely accepted, there is convincing evidence of local vertical movements of great magnitude, and some geologists* prefer to interpret the faults in terms of predominantly dip-slip movement. The San Andreas fault is now often interpreted as a transform fault which connects the Mendocino fault with an oceanic ridge crest in the Gulf of California. This interpretation is set forth in Chap. 24 on the tectonics of the Cordilleran region.

Evidence of Displacement on the Great Glen Fault

Most geologists considered the Great Glen fault to be a normal fault until 1946, when W. Q. Kennedy (1946) summarized the evidence for lateral slip as follows:

1. The dislocation possesses physical characters unlike those of most normal faults but similar to the great strike-slip shears of the California Coast Range.
2. It belongs to the same system as the Strathconon, Ericht-Laidon and Loch Tay faults, all of which have proved lateral displacements of up to 5 miles [8 km].

* One of the best discussions of this is Paschall and Off (1961).

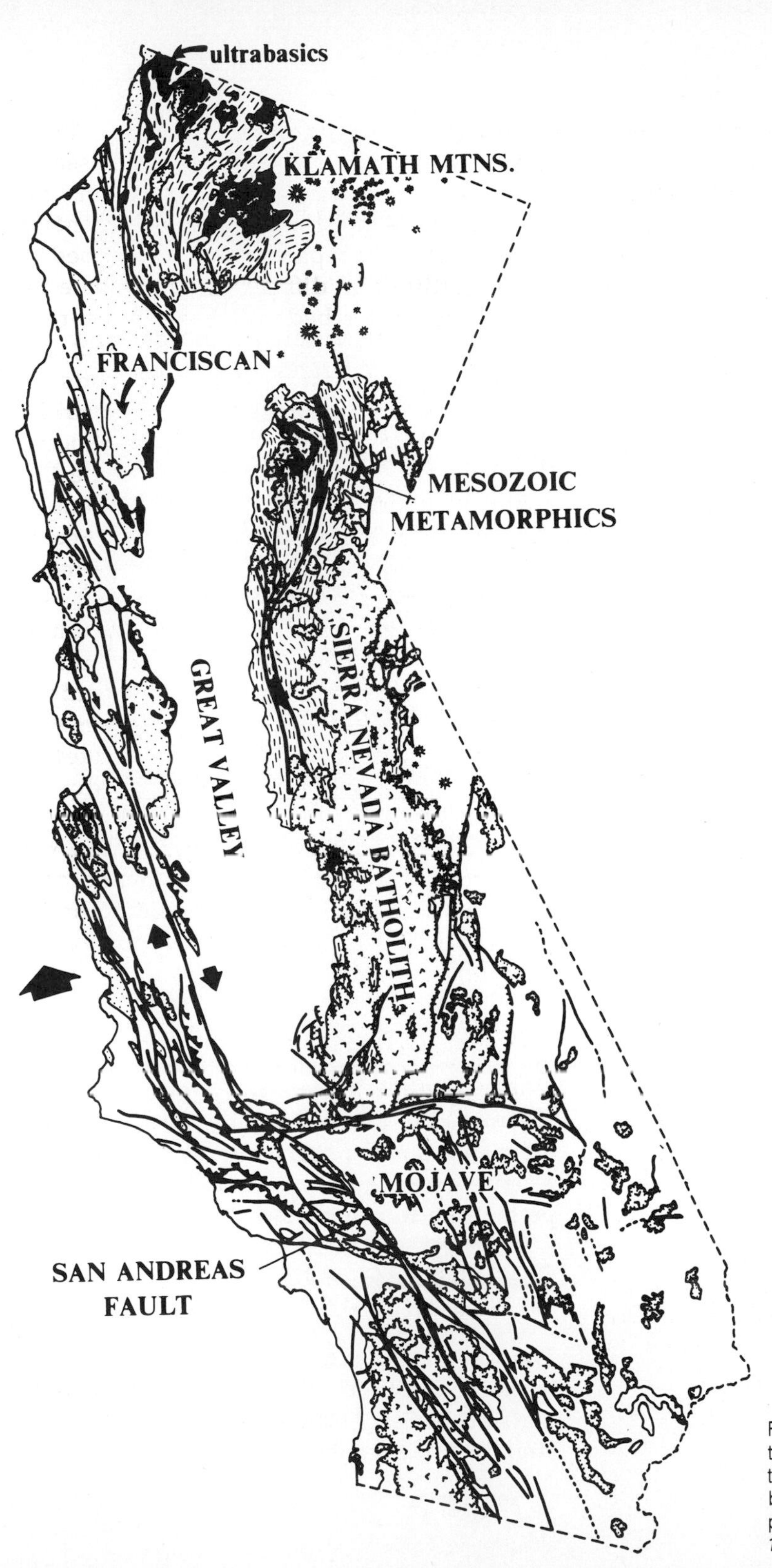

FIGURE 10-1 Sketch map of major tectonic elements in California showing the major faults, the Sierra Nevada batholith, and the metamorphic complexes. (*After the U.S. Tectonic Map, 1961.*)

3. It displaces the great belt of regional injection which affects the Moine Schists of the northern and Grampian Highlands, the nature and amount of the displacement being consistent with lateral shift but not with vertical downthrow.
4. It similarly displaces the metamorphic zones of the Highlands in an equally significant manner.
5. It truncates the Strontian Granite, the southern portion of which, according to the detailed structural evidence, is missing. The missing portion, moreover, can be identified in the Foyers mass which outcrops on the other side of the fault-line some 65 miles [about 104 km] to the north-east and is similarly truncated by the fault. These two major Caledonian intrusions consist of identical rock types and are structurally homologous.
6. Finally, the occurrence of Lewisian and Torridonian rocks in Islay and Colonsay and the presence of the Moine Thrust-plane in the former island are more readily explained on the assumption of a lateral rather than a vertical displacement along the fault.

FIGURE 10-2 Examples of *"en échelon"* patterns of most recently active breaks along the San Andreas fault. A–C are left-stepping; D–G are right-stepping. From strip maps by members of the U.S. Geological Survey showing the most recently active breaks along the San Andreas fault. (*From R. E. Wallace, in Kovach and Nur, 1973.*)

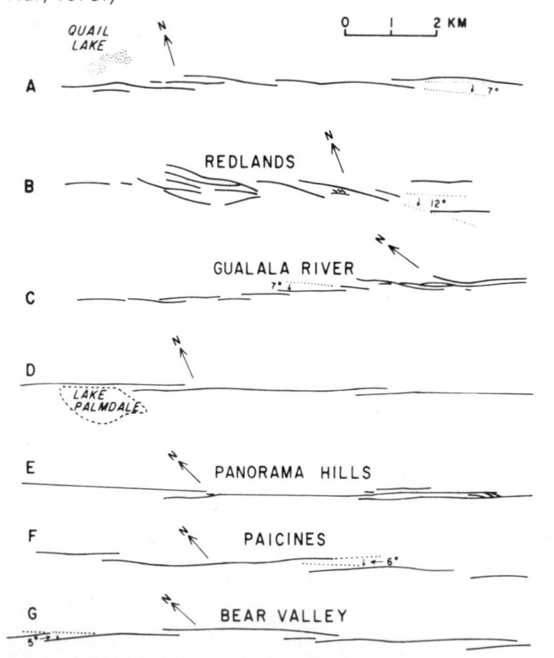

The Alpine Fault of New Zealand

The Alpine fault is a major fault which may be traced from Milford Sound on the southwestern coast of New Zealand to the northern end of the South Island, where it passes into Cook Strait. The fault has a straight trace, and the fault zone separates rocks of very different age and lithology on either side. The primary evidence for the amount of displacement is obtained by matching the rocks in the nose of a large pinched synclinal structure in the south with similar rocks in Nelson in the north, Fig. 10-3. The Upper Paleozoic units in the north and south are very similar in character and thickness across the fault but they are displaced nearly 500 km. Strike-slip displacement along the Alpine fault is suggested by the unusual fold pattern mapped by Arnold Lillie in the area near Mount Cook, which shows a system of folds which exhibit steep plunges which may have been formed as a result of drag on faults parallel to the Alpine fault, Fig. 10-4. The New Zealand Alps, located on the eastern side of the fault, stand thousands of meters above the topography west of the fault, suggesting a large component of vertical movement on the fault which has been interpreted by some geologists as a high-angle reverse. Continuation of the Alpine fault on the North Island is uncertain, because although several faults occur along the strike, none of them shows major strike-slip displacement. In recent years the Alpine fault has been interpreted as a transform fault.

STRUCTURAL PATTERN ALONG STRIKE-SLIP FAULTS*

The straight character of many strike-slip faults is probably their most distinctive feature. Parts of the San Andreas are made up of many straight-line segments arranged in parallel but

* See Bishop (1968) for a discussion of New Zealand faults. See Tchalenko (1970) for a comparison of features along faults of different sizes.

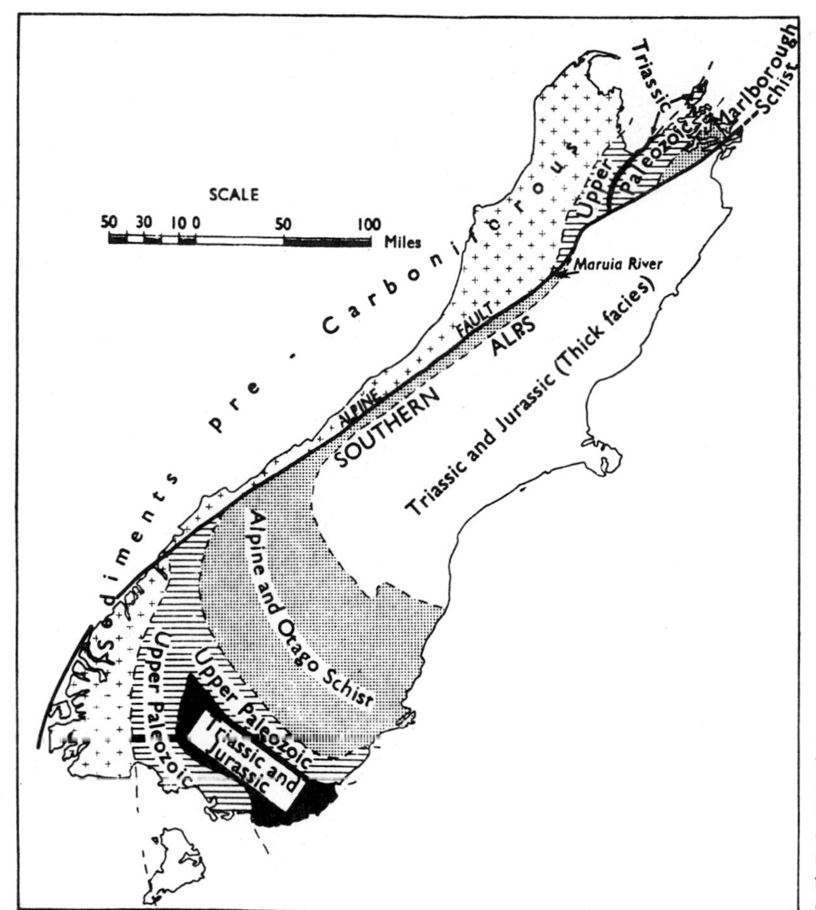

FIGURE 10-3 Sketch map showing the Alpine fault and pre-Cretaceous rocks of South Island. The 300-mile (483-km) lateral shift is based on the correspondence of regional sequences on opposite sides of the Alpine fault in the north and south of the island. (*From Wellman, 1952.*)

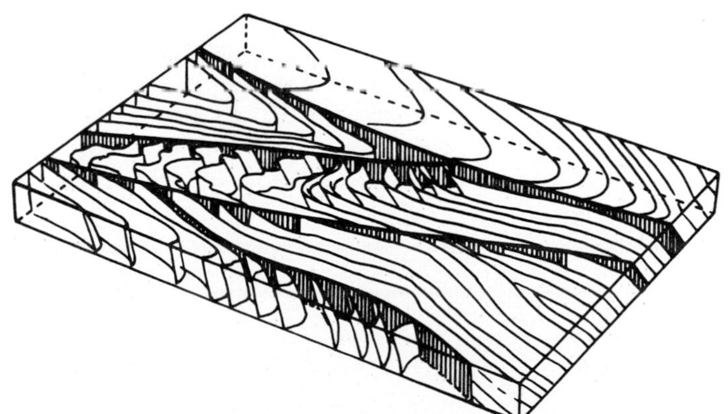

FIGURE 10-4 Steeply plunging folds located along vertical strike-slip faults in the New Zealand Alps. (*Redrawn after Lillie, 1964.*)

slightly offset lines. Other zones, like the Calaveras zone, Fig. 10-5, contain many fault segments which form an irregular braided pattern. Faults split and rejoin in a pattern somewhat like that formed by slices along a thrust. Many are discontinuous or break up into several branches, as does the Dasht E Bayaz, Fig. 10-6.

Depressions and Upbulges

Tectonic depressions and upbulges characterize some strike-slip faults (e.g., the Hope fault, Fig. 10-7). Movements have taken place along the Hope fault during the Pleistocene, and the depressions are expressed by modern topographic features. The origin of the depressions is thought to be related to *en échelon* offsets in the fault trace. A depression is formed where the offset occurs as the strike-slip movement continues. Note that depressions form when one particular combination of offset direction and sense of strike-slip movement occurs, while an upbulge with associated minor thrust faulting forms if the offset is in the opposite direction (Clayton, 1966).

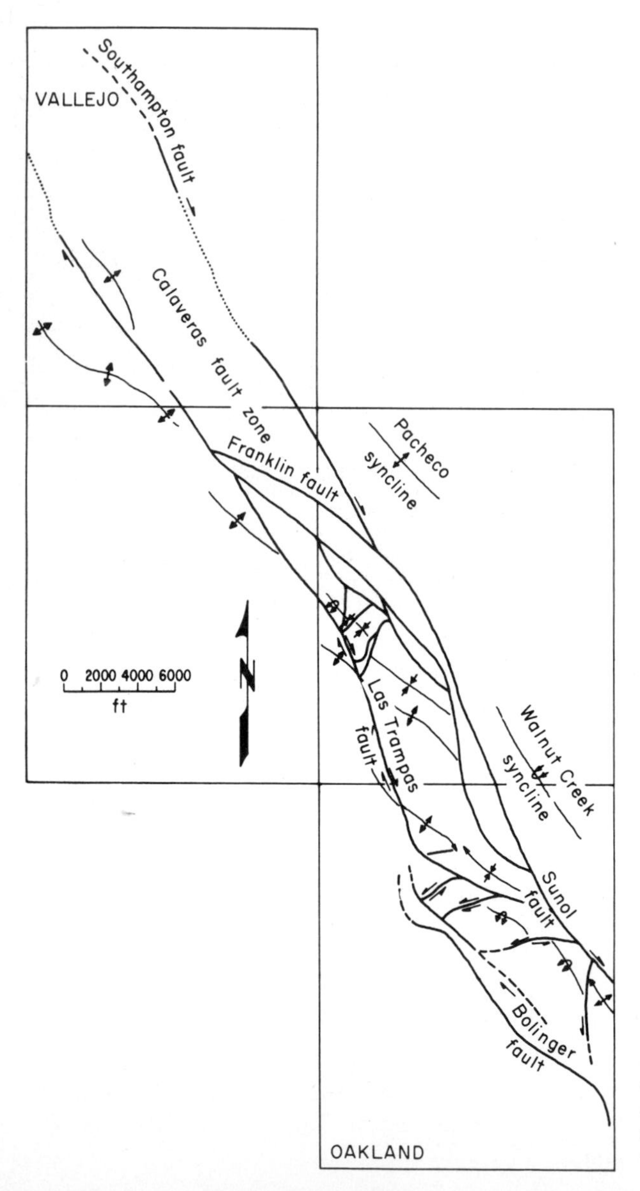

FIGURE 10-5 Part of the Calaveras fault zone of California showing bifurcations along the fault and folds caused by movements along the fault zone. Compare with the strain ellipse in Fig. 10-9. (*Adapted from Saul, 1967.*)

En Échelon Folds, Faults, and Fractures

Belts of doubly plunging anticlines and synclines are associated with many strike-slip faults. These folds may appear within a belt bounded by strike-slip faults on either side as in the Calaveras zone; they may be located to either side of the main fault, Fig. 10-8 (these folds often exhibit curved axes); or they may be confined to the sedimentary cover over a deep-seated strike-slip zone that does not break through to the ground surface. These folds are oriented as shown, with axes aligned obliquely to the trend of the major faults, Fig. 10-9.

The folds described above are often cut by transverse normal faults that appear to be extensional features. Fractures and faults with *en échelon* arrangement are characteristic features of strike-slip faults of all scales (Tchalenko, 1970). These failures are prominent within strike-slip zones, and they are often present in the sedimentary veneer over proven or suspected strike-slip zones that do not break the surface, such as the Lake Basin zone in Montana and the zones in eastern Oklahoma.

Experimentally Produced Strike-Slip Faults

Many of the structural features found along strike-slip fault zones have been successfully produced experimentally in clay. These experiments have the advantage of allowing us to see the progressive development of the features. Wilcox and others (1973) use a series of circles on the surface of the clay cake and over the zone of displacement to show the strain, Fig. 10-10. In the initial stages the circles are distorted into strain ellipses, but as strain progresses an *en échelon* zone of fractures form at a small angle to the displacement. In later stages the displacements occur in the fractures offsetting the strain ellipses.

In a second experiment with clay, Lowell (1972) produces a strike-slip movement but with a component of shortening across the fault zone. A zone of *en échelon* fractures similar to those described above form, but the material which moves up out of the fault zone as a result of the shortening forms a welt which is bounded by thrusts on both sides and a series of *en échelon* folds, Figs. 10-11 and 10-12.

MECHANICAL CONSIDERATIONS IN STRIKE-SLIP FAULTS

Moody and Hill (1956) extended Anderson's ideas concerning fault orientations to explain the origin of structures associated with strike-slip faults.

If the principal stress direction is north-south, a set of conjugate shears (first-order shears) form when the rock fails, Fig. 10-13. These are oriented at angles of about 30° to the principal stress. Once the first shears form, local stress conditions are reoriented (this is attributed to body forces developed as a result of movement on the fault), and a second set of shears becomes possible. It follows that a third-order set of shears could develop along the second-order structures; fourth-order fea-

FIGURE 10-6 Characteristic pattern of a major strike-slip fault, the Dasht E Bayaz fault in Iran. Note the splaying of the fault and its discontinuous character just east of Khezri. (*From Tchalenko and Berberiau, 1975.*)

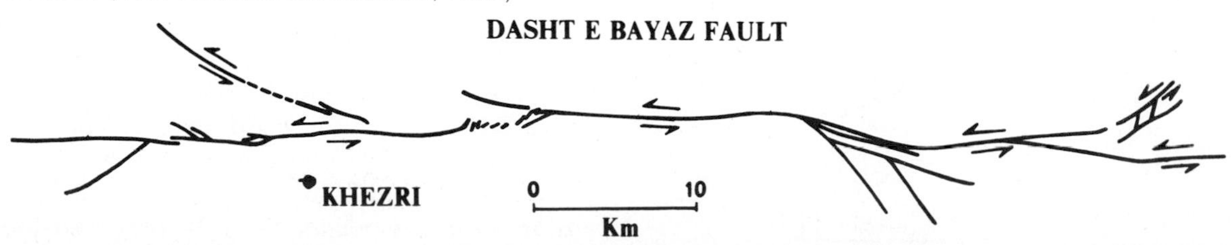

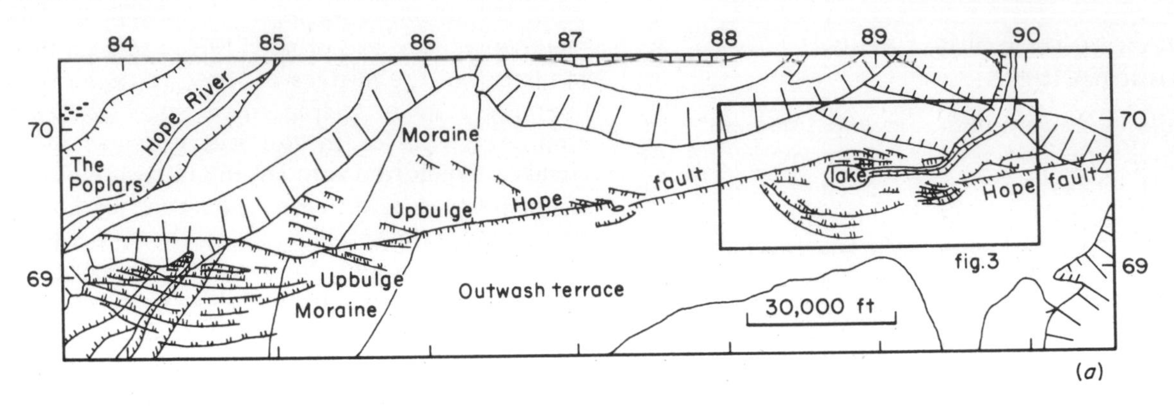

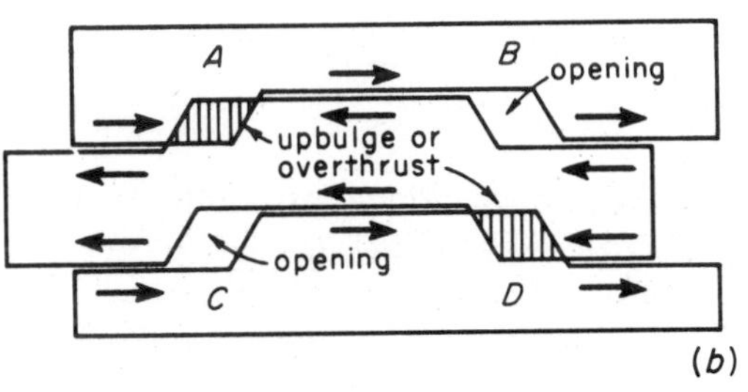

FIGURE 10-7 (*a*) Map of the lower Hope Valley showing fluvial terraces (single tick marks) and fault or slump scarps (double tick marks); (*b*) diagrammatic plan view of openings and upbulges or overthrusts formed between echelon segments of a right-lateral and a left-lateral fault. (*From Clayton, 1966.*)

FIGURE 10-8 Drag folds along San Andreas fault, California. Compare with the strain ellipse shown in Fig. 10-9. (*From Moody and Hill, 1956.*)

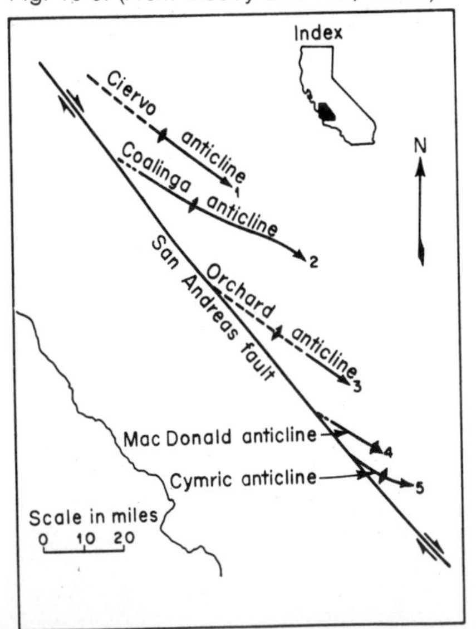

tures would parallel preexisting lower order structures. A total of eight shear directions and three directions of compression is postulated. This theory is applied to many of the large strike-slip faults, and indeed many of the small faults and folds, as in Fig. 10-13, are aligned as predicted by the theory. It is difficult to test the theory adequately because the predicted directions of shear and folding are so numerous. They box the compass so completely that almost any fold or fault lies close to one or the other of the predicted directions.

Structures the size of the faults discussed here may best be considered to result from simple shear with a rotational couple rather than from pure shear. The large strike-slip faults must extend deeply, possibly through the lithosphere, and though the rock may behave as a brittle substance near the surface, it must behave as a plastic at depth due to lithostatic pressure and elevated temperatures. Thus it is not valid to consider the lithosphere as a homogeneous layer of material. The orientation of the folds and subsidiary faults associated with the major strike-slip faults can be equally as well explained in terms of simple shear. The major fault is a zone of concentrated shear, but a couple acts over a broader zone, causing the material in this zone to be strained. If the strain is simple shear completely across this zone, it is characterized by parallel slip planes, but if any compressional component acts across the zone, then some shortening across the zone will occur producing a strain rhomb, as shown in Fig. 10-14, with potential fold orientation *F*, shear fractures *S*, and tensional fracture *T*. It is suggested that the conditions at depth in the zones of plastic flow are like those shown in the lower part of the diagram, while surface conditions are like those shown at the top.

The *en échelon* belts of faults and folds in sedimentary units in Oklahoma and in the Nye-Bowler and Lake Basin fault zones of Montana have been interpreted as resulting from movements on deep faults that do not break directly through the sedimentary veneer. A mechanism for the development of such features can be envisioned as follows: Assume that a strike-slip movement occurs in the basement but that it does not break directly through the covering sedimentary units. As lateral movements take place in the basement, the overlying sediments are subjected to drag. At some distance to either side of the fault, the sediment moves passively with the basement. The material near the fault zone is subjected to a rotational couple which has the effect of extending the material in one direction, possibly inducing extension fractures, and of shortening the material at right angles

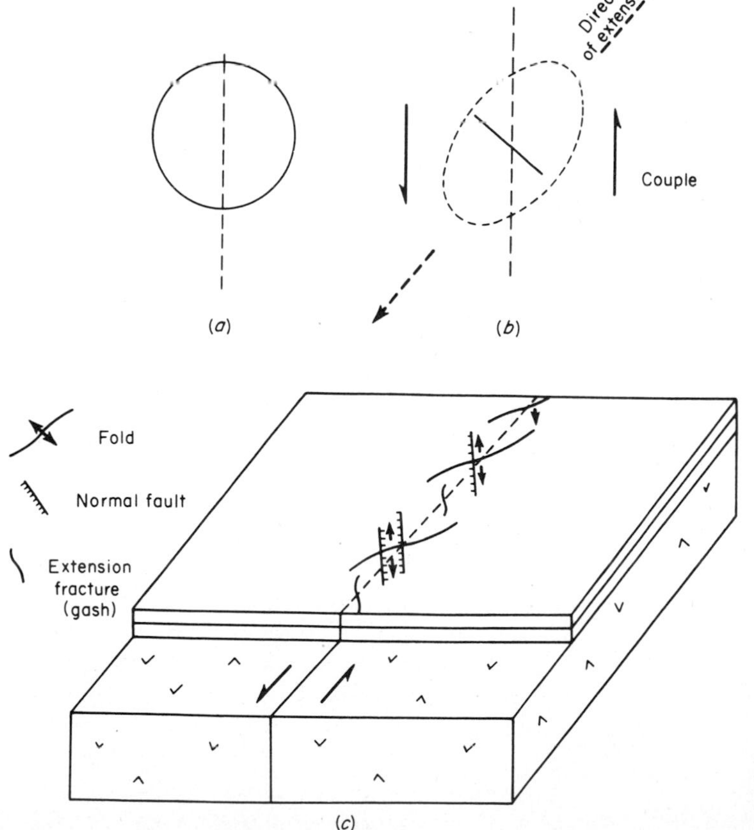

FIGURE 10-9 (*a*) Diagrammatic illustration showing an imaginary circular area of sediment over a deep-seated strike-slip fault; (*b*) the same area distorted as a result of movement on the fault; (*c*) the associated structural features.

FIGURE 10-10 Progressive development of faults and strain along a left-lateral strike-slip fault formed in a clay model. (*From Wilcox, Harding, and Seely, 1973.*)

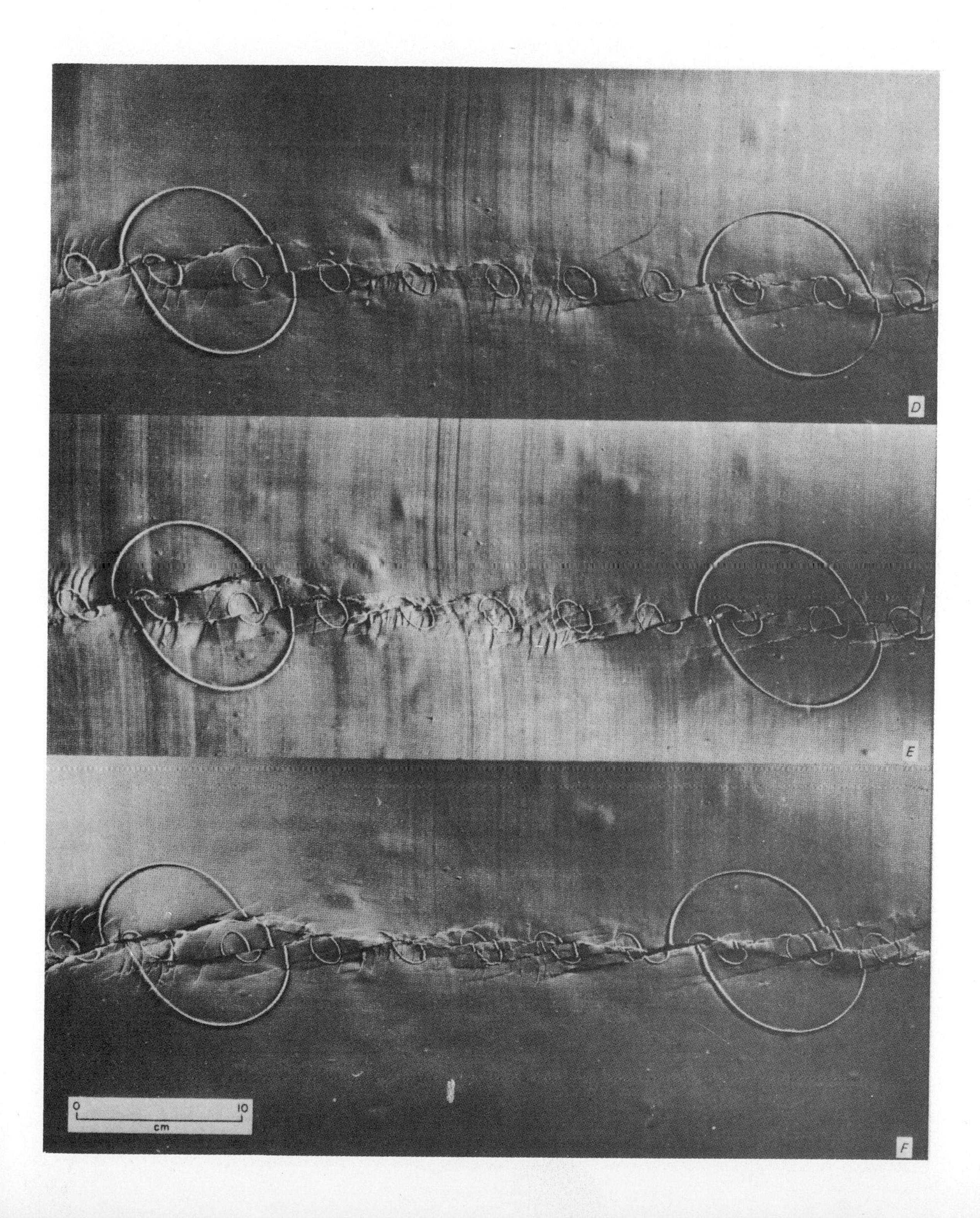
D
E
0
10
cm
F

to the extension, causing folds, domes, or doubly plunging anticlines. An original circular area of sediment over the fault is distorted into an elliptical shaped area, Fig. 10-9, in response to the rotational couple. The sedimentary veneer fails in response to the extension by the development of normal faults oriented with their strike direction normal to the direction of elongation.

VERTICAL DISPLACEMENT ALONG STRIKE-SLIP FAULTS

The topographic relief across major strike-slip faults is often great. The New Zealand Alps

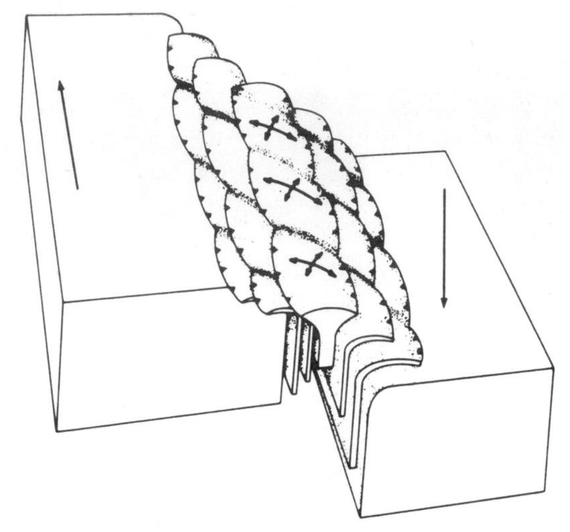

FIGURE 10-12 Conceptual diagram of an upthrust-bounded welt created by convergent strike-slip or transform motion. Portions of two plates or blocks are shown, the one moving at a low convergent angle into the other in the same plane, causing each segment to move a small distance successively past the other in a right-slip direction. A space problem is created, and the easiest direction of relief for the crowded material is upward. A welt with downward-tapering wedges and upthrust margins is created. The upthrusts are not necessarily as symmetrically disposed as shown, and the deeper faults may tend to coalesce and braid rather than be parallel. (*From Lowell, 1972.*)

FIGURE 10-11 Plan view of clay model of a thrust-bounded welt caused by and paralleling the trend of an underlying strike-slip zone. Strike-slip structures in the clay were created by convergent right-lateral motion between two underlying tin sheets; the lower tin was pulled to the left beneath the stationary upper tin, which was canted at an angle of 15°. A compressional component additional to that inherent in any strike-slip was thus introduced. First formed on the top of the originally smooth clay surface were *en échelon* anticlines approximately parallel to the long axes of the deformed ellipses. Shortly thereafter, and as the anticlines continued to grow, synthetic right-lateral shears trending at low angles to the underlying wrench zone developed along with antithetic left-lateral shears trending at high angles to the underlying wrench. In the more advanced stages of lateral movement, an uplift (or welt) formed parallel to the underlying wrench zone. The edges of the uplift are clearly thrusted. (*From Lowell, 1972.*)

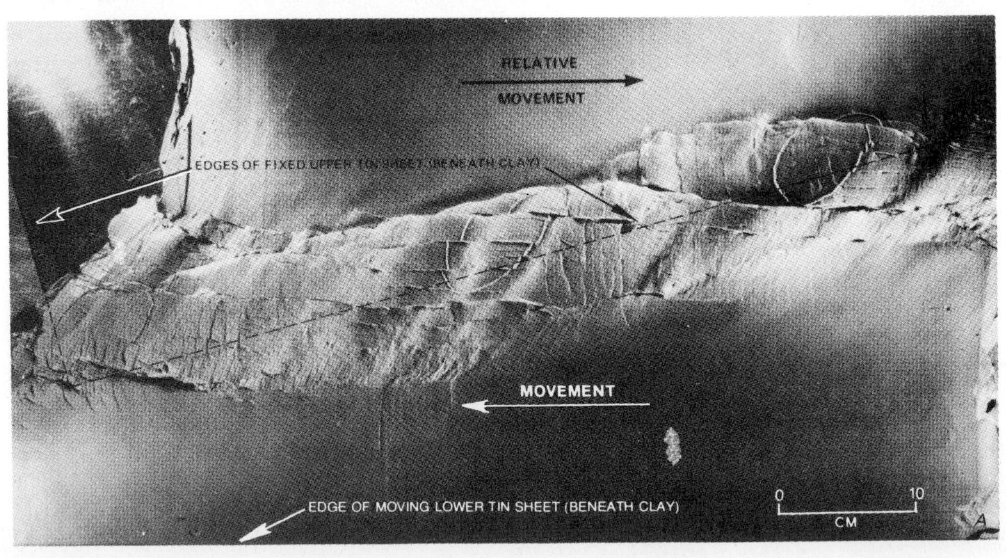

rise to 3,600 m east of the Alpine fault; deep sedimentary basins have formed adjacent to the San Andreas; and dip-slip components are common along many recent faults showing predominantly strike-slip movement. A 40-cm vertical movement accompanied a horizontal movement of several meters during the Tango and Idu earthquakes in Japan (Chinnery, 1961). Dip-slip movement also accompanied the San Fernando earthquake in 1971. The presence of demonstrable dip-slip movements and the ambiguities so often present in estimates of strike-slip displacements have led to considerable debate regarding the nature of these faults. If strike-slip faults die out, vertical displacements must accompany the horizontal movements.

An interesting approach to the problem has been made by Chinnery (1961, 1965) and Press (1965). Mathematical analyses of effects of transcurrent faulting on the ground surface

FIGURE 10-13 Orientations of second-order shears which result from a reorientation of the stresses after the first-order shear forms. (*After Moody and Hill, 1956.*)

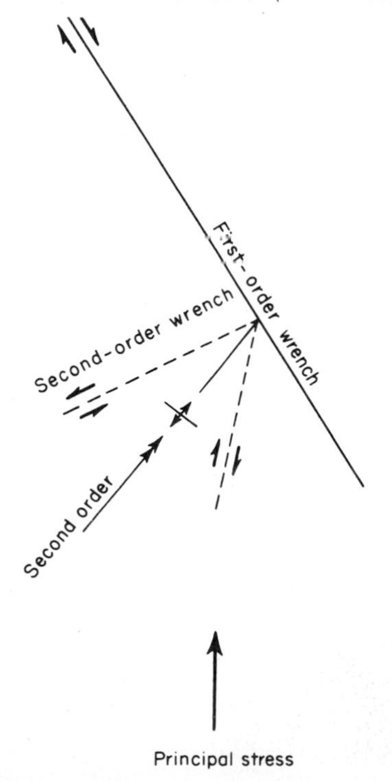

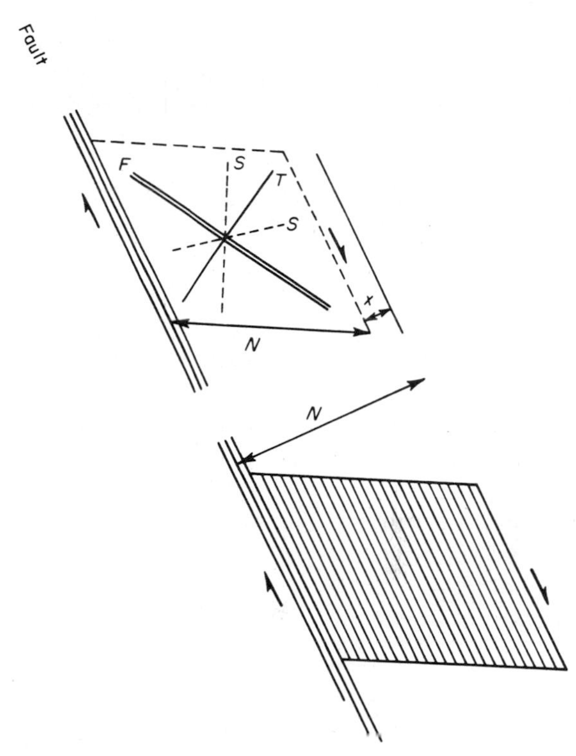

FIGURE 10-14 Simple shear (bottom) and a rotational couple (top) with the predicted fold orientation *F*, tension fracture *T*, and shears *S*.

have been based on the elastic dislocation theory of Staketee. Assumptions are that the fault is vertical; that movement is strike-slip; that the crust behaves as a semi-infinite elastic solid; that the ground surface is horizontal; and that displacement is constant over the fault surface and falls to zero in a short distance from the edge of the fault. The vertical displacements predicted near a fault in which the depth of the fault is one-twentieth of the length are shown in Fig. 10-15. Solid lines represent uplift; dashed lines are depressions. The contours are in units of 10^{-4} times the amount of the displacement. If vertical movements can be accumulated over a long time interval in this way, great vertical displacements can result. This is significant because the major strike-slip faults are so very long and the horizontal displacements estimated are great. The theory predicts scissor-like displacements along the fault and could account for large regional vertical movements near the faults and smaller regional uplift hundreds of kilometers away.

TECTONIC SETTINGS OF STRIKE-SLIP FAULTS

Strike-slip faults occur in a great variety of tectonic settings and on many scales. Small-scale strike-slip displacements are common as tear faults along the leading edge of both low-angle thrust faults in foreland fold belts and along upthrust where block faulting has taken place. On such faults one section of the fault block moves farther than another as shown in Fig. 8-3. *En échelon* zones of faults with a component of strike-slip movement are well known in the cover over deep shear zones within continental crust. Examples are found in the Lake Basin fault zone in Montana and in similar zones in Oklahoma.

FIGURE 10-15 Vertical displacements in the plane $y_1 = L$ for a hypothetical fault of length 2,000 km, depth 100 km, and displacement 200 km; (*a*) at the ground surface, (*b*) halfway down the fault, and (*c*) at the lower edge of the fault. (*Chinnery, 1965.*)

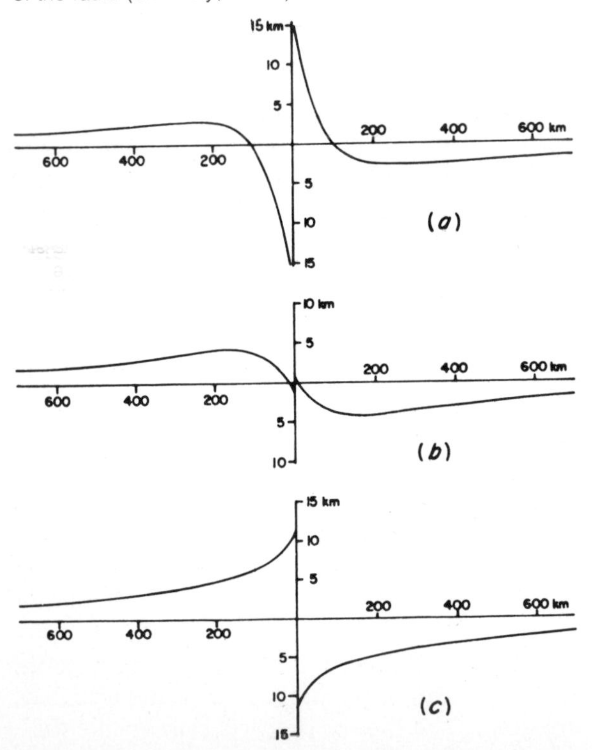

Many of the largest strike-slip faults play an important role in plate-tectonic theory in which they form one of three types of plates boundaries. As such they must be very deep-seated breaks which extend deep into or possibly through the lithosphere. The direction of movement on some of these is the reverse of the apparent displacement as is expected on transform faults as described in Chap. 17. The best documented of these are the faults which cut across the oceanic ridges (see Chap. 20), but other major strike-slip faults such as the San Andreas fault and the Alpine fault are also now interpreted as transforms.*

The Levant

The fault system at the northern end of the Red Sea is of special interest in terms of plate tectonics. Does the long fault system along the Gulf of Eilat, the Dead Sea, and the Jordan valley constitute a zone of predominantly right lateral shift which would be consistent with the movement of the Arabian Peninsula to the northeast, or is the zone one of block faulting associated with vertical movements alone? A case can be made for both types of movement. Certainly many parts of the Levant are grabens. The Dead Sea valley is estimated to contain more than 10,000 m of sediment, and negative gravity anomalies lie over the Gulf of Elath. A system of transverse faults and folds is superimposed on this graben system, Fig. 10-16. Many of these have proven strike-slip movements which offset even recent sediments. Additional evidence of offset is found at the mouth of the Gulf of Elath, where the Precambrian shield rocks of the Arabian Peninsula side are apparently offset to the northeast relative to those on the Sinai side of the gulf. Lees (1954) and Quennell (1957) ex-

* See Chaps. 17 and 24 for discussion.

plain the fault pattern in terms of simple shear parallel to the Levant accompanied by some extension normal to the valleys. The sides of the rifts, interpreted as shears, east of the valley pass northward into thrusts; those on the west pass into tensional features; folds are viewed as related to a rotational couple.

FIGURE 10-16 Generalized structural features of the Levant, including a major strike-slip fault zone which extends from the Gulf of Akaba through the Dead Sea to Turkey. Numerous folds (heavy lines) and grabens are found along and on both sides of the zone. These are oriented as might be expected if the zone if a left-lateral fault. (*After Picard, 1965.*)

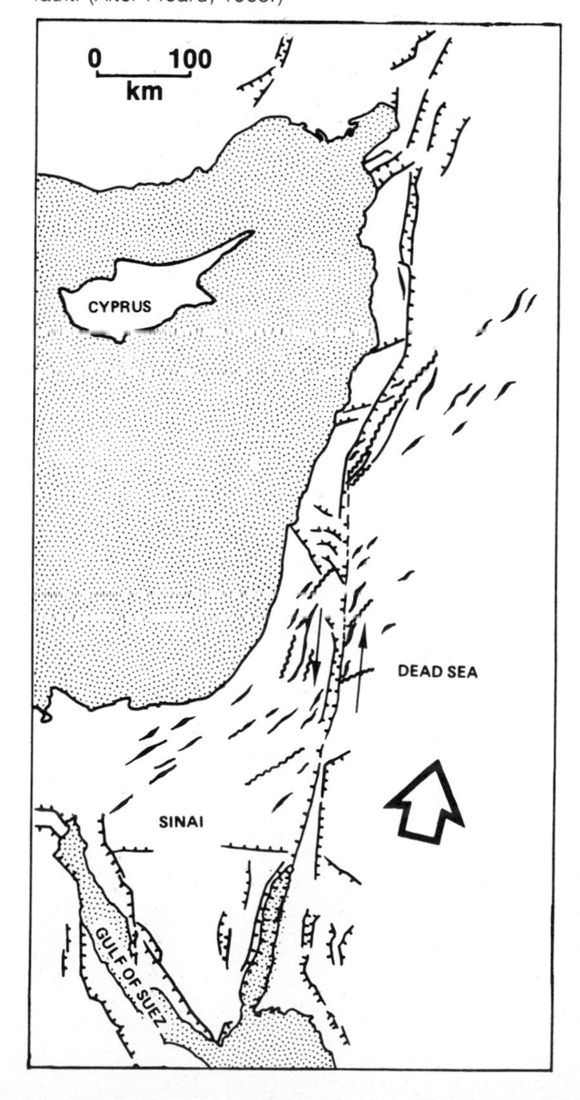

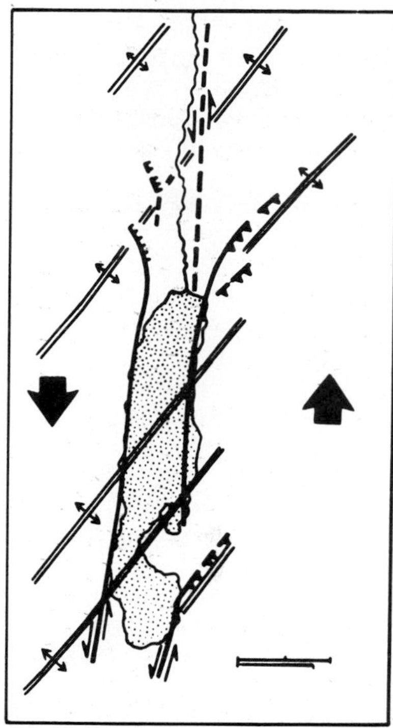

FIGURE 10-17 Faults and folds near the Dead Sea. The left-lateral strike-slip fault on the east side of the Sea turns into a thrust in the north. Note the normal faults at the north end of the Dead Sea on the west, where regional movement is south. (*After Lees, 1954.*)

Carey as early as 1958 reconciled this with the extensional origin of the Red Sea and Gulf of Aden by viewing the origin of the entire system as a result of rotation of East Africa away from Arabia.

REFERENCES

Achenbach, Abo-Zena, 1973, Analysis of the dynamics of strike-slip faulting: Jour. Geophys. Res., v. 78, no. 5, p. 866–875.

Anderson, E. M., 1951, The dynamics of faulting, 2d ed.: Edinburgh, Oliver & Boyd.

Arthaud, F. et Matte, Ph., 1975, Les décrochements tardo-hercyniens du sud-ouest de l'Europe. Géométrie et essai de reconstitution des conditions de la déformation (Late Hercynian wrench-faults in southwestern Europe. Geometry and nature of the deformation): Tectonophysics, v. 25, p. 139–171.

Bishop, D. G., 1968, The geometric relationships of structural features associated with major strike-slip faults of New Zealand: New Zealand Jour. Geo. Geophys., v. 11, no. 2, p. 405–417.

Carey, S. W., 1958, The tectonic approach to continental drift, *in* Continental drift—a symposium: Geol. Dept. Univ. of Tasmania, p. 177–355.

Chinnery, M. A., 1961, The deformation of the ground around surface faults: Seismol. Soc. America Bull., v. 51, p. 355–372.

——— 1965, The vertical displacements associated with transcurrent faulting: Jour. Geophys. Res., v. 70, no. 18, p. 4627–4632.

Clayton, Lee, 1966, Tectonic depressions along the Hope fault, a transcurrent fault in North Canterbury, New Zealand: New Zealand Jour. Geol. Geophys., v. 9, nos. 1 and 2, p. 95–104.

Crowell, J. C., 1952, Probable large lateral displacement on San Gabriel fault, Southern California: Am. Assoc. Petroleum Geologists Bull., v. 36, p. 2026–2035.

——— 1962, Displacement along the San Andreas fault, California: Geol. Soc. America Spec. Paper 71.

Cummings, David, 1976, Theory of plasticity applied to faulting, Mojave Desert, southern California: Geol. Soc. America Bull., v. 87, no. 5.

Freund, Raphael, 1970, Rotation of strike slip faults in Sistan, southeast Iran: Jour. Geology, v. 78, no. 2, p. 188–200.

——— 1974, Kinematics of transform and transcurrent faults: Tectonophysics, v. 21, p. 93–134.

Garfunkel, Z., Transcurrent and transform faults; a problem of terminology: Geol. Soc. America Bull., v. 83, no. 11, p. 3491–3496.

Kennedy, W. Q., 1946, The Great Glen fault: Quart. Jour. Geol. Soc. London, v. 102, p. 41–76.

Kovach, R. L., and **Nur, A.,** 1973, Proceedings of the conference on tectonic problems of the San Andreas fault system: Stanford University Pub., Geological Sciences, v. XIII.

Lajtai, E. Z., 1969, Mechanics of second order faults and tension gashes: Geol. Soc. America Bull., v. 80, no. 11, p. 2253–2272.

Lees, G. M., 1954, The geological evidence of the nature of the ocean floor: Roy. Soc. London Proc., v. 22, p. 400–402.

Lenson, G. J., 1968, Analysis of progressive fault displacement during downcutting at the Branch River terraces, South Island, New Zealand: Geol. Soc. America Bull., v. 79, no. 5, p. 545–556.

Lillie, A. R., 1964, Steeply plunging folds in the Sealy Range, southern Alps: New Zealand Jour. Geol. and Geophys., v. 7, no. 3.

——— and **Gunn, B. M.,** 1963, Steeply plunging folds in the Sealy Range, southern Alps: New Zealand Jour. Geol. Geophys., v. 7, p. 403–423.

Loczy, Louis de, 1970, Role of transcurrent faulting in South American framework: Amer. Assoc. Petroleum Geologists Bull., v. 54, no. 11, p. 2111–2119.

Lowell, J. D., 1972, Spitsbergen Tertiary orogenic belt and the Spitsbergen fracture zone: Bull. Geol. Soc. America, v. 83.

McKinstry, H. E., 1953, Shears of the second order: Am. Jour. Sci., v. 251, no. 6, p. 401–414.

Maxwell, J. C., and **Wise, D. U.,** 1958, Wrench-fault tectonics—A discussion: Geol. Soc. America Bull., v. 69, p. 927–928.

Moody, J. D., and **Hill, M. J.,** 1956, Wrench fault tectonics: Geol. Soc. America Bull., v. 67, p. 1207–1246.

Paschall, R. H., and **Off, Theodore,** 1961, Dip-slip versus strike-slip movement on San Gabriel fault, southern California: Am. Assoc. Petroleum Geologists Bull., v. 45, p. 1941–1956.

Picard, L., 1965, Thoughts on the graben system in the Levant, *in* Irvine, T. N., ed., The world rift system: Geol. Survey of Canada, paper 66-14.

Press, F., 1965, Displacements, strains, and tilts at teleseismic distances: Jour. Geophys. Research, v. 70, p. 2395–2412.

Quennell, A. M., 1957, The structural and geomorphic evolution of the Dead Sea rift: Geol. Soc. London Proc., no. 1544, p. 14–20.

Saul, R. B., 1967, The Calaveras fault zone: Mineral Info. Service, v. 20, no. 3.

Shawe, D. R., 1965, Strike-slip control of basin and range structure indicated by historical faults in western Nevada: Geol. Soc. America Bull., v. 76, p. 1361–1378.

Smith, G. I., 1962, Large lateral displacement on Garlock fault, California, as measured from offset dike swarm: Am. Assoc. Petroleum Geologists Bull., v. 46, p. 85–104.

Suggate, R. P., 1963, The Alpine fault: Roy. Soc. New Zealand Geol. Trans., v. 2, p. 105–129.

Tanner, W. F., 1962, Surface structural patterns obtained from strike-slip models: Jour. Geology, v. 70, p. 101–107.

Tchalenko, J. S., 1970, Similarities between shear

zones of different magnitudes: Geol. Soc. America Bull., v. 81, no. 6, p. 1625–1640.

——— and **Berberian, M.**, 1975, Dasht E Bayaz fault, Iran: Earthquake and earlier related structures in bed rock: Geol. Soc. America Bull., v. 86, p. 703–709.

Thom, W. T., Jr., 1955, Wedge uplifts and their tectonic significance: Geol. Soc. America Spec. Paper 62, p. 369–376.

Walsh, J. B., 1968, Mechanics of strike-slip faulting with friction: Jour. Geophys. Res., v. 73, no. 2, p. 761–776.

Wellman, H. W., 1952, The Permian-Jurassic stratified rocks, New Zealand. Symposium on Gondwana Series: Proc. 19th Int. Geol. Congress.

——— 1954, Active transcurrent faulting in New Zealand: Geol. Soc. America Bull., v. 65, p. 1322.

Wilcox, R. E., Harding, T. P., and **Seely, D. R.**, 1973, Basic wrench tectonics: Am. Assoc. Petroleum Geologists Bull., v. 57, no. 1, p. 74–96.

Wilson, C. W., Jr., 1936, Geology of Nye-Bowler lineament, Stillwater and Carbon Counties, Montana: Am. Assoc. Petroleum Geologists Bull., v. 20, p. 1161–1168.

Wilson, J. T., 1965, A new class of faults and their bearing on continental drift: Nature, v. 207, no. 4995, p. 343–347.

11

Geometry of Folded Rocks

The term *fold* is applied to a curve or bend developed in a preexisting surface, most frequently stratification. According to this definition a fold is a secondary structure; however, some folds form contemporaneously with the folded surfaces, (e.g., syndepositional deformation).

Folds and flexures in crustal rocks are known on every scale from microscopic to the great basins and geosynclinal belts. They are among the most fascinating and highly varied structures with which we deal, and they form under a wide range of physical conditions, in all types of rocks, and as a result of numerous applied stress conditions. Some can be readily described by simple geometrical techniques, Fig. 11-1; others defy description by analogy to regular geometric forms.

OCCURRENCE OF FOLDS

Folds and fold systems are found in a variety of geological circumstances, and fold geometry, rock type, and particularly the relation of the fold to adjacent structural features provide clues to the way folding was generated.

Compression is one of the most obvious potential causes of folding, and it has been examined thoroughly both experimentally and theoretically. From the standpoint of tectonics it is important to keep in mind the many ways compression can arise. In some experiments material laid in a box with rigid sides and bottom are squeezed between the jaws of a vise. The fixed bottom and sides and the open top are important constaints. Compressional forces are applied externally in this model. Such a model has considerable application in the earth if the continents and ocean basins are considered to be rigid plates between which geosynclines form and are subsequently compressed. Early students of folds thought that compression was generated by crustal shortening resulting from cooling and contraction of the earth. Folded mountain belts were explained in this way. The validity of this idea is being challenged on a number of grounds, and satisfactory criteria proving that any particular fold or fold system is a product of crustal shortening are difficult to establish. Compression remains an attractive explanation, but the connotation of crustal shortening does not necessarily follow. Regional compression is now explained most frequently as a product of collision or near collision between plates, but compression can arise from crowding of sediments within a basin bordered by rigid blocks that move horizontally or vertically relatively to one another, or it can arise through gravity gliding or spreading. Gravity acting as a body force in an inclined layer may cause surficial folds in the layer, and folds may be produced ahead of a block that is free in slide.

Folds and faults often occur together, and small folds are often related to drag effects along faults. Sedimentary layers may occur draped over vertical faults in areas isolated from any other folds or evidences of compression. Where reverse faults or thrusts are associated with folds, it is much more difficult to determine whether the folds formed first and were followed by progressive development of asymmetry and ultimately failure and faulting or whether the two were more or less synchronous. That folding can occur after thrusting is proven by folded low-angle thrusts.

Nondiastrophic conditions may also lead to formation of folds, such as the steeply plunging folds found within salt domes. Folds may be caused by differential compaction of sediment, gravity sliding within layers of plastic sediment or within layers associated with high pore pressures, flow and movements within and close to plutons, the flow of glacier ice, and rock and soil creep or flow.

Force couples may produce folds, such as the small folds caused by drag between layers as they move relative to one another. The folds may form in connection with horizontal couples as exemplified by *en échelon* folds in the sedimentary cover over strike-slip faults, between faults, and within zones of lateral displacement even where no faults are recognized, Fig. 10-9.

RECOGNITION OF FOLDS

The methods used to study and analyze folds depend to some extent on the size of the folds and the presence or absence of good marker horizons suitable for mapping. Large folds which contain good marker horizons are usually defined by geologic mapping techniques, and their shape is described through use of graphic representations such as structure contour maps, isopach (thickness) maps, and constructed cross sections.

Increasingly, attention has been given to analysis of mesoscopic and microscopic folds in metamorphic terrane and in regions of thick sedimentary sections lacking good mapping horizons. Often the deformational history is complex and involves repeated deformations. Conventional mapping techniques have lacked the resolution needed to define these structural features, and new geometrical techniques have been employed. Basically, the new methods involve collection of large samples of geometrical data on the attitude of *s* surfaces,* lineations, and detailed notation of the relations of the various geometrical features to one another. The data are compared with the statistical distribution of attitudes of points on selected simple hypothetical geometrical models, and the structural evolution of the deformed rock is inferred from these structural and fabric relationships.

IDEALIZED MODELS OF FOLD GEOMETRY

Cylindrical and Conical Folds

The form of many folds can be closely approximated by cylindrical or conic models, Fig. 11-1. These two shapes can be formed by

* See Appendix E for structural notations.

the rotation of a line; for a cylindrical fold, the line is rotated parallel to itself. A *conical fold* is generated by rotating a line, one end of which is fixed. In map view a plunging cylindrical fold and a conical fold are similar, but the two can be distinguished by geometrical data. The most convenient method is to plot strike and dip data from the fold on a stereographic or an equal-area projection as a point diagram. If the sample of bedding attitudes is taken from all parts of a cylindrical fold, the points will fall approximately along a great circle, and a line can be drawn through the clustering of points to define that great circle, a *pi* (π) *circle*. The *axis* of the fold (a *B* lineation) is the line perpendicular to the pi circle, and the attitude, bearing, and plunge of the axis can be measured directly from the stereographic plot, Fig. 11-2. Comparable data from a conical fold form a small circle on a point diagram.

Representation of Cylindrical Folds on the Stereographic Projection*

Much of the recent analysis of microscopic and mesoscopic fold geometry has been done with the use of stereographic and equal-area projections. Angular relationships are readily

* The definition of fold types and fold geometry is given in the glossary at the end of this chapter; structural notation is explained in Appendix E; and use of the stereographic projection is discussed in Appendix D.

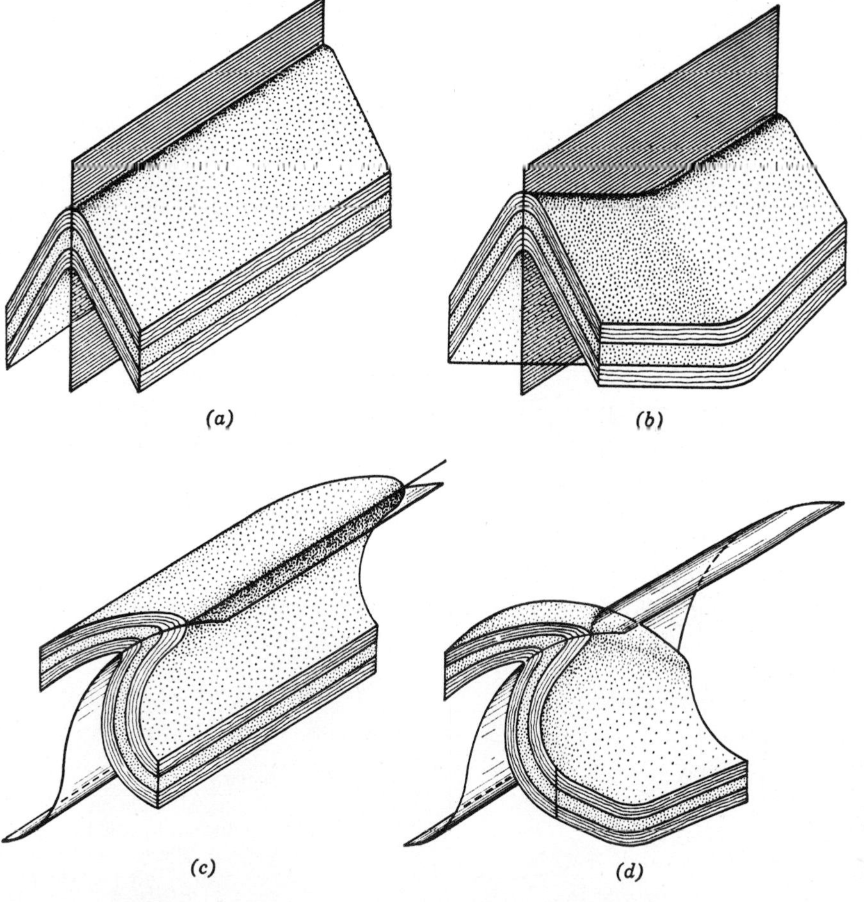

FIGURE 11-1 Geometric properties of folds. (*a*) Cylindrical plane fold; (*b*) noncylindrical plane fold; (*c*) nonplane cylindrical fold; (*d*) nonplane noncylindrical fold with cylindrical axial surface. (*From Turner and Weiss, 1963.*)

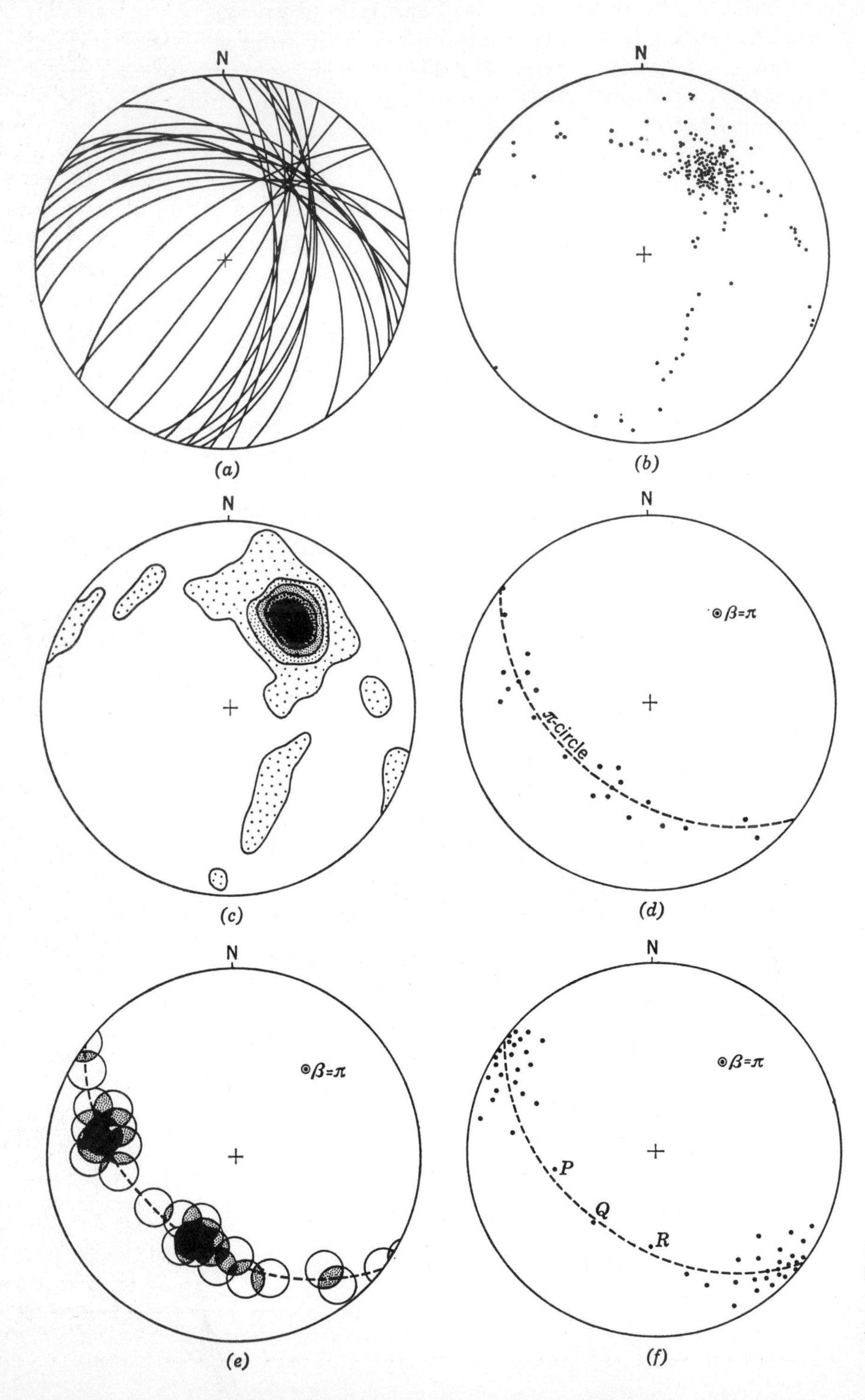

FIGURE 11-2 Stereographic plot of 25 planar segments of a cylindrically folded *s* surface. See text for explanation. (*From Turner and Weiss, 1963.*)

depicted on the stereographic projection, but the equal-area projection should be used for statistical analysis of large amounts of data in which contouring of the diagram is done.

The types of data most frequently collected for analysis are strike and dip measurements on the limbs and crests of folded rocks, fold axes, and axial planes of folds. If the surface measured is a bedding plane it is usually designated S_0, but other types of surfaces may be folded, especially where superposed folds occur.

The bedding surfaces measured on a fold may be represented on an equal-area projection as shown in Fig. 11-2a. Note that the planes always intersect in a line. This line is designated as β (beta), and it is parallel to the fold axis *B* if the fold is a cylindrical fold. The lines of intersection are represented as points on the stereographic projection, Fig. 11-2*b*, and the points yield a clear maximum when they are contoured, Fig. 11-2*c*. The great circle which is normal to β defines a pi circle. The pi circle may also be obtained by making a plot of poles to bedding planes with the original data and determining the best fit to that data, Fig. 11-2*e* and *f*.

The axial plane of a fold is the plane which comes closest to dividing the fold into two symmetrical halves, Fig. 11-1, and it may be represented as a plane in the stereographic projection or as a pole to that plane, Fig. 11-3*a*. If the axial plane is vertical, as in an upright cylindrical fold, it appears as a line across the projection, a pole to that plane appears on the edge of the circle, and the trend of the line is the strike of the axial plane, *A* in Fig. 11-3*b*. If the axial plane is horizontal it appears as the outside of the projection, a circle, but if the axial plane is inclined it appears as a great circle in the projection and its pole is 90° away, *B* in Fig. 11-3*b*.

The axis *B* of a cylindrical fold lies within the axial plane, but it may have any orientation within that plane, depending on the plunge of the fold. Its actual position may be measured in the field, or it may be determined statistically by determination of beta points as explained above. The plunge of the axis of a fold is measured in a vertical plane that includes the axis, and this should be carefully distinguished from the pitch of the axis in the axial plane, which may also be measured in the field. A number of different pitches are shown in Fig. 11-3*c* in which axis *A* has no pitch or plunge and represents the axis of a recumbent fold; *B*, *C*, and *D* correspond to

FIGURE 11-3 (*a*) Poles to axial plane of folds. *A* represents an upright fold, *B* is an inclined or overturned fold, and *C* is a recumbent fold. (*b*) Stereographic plot showing an axial surface that is vertical (*A*) and its pole, and an axial surface dipping east and its pole (*B*). (*c*) Stereographic plot showing an axial plane position with N-S strike and low dip to the east. *A*, *B*, *C*, and *D* represent some of the possible positions for fold axes. *A* is a recumbent fold, and *D* is what is sometimes referred to as a reclined fold.

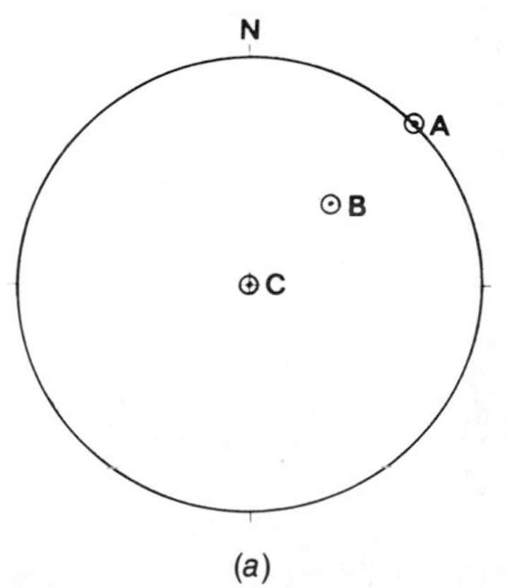

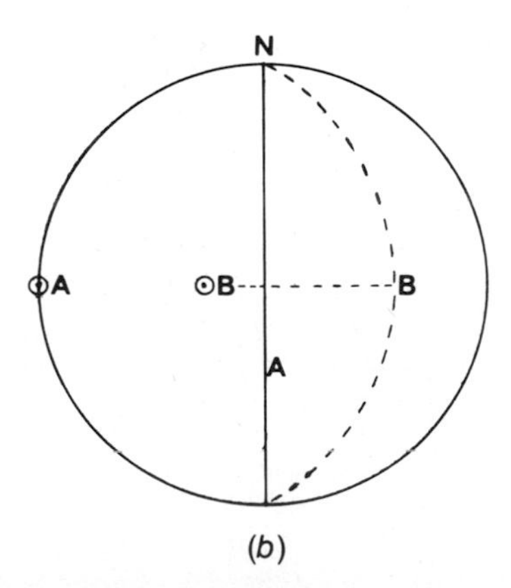

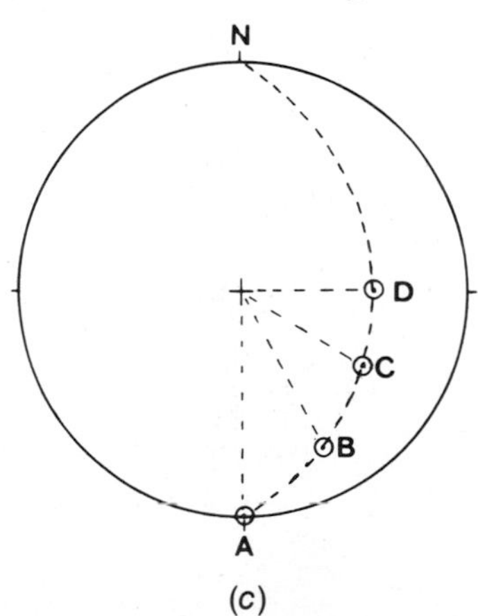

folds of increased pitch. (*Note:* pitch is the angle *AOC* measured in the axial plane for axis *C*.) The position of the axis, the axial plane, and girdles obtained by plotting and contouring poles to *s* surfaces are shown for a number of different types of folds, Fig. 11-4.

The distribution of the sample of data points on the fold is of critical importance in interpretation of point diagrams. It is generally assumed that the sample is evenly distributed across the fold. If this condition is met in the case of a cylindrical fold, the great circle defined by attitude measurements will be uniformly developed, but if a large part of the sample is collected on any one part of the fold, the corresponding part of the great circle will appear as a maximum. Thus a large sample collected from all parts of the fold should be used. Point diagrams can be interpreted as follows:

1. Curvilinear folds are represented by a uniform distribution of points along the great or small circle, and angular folds are represented by isolated maxima.

2. If the radius of hinge curvature is great, more points are located on the hinge region, and it is better defined.

3. The axial surface of an ideal cylindrical fold plots on a great circle which contains the axial line as a point on the pi circle within the crestal scatter pattern between the limb maxima.

4. The position of the axis (using the concept of the axis as a generator line) is defined by the line of intersection β of any two planes tangent to a cylinder or cone. The strike and dip measured at any point on a folded bed defines such a tangent plane. Because such measurements are liable to be in error, the line β is most reliable when it is determined by use of data from many points on the fold. If the positions of the successively determined β points (the point of intersection of the line with the hemisphere on the projection is called a β point) are not close, the fold is noncylindrical.

5. The appearances of point diagrams of ideal cylindrical folds are shown in Fig. 11-4 for cases in which the folds are open, isoclinal, asymmetric, overturned, and recumbent.

Similar and Concentric Folds

An ideal *concentric* (syn.: *parallel*) *fold,* Figs. 11-5 and 11-6, is one in which adjacent beds remain parallel to one another and is characterized by the following geometry: (1) The *true thickness* of the beds, measured perpendicular

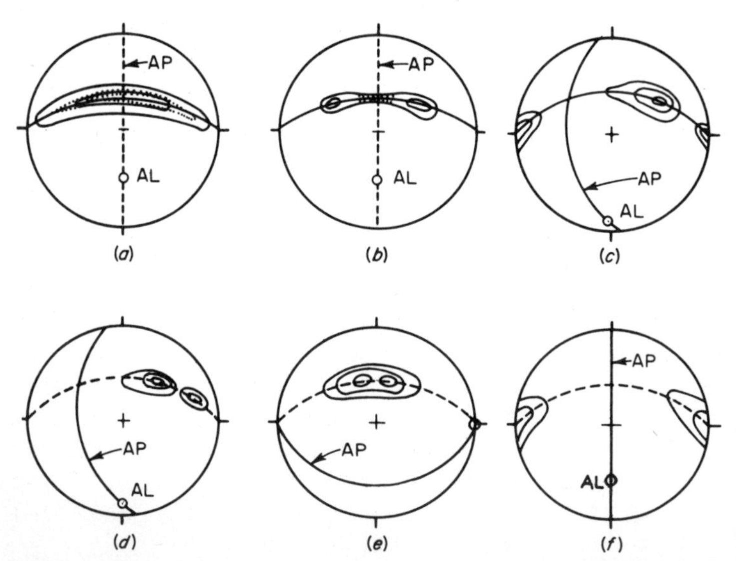

FIGURE 11-4 Geometry of some ideal cylindrical folds. AP = axial plane, AL = axial line. Contours (schematic) are of points (strike and dip data collected on the folded surface) plotted on lower hemisphere projection. (*a*) Open fold of uniform curvature; (*b*) open fold with plane limbs; (*c*) asymmetrical fold; (*d*) overturned fold; (*e*) recumbent fold; (*f*) isoclinal fold. (*After Dahlstrom, 1954.*)

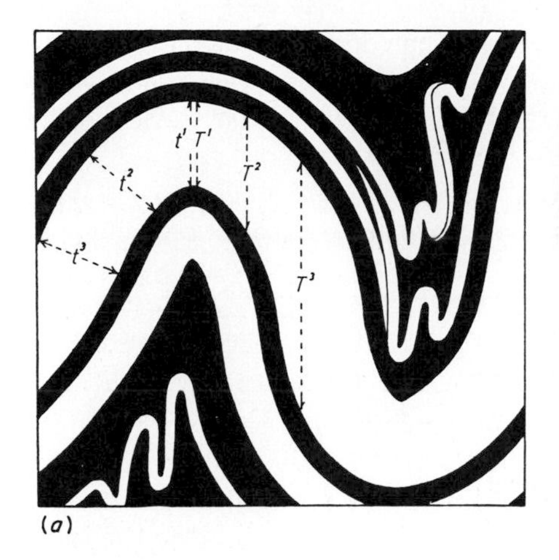

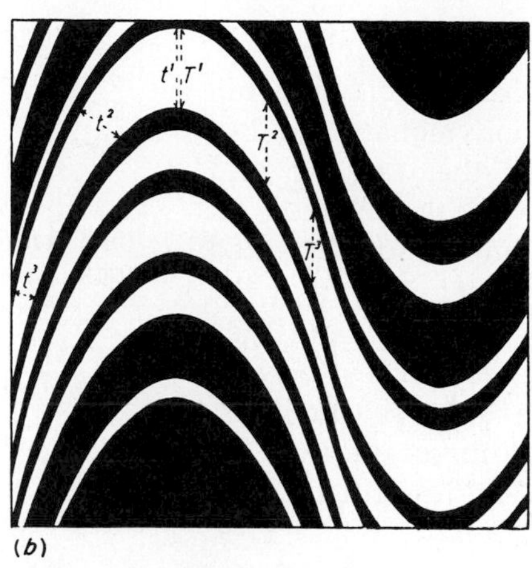

FIGURE 11-5 Comparison of the geometry of (*a*) ideal concentric (parallel) and (*b*) similar folds. Note thickness relationships described in the text and the crowding effects in the core of the parallel fold. (*From Ramsay, 1962.*)

to bedding, remains constant. (2) The thickness of beds varies considerably with position on the fold when measured parallel to the axial plane. One consequence of the uniform-thickness requirement is that the fold shape must change with depth. A cylindrical anticline at one level is first transformed into a cusp and then dies out at depth.

Ideal *similar folds*, Fig. 11-7, maintain their shape with depth, but the true thickness of the beds is highly variable from point to point along the fold. Most similar folds have strongly developed planar structures oriented approximately parallel to the axial surface. These planes are due to mineral orientation, foliation, or fractures and cleavages. These foliations often penetrate through successive layers in the fold, indicating that the individual layers were passive and did not control development of the fold form. Concentric folds, on the other hand, are usually found in sedimentary piles of varying lithology in which the individual layers are effective in controlling the deformation.

The restoration of layers folded into similar form is simple. The flow lines are parallel to the axial surface of the fold, and the volume of material of a given layer between any two flow lines is the same as that in the restored layer between the same two lines. Restoration is achieved by reversing the slip.

While individual beds in concentric folding may be isotropic, the folded pile usually contains layers of markedly different character—different layers deform by different mechanisms, the folds die out at depth, and movements parallel to bedding are dominant. The less ductile beds control the folding, and the more ductile layers may have an intricate fold pattern, not reflected above or below, that results from intralayer movements.

If the folded layers in concentric folds are envisioned as compressional features in which each bed is shortened by the same amount, the bed thicknesses and lengths may be maintained by intricate crumpling and faulting within the cores of the anticlines combined with detachment of the folded beds from the underlying unfolded strata. The volume of the shortened mass is then present as an elevated volume over the anticline.

FIGURE 11-6 Lines of discontinuity in ideal concentric folds. Curvatures and radii of curvature of bedding planes are discontinuous across the lines of discontinuity. (*From Johnson and Ellen, 1974.*)

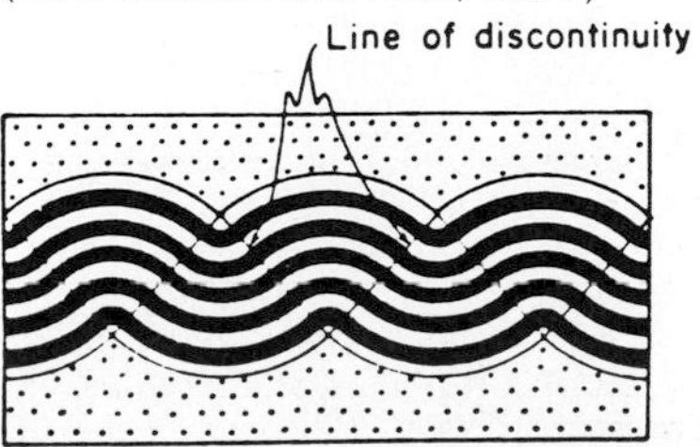

Buckling and Bending

The terms *buckle fold* and *bending fold* are often used in an essentially synonymous sense, but Hills (1953) distinguished them as two fold types. *Buckle folds* occur when the layered rock mass is subjected to compression oriented parallel to the layering. Folding occurs if the layers are not physically isotropic, and they shear along lines that are more or less parallel to the compression. In the case of *bending folds*, anisotropy of layers is unimportant, and the force causing the bend acts essentially perpendicular to the layers, Figs. 11-8 and 11-9.

Bends may be distinguished from buckles as follows (Ramberg, 1963):

1. Shortening of a buckled layer tends to form in a direction parallel to the original position of the layer and normal to the fold axis. This shortening may be recognized by such features as strained fossils, boulders, oolites, schistosity, rotated minerals, and boudins.

2. Contact strain in a buckle fold tends to die out rapidly up and down.

3. Before buckling can occur in a single enclosed layer, the competency of that layer must be greater than that of the enclosing layers. Ductility contrast is not important in bending.

4. The changes in thickness of originally evenly thick layers is opposite in the cases of bending and buckling (Fig. 11-8).

Chevron Folds

Many naturally formed folds fail to fit any of the ideal models previously described, because the fold limbs are straight and most of the curvature in the fold is confined to the fold hinge. Such folds are called *chevron folds*, and they have been described and analyzed by Ramsay (1974). Chevron folds are usually found in regularly bedded sequences which contain two alternating rock types, commonly sandstone and shale or limestone and shale. The more competent layer exhibits constant thickness even in the fold hinges, suggesting that chevron folds are a special variety of concentric folds, but the incompetent layers show

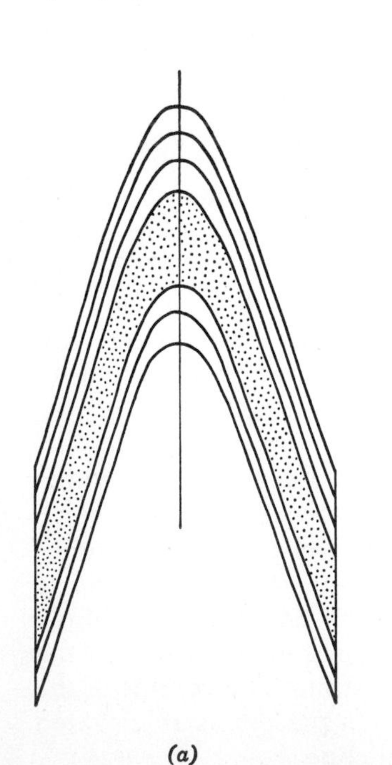

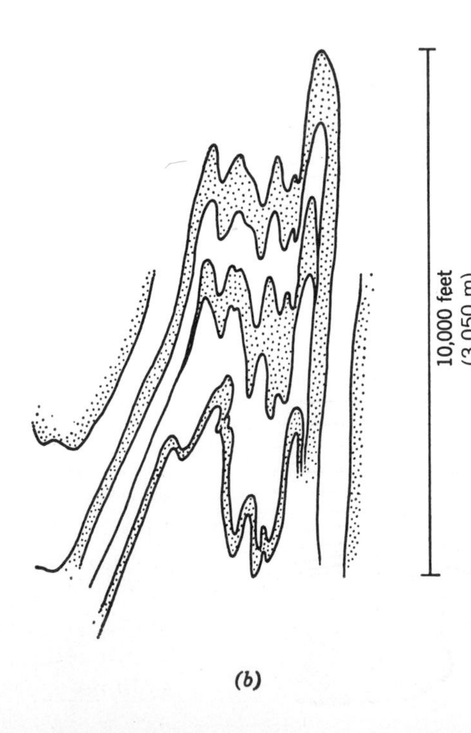

FIGURE 11-7 Features of similar folds in sections normal to fold axes. (*a*) General form of a similar fold. (*b*) Large macroscopic similar folds at Broken Hill, Australia. (*After J. K. Gustafson, H. C. Burrell, and M. D. Garretty. From Turner and Weiss, 1963.*)

pronounced variation in thickness especially in the fold hinge. An ideal geometrical model of a chevron fold is shown, Fig. 11-10, which should be compared with a schematic sketch of a field example, Fig. 11-11, showing mesoscopic features described in the following section and in Chap. 12.

Geometry of Real Folds and Clues to Their Origin

Shortening of the folded layers is an inherent part of concentric folding, and although similar folding can occur without any shortening of layers, it rarely does, Figs. 11-12 and 11-13. Among real folds intermediate fold types are common, for example, similar-type folds in which the fold form is not exactly reproduced from layer to layer. When layers of different ductility are involved, slip surfaces (cleavages) may be well developed in the shales (phyllites, schists, etc.) but not in interbedded sandstones (quartzites, granitic layers, etc.). These folds have a geometry which is suggestive of concentric folding of the more competent layers accompanied by movement within the less competent layers to produce thickening of hinges, thinning of limbs, and at some stage, foliation subparallel to the axial surface.

Classic examples of concentric folds are frequently chosen from the Appalachian Valley and Ridge province, where thick Paleozoic sequences of sandstone and quartzite alternating with shales are thrown into folds tens of

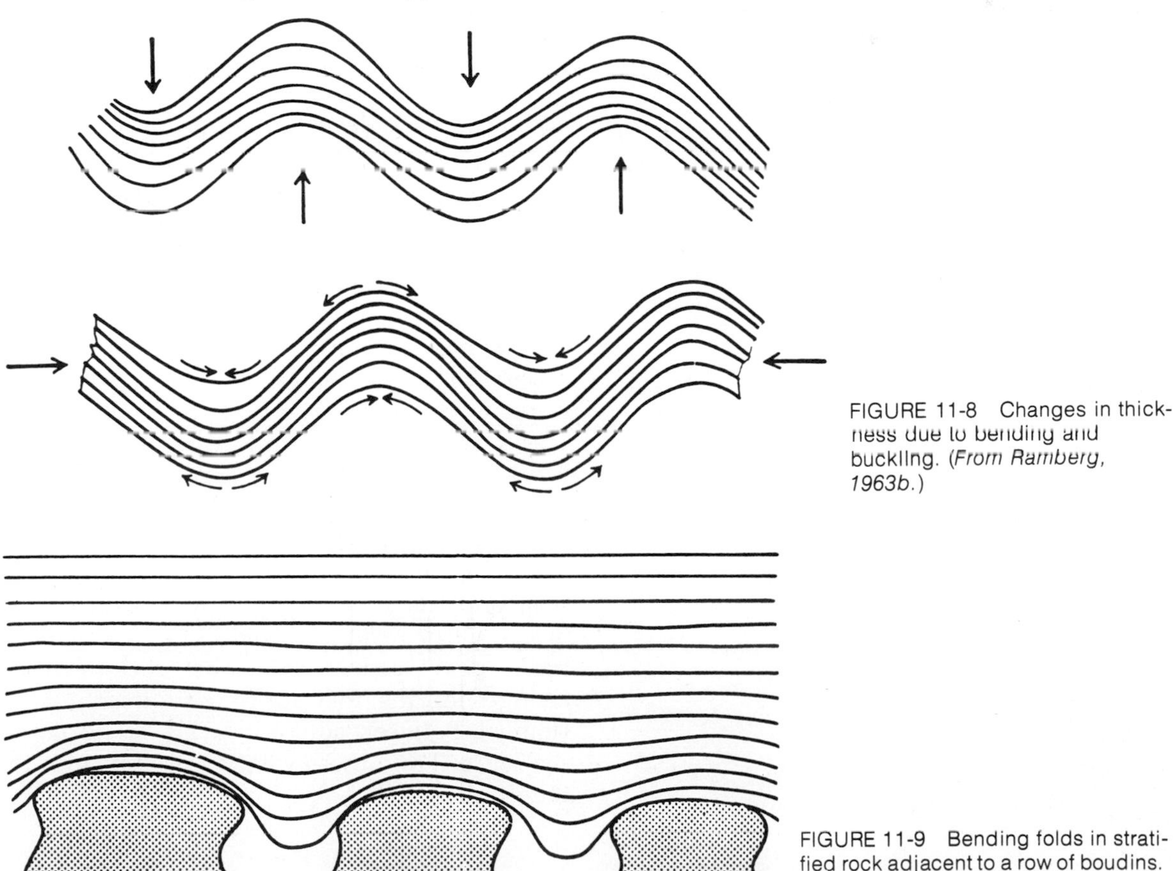

FIGURE 11-8 Changes in thickness due to bending and buckling. (*From Ramberg, 1963b.*)

FIGURE 11-9 Bending folds in stratified rock adjacent to a row of boudins. Only half of the symmetric structure is shown. (*From Ramberg, 1963b.*)

kilometers long and with hundreds of meters of wavelength and wave height (see Chaps. 14 and 22). However, some of these folds exhibit geometry like that of kinks whereas others are clearly disharmonic.

The best examples of similar folds are found in metamorphic rocks, in salt structures, and in clays. They are most likely to form when the rock behaves as a relatively isotropic material as in the ideal model, Fig. 11-14. This is most likely to occur either when the rock is uniform and ductile or when conditions of temperature and confining pressure reduce the contrast between layers. Even under conditions of high-grade regional metamorphism, different lithologies are apt to retain rheological differences, which may explain why so many folds classed as similar do not exhibit the ideal geometry. A fold shape cannot often be precisely traced downward more than a few meters in folds of similar type.

All rocks yield by folding, but the range of conditions and modes of occurrence are great. We may wonder whether we should approach folding processes from the point of view of folds generated in each rock type. Should we treat surficial folds in unconsolidated sediment separately from folds generated in gneisses? The movements and strains which accompany the development of folds are very much alike in folds obviously formed under quite different conditions. This suggests that while many different modes of occurrence may be found, the mechanisms of folding are more limited in number.

FIGURE 11-10 Chevron-fold model in multilayered sequence of layers of different properties and thicknesses. (*From Ramsay, 1974.*)

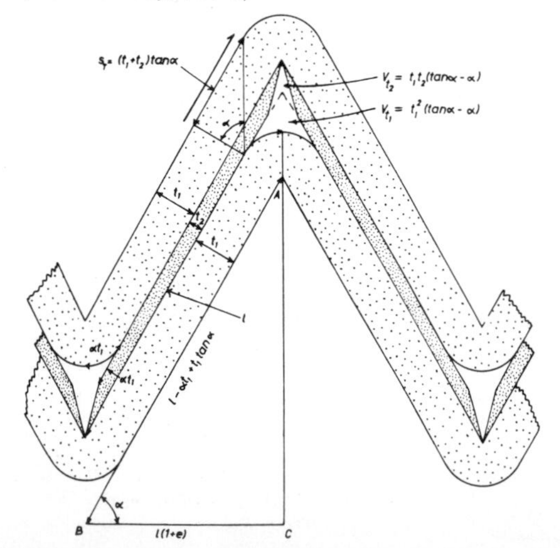

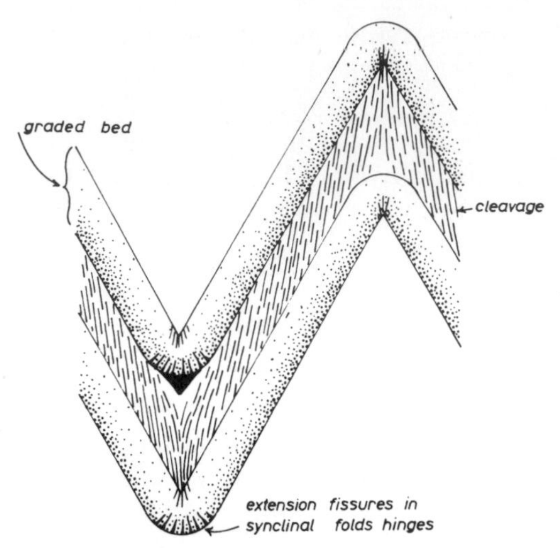

FIGURE 11-11 Appearance of small-scale structures developed in chevron-folded graded beds. (*From Ramsay, 1974.*)

Experiments in which folds are artificially created provide one of the few ways of actually observing the formation of folds. What we normally see is the result of the folding process—the resultant strain in the rock—but the way stresses were applied, and the conditions under which folding took place can only be inferred approximately.

We can observe, describe, and analyze the scale and shape of folds, the types and thicknesses of the rocks involved in the fold, structures developed within the rock layers, changes of the fold form with depth, and changes between adjacent layers in a folded sequence. From these we may obtain clues to the origin of folds.

1. *Thickness of beds involved in folding.* Thickness measured normal to contacts remains essentially constant in some folds; in others, layers are

greatly thickened in the hinge areas of the folds and thinned on the limbs. This can happen in at least two ways. Either material within the layers can move laterally, so material migrates from the limbs toward the hinges, or the effect may be caused by slips in planes parallel to the axial plane, Fig. 11-13, as already described for similar folds.

2. *Thinning in the hinge.* Thinning may occur in bending or as a result of lateral movement of material away from the fold hinge, as may occur when the core of a fold rises into a ductile layer, stretching it over the hinge.

3. *Flattening of folds.* Some departures from ideal thickness can be explained in terms of a flattening of the fold. "Flattening is defined as the process of deformation whereby the original rock shape is plastically changed by compression. This compression results in contraction in a direction parallel to the principal compressive stress and in expansion at right angles to this. . . ." (Ramsay, 1962a). If a concentric fold is flattened as shown in Fig. 11-15, both the thickness and the dip of the bed at any corresponding points on the fold change. The amount of change is systematic over the fold, and the changes are a measure of the flattening. Flattening curves defined by the relationship between angle of dip and thickness have been calculated (Fig. 11-17) for various percentages of flattening. It is possible by making several measurements of the angle and thickness at different points on a profile of a fold to plot the points on the graph, see which curve they fit, and thereby estimate the percentage of flattening.

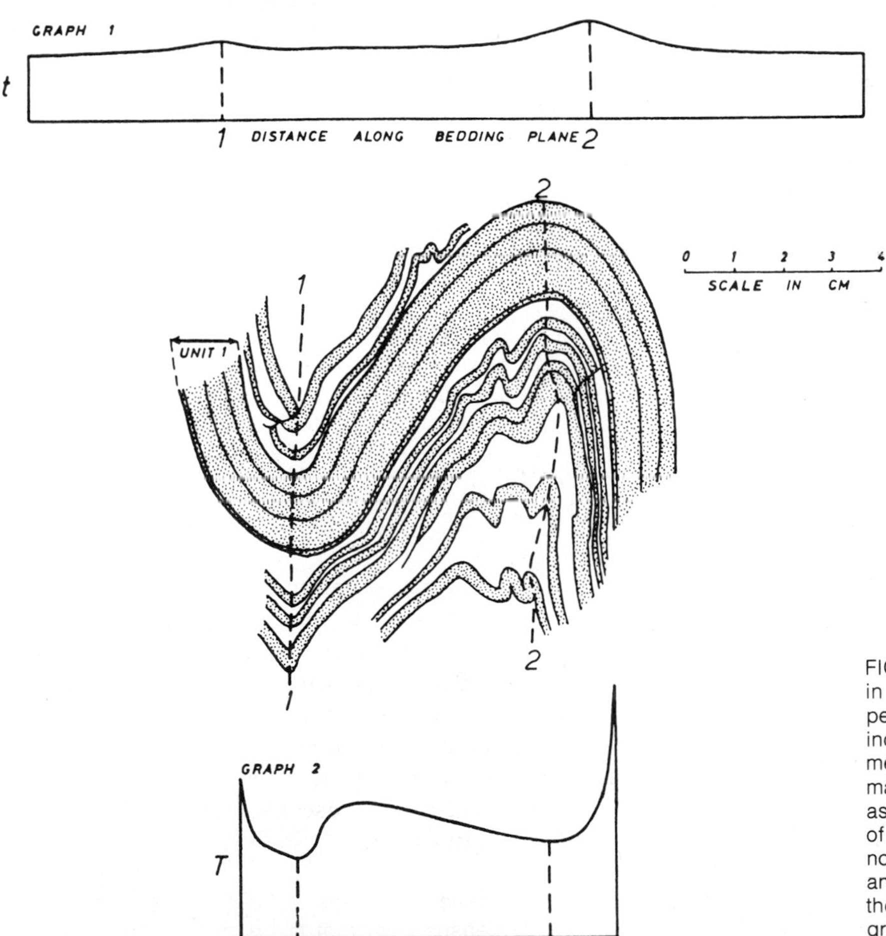

FIGURE 11-12 Flexure folds in siltstone, with coarse competent bands stippled and fine incompetent bands unornamented. The axial planes of the main folds (1 and 2) are shown as dashed lines. Measurements of the thickness of unit 1, t, normal to the bedding planes and its thickness T parallel to the axial plane have been graphically recorded. (*From Ramsay, 1962a.*)

4. *Slippage in folding.* Slip along essentially parallel planes can be identified in many folds. The two common orientations of such slippages are parallel to the bedding surfaces and parallel to the axial surface; however, slips are also found at other angles. In some folds these surfaces are widely enough spaced and have undergone so much movement that they can be identified megascopically (e.g., they cause offsets in bedding surfaces). In other instances the slips are so small that they can be detected only in thin section. In detail, whether viewed through a microscope or by the unaided eye, slight changes in orientation of the material between slips, called *microlithons* by De Sitter (1954), are usually present.

FIGURE 11-13 A group of minor folds of similar type in metamorphosed strata from the Moine series of the Scottish Highlands; changes in thickness T in the various folded rock units. (*From Ramsay, 1962a.*)

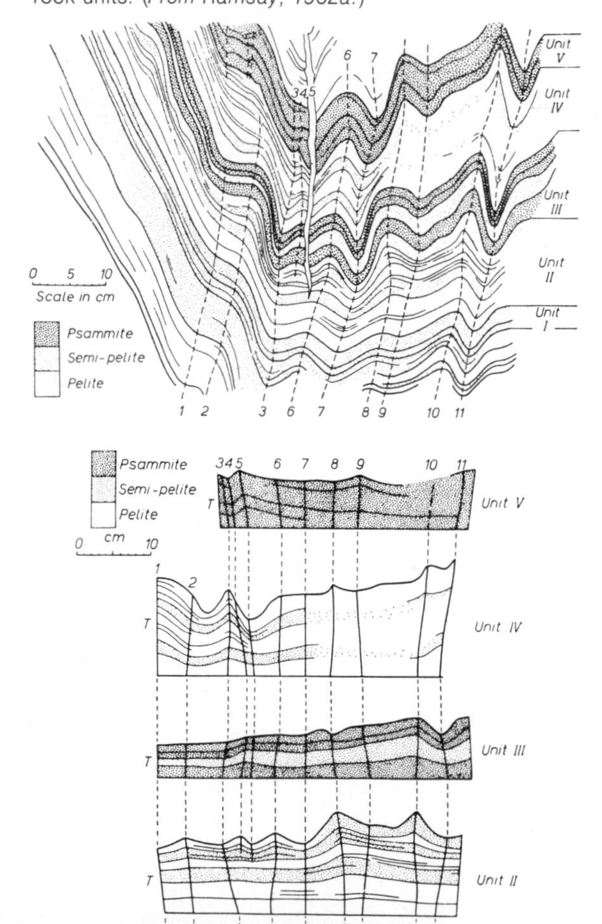

FIGURE 11-14 The formation of rheid folds. (*After Carey, 1954.*)

5. *Deformed primary features of the sediments.* If any feature of known primary shape can be identified in a deformed sequence, a comparison can be drawn between the original and the deformed shape of the feature. Particularly important in this regard are fossils, oolites, and other primary features. Valuable references to the application of this principle will be found for deformed oolites in the work of Ernst Cloos (1947, 1971), Fig. 11-18, for deformed fossils in Bucher (1953), and for deformed primary structures in Ramsay (1961). Two principal types of strains seem to emerge—(1) extension approximately in the plane of the axial surface of the fold in which they occur; (2) strains which vary in amount and orientation from bottom to top of the folded layer in which they occur. The first type of folding results from actual extension taking place within the axial surface either during initial folding or as a result of subsequent flattening. In the second type the variation of the sense of the distortion indicates a change from extension across the hinge on the upper portion of the unit to a neutral surface at some level and then to compression along the lower portion of the unit over the hinge (Fig. 13-10).

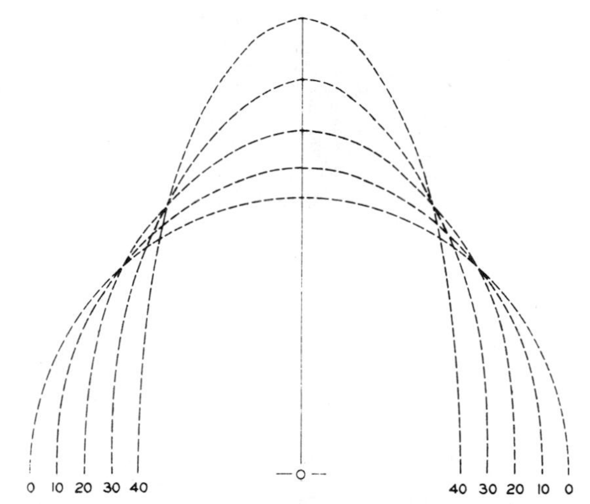

FIGURE 11-15 Semicircular area flattened by 10, 20, 30, and 40 percent. Corresponding points on the curves are indicated by breaks in the line. (*From Williams, 1967.*)

6. *Slickensides.* The striations, smeared crystals, and fibrous aligned mineral growths often confused with slickensides indicate movement between the planes on which they lie; their orientation marks the direction of that movement, and their presence indicates that the surfaces on which they are located were active as boundaries during the folding. Slickensides usually occur either on the bedding surfaces or on fractures cutting across bedding. Slickensides and mineral growths on the bedding and oriented in a plane that is perpendicular to the fold axis (the *ac* plane) suggest movement such as might be expected in a flexed pile of layers.

7. *Saddle reef.* This name is applied to ore deposits formed between bedding surfaces in folds. The shape of the saddle reef (Fig. 11-19) indicates that the rock units on either side moved relative to one another so that the upper unit folded around a smaller radius of curvature than the lower one, allowing a space to form between the layers. This separation between layers may be progressively filled as the folding proceeds, or a void may exist for a time. In either case, the beds involved are active in determining the fold form, and they must maintain sufficient strength to support part of the weight over the saddle reef.

8. *Detachment surfaces or décollement.* The shape of bedding surfaces in parallel folds changes at successively greater depths. One consequence of this change is that at a certain depth the folding dies out. The top surface of the undeformed layers is usually a plane of detachment above which folding has taken place. Material in the bed directly above the detachment moves into the anticlines and away from the synclines (Fig. 11-20). *Décollements* of this type are known in the foreland fold belts of many orogenic belts. Usually the detachment takes place in ductile rock type such as salt, shale, mudstone, or gypsum.

9. *Change in structure with depth.* The upper sur-

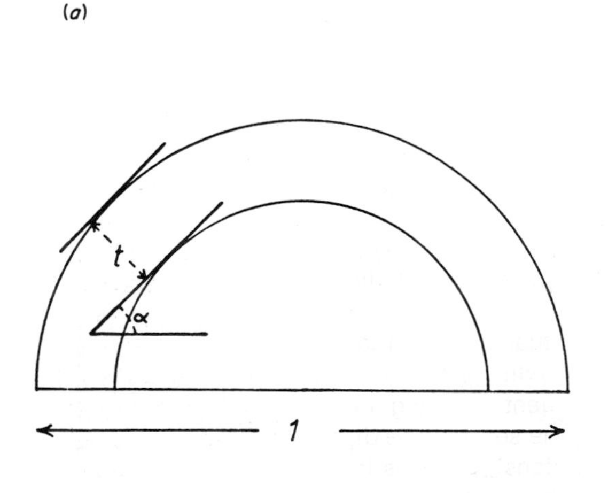

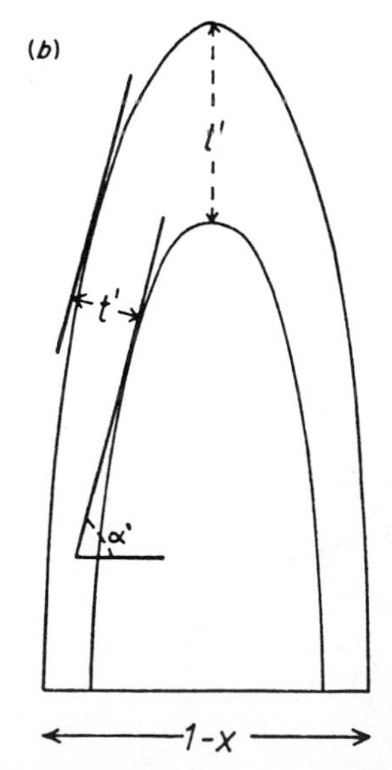

FIGURE 11-16 (*a*) A bed of thickness t folded by a concentric flexure. (*b*) The geometry of the bed is modified by flattening (x): the angle of dip at any point within the flexure is increased (α to α'), and the thickness of the beds is altered (t to t'). (*From Ramsay, 1962a.*)

face of some folded layers show evidence of extention, but the lower surface may show features arising from compression, from forcing material together in the hinge zone. This portion of such a fold is especially prone to complicated structure if a competent unit, one which has tended to maintain its coherence through the deformation, is underlain by ductile units, or especially by thin bedded units of alternating lithology. Small thrust faults and complicated folding may result from movement of material toward the hinge of the larger fold, Fig. 13-2.

10. *Wedging of beds on the limbs and in the hinge of folds.* Wedging (E. Cloos, 1964) or telescoping of beds (Fig. 11-21) arises through failure of a bed along a fracture, followed by displacements such that part of a unit is forced along its former lateral continuation. This is most commonly found in sequences composed of beds of alternating competence such as sandstones or quartzites interbedded with shale. This feature clearly indicates that the sense of movement in the portion of the fold on which it is found is parallel to the bedding surfaces.

11. *Drag.* Large-scale folds are frequently accompanied by smaller folds developed in the more ductile layers as a result of drag produced by relative movements of one layer relative to the adjacent layer. Drag folds occur in concentric folds where the layers are flexed so that couples are set up between layers (Fig. 11-22). Drags associated with parallel folds are asymmetrical toward the fold hinges on either side of the larger concentric fold, and the axial planes of the drags are inclined toward the hinges of the major fold, Fig. 11-23.

Reverse drag in which the asymmetry of the drags is reversed (Fig. 11-24) may occur on folds if the core of an anticline moves upward relative to adjacent synclines. This type of drag is found in salt structures, where plastic cores of domes or anticlines are involved, and it is also likely under metamorphic conditions (Fig. 11-25), where plasticity is more pronounced at depth, or where a clay or shale breaks upward and is injected through more brittle layers. These same conditions are often associated with thinning of beds over an anticline.

FIGURE 11-17 Graphical representation of the change of α' and t' with increasing flattening x; the original thickness of the beds before flattening, t, is taken as unity. (*From Ramsay, 1962a.*)

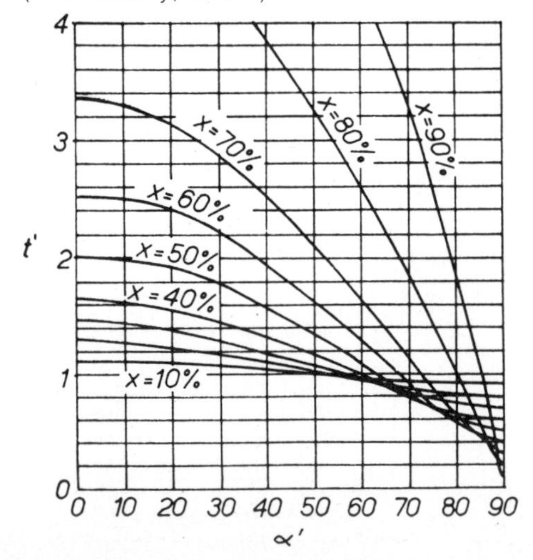

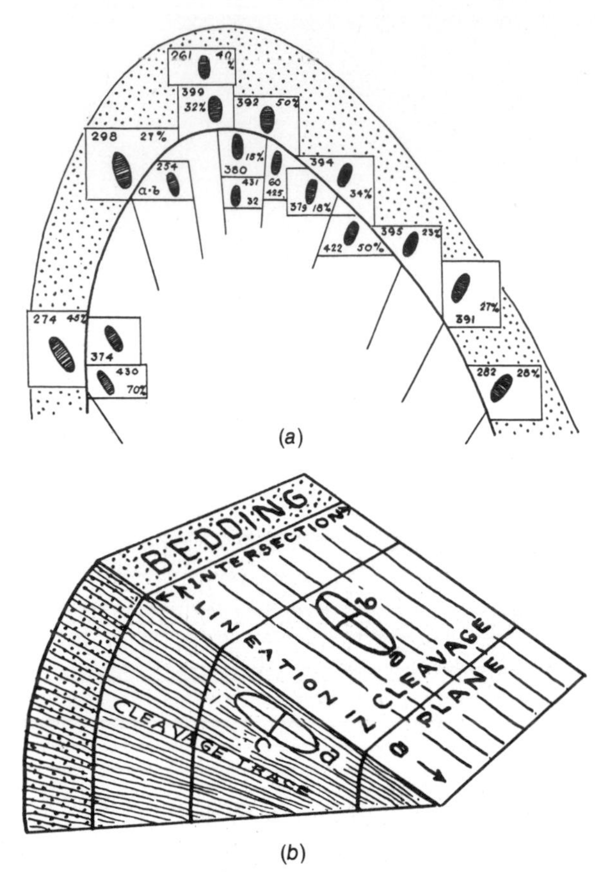

FIGURE 11-18 (*a*) Cross section through a fold showing fanning slip surfaces (cleavage) and deformed oolites, South Mountain, Maryland. (*b*) Deformation plan in South Mountain. Axial portion of overturned fold shows mutual relations of bedding, cleavage, ooid extension, and lineation. (*From E. Cloos, 1947.*)

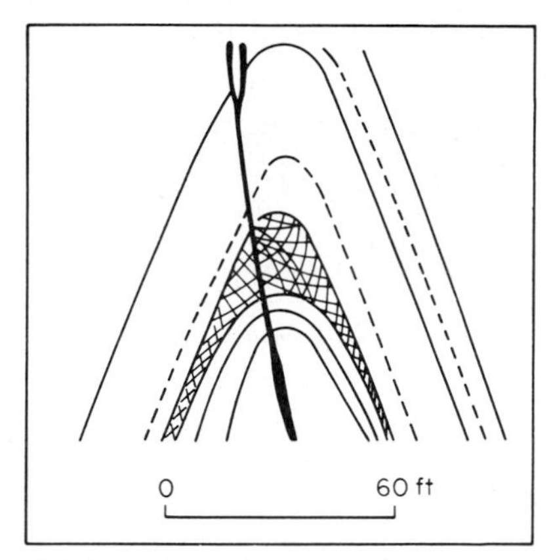

FIGURE 11-19 Saddle reef, Deborah mine, Bendigo, Australia. (*After Stillwell, 1953.*)

12. *Fractures.* The fractures found in the more brittle members of folded sequences often include a set parallel to the fold hinge and disposed in a fan shape. These fractures may completely penetrate a unit, or they may be confined to the outer portion as shown in Fig. 11-11. Some exhibit wedge-shaped fracture fillings. The orientation and shape of the fracture is indicative of extension in the hinge, with extension taking place as shown by the arrows causing failure through tension fracturing. Other fracture sets found on folds are described in Chap. 5.

13. *Angular folds.* A single set of fractures may form in the axial surface of angular folds, a special type of parallel fold in which the limbs are nearly planar and intersect at a sharp angle, Fig. 11-11.

14. *Recrystallization.* Recrystallization of minerals in folds is associated with folds that have formed under high-grade metamorphic conditions. It is often difficult to distinguish premetamorphic and synmetamorphic structures from one another because metamorphism is likely to destroy part of the premetamorphic structures through recrystallization. Postmetamorphic structures can be identified if the mechanism of the folding is incompatible with metamorphic conditions. For example, open fractures and other signs of brittle behavior are incompatible with high-grade metamorphic conditions.

The preceding observations may be summarized as follows. Structures, notably slip surfaces, tend to form across bedding planes where the bedding has not been active in controlling development of the fold. This condition is found when the rock mass behaved as a relatively isotropic material during folding. Isotropic conditions may be expected in thick masses of salt or clay and in any rock type, even in sections of greatly different lithologies, under high-grade metamorphic conditions. It should be emphasized that the condition of isotropy is not fulfilled if planes of weakness exist in the rock mass. Thus a succession of beds of a single rock type will not behave as an isotropic mass if the bedding

FIGURE 11-20 Disharmonic folds with bedding thrusts that rise into the crest of the anticlinal folds above a *décollement* zone. (*From Fitzgerald and Braun, 1965.*)

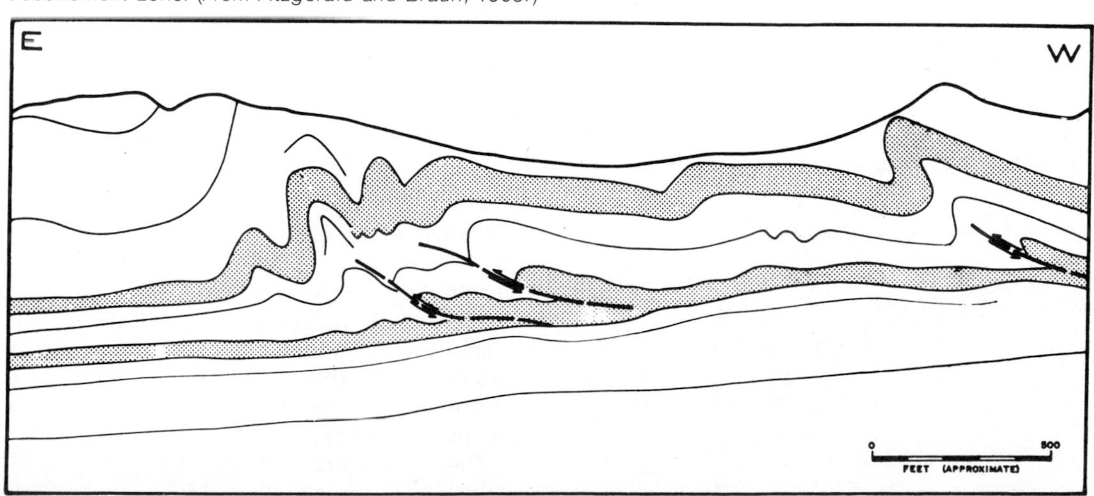

FIGURE 11-21 Folded wedge in the limb of a fold. The sandstone bed is contained in slates. (*From Marcos and Arboleya, 1975.*)

planes are open or if the strength of the rock across these planes is greatly different from that within the layers.

When layering controls fold development, movements parallel to the layering are most prominent. Wedging and drag are likely to occur when the section contains rocks of greatly different physical character. Systematic changes occur within layers from top to bottom. Fractures form in brittle materials over the hinge; thicknesses of more brittle layers tend to remain nearly constant; thickness in ductile layers varies with migration of material toward fold hinges; saddle reets may occur; marked changes in shape of the fold with depth are to be expected; and detachment surfaces are likely at depth.

Ptygmatic Folds

The name *ptygmatic* is applied to a variety of folds which have in common primarily a meandering cross-sectional pattern. Sederholm (1907, 1913) first described them as resulting from movements in rock mobilized by partial melting. They are found in metamorphic or "granitized" complexes in which at least partial remobilization has occurred. Usually ptygmatic folds are in veins composed largely of quartz and feldspar which do not show internal evidence of brittle fracture or cataclasis, Fig. 11-26.

The question whether all ptygmatic veins originate as a result of the same structural and metamorphic conditions or whether they form through a variety of circumstances is a difficult one to answer. Several ideas with corroborating field observations, theory, and experimental results have been advanced to explain the mode of origin. Two main classes of mechanism have been suggested. The first attributes the folds to a passive response of a vein to movements within the country rock at some time after the vein has been emplaced; in this case the folding is secondary. In the second class, folds are attributed to processes acting during emplacement of the vein.

Read (1928) states that the "tortuous form results from the resistance to plane fissuring

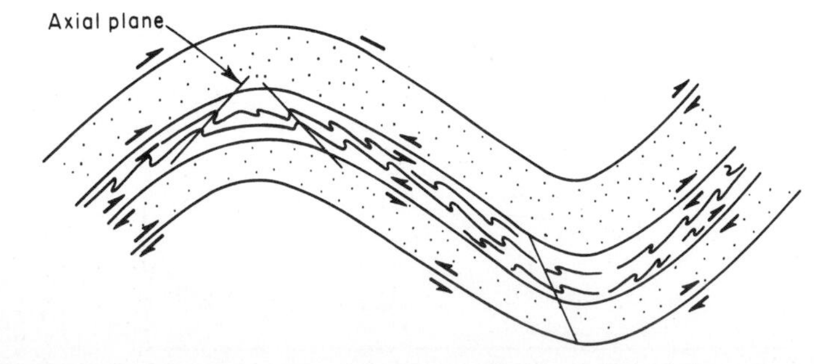

FIGURE 11-22 Drag folds in concentrically folded sequence. The arrows show bedding slip. The axial surface of the drags are inclined toward fold hinges.

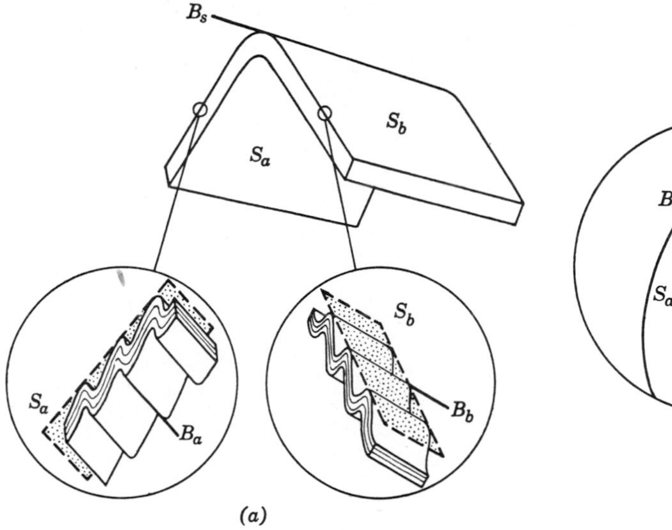

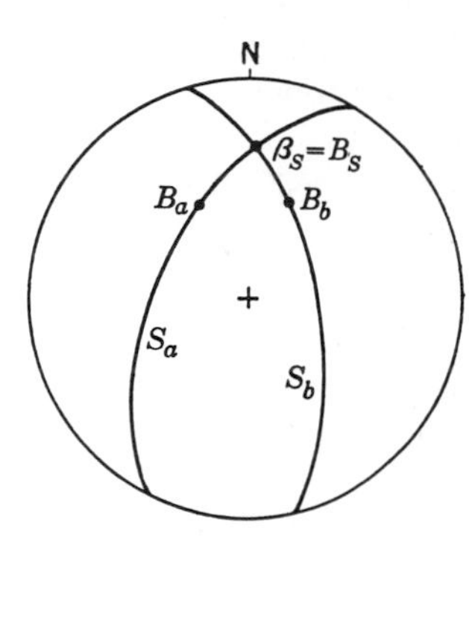

FIGURE 11-23 Use of enveloping surface of mesoscopic folds to determine axis of macroscopic fold. (*From Turner and Weiss, 1963.*)

of the country-rock . . . the veins have their original form; they were never plane . . ." and they are "true igneous injections, showing no metamorphic characters." Thus he concludes that the essential physical condition for the formation of ptygmatically folded veins is the "injection into structureless country-rock not readily giving plane fissures."

Buddington (1939) found two types of ptygmatic veins in his study of the Adirondacks—one formed as described by Read and the other conformable with plications in the containing schist, for which he concluded that "the injection of the pegmatite occurred in folded rocks during a period of active deformation. Plastic flowage of country-rock and both plastic and magmatic flowage of pegmatite is involved."

FIGURE 11-24 Reverse drag resulting from the upward movement of a mobile core (e.g., salt, clay, ductile rock).

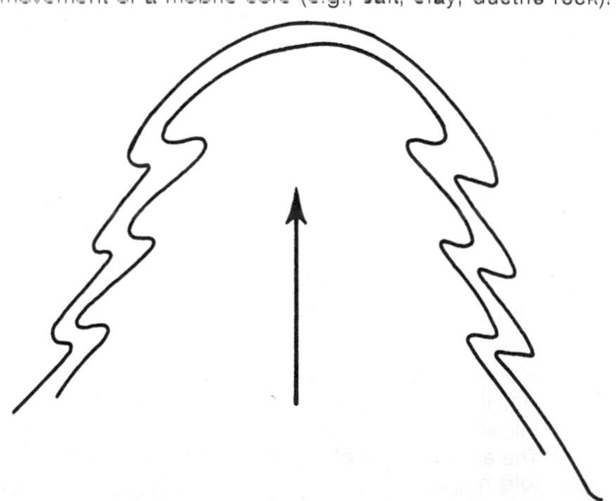

Kuenen (1938), Ramberg (1959), and others have held the view that ptygmatic folds are secondary and develop under compressive stresses. Ramberg (1959, p. 100) states his reasoning as follows:

This structure . . . is caused by a component of shortening in host rocks parallel to originally more or less planar veins. The vein can only adjust to plastic compressive strain in the host rock by a folding mechanism. . . . Plastic compression along some directions in rock complexes must be associated with simultaneous extensions in directions that make large angles to the compression, provided that volume remains essentially constant during strain. Relatively competent sheets of rocks which make large angles with direction of maximum compression will therefore develop tension fractures or necked-down regions. One should consequently expect boudinage and pinch-and-swell structures to be associated with ptygmatic structures in the field as unseparably as extensive strain is associated with compressive strain in plastic deformation at constant volume.

FOLD CLASSIFICATION

Ideally, terms which describe shape should not be used to connote the mechanism of folding, but some of them have become so closely associated with mechanisms of folding that they are used nearly synonymously. For example, the terms *parallel folding* and *concentric folding* connote folding by flexure in which the beds exercise a controlling influence on the development of the folds. *Similar folding* is often taken to mean folding in which the development of and movement along slip surfaces which cut across the beds are the dominant processes of folding. Such folds are also known as *shear folds* because the mechanism of slippage involves the development of couples which bring about shear or slip along closely spaced cleavages. However, many folds of similar geometry do not show such cleavages or slip surfaces, and frequently, slip surfaces are well developed in incompetent layers but do not cross competent layers. The mechanism seems to consist of buckling of the competent layers, with development of slip oriented to bring about thickening in the hinge of incompetent layers.

A third category of folds forms under conditions which permit flow of the rock, but flowage is not laminar and is not accomplished by movements on closely spaced planar slip surfaces. Highly irregular forms called *irregular flow folds* may resemble the flow of viscous substances.

To avoid some of the problems involved in the use of such genetic terms as shear, concentric, flow, and ptygmatic, Donath and Parker (1964) proposed a classification of folds based on mechanisms of folding and consisting of three major classes—flexural, passive, and quasi-flexural. The first two of these classes are subdivided into flexural slip and flow and passive slip and flow. The basic dis-

FIGURE 11-26 Ptygmatic fold formed in a quartz-feldspar pegmatite vein in granite.

FIGURE 11-25 Reverse drag. Simplified from a cross section across the Connecticut Valley synclinorium on the Geologic Map of Vermont, 1961. (*After Doll and others, 1961.*)

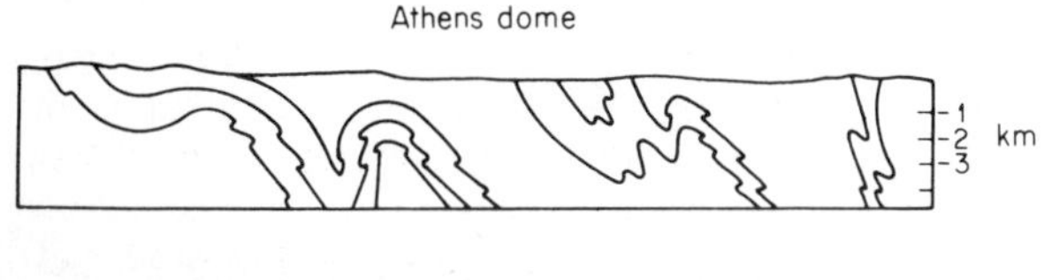

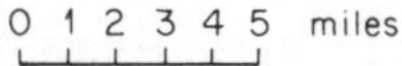

tinction between flexural and passive folds is whether or not the bedding actively controls the development of the folds. The types within each of these categories are then distinguished by whether the basic mechanism, within layers for flexural folds or through layers for passive folds, was slip along macroscopically discontinuous surfaces or flow movements that do not cause permanent loss of cohesion. The third class, quasi-flexural, is described as follows.

Under certain conditions individual layers within a folded sequence may be flexed in response to passive behavior in the associated rocks. This relationship represents a gradational class of folding, particularly characteristic in rocks of moderate to high ductility, that is here called *quasi-flexural;* the geometry and more obvious features of the fold are flexural in general aspect, but the overall behavior is predominantly passive.*

Physical Basis for the Donath-Parker Fold Classification

The Donath-Parker (1964) fold classification is based on the idea that variation in the response of layered rocks to deformation is closely related to the average ductility of the folded rock and the degree of ductility contrast between layers.

Ductility is a function of the physical conditions during folding. These conditions include the effects of temperature, confining pressure, the presence or absence of fluids, the rate of strain, and the manner of stress application. If the ductility of the folded sequence is uniformly high, the individual layers should not play a major role in determining the deformation (*passive folding*), but if the ductility of layers varies greatly, then layers of low ductility may deform by flexure, whereas those of high ductility accommodate themselves within the space available between the more competent layers. Fields of different folding response can be identified on a plot of ductility contrast against mean ductility, Fig. 11-27.

* Donath and Parker (1964).

1. A sequence of low mean ductility deforms through flexural slip.

2. A sequence of high mean ductility deforms by passive folding if the ductility contrast is low, but if the ductility contrast is high, then flexure of the low ductility members may dominate, producing quasi-flexural folds.

3. If *ductility contrast* is low, the fold type depends on the ductility of the sequence.

SUPERPOSED FOLDS

Recognition of many superposed folds has been one result of the application of the new geometrical approaches to fold analysis, especially in metamorphic rocks. These superposed structural features are sometimes the result of two distinctly different events. However, superposed deformations under similar conditions are common and may represent two stages in a single major tectonic event.

The presence of two fold systems of different character, different geometry, or different trend does not necessarily mean that two separate deformational phases took place. The two may form simultaneously. Drag folds may form on the limbs of larger folds during concentric folding, and reverse drags are often associated with diapiric structures. Superposed folds may also result from a single deformational event if the deformation is nonaffine—that is, if deformation proceeds in such a way that lines and planes that existed before deformation are transformed into new curved lines and planes. Nonaffine deformation may occur in a sedimentary basin deformed by lateral pressure where an obstacle in the basin could cause retardation and breaking up the orderly fold pattern, or as a result of lateral pressures acting from more than one direction simultaneously. Cross folding can be caused in metamorphic rocks by variations in lithologies, thermal gradients, or as a result of the presence of fluids which cause local variations in the ductility of the rock. No doubt many superposed folds are a product of changes in the direction of movement in response to changes in the orientation of the stress field, either locally or regionally, during a single period of deformation.

If the type of folding and the orientation of directions of movement alone are considered, a large number of possible types of superposition are possible. A partial list will serve to illustrate:

1. Similar-type folds superposed on similar folds, with common orientation of slip planes and slip direction
2. Similar folds superposed on older similar folds, with different orientation of slip planes
3. Concentric folds superposed on concentric folds of the same and different orientation of axial surfaces
4. Similar folds superposed on concentric folds, with the same and different orientation of avial surfaces
5. Concentric folds superposed on similar folds
6. Kink folds superposed on similar or concentric folds, etc.

It is not difficult to recognize superposed deformation, but it may be difficult to establish the sequence of events or to distinguish the products of two separated phases of deformation from those of a single event. Superposed deformations may be recognized where folds of different trends occur in the same area. Superposed deformation may be detected from study of point diagrams of bedding, foliation, lineations, and fold-axis data. The pattern of points representing foliations and lineations within a field of superposed folds depends on geometry, size, and orientations of the two fold systems.

As an example of superposed fold systems, consider the earlier generation of cylindrical folds (F_1) with horizontal axes and vertical axial surfaces S_B, which is shown distorted by a second-generation fold system (F_2) in Fig. 11-28. The original slip surfaces S, the line of

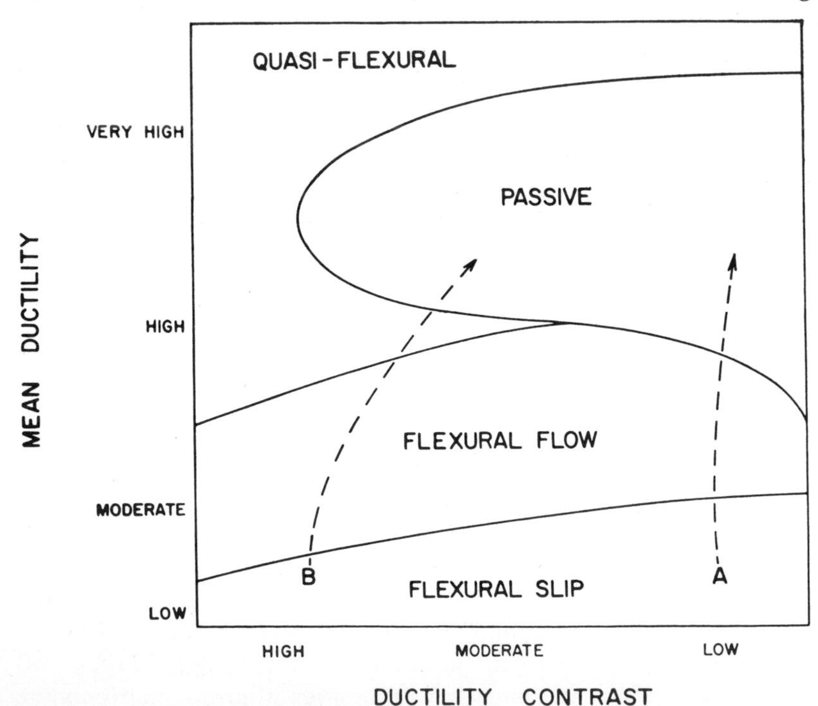

FIGURE 11-27 Fields of folding related to mean ductility and ductility contrast. (*From Donath and Parker, 1964.*)

their intersection β, and the fold axes of the first-generation folds are shown in a stereographic diagram. The axial surfaces S' of the second generation of folds are subparallel, but the resulting axes of the second-generation folds B' do not intersect in a point. They do, however, lie within the axial surfaces S' of the second-generation folds, and they are oriented parallel to the intersection of S and S' in any given part of the field of folding. The axes of the second-generation folds tend to be concentrated in two clusters, Fig. 11-28, because the limbs of the first-generation folds had been more planar. The axes of the first-generation folds are so dispersed as a result of the rotation which took place during the second folding that they do not show up at all in the plot.

The geometry of superposed fold systems is reviewed in Weiss and McIntyre (1957), Weiss (1959), Turner and Weiss (1963), Whitten (1966), and Ramsay (1967).

Determination of Sequence in Superposed Deformations

Evidence of sequence in superposed folds may be found in such key observations as the following:

1. If small-scale folds occurring on larger folds have similar geometrics (axes are parallel, etc.), their development was probably synchronous.

2. If slip surfaces are folded into angular folds (i.e., kink folds) the angular folds are younger.

3. If similar folds are developed through a metamorphic rock before a later phase of concentric folding, the direction of asymmetry of the small folds will not change on passing over the hinge of the later folds, but the direction of asymmetry will reverse over hinges of major folds if the minor folds are related to drag movements accompanying formation of the large folds.

4. Slip surfaces are very important indicators. If two generations of similar folds formed by movement on slip surfaces are superposed, the slip surfaces of the older movement will be folded, and the younger slips will be straight. If slip surfaces are parallel to the form of a concentric fold, then either the slips are earlier or they developed as part of the movement accompanying development of the concentric fold (see item 3). If similar folds follow concentric folds, the slip surfaces are unfolded.

5. If metamorphism is involved at some stage, the folds may be identified as pre-, syn-, or postmetamorphic, and other conventional means of establishing sequence, such as intrusions or vein filling, may be useful.

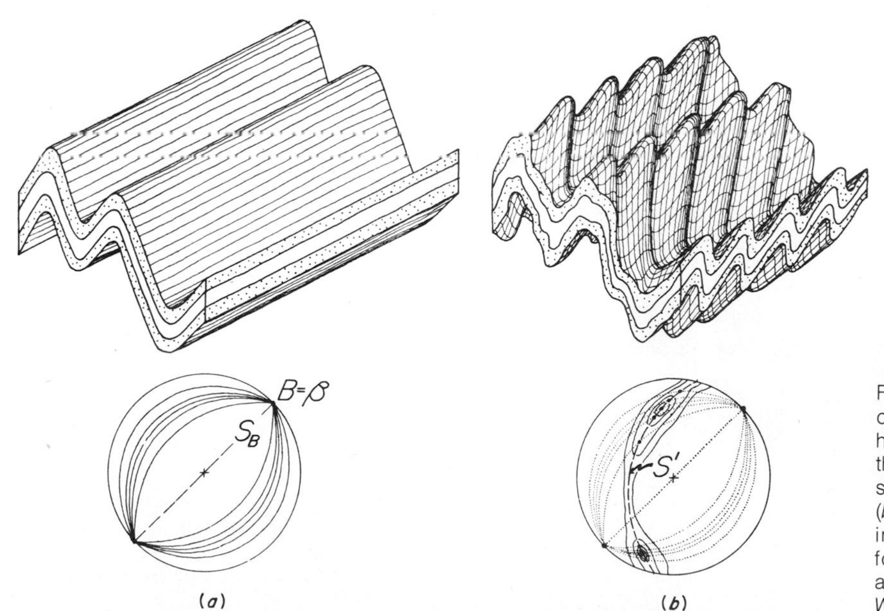

FIGURE 11-28 (*a*) A set of cylindrical folds with horizontal axes. The axes of these folds B are shown in a stereographic projection. (*b*) The result of superimposing a second set of folds with axes that trend across the first set. (*From Weiss, 1959.*)

Carey (1962) analyzed the superposition of one type of similar fold on another. Similar folds characterized by slip on closely spaced slip surfaces are the most common type of fold in metamorphic rocks when they are behaving as nearly isotropic masses. A set of similar folds is shown modified by later slip movement, Fig. 11-30. The two sets of folds have the same strike but different attitudes. A cross-sectional sketch is shown at the top. In making restoration the individual layers are identified and traced out, and the axes of symmetry of the

FIGURE 11-29 Successive stages of the development of folded folds in unsteady flow by a change in the direction of movement. (*From Wynne-Edwards, 1963.*)

FIGURE 11-30 Fold pattern (top, right) formed by the superposition of two generations of folds produced by slip at right angles to marker beds, followed by slip at right angles to first fold axes. (*From Carey, 1962.*)

most obvious folds are determined. One set of the axes of symmetry is subparallel (about vertical in the drawing), and the second set is a sinuous set extending across the diagram. By straightening the axes of symmetry of the folded folds (accomplished by moving points parallel to the axial surfaces of the second set of axes of symmetry), the effects of the second movement are removed; and by doing the same along the now restored first set of axes of symmetry, the bedding is restored to its original form.

GLOSSARY OF FOLD NOMENCLATURE

Definitions of terms most frequently used to describe folds are cited here along with some genetic terms. See Fleuty (1964) for a more comprehensive discussion of fold nomenclature.

I. Parts of a Fold

Apex: The point representing the hinge in any cross section. (Syn.: **vertex, apical region.**)

Apical plane: Syn. of **axial surface.**

Apical region: Syn. of **apex.**

Axial surface: The locus of the hinges of all beds (*s* surfaces) forming the fold. (Clark and McIntyre, 1951). [Syn.: **apical plane** (Hills, 1953); also called **axial plane**, although it is rarely plane.] Axial plane is used in several somewhat different ways, including "a plane that intersects the crest or trough in such a manner that the limbs or sides of the fold are more or less symmetrically arranged with reference to it." (Leith, 1913).

Axial trace: The line formed by the intersection of the axial surface with any other surface or plane (commonly the ground).

Axis: Multiple definitions are in use. (1) Used in the same sense as **hinge.** (2) The line of intersection of the axial surface and any bed or *s* surface. (Billings, 1954.*) (3) The nearest approximation to the line which, moved parallel to itself, generates the folded surface. (McIntyre, 1950.)

Crest: The highest point in a given *s* surface on an anticline in a given cross section.

Crest line: The line connecting the highest points in an infinite number of cross sections.

Crestal surface: A plane formed by joining the crests of all beds in a single fold.

Core: The inner part of a fold; the part located nearest to the axial surface.

Curvature: A measure of the degree of bending of a line or surface. Curvature of a portion of a line is defined by the radius of a circle which would coincide with that portion of the line. Any point on an *s* surface has a curvature which is defined by a maximum and a minimum radius of curvature. The two are equal in the case of spheres, and both are infinite for planes.

Envelope: The outer part of a fold; the part located farthest from the axial surface.

Fold. A deformation of a preexisting rock surface to a continuous curved surface that is convex in a single sense. (Carey, 1962.)

Hinge: A continuous line on a folded *s* surface connecting points of maximum curvature (minimum radius of curvature). The amount of curvature may vary along the hinge. (Syn.: **hinge line, fold hinge, axis,** or **fold axis.**)

Hinge line: Syn. of **hinge.**

Limb: The portion of a folded *s* surface located between two adjacent hinges. If the two limbs are not of equal length they may be called the *short limb* and the *long limb*. (White and Jahns, 1950.)

Normal cross section: Syn. of **profile plane.**

Profile plane: A cross section of a fold drawn normal to the hinge (McIntyre, 1950.) (Syn.: **normal cross section.**)

Trough: The lowest point in a given *s* surface on an anticline in a given cross section.

Trough line: A line connecting the lowest points in an infinite number of cross sections of a syncline.

* Indicates definitions based on the Glossary of geology and related sciences, published by the American Geological Institute.

Trough surface: A plane formed by joining the crests of all beds in a syncline.
Vertex: Syn. of **apex**.

II. Fold Dimensions

Height: The vertical distance between the crest and the trough of a single marker horizon.
Interlimb angle: The minimum angle between the limbs as measured in the profile plane. (Dahlstrom, 1954.)
Wavelength: The distance between two adjacent anticlinal or synclinal axes measured normal to those axes.

III. Shape of a Fold in Profile Plane (Right Normal Cross Section)

A. SYMMETRY

Asymmetrical: A fold in which the limbs are not symmetrically disposed about the axial surface. (Stoces and White, 1935.)
Symmetrical: A fold in which the limbs are symmetrically disposed about the axial surface. (Stoces and White, 1935.)

B. ATTITUDE OF AXIAL SURFACE

Inclined fold: A fold with an inclined axial surface.
Overfold: Syn. of **overturned fold**.
Overthrown fold: Syn. of **overturned fold**.
Overturned fold: A fold in which the beds on one limb are overturned. (Syn.: **overfold, overthrown fold**.)
Recumbent fold: A fold in which the axial surface is more or less horizontal.
Upright fold: A fold with a vertical (or near-vertical) axial plane surface. (Structural Committee.*)

C. DIVERGENCE OF LIMBS

The terms *gentle, broad, open, close, closed, tight,* and *isoclinal* have all been used to describe various interlimb angles. (Fleuty, 1964.)
Isoclinal: A fold the limbs of which have parallel dips.

D. FOLD SHAPE

Acute: Syn. of **angular**.
Accordion: Syn. of **chevron**.
Angular: Having an angle or angles; forming an angle; sharp-cornered. Fold hinges are angular when the limbs are plane. (Syn.: **sharp, acute**.)
Anticline: A fold with older rocks in its core. (Bailey and McCallien, 1937, *in* Fleuty, 1964.)
Antiform: When beds are arched so as to incline away from each other, they form an antiform. (Bailey and McCallien, 1937.)
Box: An angular fold in which the crest is flat and the limbs are steep. The fold resembles a box in profile.
Chevron: Angular folds with plane or nearly plane limbs. When distinctions among these terms are made, chevron folds have limbs of approximately equal length, alternate limbs parallel; zigzag folds have limbs of unequal length; kink folds are very small. (Syn.: **zigzag, accordion**.)
Concentric: Syn. of **parallel fold**.
Conical fold: A conical surface is a surface generated by a straight line that always passes through a fixed point and always touches a fixed curve. (Stockwell, 1950.)
Conjugate folds: Sets of paired, reversed folds whose axial planes are inclined toward one another. (Ramsay, 1962b.)
Curvilinear: Consisting of, or bounded by, curved lines. Fold hinges are described as curvilinear or rounded, meaning that they are connected by a surface with nearly constant radius of curvature. (Fleuty, 1964.)
Cylindrical fold: A fold the poles of whose bedding planes when plotted on a stereogram lie close to a great circle. (McIntyre, 1950.*)
Disharmonic fold: A fold in which pronounced changes in shape occur from layer to layer.
Fan: A curvilinear fold which resembles a fan in profile.
Irregular fold: A polyclinal fold characterized by irregularity of axial plane(s) and discontinuities and rapid variation in the thickness

of bands. (Fleuty, 1964.) This term was proposed as a replacement for such terms as **flow fold, wild fold, turbulent flow fold.**

Parallel fold: A fold in which each bed maintains the same thickness (assuming it was initially of uniform thickness) throughout all parts of the fold. (Structural Committee.*) (Syn.: **concentric.**)

Sharp: Syn. of **angular.**

Similar fold: A type of folding in which each successively lower bed shows the same geometrical form as the bed above. The thickness of beds measured parallel to the axial surface is constant.

Syncline: A fold with younger rocks in its core. (Bailey and McCallien, 1937.)

Synform: A fold in rocks in which the strata dip inward from both sides toward the axis.

Zigzag: Syn. of **chevron.**

E. FLEXURES

Homoclines: "A general name for any block of bedded rocks all dipping in the same direction. It may be a monocline, an isocline, a tilted fault block, or one limb of an anticline or syncline." (Daly, 1915.*)

Monocline: (1) Beds inclined in a single direction. (Chamberlin and Salisbury, 1927.) (2) A steplike bend in otherwise horizontal or gently dipping beds. (Lahee, 1952.)

Structural terrace: An area where beds which generally have a regional dip are locally horizontal.

IV. Terms Based on Map and Profile Shape

Basin (Structural): A depressed area with the strata dipping inward. (Emmons, 1863.*)

Closure: In an anticline, a dome, or a swell, the vertical height between the highest point on a given *s* surface and the lowest horizontal plane which gives a closed trace for that *s* surface.

Dome (Structural): A roughly symmetrical upfold, the beds dipping in all directions, more or less equally from a point. (Nevins, 1936.*)

Sag: (1) A broad gentle shallow basin, e.g., the Michigan and Illinois basins (Bucher, 1933.) (2) Downwarping of beds near a fault that is opposite of that of frictional drag. (Structural Committee.*)

Swell: (1) A low dome or quaquaversal anticline of considerable areal extent. (La Forge, 1920.*) (2) An essentially equidimensional uplift without connotation of size or origin. (Bucher, 1933.)

V. Relations among Folds

Congruous folds: Folds which conform with each other in attitude of axial surface and hinge, or vary systematically in attitudes of these structures as an original geometrical feature of the fold system. (Fleuty, 1964.)

Cross folds: Informally used for folds with axial surface or axial directions, or both, at a high angle to the direction of comparable structures related to the main regional folds. (Fleuty, 1964.) (Syn.: **transverse folds, oblique folds.**)

Incongruous folds: Folds which do not conform with each other in the ways mentioned above. (Fleuty, 1964.)

Oblique folds: Syn. of **cross folds.**

Refolded fold: Syn. of **superposed folds.**

Superposed folds: Folds formed in a rock that has been previously folded. (Syn.: **refolded fold.**)

Transverse Fold: Syn. of **cross folds.**

VI. Genetic Classification of Fold Types

Diapiric fold: A piercement structure in which a fold formed in a highly mobile material breaks through, or pierces, less mobile layers.

Drag: Generally a small-scale asymmetric fold formed as a result of a localized couple caused by the relative movement of layers in a larger flexural fold, or along a fault.

Flexural folds: Where flow or slip is restricted by layer boundaries, the layering exercises an active control on the deformation, and the resulting folds represent a true bending of layers. (Donath and Parker, 1963.)
Flow fold: A fold formed by rock flowage in which the rocks behave as fluids.
Passive folds: Where flow or slip crosses the layer boundaries, the layering exercises little or no control on the deformation (the layering is passive), and layer boundaries serve as markers, parts of which are displaced relative to other parts to produce an apparent bending. (Donath and Parker, 1963.)
Quasi-flexural: A gradational class of folding, particularly characteristic in rocks of moderate to high ductility, in which individual layers within a folded sequence are flexed in response to passive behavior in the associated rocks. (Donath and Parker, 1963.)
Rheid folding: Flow folding in which the rocks have remained solid or crystalline, but have deformed as fluids because the duration of the loading was much longer than the relevant deformation time constant (i.e., the rheidity). (Carey, 1954.)
Shear fold: A fold formed as a result of the minute displacement of beds along closely spaced fractures or cleavage planes. (Billings, 1954.*) (Syn.: **slip fold**.)
Slip fold: Syn. of **shear fold**.
Supratenuous: (1) A fold in which the beds thicken toward the syncline because the basin subsided during sedimentation. (After Nevin, 1931.*) (2) A fold which shows a thinning of the formations upward above the crest of the fold. (Hills, 1953.*)

VII. Fold Systems and Features of Folded Systems

Anticlinorium: A series of anticlines and synclines so arranged structurally that together they form a general arch or anticline. (Dana, 1873.*) (Syn.: **geanticline**.)
Culmination: A portion of a fold system, generally more or less at right angles to the folds, away from which the folds plunge. (Structural Committee.*)
En échelon folds: Parallel structural features that are offset as are the edges of shingles on a roof when viewed from the side. (Structural Committee.*)
Fold generation: A group of cognate folds. (Turner and Weiss, 1963.)
Fold system: A group of folds which occur together.
Geanticline: A broad uplift, generally referring to the land mass from which sediments in a geosyncline are derived. (Structural Committee.*)
Geosyncline: A surface of regional extent subsiding through a long time while contained sedimentary and volcanic rocks are accumulating; great thickness of these rocks are almost invariably the evidence of subsidence, but not a necessary requisite. Geosynclines are prevalently linear, but nonlinear depressions can have properties that are essentially geosynclinal. (Kay, 1951.)
Orocline: A fold where the deformed unit is the orogen itself. (Carey, 1955.)
Recess: The part of an orogenic belt where the axial traces of the folds are concave toward the outer part of the belt. (Billings, 1954.*)
Salient: That part of an orogenic belt that is convex toward the foreland, i.e., is concave toward the orogenic belt. (Billings, 1954.*)
Synclinorium: A compound syncline; a closely folded belt, the broad general structure of which is synclinal. (Structural Committee.*)

VIII. Nappe Structures

Allochthonous: A term originated by Gümbel and applied to rocks of which the dominant constituents have not been formed in situ. (Holmes, 1920.*) The term is applied to rock masses which have moved considerably from their point of origin.
Autochthon: In Alpine geology, a succession of beds that have been moved comparatively

little from their original site of formation, although they may be intensely folded and faulted. (Heritsch, 1929.)
Nappe: A sheet-like mass of rock transported laterally for great distances by recumbent folding and/or thrusting. (*Note:* Usage of the term is highly varied; see text.)
Para-autochthonous: In Alpine geology, a term applied to folds and nappe structures which can be connected by their facies and tectonic features with the **autochthon,** q.v. (Heritsch, 1929.)
Root: (1) The core of a geanticline within a geosyncline, which, after the forward drive of the geosynclinal sediments, became the recumbent fold or nappe (Heritsch, 1929.) (2) The back-remaining, steep part of a nappe. (Heritsch, 1923.)

REFERENCES

Bail, T. K., 1960, A petrofabric analysis of a fold: Am. Jour. Sci., v. 258, no. 4, p. 274–281.
Bayly, M. B., 1971, Similar folds, buckling and great-circle patterns: Jour. Geology, v. 79, no. 1, p. 110–118.
Billings, M. P., 1972, Structural geology, 3d ed.: Englewood Cliffs, N.J., Prentice-Hall.
Bucher, W. H., 1953, Fossils in metamorphic rocks: Geol. Soc. America Bull., v. 64, p. 275–300.
Buddington, A. F., 1939, Adirondack igneous rocks and their metamorphism: Geol. Soc. America Mem. 7.
Busk, H. G., 1929, Earth flexures: London, Cambridge Univ.
Cambell, J. D., 1958, En échelon folding: Econ. Geology, v. 53, p. 448–472.
Campbell, J. W., 1951, Some aspects of rock folding by shear deformation: Am. Jour. Sci., v. 249, p. 625–639.
Carey, S. W., 1954, The Rheid concept in geotectonics: Geol. Soc. Australia Jour., v. 1, p. 67–117.
——— 1962, Folding: Alberta Soc. Petroleum Geologists Jour., v. 10.
Chaudhuri, A. K., 1972, Concise description of fold orientations: Geol. Mag., v. 109, no. 3, p. 231–233.
Clark, R. H., and **McIntyre, D. B.,** 1951, The use of the terms pitch and plunge: Am. Jour. Sci., v. 249, p. 591–599.
Clifford, P., and others, 1957, The development of lineation in complex fold systems: Geol. Mag., v. 94, no. 1, p. 1–24.
Cloos, Ernst, 1947, Oolite deformation in the South Mountain fold, Maryland: Geol. Soc. America Bull., v. 58, p. 843–918.
——— 1964, Wedging, bedding plane slips and gravity tectonics in the Appalachians, *in* Tectonics of the southern Appalachians: Blacksburg, Va., V.P.I. Dept. of Geol. Sci. Mem. 1.
——— 1971, Microtectonics: Studies in geology, no. 20, Johns Hopkins University.
Cloos, Hans, 1930, Zur Experimentellan Tektonik: Naturwissenschaften, Jhg. 18.
Currie, J. B., Patnode, H. B., and **Trump, R. P.,** 1962, Development of folds in sedimentary strata: Geol. Soc. America Bull., v. 73, p. 655–674.
Dahlstrom, C. D. A., 1954, Statistical analysis of cylindrical folds: Canadian Mining Metall. Bull., no. 504, p. 234–239.
De Sitter, L. U., 1939, The principle of concentric folding and the dependence of tectonic structure on original sedimentary structure: Amsterdam, Kininkl. Nederlandse Akad. Wetensch, Proc., v. 42, no. 5, p. 412–430.
——— 1954, Gravitational gliding tectonics—An essay on comparative structural geology: Am. Jour. Sci., v. 252, p. 321–344.
Doll, C. G., and others, 1961, Geologic map of Vermont: Vermont.
Donath, F. A., 1963, Folds and folding: Geol. Soc. America Spec. Paper 68.
——— and **Parker, R. B.,** 1964, Folds and folding: Geol. Soc. America Bull., v. 75.
Duska, Leslie, 1961, Depth of the basal shearing plane in cases of simple concentric folding: Alberta Soc. Petroleum Geologists Jour., v. 9, no. 1, p. 20–24.
Elliott, David, 1968, Interpretation of fold geometry from lineation isogonic maps: Jour. Geology, v. 76, no. 2, p. 171–190.
Fitzgerald, E. L., and **Braun, L. T.,** 1965, Disharmonic folds in Besa River formation, northeastern British Columbia, Canada: Am. Assoc. Petroleum Geologists Bull., v. 49, no. 4.
Fleuty, M. J., 1964, The description of folds: Geol. Assoc. Proc., v. 75, p. 461–492.
Fyson, W. K., 1971, Fold attitudes in metamorphic rocks: Amer. Jour. Sci., v. 270, no. 5, p. 373–382.
Ghosh, S. K., 1970, A theoretical study of inter-

secting fold patterns: Tectonophysics, v. 9, no. 6, p. 559–569.

Godfrey, J. D., 1954, The origin of the ptygmatic structures: Jour. Geology, v. 62, no. 4, p. 375–387.

Hara, Ikuo, 1970, A note on "concentric" folding of multilayered rocks: Hiroshima Univ., J. Sci. Ser. C., v. 5, no. 3, p. 217–239.

Hills, E. S., 1953, Tectonic setting of Australian ore deposits, *in* Geology of Australian ore deposits: Empire Mining Metall. Cong. (Melbourne), 5th, v. 1, p. 41–61.

Hudleston, P. J., 1973, Fold morphology and some geometrical implications of theories of fold development: Tectonophysics, v. 16, no. 1–2, p. 1–46.

Johnson, A. M., and **Ellen, S. D.**, 1974, A theory of concentric, kink, and sinusoidal folding and of monoclinal flexuring of compressible, elastic multilayers. I. Introduction: Tectonophysics, v. 21, p. 301–339.

Kuenen, P. H., 1938, Observations and experiments in ptygmatic folding: Soc. Geol. Finlande, Comptes rendus, no. 2, p. 11.

Marcos, A., and **Arboleva, M. L.**, 1975, Evidence of progressive deformation in minor structures in Western Asturias (N.W. Spain): Sonderdruck aus der Geologischen Rundschau Band 64, 1975.

Mertie, J. B., 1959, Classification, delineation and measurement of non-parallel folds: U.S. Geol. Survey Prof. Paper 314-E, p. 91–94.

O'Driscoll, E. S., 1964, Cross fold deformation by simple shear: Econ. Geology, v. 59, p. 1061–1093.

Oertel, Gerhard, 1974, Unfolding of an antiform by the reversal of observed strains: Geol. Soc. America Bull., v. 85, no. 3, p. 445–450.

Ramberg, Hans, 1959, Evolution of ptygmatic folding: Norsk Geol. Tidsskr., v. 39, p. 99–131.

——— 1963a, Fluid dynamics of viscous buckling applicable to folding of layered rocks: Am. Assoc. Petroleum Geologists Bull., v. 47.

——— 1963b, Strain distribution and geometry of folds: Geol. Inst. Univ. Uppsala Bull., v. 42, p. 3–20.

Ramsay, J. G., 1960, The deformation of early linear structures in areas of repeated folding: Jour. Geology, v. 68.

——— 1961, The effects of folding upon the orientation of sedimentation structures: Jour. Geology, v. 69, no. 1.

——— 1962a, The geometry and mechanics of formation of similar type folds: Jour. Geology, v. 70, p. 309–327.

——— 1962b, The geometry of conjugate fold systems: Geol. Mag. (Great Britain), v. 99, p. 516–526.

——— 1964, The uses and limitations of beta-diagrams and pi-diagrams in the geometrical analysis of folds: Geol. Soc. London Quart. Jour., v. 120.

——— 1967, Folding and fracturing of rocks: New York, McGraw-Hill.

——— 1974, Development of chevron folds: Geol. Soc. America Bull., v. 85, p. 1741–1754.

Read, H. H., 1928, A note on ptygmatic folds in the Suterland granite complex: Summ. Prog. Geol. Survey for 1927, pt. II, p. 72.

Rickard, M. J., 1970, A classification diagram for fold orientations: Geol. Mag., v. 108, no. 1, p. 23–26.

Rowan, L. C., and **Mueller, P. A.**, 1971, Relations of folded dikes and Precambrian polyphase deformation, Gardner Lake area, Beartooth Mtns., Wyoming: Geol. Soc. America Bull., v. 82, no. 8, p. 2177ff.

Sander, B., 1930, Gefügekunde der Gesteine: Vienna, Springer-Verlag OHG.

——— 1950, Einführung in die Gefügekunde der Geologischen Körper: v. I & II, Vienna, Springer-Verlag OHG.

Sederholm, J. J., 1907, Om Granit Och Gneis: Comm. Geol. Finlande Bull., no. 23.

——— 1913, Über Ptygmatische Faltungen: Neues Jahrbuch Mineralogie, v. 36.

Stauffer, Mel R., 1973, New method for mapping fold axial surfaces: Geol. Soc. America Bull., v. 84, no. 7, p. 2307–2318.

Stillwell, F. L., 1953, Geology of Australian ore deposits: Melbourne, Australian Inst. Mining and Metallurgy.

Tobisch, Othmar T., 1966, Large-scale basin-and-dome pattern resulting from the interference of major folds: Geol. Soc. America Bull., v. 77, no. 4, p. 393–408.

Turner, F. J., and **Weiss, L. E.**, 1963, Structural analysis of metamorphic tectonites: New York, McGraw-Hill.

Weiss, L. E., 1955, Fabric analysis of a triclinic tectonite and its bearing on the geometry of flow in rocks: Am. Jour. Sci., v. 253, p. 225–236.

——— 1959, Geometry of superposed folding: Geol. Soc. America Bull., v. 70, p. 91–106.

——— and **McIntyre, D. B.**, 1957, Structural geometry of Dolradian Rocks at Loch Leven, Scottish Highlands: Jour. Geology, v. 65, p. 575–602.

Whitten, E. H. T., 1959, A study of two directions of

folding: The structural geology of the Monadhliath and Mid-Strathspey: Jour. Geology, v. 67.

——— 1966, Structural geology of folded rocks: Chicago, Rand McNally.

Willis, Bailey, 1893, Mechanics of Appalachian structure: U.S. Geol. Survey Ann. Rept. 13.

Williams, Emyr, 1965, The deformation of competent granular layers in folding: Am. Jour. Sci., v. 263, p. 229–237.

——— 1967, Notes on the determination of shortening by flexure folding modified by flattening: Papers and Proc. Royal Soc. Tasmania, v. 101.

Wilson, Gilbert, 1952, Ptygmatic structures and their formation: Geol. Mag. (Great Britain), v. 89, p. 1–52.

——— 1969, The geometry of cylindrical and conical folds, Geol. Assoc., London, Proc., v. 78, part 1, p. 179–209.

Wynne-Edwards, H. R., 1963, Flow folding: Am. Jour. Sci., v. 261, p. 793–814.

12

Mesoscopic Features Commonly Associated with Folded Rocks

CLEAVAGE AND FOLIATION*

Many terms (e.g., foliation, flow cleavage, fracture cleavage, schistosity, strain-slip cleavage, slaty cleavage) are applied to closely spaced fractures or strongly aligned planar structures of a rock along which the rock tends to break, Fig. 12-1. Some of the terms are genetic, but the mode of origin of all these features is clearly not the same.

Some planar fissility is due to breaks or planes of weakness in the rock that are unrelated to the orientation of the rock fabric. Others are indeed due to the strong alignment of fabric components, especially to the alignment of platy minerals—notably the clays and micas. Planar fissility is found in metamorphic rocks (where the fabric is due to recrystallization), in sedimentary rocks, and in unconsolidated sediments.

The term *foliation* is a general term sometimes used as essentially synonymous with cleavage, but it is applied most generally to mineral alignment in metamorphic rocks. Thus "slaty" cleavage and "schistosity" are special types of metamorphic foliation characterized by the types of fabric or mineral arrangement commonly found in slates and schists.

Sander (1930) avoided the problem of genetic connotations by designating sets of planes of mechanical inhomogeneity as *s* surfaces or *s* planes. Subscripts are used for rocks with more than one plane structure. If bedding can be identified, it is one *s* surface designated S_0 generally; others are designated S_1, S_2, S_3, etc. The intersection of S_1 with S_0 forms a lineation, l_1, etc. This procedure evades the problem of deciding how each plane formed, but the geometrical elements thus defined can be very useful in describing and analyzing the structure.

Two major classes of cleavage, fracture cleavage and slaty or flow cleavage, are readily differentiated as follows:

Fracture cleavage (also called *false cleavage* and *strain-slip cleavage*) is conditioned by the

* See review by Siddens (1972), and see structural notation in Appendix E.

existence of incipient, cemented, or welded parallel fractures and is nonpenetrative; it is independent of a parallel arrangement of the mineral constituents, Fig. 12-2.

Slaty (flow) *cleavage* (also called *penetrative cleavage* and *schistosity*) is dependent on the parallel arrangement of the mineral constituents of the rock. It should be noted, however, that a much more rigorous parallelism of fabric is present in slates than in schists or sediment.

These definitions by Leith (1905) provide a clear way to differentiate cleavages into two categories. However, the term *flow* used by Leith implies a mode of origin and is better replaced by the term *slaty*.

FIGURE 12-1 A strongly developed cleavage cuts across the primary stratification. Small-scale crenulations occur between the cleavage planes. (*From Marcos, 1973.*)

Fracture Cleavage

No clear distinction can be drawn between fracture cleavage and other fractures and joints which are brittle failure phenomena. The name cleavage is applied only when the fractures are very closely spaced, usually several per centimeter. The cleavage surfaces are often distinct cracks separated from one another by thin plates of rock showing no alignment of minerals parallel to the cleavage. The fracture cleavages are often only subparallel, and their surfaces may be rough, irregular, and discontinuous. The term *fracture cleavage* has been applied to closely spaced fractures located in a variety of structural positions, and their origin has been attributed to a number of mechanisms. For example:

1. Fractures parallel to a fault and attributed to shear.

2. Fractures along which movements have occurred, giving rise to passive-slip (similar-type) folds.

3. Fractures parallel to axial planes of chevron- or accordion-type folds.

4. Fractures parallel to axial planes of kinks in foliation of schists and slate.

5. Fractures in the hinge zones of competent layers in concentric folds and attributed to extension fracturing. Fractures much farther down on the limbs and approximately normal to bedding have similarly been attributed to extension as the fold amplitude has increased.

6. Fractures originating from interbed shearing during folding as a result of couples generated by movements along bedding surfaces.

If all these fractures are to be considered fracture cleavages, several mechanisms are needed to generate them. The three most commonly employed are shearing, extension, and flattening. De Sitter (1964) emphasizes the role of flattening in development of cleavages of both types. He calls the rock slivers between two cleavages *microlithons* and points convincingly to the folding and kinking within the microlithons as evidence of the shortening of the rock normal to the cleavages. Thus these cleavages arise as a result of flattening accompanied by movements in planes normal to the maximum stress direction.

In summary, some fracture cleavages are nothing more than closely spaced fractures—breaks across which the rock has lost cohesion but has not been displaced. In other instances movements have occurred along the cleavage (a slip takes place parallel to the cleavage), giving rise to offsets or small folds called *crenulations* in any *s* surface formed before the cleavage. This variety is called *strain-slip cleavage.*

Superposed Cleavages

Two generations of cleavage, usually one with the characteristics of slaty or flow cleavage and a later one involving rupture of the earlier cleavage, are often encountered, Fig. 12-3. The two may be clearly differentiated, or one may appear to grade into the other. This type of superposition of *s* surfaces may lead to progressive deformation of the earlier surface as a result of movements on the second, as shown in Fig. 12-4, and is called strain-slip cleavage. Two cleavages, one the shistosity of the rock and the second a fracture cleavage, occur in metasedimentary rocks of east-central Vermont (White, 1949). The fracture cleavage cuts the earlier schistosity and appears as parallel fractures along which local mica flakes are aligned. The fractures lie in the axial plane of small-scale "crinkles" or chevron-shaped folds in the schistosity. The fracture cleavage becomes increasingly developed toward the west until it is aligned parallel with the schistosity. Progressive rotation of the earlier schistosity by movements along the fracture cleavages apparently took place until the old schistosity disappeared into the plane of the new schistosity. This was accompanied by growth of new micas parallel to the new schistosity.

The types of superposition of one cleavage on another just described often affects rocks that have been folded during the formation of an early cleavage. As the second deformation is imposed on the first, the early structures may be largely transposed and become difficult to recognize, Fig. 12-5.

Schistosity

Schistosity of some metamorphic rocks is formed parallel to the original bedding of the sedimentary rocks from which they were derived. Schistosity can form in the absence of strong deformation as shown by the presence of undeformed primary structures in some schists. In other cases, the schistosity cuts across primary bedding; it may be aligned parallel to the long axis of the strain ellipsoid as indicated by deformed primary features; and in still other cases, two or more superimposed schistosities in the same rock.

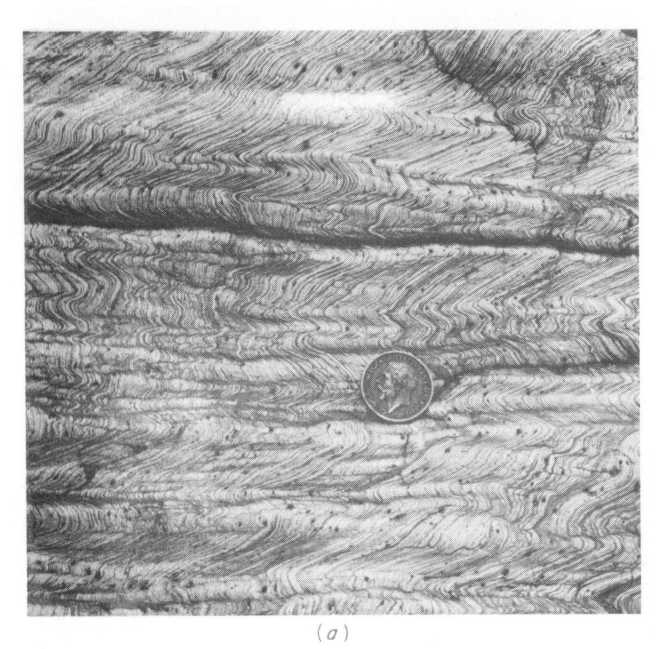

(a)

FIGURE 12-2 Examples of cleavage. (*a*) Crenulation cleavage cuts across primary layering in Dartmouth slates, Devon (*Crown copyright*); (*b*) cleavage developed in shale interbedded with uncleaved limestone of the Edinburg formation. Virginia; (*c*) fracture cleavage in argillaceous limestone of Edinburg formation, Virginia; (*d*) fracture cleavage in schistose grits (*Crown copyright*).

(b)

(c)

(*d*)

FIGURE 12-2 (Continued.)

Hypotheses of the Formation of Slaty Cleavage and Schistosity

A basic question regarding the origin of cleavage is whether all types of slaty cleavage originate in the same way. Slate, phyllite, and schist are all products of metamorphism of shale. Slate is generally viewed as a product of dynamic metamorphism, while phyllite and schist involve a higher degree of recrystallization. The orientation of minerals in these metamorphic rocks is generally ascribed to:

1. Mechanical rotation of existing minerals
2. Mechanical flattening of existing minerals
3. Recrystallization and growth of new minerals with preferred orientation
4. Combinations of the above

The first of these ideas was suggested by Sorby (1853) on the basis of petrographic studies of slate and artificially compressed clay. He concluded that the mica flakes and other planar minerals were rotated into parallel positions normal to the compression. The rotations were envisioned as being those of rigid grains in a plastically flowing matrix. This idea, especially in combination with rupturing and extension of grains and recrystallization, has been popular. Recent experimental work has shown more clearly the process by which plastic deformation in rocks occurs. Mechanical flattening of grains occurs through a process of brittle fracture, by granulation, by gliding within mineral grains, by slip between layers of atoms, by twinning, and by recrystallization.

Crystallization contemporaneous with deformation has long been recognized as an important process in the development of schistosity. The rolled garnets found in some schists show clearly that recrystallization was contemporaneous with movements parallel to cleavage. But the presence of undeformed fossils in schistose rocks indicates that recrystallization without strong distortion of the rocks also takes place. The mechanisms which have been suggested for syntectonic recrystallization include solution and redeposition according to *Riecke's principle,* which holds that the solubility of a stressed face of a crystal is increased. Thus material tends to dissolve most rapidly from points of stress, to be transported by solution, and to be redeposited elsewhere in the fabric where stress is less. The rock fabric tends toward the development of a foliated structure, with the foliation normal to the maximum principal stress direction due to flattening and to deposition of dissolved material in unstressed positions.

Experiments have shown that marbles strained slowly at temperatures from 400 to 800°C undergo some recrystallization. New crystals have a strongly developed preferred orientation of the *c* crystallographic axis in a position normal to the maximum stress direction. A similar effect is seen in recrystallized quartz.

FIGURE 12-3 A greatly enlarged photograph of strain-slip cleavage showing small-scale crenulations. (*From Marcos, 1973.*)

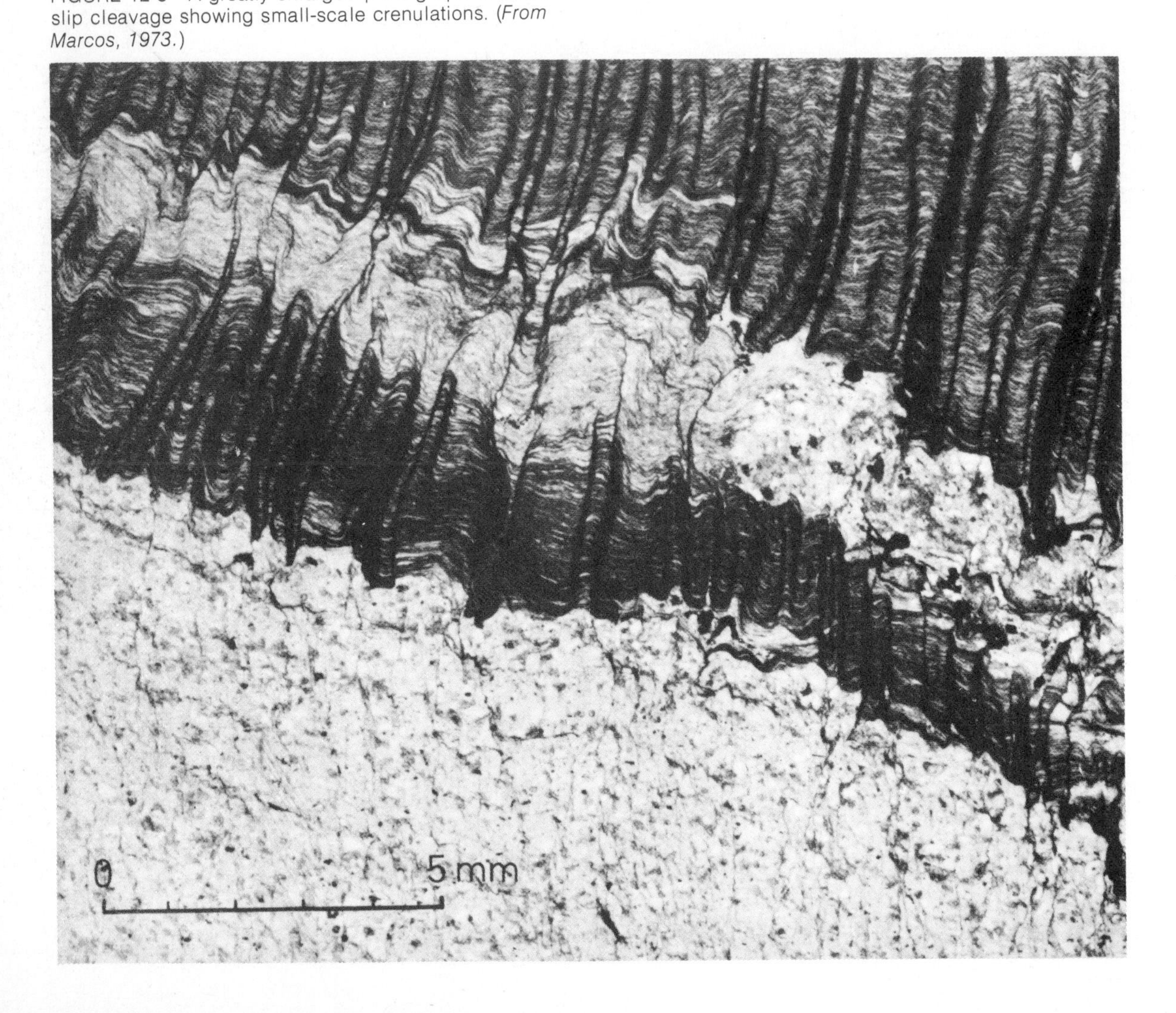

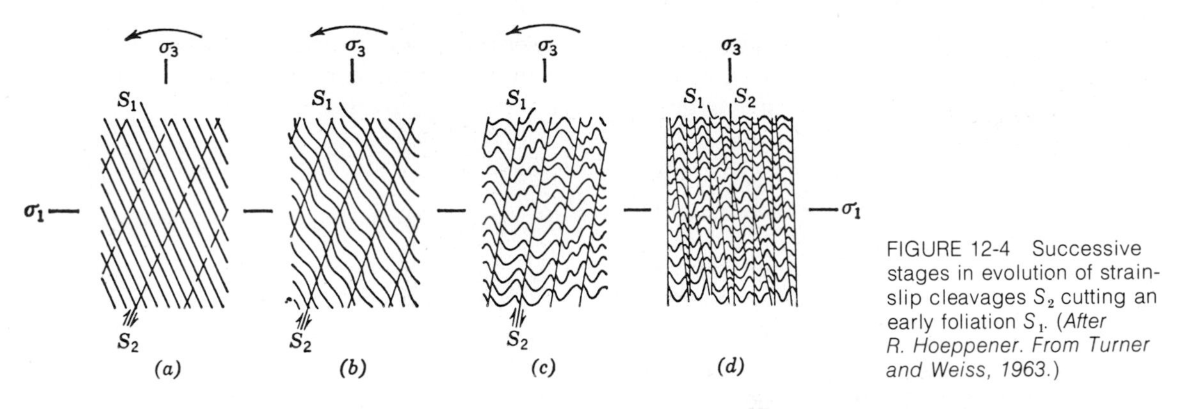

FIGURE 12-4 Successive stages in evolution of strain-slip cleavages S_2 cutting an early foliation S_1. (*After R. Hoeppener. From Turner and Weiss, 1963.*)

Slaty Cleavage of Eastern Pennsylvania

The slates of Pennsylvania are especially well known (Behre, 1933; Maxwell, 1962; Epstein, 1974). Slaty cleavage is found in the fine-grained portions of the Ordovician Martinsburg formation in the folded Appalachians. Fine micaceous minerals, muscovite, sericite,

FIGURE 12-5 Diagrammatic geologic map (plane horizontal section) of a body containing superposed folds involving transposition. The four fomains I to IV represent progressive stages of superposition of folds with axial planes striking NW on earlier folds with axial planes striking NE. Transposition of the initial s surface S_1 (given by the stippled layer) into the secondary foliation S_2 is complete on a mesoscopic scale in domain IV. (*From Turner and Weiss, 1963.*)

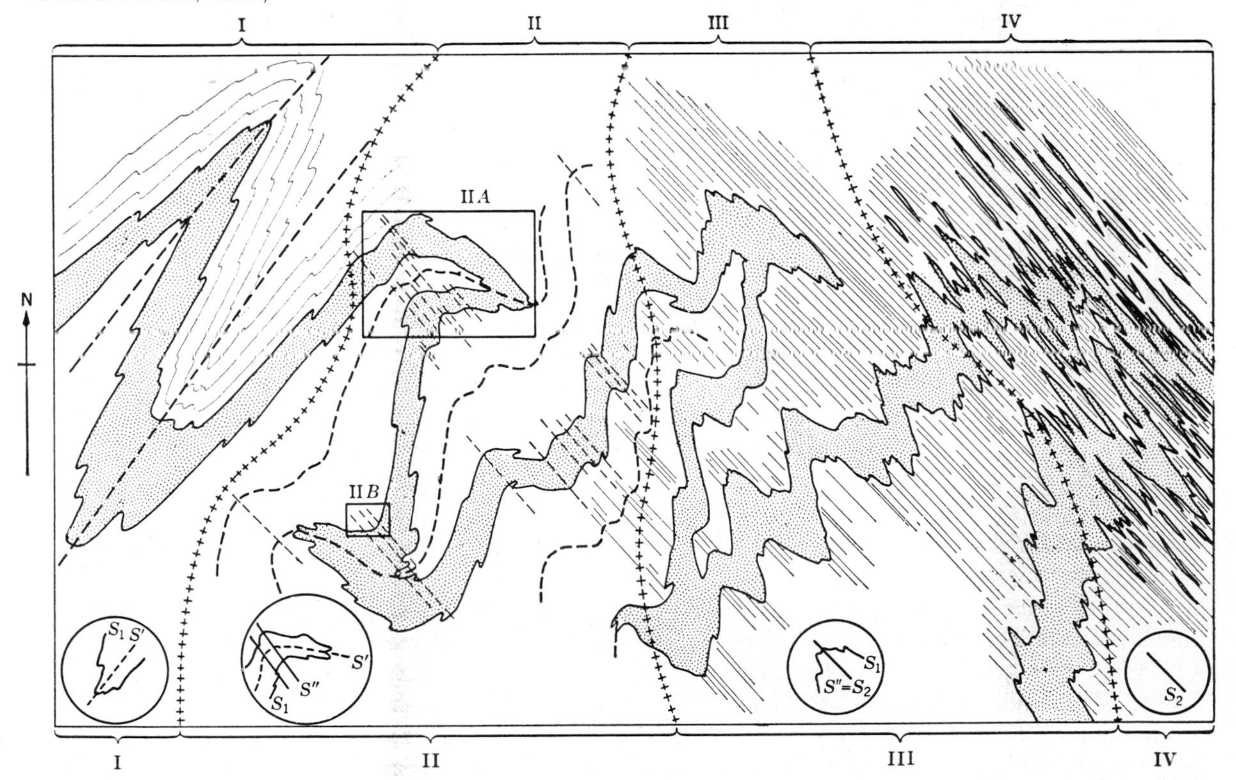

quartz fragments, and some carbonate, pyrite, graphite, and rutile needles occur in the slate. The micaceous minerals are so strongly oriented in the cleavage that bedding is often obscured. The cleavages lie approximately parallel to the axial planes of the folds, suggesting that they formed contemporaneously. Folding continued after the slaty cleavage formed, and in some places a fracture cleavage is imposed on the slaty cleavage. Despite these later deformations, the slaty cleavage is of regional scope and is roughly parallel over the region.

The Case for Formation of Cleavage in Unconsolidated Sediment

Maxwell (1962) suggests that the slaty cleavage in the Pennsylvania slates originated while the sediment was unconsolidated. Evidence of the unconsolidated nature of the sediment at the time of cleavage formation includes:

1. Sandstones interbedded with the slate were injected along cleavage directions as dikes, a process associated with water-saturated unconsolidated sand, Fig. 12-6.

2. The sands and slates are folded into similar-type patterns. Apparently the bedding exercised little control in folding; the folds are passive, indicating that the sands were not brittle materials.

3. The cleavage formed in Late Orodovician time; cleavage of the same orientation and character does not occur in younger units. It is unlikely that the sediment was deeply buried.

4. The main micaceous mineral in the slate is illite, and recent studies indicate that illite and even

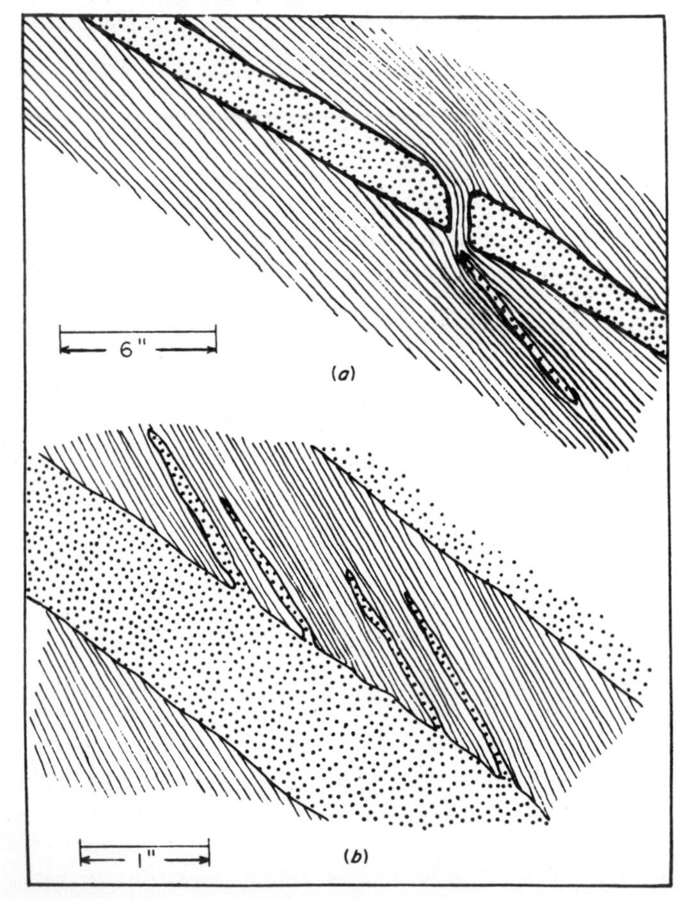

FIGURE 12-6 Sandstone dikes injected parallel to slaty cleavage in the Martinsburg formation. (*From Maxwell, 1962.*)

chlorite can be generated in marine environments at low temperatures and low pressures. Thus the usual metamorphic conditions of high temperature and pressure are unnecessary to account for micaceous minerals of slate.

Maxwell (1962, p. 300) summarizes his hypothesis as follows:

In the early stages of compaction a clay may be visualized as an open network of clay plates and fibers with water-filled interstices. As compaction proceeds, water is expelled, and the clay particles rotate, approaching parallelism as more and more water is expelled (Buessem and Nagy, 1954). If the process proceeds slowly and water pressure remains essentially hydrostatic, the load is carried by the clay particles, friction and bearing strength increase, and in later stages compaction occurs only at the expense of considerable deformation of the clay particles. In the process proposed in this paper, it is assumed that a thick sequence of impermeable shaly sediments accumulated just prior to deformation and that escape of pore water was so slow that abnormally high pore pressures characterized the sequence when deformation began. The resulting compression tended to increase the pore-water pressure which may very well have approximated the lithostatic pressure through much of the formation; perhaps even locally lithostatic pressure was exceeded, with resulting instabilities giving rise to fluid flow perpendicular to the maximum pressure. Flowage parallel to cleavage . . . may well be of this origin, as may be the peculiar downward-injected sandstone dike of [Fig. 12-6].

During deformation, flattening, rotation, and extensive flowage parallel to cleavage resulted in a near-parallel orientation of clay particles. This in turn destroyed the original network of clay particles which held the water so tenaciously; permeability increased in the plane of the cleavage and in the direction of the grain, facilitating the rapid dewatering of the slaty rocks as the last stages of the slaty cleavage episode. Thereafter the rocks were relatively brittle, as indicated by the post-slaty cleavage deformation.

Slaty cleavage has now been reported in highly deformed Pleistocene mudstones cored from the inner wall of the Aleutian Trench and from the continental rise in the Gulf of Mexico (Moore and Geigle, 1974). In both cases an incipient cleavage is observed in the unconsolidated sediments. Platy minerals are aligned subparallel to the axial surfaces of the folds.

The Case for a Metamorphic Origin of Cleavage in Pennsylvania

Epstein (1974) renews the dynamic metamorphic argument for the origin of the Martinsburg slates. Based on study of mineral fabric relationships, he concludes that

Cleavage formed under conditions at and just below those of greenschist facies regional metamorphism by (1) tectonic compaction perpendicular to cleavage resulting in pressure solution (elongation and removal of quartz by corrosion from the folia and concentration of less soluble minerals in the folia), (2) mechanical mineral reorientation parallel to the folia aided by volume loss and accompanied by some grain diminution (3) laminar flow (mineral migration) and injection of pelitic and some sandy material along the folia aided by silica migration, by connate water squeezed out of the rocks, or by release of water from hydrous minerals, and (4) some new mineral growth. Numerous lines of evidence indicate that cleavage developed after the rocks were indurated.

While this hypothesis clearly differs from the dewatering hypothesis regarding the degree of consolidation of the rock at the time of cleavage formation, they agree that fluids and high pore pressures facilitate cleavage development.

Relationship of Cleavage to Folds

As a generalization it is correct to say that most slaty cleavage is found aligned parallel to the axial surface of the folds with which it is associated, Fig. 12-7, but it is not uncommon for cleavage to show a fan-shaped disposition also. When the folded layers differ in grain size the cleavage may be confined to the finer grained layers, Fig. 12-7.

Where the folded layers are strongly contrasted in physical properties, e.g., quartzite and shale or schist, the cleavage or schistosity orientation may show strong departures from parallelism with the axial surfaces; the patterns, Figs. 12-8 and 12-9, are typical. These orientations bear a strong resemblance to the stress trajectory patterns found in experimental work and suggest again the importance of flattening effects in cleavage development.

FIGURE 12-7 Axial-plane cleavage well developed in the finer grained layers but not apparent in the coarser sands. Notice the prominent thickening of the sand layers in the fold crest.

Summary of Observation on the Occurrence of Cleavage

1. *Spacing.* Cleavage tends to pervade the rock in slates, but in schists, spacing of the planes of schistosity may vary from a few millimeters to fractions of an inch apart. Fracture cleavages vary in spacing from millimeters to centimeters.

2. *Rock type.* Cleavage is most common in fine-grained rocks and tends to be less developed in sandstones than in silts, shales, or limestones.

3. *Relation of cleavage to folds.* Cleavages often are subparallel to axial planes and exhibit a fan-shape disposition, with dips slightly toward the axial planes of the anticlines. Fracture cleavages tend to strike in the same direction as the strike of the axial planes of folds on which they occur, but their dips show greater divergence from the axial plane as they fan across fold axes.

4. *Variations in degree of development, spacing, and attitude where cleavages cross different lithologies.* The cleavages often appear to be "refracted" on passing from one layer to another of different character.

FIGURE 12-8 Part of folded quartzite layer sandwiched between mica schist. Schistosity in mica schist is indicated by short lines. The short conformable lines in the quartzite represent original layering as indicated by concentrations of mica. Compare this with Fig. 12-9.

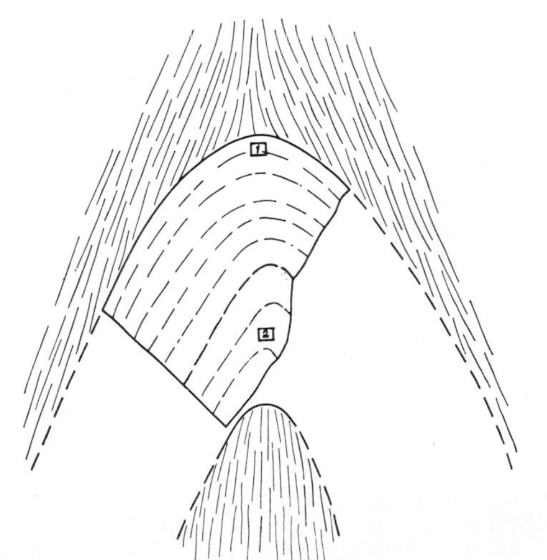

FIGURE 12-9 S_1 cleavage pattern in multilayered pelite/psammite, showing the development of arcuate hinge-cleavage in the inner arcs of folded pelite layers. Kongsfjord formation, southwest Risfjord. (*From Roberts, 1972.*)

5. *Movement along and between cleavages.* Movements occur parallel to cleavage, giving rise to a variety of structures, i.e., small folds, kinking, or shear offsets. De Sitter (1964) calls the thin slivers of rock between cleavages *microlithons,* Fig. 12-3. When folding or kinking occurs between cleavages, a shortening normal to the cleavages is indicated.

6. *Fracture and slaty cleavage occurring together in a rock mass.* Progressive movements along fracture cleavage can lead to the formation of a new cleavage with penetrative characteristics.

7. *Recrystallization.* The mineral constituents of many schists include some components (e.g., garnet, kyanite, etc.) formed only under metamorphic conditions and others which are rarely large in sediment (e.g., mica). The mineral constituents owe their large size to processes of re-

crystallization. In some schists and phyllites recrystallization is contemporaneous with movements parallel to the cleavage. This is shown by the presence of "rolled or snowball" garnets which show rotation during growth.

8. *Shortening normal to cleavages.* Deformed oolites, fossils, pebbles, and other primary geological objects in rocks with slaty cleavage show clearly that shortening normal to the principal cleavage plane often accompanies the development of cleavage, and that elongation is parallel to the cleavage.

KINKS—KINK BANDING

The term *kink* is used to describe the abrupt bending or rotation of cleavage, foliation, or bedding planes, Fig. 12-10. It was first described by Clough (1897), who called the bands produced by kinks *strain bands.* Kinks are common features of deformed, thinly foliated, fine-grained rocks such as slates and phyllites. The foliation between kinks is usually planar, and the kink itself may be a very abrupt flexure or a fracture. Two zones, inclined to one another and showing opposite senses of rotation, called *conjugate sets,* often occur together.

Four different types of kink bands (see Fig. 12-11) are distinguished by Dewey (1965):

1. *Joint-drag kink bands* are those in which the kink plane is a fracture along which the foliation is externally rotated. Within the foliation, internal rotation occurs by slip or gliding along foliation planes. This type of deformation is thought to occur in brittle materials and at a rapid rate.

2. *Segregation kink bands* are similar to joint drags except that the foliation is pulled apart and the voids are filled with quartz or calcite. The development of voids indicates shallow deformation.

3. *Pelitic strain bands* are characterized by pelitic kink bands from which quartz has migrated and by semipelitic limbs. The kink plane contains sericite and muscovite.

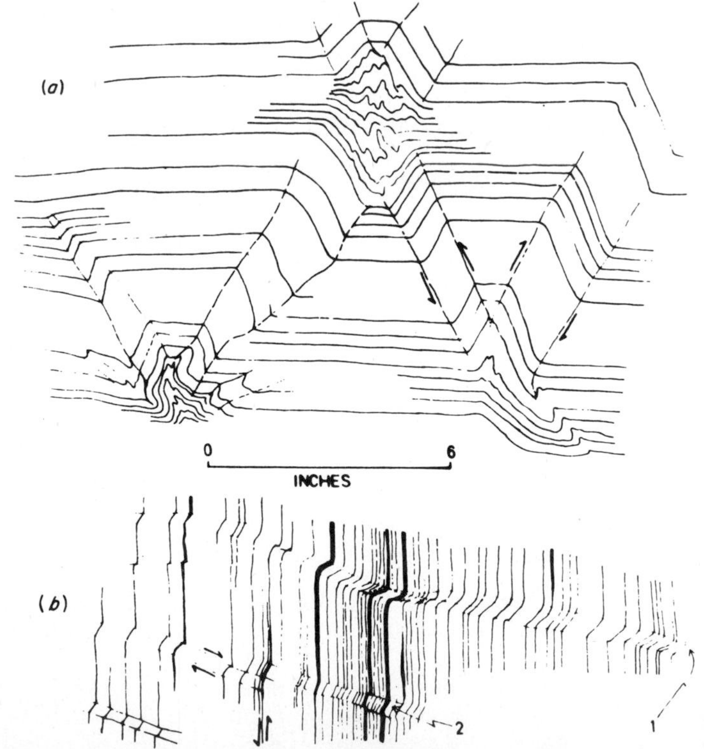

FIGURE 12-10 Late folds in black Carboniferous slates, County Cork, Ireland. (*a*) Conjugate folds at the Slath; (*b*) kink bands at Black Ball head. (*From Dewey, 1965.*)

4. *Shear kink bands* are abrupt changes in orientation of foliation due to continuous shear. Note the absence of kink planes.

Kinks are commonly attributed to late-stage deformation, and particularly, the joint drags and segregation kink bands seem to fall into this category. The mechanism of formation involves primarily slip between foliation layers. Experimental work (Chap. 4) has revealed much about the origin of kinks, and we will return to them again in connection with folds.

MULLION AND RODDING

The terms *mullion* and *rodding* are applied to columnar or rod-like structures found in folded sedimentary and metamorphic rocks. Mullion structure has the external appearance of columns and occurs in many rock types. Rods have a more distinct cylindrical form and usually are composed of quartz. The origin of mullions and rods has been interpreted in quite different ways, and it is probable that similar features can originate by different means. Many arguments concern whether the rods and columnar forms are developed parallel or perpendicular to the direction of movement. Terms such as *slickensides grooving* have been applied to "parallel striated and grooved prisms suggesting logs of wood," and the elongate forms seen on fault planes are well known. Such structures have almost certainly formed by movements parallel to the grooves.

FIGURE 12-11 Idealized geometry of four types of kink band (*a*) Joint-drag band; (*b*) segregation kink band; (*c*) pelitic strain band; (*d*) shear kink band. (*From Dewey, 1965.*)

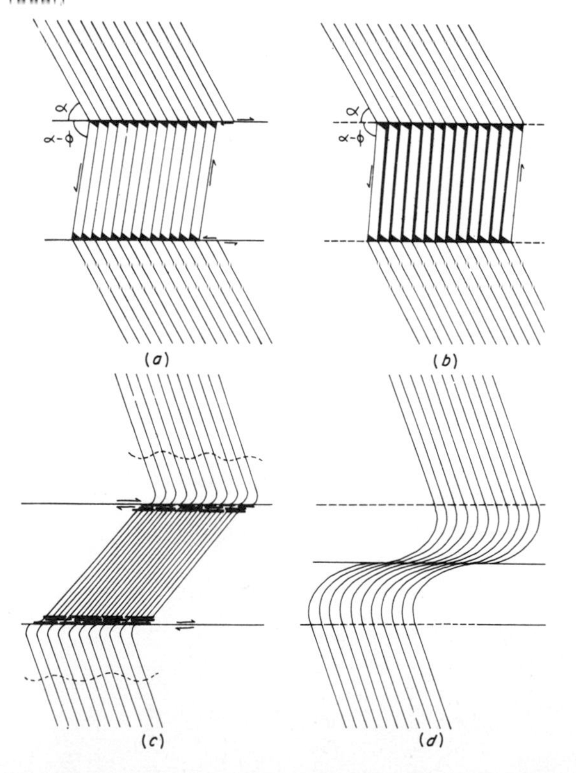

A very different mode of origin is proposed for the mullion and rods described by G. Wilson (1953) in the Scottish Highlands. He recognized three types of mullions:

1. Fold mullions
2. Cleavage mullions
3. Irregular mullions

Fold mullions are cylindrical undulations of bedding, as shown by the conformable laminations of bedding within the mullions. Cleavage mullions are formed by the intersection of cleavage with bedding; many are angular prisms for this reason, Fig. 12-12. The contacts between thin and massive beds are corrugated by asymmetrical rolls. A third variety of mullion characterized by highly irregular outline in cross section is described but not explained by Wilson. The mullions are remarkably constant along their lengths and are parallel to the fold axes of the larger folds on which they are located.

Much of the quartz in rods is a product of quartz segregation during metamorphism. This metamorphism is often accompanied by folding, shearing, and penetrative deformation. The segregated quartz tends to be concentrated (1) along bedding, (2) along a cleavage, or (3) in open fractures, Fig. 12-13.

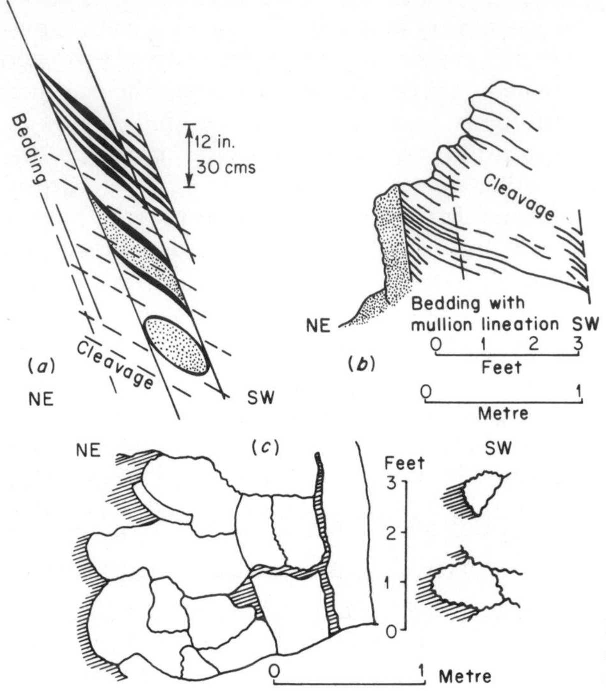

FIGURE 12-12 (*a*) and (*b*) Examples of the development of cleavage mullions; (*c*) profiles of irregular mullions. (*From Gilbert Wilson, 1953.*)

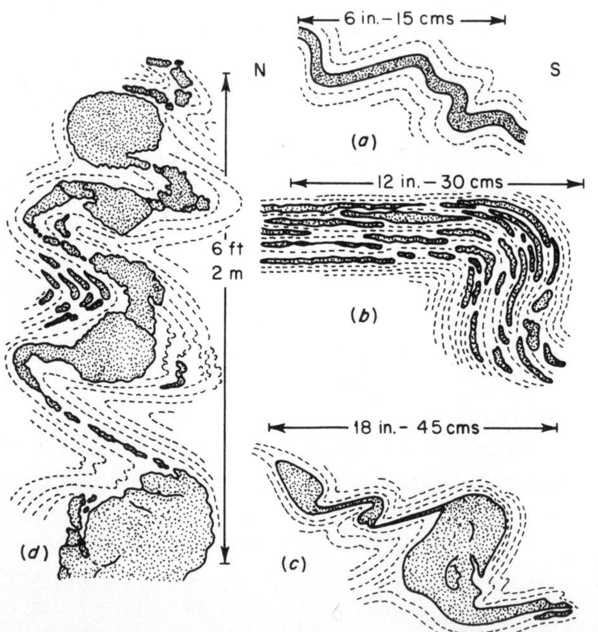

FIGURE 12-13 (*a*) Slightly folded quartz vein parallel to the bedding; (*b*) stretched and folded quartz lenses parallel to the bedding; (*c*) and (*d*) profiles of quartz rods developed in the apices of small-scale folds. (*From Gilbert Wilson, 1953.*)

Quartz rods may then be produced:

1. By movement in the plane of a quartz vein (e.g., slickensides).

2. As a result of intersection of cleavage with a thin quartz layer (bed or vein) followed by enough movement on the cleavage to displace the quartz. Rods parallel to the fold axes are produced in this way.

3. By folding and rolling of thin quartz veins or beds on the limbs of folds produces rods essentially parallel to fold axes.

BOUDINAGE*

Lohest (1909) applied the term *boudinage structure* to sausage-shaped bodies of one rock layer sandwiched between layers of different rock type. Boudins occur in layered sequences ranging from semiconsolidated sediments to high-grade metamorphic rocks.

They may be separated from one another, but generally they occur as a linked chain, and the individual bodies are usually elongate in the plane of the layering.

Boudins may be sharp and retangular blocks which have separated, but boudins with a barrel-shaped cross section and pinches and swells are far more common. Generally, boudins appear to have formed from a layer which was uniform in thickness before deformation.

Most students of boudinage agree that it results from extension and elongation parallel to the layering in a sequence of materials of varied ductility in which the more brittle material breaks while the more ductile rock flows. Boudins are sometimes rotated or deformed by shearing, but such cases appear to be due to superposition of deformations and are not an integral part of the process by which most boudins originate.

The shapes arise from plastic flowage and lateral elongation of the layer. Lens-shaped bodies or pinch and swell shapes may be produced in extreme cases of plastic flowage. Where the plastic flowage took place after rupture, the upper and lower edges of the boudin are drawn out farther than the central portion of the boudin. Structures within the layers which enclose boudins support the idea that plastic flowage parallel to the layering takes place within them and is directed toward the places where the boudins separated. Sometimes the layering curls around the edge of a boudin. Such zones of separation are sites of secondary mineral growth in metamorphic rocks, and in cases of extreme replacement the structure within the zone is totally obscured, Figs. 12-14 and 12-15.

Elongation in a rock sequence may result from a variety of conditions—stretching as a result of tensile stress, compression perpendicular to the direction of elongation, or a stress couple. Ramberg (1955) favors the second of these alternatives as the most general condition. The evolution of the structure is viewed in terms of a model consisting of a brittle layer sandwiched between two ductile layers. As the sequence is compressed be-

* Excellent detailed descriptions of boudinage are given by Wegmann (1932), Ernst Cloos (1947), and Ramberg (1955).

FIGURE 12-14 Sketches showing natural boudinage. (*Drawn from photographs in Ramberg, 1955.*)

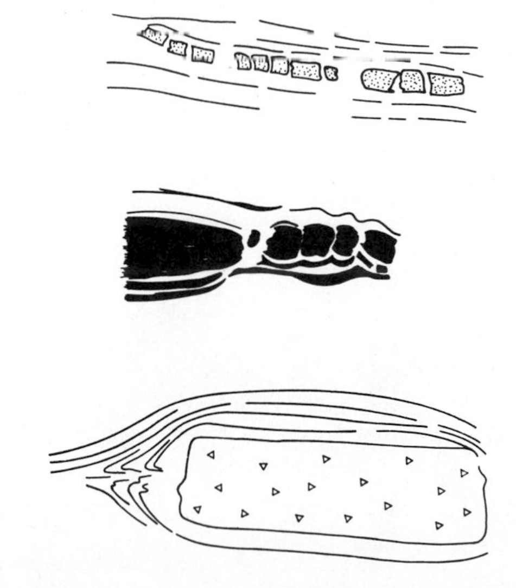

tween two rigid plates, plastic flowage occurs within the ductile layers from the center toward the margins. Depending on the boundary conditions, flowage is possible in a single direction (giving rise to rod-shaped boudins) or in several directions within the layering (giving rise to irregular shapes in plan). Flowage exerts drag on the brittle layer, fracturing it or at least separating older fractured blocks, Fig. 12-16. The magnitude of the tensile stress built up in the brittle layer depends on such factors as the magnitude of compressive stress, rate of flow, viscosity coefficients, boundary conditions between the layers, area of contact, and thickness.

LINEATIONS

The term *lineation* refers to all linear types of structures regardless of their origin,* but many lineations are associated with folded structure and owe their origin to the folding process. Lineations may be caused by flowage in igneous rocks, by the intersection of planar structures, by growth of elongate minerals, or by slippage which causes streaks or scratches. Thus slickensides formed by movement on a fault or bedding-plane slippage in concentric folding form lineations, the intersection of fractures or the intersection of a fracture with a bed result in lineations, and the long axis of a deformed oolite forms a lineation. Thus lineations may form in a great many quite different

* See Cloos (1946) for a definitive treatment of lineations.

FIGURE 12-15 Boudins in a folded quartz-rich layer of gneiss located in the Madison Mountains, Montana.

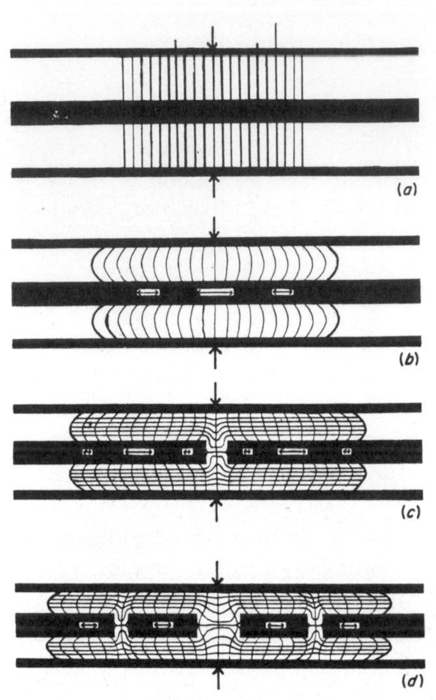

FIGURE 12-16 Successive steps during formation of boudinage structures. (*a*) State prior to compression. The black layer in the middle is the competent body on either side of which incompetent bodies exist (vertically lines regions). Compressive stress is transmitted by means of two stiff sheets (black). (*b*) Compression has started; the plastic flowage in the incompetent layers is indicated by the distortion of the originally vertical lines. Arrows in the competent black layer indicate tensile stress. (*c*) A more advanced step than (*b*). Competent layer is now ruptured in the middle where tension was greatest. The network in the incompetent layer indicates pattern of flowage. This network is not a further evolution of the deformed network in (*b*), but rather the evolution of an imaginary rectangular network not shown in (*b*). Horizontal arrows show tensile stresses in the competent layer. (*d*) A stage more advanced than (*c*). Competent layer has ruptured at two new places. The pattern of the network indicates plastic flowage in the incompetent layers during evolution from (*c*) to (*d*). Again, the deformed network in (*d*) is supposed to have developed from a rectangular imaginary network not shown in (*c*). (*From Ramberg, 1955.*)

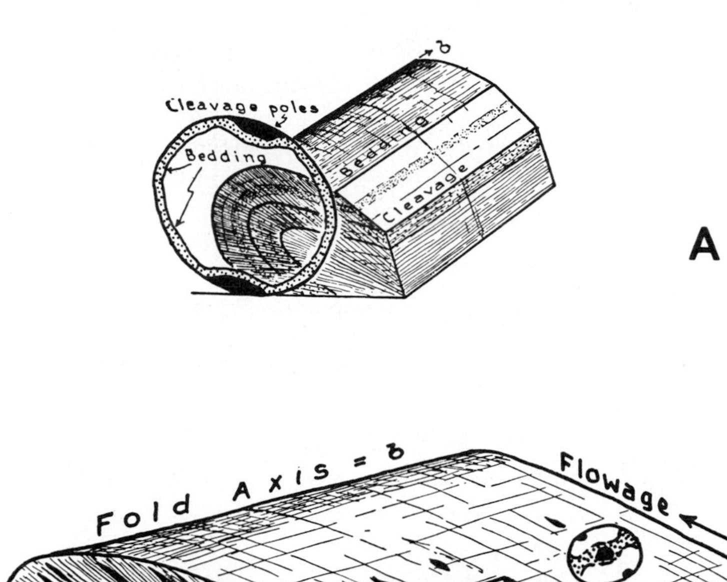

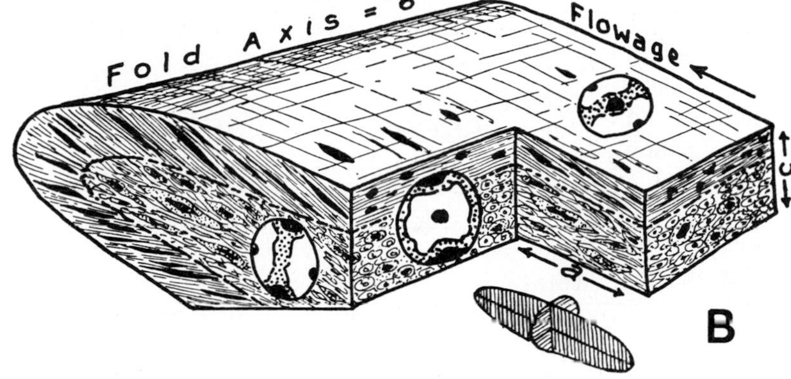

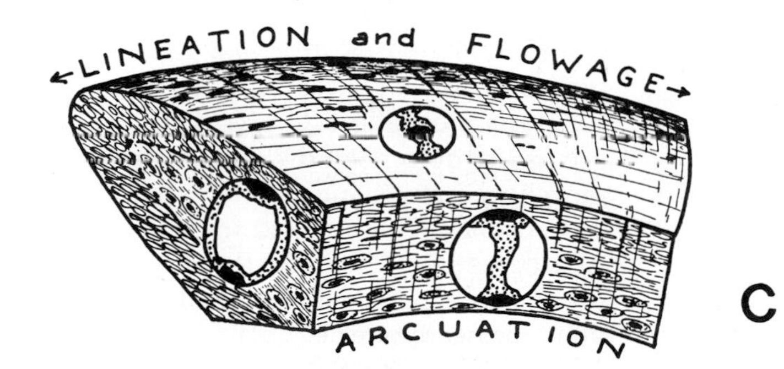

FIGURE 12-17 Lineations commonly associated with folds. (*a*) A fold showing cleavage fanning from the axial plane. The direction of the fold axis, *b*, is shown, and a fabric diagram shows both plots of bedding and cleavage contoured. The fabric diagrams are oriented to represent the data as shown on the surface on which they appear. (*b*) This overturned fold shows flowage up the dip of the cleavage. The dark elongate shapes are pebbles or ooids. Mica diagrams are forming the girdles around lineations in the *a* direction. This is based on a study at South Mountain, Maryland. (*c*) This overturned fold has been stretched parallel to the *b* direction. Note the elongation of ooids parallel to *b*. (*d*) This overturned fold has been deformed in such a way that drags are formed on the limbs, and s sandstone layer has been rolled to form rods or pencil-shaped linear features in the fold's hinge. (*From Cloos, 1946.*)

ways; so, great care should be taken in observing lineations in the field to distinguish the exact cause of the lineation as well as its orientation. The orientation of lineations is conveniently measured in either of two ways—the bearing and plunge may be measured, or the pitch of the lineation within some planar structure such as bedding or fracture in which it occurs may be used. It is often easier to obtain a more accurate measurement by use of pitch.

Only a few of the more important and frequently encountered types of lineations found in deformed rocks are listed below, Fig. 12-17. *B* is used to designate the direction of the fold axis, *a* is the direction of tectonic transport and is at right angle to *b* (*B*), and *c* is the direction normal to the plane that contains *a* and *b*, the *ab plane*.

1. Cleavage bedding lineation—cleavage often occurs parallel to the axial planes of folds (the *ab* plane), and almost invariably it contains the *b* axis even when the cleavages are not rigorously parallel (note the way cleavage fans around fold hinges, Fig. 12-17*a*).

2. *Rods, mullions*, and *pencil structures* are most frequently found oriented in the *b* direction. Some of these are formed as a result of cleavage-bedding intersections, and they are often accentuated by a rolling-type motion which may separate the features and make them more distinct.

3. *Axes of drag folds* are usually in the *b* direction and result from the action of a couple that may cause rotation of the rods or pencils.

4. Elongation in *b* is found when the folds are stretched in the *b* direction as shown in Fig. 12-17*b*. This stretching may cause oolites, etc., to become elongated parallel to the fold axis instead of perpendicular to it.

5. *Deformed geological objects* such as fossils, pebbles, or oolites are frequently aligned so that their long axis is oriented in the *a* direction. This lineation may also be formed by elongate minerals, deformed cavities, or elongate blobs of mineral aggregates.

Analysis of lineations has become a very powerful tool for identification of superposed deformation. The lineations and planes formed during an early deformation may become transformed into curved lines or surfaces during a second deformation; therefore, the initial problem is to determine the sequence of formation of the different *s* surfaces (S_0, S_1, S_2, etc.) and lineations (L_1, L_2, etc.) and then to relate each of the geometrical elements to the corresponding folding phase (F_1, F_2, etc.) Excellent treatments of this problem are available in Turner and Weiss (1963) and in Ramsay (1967).

GLOSSARY FOR MESOSCOPIC FEATURES

Cleavage: The property or tendency of a rock to split along secondary, aligned fractures or other closely spaced, planar structures or textures.

Fabric: The sum of all the structural and textural features of a rock.

Flow cleavage: A cleavage in which it is assumed recrystallization of the platy minerals is accompanied by rock flowage.

Foliation: A general term for a planar arrangement of textural or structural features in any type of rock, e.g., cleavage, schistosity, etc.

Fracture cleavage: A type of cleavage that occurs in deformed but only slightly metamorphosed rocks and that is based on closely spaced, parallel joints and fractures.

Hydroplasticity: Plasticity that results from the presence of pore water and absorbed water films in a sediment so that it yields easily to changes of pressure.

Kink band: A type of deformation band in which the orientation of the lattice or of the foliation is changed or deflected by gliding or slippage and shortening along slippage planes. (Syn.: **strain band**.)

Microlithon: The rock material between cleavage planes that is folded or kinked. (De Sitter, 1954.)

Mullion: A columnar structure in folded sedimentary and metamorphic rocks in which columns of rock appear to intersect.

Pore pressure (neutral stress): The stress transmitted through the fluid that fills the voids between particles of a soil or rock mass.

Riecke's principle: The statement that solution of a mineral tends to occur most readily at points where external pressure is greatest, and that crystallization occurs most readily at points where external pressure is least.

Rodding: In metamorphic rocks, a linear structure in which the stronger parts, such as vein quartz or quartz pebbles, have been shaped into parallel rods.

Schistosity: The foliation in schist or other coarse-grained crystalline rock due to the parallel, planar arrangement of mineral grains of the platy, prismatic, or ellipsoidal types, usually mica.

Slaty cleavage: Flow cleavage as it occurs in slate or other homogeneous sedimentary to low-grade metamorphic rock.

Strain band: Syn. of **kink band**.

Strain-slip cleavage: A type of cleavage that is superposed on slaty cleavage or schistosity and is characterized by finite spacing of cleavage planes between which there occurs thin, tabular bodies of rock displaying a crenulated cross lamination.

REFERENCES

Anderson, E. M., 1948, On lineation and petrofabric structure: Geol. Soc. London Quart. Jour., v. 104, p. 99–132.

Bader, H., 1951, Introduction to ice petrofabrics: Jour. geology, v. 59, no. 6.

Bailey, E. B., 1935, Tectonic essays: Oxford, Clarendon.

Balk, Robert, 1936, Structural and petrologic studies in Dutchess County, New York: Geol. Soc. America Bull., v. 47, pt. 1.

Behre, C. H., Jr., 1933, Slate in Pennsylvania: Pennsylvania Geol. Survey Bull. no. 16.

Cloos, Ernst, 1946, Lineation, a critical review and annotated bibliography: Geol. Soc. America Mem. 18.

——— 1947a, Oolite deformation in the South Mountain fold, Maryland: Geol. Soc. America Bull., v. 58, p. 843–918.

——— 1947b, Boudinage: Am. Geophys. Union Trans., v. 28, p. 626–632.

Cooper, B. N., 1961, Grand Appalachian field excursion: Virginia Polytechnic Institute Engineering Extension Series, Geological Guidebook No. 1.

Crook, K. A. W., 1964, Cleavage in weakly deformed mudstones: Am. Jour. Sci., v. 262, p. 523–531.

Crosby, G. W., 1963, Structural evolution of the Middlebury synclinorium, west-central Vermont: Ph.D. thesis, Columbia Univ., New York.

Dale, T. N., 1892, On plicated cleavage-foliation: Am. Jour. Sci., 3d ser., v. 43, p. 317–319.

Dalziel, I. W. D., and **Stirewalt, G. L.**, 1975, Stress history of folding and cleavage development, Baraboo syncline, Wisconsin: Geol. Soc. America Bull., v. 86, p. 1671–1690.

De Sitter, L. U., 1964, Structural geology, 2d ed.: New York, McGraw-Hill.

Dewey, J. F., 1965, Nature and origin of kink bands: Tectonophysics, v. 1, p. 459–494.

Epstein, J. B., 1974, Metamorphic origin of slaty cleavage in eastern Pennsylvania (abs.): Geol. Soc. America, 27th annual meeting, Miami.

Etheridge, M. A., and **Lee, M. F.**, 1975, Microstructure of slate from Lady Loretta, Queensland, Australia: Geol Soc. America Bull., v. 86, p. 13–22.

Gonzaley-Bonorino, F., 1960, The mechanical factor in the formation of schistosity: Internat. Geol. Congr., 21st, Norden, v. 18, p. 303–316.

Hancock, P. L., 1965, Axial-tract-fractures and deformed concretionary rods in South Pembrokeshire: Geol. Mag. (Great Britain), v. 102, p. 143–163.

Haughton, S., 1856, On slaty cleavage and distortion of fossils: Philos. Mag., v. 12, p. 409–421.

Holinquist, P. J., 1931, On the relations of the boudinage structure: Geol. Fören, Stockholm, v. 53, p. 193–208.

Leith, C. K., 1905, Rock cleavage: U.S. Geol. Survey Bull., v. 239.

——— 1913, Structural geology: New York, Holt.

Lohest, M., 1909, De L'origine des Veines et des Geodes des Terrains Primares de Belgique: Soc. Geol. Belgique Annales, v. 36b, p. 275–282.

Marcos, Alberto, 1973, Las series del Paleozoico inferior y la estructura Herciniana del occidente de Asturias (NW de Espana): Trabajos de Geologia, no. 6.

——— and **Arboleya, M. L.**, 1975, Evidence of progressive deformation in minor structures in western Asturian (NW Spain): Sonderabdruck aus der Geol. Rundsch., Band 64, p. 278–287.

Maxwell, J. C., 1962, Origin of slaty and fracture cleavage in the Delaware Water Gap area, New Jersey and Pennsylvania, *in* Petrologic studies—A volume to honor A. F. Buddington: Boulder, Colo., Geol. Soc. America.

Mead, W. J., 1940, Folding rock flowage and foliate structures: Jour. Geology, v. 48, p. 1007–1021.

Moore, J. C., and **Geigle, J. E.**, 1974, Slaty cleavage: Incipient occurrences in the deep sea: Science, v. 183.

Oertel, Gerhard, 1962, Extrapolation in geologic fabrics: Geol. Soc. America Bull., v. 73, p. 325–342.

Paterson, M. S., and **Weiss, L. E.**, 1961, Symmetry concepts in the structural analysis of deformed rocks: Geol. Soc. America Bull., v. 72, p. 841–882.

——— 1966, Experimental deformation and folding in phyllite: Geol. Soc. America Bull., v. 77, no. 4, p. 343–374.

Powell, C. McA., 1974, Timing of slaty cleavage during folding of Precambrian rocks, northwest Tasmania: Geol. Soc. America Bull., v. 85, p. 1043–1060.

Ramberg, Hans, 1955, Natural and experimental boundinage and pinch and swell structures: Jour. Geology, v. 63, no. 6.

——— 1960, Relationships between length of arch and thickness of ptygmatically folded veins: Am. Jour. Sci., v. 258, no. 1, p. 36–46.

Ramsay, J. G., 1967, Folding and fracturing of rocks: New York, McGraw-Hill.

Roberts, David, 1971, Abnormal cleavage patterns in fold hinge zones from Varanger Peninsula, northern Norway: Amer. Jour. Sci., v. 271, no. 2, p. 170–180.

——— 1972, Tectonic deformation in the Barents Sea region of Varanger Peninsula, Finnmark: Norges Geologiske Undersøkelse.

——— and **Strömgård, Karl-Erik**, 1971, A comparison of natural and experimental strain patterns around fold hinge zones: Tectonophysics, p. 105–120.

Sander, B., 1930, Gefugekunde der Gesteine: Vienna, Springer.

Sanderson, D. J., 1974, Patterns of boudinage and apparent stretching lineation developed in folded rocks: Jour. Geology, v. 82, no. 5, p. 651.

Sharpe, D., 1849, On slaty cleavage: Geol. Soc. London Quart. Jour., v. 5, p. 111–129.

Siddens, A. W. B., 1972, Slaty cleavage—A review of research since 1815: Earth-Science Revs., v. 8, no. 2, p. 205ff.

Smith, R. B., 1975, Unified theory of the onset of folding, boundinage, and mullion structure: Geol. Soc. America Bull., v. 86, p. 1601–1609.

Sorby, H. C., 1853, On the origin of slaty cleavage: Edinburgh New Phil. Jour., v. 10, p. 137–147.

——— 1856, On the theory of slaty cleavage: Philos. Mag., v. 12, p. 127–129.

Spry, Allen, 1964, The origin and significance of snowball structures in garnet: Jour. Petrology, v. 4.

Tobisch, O. T., 1971, Nappe formation in part of the southern Appalachian Piedmont: Geol. Soc. America Bull., v. 82, p. 2209–2230.

Tullis, Terry E., 1976, Experiments on the origin of slaty cleavage and schistosity: Geol. Soc. America Bull., v. 87, no. 5.

Turner, F. J., 1942, Current views on the origin and tectonic significance of schistosity: Royal Soc. New Zealand Trans., v. 72, pt. 20.

——— 1948, Mineralogical and structural evolution of metamorphic rocks: Geol. Soc. America Mem. 30.

——— and **Weiss, L. E.**, 1963, Structural analysis of metamorphic tectonites: New York, McGraw-Hill.

Walls, R., 1937, A new record of boudinage structure from Scotland: Geol. Mag. (Great Britain), v. 74, p. 325–332.

Wegmann, C. E., 1932, Note sur le boudinage: Soc. Geol. France Comptes rendus, v. 5, pt. 2, p. 477–489.

Weiss, L. E., and **McIntyre, D. B.**, 1957, Structural geometry of Dolradian Rocks at Loch Leven, Scottish Highlands: Jour. geology, v. 65, no. 6.

White, W. S., 1949, Cleavage in east-central Vermont: Am. Geophys. Union Trans., v. 30, p. 587–594.

Williams, P. F., 1972, Development of metamorphic layering and cleavage in low grade metamorphic rocks at Bermagui, Australia: Amer. Jour. Sci., v. 272, no. 1, p. 1–47.

Wilson, Gilbert, 1946, The relationship of slaty cleavage and kindred structures to tectonics: Geol. Assoc. London Proc., v. 57, p. 263–302.

——— 1953, Mullion and rodding structures in the

Moine Series of Scotland: Geol. Assoc. Proc., v. 64, p. 118–151.

——— 1961, The tectonic significance of small scale structures, and their importance to the geologist in the field: Ann. Soc. Geol. Belgium, v. 84, p. 423–548.

Wilson, M. E., 1953, Early Precambrian rocks of Western Quebec: Geoi. Soc. America Bull., v. 64, p. 1492.

Wynne-Edwards, H. R., 1963, Flow folding: Am. Jour. Sci., v. 261, p. 793–814.

13

Folding in Principle and Experiment

THEORY AND EXPERIMENTS IN FOLDING*

Theoretical analysis of folding grows out of the physical theory of elasticity and fluid dynamics. To a first approximation rock layers are idealized in terms of mechanically simple models—e.g., the rock is assumed to behave like an ideal elastic, viscous, or elasticoviscous substance. Other common simplifying assumptions are:

1. Conditions of plane stress and strain apply.

2. Viscous materials are treated as being incompressible.

3. Viscous materials exhibit linear viscosity (viscosity is independent of the strain rate), and simple stress configuration is postulated to initiate folding. Usually the principal stress is applied in a direction parallel to the layering.

The models analyzed theoretically are of the following general types:

1. An elastic layer surrounded by air.

2. A pile of elastic layers with variations in layer thickness.

3. An elastic layer embedded between viscous or elasticoviscous layers.

4. Interbedded elastic and viscous layers.

5. A pile of viscous layers. This may be varied by assuming that a strong viscosity contrast exists among the layers, that all layers have the same viscosity, and that the number of layers and their thickness may be varied.

Physical modeling has become increasingly sophisticated. Early models were designed to

* The basic derivation of theories now widely used in theoretical and experimental work with folds is found in the writings of Maurice Biot (1957, 1964, 1965) for buckling stability of elastic, viscoelastic, and viscous layered media and Hans Ramberg (1967, 1968) for a fluid dynamics approach. Ramsay (1967) presents one of the most complete accounts of the use of finite strain analysis. Refinement and application of these approaches is found in the writings of Chapple (1968, 1969), Bayly (1970, 1971), Price (1967), Dieterich and Carter (1969), S. H. Treagus (1973), Smythe (1971), Sherwin and Chapple (1968), and Hobbs (1971).

approximate rocks only crudely. Later attempts were made to use scale-model experiments in which time, dimensions, viscosity, and elastic properties of materials in the models were related to one another in the model in ratios similar to those that exist in the earth. Modern models are constructed to duplicate experimentally the conditions postulated in theoretical treatments, and these results are compared with actual field examples which they approximate most closely.

Variables in Folding

Many factors are important in folding processes:

1. The physical conditions under which folding takes place and which affect material behavior:
a. Temperature
b. Confining pressure
c. Solutions
d. Strain rate
e. Pore pressure

2. The character of the folded rocks:
a. The physical properties of the strata at the time of deformation.
b. The homogeneity of physical properties within each layer.
c. Variations within the stratified section of thickness and rock types.

3. The application of the deforming stresses:
a. The direction of application of the principal stresses.
b. The rate at which deformation proceeds.
c. The duration of the applied stress.
d. Uniformity of direction or stress and rate of deformation.

Several of these major categories have already been examined as they apply to small samples of rock (Chap. 4). Among the significant conclusions reached were:

1. Most rocks deform as brittle elastic substances at low temperatures and low confining pressures and high strain rates.

2. At higher temperatures and confining pressures, most rocks tend to exhibit plastic deformation or elasticoviscous flow.

3. Even under relatively low loads, rocks tend to creep when the directed stress is sustained over long periods of time.

These conclusions suggest that different mechanisms of folding are related to the physical conditions under which deformation occurs. Thus superficial folding of crustal rocks at relatively shallow depth and at low temperatures may best be explained in terms of elastic deformation, especially if the folding is of short duration, while folding of rocks at depth, under conditions of metamorphism, or over long time intervals will likely be accounted for in terms of viscous behavior.

For the purposes of experimental analysis the character of the folded rocks is greatly simplified. Generally the rock units are treated as a stratified sequence in which each layer is considered to be homogeneous, of uniform thickness, and of uniform elastic moduli or viscosity.

Scale-Model Experiments*

Although the validity of conclusions reached from model studies is debated, features formed experimentally often closely resemble those seen in the field. Moreover, results of model experiments have been influential in the development of structural interpretations. Varying degrees of care have been exercised in the setting up of models. Some have been carefully designed so that the dimensions and physical properties of the materials used are accurately scaled. Other experiments are made using materials which approximate only in a general way the rocks they represent.

Scale models are designed by selecting appropriate dimensions for the model, choosing a suitable length of time for the experiment to run, and obtaining a material with appropri-

* See Hubbert (1937).

ately scaled density and viscosity. Model ratios for time, dimensions, and density are established as follows:

$$\frac{L_1}{L_2} = \lambda \qquad L_1 = \text{length in nature}, \quad L_2 = \text{length in model}$$

$$\frac{t_1}{t_2} = \tau \qquad t_1 = \text{time in nature}, \quad t_2 = \text{time in model}$$

$$\frac{\rho_1}{\rho_2} = \delta \qquad \rho_1 = \text{density in nature}, \quad \rho_2 = \text{density in model}$$

Model ratios for such mechanical quantities as area, volume, frequency, velocity, acceleration, momentum, force, work, power, stress, elastic modulus, and viscosity can be expressed in terms of dimension, time, and density. Such characteristics as viscosity and strength are dependent on the values selected for length and duration of experiment; thus the selection of materials cannot be random if properly scaled results are to be obtained. The basic criterion used in scale-model theory is that of dynamic similarity: that all forces of like kinds be proportional. The conditions are approximately satisfied, provided inertial forces are negligible, which they are for deformation of plastic or elasticoviscous deformation, and provided the gravitational constant has remained constant. The application of this scale-model theory is found in the experiments of Nettleton (1943) for salt dome formation and Bucher (1956) for formation of fold structures.

Classic Model Experiments

Bailey Willis (1893) produced folds resembling those of the Appalachian region long before the principles of scale-model design were formulated. His models were constructed of layers composed of mixtures of wax, plaster, and turpentine. Some of the layers were much too hard to be suitable for use in scale models, but this was overcome in part by the use of heavy loads of lead shot which built up confining pressure. The models were compressed by a lateral pressure, and results such as those illustrated in Fig. 13-1 were obtained. The box-shaped folds produced in the experiment resemble folds of the Jura Mountains. Folds in the model developed over a rigid basement and by lateral compression accompanied by flow of the soft material into the cores of anticlines from adjacent synclines. A similar hypothesis has been advanced to explain the Jura folds.

In another experiment, Willis modeled the development of a thrust fault on the forelimb of an asymmetric anticline, Fig. 13-2. As the layers arched upward, faults formed in the ductile layers, and material moved into the arch formed by the brittle layers along these faults. As shortening continued the strata in the inversion first stretched and then faulted to produce an overthrust similar to those in the Alps described by Heim.

Bucher (1956) used scale models to investigate the role of gravity in foreland folding. He used stitching wax, a temperature-sensitive elasticoviscous material with a viscosity of 10^6 poises, and petrolatum. These materials provide excellent scaled-down equivalents of estimated rock viscosities.

One set of experiments was designed to simulate deformation due to gravity in pseudoviscous layers tilted into an inclined position. A model consisting of interbedded layers of stitching wax and grease was con-

FIGURE 13-1 Concentric folds produced in a pressure box by lateral compression. (*Redrawn from Willis, 1934.*)

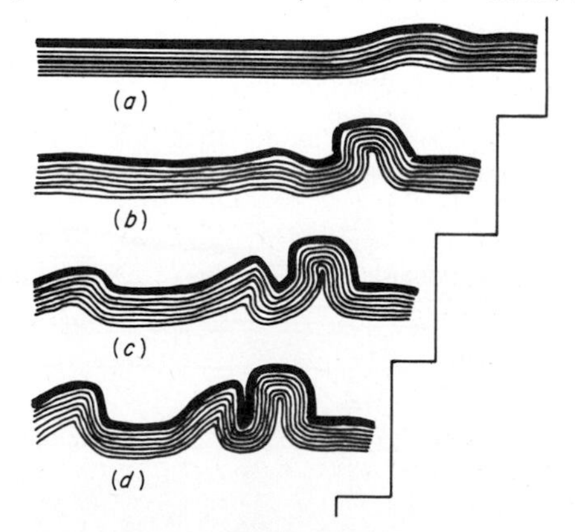

structed so that the top surface at one end sloped 23°. As the wax at the high end of the mass spread under the force of gravity, the layers corresponding to the foreland were thrown into a system of disharmonic folds resembling those in mountain belt forelands.

A second set of experiments was designed to simulate the formation of nappe-like structures of the Alpine type. A wooden piston was driven at a slow constant rate into a thickened wax layer at one end of the model. Wax at the thicker end was forced out and upward. Since the piston moved several times faster than the creep rate of the wax, a pile of wax built up. As the wax spread out, a larger recumbent fold and root formed, Fig. 13-3. The forelimb of the fold is greatly thinned and drawn out; digitations similar to those of Alpine recumbent folds developed, as did surficial folds in front of the nappe. Close to the edge of the nappe the folds are overturned, but farther away from this leading edge the folding dies out in a manner analogous to that seen in many orogenic belts. The actual shortening which took place was 20 cm, but measurement of the bed's length along the folds, including the nappe, leads to estimates of 80 to 100 cm, or nearly five times the actual shortening and raises questions about the amounts of shortening reported in orogenic belts.

Photoelastic Strain Analysis

Photoelastic techniques* afford a method by which the pattern of stress in a transparent medium can be directly observed. The method is based upon the phenomenon of double refraction or birefringence. David Brewster, in

* Refer to Frocht (1964) for a complete discussion of the methods.

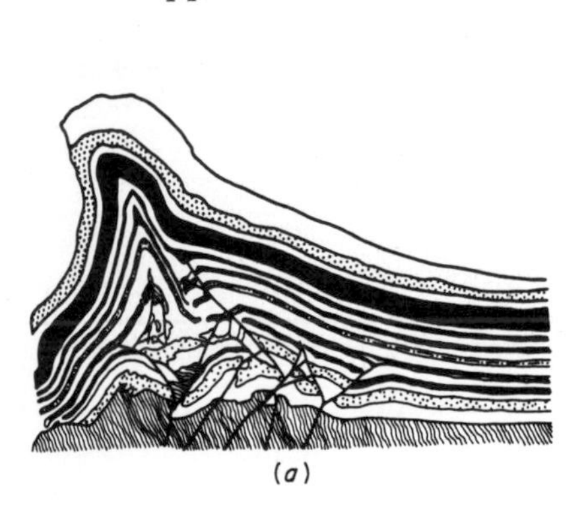

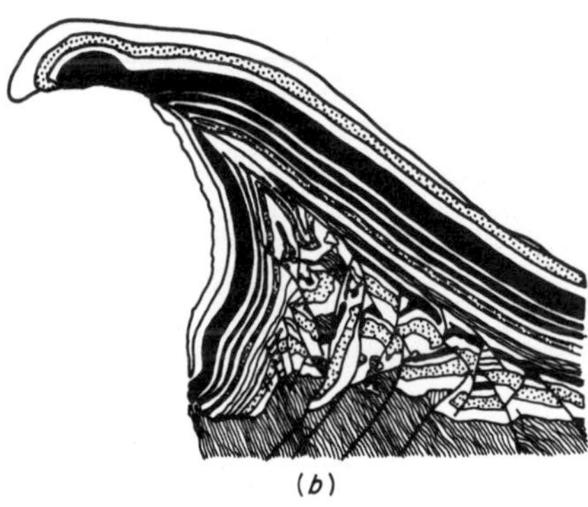

FIGURE 13-2 Folds formed in pressure-box experiments. The rise of the competent anticline continued until it overtopped the piston by which pressure was applied. Then the nearer limb of the fold was pushed under the further limb and the strata in the inversion were stretched, producing an overthrust of the alpine type described by Heim—a stretch thrust. (*Redrawn from Willis, 1934.*)

FIGURE 13-3 Folds and nappes formed experimentally in stitching wax. Layers of wax were laid horizontally between a barrier at the right and a wooden block, and other wax layers covered the block. The block (diagonal lines) was moved laterally into the wax, which bulged upward initially and then flowed laterally, folding the wax above the block as a result of gravitative spreading. (*From Bucher, 1956.*)

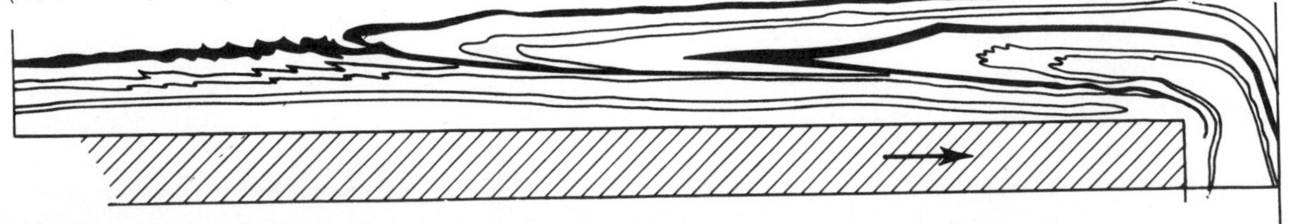

1815, discovered that a distinctive color pattern due to stress can be seen when a piece of stressed glass is viewed by polarized light transmitted through it. Monochromatic light is now used to produce a black and white fringe pattern such as that in Fig. 13-4. The method is used extensively in modeling of engineering structures and is being increasingly applied to geological problems.* For fold studies the glass is replaced by gelatin, which exhibits similar behavior.

The birefringence is caused by effects of the principal stresses oriented in the plane transverse to the axis of propagation of the polarized light. The stress pattern consists of colored bands in cycles of yellow, red, and green. Areas of the same color are called *isochromatics*. The fringes represent the loci of points of equal stress differences between the maximum and least principal stresses oriented in the plane transverse to the propagation of the light.

As an example of this technique consider the model, Fig. 13-4, of a rubber strip embedded in gelatin which has been shortened 18 percent by compression. The black dots with first-order fringes are points of no finite strain. Trajectories connecting points of equal strain can be constructed from this pattern of isochromatics.

* Currie, Patnode, and Trump (1962); Ramberg and Strömgård (1971).

Folding by Elastic Buckling

As defined in physics, buckling of a structural system occurs when the system is stressed in such a way that the original configuration becomes unstable and the system assumes some other more stable shape. Flexural buckling is a special type in which the bar (plate, etc.) is loaded so that it is compressed at the ends (edges) and parallel to the initial bar length, Fig. 13-5. Ideally the bar is hinged at each end so that rotations are unrestrained. When a plate (layer) is subjected to a uniform compression, it remains straight and stable until the stress obtains a critical value:

$$\sigma_{crit} = \frac{\pi^2 EI}{L^2} \qquad \text{(Euler equation)}$$

where E = Young's modulus
I = moment of inertia of the plate
L = wavelength of the buckle

When this critical value is reached, the plate flexes about its weakest axis and assumes a

FIGURE 13-4 Isochromatic fringes produced by buckling rubber strips embedded in gelatin. This model shows offsets after an 18 percent bulk compression. (*From Roberts and Strömgård, 1972.*)

sinusoidal form with greatest amplitude at the midlength. Buckling is elastic when all regions of the material of the structure remain elastic during buckling. (The deformation is completely recoverable upon removal of the loads.) (Rocks exhibit elastic rigidity in the range of 10^{10} to 10^{12} dynes/cm^2.)

Not all geologists use this definition precisely. Bayly (1971) adopts a more general definition that does not imply specific movements such as flexural slip. He defines buckling as a process that (1) involves a slab or slabs, more or less planar initially, with compressive stresses applied parallel to the initial attitude of the slabs; and (2) results in conversion of the slabs to a geometry where departures from planarity are much greater than before, and where the overall length of the slabs measured in the general direction of the compressive stresses is less than before.

Both field studies and laboratory experiments indicate that stratification, its thickness and arrangement of layers, is significant in determining how a sequence of rocks responds to deformation. The behavior of the units is dependent on the arrangement of units of different ductility or elasticity in the stratigraphic column. The physical properties and thickness of a dominant member are generally found to control the fold wavelength that develops in the early stages of deformation in the buckling process.

Three special cases of elastic buckling have been analyzed by Biot (1961) and by Currie, Patnode, and Trump (1962), Figs. 13-6 and 13-7:

Case I involves a single competent layer (one that has low ductility) in air.

Case II involves a dominant layer of thickness T located between two elastic layers of thickness J which have rigid outer boundaries.

Case III involves a large number n of thin elastic layers having individual thicknesses t and a total thickness T.

In all these cases the load is applied parallel to the layering; the layer buckles rapidly and assumes a sinusoidal form. Both the load necessary to cause buckling and the wavelength are found to depend on the elasticity of the layers and the gravity buoyancy of the fluid around the plate.

Case I. The Euler equation applies in this case. This equation was derived from engineering problems carried out under 1 atm of pressure, but most deformation in the earth takes place at some depth; so, the critical stress, σ_{crit}, under which buckling is initiated in the earth is equal to the difference between the maximum principal stress and the confining pressure.

The moment of inertia I of a buckled layer is a function of the width w and thickness T of the layer:

$$I = \frac{T^3 w}{12}$$

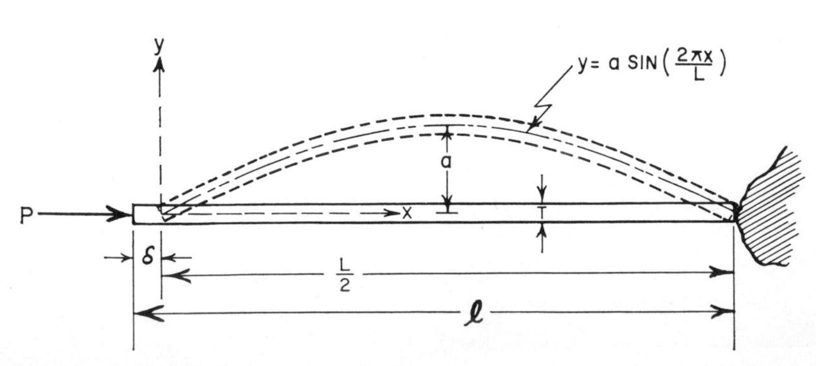

FIGURE 13-5 Sketch of the buckling of a weightless beam with hinged ends. The original length of the beam (l) is shown buckled into a fold with wavelength (L). (*From Currie, Patnode, and Trump, 1962.*)

If the width of the layer is taken as unity, $I = T^3/12$ and $\sigma_{crit} = \pi^2 E T^3/12L^2$. Thus the critical stress is a function of the unit thickness as well as the elasticity of the layer. A sinusoidal wave of wavelength L is produced as a result of the buckling (Fig. 13-5), and that wavelength is given by the following expression:

$$L = 2\pi T \sqrt[3]{\frac{E}{6E_0}}$$

where E = Young's modulus of embedded unit
E_0 = Young's modulus of medium in which it is embedded

The medium in which the competent layer is embedded has to have an elastic modulus less than one-hundredth that of the embedded member to allow elastic instability. Such a marked difference in elastic moduli in nature would exist in the case of a cemented sandstone or quartzite interbedded with thick, weak materials such as shale, salt, gypsum, clay, and mudstone. Price (1967) cites the ratio of shear moduli for cemented sandstone embedded in mudstone as 1,000:1. Much more common ratios are in the range of 5:1 to 50:1. The corresponding values for the ratio L/T of the dominant member are 6:1 to 12:1.

If a ratio of $E/E_0 = 100$ is selected, the equation above indicates that the lowest possible wavelength-to-bed thickness ratio for this case is 16:1.

Case II. The wavelength of the competent layer under these circumstances becomes

$$L = \sqrt[4]{\frac{2\pi^4 E T^3 J}{3E_0}}$$

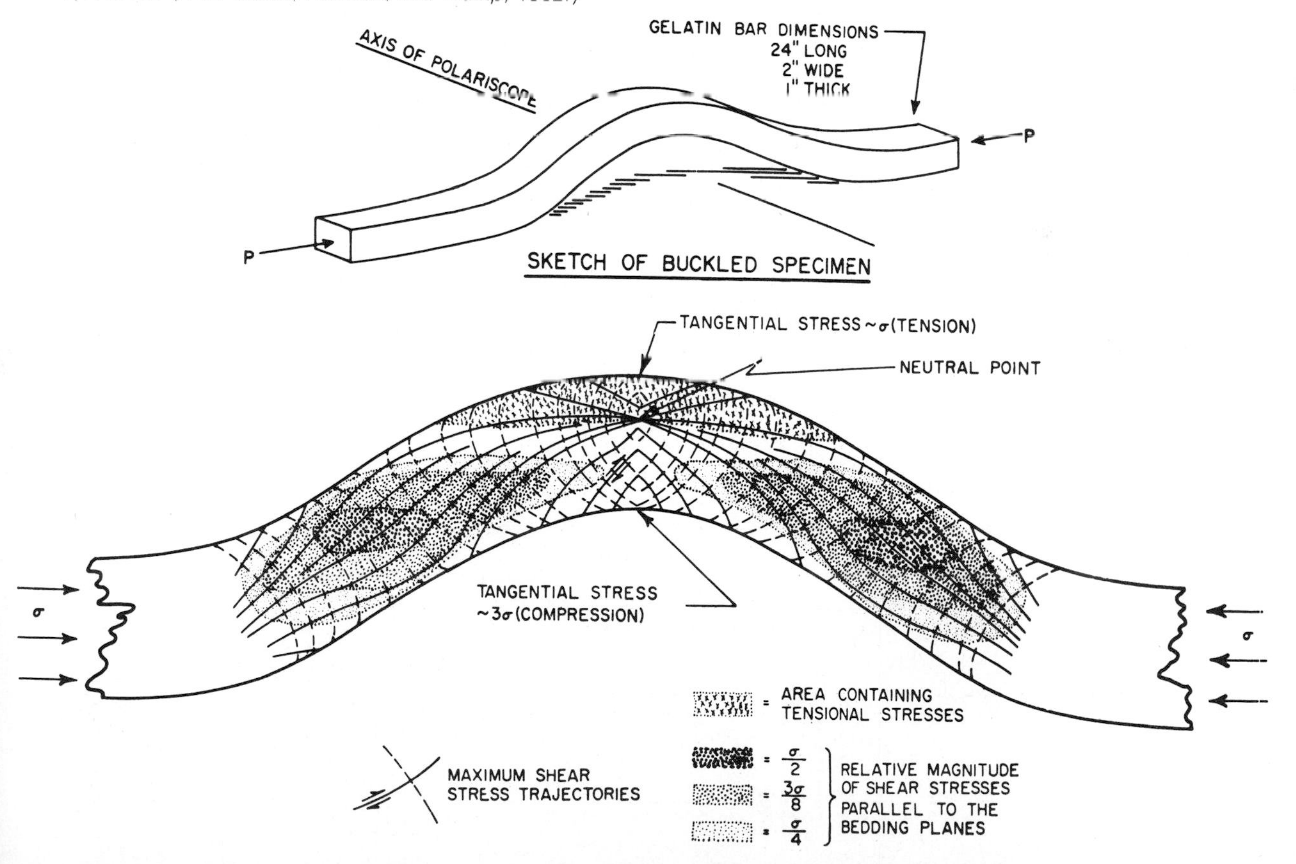

FIGURE 13-6 A qualitative photoelastic study of the crest and flanks of a buckled gelatin beam. Note the stress trajectories. (*From Currie, Patnode, and Trump, 1962.*)

This equation reduces to a value that is close to that in case I when J is equal to or greater than the wavelength.

Case III. The dominant wavelength becomes

$$L = 2\pi T \sqrt[3]{\frac{E}{6n^2 E_0}}$$

where E_0 is the elasticity of an infinite medium in which the layers are embedded. This case is most favorable to elastic instability, but it is dependent on the assumption that the boundaries of individual competent members are frictionless.

To test the above formulas, Currie and others (1962) performed a series of experiments using gelatin layers for incompetent units and gum rubber strips for more competent units. The Young's moduli of the two are 10 and 100 psi, respectively. Photoelastic experiments confirm the theoretical predictions of fold development through buckling of a dominant member within a relatively less competent medium.

The theoretical values for wavelengths were compared with actual field examples of folded sequences by Currie and others. The fold wavelength of a number of folds in sedimentary rocks is plotted against the thickness of the dominant member, Fig. 13-8. They found a linear relationship between fold wavelength and dominant-member thickness. For this type of comparison particular care must be exercised in dividing the sedimentary section into competent and incompetent units. Formational boundaries do not always correspond to boundaries separating rock layers of different structural properties. In summary, this study indicates that the thickness of a dominant member and the relative physical properties of the layers control the wavelength that develops in early stages of buckling.

Folding of Viscoelastic Materials

Many rocks behave as elastic substances at least initially when they are deformed at low temperatures and low confining pressures—at shallow depths. They are elastic and brittle even at greater depths when the strain rate is high, but when strain rates are low and tem-

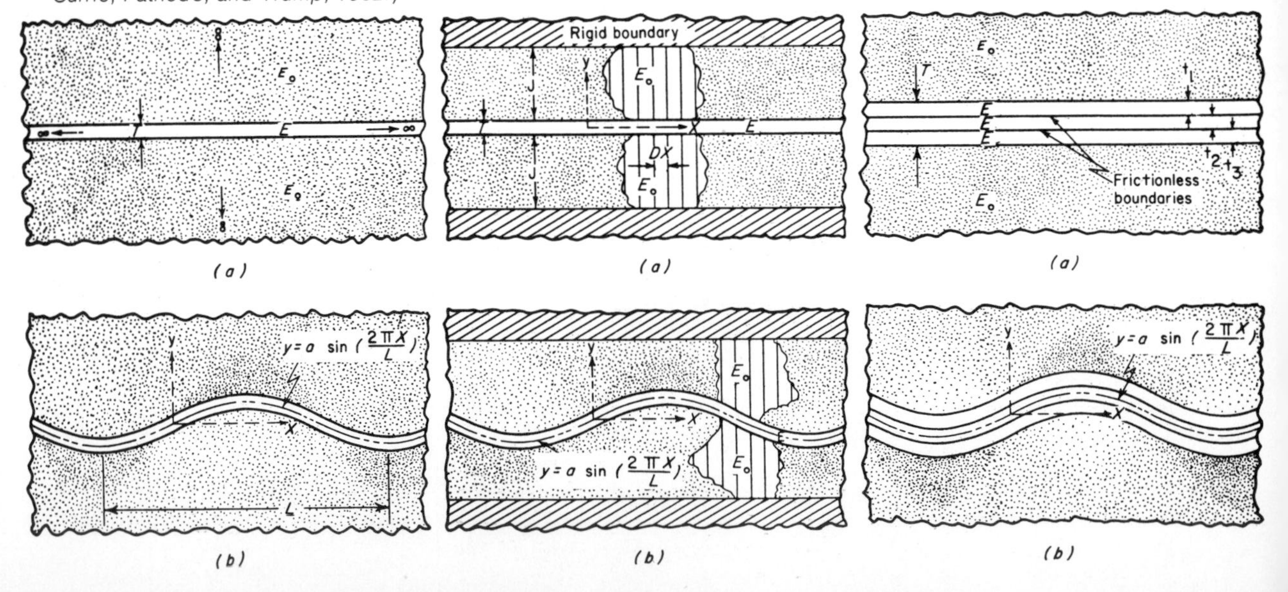

FIGURE 13-7 Sketch of the buckling of a beam in a continuous, infinite medium (left); sketch of the buckling of a beam in a medium made up of vertical columns bounded by rigid members (center); sketch of the buckling of multiple members in an infinite medium (right). (*From Currie, Patnode, and Trump, 1962.*)

perature and pressures are high, rock behavior is much more that of the ideal elasticoviscous material. The elasticoviscous model is applicable to deformation that takes place under conditions of high-grade metamorphism for all rocks, and also for unconsolidated sediments, salt, and clay even under near-surface conditions. Biot (1961, 1965) and Ramberg (1963) laid the mathematical basis for analysis of folding of viscous layers. Three models analyzed by Biot are described below. In each case folds of a certain wavelength grow much more rapidly than others and become the dominant wavelengths.

CASE I: BUCKLING ELASTIC PLATE IN A VISCOUS MEDIUM

This case approximates the folding of a competent bed in an incompetent medium (e.g., sandstone in shale, etc.). A dominant wavelength forms which is given by the following equation:

$$L_D = \pi T \sqrt{\frac{E}{(1 - v)^2 P}}$$

where v = Poisson's ratio

The wavelength in this case is dependent on the compression load P as well as the thickness of the elastic plate (T).

CASE II: VISCOUS PLATE IN A VISCOUS MEDIUM

In elasticoviscous and viscous deformation the sudden instability and rapid buckling of flexural buckling does not occur. Instead the deformation is continuous; it grows as a function of time, and folding depends on initial imperfections and departures from plane. The rate of growth of a deflection is a function of the wavelength of the folds. Some wavelengths are found to grow faster than others; one will be the fastest growing, and it is called

FIGURE 13-8 Log-log plot of the wavelength to dominant member thickness for field examples. *(From Currie, Patnode, and Trump, 1962.)*

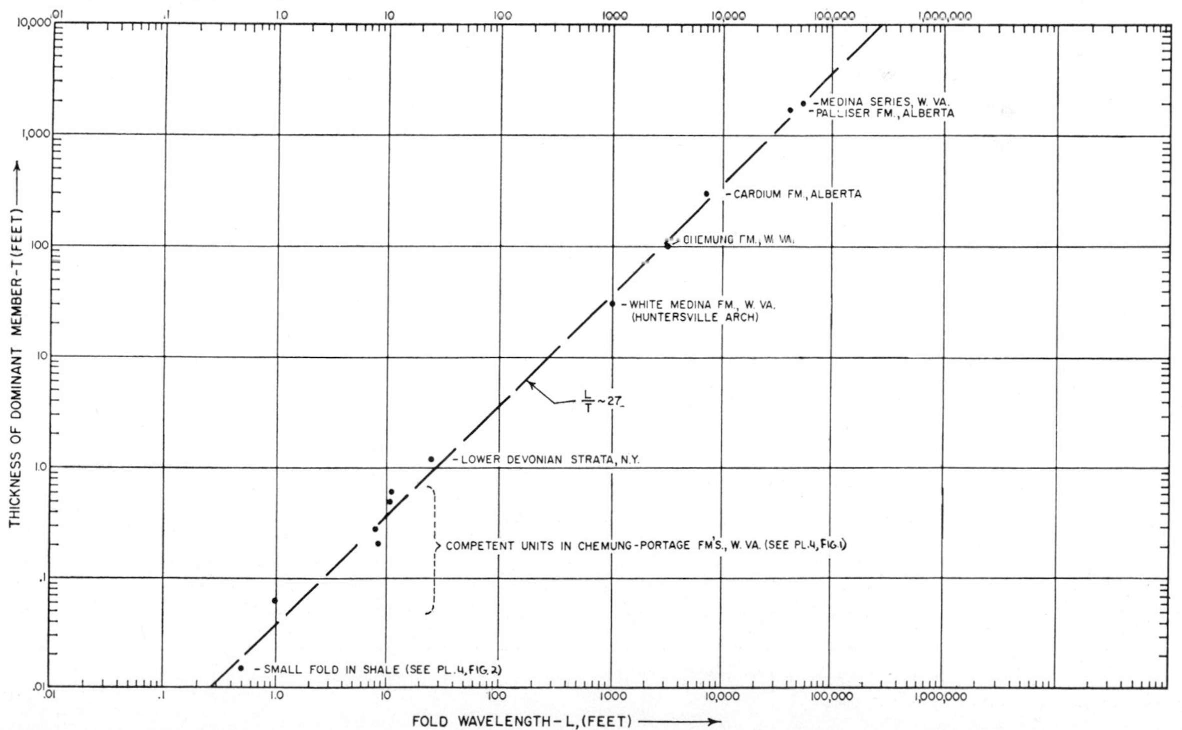

the *dominant wavelength* L_D. After some lateral compression this wavelength appears as a more or less regular sinusoidal wave. Folding of a compressed viscous plate in a viscous medium is thought to approximate conditions of the folding which takes place during high-grade metamorphism. An equation for the dominant wavelength in this case is

$$L_D = 2\pi T \sqrt[3]{\frac{\eta}{6\eta_1}}$$

where η = viscosity of layer
η_1 = viscosity of medium

Note that the dominant wavelength is not dependent on the compressive stress in this case. The stress does effect the rate of folding but not the dominant wavelength, which is 20 to 50 times the layer thickness for η/η_1 ratios expected to occur in rocks.

Biot extends this theory to define the rate at which an initial wavelength present in the viscoelastic layer is amplified. The amplification factor is a function of the viscosity ratio of the layer and medium and time. When the amplitude of the folds reaches a certain point, it increases at an explosive rate. The time required for this to happen is a function only of the viscosity ratios. For example, if a layer of viscosity 10^{21} poises is interbedded in other layers of viscosity 10^{18} poises and is subjected to a compressive load of 10^8 dynes/cm^2 (about 1,450 psi), the time required for the layer to shorten by 25 percent is

$$\text{Time} = \frac{\eta}{P} = \frac{10^{21}}{10^8} = 10^{13} \text{ sec} = 317{,}000 \text{ years}$$

and the dominant wavelength is $L_D = 37.5 \times$ layer thickness. These values fall within the range of geologically feasible values.

CASE III: MULTIPLE VISCOUS LAYERS

The dominant wavelength, when multiple layers N of identical thickness T and viscosity η (with lubricated boundaries between layers) are embedded in a viscous medium η_1 and subjected to a horizontal compression P, is

$$L_D = 2\pi T \sqrt[3]{\frac{N\eta}{6\eta_1}}$$

and all conclusions for the single layers are applicable to the multiple-layer case except that the values are affected by the value of N.

Application of Theory to Field Studies

These studies can be applied to field studies of folds and contribute significantly to our understanding of the folding process. The following are among the most significant of Biot's conclusions:

1. Rocks are elastic for fast deformation. They tend to behave like viscous materials with viscosities in the range of 10^{17} to 10^{22} poises for slow deformation. When compressive stresses act over a long period of time, on the order of a million years, even relatively small loads may produce explosive folding in rock of great strength. Much more time is required for elastic folding at low load values.

2. Viscous behavior dominates over elastic for viscoelastic deformation at low loads, but rupture and plastic deformation dominate, especially at fold hinges, where bending stresses are great for high loads.

3. Maximum stresses occur at the points of maximum amplitude. They are compressions in troughs, tension over crests. When the maximum stress reaches failure levels, creep rates increase, cracks form, and local weakening of the layer occurs. Further deformation occurs as if the layer were hinged at these points and the bends are sharpened. At high stresses, fracture and plastic behavior are almost instantaneous. Therefore, fracture depends on strain rate.

4. Viscous behavior often predominates in tectonic folding, and this leads to large deformations without fracture.

5. For purely viscous deformations and when the influence of gravity is not important, the dominant wavelength is independent of the tectonic stress.

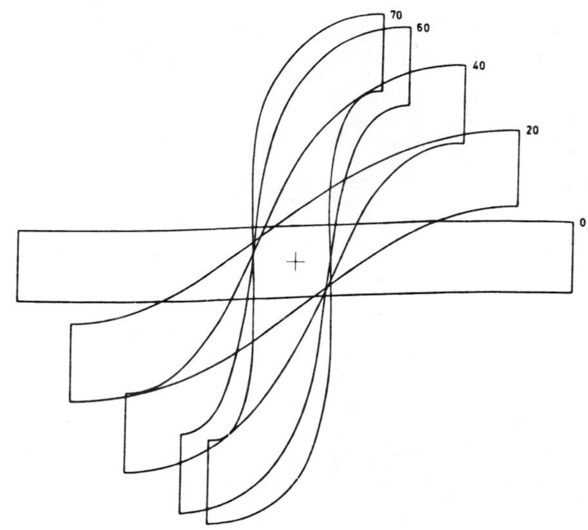

FIGURE 13-9 Single-layer buckle folds produced by finite-element analysis. Progressive development of fold shape with overall shortening. The initial wavelength/thickness ratio is taken from the basic equation of Biot (1961) and Ramberg (1961). Viscosity ratio $\mu_1/\mu_2 = 100$. (*From Hudleston and Stephansson, 1973.*)

6. The *dominant wavelength* depends on the cubic root of the viscosity ratio, with the consequence that it is not sensitive to large variations in viscosity contrast.

Analysis of Strain in Folding

Analysis of folding has grown substantially following the pioneering work of Biot and Ramberg. This has been aided by the use of computer techniques which allow rapid calculation of stresses and strains on a point by point basis throughout a folded layer. It is possible to trace the path of any given point as folding proceeds.

An example of this type of finite element analysis in the case of a single-layer buckle fold is illustrated in Fig. 13-9, which depicts the changes in fold form that accompany progressive shortening.

Mathematical analysis allows calculation of strain ellipses from point to point in a bed folded to form parallel or similar folds, as shown for a buckle fold, Fig. 13-10. A dashed line connects points along a *neutral surface* in which the strain ellipses show little strain. Tensional stress causes extension in the layer above the neutral zone over the crest of the fold and the lower part of the folded layer is compressed. Compare this with the section of a folded layer of rubber experimentally deformed, Fig. 13-11. Strain analysis for a similar fold, Fig. 13-12, emphasizes the difference between similar and parallel folds. Strain does not vary significantly from top to bottom of a similar fold as it does in parallel folds. Minimal strain occurs over the crest of a similar fold, but strain increases from the crest onto the limbs. Note that the strain ellipses are inclined toward the fold hinge.

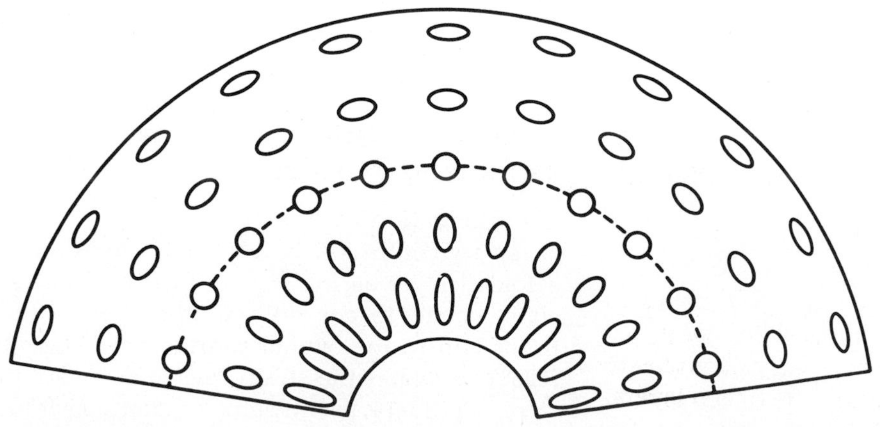

FIGURE 13-10 Distribution of strain in a bar deformed by buckling. The surface of no-finite strain is shown dotted. (*From Hobbs, 1971.*)

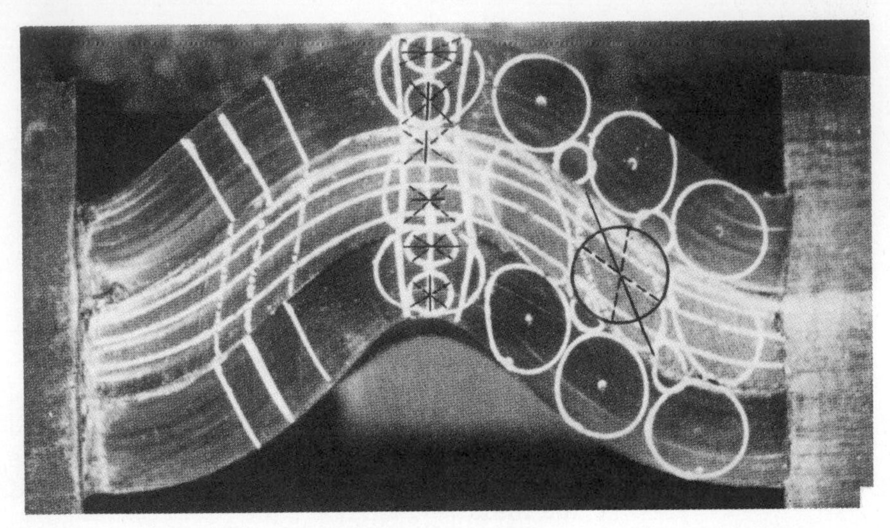

FIGURE 13-11 A layer of soft rubber flanked by two layers of stiffer rubber about 12 mm thick. Stippled black lines indicate finite maximum shear directions; full black lines are parallel to finite maximum extensive strain. The black circle on the flank of the fold is identical in size to the original unstrained circular markers. Lines between intersections of this circle with the strained ellipse that has the same center are almost parallel to the two directions of finite maximum shearing strain. (*From Ramberg, 1963b.*)

FIGURE 13-12 Distribution of strain in ideally similar, sine-curve folds. (*a*) No shortening; (*b*) 20 percent shortening; and (*c*) 50 percent shortening. (*From Hobbs, 1971.*)

A

B

C

We have good reason to believe that some essentially homogeneous shortening accompanies much folding. This is responsible for the numerous flattening effects seen best in studies of deformed primary features (e.g., oolites). Shortening effects on strain ellipses are found for both similar and parallel folds, Figs. 13-12 and 13-13. In both instances this flattening produces a strong alignment of flattened features subparallel to the axial surface of the fold. Note that the long axes of ellipses in flattened similar folds become rotated from a position of being inclined toward the axial surface to a position approaching parallelism with the axial surface as flattening increases, Fig. 13-12. The axes of flattened ellipses in parallel folds vary in inclination with position on the fold.

Variation in Stress and Strain as Folding Proceeds

The orientation of principal stress and strain axes vary continuously from point to point in a folded layer as folding continues. Progressive changes occur both in the dip of fold limbs and in the overall shortening. Chapple (1968) calculates the strain axes and the strain rate, Fig. 13-14, from point to point in folds characterized by limb dip. [*Note:* The small symbols to the right of the figures show the strain axes (crosses) and strain rate (vertical line) calculated for uniform homogeneous compression equivalent in magnitude to that in the folded rock mass.]

The variation in stresses at four points in a folded viscous layer in a less viscous matrix as folding proceeds is shown in Fig. 13-15. Total percent strain and accompanying limb dip are plotted on the abscissa, and stress amounts are plotted on the ordinate. The position of the four points is shown in the insert. This study represents a significant advance over the strain ellipses on the flexed rubber sheet in that relative stress magnitudes as well as the way these change is indicated. Note that the stress at a point on the fold trough reach a maximum value when the limbs dip about 20°, while compressive stresses are a maximum initially for all other cells considered. Points above the mid-line of the layer experience compressive stresses initially but these become tensile stresses as the folding takes place. The upper part of this figure shows the orientation of the principal stress direction. Initially these were all parallel and oriented in the plane of the layering. As soon as the layer deflected from the initial plane the stresses became reoriented. This reorientation continues as the fold form changes and are shown here after 40 percent shortening.

FIGURE 13-13 Folds formed by buckling together with various amounts of homogeneous shortening normal to the axial plane. (*a*) 20 percent shortening. (*b*) 50 percent shortening. These are parallel folds with a superimposed homogeneous shortening (i.e., inhomogeneous strain + homogeneous strain = total strain). (*From Hobbs, 1971.*)

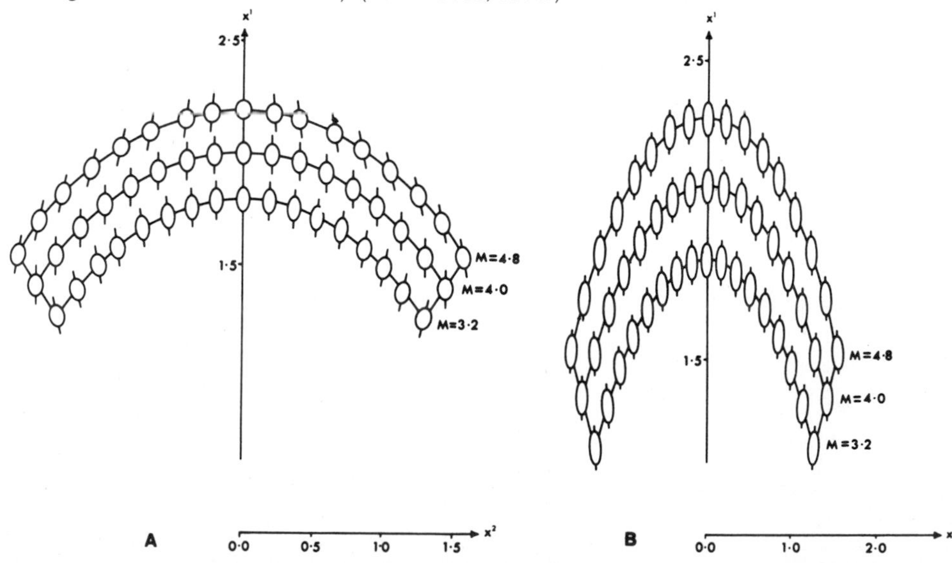

Development of Asymmetric Folds

The theories and experiments described in preceding sections generally postulate that the maximum principal stress acts parallel to the layering. Folds formed in this way are symmetric in section, and their axial planes are normal to the principal shortening of the system. However, many real folds are asymmetric in section and have inclined axial surfaces. The formation of this configuration has been investigated by Ramberg (1963), Ghosh (1966), Price (1967), Smythe (1971), and Treagus (1973).

Ramberg (1963) and Ghosh (1966) lay a theoretical basis for asymmetric folding which is

FIGURE 13-14 Orientations of the axes of strain ellipses are shown by crosses, and extension rates are shown by the length of the solid dashes for dominant wavelength folds with limb dips of (1) 23, (2) 46, (3) 66, and (4) 89°. In each case the axes of the strain ellipse and strain rate for an equivalent uniform compression are shown along the right edge of the figure. (*From Chapple, 1968.*)

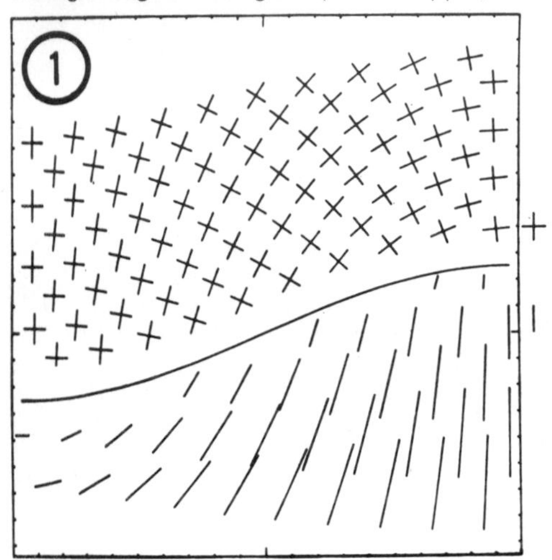

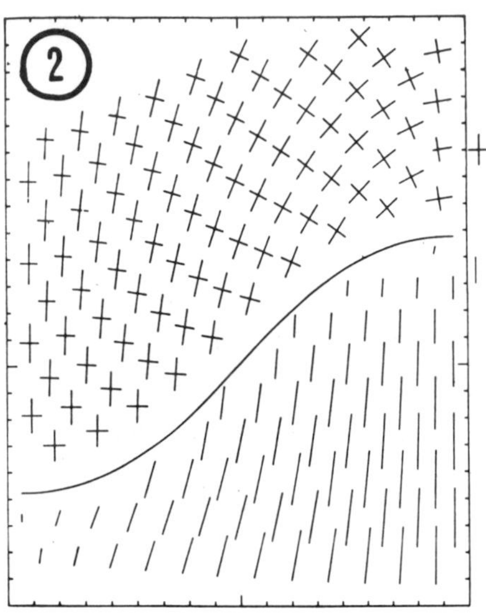

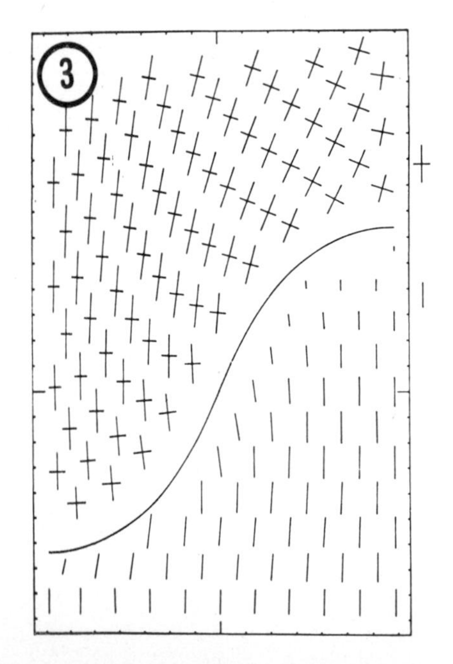

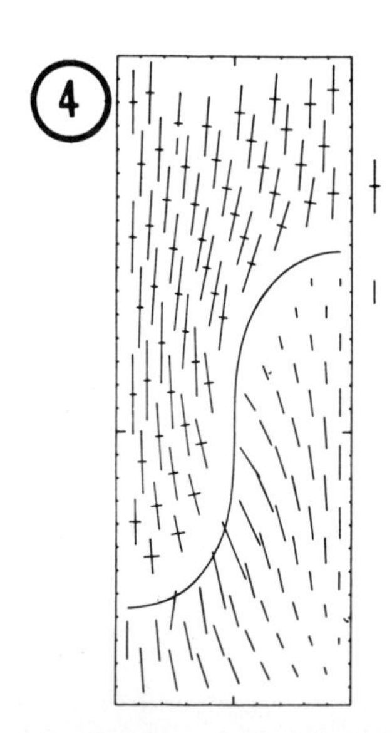

FIGURE 13-15 Theoretical stress history of folding a viscous layer in a less viscous matrix (*Dieterich and Carter, 1969*). (*a*) Orientations of σ_1 after 40 percent shortening of the layer. (*b*) Variation with total shortening (strain within matrix) of relative magnitudes of principal stresses for various localities, within fold (see inset). Numbers at triangles for cell 29 indicate inclination of σ_1 to dip of bed. (*From Carter and Raleigh, 1969.*)

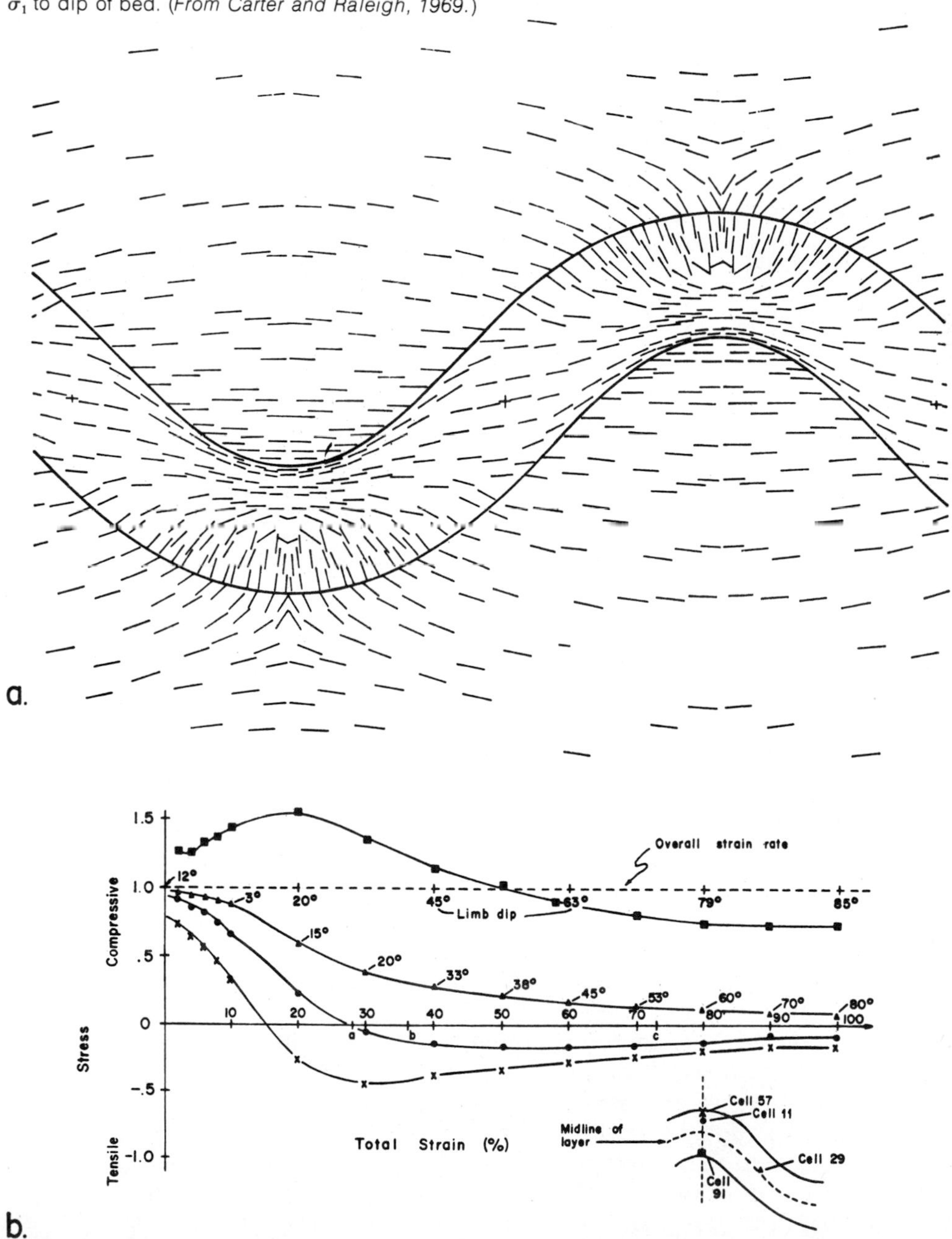

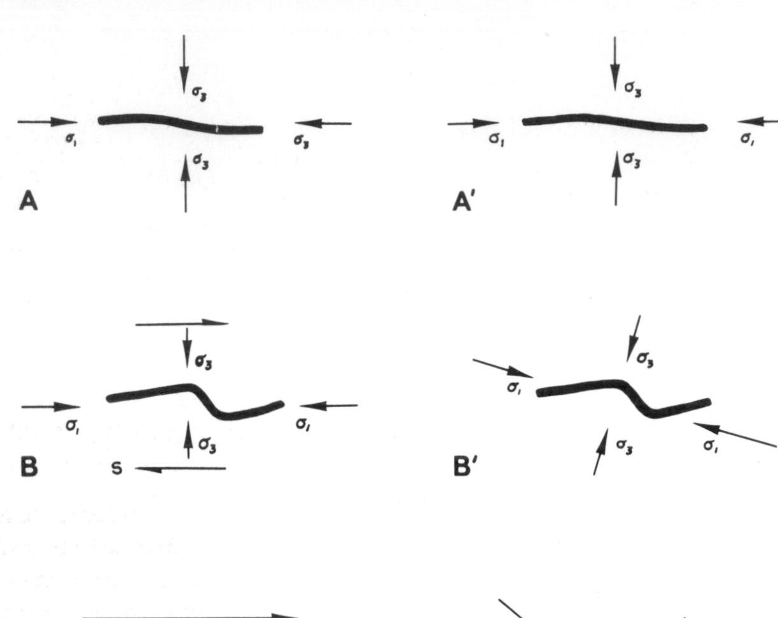

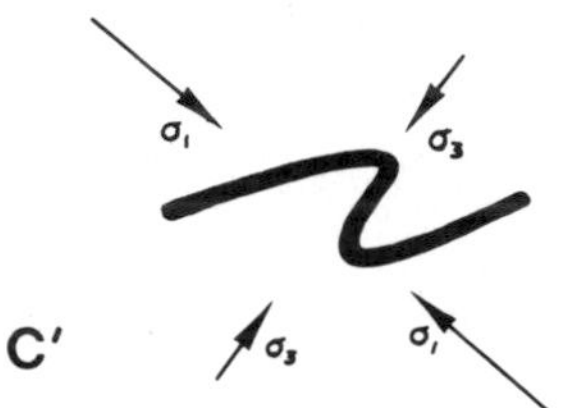

FIGURE 13-16 Schematic development of asymmetrical folds. (*a*) Initial buckling of a horizontal layer under horizontal maximum compressive stress σ_1 and minimum compressive stress σ_3. (*b*) Superimposition of small shear couple *S* parallel to σ_1, resulting in *B'* with new maximum and minimum stresses σ_1, and σ'_3, rotated slightly in the same sense as *S*. (*c*) Large shear couple *S* resulting in large rotation of orthogonal compressive stresses and development of an asymmetrical fold. (*From Smythe, 1971.*)

derived by combining both shear and compressive stresses parallel to the original layering. Small drag folds are commonly associated with flexural slip which involves localized couples as well as the overall compression which causes shortening. Presumably, large folds become asymmetric as a result of being initially inclined to compressive stresses or as a result of shear generated by differences in the amount of tectonic transport at different levels in the folded sequence.

The degree of asymmetry of a fold can be expressed as the ratio of the length of the long limb to the short limb, a ratio that is rarely less than 1.5:1 and is sometimes as much as 6:1. Probably the most consistent feature of a system of folds is the attitude of the axial planes, which usually are subparallel. The axial planes are perpendicular to the direction of compression or initiating stress in cases of upright symmetrical folds. Price (1967) shows that this is also true of asymmetrical folds which are initiated when the axis of the maximum principal stress is inclined to the unfolded layer (Fig. 13-16). Under this condition a shearing stress is set up which in turn generates a bending moment. The total moment curve is asymmetrical in shape, and the resulting fold is asymmetrical. After the fold is initiated and the position of the fold axes determined by elastic failure, the later development of the fold takes place largely by rotation of the limbs. Price hypothesizes that the amount of work done in rotating each limb is equal; therefore, the shorter limb rotates through a larger angle.

REFERENCES

Bayly, M. B., 1964, A theory of similar folding in viscous materials: Am. Jour. Sci., v. 262, p. 753–766.

——— 1970, Viscosity and anisotropy estimates

from measurements on chevron folds: Tectonophysics, v. 9, no. 5, p. 459–474.

——— 1971, Similar folds, buckling, and great-circle patterns: Jour. Geol., v. 79.

Biot, M. A., 1957, Folding instability of a layered viscoelastic medium under compression: Royal Soc. London Proc., A., v. 242.

——— 1961, Theory of folding of stratified viscoelastic media and its implications in tectonics and orogenesis: Geol. Soc. America Bull., v. 72, p. 1595–1632.

——— 1964, Theory of internal buckling of a confined multilayered structure: Bull. Geol. Soc. America, v. 75.

——— 1965, Mechanics of incremental deformation: New York, Wiley.

——— and others, 1961, Experimental verification of the theory of folding of stratified viscoelastic media: Geol. Soc. America Bull., v. 72, p. 1621–1632.

Bucher, W. H., 1956, Role of gravity in orogenesis: Geol. Soc. America Bull., v. 67, p. 1295–1318.

Chapple, W. M., 1968, A mathematical theory of finite-amplitude rock-folding: Geol. Soc. America Bull., v. 79, p. 47–68.

——— 1969, Fold shape and rheology; the folding of an isolated viscous-plastic layer: Tectonophysics, v. 7, no. 2, p. 97–116.

——— and **Spang, J. H.,** 1974, Significance of layer-parallel slip during folding of layered sedimentary rocks: Geol. Soc. America Bull., v. 85, no. 10, p. 1523.

Currie, J. B., Patnode, H. W., and **Trump, R. P.,** 1962, Development of folds in sedimentary strata: Geol. Soc. America Bull., v. 73, p. 655–674.

Dieterich, J. H., and **Carter, N. L.,** 1969, Stress-history of folding: Am. Jour. Sci., v. 267, no. 2.

Donath, F. A., 1962, Role of layering in geologic deformation: New York Acad, Sci. Trans., v. 24, p. 236–249.

——— 1964, Folds and folding: Geol. Soc. America Bull., v. 75.

——— and **Parker, R. B.,** 1961, Folds and folding (abs): Geol. Soc. America Spec. Paper 68, Abstracts for 1961, p. 87–88.

Frocht, M. M., 1964, Photoelasticity: New York, Wiley.

Gay, N. C., and **Weiss, L. E.,** 1974, The relationship between principal stress directions and the geometry of kinks in foliated rocks: Tectonics, v. 21, p. 287–300.

Ghosh, S. K., 1966, Experimental tests of buckling folds in relation to strain ellipsoid in simple shear deformation: Tectonophysics, v. 3.

——— 1968, Experiments of buckling of multilayers which permit interlayer gliding: Tectonophysics, v. 6, no. 3, p. 207–249.

Handin, J. M., Friedman, M., and **Logan, J. M.,** 1973, Experimental folding of rocks under confining pressure; buckling of single-layer rock beams: Am. Geophys. Union, Geophys. Monogr. no. 16, p. 1–28.

Hobbs, B. E., 1971, The analysis of strain in folded layers: Tectonophysics, v. 2, no. 5, p. 329–375.

Hubbert, M. K., 1937, Scale models and geologic structures: Geol. Soc. America Bull., v. 48, p. 1459.

Hudleston, P. J., 1973, The analysis and interpretation of minor folds developed in the Moine rocks of Monar, Scotland: Tectonophysics, v. 17, p. 89–132.

——— and **Stephansson, O.,** 1973, Layer shortening and fold-shape development in the buckling of single layers: Tectonophysics, v. 17, no. 4, p. 299–321.

Kuenen, P. H., and **De Sitter, L. U.,** 1938, Experimental investigations into the mechanism of folding: Leidse Geol. Med., v. 10, p. 271–240.

McBirney, A. R., 1961, Experimental deformation of viscous layers in oblique stress fields: Geol. Soc. America Bull., v. 72, no. 3, p. 495–498.

Nettleton, L. L., 1943, Recent experimental and geophysical evidence of salt dome formation: Am. Assoc. Petroleum Geologists Bull., v. 27, p. 51–63.

Paterson, M. S., and **Weiss, L. E.,** 1962, Experimental folding in rocks: Nature, v. 195.

Price, N. J., 1967, The initiation and development of asymmetrical buckle folds in non-metamorphosed competent sediments: Tectonophysics, v. 4, p. 173–201.

Ramberg, Hans, 1961, Relationship between concentric longitudinal strain and concentric shearing during folding of homogeneous sheets of rocks: Am. Jour. Sci., v. 259, p. 382–390.

——— 1963a, Fluid dynamics of viscous buckling applicable to folding of layered rocks: Am. Assoc. Petroleum Geologists Bull., v. 47.

——— 1963b, Strain distribution and geometry of folds: Geol. Inst. Univ. Uppsala Bull., v. 42, p. 3–20.

——— 1967, Gravity deformation and the earth's crust as studied by centrifuged models: New York, Academic.

——— 1968, Instability of layered systems in the field of gravity: Phys. Earth Planet. Interiors, v. 1. no. 7.

——— and **Strömgård, K. E.**, 1971, Experimental tests on modern buckling theory applied to multilayered media: Tectonophysics, v. 11.

Ramsay, J. G., 1967, Folding and fracturing of rocks: New York, McGraw-Hill.

——— 1974, Development of chevron folds: Geol. Soc. America Bull., v. 85, p. 1741–1754.

Roberts, David, and **Strömgård, K. E.**, 1972, A comparison of natural and experimental strain patterns around fold hinge zones: Tectonophysics, v. 14.

Sanford, A. R., 1959, Analytical and experimental study of simple geologic structures: Geol. Soc. America Bull., v. 70, p. 19–52.

Sherwin, Jo-Ann, and **Chapple, W. M.**, 1968, Wavelengths of single layer folds: A comparison between theory and observation: Am Jour. Sci., v. 266, no. 3, p. 167–179.

Smythe, D. K., 1971, Viscous theory of angular folding by flexural flow: Tectonophysics, v. 12.

——— 1972, Viscous theory of angular folding by flexural flow: Tectonophysics, v. 12, no. 5, p. 415–430.

Strömgård, Karl-Erik, 1973, Stress distribution during formation of boudinage and pressure shadows: Tectonophysics, v. 16, p. 215–248.

Timoshenko, S. P., and **Gere, J. M.**, 1961, Theory of elastic stability: 2d ed., New York, McGraw-Hill.

Treagus, S. H., 1972, Similar folds and anistropy: Jour. Geology, v. 80, no. 1, p. 122–123.

——— 1973, Buckling stability of a viscous single-layer system, oblique to the principal compression: Tectonophysics, v. 19.

——— 1974, Buckling stability of a viscous single layer system, oblique to the principal compression: Tectonophysics, v. 19, no. 3, p. 271–289.

Vikhert, A. V., and **Kurbatova, N. S.**, 1969, Folds in tectonic model experiments arising as a result of an increase in volume: Geotectonics, no. 2, p. 137.

Willis, Bailey, 1893, Mechanics of Appalachian structure: U.S. Geol. Survey Ann. Rept. 13.

——— 1934, Geologic structures, 3d ed., New York, McGraw-Hill.

14

Large-Scale Folds and Fold Systems

OCCURRENCE AND TECTONIC ASSOCIATIONS OF LARGE FOLDS AND FOLD SYSTEMS

Although much of the information developed about the geometry of mesoscopic folds in the preceding chapter is applicable to larger folds, it is often easier to see the relationship of big folds to the tectonic framework of which they are a part. Otherwise the primary distinction between mesoscopic and larger folds lies in the techniques used to describe and analyze the two groups of features. While detailed description of *s* surfaces and fabric relationships may be most fruitful with mesocopic folds, the larger features are more often described by use of structure contour and geologic maps and by construction of cross sections based on these and other types of subsurface data.

Many large folds such as compaction folds, salt domes, and folded slump and slide masses form in the absence of any strong directed stress other than gravity, and they are clearly nontectonic folds. It is equally clear in other instances that folds are part of large orogenic belts in which significant parts of the crust are strongly and systematically deformed as a result of lateral compression and gravity-induced spreading and sliding of uplifted areas. Still other folds, both in and out of orogenic belts, are formed as part of the crustal response to movements along faults. A third important group of folds are related to vertical movements, some of which are directly caused by lateral movements, but many others show no such connection.

Nontectonic Folds

Most of the nontectonic folds are described in other parts of this text—compaction or supratenuous folds are described in Chap. 2, domal structures over salt diapirs are discussed in Chap. 15, and folds formed as part of gravity glides are covered in Chap. 8. In addition, domal uplifts may be formed over igneous intrusions. Supratenuous folds may be found in any sedimentary basin where sediments of different compactability are deposited unevenly, and they are especially common where sediment is deposited over an irregular base such as preexisting topography. Such folds are characterized by thinning of sedimentary layers over the fold crust. Salt domes are confined to the sedimentary basins in which great thicknesses of both salt and other sediment have accumulated. Salt has a high ductility and has played an important role as a lubricant for gravity gliding and as a layer over which other strata might slip and fold as in the Jura Mountains (Chap. 23), but most of the salt domes are located in areas where effects of lateral compression are absent, as in the Gulf of Mexico and northern Germany. Folded slump and slide masses are well known both in and out of orogenic belts, and it is sometimes difficult to determine their relationship to external stresses. Many slides occur in marine environments where water lubricates and facilitates movement of rock masses down the continental slope into a trench, where the resulting deposits may form a chaotic melange or mixture of sediment and more or less isolated, disoriented, folded rock masses. Sliding may also be important following vertical uplift if potential glide planes are present.

Syndepositional Folding

Large-scale folds can form as a result of essentially vertical movements in a depositional basin during sedimentation, and Cooper (1964) suggests that this may be the primary origin of many folds in the Appalachian foreland fold belt, Fig. 14-1, although he fully recognizes that these folds have been accentuated and modified by later lateral compression.

The evidence for syndepositional folding is essentially of two types: first, the localization and restriction of certain types of sedimentary facies on synclines and anticlines. Second, variation in thickness which are related to position on the folds; synclines have thicker sections than anticlines. Both of these lines of evidence indicate that the structures were in existence during the time of sedimentation. The persistence of these relations for different times on a given structure indicates that the folds continued to form over a long period of time. In the case of the folds described by Cooper, a section that is 838 m (2,750 ft) thick on an anticline is 2,853 m (9,360 ft) thick in the center of the syncline (Price Mountain, Va.). Not only are the synclines characterized by greater thicknesses, they contain (Cooper, 1964):

> Local segregations of special facies including black shales (which suggest stagnant water conditions), green shales, polymicitc conglomerates, and poorly sorted coarse clastics. . . . Anticlinal crests commonly exhibit naturally thinner sections characterized by prevalence of biostromes and bioherms, abundant desiccation jointing, and thinner well washed and more oxidized sediments.

Regional Features due to Vertical Movements

Syndepositional folds of the type described in the preceding section have generally been attributed to lateral compression, but the stratigraphic evidence for localized vertical uplift within a large sedimentary basin seems extremely strong in the case of a number of such folds, and many other broad fold structures lie outside orogenic belts where the question of folding by lateral compression is not raised. Such broad structures constitute some of the most prominent features of continental platforms, and excellent examples are found in North America. These include regional basins such as the Powder River, Denver, and Delaware basins just east of the Rocky Mountain front, and others in the interior such as the Michigan and Illinois basins. The cause of these basins, the broad arches (e.g., Findley, Cincinnati, Cape Fear, etc.), and domal uplifts like the Nashville, Ozark, and Adirondack

FIGURE 14-1 Syndepositional deformation as proposed in the evolution of the Hurricane Ridge syncline by Byron Cooper. This model depicts the theory of basement involvement in foreland folding. (*From Cooper, 1961.*)

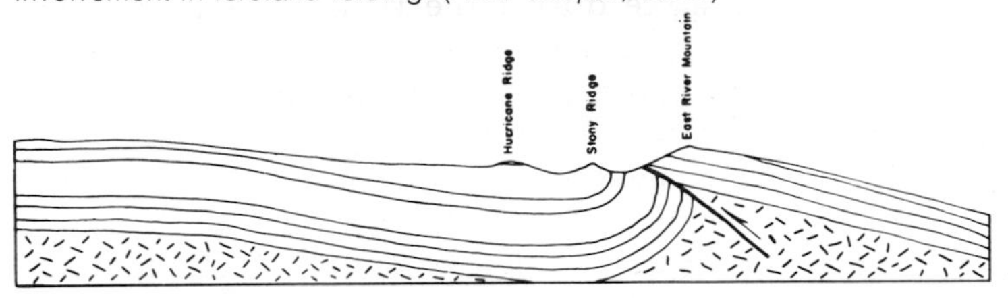

Erosion removes leading edge of overriding block and fault trace recedes down dip.

Localized subsidence controlled by basement ultimately produces a break along which upper block rides out over subsiding syncline.

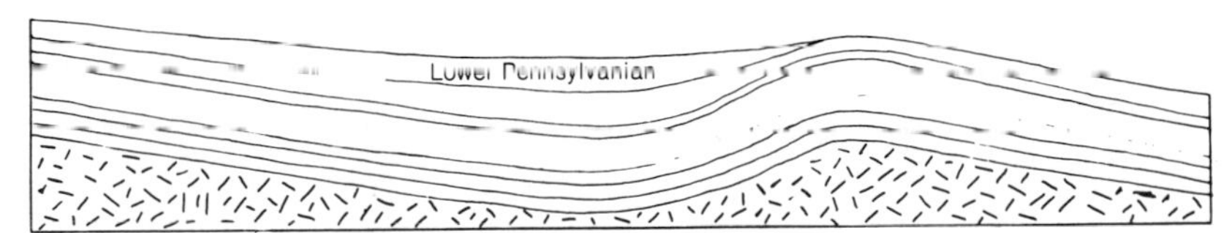

Localized downward warping and thicker deposition continues in early Pennsylvanian time.

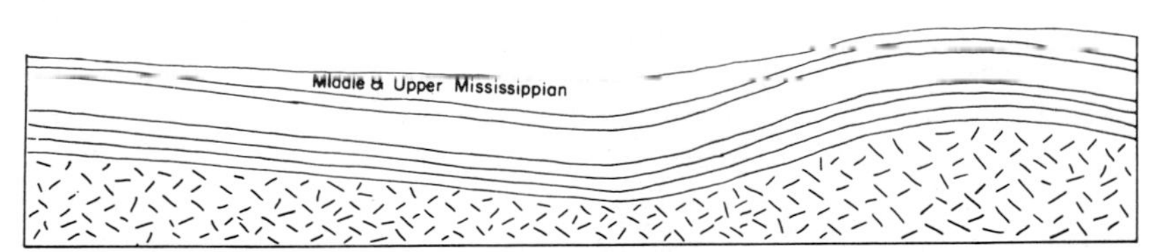

Localized downward warping of portion of foreland shelf in Meramecian time creates depositional syncline; contemporaneous deformation in thick Mississippian clastics.

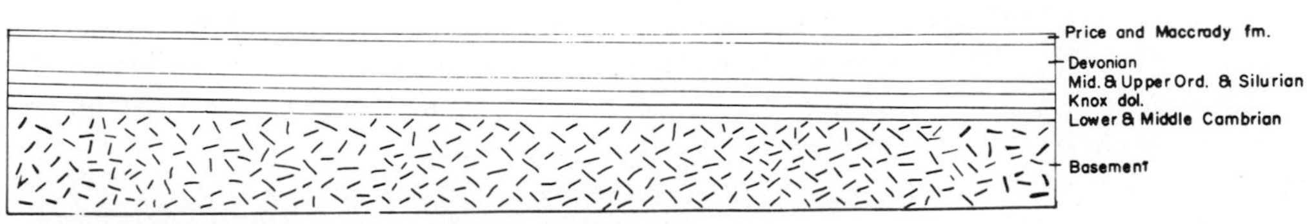

Uniform depositional conditions when area was part of foreland shelf

EVOLUTION OF HURRICANE RIDGE SYNCLINE

uplifts remain obscure, but they must have been caused by essentially vertical uplift or subsidence, and their cause must be deep seated because the relief of some of them is great as is shown in the Michigan basin, where the Precambrian-Cambrian contact is over 3,050 m (10,000 ft) below sea level in the basin and over 152 m (500 ft) above sea level nearby in the Wisconsin Highlands.*

Most of the larger regional features are described as domes, basins, arches, downwarps, or uplifts, but the dips observed on these features are generally so low that they are identified best on regional geologic or structure contour maps, Fig. 14-2. Locally these may have the form of structural terraces (flat-lying strata), homoclines (uniformly dipping), or monoclines (a flexure in flat or otherwise homoclinal strata), Fig. 14-3.

* See Chap. 18.

FOLDS OF TECTONIC ORIGIN

Folds are certainly one of the most prominent features of the orogenic belts, where they are usually associated with thrust faults and often with strongly tectonized and metamorphosed rock. The best known of these orogens, which are briefly described later in this book, are the Alpine-Himalayan system of Europe and Asia, the Appalachian-Caledonian system of eastern North America and western Great Britain and Scandinavia, and the Cordilleran system of western North America. Older orogenic belts, characterized by folds, are recognized in many other parts of the world also, but the structure

FIGURE 14-2 The Michigan basin is one of a number of large basins found in continental platforms. The contours show depth to basement in feet. (*From the North American Tectonic Map, 1969.*)

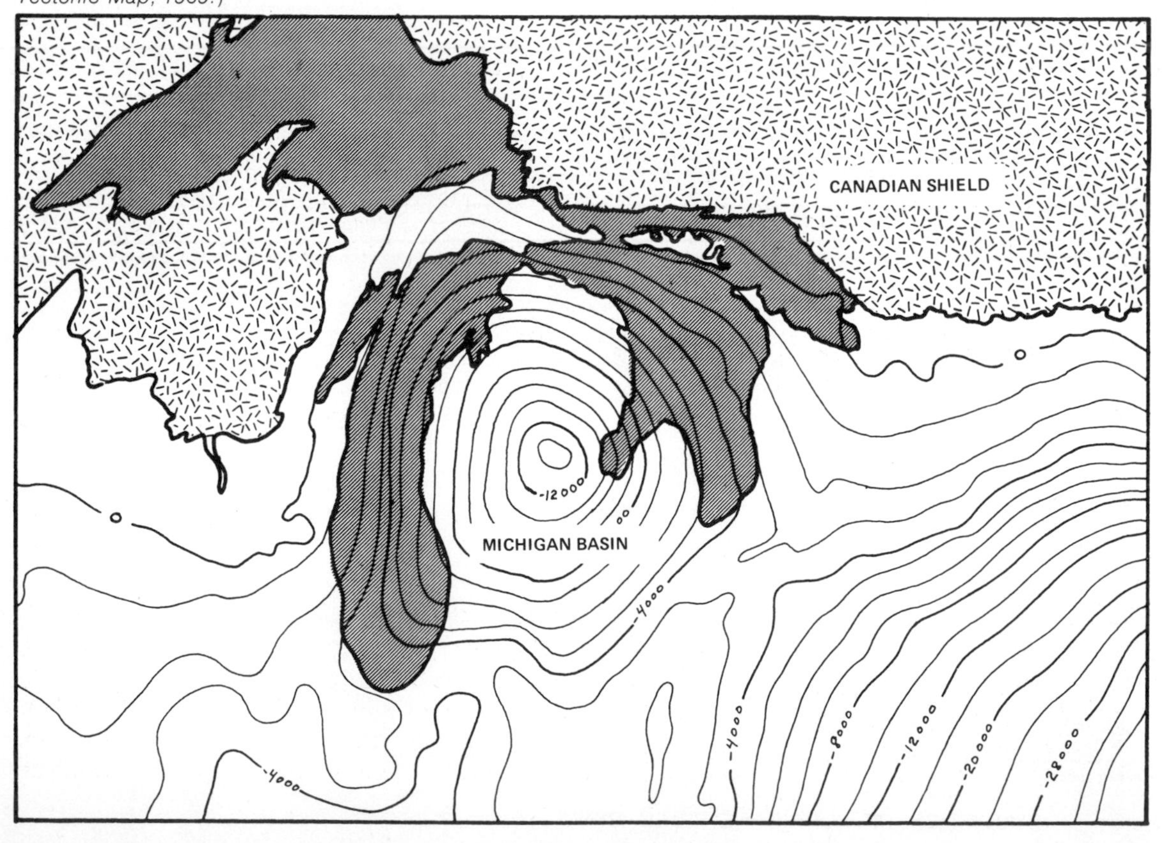

of the older orogens is often obscure because of later metamorphism, a cover of superimposed deformation. Distinctive types of folds are found in different parts of a typical orogen, which may be thought of as possessing an internal or *core zone* characterized by rocks that have been regionally and dynamically metamorphosed to a high grade and a more external belt of folded but unmetamorphosed sedimentary rocks which composed what are called *foreland fold belts,* Fig. 14-4.* Some geologists have described orogens as consisting of a deeper seated *infrastructure,* generally a metamorphic-plutonic core, and an overlying *suprastructure,* consisting of rocks deformed closer to the surface and often consisting of gravity-driven sliding and folding, Fig. 14-5. In both of these divisions the direction of tectonic transport in the suprastructure or foreland is almost always directed externally and outward away from the exposed internal core of infrastructure. This is expressed in the direction of displacement on thrusts and in the direction of overturning of most folds.

*The Appalachian foreland fold belt is described in Chap. 22, and the Jura Mountains are covered in Chap. 23.

FIGURE 14-3 Schematic cross sections of flat-lying strata, top; a monocline, middle; and a homocline, bottom.

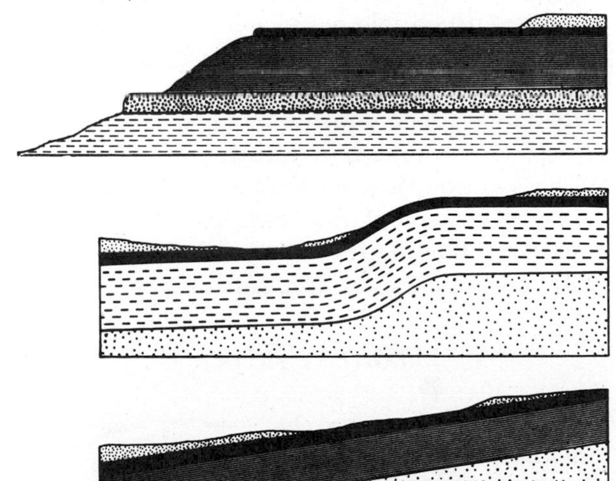

Most orogenic zones appear to have formed in long narrow belts that were originally along continental margins and sites of accumulation of sedimentary deposits—in many cases of great thickness. These preorogenic sedimentary belts are called *geosynclines** where the stratigraphic record proves long-term subsidence and sediment accumulation. Subsidence during the depositional phase of a geosyncline is not uniform, and some parts of the geosyncline are subject to uplift, creating "*swells*" or localized uplifts. When orogenic deformation occurs, large areas of the crust (often part of the geosyncline) are uplifted as large elongate anticlinal features, sometimes called *geanticlines,* or if the uplifted crust contains many small folds, it may be termed an *anticlinorium.* Similarly a large region of folded strata that has an overall synclinal form is called a *synclinorium.*

Many orogenic belts are curved, and some, like the Alpine system of southern Europe, Fig. 23-1, exhibit meandering patterns on maps. These sweeping arcs and more gentle curves such as those in the Appalachians have long attracted the attention of geologists. Some have suggested that they are primary features, that the geosynclines were curved even before orogenic deformation took place, but most have argued that they are forms resulting from large-scale deformation. This subsequent deformation may consist of bending or folding of the orogen as a whole to form what is called an *orocline,* as originally suggested by S. W. Carey, but others think that the bends are due to complex movements related mainly to strike-slip offsets within the orogen as described for the Alps in Chap. 23.

Crystalline igneous and metamorphic rocks can become sufficiently mobilized to fold and to even undergo plastic flowage as is shown in gneiss domes and broad anticlinal structures in many orogenic belts and in the crystalline

* See Chap. 17.

nappes in some orogens. These features are usually found only in the core or infrastructure, where temperatures and depth of burial were great. The crystalline basement under most foreland fold belts is not strongly deformed and may be essentially flat as is indicated by seismic profiles across the Appalachians and Northern Rocky foreland fold belts. The folds must thus be confined to the more surficial sedimentary veneer and result from uncoupling, *décollement,* from the underlying rocks as is postulated to occur in salt beds in the Jura Mountains and in shales in the Appalachians. The shapes of the folds in the sedimentary veneer or foreland fold belts is most often described as varying from strongly overturned, asymmetric, and often broken by forelimb or backlimb thrust in the more internal part of the belt to open, upright, symmetrical parallel (concentric) folds farther out. The model of parallel folds has been widely used to describe foreland folds, but box-shaped folds are well known, especially in the Jura Mountains, and folds that have characteristics of *kink bands* have been described in Pennsylvania* and in other foreland fold areas, Figs. 14-6 and 14-7. Many other foreland folds are well enough exposed to make it evident that they are *disharmonic* and cannot be represented by a simple geometric model.

Foreland belts show variations in form longitudinally and vertically as well as in cross section. Curves in the fold belts sometimes coincide with areas where the folds extend farther out into the foreland region to form a convex arcuate shape called a *salient* or with places where the belt has a concave section called a *recess.* Somewhat analogous variations are found in the level of fold crests, many of which may be higher in one section of a fold belt and lower in another. The area where crests are at a high level is called a *culmination,* and areas where a number of folds plunge or are low are called *depressions.*

* Faill (1969).

Fault-related Folds

We have seen in earlier chapters several examples of the way folds may be formed in connection with faults. Notable among these are: (1) drag folds formed along thrust faults, (2) distinctive vertical folds formed along major strike-slip faults such as folds along the Alpine fault, Fig. 10-4, (3) *en échelon* folds formed in the sedimentary cover over strike-slip faults and between strike-slip faults, as is common along the San Andreas fault system, Fig. 10-8, and (4) drape folds formed in sedimentary cover over high-angle faults, which are so beautifully exposed in the Middle Rocky Mountains, Fig. 14-8.

It is often difficult to determine in the field if an overturned fold associated with a thrust is a drag fold or if the folding preceded development of the thrust. This is especially true of the larger folds, which may be cut by faults on either the forelimb (overturned limb) or on the back limb. Most geologists seem to favor the development of the folds as proceeding from and following the break and fault movement in the case of forelimb thrusts, but

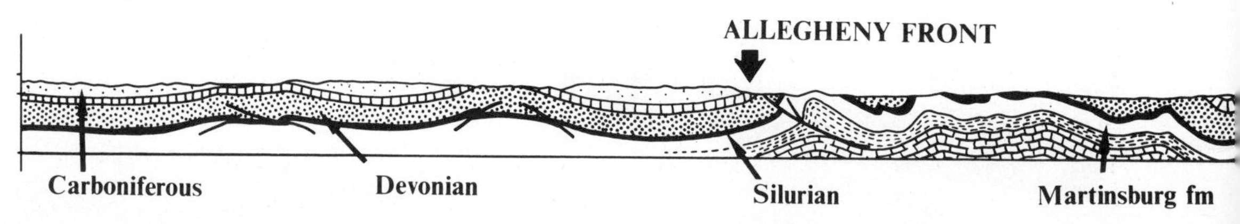

FIGURE 14-4 Cross section across the Appalachian foreland fold belt in Maryland. (*From the Maryland State Geologic Map, 1968.*)

thrusting is usually thought to follow folding when the thrust cuts the back limb of the fold.

Nappe Structure

The term *nappe* is a French word meaning sheet which is applied to large, tectonically emplaced sheets of rock that are far removed from their place of origin.* The name has been applied to rock masses transported laterally by both recumbent folding and thrusting. Some geologists have confined the use of the term to recumbent folds which have been moved laterally a great distance, usually resulting in shearing out of the lower, overturned limb (Fig. 14-9), but this restricted usage is not accepted by most European geologists. Others restrict the term to situations where the movement has been so great that the nappe has become detached from the beds with which it was originally connected, in a place known as the *root* of the nappe. The term is applied to a number of the great thrust masses, thrust sheets, drawn-out recumbent folds, and overturned recumbent folds which are characteristic of the western Alps, where the term was first applied to structures which were interpreted as having moved some distance laterally. Much of the early debate about the Alps revolved around the validity of this concept of lateral transport. From the original question of whether or not such movement had taken place, the argument progressed to how much movement and what direction of movement, and to the location of original source material, the roots, and the important tectonic question of the mechanics of the movement. Many alternatives have been suggested, including low-angle thrusting resulting from horizontal compression, flow of the rock at depth followed by uplift, uplift followed by flow at shallow depths, and vertical uplift followed by gravity sliding.

The theory of nappes has played a major role in the interpretation of the complex tectonic structures of the Alps. The importance of nappes was aptly indicated by Heritsch (1929): "The fundamental question of Alpine tectonics is the extent to which the Nappe Theory is true." At the time European geologists were embroiled in a controversy over the application of the theory to Alpine structures. It would be unfair to say that the argument is completely resolved even now, but the view that large-scale complex nappes do exist and that they are one of the most significant features of Alpine geology is universally held, and nappes have been described in most of the world's orogenic belts.

Arnold Escher Van der Linth (1841) first proposed large-scale overthrusting to explain Alpine structure. He had studied the geology of the region near Glarus, Switzerland, where he found Permian conglomerates and sandstones overlying Tertiary sediments known as *flysch*, a marine sequence of interbedded sands, silts, shales, marls, conglomerates, and breccias, much of which is highly contorted and contains exotic blocks. The flysch is highly folded and faulted in the Glarus area. It was

Allochthon: A mass of rock which has been moved from its place of origin. Adj.: allochthonous.

Autochthon: A mass of rock that is still located at its place of origin. Adj.: autochthonous.

Para-autochthon: A mass of rock which has been moved only slightly from its place of origin. Adj.: para autochthonous.

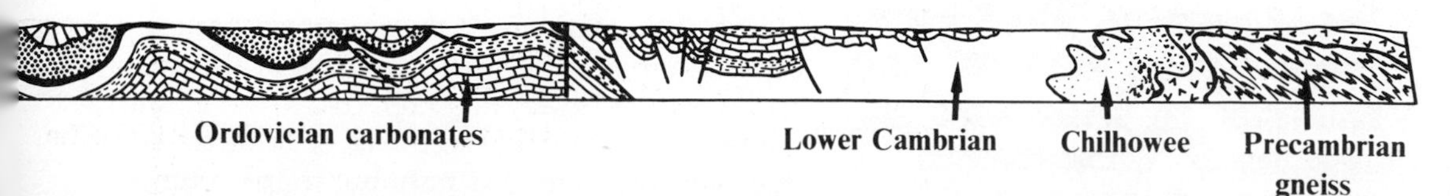

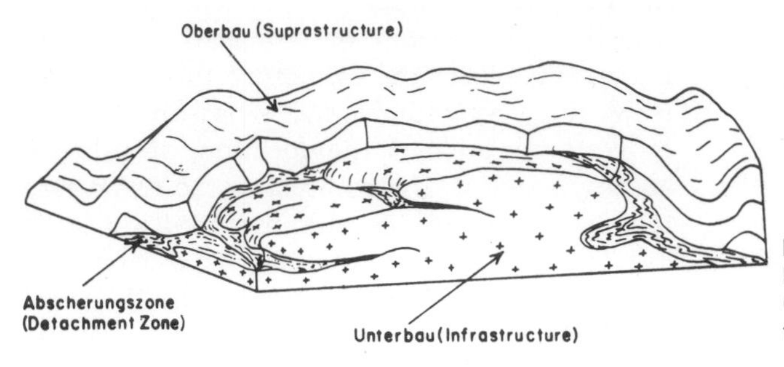

FIGURE 14-5 A schematic stockwork folding model based on the tectonic relationships in the east Greenland caledonides [modified after Haller (1956), fig. 7]. (*From Griffin, 1970.*)

this flysch that Escher found in all the valley bottoms for a distance of about 50 km, while the crest and peaks in the same area consisted of horizontal schist overlain by limestones containing Liassic fossils.

Apparently Escher was anxious to minimize movement on this nearly horizontal fault. This was a new type of structure, and perhaps he was anxious to keep the scale of the overthrust as reasonable as possible. In any case he interpreted the geology of the area as a double fold, two large recumbent folds of opposite orientation. It remained for Marcel Bertrand to replace Escher's double fold with a single northward-transported structure (Fig. 14-9). In the process, however, it became necessary to interpret the existence of an additional overthrust of gigantic size. The origin of the structure of the region was argued, bitterly at times, for a period of 50 years. These arguments centered on questions regarding the concept of the double fold and Bertrand's proposed substitute, the direction of thrusting along each fault, the existence or nonexistence of a recumbent fold with its inverted limb thinned and mylonitized, and the amount of movement. Gradually Bertrand's interpretation of the structure as a great nappe became the accepted view of the Glarus fold.

FIGURE 14-6 Lines of discontinuity associated with ideal, conjugate, sharp-kink folds. (*From Johnson, 1969.*)

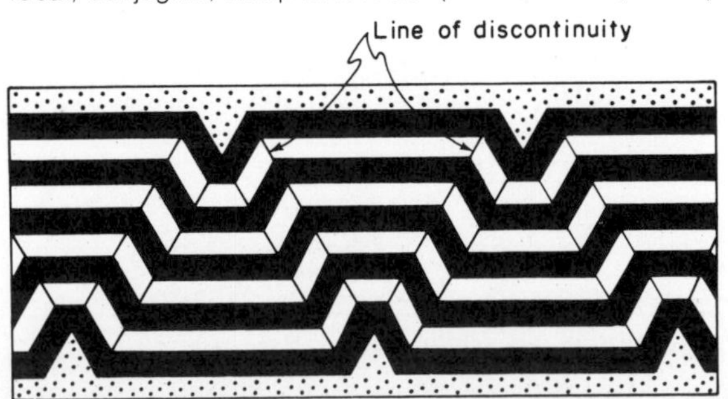

The appearance of a nappe on a geologic map depends on the relief and topography as well as on the complexity of shape of the nappe sheet itself, but an area in the Canadian Rockies of British Columbia (Fig. 14-10) will serve as an example. Major horizontal transport was involved in this case to bring Precambrian units over Triassic. The sinuous trace of the major fault outcrop is characteristic of low-angle faulting in areas of high relief, as is the klippe at Mount Hosmer.

ANALYSIS AND REPRESENTATION OF LARGE-SCALE FOLDS

When the size of a fold exceeds the size of an outcrop, geological mapping in combination with preparation of cross sections and structure contour maps provides the best means of defining the fold shape. The appearance of a fold on a geologic map depends on the topography as well as on the geometry of the fold, so maps of real folds may appear quite different from schematic illustrations which do not show relief effects. Actual geological map patterns may closely resemble the schematic patterns of illustrations if the mapped area either has subdued relief or if the size of the structural feature is very large relative to the amount of relief. For this reason the large folds that can be seen on state and national maps often resemble schematic maps more than those on 7½ minute quadrangle maps.

Several very good sources of information are available on the interpretation and geometrical analysis of folds structures on geologic maps.* Only a few of the basic principles of interpretation are described here. Many geological maps are drawn on topographic map bases, and these are much easier to interpret accurately and are especially helpful in determining if an unusual map pattern is due to the form of the structure or the topography.

* Simpson (1968), Dennison (1964).

FIGURE 14-7 Superposition of a modified kink-band structure on a block diagram of the Cocolamus fold structure in the Valley and Ridge province in Pennsylvania. The bed rotation and fold geometry resulting from the development of the kink-band structure correspond closely to the field data. (*From Faill, 1969.*)

Strike and dip of beds are often shown on maps, and they can be determined if the map is on a topographic base by solution of a "three-point" problem (Appendix C) using the orthographic projection. If the ground surface in the map area is horizontal, then, of course, the strike of a bed at any point along a contact is the tangent to that contact. Strike and dip data collected at the ground surface are critical information in map interpretation because they usually provide the best basis for projection of the contacts below ground surface; thus they are the basis for cross-section construction which is necessary for development of a three-dimensional model of the fold. Dip of the strata, layer thickness, and ground-surface slope are the three factors that determine the width of an outcrop belt on a map of

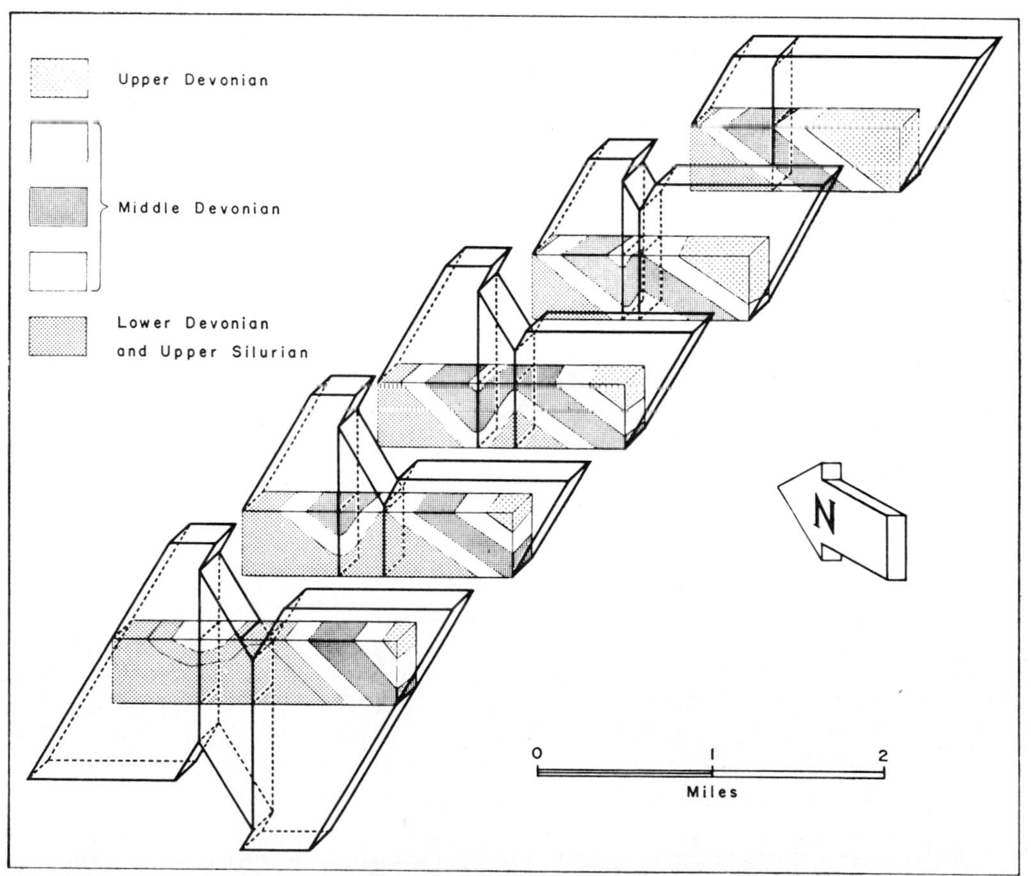

FIGURE 14-8 Interpretative cross section of the flank of Rattlesnake Mountain, Wyoming, showing details of the structure of a drape fold in the sedimentary cover over an uplifted basement block. (*After Stearns and others, 1974.*)

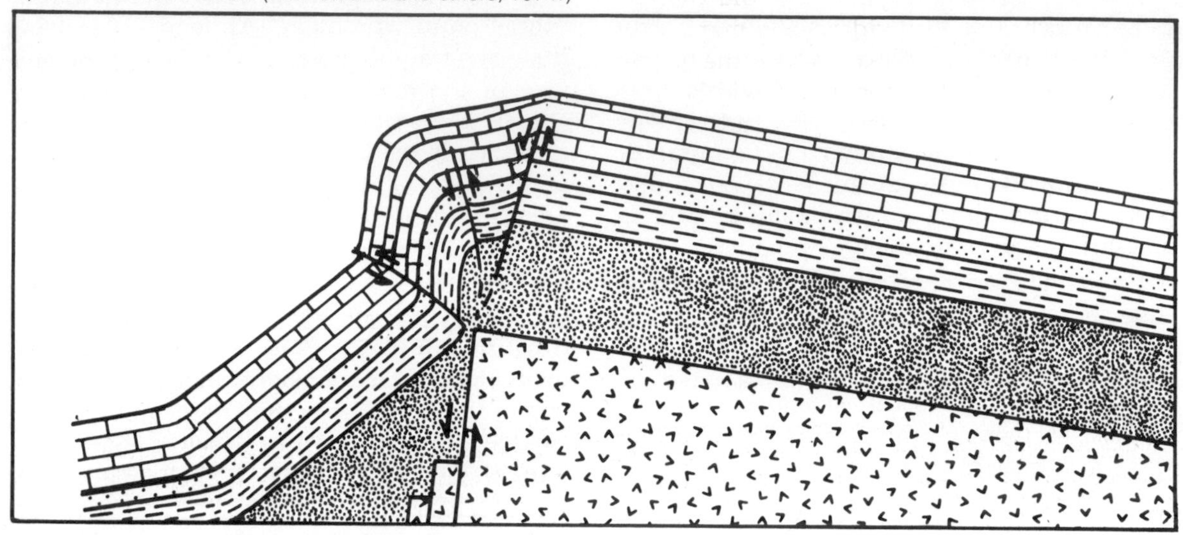

FIGURE 14-9 Interpretative cross sections across the Glarus nappes, Switzerland, where the nappe theory originated. The section at top depicts the original double-fold concept. Large-scale lateral movement is required by the interpretation at bottom. m = molasse, f = flysch, c = Cretaceous, j = Jurassic, t = Trias. (*From Bailey, 1935.*)

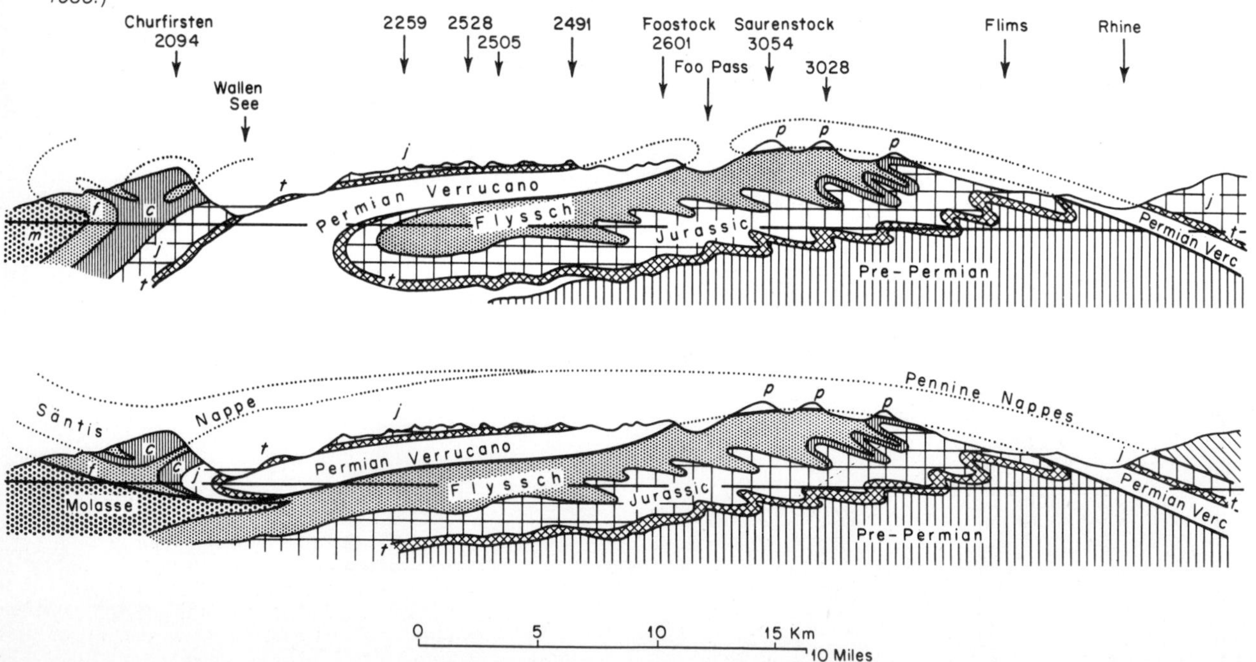

any given scale (see Appendix B). Lower dips generally produce wider outcrop patterns; so, the two limbs of an asymmetric fold will appear to have different widths on a map. If the unit is of constant thickness and the topography is subdued, the outcrop width is an excellent indication of layer dip and the fold form can thus be inferred from observing the way the outcrop width changes. Width will be a minimum where the layers are vertical and usually reaches a maximum where the fold plunges—again outcrop width in the fold hinge is an indication of the amount of plunge. The map pattern will provide an indication of fold symmetry and can be used to identify the axial trace.

The map pattern of some typical foreland folds in the Valley and Ridge province of Pennsylvania is shown on the map of the Oriskany sandstone, Fig. 14-11. The axial traces of these folds follow the gentle curved form of one of the salients in the foreland belt. These folds are typical of foreland folds in many respects—for example (1) they have shorter wavelength in the east and become broader to the northwest; (2) they are plunging folds (the regional plunge in this area is southward); and (3) small folds occur on some of the larger folds, producing the unusual pattern south of East Waterford. The broad outcrop belt at Huntingdon is an indication of the low dips in that area, and the V-shaped outcrop there shows that the Oriskany dips eastward; so, Mt. Union lies in a broad syncline that is strongly asymmetric with a steep eastern limb. The irregular broad outcrop south of

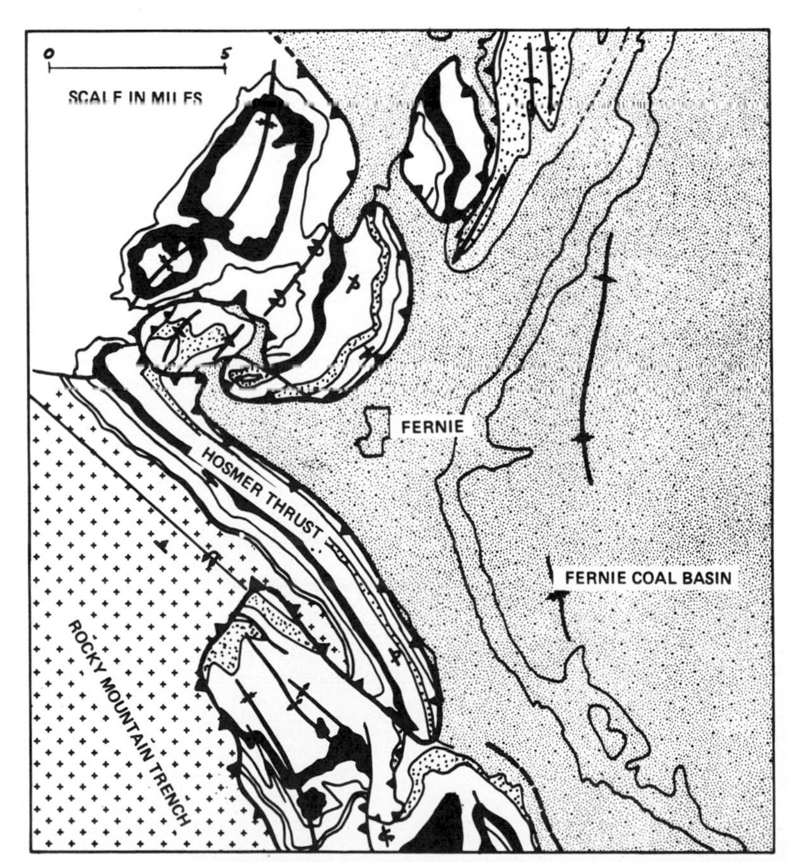

FIGURE 14-10 Overthrusts and recumbent folds in the Fernie area, British Columbia. Mesozoic rocks underlie the area shown by dots, a Mississippian unit is shown black, a Permian unit is shown by coarse dots, and the Precambrian is shown by crosses. (*From Henderson and Dahlstrom, 1959.*)

East Waterford shows not only that the Oriskany is at the surface but that it is folded in a series of small doubly plunging folds.

Down-Structure Method of Viewing Maps

A method for approximating the shape of the cross section of a structure, known as the *down-structure method* of viewing geologic maps, is described by Mackin (1950). By looking at a geologic map of a plunging fold from a position in line with the axis of the fold and from an angle that is close to the angle of the plunge of the structure, the structure may be seen approximately as it would appear in cross section. The method can be used with most types of structures. It helps to clarify probable subsurface shape as well as the form that has been removed from the structure through ero-

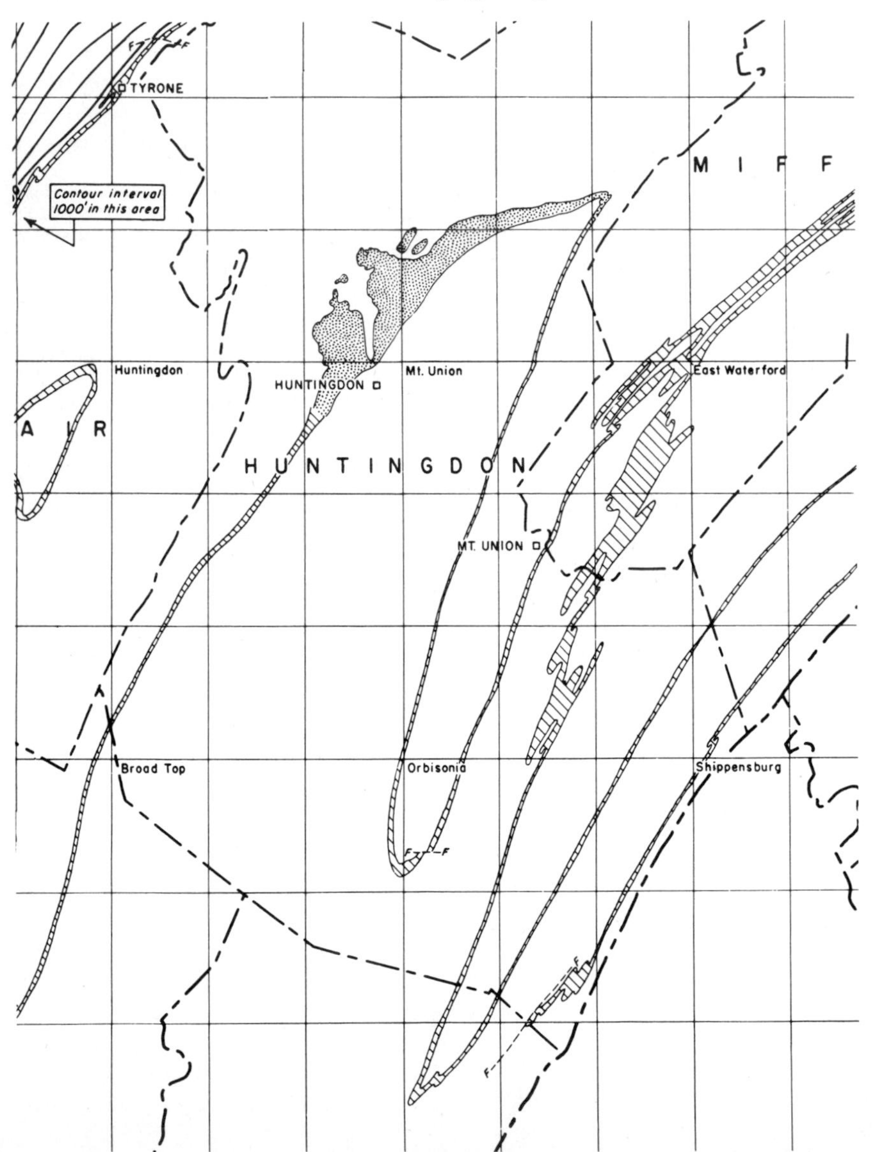

FIGURE 14-11 Outcrop pattern of the Oriskany and Helderberg formations, Pennsylvania. See text discussion. (*From Cate and others, 1961.*)

sion. It should be evident that the cross section you see when looking down plunge is not a vertical section, but a section at right angles to the line of sight. One of the other limitations to this method is that in dealing with detailed geologic maps in areas of high topography, contacts of even perfectly planar inclined surfaces may appear as an intricate pattern. The method works best in regions of low relief or on regional maps of such small scale that relief effects are minimized.

Structure Contour Maps*

Structure contour maps provide the best representation of the three-dimensional geometry of a deformed layer, but they are constructions based on limited amounts of data and they are no better than the data and the accuracy of the contouring. If the contoured horizon comes to the ground surface, the outcrop trace is a prime source of data; otherwise the depth data needed for the map is obtained from drill holes, thickness estimates, seismic lines, or models based on indirect geophysical observations. Special precautions must be used in plotting some of these data and in interpretation of structure contour maps. For example, an effort is made to drill most wells vertically; so the apparent thickness of a layer penetrated in a well is not the true thickness of that layer, and this can lead to errors in construction of cross sections such as that illustrated and described in Fig. 14-12.

If we assume that a structure contour map is a correct representation of subsurface shape, then the strike of the marker horizon is tangent to the structure contours at every place on the map, and dip of the horizon is equal to the slope of the contoured surface at that place. Plunge and axial traces of folds can be readily identified, and a cross section of the horizon can be constructed readily along any line across the structure.

Structure contour maps may be prepared on a number of different horizons in a single area, and often additional maps are prepared for other horizons on the basis of information or some hypothesis about the thickness of the intervening strata. A map showing thicknesses of a layer is called an *isopach map* if true thicknesses are known, or it is called an *isochore map* if the map is constructed on the basis of vertical drill-hole data. A structure contour map on one horizon may be combined with an isopach or isochore map to produce a second structure contour map of the lower horizon.

Cross Sections*

Generally, geologic maps are accompanied by cross sections in which the geologist presents his idea of what the structure looks like at depth. In some cases these sections may be prepared from a wealth of control data (surface exposures, well logs, seismic lines, etc.), but more commonly they are prepared from geologic maps and scattered surface control. When there is no subsurface control, cross sections must be constructed on the basis of certain assumptions; so, they usually represent but one of several possible interpretations. The selection of assumptions depends on the mapper's conception of the geometry and the origin of the structure. The mapper is likely to be influenced by current opinion regarding the structural style. Thus literal interpretation of cross sections prepared without subsurface control is unwarranted. It is wise to regard them as interpretations, with the understanding that interpretations change. Generally, cross sections lack the degree of accuracy found in surface geologic maps. In spite of these shortcomings, cross sections are very useful. They are much more easily visualized than surface geologic maps; they reflect what the

* See Appendix C. Refer to the North American Tectonic Map (1969) and the United States Tectonic Map (1961) for good examples of very-large-scale structure contour maps.

* See Appendix C.

FIGURE 14-12 Structure contour map drawn on the Oriskany sandstone in the folded Appalachians of Pennsylvania. Doubly plunging folds trend northeast. (*From Cate and others, 1961.*)

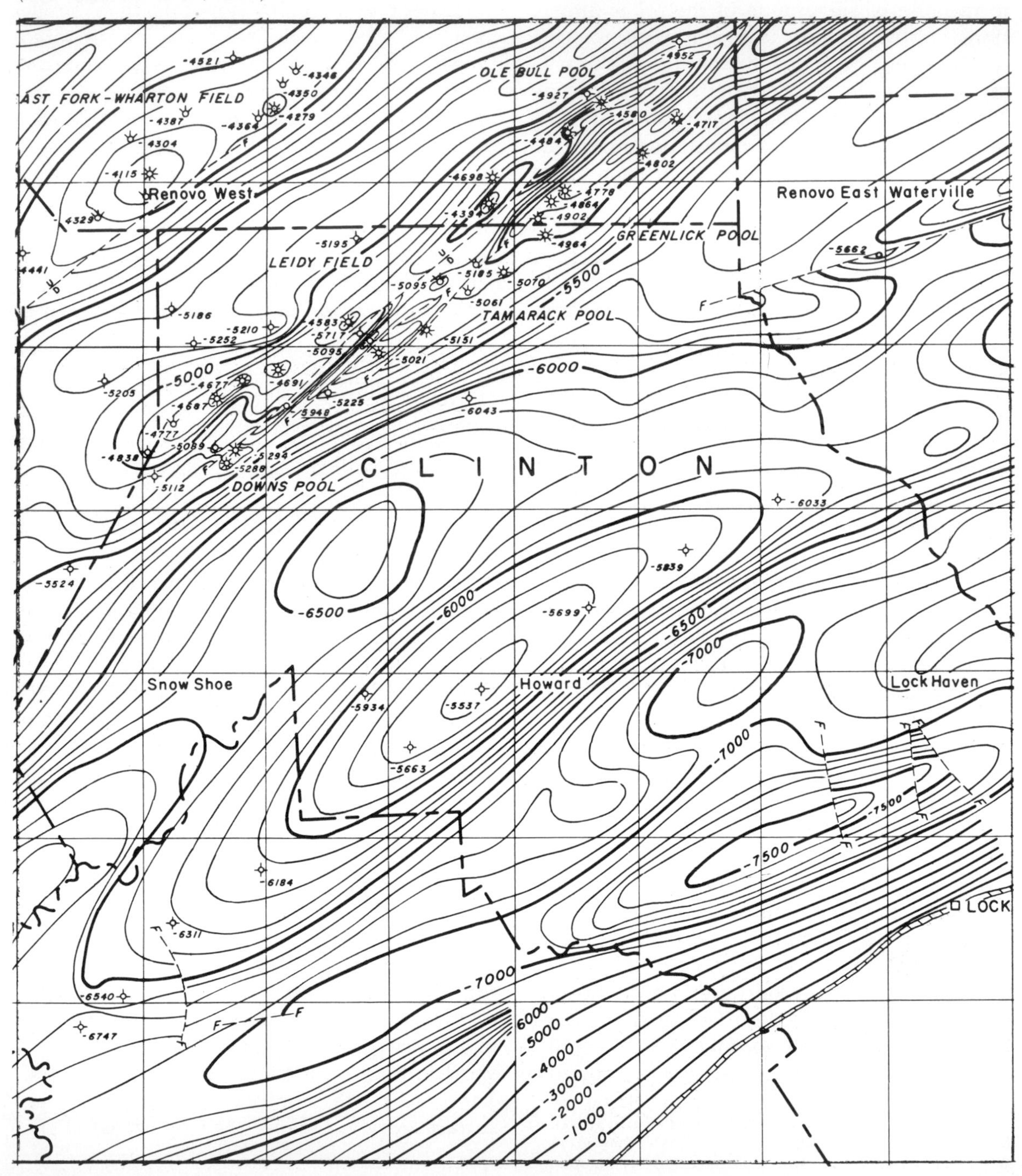

person who draws them thinks about the subsurface configuration. Drawing cross sections helps point up various strengths and weaknesses in particular map interpretations.

LARGE-SCALE SUPERIMPOSED FOLDS

Oroclines (bent or folded orogenic belts) constitute the largest superimposed folds. Although their mode of origin is still debated, many curves in orogens have been attributed to such bending, including the drag-like bend in the northwestern United States, Fig. 24-15, the bends in the Cordilleran system in Alaska and the Yukon, and the meandering loops in the European Alps, Fig. 23-1. Very abrupt bends occur in the Himalayas, where the Indian plate has apparently moved into the orogenic belt, deflecting the fold trends, Fig. 23-19.

An unusually good example of large-scale superimposed structures is found in the northern part of Spain in the Cantabrian zone, which is a part of the Spanish Hercynian orogenic belt.* The most prominent set of folds, which is paralleled by a number of major thrust faults, define a great arc (called the Asturian arc) which swings from an east-west trend along the southern margin of the fold belt through about 120° to trend northeast at the coast, Fig. 14-13. The second set of folds trends east-west but tends to be some-

* See Julivert and Marcos (1973).

FIGURE 14-13 . Superimposed folds in the Cantabrian zone in northern Spain. (*From Julivert, 1971.*)

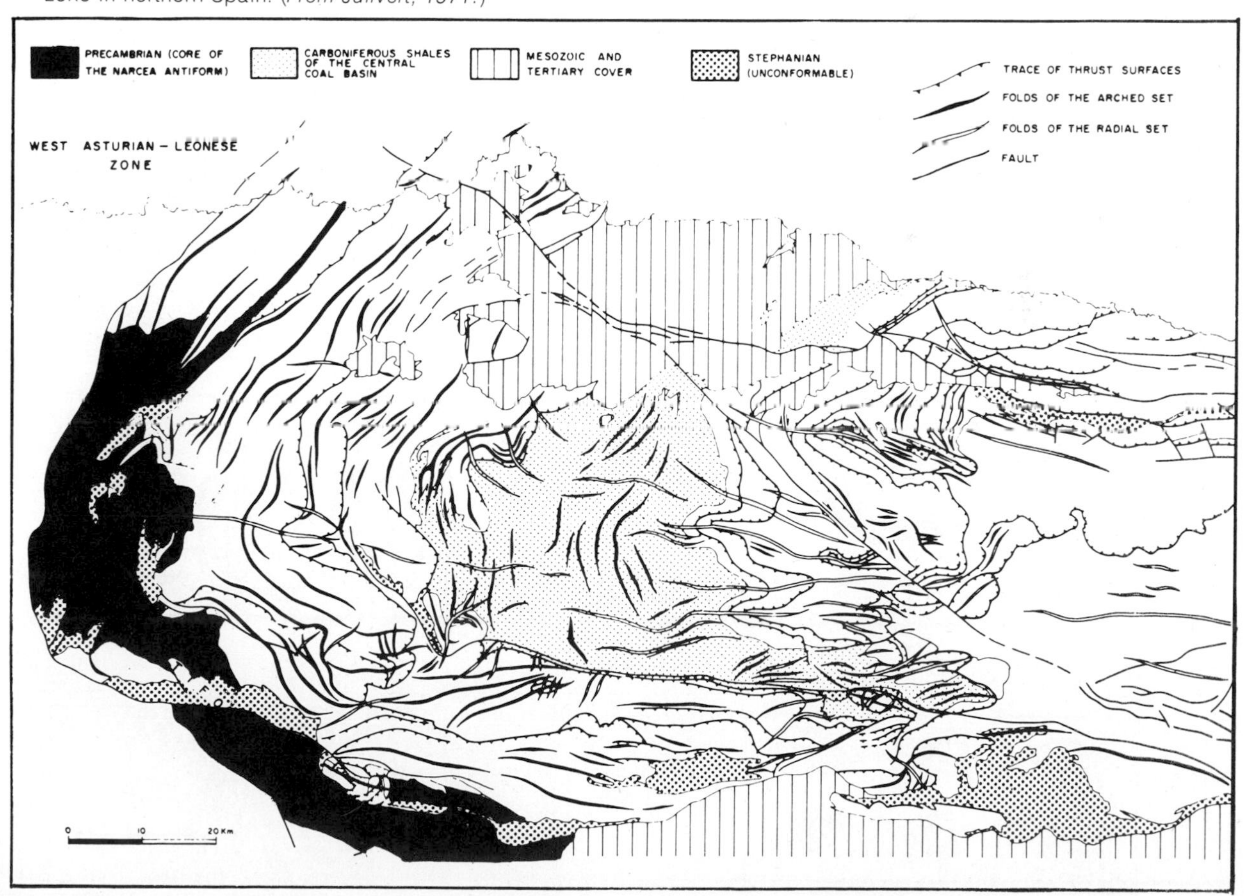

what radially disposed. Most of the examples of superimposed folds which have been described are mesoscopic in size and are passive or similar folds, but these Spanish folds formed as flexural folds. The stratigraphic section represented in the folded features here include both a Precambrian metamorphic core zone in the west and a thick Paleozoic section of sedimentary rock of great ductility contrast.

Rocks in the west are more intensely tectonized, metamorphosed, and have a well-developed flow cleavage that is missing in the east, where *décollement*-type flexural folding is the main style of deformation. Three types of structures are superimposed in this area, the first and earliest of which consist of *décollement* nappes and thrust slices, the second of arched folds, and the last and youngest of a set of radial folds. The gradual unroofing of superimposed folds gives rise to a progression of changing but generally complex map patterns such as those depicted in Fig. 14-14, which serve as a guide to recognition of cross folds on geologic maps.

Superimposed folds are more commonly found in metamorphic rocks, but it is often difficult to trace out the large-scale features as

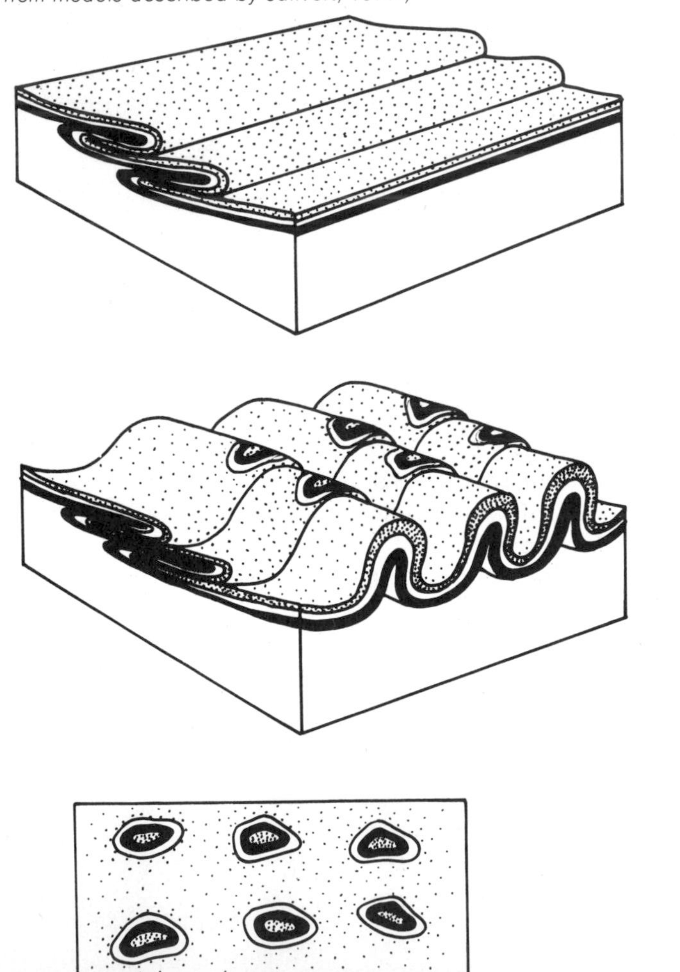

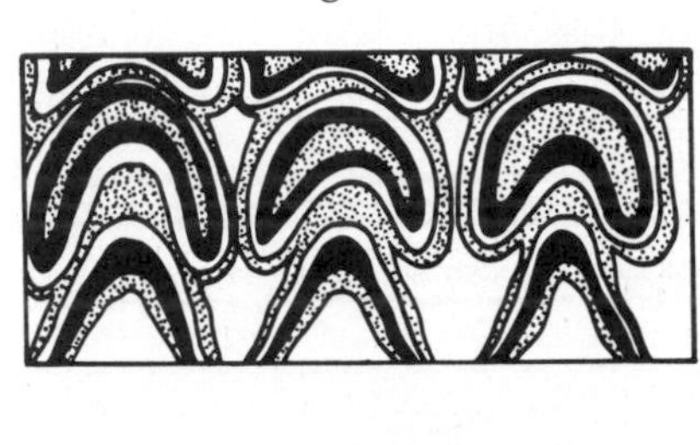

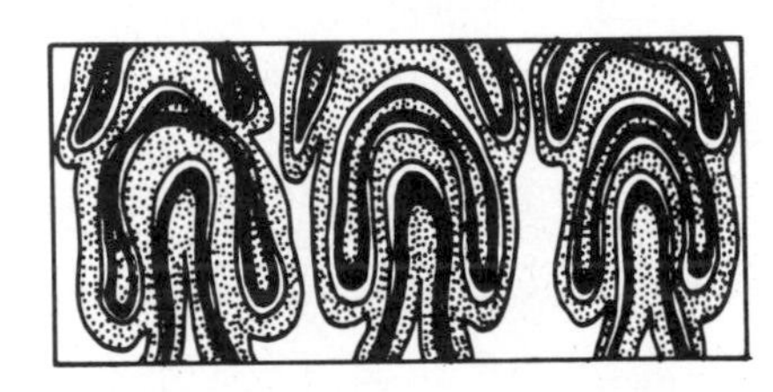

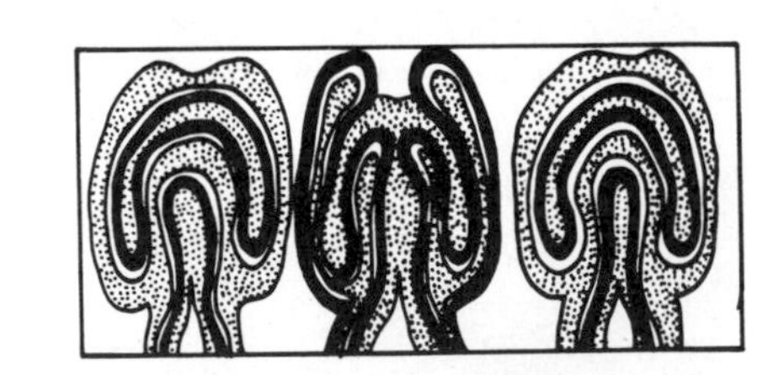

FIGURE 14-14 Map patterns formed as a result of erosion of refolded folds to successively deeper levels. (*Drawn from models described by Julivert, 1971.*)

has been done in the case the large fold shown in Fig. 14-15 which occurs in the Piedmont of Virginia and North Carolina.* The earlier form here, as in the Spanish example, was a large antiformal nappe which preceded the metamorphism and probably was formed from what were initially buckles. With the onset of high temperatures, a metamorphic flow cleavage appeared, the nappe was emplaced, and the preexisting fold forms were folded as new folds developed. The nappe was probably arched by upwelling of hot ductile material in the area of high-metamorphic-grade rocks.

Refolded folds are relatively common features in metamorphic terranes of shield areas and in the infrastructure of young orogenic belts. Such folds are readily recognized on both large- and small-scale maps once the diagnostic map patterns are known. A fine example of large-scale refolding in the Canadian Shield is found in the Westport area, Ontario, where large folds with a plunge of about 35° have been refolded about the same axis as the earlier folding, Fig. 14-16. Compare this with Fig. 11-29, which shows a schematic drawing of this type of feature.

* Tobisch and Glover (1971).

FIGURE 14-15 Refolded fold mapped in the Appalachian Piedmont. (*From Tobisch and Glover, 1975.*)

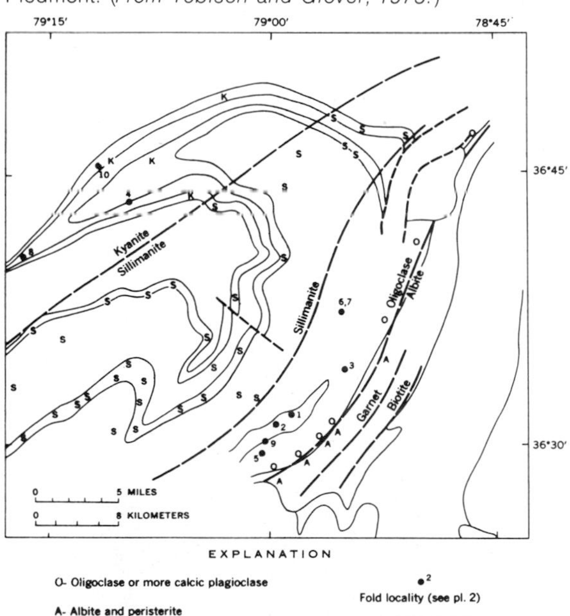

GLOSSARY OF FOLD SYSTEMS AND FEATURES OF FOLDED SYSTEMS

Allochthon: A term originated by Gümbel and applied to rocks of which the dominant constituents have not been formed in situ. (Holmes, 1920.*) The term is applied to rock masses which have moved considerably from their point of origin.

Anticlinorium: A series of anticlines and synclines so arranged structurally that together they form a general arch or anticline. (Dana, 1873.*) (Syn.: geanticline.)

Autochthon: In Alpine geology, a succession of beds that have been moved comparatively little from their original site of formation, although they may be intensely folded and faulted. (Heritsch, 1929.)

Culmination: A portion of a fold system, generally more or less at right angles to the folds, away from which the folds plunge. (Structural Committee.*)

En échelon folds: Parallel structural features that are offset as are the edges of shingles on a roof when viewed from the side. (Structural Committee.*)

Fold generation: A group of cognate folds. (Turner and Weiss, 1963.)

Fold system: A group of folds which occur together.

Geanticline: A broad uplift, generally referring to the land mass from which sediments in a geosyncline are derived. (Structural Committee.*)

Geosyncline: A surface of regional extent subsiding through a long time while contained sedimentary and volcanic rocks are accumulating; great thickness of these rocks are almost invariably the evidence of subsidence, but not a necessary requisite. Geosynclines are prevalently linear, but nonlinear depressions can have properties that are essentially geosynclinal. (Kay, 1951.)

Nappe: A sheet-like mass of rock transported laterally for great distances by recumbent folding and/or thrusting. (*Note:* Usage of the term is highly varied; see text.)

Orocline: A fold where the deformed unit is the orogen itself. (Carey, 1955.)

Para-autochthonous: In Alpine geology, a term applied to folds and nappe structures which can be connected by their facies and tectonic features with the **autochthon**, q.v. (Heritsch, 1929.)

Recess: The part of an orogenic belt where the axial traces of the folds are concave toward the outer part of the belt. (Billings, 1954.*)

Root: (1) The core of a geanticline within a geosyncline, which, after the forward drive of the geosynclinal sediments, became the recumbent fold or nappe. (Heritsch, 1929.) (2) The back-remaining, steep part of a nappe. (Heritsch, 1923.)

Salient: That part of an orogenic belt that is convex toward the foreland, i.e., is concave toward the orogenic belt. (Billings, 1954.*)

Synclinorium: A compound syncline; a closely folded belt, the broad general structure of which is synclinal. (Structural Committee.*)

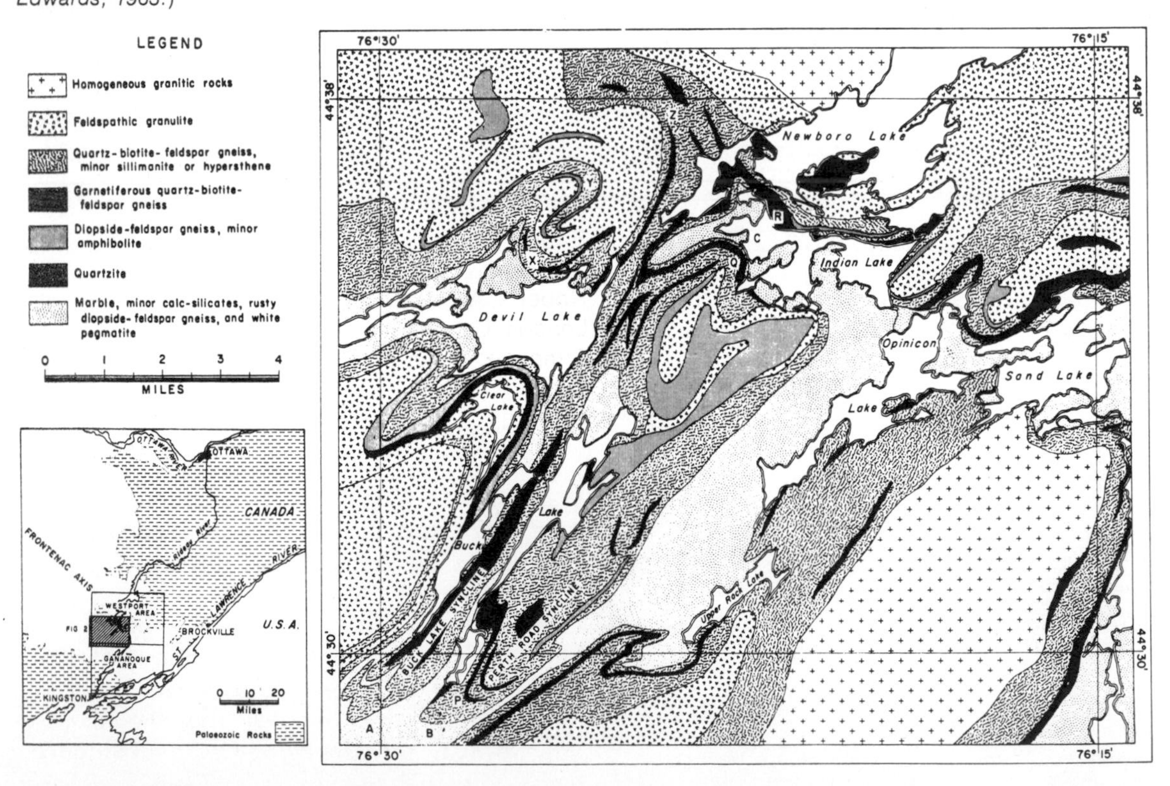

FIGURE 14-16 Geology of part of the Westport map area, Ontario. The folds plunge northeast at an average of 35° so that the map is an oblique section through the structure. Stratigraphic units are repeated at successive structural levels B, C (marble); X, Y, Z (gneiss); and P, Q, R (gneiss) by refolding about the same axis. (*From Wynne-Edwards, 1963.*)

REFERENCES

Anderson, D. E., 1971, Kink bands and major folds, Broken Hill, Australia: Geol. Soc. America Bull., v. 82, no. 7, p. 1842ff.

Bailey, E. B., 1935, Tectonic essays: Oxford, Clarendon.

Bayly, M. B., 1974, An energy calculation concerning the roundness of folds: Tectonophysics, v. 24, p. 291–316.

Bennison, G. M., 1964, Introduction to geological structures and maps: London, Edward Arnold.

Bucher, W. H., 1956, Role of gravity in orogenesis: Geol. Soc. America Bull., v. 67, p. 1295–1318.

Carey, S. W., 1955, The orocline concept in geotectonics: Royal Soc. Tasmania Proc., v. 89, p. 255–288.

Cate, A. S., and others, 1961, Subsurface structure of plateau region north-central and western Pennsylvanian on top of Oriskany formation: Pennsylvania Geol. Survey, 4th ser., map.

Chalmers, R. M., 1926, Geological maps: London, Oxford.

Cooper, B. N., 1961, Grand Appalachian field excursion: Virginia Polytechnic Institute Engineering Extension Series, Geological Guidebook No. 1.

——— 1964, Relation of stratigraphy to structure in the southern Appalachians, *in* Tectonics of the southern Appalachians: Blacksburg, Va., Dept. of Geol. Sci. Mem. 1.

Currie, J. B., Patnode, H. W., and **Trump, R. P.**, 1962, Development of folds in sedimentary strata: Geol. Soc. America Bull., v. 73, p. 655–674.

Dennison, J. M., 1968, Analysis of geologic structures: New York, W. H. Norton.

Dwerryhouse, A. R., 1942, Geological and topographical maps: London, Edward Arnold.

Elles, G. L., 1931, The study of geological maps: London, Cambridge.

Escher van der Linth, A., 1841, *in* Verhandel. Schweiz. naturforsch. Gesellschaft, Zürich.

Faill, Rodger T., 1969, Kink band structures in the Valley and Ridge province, central Pennsylvania: Geol. Soc. America Bull., v. 80, no. 12, p. 2539–2550.

Fitzgerald, E. L., and **Braun, L. T.**, 1965, Disharmonic folds in Besa River formation, northeastern British Columbia, Canada: Am. Assoc. Petroleum Geologists Bull., v. 49, no. 4.

Griffin, V. S., Jr., 1970, Relevancy of the Dewey-Bird hypothesis of cordilleran-type mountain belts and the Wegmann stockwork concept: Jour. Geophys. Res., v. 75, p. 7504–7507.

Haller, J., 1962, Probleme der Tiefentektonik: Bauformen im Migmatit-Stockwerk der ostgrönländischen Kaledoniden: Geol. Rundsch., v. 45, p. 159–167.

Harker, Alfred, 1926, Notes on geological map reading: Cambridge, W. Heffer & Sons.

Henderson, G. G. L., and **Dahlstrom, C. D. A.**, 1959, First-order nappe in Canadian Rockies: Am. Assoc. Petroleum Geologists Bull., v. 43, p. 641–653.

Heritsch, F., 1929, The nappe theory in the Alps: London.

Johnson, Arvid M., 1969, Development of folds within Carmel formation, Arches National Monument, Utah: Tectonophysics, v. 8, no. 1, p. 31–77.

——— and **Ellen, S. D.**, 1974, A theory of concentric, kink, and sinusoidal folding and of monoclinal flexuring of compressible, elastic multilayers. I. Introduction: Tectonophysics, v. 21, p. 301–339.

Julivert, Manuel, 1971, Décollement tectonics in the Hercynian Cordillera of northwest Spain: Am. Jour. Sci., v. 270, p. 1–29.

——— and **Marcos, Alberto**, 1973, Superimposed folding under flexural conditions in the Cantabrian zone: Am. Jour. Sci., v. 273, p. 353–375.

Lebedeva, N. B., 1969, A model of a folded zone: Tectonophysics, v. 7, no. 4, p. 339–351.

Lillie, A. R., 1964, Steeply plunging folds in the Sealy Range, Southern Alps: New Zealand Jour. Geol. Geophys., v. 7, no. 3, p. 406.

Mackin, J. H., 1950, The down-structure method of viewing geologic maps: Jour. Geology, v. 58, p. 55–72.

Nelson, A. (no date), Geological maps—Their study and use: Colliery Guardian Co.

Ragan, D. M., 1973, Structural geology: 2d. ed., New York, Wiley.

Roberts, A., 1958, Geological structures and maps: London, Cleaver-Hume.

Simpson, Brain M., 1968, Geological maps: New York, Pergamon.

Stearns, D. W., Logan, J. M., and **Friedman, M.**, 1974, Structure of Rattlesnake Mountain and related rock mechanics investigations: Rock mechanics, p. 18–25.

Tobisch, Othmar T., and **Glover, Lynn III**, 1971, Nappe formation in part of the southern Appalachian Piedmont: Geol. Soc. America Bull., v. 82, no. 8, p. 2209ff.

Willis, Bailey, and **Willis, Robin**, 1934, Geologic structures, 3d. ed.: New York, McGraw-Hill.

Wynne-Edwards, H. R., 1963, Flow folding: Am. Jour. Sci., v. 261, p. 793–814.

15

Diapirs and Salt Domes

Diapirs are piercement structures in which a solid but mobile core material is injected through overlying, less mobile layers. The necessary conditions of high mobility are encountered under metamorphic conditions, where ductility of most rocks is increased by elevated temperatures and confining pressure, but diapirs are perhaps better known in the case of certain sediments—particularly salt, gypsum, and clay—which are highly ductile at relatively low temperatures and pressures.

The ductile material in diapirs may be mobilized as a result of stresses arising in connection with folding or faulting, but many diapirs, especially salt structures, form apparently as a result of a gravitative instability arising from the burial of low-density salt deposits under normal-density sediments. Such diapirs are well known in the Gulf Coast, in northern Germany, and elsewhere.

SALT STRUCTURES

Structures related to salt in and around the Gulf of Mexico have been responsible for entrapment of much of the oil and gas of that region. Exploration in the region and interest in the formation of these salt bodies have been intense since the discovery in 1901 of Spindletop near Beaumont, Texas, the first salt-dome oil field. Many ideas concerning the origin and emplacement of the salt have been advanced (see De Golyer, 1925), but our present conception, based on thousands of wells around salt structures and a vast amount of geophysical data, is that the salt was deposited as an extensive sedimentary unit in a large basin and was later buried under increasingly greater loads of unconsolidated Mesozoic and Cenozoic sediment which caused the salt to flow and locally to force its way up through the overlying higher density sediments. The overlying sediments are domed, faulted, and forced apart as the salt moves upward. The salt movements and structures related to it are thought to have formed independently of tectonic activity.

Gulf Coast Salt Basin

A thick layer of bedded salt overlying anhydrite and associated with red beds and other evaporites has been penetrated by wells in Arkansas, northern Louisiana, and east Texas, Fig. 15-1. These beds, known as the Louann salt, are thought to be the source of the salt found in domes, plugs, and pillars penetrating much younger strata. The Louann salt is gray, nearly clear, coarsely crystalline, and contains streaks of anhydrite. It is underlain by anhydrite, with minor red beds and gravel (Werner formation), and it is overlain by a sequence of red beds (Norphlet formation). These units lie above an unconformity developed on the Paleozoic basement. Palynological evidence indicates that the salt is of Late Triassic to Early Jurassic age.

The shape and extent of the basin containing Louann salt and the source of the salt are both unsolved problems, but the salt extends at least as far as the edge of the present continental shelf, and it is probable that the salt covered most of the Gulf region and that it is the same salt deposit penetrated in Cuba and from which the salt domes in the Tehuantepec region of Mexico originate. Several closed depressions in the central part of the Gulf mark sites where salt has risen to the surface, broken into the water, and been dissolved, and salt plugs underlie the Sigbee knolls.

Certainly the salt deposit was not initially of uniform thickness. The thinned northern edge of the salt is known in southern Arkansas.

FIGURE 15-1 Offshore distribution of salt within Gulf of Mexico basin (shaded area) and onshore distribution as shown by Murray (1966) (cross-hatched area). Dashed area represents boundary of knolls and domes on abyssal plain and in Bay of Campeche as surveyed by Worzel and others (1968). (*From Antoine and Bryant, 1969.*)

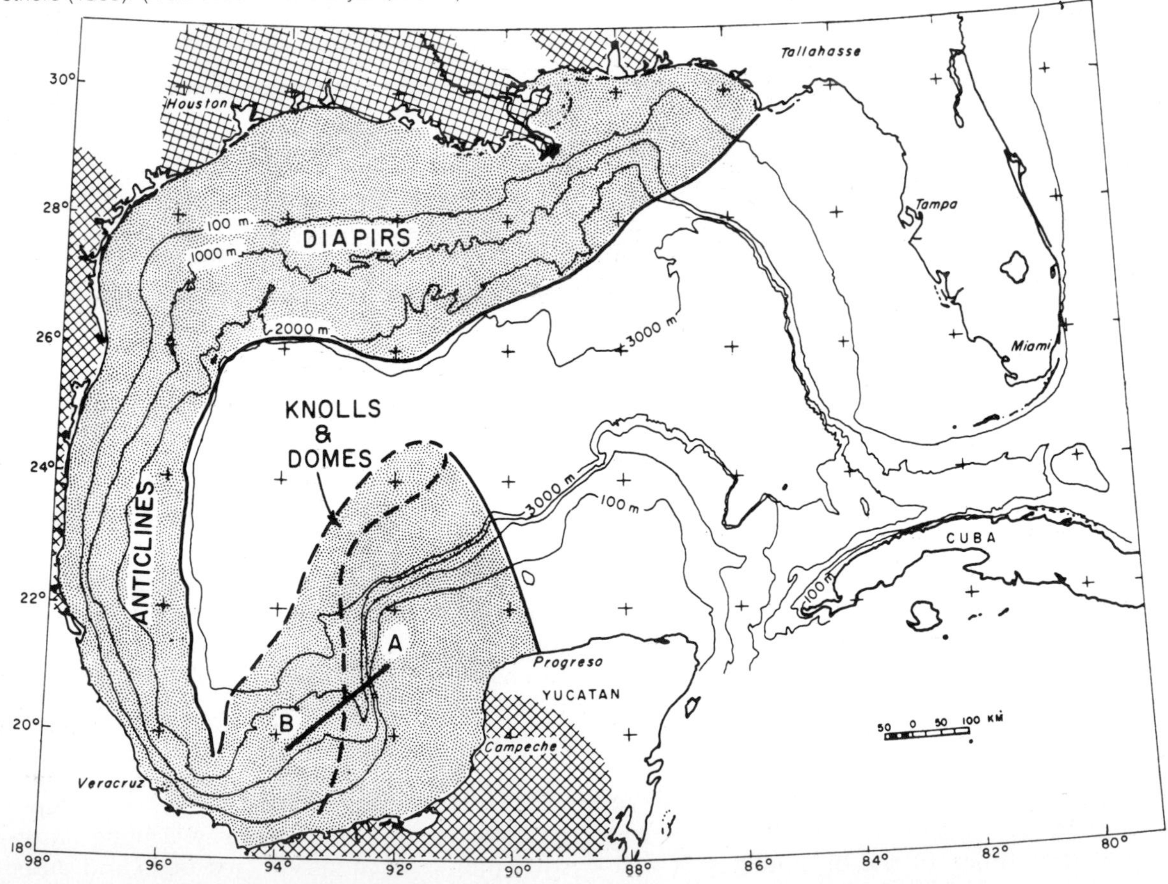

Wells into the salt show its thickness to be more than 396 m (1,300 ft) in south Arkansas, 366 m (1,200 ft) in northwest Louisiana, and 244 m (800 ft) in east Texas—and all these are thicknesses penetrated near the edge of the salt basin. The salt is covered by such great thicknesses of sediments farther south that no wells have reached it. Thickness estimates for the salt range from 1,219 to over 5,182 m (4,000 to over 17,000 ft), with 1,524 m (5,000 ft) as an average thickness. These variations are caused in part by differences in original thickness, but the salt has also undergone considerable postdepositional movement that has affected its thickness.

The origin of such thick accumulations of salt is a difficult problem, as is the thickness of salt relative to the amount of underlying anhydrite. The proportion of the two does not correspond to the relative amounts that can be precipitated from sea water today. Because the anhydrite would be precipitated first, it has been suggested that highly concentrated brines were introduced into the Louann salt basin during deposition. These could possibly have come from the Permian basins of western Texas or from the Sabinas basin of Mexico. Another idea is that concentration took place in an environment where bars or barriers of some type allowed water into the basin or parts of it only after the water was partially evaporated and part of the calcium sulfate precipitated. The shape of this basin is much disputed. Some propose a lagoonal environment, others suggest local basins, and still others consider that the Gulf may have been a vast closed sea at the time.

One of the postulated development histories of the basins is illustrated, Figs. 15-2 and 15-3. The initial stage represents deposition of the salt over an extensive area, with some possible highs over which deposition was thin. The second stage, close to the end of the Cretaceous, depicts subsidence of the basin and accumulation of Jurassic and Cretaceous sediments, with the consequent formation of some salt domes. The last two stages depict the gradual extension of the areas of subsidence toward the coast, with consequent accumulation of sediments and activation of salt intrusion. Because salt movement is thought to be triggered by the weight of overlying sediment, it is probable that salt domes grow most rapidly during and shortly after deposition of sediment, and growth should be related to rates of accumulation. Movements of salt in the Paradox basin in the Colorado Plateau region have also been shown to be associated with basin filling.

Shapes of Gulf Coast Salt Features*

1. *Salt swells along faults at the edge of the basin.* These are best seen in the zone of normal faults (e.g., Mexia-Talco zone). The salt wells up along the upthrown sides of the faults, forming domes or folds with as much as 200 m of closure at the depth of the salt.

2. *Salt pillows.* These are pillow-shaped bodies that resemble salt swells, but they are not associated with faults, although they are often aligned. Similar features occur in the salt basin of Germany, Fig. 15-4.

3. *Deep-seated salt domes, buried at depths of 2 km or more.* Sediments are domed and faulted over these bodies, but piercement of the sediments by the salt is minor. The shallow domes appear to have more faulting associated with them than the deep ones. Domes below 4,000 m appear to have little associated faulting.

4. *Piercement domes.* The salt has forcefully intruded the overlying sediment by moving up along complex faults in some instances and by displacement of the unconsolidated sediments.

5. *Residual salt features called turtle structures in northern Germany.* These are structures which result from the withdrawal of salt from around the salt bed into piercement domes, leaving a body with thinned edges.

* See the tectonic map of the Gulf of Mexico, 1972 (A.A.P.G.).

Basement Involvement in Salt Structures

The role of the Paleozoic "basement" underlying the salt in the development of the salt structures is one of the important unsolved problems. Few wells are deep enough to penetrate this basement. Where data are available, the surface on which the salt was deposited is underformed, but faults offsetting the basement may be present. Movements on these faults could initiate the development of salt ridges from which the domes and swells have originated. Support for this argument is found in the apparent alignment of many domes and in the salt swells located along faults at the edges of the salt basin. In any case the dominant process in the development of structures appears to be directly related to the mobility of the salt.

Avery Island Dome

Salt has been mined at Avery Island since the Civil War in the top of a dome which has intruded so close to the surface that its shape is reflected in the topography of the island. It has a roughly elliptical shape in map views, Fig. 15-5, but salt has flowed laterally near the surface overhanging part of the sediments, Fig. 15-6.

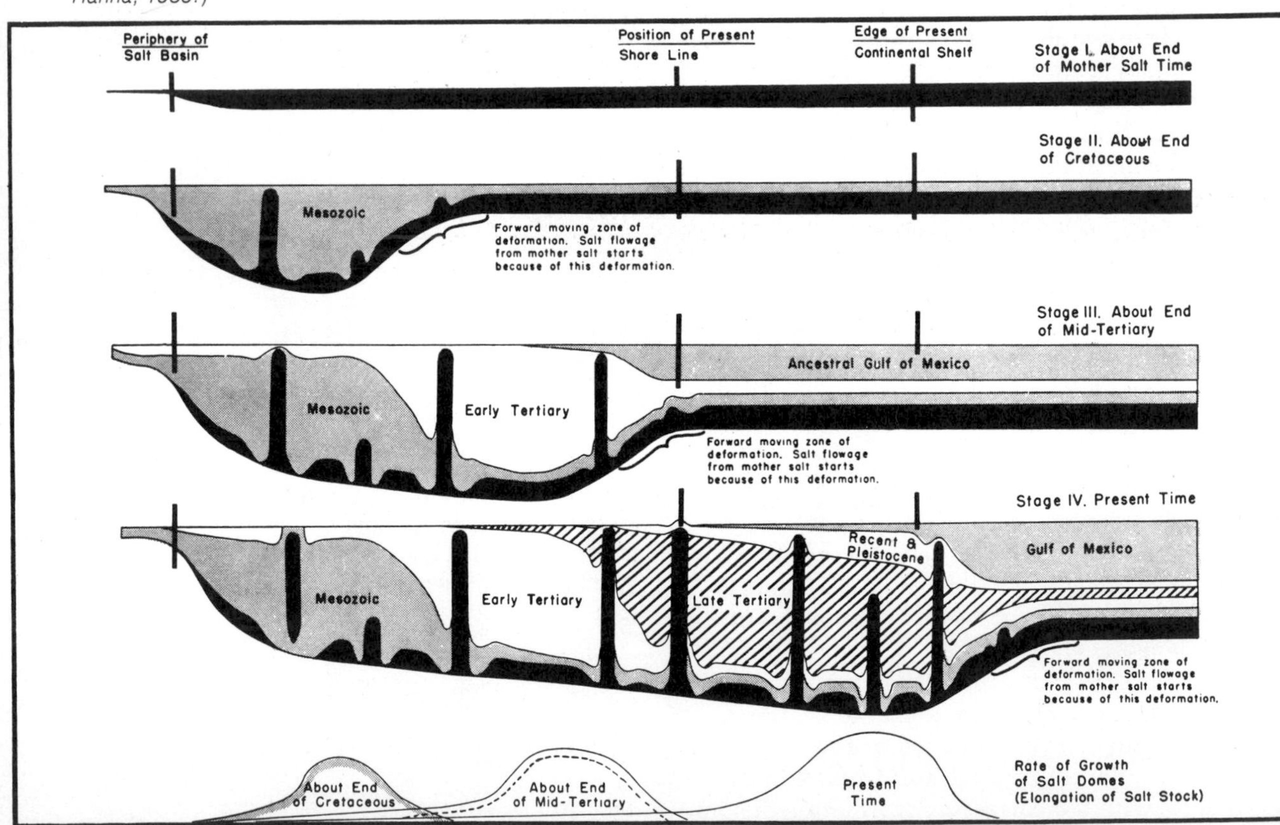

FIGURE 15-2 Rise of Gulf Coast domes through geologic time shows in this cross-sectional account (top to bottom) of four stages in deposition of sediments atop mother salt bed by rivers, streams, and prehistoric seas. Domes were supposedly triggered by difference in static load of sediments at, and ahead of, area of deposition. (*From Hanna, 1959.*)

Fault Patterns Associated with Salt Structures*

A characteristic pattern of normal faulting is found over salt domes. The pattern may be complex in detail, but in general, two or more major normal faults form a graben over the dome, and radial high-angle normal faults are frequently present. The strike of these faults parallels the long axis of the dome when the dome is elongate. Frequently the major fault bifurcates, splitting into two or more normal faults which form a *step-fault* pattern, with the downthrown side of the fault toward the center of the dome. Other small subsidiary faults, called *antithetic faults,* have strikes parallel to the major faults but with opposite dip direction. The fault patterns have been closely duplicated through model experiments such as those performed by Currie (1956), Fig. 15-7.

* See Chap. 9.

Growth Faults

Changes of dip and amount of throw with depth are especially interesting aspects of these faults. These changes are reflected in the cross section, Fig. 15-8. Although not all faults show the same type of change in dip, many tend to increase in throw and decrease in dip with depth (Hughes, 1960). This is interpreted as meaning that the amount of throw on the faults has increased with time. This growth is caused by the slow upward movement of the salt. Stages in the growth of the fault through the overlying sediments as they were deposited is shown in Fig. 15-9. Notice that the *stratigraphic throw* has increased on the fault from top to bottom. This is due to lengthening of the section in the downthrown beds and to

FIGURE 15-3 Diagrammatic cross section of northern gulf margin southeastward from the Bend arch to the Sigsbee escarpment, showing the Gulf Coast geosyncline and its relationship between the young geosynclinal deposits and the Paleozoic structures of the Ouachita system. (*From Martin and Case, 1975.*)

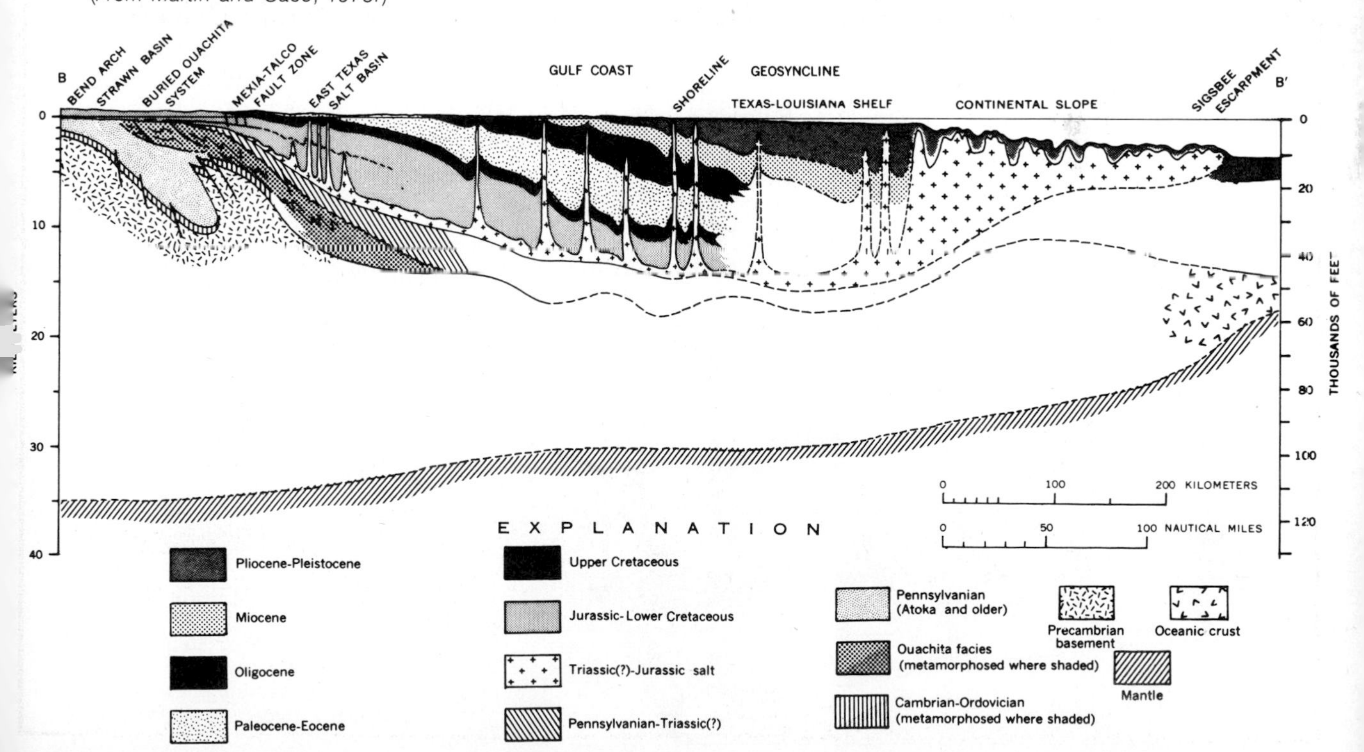

the presence of wedges below unconformities that are preserved in the downthrown side but not on the upthrown block.

Deformational Pattern within Salt Domes

Mines located in the top of near-surface salt domes have afforded an unusual opportunity to study the structure of the salt body and to determine directly the character of the processes by which the salt has been intruded. A classic study of the megascopic features in a salt dome was made by Balk (1949)* in the mines at the Grand Saline salt dome, Texas. Layers of salt and anhydrite are visible throughout the mined parts of the dome. These layers dip very steeply near the margins and presumably parallel to the margins of the dome. The layers are intricately folded elsewhere in the dome, and the axes of nearly all the folds plunge almost vertically. Many of these are isoclinal folds in which the beds show slip planes parallel to the axial planes; however, none of the slippages causes a complete break in the bedding, Fig. 15-10. Balk (1949, p. 1803) describes the folds as follows:

In contrast to the monotonous sequence of parallel layers of salt on vertical walls, the ceilings of rooms and tunnels exhibit an impressive array of folds. Few structural features express the remarkable evolution of a salt dome through deformation as tellingly as the wonderful sweep and beauty of single beds, or swarms of salt layers, trending in broad, smooth curves across the spacious ceilings of large, electrically lighted rooms, forming here a large isoclinal fold, there grouped in numerous smaller flexures that reconcile diverging trends of larger folds.

. . . It is probably significant that only in this section (one to two hundred feet from the border of the dome) are folds almost lacking. The salt here displays straight layers. . . .

Salt folds vary greatly in size. Some are small, insignificant features, involving only a few, or one single layer. . . . There are gently curved, open folds as well as tightly compressed, isoclinal folds. The innumerable folds which vary the orientation of their limbs in a bewildering manner, have one element in common; all their axes are parallel, in nearly vertical attitudes. . . .

There are literally thousands of smaller folds, superposed on the limbs of the larger folds. . . . The limbs of the shear folds may be isoclinal, or may diverge in directions, causing rows of zigzags, or chevrons. . . . Salt layers whose cross sections approach the axial plane zone as smooth curves are abruptly thrown into scores of isoclinal contortions.

The axial planes of these small shear folds are without exception parallel with the axial planes of the folds on which they are found. Followed through the mine, their directions vary over short distances, and nowhere have shearing movements cut through the limbs of folds. . . .

It seems, therefore, that the shear folds developed in response to the same stresses that governed the direction of compression of the larger folds.

* See also Muehlberger (1959).

Lineations formed through the alignment of minerals (e.g., pencil-shaped aggregates of anhydrite) are found throughout the mine, and most of these are vertically oriented.

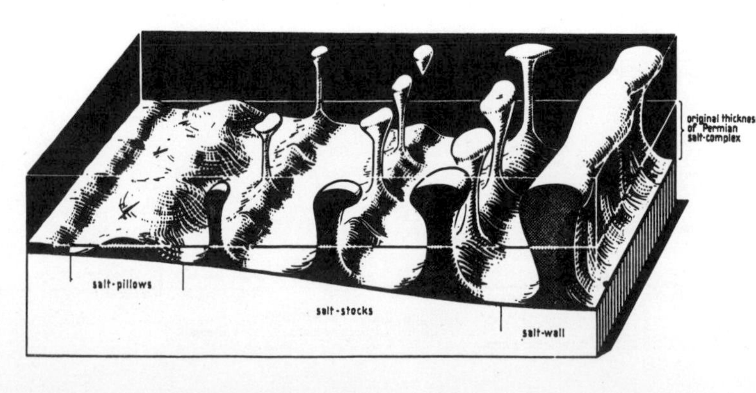

FIGURE 15-4 Diagram of different types of salt structures in relation to original thickness of Permian salt complex of northwest Germany. (*From Trusheim, 1960.*)

Halite crystals also tend to show a vertical elongation. A number of other observations also point to solid flowage as the mechanism by which the salt plug was emplaced. Neither fractures nor faults occur within the salt. Salt layers are nowhere crossed by others, and there are no unconformable contacts between layers. Where groups of salt layers approach each other with contrasting strikes, a zone of compact salt intervenes.

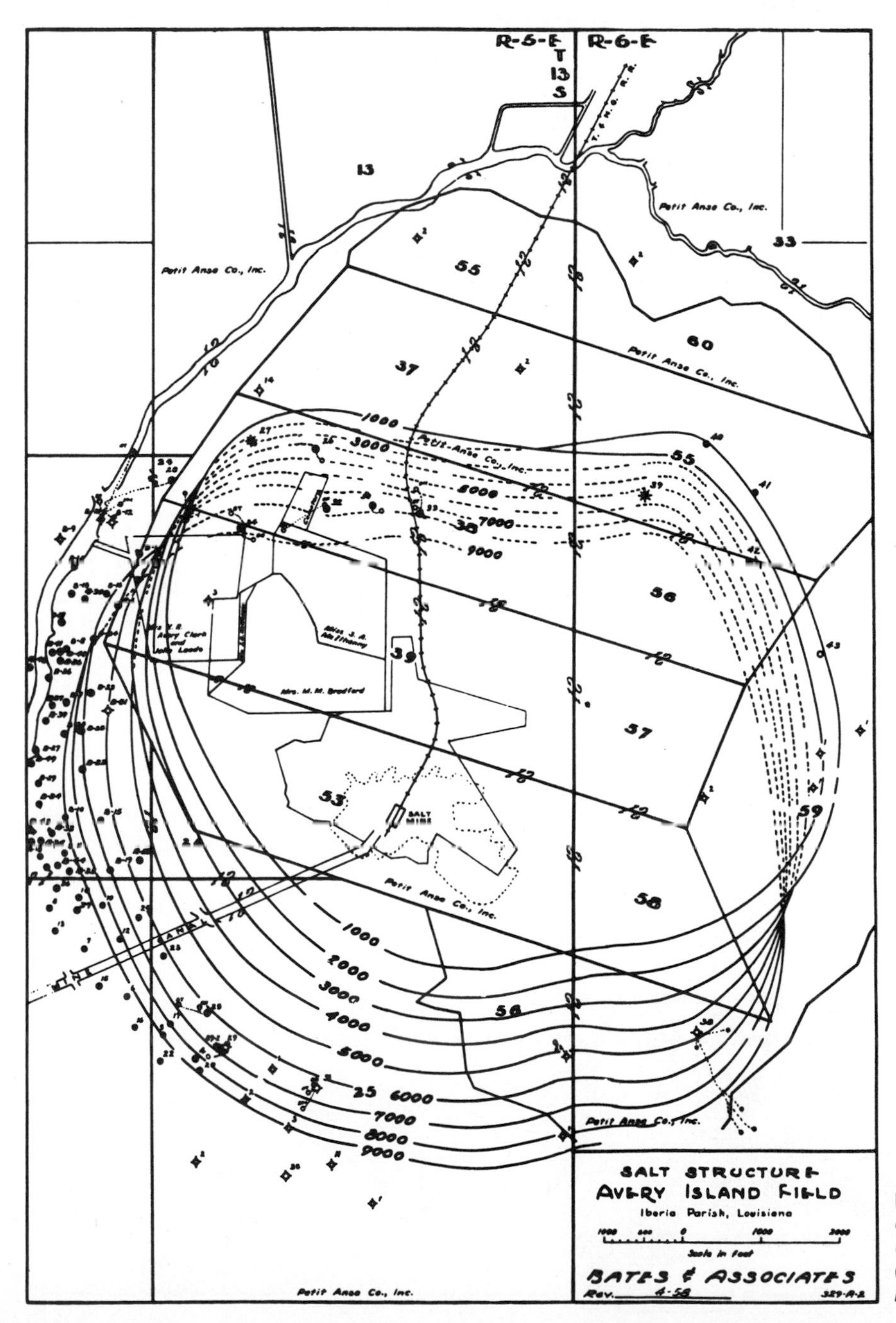

FIGURE 15-5 Structure contours of Avery Island dome on top of the salt. (*From Bates, Copeland, and Dixon, 1959.*)

The movement of salt toward and into a salt dome is envisioned by Escher and Kuenen (1929) as follows:

While the direction of principal propagation of the whole material is centripetally inward and upward, the converging motion of all particles generates peripheral, tangential stresses. Their varying directions would be represented by horizontal lines, drawn on the surface of the cone. As these stresses act parallel with the layers, they constitute shearing stresses. If the material chosen is homogeneous, there is a continuous gradation of the rate of yielding from one small area to another, but if the mobility of adjacent layers varies abruptly, the rates of yielding will also vary abruptly so that one layer, or one group of layers, must slip over or under adjacent layers. . . . Thus there originates around the periphery, at the base of the dome, numerous folds whose axes trend radially, and plunge outward from the dome in all directions.

A simple and familiar way of reproducing this fold pattern is to raise several horizontal sheets through a horizontal ring. . . . As each mass of salt approaches the axis of the dome, the direction of principal propagation becomes steeper. Therefore, the axes of the folds turn also into progressively steeper directions until, at high levels in the dome, there should remain only a multitude of folds with nearly vertical axes.

Structural features similar to those in the Grand Saline dome have been described in the

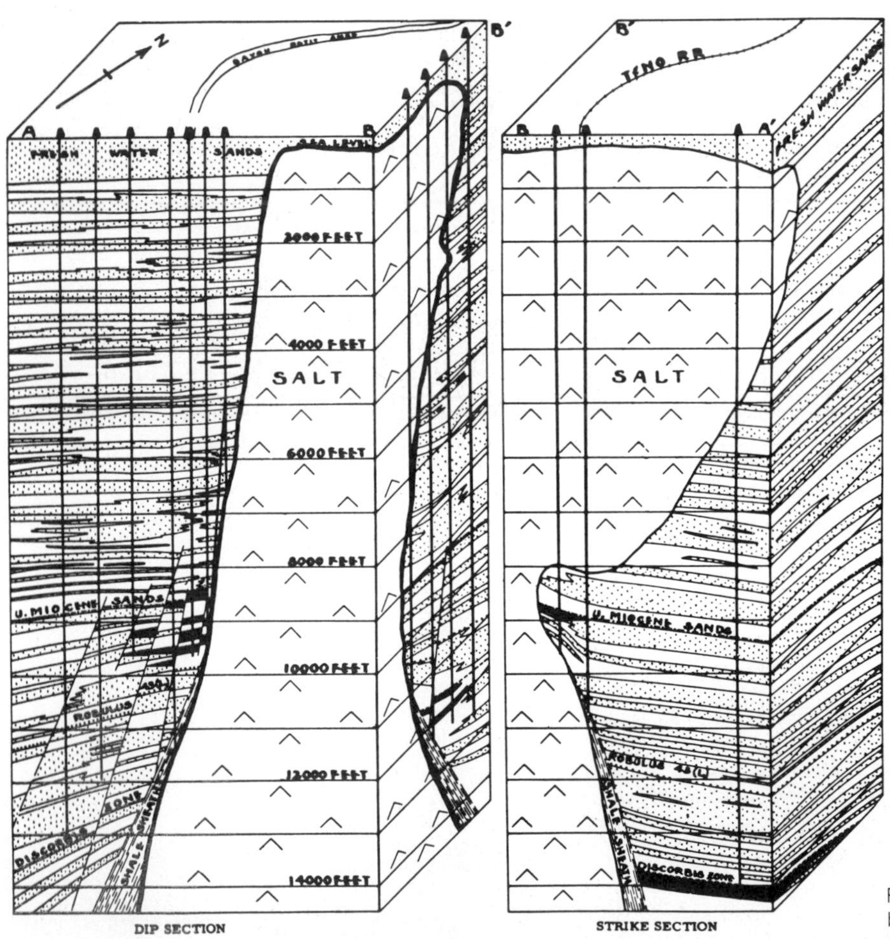

FIGURE 15-6 Generalized block diagram of Avery Island dome. (*From Bates, Copeland, and Dixon, 1959.*)

Winnfield dome (Hoy, Foose, and O'Neill, 1962) and in the Weeks Island dome (Kupfer, 1962), Fig. 15-11.

Formation of Salt Domes

Nettleton (1934) demonstrated that salt can rise from an original bedded deposit of salt as a result of the instability which exists when salt is overlain by a thick sequence of higher density sediments. This theory has gained acceptance because it accounts for the formation of the observed features by a process which does not require lateral compression. Nettleton concluded that salt domes formed as a result of movement by solid flowage under gravitational forces because salt is lighter than the surrounding sediments. The domes have a nearly circular form, suggesting that their shape is controlled by yielding and flow and not as a result of fracturing or faulting.

The contrast between the density of salt and the surrounding sediments provides a gravitational motive force which is sufficient to cause the salt flow. The density of the salt (2.2) does not vary with depth, but the density of the surrounding sediments increases with depth as a result of increasing compaction. A plot showing the predicted change in density of sediments with depth, Fig. 15-12, also has a curve showing the differential hydrostatic pressure to be expected at the bottom of the salt column. Below a depth of 600 m (2,000 ft) the density of the sediment exceeds that of the salt. The differential hydrostatic pressure ranges from a few tens of kilograms per square centimeter to nearly 98 kg/cm^2 (1,400 lb/in.^{2}), certainly enough to cause solid flowage of the

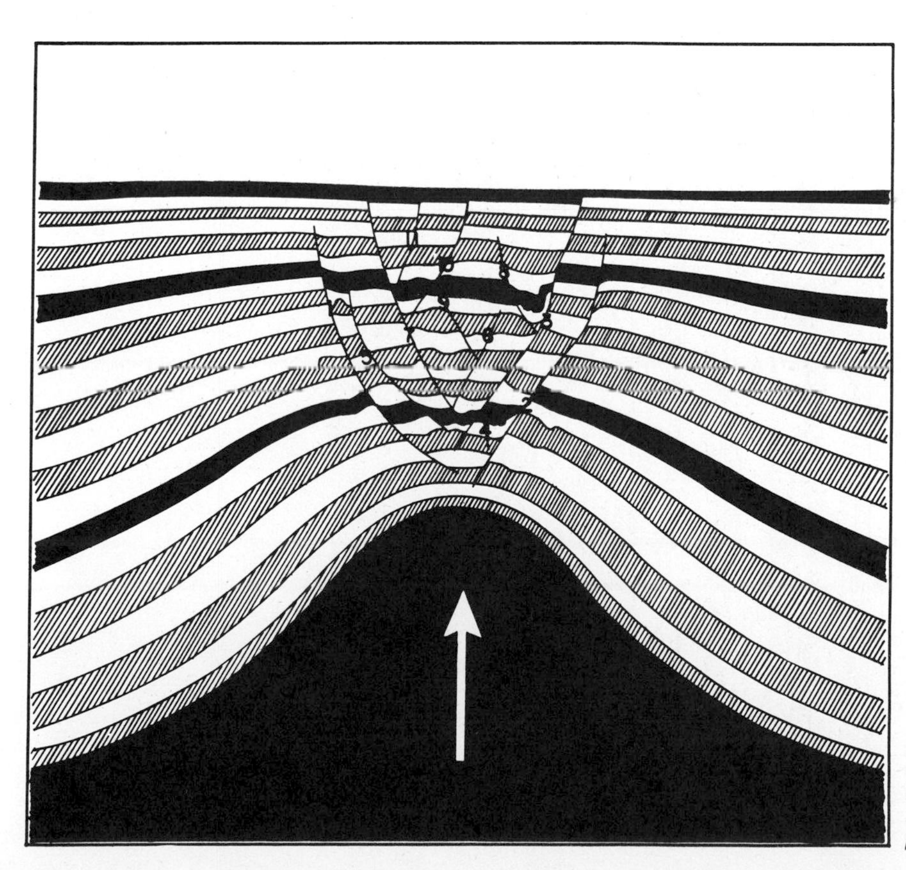

FIGURE 15-7 Graben fault pattern formed in model experiment. (*Modified after Currie, 1956; redrawn from Hughes, 1960.*)

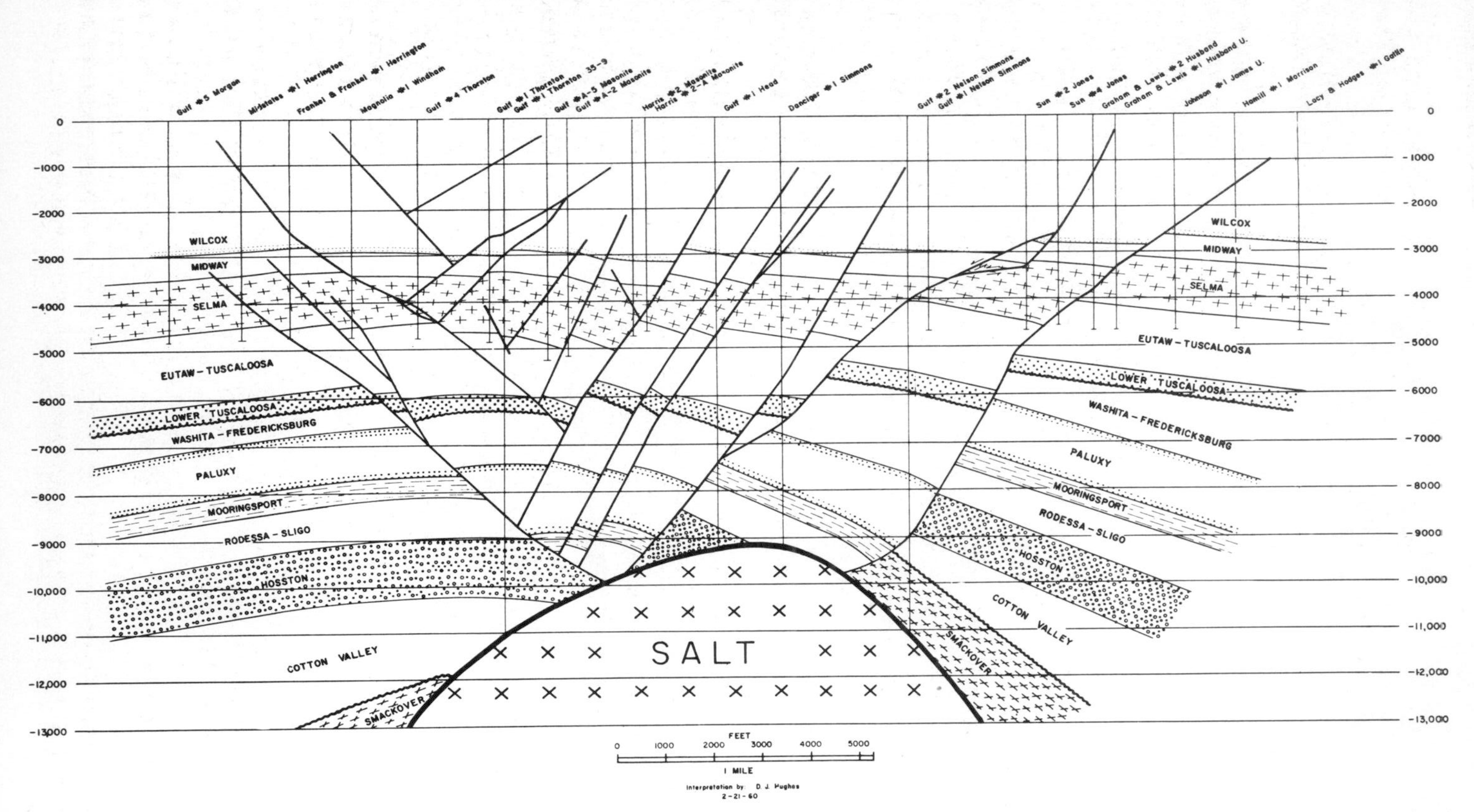

FIGURE 15-8 East-west cross section of the Heidelberg structure, Jasper County, Mississippi, illustrating actual fault pattern over deep-seated salt dome. (*From Hughes, 1960.*)

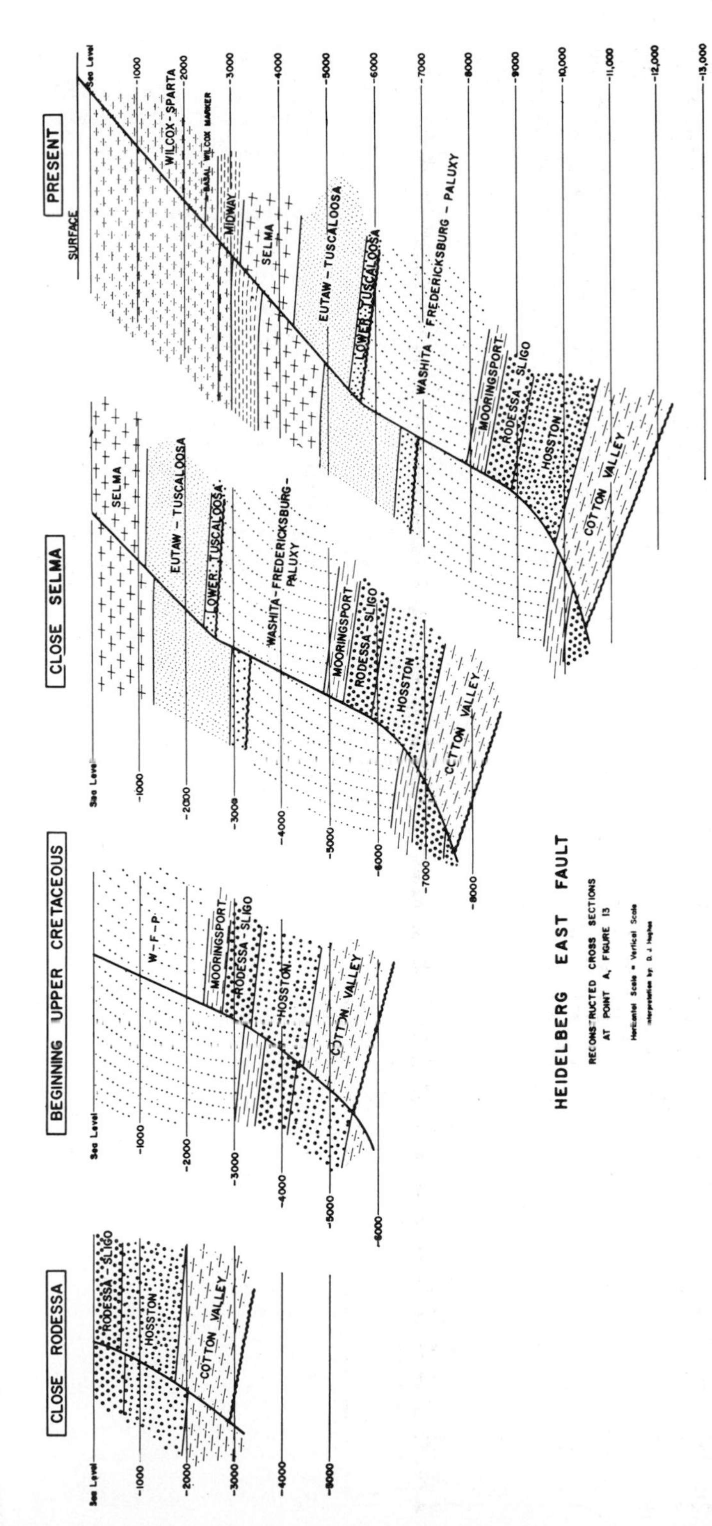

FIGURE 15-9 Stages in growth of the East Heidelberg fault, Jasper County, Mississippi. Fault is reconstructed at various times, assuming zero throw on each horizon at time of deposition of that horizon. *(From Hughes, 1960.)*

salt. Experiments have shown that salt will flow at differential pressures of this order of magnitude. It should be pointed out that some salt domes have apparently formed under overburdens of as little as 1,000 m.

Nettleton (1943) and others, notably Parker and McDowell (1955) and Dobrin (1941), have conducted scale-model experiments to simulate the formation of salt domes. Stages in the development of three of these experiments by Nettleton are shown in Fig. 15-13.

Salt Structures of Northern Germany

More than 200 salt stocks and similar features are now known in the oil- and gas-producing region of northern Germany, Fig. 15-14. The mechanics of formation of these bodies is thought to be similar to that in the Gulf Coast (Trusheim, 1960), although the general geologic setting and the shapes of many of the salt structures differ from those normally associated with American Gulf Coast domes. The salt in the German basin is shallower than the salt of the Gulf Coast, and the structure underlying the German basin is better known. Stock-like masses resemble American salt domes; other bodies resemble great salt pillows and salt-walls, Fig. 15-14, most of which are oriented roughly parallel to one another, but a few have almost meandering courses. Some of these walls are nearly 97 km (60 miles) long.

The "mother" salt was deposited in northern Germany in the Permian on a basement surface of low relief. The thickness of the salt now varies from a few meters at the edges to more than 1,000 m in the deepest portions of the basin. The sedimentary cover overlying the salt attains a maximum thickness of between 3,000 and 4,000 m. The salt stocks and walls have risen from the deepest portions of the basin, and the salt pillows are concentrated in the shallower portions, suggesting that the salt thickness and the weight of overlying strata determine the form of the salt structures.

The basement shows little post-Permian deformation except for a few faults and fractures, but the overlying section is strongly deformed. Folds, faults, and flowage features have formed independently of the underlying basement, and the intensity of deformation decreases with depth. A thickness of about 1,000 m of sediment is required to start salt flowage and migration. Various ideas have been put forth in Germany to explain the

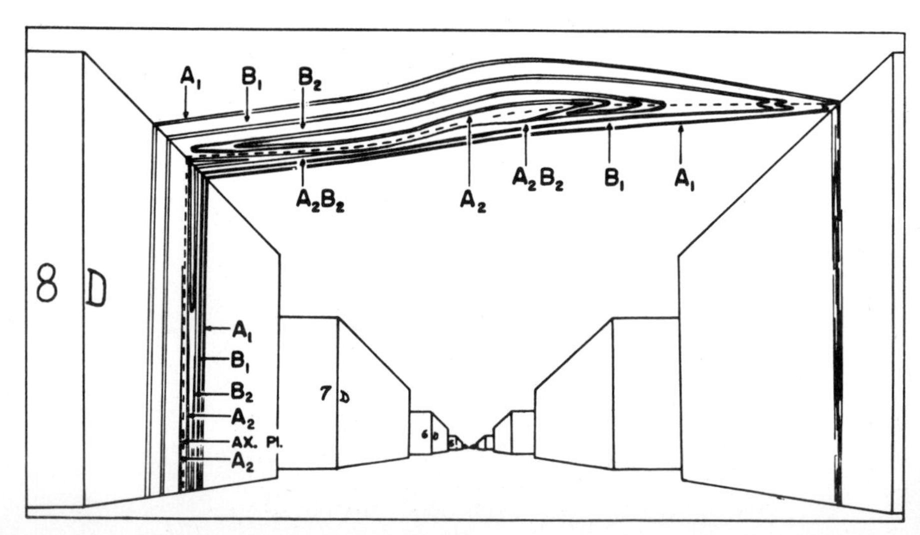

FIGURE 15-10 Block diagram of refolded fold, exposed in the walls and ceilings of the salt mine in the Weeks Island salt dome. If salt structure is unfolded along dashed line representing trace of axial plane, a second axial plane is revealed between B_1 and B_2. (*From Kupfer, 1962.*)

initial movements. These range from attributing the initial movements to faulting, to fracture zones in the basement, and to initial difference in thickness of the salt.

Trusheim suggests "that the development of a salt structure normally began with the *salt pillow* stage. The salt pillow consists of a salt accumulation, which is plani-convex, at first hourglass-shape, later dome-shape, and usually almost symmetrical. . . ." These features as well as the salt stocks have *peripheral sinks.* If the supply of salt is maintained, the pillow structure swells, its flanks become steeper and eventually form shearing cracks in the overlying sediment, and the diapir stage begins in which vertical salt movement dominates the development of the structure.

Some of the American domes come close to the surface, and apparently some of these domes in Germany broke through the surface and flowed out as salt extrusions (possibly under unconsolidated sediment) or *"salt glaciers"* flowing over and around brecciated rock and sediment.

Some typical structures associated with these salt intrusions are illustrated in Fig. 15-15. Trusheim (1960, p. 1527) described their formation as follows:

> Many salt structures in Northern Germany are asymmetric especially in the deeper levels. . . . The salt

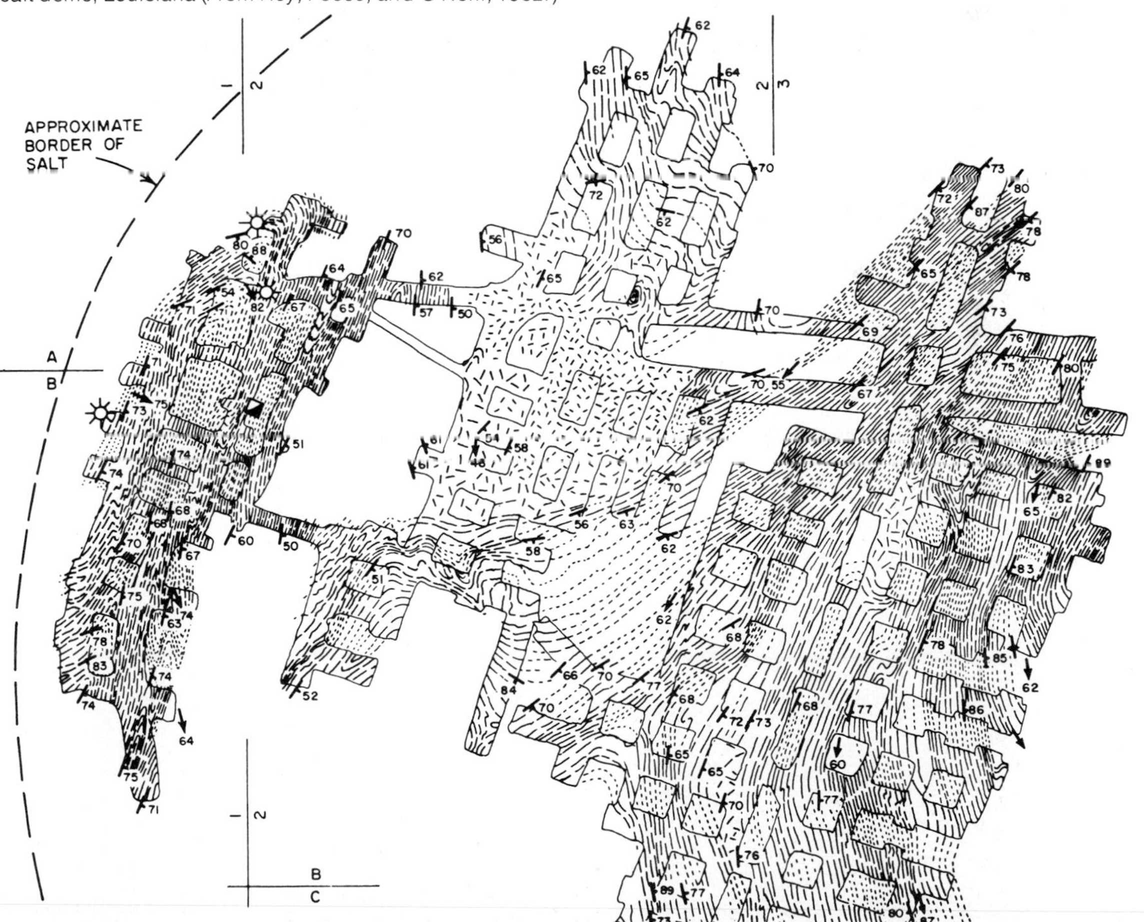

FIGURE 15-11 Structure of salt as exposed on 811-ft (247-m) level of Carey Salt Company mine, Winnfield salt dome, Louisiana (*From Hoy, Foose, and O'Neill, 1962.*)

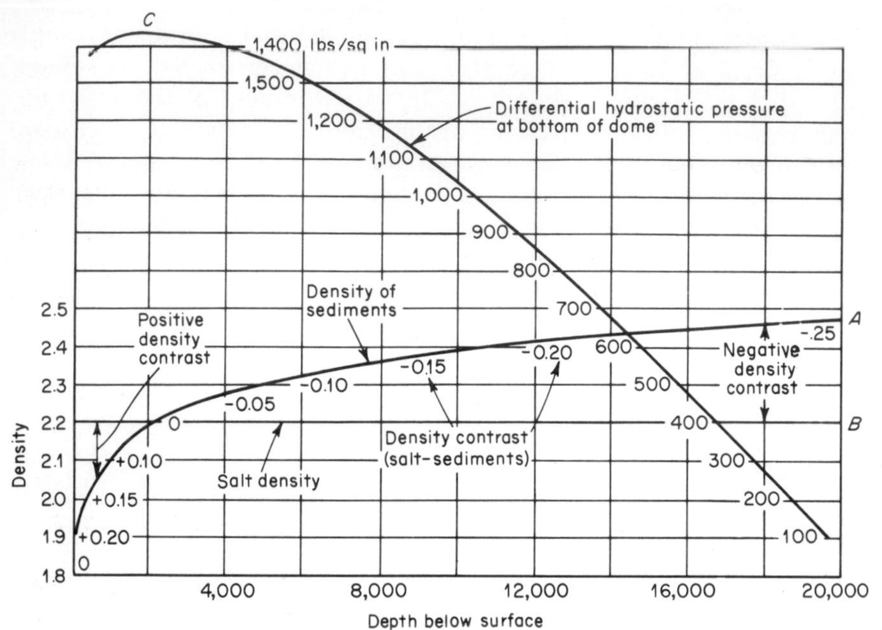

FIGURE 15-12 Density contrast and differential pressure at bottom of salt column, Texas Gulf Coast. The lower curve is a plot of the density of sediments as a function of depth of burial (note that salt has a density of 2.2). Values for density contrast between the sediment and salt are shown along curve. (*Redrawn from Nettleton, 1934.*)

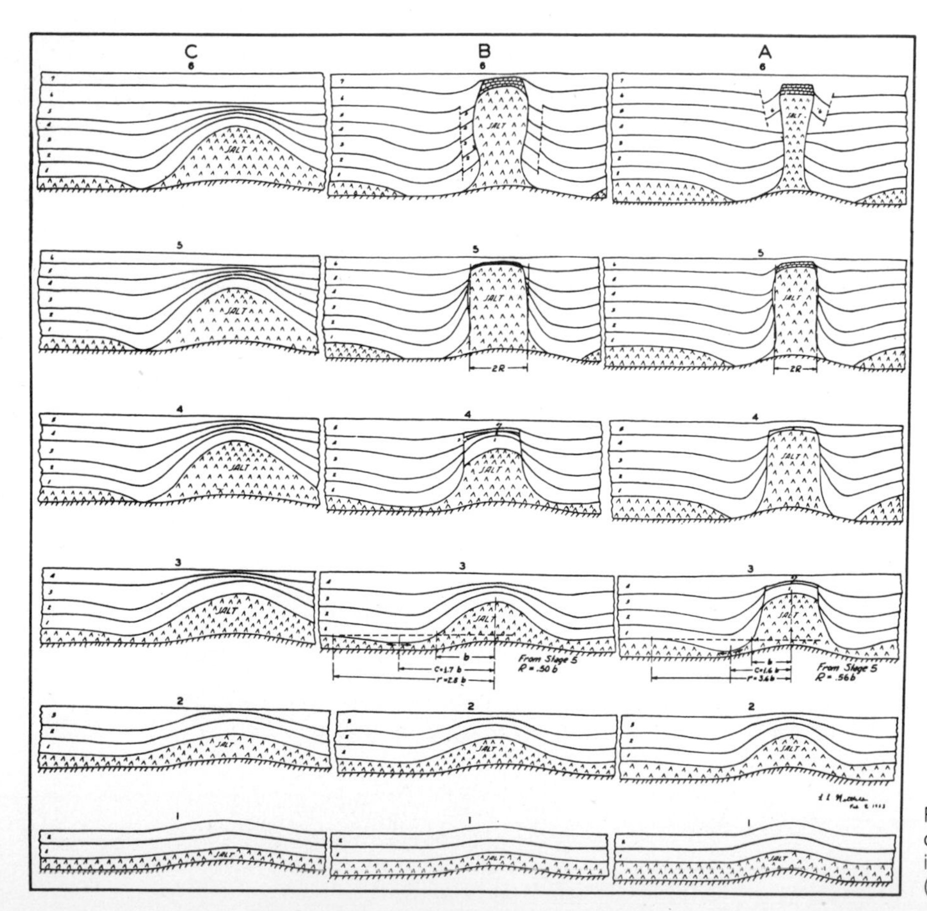

FIGURE 15-13 Formation of salt-dome structural form in scale-model experiments. (*From Nettleton, 1934.*)

has been forced obliquely upward, through a half-open "trapdoor" so that in cross section the salt body has assumed the appearance of a duck's head. The salt paste, during the upward thrust, probably intruded along the inclined shearing planes, and consequently initiated the asymmetric development of the diapir. In this way one of the German salt plugs even becomes detached from its source of supply, and rises into higher levels as a rootless, drop-shape salt body. In general, it may be stated that even neighboring salt stocks can be of very different construction. No two salt stocks are alike.

ROLE OF SALT IN STRUCTURES OF TECTONIC ORIGIN*

Salt deposits occur in many sedimentary basins throughout the world. Where the thickness of the salt and overlying sediments is great, and where tectonic activity followed salt deposition, the salt has exhibited a high degree of mobility and has consequently been an important factor in the determination of the structures formed. In the Gulf Coast, Germany, and some other basins, behavior of the underlying rocks does not appear to have played an important part in the movements which caused the salt to penetrate overlying rock layers and to form domes and other structures, although it has been suggested that movements in the rocks beneath the salt may have been important in initiating the salt movement.

In other areas salt, gypsum, and anhydrite have played an important role in the development of large-scale structural features when external stresses have been present. Massive gypsum beds are involved in the folds and faults of the Jura Mountains of France and in

* For other studies see Carter and Elston (1963)—salt anticlines in Colorado; Small (1959)—salt folds in the Carpathians.

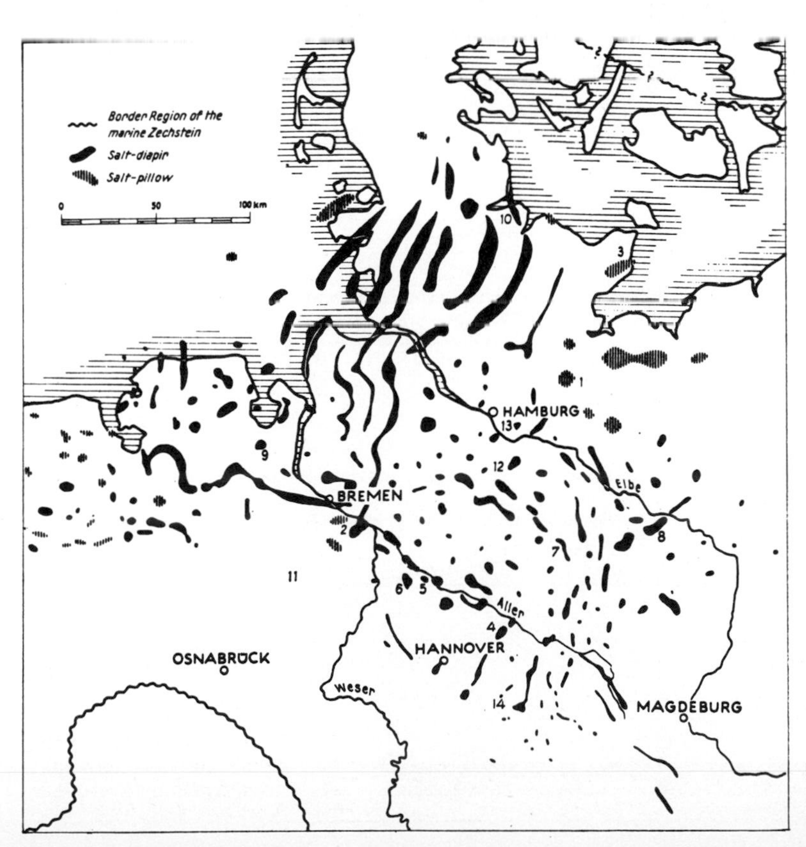

FIGURE 15-14 Map of salt structures in northwest Germany. (*From Trusheim, 1960.*)

the Alps (Chap. 23). The salt beds in the Jura Mountains provided a gliding surface over which the sedimentary veneer is folded and faulted and below which the basement is relatively undeformed. Salt and shale beds function in a similar manner in the fold and fault belt of the Appalachian Mountains.

In southern France a salt basin formed during the Triassic over the area immediately north of the Pyrenees Mountains. Approximately 2 km of Triassic sediment accumulated and was followed by an additional 3 to 7 km of sediment. This sequence was deformed during several phases of tectonic activity in the

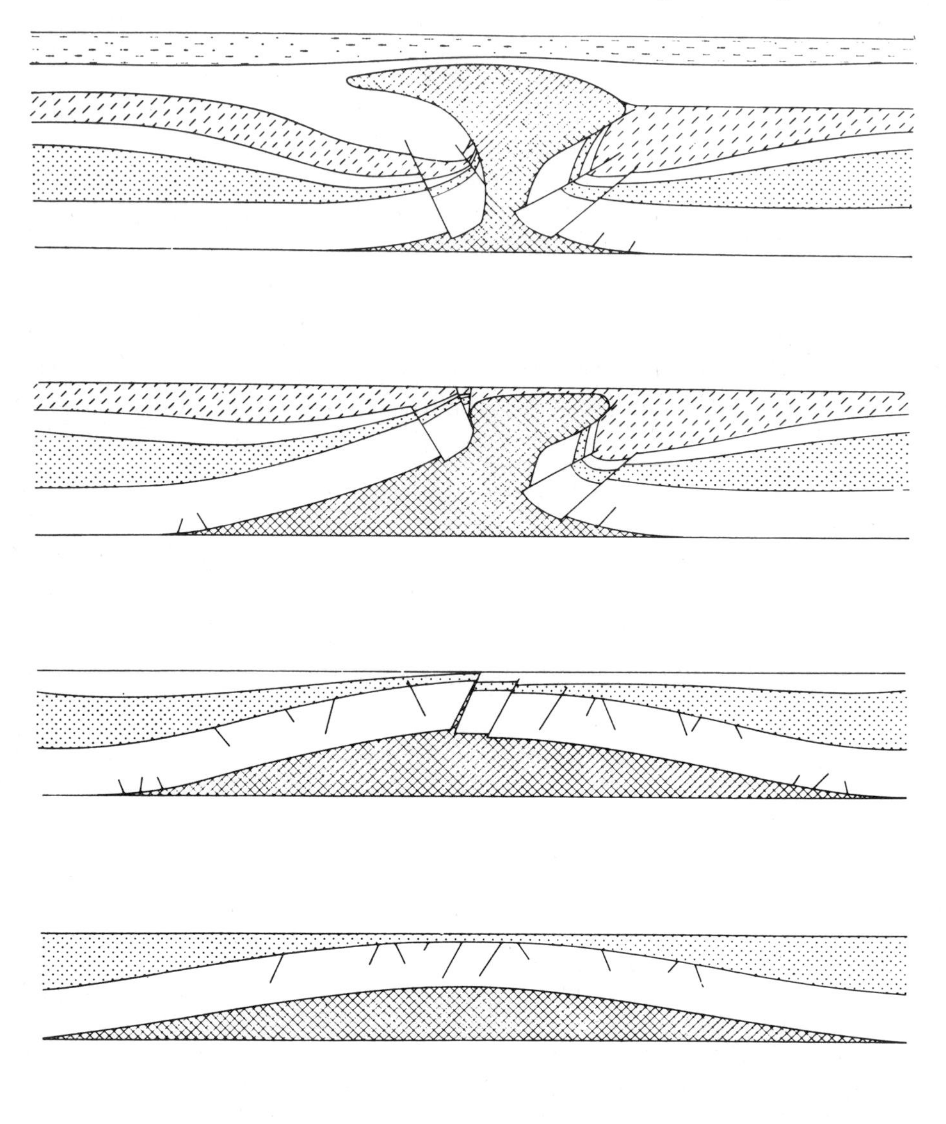

FIGURE 15-15 Diagrammatic development of asymmetrical Zechstein salt stock. (*From Trusheim, 1960.*)

Pyrenees. Salt is found as isolated domes, piercement folds with salt plug cores, anticlines over deep diapiric plugs, salt dikes along faults, and erratic salt outcrops probably connected with unexposed faults. Many of these structures are related to deep fractures and faults in the underlying basement rocks (Dupouy-Camet, 1953), and both tectonic forces and density differences have been important in their formation.

Collapse Structures in Salt Basins

A very different type of feature from the salt domes and piercements and the high mobility of salt in tectonically deformed areas is found in the great Silurian salt basin of central North America, where removal of salt has promoted subsidence and collapse of overlying rock units. A large area of probable collapse occurs above the Salina salt under Lake Michigan and Lake Huron (Landes, 1945). Surface evidence is seen in the vicinity of the straits of MacKinac, where great masses of brecciated rocks, some up to 400 m in vertical dimension, are found in the otherwise slightly deformed Silurian and Devonian section. Some of the blocks have moved down hundreds of feet in the collapse features, but others are only slightly tilted.

REFERENCES

Ala, M. A., 1974, Salt diapirism in southern Iran: Am. Assoc. Petroleum Geologists Bull., v. 58, no. 9, p. 1758–1770.

Andrews, Donald I., 1960, The Louann salt and its relationship to Gulf Coast salt domes: Gulf Coast Assoc. Geol. Soc. Trans., v. 10, p. 215–240.

Antoine, J. W., and **Bryant, W. R.**, 1969, Distribution of salt and salt structures in Gulf of Mexico: Amer. Assoc. Petroleum Geologists Bull., v. 53, no. 12, p. 2543–2550.

Applin, P. L., 1951, Preliminary report on buried pre-Mesozoic rocks in Florida and adjacent states: U.S. Geol. Survey Circ. 91, p. 1–27.

Atwater G. I., and **Forman, M. J.**, 1959, Nature of growth of southern Louisiana salt domes and its effect on petroleum accumulation: Am. Assoc. Petroleum Geologists Bull., v. 43, p. 2592–2623.

Balk, Robert, 1949, Structure of Grand Saline salt dome, Van Zandt County, Texas: Am. Assoc. Petroleum Geologists Bull., v. 33, p. 1791–1829.

Barton, D. C., 1933, Mechanics of formation of salt domes, with special reference to Gulf Coast salt domes of Texas and Louisiana: Am. Assoc. Petroleum Geologists Bull., v. 17, p. 1025–1083.

Bates, F. W., Copeland, R. R. Jr., and **Dixon, K. P.**, 1959, Geology of Avery Island salt dome, Iberia Parish, Louisiana: Am. Assoc. Petroleum Geologists Bull., v. 43, p. 944–958.

Berry, G. F., Jr., and **Harper, P. A.**, 1948, Augusta Field, Butler County, Kansas, *in* Structure of typical American oil fields, v. 3: Am. Assoc. Petroleum Geologists.

Bornhauser, Max, 1958, Gulf Coast tectonics: Am. Assoc. Petroleum Geologists Bull., v. 42, p. 339–370.

Braunstein, Jules, and **O'Brien, G. D.**, 1968, Diapirism and diapirs, a symposium: Am. Assoc. Petroleum Geologists, Tulsa, Oklahoma.

Bryant, W. R., and others, 1969, Escarpments, reef trends, and diapiric structures, eastern Gulf of Mexico: Am. Assoc. Petroleum Geologists Bull., v. 53, no. 12, p. 2506–2542.

Carter, F. W., and **Elston, D. P.**, 1963, Structural development of salt anticlines of Colorado and Utah, *in* Backbone of the Americas: Am. Assoc. Petroleum Geologists Mem. 2.

Coats, R. P., 1964, The geology and mineralization of the Blinman dome diapir: South Australia Geol. Survey, Rept. Inv. no. 26.

Currie, J. B., 1956, Role of concurrent deposition and deformation of sediments in development of salt-dome graben structures (Mississippi): Am Assoc. Petroleum Geologists Bull., v. 40, p. 1–16.

De Golyer, E., 1925, Origin of North American salt domes, problems of petroleum geology: Am. Assoc. Petroleum Geologists Bull., v. 9.

Dixon, J. M., 1975, Finite strain and progressive deformation in models of diapiric structures: Tectonophysics, v. 28, p. 89–124.

Dobrin, M. B., 1941, Some quantitative experiments on a fluid salt dome model and their geologic implications: Am. Geophys. Union Trans., Ann. Mtg., 22d, p. 528–542.

Dupouy-Camet, J., 1953, Triassic diapiric salt structures, southwestern Aquitaine basin, France: Am. Assoc. Petroleum Geologists Bull., v. 37, p. 2348.

Escher, B. G., and **Kuenen, P. H.**, 1929, Experiments in connection with salt domes: Leidse Geol. Med., v. III, pt. 3, p. 151–182.

Hanna, M. A., 1959, Salt domes: Favorite home for oil: Oil and Gas Jour., v. 57, p. 138–142.

Hoy, R. B., Foose, R. M., and **O'Neill, B. J., Jr.**, 1962, Structure of Winnfield salt dome, Winn Parish, Louisiana: Am. Assoc. Petroleum Geologists Bull., v. 46, p. 1444–1460.

Hughes, Dudley J., 1960, Faulting associated with deepseated salt domes in the northeast portion of the Mississippi salt basin: Gulf Coast Assoc. of Geol. Soc. Trans., v. 10, p. 155–173.

Kupfer, D. H., 1962, Structure of Morton Salt Company mine, Weeks Island salt dome, Louisiana: Am. Assoc. Petroleum Geologists Bull., v. 46, no. 8, p. 1460–1467.

Landes, K. K., 1945, Mackinac breccia, subsurface stratigraphy and regional structure, *in* Geology of the Mackinac Straits region: Michigan Geol. Survey Div. Pub. 44, Geol. Ser. 37.

Lehner, Peter, 1969, Salt tectonics and Pleistocene stratigraphy on continental slope of northern Gulf of Mexico: Am. Assoc. Petroleum Geologists Bull., v. 53, no. 12, p. 2431–2479.

Leyden, R., Asmus, H., Zembruscki, S., and **Bryan, G.**, 1976, South Atlantic diapiric structures: Am. Assoc. Petroleum Geologists Bull., v. 60, no. 2, p. 196–212.

Martin, R. G., and **Case, J. E.**, 1975, Geophysical studies in the Gulf of Mexico, *in* Nairn, A. E. M., and Stelhi, F. G., eds., The ocean basins and margins, v. 3: New York, Plenum.

Martinez, J. D., 1969, The impact of salt on man's environment: Transactions, Gulf Coast Assoc. of Geol. Soc., v. XIX, 1969, p. 49–62.

Mattox, R. B., ed., and others, 1968, Saline deposits: Geol. Soc. America Spec. Paper 88.

Muehlberger, W. R., 1959, Internal structure of the Grand Saline salt dome, Van Zandt County Texas: Texas Univ. Bur. Econ. Geology, Rept. Inc., no. 38.

Murray, G. E., 1966, Salt structures of Gulf of Mexico Basin: Am. Assoc. Petroleum Geologists Bull., v. 50, p. 439–478.

——— 1967, Salt structures of Gulf of Mexico basin —A review: Am. Assoc. Petroleum Geologists Bull., v. 50, p. 440.

Nettleton, L. L., 1934, Fluid mechanics of salt domes: Am. Assoc. Petroleum Geologists Bull., v. 18, p. 1175–1204.

——— 1943, Recent experimental and geophysical evidence of mechanics of salt-dome formation: Am. Assoc. Petroleum Geologists Bull., v. 27, p. 51–63.

——— 1955, History of concepts of Gulf Coast salt dome formation: Am. Assoc. Petroleum Geologists Bull., v. 39, p. 2373–2383.

Omara, S., 1964, Diapiric structures in Egypt and Syria: Am. Assoc. Petroleum Geologists Bull., v. 48, p. 1116.

Parker, Travis J., 1951, Scale models as guide to interpretation of salt-dome faulting: Am. Assoc. Petroleum Geologists Bull., v. 35, no. 9, p. 2076–2086.

——— and **McDowell, A. N.**, 1955, Model studies of salt dome tectonics: Am. Assoc. Petroleum Geologists Bull., v. 39, no. 12.

Prucha, J. J., 1968, Salt deformation and décollement in the Firtree Point anticline of central New York: Tectonophysics, v. 6, no. 4, p. 273–299.

Read, J. L., 1959, Geologic case history of Slocum dome, Anderson County, Texas: Am. Assoc. Petroleum Geologists Bull., v. 43, p. 958–973.

Richter-Bernberg, G., and **Schott, Wolfgang**, 1959, The structural development of northwest German salt domes and their importance for oil accumulation: World Petroleum Cong., 5th, sec. I, Paper 4.

Small, W. M., 1959, Thrust faults and ruptured folds in Roumanian oil fields: Am. Assoc. Petroleum Geologists Bull., v. 43, p. 455–472.

Smith, Derrell A., 1961, Geology of South Pass Block 27 oil field, offshore, Plaquemines Parish, Louisiana: Am. Assoc. Petroleum Geologists Bull., v. 45.

Trusheim, F., 1960, Mechanism of salt migration in Northern Germany: Am. Assoc. Petroleum Geologists Bull., v. 44, p. 1519–1541.

Woods, R. D., 1956, The northern structural rim of the Gulf basin: Gulf Coast Assoc. Geol. Soc. Trans., v. 6, p. 3–9.

TWO

TECTONICS

16

Major Structural Elements of the Lithosphere

The character of the earth's interior and the processes functioning there are of fundamental importance in the origin and development of the structural features we can examine directly at and near the surface. All modern tectonic theories involve at least the upper parts of the mantle, and it is in the mantle that we must seek the processes basically responsible for continental drift, sea-floor spreading, and the origins of stresses responsible for the formation of mountains and the small-scale features associated with them.

Knowledge of the earth's interior is derived primarily from indirect geophysical observations and from experimental petrology. Direct observations can be obtained from well samples, from study of volcanic and igneous products erupted from depth, and from examination of rocks once deeply buried but subsequently uplifted and laid bare by erosion. A few oil wells have been drilled to depths of 8 km, but these are located on continents nearly 40 km thick, and all these wells are terminated in crustal rocks, generally in sedimentary rocks. A number of the much shallower wells (one to several hundred meters) drilled in the oceans have been drilled through a thin sedimentary section and into a basalt (often glassy) layer below. These wells do afford us proof that the sedimentary veneer in the oceans is remarkably thin in many places.

Direct evidence for the composition of the upper mantle is obtained from xenoliths in the diamond pipes of South Africa and from blocks found in lavas from deep-seated magma chambers. These materials are all ultramafic rocks (e.g., dunite, peridotite, and eclogite). Similar ultramafic rocks, often altered to serpentine, are found in the core or central parts of high folded mountains and in the island arcs, where it is thought they may represent portions of the mantle that have been squeezed or faulted into the deformed belt.

Once-deep portions of the crust are now exposed in the uplifted and deeply denuded mountain belts and Precambrian shield areas.

Here we can examine rocks once buried to depths of 10 to 20 km—depths that are too shallow to expose the mantle. Thus, methods of direct observation provide only limited insight to the character of the earth's interior and then only of the uppermost parts of the mantle.

PROBING THE INTERIOR

An impressive array of geophysical techniques have been applied to the problems of the internal structure of the earth. Seismic studies are among the most important of these.

1. *Earthquake seismology:* Analysis is made of travel-time data obtained for seismic waves generated by earthquakes and, more recently, atomic explosions. Models of the interior are constructed which locate seismic discontinuities (i.e., changes in elastic properties of the materials) at depths so that the times needed for the seismic waves to travel from a source to the discontinuity, to be reflected or refracted there and return to the surface, are compatible with observations made at seismic observatories. Our large-scale models of the earth (e.g., Fig. 16-1 and Table 16-1) are derived largely from earthquake seismology. Refer to Richter (1958).

2. *Surface-wave dispersion:* Special attention has been given to those seismic waves which travel at the surface and within a relatively thin outer shell of the earth. The velocity and attenuation of these waves is related to the character of the shell through which they travel. Thus, they have been useful in analyzing the average characteristics of the earth along their paths. Refer to Press and Ewing (1955).

3. *Reflection and refraction prospecting methods:* Seismic methods developed primarily for petroleum prospecting have been applied with success to study of the crust and upper mantle. Reflection techniques involve measuring the time required for an artificially generated seismic wave (e.g., TNT or atomic explosion) to travel from its source to a reflecting discontinuity and back to recorders. Depth to the reflecting discontinuity can be calculated if seismic velocity in the layers is known.

Seismic refraction techniques involve measuring the time for an artificial signal to travel to the discontinuity, be refracted along it, and return to a set of detectors at the surface. Seismic velocities as well as depth can be calculated by this method.

These methods have been valuable in studies of near-surface crustal structure, but generally, the strength of the artificial signal is insufficient to probe below the crust. Refer to Dobrin (1976), Nettleton (1940), or Howell (1959).

TABLE 16-1
BULLEN'S (1963) MODEL OF THE EARTH'S INTERIOR SHOWING REGIONS, DEPTH AND FEATURES OF THE WAVE VELOCITY IN EACH REGION

Region	Depth (km)	Features of region
	—	
A		Crustal layers
	33	
B		Steady positive P- and S-velocity gradients
	413	
C		Transition region
	984	
D		Steady positive P- and S-velocity gradients
	2,898	
E		Steady positive P-velocity gradient
	4,982	
F		Negative P-velocity gradient
	5,121	
G		Small positive P-velocity gradient
	6,371	

FIGURE 16-1 Division of the interior of the earth on the basis of seismic evidence. See Table 16-1 for characteristics of the regions labeled A–G. (*After Bullen, 1963.*)

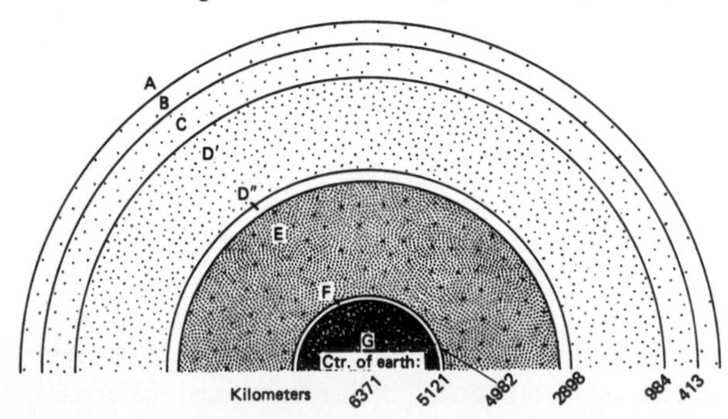

SEISMIC DISCONTINUITIES

Seismologists have long recognized a three-part division of the earth's interior: the crust, the mantle, and the core. These were discovered through study of travel times of earthquake shock waves. The crust-mantle discontinuity is marked by the sudden velocity increase of *P* waves. The mantle-core break is marked by the disappearance of shear waves, suggesting a liquid core. Subsequently the core was divided into an inner and outer core, the inner core being characterized by somewhat higher *P*-wave velocities (Lehmann, 1936). Velocities of compressional, *P*, and shear, *S*, waves at various depths in the interior are illustrated in Fig. 16-2. At least two additional breaks occur within the mantle. The first is a zone between 50 and 200 km in which velocities decrease, called the *Gutenberg low-velocity zone*, and the second is a break at about 900 km in which the rate of increase of velocities of both *P* and *S* decreases with depth. Anderson, Sammis, and Jordan (1971) have summarized much of the recent knowledge of the interior.

The Lithosphere

The *lithosphere* is the relatively brittle "lithified" outer shell of the earth which extends from the surface to the Gutenberg low-velocity zone. It includes the crust and upper portions of the mantle where most earthquakes originate. The lower boundary of the lithosphere is formed by a low-velocity zone in the mantle at a depth of 50 to 200 km. This zone exhibits less strength than the lithosphere, is more ductile or plastic, and is often referred to as the *asthenosphere*, Fig. 16-3. This zone, marked by low-viscosity materials, coincides in depth with a pronounced change in temperature gradient, Fig. 16-4. This change in physical properties is certainly of great importance in the development of near-surface structure and probably of greater significance than the M discontinuity.

The Crust

The *crust* of the earth is defined by a seismic discontinuity discovered by Mohorovicic and named the Moho or M discontinuity in his honor. The M discontinuity is recognized throughout most of the earth by a sharp increase in the velocity of compressional waves from about 6.0 to 8.0 km/sec. The crust so defined has various marked characteristics:

1. The continents and ocean basins have different thicknesses of crust, Fig. 16-5. The average continental thickness is 35 km, but considerable variation

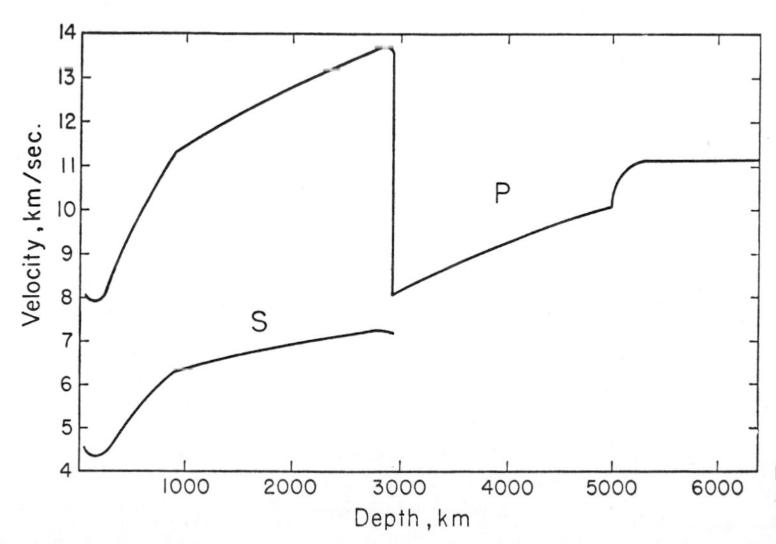

FIGURE 16-2 Velocity of *P* and *S* waves as a function of depth as determined by Gutenberg.

is found, and oceanic crust thickness ranges between 5 and 10 km.

2. The main areas of departure from the average continental and oceanic thickness lie in long narrow belts in the regions of island arcs, in folded mountain belts where crustal thicknesses of 50 to 70 km have been observed, and in mid-oceanic ridges, where no sharp M discontinuity is observed.

Most of the crust has a sedimentary veneer with a bulk chemical composition similar to granite. Shield areas and other exposures of deep continental rocks are predominantly granites and granitic gneisses. Basic materials, primarily andesite and basalt, make up a small fraction of the exposed continental rocks. Basalt and other basic rocks are exposed as dikes and sills, and in a few places, e.g., the Columbia River Plateau and in the Deccan Plateau of India, large volumes of basalt occur on continents. Thus the continents are primarily granitic, but a source of basalt must be present at depth under continents. The oceans, on the other hand, are underlain by extensive basaltic materials, judging from the

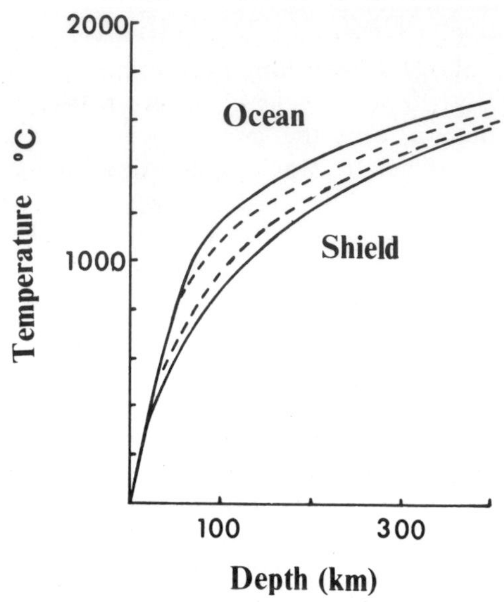

FIGURE 16-4 Model of the variation of temperature with depth under Precambrian shields, continents, and oceans. (*After Clark and Ringwood, 1964.*)

FIGURE 16-3 Division of upper mantle into tectosphere, asthenosphere, and mesosphere. This plot shows variation of the log of the viscosity of the material against depth. (*After Elsasser, 1969.*)

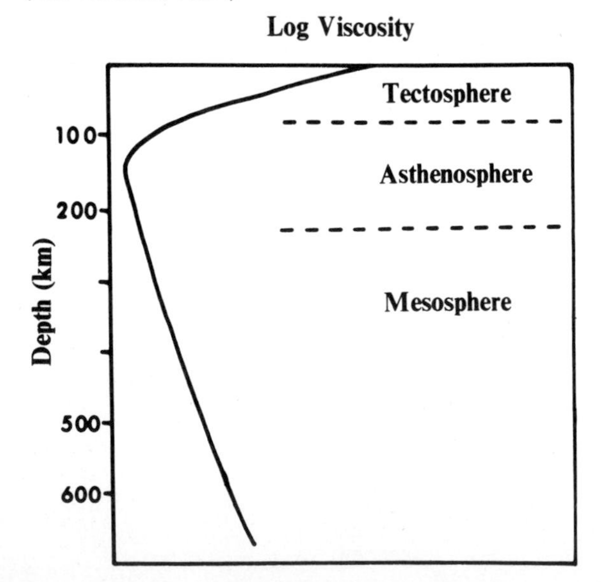

FIGURE 16-5 Seismic indications of crustal layering in continental shields and ocean basins. Compressional and shear velocities (in kilometers per second) are shown. Shear velocities are in parentheses. (*From Press, 1961.*)

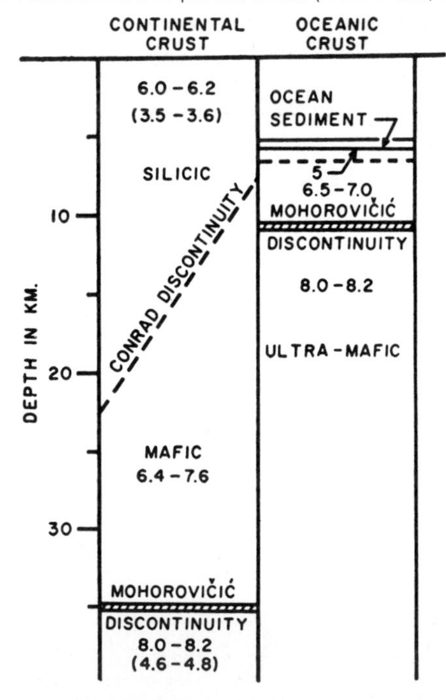

predominence of basaltic volcanic products on the mid-Atlantic ridge, the basalt dredged from seamounts, and lava coming from active volcanoes in all oceans.

Seismic velocities of the subcrustal materials (P = 8.1 km/sec) are similar to those indicated by experimental work for ultramafic rocks such as dunite and peridotite at the temperatures and pressures which prevail at the level of subcrustal rocks. Thus the composition of the mantle is interpreted to be one of the ultrabasic rock types. Velocities in the higher portions of the continental crust (P = 6.0 to 6.2 km/sec) are velocities found experimentally to be characteristic of granite, while the lower parts of the continental crust and most of the standard oceanic crust have velocities (P = 6.5 to 7.6 km/sec) of basalt. This twofold division of the continental crust is not confirmed everywhere.

Defining the Crust by Gravity Anomalies

Because the materials of the continental crust have lower densities than those of the mantle, Bouguer gravity anomalies are related to the thickness of the crust. Woollard (1959) compiled records which illustrate the relations between gravity anomalies, topography, and crustal thicknesses. Figure 16-6 shows a plot of crustal thicknesses measured by means of seismic refraction data, data on thickness derived from phase velocities of surface waves, and Bouguer anomalies. The anomalies over deeper parts of the continental crust are greater than they would be if a straight-line fit were made. It appears that the density of the crustal rocks increases in deeper parts of the continental crust. This is thought to be due to thickening of an intermediate, basaltic layer rather than to change in the upper sialic layer. This is confirmed by seismic studies. Figure 16-7 shows the relation between topography and Bouguer anomalies. In both cases a positive correlation is seen. Gravity anomalies and seismically determined crustal thicknesses may be used to construct contour maps of the crust-mantle contact, Fig. 16-8.

The Upper Mantle

All rocks which have apparently come from the upper mantle are ultrabasic. They occur as inclusions in basalts, as kimberlites, and as associated ultrabasic xenoliths. The main

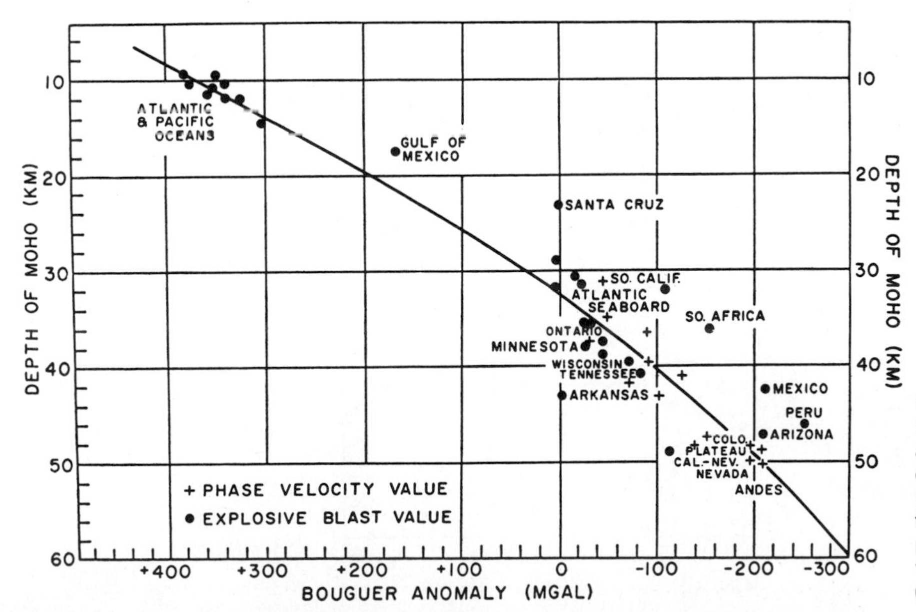

FIGURE 16-6 Relation between the Bouguer gravity anomaly and the thickness of the crust. Circles = points plotted from seismic refraction data; crosses = points plotted from data on phase velocity of surface waves. (*After Woollard, 1959.*)

minerals which occur in the ultrabasic inclusions are olivine, enstatite, and augite. They occur in a rock with density in the range of 3.30 to 3.33 g/cm^3. The proportions of these minerals are highly variable, ranging from dunite, 100 percent olivine, to rock without olivine. The major xenoliths from kimberlite (a peridotite) are eclogites which contain garnet and mica as well as olivine. Thus subcrustal rocks show mineralogical variation, indicating that the upper mantle is not a uniform homogeneous layer. Unfortunately it is impossible to obtain an accurate idea of the distribution of such variations as may exist without drilling.

Lateral variation in the upper mantle is also suggested by seismic velocities (P_n and S_n) of phases propagated in the upper mantle. P_n varies from 7.9 to 8.3 km/sec under the continental United States, Fig. 16-9. In general, both P_n and S_n are lower beneath mountain belts, on the concave side of island arcs, and on mid-ocean ridges and are higher beneath the continental shield areas and deep-ocean basins, Table 16-2. High velocities are thought to be related to the rigidity or strength of the mantle. This suggests that the upper mantle has a higher rigidity under the more tectonically stable crustal elements and that the regions under which these phases are attenuated are weaker (Molnar and Oliver, 1969).

Surface-Wave Interpretation

Unlike body waves, which arrive suddenly and pass through a station quickly, surface waves travel as a wave train. Once wave motion starts it continues for a long period of time. The seismograph traces are set in long-period motion. Periods of surface waves fall in a range of 15 to over 3,000 sec; waves used in most studies, however, have periods in the range of 15 to 75 sec. Generally the earlier arrivals are of longer period than the later arrivals. This suggests that many different waves originate at about the same time but travel with different velocities—velocities related to the period of the wave motion. This

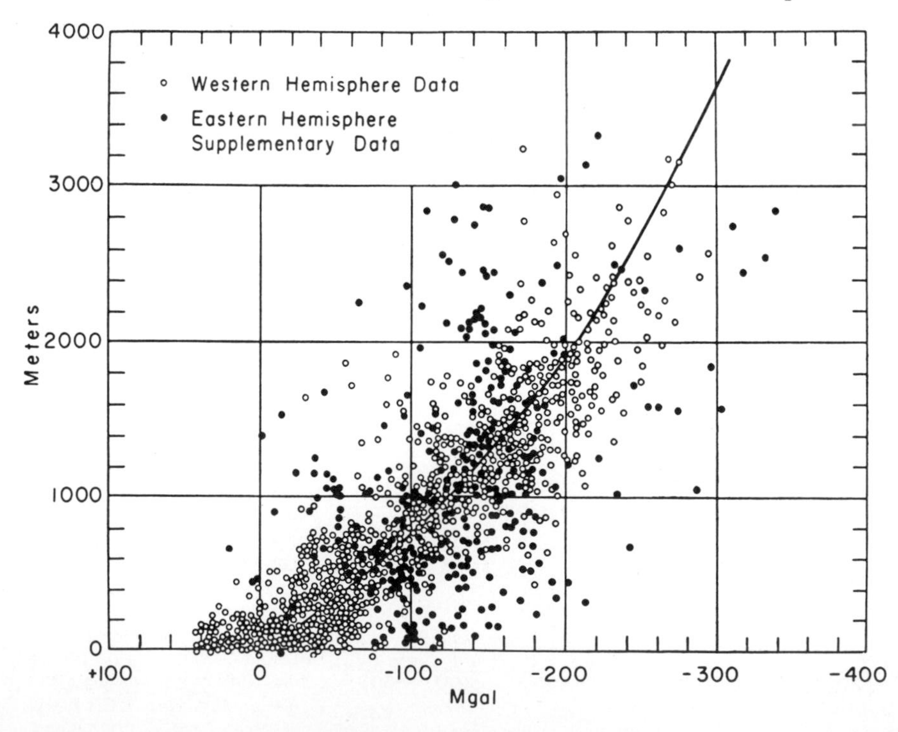

FIGURE 16-7 Relation between the Bouguer gravity anomaly and topographic elevation. (*After Woollard, 1959.*)

continual spreading out of an initial disturbance into a train of waves is called *dispersion*. The original motion is sorted out into *groups* of simple harmonic waves, and each of these groups has a particular wavelength, period, and travels forward with a particular *group velocity*. It is possible to identify each group by measuring the period of each surface wave, and, if the time and place at which the earthquake originated are known, the velocity of each group can be calculated. Group velocity versus group period is plotted as shown in Fig. 16-10. (It turns out to be more convenient to plot this on logarithmic graph paper because the range of periods is so great.) When plots of this type are made at one station for several different epicentral locations, it becomes clear that the *dispersion curves* are not identical—surface-wave dispersion is related to the particular path followed by the waves, an observation suggesting this is a tool that may be used to interpret lithospheric structure.

The key to the use of surface waves for interpretation lies in the fact that the depth to which a wave penetrates the medium through which it moves is a function of its wavelength. Long-period waves penetrate more deeply. Those waves that penetrate deeply into the upper mantle are influenced by its higher velocities. Thus the longer period waves arrive first—they have traveled with greater velocity. It follows that the dispersion curves should reflect differences in the distribution of physical properties along the path. The prominent differences between the structure of the crust (depth to M discontinuity) under oceans and continents is clearly defined by dispersion curves; the curves join to form a single curve for groups with periods of about 100 sec. The longer period groups are influenced by the mantle, in which difference in properties along different paths is not as great as the dif-

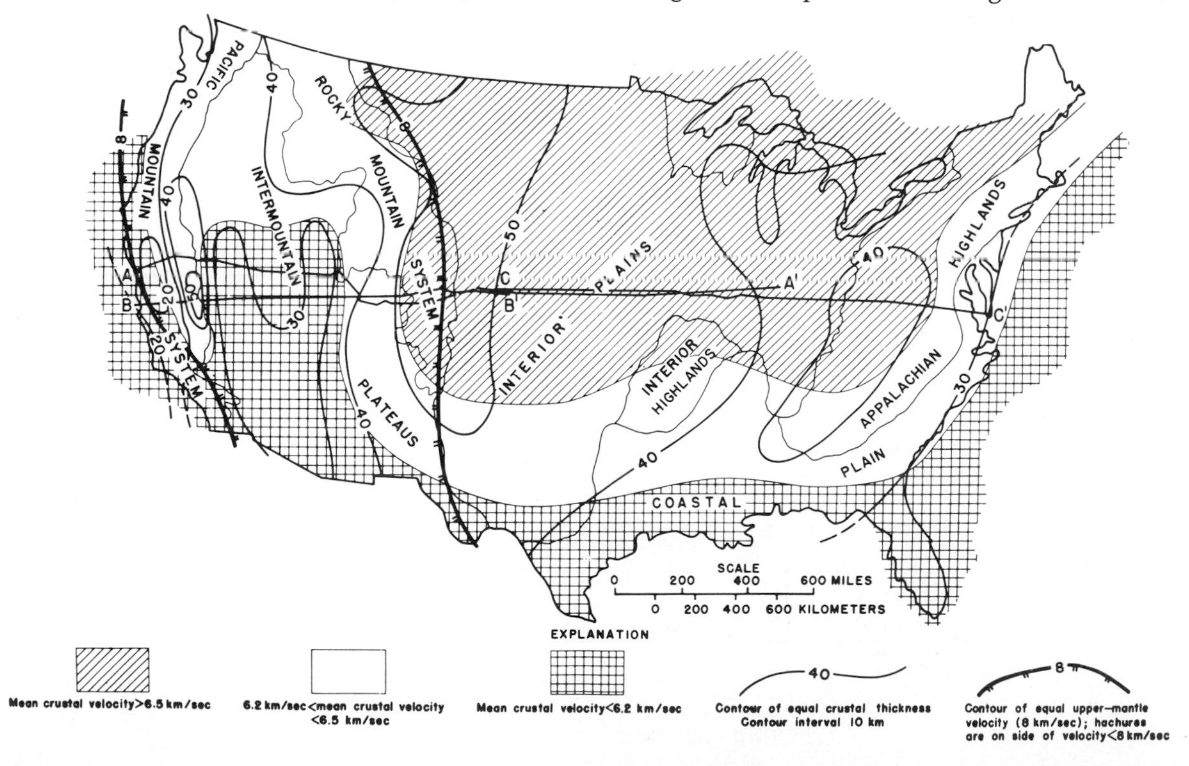

FIGURE 16-8 Crustal thicknesses in the coterminous United States. Crustal depths are taken from seismic refraction data. (*After Pakiser and Zietz, 1965.*)

ferences between oceanic and continental plates.

Additional information can be obtained from the analysis of *phase-velocity* curves for Rayleigh waves (Press, 1956). Phase velocity is the velocity of a particular spectral component of a wave group. This technique may be used where stations are so closely spaced that it is possible to recognize and correlate individual waves from station to station. When this is possible, the velocity of a given wave of specific period can be calculated between stations. A phase-velocity curve may be prepared by plotting the phase velocity of a number of waves of different period so that a plot of velocity versus period is obtained. Interpretation of this in terms of crustal structure is then possible by calculating what the phase-velocity curve would be for specific models of crustal structure. The variables in such models are layer thickness and elastic moduli. A correct model should yield a phase-velocity curve that coincides with the observed curve.

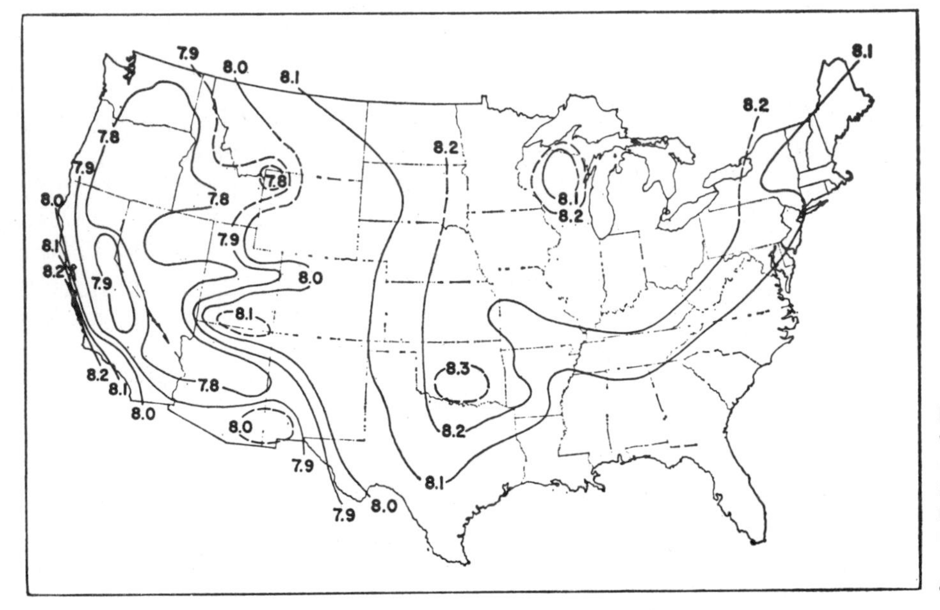

FIGURE 16-9 Estimated velocity of P_n in the upper mantle under the United States. Variations may reflect differences in materials of the upper mantle. (*From Pakiser and Zietz, 1965.*)

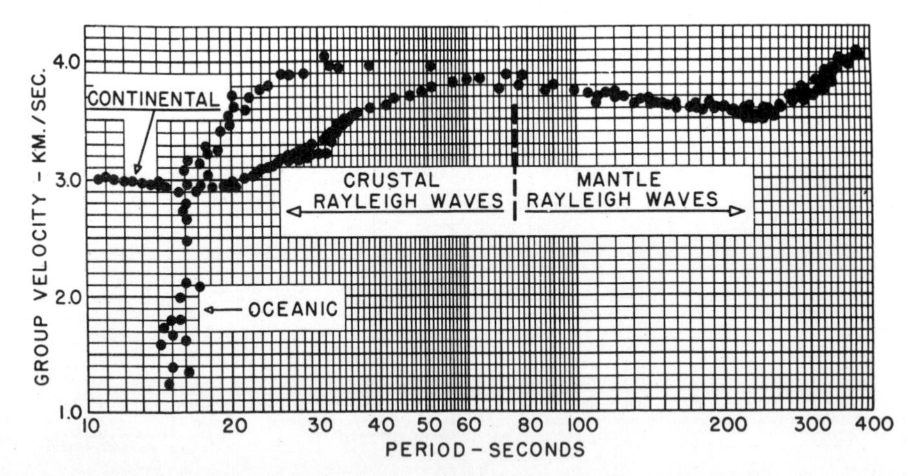

FIGURE 16-10 Observed dispersion of Rayleigh waves in the period range 10 to 400 sec for continental and oceanic paths. (*After Ewing and Press, 1956.*)

The Gutenberg Low-Velocity Zone

The discovery of a low-velocity zone near the top of the mantle must be ranked as one of the most important recent discoveries in earth science, Fig. 16-11. Gutenberg (1948) and Gutenberg and Richter (1954) postulated the existence of such a low-velocity zone at a depth of 60 to 150 km from a shadow-zone effect close to earthquake epicenters.

Low-velocity layers are difficult to detect by seismic methods because the lower velocity causes waves refracted into the layer from above to be bent downward. If velocity increases steadily with depth, a shock wave travels out from the focus as a continuous wave front, but if a low-velocity zone occurs at depth, the waves become channeled into it and are not detected close to the focus. Effects vary, depending on the relative depth of the focus and the low-velocity zone. If the focus is in the low-velocity zone, then a larger part of the energy remains in the channel formed by the low-velocity zone. If the focus is above the zone, waves passing downward into the zone

FIGURE 16-11 Generalized structure for the Pacific basin. (*From Ewing and others, 1962.*)

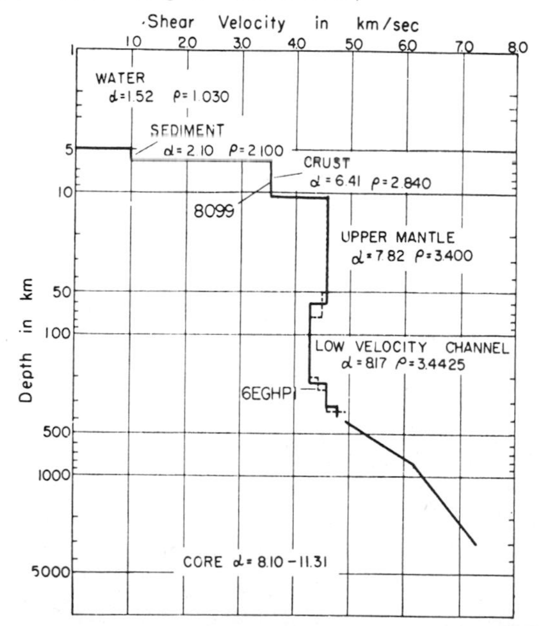

must first be refracted through the zone and into high-velocity materials before they can pass back through the zone to the surface. In the time taken for these movements, the waves are traveling laterally as well; thus, a shadow zone is created. The existence of the low-velocity layer has been confirmed from other sources:

1. Underground nuclear explosions provide the carefully controlled conditions needed to test this idea. These confirm Gutenberg's findings but indicate that the zone lies at depths between 60 and 250 km.

2. Analysis of the dispersion of surface waves indicates that the material in the low-velocity zone is less rigid than that above and below. Surface-wave studies show that the layer is worldwide in its extent but that it differs under continents and ocean basins.

3. Plots of focal depths of large numbers of earthquakes show a gap within or below the depth range of the low-velocity zone.

4. A layer of lower rigidity is necessary to explain free oscillations (vibrations) of the earth as a whole, such as those that are caused by the most intense earthquakes.

The significance of the low-velocity layer depends on the physical conditions responsible for it. The depth of the zone is about that predicted for the generation of primary basaltic magma. It is in this depth range that temperatures come close to those required for partial melting of basalt. If the zone is one of high temperature and increased plasticity or low shear strength, it is certainly of critical importance in regard to continental drift, apparent polar wandering, and crustal deformation.

STRUCTURAL DIVISIONS OF THE CRUST

Only a few decades ago the outer shell of the earth was often regarded as a relatively uniform layer made up of the lighter minerals which segregated as the once-molten earth solidified. The continents and ocean basins were regarded as elevated and depressed segments of this shell and the mountains as wrinkled and thickened portions of this crust. The development and application of geophysical techniques to crustal structure gradually destroyed this overly simplified model, and though a great deal remains to be clarified, a much more accurate and sophisticated model now exists.

A number of criteria have proven valuable in recognizing major types of crustal elements. Among these, topography, surface and subsurface geology, gravity and magnetic anomalies, seismicity, seismically determined crustal structure, and heat-flow studies are especially valuable.

The best-defined division of the crust is between continents and ocean basins, with a third, highly varied division where the transition from continents to ocean basins occurs.

1. Topographically the surfaces of the continents stand high. The level of the top of the oceanic crust is low. This is evident in the hypsographic curve of the earth's surface, Fig. 16-12, which shows two distinct levels nearly 5 km apart on the average, Fig. 16-13.

TABLE 16-2
SUMMARY OF GEOPHYSICAL EVIDENCE COMPARING MANTLE PROPERTIES BENEATH CONTINENTS AND OCEAN BASINS. (*MODIFIED AFTER Alexander and Sammis, 1975*)

Geophysical data	Conclusions: Continent (shield)	Conclusions: Ocean basin
SEISMIC		
Surface waves: (Love and Rayleigh; fundamental mode primarily)	High-compressional (*P*) and shear (*S*) velocity to 400 km or more; *no low-velocity zone* between 100 and 200 km depth All shields give the same response	Low-velocity zone at about 100-km depth Lateral variations at depth
Body waves: *P*- and *S*-wave travel times vs. distance	High average velocity; no mantle low-velocity zone	Lower average velocity; low velocity inferred
P and *S* delays	Early with respect to world average, implying high average mantle velocity	Late with respect to world average, implying low average mantle velocity
Attenuation (Q): Surface waves	High average *Q* throughout	Low average *Q* below 100 km
Body waves	*Q* average greater than 500 for upper 400 km with no minimum above 400; transmits high-frequency signals	*Q* average approx. 300 with minimum of approx. 50–100 near the low velocity zone; attenuates high-frequency signals.
Seismicity	Aseismic (except at plate boundaries)	Aseismic (except at plate boundaries)
HEAT FLOW (at Moho)	Low (mean approx. 0.8 heat-flow units) and very consistent from continent to continent	High (mean approx. 1.2 heat-flow units)
ELECTRICAL CONDUCTIVITY (Magnetotellurics, geomagnetic and resistivity surveys)	Low throughout upper 400 km; no pronounced high-conductivity zone	Low in top 100 km, but increases sharply by several orders of magnitude at greater depths
TILTS AND STRAINS (Response to surface loading or unloading, e.g., postglacial uplift)	High average upper mantle viscosity (approx. 10^{24} poises)	Low viscosity at shallow (approx. 100 km) depths with high-viscosity lid

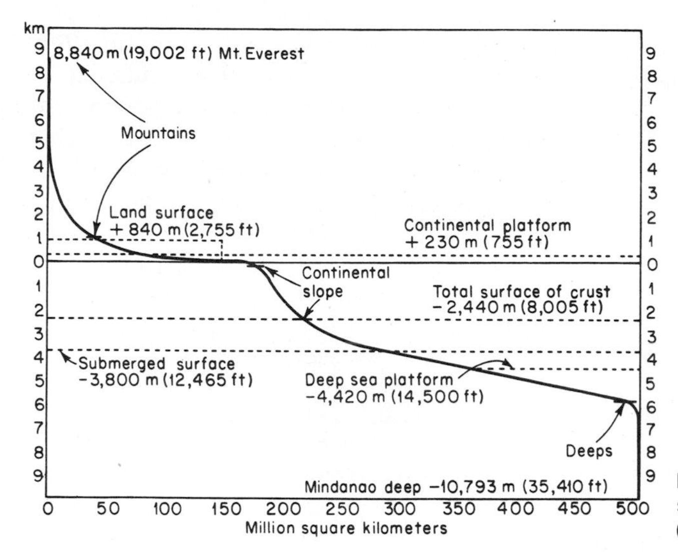

FIGURE 16-12 Hypsographic curve of the earth's surface (mean elevations given by dashed lines). (*Redrawn from Bucher, 1933.*)

2. The bulk chemical composition of the continental crust is granitic. This includes the thick sedimentary sections and gneissic basement rocks as well as granites and their eruptive equivalents, all of which are similar in bulk composition. In contrast, rocks found under the ocean basins are, with the exception of the thin sedimentary veneer, basic (basalt or gabbro) or ultrabasic.

3. Structurally, a most significant difference is that young folded mountains of the types found in the Alpine-Himalayan, Appalachian, Caledonian, and Cordilleran belts are absent in the oceanic crust. Thus the processes and conditions which give rise to folded belts appear not to exist in the oceans. On the other hand the mid-oceanic ridges have no counterparts of comparable dimensions on the continents.

4. Seismic studies indicate prominent differences between continents and ocean basins as illustrated in Fig. 16-5. The differences in level of the M discontinuity are most striking. A typical oceanic crustal section consists of water, sediment, a crust with velocity and density comparable to basalt, and the low-velocity channel in the mantle. The twofold division of the continental crust into an upper sedimentary and granitic layer separated by the Conrad discontinuity from a lower basaltic-gabbroic layer shows up prominently in Fig. 16-14.

A less-well-defined difference between continents and oceans is found in the character of the low-velocity layer of the upper mantle discussed earlier. Either the layer is at a different level or the velocity reduction is different under continents and oceans.

5. Average heat flow from continental crustal plates is almost as high as that from the ocean basins. Departures from this background are localized primarily in the same areas that are seismically active (Langseth, 1969).

The differences in heat production of different rocks are great (e.g., granite 285 ergs/g year^{-1}, basalt 58 ergs/g year^{-1}, dunite 0.08 erg/g year^{-1}).

FIGURE 16-13 Graph showing the two frequency maxima of elevation on the earth's surface. Solid line = actual frequency of elevations based on a contour interval of 100 m. (*From Wegener, 1922.*)

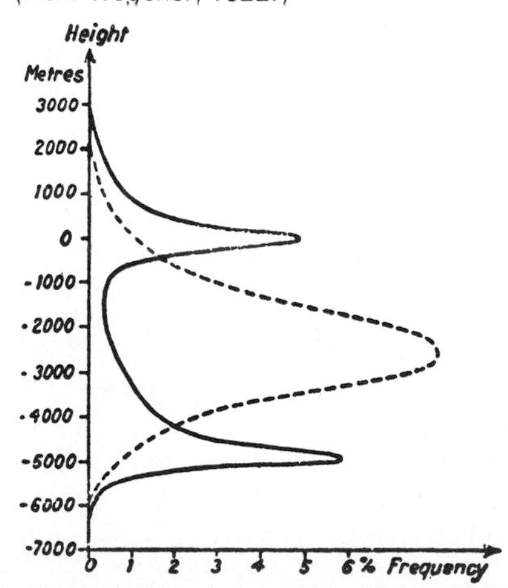

Despite this difference the granitic continental crust fails to be clearly differentiated from the basaltic oceanic crust on the basis of heat-flow measurements. This is interpreted by MacDonald (1964) as indicating that the differences in the continents and ocean basins extend deep into the mantle. But it may simply reflect the relative age of these plates—the oceanic plates being young and still relatively warm.

6. Gravity data also tend to support the idea that differences between oceans and continents extend into the mantle. The gravity field of the earth shows that average values are the same over continents and oceans. This means that the mass per unit area is the same beneath oceans and continents despite the prominent differences in the density of the rocks in the two. Thus differences in the density of deeply buried material must compensate for surface inequalities in density that are (MacDonald, 1964) not compensated for by differences in elevation.

Classification of Crustal Elements

A classification of the major crustal elements based on tectonics and geophysical observations proposed by Brune (1969) is most useful for purposes of structural study, Table 16-3. Crustal elements are first distinguished as continental or oceanic on the basis of the depth to the Moho (crustal thickness). Within this framework elements are identified as stable or unstable and finally characterized by topography, form, and structure.

1. *Continental shields:* Every continent is composed in part of a shield area. Although these areas were the sites of widespread orogenic activity in the Precambrian and much of the rock is recrystallized and fused by igneous intrusions, they have been relatively unaffected by post-Precambrian orogeny. Post-Precambrian deformation effects take the form of block faulting and brittle fracturing, accompanied by shallow earthquakes, and in a number of areas Cenozoic lavas have been erupted along these high angle faults. The shields are generally of low relief and moderate to low elevation.

2. *Mid-continent or shield extensions:* The shield areas may be drawn to include stable platform-like areas which are covered by relatively thin undeformed post-Precambrian sedimentary veneers. The central United States is essentially of this character and may be conveniently thought of as an extension of the Canadian Shield.

Although none of these areas has been involved

FIGURE 16-14 Crustal cross section across North America showing thickness, seismic velocities, gravity, and magnetic fields. (*After Pakiser and Zietz, 1965.*)

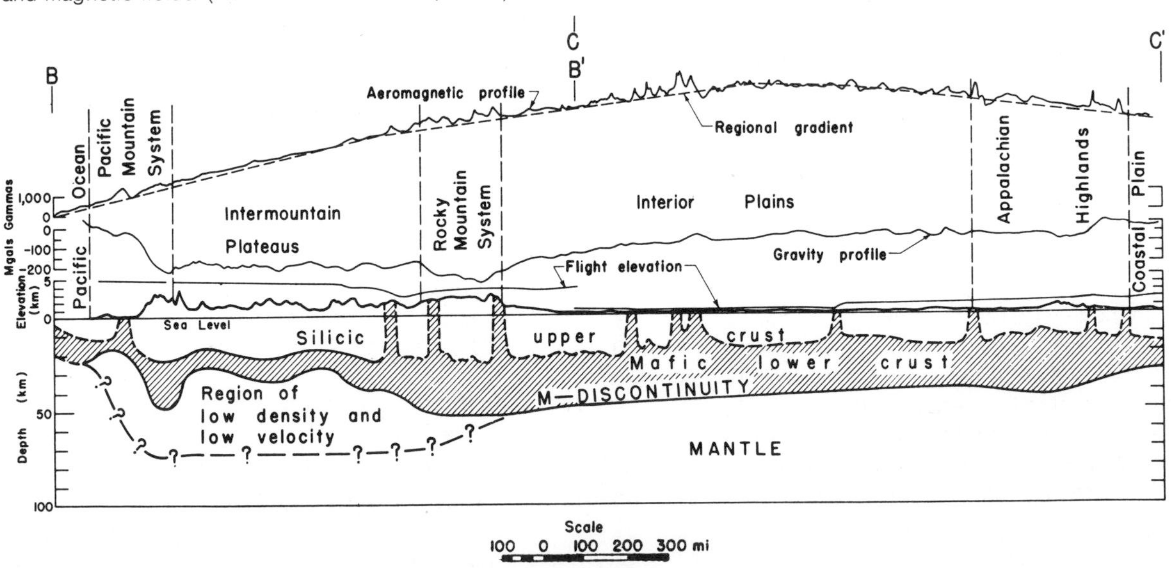

in orogenic deformations, deep subsidence and consequent accumulation of thick sedimentary cover have occurred in a number of large basins (e.g., Anadarko, Michigan, and Illinois basins).

3. *Basin and range:* This type of crust, named after the physiographic province in the western United States (Nevada, eastern California, and parts of Arizona, New Mexico, Idaho, and Oregon), is characterized by Cenozoic block faulting, volcanism and intrusion, high average elevation, thin crust, and high average heat flow.

4. *Alpine mountains—or young folded mountain belts:* These are the narrow, elongate, belts in which the crust has exhibited great mobility, yielding to the formation of high mountains by uplift, folding, and thrusting in the sedimentary veneer and often even in the underlying crystalline complexes. Many of these belts are sites of long-term instability passing initially through a geosynclinal phase and later one or more episodes of crustal shortening and uplift.

5. *Plateau:* An example of this type of crustal element is found in the Colorado Plateau—a large square block of continental crust which lies within the Rocky Mountain region and has been uplifted by nearly a mile (1.6 km) but lacks the effects of folding and thrusting which characterize the folded mountain belts.

6. *Island arcs:* Arcuate rows of islands form an interesting series of chains from the Aleutians to Australia across the western Pacific basin, and other arcs are present in the Caribbean and the Scotia Sea. The arcuate form of the andesitic volcanic islands, the associated deep-sea trenches, active volcanism, and high seismicity combine to make the arcs an especially active and interesting type of crustal element. The crustal structure is intermediate between oceanic and continental types in thickness, and the seismic velocities in the upper mantle are abnormally low in the arcs.

7. *Ocean basin:* The physiography of the ocean basins is slowly being revealed. A high degree of variation is already known, as is the underlying crustal structure. The type crust included in this category is best exemplified by the vast abyssal plains of the ocean, where only a thin veneer of sediment overlies the basaltic substrata. These plains have often been compared with the shields in terms of their stability. They are broken by long, straight faults or fractures along which basaltic lava has

TABLE 16-3
TECTONIC CLASSIFICATION OF THE CRUST SHOWING CHARACTERISTIC CRUSTAL THICKNESSES, P_n VELOCITIES, HEAT FLOW, AND GRAVITY ANOMALIES
(*After Brune, 1969*)

	Crust *T* (km)	P_n (km/sec)	Heat flow (μcal/cm² sec)	Bouguer anomalies (mgals)
CONTINENTAL CRUST				
Stable Mantle				
A. Shields	35	8.3	0.7–0.9	−10 – −30
B. Mid-continent	38	8.2	0.8–1.2	−10 – −40
Unstable Mantle				
C. Basin and range	30	7.8	1.7–2.5	−200 – −250
D. Alpine mountains	55	8.0	0.7–2.0	−200 – −300
E. Plateau				
F. Island arcs	30	7.6	0.7–4.0	−50 – +100
OCEANIC CRUST				
Stable Mantle				
G. Ocean basin	11	8.1	1.3	+250 – +300
Unstable Mantle				
H. Ocean ridge	10±	7.5	1–8	+200 – +250
I. Ocean trench				

risen to form seamounts, but folded mountains are absent.

8. *Ocean ridges:* The mid-Atlantic ridge and the East Pacific rise are the best known examples of this type of element. These ridges are composed almost entirely of great arched piles of lava, dikes, and sills of basaltic composition. High heat flow, shallow seismic activity, high-angle faults, and a central rift valley (a graben) are characteristic features. These ridges can be traced into continental crust in the Gulf of California and in the Gulf of Aden.

9. *Ocean trenches:* The deep-sea trenches are so closely associated with the island arcs that they might best be considered an integral part of them. The trenches are narrow, elongate, deep (up to 10 km), and the sites of large negative gravity anomalies.

REFERENCES

Aki, K., 1972, Earthquake mechanism, *in* Ritsema, A. R., ed., The upper mantle: Tectonophysics, v. 13, p. 423–446.

Alexander, S. S., and **Sammis, C. G.,** 1975, New geophysical evidence on the driving mechanisms for continental drift: Earth Miner. Sci., v. 44, no. 4.

Anderson, D. L., 1962, The plastic layer of the earth's mantle: Sci. Am., July.

———, **Sammis, Charles,** and **Jordan, Tom,** 1971, Composition of the mantle and core, *in* E. Robertson and others, eds., The nature of the solid earth: New York, McGraw-Hill.

Benioff, Hugo, 1954, Orogenesis and deep crustal structure—Additional evidence from seismology: Geol. Soc. America Bull., v. 65, no. 4.

Birch, A. F., 1952, Elasticity and constitution of the earth's interior: Jour. Geophys. Res. v. 57, p. 227–286.

——— 1954, The earth's mantle—Elasticity and constitution, *in* Bucher, W. H., ed., Symposium on the interior of the earth: Am. Geophys. Union Trans., v. 35, p. 79–85.

Brune, J. N., 1969, Surface waves and crustal structure, *in* The earth's crust and upper mantle, Am. Geophys. Union Mon. 13.

Bucher, W. H., 1933, The deformation of the earth's crust: Princeton, N.J., Princeton Univ.

——— ed., 1954, Symposium on the interior of the earth: Am. Geophys. Union Trans., v. 35, p. 48–49.

Bullen, K. E., 1963, An introduction to the theory of seismology, 3rd ed.: Cambridge, Cambridge Univ. Press.

Clark, S. P., Jr., 1963, Variation of density in the earth and the melting curve in the mantle, *in* Donnelly, T. W., ed., The earth sciences: Chicago, Univ. of Chicago.

——— and **Ringwood, A. E.,** 1964, Some properties of the earth and other terrestrial planets (abs.): Am. Geophys. Union Trans., v. 45, no. 1, p. 105.

Dobrin, M. B., 1976, Introduction to Geophysical Prospecting, 3rd ed.: New York, McGraw-Hill.

Elsasser, W. M., 1969, Convection and stress propagation in the upper mantle: *in* Runcorn, S. K., ed., The application of modern physics to the earth and planetary interiors: London, Wiley-Interscience.

Ewing, Maurice, and **Press, Frank,** 1956, The long-period nature of *S* waves (abs.): Am. Geophys. Union Trans, v. 37, no. 3, p. 343.

——— and others, 1962, Surface wave studies of the Pacific crust and mantle, *in* Crust of the Pacific basin: Am. Geophys. Union Mon. 6.

Fox, P. J., Schreiber E., and **Peterson, J. J.,** 1973, The geology of the oceanic crust: Compressional wave velocities in oceanic rocks: J. Geophys. Res., v. 78, p. 5155–5172.

Gutenberg, Beno, 1948, On the layer of relatively low wave velocity at a depth of about 80 kilometers: Seis. Soc. America Bull., v. 38, p. 121–148.

——— 1951, Internal constitution of the earth: Princeton, N.J., Princeton Univ.

——— 1954, Low-velocity layers in the earth's mantle: Geol. Soc. America Bull., v. 65.

——— and **Richter, C. F.,** 1954, Seismicity of the earth, 2d ed.: Princeton, N.J., Princeton Univ.

Hodgson, J. H., 1962, Movement in the earth's crust as indicated by earthquakes, *in* Runcorn, S. K., ed., Continental drift: New York, Academic, p. 67–102.

Horai, Ki-iti, and **Nur, Amos,** 1970, Relationship among terrestrial heat flow, thermal conductivity, and geothermal gradient: Jour. Geophys. Res., v. 75, no. 11, p. 1985–1991.

Howell, B. F., Jr., 1959, Introduction to Geophysics: New York, McGraw-Hill.

Hudleston, P. J., 1973, An analysis of "single layer" folds developed experimentally in viscous media: Tectonophysics, v. 16, p. 189–214.

Kanamori, Hiroo, 1972, Relation between tectonic stress, great earthquakes and earthquake swarms: Tectonophysics, v. 14, p. 1–12.

Kaula, W. M., 1969, A tectonic classification of the main features of the earth's gravitational field: Jour. Geophys. Res., v. 74, no. 20, p. 4807–4826.

——— 1972, Global gravity and mantle convection, *in* Ritsema, A. R., ed., The upper mantle: Tectonophysics, v. 13, p. 341–359.

Langseth, M. G., Jr., 1969, The heat flow through the surface of the oceanic lithosphere, *in* Deep-seated foundations of geological phenomena: Tectonophysics, v. 7, no. 5–6.

Lee, W. H. K., ed., 1965, Terrestrial heat flow: Baltimore, Am. Geophys. Union Pub. 1288.

Lehmann, I., 1936, P'': Pubs. Bur. Central Seismol. Intern. Trav. Scie., v. 14.

MacDonald, G. J. F., 1964, The deep structure of continents: Science, v. 143, p. 921–930.

McKerrow, W. S., and **Lanbert, R. St. J.,** 1973, Deep earthquakes, surface subsidence, and mantle phase changes: Jour. Geology, v. 81, no. 2.

Molnar, Peter, and **Oliver, Jack,** 1969, Lateral variations of attenuation in the upper mantle and discontinuities in the lithosphere: Jour. Geophys. Res. v. 74, no. 10.

Mueller, Stephan, The structure of the earth's crust, *in* Developments in geotectonics 8: Amsterdam-London-New York, Elsevier.

Nettleton, L. L., 1940, Geophysical prospecting for oil: New York, McGraw-Hill.

Oliver, J., and **Isacks, B.,** 1967, Deep earthquake zones, anomalous structures in the upper mantle, and the lithosphere: Jour. Geophys. Res., v. 72, no. 16, p. 4259–4275.

Pakiser, L. C., 1963, Structure of the crust and upper mantle in the western United States: Jour. Geophys. Res., v. 68, no. 20, p. 5747–5756.

——— and **Zietz, Isidore,** 1965, Transcontinental crustal and upper mantle structure, *in* U.S. Program for the Internat. Upper Mantle Project: Prog. Rept. 1965: Natl. Acad. Sci., Natl. Research Council, Washington.

Parmentier, E. M., Turcotte, D. L., and **Torrence, K. E.,** 1975, Numerical experiments on the structure of mantle plumes: Jour. Geophys. Res., v. 80, no. 32, p. 4417–4424.

Poldervaart, Arie, 1955, Chemistry of the earth's crust, *in* Crust of the earth: Geol. Soc. America Spec. Paper 62.

Press, Frank, 1956, Southern California, pt. 1 of Determination of crustal structure from phase velocity of Rayleigh waves: Geol. Soc. America Bull., v. 67, no. 12.

——— 1959, Some implications on mantle and crustal structure from G waves and Love waves: Jour. Geophys. Res., v. 64.

——— 1961, The earth's crust and upper mantle: Science, v. 133, p. 1455–1463.

——— and **Ewing, Maurice,** 1955, Earthquake surface waves and crustal structure, *in* Geol. Soc. America Spec. Paper 62, p. 51–60.

Richter, C. F., 1958, Elementary seismology: San Francisco, Freeman.

Ringwood, A. E., 1969, Composition and evolution of the upper mantle, *in* The earth's crust and upper mantle, Am. Geophys. Union Mon. 13.

Shoemaker, E. M., 1974, Continental drilling: Carnegie Institution of Washington, p. 1–56.

Turcotte, D. L., and **Oxburgh, E. R.,** 1969, Convection in a mantle with variable physical properties: Jour. Geophys. Res., v. 74, no. 6, p. 1458–1474.

Walcott, R. I., 1970, Flexural rigidity, thickness, and viscosity in the lithosphere: Jour. Geophys. Res., v. 75, no. 20, p. 3941ff.

Warren, D. H., and **Healy, J. H.,** 1973, Structure of the crust in the conterminous United States, *in* The structure of the earth's crust based on seismic data: Tectonophysics, v. 20, no. 1–4, p. 203–213.

Weertman, J., 1970, The creep strength of the earth's mantle: Reviews of Geophysics and Space Physics, v. 8, no. 1, p. 145–168.

Wegener, Alfred, 1922, The origin of continents and oceans: London, Methuen.

Woollard, G. P., 1959, Crustal structure from gravity and seismic measurements: Jour. Geophys. Res., v. 64.

17

Global Tectonics

CONCEPTS OF GLOBAL TECTONICS AND THE ORIGIN OF STRESS IN THE LITHOSPHERE

Examination of the rocks near the top of Mount Everest reveals that they are sedimentary rocks containing marine Jurassic ammonites. Such an observation leads us to conclude that the crust of the earth is subject to great uplift—in this case more than 9,000 m. When the rock units are traced and the construction of a cross section completed, Fig. 23-20, an even more startling realization of the extent of crustal deformation is forced on us. One is inevitably drawn from this realization to consider the causes of the uplift and deformation and in turn the sources of stress in the lithosphere. Students of the earth have never wanted for grand schemes to explain the causes of crustal movements. Ideas to explain the origin of mountains had been advanced long before the field of geology was recognized as a scientific discipline. Evaluation of tectonic concepts is especially difficult because these ideas are almost invariably based on assumptions about processes taking place in the earth's interior at depths well below our deepest observations. It should be useful here to review some of these ideas about the origin of stresses in the earth before turning to a more detailed account of the resulting strains.

Compelling evidence can be marshalled to prove that the crust has exhibited and, in fact, exhibits today, both vertical and lateral instability. The presence of marine strata thousands of meters above sea level and the great thicknesses of shallow-water sediment in some old basins convince us that vertical movements (at least 9 km above and 10 to 15 km below sea level) have taken place. Much smaller vertical movements related to glacial rebound occur at measurable rates today. Several types of movements of the crust can also be documented. Deep-seated, nearly vertical faults (e.g., the San Andreas, Great Glen, Alpine) have undergone great lateral displacements parallel to the strike of the faults. Folded mountain systems embodying intricate folding and thrusts piling up crustal elements provide convincing evidence of strong lateral compression. Fault systems allowing the formation of grabens and the separation of continents require us to recognize large-scale crustal extension. Thus a satisfactory system for generation of lithospheric stresses must allow large-scale compression, extension, and lateral displacement as well as vertical movements. A brief review of tectonic hypotheses follows here, and some of these will be discussed as they apply in the explanation of major crustal elements later.

CLASSICAL TECTONIC THEORIES

The Contraction Hypothesis

Lord Kelvin developed the view that the earth is like a heat engine slowly dissipating the heat energy originally present in a primitive molten earth through processes of volcanism and diastrophism. Elie de Beaumont (1829) proposed that folds and faults develop in the outer crust of the earth as it accommodates to a cooling and shrinking interior. Thus the outer shell of the earth is under all-sided compression. Sir Harold Jeffreys (1952) supported this view on geophysical grounds, suggesting that the outer 600 km of the earth is brittle. Jeffreys pointed out that few earthquakes occur below this depth, but this idea is incompatible with the more recent concept that the Gutenberg low-velocity layer is plastic or partial melt.

This once-popular hypothesis now seems incompatible with a vast amount of observational data. Extensional features abound on the crest of oceanic ridges and in the rift valleys of East Africa, the Red Sea, the Gulf of California, and elsewhere. Extension of this order of magnitude has not been explained in terms of the contraction hypothesis, nor does it offer the flexibility to accommodate continental drift. Finally, we must call into question the basic premise that the earth is cooling and contracting. The thermal history of the earth is poorly known at best, but we do now know that heat is liberated by radioactive decay and a significant amount of heat may have been liberated in this way, maintaining the temperature of the interior or even possibly increasing it as some levels. Among various techniques used to establish the size of the earth in the past (e.g., length of the day, rate of rotation, paleomagnetic indicators, etc.), at least as many point to an earth of smaller radius in the past as to one of larger radius.

Continental Drift Hypothesis

The idea that the continents of today are the fragments of one or possibly two massive original continents as suggested by Snider (1858), Taylor (1910), Wegener (1915), and later a host of others was widely discredited in the United States only two decades ago. A considerable body of geological evidence supported the idea of drift, but the deep crust under continents appeared to offer an insurmountable obstacle to movement of the continental plates through a strong and near-surface oceanic crust.

The near-perfect fit of the continental margins of Africa and South America (best at the 2,000-m isobath; Carey, 1958) has long provided the intriguing and perhaps strongest evidence for continental drift, Fig. 17-1, and this fit is borne out by impressive corroborating evidence:

A continental glaciation affected South America, Africa, India, Australia, and Antarctica in the late Paleozoic. The pattern of glacial striations, the composition of tills and erratics, and other details of glaciation are compatible with the idea that the above continents were contiguous and located in a polar position.

A large number of striking similarities are found in the fossil flora and fauna from the Paleozoic Era, especially for the southern continents and to a lesser extent between North America and Europe. Many of these similarities are among plants but include amphibians as well. Movement of plant spores and animals among these continents was apparently free until the end of the Paleozoic, after which the development of life followed different lines of evolution on the various continents.

Numerous efforts to match the geology of one continent with that of others have been made. Many of these are not surprisingly unsuccessful when the varied character of the geology and the width of the continental shelf and slope are considered. The fit of Precambrian orogenic belts established by geochronology and structural trends between South America and Africa, Figs. 18-17 and 18-19, exemplifies the most successful of these efforts.

Interest in the drift hypothesis was reawakened in the United States with the emergence of paleomagnetism as a means of documenting paleopole positions. When these studies established that the magnetic poles had apparently wandered through geologic time, the idea of a fixed crust was weakened, and when it emerged that the paths of polar wandering for North America and Europe were different and could best be made to coincide by restoring the two plates to a contiguous position, continental drift was essentially reborn.

Some drift is now accepted as a corollary or an essential part of most geotectonic theories. Mechanisms of drift differ but the effects are similar. Some move the continents by drag on the base of the lithosphere caused by convective flow in the mantle. Others move the plates as new sea floor is formed along the oceanic ridges. Or a combination of convection and sea-floor spreading is postulated. Drift is also effected on an expanding earth even as dots are dispersed on an inflating balloon. Drift is also compatible with the idea of continental plates sliding down off the flanks of large swells in the mantle. Thus a theory once dead for lack of a suitable mechanism is restored and with a number of feasible mechanisms.

FIGURE 17-1 One of the earliest reconstructions of continents in their predrift configuration as proposed by Alfred Wegener (1915).

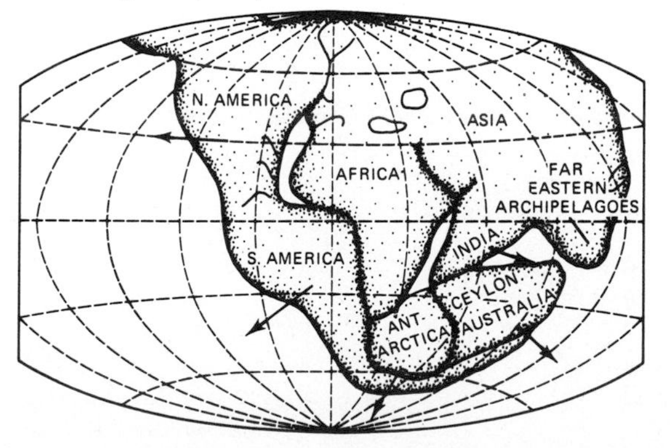

Classical Geosynclinal Theory*

The idea that orogenic belts form in long, relatively narrow zones of the crust where subsidence and long-term sedimentation produce great thicknesses of sediment was first proposed by James Hall (1859), an early American geologist. Hall studied the central part of the Appalachian Mountains in New York, Pennsylvania, and Virginia, where the thickness of the Paleozoic sequences is on the order of 9,000 to 12,000 m. Such long narrow belts of subsidence, sedimentation, and folded strata are called *geosynclines*, and these characteristics have been found in many other, but by no means all, folded mountain belts. Hall noticed that the present Appalachian Mountains are located in an area that had been the site of thick sedimentary accumulations in the Paleozoic and that almost all the sediment had accumulated in shallow water. The water could remain shallow only if the geosyncline subsided as sediment accumulated. Thus the geosyncline was not a deep trough that became filled gradually; it was a sedimentary basin in which subsidence and sedimentation took place together. This led Hall to think that the weight of the sediment caused the subsidence. Later studies show that the crust is strong enough to support thick sedimentary piles; thus the cause of the subsidence lies

* An excellent account of classical geosynclinal theory is found in Aubouin (1965) and in Kay (1951).

primarily in some process that forces the crust down or causes it to subside.

The term *geosyncline* is applied to long narrow sedimentary basins which contain thick accumulations of sediment. Some such basins have been subsequently deformed into folded mountain belts, and it has been suggested that most orogenic belts pass through a geosynclinal phase before they are deformed and uplifted. Many of the geosynclines, both modern and ancient, are or were situated along continental margins and may best be thought of as the prisms of sediment which accumulate in the transitional zone between continent and ocean. Geosynclines have often been described as trough-like, but we should remember that the nature of the oceanward side is usually obscure in orogens because that margin is severely deformed and altered.

Hall also suggested that the folds in the folded mountains are caused by the sliding and slumping of the sediment toward the center of the trough during sedimentation. Now we know that the folding and faulting came much later in most instances. He did not explain why mountains are high, and for this reason his theory has been called a theory of mountain building with the mountains left out. Typically geosynclinal thicknesses are 10 to 20 times as great as the sediment accumulated on an adjacent platform in the same time interval. Relatively undeformed sediment accumulations of geosynclinal proportions lie off the eastern and Gulf coasts of the United States under the continental shelves and continental slopes today.

Hans Stille (1941) suggested that many geosynclines consist of two subparallel troughs that can be distinguished by the types of sediments they contain. The inner trough, closer to the continent, he called the *miogeosyncline*. The miogeosyncline typically contains thick sequences of sandstone, shale, and limestone. The outer belt, called a *eugeosyncline*, contains the same sedimentary types as the miogeosyncline, but in addition it usually has thick deposits of graywacke (an arenaceous sediment with fragments of other rocks, often of igneous or volcanic origin). Eugeosynclines also contain considerable quantities of volcanic debris and lava flows, even including some ultrabasic rocks and their derivative serpentine. Many of the flows are pillow-shaped masses of lava, formed when lava is extruded underwater. Some deep-sea sediments, notably radiolarian oozes, which now form only in very deep water, are also found.

Marshall Kay (1951) applied this concept of geosynclines to North America and concluded that the origin of the sedimentary rocks in both the Appalachian-Ouachita geosyncline and the Cordilleran geosyncline of the western United States could be best explained in terms of an inner miogeosyncline such as the Appalachian Valley and Ridge area, and an outer eugeosyncline, the Appalachian piedmont and New England highlands, Fig. 17-2. Kay (1951) recognized a great variety of different types of geosynclines and likened the eugeosyncline to modern volcanic island arcs, and it is in this context that the term is still used.

Concept of the Sedimentary Orogenic Cycle

Because sediments provide a key to the environment in which they form, it is possible

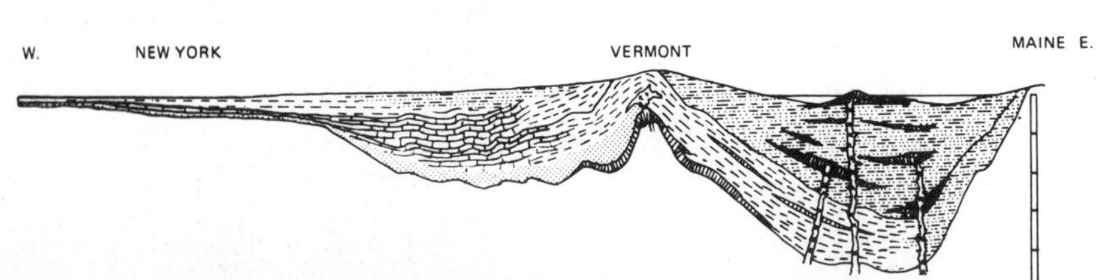

FIGURE 17-2 One of the best known cross-sectional representations of the geosynclinal interpretation of the Appalachians. The section is across both the miogeosyncline and eugeosyncline as restored to a time in the Ordovician Period by Marshall Kay (1951).

to relate the types of sediment in a sedimentary basin to adjacent orogenic activity.

Pettijohn (1957) subdivided the sedimentary pile in the Appalachian miogeosyncline into a series of sedimentary cycles, each of which can be subdivided into four stages reflecting the tectonic activity in the area. Each group of stages makes up an orogenic cycle—a concept which has been successfully applied in many orogens. The four stages are:

Preorogenic stage: Sediments deposited in the miogeosyncline are clean sandstones and limestone.

Euxinic stage: The initial indications of orogenic activity are the development of deeper basins in which black shales and limestones are deposited because circulation of water is not good.

Flysch stage: At this stage uplift in the core (eugeosyncline) of the orogenic belt has started, and deposits of sandstone and shale are deposited by turbidity currents in deep water.

Molasse stage: In this final stage the mountains are high, and coarse sediment is shed into the sedimentary basins. These sediments may contain abundant feldspar, as in arkose; they may be extremely coarse; the iron may be oxidized; and they can often be traced into terrestrial deposits.

Of course the type of sediment deposited at any given time is a function of location in the orogen as well as stage in the orogenic cycle. Thus significant differences occur contemporaneously in miogeosynclines and eugeosynclines.

Convection Current Hypothesis

Convective overturn of materials within the earth's core and mantle has been postulated for many years, Fig. 17-3. Convection currents form in fluids subject to strong temperature gradients. Although estimates of temperature in the earth's interior, Fig. 16-4, embody many assumptions, it is clear that strong temperature gradients between the surface and the deep mantle do exist. Convection cannot exist

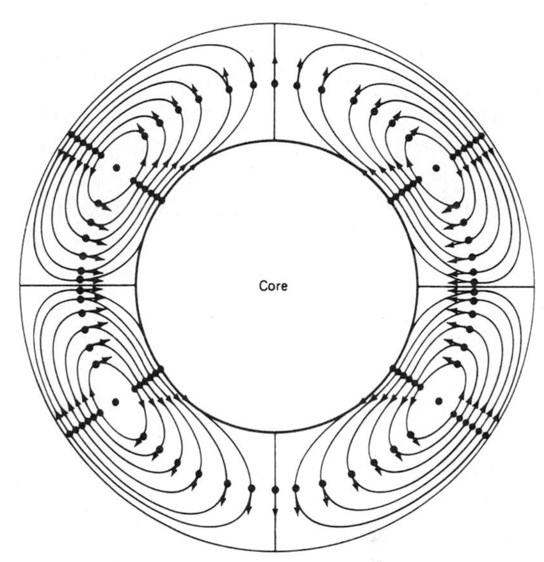

FIGURE 17-3 One of the early models of convection in the earth. The length of the arrows indicates relative rate of movement. (*From Pekeris, 1935.*)

in the earth's crust because it is too rigid; even the upper mantle down to the level of the Gutenberg low velocity channel is probably too rigid to allow convection, but deeper within the upper mantle, where the mantle is pseudoviscous as a result of high temperature and confining pressure, convection is possible.

Proof of mantle convection rests primarily on theoretical grounds, but the theory finds support in many near-surface phenomena attributed to convection, and it has been called on repeatedly in various geotectonic hypotheses as a driving mechanism of crustal tectonism.

Seismic evidence indicates that the outer core is liquid, and convection there is highly probable—a conclusion supported by the character of the earth's magnetic field; however, movements within the core probably have no direct effect on the lithosphere.

The Tectogene Hypothesis

A Dutch geophysicist, Vening Meinesz (1948), discovered large negative gravity anomalies over the Java trench. Later, such anomalies were found to be located over most trenches and over some islands and submarine ridges that are aligned with the trenches. These large negative gravity anomalies remain even after isostatic reductions are made. Vening Meinesz concluded that these anomalies resulted from projection of low-density rocks down into the denser subcrust under the island arcs. He postulated that lower density rocks, the sialic part of the crust, had been pulled into a downbuckle, later called a *tectogene,* by convection currents. According to his hypothesis, convection within the mantle exerts a frictional drag on the crust, and in areas where two cells converge and move down, the crust is pulled down into the mantle. Initial downwarping of the crust produces a geosyncline in which great quantities of sediment may accumulate. As the downbuckle is accentuated, the deeper portions are heated, deformed, and metamorphosed; magmas are created and intrude the mass. Eventually the downbuckle may be pressed together so that part of the material in it is forced upward to become a folded mountain range. Uplift is sustained by isostatic adjustments that arise from the low-density crust's having been pulled down into a high-density mantle. Isostatically driven uplift is most pronounced when convection slows or stops, as postulated by Vening Meinesz. As the uplift occurs, a metamorphic complex is brought close to the surface where it may be exposed later by erosion in the central part of the orogenic belt, and foreland folding takes place. Possibly the marginal folding is derived from the upward and outward push of the mountain root, or it may be the result of gravity sliding from the uplifted core. Thus Vening Meinesz's hypothesis successfully explained many observed features of folded orogenic belts, and it became the most popular tectonic hypothesis during the 1950s. (See Vening Meinesz, 1964, for the most complete account of this theory.)

The tectogene hypothesis was applied to many aspects of global tectonics with a great deal of success. Areas of crustal rifting such as the oceanic ridge system and continental rift zones could be explained as near-surface manifestations of rising convection currents in the mantle. The island arcs were viewed as representing an early stage in the evolution of orogenic belts. The application of this hypothesis finds many parallels in modern plate theory.

Some of the initial ideas of the tectogene hypothesis had to be modified as a result of discoveries in the ocean basins. The idea of a granitic crust under the oceans was disproved by seismic studies. Thus, the tectogene could not be a downbuckle of sialic crust as Vening Meinesz originally thought.* Subsequently it was suggested that the downbuckle takes place in the zone of transition between continental- and oceanic-type crusts. Fisher and Hess (1963) suggested that oceanic crust consisting of basalts and water-laden sediment on mantle of peridotite might be carried into the mantle by convection. The descending mafic rocks would be altered to serpentine by the rising temperature and presence of water. Eventually partial melts would form and rise to the surface to cause volcanic activity as in the island arcs.

A new concept of tectonics in which Hess' ideas were to be an important part were evolving rapidly even as attempts were being made to modernize the tectogene hypothesis. These were the concepts of sea-floor spreading and plate tectonics.

MODERN TECTONIC HYPOTHESES

Continental drift is now almost universally accepted among geologists, and the mechanisms for this movement most generally adopted are those associated with the ideas that new sea floor is created along oceanic

* The mechanical problem becomes especially difficult if the lithosphere is of greater density than the upper mantle, as suggested by Clark and Ringwood, because isostatic uplift would not occur.

ridges (sea-floor spreading), that the pattern of drift is governed by movements of relatively rigid lithospheric plates, that oceanic crust subsides into the mantle in island arcs, and that orogenic systems are formed along these plate boundaries by several different types of plate interactions (plate tectonics). All modern theories place some emphasis on gravity tectonics, though some would restrict it primarily to surficial downslope movements, while others stress the importance of gravitative spreading of large uplifted masses of material which yields by pseudoviscous flow as well as sliding. All modern theories encompass some vertical uplifts, but again these are secondary effects in some theories such as plate tectonics, while it is the vertical uplift which primarily motivates continental drift in others. One type of vertical movement, due to rise of diapiric masses from the mantle, is employed both as a primary cause of uplift by some and as an additional component in the overall tectonic process by others.

The convection current hypothesis remains important as a primary cause of both vertical and lateral movements in many modern models, and the theory of global expansion which originated long ago remains enigmatic.

The Roles of Gravity in Geotectonic Hypotheses

Gravitative forces account for the existence of body stresses throughout the earth. The types of effects we should expect from these stresses are predictable, and little doubt of their importance in crustal movement remains. The concept of isostasy*—the idea that the outer shell of the earth is essentially in a state of flotational equilibrium on the interior—is of fundamental importance. The strongest evidence for isostasy derives from studies of gravity which confirm that gravity anomalies are vastly reduced or eliminated over most of the earth if corrections are made for the observed near-surface density variations. The effects of these variations are compensated at a depth of about 100 km. Also, the lower density sialic continents are forced to protrude deeper into the more dense substrata than the oceanic crust. Confirmation of isostasy is found in the ease with which it can be used to explain isostatic anomalies where compensation or flotational equilibrium does not exist. Such areas include the sites of Pleistocene ice caps (e.g., Canadian and Fennoscandian Shields), where negative anomalies occur because the rate of removal of the ice cover has been more rapid than the rate of subcrustal compensation now in progress as shown by modern uplift. Isostatic anomalies also exist over some mountains, island arcs, and other areas of unstable crust.

Seismic evidence indicates that the lithosphere has sufficient strength and rigidity to be considered as a viscous material only in the context of very long time spans. The low-velocity zone, however, may include some melt and be hot enough for viscous flow of considerable magnitude over much shorter intervals. Figure 16-3 indicates the way viscosity varies with depth. A condition of vertical stability would exist if the density of the mantle is greater than that of the lithosphere, but upper mantle models formulated by Clark and Ringwood (1964) on the basis of temperature and compositional estimates for the mantle portray a condition of quasi-equilibrium in which the density is actually somewhat lower in the low-velocity zone than in the overlying lithosphere—especially over oceanic crust.* Thus, much of the lithosphere is potentially subject to sinking into the mantle simply under its own weight. Presumably the rigidity, strength, and continuity of the lithosphere and the small density contrast between the lithosphere and mantle prevent its rapid breakup and floundering into the mantle.

* Refer to Daly (1940) and Heiskanen and Vening Meinesz (1958) for a history of the theory and details of its application.

* A good summary of the literature on this subject is presented in Wyllie (1971, pp. 63–136, 233–255).

Effects of gravitative forces in the crust must be considered in yet another context—that of the components of gravity acting down an inclined plane. Such forces act within most crustal materials and may be an important factor in spreading of the sea floor which slopes from oceanic ridge crests to subduction zones. The situation is similar in principle to the inclined plane problem of elementary physics. A block located on an inclined plane is acted on by a component of gravity which tends to move the block down the plane and will do so if the frictional resistance on the contact is insufficient to hold it in place. By analogy, uplifted masses of rock are subjected to gravitative stresses which may cause them to flow or even slide (called a *gravity-glide fault*) downslope. Van Bemmelen (1965) applies this principle to the movement of whole continents over the upper mantle.

Gravity is called on in most geotectonic hypotheses to cause upward movement of low-density material at depth, to prevent downward movement of low-density material into more dense substrates, and to cause lateral spreading at depth of uplifted pseudoviscous material, to cause lateral pressures responsible for shallow folding and faulting of sedimentary veneers, and to cause surficial slumping and sliding in both marine and continental environments.

Ultrabasic Rocks—Proof of Mantle Involvement in Orogeny*

The unusual association of coarse-grained ultrabasic rocks such as peridotite, and its hydrothermal alteration product serpentinite, with extrusive basaltic rocks and frequently radiolarian cherts, often called Steinmann's trinity, has long been recognized, and the origin of this assemblage and its mode of emplacement in orogenic belts has been a topic of considerable controversy and debate.* These rocks play an important role in both plate-tectonic and mantle diapirism theories, because the composition of this assemblage is highly suggestive of what we know of the deep oceanic crust and upper mantle. All parties to the debate agree that the ultrabasic materials, at least, are part of the upper mantle.

Hess had been interested in these ultrabasic masses long before he first proposed sea-floor spreading. He pointed out the presence of peridotites in folded mountain systems in 1937 and concluded that these "alpine ultramafics" were emplaced early during orogeny, in the internal portions of the orogens, and only once in the history of each orogen—a conclusion now shared on the basis of detailed field work by many others (Maxwell, 1973). Ultrabasic bodies occur mainly in only one other tectonic setting—the island arcs. The question then is how did these pieces of the deep oceanic crust and upper mantle become emplaced—What was the mechanism? Do all such bodies originate in the same way? This is an important question which each of the modern tectonic theories must address. We will consider the explanations in the following sections.

Sea-Floor Spreading and Plate Tectonics†

The basic concepts of plate tectonics are that the lithosphere can be subdivided into a mosaic of relatively rigid plate-like masses and that the dynamics of the lithosphere can be

* See Maxwell (1973), Coleman (1971), and Chidester and Cady (1972) for discussions of this problem.

* The lithology of ophiolite sequences in the Alps and Appenines where they were first described is as follows (Maxwell, 1973). The base consists of dunite, serpentinized peridotite, and serpentinite overlain by gabbro, diabase, pillow basalt and volcanic breccia, radiolarian cherts, and calcareous oozes. Thickness of the sequence is 1 km±.

† Two excellent books of readings have been prepared which contain many of the important original papers written on these subjects. Refer to Cox (1973) and Bird and Isacks (1972).

analyzed in terms of the relative movements of these plates.

It is not difficult to identify some of the discoveries which formed the basis for the development of the theory of plate tectonics, but it obviously grew during a period of exceptionally fast discovery in many aspects of earth science and especially in the fields of geophysics and oceanography. At the time of its formulation and expression in print, the theory of continental drift had been revitalized by a number of paleomagnetic discoveries, the crustal structure of many parts of the earth was well known, the Gutenberg low-velocity zone had become widely accepted and interpreted as a zone of plasticity in the mantle, the mid-oceanic-ridge system had been charted and the extensional character of its rift valley recognized, the seismicity of the world had been established and the planes of inclined seismicity under trench systems—Benioff zones—were known. Studies of heat flow had revealed that the crest of the oceanic ridges are abnormally hot. Convection was favored as the primary cause of crustal movements, and both the island arcs and oceanic ridges came under intensive study. Reversals in the polarity of the earth's magnetic field had been recognized and provided the basis for establishing a geomagnetic time scale.

It is against this background of discovery that we should view the now famous papers of Hess (1960, 1962) in which he suggests the idea of a mobile sea floor involving spreading of the oceans away from the oceanic ridges. Shortly afterward, Vine and Matthews (1963) devised a way to test the idea of sea-floor spreading based on polarity reversals. If new material in the form of basaltic intrusions is being added along the crest of oceanic ridges while older solid oceanic crust is moving aside, and if the polarity of the magnetic field is essentially "frozen in" to the basalt at the time the melt cools to the Curie point, then anomalies related to the polarity reversals should be found across the ridge. These anomalies should occur as bands essentially parallel to the ridge and symmetrical about the axis of the ridge. Positive anomalies in the Northern Hemisphere should coincide with periods of present-day polarity, and negative anomalies would identify reversals. Vine and Matthews (1963) published the results of their geomagnetic study of the Carlsberg Ridge in the Indian Ocean and a short time later a study over the mid-Atlantic ridge south of Iceland showing strong confirmation of the pattern as predicted by sea-floor spreading, Fig. 17-4.

The idea of sea-floor spreading provided a simple clear-cut answer to one of the most perplexing questions of marine geology raised by early seismic studies: why is the sedimentary layer so thin in ocean basins? The answer: because the ocean floor is relatively young.

The decade of the sixties is already described as the time of a revolution in the

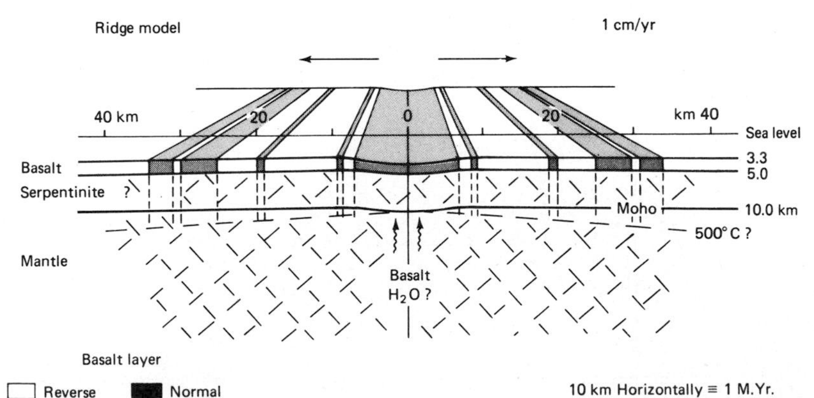

FIGURE 17-4 Diagrammatic representation of the oceanic crust at a mid-ocean-ridge crest, assuming active spreading at a rate of 1 cm per year per ridge flank, and the geomagnetic reversal time scale of Cox and his associates (1967). (*From Vine, 1969.*)

earth sciences. New discoveries followed one another in rapid succession, and these were quickly synthesized into the theory best known as *plate tectonics.* The globe-encircling oceanic ridge system became accepted as the site where new sea floor is created and as the boundary between diverging lithospheric plates. This boundary can be traced into continental crust in the Gulf of California and in the Gulf of Aden—both are areas characterized by extensional tectonics.

The earth's lithosphere was depicted as being composed of six major lithospheric plates by Le Pichon (1968), Fig. 17-5. The boundaries between these plates are well outlined by the zones of high seismicity and are defined by oceanic ridge crests, modern orogenic belts, island arcs, and major strike-slip faults—the transform faults described by J. T. Wilson (1965).

The subsequent trend of plate-tectonic interpretations has led to the subdivision of these six major plates into numerous smaller plates. Some of these new plates result from the discovery of additional spreading centers defined by magnetic anomalies. The divergent junctions are generally long, essentially straight boundaries between oceanic plates, but a number of junctions are now recognized where a single boundary splits. Such Y-shaped junctions formed where three plates share a corner are called *triple junctions.*

FIGURE 17-5 Division of the earth's crust into plates. Borders are seismically active. Directions of movement are indicated by arrows. Plates are bounded by trenches, oceanic ridge crests, and orogenic belts. (*Modified after Le Pichon, 1968.*)

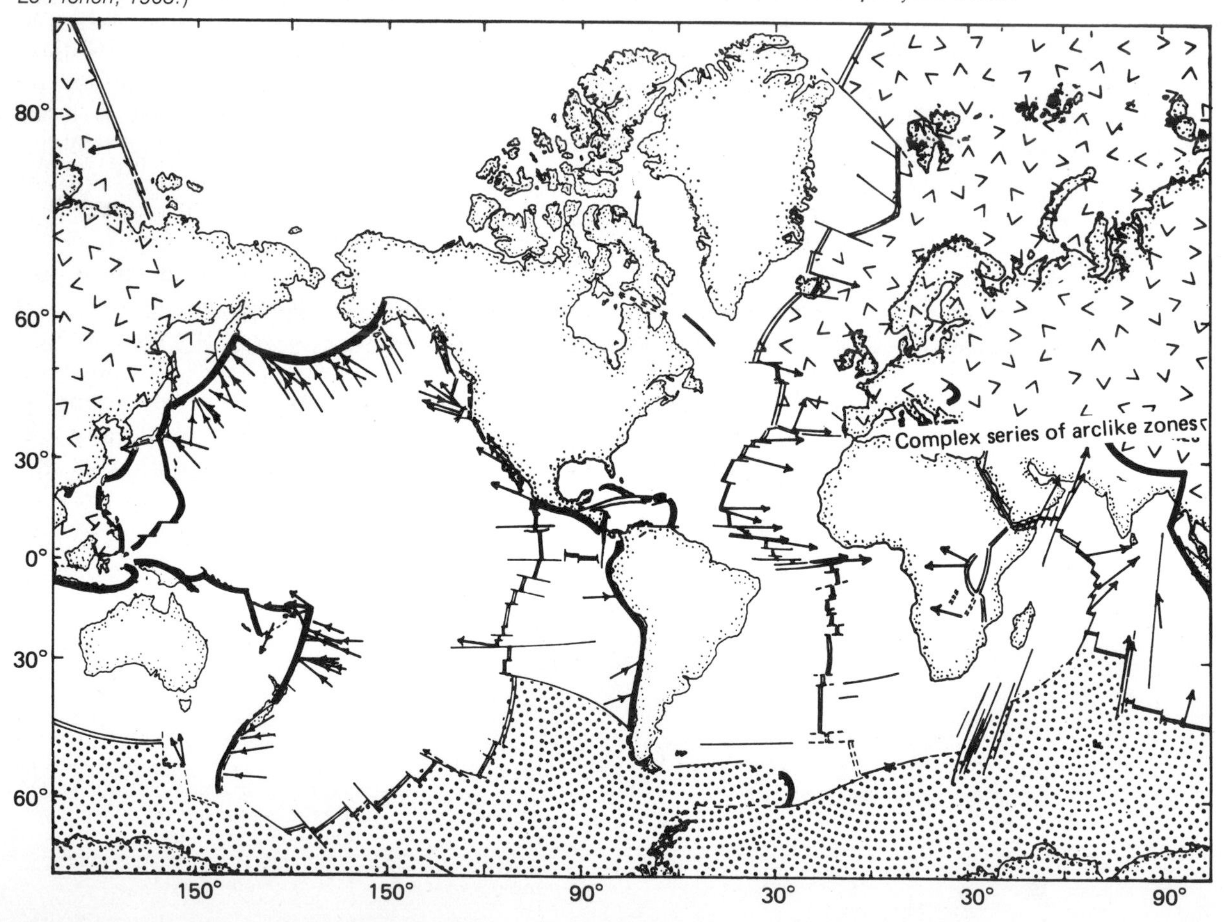

Acceptance of the sea-floor-spreading idea leads directly to another conclusion—either the earth is expanding or the lithosphere is being destroyed, most probably by becoming reincorporated in the mantle. Vening Meinesz had suggested that the island arcs are locations where the crust is dragged down by convection. Benioff and others had earlier demonstrated that earthquake foci define inclined seismic shear zones under the arcs and along the Central and South American margins. Oliver and Isacks (1968) demonstrated the probable existence of a slab of lithosphere under the Benioff zones by studying the attenuation of surface waves from deep earthquakes. This slab appears to plunge down into the mantle where it could be assimilated into the mantle, Fig. 17-6. The direction of movement of the slab is indicated by analysis of the initial motion associated with earthquakes in the seismic shear zone (Isacks, Oliver, and Sykes, 1968). These studies indicate that the slab is moving down under the volcanic islands. Plate boundaries where lithosphere is consumed are known as *subduction zones*. The island arcs and the trenches along the coast of Central and South America are the prototypes.

FIGURE 17-6 Block diagram illustrating schematically the configurations and roles of the lithosphere, asthenosphere, and mesosphere in a version of the new global tectonics in which the lithosphere, a layer of strength, plays a key role. Arrows in asthenosphere represent possible compensating flow in response to downward movement of segments of lithosphere. One arc-to-arc transform fault appears at left between opposing zones of convergence (island arcs), two ridge-to-ridge transform faults along ocean ridge at center, and simple arc structure at right. (*From Isacks, Oliver, and Sykes, 1968.*)

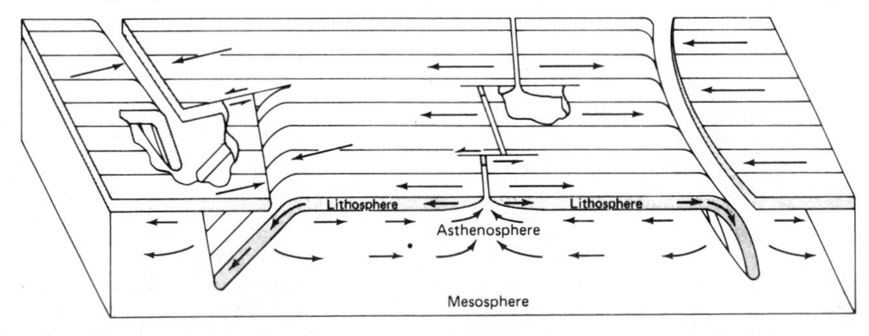

Geometry of Plate Movements

In its simplest version plate theory treats the movements of essentially rigid thin plates on the surface of a sphere. The plates exhibit three basic types of boundaries—spreading centers (divergent junctions) where new oceanic crust is formed, subduction zones where oceanic crust sinks back into the mantle, and transform faults, a special type of strike-slip fault connecting other plate boundaries. Plate boundaries may or may not coincide with the edges of continents; for example, the west coast of South America coincides with a subduction zone, but the eastern margin does not. The continents are high because they are composed of lower density materials, and for this same reason, large quantities of continental crust cannot be subducted.

Seismicity maps provide one of the clearest indications of the location of modern plate boundaries, and the subduction zones are well defined on maps showing intermediate- and deep-focus earthquakes. In addition, most modern subduction zones are characterized by trenches and volcanic activity.

Plate movements can be discerned in a number of ways. Solutions to earthquake focal mechanisms (first motion studies) for earthquakes which occur along plate boundaries provide a direct indication of localized motion on faults at boundaries, and from these movements vectors for the plates can be determined successfully in subduction zones and along transform faults, Fig. 17-7. A number of geological features also provide guides to movements. The magnetic anomaly patterns arising from polarity reversals during sea-floor spreading are especially important because they not only show by their orientation the direction of growth at the spreading center, but the spacing is a guide to the rate of spreading. Once the anomalies are correlated and dated, the ocean can be restored to its form at any given time by closing the oceans—a process involving bringing the two anomaly stripes that were formed at that time back into coincidence at the spreading center. If the spreading has not been uniform (as is usually the case), then the plates on either side of a spreading center will rotate so that direction to the magnetic poles will change. Since the direction to the magnetic poles (paleopole positions) can be determined through paleomagnetic studies, it is possible to check the rotations indicated by restoration of magnetic anomaly belts. Zones of different spreading rates are separated by transform faults within the oceanic plate, and these two provide a guide to the direction of localized plate movements. Finally, in fitting continental blocks that have been split and separated back together, it is sometimes possible to use the shape of the continental margin, but these are sites of rapid change due to both sedimentation and tectonic activity; so, the fit usually will not be very precise. The most effective topographic fits between continents have been obtained by use of the 1,000- to 2,000-m isobaths. In a number of instances the geology on the continents can also be used to check a fit.

Because the earth is a sphere, the motion of any two adjacent rigid plates can be related to a *pole of relative motion*—a point on the globe that does not move relative to either plate and that remains relatively fixed for an extended period of geologic time, Fig. 17-8. From such a premise we can infer several important characteristics of plate motion. For example, transform faults on the plates will lie along segments of concentric circles drawn from the pole. Actually the pole can be found by extending normals to the traces of transform faults. It also follows from the concept of the pole of relative motion that the width of new crust formed along a spreading center over a given period of time varies from zero at the pole of relative motion to a maximum amount at a distance 90° from the pole.

Tracing Plate Motion

Morgan (1968) points out that the motion of crustal plates relative to one another may be

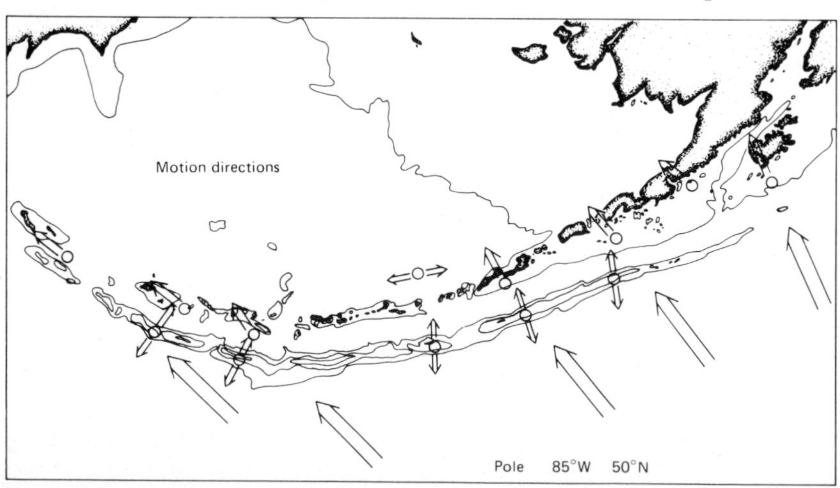

FIGURE 17-7 Motion directions in the Aleutians. The Pacific plate is moving northwesterly with respect to Alaska and the Bering Sea. These motions are inferred from the mechanisms of earthquakes located along the inner margin of the trench. Events in the deep part of the trench are characterized by large components of normal faulting, with the axes of tension nearly perpendicular to the local strike of the trench. (*From Stauder, 1968.*)

described by a rotation of one block relative to another. In this model the plates are treated as though they are rigid and not given to stretching, folding, or distortion within the block. The rotation can be described in terms of the position of a pole or the relative rotation and angular velocity. The application of this idea is best seen in the case of the opening of the Atlantic, where a number of transform faults lie nearly perpendicular to the boundary between the eastern and western sides of the ocean. Morgan constructed lines normal to the various transform faults along this boundary and found that they intersect in a pole located at about 62° N. In other words the transform faults lie on "circles of latitude" about the pole of rotation. The faults along the mid-Atlantic ridge do lie very closely along such circles. A good fit is obtained for the spreading of the Pacific-Antarctic ridge using a pole at 71° S. This study provides a further insight to the application of transform faults to global tectonics.

Transform Faults

Transform faults are one of the most important features of plate-tectonic models; yet, their geometry is often difficult to understand because the faults often involve displacements which are in the opposite direction from the apparent offsets Fig. 17-9. The name transform fault was proposed by J. Tuzo Wilson (1965) for strike-slip faults along which the displacement suddenly stops or changes form. We earlier considered one type of transform fault which cuts across spreading centers on oceanic ridges separating ridge crust segments which are spreading at different rates or which were initially offset when spreading started. Other transforms connect a ridge spreading center with a subduction zone or two subduction zones, Fig. 17-9. Examples of some of these and other geometric features of plate-tectonic theory are applied to the problems of regional geology later in the book.

FIGURE 17-8 On a sphere, the motion of block 2 relative to block 1 must be rotation about some pole. All faults on the boundary between 1 and 2 must be small circles concentric about the pole, *A*. (*From Morgan, 1968.*)

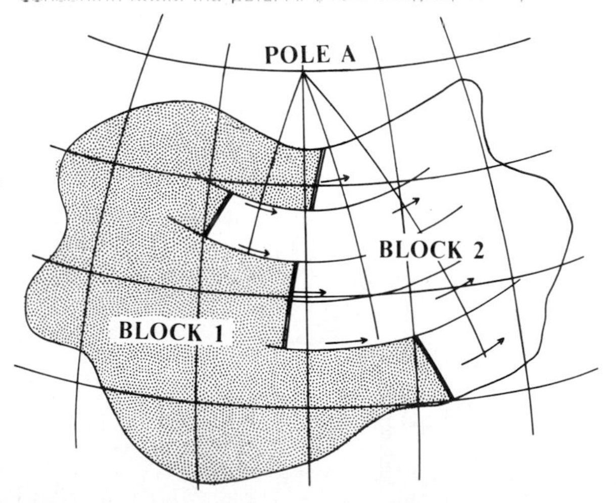

Plate Tectonics and the Causes of Orogeny

Plate tectonics grew primarily out of discoveries in the ocean basins, but it has been successfully applied to explain the cause of orogeny and many features of orogenic belts. The Alpine-Himalayan orogen, the orogenic belt along the western margins of the Americas, and the active zones in and near modern island arcs are the sites of most Cenozoic orogeny. Tectonism in these three belts occurs under very different circumstances. The Alpine-Himalayan orogen is situated between the Eurasian craton and the African and Indian cratons. The Himalayas especially appear to be formed where the Indian craton has collided with and been driven into the edge of the Eurasian craton. The mountain system of western North America occurs along a margin which appears to be overriding the spreading center in the eastern Pacific. The Andes and the island arcs appear similarly located over subduction zones associated with adjacent trenches.

The most comprehensive synthesis of orogeny in terms of plate tectonics is that of Dewey and Bird (1970). They recognize two fundamentally different mechanisms of orogeny—one is a mechanically driven orogeny resulting from the collision of two continents or a continent and an island arc, Fig. 17-10 and 17-11. The second is basically a thermally driven orogeny caused by the subduction of one plate under another, called an *island-arc-type* or a *cordilleran-type mountain belt,* Fig. 17-12. Plate-tectonic interpretation of features in major tectonically active crustal elements is as follows (Dewey and Bird, 1970):

Island arcs: A trench forms as the oceanic plate begins to descend. This is accompanied by thrusting of wedges of oceanic crust and mantle in the inner side of the trench and formation of a submarine ridge. Oceanic chert, argillite, and carbonates may become involved in large submarine gravity slides into the newly forming trench thus

FIGURE 17-9 Transform faults as originally conceived by J. Tuzo Wilson. (*a*) A ridge-ridge transform; (*b*) a ridge-trench transform; (*c*) a trench-trench transform. The figure at the right shows the migration of the features shown at left after some time interval. A marker is added for reference. (*Modified after J. Tuzo Wilson, 1965.*)

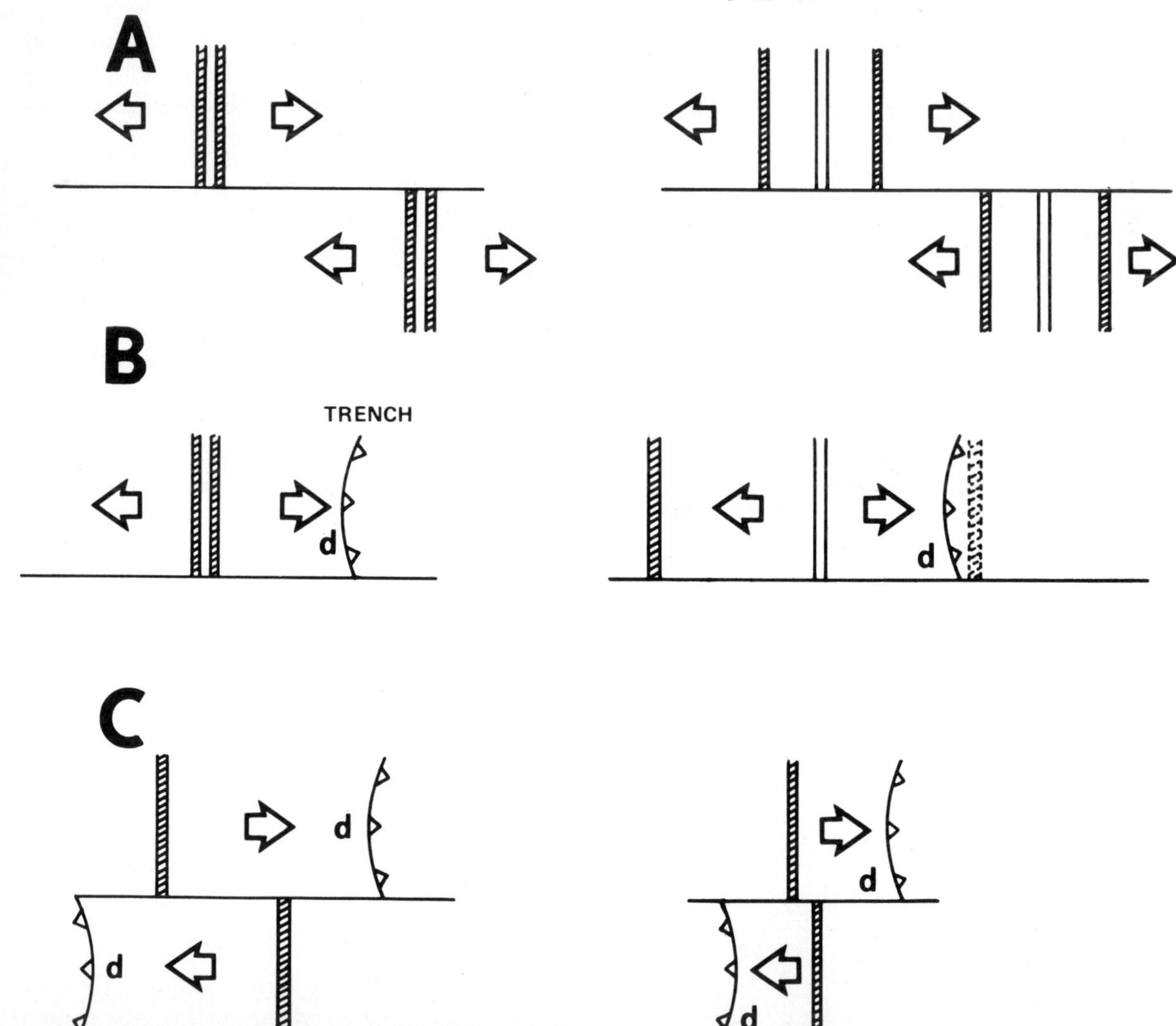

creating the ophiolitic sequences previously discussed. These slides may also carry basic and ultrabasic blocks derived from the upper edges of the early thrust wedges. This material moves into a trench located where the oceanic slab is going down. Thus the material in the slide masses tends to be carried down and under the trench where it is at least partially scraped off the descending plate and is strongly deformed in blueschist metamorphic facies [a relatively low-temperature (100 to 300°C) high-pressure (1- to-10-kbar) assemblage]. When the descending plate reaches depths over 100 km, the amphibolite and quartz eclogite crust begins to undergo partial melting, producing calcalkaline magmas that undergo differentiation as they rise, Fig. 17-13. Basalt magmas form at shallow depths, and a volcanic pile begins to form at the surface. When this pile grows thick enough to rise above the surface, it is eroded and a thick wedge-shaped sedimentary accumulation derived from the volcanoes develops between the trench and the volcanic front. The result is formation of a volcanic arc with fair metamorphic zonation.

Cordilleran-type orogen. This type of orogen is formed where a trench and subduction zone develop along the edge of a continent. The Andes and North American Cordilleran are cited as modern examples. The evolution of this type of orogen starts with formation of a consuming plate margin near the continental margin. Oceanic crust is driven toward the newly formed trench. Flysch accumulations thicken toward the trench, and blueschist mélanges (gravity slides composed of mixtures of materials) are formed. As the oceanic slab descends under the continental rise it is heated, and when it reaches approximately 100 km in depth, magma is generated and volcanics are erupted to form a chain of volcanoes, Fig. 17-13. Magmas and the heat from them rises and an orogenic welt forms above the expanding dome, the core of which is made up of gabbroic and granodioritic magmas. As the core of this dome grows in size, the overlying sedimentary pile which was formed on the continental rise is affected by high-temperature deformation and metamorphism. When this growing welt rises above sea level, sediment derived from it is transported both toward the ocean as a flysch wedge and toward the continent. Eventually this wave of deformation reaches the edge of the continent and the continental shelf subsides, forming a trough into which flysch and gravity slides from the welt accumulate. Initially, the gravity slides are composed of flysch and sedimentary rocks, but materials derived from

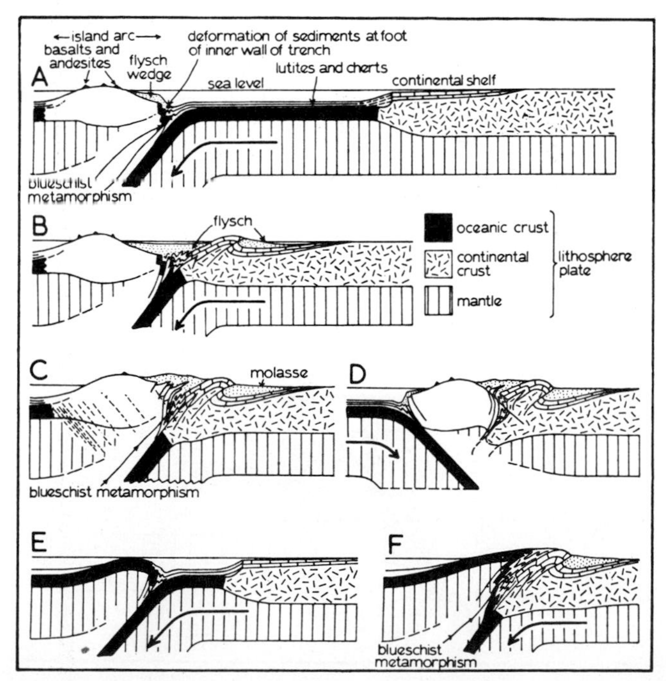

FIGURE 17-10 (A–D) Schematic sequence of sections illustrating the collision of a continental margin of Atlantic type with an island arc, followed by change in the direction of plate descent (D–F). (*From Dewey and Bird, 1970.*)

progressively deeper levels are involved until finally metamorphic rocks are involved in the slides which move toward the continent. Even the cratonic basement may become involved. Lateral pressures are generated both by spreading of the mobile core of the orogen and by gravity-induced movements of materials over the uplift. About the time the metamorphic rocks begin to reach the surface, coarse fluviatile (molasse) deposits are carried into basins and troughs located farther inland. Finally the system becomes resistent to additional contraction. Granitic plutons are emplaced at high levels in the welt at this stage. The crest of the welt becomes broken by extensional block faulting as it grows.

Continent–island-arc collision: An ocean initially separates these two crustal elements, but as the oceanic crust is consumed in the subduction zone, the continental margin is drawn toward and into the subduction zone. Flysch floods into and fills the ocean just before collision. The low density of the continental rocks makes it difficult for them to be drawn into the mantle. Instead, thrust slices from the leading edge of the arc are driven into the sediment accumulation along the edge of the continental plate. These thrust wedges carry flysch, blueschist-facies rocks, and slices of oceanic crust (true ophiolite sequences composed of oceanic chert, ultramafic rocks, and pillow basalts). These rocks are forced onto the rocks of the continental shelf and rise, Fig. 17-14. The consolidation of the island arc and the continental margin eventually forms a resistant mass, and this may be followed by development of a new subduction zone on the opposite side of the island arc, in which the direction of subduction flips.

Continent-continent collision: The collision of two continents, one of which is bordered by a subduction zone and the other bordered by a tectonically inactive continental margin like that on the east coast of North America, might proceed as follows. The initial stages of collision occur as the inactive

FIGURE 17-11 Schematic sequence of sections illustrating the collision of two continents. See description in text. (*From Dewey and Bird, 1970.*)

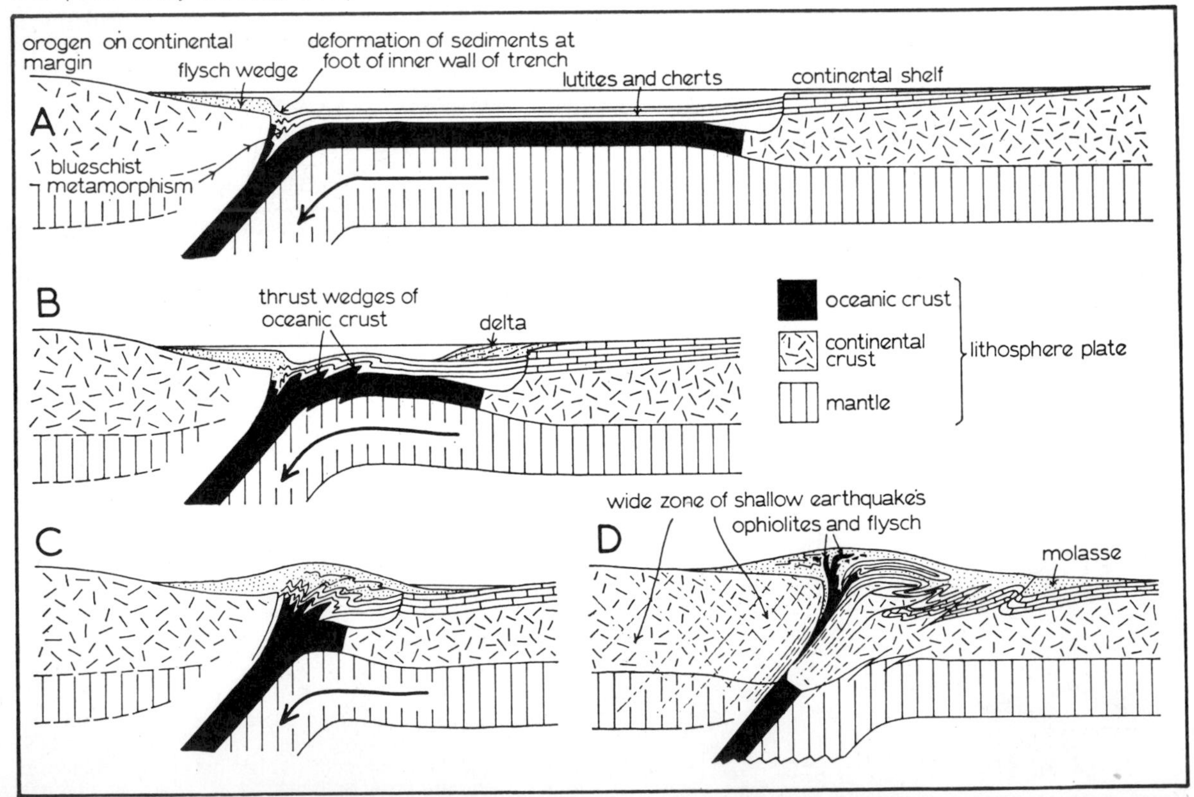

FIGURE 17-12 Schematic sequence of sections illustrating a model for the evolution of a cordilleran-type mountain belt developed by the underthrusting of a continent by an oceanic plate. (*From Dewey and Bird, 1970.*)

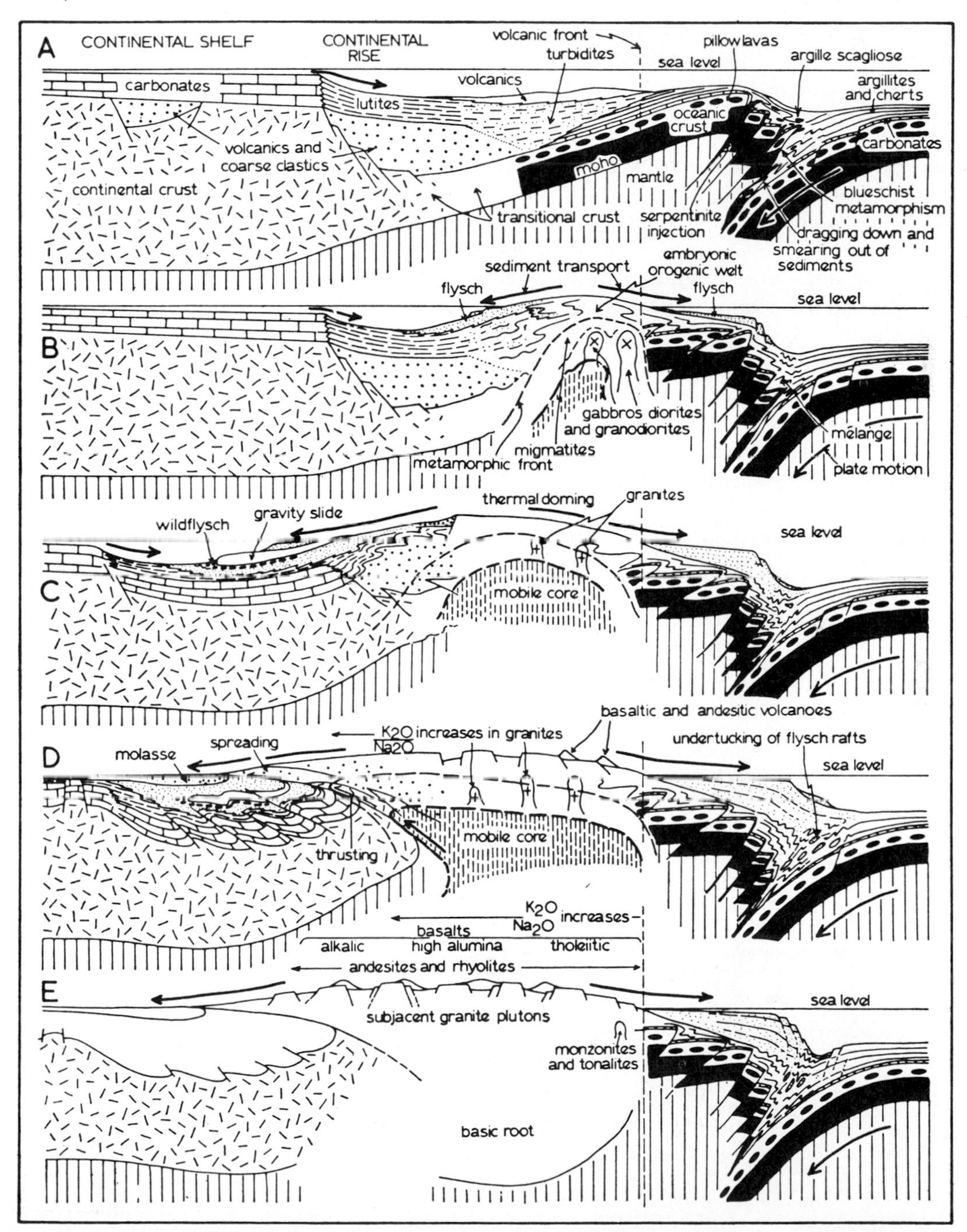

margin moves into the subduction zone, and these stages are similar to the arc-continent collision, including splintering and thrusting of the continental basement to form the cores of nappes. The oceanic crust (ophiolites) and flysch are squeezed and ultimately thrust over the lower thrust sheets. The low density of the continental rocks and consequent buoyancy prevent their subduction, and this minimized the generation of magma and thermal effects. It is suggested that the descending oceanic plate may even break off and sink. The trench-like subduction zone is replaced by a wide zone in which the continental crust is splintered.

The concepts of sea-floor spreading and movement of lithospheric plates developed in the early 1960s (Hess, 1960, 1962; Dietz, 1961) and have become so widely accepted that this model of global tectonics is the primary working hypothesis against which other ideas and new observational data are compared. The use of multiple working hypotheses (Chamberlin, 1897) has not been abandoned in tectonics, but with a few notable exceptions, most geologists and geophysicists favor "plate tectonics" as the best model available today. It is compatible with a large body of observational data, and it has withstood many tests devised to establish its validity.

Mantle Convection in Plate-Tectonic Theory

Mantle convection has been and continues to be the favored mechanism for driving the movements in the lithosphere, but convection, because it cannot be measured directly, must always be inferred from the lithospheric movements it is called on to explain or from other indirect observations such as gravity or heat-flow measurements thought to be directly related to mantle properties. In addition, theoretical models of convection can be constructed and compared with the proposed tectonic models to see if they are compatible. A considerable effort has been and is being mounted by geophysicists in this direction.

Heat-flow observations show that the oceanic ridge crests are sites of high heat flux from the mantle and that the subduction zones are sites of abnormally low heat flux—suggesting that the ridges are sites of upwelling in the mantle, while subduction areas may well be sites where sinking of cooled mantle material occurs. If lithospheric plates are moving laterally as a result of convection, the question

FIGURE 17-13 Hypothetical cross section of the ocean showing a spreading ridge at left, some oceanic islands, and a subduction zone located along an island-arc or Andean-type margin. Basaltic magma is erupted at the oceanic ridge, and that is covered by sediment as the sea floor spreads toward the subduction zone. A trench marks the edge of the subduction zone, deep-water sediment accumulates, and sediment is scraped off the descending crust. This sediment becomes tightly folded and faulted. The high pressure caused blueschist-facies metamorphism to occur beneath the trench. At greater depth the basalt undergoes a transformation to amphibolite and then to eclogite. At depths at about 100 km, partial melts begin to form and rise to the surface to form andesitic volcanoes and the island arc. The drawing is based on earlier works of Hess, Ringwood, Green, and Miyashiro.

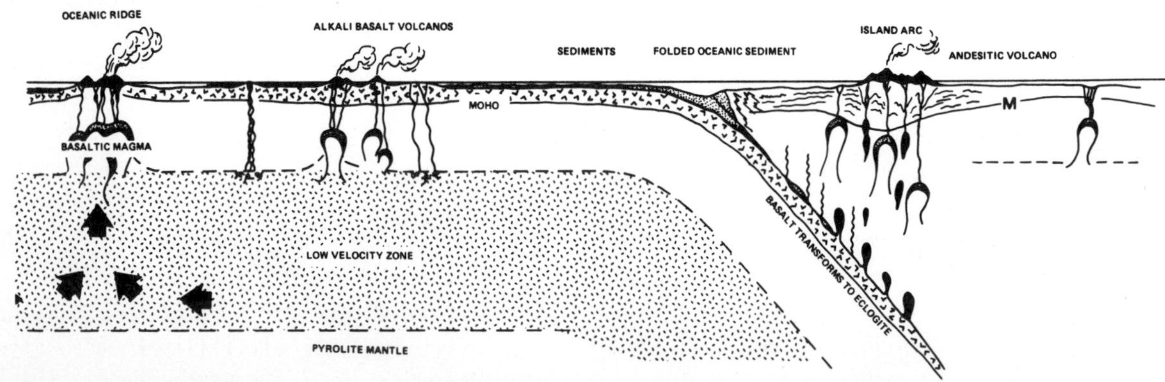

of to what extent those very thin, broad plates are coupled with the underlying mantle must be considered.*

Elsasser (1967, 1969) devised one of the first models in which a cold dense oceanic lithospheric plate sinks into the underlying hot and pseudoviscous mantle, pulling the rest of the plate toward the subduction zone. He shows that in this model the lithosphere acts as a stress guide capable of transmitting stresses over long distances and that the model is independent of any viscous coupling with mantle convection. A viscous drag would be exerted on the base of the lithosphere, however, and we should therefore expect to see small plates move faster than large plates. This is not the case; so the convection model must be more complex.

The methods used to approach this problem are of two main types—in one it is assumed that a lithospheric slab is being driven into the subduction zone at observed rates, and the thermal and mechanical consequences of this are compared with observations (Griggs, 1972; McKenzie, 1969). The other is a more purely theoretical approach based on the classical model of convection in a viscous medium in which temperature gradients, viscosity, strain rate, and internal changes are varied to produce models (Turcotte and Oxburgh, 1968, 1969).

* Garfunkel (1975) discusses additional restraints placed on flow in the mantle by plate theory.

FIGURE 17-14 Possible mechanisms for the obduction of ophiolite sheets onto continental margins (*Dewey and Bird, 1971.*)

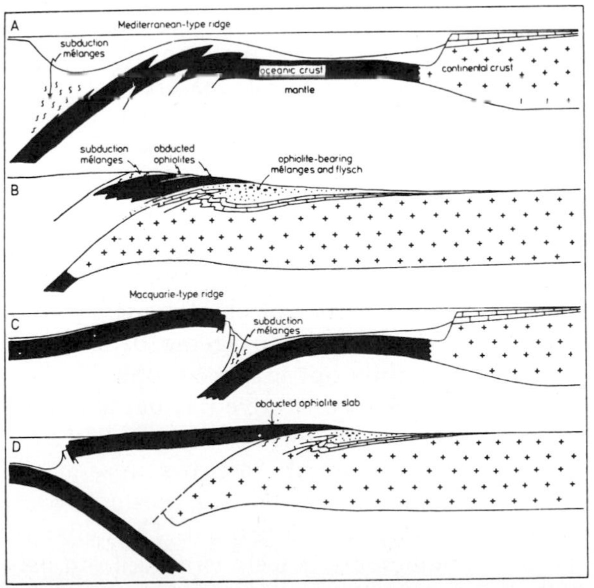

Mantle Plumes

J. T. Wilson (1965) suggested that the Hawaiian Islands could be explained as having formed as a result of the lateral movement of the oceanic crust over a hot spot in the mantle —the site of upwelling of hot magma possibly from deep in the mantle more or less resembling a convection cell associated with a thunderhead in the atmosphere. A number of other hot spots have been subsequently identified, and plate movements deduced from spreading rates agree well with the idea of fixed hot spots (Morgan, 1971). This concept is extended by Morgan, who proposes that the hot spots are a near-surface manifestation of convection in the lower mantle which provide the motive force for plate movements.* If the convection is plume-like rather than cellular as proposed in most earlier models, then the interaction of plates becomes an important determinant of the resultant forces which govern plate movements. Evidence supporting this concept is (1) that hot spots occur most frequently on oceanic rise crests and at each triple junction, and (2) world gravity maps reveal gravity highs over the hot spots, indicating deep-seated high-density material and topographic highs.

*Shaw (1973) reports evidence which indicates that the deep-seated thermal plume is not required in interpretations of oceanic chain volcanism, and Wesson (1972) proposes an earth model in which the lower mantle is nonconvective. He concludes that an impassable boundary due to phase changes occurs at 700 km depth, which separates the nonconvective lower mantle from an irregularly convective upper mantle.

Thus two strongly contrasted mechanisms of convection are considered to explain plate movements—(1) the rise of plumes from deep in the mantle to the base of the lithosphere, where they spread laterally, exerting viscous drag on the overlying plates with resultant opening along divergent junctions which are filled by upwelling of the mantle and or injection of dike complexes into the cracks, and (2) the oceanic rises and trench systems delineate the edges of great convective cells and indicate the sites of upwelling and downwelling in the asthenosphere.

VERTICAL TECTONIC MODELS

Much of the debate about lithospheric dynamics today concerns the relative importance of vertical movements compared with horizontal movement. Modern plate-tectonic theory involves some vertical movement—upwelling of mantle at the oceanic ridge crests, subduction of crust in trenches, and uplift in orogenic belts over thermal zones created by subduction. Nevertheless, the dominant process is lateral movement and lateral compression.

Those who emphasize the importance of vertical tectonic movements do not agree on the question of how much lateral movement (e.g., continental drift) has taken place. Some, exemplified by the renowned Russian geologist Beloussov (1952, 1962), question the validity of most of the evidence for continental drift and prefer to interpret most lithospheric dynamics as involving primarily vertical movement, with horizontal spreading under gravity as a subsidiary process. Others, like the prominent Dutch geologist Van Bemmelen (1973), accept continental drift and suggest that drift as well as orogenic processes are caused by vertical tectonics.

Mantle Diapirism

The mechanisms which have been suggested as a cause of vertical tectonics are of several types. Convective overturn driven by the thermal properties of the mantle and lithosphere provide one type of vertically directed movement. A second is gravitatively driven segregation of pseudoviscous mantle materials which become unstable as a result of physical and chemical processes which accompany bulk changes of density of mantle materials.

Maxwell (1968, 1973) suggests diapirism as a cause of tectonic activity. The plastic nature of the material in the low-velocity zone and the high temperatures should make this layer unstable. If the material in the soft layer becomes sufficiently unstable and starts to rise in a manner analogous to the rise of salt in the salt basins of Germany and the Gulf Coast, the material would melt as it rises in response to the drop in confining pressure. Thus the column of material would become less dense, accelerating the process. This material would be driven upward by the sinking of cool, more dense lithosphere outside the zone of upward movement.

Both the mid-oceanic ridges and the young folded mountain systems are locations of high thermal energy, and both are sites where mantle materials, ultrabasic rocks, have risen. Thus Maxwell postulates that the uplift, the thermal activity in the form of magmas produced, volcanic activity, metamorphism, and the associated structural features of both oceanic ridges and mountain systems may be caused by the same mantle processes. The differences arise primarily from the differences in the types of crusts (continental and oceanic) which lie above the zones of diapirism.

The arguments for mantle diapirism rest in part on the mechanics of emplacement of the ophiolite sequences in orogenic belts. Maxwell (1973) points out that most ophiolite sequences can be characterized as being one of two types—very extensive and thick sheets lying on continental margins (examples: Papua, Omau, New Caledonia, western Newfoundland, Cyprus, Vourinos). The other of broken fragments of sheets varying from fist-sized chunks to pieces of mountainous proportions enclosed in more or less chaotic flysch-type rocks (examples: Apennines, Alps, Coast Ranges of California and southwestern Oregon, southwestern Greece, Turkey). In all cases the ophiolites have been moved from their site of origin to a more external (continentward) position. Maxwell concludes that the initial stage is emplacement of the ophiolites on top of relatively undisturbed flysch followed by gravity sliding and separation into detached blocks. The initial movement is interpreted as being due to diapiric intrusion.

Some of these mantle diapirs are apparently still deeply rooted in the mantle as is suggested in the case of the Rhonda massif (Chap. 23) on the basis of large positive gravity anomalies over them. Evidence of their intrusion as hot masses is supported by thermal metamorphic aureoles reported around some of them. The initial flysch sedimentation may be caused by spreading and thinning of the crust which precedes the upward movement of the diapir.

Vertical Movements Due to Differentiation in the Mantle

Beloussov (1951, 1962) stresses the importance of differentiation at various levels in the earth as a cause of vertical instability. Differentiation occurs in response to unidentified changes which affect the physicochemical equilibrium of the system. This system lies in a gravitational field which has mechanical effects on unstable arrangements. For example, if a "bubble" of acidic material rises upward, it is heated by its radioactive elements, and it expands. The expansion exerts a mechanical pressure on the surrounding masses, causing compression and wave-like oscillatory movements of the crust—the uplifts and subsidence of geosynclines. Thus, uplifts are more frequently associated with acidic magmas and subsidence with basic magmas.

Beloussov views the sialic continents as being made up primarily of multiple granitic intrusions and the products of their erosion. Geosynclines are envisioned as evolving through a cycle that has an early stage marked by basaltic flows, followed by intermediate and acid lavas. Acidic magmas are injected with the formation of the new central uplift, and surrounding rocks are granitized. Later, magmas of various compositions rise, and finally, when the differentiation products are exhausted, a homogeneous primary basalt pours out.

Just as a change in equilibrium of the physicochemical system may cause differentiation of basalt to form acidic magma, a change in the opposite direction may lead to the reverse reaction and granitic rocks may undergo basification. Basification would lead to an increase in density and subsidence. This could provide a process by which continents are converted into ocean basins—a process referred to as *oceanization*.

The Undation Hypothesis of Van Bemmelen

Another global view which, like Beloussov's, embodies physiochemical processes deep in the earth but with very different consequences in the crust has been advanced by Van Bemmelen (1964, 1965, 1973). He postulates that crustal movements result from the development of various size undations of the earth, Figs. 17-15 and 17-16. Five classes of undations, based on their diameters, are suggested: local, 1 km; minor, 10 km; meso, 100 km, geo, 1,000 km; and mega, 10,000 km. When heavier constituents in the lower mantle are segregated toward the core, the density of the residual matter is reduced and rises. Currents in the inner mantle cause bulges in the boundary between the inner and outer mantle which are transmitted vertically by plastic flow to the surface to form megaundations.

The undations create potential energy which can be carried off by (1) volcanic activity, (2) spreading of matter by erosion and sedimentation, or (3) gravity tectonics. The

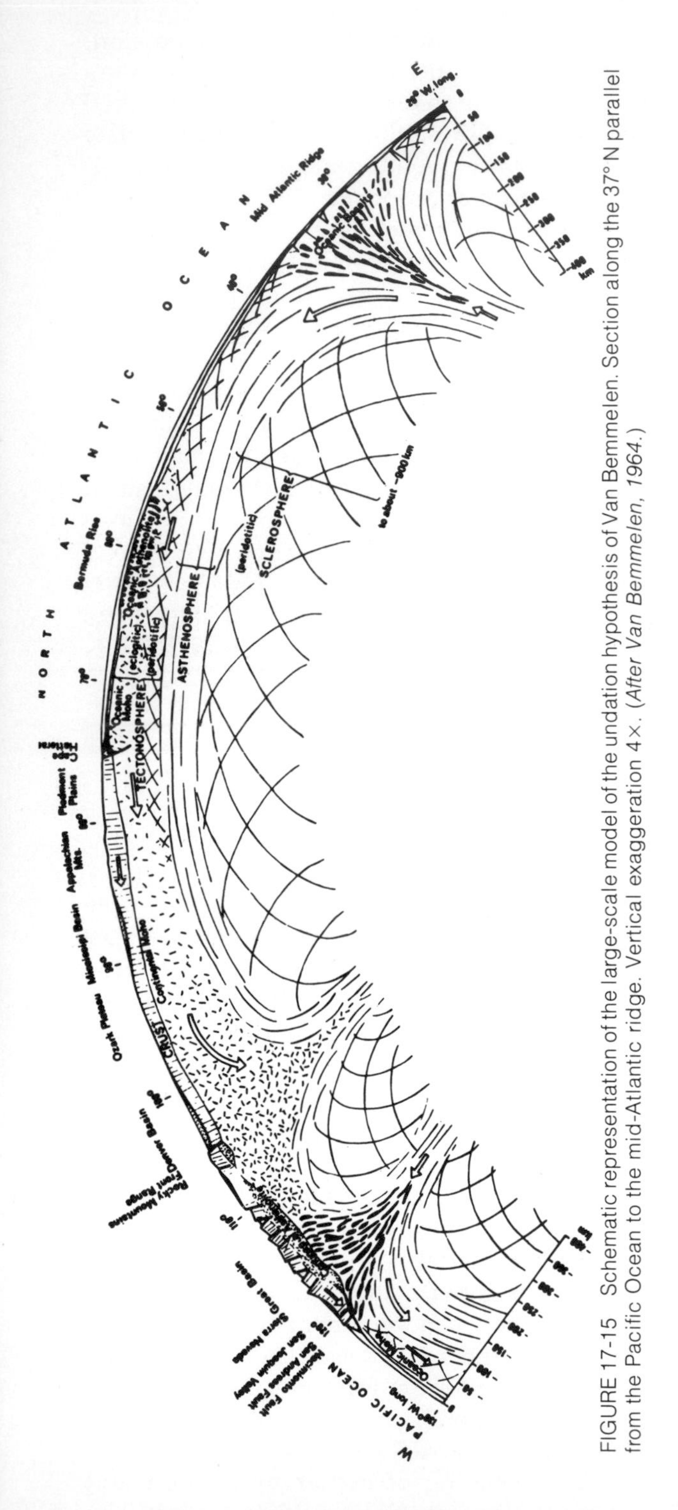

FIGURE 17-15 Schematic representation of the large-scale model of the undation hypothesis of Van Bemmelen. Section along the 37° N parallel from the Pacific Ocean to the mid-Atlantic ridge. Vertical exaggeration 4×. (*After Van Bemmelen, 1964.*)

maximum height attained by an undation depends on the character of the rock and the rate of uplift. Maximum height of about 9 km is reached in mesoundations. In Van Bemmelen's model, volcanic spreading was most important in the early phases of earth history, followed by the effects of erosion and sedimentation accompanied by continental crustal development by means of zonal accretion around continental nuclei as geosynclinal belts formed, became the site of sediment accumulation, and were later transformed into metamorphic and igneous rocks which became the cratonic centers of Gondwana and Laurasia. Breakup and drift of these early continental plates, driven by gravity tectonics, followed.

The mechanism for continental drift is uplift, and downwarping of the geoid. The highest unit, the lithosphere, glides over the next deeper one, the *asthenosphere*. The theory explains the origin of geosynclines as a product of continental drift—depression of the forward margin of the moving plate. Later the geosynclinal area is pushed up as a result of the buoyancy of an orogenic asthenolithic root. Uplifted strata glide toward the basin from the emerging center; and the mobilized portions of the crystalline basement and mantle spread. Extension phenomena occur on the rear side; and a system of transcurrent, or strike-slip, faults occurs along the sides of the plate.

If a megaundation occurs under a continental area, the continent can be split as tensional structures develop over the undation. If parts of the continent then lie on either side of the crest of the undation, they may slip apart, allowing the opening of a new ocean basin, and voluminous extrusions of basalts produced as a result of the reduction of pressure on the underlying peridotite.

Oceanization

The transformation of portions of the continental lithosphere to oceanic crust—a phe-

FIGURE 17-16 Undation hypothesis models of two stages in formation of the European Alps. See Chap. 23 for discussion of the Alps. (*From Van Bemmelen, 1973.*)

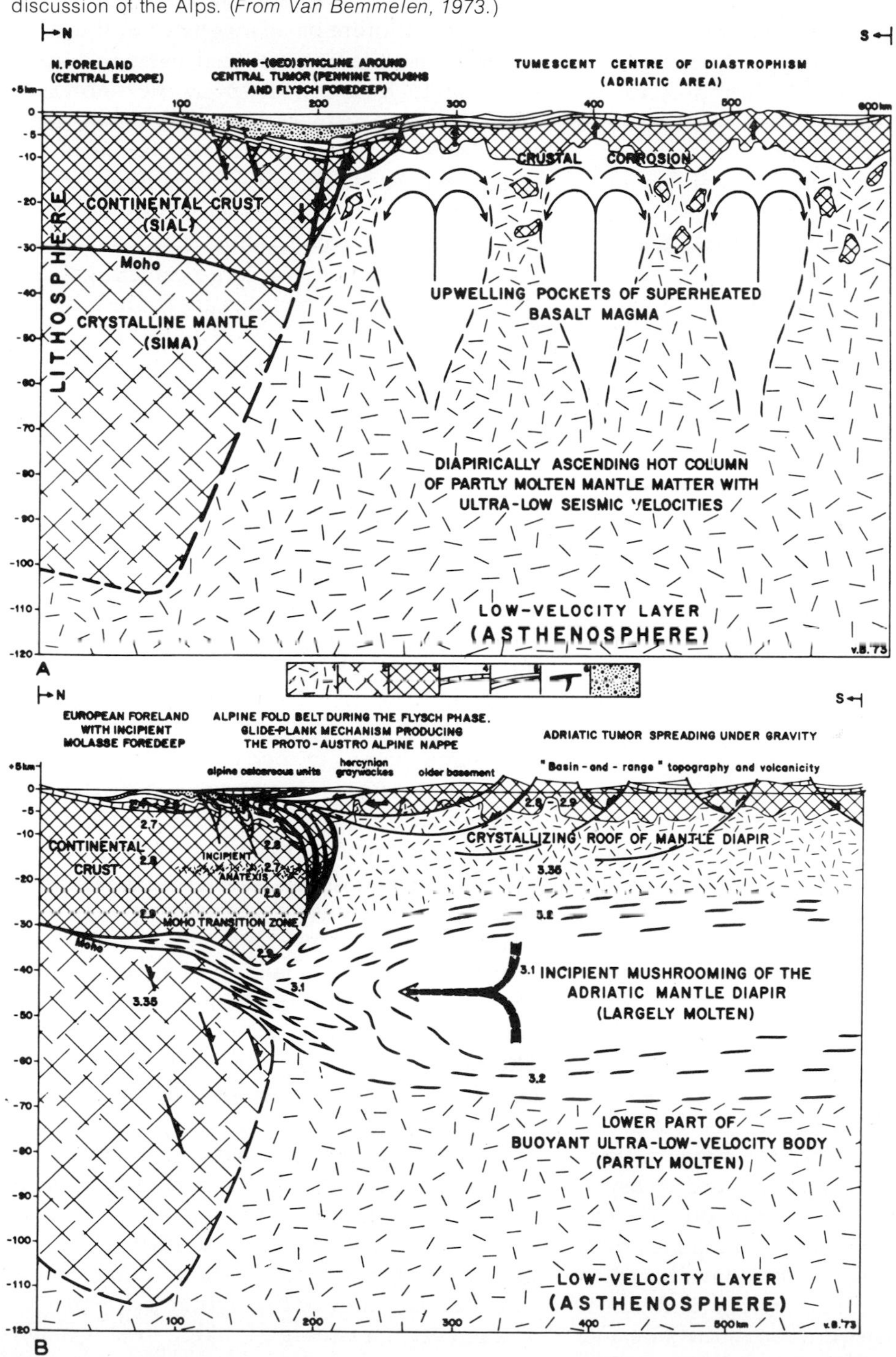

nomenon known as oceanization—is a common feature of vertical tectonic or mantle-diapiric theories. This idea has appealed to many as a means of explaining the appearance of oceanic crust in areas which appear to be probable source areas for rock masses now located in surrounding regions. This idea has been especially important in many interpretations in the Mediterranean region.

Van Bemmelen (1973) describes two types of oceanization. One is essentially a mechanical oceanization resulting from the floundering of a more dense lithospheric section into a less dense substrate. The second type is called *geochemical* oceanization. It is this type which he thinks is at work in the Mediterranean region.

During the process of Mediterranean oceanization, mantle diapirs corrode the overlying crust by means of overhead stoping. The detached blocks are composed of mixtures of crystallized ultramafic rocks, belonging to the cooling roof of the diapir, and katametamorphic rocks of the lower part of the continental crust, which have lost their more volatile constituents (H_2O, CO_2).

These blocks of mixed composition and high density "fall" through the ultralow-velocity diapir and the low-velocity layer, both having a relatively low viscosity and presumably a lower density than the crust. Thus parts of the sialic crust will be transported to the lower part of the upper mantle, possibly some hundreds of kilometers in depth, where they ultimately find a position of physicochemical equilibrium as high-temperature and high-density mineral phases.

Ramberg (1945, 1964, 1967) has carried out an important series of experiments to test his ideas on vertical tectonics. He suggests that subsidence of the ocean floor and geosynclines may be caused by extrusion of large quantities of basic lava in and on sialic crust. The region invaded by the basic lava becomes heavier than surrounding regions and sinks, and sialic material is squeezed to the side and appears as batholiths around the subsiding basin. Such subsiding areas are likely to become sites of more basic magma, but this sima is less dense than crystallized basic rocks and will rise to form the floor of the depressed area. Ramberg uses scale-model experiments with layers of pseudoviscous materials of different densities. Density inversions can be built into the model, and long-term effects of gravity are produced in short time intervals by placing the models in a large centrifuge. The resulting diapiric structures are represented in Fig. 17-17.

Orogenic Belts and Mantle Diapirism

Many theories of orogeny now rely on vertical uplift as the primary cause of orogeny, with the folding and thrusting which are so prominent in orogenic belts being the result of lateral spreading of the rising mass. Vertical displacements of large masses of matter as much as 10 to 15 km can often be proven in orogenic belts. The energy needed to elevate such large masses of rock by this amount is so great that most geologists consider it almost certain that such uplift is a result of changes in the average density of the crustal zones involved, followed by isostatic adjustment. Density of lithospheric sections may be altered in a number of ways. One method, called on in plate theory especially, is the compression of continental marginal (geosynclinal) deposits, causing thickening of the sialic crust. The collision of continents would have a similar effect. The subduction of water-laden sediments into the mantle would cause hydration of mantle materials with a similar effect. It has also been suggested that the granitization of the simatic basement under a eugeosyncline might cause the density alteration. The processes of magmatic differentiation may also be called on to produce the lighter materials.

Ramberg considers it likely that geosynclines and orogens are generated by lateral spreading along continental margins. In periods of intense magmatic activity, oceanic regions spread and expand at the expense of continents, but during periods of quiet the continents spread overriding the ocean margin.

Sections of some of the famous experiments of Ramberg, Fig. 17-17, illustrate the types of patterns formed as a result of gravitatively driven separation of psuedoviscous layers of different density. The strong resemblance of these models to sections across orogenic belts is striking.

Global Expansion Hypothesis

Extension appears necessary to explain the structure of the Basin and Range province in the western United States, the great rift system of East Africa, the Triassic basins of the Appalachians, and numerous other large graben systems. Block faulting is prominent in sonic profiles of the Japan trench, Fig. 21-10, and rift systems are now widely recognized along the crest of the globe-encircling oceanic ridges. Such rifting may be a result of local extension, but the number and size of extensional features in the crust have been interpreted as evidence for global expansion.

Hilgenberg (in the 1930s) pointed out that continental drift could be accomplished by means of expansion of the earth's interior. Continental blocks would be dispersed as the zones between the continental plates, the oceanic crust, grew. Recent work in the oceans suggests just such growth, with new oceanic crust being added along the oceanic ridges and older oceanic crust being spread farther apart. One marked advantage of this idea is that it is unnecessary to seek forces to move the thick continental plates over or through oceanic crust in order to explain continental drift. The continental crust remains essentially in place and rides passively on the mantle.

Several lines of evidence support global expansion.* Egyed (1956) sought to prove expansion by means of a paleogeographical argument in order to support his theory that the earth's core is composed of an unstable high-pressure-phase material that is undergoing a steady transformation to a lower density phase, resulting in a volume increase for the earth. Egyed selected two sets of paleogeographic maps (Termier and Strakhov) and measured the areas of the continental plates, including shelf areas that are shown covered by water, for each period. He plotted these against time, and it appears in both plots that the areas of continents covered by water have decreased steadily since the start of the Paleozoic. This has apparently happened in spite of an increase in the volume of water on earth through time. He interprets the curve as meaning that the oceans have grown in size relative to the continents as the earth has expanded through time. More recent analysis of

* See Holmes (1965) for an extended discussion of the hypothesis. See a review article by Carey (1975).

FIGURE 17-17 Model designed to illustrate features of mantle diapirism using silicone, wax, and putty. Layers were originally laid down horizontally. The layer shown as white was originally thicker in the center of the section. (*After Ramberg, 1973.*)

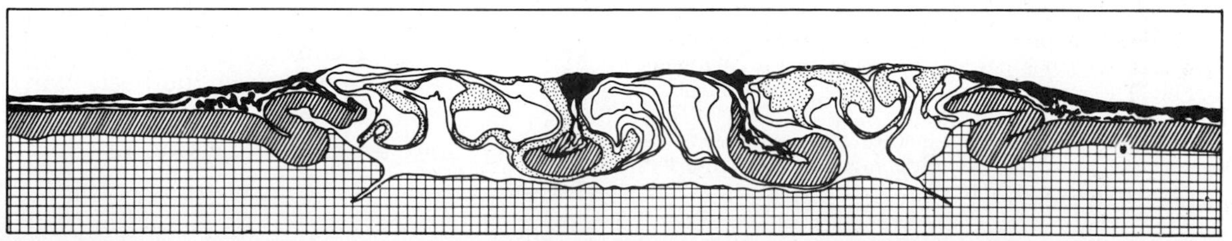

this data by Wise (1974) cast doubt on the validity of these results.*

Carey (1958) reached the conclusion that the earth is expanding through a totally different approach. He attempted to synthesize all crustal structure and restore the continents to a predrift pattern by removing all subsequent deformation. After trying various assemblies he concluded that a reasonable fit could be obtained only on an earth of smaller radius (¾ of the present diameter for late Paleozoic time).

A second argument for expansion grows out of Carey's reconstructions. The continents appear to be dispersing. North America is moving apart from Europe and Asia, Australia and Antarctica away from Africa and India. All the continental blocks surrounding the Pacific seem to be moving toward it; yet the Pacific is still by far the largest ocean basin of all.

The expansion hypothesis has not been widely accepted, although proponents continue to argue their case (see Jordan, 1971, and Carey, 1975). Paleomagnetic arguments have been used both pro and con, but like the tectonic reconstructions, the margin of possible error remains too large for a conclusive answer.

More conclusive answers should be forthcoming as a result of experiments now in progress. One of these involves precise observation using laser beams of the distance from astronomical observatories on different continents to a reflector placed on the moon. According to plate tectonics some observatories are approaching one another at rates of several centimeters per year; the expansion theory predicts they are moving apart. Repetition of measurements over a period of a few years should resolve the question.

In some respects the theory of global expansion employs explanations that are similar to those of plate theory. Both subdivide lithosphere into slab-like plates, both accept essentially the same interpretations of spreading centers—the divergent plate junctions, but they differ dramatically in other respects, especially with regard to the deep-sea trenches. These are zones of crustal shortening, convergence, and subduction in plate theory. They are zones of divergence and extension in the expansion hypothesis. Tanner (1973), after reviewing the characteristics of trenches, concludes that they are caused by regional extension, that they exhibit few compressional features, that there is no downgoing slab, and that most of the motions in trench areas are extensional or strike-slip. The presence of undisturbed sediments in the trench floors appears to deny effective "scraping off" or local compression, and the age of some sediments in trenches seems inconsistent with the proposed rates of subduction. The high heat flux in the island arcs is consistent with processes bringing isotherms upward, such as a geotumor. Proponents of expansion question the likelihood of this over a downgoing slab (Sychev, 1973). Orogeny on an expanding earth is readily explained in terms of the types of mantle diapiric hypotheses described in the preceding section.

The pendulum of scientific opinion has certainly not swung in the direction of acceptance of global expansion, but the arguments advanced are strong enough to warrant a more serious consideration of the hypothesis than it has thus far received and caution us against a too hasty and uncritical rejection of the hypothesis.

* Wise (1972, 1974) has reexamined the maps used by Egyed and compared them with a more detailed set of paleogeographic maps prepared for North America by Schuchert (1955). The plots of the percentage of the North America continent that is flooded against time do not agree with similar plots of Egyed, and Wise believes this is due to biases inherent in the paleogeographic maps used. He concludes that for over 80 percent of post-Precambrian time, sea level has been within 60 m of a normal value that is 20 m above modern sea level. If this conclusion is true, one of the original arguments for global expansion is invalid, and it is implied that the quasi-balanced state exists in which continents are rebuilt about as fast as they are eroded and degraded.

REFERENCES

Alexander, S. S., and **Sammis, C. G.**, 1975, New geophysical evidence on the driving mechanisms for continental drift: Earth Miner. Sci. v. 44, no. 4.

Atwater, Tanya, 1970, Implications of plate tectonics for the Cenozoic tectonic evolution of western North America: Geol. Soc. America Bull., v. 81, p. 3513–3536.

Aubouin, Jean, 1965, Geosynclines: Developments in geotectonics I: Amsterdam, Elsevier.

Beloussov, V. V., 1951, The problems of inner structure development of the earth: Izvest. Akad. Nauk SSSR Ser. Geograf. Geofiz., no. 7, p. 2, 4–16, 3–19.

——— 1962, Basic problems in geotectonics: New York, McGraw-Hill.

——— 1972, Basic trends in the evolution of continents, *in* Ritsema, A. R., ed., The upper mantle: Tectonophysics, v. 13, p. 95–117.

Benioff, Hugo, 1954, Orogenesis and deep crustal structure: Additional evidence from seismology: Geol. Soc. America Bull., v. 65, p. 385–400.

——— 1955, Seismic evidence for crustal structure and tectonic activity: Geol. Soc. America Spec. Paper 62, p. 61–74.

Bird, J. M., and **Isacks, Bryan,** eds., 1972, Plate Tectonics: Selected Papers from the Journal of Geophysical Research: Am. Geophys. Union, Washington, D.C.

Bird, Peter, Toksöz, M. N., and **Sleep, N. H.,** 1975, Thermal and mechanical models of continent-continent convergence zones: Jour. Geophys. Res., v. 80, no. 32, p. 4405–4416.

Blake, M. C., Jr., Irwin, W. P., and **Coleman, R. G.,** 1969, Blueschist facies metamorphism related to regional thrust faulting: Tectonophysics, v. 8, p. 237–246.

Bucher, W. H., 1933, The deformation of the earth's crust: Princeton, N.J., Princeton Univ.

Carey, S. W., 1954, The Rheid concept in geotectonics: Geol. Soc. Australia Jour., v. 1, p. 67–117.

——— 1958, The tectonic approach to continental drift, *in* Continental drift—A symposium: Geol. Dept. Univ. Tasmania, p. 177–355.

Carey, S. W., 1975, The expanding earth—An essay review: Earth-Sci. Rev., v. 11, no. 2, p. 105–143.

Chamberlin, T. C., 1897, The method of multiple working hypotheses: Jour. Geology, v. 5.

Chidester, A. H., and **Cady, W. M.,** 1972, Origin and emplacement of Alpine-type ultramafic rocks: Nature, Phys. Sci., v. 240, no. 98.

Clark, S. P., Jr., and **Ringwood, A. E.,** 1964, Density distribution and constitution of the mantle: Revs. Geophys., v. 2.

Coleman, R. G., 1971a, Petrologic and geophysical nature of serpentinite: Geol. Soc. America Bull. v. 82.

——— 1971b, Plate tectonic emplacement of upper mantle peridotites along continental edges: J. Geophys. Res., v. 76, no. 5.

Coney, P. J., 1970, The geotectonic cycle and the new global tectonics: Geol. Soc. America Bull., v. 81, p. 739–748.

Cox, Allan, 1969, Geomagnetic reversals: Science, v. 163, p. 237–245.

——— 1973, Plate tectonics and geomagnetic reversals: San Francisco, Freeman.

———, **Doell, R. R.,** and **Dalrymple, G. B.,** 1964, Reversals of the earth's magnetic field: Science, v. 144, p. 1537–1543.

Daly, R. A., 1940, Strength and structure of the earth: Englewood Cliffs, N.J., Prentice-Hall.

Dana, J. D., 1873, On some results of the earth's contraction from cooling: Am. Jour. Sci., Ser. 4, vol. v., p. 423.

Dearnley, R., 1966, Orogenic fold-belts and a hypothesis of earth evolution, *in* Ahrens, L. H., and others, eds., Physics and chemistry of the earth (7): New York, Pergamon.

De Beaumont, Elie, 1829–1830, Recherches sur quelques unes des revolutions de la surface du globe: Annales Sci. Nat., v. xviii, xix.

——— 1832, Observations et memoires geologiques publies par M. Elie de Beaumont, Professeur Adjoint de Geologie a l'École des Mines, In–4°, 4 p. autogr., s.l.n.d.

DeJong, Kees A., and **Scholten, Robert,** 1973, Gravity and tectonics: New York, Wiley.

Dewey, John F., 1975, Finite plate implications: Some implications for the evolution of rock masses at plate margins: Am. Jour. Sci., v. 275-A, p. 260–284.

——— and **Bird, J. M.,** 1970, Mountain belts and the new global tectonics: Jour. Geophys. Res., v. 75, no. 14, p. 2625ff.

——— 1971, Origin and emplacement of the ophiolite suite: Appalachian ophiolites in Newfoundland: Jour. Geophys. Res., v. 76, no. 14, p. 441–470.

Dickinson, W. R., 1972, Evidence for plate-tectonic regimes in the rock record: Am. Jour. Sci., v. 272, no. 7, p. 551–576.

——— 1974, Plate tectonics and sedimentation: Soc. Econ. Paleont. Mineralog. Special Pub. No. 22.

Dietz, R. S., 1961, Continental and ocean basin evolution by spreading: Nature, v. 189.

——— 1963, Collapsing continental rises: An actualistic concept of geosynclines and mountain building: Jour. Geology, v. 71.

——— and **Holden, J. C.,** 1966, Miogeoclines (miogeosynclines) in space and time: Jour. Geology, v. 75, no. 5, pt. 1, p. 566–583.

——— 1970, Reconstruction of Pangaea: Breakup and dispersion of continents, Permian to Present: Jour. Geophys. Res., v. 75, no. 26, p. 4939ff.

Egyed, L. 1956, Determination of changes in the dimensions of the earth from paleogeographic data: Nature, v. 178, p. 534.

——— 1957, A new dynamic conception of the internal constitution of the earth: Geol. Rundsch. v. 46, no. 1, p. 101–121.

Elsasser, W. M., 1967, Convection and stress propagation in the upper mantle: Princeton Univ. Tech. Rep. 5.

——— 1969, The mechanics of continental drift, *in* Gondwanaland revisited; new evidence for continental drift: Amer. Phil. Soc., Proc., v. 112, no. 5, p. 344–353.

Engel, A. E. J., and **Kelm, D. L.,** 1972, Pre-Permian global tectonics: A tectonic test: Geol. Soc. America Bull., v. 83, no. 8, p. 2325ff.

Ernst, W. G., 1975, Systematics of large-scale tectonics and age progressions in Alpine and circum-Pacific blueschist belts: Tectonophysics, v. 26, p. 229–246.

Fisher, R. L., and **Hess, H.,** 1963, Deep-sea trenches, *in* The sea: New York, Wiley.

Fyfe, W. S., and **McBirney, A. R.,** 1975, Subduction and the structure of andesitic volcano belts: Am. Jour. Sci., v. 275–A, p. 285–297.

Garfunkel, Zvi, 1975, Growth, shrinking, and long term evolution of plates and their implications for the flow pattern in the mantle: Jour. Geophys. Res., v. 80, no. 32.

Gilluly, James, 1949, Distribution of mountain building in geologic time: Geol. Soc. America Bull., v. 60, no. 4, p. 561–590.

——— 1967, Chronology of tectonic movements in the western United States: Am. Jour. Sci., v. 265, p. 306–331.

Griggs, D. T., 1972, The sinking lithosphere and the focal mechanism of deep earthquakes, *in* Robertson, E., and others, eds., The nature of the solid earth: New York, McGraw-Hill, p. 361–384.

Hall, James, 1859, Geological survey of New York: Paleontology, v. iii, introd.

——— 1859, Description and figures of the organic remains of the Lower Helderberg group and the Oriskany sandstone: N.Y. Geol. Survey, Paleont., v. 3.

Hatherton, Trevor, and **Dickinson, W. R.,** 1969, The relationship between andesitic volcanism and seismicity in Indonesia, the Lesser Antilles, and other islands arcs: Jour. Geophys. Res., p. 5301–5310.

Heirtzler, J. R., Le Pichon, X., and **Baron, J. G.,** 1965, Magnetic anomalies over the Reykjanes ridge: Deep Sea Res., v. 13, p. 427.

——— and **Hayes, D. E.,** 1967, Magnetic boundaries in the North Atlantic Ocean: Science, v. 157, no. 3785, p. 185–187.

———, **Dickson, G. O., Herron, E. M., Pitman, W. C. III,** and **Le Pichon, X.,** 1968, Marine magnetic anomalies, geomagnetic field reversals, and motions of the ocean floor and continents: Jour. Geophys. Res., v. 73, no. 6.

Heiskanen, W. A., and **Vening Meinesz, F. A.,** 1958, The earth and its gravity field: New York, McGraw-Hill.

Hess, H. H., 1948, Report of the chairman of the special committee on geophysical and geological study of ocean basins, 1947–1948: Am. Geophys. Union Trans., v. 29, no. 6.

——— 1955, Serpentines, orogeny and epeirogeny: Geol. Soc. America Spec. Paper 62, p. 391–407.

——— 1960, History of ocean basins: Preprint.

——— 1962, History of the ocean basins, *in* Petrologic studies: New York, Geol. Soc. America, p. 599—620.

Hill, M. L., 1971, A test of new global tectonics: Comparisons of NE Pacific and California structures: Am. Assoc. Petroleum Geologists Bull., v. 55, no. 1, p. 3–9.

Holmes, A., 1965, Principles of physical geology: Ronald, New York.

Hsü, K. J., and **Ryan, W. B. F.**, 1972, Summary of the evidence for extensional and compressional tectonics in the Mediterranean, *in* Initial reports of the deep sea drilling project, v. 13, ch. 37, U.S. Gov't. Printing Office, p. 1011–1019.

Isacks, Bryan, Oliver, Jack, and **Sykes, L. R.,** 1968, Seismology and the new global tectonics: Jour. Geophys. Res., v. 73, p. 5855–5899.

——— and **Molnar, Peter,** 1969, Mantle earthquake mechanisms and the sinking of the lithosphere: Nature, v. 223, p. 1121–1124.

——— 1971, Distribution of stresses in the descending lithosphere from a global survey of focal-mechanism solution of mantle earthquakes: Rev. Geophys. Space Physics, v. 9, no. 1, p. 103ff.

Jeffreys, H., 1952, The earth: Cambridge, Mass., Cambridge Univ.

Jordon, Pascual, 1971, The expanding earth; some consequences of Dirac's gravitation hypothesis: Oxford, Pergamon.

Kaula, W. M., 1972, Global gravity and tectonics, *in* Robertson, E., and others, eds., The nature of the solid earth, New York, McGraw-Hill, p. 385–405.

Kay, G. M., 1951, North American geosynclines: Geol. Soc. America Mem. 48.

King, P. B., 1955, Orogeny and epeirogeny through time: Geol. Soc. America Spec. Paper 62, p. 723–739.

Larson, R. L., and **Pitman, W. C. III,** 1972, Worldwide correlation of Mesozoic magnetic anomalies, and its implications: Geol. Soc. America Bull., v. 83, no. 12, p. 3645–3662.

Le Pichon, X., 1968, Sea-floor spreading and continental drift: Jour. Geophys. Res., v. 73, no. 12, p. 3661–3697.

Lowry, W. D., 1974, North America geosynclines—Test of continental-drift theory: Am. Assoc. Petroleum Geologists Bull., v. 58, no. 4, p. 575–620.

McKenzie, D. P., 1969, Speculations on the consequences and causes of plate motions: Geophys. Jour. Royal Astron. Soc., v. 18, p. 1–32.

——— and **Morgan, W. J.,** 1969, Evolution of triple junctions: Nature, v. 224, p. 125–133.

——— and **Parker, R. L.,** 1967, The North Pacific: An example of tectonics on a sphere: Nature, v. 216, p. 1276–1280.

Mantura, A. J., 1972, Geophysical illusions of continental drift: Amer. Assoc. Petroleum Geologists Bull., v. 56/8, p. 1552ff.

Mason, R. C., and **Raff, A. D.,** 1961, A magnetic survey off the west coast of North America, 32° N to 42° N: Geol. Soc. America Bull., v. 72, p. 1259–1265.

Matsuda, Tykihiko, and **Uyeda, Seiya,** 1971, On the Pacific-type orogeny and its model: Extension of the paired belts concept and possible origin of marginal seas: Tectonophysics, v. 11, p. 5–27.

Maxwell, J. C., 1968, Continental drift and a dynamic earth: Am. Scientist, v. 56, no. 1.

——— 1973, Symposium on "Ophiolites in the earth's crust": Preprint, Moscow, May 31–June 2, 1973.

Menard, H. W., 1958, Development of median elevations in ocean basins: Geol. Soc. America Bull., v. 69, p. 1179–1186.

——— 1967, Sea-floor spreading, topography, and the second layer: Science, v. 157.

——— 1973a, Does Mesozoic mantle convection still persist? Earth Planet. Sci. Letters, v. 20, p. 237–241.

——— 1973b, Depth anomalies and the bobbing motion of drifting islands: Jour. Geophys. Res., v. 78, no. 23, p. 5128ff.

Mikhaylov, A. Y., 1971, The development of geosynclines and folding: Int. Geol. Rev., v. 12, no. 12, p. 1490–1495.

Minear, J. W., and **Toksöz, M. N.,** 1970, Thermal regime of a downgoing slab and new global tectonics: Jour. Geophys. Res., v. 75, no. 8, p. 1397–1419.

Mitchell, A. H., and **Reading, H. G.,** 1969, Continental margins, geosynclines, and ocean flood spreading: Jour. Geology, v. 77, no. 6, p. 629–646.

Miyashiro, Akiho, 1972, Metamorphism and related magnetism in plate tectonics: Am. Jour. Sci., v. 272, p. 629–656.

Morgan, W. J., 1968, Rises, trenches, great faults and crustal blocks: Jour. Geophys. Res., v. 73, no. 6, p. 1959ff.

——— 1971, Convection plumes in the lower mantle: Nature, v. 230, p. 42.

Orowan, E., 1964, Continental drift and the origin of mountains: Science, v. 146, p. 1003–1010.

Pekeris, C. L., 1935, Thermal convection in the interior of the earth: Royal Astron. Soc. Monthly Notices Geophys. Supp., v. 3.

Phillips, J. D., and **Forsyth, D.,** 1972, Plate tectonics, paleomagnetism, and the opening of the

Atlantic: Geol. Soc. America Bull., v. 83, no. 6, p. 1579ff.

Pitman, W. C. III, and **Talwani, Manik,** 1972, Sea-floor spreading in the North Atlantic: Geol. Soc. America Bull., v. 83, no. 3, p. 619–646.

Poole, W. H., 1967, Tectonic evolution of Appalachian region of Canada, *in* Geology of the Atlantic region: Geol. Assoc. Canada Spec. Paper no. 4, p. 9–51.

Press, Frank, 1969, The suboceanic mantle: Science, v. 165, p. 174–176.

Ramberg, Hans, 1945, The theormodynamics of the earth's crust II: Norsk Geol. Tidsskr., v. 25.

——— 1964, A model for the evolution of continents, oceans, and orogens; Tectonophysics, v. 2.

——— 1967, Gravity, deformation and the earth's crust: New York, Academic Press, p. 1–214.

——— 1973, Model studies of gravity-controlled tectonics by the centrifuge technique: Gravity and tectonics: New York, Wiley, p. 49–66.

——— and **Sjöström, Håkan,** 1973, Experimental geodynamical models relating to continental drift and orogenesis: Tectonophysics, v. 19, p. 105–132.

Rodgers, John, 1967, Chronology of tectonic movements in the Appalachian region of eastern North America: Am. Jour. Sci., v. 265, p. 408–427.

Roeder, D. H., 1973, Subduction and orogeny: Jour. Geophys. Res., v. 78, no. 23, p. 5005ff.

Roper, Paul J., 1974, Plate tectonics: A plastic as opposed to a rigid body model: Geology, p. 247–250.

Schuchert, Charles, 1955, Atlas of paleogeographic maps of North America: New York, Wiley.

Sclater, J. G., and **Francheteau, Jean,** 1970, The implications of terrestrial heat flow observations on current tectonic and geochemical models of the crust and upper mantle of the earth: Geophys. Jour. Royal Astron. Soc., v. 20, p. 509–542.

Shaw, H. R., 1973, Mantle convection and volcanic periodicity in the Pacific; evidence from Hawaii: Geol. Soc. America Bull., v. 84, p. 1505–1526.

Sloss, L. L., 1966, Orogeny and epeirogeny: The views from the craton: New York Acad. Sci. Trans., set 2, v. 28, p. 579–587.

Stauder, William, 1968, Tensional character of earthquake foci beneath the Aleutian trench with relation to sea-floor spreading: Jour. Geophys. Res., v. 73.

Stille, Hans, 1936, The present tectonic state of the earth: Am. Assoc. Petroleum Geologists Bull., v. 20, p. 849–880.

——— 1941, Einführung in den Bau Amerikas: Borntraegar, Berlin.

Suess, Edward, 1904, The face of the earth: Oxford, Clarendon.

Sychev, P. M., 1973, Upper-mantle structure and nature of deep processes in island arcs and trench systems: Tectonophysics, v. 19, no. 4.

Sykes, L. R., 1967, Mechanism of earthquakes and nature of faulting on the mid-ocean ridges: Jour. Geophys. Res., v. 72, p. 2131–2153.

Talbot, C. J., 1974, Fold nappes as asymmetric mantled gneiss domes and ensialic orogeny: Tectonophysics, v. 24, p. 259–276.

Tanner, W. F., 1973, Deep-sea trenches and the compression assumption: Am. Assoc. Pet. Geol., Bull., v. 57, no. 11.

Tarling, D. H., and **Runcorn, S. K.,** 1973, Implications of continental drift to the earth sciences, v. 2: New York, Academic.

Toksöz, M. N., Minear, J. W., and **Julian, B. R.,** 1971, Temperature field and geophysical effects of a downgoing slab: Jour. Geophys. Res., v. 76, no. 5, p. 391–416.

Turcotte, D. L., and **Oxburgh, E. R.,** 1968, A fluid theory for the deep structure of dip-slip fault zone: Phys. Earth Planet. Interiors, v. 1.

——— and ———, 1969, Convection in a mantle with variable physical properties: J. Geophys. Res., v. 74.

———and **Schubert, G.,** 1973, Frictional heating of the descending lithosphere: Jour. Geophys. Res., v. 78, no. 26, p. 5876ff.

Umbgrove, J. H. F., 1947, The pulse of the earth: The Hague, Nijhoff.

Van Bemmelen, R. W., 1964, The evolution of the Atlantic megaundation: Tectonophysics, v. 1, no. 5, p. 385–430.

——— 1965, The evolution of the Indian Ocean mega-undation: Tectonophysics, v. 2, no. 1, p. 29–57.

——— 1973, Geodynamic models for the Alpine type of orogeny (Test case II: The Alps in central Europe): Tectonophysics, v. 18, no. 1–2, p. 33–79.

——— 1974, Driving forces of orogeny, with emphasis on blue-schist facies of metamorphism (Test case III: The Japan arc): Tectonophysics, v. 22, p. 83–125.

Van Hilten, D., 1964, Evaluation of some geotectonic hypotheses by paleomagnetism: Tectonophysics, v. 1, no. 1, p. 3–71.

Vening Meinesz, F. A., 1948, Gravity expeditions at sea: Pub. Netherlands Geodetic Comm., v. 4, p. 1–233.

——— 1964, The earth's crust and mantle: New York, Elsevier.

Vilas, J. F., and **Valencio, D. A.,** 1970, Paleogeographic reconstructions of the Gondwanic continents based on paleomagnetic and sea floor spreading data: Earth Planet. Sci. Letters, v. 7, no. 5, p. 397–405.

Vine, F. J., 1969, Sea-floor spreading—New evidence: Jour. Geol. Educ., v. xvii, no. 1, p. 6–16.

——— and **Matthews, D. H.,** 1963, Magnetic anomalies over oceanic ridges: Nature, v. 199, p. 947–949.

Wegener, Alfred, 1924, The origin of continents and oceans: London, Methuen.

Wellman, H. W., 1972, Recent crustal movements: Techniques and achievements, *in* Ritsema, A. R., ed., The upper mantle: Tectonophysics, v. 13, p. 373–392.

Wesson, P. S., 1972, Mantle creep: Elasticoviscous vs. modified Lomnitz law, and problems of "The new global tectonics": Am. Assoc. Petroleum Geologists Bull., v. 56, no. 11, p. 2127ff.

Wilson, J. T., 1965, A new class of faults and their bearing on continental drift: Nature, v. 207, p. 343–347.

——— 1973, Mantle plumes and plate motions, *in* Irving, E. ed., Mechanism of plate tectonics: Tectonophysics, v. 19, p. 149–164.

Wise, D. U., 1972, Freeboard of continents through time, *in* Studies in earth and space sciences, Geol. Soc. America, Mem. no. 132.

——— 1974, Continental margins, freeboard and the volumes of continents and oceans through time: Geol. Continental Margins, p. 45–58.

Woollard, G. P., 1962, Gravity anomalies and the crust of the earth in the Pacific Basin, *in* The crust of the pacific basin: Am. Geophys. Union Mon. 6, p. 60–80.

Wyllie, P. J., 1965, A modification of the geosyncline and tectogene hypothesis: Geol. Mag., v. 102, no. 3, p. 231–245.

——— 1971, The dynamic earth: Textbook in geosciences: New York, Wiley.

Zen, E-An, 1967, Time and space relationships of the Taconic allochthon and autochthon: Geol. Soc. America Spec. Paper 97, p. 107.

18

Continental Cratons

Each continent contains one or more large areas, called *cratons,* which have exhibited long-term crustal stability. The term was also applied to parts of the deep-sea floor until the mobility and young age of the present oceanic crust became widely accepted. Some restrict the term craton to the oldest portions of the Precambrian shields, but it is widely applied as in the following discussion to the shields and those parts of the continents which are covered by relatively undeformed sedimentary veneers. Thus the cratons can be thought of as those continental areas which have not been involved in orogenesis since at least the start of the Paleozoic, and in most cratons the youngest thermal event is dated at about 1 billion years. The best exposures of the underlying cratonic rocks are in the shields, and of these the Canadian Shield, the Baltic Shield, and the shield areas of southern Africa are best known.

Deciphering the history and structure of the shields has been slow because the rocks are largely igneous or metamorphic, they contain few fossils, and the conventional techniques of stratigraphic correlation have not worked well. Conventional mapping is very slow, and large areas of the shields have been subjected to multiple deformations, making the determination of structural evolution especially difficult. Gradually, detailed mapping is being achieved and important large-scale features have emerged as a result of studies of radiometric age determinations and the application of aeromagnetic geophysical studies. As the shortage of metals grows more intense, investigation of the shields will be spurred by economic incentives because most of the major ore deposits are found in the cratons.

Several important syntheses of cratons have been reported in recent years by Hurley and Rand (1969), Anhaeusser and others (1969), and Engel and others (1974). Anhaeusser compares the South African shields* with others. He finds a "clear and well-defined pattern of events" in the early evolution of the South African craton and believes a similar pattern can be found in all shields. That pattern is described in the following paragraph.

The typical shield consists of one or more cratonic nuclei composed of ancient (2.4+ billion years old) granitic rocks and belts of metabasalts (greenstone belts) surrounded by younger linear belts of high-grade and often highly tectonized metamorphic rocks. The evolutionary model starts with formation of cratonic nuclei, Fig. 18-1. The cratonic nuclei are thought to have been derived from the mantle by magmatic differentiation and sedimentary reworking of the primitive crust. Younger granitic material has caused thickening of the earlier primitive crust. That crust has been intruded, migmitized, and gran-

* The following definitions are used by Anhaeusser and others (1969):

Shield: A continental or subcontinental area of exposed crystalline rocks of Precambrian age.

Craton: A stable nuclei within a shield area which consists of complex granitic terranes incorporating early Precambrian greenstone belts. It is unaffected by tectothermal events younger than 2.4 billion years.

Greenstone belts: Distinctive metavolcanic and sedimentary assemblages which occur as scattered remnants on the cratons. Greenstones (metabasalts and andesites) generally predominate in these belts. Low-grade metamorphism and tight folding are also major features.

Mobile belts: Younger, linear, metamorphic belts which tend to surround the cratonic nuclei. The mobile belts are characterized by high-grade metamorphism, granitization, and faults. (*Mobile belt* is used in preference to *orogenic belt* to avoid a connotation that these belts resembled Alpine- or Appalachian-type orogenic belts.)

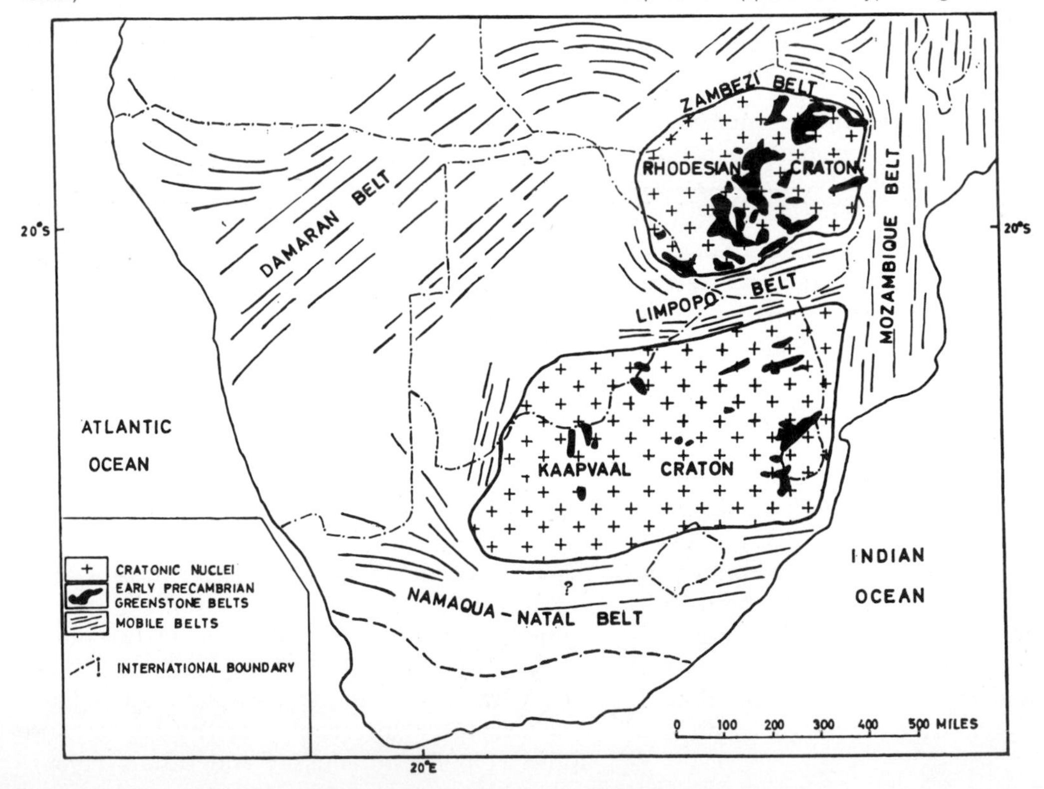

FIGURE 18-1 The southern African crystalline shield stripped of its younger cover and showing the ancient greenstone-granite cratons and the encircling, younger, mobile metamorphic belts. (*From Anhaeusser and others, 1969.*)

itized by later events which have largely obliterated original detail, Fig. 18-2. Early Precambrian greenstone belts evolved on this thin granitic crust. The thin crust was extended, forming trough-like features which became sites of sedimentation from above and intrusion from below, Fig. 18-3. The initial phase of volcanic activity is characterized by ultramafic rocks in South Africa. These lavas are interbedded with chemically deposited sediments. Later basalts, andesites, dacites, and rhyolites are emplaced in detrital and chemical sediments. A sedimentary sequence is then deposited unconformably over the volcanic-rich sequences as volcanic activity subsides. The rocks of the greenstone belt are deformed and affected by a low-grade metamorphism, and a steep linear schistosity and crenulation cleavage are developed. This deformation and thermal event is thought to be caused by upwelling of granites and sagging of the greenstone belt. Movements at depth are predominantly vertical, but at higher levels lateral compression is caused by spreading of the plutons.

Later, linear mobile belts develop across and along the margins of the cratonic nuclei. Their cause is still uncertain. Many have taken the view that these evolved in a fashion similar to younger folded mountain belts such as the Appalachians or the Alps. Anhaeusser prefers an interpretation in which the sites of accumulation of new sediment are initiated along fracture zones and zones of large-scale transcurrent motion. Once the location of the new mobile belts becomes established, the belts become sites of accumulation of younger Precambrian volcanics and sediments derived from the cratonic nuclei and greenstone belts. When these mobile belts are tectonized, the pattern of deformation and metamorphism is distinctly different from the earlier tectonothermal event. Potassium metasomatism leading to granitization is widespread. High-grade metamorphism occurs and the rocks are highly tectonized, with development of complex fold patterns. This appears to be related to nonuniform laminar flow while the rocks are ductile.

Coward and others (1976) provide a detailed description of the evolution of one of the greenstone belts found in the southern part of the Rhodesian craton in southern Africa.

These greenstone belts, Fig. 18-4, contain 2.6-billion-year-old volcanic and sedimentary rocks deposited on an Archean gneissic basement 3.5 billion years old. The greenstone belt exhibits early isoclinal and recumbent folds that were deformed before being intruded by large diapiric masses of granite. Later deformation and formation of two superimposed fold systems is apparently due to intracratonic block tectonics.

FIGURE 18-2 Schematic cross section through a portion of a granite-greenstone craton. (*From Anhaeusser and others, 1969.*)

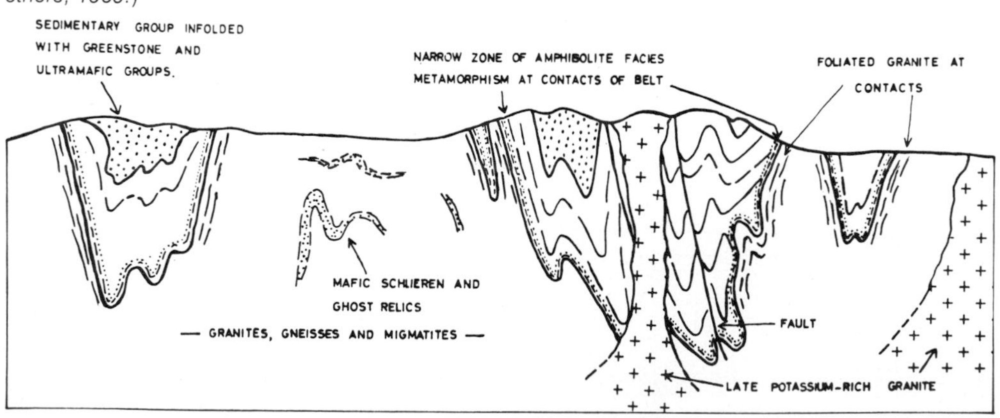

A Petrochemical Approach to Precambrian Evolution of the Cratons

Engel and others (1974) have synthesized information on chemical compositions of rocks in the shield areas and compared them with rocks from other major crustal elements. The ratio K_2O/Na_2O is used as a petrochemical index which indicates the degree of differentiation. The mantle and oceanic crust are little differentiated chemically. Progressively greater differentiation is seen in the bulk composition of island arcs, continental borderlands, and shields, Fig. 18-5. These crustal elements are progressively enriched in group 1 elements (e.g., K, U, Th, Pb, Ba, etc.). This is seen in the change in the K_2O/Na_2O ratio in the differentiation sequence basalt—granite or graywacke—standstone, and shale. Other ratios could be used, but K_2O/Na_2O is established as a standard for comparison and is

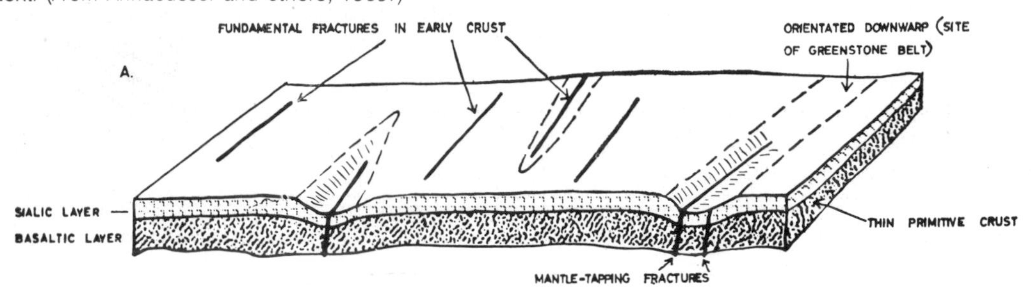

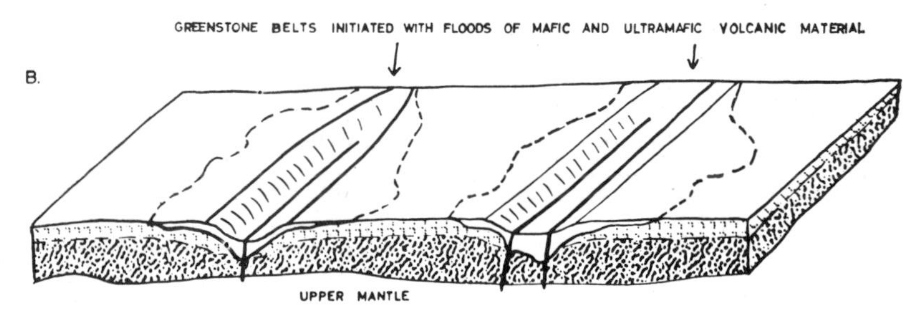

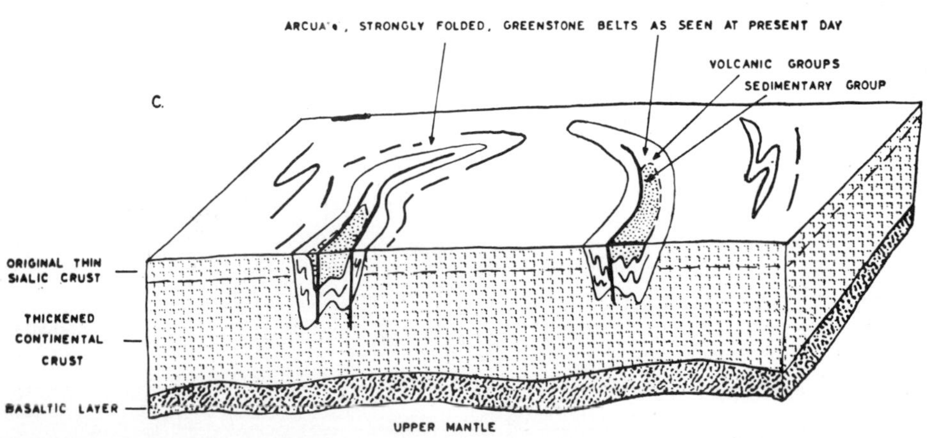

FIGURE 18-3 Diagrammatic illustrations showing the suggested evolution of greenstone belts described in the text. (*From Anhaeusser and others, 1969.*)

widely used (see Wyllie, 1971). A plot of potassium/sodium ratios against ages reveals some long-term changes of the differentiation index which can be correlated with major events in crustal evolution, Fig. 18-6. The differentiation index remained less than 1 until about 2.5 billion years ago, when it rose to values greater than 1 and even as high as 2 for some sediments. It remained high until about the middle of the Paleozoic, when it dropped. Engel argues that the Archean low differentiation ratio was related to the early formation of continental crust. The second stage was accre-

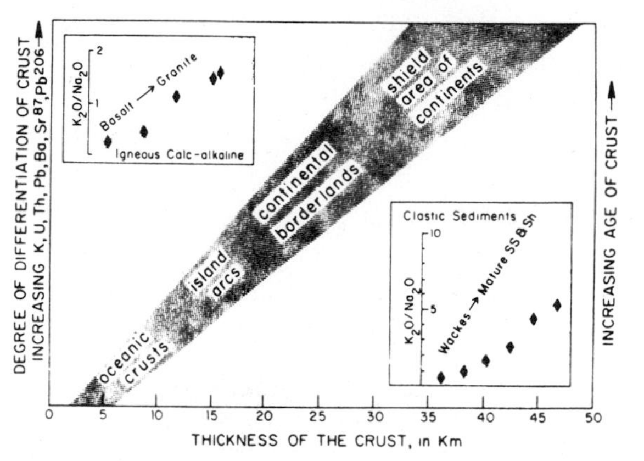

FIGURE 18-5 Interrelations in age, thickness, and composition of the earth's crust, and the complementary variations in the petrogenic index K_2O/Na_2O for major crustal rocks. Diamond "points" in smaller graphs represent weighted "average" compositions of common rock types. (*From Engel and others, 1974.*)

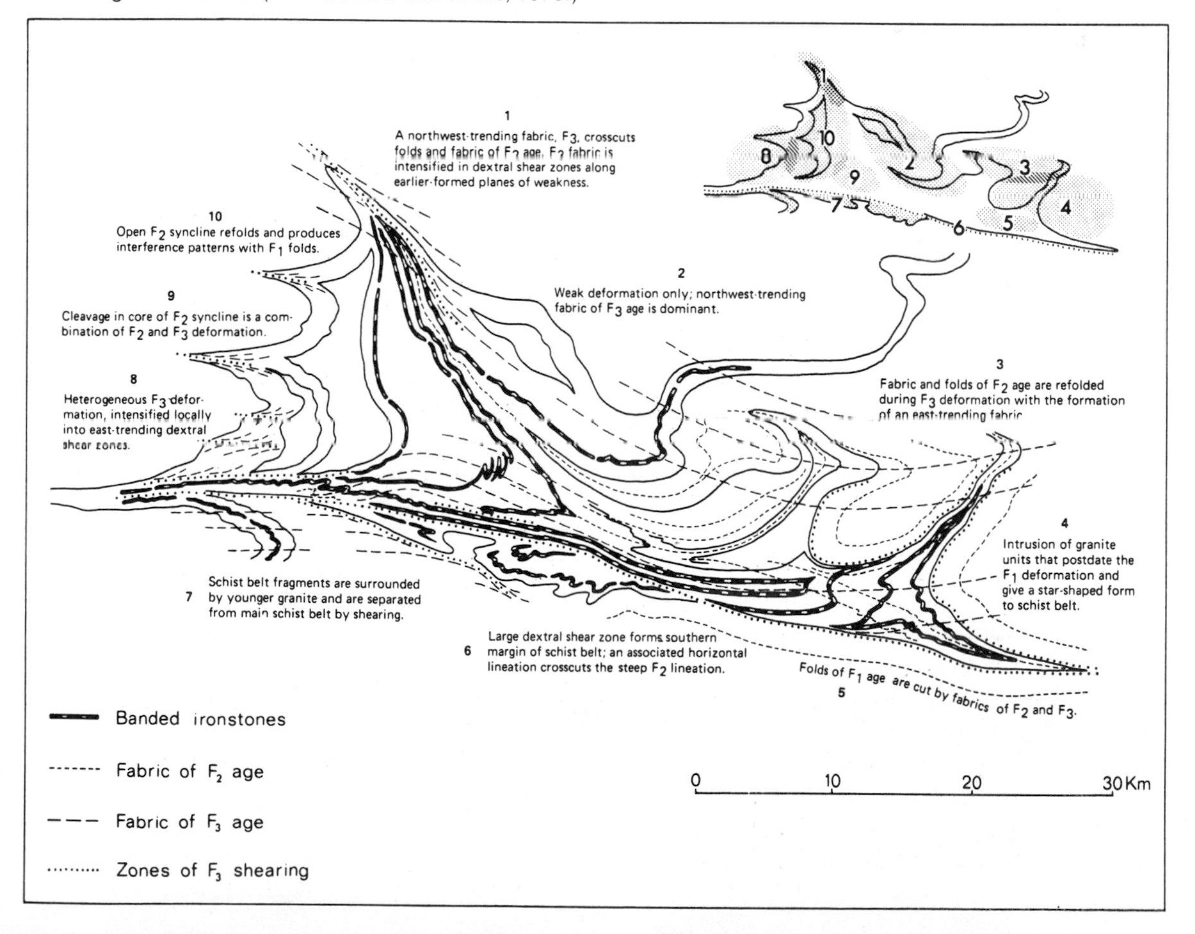

FIGURE 18-4 Map of a small part of the Rhodesian craton, the Gwanda greenstone belt, showing complex superimposed folds formed in bedded ironstones and schist. The surrounding rock is largely Precambrian intrusive granite bodies. (*From Coward and others, 1976.*)

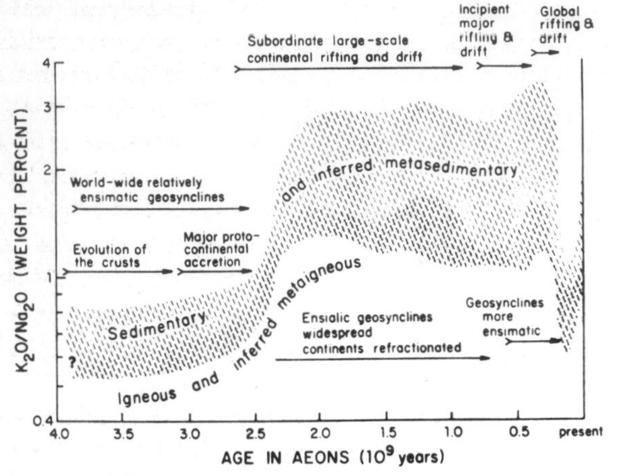

FIGURE 18-6 Variations in the petrogenic index K_2O/Na_2O of igneous, sedimentary, and metamorphic rock complexes plotted as a function of geologic time. Variations in crustal environments of the constituent rocks, as indicated by the index, and by related lithologic features, rock associations, and relative abundances, suggest a crude first-order sequence of crustal evolution and global tectonics. (*From Engel and others, 1974.*)

tion of this primitive crust into one or possible two large protocontinents. This accretion was accompanied by orogenic episodes and high thermal activity which culminated and began to subside about 2.5 billion years ago, leaving most Archean rocks with radiometric ages between 2.4 and 2.6 billion years.

In the following early Proterozoic (1.7 to 2.3 billion years) the protocontinents became deformed internally. Orogenic areas formed along the thinner portions of the granitic crust, and this led to greater differentiation of largely reworked older crustal materials formed at that time. Two major episodes of Proterozoic orogeny are apparent, one of which occurred at 1.7 billion years and the youngest at 1.0 billion years, known as the *Grenville event*—an episode in which distinctive granulite facies rocks and anorthositic-charnokitic complexes formed.

A long quiet period of nearly half a billion years followed the Grenville event. Few magmas formed in this time. This quiet preceded the great Paleozoic orogenies—Pan-African, Appalachian, Hercynian, and Caledonian—which affected most of the continents in the middle and late Paleozoic. After these, the older cratons have remained relatively passive as the protocontinents fragmented and drifted apart as large, thick, cool, fractionated fragments.

THE CANADIAN SHIELD

The Precambrian rocks of the Canadian Shield consist largely of granites, granitic gneisses, and schists, although other rocks are represented as well. The scarcity of fossils and lack of marker horizons of regional extent made synthesis of the structure a difficult task, as has the vast size, the extensive cover of glacial lakes and swamps, and the cold climate. That the Shield consists of a number of Precambrian orogenic belts superimposed and welded together has long been recognized, but the exact character of these belts, precise correlation of events within the belts, and the delineation of their boundaries has varied with time and author. Gastil (1960) has delineated the date provinces as illustrated in Fig. 18-7, and major orogenic belts are identified on the Tectonic Map of Canada, Figs. 18-8 and 18-9.

Four major periods of orogeny are identified in the Shield—Kenoran (2,480 million years), Hudsonian (1,735 million years), Elsonian (1,370 million years), and Grenville (955 million years). The establishment of boundaries of the belts is complicated because some of the orogenies affected older orogenic belts and consequently reset the radiometric "clocks" and obliterated the earlier structures. The long and straight trend lines on the map represent traces of long tight folds; the curvilinear features are open folds or tight folds bent by batholithic intrusions or around mantled gneiss domes, Fig. 18-8.

Structural Provinces of the Canadian Shield

The Canadian Shield is subdivided into seven structural provinces on the basis of the trends and structural style of the folding, Fig. 18-9.

Superior province: This is characterized by easterly trending folds and faults throughout much of its area, although variations occur east of Hudson Bay and around a number of large batholithic intrusions. These highly tectonized rocks are overlain by slightly folded units in place. This was the main site of the Kenoran orogeny (2,480 million years), with thermal events continuing into the Hudsonian orogeny (1,735 million years).

Slave province: This is characterized by northerly trending and curved folds which were intruded by syenites after the orogenic deformation. Its thermal history is similar to that of the Superior province.

Bear province: This is separated from the Slave province by an angular unconformity, but the rocks above the unconformity in the Bear province are also folded away from the contact, and both were affected by the Hudsonian orogeny.

Churchill province: This is characterized by curved structures which trend northeast and form a broad arc around Hudson Bay. It was the main site of the Hudsonian orogeny (1,735 million years).

Southern province: This includes the easterly trending Penokean fold belt which is overlain by gently folded cover rocks of the Lake Superior basin.

Nain province: This is bounded on the west by northerly trends of structures which truncate northeast trends in the Churchill province. The Nain province was the site of an orogeny about 1,370 million years ago known as the Elsonian orogeny.

Grenville province: This is bounded for most of its length by a northeasterly trending orogenic front which truncates the provinces to the west. In places the front is a fault, and in other places it is a sharp metamorphic front. Folding within the Grenville province is extremely complex and on a small scale,

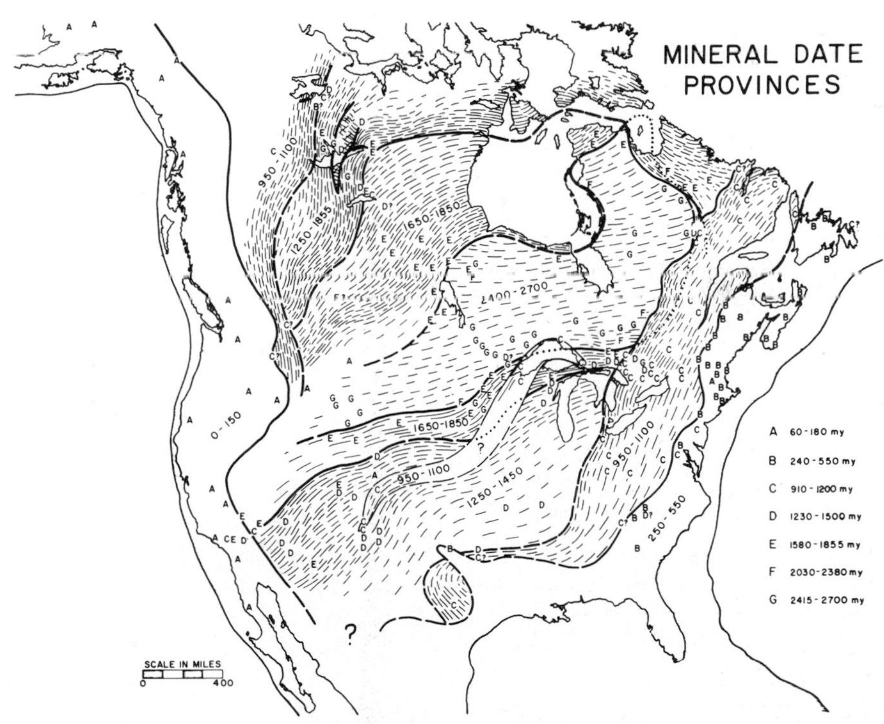

FIGURE 18-7 Precambrian orogenic belts of North America. Age ranges for each belt, obtained by radioactive dating methods, are indicated in millions of years. (*Courtesy Dr. Gordon Gastil.*)

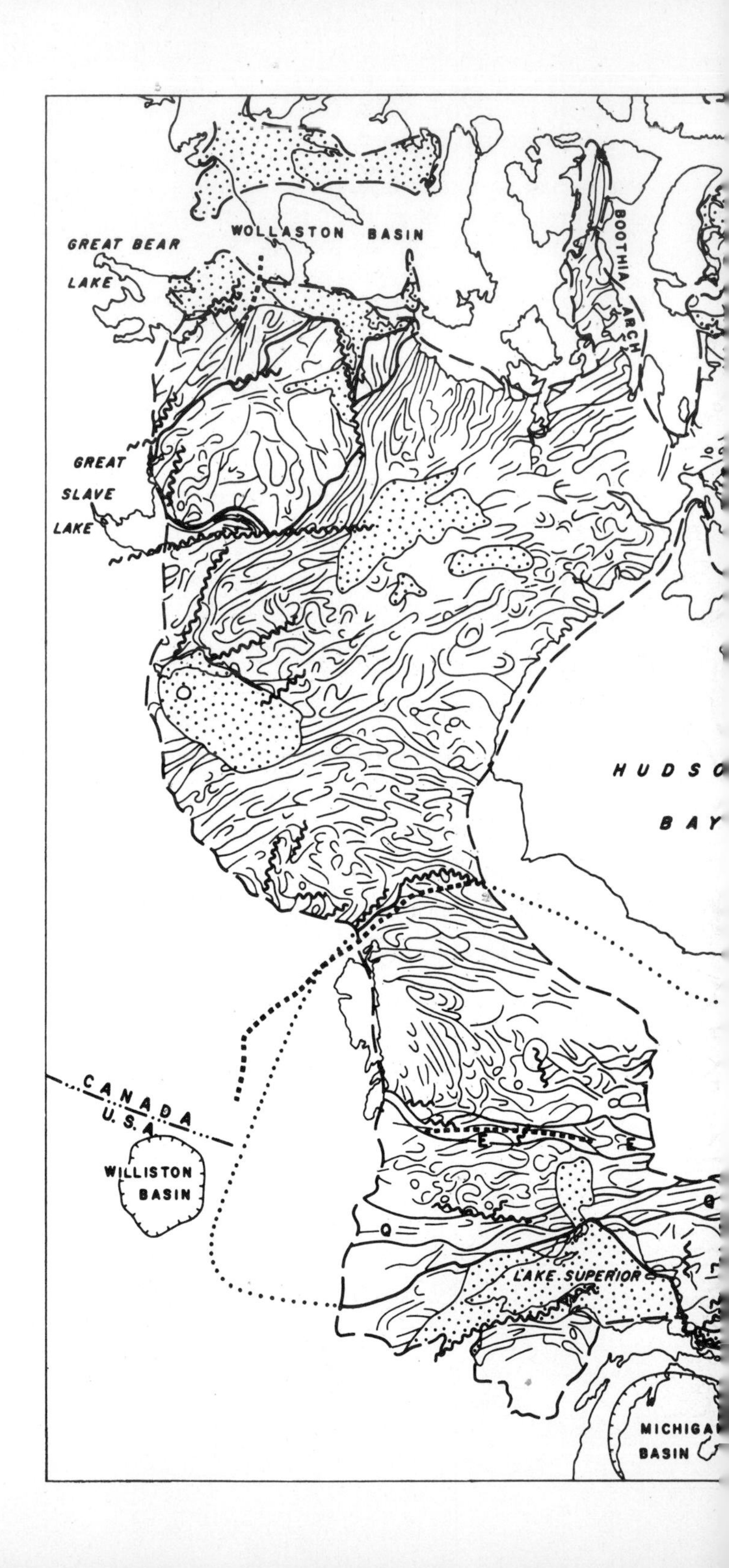
WOLLASTON BASIN
GREAT BEAR
LAKE
BOOTHIA
ARCH
GREAT
SLAVE
LAKE
HUDSO
BAY
CANADA
U. S. A.
WILLISTON
BASIN
E
Q
LAKE SUPERIOR
MICHIGA
BASIN

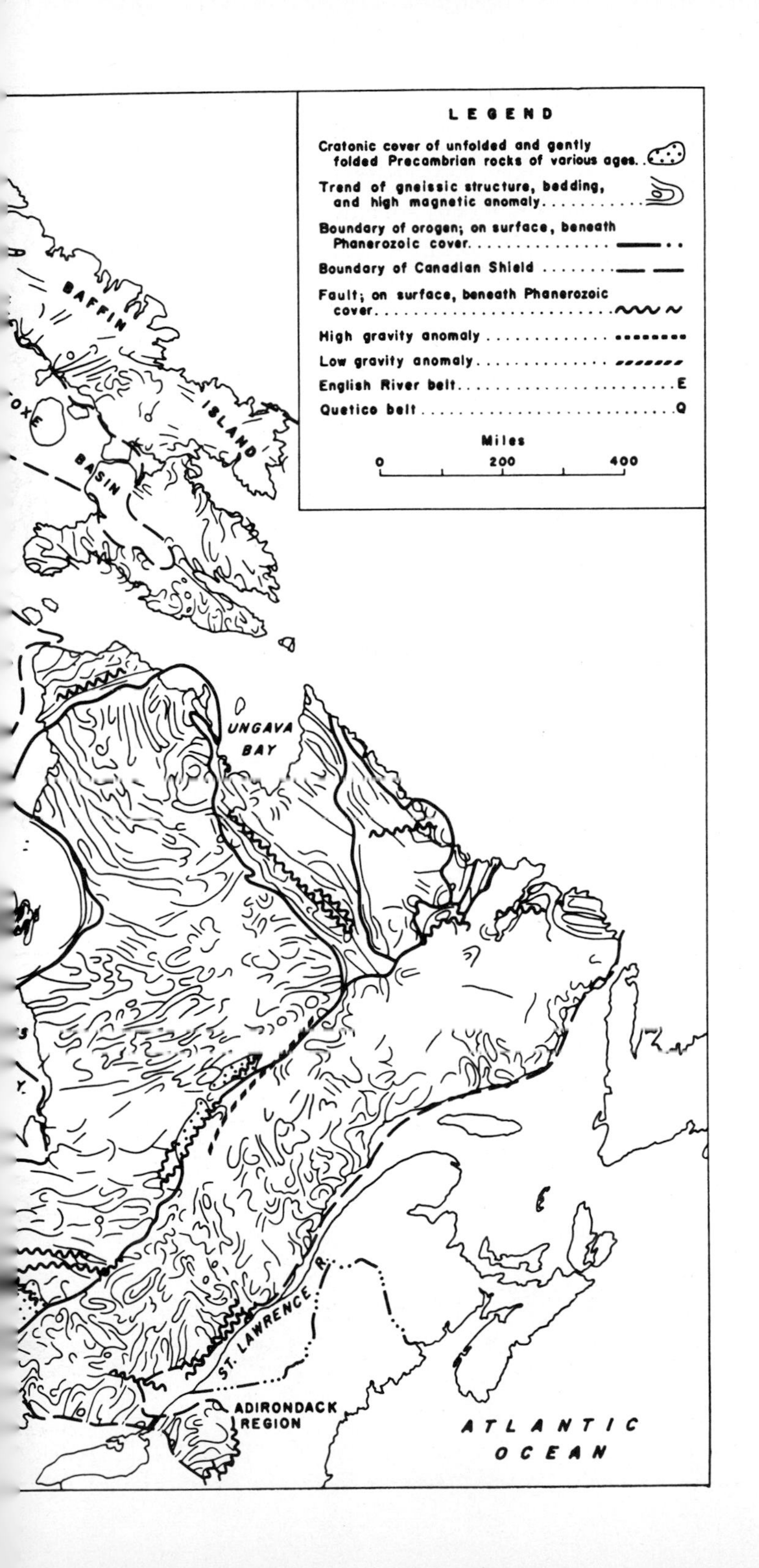

FIGURE 18-8 Structural trends in the Canadian Shield. (*From Stockwell, 1965.*)

and numerous gneiss domes are found in the province. The Grenvillian orogeny is dated at 955 million years.

Pettijohn (1972) synthesizes the Archean of the Shield, describing the early crust and the Kenoran belts. He describes the Archean as

Island-like areas of supracrustal rocks in a sea of invasive granite (all over 2.5 b.y. old). These "islands" consist of metavolcanic rocks that are mainly andesitic and basaltic greenstones . . . altered graywacke and slate with locally interbedded conglomerates and iron bearing formation. . . .* The greenstones are cut by dome-like diapiric stocks and batholiths of granitoid rocks . . . sediments outside the greenstone belts are intimately interleaved with granite sheets and are in places much granitized and engulfed [Fig. 18-10] . . . the greenstones are organized into broad belts . . . separated by equally broad and continuous belts of sediments that were apparently deposited in protogeosynclinal tracts.

* See Drever (1974) for origin of iron formations.

As in South Africa, these sags were filled in from both sides and are thus unlike many younger geosynclines.

Character of Structures in the Grenville Belt

The Grenville belt is located along the eastern margin of the North American craton and in-

FIGURE 18-9 Structural provinces of the Canadian Shield.

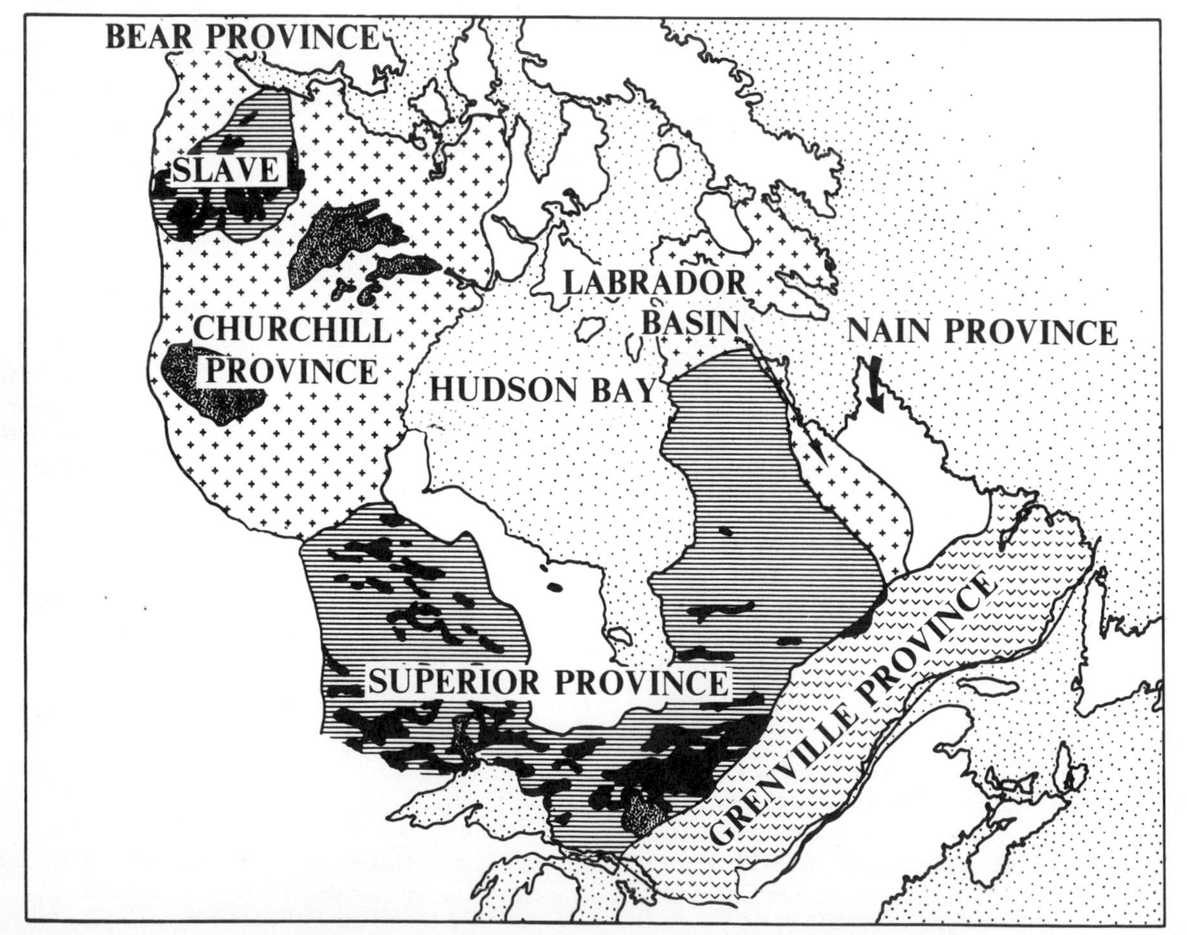

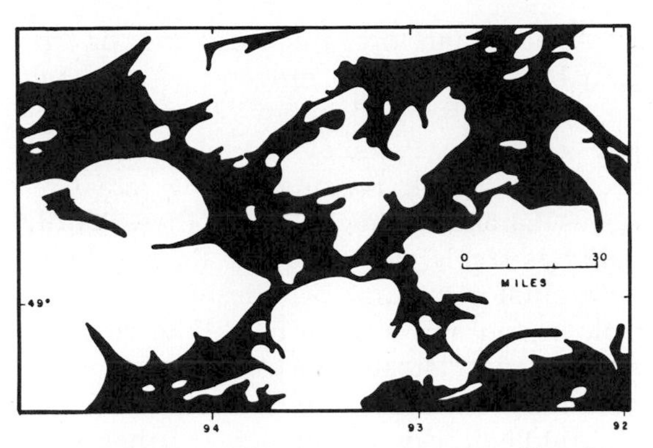

FIGURE 18-10 Areal pattern of schists (black) and granites (white) within Wabigoon greenstone belt in the Canadian Shield. Schists are largely metavolcanic rocks but include some metasediments and basic intrusives. *(From Pettijohn, 1972.)*

cludes rocks exposed in the Adirondack Mountains and the basement of the Appalachian Mountains. Both mantled gneiss domes and discordant plutons are abundant, and map patterns frequently suggest refolding of earlier structures. The full range of metamorphic and igneous structures is found, but the origin of much of the granitic gneiss remains to be determined. Representative examples of the structural styles are found in the French River, Harvey-Cardiff arch, and Stark complex areas described below.

The French River area of Ontario consists of a system of plunging folds formed in granitized metasedimentary rocks, unlabeled in Fig. 18-11. Evidence that the gneisses here are products of granitization is found in the partial conversion of the quartzite to granite, the well-developed fold structure in the foliated and layered gneisses, and the abundance of replacement of other minerals by microcline. The main structural feature is a southward plunging isoclinal syncline.

The geology of the Harvey-Cardiff arch reveals a very different type of structure. Here a line of granite gneiss domes forms an arch over 64 km (40 miles) long, Fig. 18-12. Anticlines and synclines in the well-foliated gneisses of the Burleigh dome indicate its metasedimentary origin. The Anstruther dome has a migmatite border and is interpreted as having formed under conditions which promoted mobility of the domes and their mantling rocks. Hewitt (1956) thinks it probable that the Cheddar dome differs from the Anstruther dome in that a magma formed in the former and migrated farther upward, while in the latter the rock reached only a stage of incipient mobility with some plastic flowage. The Centre Lake (Cardiff) pluton represents still a higher level of intrusion. It possesses both a center of upward intrusion and flank structures where magma has invaded the country rock. The plutons in this region seem to form a progressive sequence from granitic gneisses formed in place and showing structures associated with metasedimentary rocks in the south to intrusive plutons in which magma formed and intruded the country rock in the north.

The map pattern of the third example indicates a complicated configuration of bodies of granite, gneisses, marbles, and quartzites, Fig. 18-13. The mixed gneisses are a combination of altered metasedimentary rocks and thin intrusive sheets. A major fault separates an area of predominantly metasedimentary rocks from one of mainly igneous and metaigneous origin. The Stark complex is a tight anticline, the western limb of which was later cut out by intrusion of granite. The rocks west of the Stark complex are involved in a large isoclinal syncline overturned toward the southeast and dipping 25 to 50° N.

Complex fault and regional fracture systems which appear on aerial photographs as multidirectional patterns are characteristic of Precambrian rocks throughout the Shield. Many of these lineaments are wide shatter zones, some are occupied by dikes, and many are known as faults of considerable displacement. They are not related to the fold structures of the metamorphic rocks in which they occur, or if they are, the relationship is obscured by the multiplicity of directions. Displacements along such zones have played a major role in later development of structures in the American mid-continent and in the Rocky Mountains.

THE NORTH AMERICAN PLATFORM

The craton of North America may be thought of as including the vast area of the continental interior covered by thin platform sediments which lies south and west of the Shield, Fig. 18-14. The platform is bounded on the east by the Appalachian orogenic belt and on the south by the Ouachita orogen and the Gulf Coast geosyncline, but the western margin is more complex. Parts of the platform were affected by block faulting in the late Cretaceous Laramide orogeny, and these hardly qualify as stable elements. The Rocky Mountain frontal structural belt lies between the stable craton and a large block-like element, the Colorado Plateau. Blocks of the basement come to the surface in the front ranges and in the Middle Rocky Mountains. Farther west

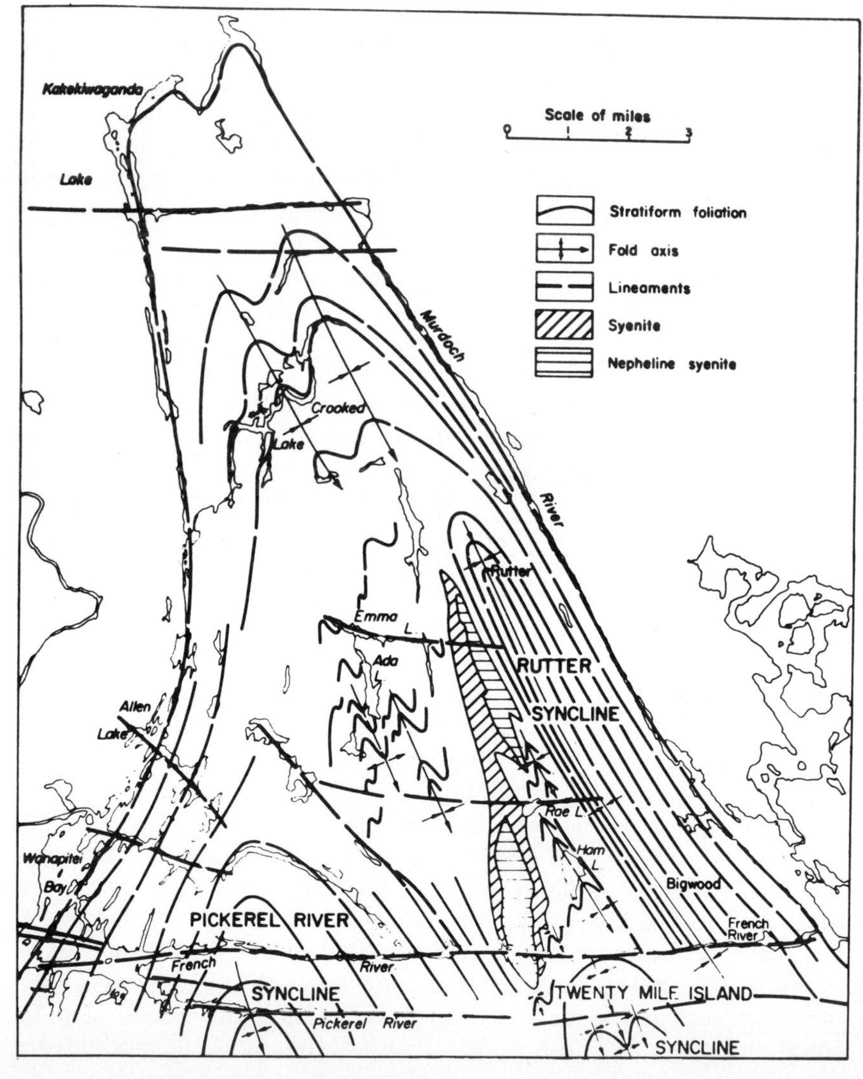

FIGURE 18-11 An example of the structural style exhibited by the crystalline rocks of the Canadian Shield in the French River area. (*From Hewitt, 1956.*)

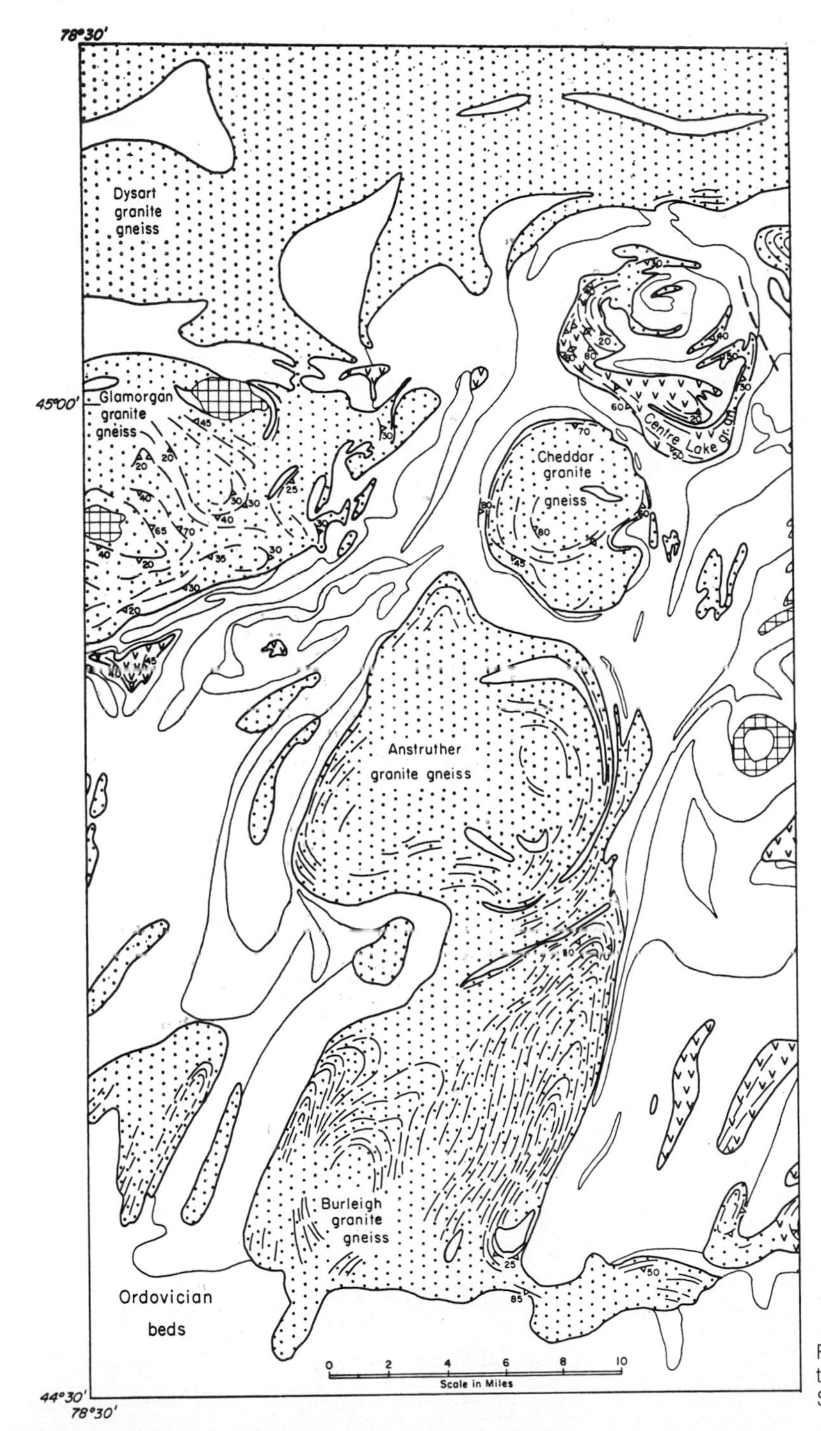

FIGURE 18-12 Gneiss domes along the Harvey-Cardiff arch, Canadian Shield. (*From Buddington, 1959.*)

the continent is made up of a succession of complex young folded mountain belts.

The cratonic platform has not been uniformly stable since the Precambrian. The edges of the craton have subsided in the Appalachians, Ouachitas, Marathon region, and along the western Cordilleran margin. Numerous smaller basins occur in the Rocky Mountain region of the United States. Some deeply subsiding basins also occur within the platform. Notable among these are the Michigan, Illinois, Anadarko, West Texas, and Williston basins. These are broad deep sags in the craton that must also affect the underlying Precambrian crust, and a few like the Anadarko basin are bounded by faults.

Steep faults penerate the basement within the platform and many also affect the cover. A major fault zone (Rough Creek) extends from the Ozark Mountains to the central Appalachians. Faults are associated with the north-south-trending buried Nemaha Mountains of Oklahoma, Kansas, and Nebraska, and faults are associated with the mid-continent gravity high.

Mid-Continent Gravity High

The Bouguer anomaly gravity map of the United States shows a very pronounced positive anomaly with marginal negative anomalies situated over the Duluth lopolith along the western edge of Lake Superior. It is offset to the east at the south end of Lake Superior but continues as a broad flat high to Minneapolis, where a second eastward offset occurs. The high continues southwest across Iowa and Nebraska, Fig. 18-15. Negative anomalies of similar trend can be followed well into Kansas. The maximum coincides with a belt of Keweenawan basic igneous rocks estimated to be

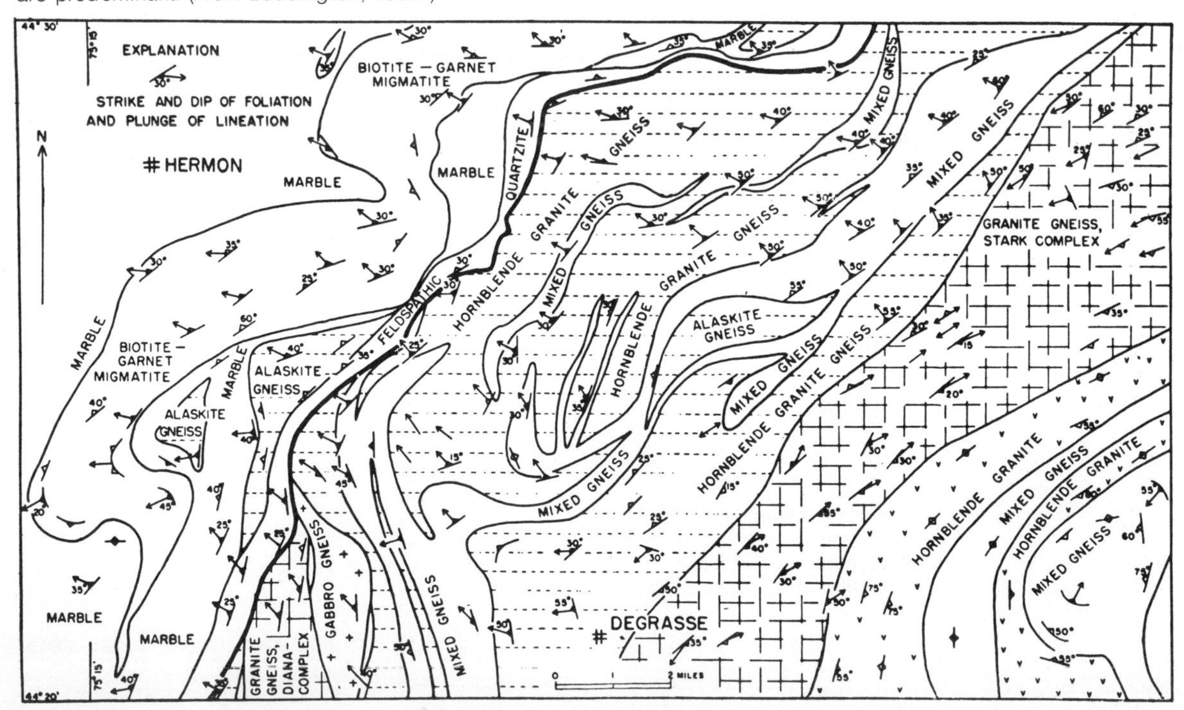

FIGURE 18-13 Details of overturned isoclinal syncline on northwest flank of relatively rigid unit of Stark complex. The lineation is subperpendicular to the trend of the major fold axes, northwest Adirondacks. The heavy line marks the boundary between the belt on the northwest in which rocks of the Grenville series predominate and the belt on the southeast in which igneous rocks and orthogneisses are predominant. (*From Buddington, 1956.*)

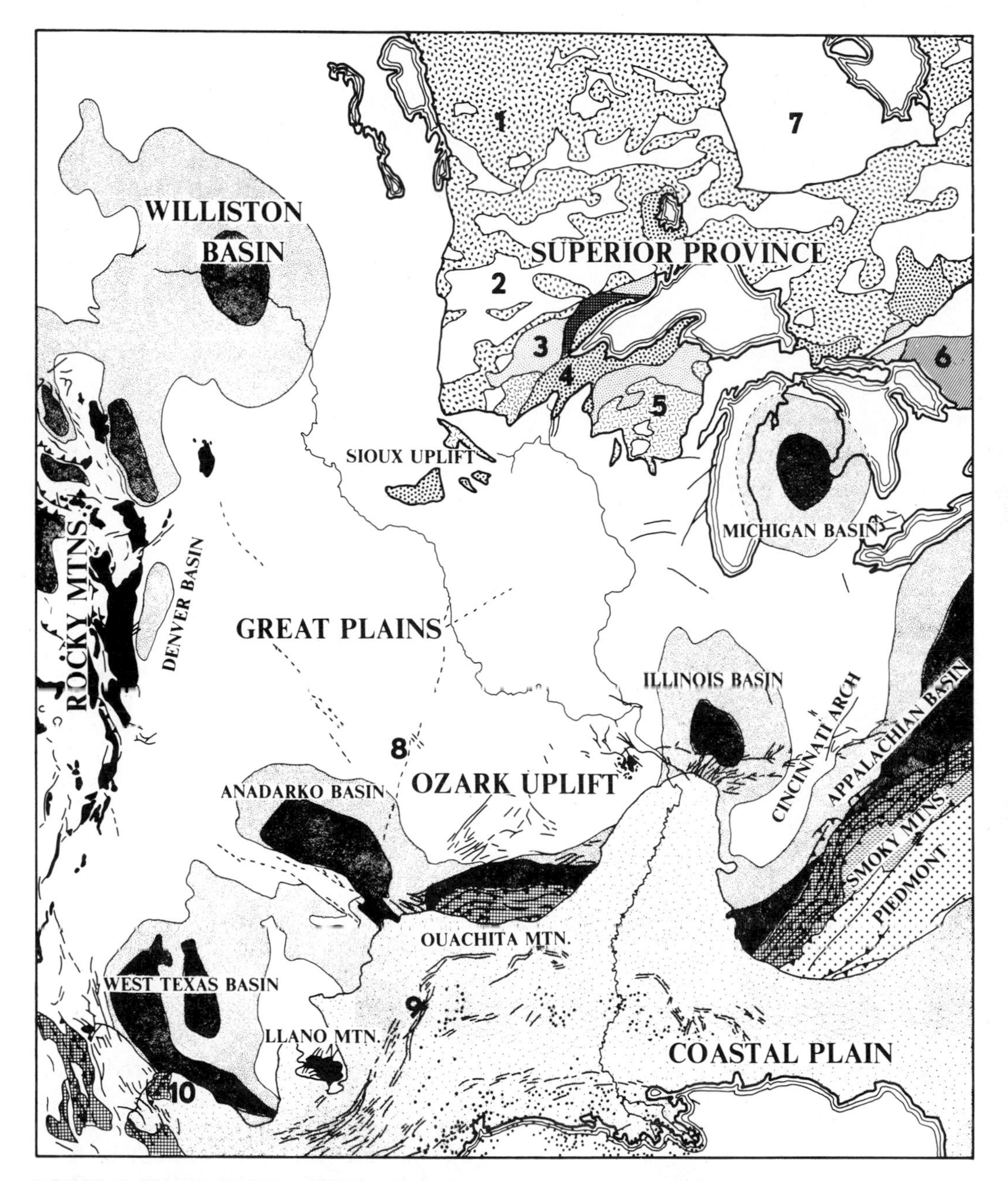

FIGURE 18-14 The North American platform showing the surrounding orogenic belts and the southern part of the Canadian Shield. The structure contours which bound shaded areas are for 2,000 and 1,000 m below sea level. The numbered references are:

1. Archean granitic gneisses
2. Archean sedimentary rocks
3. Lower Paleozoic
4. Lake Superior syncline
5. Archean gneisses reworked by Hudsonian events
6. Grenville metamorphosed Proterozoic
7. Paleozoic platform deposits.
8. Nemaha uplift
9. Mexia-Taco fault zone
10. Marathon Mountains

(*After the tectonic map of North America compiled by King, 1969.*)

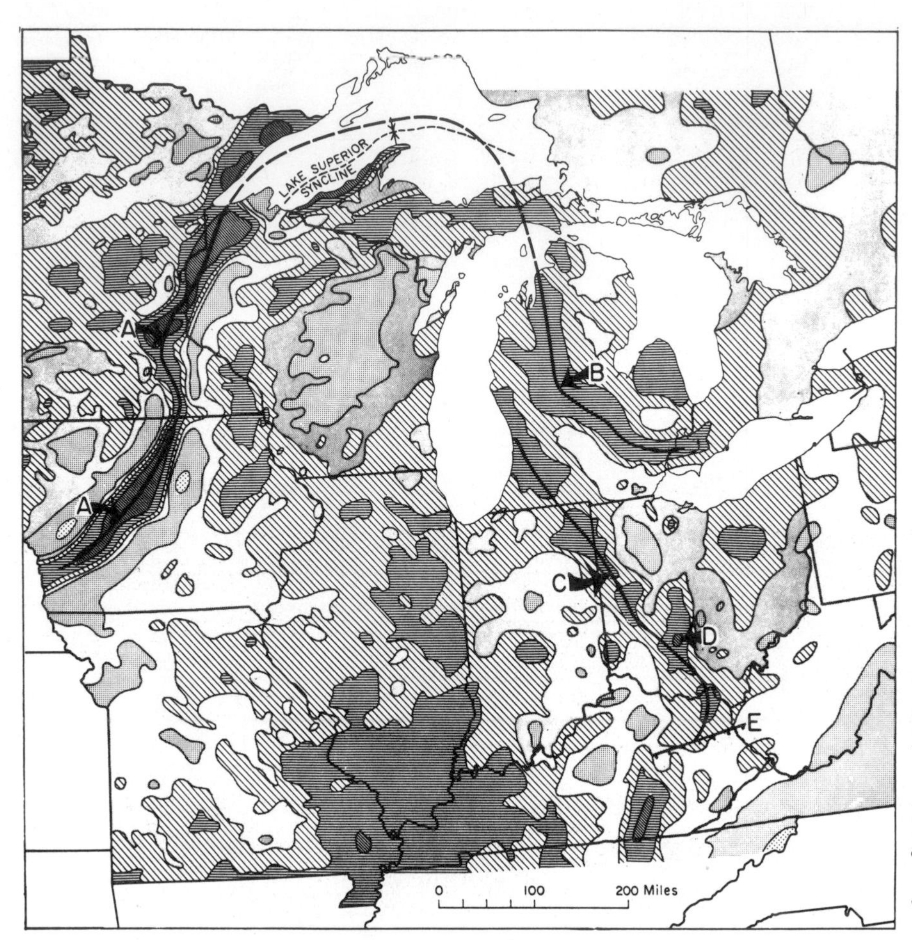

FIGURE 18-15 Bouguer gravity anomaly map of the midwestern United States. The darker patterns correspond to positive anomalies. (*Modified from Woollard and Rose, 1963, and Thompson and Miller, 1958; from Rudman, Summerson, and Hinze, 1965.*)

8,840 m (29,000 ft) thick where the basement is exposed near Minneapolis (Craddock and others, 1963). Craddock interprets the high as being due to a southward continuation of these thick igneous rocks, and the lows as being due to adjacent Keweenawan sedimentary sequences estimated to be 3,353 m (11,000 ft) thick.

The positive anomaly also parallels the St. Croix horst and is characterized in this region by a broad flat over the crest of the horst and by bands of steep gravity gradient over the basement faults which bound the horst. This horst was elevated in late Precambrian time.

Southwest of Minneapolis the gravity anomaly bends sharply to the southeast before resuming a southwesterly trend. This could be due to the primary shape of a volcanic fissure along which basic rocks are localized, or it could reflect an offset in the basement along a transcurrent fault. A northwest-trending dip-slip fault has been described in the Paleozoic rocks along this trend (Sloan and Danes, 1962).

The continuation of the mid-continent gravity high across Iowa is also postulated as representing a continuation of the Precambrian Keweenawan basalt flows. This high has been tentatively connected with another high which extends across the Michigan basin (Rudman, Summerson, and Hinze, 1965), Fig. 18-15. Magnetic data show highs coincident with this anomaly. Still another high may be traced across Michigan and Iowa and into Kentucky, and it too has been interpreted as resulting from basement basalts or ultramafics. Wells which have penetrated to the Precambrian have revealed basic rocks in central Iowa and in Ohio, adding strength to the argument that extensive volcanic flows occur in the midwestern continental interior.

The belt of gravity highs has been reinterpreted (Ocola and Meyer, 1973) on the basis of

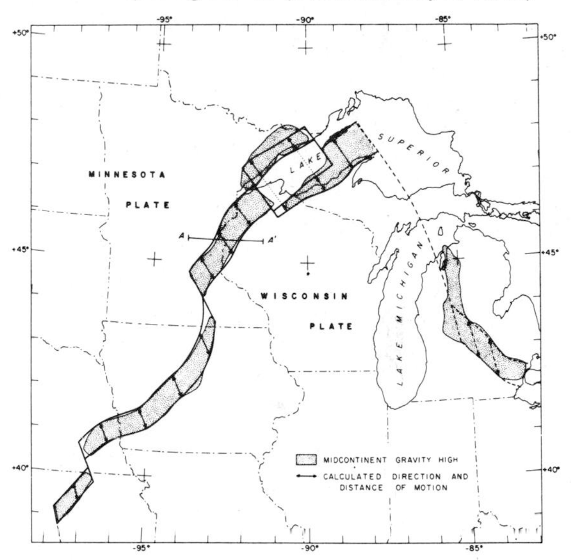

FIGURE 18-16 Interpretation of the mid-continent gravity high as a Precambrian intracratonic rift filled by basic materials. The arrows indicate the direction of movement involved in opening the rift. (*Ocola and Meyer, 1973.*)

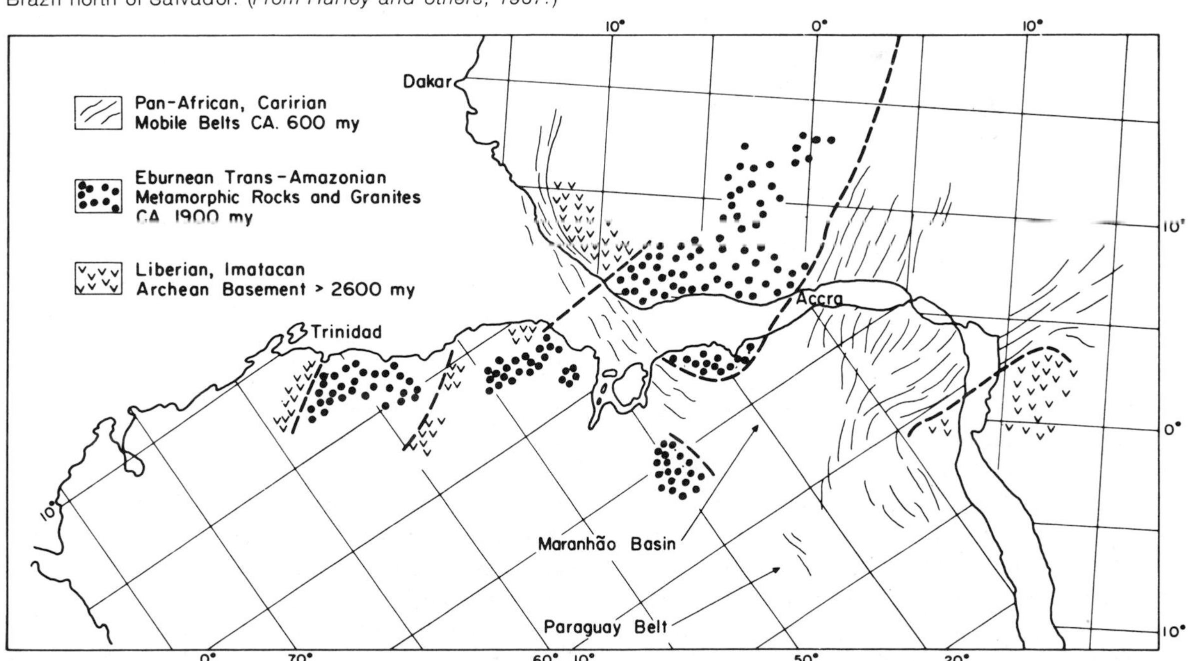

FIGURE 18-17 West Africa and South America shown fitted together according to the reconstruction of Bullard and others (1965). In West Africa the 2,000-million-year Eburnean age province (solid circles) adjoins the 550-million-year Pan-African age province (open circles); the boundary between them is shown by the heavy dashed line. If Africa and South America were once joined together, this line would have entered Brazil near São Luis. The age measurements for Brazil appear to show the same age provinces as those in West Africa, with the boundary at the predicted location. There may be a similar correlation between West Africa and the east coast of Brazil north of Salvador. (*From Hurley and others, 1967.*)

FIGURE 18-18 Continents reassembled in a predrift reconstruction. Lighter hatching—regions underlain by rocks having apparent ages in the range 800 to 1,700 million years; heavier hatching—regions having apparent ages > 1700 million years. It appears that there are two (or one) central regions of older rocks, transected and totally surrounded by belts of younger rocks. (*From Hurley and others, 1967. Copyright Am. Assoc. Adv. of Science.*)

new seismic data. They conclude that the zone defined by the gravity high most closely resembles the Red Sea rift system in terms of its geophysical characteristics, and that it may represent a long graben-like rift zone along which the craton has been pulled apart, allowing the emplacement from below of high-density mafic intrusive rocks, Fig. 18-16.

RESTORATION OF CRATONS TO PREDRIFT CONFIGURATIONS

The strong similarities in the thermal and orogenic histories, pattern and styles of deformation, and stratigraphy of the shield areas suggest the possibility of a closer physical connection than now exists. The theory of continental drift was not formulated on the basis of comparing shield areas; indeed, little was known about the shields and Precambrian history during the early development of continental drift theory. Models showing postulated fits between continents were initially developed on the basis of shape, floral and faunal similarities, and matching of major geological provinces. As details of the structure of the shields became better defined and especially as radiometric dates became sufficiently numerous to allow identification of belts of distinctive thermal history, matching of the internal structure of shields became possible. Parts of some shields are very near the coast. Among these are parts of the shields of Africa and South America in the area where those two continents are thought to have fit together.

Hurley and others, (1967) carried out a comprehensive field study of structural and radiometric dates in areas near the coast of these two shields. Their results seem to give conclusive confirmation that these two shields were, in fact, contiguous during the Precambrian, Fig. 18-17. In a later paper, Hurley and Rand (1969) compare structural trends and ages for all shields and show that the best predrift reconstructions allow interpretation of the Precambrian cratons as either one or two major units of older rock crossed and surrounded by younger rocks, Fig. 18-18. A somewhat similar, but more detailed, pattern is defined for the southern continents by Engle and others (1974), Fig. 18-19, which shows that the orogenic belts are systematically younger toward the south.

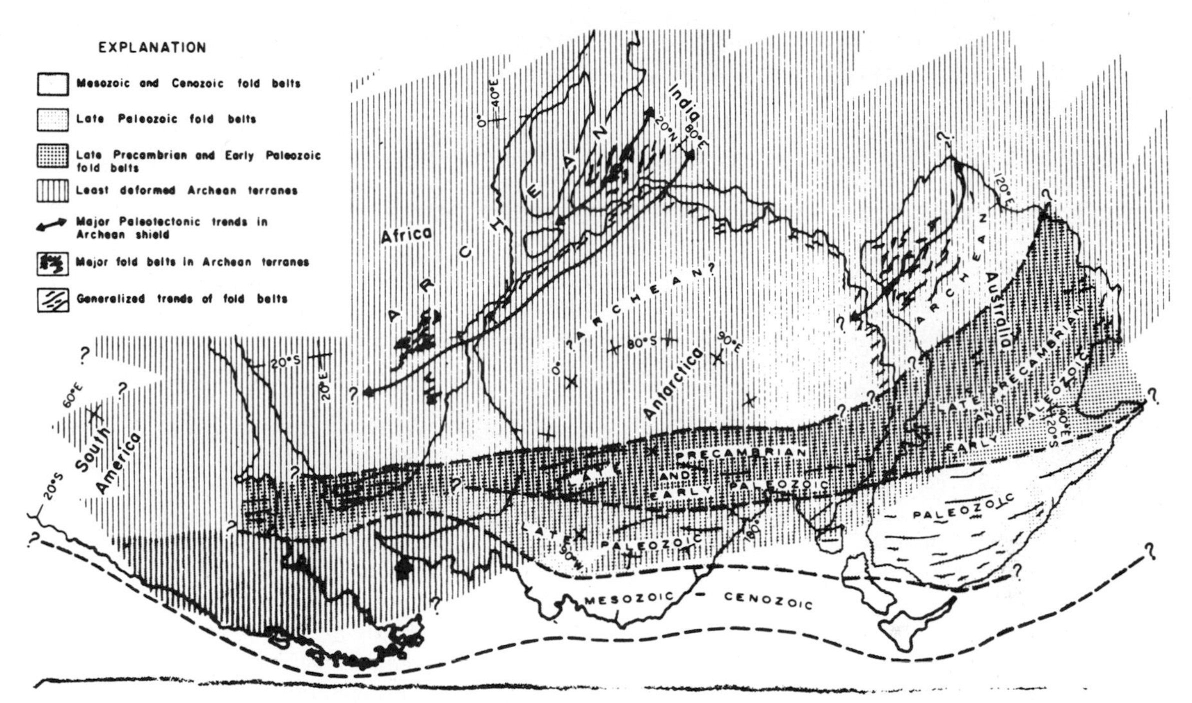

FIGURE 18-19 Subparallel alignment of major fold belts ranging in age from Archean to Cenozoic in Africa, India, Australia, and South America that appear when these continents are clustered into the classical late Paleozoic Gondwana. The grossly accordant Limpopo, Zambesi, and Grenville fold belts of Proterozoic age in South Africa are omitted in this oversimplified diagram. The gross trend from north to south of progressively younger orogenies suggests the progressive secular thickening and refractionation of Gondwana in this direction. (*From Engel and others, 1972.*)

REFERENCES

Anhaeusser, C. R., Mason, Robert, Viljoen, M. J., and **Viljoen, R. P.,** 1969, A reappraisal of some aspects of Precambrian shield geology: Geol. Soc. America Bull., v. 80, no. 11, p. 2175–2200.

Berry, M. J., and **Fuchs, Karl,** 1974, Crustal structure of the Superior and Grenville provinces of the northeastern Canadian Shield: Seis. Soc. Bull., v. 63, no. 4, p. 1393–1432.

Buddington, A. F., 1956, Correlation of rigid units, types of folds, and lineation in a Grenville belt (New York and New Jersey), *in* Thompson, J. E., ed., The Grenville problem: Royal Soc. Canada Spec. Pub. no. 1, p. 99–118.

——— 1959, Granite emplacement with special reference to North America: Geol. Soc. America Bull., v. 70, p. 671–747.

Chase, C. G., and **Gilmer, T. H.,** 1973, Precambrian plate tectonics: The midcontinent gravity high: Earth Planet. Sci. Letters, v. 21, p. 70–78.

Condie, K. C., 1973, Archean magmatism and crustal thickening, Geol. Soc. America Bull., v. 84, no. 9, p. 2981–2991.

Cooke, H. C., 1947, The Canadian Shield: Canada Geol. Survey Econ. Geol. Ser. no. 1, p. 11–97.

Coons, R. L., Woollard, G. P., and **Hershey, Garland,** 1967, Structural significance and analysis of mid-continent gravity high: Am. Assoc. Petroleum Geologists Bull., v. 51, no. 12, p. 2381–2399.

Coward, M. P., James, P. R., and **Wright, L.,** 1976, Northern margin of the Limpopo mobile belt, southern Africa: Geol. Soc. America Bull., v. 87, p. 601–611.

Craddock, Campbell, Thiel, E. C., and **Gross, Barton,** 1963, A gravity investigation of the Precambrian of southeastern Minnesota and western Wisconsin: Jour. Geophys. Res., v. 68, p. 6015–6032.

Dolginov, Ye. A., and **Ponikarov, V. P.,** 1974, Types

of early Precambrian structures (Hindustan, Australian, Antarctic, and African platforms): Geotectonics, no. 2, p. 64–69.
Drever, J. I., 1974, Geochemical model for the origin of Precambrian banded iron formation: Geol. Soc. America Bull., v. 85, no. 7.
Engel, A. E. J., and **Kelm, D. L.**, 1972, Pre-Permian global tectonics; a tectonic test: Geol. Soc. America Bull., v. 83, no. 8.
———, **Itson, S. P., Engel, C. G., Stickney, D. M.**, and **Cray, E. J., Jr.**, 1974, Crustal evolution and global tectonics: A petrogenic view: Geol. Soc. America Bull., v. 85, no. 6, p. 843–858.
Gastil, G., 1960, Continents and mobile belts in the light of mineral dating: Internat. Geol. Cong., 21st, Norden, p. 162–169.
Gill, J. E., 1948, The Canadian Pre-cambrian shield: Structural geology of Canadian ore deposits: Canadian Inst. Mining Metallurgy.
——— 1949, Natural divisions of the Canadian Shield: Royal Soc. Canada Trans., s. 3, sec. 4, v. 43, p. 61–69.
——— 1952, Mountain-building in the Pre-cambrian shield: Internat. Geol. Cong., 18th, London, v. 13, p. 97–104.
Glikson, A. Y., 1972, Early Precambrian evidence of a primitive ocean crust and island nuclei of sodic granite: Geol. Soc. America Bull., v. 83, no. 11, p. 3323ff.
Goldich, S. S., and others, 1961, The Precambrian geology and geochronology of Minnesota: Univ. Minnesota and Minnesota Geol. Survey Bull., v. 41.
Hewitt, D. F., 1956, The Grenville region of Ontario, *in* The Grenville problem: Royal Soc. Canada Spec. Pub. 1, p. 23–41.
Hurley, P. M., 1972, Can the subduction process of mountain building be extended to Pan-African and similar orogenic belts?: Earth Planet. Sci. Letters, v. 15, p. 305–314.
——— and **Rand, J. R.**, 1969, Pre-drift continental nuclei: Science, v. 164.
——— and others, 1967, Test of continental drift by comparison of radiometric ages: Science, v. 157, no. 3788.
Illies, J. H., 1969, An intercontinental belt of the world rift system: Tectonophysics, v. 8, no. 1, p. 5–29.
Krutikhoskaya, Z. A., Pashkevich, I. K., and **Simonenko, T. N.**, 1973, Magnetic anomalies of Precambrian shields and some problems of their geological interpretation: Can. Jour. Earth Sci., v. 10, no. 5, p. 629–636.
Lowdon, J. A., and others, 1963, Age determinations and geological studies, including isotopic ages: Geol. Survey Canada, Rept. 3, Paper 62–17, p. 5–120.
MacGregor, A. M., 1951, Some milestones in the Precambrian of southern Rhodesia: Geol. Soc. South Africa Trans. Proc., v. 54, p. xxvii.
Magnusson, N. H., 1965, The Precambrian history of Sweden: Geol. Soc. London Quart. Jour., v. 121, p. 1–30.
Muehlberger, W. R., and others, 1967, Basement of the continental interior of the United States: Am. Assoc. Petroleum Geologists Bull, v. 51, no. 12.
———, **Denison, R. E.**, and **Lidiak, E. G.**, 1967, Basement rocks in continental interior of U.S.: Amer. Assoc. Petroleum Geologists Bull., v. 51, no. 12, p. 2351–2380.
Ocola, L. C., and **Meyer, R. P.**, 1973, Central North American rift system: 1. Structure of the axial zone from seismic and gravimetric data: Jour. Geophys. Res., v. 78, no. 23, p. 5173ff.
Pettijohn, F. J., 1972, The Archean of the Canadian Shield; a resume, *in* Studies in mineralogy and Precambrian geology: Geol. Soc. America Mem. No. 135, p. 131–149.
Ravich, M. G., 1973, Regional metamorphism of the Antarctic platform crystalline basement, *in* Antarctic geology and geophysics: Int. Union Geol. Sci., Ser. B, no. 1, p. 505–515.
Rudman, A. J., Summerson, C. H., and **Hinze, W. J.**, 1965, Geology of basement in midwestern United States: Am. Assoc. Petroleum Geologists Bull., v. 49, p. 894–905.
Sloan, R. E.., and **Danes, Z. F.**, 1962, A geologic and gravity survey of the Belle Plaine area, Minnesota: Minnesota Acad. Sci. Proc., v. 30, no. 1.
Stevenson, J. S., ed., 1962, The tectonics of the Canadian Shield: Royal Soc. Canada Spec. Pub. 4.
Stockwell, C. H., 1965, Structural trends in the Canadian Shield: Am. Assoc. Petroleum Geologists Bull., v. 49, p. 887–894.
Sutton, J., and **Windley, B. F.**, 1973, A discussion on the evolution of the Precambrian crust: Phil. Trans. Roy. Soc. London, v. 273, p. 315–581.
Thompson, J. E., ed., 1956, The Grenville problem: Royal Soc. Canada Spec. Pub. 1.

Wegmann, C. E., 1929, Biespiele Tektonischer Analysen des Grundgebirges in Finland: Comm. Geol. Finlande Bull., v. 87, no. 8, p. 98–127.

Wilson, M. E., 1939, The Canadian Shield, Geologie der Eerde, *in* Ruedeman, R., and Balk, R., eds. Geology of North America: Berlin, Borntraeger.

——— 1948, An approach to the structure of the Canadian Shield: Am. Geophys. Union Trans., v. 29, p. 691–726.

——— 1949, Some major structures of the Canadian Shield: Canadian Inst. Mining Metallurgy Trans., v. 52, p. 231–242.

Wyllie, P. J., 1971, The dynamic earth: Textbook in geosciences: New York, Wiley.

Zietz, I., and others, 1966, Crustal study of a continental strip from the Atlantic Ocean to the Rocky Mountains: Geol. Soc. America Bull., v. 77, p. 1427–1448.

19

Atlantic-Type Continental Margins

The marginal zones between continents and the deep-sea floor exhibit many diverse structural configurations. Because the seaward side of these margins is so often the site of rapid sedimentation, the topography reflects the effects of marine agents—erosion and sedimentation—more than the underlying structure. This is especially true along margins where the structural configurations are old. Most classifications of continental margins reflect this dual control of diastrophism and sedimentation.

For tectonic classification the margins may be best subdivided according to gross structural characteristics as follows (see later chapters for discussion):

Atlantic-type stable margins

Cordilleran-type unstable margins

Andean-type unstable margins

Island-arc type unstable margins

The Atlantic-type margins are almost aseismic; they have little or no modern volcanic activity; mountains near the coast are old and tectonically inactive, many of these margins have broad coastal plains or continental shelves; they are much more stable than any other type.

Atlantic-type margins do not uniformly surround the Atlantic—notable departures are present in the island arcs of the Caribbean region and the Scotia arc. Nor are Atlantic-type margins confined to the Atlantic—similar margins occur in the Arctic Ocean and around Australia, Antarctica, Africa, India, and the eastern margin of South America, all of which were part of Gondawana in the late Paleozoic. Portions of many of these margins have been studied, but the deep structure is nowhere known in detail comparable to that along the Atlantic margin of North America, which is described in the following section.

THE NORTH AMERICAN CONTINENTAL MARGIN

Surface Configuration

The continental margin south of Florida is obscured by the Bahama Banks, vast areas of thick carbonate sediments which are now covered by shallow water. South of the Banks the margin is complicated by the Antilles island arc, and that portion of the continental margin is not Atlantic-type. From Florida northward, the edge of the continent is covered by a broad veneer of Cenozoic sediments in the Gulf and Atlantic Coastal Plains. These sediments and sedimentary rocks thicken toward the Gulf and the ocean, and they are gently tilted toward the ocean. They are warped in some places, as at Cape Fear; otherwise they show little sign of deformation. Northward this sedimentary veneer becomes a narrow belt, and north of Long Island the sedimentary units lie submerged on the continental shelf, and the much older, strongly deformed, and metamorphosed rocks of the Appalachian orogen are exposed at the shore from New England to Newfoundland. Even in the south, wells have penetrated the Coastal Plain sediments into the Paleozoic rocks which lie with profound unconformity beneath them. This highly tectonized and altered complex, together with several isolated fault basins which contain unfolded and unmetamorphosed Pennsylvanian and Triassic sediments, make up the basement along this Atlantic margin.

Configuration of the Submerged Margin

The topography of Atlantic-type margins is relatively simple compared with other types. The North American margin can be subdivided into three major parts—a broad, flat coastal plain or continental shelf (the shelf is essentially a submerged equivalent of the coastal plain), a continental slope, and a continental rise, Figs. 19-1 and 19-2. This pattern is well developed toward the north, but a more complicated pattern occurs south of Cape Hatteras, where the seaward side of the shelf is a narrow and short slope which leads onto a broad, flat, deeply submerged plateau, the Blake Plateau. The edge of this plateau is an escarpment which extends to great depths characteristic of the oceanic crust, Fig. 19-3.

Shallow Structure of the Atlantic Margin

Geologic mapping provides insight to the landward portions of the margin, but the Cenozoic sedimentary cover on the coastal plain and the continental shelf limit access to deeper features. A number of water wells have been drilled in the Coastal Plain, but these are generally of shallow depth, and even these are not drilled in water-covered areas. Deeper exploratory drilling for petroleum on the continental shelf has started and will undoubtedly be an important source of data in the future; in addition a number of wells have been drilled as part of the Joides program, but the Joides wells are widely spaced and provide mainly data on stratigraphy rather than structure.

At the present time the results of geophysical surveys, and especially continuous seismic reflection profiling, are the most important source of shallow structural data for this margin. Results of representative profiles are reproduced, Fig. 19-4. These surveys show that the Atlantic margin has been shaped in large part by the building up of sedimentary deposits. The shelf has been built up by deposition of between 200 and 1,000 m of Cenozoic sediment. The slope has been extended seaward by 5 to 35 km during the same interval, with the most rapid construction taking place where the slope is flanked by the Blake Plateau. The Cenozoic sedimentary beds can be traced continuously from the continental slope to the continental rise off Nova Scotia, but south of New England the beds of the slope appear to cover those of the rise and the Blake Plateau (Uchupi and Emery, 1967).

Several variations in shallow structure are apparent on the cross sections, Fig. 19-4. Seamounts which rise through the sedimentary cover high into the water are seen in the sections off Georges Bank and off Cape Cod. Strong positive magnetic anomalies and Bouguer gravity anomalies identify them and similar buried seamounts as former volcanic centers. These volcanos are found in various stages of burial by sediments which can be traced great distances away from them. Some of the seamounts (e.g., Mytilus) are capped by carbonate caps. Many of these cross sections are interpreted as showing surficial layers of undisturbed turbidites which cover thick

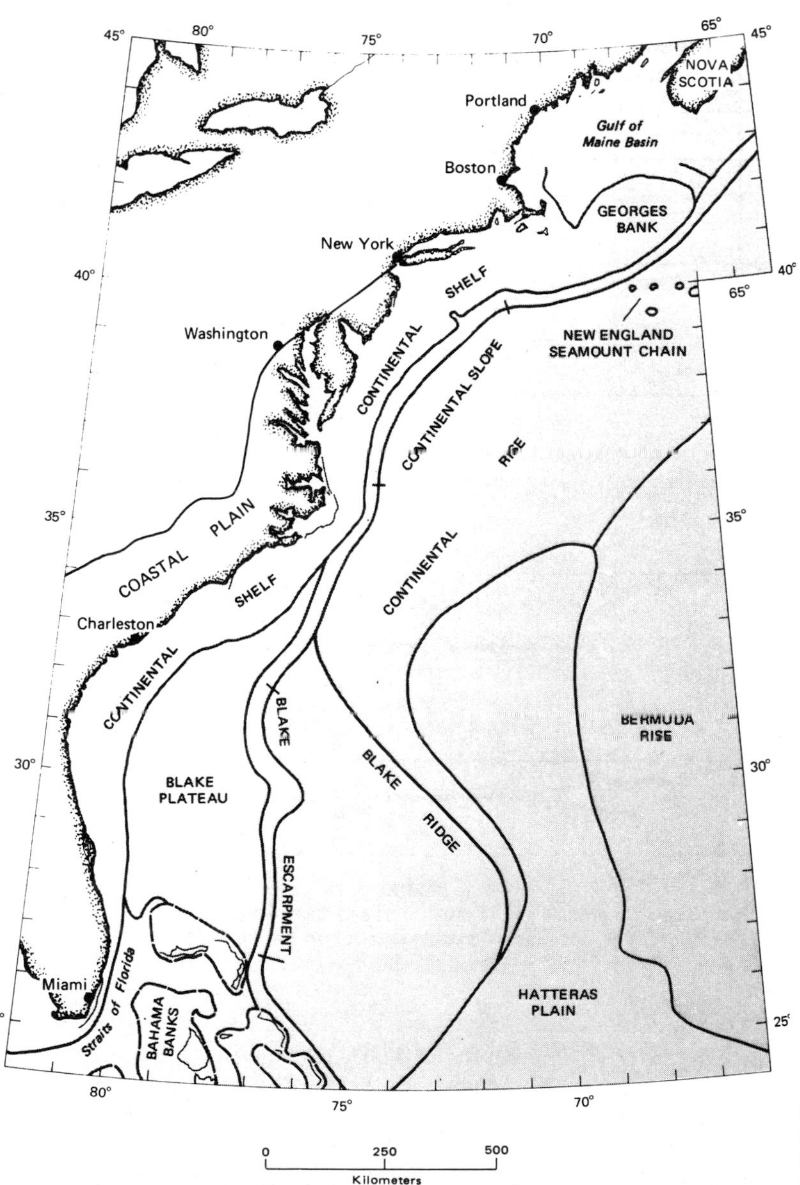

FIGURE 19-1 Major physiographic divisions off the eastern margin of the United States. (*From Emery and others, 1966.*)

wedges of slumps and slides. Another layer of undisturbed turbidites occurs under these. The fourth and lowest sedimentary layer is composed of undisturbed pelagic sediments resting on oceanic crust. Important marker horizons are identified as A and B in these cross sections.

The A horizon is a strong acoustic reflector known in sounding profiles long before its stratigraphic identity was established as being due to a blanket-like deposit of chert thought to have formed very widely in the deep-ocean basins during the time interval between the Late Cretaceous and the middle Eocene. The B horizon is the contact of the sedimentary veneer with oceanic crustal basement.

FIGURE 19-2 Bathymetric map of the region off eastern North America compiled by Emery and others (1970).

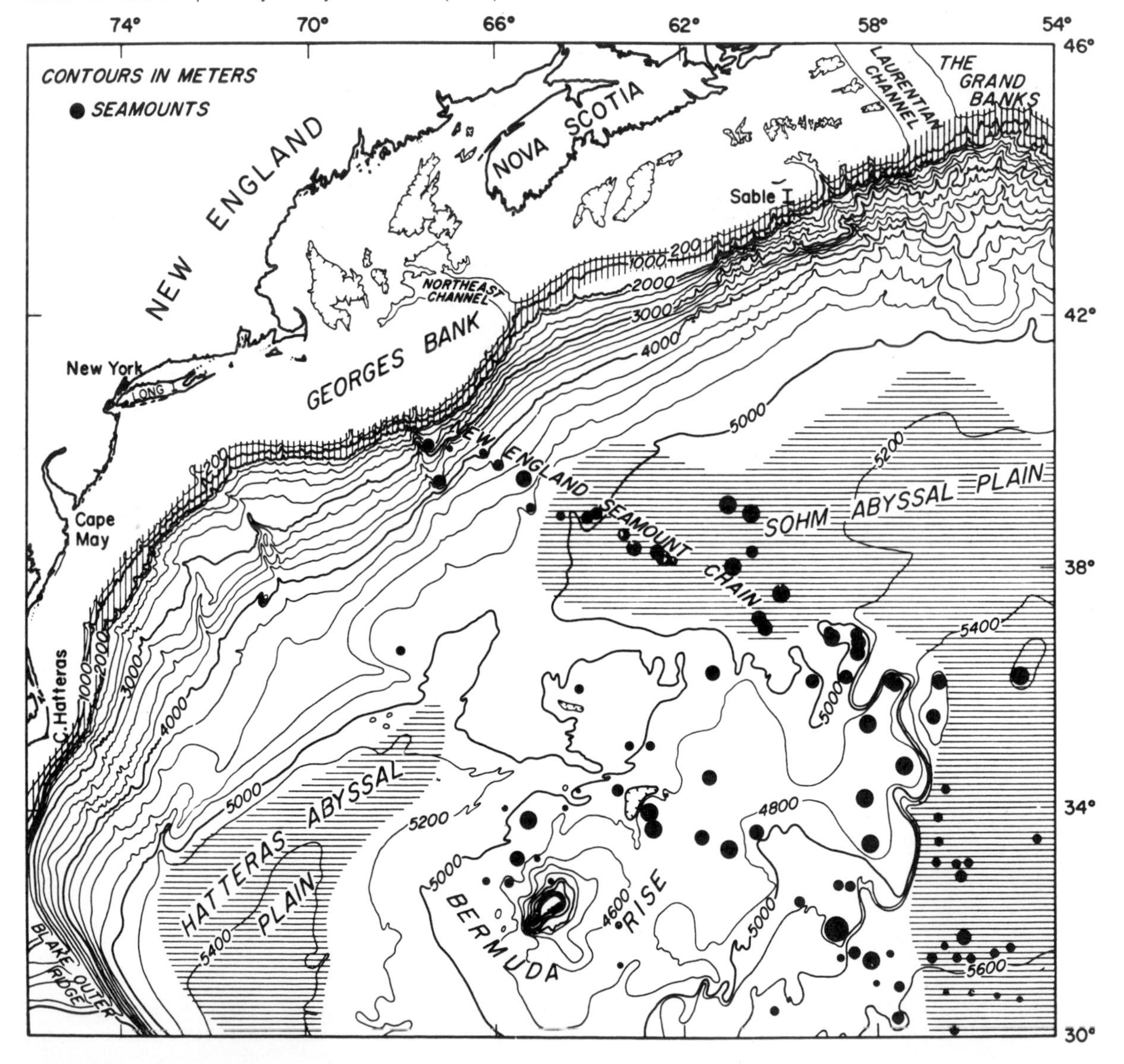

Deeper Structure of the Continental Margin

Our first major insights into the deeper structure of the North American margin came as a result of a study of refraction profiles done by the Lamont Observatory (Drake, Ewing, and Sutton, 1959). These profiles were the beginning of the resolution of what had been a highly generalized model of the continent-ocean transition, Fig. 19-5. The refraction profiles reveal the presence of a great thickness of low-velocity sedimentary rock buried beneath the continental slope and continental rise. The M discontinuity appeared in only two of these sections, and in both it appears as a steeply inclined surface as it drops from oceanic to continental levels. Subsequent geophysical studies and results of Joides drilling have provided a more complete picture of the subsurface along this margin.

The depth to the top of the mantle (to seismic velocity 7.8 km/sec) on the continental margin has been mapped through seismic refraction methods, Fig. 19-6. These indicate that the M discontinuity drops off rapidly toward the continent under the continental slope and rise all the way from the Grand Banks to Florida. Good records could not be obtained where the Kelvin seamount belt crosses the margin, and little data is available south of Florida. Mantle highs appear as a ridge extending from just north of Puerto Rico along the eastern edge of the Bahama Banks and the Blake Plateau. A second high with a central depression occurs under the area between the Hatteras abyssal plain and the Sohm abyssal plain.

The shape of the sedimentary deposits along the continental margin is depicted by structure contour maps showing the depth to various horizons and by isopach maps. Emery and others (1970) prepared maps showing the depth to the basement and the depth to horizon A for the same area depicted in Fig. 19-6. The map showing the total thickness of all sediment reveals that the sediments along the margin lie in a sinuous relatively narrow belt that closely follows the continental slope and rise, Fig. 19-7. The greatest thicknesses are in general located along the slope, but several deep sedimentary basins occur west of the Grand Banks, off the Florida and South Carolina coasts, and in the eastern Gulf of Mexico. These sedimentary accumulations are great (more than 9 km in places) and have been compared in thickness and extent with the geosynclinal accumulations in the Appalachian geosyncline.

FIGURE 19-3 Structure cross section across the Florida peninsula and Blake plateau based on drill hole and seismic data. Data at east is from Pratt and Heezen (1961). (*From Sheridan and others, 1966.*)

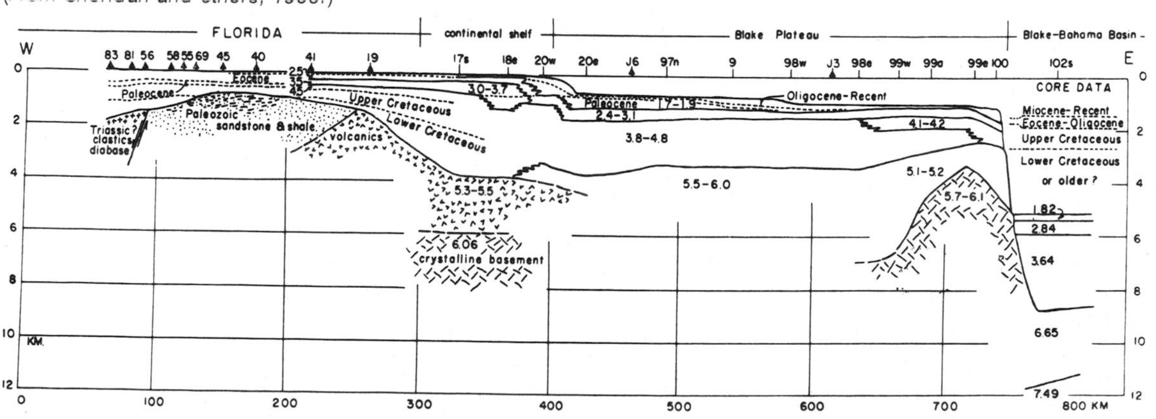

By using horizon A as the approximate base of the Cenozoic system, it is possible to prepare structure contour maps on this strong reflector and to determine approximately the thickness of Cenozoic sediments along the margin, Fig. 19-8. These show thick sedimentary accumulations along the continental slope from the Grand Banks to the northern end of the Blake Plateau. Sedimentary cover on the Blake Plateau is thin, but thick covers are found under the Bahama Banks and northeast of the Blake Plateau.

FIGURE 19-4 Cross sections across the eastern margin of North America based on continuous seismic reflection profiles records. (*From Emery and others, 1970.*)

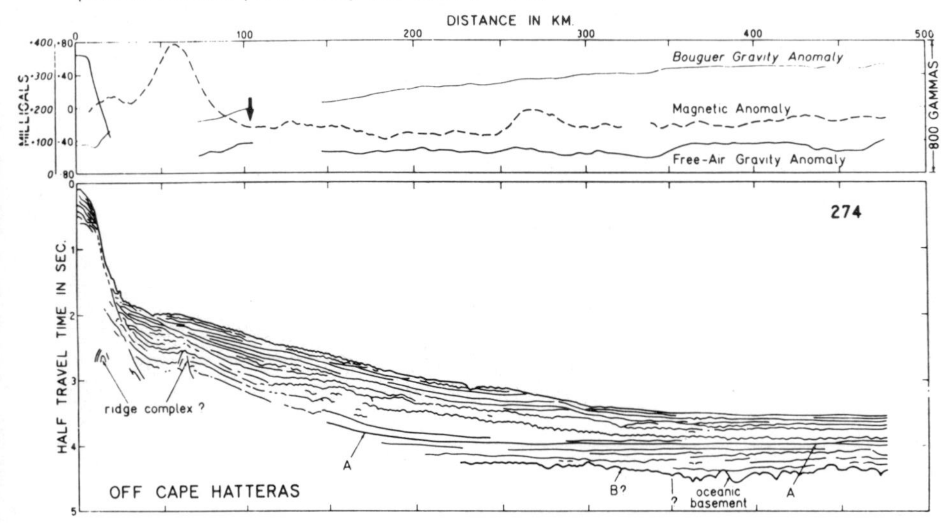

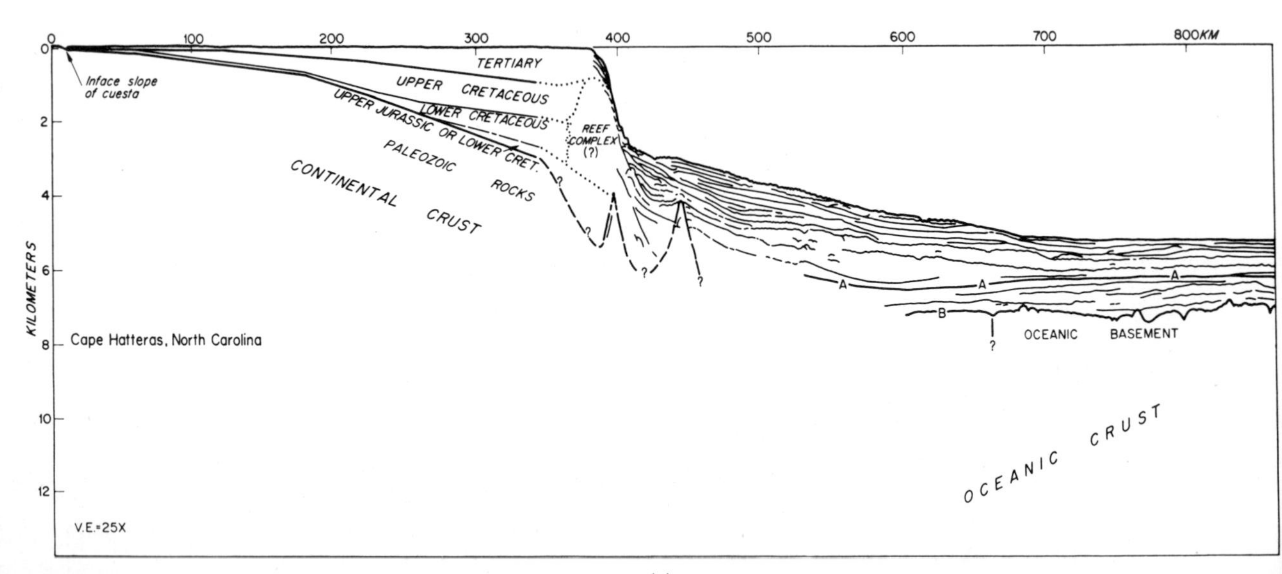

(*a*)

These great prisms of sediment are potential reservoirs of oil and gas. The rocks are young enough to have petroleum potential, they are the right kinds of rocks, the thickness of the pile is great enough, and potential structural and stratigraphic traps appear to be present. Water depth is the most important physical restraint at present.

Both the total sediment and the Cenozoic sediment thicknesses show some complications in distribution patterns which must be related to major structural features along the margin. The shape of the trough is anomalous at about 40° N latitude. The trend of the accumulations swing into an east-west line along which the Kelvin seamount group is situated to the east and one of the faulted Triassic basins to the west. The great bend of the Appalachian fold system in Pennsylvania occurs along this same line, and it has been suggested that this line is a major fault or fracture zone at depth (Drake and Woodward,

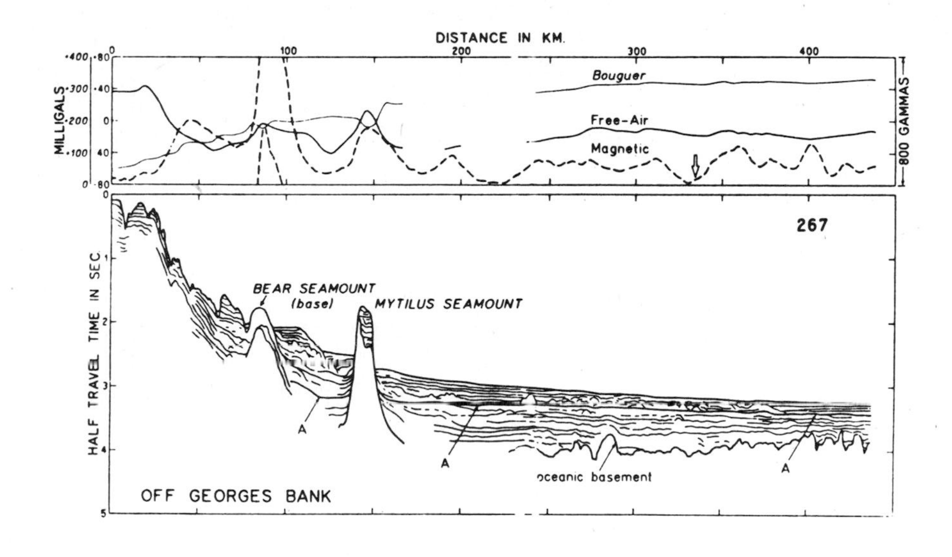

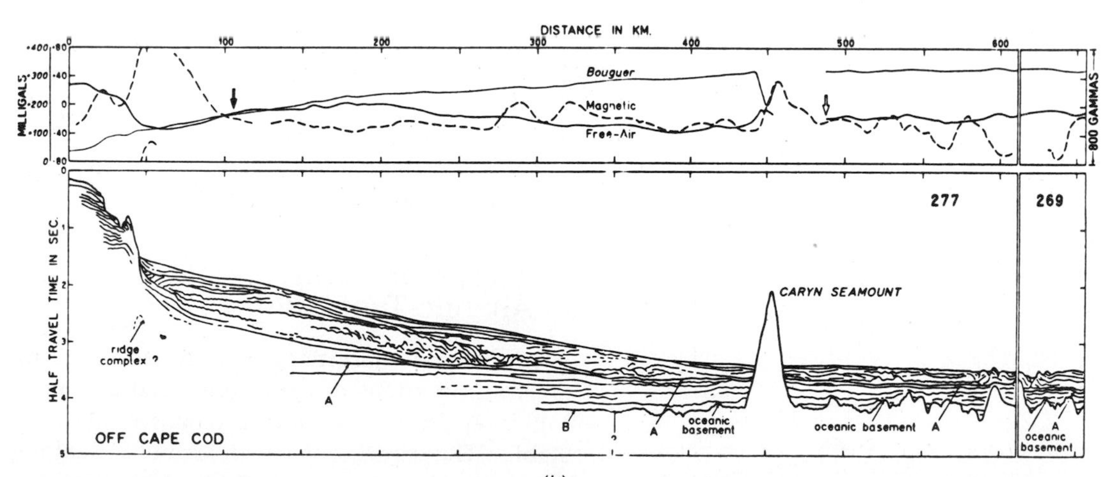

(b)

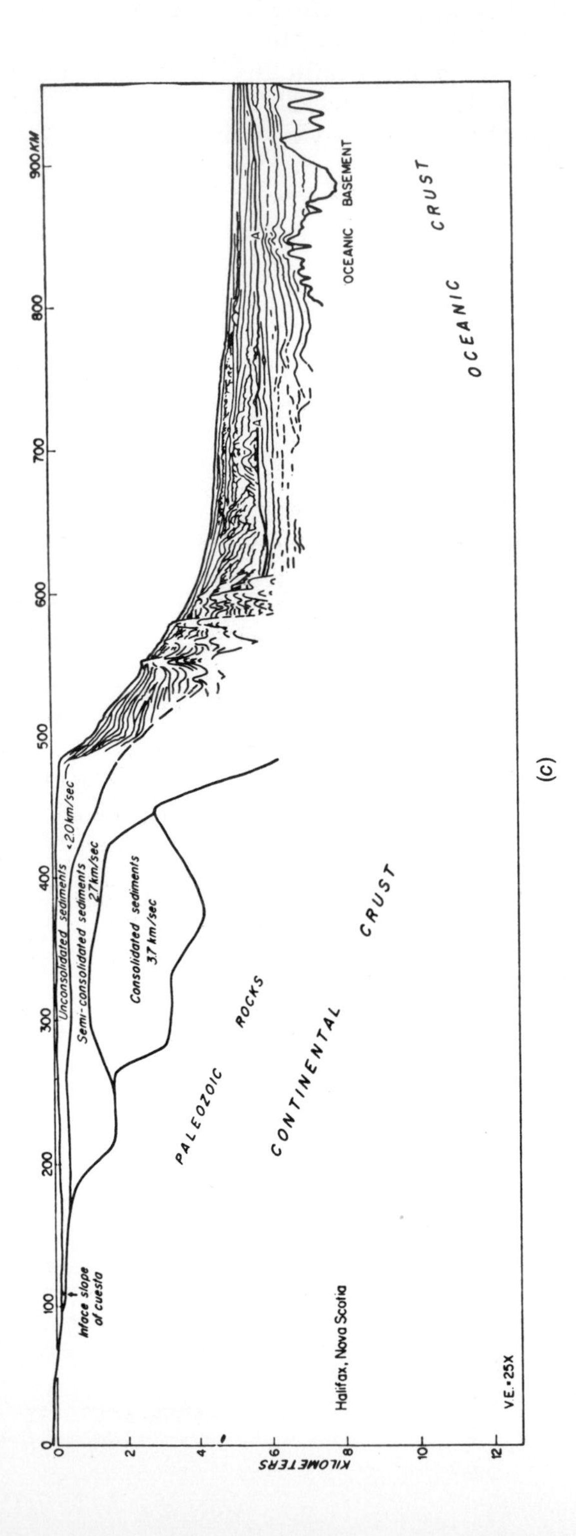

1963) oriented transverse to the structural trends of the older mountain system and the modern continental margin structure. A basement high lies under the Florida Peninsula and the geosyncline-like accumulation does not cross northern Florida into the Gulf of Mexico. Seismic and magnetic studies east of Florida in the region of the Blake Plateau and the Bahama basin have demonstrated that this is an area of complex structure, unlike that farther north. Several major faults have been postulated, and the basement may be complexly block-faulted (Fig. 19-3). Block faulting developed on a regional scale in the Appalachians during the Triassic. This block faulting is thought to be an expression of the extension as the Atlantic Ocean began to open. Block faulting may thus be widely developed in basement rocks along Atlantic continental margins. One fault of northerly trend is mapped along the steep slope from the Blake Plateau to the Blake Bahama basin. Other faults of eastern trend cross the Blake Plateau and strike more nearly parallel with the northern coast of the Gulf of Mexico.

The basement structure in southern Florida is known from deep wells. Rocks of Precambrian to Silurian age are found in wells deep enough to penetrate the Cretaceous Coastal Plain sediments (Applin, 1951). Magnetic anomalies in this region and on the Blake Plateau trend toward the west, across the Appalachian structural trend and toward athe Ouachitas (King, 1959). It has been suggested that the Ouachita structures may continue east toward the Bahamas (Drake, Heirtzler, and Hirshman, 1963).

Plate-Tectonic Interpretation of Atlantic-Type Margins

Atlantic-type margins generally occur along the trailing edges of continents that are moving away from a spreading center or that are stable. The continent-ocean transition along these margins is passive because there is little

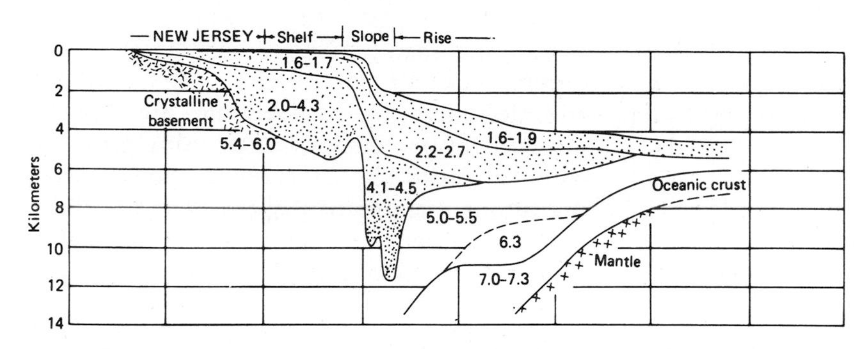

FIGURE 19-5 Deep seismic refraction section across the continental margin east of Cape May, New Jersey. (*From Drake, Ewing, and Stockard, 1968.*)

FIGURE 19-6 Depth to mantle in kilometers based on seismic refraction measurements. Contours show depth to seismic velocities of 7.8 km/sec, except in stippled areas where velocities are 7.2 to 7.7 km/sec. Dashed line near land is 200-m depth contour. (*Adapted from Drake and others, 1968, Emery and others, 1970.*)

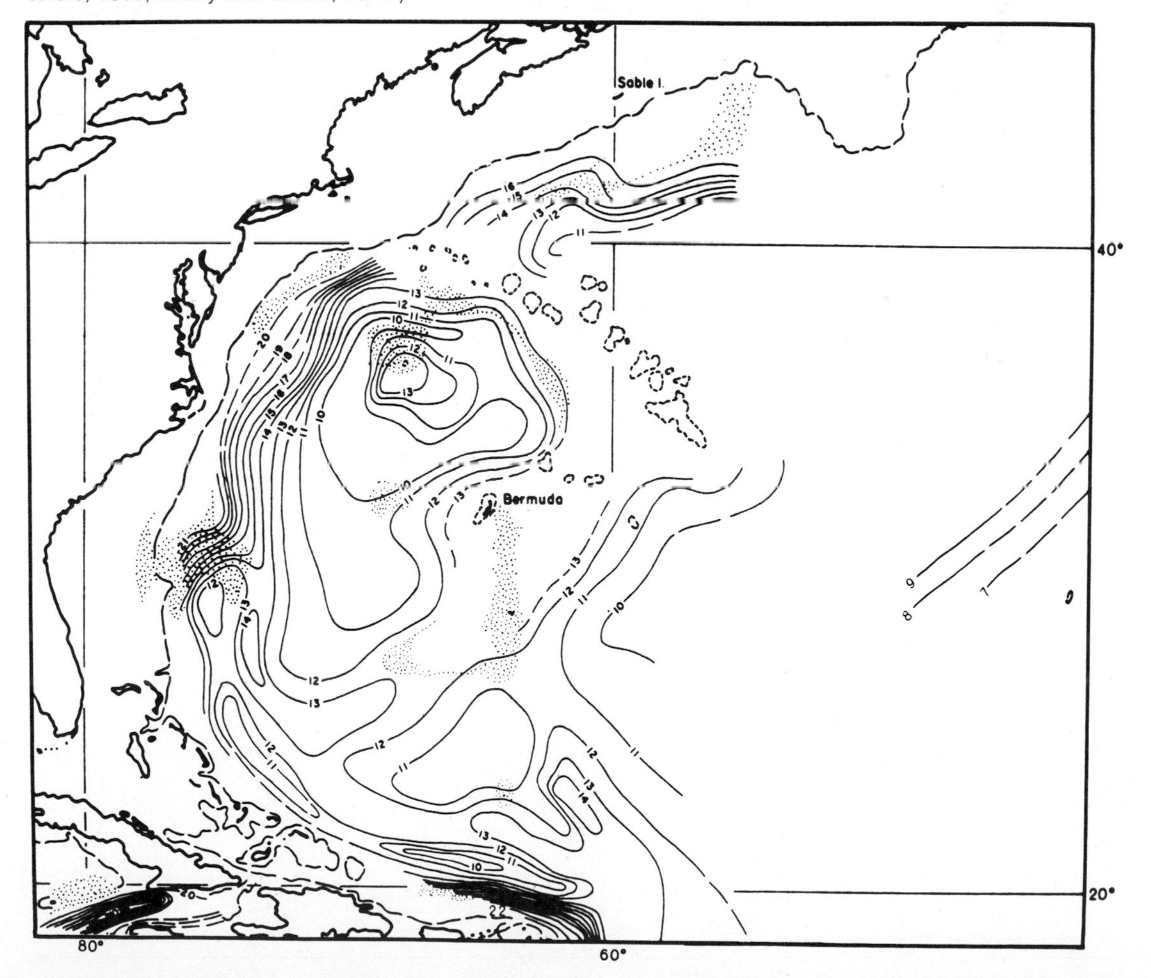

or no relative movement between the continental plate and the immediately underlying mantle. The continent is moving with the underlying mantle. Thus these margins are neither sites of subduction, the associated seismic and volcanic activity, nor are the margins sites of immediately impending plate collision. Refer to Fig. 22-1 which shows one of the reconstructions of Pangea. Almost all the margins which originally were located within the reconstruction assembly are now consid-

FIGURE 19-7 Isopach map (in kilometers) of thickness of all sediments, based essentially on difference in depths for topography of sea floor and of basement. Series of seamounts and ridge complexes are contoured as continuous features. (*From Emery and others, 1970.*)

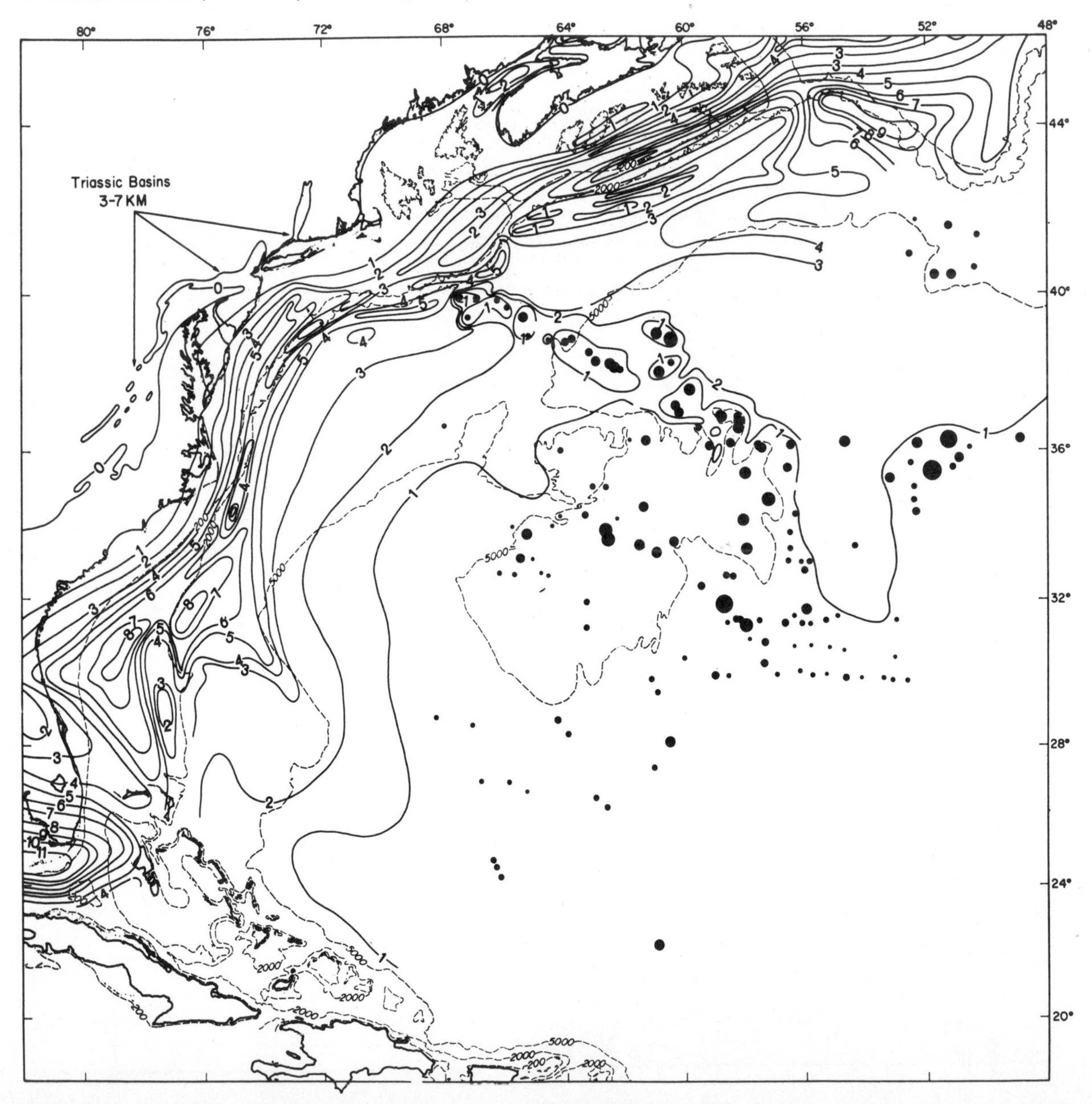

ered Atlantic-type. The outside margins of the assembly and the marginal zones which now connect adjacent continents are unstable types.

Long segments of the Atlantic-type margins are characterized by close proximity of Precambrian shields or stable platforms to the ocean basin. This is especially true for South America, Africa, India, and Australia. However, the margin between North America, Africa, and Europe is largely oriented along a Paleozoic orogenic belt which was inactive by the time the last break up of Pangea began.

FIGURE 19-8 Isopach map (in meters) of thickness of Cenozoic sediment, based essentially on difference in depths for topography of sea floor and of horizon A. (*From Emery and others, 1970.*)

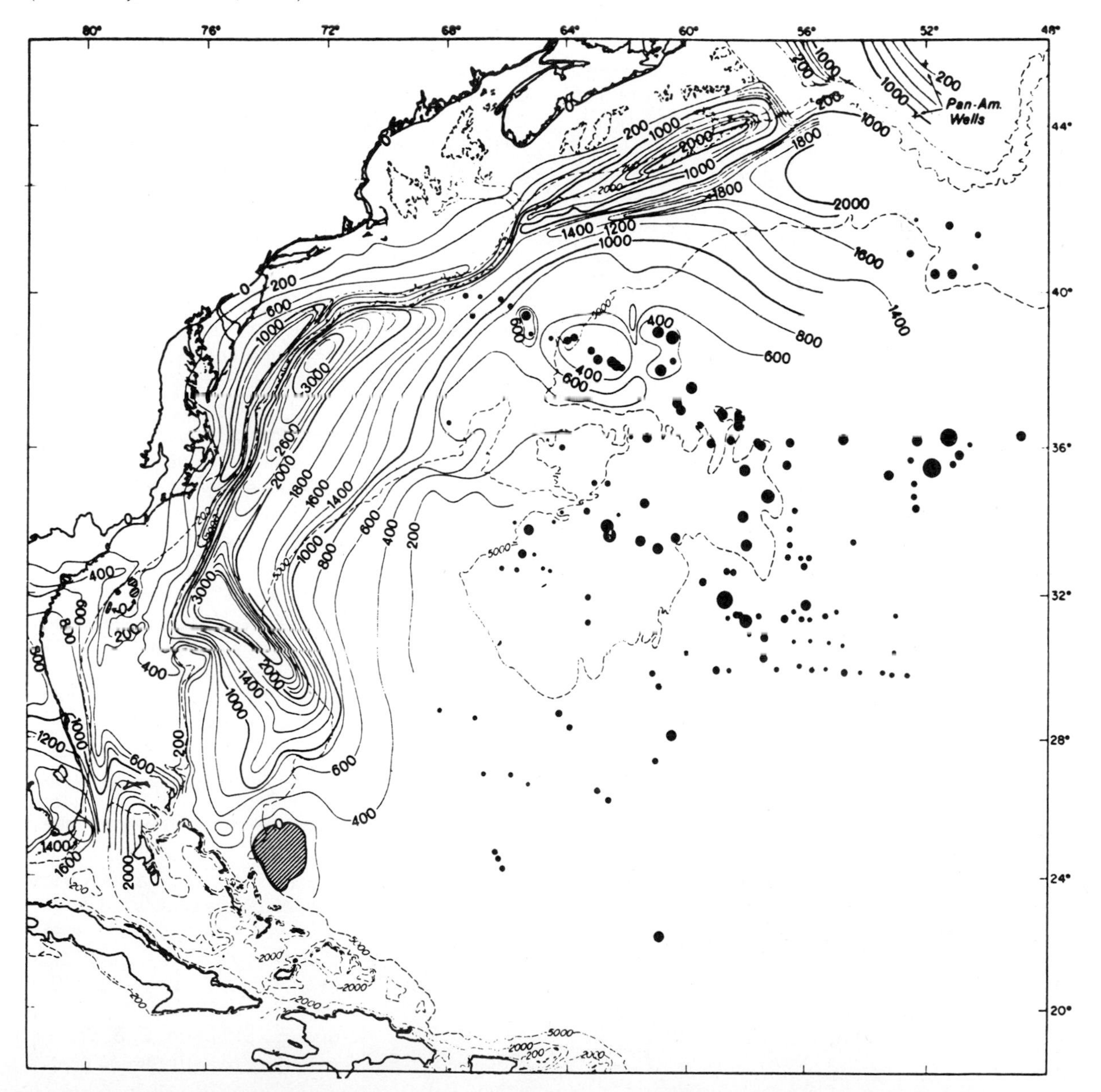

REFERENCES

Amery, G. B., 1969, Structure of Sigsbee scarp, Gulf of Mexico: Am. Assoc. Petroleum Geologists Bull., v. 53, no. 12, p. 2480–2482.

Antoine, J. W., and **Ewing, John**, 1963, Seismic refraction measurements on the margin of the Gulf of Mexico: Jour. Geophys. Res., v. 68, p. 1765–1996.

——— and **Harding, J. L.**, 1965, Structure of the continental shelf, northeastern Gulf of Mexico: Am. Assoc. Petroleum Geologists Bull., v. 49, p. 157–171.

Applin, P. L., 1951, Preliminary report on buried pre-Mesozoic rocks in Florida and adjacent states: U.S. Geol. Surv. Circ. 91.

Behrendt, J. C., Schlee, John, Robb, J. M., and **Silverstein, K.**, 1974, Structure of the continental margin of Liberia, West Africa: Geol. Soc. America Bull., v. 85, no. 7, p. 1143–1158.

Bryant, W. R., Antoine, J. W., Ewing, M., and **Jones, Bill**, 1968, Structure of Mexican continental shelf and slope, Gulf of Mexico: Amer. Assoc. Petroleum Geologists Bull., v. 52, no. 7, p. 1204–1228.

Butler, L. W., 1970, Shallow structure of the continental margin, southern Brazil and Uruguay: Geol. Soc. America Bull., v. 81, no. 4, p. 1079–1096.

Crosby, Gary W., 1971, Gravity and mechanical study of the Great Bend in the Mexia-Talco fault zone, Texas: Jour. Geophys. Res., v. 76, no. 11, p. 2690ff.

Drake, C. L., and **Woodward, H. P.**, 1963, Appalachian curvature, wrench faulting, and off-shore structures: New York Acad. Sci. Trans., ser. II, v. 26, p. 48–63.

———, **Ewing, J. I.**, and **Stockard, Henry**, 1968, The continental margin of the eastern United States, *in* Symposium on continental margins and island arcs, 3rd, Zurich, 1967: Can. Jour. Earth Sci., v. 5, no. 4, pt. 2, p. 993–1010.

———, **Ewing, M.**, and **Sutton, G. H.**, 1959, Continental margins and geosynclines: The east coast of North America, north of Cape Hatteras: Phys. Chem. Earth, v. 3, p. 110–198.

———, **Heirtzler, J.**, and **Hirshman, J.**, 1963, Magnetic anomalies off eastern North America: Jour. Geophys. Res., v. 68, no. 18.

Emery, K. O., 1966, Atlantic continental shelf and slope of the U.S.: U.S. Geol. Survey Prof. Paper 529-A.

——— 1968, Shallow structure of continental shelves and slopes: Southeastern Geol., v. 9, no. 4, p. 173–194.

——— and **Zarudzki, E. F. K.**, 1967, Seismic reflection profiles along the drill holes on the continental margin off Florida: Geol. Soc. America Prof. Paper 581-A.

——— and others, 1970, Continental rise off eastern North America: Amer. Assoc. Petroleum Geologists Bull., v. 54, no. 1, p. 44–108.

——— 1975, Continental margin off western Africa: Angola to Sierra Leone: Am. Assoc. Petroleum Geologists Bull., v. 59, no. 12, p. 2209–2265.

Ewing, John, Worzel, J. L., and **Ewing, Maurice**, 1962, Sediments and oceanic structural history of the Gulf Mexico: Jour. Geophys. Res., v. 67, p. 2509–2527.

Ewing, Maurice, 1972, Geology and history of the Gulf of Mexico: Geol. Soc. America Bull., v. 83.

——— and **Antoine, J. W.**, 1966, New seismic data concerning sediments and diapiric structures in Sigsbee deep and upper continental slope, Gulf of Mexico: Am. Assoc. Petroleum Geologists Bull., v. 50, no. 3, p. 479–504.

Francheteau, Jean, and **Le Pichon, X.**, 1972, Marginal fracture zones as structural framework of continental margins in South Atlantic Ocean: Am. Assoc. Petroleum Geologists Bull., v. 56/6, p. 991ff.

Garrison, L. E., 1970, Development of continental shelf south of New England: Am. Assoc. Petroleum Geologists Bull., v. 54, no. 1, p. 109–124.

Gough, D. I., 1967, Magnetic anomalies and crustal structure in eastern Gulf of Mexico: Am. Assoc. Petroleum Geologists Bull., v. 50, no. 2, p. 200–211.

Hedberg, H. D., 1970, Continental margins from viewpoint of the petroleum geologist: Am. Assoc. Petroleum Geologists Bull., v. 54, no. 1, p. 3–43.

Hersey, J. B., Bunce, E. T., Wyrick, R. F., and **Dietz, F. T.**, 1959, Geophysical investigation of the continental margin between Cape Henry, Virginia, and Jacksonville, Florida: Geol. Soc. America Bull., v. 70.

King, E. R., 1959, Regional magnetic map of Florida: Am. Assoc. Petroleum Geologists Bull., v. 43, no. 12.

Krause, D. C., 1966, Seismic profile showing Cenozoic development of the New England continental margin: Jour. Geophys. Res., v. 71, no. 18, p. 4327–4332.

Ludwig, W. J., Nafe, J. E., Simpson, E. S. W., and **Sacks, S.**, 1968, Seismic-refraction measurements on the southeast African continental margin: Jour. Geophys. Res., v. 73, no. 12.

McMaster, Robert L., DeBoer, Jelle, and **Ashraf, Asaf,** 1970, Magnetic and seismic reflection studies on continental shelf off Portuguese Guinea, Guinea, and Sierra Leone, West Africa: Am. Assoc. Petroleum Geologists Bull., v. 54, no. 1, p. 158–167.

Massingill, J. V., Bergantino, R. N., Fleming, H. S., and **Feden, R. H.,** 1973, Geology and genesis of the Mexican ridges: Western Gulf of Mexico: Jour. Geophys. Res., v. 78, no. 14, p. 2498ff.

Rona, Peter A., 1970, Comparison of continental margins of eastern North America at Cape Hatteras and northwestern Africa at Cap Blanc: Am. Assoc. Petroleum Geologists Bull., v. 54, no. 1, p. 129–157.

Sheridan, R. E., and others, 1966, Seismic-refraction study of continental margin east of Florida: Am. Assoc. Petroleum Geologists Bull., v. 50, no. 9, p. 1972–1991.

Uchupi, Elazar, and **Emery, K. O.,** 1967, Structure of continental margin off Atlantic coast of U.S.: Am. Assoc. Petroleum Geologists Bull., v. 51, no. 2, p. 223–234.

Wilhelm, Oscar, and **Ewing, Maurice,** 1972, Geology and history of the Gulf of Mexico: Geol. Soc. America Bull., v. 83, no. 3, p. 575ff.

20

Divergent Junctions

Divergent Plate Boundaries

According to plate-tectonic theory the earth's lithosphere can be subdivided into plates or relatively stable blocks which are bound by various types of structural zones. The boundaries include constructive zones wherein new oceanic crust is formed and from which plates diverge, subduction zones in which oceanic crust sinks to become destroyed as it is assimilated in the north, and fault zones along which lateral movement of plates occurs. This chapter is devoted to divergent zones—the oceanic ridge system where sea-floor spreading is taking place, and to the landward continuation and expression of these ridges. We will also consider some zones within continental plates which exhibit extensional features suggestive of spreading, the rift systems.

Methods Used to Study the Ocean Basins

Their size alone is enough to emphasize the importance of the structure of the ocean basins in any global analysis of tectonics. The conventional methods used on land cannot be applied in the oceans except in the marginal zones and on islands, and many of these islands are volcanic. Thus most of what we know about the oceans is derived from indirect methods of study. Extensive programs of drilling are carried on by petroleum companies on continental shelves, and many shallow wells have been drilled in the oceans as part of the Joides Project. Seismic reflection and refraction techniques are used to provide cross sections of the oceanic crust. Gravity measurements, now made at sea both in submarines and on surface ships, provide data on density distribution in the crust and upper mantle. Magnetic-field measurements are used to detect anomalies related to near-surface distribution of rocks containing different amounts of magnetic minerals, differences in field intensity, and polarity. Earthquake seismology is used to outline seismic and aseismic areas, to determine apparent initial directions of movement on faults, and to analyze crustal structure through surface-wave dispersion.

Heat-flow measurements give an indication of the crustal temperature and heat flux through various types of oceanic crustal elements. Analysis of submarine topography has been one of the most valuable tools. Because processes of erosion are ineffective in the oceans, the topography often faithfully reflects the underlying rock structure.

THE OCEANIC RIDGE SYSTEM

The existence of the mid-Atlantic ridge, a largely submerged mountainous ridge located along the center line of the North and South Atlantic Ocean, has long been known, but the discovery in the 1950s that this ridge is only a small part of a much longer and essentially continuous ridge system had momentous implications for tectonic theories. The system extends almost exactly down the center of the North and South Atlantic Oceans and then passes around the southern tip of Africa into the Indian Ocean, where it splits, Fig. 20-1. One section goes north into the Gulf of Aden and the Red Sea. The other branch trends southeast, south of Australia, across the South Pacific, then northwest by Easter Island into the Gulf of California. Subsidiary ridges run from the vicinity of Easter Island toward the Cape of Good Hope, from the East Pacific Rise to Chile, and from the southwest Pacific toward New Zealand. The north end of the mid-Atlantic ridge in Iceland is traced into the Arctic near Spitzbergen across the Arctic Ocean.

FIGURE 20-1 Location of plate boundaries and spreading oceanic ridges (after Le Pichon, 1968). Computed movements were derived from rates of spreading determined from magnetic data and from orientations of fracture zones along features indicated by double lines. The extensional and compressional symbols in the legend represent rates of 10 cm/year; other similar symbols are scaled proportionally. Historically active volcanoes (Gutenberg and Richter, 1954) are denoted by crosses. Open circles represent earthquakes that generated tsunamis (seismic sea waves) that were detected at distances of 1,000 km or more from the source (from Isacks and others, 1968). *(From Isacks, Oliver, and Sykes, 1968.)*

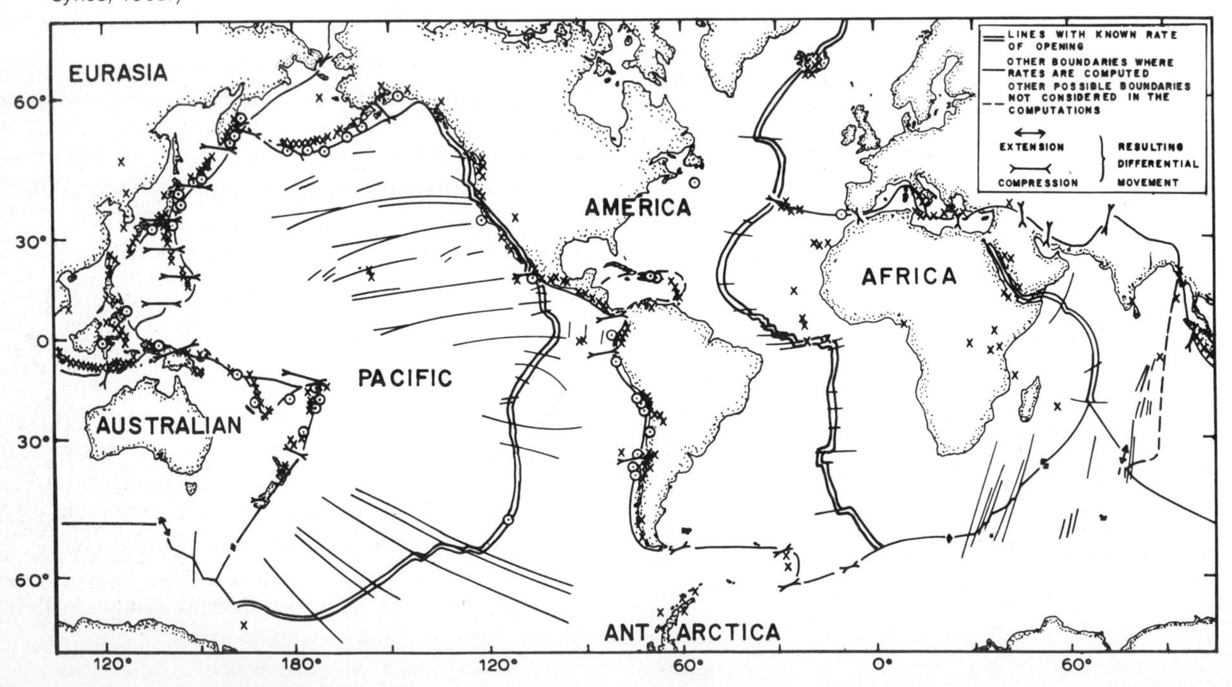

Mid-Atlantic Ridge

The mid-Atlantic ridge rises from abyssal plains on either side to a crest that comes to the surface as volcanic islands at a number of places. Thus the relief is over 6,000 m, and the width of the whole ridge province is over 2,000 km. Heezen has identified plateau-like features on the flanks of the ridge, called *steps,* which are interpreted as tilted fault blocks. High mountains near the ridge crest ± 800 m above the adjacent plateaus often have relief of several hundred meters. The basins within and between the step-like plateaus are flat-floored, sedimentary basins interpreted as tilted or downdropped fault blocks.

Iceland is located near the northern end of the mid-Atlantic ridge, and a major graben of the same trend is found in central Iceland, where modern earthquakes and volcanism are associated with the graben (Fig. 20-2). That this is an active fault zone is also shown by the open fissures found there. Topography north of Iceland is not as high as that to the south, but a ridge with local rift valleys and on which shallow-focus earthquakes are concentrated can be traced toward Spitzbergen and into the Arctic, where an ocean ridge and seismic belt is again identified.

The oceanic ridge system is characterized over most of its length by a distinct topo-

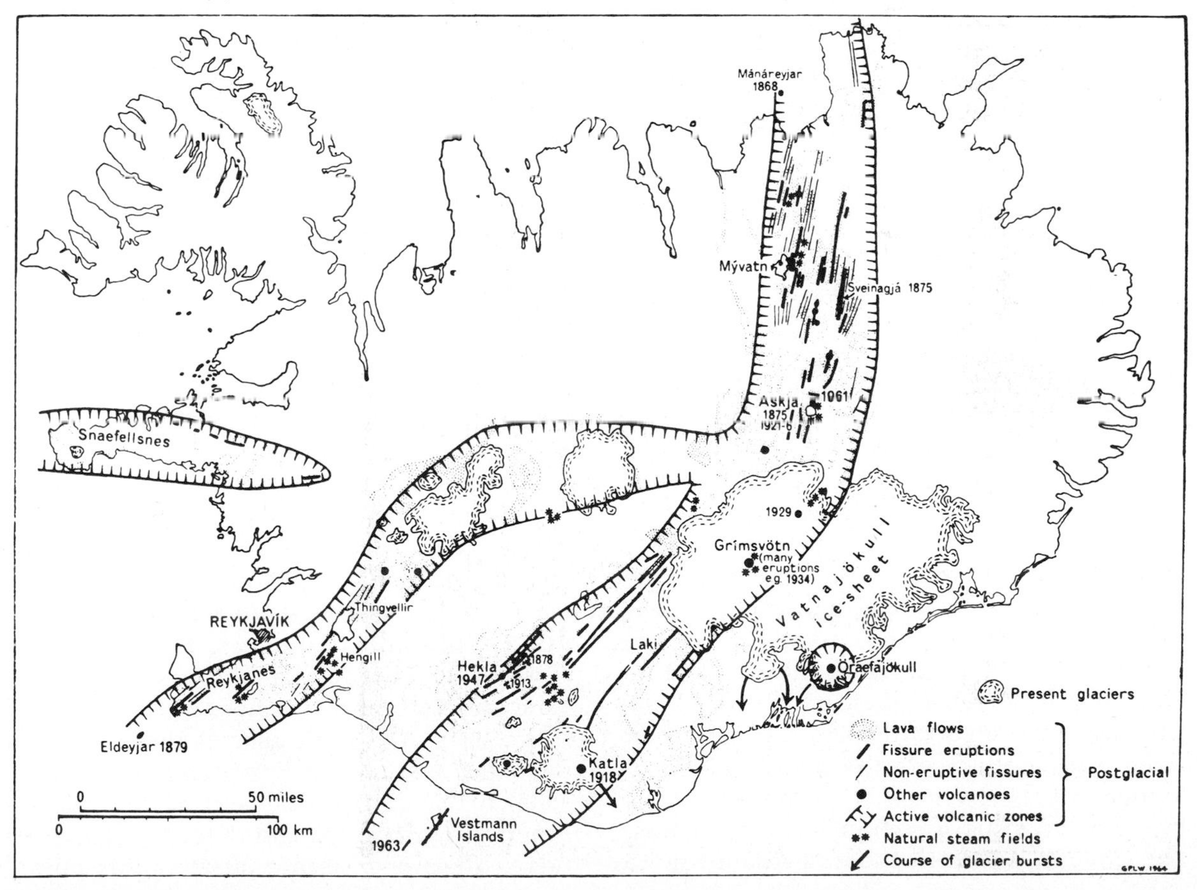

FIGURE 20-2 Map of Iceland showing the postglacially active volcanic zones and the distribution of postglacial volcanoes and fissures. (*From Gibson and Walker, 1964, in Thorarinsson, 1965.*)

graphic rise, but the character of that rise varies from place to place. Volcanic seamounts, some of which rise high above sea level, are a common feature along some sections of the ridge system. Some of these are still active. The ridge system is generally characterized by a broad upwarp of the ocean floor. Some sections of the sea floor are smooth, especially on the flanks of the ridge, but many sections of the ridge are affected by block faulting. These faults are expressed by sharp topographic breaks, and many of these scarps face toward the ridge crest. The crestal portion is marked by a central valley bounded on both sides by in-facing scarps, and this valley is interpreted as a graben. It is as much as 1,000 m deep in places and up to 10 km wide. The central graben is discontinuous along the ridge crest, is a compound feature in places, and is missing altogether in others. A number of volcanoes are located close to the edge of the valley, suggesting that the lava found its way to the surface along a fault which was open as a result of extensional movements.

The best documented of these transverse zones in the Atlantic lie just north of the equator, where they cut and displace the mid-Atlantic ridge (Heezen and others, 1964). The Chain and Romanche fracture zones are two of these, all of which show left-lateral displacements. The displacements in this case can be shown by means of detailed topography, Fig. 20-3. The Romanche zone appears as a deep, narrow trench which is about 20 km wide at a depth of 1,000 m. The walls are very steep, 45° slopes in places, and the floor of the trench is broken by a longitudinal ridge. The maximum depth in the Romanche zone, the Vema deep, is 2,000 m, and thus the fracture zone has local relief of nearly 3.3 km. The Chain zone located to the south is similar in trend and depth. From the displacements of topography, it appears that the movement has been in the same direction along the faults. The Chain zone offsets the mid-Atlantic ridge about 300 km; the Romanche zone offsets it about 500 km.

The Great Fault Zones of the Pacific Basin

Islands in the Pacific show a strong alignment. Some lie on arcs, but those in mid-ocean lie along more or less straight lines. As submarine topography has gradually become better known, many seamounts and guyots are also found to be aligned, and long narrow ridges and depressions in the deep-sea floor have been defined. Some volcanic islands, seamounts, and guyots probably owe their alignment to faults or fracture systems in the oceanic crust through which lavas have found their way to the surface. Other faults and fracture zones are not covered by volcanic products and are still clearly defined by offsets in the submarine topography.

The Clipperton zone (Fig. 20-1) was one of the first major fracture zones to be described (Menard, 1955, 1958). It was recognized by the anomalous topography associated with it. The Guatemala basin, situated at its eastern end, is bounded on the north by a sharp ridge, the Tehuantepec ridge, and on the south by a sharply defined depression which lines up with the Clipperton zone. To the west the zone is represented by a broad rise, with a central trough between 400 and 800 m deep where the zone cuts across the East Pacific rise. A volcano-studded ridge, 70 km wide and 500 km long with a narrow trough on the north side, lies west of this. Local relief along this trough reaches 6,000 m. This lineament can be traced east-west for nearly 5,000 km; thus it is one of the longest features in the crust. Submarine topography has been offset along the zone, and volcanoes occur along one side of it for a long distance. It cuts across the still larger feature, the East Pacific rise, and it ends in the east by intersecting the middle America trench at the end of the Tehuantepec ridge. The trench is not offset at this point, but it is deflected, the depth to the M discontinuity is altered, the depth of the trench changes, and the character of volcanism is altered (Fisher and Shor, 1958).

The Clipperton is but one of a number of similar east-west structures now known in the eastern Pacific, Fig. 20-1. The offset on the zones has been inferred by offsets of topography, by seismic studies, and by the use of offsets in contours of magnetic anomalies (Fig. 20-4). The Mendocino zone is one of the most spectacular of these zones. The elevation of the sea floor is about 1 km higher on the north side of the fault scarp, an asymmetrical ridge with a south-facing escarpment between 1.5 and 3 km high. The fault separates two provinces of different physiographic characteristics. Narrow northeast-trending ridges and troughs occur north of it; a relatively smooth abyssal plain lies south of it. Seismic work has shown the region south of the fault to be normal oceanic crust, while that to the north is abnormal in that the velocities are not those found normally at the M discontinuity and in that gravity surveys along the fault reveal a thick low-density upper mantle layer north of the fault which probably is

FIGURE 20-3 Bathymetric sketch of portions of the Chain and Romanche fracture zones. Arrows indicate a suggested pattern of flow of the coldest bottom water based on interpretation of bottom temperatures, echo soundings, and topographic trends. Basic contour interval is 200 fathoms, except below 3,000 fathoms, where the slopes are too steep for contour portrayal at this scale, and between 1,400 and 2,000 fathoms, where topography is too irregular to permit detailed contours with the present control. (Survey of chain fracture zone by Bunce, contours by Heezen.) (*From Heezen and others, 1964.*)

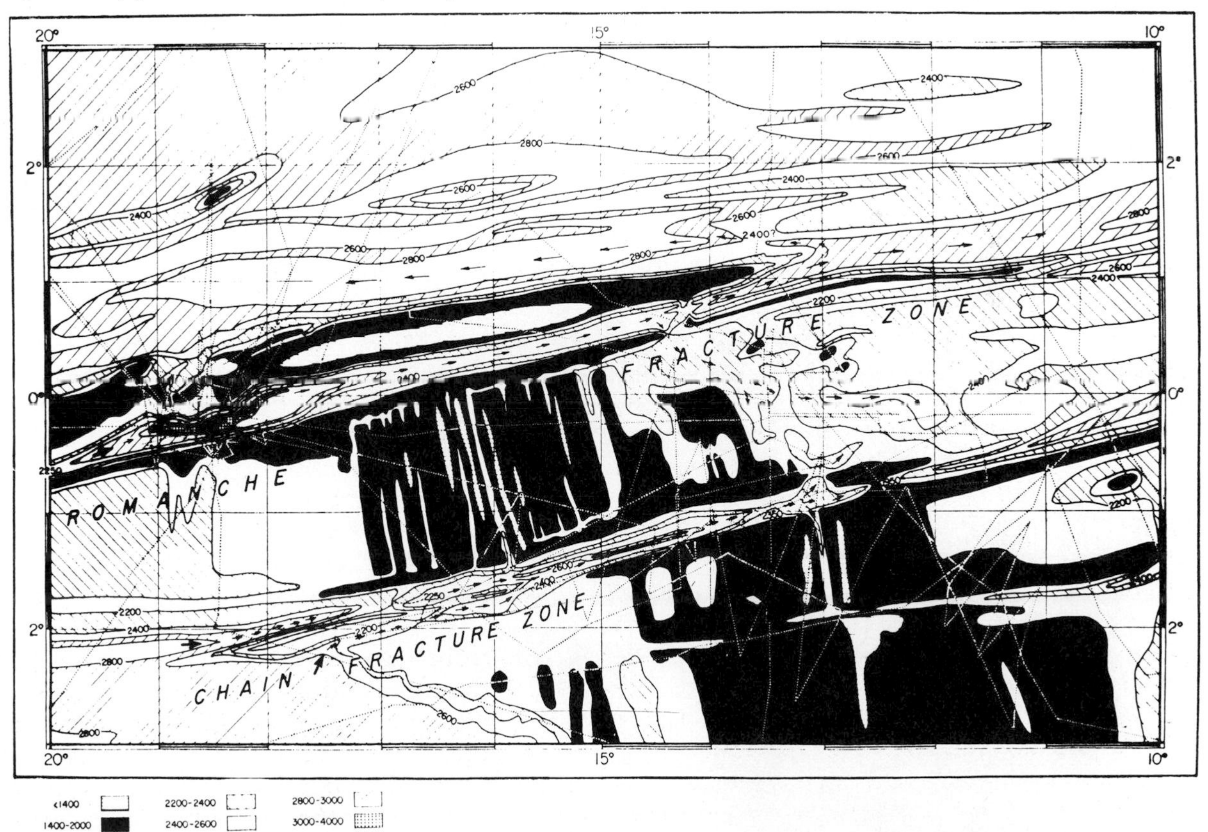

chemically different from that south of the fault. The high relief and abnormal mantle across the fault are found most prominent toward the eastern end of the fault. West of longitude 130° W these effects die out (Dehlinger, Couch, and Gemperle, 1967).

The north-south trend of the magnetic anomalies is striking, as are the offsets in it. This pattern has been traced for over 2,500 km, and it may extend even farther west as well. To determine displacements, the anomalies are shifted along each zone until a best fit is obtained.

Seismic Activity along the Oceanic Rises

Studies of the regional distribution of earthquakes in the oceans reveal that the oceanic ridge system is the site of frequent shallow-focus earthquakes of low magnitude. The activity is generated near the surface and is closely associated with the central graben and with the transverse fault zones. In contrast, large sections of the ridge flanks and the abyssal plains are almost aseismic. Focal-plane solutions* for first motions of earthquakes which occur along the transverse faults has been one of the most important proofs of the transform nature of these faults. The focal-plane solutions indicate that the sense of motion is opposite from that which might be expected for a strike-slip fault, with the apparent offset observed in the alignment of the ridge crest. This is precisely what Wilson predicted the movement should be on a transform fault.

Seismic Evidence for Crustal Structure in the Ocean Basin

Information about crustal structure comes from seismic refraction and reflection

FIGURE 20-4 Map of the total magnetic intensity from 32 to 42° N after removal of the regional field. The contour interval is 50γ, and spot values are rounded off to the nearest 5γ beyond the extreme value recorded. (*From Mason and Raff, 1961.*)

* Focal-plane solutions are based on the observed direction of initial motion at stations located in different quadrants from the earthquake. The method is explained by Hodgson (1964), and it has been extensively applied by Sykes (1967) and others.

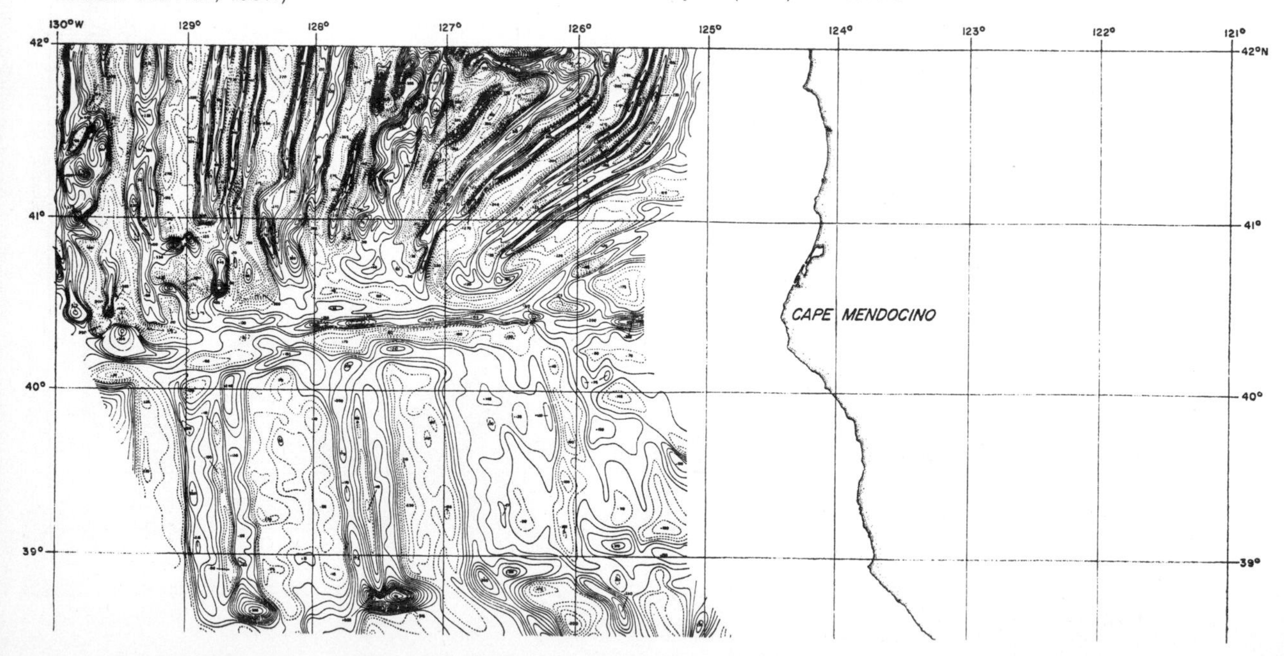

methods, from gravity, heat flow and magnetic studies, and for the upper parts, from drilling. The velocity structure and inferred rock types typical of the oceanic crust, Fig. 20-5, consist of a simple widely representative structure. The model consists of an upper layer composed of unconsolidated sediment underlain by basalt flows which often show pillow-lava characteristics. The lower parts of the basalt are altered and underlain by gabbro and possibly serpentine. The Moho occurs at a depth of from 10 to 12 km below sea level and is underlain by ultrabasic rocks.

Representative seismic sections, Fig. 20-6, across the oceanic ridge system show that the mantle boundary tends to rise slightly toward the ridge crest. Normal mantle seismic velocities of about 8 km/sec on the flanks give way to velocities of 7.2 to 7.7 km/sec under the ridge crest. It appears that altered mantle material of both lower velocity and lower density swells upward against the oceanic crustal layer under the ridge crest. This would account for the upbulge and the extension across the ridge crest. Sediment layers tend to be thin or missing over the ridge crest and thicken toward the flanks.

East Pacific Rise

That portion of the oceanic ridge and seismic zone that lies in the eastern Pacific is of particular interest because it may be traced into the North American continent. It is also significant that this portion of the ridge system is simpler than that of the Atlantic. The width of the rise, approximately 800 km, is comparable, but the topography of the East Pacific rise is smooth by comparison with the mid-Atlantic ridge, suggesting a simpler crustal structure. This conclusion is born out by the seismic section across it (Fig. 20-7). The East Pacific rise passes into the Gulf of California, where

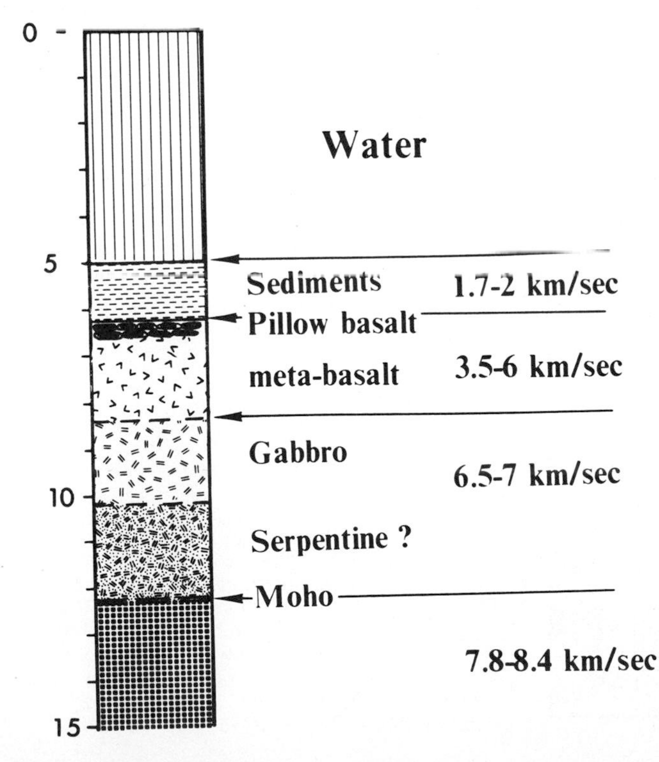

FIGURE 20-5 Velocity structure of the oceanic crust and the inferred composition of the velocity layers based on a correlation between the compressional wave velocities of oceanic rocks and seismic refraction results. (*From Fox and others, 1973.*)

it appears to pass under western North America. It is identified to the north in the Pacific north of Cape Mendocino.

Like other portions of the ridge system, shallow-focus earthquakes occur along the crest of the East Pacific rise. Heat-flow measurements show abnormally high values in a narrow zone over the crest (Fig. 20-7), the crust is thinned over the crest of the rise, and abnormally low seismic velocities are found in the mantle in a narrow zone below the crest (Menard, 1960).

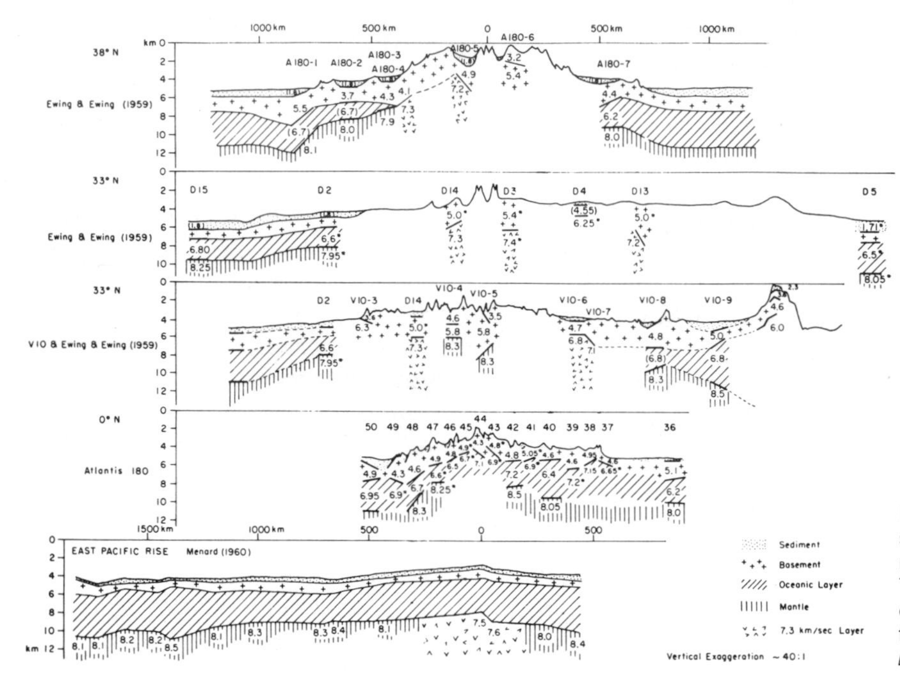

FIGURE 20-6 Seismic cross sections of the mid-Atlantic ridge and the East Pacific rise. The topography is schematic, especially between the different stations. (*From Le Pichon and others, 1965.*)

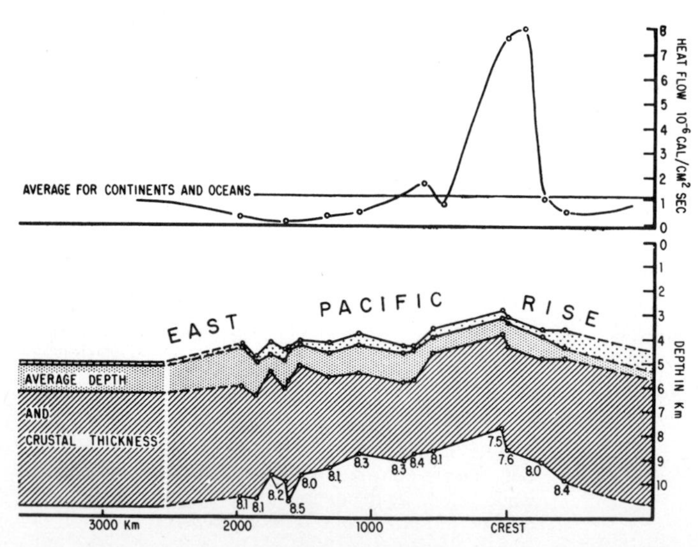

FIGURE 20-7 Crustal section and heat-flow profiles of the East Pacific rise. Velocities below the "crust" are in kilometers per second. The crust is thinned under most of the rise, but the region of very high heat flow and anomalous "mantle" velocities is confined to a narrow band on the crest. (*From Menard, 1960. Copyright 1960 by Am. Assoc. Adv. Sci.*)

Gravity Measurements over the Ridges

Measurements of the Bouguer anomalies in sections over the oceanic ridges consistently reveal a pronounced negative gravity anomaly closely associated with the ridge crest. Models have been constructed which are consistent with the seismic sections and which also allow good fits with the gravity data, Fig. 20-8 (Talwani and others, 1965). The necessary compensation for the crestal portion of the ridge can be explained if the anomalous mantle material extends laterally under the normal mantle as shown in three of the possible models. Deeper extensions of the abnormal mantle material under the ridge flanks were ruled out by the steep gradient of the Bouguer anomaly.

Heat-Flow Measurements over Oceanic Ridges

Several thousand measurements have now been made of the heat flux in the vicinity of oceanic ridges (results are summarized by Anderson, 1972). The highest values appear along the ridge crest, and some of these are as much as eight times as high as the average values for ridge flanks and deep ocean basins. Considerable variability is found along the crest, but the basins are relatively uniform

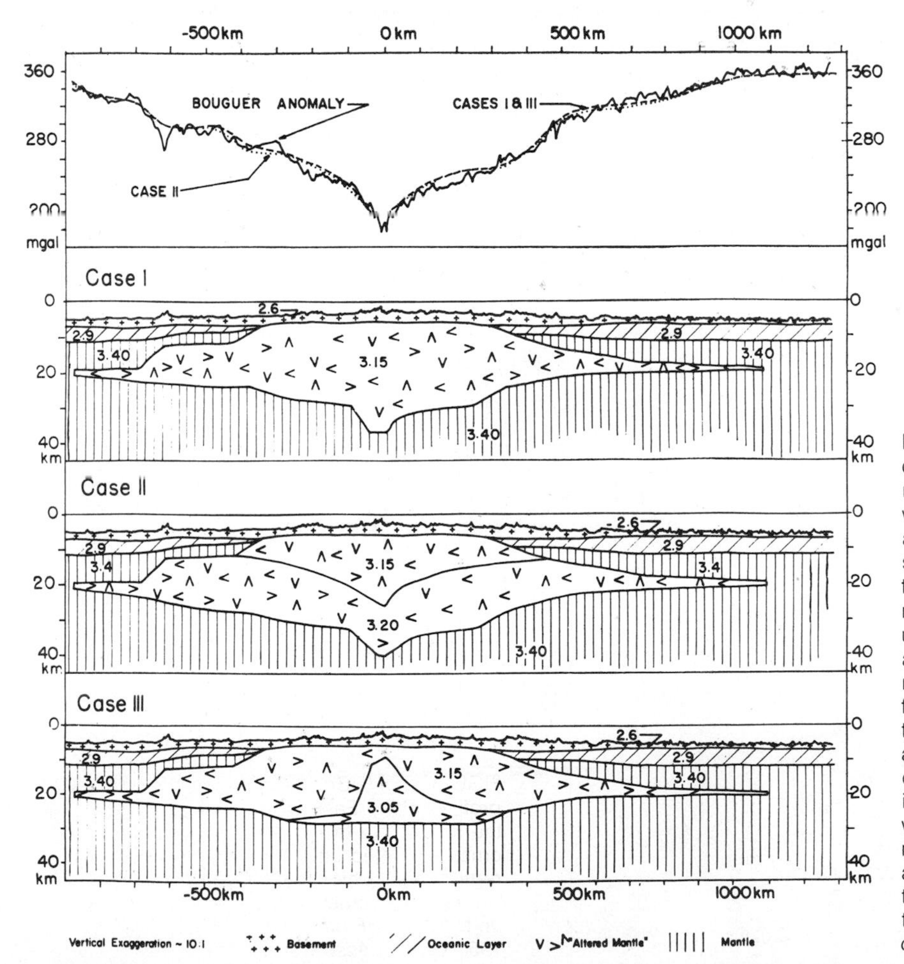

FIGURE 20-8 Three possible crustal models across the north mid-Atlantic ridge which satisfy gravity anomalies and are in accord with seismic refraction data. In all three models the anomalous mantle found seismically under the crest of the ridge is assumed to underlie the normal mantle under the flanks of the ridge. In case I the anomalous mantle is assumed to have a uniform density, in case II its density is assumed to increase downward, and in case III the material constituting the anomalous mantle is assumed to be lighter near the axis of the ridge. (*From Talwani and others, 1965.*)

in heat flux. Sections showing profiles of the heat-flux variation with distance from ridge crest, Fig. 20-9, show some similarities from one ridge to another, but differences appear as well. Commonly the curves have their peaks displaced from the ridge crest to a position some tens of kilometers off the crest. Farther away, the heat flux decreases until it reaches a minimum value on the flanks, and then it rises slightly before leveling off in the ocean basins.

Anderson (1972) suggests that the low values over the crest where we might expect the highest flux are due to the presence of hydrothermal activity, the circulation of water through this portion. The high values observed occur where a thin cover of sediment seals the underlying lavas, but because the water circulation is less here the heat flux is high. The trough is explained as being due to dehydration of the underlying crustal materials and associated endothermic reactions.

Oceanic rises which are spreading rapidly, like the East Pacific rise and the Galapagos ridge, have heat-flow profiles which closely approach the theoretical form expected, but the position of highest heat flux is displaced from the ridge crest over the slower spreading ridges such as the mid-Atlantic ridge, Fig. 20-9.

Magnetic Anomalies and Spreading Rates

Few geologists in the 1950s suspected that studies of variation in intensity of the magnetic field would yield such dramatic results and lead to the formulation of new tectonic theories. The results of these studies began to appear in the early 1960s with publication of

FIGURE 20-9 Composite profile normalizing to age all published heat-flow profiles with sufficient resolution to distinguish the crestal distribution in 0-to-10-million-year-old crust. Mean curves are shown. (*From Anderson, 1972.*)

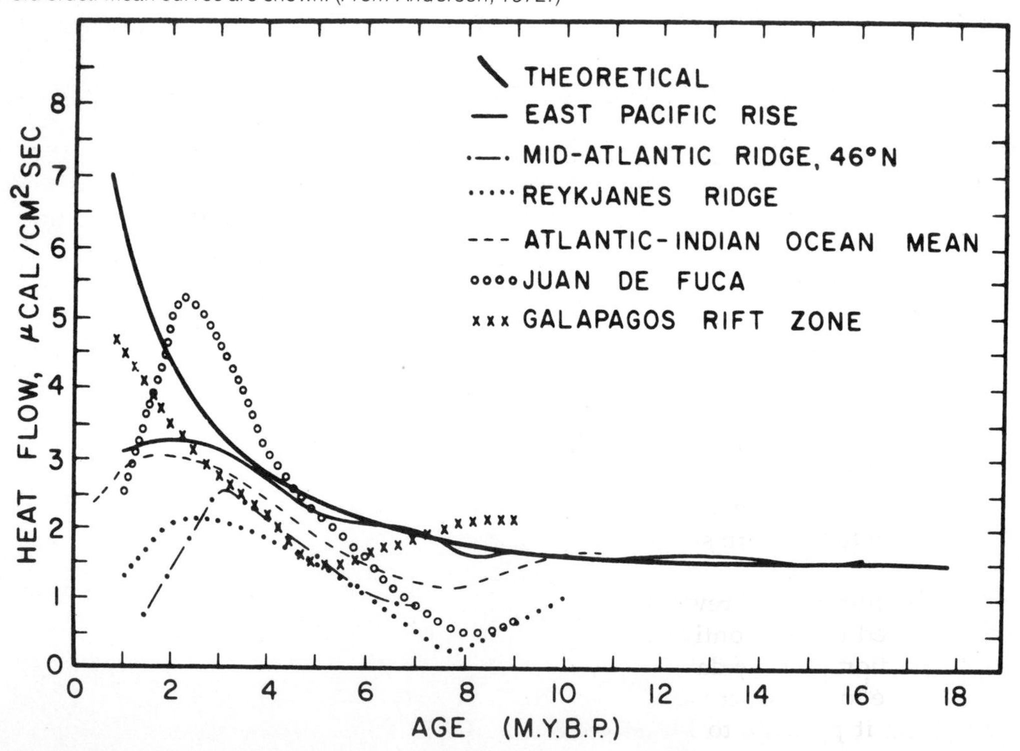

maps of the Pacific Ocean just off the North American continent, Fig. 20-4 (Raff and Mason, 1961). The contoured intensity values revealed a distinctive pattern, linear belts of high intensity alternating with belts of low magnetic intensity. The pattern revealed distinct offsets in the anomaly belts which would have been suggestive of faulting by themselves, but because these offsets coincided with known topographic lineaments in the sea floor, the anomaly pattern simply confirmed the fault interpretation.

New ideas and discoveries developed very rapidly in the early sixties, and it was soon recognized that these belts of high and low magnetic intensity might be caused by reversals in the polarity of the earth's magnetic field, details of which were being resolved on continents. Hess (1962) and Dietz (1961) were formulating the theory of sea-floor spreading in which it was postulated that new crust is forming in the oceans along the oceanic ridges by emplacement of hot mantle materials from depth. Thus the reversals in magnetic polarity could be frozen into the new crust as it cooled and passed through the Curie point. If this were the case, the belts of low magnetic intensity would correspond to times of reversed magnetic polarity and the highs to periods of normal polarity. Vine and Matthews (1963) conducted their now-famous experiment on the Reykjanes ridge south of Iceland to test this hypothesis, and they found that the crest of the ridge is a high-intensity belt and that a pattern of anomaly belts symmetrical with the axis of the ridge is present. This has been taken as one of the most significant confirmations of the sea-floor-spreading hypothesis. Subsequently, magnetic surveys have been widely conducted with similar results.

Because the epochs of normal and reversed polarity have been worked out on continents where it is possible to radiometrically date the rocks, it has been possible to construct a magnetic time scale, making it possible to know the age of each reversal if the sequence starts with the present (Heirtzler and others, 1968).

Because the age of the polarity reversals can be inferred from the sequence of belts starting with the crest of the oceanic ridges, and because the distance of a particular belt from the ridge crest can be measured, it becomes possible to determine the rate of spreading for any portion of a ridge. Spreading rates are found to vary both from place to place today, and they have not remained constant through time, Fig. 20-10. Average spreading along the mid-Atlantic ridge has been about 1.0 to 2.0 cm/year during the Cenozoic, while parts of the Pacific ridge have spread at considerably higher rates. Comparison of spreading rate with topography along various segments of the ridge system bring out a striking relationship. High relief and abundant block faulting are associated with areas that are spreading slowly, while the more rapidly spreading segments have subdued topography and often no central rift valley.

Magnetic anomaly belts like those found near the mid-Atlantic ridge do not extend to the continental margins. A rather distinct boundary more or less parallel to the continental slope separates the magnetically disturbed from the undisturbed regions of the ocean on both sides of the North Atlantic Ocean. The boundary lies between 2,000 and 2,500 km from the axis of the mid-Atlantic ridge and roughly equidistant from it (Heirtzler and Hayes, 1967). The wide tracts of uniformly magnetized crust could represent a long period during which no reversals took place, possibly late Paleozoic, or thick accumulations of sediment may mask the anomalies.

Sediment Distribution and Sea-Floor Spreading

The distribution of sediments of various ages in the ocean basins is an independent way of determining the age of various parts of the sea floor. If various sections of sea floor were formed at a particular time, then those sections could not have older sediments deposited on them. The main problem in answering the question here is that so little deep drilling has been done. Coring of the upper tens of meters of the sea-floor sediment has been done in many places, but deep drilling has been carried out in only a few carefully selected places as part of the Joides program. One of the earlier studies, on the East Pacific rise, showed that only Pleistocene sediments occur along the crest of that rise, but pre-Pleistocene sediments are penetrated on the flanks (Burckle and others, 1967). More recently, Fischer and others (1970) summarize the basement ages determined from drilling in the Pacific. Although little data are available for the western Pacific, the general pattern of increasing age and thickness of sediments away from the east Pacific rise is clear.

FIGURE 20-10 Average spreading rates in the Atlantic and Pacific Oceans as a function of geologic time. Note the marked increases in spreading rates of all spreading systems at 85 to 100 million years B.P. (*From Larson and Pitman, 1972.*)

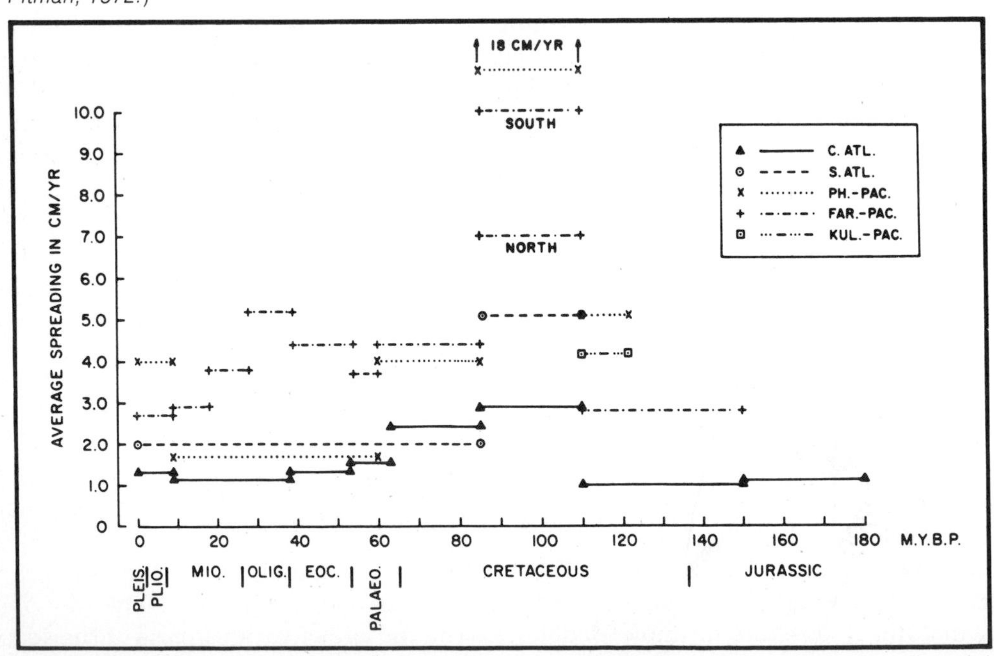

Petrology of the Oceanic Ridges*

Most of what we know about the petrology of the oceanic ridge system comes from examination of samples dredged up, a few cores, and the products of the volcanic activity along the ridge crest.

Most of the dredge hauls from the ridge crest of the mid-Atlantic ridge have contained basalt with some metabasalt altered to the greenschist facies. A more varied assemblage has been found in hauls from the transverse fault zones. These contain metabasalt, gabbro, and serpentinite as well as basalt. A few blocks of ultrabasic rocks have been found in some of the volcanic debris, and ultrabasic rock is thought to make up the mantle.

Since mantle rock is thought to be coming up toward the surface along the ridge crest, it is important to understand what reactions that rock might undergo as it encounters

* See Dickinson (1971); Hess (1962); Vogt, Schneider, and Johnson (1969).

water. The hydration reactions involving common minerals in peridotite are as follows:

Enstatite + water→
forsterite (olivine) + talc at 600–700°C

Forsterite + talc + water→
serpentine at 500–550°C

Forsterite + water→
serpentine + brucite at 350–450°C

Thus we might expect a variety of rocks, especially serpentine, to be present where the mantle rises and encounters water. When conditions are reached at which the hydration process begins to reverse, the serpentine may be altered again through dehydration which occurs at higher temperatures.

Mantle Convection and Oceanic Ridges

Thermal convection in the upper mantle is the process most frequently called on to explain the cause of sea-floor spreading. Heat-flow measurements provide the most direct indication of this convective movement. The high heat flow plus the fact that the ridges are uplifted and are sites of volcanic activity appear to substantiate the convection hypothesis. Hypothetical models of the convective plumes have been constructed, and study of the associated problems remains one of the most active topics of investigation in tectonics.* Two schematic models showing isotherms and probable movement vectors are illustrated, Fig. 20-11.

* See Hess (1962), Shaw (1973), Turcotte and Oxburgh (1969), Wesson (1972), Langseth (1969).

FIGURE 20-11 Detail of the top of the ascending plume of a convection cell. Isotherms are shown as solid lines (values in degrees Celsius in heavy numerals); liquidus temperatures for olivine tholeiite are indicated by horizontal dashed lines, temperatures in degrees Celsius. The outline of the left half of the zone of fusion is indicated by the dashed line +0, and the zone within which the predicted temperature exceeds the liquidus temperature by 200°C or more by the dashed line +200. Small arrows are velocity vectors for particle motion. (*After Oxburgh and Turcotte, 1968.*)

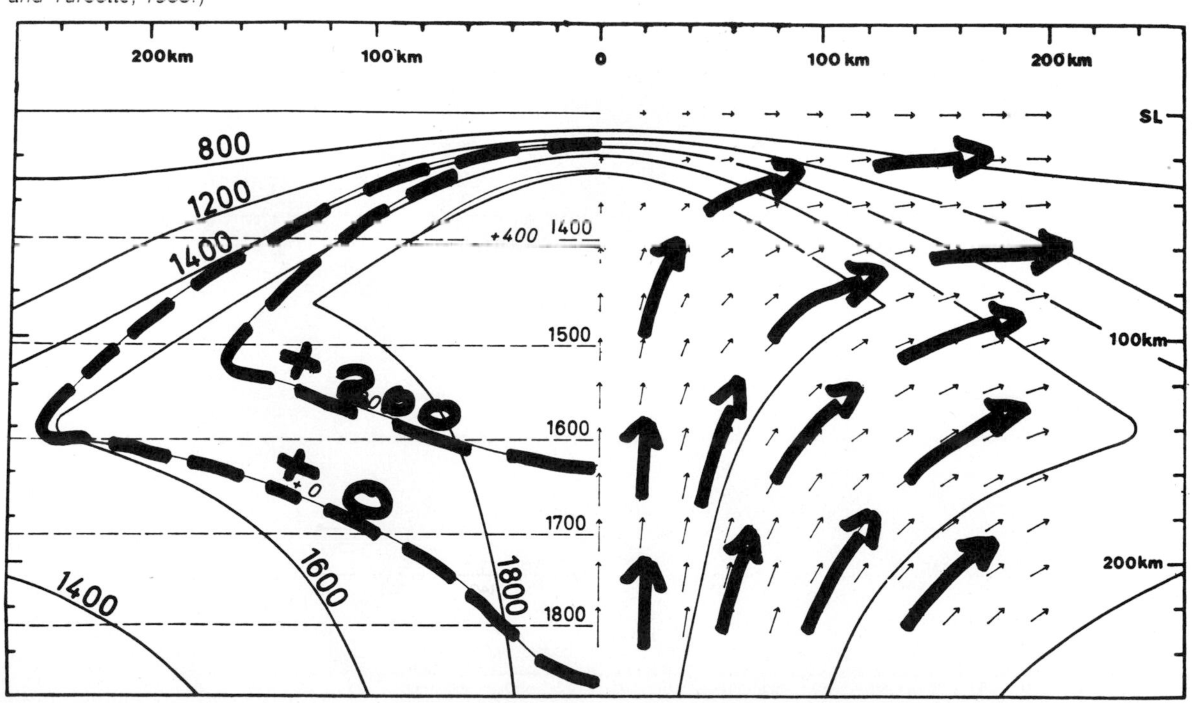

Triple Junctions and the Pacific Plates

The Atlantic Ocean has a much simpler plate model than that of the Pacific. The mid-Atlantic ridge is a divergent plate junction which divides the Atlantic into plates which are moving apart relative to the spreading center.

The pattern of magnetic anomaly belts in the Pacific is not only much more broken up by faults, but the spreading ridge is not centrally located in the ocean, and the ridge passes into the Gulf of California so that its northern continuation is less apparent. Some geologists think that the San Andreas fault is a transform fault and that the ridge continues after that offset just north of the Mendocino fault. Others think that the ridge lies under the western United States and is overridden by the continental plate. In either case the magnetic anomaly pattern in the Pacific is additionally complicated by the unusual changes of trend both in the northern Pacific and along the ridges which extend from the East Pacific rise to the northern coast of South America, Fig. 20-12. The favored interpretation now is that the Pacific Ocean should be subdivided into several plates and that three of these join at the corners to form a *triple junction*. The plate geometry (Pacific, Kula, Farallon, and Phoenix) is now partially obscured as a result of the total subduction of the Kula plate and the Kula ridge into the trenches along the northern and western side of the Pacific, and the Farallon plate is partially lost due to the overriding by North America. The plates are shown restored to a position estimated for 110 million years ago, Fig. 20-12.

FIGURE 20-12 Pacific Ocean plates and plate boundaries 110 million years ago (after Larson and Chase, 1972) plotted relative to the Atlantic Ocean continents at that time. Dotted plate boundaries show probable extensions of known features. Relative configuration accomplished by superimposing the paleomagnetic poles for the Pacific plate and North America at 110 million years B.P. and by reconstructing the Atlantic continents relative to North America. Plot construction based on the present-day location of North America, so the paleoquator appears as a curved line. (*From Larson and Pitman, 1972.*)

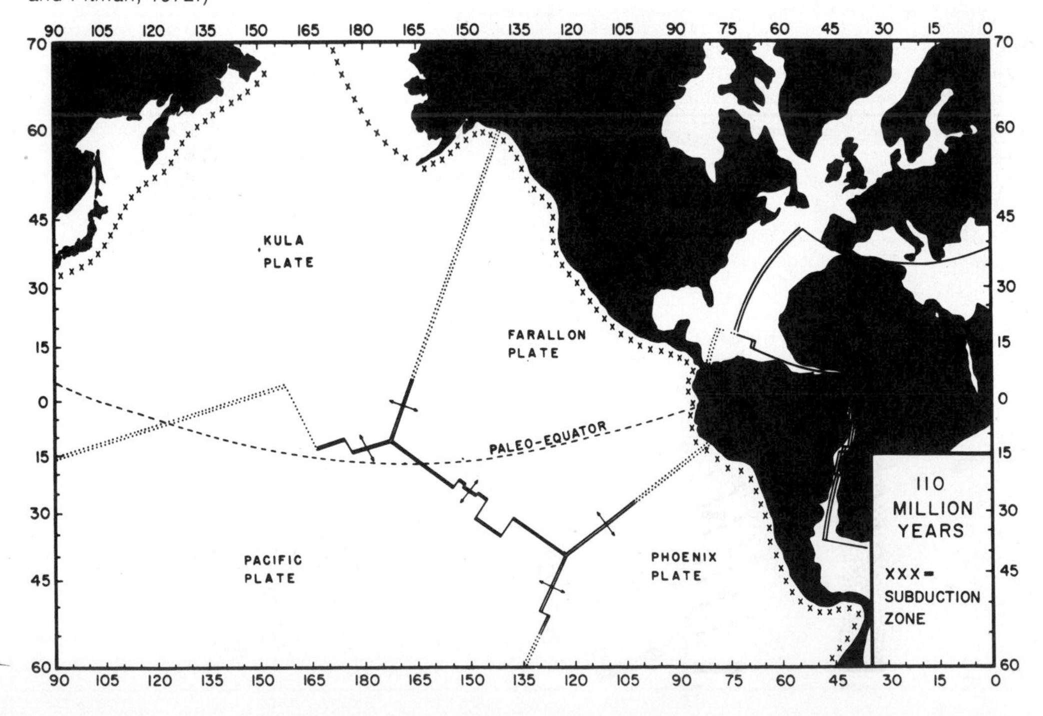

The growth of the sea floor around a triple junction is quite different from that across an ordinary two-plate boundary. The position of the triple junction may migrate as a result of differences in spreading rates of the three plate-boundary zones.

Opening and Growth of the Atlantic

The history of growth of the oceans can be traced on the assumption that the magnetic anomaly belts were formed at certain times. Analysis of the history of the Atlantic has been undertaken by several authors. The position of the continents at various times can be reconstructed from the anomaly belts as illustrated, Fig. 20-13. The opening is thought to have begun in the Triassic Period. Evidence for this comes largely from the structural features along the continental margins, where Triassic grabens and step faults constitute evidence of widespread extension. The magnetic quiet zone located on the continental rises may represent the edge of the continental plates. The major phase of opening of the Atlantic took place in the Late Cretaceous when spreading rates of nearly 4 cm/year prevailed. The spreading rate slowed down after the Cretaceous and has been between 2 and 3 cm/year during most of the Cenozoic.

The Arctic Ocean apparently did not begin to open until the early Cenozoic, and Greenland did not separate from North America until the Eocene.

Many of the seamounts in the Atlantic are located along the great fracture zones and transform faults. It is probable that the New England seamounts lie on the same zone as the

FIGURE 20-13 The relative position of Europe and Africa with respect to North America as the Atlantic opened. Blacked-in continents represent present positions. Dates for earlier positions are indicated in the map. The arrows show the path of drift of Africa and Europe away from North America. The line *AB* which joins a point in Spain and a point in Africa is shown as a guide to seeing the relative motion of Eurasia and Africa. (*From Pitman and Talwani, 1972.*)

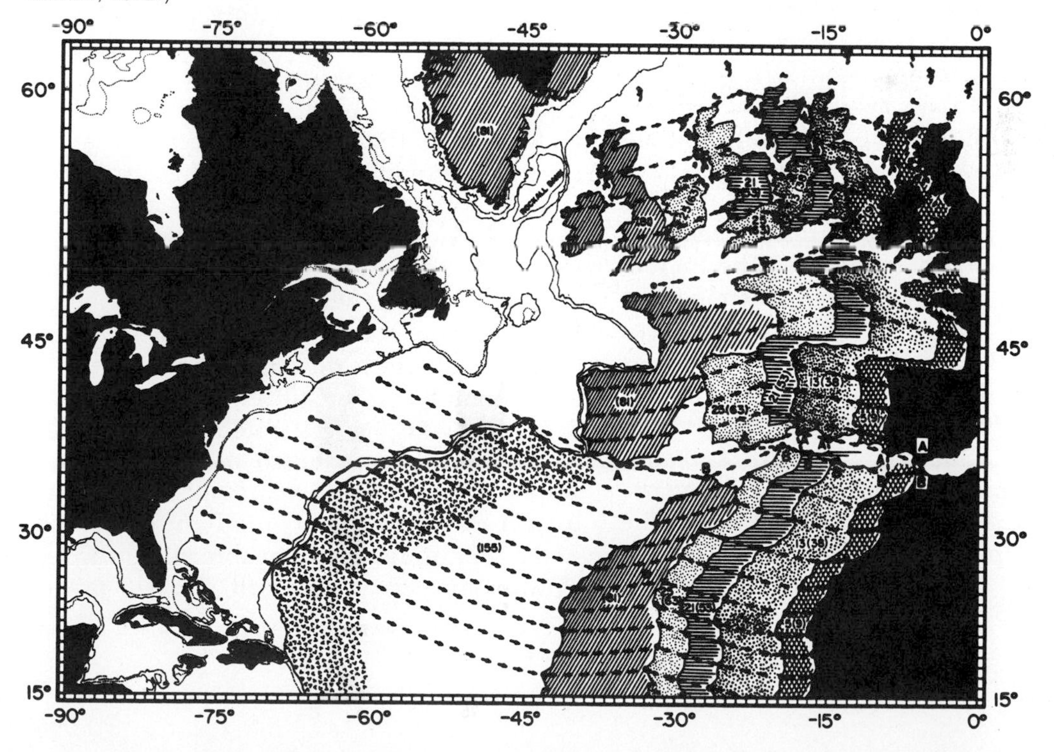

Canary Islands. Thus the faults serve as guides to the trace of movements.

Divergent Junctions in the Continental Crust

The oceanic ridge system can be traced into land areas in several places, notably the Gulf of California, the Gulf of Aden and the Red Sea, and Iceland. Extensional tectonics is the primary mode of crustal deformation in all three of these places. Iceland differs from the other two in that it is not underlain by continental type crust but is entirely composed of volcanic products and sediments. The Gulf of California and the whole region of East Africa have been studied in considerable detail and provide good examples of the changes in deformational pattern where the oceanic ridges pass into or under continental crust.

Gulf of California

The Gulf of California is one of the most intensively studied pieces of sea floor in the world. Its physiography has been charted with sonic equipment, numerous seismic refraction and reflection sections are available, the gravity field, magnetic anomalies, and heat flow are known, and the surrounding land areas have been mapped.*

The eastern part of the Pacific Ocean presents a complicated plate picture. The Pacific-Antarctic ridge passes between Australia and Antarctica and then bends northward to become the East Pacific rise, which then passes northward through the Galapagos Islands and into the Gulf of California. Several ridges connect the East Pacific rise with parts of South and Central America, and these break the sea floor east of the rise into smaller plates. The larger of these plates, called the Nazca plate, is located between the Chile ridge, which runs from southern Chile to the East Pacific rise, and the Carnegie ridge, which trends almost east west from the Galapagos Islands to Ecuador. The Nazca plate is bounded on the east by the Chile-Peru trench. The area north of the Carnegie ridge and east of the East Pacific rise is called the Cocos plate, although it may be subdivided into two smaller plates by a ridge, the Cocos ridge, which runs from the Galapagos Islands to Costa Rica.

The East Pacific rise is severely fragmented by transform faults toward its northern end. The ridge crest trends are consistently northeast, but faults break it into sections, especially within the Gulf of California, Fig. 20-14. The ridge crest is identified by the symmetrically disposed magnetic anomalies on either side of the larger segments. High heat-flow values and positive gravity anomalies are found within the Gulf, and a large number of shallow-focus earthquakes occur along the ridge and the transform faults.

The transform faults within the Gulf show up in the topography and many displace recent sediments, proving that the Gulf is still tectonically active, Fig. 20-15. The transform faults within the Gulf have trends that are close to the trend of the San Andreas and other faults* located at the northern end of the Gulf, but the faults in the Gulf do not pass directly into faults on continental crust on either side.

Spreading rates within the Gulf can be determined from the magnetic anomaly pattern. These calculations indicate that the spreading at the mouth of the Gulf is at a rate of about 3 cm/year. At the estimated rates the Gulf would have been a much narrower body of water some 4 million years ago. A thick sequence of sediments is found on the margins of the Gulf, indicating that spreading was very slow, if any occurred at all, 4 to 10 million

* Elders and others (1972), Henyey and Bischoff (1973), Moore (1973), Larson and others (1968), Gastil and Jensky (1973), map of Baja.

* The relation between the San Andreas and the transforms is discussed further in Chap. 24.

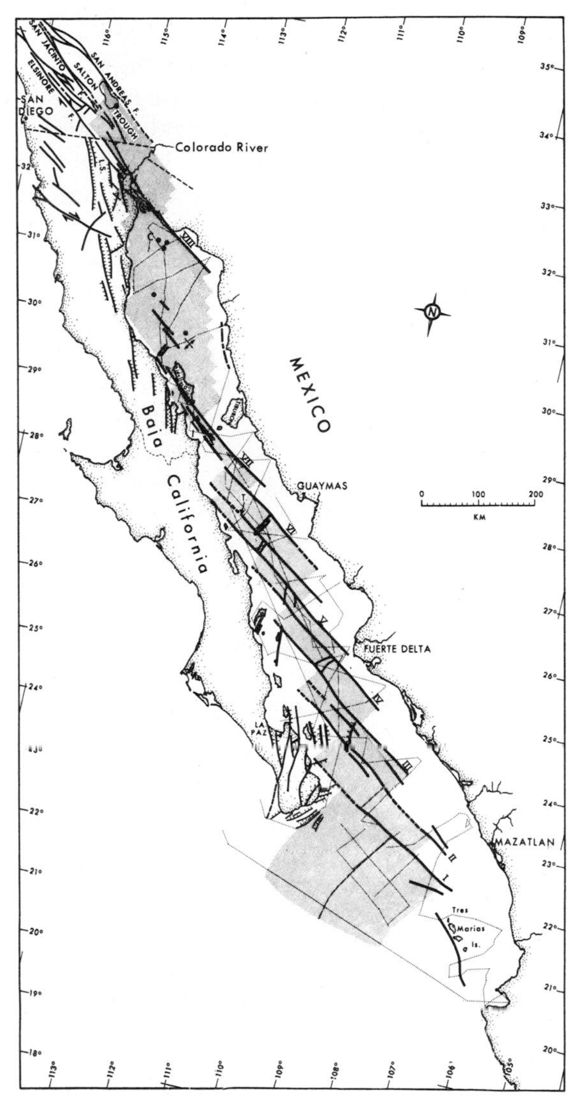

FIGURE 20-14 Fracture zones and other significant faults in the Gulf of California. Dashed lines indicate inferred fracture zones and faults. Shading indicates areas interpreted as new crust. Light dashed lines show data control of ship's tracks. Major fracture zones are numbered I through VIII. (*From Moore, 1973.*)

years ago. Gastil and Jensky (1973) suggest a restoration based on reconstruction of presently displaced old belts of various types of intrusions. He concludes that the Gulf was completely closed 30 million years ago but that the northwestward translation of slices of crust had begun by that time.

Most students of the region conclude that the protogulf began to broaden about 4 million years ago and that the movements occurred on transform faults which caused offsets of the spreading centers. The effect of these offsets has been to bring about a northwestward translation of the Baja Peninsula.

Cross sections of the Gulf are based on both gravity and seismic refraction studies. Seismic studies indicate that the mantle rises under the Gulf but that the depth to mantle and the thickness of the sedimentary and oceanic layers vary considerably. This variation is due in part to the presence of a thick sediment fill near the mouth of the Colorado River. Water deepens from north to south in the Gulf, and the sediment thickness generally decreases, but the depth to mantle and the thickness of the oceanic layer are less toward the south. Mantle and isotherms apparently rose into the space between the spreading sections of continental crust to either side. The elevated temperatures would account for magma generation and metamorphism in places along the rifted continental plate, and the presence of the dense mantle near the surface gives rise to the positive anomalies.

The Salton trough is located at the head of the Gulf, and it is essentially a landward continuation of the Gulf. It is a structural depression, similar in size to some of the basins within the Gulf, but here sediments from the Colorado River have filled in the depression. A model for the structural development of this area has been proposed by Elders and others (1972). They propose "a model in which the continental crust is being thinned beneath a deepening and widening rift. . . . The trough forms as successive sections of the crust are sliced off along strike-slip faults. These slices move northwest and are transferred from the

North American to the East Pacific plate." This model is consistent with much of the observational data concerning the Gulf and the Salton area.

Rift Systems of East Africa*

The East African system can be traced into the junction of the Gulf of Aden and the Red Sea, and the fault system at the north end of the Red Sea continues into the Dead Sea rift, Fig. 20-16. The fault system thus defined extends from Lebanon to the Zambesi River of South Africa, a distance more than one-sixth the circumference of the earth. Over a large portion of this region, Precambrian rocks of the shield are covered to shallow depths by Cenozoic volcanics and other sedimentary deposits.

* See McConnell (1972).

FIGURE 20-15 Reflection profile crossing San Pedro Martir basin and shelf south of Tiburon Island. Note erosional unconformity and upturned strata on east side of basin. Deep part of basin is bounded by fracture zones VI and VII. Probable turbidite channel occupies center of basin, and small patch of ponded turbidites can be seen beneath channel-flank deposits at base of east wall. (*From Moore, 1973.*)

The valleys of East Africa which make up part of this system extend from the southern end of the Red Sea as the Abyssinian rift to Lake Victoria, where the central Tanganyika Plateau is largely surrounded by rift valleys which join at the southern end and continue south. The rift valleys separating the high plateaus have steep borders; some are precipitous drops of over 1,000 m, and they have been interpreted as fault scarps of relatively recent age by all who have studied them.

Cenozoic volcanic activity is associated with the system as a whole. Some of the highest volcanos in Africa (Kilimanjaro, Kenya, and Elgon) are located along but outside the rifts. Many smaller volcanos occur in and along the rifts, but the arrangement of the volcanic centers is not systematic, and the activity does not coincide exactly with the faulting. Both normal and reverse faults are known along the edges of the rifts, and both strike parallel to the rift. Regionally, the structure is that of a broad low dome; folding is not commonly associated with the sediments in the rift valleys or the adjacent bordering plateaus. Negative-gravity anomalies lie over the rift valleys, indicating a mass deficiency at depth but also raising the question of why an upward adjust-

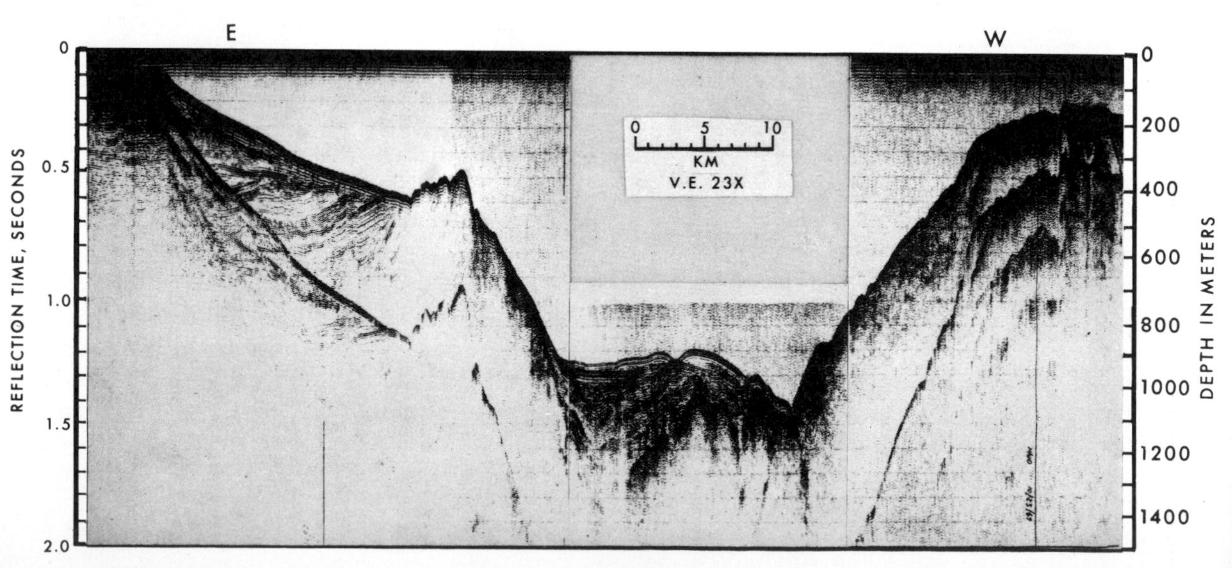

FIGURE 20-16 East Africa, Red Sea, Gulf of Aden rift system.

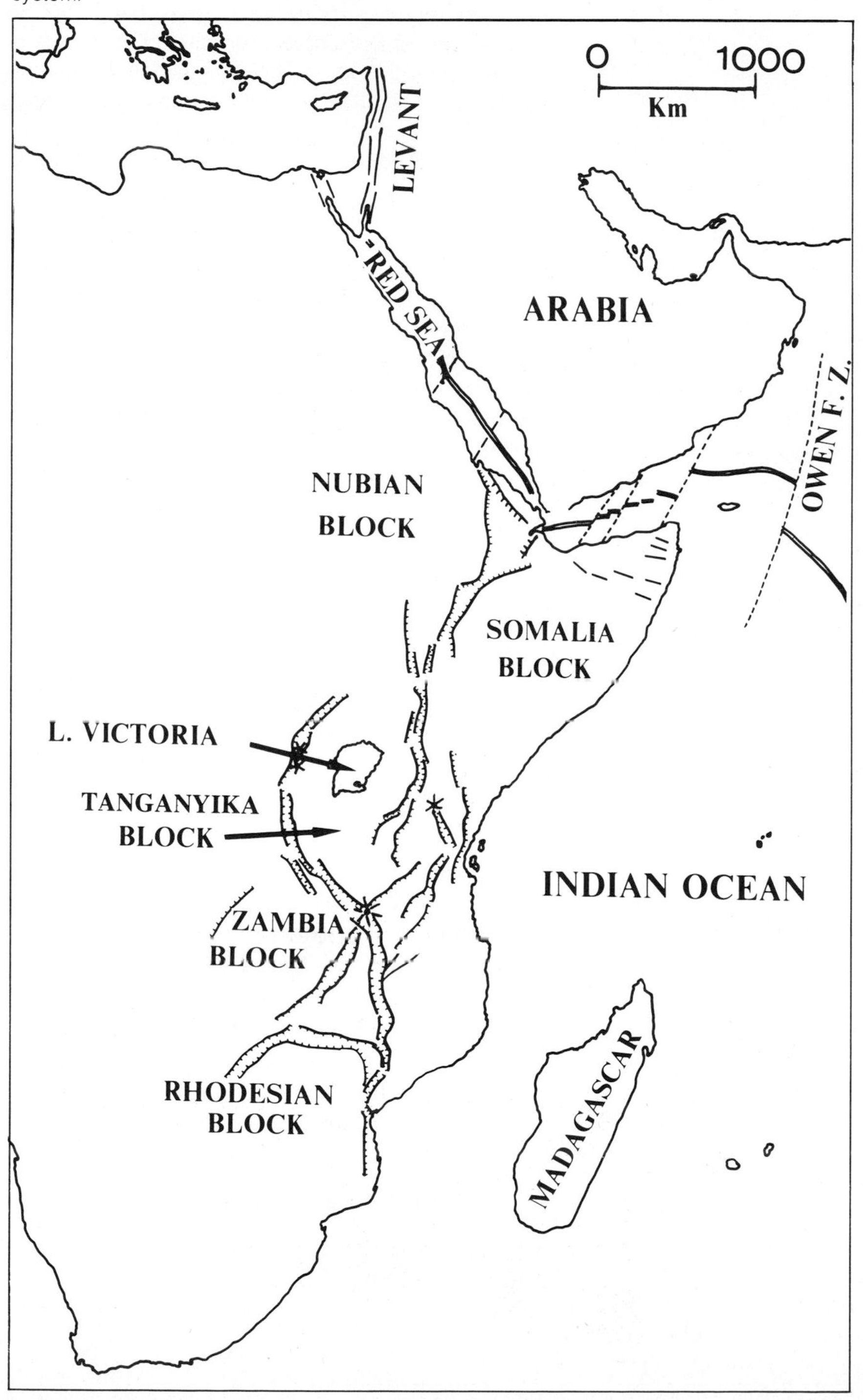

ment has not occurred over the long time the rifts have been forming.

Hypotheses on the Formation of the Rift System

An argument over the origin of the valleys developed early in the history of the study of rift systems (Willis, 1936). Some of the first workers concluded that the rifts are due to extension over an uplifted area. This view was expressed by the German geologist Abendanon (1914):

> The increase of dimension in consequence of the anticlinal stretching is made evident by the numerous graben or depressions. . . . Throughout the length of the Gross Falte we observe as direct and indirect effects of the extension in the anticlinal zone the occurrence of seismotectonic and volcanic phenomena.

Others concluded that the rifts resulted from thrusting, that the downdropped valley centers are actually forceably depressed by overriding thrust sheets (Uhlig, 1909):

> (1) In certain districts where great vertical displacements have been produced, volcanic eruptions, such as would presumably occur along tension faults, are lacking; (2) ancient mica schist overlies young lavas east of the great escarpment on the western side of Lake Natron indicates an overthrust of 2½ km.; (3) the notable elevation of the plateau between the Great Rift Valley and the Indian Ocean appears to be of the nature of uparching due to horizontal pressure; (4) the elevated margins of the rift valleys suggest uparching along their trends in a manner consistent with the hypothesis of overthrusting.

Willis (1936) regarded the thrust faults as defining wedges of the crust and explained the steep valley walls as slumping of the edges of the thrusts. Gregory (1920) and Krenkle (1922) argued against thrusting. Krenkle stated his argument as follows:

> The tectonic setting of the East African fault zones admits of only one explanation: they are zones of tearing apart of the crust, produced by a directed tension. Only as tension phenomena can the wealth of observed structural facts and changes be brought into orderly relations. Accompanying these deep-seated developments as inevitable accompaniments are disturbances of gravity, earthquakes, and volcanic activity. Forces that produce partings alone produce structures like the rift valleys with their series of splits, that extending to great depths converge downward and are unequally filled with light-weight superficial rocks. The separation is, however, most evident in those gaping rifts whose depths are filled with water. The action of compressive forces is nowhere recognizable. Overthrusts of significant displacement toward the axes of splitting are everywhere lacking. The tearing apart has resulted in a certain areal expansion of the central African landmass, which has been most strongly affected.

Almost without exception the more recent students of East Africa conclude that the rift system is an expression of extension of the ancient cratonized African region. McConnell (1972) summarized much of the more recent work. He concludes that the modern rift valleys follow ancient mobile belts which were active before Silurian time and which developed between Precambrian cratons. The rifted belts show evidence of repeated reactivation and probable control by mantle processes. Unlike the Red Sea and Gulf of Aden, however, no evidence of major spreading with accompanying rise of mantle material is found under the East African rifts. The rift volcanism is of continental type and largely confined to the northeastern section of the system.

The Red Sea–Gulf of Aden*

An oceanic ridge similar to the mid-Atlantic ridge passes from the central part of the Indian Ocean into the mouth of the Gulf of Aden,

* Refer to Hutchinson and Engels (1972); Ross and Schlee (1973); Carey (1958); Illies (1969); and Picard (1966).

FIGURE 20-17 Diagrammatic cross sections illustrating tectonic history and geologic evolution of the southern Red Sea. (*a*) Jurassic-Paleocene; (*b*) Eocene-Oligocene; (*c*) early Miocene; (*d*) late Miocene-Quaternary. (*After Hutchinson and Engels, 1970.*)

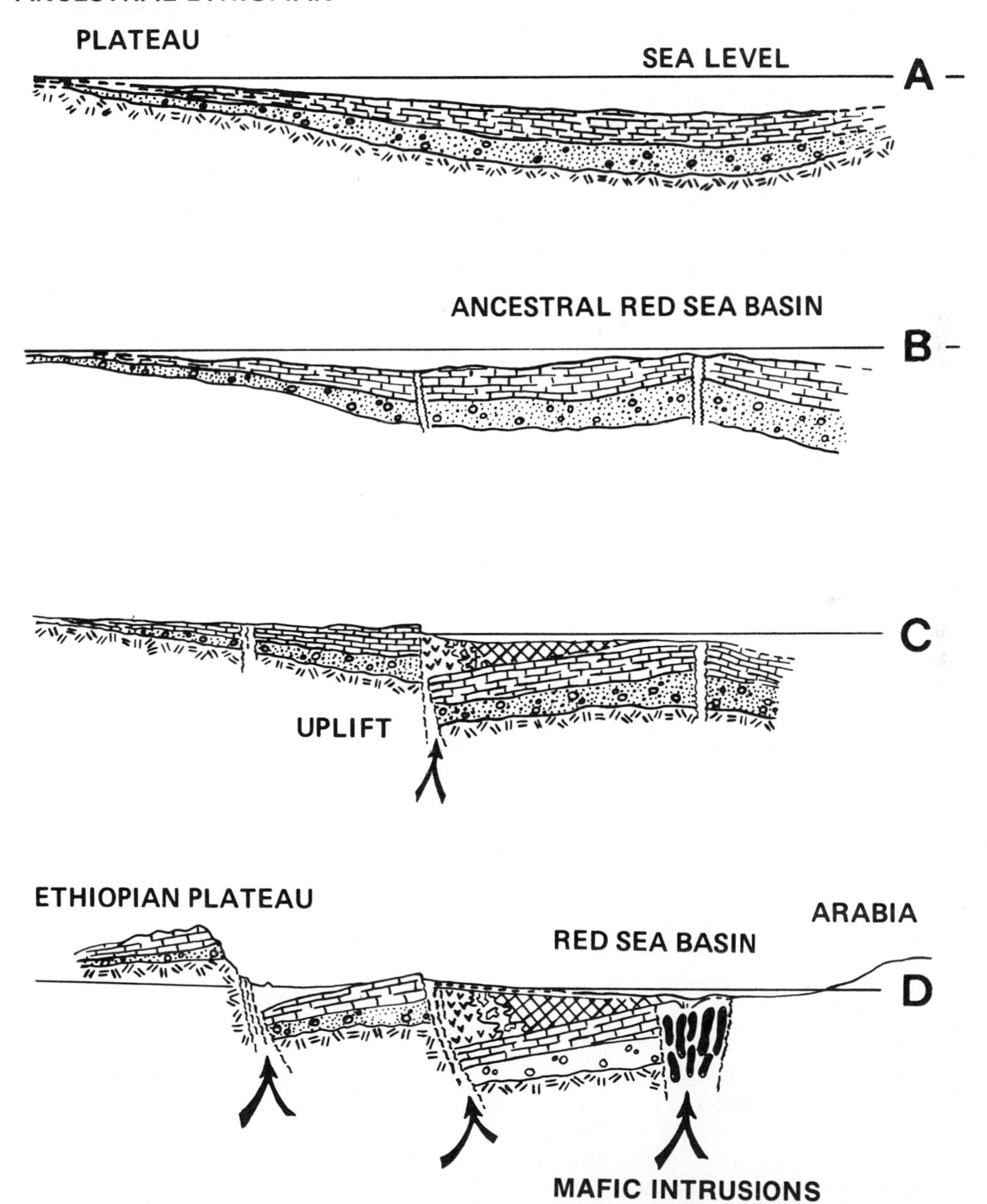

Fig. 20-16. The ridge can be traced into the Gulf by the magnetic anomaly patterns, and it is offset and segmented by transcurrent faults within the Gulf which cut the ridge obliquely as do those at the Gulf of California. Here also the faults are well defined by topographic lineaments in the floor of the Gulf. The center of the Red Sea is marked by a deep rift valley, and abnormaly large magnetic anomalies are found in the southern part, but the ridge crest and symmetrically disposed magnetic anomalies found over oceanic ridge crests have not been identified here.

The Gulf of Aden and the Red Sea resemble the Gulf of California in many respects. The relatively angular borders of the seas suggest fault control; abnormally high heat-flow values are associated with the axial portions of the gulfs; positive Bouguer anomalies lie over the axes; positive magnetic anomalies are associated with the axes; and seismic refraction studies show that the crustal sections are not continental and that the mantle is higher than normal under these seas. All are thought to be places where rifting of the continental crust is taking place and where mantle material has risen closer to the surface as isostatic compensation occurs.

The regional structural setting of the Red Sea–Gulf of Aden area is quite different from that of the Gulf of California. The Red Sea–Gulf of Aden–East African rift valleys form a three-pronged system of extensional tectonics centered in a region of Precambrian continental crust, Fig. 20-17. Each of the three prongs is different. The Gulf of Aden seems clearly to be formed by spreading of the Arabian Peninsula away from the Somali Peninsula of Africa along trends indicated by the orientation of the faults in the floor of the Gulf of Aden. Continuation of the oceanic ridge beyond the junction of the Gulf of Aden and the Red Sea is not in evidence.

The Red Sea is oriented more nearly at right angles to the direction of spreading than the Gulf of Aden, but the width of the Red Sea increases toward the south. The western edge of the Red Sea rift runs along the west bank of the Sea in the north but includes land area at the southern end. At its northern end the Red Sea narrows to form the Gulf of Suez, and a major zone of faults called the Levant trends to the north along the Gulf of Eilat and the Dead Sea. This fault zone is complicated and shows structural styles associated both with extension and left-lateral strike-slip movement that would be consistent with northeastern movement of the Arabian Peninsula. The structure of the Levant is discussed in greater detail in Chap. 23.

The third prong of the system, the East African rift system, is an extensional fault system situated entirely within the African continent.

REFERENCES

Abendanon, E. C., 1914, Die grossfalten der Erdrende: Leiden.

Allen, T. D., 1969, A review of marine geomagnetism: Earth-Sci. Rev., v. 5, no. 4, p. 217–254.

Anderson, Roger N., 1972, Petrologic significance of low heat flow on the flanks of slow-spreading mid-ocean ridges: Geol. Soc. America Bull., v. 83, no. 10, p. 2947ff.

——— 1974, Cenozoic motion of the Cocos plate relative to the asthenosphere and cold spots: Geol. Soc. America Bull., v. 85, no. 2, p. 175–180.

Bonatti, Enrico, 1973, Origin of offsets of the mid-Atlantic ridge in fracture zones: Jour. Geology, v. 81, no. 2, p. 144ff.

Burckle, L. H., Ewing, M., Saito, T., and **Leyden, R.**, 1967, Tertiary sediment from the East Pacific rise: Science, v. 157, no. 3788, p. 537–540.

Carey, S. W., 1958, The tectonic approach to continental drift, *in* Continental drift—A symposium: Geol. Dept. Univ. Tasmania, p. 177–355.

Dehlinger, P., Couch, R. W., and **Gemperle, M.**, 1967, Gravity and structure of the eastern part of the Mendocino escarpment: Jour. Geophys. Res., v. 72, no. 4, p. 1233–1247.

Dietz, R. S., 1961, Continental and ocean basin evolution by spreading: Nature, v. 189.

Dickinson, W. R., 1971, Plate tectonics in geologic history: Science, v. 174.

Dickson, G. O., Pitman, W. C. III, and **Heirtzler, J. R.,** 1968, Magnetic anomalies in the South Atlantic and ocean floor spreading: Jour. Geophys. Res., v. 73, no. 6, p. 2087–2100.

Elders, W. A., and others, 1972, Crustal spreading in Southern California: Science, v. 178, no. 4056.

Ewing, Maurice, Le Pichon, X., and **Ewing, John,** 1966, Crustal structure of the mid-ocean ridges: Jour. Geophys. Res., v. 71, no. 6, p. 1611–1636.

Fisher, R. L., and **Shor, G. G.,** 1958, Topography and structure of the Middle America trench: XX Internat. Geol. Cong. Proc.

Fox, P. J., Schreiber, Edward, and **Peterson, J. J.,** 1973, Geology of the oceanic crust; compressional wave velocities of oceanic rocks: Jour. Geophys. Res., v. 78, no. 23.

Gastil, R. G., and **Jensky, Wallace,** 1973, Evidence for strike-slip displacement beneath the Trans-Mexican volcanic belt, *in* Conference on tectonic problems of the San Andreas fault system, Proceedings, Stanford Univ. Publ. Geol. Sci., v. 13.

Gibson, I. L., and **Walker, G. P. L.,** 1964, Some composite rhyolite-basalt lavas and related composite dykes in eastern Iceland: Geologists' Assoc., London, v. 74, p. 301–318.

Gregory, J. W., 1920, The African rift valleys: Geogr. J., v. 56.

Gzovsky, M. V., and others, 1973, Problems of the tectonophysical characteristics, stresses, deformations, fractures and deformation mechanisms of the earth's crust: Tectonophysics, v. 18, p. 167–205.

Harrison, C. G. A., and **Ball, Mahlon M.,** 1973, The role of fracture zones in sea floor spreading: Jour Geophys. Res., v. 78, no. 32, p. 7776ff.

Heezen, B. C., and others, 1964, Chain and Romanche fracture zone: Deep-Sea Research, v. 11, p. 11–33.

Heirtzler, J. R., and **Hayes, D. E.,** 1967, Magnetic boundaries in the North Atlantic Ocean: Science, v. 157, no. 3785, p. 185–187.

——— and **Le Pichon, X.,** 1965, Crustal structure of the mid-ocean ridges, 3, Magnetic anomalies over the mid-Atlantic ridge: Jour. Geophys. Res., v. 70.

——— and others, 1968, Marine magnetic anomalies, geomagnetic field reversals, and motions of the ocean floor and continents: Jour. Geophys. Res., v. 73, no. 6, p. 2119–2136.

Henyey, T. L., and **Bischoff, J. L.,** 1973, Tectonic elements of the northern part of the Gulf of California: Geol. Soc. America Bull., v. 84, no. 1.

Hess, H. H., 1962, History of ocean basins, *in* Petrologic studies: A volume in honor of A. F. Buddington, Geol. Soc. America, New York.

Hodgson, J. H., 1964, Earthquakes and earth structure: Englewood Cliffs, N.J., Prentice-Hall.

Hutchinson, R. W., and **Engels, G. G.,** 1970, Tectonic significance of regional geology and evaporite lithofacies in northeastern Ethiopia, *in* A discussion on the structure and evolution of the Red Sea and the nature of the Red Sea, Gulf of Aden and Ethiopia rift junction: Roy. Soc. London, Phil. Trans., Ser. A., v. 267, no. 1181, p. 313–329.

——— 1972, Tectonic evolution in the southern Red Sea and its possible significance to older rifted continental margins: Geol. Soc. America Bull., v. 83, no. 10, p. 2989ff.

Illies, J. H., 1969, An intercontinental belt of world rift system: Tectonophysics, v. 8, no. 1.

Isacks, B., Oliver, J., and **Sykes, L. R.,** 1968, Seismology and the new global tectonics: Jour. Geophys. Research, v. 73, no. 18.

Kay, R., Hubbard, N. J., and **Gast, P. W.,** 1970, Chemical characteristics and origin of oceanic ridge volcanic rocks: Jour. Geophys. Res., v. 75, no. 8, p. 1585–1613.

King, E. R., Zietz, I., and **Alldredge, L. R.,** 1966, Magnetic data on the structure of the central Arctic region: Geol. Soc. America Bull., v. 77, no. 12, p. 619–646.

Krenkle, E., 1922, Die bruchzonen Ostrafrikas: Berlin.

Langseth, M. G., Jr., 1969, The heat flow through the surface of the oceanic lithosphere (abs.), *in* Deep seated foundations of geological phenomena: Tectonophysics, v. 7, no. 5–6.

———, **Le Pichon, X.,** and **Ewing, M.,** 1966, Crustal structure of the mid-ocean ridges: Jour. Geophys. Res., v. 71, no. 22, p. 5321–5355.

Larson, R. L., and **Pitman, W. C. III,** 1972, World-wide correlation of Mesozoic magnetic anomalies, and its implications: Geol. Soc. America Bull., v. 83, no. 12.

———, **Menard, H. W.,** and **Smith, S. M.,** 1968, Gulf of California, a result of ocean-floor spreading and transform faulting: Science, v. 161.

Le Pichon, X., and **Heirtzler, J. R.,** 1968, Magnetic anomalies in the Indian Ocean and sea-floor spreading: Jour. Geophys. Res., v. 73, no. 6, p. 2101–2117.

——— and others, 1965, Crustal structure of the mid-ocean ridges, 1, Seismic refraction measurements: Jour. Geophys. Res., v. 70.

Lowell, J. D., 1972, Spitsbergen Tertiary orogenic belt and the Spitsbergen fracture zone: Geol. Soc. America Bull., v. 83, no. 10, p. 3091ff.

McConnell, R. B., 1972, Geological development of the rift system of Eastern Africa: Geol. Soc. America Bull., v. 83, no. 9, p. 2549ff.

Marauchi, S., and others, 1968, Crustal structure of the Philippine Sea: Jour. Geophys. Res., v. 73, no. 10.

Mason, R. C., and **Raff, A. D.**, 1961, A magnetic survey off the West Coast of North America, 32° N to 42° N: Geol. Soc. America Bull., v. 72, p. 1259–1265.

Menard, H. W., 1955, Deformation of the northeastern Pacific basin and the West Coast of North America: Geol. Soc. America Bull., v. 66.

——— 1958, Development of median elevations in ocean basins: Geol. Soc. America Bull., v. 69.

——— 1960, The East Pacific rise: Science, v. 132, p. 1737–1746.

——— 1964, Marine geology of the Pacific: New York, McGraw-Hill.

——— and **Fisher, R. L.**, 1958, Clipperton fracture zone in the northeastern equatorial Pacific: Jour. Geology, v. 66, p. 239–253.

——— and **Smith, S. M.**, 1966, Hypsometry of ocean basin provinces: Jour. Geophys. Res., v. 71, no. 18, p. 4305–4325.

Milanovsky, E. E., 1972, Continental rift zones; their arrangement and development, *in* Girdler, R. W., ed., East African rifts: Tectonophysics, v. 5 (1/2), p. 65–70.

Moore, D. G., 1973, Plate-edge deformation and crustal growth, Gulf of California structural province: Geol. Soc. America Bull., v. 84, no. 6, p. 1883–1905.

Morgan, W. J., 1968, Rises, trenches, great faults, and crustal blocks: Jour. Geophys. Res., v. 73, no. 6, p. 1959–1982.

O'Bryan, J. W., **Cohen, R.**, and **Gilliland, W. N.**, 1975, Experimental origin of transform faults and straight spreading-center segments: Geol. Soc. America Bull., v. 86, p. 793–796.

Oxburgh, E. R., and **Turcotte, D. L.**, 1968, Mid-ocean ridges and geotherm distribution during mantle convection: Jour. Geophys. Res., v. 73, no. 8, p. 2643–2661.

Phillips, J. D., 1967, Magnetic anomalies over the mid-Atlantic ridge near 27° N: Science, v. 157.

Picard, L., 1966, Thoughts on the graben system in the Levant, *in* Irvine, T. N., ed., The world rift system: Geol. Survey Canada Paper 66–14.

Pitman, W. C. III, and **Heirtzler, J. R.**, 1966, Magnetic anomalies over the Pacific-Antarctic ridge: Science, v. 154.

——— and **Talwani, M.**, 1972, Sea floor spreading in the North Atlantic: Geol. Soc. America Bull., v. 83.

———, **Herron, E. M.**, and **Heirtzler, J. R.**, 1968, Magnetic anomalies in the Pacific and sea floor spreading: Jour. Geophys. Res., v. 73, no. 6, p. 2069–2085.

Raff, A. D., 1962, Further magnetic measurements along the Murray fault: Jour. Geophys. Res., v. 67, p. 417–418.

——— and **Mason, R. G.**, 1961, Magnetic survey off the west coast of North America 40° N latitude to 50° N latitude: Geol. Soc. America Bull., v. 72.

Ross, D. A., and **Schlee, John**, 1973, Shallow structure and geologic development of the southern Red Sea: Geol. Soc. America Bull., v. 84, no. 12, p. 3827ff.

Saemundsson, Kristjan, 1974, Evolution of the axial rifting zone in northern Iceland and the Tjornes fracture zone: Geol. Soc. America Bull., v. 85, no. 4, p. 495–504.

Sclater, J. G., and **Fisher, R. L.**, 1974, Evolution of the east central Indian Ocean, with emphasis on the tectonic setting of the Ninety East ridge: Geol. Soc. America Bull., v. 85, no. 5, p. 683–702.

Shaw, E. W., 1963, Canadian Rockies—orientation in time and space, *in* Childs, O. E., and Beebe, B. W., eds., Backbone of the Americas: Tectonic history from pole to pole: Tulsa, American Association of Petroleum Geologists.

Stover, C. W., 1968, Seismicity of the South Atlantic Ocean: Jour. Geophys. Res., v. 73, no. 12, p. 3807–3820.

Sykes, L. R., 1967, Mechanism of earthquakes and nature of faulting on the mid-oceanic ridges: Jour. Geophys. Res., v. 72, no. 8, p. 2131–2153.

Talwani, Manik, 1964, A review of marine geophysics: Marine Geol., v. 2, p. 29–80.

———, **Windisch, C. C.**, and **Langseth, M. G., Jr.**, 1971, Reykjanes ridge crest: A detailed geophysical study: Jour. Geophys. Res., v. 76, no. 2, p. 473ff.

——— and others, 1965. Crustal structure of the mid-ocean ridges, 2, Computed model from gravity and seismic refraction data: Jour. Geophys. Res., v. 70, p. 341–353.

Thorarisson, Sigurdur, 1965, The Median zone of Iceland, *in* The world rift system: Geol. Survey Canada Paper 66–14, p. 187–211.

Turcotte, D. L., and **Oxburgh, E. R.,** 1969, Convection in a mantle with variable physical properties J. Geophys. Res., v. 74.

Uhlig, Johannes, 1909, Untersuchung einiger Gesteine aus dem mordostlichsten Labrador: Ver Erdk Dresden, v. 8.

Vacquier, V., 1959, Measurement of horizontal displacement along faults in the ocean floor: Nature, v. 183.

———, **Raff, A. D.,** and **Warren, R. E.,** 1961, Horizontal displacements in the floor of the northeast Pacific Ocean: Geol. Soc. America Bull., v. 72.

Van Andel, Theerd H., and Bowin, C. O., 1968, Mid-Atlantic ridge between 22° and 23° N. latitude and the tectonics of mid-ocean rises: Jour. Geophys. Res., v. 73, no. 4, p. 1279–1298.

Vine, F. J., 1966, Spreading of the ocean floor: New evidence: Science, v. 154, no. 3755, p. 1405–1415.

——— and **Matthews, D. H.,** 1963, Magnetic anomalies over oceanic ridges: Nature, v. 199.

Vogt, P. R., and **Johnson, G. L.,** 1975, Transform faults and longitudinal flow below the mid-oceanic ridge: Jour. Geophys. Res., v. 80, p. 1399–1428.

———, **Ostenso, N. A.,** and **Johnson, G. L.,** 1970, Magnetic and bathymetric data bearing on sea-floor spreading north of Iceland: Jour. Geophys. Res., v. 75, no. 5, p. 903–920.

———, **Schneider, E. D.,** and **Johnson, G. L.,** 1969, The crust and upper mantle beneath the sea, *in* The earth's crust and upper mantle, Am. Geophys. Union Mon. 13.

———, **Anderson, C. N., Bracey, D. R.,** and **Schneider, E. D.,** 1970, North Atlantic magnetic smooth zones: Jour. Geophys. Res., v. 75, no. 20, p. 3955ff.

Von Herzen, R. P., and **Uyeda, S.,** 1963, Heat flow through the east Pacific Ocean floor: Jour. Geophys. Res., v. 68, no. 14.

Wesson, P. S., 1972, Objections to continental drift and plate tectonics: Jour. Geol., v. 80, no. 2.

Willis, Bailey, 1936, East-African plateaus and rift valleys: Carnegie Inst. Washington Pub. 470.

Wilson, J. T., 1965, A new class of faults and their bearing on continental drift: Nature, v. 207, no. 4995, p. 343–347.

21

Island Arcs

The arcuate island festoons of the western Pacific have intrigued geologists ever since their form was first recognized. Suess (1885) in his monumental work "The Face of the Earth" described the nearly perfect arcuate form of some of these island chains, Fig. 21-1.* Even at that early time, andesite volcanism, earthquakes, elevated coral reefs, and deep depressions in the sea floor were associated with the island arcs. Other closely related island chains in the Pacific between New Guinea and Samoa have most of these characteristics except that they are not arcuate in form.

The island arcs are distinctly different from the numerous oceanic island groups and seamounts of the central and eastern Pacific. This distinction was initially pointed out by Marshall in 1912 when he defined what is known as the "andesite line"—the boundary line between volcanoes of the island arcs with their andesite volcanism and oceanic islands which exhibit basaltic volcanism. The oceanic islands are also quite different in their deformational and geophysical characteristics from the island arcs.

* Several important summaries on island-arc geophysics and tectonics have been published in recent years. Notable among these are Sugimura and Uyeda (1973); Coleman, P. J., ed., 1973, The Western Pacific: New York, Crane-Russak; and Spencer, A. M., ed., 1974, Mesozoic-Cenozoic orogenic belts.

Distribution

The island arcs are located primarily along the western margin of the Pacific Ocean, although two occur on the western side of the Atlantic. The West Indies constitute an island arc between North and South America, and the Scotia arc connects South America and Antarctica.

The Pacific island arcs fall clearly into those with distinct arcuate form and those without arcuate form, Fig. 21-2. The arcuate island arcs form a nearly continuous chain from north to south. These are the Aleutian, Kurile, Japan, Bonin, Mariana, Yap, and Palau arcs. Of the above, those south of Japan form the eastern margin of the Philippine Sea,

FIGURE 21-1 The western Pacific Ocean. Active volcanoes are shown as solid circles, most of the deep-sea trenches are solid black and their names are shown. Areas behind arcs where heat flow is unusually high are shaded (data from Karig, 1973). Some large transform faults are shown and a few magnetic anomalies are indicated south of the Aleutians.

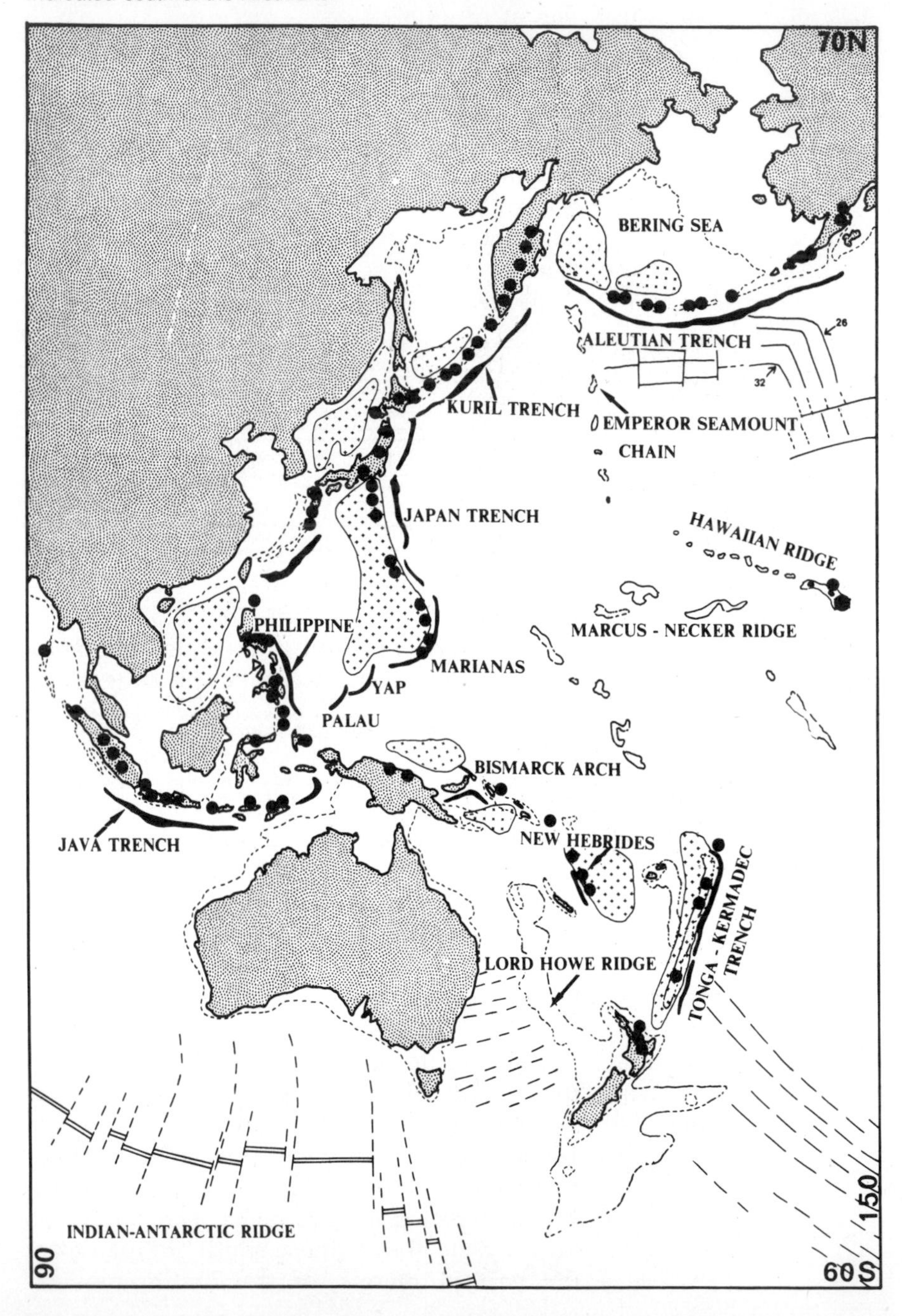

which is bounded by the Ryukyu and Philippine arcs on its western side. An additional nearly perfect arc lies along the southern border of Indonesia. This arc swings between New Guinea and Borneo at its eastern end, where a strongly curved segment is called the Banda arc, and the Andaman and Nicobar Islands south of Burma.

The island chains east of New Guinea do not exhibit the arcuate form of those to the north, but these southern Pacific islands are so similar in other ways that they must be closely related in origin to the arcuate islands. The island groups in this southern sector were called *strewn islands* by Wegner. Included here are the islands of the Bismarck arch, the Solomons, and the New Hebrides. At the eastern end of these islands lie the long nearly straight chains of the Tonga and Kermadec Islands which extend north-south between New Zealand and Samoa.

The pattern formed by these island groups is varied and complex. The Aleutian, Kurile, and Japan arcs seem simple by comparison

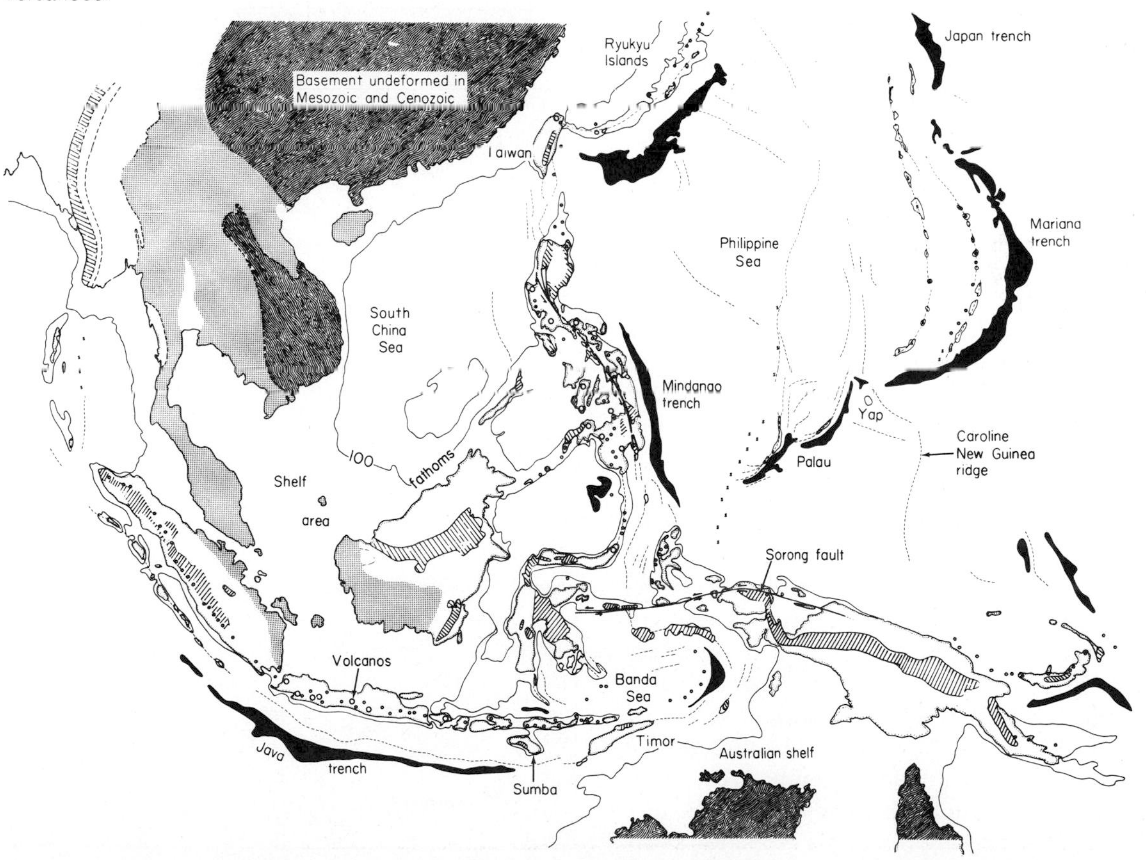

FIGURE 21-2 Tectonic sketch map of island arcs of the southwestern Pacific. The margins of the trenches are drawn approximately on the 3,000-fathom contour of the U.S. Hydrographic Office map of the world, 1961. The 100-fathom contour is drawn to outline shelf areas. Areas where pre-Cenozoic rocks were undeformed during the Mesozoic and Cenozoic are shown by a wavy pattern. Areas involved in Mesozoic folding are covered by a dot pattern. Areas where pre-Cenozoic rocks were deformed in Cenozoic orogeny, particularly Miocene, are shown by diagonal patterns. Deformed and undeformed Cenozoic sediments are shown blank. The dashed lines mark the location of ridges in the submarine topography; s designates seamounts; small circles indicate location of volcanoes.

with those to the south. The northern arcs are made up of single chains of mainly volcanic islands which intersect at high angles. They are curved, with their concave side toward the continents. Each is separated from continents by a marginal sea and from the ocean basin by a deep-sea trench.

South of Japan two chains of arcs separate and bound the Philippine Sea, but note that the arcs in both chains are concave toward Asia. If we view the entire area of Malaysia, Indonesia, Borneo, and surrounding seas as a unit, then this unit is bounded all along its southeastern border by island arcs—the Indonesian arc, the Banda arc, and the Philippine arc. These are much more complicated in that the geology of the larger islands is much more complex than that of the Aleutians and Kuriles, and the physiography and structural configuration is also more complex.

Physiography—Bathymetry of Island Arcs

Despite the considerable variety of arc form, some features recur frequently enough to constitute general characteristics. Prominent among these is the presence of a deep-sea trench with the arc. The trenches are usually long, narrow, and conform in trend with the volcanic islands. The trenches generally lie on the oceanward side of the islands, and they contain the greatest depths of the oceans, the deepest point being in the Marianas trench [over 11,600 m (about 38,000 ft) deep]. Many trenches are asymmetric in transverse profile. The oceanward side tends to be smoother and to have a gentle curved slope as in the Peru-Chile trench, Fig. 21-3. The island side is steeper; some are highly irregular, and terraces may be present. These trenches are sediment traps; where they are close to a source of sediment they may be shaped to some degree by the deposits. The eastern end of the Aleutian trench is much shallower and has less relief for this reason. In some trenches sediment fill has created a relatively flat floor in the trench.

The zone between the trench and the volcanic islands is called the *arc-trench gap*, Fig. 21-4. This gap is much narrower in some arcs (e.g., the Aleutian, Kurile, Japan, and Ryukyu Islands) than in others. Where it is narrow the sea floor rises from the trench to the volcanic islands. In other arcs, however, a nonvolcanic arcuate ridge lies between the trench and the volcanic arc. In these arcs the nonvolcanic arc is often separated from the volcanic arc by an

FIGURE 21-3 Schematic seismic reflection profiles across Peru-Chile trench and apparent fracture zone (22–23). Sediment reflectors drawn from original seismic records to show major structures. Fault locations were picked from seismic reflection records, but their orientation is schematic. Dotted pattern indicates acoustic basement; gaps in pattern show that nature of subbottom is undetermined. (*From Prince and Kulm, 1975.*)

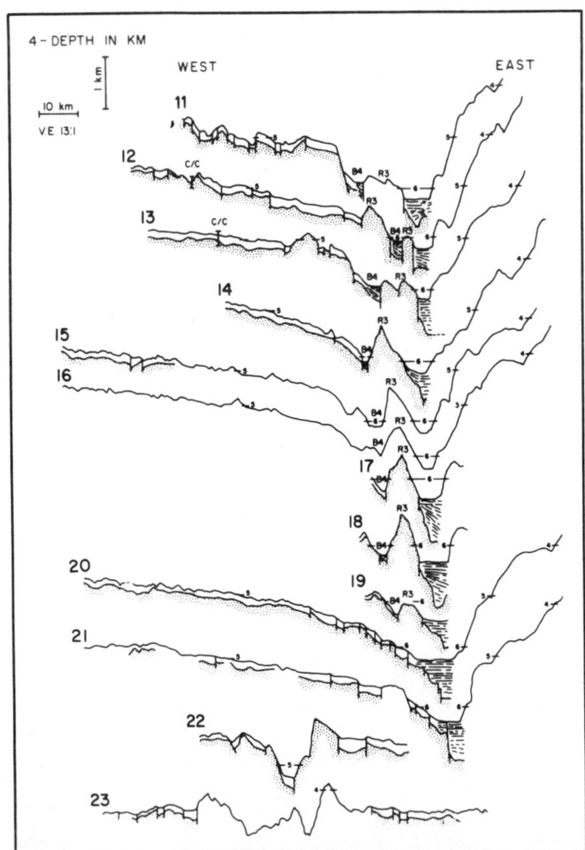

interdeep, as in the Indonesian arc. The outer nonvolcanic arc may appear as a submerged ridge, or it may include some islands.

The most prominent part of most arcs is made up of the active and dormant volcanoes which may occur as a chain of isolated volcanoes, as in the Aleutians, or as a prominent chain of volcanic centers joined to form a long and larger island complex, as in Japan, Sumatra, and the Philippines.

Several arcs exhibit one or more arcuate submerged ridges or swells located behind the volcanic arcs and separated from them by an interarc basin, Fig. 21-2. These interarc basins and the seas which separate them from the continents are generally areas of deep water, and in the case of the Philippine Sea and the South Fiji basin the water approaches abyssal-plain depths.

GEOPHYSICAL CHARACTERISTICS OF ISLAND ARCS

Seismicity

The seismic character of the island arcs has long been known. They are the sites of concentrated earthquake activity of both shallow and deep focal depths. The distribution of epicenters was cataloged for many of these areas by 1938 (Gutenberg and Ritcher), and a well-defined pattern had emerged by the time Umbgrove compiled his notable study of the "Structural History of the East Indies" (1949). The island arcs stand in stark contrast with the essentially aseismic character of the deep-sea floor. It was soon recognized that the shallow-focus earthquakes are concentrated under the trenches and that deeper focus earthquakes are located progressively farther behind the volcanic islands, with the deepest 600-to-700-km focal depths occurring well behind chains of volcanoes. This arrangement is evident both in the early studies of the Indonesian arc and in the more recent study of Japan, Fig. 21-5.

Benioff (1954)* made a now classic study of seismicity in island arcs in which he compiled both the areal distribution and the focal depths in most of the island arcs. He found that a clear cross-sectional pattern of focal distribution emerged when he projected foci along the trend of the arc into a cross-sectional plane, Fig. 21-6. In most arcs the foci define a steeply inclined seismic zone which emerges near the trench and slopes under the volcanic arc. These zones are now commonly called *Benioff zones* or *seismic shear zones*.

Many of the Benioff zones contain intervals, especially between 300 to 500-km depths, where relatively fewer earthquakes occur.

More recently efforts have been made to draw contour maps on the planes defined by foci. Stoiber and Carr (1971) found that most foci occur on one or two undulating surfaces that are conformable and separated by 50 to 70 km. They believe these may represent the top and bottom of a sinking lithospheric slab. Closer studies of foci planes under the Japanese arcs reveal that these planes are offset in places by what appear to be transverse breaks of as much as 50 km. These transverse zones can be traced to surface features and identified with transverse faults, lines of volcanic cen-

* More recent detailed summaries are made by Sykes (1966, 1967, 1969, 1971) and others.

FIGURE 21-4 Typical morphology over Tonga, Marianas, and Bonin arcs after Karig (1970). The morphology of the central Aleutian arc between the volcanoes and trench is very similar to Karig's frontal arc. (*From Grow, 1973.*)

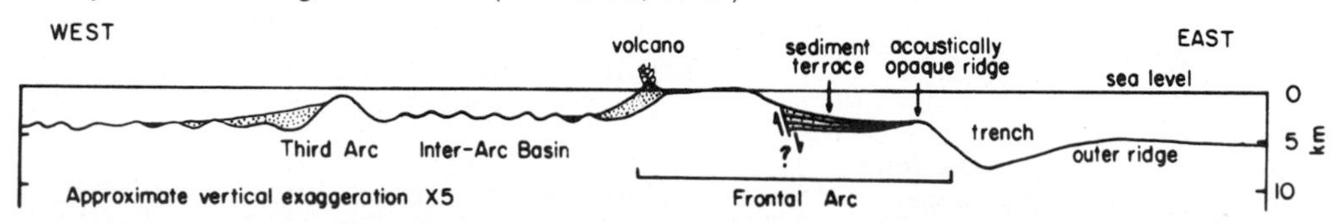

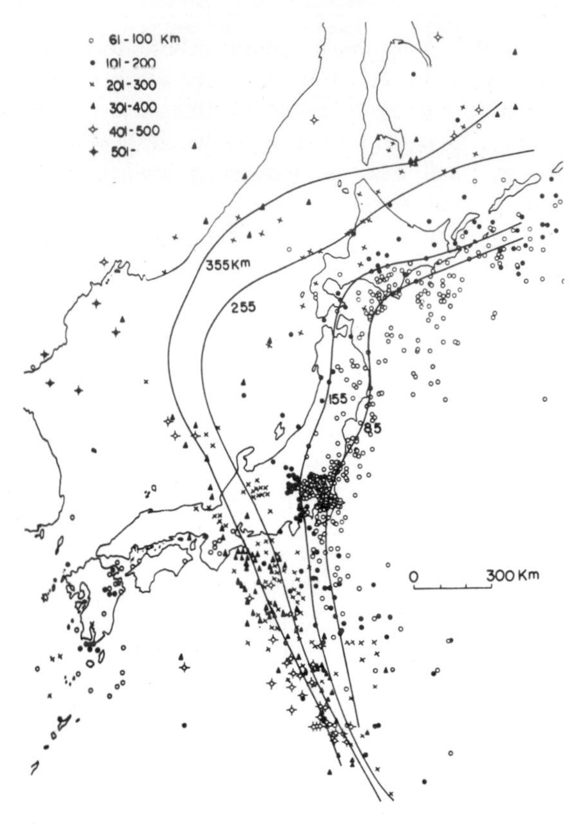

FIGURE 21-5 Epicenters of the deep- and intermediate-focus earthquakes during 1926–1956 as reported by the Japan Meteorological Agency (contours by Sugimura and Uyeda, 1970. (*From Watanabe, 1970.*)

ters, and other structural zones oriented transverse to the arc trend (Carr and others, 1973).

Gravity*

Determination of gravity in island arcs was restricted to measurements made on land until F. A. Vening Meinesz (1948) devised equipment which could be used aboard a submarine. These early studies by Vening Meinesz, Umbgrove, and Kuenen revealed very strong negative gravity anomalies located along and over the trench off Java and Sumatra. Later work by others has shown that this type of anomaly is characteristic of trenches throughout the oceans.

It might seem that the negative gravity anomalies could arise from the fact that the trenches are filled with water and are deep enough to have high-density rock located on both sides; however, calculations show that negative anomalies exist even if the trenches are assumed to be filled with rock. Furthermore, the negative anomalies pass over a number of islands (Andaman-Nicobar Islands, Sumba, Timor, Tanimbar, and Ceram), Fig. 21-7. This can only mean that the negative values arise from some deep-seated cause and that the density of the rocks under the trenches and islands deep into the mantle must be less than that to either the oceanward or the landward side.

The negative anomalies persist regardless of the type of corrections made to the observed values. The free-air, Bouguer, and both regional and local isostatic anomalies are all negative. These studies lead Vening Meinesz to formulate the tectogene hypothesis described in Chap. 17.

In a typical gravity profile across an arc, the axis of the negative anomaly lies on the inner side of the trench. From this low the values increase toward the volcanic arc and generally reach a maximum positive value in the arc-trench gap. Values decline across the arc and into the back-arc area.

The technique of modeling is one of the most valuable methods used to analyze gravity data. A density model is devised which conforms as closely as possible to all subsurface information (generally seismic sections) and the gravity values with observed values over an actual arc. This is done in the case of the Aleutian Islands in Fig. 21-8. The model has been adjusted to produce a perfect fit with the observed data. In this instance the seismic section for the sea floor at the left is known, and density and thickness values for rocks near the surface are known. Also, seismic sections have been made under the arc and in the Bering

* Other gravity studies in island areas are reported by Hess (1938) and Talwani, Worzel, and Ewing (1961).

Sea. This model depicts a large slab of lithosphere bent down under the arc into the mantle as is suggested by other geophysical data. This model can be translated into a hypothetical geological cross section as shown in Fig. 21-9.

Magnetic Studies of Island Arcs

Several types of magnetic studies have important implications for island-arc tectonics. These are (1) magnetic anomaly patterns over the trenches and arcs which reflect the localized distribution of magnetic materials in the crust and mantle; (2) analysis of the relationship of the magnetic anomaly patterns in the oceanic crust to island arcs; and (3) analysis of the relationship of island arcs to paleopole positions.

Local magnetic anomalies do not show a consistently strong correlation with the topography of island arcs, but in the Aleutians the magnetic anomalies do parallel the trend of the

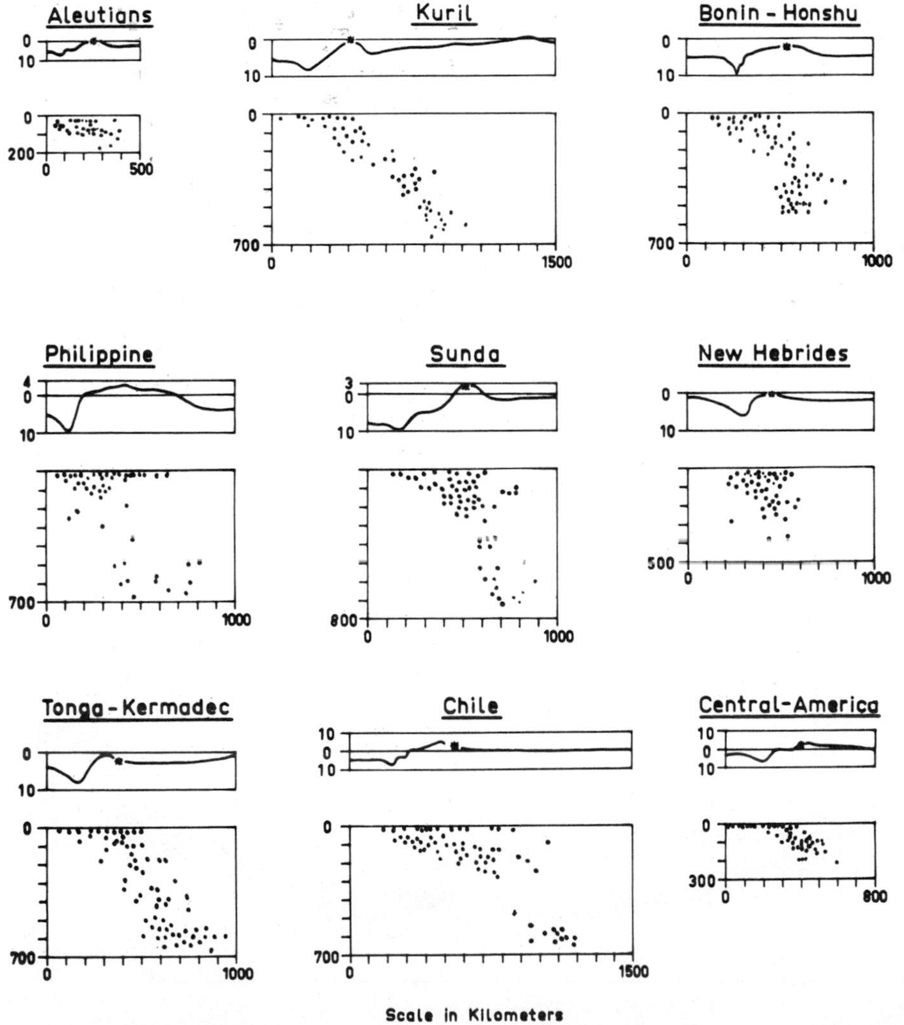

FIGURE 21-6 Foci patterns across island arcs and margins of Pacific. Foci from long sections parallel to the arc have been projected into a single section. (*Modified after Benioff, 1954.*)

trench. A narrow positive anomaly lies over the trench, and a second positive anomaly lies along the south rim of the trench. This has been interpreted (Peter and others, 1965) as being due to a fissure in the floor of the trench that is filled with basalt. It seems probable that the magnetic anomalies in arcs are primarily related to the distribution of volcanic intrusions and to faults that bring rocks of varying susceptibility in contact to produce irregular and difficult-to-interpret patterns.

One of the most interesting results from magnetic studies is the relationship between the strip-like anomaly patterns found in the sea floor and the trenches. These anomaly belts, which are generally thought to have formed at spreading centers along the oceanic ridge system, can be identified over the oceanic crust immediately in front of the trenches; the pattern either changes trend or is lost in the trench and no pattern is identified behind the arc. It is also notable that the anomaly patterns exhibit many divergent trends relative to the orientation of the trench. The pattern in the sea floor south of the Aleutians trends nearly east-west, Fig. 21-1, while others are nearly parallel in trend to trenches (Hayes and Heirtzler, 1968).

These relationships are easily explained according to plate-tectonic theory. The patterns are "frozen in" the sea floor; the pattern approaches the trench as the sea floor spreads; and finally the oceanic crust is subducted in the vicinity of the trench, and the anomaly pattern is erased as the lithospheric plate is heated to the Curie point as it descends into the mantle.

Seismic Cross Sections*

Many cross sections have been prepared across island arcs using both shallow continu-

* See the following references: *Aleutians*—Grow (1973), Stone (1968); *Philippines*—Ludwig (1970), Ludwig and others (1973); *Mariana*—Bracey and Ogden (1972); *Japan*—Yoshii (1973), Ludwig and others (1966, 1973); *Indonesia*—Murauchi and others (1973), Weeks and others (1967), Rodolfo (1969).

FIGURE 21-7 Regional isostatic anomalies over the Indonesian island arc area (−200 to −250 mgals, black; −50 to −200 mgals, vertical lines). (*After Vening Meinesz, 1948.*)

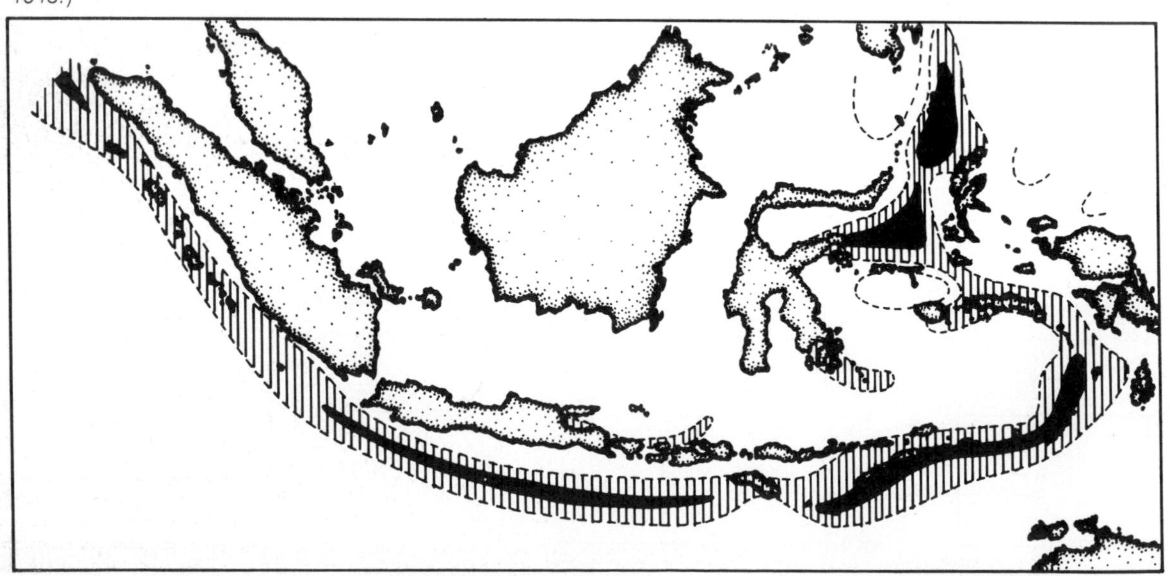

ous seismic sounding and deep penetrating seismic reflection and refraction techniques.

Results of some of the shallow soundings in arcs are illustrated, Figs. 21-10 and 21-11, which show some detail in the thin sedimentary layers over the sea floor at the lip of a trench and in the arc-trench gap. Records from the trench lip indicate extensive high-angle faulting in the sediment which can be explained as an extension of that sediment. Most geologists interpret that as being due to bending of the oceanic crust as it sinks under the trench, but others who favor an extensional origin for the arcs point to this evidence of extension. Thrust faults are suggested in the records from the arc-trench gap, Fig. 21-11, and an interpretation of this is shown for the Aleutians in the geological section, Fig. 21-9.

The deeper cross sections across arcs, Fig. 21-12, have similar features in many arcs. The common characteristics are that the M discontinuity, which is at about a 10-km depth in the oceans, begins to drop lower at the trench and generally extends down to a 20- to 40-km depth under the arc-trench gap and the volcanic arc. These crustal sections often have a threefold subdivision—an upper sedimentary layer, a second layer with a velocity of 5.5 km/sec, and lower layer with a velocity of 6.5 to 7.5 km/sec. The lower layer has a velocity often associated with that due to altered oceanic crust, serpentine.

Anomalous Zones in the Mantle under Island Arcs

Discovery of anomalous patterns of surface-wave dispersion in the vicinity of island arcs was another of the important developments of the plate-tectonic theory (Oliver and Isacks, 1967). The initial study took place in the region of the Tonga-Kermadec arc. The character of the dispersion of seismic waves generated by earthquakes at various depths in the Benioff zone under this arc and propagated through the crust and mantle to receiving stations revealed that seismically anomalous mantle material underlies the arc. A zone approximately 100 km thick and situated below and conformable with the Benioff zone is identified. One of the primary seismic characteristics of this zone is that *S* waves propagating through it are less subject to attenuation (high *Q*)* than are the same types of waves moving through other parts of the mantle at the same depth, Fig. 21-13.

This anomalous zone is now widely interpreted as being a slab of the lithosphere which is sinking beneath the arc. It has seismic characteristics similar to those of the lithosphere and can be traced to depths of 500 to 700 km under the arcs, where it presumably is assimilated into mantle.

Heat Flow

Measurement of heat flow in the vicinity of island arcs shows that they are strongly anomalous compared with the ocean basins in general, exhibiting both anomalously high and low values, Fig. 21-14. Values on the oceanward side are about average for oceanic crust, but a distinct low associated with subduction occurs over the trench, and unusually high values occur both in the volcanic arc and in the back-arc areas.

A good example of the back-arc anomalies is found in the Mariana arc, which consists of a series of concentric topographic features convex to the east and including from east to west the Mariana trench, a frontal arc on which the island of Guam is located, a broad shallow trough, the Mariana trough, and a third arcuate ridge. Heat-flow measurements in the Mariana trough show that a zone of high heat flow closely corresponds with the axis of a topographically high ridge in the trough and is flanked by low values on either side. The ridge

* High *Q* values indicate high rigidity.

is thought to be the site of magmatic intrusion and sea-floor spreading similar to that on the oceanic ridges, and the low values may be due to circulation of water in the oceanic crust (Anderson, 1975).

High heat flow can be caused by a number of processes, the most obvious of which is general heating due to the presence of magmas. A large region could be hot as a result of either numerous intrusions or the close proximity of a large cooling pluton, and this presumably is the cause of much of the high heat flow in the vicinity of the volcanic arcs and probably in interarc areas as well. It is more difficult to prove that the high heat flow in the interarc basins is due to magma, because so few active volcanos occur here, especially as compared with the frontal arcs which are usually volcanically active. It is possible that the high heat flow is due to the regional distribution of radioactive materials in the mantle, but a more generally favored interpretation is that these basins are tectonically active, that intrusions are present at depth, and that these are

FIGURE 21-8 Gravity values observed in a section across the Aleutian Trench area (top), and a model (below) constructed on the basis of seismic data and adjusted to give the calculated gravity values (also shown above). (*From Grow, 1973.*)

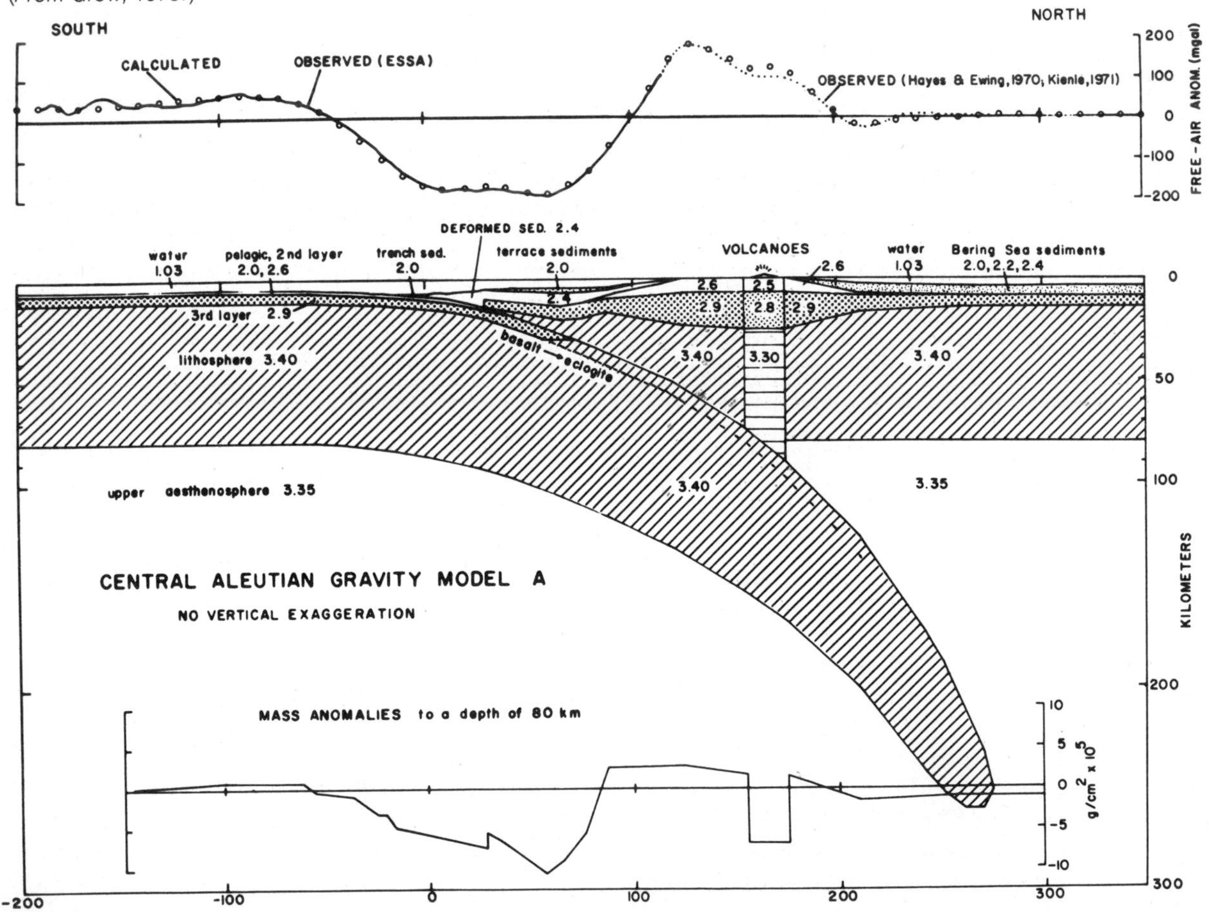

areas which are now or have been recently spreading apart, with the ascent of hot mantle material as the crust is pulled apart (Karig, 1971b). This may constitute a different kind of spreading from that associated with the oceanic ridges, since broad zones of long linear, symmetrically disposed magnetic anomalies like these seen flanking oceanic ridges are absent. Discordant magnetic anomalies have been found in the Mariana trough area, but the tectonics of the interarc areas remain obscure.

Heat could also be generated along the Benioff zone as a result of friction along the margins of the downgoing slab. Efforts have been made to postulate the thermal character of the downgoing slab on the basis of the observed heat flow and other features of island arcs (Oxburgh and Turcotte, 1970; Minear and Toksöz, 1970). Somewhat different models of temperature distribution are postulated in these two studies, but other features are common to both. Both postulate that a relatively cool lithospheric slab is gradually warmed both by frictional heating and by thermal conduction from the adjacent much hotter mantle rock as it sinks or is pulled down under the trench along the Benioff zone. Low-temperature iso-

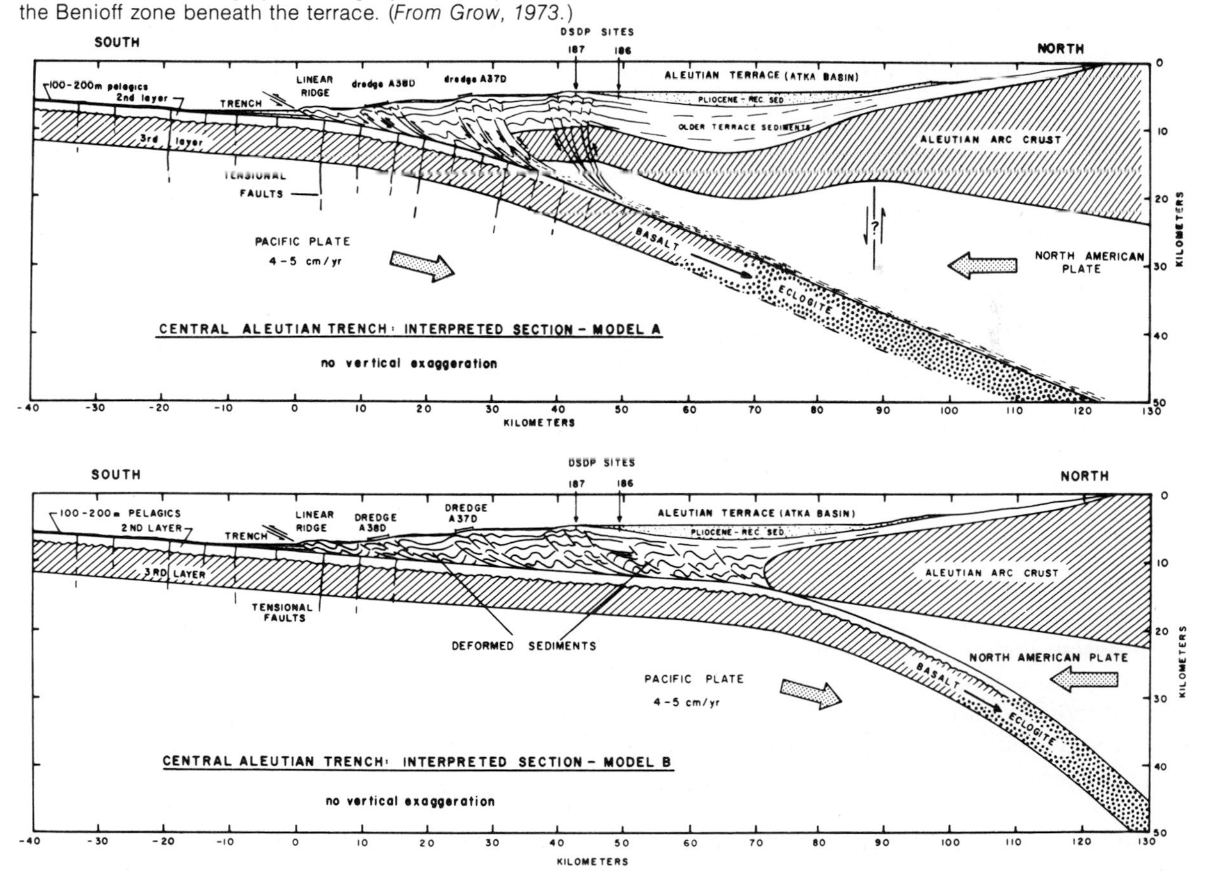

FIGURE 21-9 Interpreted structural sections of trench and terrace. Both models A and B are compatible with essentially all the available data. Uncertainty between the two models is due largely to ambiguity in the depth to the Benioff zone beneath the terrace. (*From Grow, 1973.*)

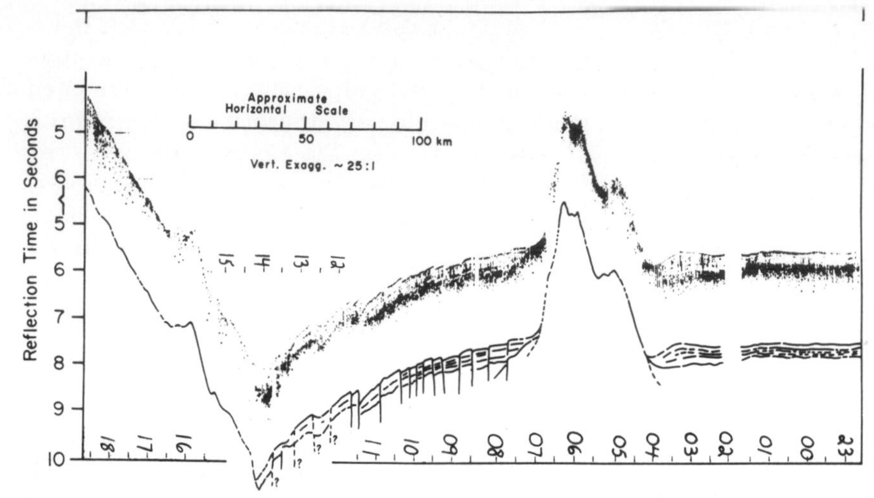

FIGURE 21-10 Representative seismic profiler record and line drawing across the Japan trench. The vertical scale represents two-way reflection time in seconds. (*From Ludwig and others, 1966.*)

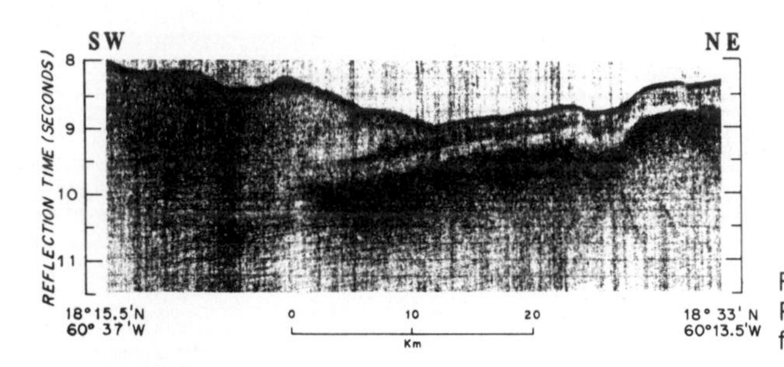

FIGURE 21-11 Seismic reflection profile across the Puerto Rico trench showing the downgoing slab of sea floor at right. (*From Chase and Bunce, 1969.*)

therms are deflected downward in the slab, but the depth to which they go depends on the rate of sinking, Fig. 21-15. Cool rock can reach a much greater depth if the rate of descent is rapid.

Focal Mechanism Studies*

Analysis of the ground motion as recorded at seismograph stations throughout the world provides an indication of the initial movement at the focus of an earthquake. Most seismologists accept the elastic rebound theory regarding the release of energy in shallow-focus earthquakes—most earthquakes are thought to be caused by shear failure or slip along a fault surface. Such movement can be viewed as resulting from the action of a single or double force couple acting in opposite directions on either side of the fault. When energy release occurs, the first pulse travels out from the focus as a compression or a dilation, depending on the direction of the couple. The movement field is divided into quadrants by the fault surface and the plane at right angles to it. The initial movement direction of the *P* wave at each station can be read from the records, and the orientation of the two reference planes can be determined from a plot of initial movements at many stations.

The initial movement data are now often plotted, and the reference plans are drawn on

* See a review and bibliography in Khattri (1973).

an equal-area projection, Fig. 21-16. If the earthquake is generated as a result of shear and the fault plane is a shear plane, then it is one of the two potential shear planes as discussed in Chap. 3. The second shear plane is the plane at right angles to the fault. The principal stress axes bisect the angles between these planes. Thus the principal stress directions and the orientation of the probable planes of shearing can be obtained from a plot of dilations and compressions.

Unfortunately the solution is not unambigu-

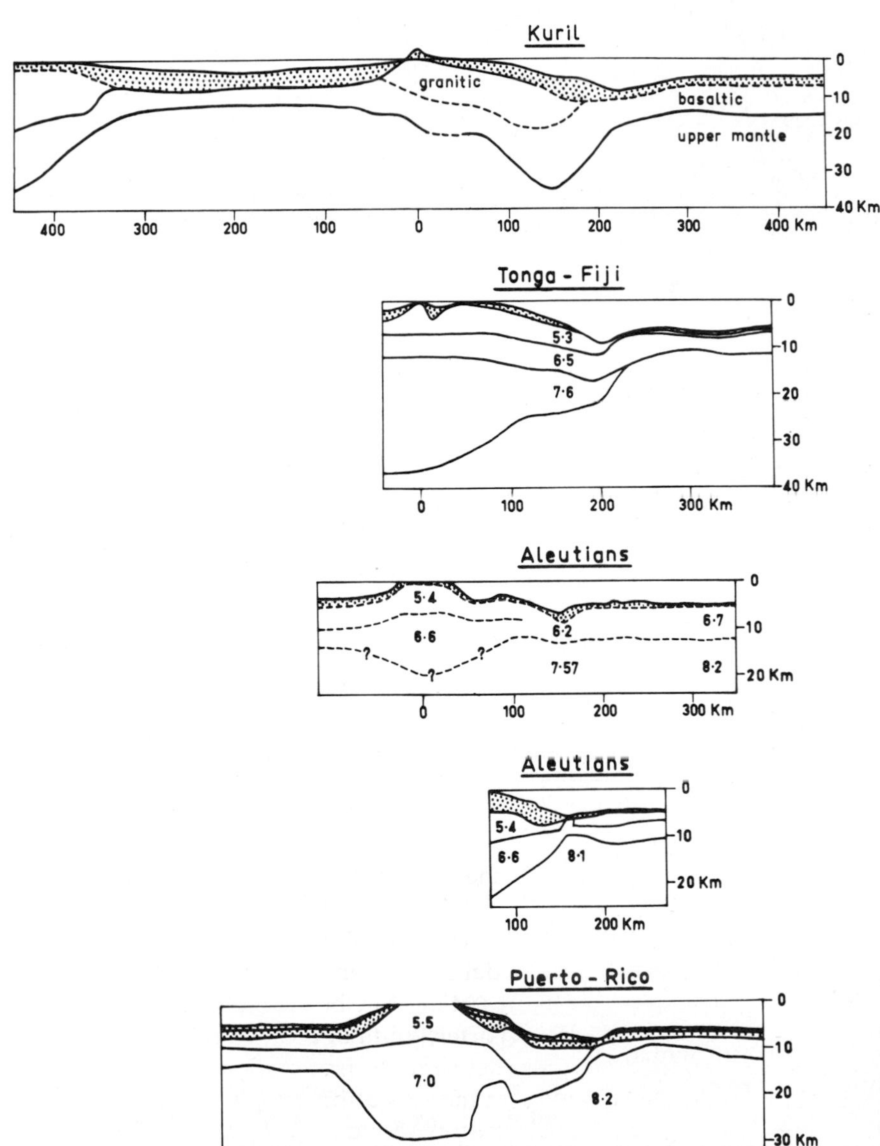

FIGURE 21-12 Crustal cross sections across island arcs. Kuriles (*from Belyayevskiy and Fedynskiy, 1962*); Tonga-Fiji (*from Talwani, Worzel, and Ewing, 1961*); Aleutians (*from Shor, 1962, and Peter and others, 1965*); Puerto Rico (*from Talwani, Worzel, and Ewing, 1959*).

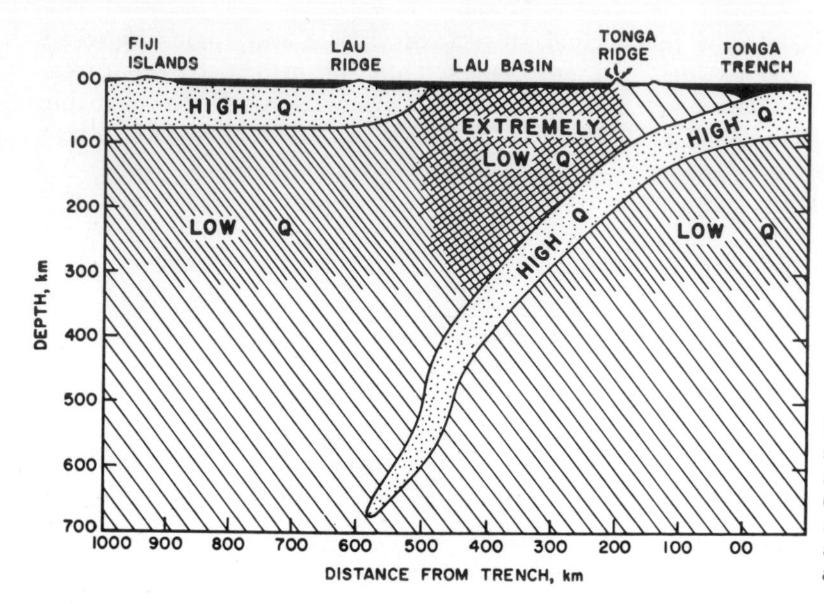

FIGURE 21-13 A schematic cross section perpendicular to the Tonga arc showing lithospheric plates (dotted) and the high- and low-attenuation zones (respectively, low and high Q values). (*From Barazangi and Isacks, 1971.*)

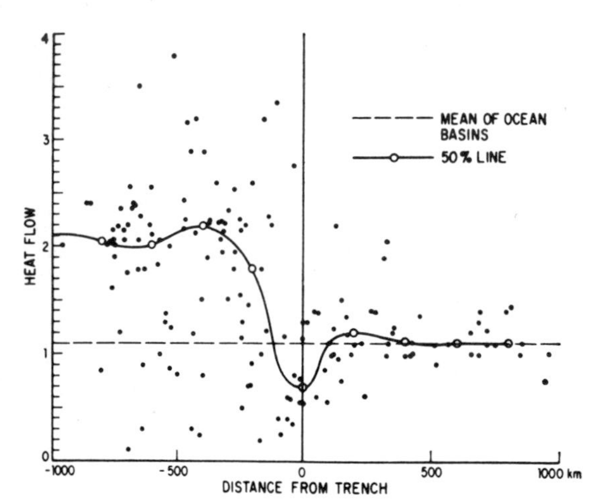

FIGURE 21-14 Compilation of heat-flow data across the Kurile, Japan, and Bonin-Marianas arc systems (after McKenzie and Sclater, 1968). Negative distance is behind the trench; open ocean is to the right and positive.

ous, because it is not always apparent which of the potential shear planes underwent displacement.

Numerous studies of focal mechanisms have been made in island arcs. In general these have led to the conclusion that the horizontal component of the compression axis is approximately perpendicular to the axis of the trench, although the slip direction and the slip planes for earthquakes in the Benioff zone are not exactly parallel to the orientation of the zone. Variations in focal mechanism solutions vary from the trench down the Benioff zone. Focal solutions for the area immediately under the trenches are generally interpreted as being due to extension, and the high-angle faults which can be seen in continuous reflection profiles there support this interpretation. Solutions under the arc-trench gap are interpreted as due to thrusting. Solutions for intermediate and deep earthquakes indicate both extension and compression as shown in Isacks and Oliver's well-known analysis, Fig. 21-17.

GEOLOGICAL FEATURES OF ISLAND ARCS

The distinction between island arcs and adjacent land areas is not always clear physiographically or, especially, in terms of the geology. Features of the Aleutians can be traced into the Alaska Peninsula, the Kurile arc passes into Kamchatka, the Tonga-Kermadec arc passes into New Zealand, etc. The structure, volcanicity, petrology, and history invariably appear much simplier where the arc consists of chains of volcanic islands and submarine features known only indirectly, as in the Bonin arc. The geology is far more complex on the larger islands such as New Zealand, New Guinea, Japan, those of the Indonesian region, and where the arcs pass into continent-connected peninsulas such as Alaska, Kamchatka, and Malaysia.

The Paleozoic history of the margins of the Pacific is obscure. A few outcrops of Paleozoic rocks are known on some of the larger islands, and these are often deformed, but the regional relationships are very difficult to decipher. The Mesozoic history of the marginal lands and larger islands is much better preserved, and these rocks reveal widespread Cretaceous tectonic activity. Typically, Mesozoic rocks form a basement as under the eastern end of the Aleutian arc, and these are eugeosynclinal assemblages intruded by Jurassic granitic plutons. However, without question the island arcs are predominantly covered near the surface by Cenozoic volcanic eruptives and associated sediments.

Many of the volcanic island arcs are made up almost entirely at the surface of late Tertiary and Quaternary andesite volcanoes. Composition of the volcanic products varies from olivine basalt to rhyolite, but most are andesitic. The older of these volcanics are intruded by plutonic rocks which range from gabbro to albite granite, but again most are

FIGURE 21-15 A possible distribution of isotherms under an island-arc system. The rate of slip on the Benioff zone is taken as 9 cm/year; the mean heat flow to the ocean basin as 1.0 μcal cm^{-2} sec^{-1}, and that behind the trench as 1.6 μcal cm^{-2} sec^{-1}. (*From Oxburgh and Turcotte, 1970.*)

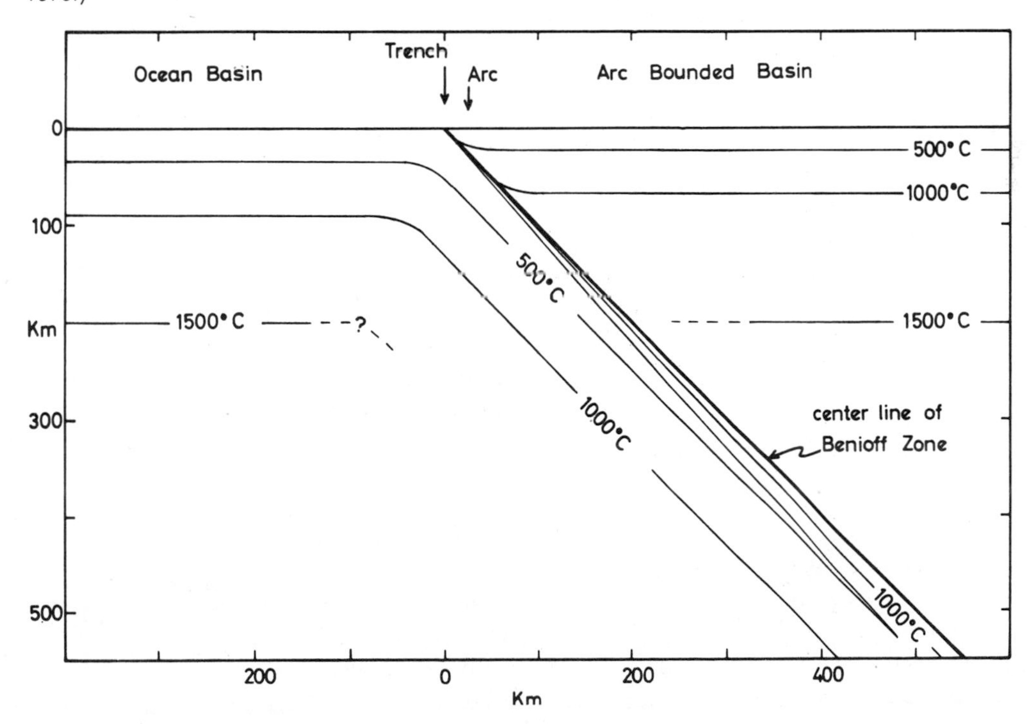

intermediate in composition. Less silicic lavas occur in intraoceanic arcs.

The volcanoes generally are aligned in a sweeping arc, but in some places as in the Aleutians, Fig. 21-18, a number of volcanic centers are aligned transverse to the trend of the arc and certainly reflect the orientation of deep-seated rupture in the crust. This volcanic activity extends into adjacent areas underlain by continental-type crust as in Alaska and New Zealand, where highly silicic lavas (ignimbrites, rhyolite) are found.

Evidence of Cenozoic tectonic activity is found in most arcs and takes a number of forms. In the Kurile arc, transverse depressions separate areas which have recently been uplifted. Emergent wave-cut terraces are present here, and uplifted reefs are found among

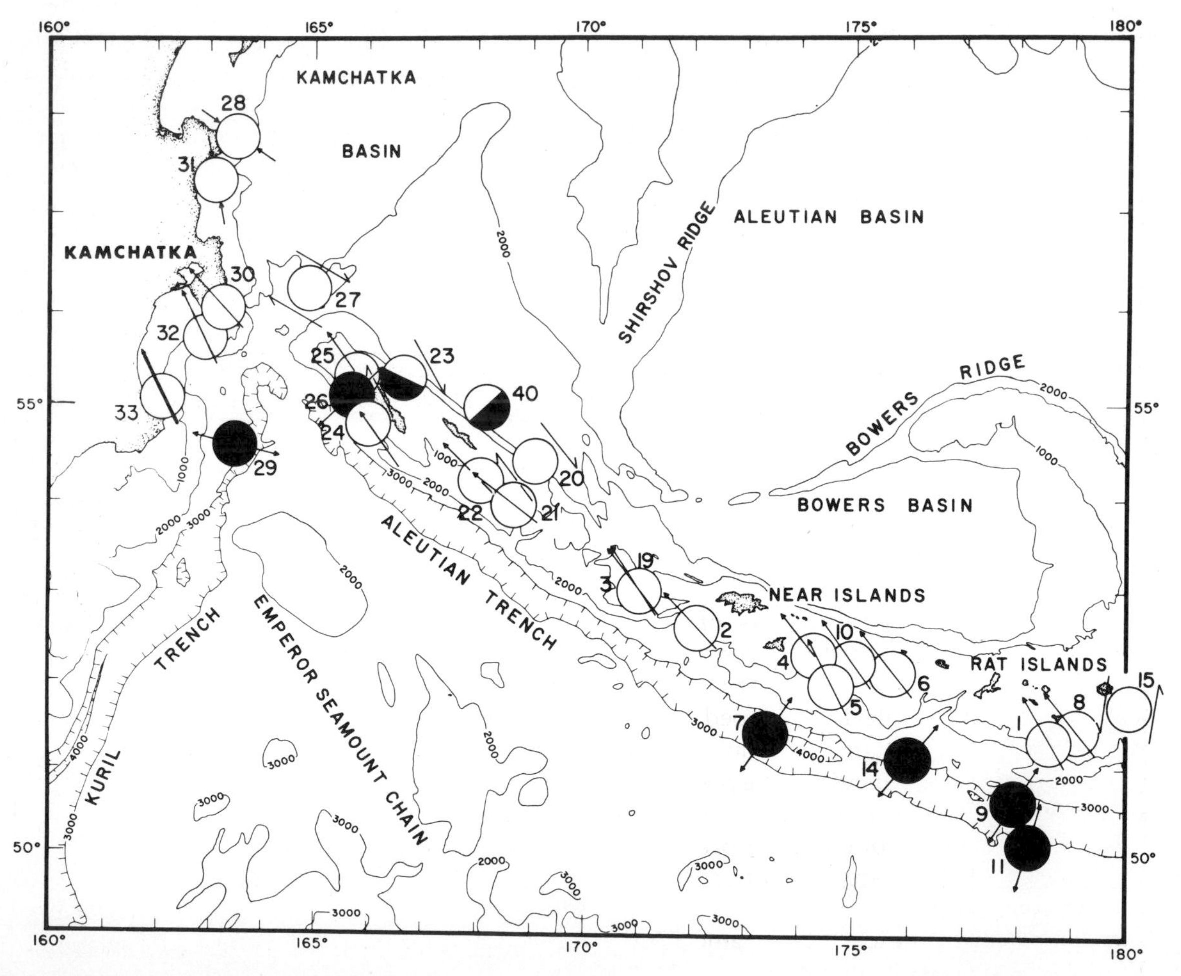

FIGURE 21-16 Azimuth of slip vectors from fault-plane solutions in western Aleutians and northeast Kamchatka. Open circle with arrow = thrust mechanism; open circle with tangent arrows pointing in opposite directions = strike-slip mechanism; half-open, half-filled circle = vertical fault plane. Fault motion at epicenter is such that side filled indicates side that moved down relative to open side. Filled circle = normal faulting mechanism. When the ambiguity of which nodal plane is the fault plane cannot be resolved, the strike of either the axis of compression or the axis of tension is shown. (*From Cormier, 1975.*)

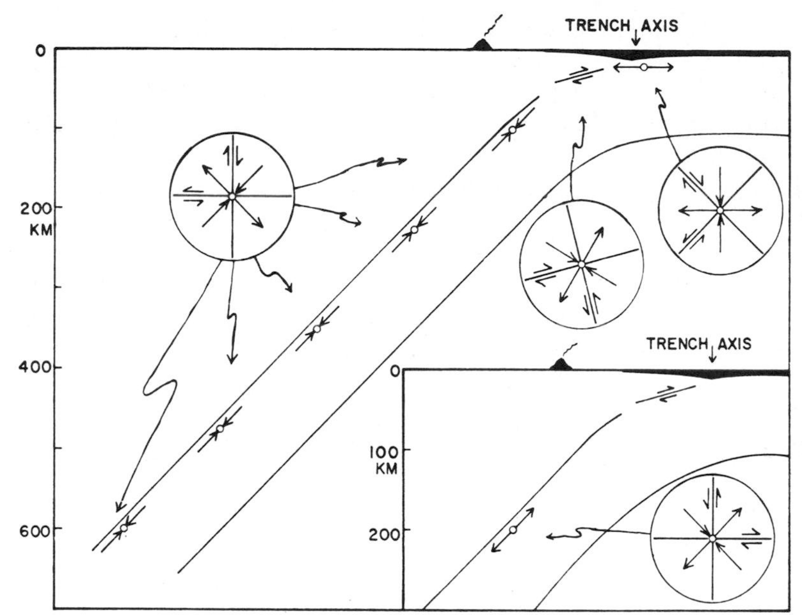

FIGURE 21-17 Vertical sections perpendicular to the strike of an island arc showing schematically typical orientations of double-couple focal mechanisms. The horizontal scale is the same as the vertical scale. The axis of compression is represented by a converging pair of arrows; the axis of tension is represented by a diverging pair; the null axis is perpendicular to the section. In the circular blowups, the sense of motion is shown for both of the two possible slip planes. The features shown in the main part of the figure are based on results from the Tonga arc and the arcs of the north Pacific. The insert shows the orientation of a focal mechanism that could indicate extension instead of compression parallel to the dip of the zone. (*From Isacks and Oliver, 1968.*)

the island arcs farther south. Both longitudinal and transverse fault and fracture patterns are found in many arcs. High angle faults appear to be more common, although thrusts and major strike-slip faults occur in some arcs. Folding is generally broad and open, often best described as warping; however, continuous reflection profiles across the arc-trench gap of many arcs reveal low-angle thrusts which dip back under the volcanic arc, and large-scale folds some of which are best described as nappes are known as in Sumatra, Fig. 21-19. How these thrust masses originated is usually uncertain. Many of them may be gravity-glide features or results of submarine slides directed toward the trench, but many geologists now prefer to think that these are the sedimentary materials scraped off a downgoing lithospheric slab. The whole problem is made even more difficult by the presence of undeformed sediments in the floors of some trenches. It is hard to reconcile their presence with the idea that these are sites of scraping and strong compression.

It is possible to predict the metamorphic conditions within the subduction zone on the basis of knowledge of the temperature, pressure, and types of materials which go into the zone. Metamorphism immediately under the trench where the cold slab begins to sink should involve recrystallization at low temperature and low pressure (i.e., in the low greenschist and blueschist facies). At greater depth the metamorphism would involve both high temperatures and high pressures, producing rocks in the greenschist to amphibolite facies. Presumably the volcanoes mark the approximate position (in a vertical plane) of partial melting in the subduction zone. All these features should be important as guides to the recognition of fossil subduction zones as the search for these goes on in land areas where the former arcs may have been uplifted and incorporated in continental masses.

THE SUBDUCTION-ZONE MODEL OF ISLAND ARCS

Island arcs are the sites of subduction of the lithosphere according to the widely accepted

Solid black dots are volcanos

Dotted areas - Jurassic sedimentary rocks

White areas - Post Jurassic-largely Tertiary and Quaternary

vvv Pattern = Tertiary granitic intrusions

= Jurassic granitic intrusions

K = Cretaceous slate and graywacke belt

U/D = High angle fault

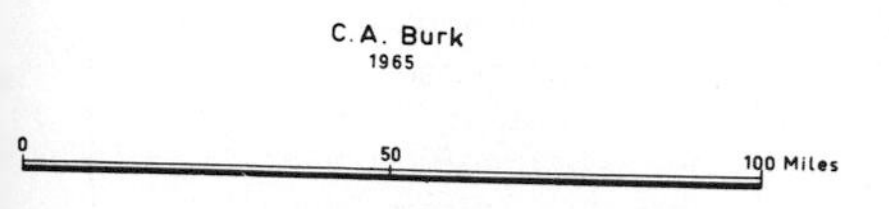

FIGURE 21-18 Geologic sketch map of the Alaska Peninsula. (*After Burk, 1965.*)

plate-tectonic theory. Most of what we know about the arcs is compatible with or can at least be reconciled with this model. That model can be summarized as follows, Fig. 21-20.

The oceanic crust bearing a surficial cover of pelagic sediment and the residual magnetic anomalies imprinted on the basaltic oceanic crust as it formed moves toward the island arcs as sea-floor spreading continues. The trenches mark sites where the lithospheric plates sink into the mantle, possibly as a result of mechanical or chemical changes. It appears likely that the lithosphere is in a state of quasi-equilibrium as a result of its density being close to that of the material in the underlying low-velocity zone. Where the downgoing slab is bent under the trench its upper surface is brought into tension and fails, forming high-angle faults. This accounts for both the focal plane solutions and the observed faults in the sediments in that location. The slab generally sinks along a zone inclined at 30 to 60°, but some are steeper and others shallower than

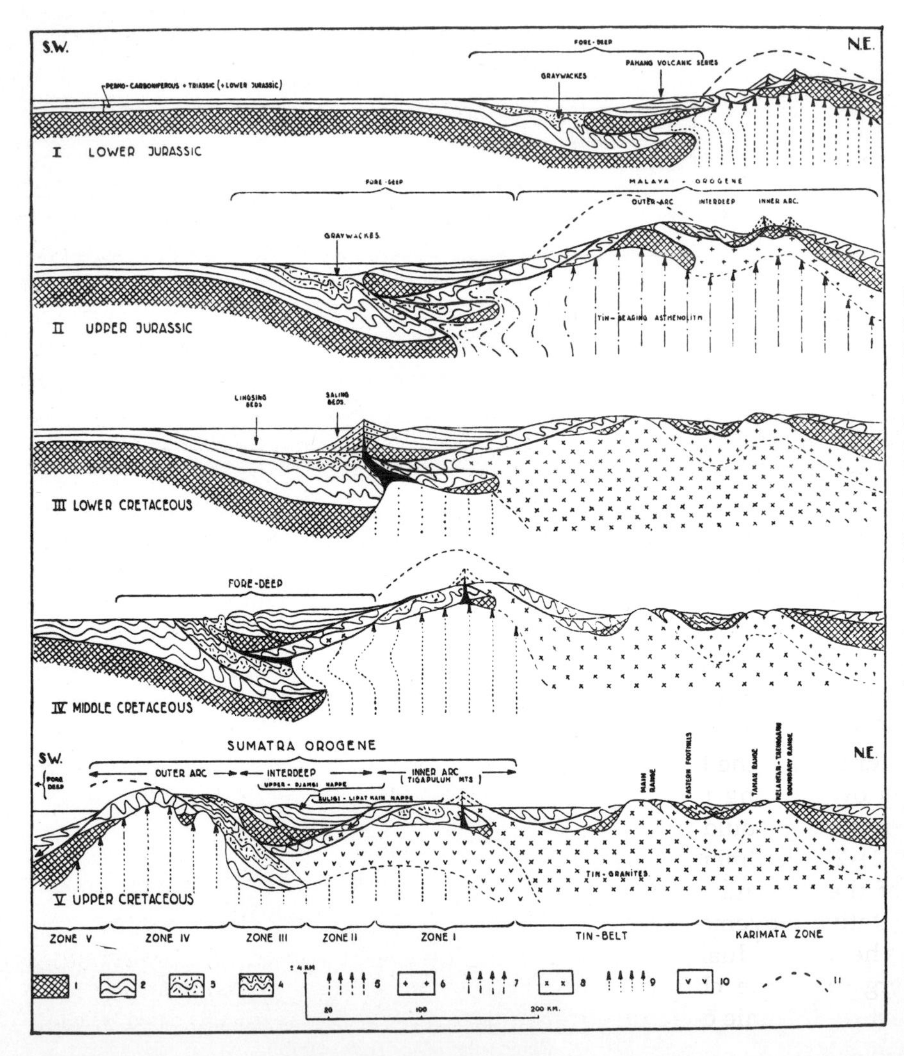

FIGURE 21-19 Five schematic sections illustrating the pre-Tertiary evolution of the Malay Peninsula and Sumatra. The symbols are numbered as follows: (1) Pre-Carboniferous and Permian basement complex. (2) Sedimentary epidermis (young Paleozoic and Mesozoic, without further distinction of age). (3) Graywacke formation (Jurassic?). (4) Saling beds (Lower Cretaceous). (5) Asthenolithic migma and magma zone of the Karimata zone. (6) Same as (5), consolidated. (7), Asthenolithic migma and magma zone of the Tin belt. (8) Same as (7), consolidated. (9) Asthenolithic migma and magma zone of the Sumatra orogen. (10) Same as (8), consolidated. (11) Outlines which the geanticlinal uplifts from the foredeeps would have reached, if no bathydermal spreading had occurred. (*From Van Bemmelen, 1949.*)

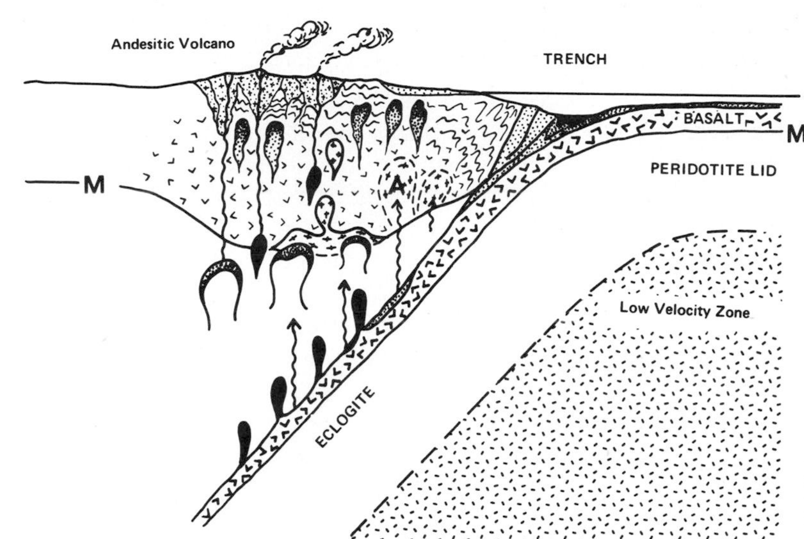

FIGURE 21-20 Schematic drawing of the ocean-floor-spreading hypothesis as modified by Ringwood (1969) after Hess (1962).

this. Movements along the top and bottom of the slab generate earthquakes, and this movement also causes frictional heating of the slab and adjacent sediments. Stresses are set up within the downgoing slab, and in some places the slab is pushed from above by overlying sinking lithosphere, causing compressive failure. In other places the slab sinks faster than new sea floor is being brought to the trench. When this happens the slab may fail as a result of extension.

Sediments being transported on the surface of the subducting plate either sink with the slab, or they are scraped off to become part of the arc-trench gap. One suggested model of the structural configuration of the arc-trench gap, Fig. 21-21, consists of stacked fault slivers of sediment and slices of oceanic crust which are sheared off at the inner side of the trench. If this process continues over a long period, the width of the arc-trench gap would likely increase as shown, Fig. 21-22. Sediment may also slip and slide off of the volcanic arc toward the trench by gravitative movements.

The downgoing slab is heated gradually as it is subducted and undergoes phase changes which cause the basalts of the oceanic crust to be changed to eclogite. At some point the temperatures finally reach the level necessary to cause partial melting of the oceanic crust, producing magmas generally of intermediate composition; however, magmatic differentiation can account for a range of compositional variation. These magmas rise vertically and emerge as andesite volcanoes or near-surface intrusions in the volcanic arc. Altered oceanic crustal materials such as serpentine and even ultrabasics may be forced to the surface in the arcs. This rising magma accounts for the high heat flows observed in the arcs, and the altered crust accounts for the abnormal seismic velocities and gravity anomalies over the arcs.

The interarc basins, Fig. 21-23, located behind the volcanic arcs are viewed as areas of modern spreading produced by the opening up of these seas such as the Philippine Sea and the area behind the Tonga-Kermadec arc.

Emplacement of Ultrabasics in Island Arcs by Obduction

Many of the largest known ultrabasic bodies occur in island arcs, and although they have long been recognized, the mechanism of their emplacement has not been obvious. The alternatives of emplacement as mantle diapirs and as thrust slices was mentioned earlier (Chap. 17). A number of variations have been proposed for emplacement in subduction zones (Dewey, 1975; Dewey and Bird, 1971). These include (1) development of imbricate thrust slices on the inner edge of the trench which may involve oceanic crust that is buckled upward or uplifted isostatically later; (2) development of a low-angle thrust when a continent reaches a subduction zone and will not go down—movement between the plates is then accommodated by thrusting of the island arc and underlying mantle mantle up onto the edge of the continental plate; and (3) opening of a rift and development of some sea floor followed by closing of that rift by resumption of subduction or continental collision could result in lateral shortening and forceful upward movement of the trapped oceanic crust and thrusting of it onto the continental margin.

Obduction by various methods has been proposed in many orogenic belts as well as in island arcs. Notable examples are found in the Newfoundland Appalachians central mobile belt (Dewey, 1975), New Guinea and New

FIGURE 21-21 Speculative models of the structure in the inner wall of a trench. Turbidite sections are sheared off along weak high-porosity pelagic sections and ride over the trench wedge. Slabs of the upper oceanic crust are sheared off when topographic irregularities enter the trench. (*From Karig and Sharman, 1975.*)

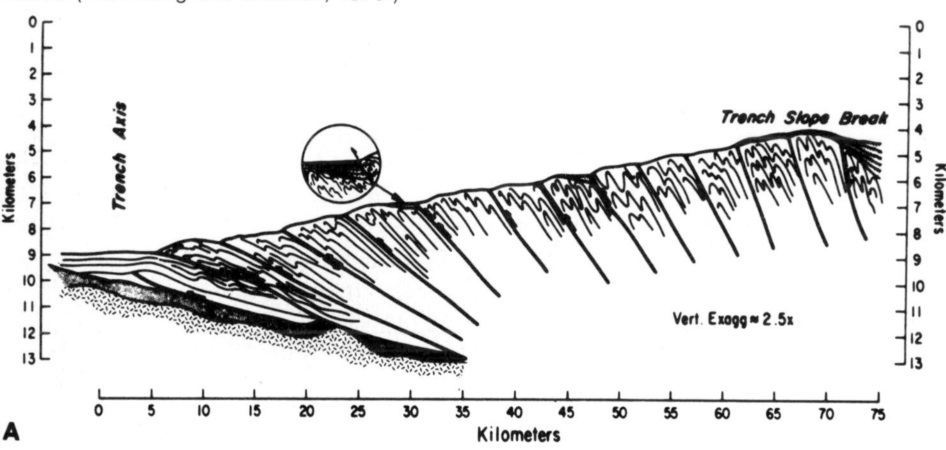

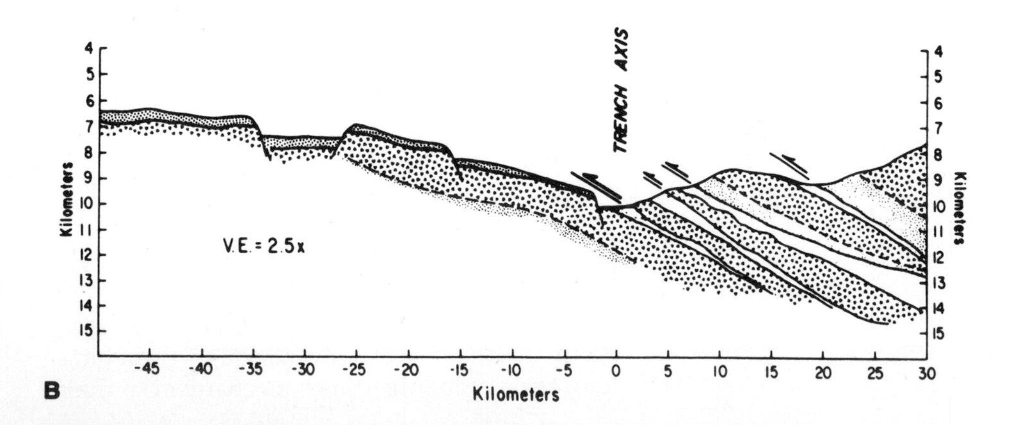

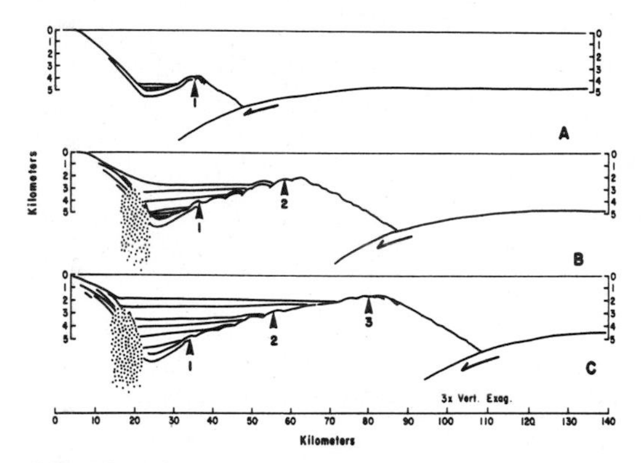

FIGURE 21-22 Hypothetical "building out" of accretionary prism, showing migration of the morphology over the accreted material. The original trench slope break becomes part of the subsiding basement of the upper slope area. Numbers mark arbitrary successive locations of the trench-slope break. *(From Karig and Sharman, 1975.)*

Caledonia (Coleman, 1971), Oman, and in the Klamath Mountains and along the Roberts Mountain thrust in Nevada (Burchfiel and Davis, 1972). The mechanics of this process are analyzed by Elliott (1976).

COMPLEXITIES OF THE SOUTHWESTERN PACIFIC ISLAND ARCS

The whole structural and topographic pattern of the region south of Japan is far more complex than that to the north. This complexity begins with the Philippine Sea, which is bounded by arcs on all sides to form a rhombic-shaped crustal element. According to the plate-tectonic model, Pacific plate oceanic crust is being subducted all along the row of arcs from Palau-Yap-Mariana to the Bonins, and crust of the Philippine Sea is being subducted along the Okinawa trench and the Philippine trench. This latter crust is being generated by spreading taking place behind the Mariana and Bonin Island groups.

The configuration in the Indonesian region is quite different. Major strike-slip faults cross both the Philippine Islands and Sumatra, Fig. 21-2, and the movements on these are such that the Eurasian plate is moving south relative to the Philippine and the Indian Ocean plates. Vening Meinesz pointed out the likelihood of this movement years ago and described the Java, Banda, and Philippine trenches as being due to the overriding of this crustal element.

Why then is the southeastern edge of the Eurasian plate so irregular? Visser and Hermes (1962) suggested the evolution of this complication, Fig. 21-24. Perhaps the major element causing this irregularity is the movement of a corner of the Pacific plate toward the west just north of New Guinea. A major strike-slip fault, the Sorong fault, trends east-west here and has left-lateral displacement. The Sorong fault is located at the western end of what many geologists consider to be one of the largest displacement zones in the earth. This zone

FIGURE 21-23 Cross section of a typical island-arc system showing tectonic units and terminology. *(From Karig and Sharman, 1975.)*

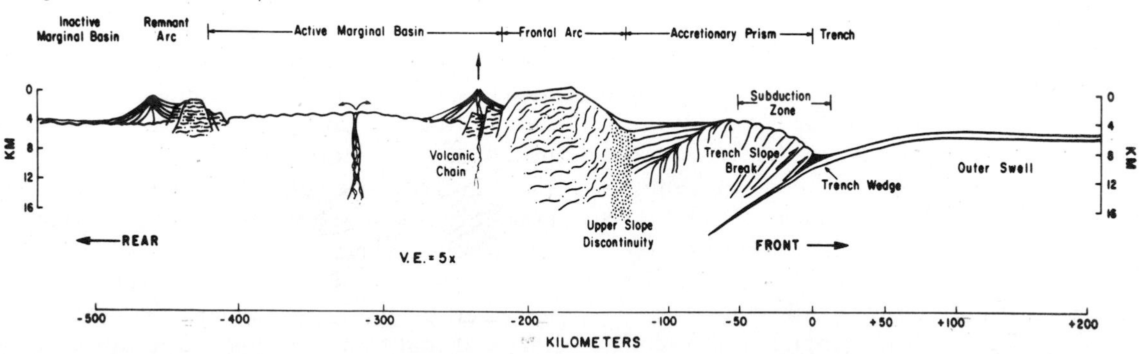

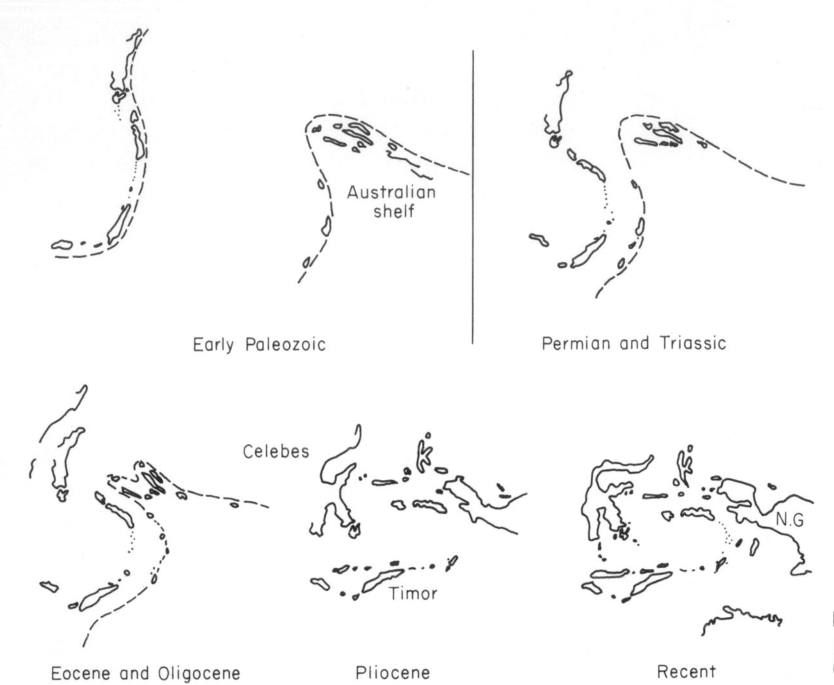

FIGURE 21-24 Evolution of the physiography of eastern Indonesia. (*After Visser and Hermes, 1962.*)

extends from north of Australia and New Guinea eastward to Samoa, just south of the andesite line. As was pointed out by Carey (1958), the whole region east of Australia appears to have opened up as New Zealand and the islands north of it moved eastward relative to Australia. Paleogeologic reconstructions support this view. Wegener pointed out long ago how the curve in New Britain suggests drag as the region to the south moved east relative to the central Pacific. This zone lies, of course, along the globe-encircling "equatorial shear zone" along which it appears that the Southern Hemisphere is rotated relative to the Northern Hemisphere.

The so-called strewn islands of the Solomon group, the Fiji Islands, and the New Hebrides pose some of the most difficult problems for interpretation in terms of relatively simple models. Here the arcuate forms tend to be curved so that the convex side of the arcs faces the Coral Sea rather than the Pacific, Fig. 21-25. The Benioff zones are inclined toward the Pacific basin; so, in general, the orientation of these arcs (New Britain, Solomon, New Hebrides) is reversed from that of all others. Thus according to the plate-tectonic model it is part of the Australian plate that is being consumed along these arcs, despite the fact that no oceanic ridge-type spreading center lies within that plate, and even the abnormally high heat-flow values associated with active basins is missing here now.

ALTERNATIVES TO THE SUBDUCTION MODEL OF ARCS

Not all geologists agree with the subduction model. Most of the observational data are not called into question, but the interpretations of those data are. Alternative models are that the island arcs are sites of crustal extension, as is favored by those who argue for global expansion such as Carey (1958, 1974), that the arcs are sites for essentially vertical tectonics, as described by the Russian geologist Beloussov (1970), or that they are sites where large unda-

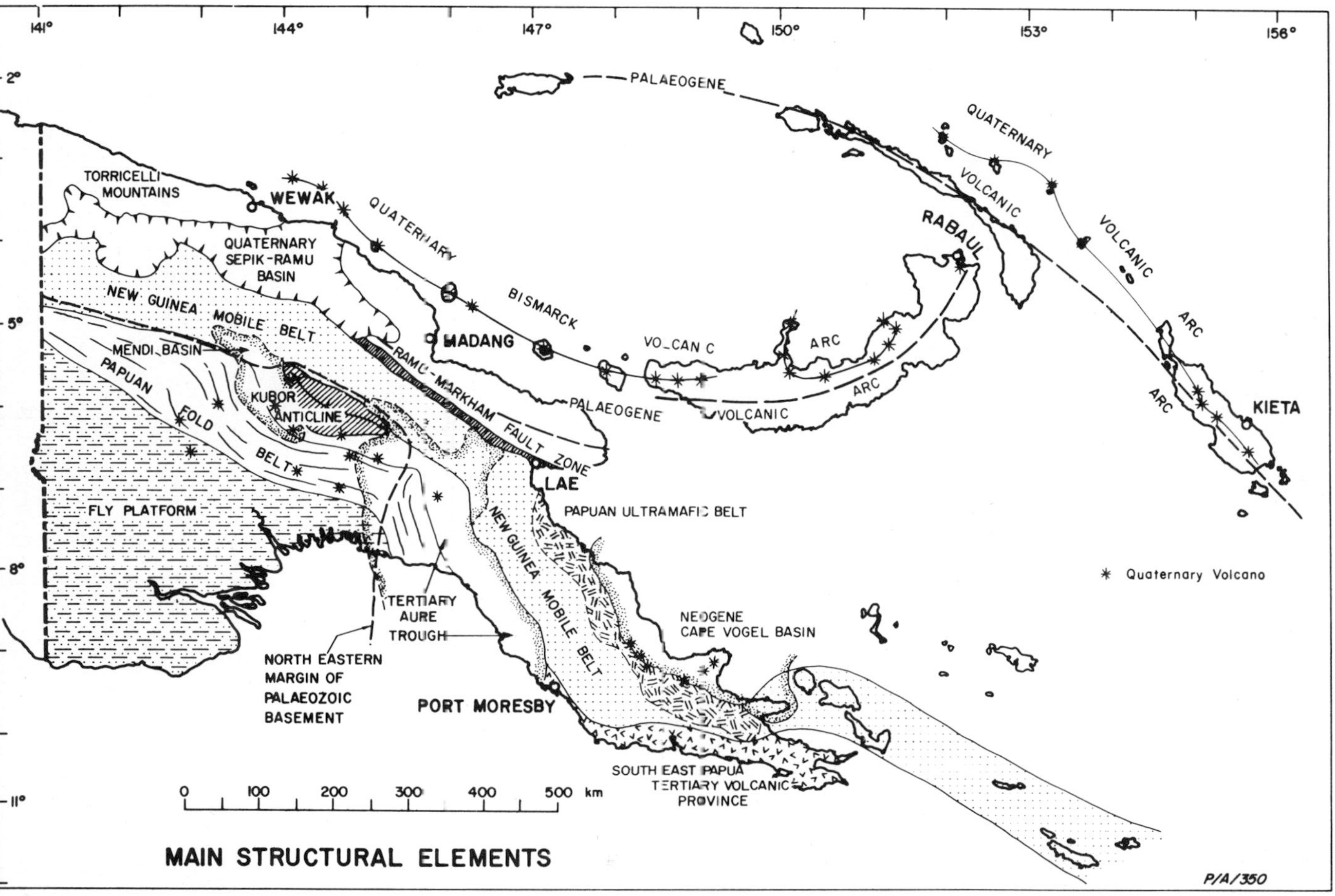

FIGURE 21-25 The primary structural tectonic elements in the area of Papau New Guinea, Louisiade, and Bismarck Archipelagos. (*From Bain, 1973. Reproduced with permission from P. J. Coleman, ed.,* The Western Pacific: Island arcs, marginal seas, geochemistry: *Perth, University of Western Australia Press, 1973.*)

tions in the earth have caused overriding at their edges as described by Van Bemmelen (1972).

The extensional character of the oceanic plate at the lip of the trench is agreed to by all, but Carey and others would attribute many of the other observations under the volcanic arc to extension as well. Deep-seated extension is compatible with the idea that heat is rising here; extension would open fractures in the crust and help the introduction of water which would serpentinize the underlying mafic rocks, reducing the density and providing an explanation for gravity anomalies; rising heat would account for the volcanic activity; even the focal plane solutions in the Benioff zone can be interpreted as extensional features. Deformed sediment in the arc-trench gap is attributed to sliding and gravitative spreading, and the undeformed sediments (such as that in the Chile trench) are more readily understood. Finally the new sea floor found behind the volcanic arcs is another expression of the overall expansion. Carey (1958) termed such areas *sphenochasms* and *rhombochasms*, depending largely on shape.

REFERENCES

Anderson, R. N., 1975, Heat flow in the Mariana marginal basin: Jour. Geophys. Res., v. 80, no. 29, p. 4043–4048.

Bain, J. H. C., 1973, A summary of the main structural elements of Papua New Guinea, *in* Coleman, P. J., ed., The Western Pacific: island arcs, marginal seas, geochemistry: New York, Crane-Russak.

Barazangi, Muawia, and **Isacks, Bryan**, 1971, Lateral variations of seismic-wave attenuation in the upper mantle above the inclined earthquake zone of the Tonga island arc: Deep anomaly in the upper mantle: Jour. Geophys. Res., v. 76, no. 35, p. 8493–8516.

Beloussov, V. V., 1970a, Deep causes of geological processes: Int. Union Geol. Sci. Geol. Newslett. v. 1970.

——— 1970b, Against the hypothesis of ocean-floor spreading: Tectonophysics, v. 9., no. 6.

Ben-Avraham, Zvi, and **Emery, K. O.**, 1973, Structural framework of Sundra shelf: Amer. Assoc. Petroleum Geologists Bull., v. 57, no. 12, p. 2323ff.

Benioff, H., 1949, Seismic evidence for fault origin of oceanic deeps: Geol. Soc. America Bull., v. 60, p. 1837–1856.

——— 1954, Orogenesis and deep crustal structure —Additional evidence from seismology: Geol. Soc. America Bull., v. 65, p. 385–400.

Bentley, C. R., and **Laudon, T. S.**, 1965, Gravity and magnetic studies in the Solomon Islands: Natl. Acad. Sci., Prog. Rept. United States Program for the Internat. Geophys. Year Upper Mantle Project.

Biq, Chingchang, 1960, Circum-Pacific tectonics in Taiwan: Internat. Geol. Cong., 21st, Norden, p. 203–214.

Bracey, D. R., and **Ogden, T. A.**, 1972, Southern Marina arc: Geophysical observations and hypothesis of evolution: Geol. Soc. America Bull., v. 83, p. 1509ff.

Brooks, J. A., 1962, Seismic wave velocities in the New Guinea–Solomon Islands region, *in* The crust of the Pacific basins: Am. Geophys. Union Mon. 6, p. 2–10.

Brouwer, H. A., 1947, Geological exploration in the island of Celebes: Amsterdam, North Holland.

Burchfiel, B. C., and **Davis, G. A.**, 1972, Structural framework and evolution of the southern part of the Cordilleran orogen, western United States: Am. J. Sci., v. 272, no. 2.

Burk, C. A., 1965, Geology of the Alaska peninsula —Island arc and continental margin, pt. 1: Geol. Soc. America Mem. 99.

Carey, S. W., 1958, The tectonic approach to continental drift, *in* Continental drift—A symposium: Geol. Dept. Univ. Tasmania, p. 177–355.

Carr, M. J., **Stoiber, R. E.**, and **Drake, C. L.**, 1973, Discontinuities in the deep seismic zones under the Japanese arcs: Geol. Soc. American Bull., v. 84, p. 2917–2930.

Chase, R. L., and **Bunce, E. T.**, 1969, Underthrusting of the eastern margin of the Antilles by the floor of the western North Atlantic Ocean, and origin of the Barbados ridge: J. Geophys. Res., v. 74.

Coleman, P. J., 1975, On island arcs: Earth-Sci. Rev., v. 11, p. 47–80.

Coleman, R. G., 1971, Plate tectonic emplacement of upper mantle peridotites along continental edges: Jour. Geophys. Res., v. 76, no. 5, p. 1212.

Cormier, V. F. (no date), Tectonics near the junction of the Aleutian and Kurile-Kamchatka arcs and a mechanism for Middle Tertiary magmatism in the Kamchatka basin: Geol. Soc. American Bull., v. 86, p. 443–453, April, 1975.

Dewey, J. F., and **Bird, J. M.,** 1971, Origin and emplacement of the ophiolite suite: Appalachian ophiolites in Newfoundland: Jour. Geophys. Research, v. 76.

Dickinson, W. R., 1970, Relations of andesites, granites, and derivative sandstones to arc-trench tectonics: Rev. Geophys. Space Physics, v. 8, no. 4.

——— 1973, Widths of modern trench-arc gaps proportional to past duration of igneous activity in associated magmatic arcs: Jour. Geophys. Res., v. 78, p. 3376ff.

Elliott, David, 1976, The motion of thrust sheets: Jour. Geophys. Res., v. 81, no. 5, p. 949–963.

Fitch, T. J., 1972, Plate convergence, transcurrent faults, and internal deformation adjacent to SE Asia and the western Pacific: Jour. Geophys. Res., v. 77, no. 23, p. 4432ff.

——— and **Molnar, Peter,** 1970, Focal mechanisms along inclined earthquake zones in the Indonesia-Philippine region: Jour. Geophys. Res., v. 75, no. 8, p. 1431–1444.

Goryatchev, A. V., 1962, Specific features of recent tectonism along the Kurile island arc: Internat. Geol. Rev., v. 4.

Grim, P. J., and **Erickson, B. H.,** 1969, Fracture zones and magnetic anomalies south of the Aleutian trench: Jour. Geophys. Res., v. 74, no. 6, p. 1488–1494.

Grow, J. A., 1973, Crustal and upper mantle structure of the central Aleutian arc: Geol. Soc. America Bull., v. 84, p. 2169–2192.

Hatherton, Trevor, and **Dickinson, W. R.,** 1969, The relationship between andesitic volcanism and seismicity in Indonesia, the Lesser Antilles, and other island arcs: Jour. Geophys. Res., v. 74, n. 22, p. 5301–5310.

Hayes, D. E., and **Heirtzler, J. R.,** 1968, Magnetic anomalies and their relation to the Aleutian island arc: Jour. Geophys. Res., v. 73, no. 14, p. 4637ff.

——— and **Ludwig, W. J.,** 1967, The Manila trench and west Luzon trough—II. Gravity and magnetics measurements: Deep-Sea Res., v. 14, p. 545–560.

Hess, H. H., 1938, Gravity anomalies and island arc structure with particular reference to the West Indies: Am. Philos. Soc. Proc., v. 70, p. 71–96.

——— 1948, Major structural features of the western North Pacific: An interpretation of H.O. 5485, bathymetric chart Korea to New Guinea: Geol. Soc. America Bull., v. 59, p. 417–548.

——— 1953b, Major structural features of the southwest Pacific: Pacific Sci. Cong. Proc., 7th, v. 2, p. 14–17.

——— 1962, History of ocean basins, *in* Petrologic studies (Buddington volume): Geol. Soc. America, p. 599–620.

Isacks, Bryan, and **Oliver, Jack,** 1968, Seismology and the new global tectonics: Jour. Geophys. Res., v. 73, no. 18, p. 5855–5867.

———, **Sykes, L. R.,** and **Oliver J.,** 1967, Spatial and temporal clustering of deep and shallow earthquakes in the Fiji-Tonga-Kermadec region: Seis. Soc. America Bull., v. 57, no. 5, p. 935–958.

Jenkins, D. A. L., 1974, Detachment tectonics in western Papua New Guinea: Geol. Soc. America Bull., v. 85, no. 4, p. 533–548.

Karig, D. E., 1970, Ridges and basins of the Tonga-Kermadec island arc system: Jour. Geophys. Res., v. 75, no. 2, p. 239–254.

——— 1971a, Origin and development of marginal basins in the western Pacific: Jour. Geophys. Res., v. 76, no. 11, p. 2542ff.

——— 1971b, Structural history of the Mariana island arc system: Geol. Soc. America Bull., v. 82, p. 323ff.

——— 1972, Remnant arcs: Geol. Soc. America Bull., v. 83, p. 1057ff.

——— and **Sharman, G. F.,** 1975, Subduction and accretion in trenches: Geol. Soc. America Bull., v. 86, p. 377–389.

Katili, J. A., 1973, Geochronology of west Indonesia and its implication on plate tectonics: Tectonophysics, v. 19, p. 195–212.

——— 1975, Volcanism and plate tectonics in the Indonesian island arcs: Tectonophysics, v. 26, p. 165–188.

Katsumata, Mamoru, and **Sykes, L. R.,** 1969, Seismicity and tectonics of the western Pacific: Izu-Mariana-Caroline and Ryukyu-Taiwan regions: Jour. Geophys. Res., v. 74, no. 25, p. 5923–5948.

Kelleher, John, Sykes, L., and **Oliver, J.,** 1973, Possible criteria for predicting earthquake locations and their application to major plate boundaries of the Pacific and the Caribbean: Jour. Geophys. Res., v. 78, no. 14, p. 2547ff.

Khattri, Kailash, 1973, Earthquake focal mechanism studies—A review: Earth-Sci. Rev. v. 9, no. 1, p. 19–63.

Kogan, M. G., 1975, Gravity field of the Kurile-Kamchatka arc and its relation to the thermal regime of the lithosphere: Jour. Geophys. Res., v. 80, p. 1381–1390.

Krause, D. C., 1966, Tectonics, marine geology and bathymetry of the Celebes Sea–Sulu Sea region: Geol. Soc. America Bull., v. 77, p. 813–832.

Krebs, Wolfgang, 1975, Formation of southwest Pacific island arc-trench and mountain systems—A plate or global-vertical tectonics?: Am. Assoc. Petroleum Geologists Bull., v. 59, no. 9, p. 1639–1666.

Lanphere, M. A., and **Reed, B. L.**, 1973, Timing of Mesozoic and Cenozoic plutonic events in circum-Pacific North America: Geol. Soc. America Bull., v. 84, p. 3773ff.

Lillie, A. R., and **Brothers, R. N.**, 1969, The geology of New Caledonia: N. Z. Jour. Geol. Geophys., v. 13, no. 1, p. 145–183.

Ludwig, W. J., 1970, The Manila trench and west Luzon trough; III, Seismic-refraction measurements: Deep-Sea Res., v. 17, no. 3.

———, **Murauchi, S.**, **Den, N.**, and others, 1973, Structure of the East China Sea–West Philippine Sea margin off southern Kyushu, Japan: Jour. Geophys. Res., v. 78, no. 14.

——— and others, 1966, Sediments and structure of the Japan trench: Jour. Geophys. Res., v. 71, no. 8, p. 2121–2137.

MacDonald, D. C., **Luyendyk, B. P.**, and **von Herzen, R. P.**, 1973, Heat flow and plate boundaries in Melanesia: Jour. Geophys. Res., v. 78, no. 14, p. 2537ff.

McKenzie, D. P., and **Sclater, J. G.**, 1968, Heat flow inside the island arcs of the northwestern Pacific: Jour. Geophys. Res., v. 73, no. 10, p. 3173–3179.

Marlow, M. S., **Scholl, D. W.**, and **Buffington, T. R. A.**, 1973, Tectonic history of the Central Aleutian arc: Geol. Soc. America Bull., v. 84, p. 1555–1574.

Marshall, P., 1912, Oceania: Handb. Regional Geology, v. 7, no. 2.

Menard, H. W., 1969, Growth of drifting volcanoes: Jour. Geophys. Res., v. 74, no. 20, p. 4827–4837.

Milson, John, 1973, Papuan ultramafic belt: Gravity anomalies and the emplacement of ophiolites: Geol. Soc. America Bull., v. 84, p. 2243–2258.

Minear, J. W., and **Toksöz, M. N.**, 1970, Thermal regime of a downgoing slab and new global tectonics: J. Geophys. Res., v. 75.

Moore, J. C., 1973, Complex deformation of Cretaceous trench deposits, southwestern Alaska: Geol. Soc. America Bull., v. 84, no. 6, p. 2005.

Murauchi, S., and others, 1968, Crustal structure of the Philippine Sea: Jour. Geophys. Res., v. 73, no. 10, p. 3143–3171.

———, **Ludwig, W. J.**, **Den, N.**, and others, 1973, Refraction measurements on the Ontong Java Plateau northeast of New Ireland: Jour. Geophys. Res., v. 78, no. 35.

Naugler, F. P., and **Wageman, J. M.**, 1973, Gulf of Alaska: Magnetic anomalies, fracture zones, and plate interaction: Geol. Soc. America Bull., v. 84, p. 1575–1584.

Oliver, J., and **Isacks, B.**, 1967, Deep earthquake zones, anomalous structures in the upper mantle, and the lithosphere: Jour. Geophys. Res., v. 72, no. 16, p. 4259ff.

———, **Isacks, Bryan**, **Barazangi, Mauwia**, and **Mitronovas, Walter**, 1973, Dynamics of the downgoing lithosphere: Tectonophysics, v. 19, p. 133–147.

Peter, G., **Elver, D.**, and **Yellin, M.**, 1965, Geological structure of the Aleutian trench southwest of Kodiak Island: Jour. Geophys. Res., v. 70, no. 2.

Oxburgh, E. R., and **Turcotte, D. L.**, 1970, Thermal structure of island arcs: Geol. Soc. America Bull., v. 81, p. 1665–1688.

Poole, W. H., 1965, Continental margins and island arcs: Geol. Survey of Canada Paper 66–15, Dept. of Mines and Technical Surveys, Ottawa, Canada.

Prince, R. A., and **Kulm, L. D.**, 1975, Crustal rupture and the initiation of imbricate thrusting in the Peru-Chile trench: Geol. Soc. America Bull., v. 86, p. 1639–1653.

Raitt, R. W., **Fisher, R. L.**, and **Mason, R. G.**, 1955, Tonga trench: Geol. Soc. America Spec. Papers, v. 62, p. 237–254.

Rodolfo, K. S., 1969, Bathymetry and marine geology of the Andaman basin, and tectonic implications for southeast Asia: Geol. Soc. America Bull., v. 80, no. 7.

St. John, V. P., 1967, The gravity field in New Guinea: Ph.D. thesis, Univ. Tasmania.

Scholl, D. W., **von Huene, R.**, and **Ridlon, J. B.**, 1968, Spreading of the ocean floor: Undeformed sediments in the Peru-Chile trench: Science, v. 159, p. 869–871.

———, **Christensen, M. N.**, **von Huene, R.**, and **Marlow, M. S.**, 1970, Peru-Chile trench sediments

and sea-floor spreading: Geol. Soc. America Bull., v. 81, no. 5, p. 1339–1360.
Sclater, J. G., 1972, Heat flow and elevation of the western Pacific: Jour. Geophys. Res., v. 77, no. 29, p. 5705ff.
———, **Karig, Dan, Lawver, L. A.**, and **Louden, Keith**, 1976, Heat flow, depth and crustal thickness of the marginal basins of the South Philippine Sea: Jour. Geophys. Res., v. 81, no. 2, p. 309–318.
Shaw, H. R., and **Jackson, E. D.**, 1973, Linear island chains in the Pacific: Result of thermal plumes or gravitational anchors?: Jour. Geophys. Res., v. 78, no. 35, p. 8634ff.
Shor, C. G., 1962, Seismic refraction studies off the coast of Alaska: Seismol. Soc. America Bull., v. 52.
———, **Kirk, H. K.**, and **Menard, H. W.**, 1971, Crustal structure of the Melanesian area: Jour. Geophys. Res., v. 76, no. 11, p. 2562ff.
Silver, E. A., 1971, Tectonics of the Mendocino triple junction: Geol. Soc. America Bull., v. 83, p. 2965ff.
Smith, J. G., 1965, Orogenesis in western Papua and New Guinea: Tectonophysics, v. 2, p. 1–27.
Stauder, Wm., 1968, Mechanism of the Rat Island earthquake sequence of Feb. 4, 1965, with relation to island arcs and sea-floor spreading: Jour. Geophys. Res., v. 73, no. 12.
——— and **Mualchin, Lalliana**, 1976, Fault motion in the larger earthquakes of the Kurile-Kamchatka arc and of the Kurile-Hokkaido corner: Jour. Geophys. Res., v. 81, no. 2, p. 297–308.
Stoiber, R. E., and **Carr, M. J.**, 1971, Lithospheric plates, Benioff zones, and volcanoes: Geol. Soc. America Bull., v. 82, no. 2.
Stone, D. B., 1968, Geophysics in the Bering Sea and surrounding areas; a review: Tectonophysics, v. 6, no. 6.
Suess, E., 1885, Das Antlitz der Erde, v. 1: F. Tempsky, Prague.
Sugimura, A., and **Vyeda, S.**, 1973, Island arcs; Japan and its environs: Amsterdam, Elsevier.
Sykes, L. R., 1966, The seismicity and deep structures of island arcs: Jour. Geophys. Res., v. 71, no. 12, p. 2981–3006.
——— 1967, Mechanism of earthquakes and nature of faulting on the mid-oceanic ridges: Jour. Geophys. Res., v. 72, no. 8.
——— 1969, Seismicity of the mid-ocean ridge system, *in* Hart, P. J., ed., The earth's crust and upper mantle: Amer. Geophys. Union Mon. 13.
——— 1971, Aftershock zones of great earthquakes, seismicity gaps, and earthquake prediction for Alaska and the Aleutians: J. Geophys. Res., v. 76.
Talwani, M., Worzel, J. L., and **Ewing, M.**, 1961, Gravity anomalies and crustal section across the Tonga trench: Jour. Geophys. Res., v. 66, no. 4, p. 1265–1278.
Tobin, D. G., 1968, Seismicity and tectonics of the northeast Pacific Ocean: Jour. Geophys. Res., v. 73, no. 12.
Umbgrove, J. H. F., 1949, Structural history of the East Indies: Cambridge, Mass., Cambridge.
Uspensky, D. G., 1972, Geological interpretation of the gravity data for a profile across the Pacific Ocean: Jour. Geophys. Res., v. 77, no. 32, p. 6316ff.
Uyeda, Seiya, and **Miyashiro, Akiho**, 1974, Plate tectonics and the Japanese Islands: A synthesis: Geol. Soc. America Bull., v. 85, no. 7, p. 1159–1170.
Van Bemmelen, R. W., 1949, The geology of Indonesia, v. I: The Hague, Govt. Printing Office.
——— 1972, Geodynamic models, an evaluation and a synthesis: Amsterdam, Elsevier.
Vening Meinesz, F. A., 1948, Gravity expeditions at sea: Pub. Netherlands Geodetic Comm., v. 4, p. 1–233.
Visser, W. A., and **Hermes, J. J.**, 1962, Geological results of the exploration for oil, Netherland, New Guinea: Holland, Staatsdrukkerij en Uitgeverij Christoffel Plantijnstratt's *Gravenhage*.
Von Huene, R., 1972, Structure of the continental margin and tectonism at the eastern Aleutian trench: Geol. Soc. America Bull., v. 83, p. 3613ff.
Wageman, J. M., and **Johnson, G. L.**, 1967, A study of part of the eastern flank of the mid-Atlantic ridge: Jour. Geophys. Res., v. 72, no. 4, p. 1175–1182.
Watanabe, Terijolp, 1970, Heat flow in the Philippine Sea: Tectonophysics, v. 10, no. 1-3, p. 205–234.
Watts, A. B., and **Talwani, M.**, 1975, Gravity effects of downgowing lithospheric slabs beneath island arcs: Geol. Soc. America Bull., v. 86, p. 1–4.
Weeks, L. A., Harbison, R. N., and **Peter, G.**, 1967, Island arc system in Andaman Sea: Amer. Assoc. Petroleum Geologists Bull., v. 51, no. 9, p. 1803–1815.
Yoshii, Toshikatsu, 1973, Upper mantle structure beneath the north Pacific and the marginal seas: J. Phys. Earth, v. 21, no. 3.

22

The Appalachian Orogen

THE TECTONIC SETTING OF THE APPALACHIAN OROGEN

Today the Appalachian Mountains are located along or near the eastern continental margin of North America, but the character of the eastern side of the Appalachians as well as both ends of the belt are obscured by the ocean and Coastal Plain sediments. Relatively undeformed Mesozoic and Cenozoic sediments of the Coastal Plain unconformably overlie the strongly folded, faulted, and metamorphosed rocks of the Appalachian orogen and cover the probable connections between the Appalachians, the Ouachitas, and the Marathon Mountains. The Coastal Plain is essentially an emergent part of the continental shelf which is submerged north of New York. It is important to realize that this continental shelf–Coastal Plain region is more than 100 km wide; so, the eastern flank of the Appalachian orogen lies buried and obscured by this cover. The character of this margin is known largely through geophysical studies, some of which are described in Chap. 19. The great width of the continental shelf makes it difficult to match regional structural features in North America with those of other continental plates which are thought to have been contiguous at times during the Paleozoic, Fig. 22-1.

The western margin of the Appalachians passes into the North American platform and craton. The northern half of this margin, from Newfoundland to New York, is sharply defined by the great thrust fault known as *Logan's line* which carries rocks of the Appalachian orogen westward against Precambrian rocks of the Grenville province of the Canadian Shield. The folded and faulted Appalachians lie on the east side of the Adirondacks, which form a southern extension of the Shield, but farther south the character of the margin changes drastically. The Precambrian basement lies deeply buried under a great thickness of Paleozoic sedimentary rocks in the Appalachian basin. The depth to the Precambrian basement is greatest along the eastern side of the Appalachian basin, where it is more than

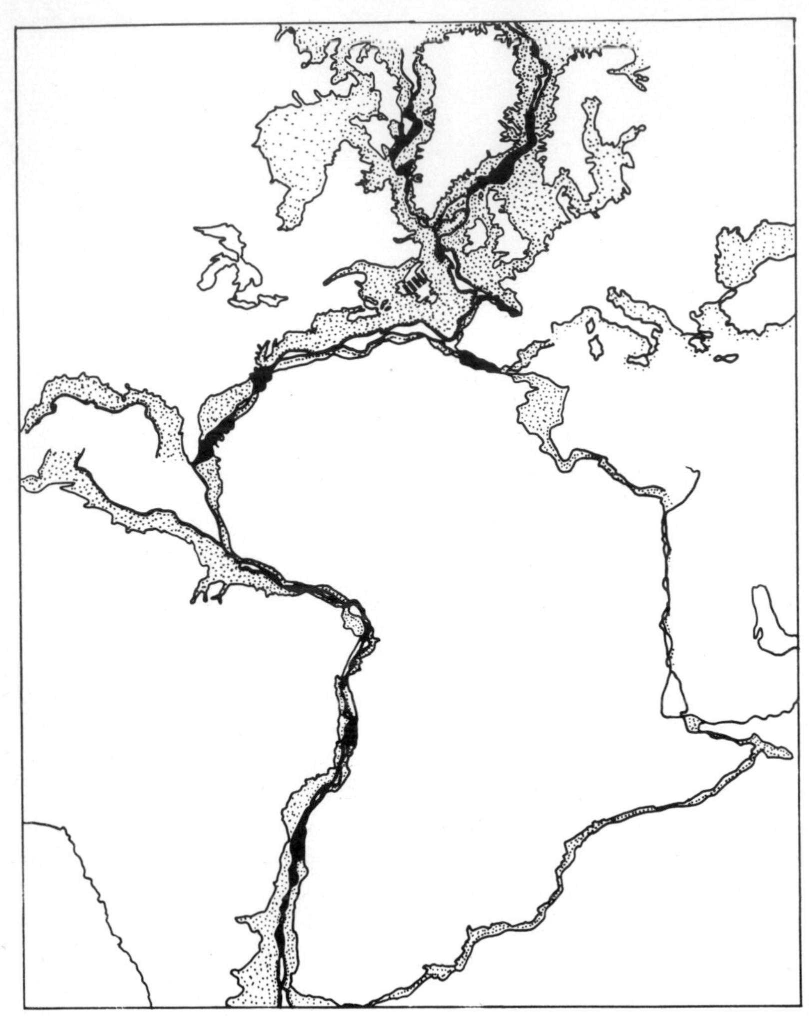

FIGURE 22-1 Reconstruction of the continents across the Atlantic to a predrift position using a computer fit at the 500-fathom level. (*From Bullard and others, 1965.*)

10 km deep, but it rises gently toward the west until basement is less than 1 km deep over the Cincinnati arch which forms the western edge of the Appalachian basin and the Appalachian orogen.

Although the Appalachian orogen is clearly in a continental marginal position today, its position during the Paleozoic when the orogenic deformation was taking place is not so clear. Predrift reconstructions, such as Bullard's fit, Fig. 22-1, would place the Appalachians in a position parallel to the coast of Morocco, and the Appalachian system would be almost perfectly aligned with Ireland, Scotland, and the Scandinavian Caledonides. This reconstruction is generally thought to represent the configuration toward the close of the Paleozoic Era. Thus one of the crucial questions in the history of this orogen is whether the sedimentary deposits and orogenic activity took place along an intracratonic trough or whether the Appalachian geosyncline was a continental margin that was deformed by one or more continental collisions as the Atlantic Ocean underwent successive openings and closures.

What Is the Basement?

The term *basement* is frequently encountered in discussions of regional tectonics and often with a variety of meanings. Some authors confine its use to Precambrian crystalline rocks—the metamorphic and igneous complexes found beneath stratified sedimentary strata. Others use the same distinction—the mechanical distinction between the behavior of crystalline rocks below and stratified sedimentary piles above—regardless of the age of the two. The latter usage is employed throughout this book; so, the "basement" in the Appalachians during the Paleozoic was the Grenville age (1 billion years old) crystalline complex, but the "basement" in the Alps is the Hercynian age (late Paleozoic) crystalline complex. It is important to note that with this usage, the age of basement rocks may vary with time within a region. Thus the Grenville rocks formed the basement under the Paleozoic sediments in the Blue Ridge area, but we might also call the Paleozoic metamorphic and igneous rocks of the Piedmont the basement under the Coastal Plain sediments.

MAJOR DIVISIONS OF THE APPALACHIANS

The structural elements of the Appalachians are so numerous that generalizations are difficult and invariably involve oversimplifications; so, with this in mind, from New England southward the Appalachian orogen can be viewed as having three main elements: a central discontinuous ridge of Precambrian rocks uplifted and faulted (the Blue Ridge–Reading prong–Berkshire highlands–Green Mountains) which lie between folded and thrust-faulted but generally unmetamorphosed Paleozoic sediments on the west, and crystalline complexes of igneous and strongly deformed metamorphic rocks to the east. The age of the metamorphic rocks continues to be a subject of debate, but they generally yield Paleozoic radiometric ages, although many show signs of superimposed deformations and thermal events indicative of a long and complex history of deformation. Fossils have been found in a few localities in these metamorphic rocks, proving a Paleozoic age for some of them; most of the unmetamorphosed plutonic rocks yield Paleozoic radiometric ages, but conventional methods of correlation between the unaltered sedimentary rocks to the west and the metamorphic rocks to the east is made difficult by the presence of the intervening ridge of Precambrian rocks, which generally are dated as being more than 1 billion years old. Elsewhere the presence of thrust faults makes correlations uncertain.

The large elongate basement uplifts do not occur north of Vermont. Deformed early and mid-Paleozoic rocks form a broad belt across Maine and eastern Canada in which many early and mid-Paleozoic intrusions occur. Newfoundland differs from all the southern parts of the Appalachians in that it has a central Paleozoic crystalline complex bounded on both sides by Precambrian rocks.

The Appalachians are conveniently divided into northern and southern parts along a line extending south of the Adirondacks and southeast to New York City. This line is chosen because some of the physiographic and structural divisions do not extend across it. For example, this is where the character of the western margin of the orogen changes. In addition, the southern region was the site of a late Paleozoic orogenic phase, the *Alleghanian orogeny*, which did not affect the northern part of the Appalachians.

Most of the large-scale structural features in the orogen, such as fold systems, fault zones, and basement uplifts, have a pronounced northeast-southwest alignment throughout the Appalachians all the way from Newfoundland to Alabama, and some of these features can be traced for long distances and allow a longitudinal subdivision of the orogen. These divisions are especially clear in the south

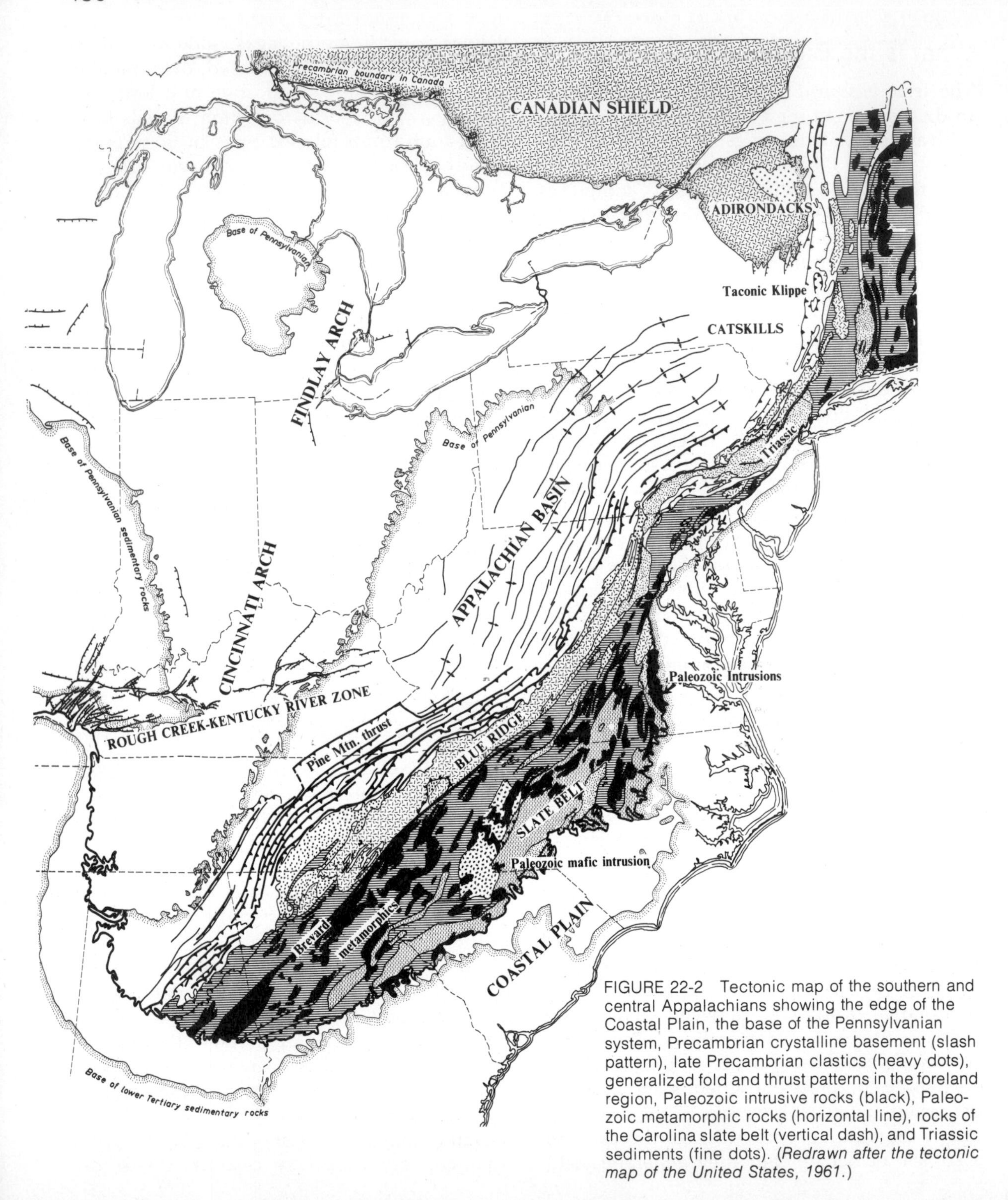

FIGURE 22-2 Tectonic map of the southern and central Appalachians showing the edge of the Coastal Plain, the base of the Pennsylvanian system, Precambrian crystalline basement (slash pattern), late Precambrian clastics (heavy dots), generalized fold and thrust patterns in the foreland region, Paleozoic intrusive rocks (black), Paleozoic metamorphic rocks (horizontal line), rocks of the Carolina slate belt (vertical dash), and Triassic sediments (fine dots). (*Redrawn after the tectonic map of the United States, 1961.*)

where the belt may be subdivided into provinces which are expressed in both structural and topographic characteristics. These divisions are:

1. *The Appalachian plateau–Allegheny synclinorium* is a large elongate structural basin in the western part of the southern Appalachians which extends north to central New York and contains flat-lying and gently folded Paleozoic sequences.

2. *The Valley and Ridge–foreland fold belt* is a belt of folded and thrust-faulted lower and middle Paleozoic sedimentary sequences exposed in a continuous belt from northern Vermont to Alabama. This province is represented in the north by the Hudson and Champlain valleys. The southern portion from Pennsylvania southward is characterized by long ridges and valleys which are caused by differential erosion of the folded Paleozoic sedimentary rocks.

3. *The Blue Ridge* is a prominent ridge of Precambrian igneous and metamorphic rocks which dies out in Maryland, but the Reading Mountains and New Jersey highlands are composed of similar rock types and lie along trend with the Blue Ridge. The Blue Ridge province included a late Precambrian and Cambrian cover of clastic rocks, especially well developed in the Smoky Mountain region.

4. *The Piedmont* is the metamorphic and igneous province in the southern Appalachians east of the Blue Ridge.

Some elements of the southern Appalachians have recognizable counterparts in the north. The Hudson and Champlain valleys are similar in physiography, age, and structure to the easternmost valley in the Valley and Ridge; the New Jersey highlands and the Green Mountains are similar to the Blue Ridge, although they lack the latter's continuity. Most of the New England highlands, Maine, and Canada east of the St. Lawrence is a crystalline complex similar in many respects to the Piedmont. One notable difference in the northern Appalachians is the presence within the crystalline complex of several major basins, the Boston basin and the New Brunswick basin, which contain relatively undeformed and unmetamorphosed upper Paleozoic sedimentary rocks lying unconformably on the crystalline complexes. In contrast upper Paleozoic rocks in the south are west of the Blue Ridge, and they are involved in folding and thrusting.

Triassic continental deposits lie unconformably and undeformed (except for block faulting) on the Paleozoic folded, thrust-faulted, intruded, and metamorphosed rocks of the Appalachian orogen. These deposits, located in relatively isolated fault-bounded basins, were laid down after the last orogenic episode in the Appalachians, and they are generally thought to mark the initial extensional deformation associated with the rifting and spreading of the Atlantic Ocean.

THE GEOSYNCLINAL MODEL

It was on the basis of studies in the western Appalachians that James Hall proposed the concept of the geosyncline. He recognized that the sedimentary units in the folded belt had originally been deposited as shallow marine sediment in seas that occupied the eastern margin of North America in the Paleozoic. These sediments accumulated slowly in a long, relatively narrow belt that gradually subsided until a great thickness of sediment, 10 to 15 km, had accumulated. Ripple marks, mud cracks, and shallow-water fossils found throughout the stratigraphic section indicate that almost all this sediment accumulated in shallow water; so, the sediment must have accumulated on subsiding crust rather than by filling in an originally deep basin. Hall recognized that the thickness of units within the belt of subsidence is far greater than that of units of comparable age in the Midwest on the craton, and he concluded that the sedimentary rocks of the geosyncline became folded during the subsidence of the geosyncline and were uplifted later.

Hall's concept of the geosyncline as a long, relatively narrow belt of mobility, character-

FIGURE 22-3 Tectonic map of the northern Appalachian region showing Precambrian basement rocks (slash pattern), Paleozoic (largely eugeosynclinal) rocks affected by the Taconic orogeny (horizontal lines widely spaced) and the Acadian orogeny (finely spaced horizontal lines), Paleozoic intrusives (black), late Paleozoic little-deformed basins (light dot), and Triassic rocks (fine dot). (*After the tectonic map of North America, by Philip King, 1969.*)

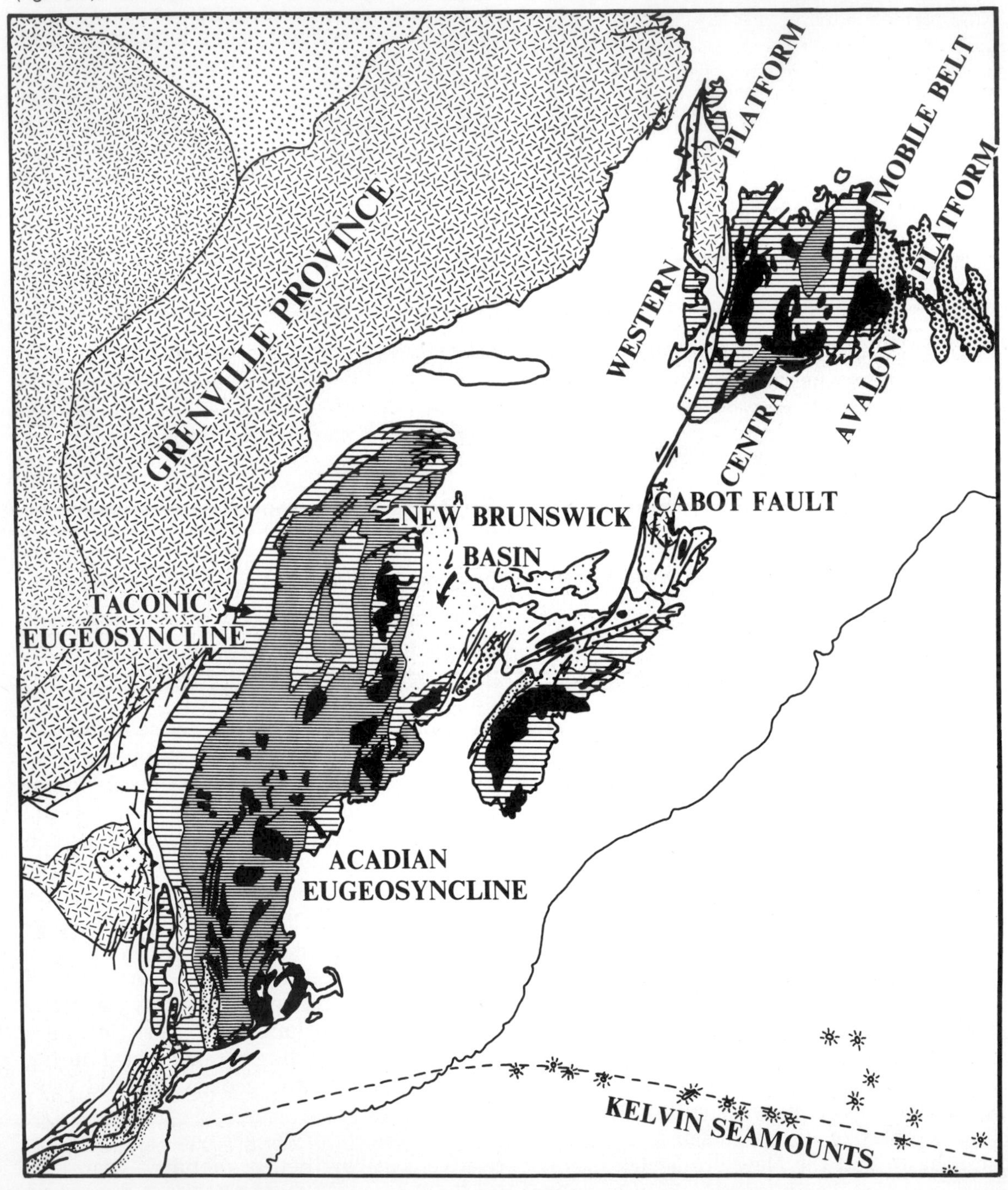

ized by thick sediment accumulations and strong deformation, has survived, although it has been considerably modified. The folds and faults are no longer generally interpreted as having formed during subsidence, and the trough envisioned by Hall is now usually viewed as a continental margin open to the sea on one side. It is not difficult to see why Hall suggested a trough-like form, because the sedimentary rocks he saw as filling the geosyncline are bounded on the east by igneous and metamorphic complexes thought at that time to be Precambrian in age. Subsequently many of the crystalline rocks proved to be metamorphic equivalents of Hall's geosynclinal filling, and the complexity of the overall pattern became more fully recognized.

The geosynclinal concept was elaborated and refined by Stille (1941) and Kay (1951). They subdivided geosynclines like the Appalachian geosyncline into a nonvolcanic interior part, called a *miogeosyncline*, and an outer belt which contains great thicknesses of volcanic materials as well as deeper water sediments, called a *eugeosyncline*. Kay suggested that island arcs are modern equivalents of the ancient eugeosynclines.

The term *miogeosyncline* (Stille, 1941) is applied to the region of sedimentary rocks west of the Blue Ridge and uplifts of Precambrian rocks in New England. Most of these units are compact and tightly cemented and some of the shales are locally altered to slate, but otherwise the rocks are unmetamorphosed; they contain few intrusions or volcanics, and they are largely composed of sandstones, shales, limestones, dolostones, and mixtures of these. The individual units often have great lateral extent, many being traced for hundreds of kilometers along the geosyncline; but quite commonly the units change character and thickness rapidly when traced across the geosyncline, and in addition, it is difficult to trace units across the geosyncline because successively younger rocks are exposed toward the west. Much of the history of Appalachian tectonics has been interpreted from the miogeosynclinal deposits.

The crystalline complexes of the Appalachians have generally been characterized as metamorphosed eugeosynclines, in which volcanic products are a common ingredient. The miogeosyncline and the eugeosyncline were contemporaneous features of the North American margin during the Paleozoic, according to the geosynclinal model of the Appalachians; however, tectonic activity was neither uniform nor synchronous throughout the geosyncline. Large portions of the crystalline complex in the northern Appalachians can be proven to be Paleozoic in age, but the age of large portions of the Piedmont is still uncertain; so, some of the "eugeosynclinal" rocks in the Piedmont may be much older than the miogeosynclinal rocks in the Valley and Ridge.

KEYS TO THE TECTONIC HISTORY OF THE APPALACHIANS

Deformation has taken many forms in the Appalachians. *Décollement* tectonics characterizes the Allegheny synclinorium and the foreland fold belt of the Valley and Ridge, parts of the Blue Ridge and Smoky Mountains are uplifted and complexly thrust toward the west, and elements which originated in the New England crystalline complex have slid westward to rest on unmetamorphosed rocks of the miogeosyncline as in the Taconic klippe, Fig. 22-4. In addition, the crystalline complexes contain a complex internal deformational and metamorphic history. For example, large bodies of crystalline rock have risen as plastic masses into their cover, forming mantled gneiss domes.

Evidence of the deformational history of the Appalachians has been deciphered by use of many relationships: (1) unconformities separate deformed from undeformed units in some places; (2) major thermal events are recognized by the resetting of radiometric "clocks" and

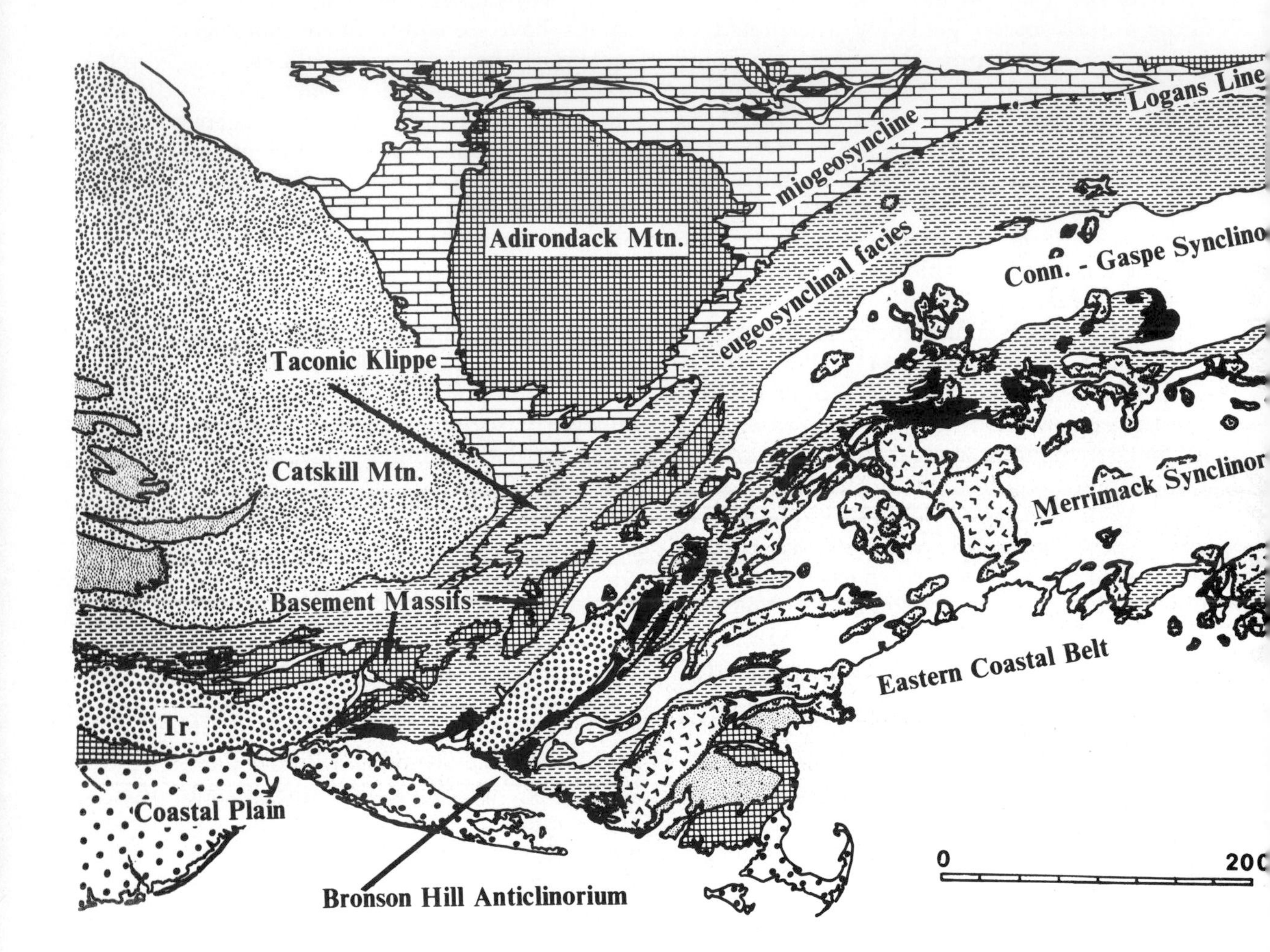
Logans Line
miogeosyncline
Adirondack Mtn.
eugeosynclinal facies
Conn. - Gaspe Synclino
Taconic Klippe
Catskill Mtn.
Merrimack Synclinor
Basement Massifs
Eastern Coastal Belt
Tr.
Coastal Plain
Bronson Hill Anticlinorium
0
200

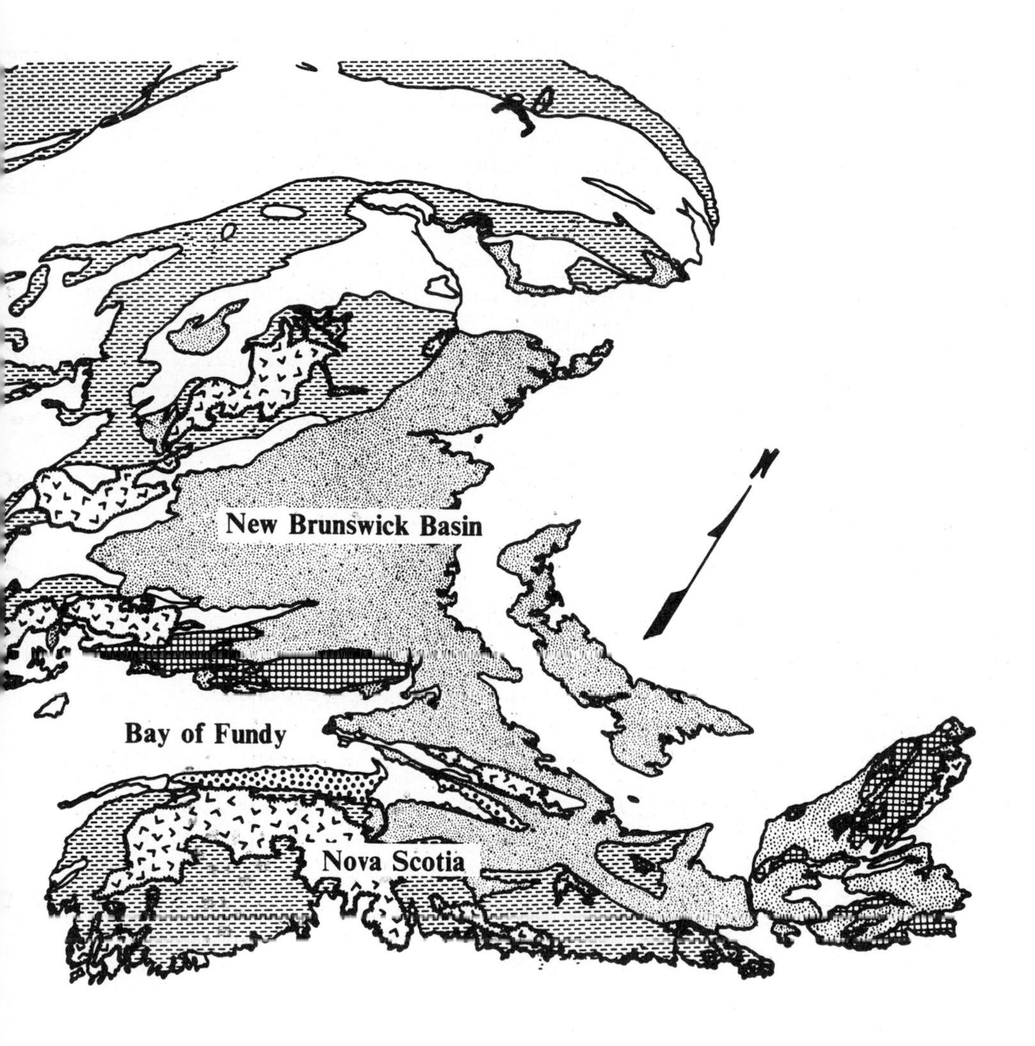

FIGURE 22-4 Sketch map of the northern Appalachians showing Precambrian basement rock (grid), early Paleozoic miogeosyncline (brick pattern), early Paleozoic eugeosynclinal deposits (dash pattern), Silurian and Devonian eugeosyncline (blank), Ordovician plutons (black), later plutons (V pattern), Carboniferous deposits (fine dots), Triassic (heavy dots), Coastal Plain deposits (open circles). (*After U.S. Geological Survey open file map, 1968, compiled by W. S. White.*)

overprinting of metamorphic mineral assemblages; (3) the thickness and pattern of distribution of clastic wedges—mainly terrigenous sediment emanating from uplifted areas—indicates some major uplifts; and (4) sedimentary orogenic cycles have been recognized in the miogeosyncline.

It is clear from these relationships that deformation was not synchronous throughout the Appalachian orogen. It is also evident that many parts of the orogen were reactivated time and again in a long and complex history of deformation which began in the Precambrian and extended through the Triassic.

The End of Appalachian Orogenic History—The Palisades Phase

The proof of the upper limit in time of deformation in the Appalachians is provided by the presence of undeformed Cretaceous sediments in the Coastal Plain (Jurassic rocks are not exposed). In addition, none of the Triassic rocks are folded or metamorphosed; although their occurrence in grabens suggest that they were deposited during extensional deformation, their terrigenous and coarse character proves that the surrounding areas were uplifted, and large intrusions of basalt confirms igneous activity. Unfolded nonmarine Triassic deposits occur in fault-bounded troughs and grabens in both the northern and southern Appalachians, proving that no folding or thrusting in the orogen occurred after the Triassic. The Palisades phase of deformation consisted of high-angle faulting, uplift at least along these faults, and intrusion and extrusion of basaltic dikes and sills such as the famous one which forms the Palisades along the west banks of the Hudson River.

Dating the Last Folding and Thrusting in the Appalachians—The Alleghanian Orogeny

None of the Mesozoic or Cenozoic rocks in the region show evidence of folding, and the only Permian units are confined in their present outcrop to a small area in the central portion of the Allegheny synclinorium (West Virginia and Ohio); these rocks show only slight warping, but Pennsylvanian rocks in the Allegheny synclinorium exhibit strong compressional-type deformation. All units in the stratigraphic column from Pennsylvanian to late Precambrian are folded and thrust-faulted in the southern Appalachian miogeosyncline. Most of this deformation is attributed to what is commonly called the *Appalachian* or *Alleghanian orogeny*. However, this deformation did not affect most of the northern Appalachian region. The northernmost folds are found in northern Pennsylvania and New York where they die out in the great clastic sequences of the Catskill Mountains. Farther south, rocks in the miogeosyncline are involved in large-scale thrusting, as is seen along the western edge of the Smoky Mountains where Carboniferous rocks are affected.

Use of Clastic Wedges and Sedimentary Rock Types to Date Uplifts

The changing character of the sediments supplied to the geosyncline has proved to be a most important key to the age of uplifts within the Appalachians. In many instances the source of these sediments appears to have been outside the area where the miogeosynclinal rocks are now exposed. Much of the great volume of clastic sediment thickens to the east, suggesting source areas in the region now occupied by the Blue Ridge, Piedmont, and New England highlands. Some sediment was derived locally from folds which rose in the miogeosyncline, but this is of minor importance compared with the external sources. Large bodies of clastic materials, now metamorphosed and showing varying degrees of internal deformation, are known in the eugeosynclinal areas also, but it is much more difficult to determine the age and thickness of these rocks than it is for similar deposits found in the miogeosyncline.

A cyclic pattern is recognized in geosynclinal sedimentation by Pettijohn (1957), who describes two cycles in the Paleozoic of the middle Appalachians—one from Early Cambrian to Middle Silurian, and a second from Late Silurian through the Carboniferous. These are characterized by a preorogenic facies, chiefly orthoquartzite and carbonates; a euxinic facies, characterized by black shales and, commonly, chert; a flysch facies in which the sandstones become coarser and cleaner upward and in which they show cross bedding and other current structures. These facies are thought to be tectonically controlled, reflecting initially slow, shallow marine sedimentation on a flooded craton or its margin; a rapid basining and starved sedimentation; then basin filling by turbidity flows and other gravity processes taking place in deep water; and later basin filling by paralic sedimentation.

Evidence of Late Precambrian Orogeny

The oldest of the clastic sequences, the late Precambrian Ocoee series, is found in the Smoky Mountains and is estimated to be as much as 10 to 15 km thick. A major source area, one of considerable height or size, is necessary to explain the origin of this sedimentary pile early in Appalachian history. A large clastic wedge of late Precambrian age is also known in the northern Appalachians (especially on the eastern side of Newfoundland) where large masses moved apparently by gravity sliding off of uplifted areas in what has been called the *Avalonian orogeny*.

Ordovician Phases—The Taconic Orogeny*

Carbonates were deposited in the miogeosyncline during most of the Cambrian and lower Ordovician. A second major clastic wedge, the Blount and Bays units, was flooding into the geosyncline in the Smoky Mountain region by the Middle Ordovician; the sands and shales of this deposit spread north and west, reaching a maximum thickness over 2,000 m. Uplift and deformation affected the northern Appalachians at the end of the middle Ordovician. A clastic wedge, the Queenstown delta, spread westward from New England and is preserved in Pennsylvania and New York, and Silurian sands and conglomerates spread widely down the geosyncline. A major unconformity, one with Silurian conglomerates resting on Precambrian gneisses, is found in this region, and marks what is called the *Taconic orogeny*. Much of the folding and thrusting including emplacement of the Taconic klippe is thought to be of this age, and igneous activity and metamorphism was widespread in the eugeosyncline which was the source area for the clastic deposits.

* See Rodgers (1971).

An orogenic phase, the *Penobscot orogeny*, which took place along the western side of the northern Appalachians, in Ordovician time, is evidenced by an angular unconformity, and it was followed in the Silurian by another uplift, the *Salinic phase*, which is indicated by an angular unconformity and small clastic wedges in the same general areas of the northwest Appalachians.

The Acadian Orogeny

A second major uplift took place in the New England area about the middle of the Devonian. The famous Catskill delta formed from the erosion products of this uplift. These clastic sequences are over 2,000 m thick in New York, and sands and shales are prominent in Devonian sequences far to the south. Unlike the earlier southern clastic wedges, the Catskill beds have not been folded or thrust-faulted by later movements in the region where they are thickest, but it is often difficult to distinguish the structures associated with this deformation, the *Acadian orogeny*, from those of the Taconic orogeny.

Many intrusions in both the northern and southern Appalachians are of mid-Paleozoic age and are probably related to thermal events associated with this orogeny.

Carboniferous sedimentary sequences are thick over much of the Appalachian geosyncline; however, they are not present east of the Valley and Ridge, presumably because they have been eroded away. Great thicknesses of Mississippian clastic sediments which thicken toward the east are found in eastern Tennessee (about 2,000 m thick) and in Pennsylvania and West Virginia. Pennsylvanian sequences are predominantly clastic and they are thick (over 3,000 m) in Alabama.

It is clear from the consideration of the sedimentary record that the common practice of recognizing only three orogenies in the Paleozoic (Taconian, Acadian, and Alleghanian or Appalachian) is misleading. Major uplifts over large regions are necessary to provide the clastic sediments which spread into the miogeosyncline in one area or another in almost every period of the Paleozoic. Rodgers (1970) recognizes 10 orogenic episodes in the Appalachians, Table 22-1.

The question of the relationship between the miogeosynclinal units and the more complex metamorphic rocks and intrusions in the Piedmont and the north has long been debated. Kay (1951) cites evidence that this eastern zone was a eugeosyncline. It was here that great intrusions were emplaced, and it was here that rocks were metamorphosed and highly deformed. This was the core of the orogenic belt in every sense, and uplifts in the

TABLE 22-1
OROGENIC MOVEMENTS IN THE APPALACHIAN REGION. (*After Rodgers, 1970*)

Orogenic episode	Area	Manifestation
Palisades (Triassic)	Along axis of Appalachians	Block faulting, basaltic lavas
Alleghanian (Pennsylvanian)	Western southern Appalachians and possibly Carolina Piedmont	Folding, metamorphism, granite intrusions
Ouachita (Mississippian)	Southernmost Appalachians	Clastic wedge
Acadian (Devonian)	Northern Appalachians	High-grade metamorphism, granites, clastic wedge
Salinic (Silurian)	NW northern Appalachians	Minor clastic wedge, angular unconformity
Taconic (Ordovician)	Early phase in south, late phase on NW side of northern Appalachians	Angular unconformity, clastic wedge, gravity slides (?), metamorphism, intrusion
Penobscot (Ordovician)	NW side of northern Appalachians	Angular unconformity, cleavage, intrusions (?)
Avalonian (Late Precambrian)	SE Newfoundland, New Brunswick, central and southern Appalachians	Uplift, clastic wedge, gravity slides (?)
_ _ _ _ (Late Precambrian) 580 m.y.	Far northern Appalachians	Metamorphism granitic intrusions
Grenville (Late Precambrian) 800–1,100 m.y.	Eastern North America	High-grade metamorphism

eugeosynclinal area could have been the source of sediments now found to the west.

The idea that a continental borderland was located east of the miogeosyncline during the Paleozoic stands in strong contrast to the concept of having a eugeosyncline to the east. That the source of sediment for the miogeosyncline was to the east during much of the Paleozoic is indicated by variations in sediment thickness and coarseness across the geosyncline and current directions interpreted from cross-bedding. The eastern source was persistent over long periods of time and throughout much of the geosyncline, although western sources are also indicated from time to time, and during long intervals of the upper Cambrian and Ordovician no land source is needed for the carbonate sequences. The location of this hypothetical eastern borderland has not been proven. Most of the Piedmont metamorphic rocks were originally thought to be Precambrian, and the Blue Ridge, Piedmont, and New England regions were thought to be the site of the borderland by many of the early American geologists. Over the years more and more of these rocks have been shown to be Ordovician, Devonian, and Carboniferous intrusions and Paleozoic metasedimentary materials. Thus the borderland of the east has become less distinct and it seems more likely that this region was an island arc. The Blue Ridge remains a possibility, the region covered by the Coastal Plain sediments is a possible location, and the metasedimentary rocks in the Piedmont may be old enough to have been part of a borderland after the Ordovician. As we place more confidence in the predrift reconstructions, it seems increasingly possible that a borderland source area could have been along the continental margin where North America and Africa were adjacent. Unfortunately the age and structure of this zone are still obscure. Several hypotheses remain possible alternatives—the geosyncline could have been located between the continents, or the eugeosyncline could have been a subduction zone along the edge of an earlier Atlantic. The sediments now found could have been derived from the volcanic arcs along the subduction zone or possibly from blocks of continental crust that were moved into the subduction zone or collided with the arcs there.

THE NORTHERN APPALACHIANS*

The northern sector of the Appalachian orogen lies between the Atlantic Ocean on the east and Precambrian rocks of the Grenville province of the Canadian Shield and the Adirondacks, which were largely unaffected by Paleozoic deformation in the geosyncline.

Intensive study of the northern Appalachians make it one of the best known sections of an orogenic belt anywhere in the world, and these studies have revealed many of the complexities of the structure and tectonic evolution of this region. The region may be subdivided into a number of longitudinal belts that can be traced for great distances along trend, although they lack the extraordinary continuity of their southern counterparts. From west to east these subdivisions are (see Zen, 1968), Fig. 22-4:

1. *The stable craton:* Rocks of the Adirondacks and Canadian Shield largely unaffected by Paleozoic deformation.

2. *A foreland belt:* A narrow belt composed of unmetamorphosed Cambrian and Ordovician limestones, sandstones, and shale that are little deformed to the west but show increasing amounts of deformation to the east where they pass under westward-directed thrusts. This belt is located along the Hudson, Champlain, and St. Lawrence valleys, on Anticosti Island, and in the western side of Newfoundland.

* Several review and summary volumes have been published including: Kay, M., 1969, North Atlantic—Geology and continental drift; Zen, E-An, and others, eds., 1968, Studies of Appalachian geology; Burk, C., A., and Drake, C. L., eds., 1974, The geology of continental margins.

3. *The western fold belt:* A zone of strongly folded, metamorphosed Cambrian and Ordovician carbonates which have been carried westward on a number of major thrusts, including the famous Logans line fault and the imbricated thrust of northern Vermont and Quebec. Folding in these rocks is intense; some are isoclinally folded and two generations of folds are common—one presumably Taconic, followed by a later Acadian deformation. Deformation within this belt is especially intense in New England, where large recumbent folds and huge allochthonous rock masses (the Taconic klippe) appear to have slid in from the east. Similar masses occur in western Newfoundland, but the intensity of deformation is less in between.

4. *Zone of basement uplifts:* A number of discrete uplifts composed of Precambrian crystalline rocks are aligned in a belt which extends from the New Jersey highlands northeast and includes the Hudson, New Milford, Housatonic, Berkshire, and Green Mountain massifs. This zone is not found through eastern Canada but may be represented by the Long Range in western Newfoundland. Structural relationships within these massifs and between them and their cover is varied. Some appear to be essentially domal uplifts of basement rock, but others are sheared and faulted especially along the western margins and some exhibit complicated internal thrusting.

5. *Connecticut-Gaspé synclinorium:* This zone lies east of the basement massifs and is very narrow in New England, but it becomes much wider to the north and is probably represented in the central part of Newfoundland. Rocks of this synclinorium are lower Paleozoic metasedimentary rocks which were derived from eugeosynclinal facies. The western flank of the syncline is much less deformed than the east, where it rises and is involved in isoclinal and recumbent folds associated with large domes. Syntectonic ultramafic rocks occur within this zone.

6. *Bronson Hill anticlinorium:* This zone may be traced from the shores of Long Island Sound along the east side of the Connecticut basin (Triassic) and to northern New Hampshire. It is characterized by a series of nappes that are composed of Paleozoic and possibly Precambrian gneisses, metavolcanics, and their altered cover which resemble the Penninic nappes of the European Alps. These nappes have been affected by a late stage of domal uplift which took place while the rocks were still ductile and which produced complex superimposed structures. The nappes appear overturned both toward the west and the east and undoubtedly represent a zone of high mobility and ductile deformation.

7. *The Merrimack synclinorium:* This belt may be an eastern part of the Connecticut-Gaspé synclinorium which is divided in New England by the Bronson Hill domes. The rocks are lower and mid-Paleozoic eugeosynclinal facies with abundant volcanics which have been intruded by numerous Acadian plutonic bodies and altered to high-grade metamorphic rocks.

8. *The eastern coastal belt:* A complex and highly varied group of rocks that is still relatively unknown. This belt extends from Newfoundland, where it is made up of late Precambrian sedimentary rocks of the Avalon platform, south through Nova Scotia to southeastern New England. It is composed of some Precambrian crystalline rocks, some eugeosynclinal rocks, and some nonmarine units. Fossils reveal not only its early Paleozoic age, but the fauna are types found in Great Britain and are distinctly different from similar fauna found in the western foreland belts. These differences have long been recognized, but now they suggest that the two regions may have been parts of two plates that were widely separated during the early Paleozoic.

This eastern zone is further complicated by considerable high-angle faulting and a number of major transcurrent faults.

The Taconic Klippe

The name Taconic klippe is applied to an approximately 300-km-long belt of mountains and associated rocks that are located in western New England and are surrounded by thrust faults, Figs. 22-1 and 22-4. The structural evolution of this area has an unusual history which reflects the way interpretations may and often do change with time.

Bucher (1957, p. 658) summarizes the history of the origin of the klippe hypothesis as follows:

The outline of the "klippe" separates two terranes of contrasting lithologies. Outside of it, the Cambrian and pre-Trentonian Ordovician rocks are largely dolomites and limestones. Inside, the same large time interval is represented by quartzites, sandstones, shales, and radiolarian cherts. . . . The so-called autochthonous carbonate sequence contains practically no argillaceous material. . . . The alleged allochthonous terrane, on the other hand, is largely somber-colored, with tints ranging from greenish gray and gray to black with green and red beds in some horizons.

The contrast between these two terranes is so great, the formational units in each maintain their characteristics over such distances, they lie so close together, and transitional beds are so few, that they could not well have been deposited side by side where they now lie.

This led Ulrich to the hypothesis that these and similar belts to the east and north were "deposits in originally distinct troughs that have since been thrust westward over each other" (Ulrich, 1911, p. 443).

Bucher goes on to point out the main flaws in this hypothesis, and propose that the two facies are separated by an unconformity and that the main structure is essentially in place although affected by thrusting and folding.

More recently Zen (1961) outlines the merits and defects of each of three possible hypotheses which might be called on to explain the structure of the Taconic Mountains.

First, that no klippe is present—that the deformed slate and phyllite units lie unconformably on typical miogeosynclinal deposits as favored by Bucher. In this case the structure of the northern end of the Taconic Mountains must be interpreted as consisting of large folds which are locally recumbent. This hypothesis has the merit of avoiding large-scale tectonic transport of this mountain mass and the problem of determining its root zone. It has the defects of requiring some eastward movement which is not observable, and it requires rapid facies changes during the deposition of Cambrian and Ordovician units.

The second hypothesis is that Taconic mountains are a klippe composed of a number of thrust slices. In this case the thrust sheets were transported from the east, possibly from over the area presently forming the Precambrian exposures in the Green Mountains. This hypothesis makes it possible to explain some of the exposures of miogeosynclinal units within the area of slate and phyllite as windows.

A third hypothesis is that the Taconic area is a klippe, but one representing a recumbent fold complex rather than a series of thrust slices. This theory, is supported by the presence of stratigraphic sections within the allochtonous mass where its western edge is in contact with the miogeosynclinal units. This theory has the advantages of explaining the westerly dip on the east flank of the Taconic Range and some of the east-west-trending structures within the central part of the area.

Several years later Zen (1967, p. 1) provided a comprehensive summary and synthesis of the work done on the Taconic region and offered a new hypothesis which is now widely accepted to explain the relationships now documented.

The geologic history of the area is reconstructed as follows: The Taconic rocks were deposited in the area of the present Precambrian massifs of the Green Mountains–Berkshire highlands belt between the eugeosynclinal sequence to the east and the miogeosynclinal synclinorium sequence to the west; they constitute the transitional facies between these two belts. Conditions were relatively stable until early Middle Ordovician time, when the Green Mountain–Berkshire highlands area began to rise and the area of the present Middlebury synclinorium began to subside. Subsidence took place largely by a series of high-angle longitudinal faults that, as a whole, step down to the west. Argillaceous sediments began to inundate the former miogeosynclinal area; because the conditions of sedimentation had become similar, the sediments resembled, in facies, the synchronous Taconic rocks that were being deposited to the east.

Continued rise of the Green Mountain–

Berkshire highlands area led in middle Trenton time to the *décollement* of the Cambrian and Ordovician sediments into the area of the present Middlebury synclinorium in a series of giant submarine slides. Sedimentation continued at the receiving site throughout the event; sedimentation may also have persisted on the moving slides. The record is found today in the turbidite-laden shale and graywacke in the upper part of the Normanskill shale of both the allochthon and the autochthon.

This concept of submarine sliding of the Taconic transitional or eugeosynclinal facies to the west is now accepted in the plate-theory interpretations, which are very similar to the model described for the Newfoundland region, Fig. 22-5. In this model, uplift in the area of basement massifs was contemporaneous with sliding of the Taconic allochthon toward the west and intrusion of the magma in the east.

Nappes and Mantled Gneiss Domes of the Bronson Hill Anticlinorium*

The gneiss domes and nappes of the Bronson Hill anticlinorium are of special interest as an example of the ductile deformation that may occur in all rocks at sufficiently high temperatures and confining pressures. Domal structures with cores of gneiss and conformable mantles of sedimentary or more frequently metasedimentary rocks and their occurrence in many orogenic belts was discussed years ago by Eskola (1949). The gneiss domes of New England provide excellent examples, although with a more complex tectonic history than many others. These domes are aligned roughly north-south and form an elongate ridge that separates the early Paleozoic eugeosynclinal facies rocks of the Connecticut-Gaspé synclinorium from similar rocks in the Merrimack synclinorium. The cores of these domes are made up of gneisses with the composition of granite or diorite, and their mantles are composed of quartzites, schists, amphibolites, and calc-silicate rocks of Paleozoic age.

* See Thompson and others, *in* Zen and others (1968).

Careful field mapping of these domes by Thompson, Robinson, Clifford, and Trask (1968) reveals that the domal shape of these features is a later structure superimposed on older nappes which formed in an early phase of recumbant folding. After the domes formed they were affected by rotation of the zone about a steep axis by a counterclockwise couple, and finally the domes were affected by uplift along a number of steep faults which probably occurred in Triassic time.

A series of three large nappes is indicated by field evidence in the Bronson Hill anticlinorium. Evidence of the recumbent fold form of at least one of these is strong but the roots are less certain, appearing to lie to the east possibly along a bedding fault. The domal structures superimposed on these nappes are broad arched features separated by narrow, deep, isoclinal synclines. The northern domes have nearly vertical flanks, but the more southerly domes have margins that are overturned as the cores appear to have mushroomed and overridden one another with overturning directed both east and west.

Movements of the nappes and later doming affect the disposition of metamorphic isograds in such a way that in some places high-grade rocks now occur at high structural levels and are surrounded at lower levels by rocks of lower grade—the isograde surfaces have been uplifted and even overturned. This led Thompson and others (1968) to conclude that successive nappes carried successively hotter rocks westward over colder rocks. High-grade rocks, kyanite to sillimanite facies, occur in this area and suggest that pressures must have been near that required to reach the kyanite-and-alucite-sillimanite triple point, which is slightly over 4 kbars based on experimental evidence and would indicate a depth of burial of about 15 km.

NEWFOUNDLAND*

The large island of Newfoundland situated at the northern end of the Appalachian orogen provides a critical link for comparisons of the Caledonide system with the Appalachians. Newfoundland may be subdivided longitudinally into three major belts—eastern and western platforms made up of Precambrian rocks covered by relatively little deformed Cambrian and Ordovician rocks, and a wide central mobile belt which exhibited greater instability during the lower Paleozoic. In addition, this central belt was the site of extensive volcanic activity and is broken into blocks separated by high-angle transcurrent faults which make reconstructions there difficult.

The Western Platform

The western platform is separated from the central mobile belt by a major fault zone with probable transcurrent movement, bounded by the Cabot and Codroy faults. West of these faults, a large massif of Precambrian Grenville basement rock is exposed in the Long Range. The lower Paleozoic cover on the Long Range is little deformed, and it seems probably that the block stood as a buttress to the deformation farther east much as the Adriondacks did farther south. Stable conditions characterize the western platform throughout the Cambrian, but Ordovician rocks suggest subsidence of the platform as clastic material flooded in from the central mobile belt. Ultimately this activity in the central belt was accompanied by sufficient uplift to allow large masses of Cambrian and Ordovician rock to become transported west, where these allochtonous masses are now found both north and southwest of Long Range. It has been suggested that these masses moved as submarine slides containing both shallow-water limestones and deeper water clastic units and including a coarse limestone conglomerate known as the Cow Head breccia. Emplacement of these transported masses was the main expression of the Taconic orogeny in the western platform.

Very few outcrops of Silurian or Devonian rocks occur in the western platform, but these and the allochthonous masses emplaced in the Taconic orogeny were slightly affected by orogenic deformation in the Acadian orogeny in mid-Devonian time. Some faulting and westward tilting were the main effects in the west, but along the eastern margin of the platform intrusion, metamorphism, and folding took place.

The Avalon Platform

This eastern platform area is made up largely of little-deformed late Precambrian volcanics, graywacke, and slates that are not present in the west where an unconformity separates the Grenville basement of the Long Range from its cover of Cambrian age. Scattered outcrops of Cambrian, Ordovician, and Silurian sedimentary rocks are found in the Avalon platform area. They are relatively thin, indicative of essentially stable platform conditions, and although they are folded, the fold patterns are simple and probably formed in a single period of deformation. In addition, some high-angle faults are present and the units are tilted, but tight folding and thrusting are absent.

The Central Mobile Belt

The Precambrian basement is exposed in only one place within the central mobile belt, and that is Grenville-age basement like that farther west. The Cambrian sedimentary record is also largely obscured by later cover and metamorphism, but clastic sequences of probable Cambrian age include sands, silts, conglomerates, and graywacke in the east. In contrast the

* Refer to review papers by Bird and Dewey (1970), Kay (1969), 1967), and Williams (1964).

Ordovician section is widely exposed and may be as much as 6 to 7 km of marine volcanics, including pillow lavas, basic pyroclastic rocks, intermediate volcanic and plutonic rocks, graywacke, chert, and conglomerates. This belt was affected by the Taconic orogeny in mid-Ordovician time, but as in so many areas to the south, it is not easy to distinguish Taconic from later (Acadian) features. The most clear-cut evidence of deformation is the allochthonous rocks, including mafic and ultramafic intrusions that were transported from this belt and now lie on the western platform.

Graywacke, conglomerates, and other clastic rocks accumulated in the central belt during the Silurian, but few sediments of Devonian age remain. The central belt was very strongly affected by the Acadian orogeny in the middle and late Devonian and became the site of large granitic plutons, regional metamorphism, and widespread folding. Folds trend northeast, are often overturned and even isoclinal in places, and are accompanied by an axial-plane cleavage that is well developed in the finer sediments.

Plate-Tneory Interpretations of Newfoundland Structure

The importance of the thrust sheets and exotic rocks (the Cow's Head breccia) found in the western platform area as indicators of the Taconic orogeny in Newfoundland have already been pointed out. A second feature, found in the central zone and consisting of a bouldery mudstone and known as the *Dunnage mélange,* is also accorded special significance in the tectonic evolution of Newfoundland. The Dunnage mélange is now interpreted as a deposit formed by submarine slumping into an oceanic trench located over an active subduction zone. The Dunnage deposits are now exposed in an area bounded by faults; so, it is difficult to establish exact relations with adjacent rocks, but the contents of the mélange are those so commonly associated with subduction—pillow lava, tuff, graywacke, and argillite. The zone subsequently became the site of many intrusions. The sequence of stages in development of the Taconic or Humberian orogeny as proposed in one plate-theory interpretation, that of Bird and Dewey (1970), is illustrated, Fig. 22-5. This model requires development of a subduction zone located under the central mobile belt. Volcanic islands form and tectonic activity begins as the crust in the arc-trench gap buckles and slices of oceanic crust are thrust to the surface, providing source rock for the Dunnage mélange. Intrusions and uplift follow this stage, and the Cow's Head allochthon slides west onto the subsiding platform in the west from the uplifted area.

It should be noted that this model requires the existence of a proto-Atlantic ocean (one that preceded the formation of the present Atlantic) as suggested by Wilson (1966).

COMPARISON OF NEWFOUNDLAND AND THE BRITISH CALEDONIDES*

The broad continental shelves off both Newfoundland and Great Britain ensure that any attempt to fit these two together will be at least somewhat imperfect. Despite this, some direct comparisons seem warranted. The Caledonide orogen sweeps across Ireland, Scotland, and northern England with a northeasterly trend like that in Newfoundland. The Caledonides orogen lies between a northwestern Precambrian basement located in the Hebrides and northwestern Scotland, Fig. 22-6, and a Precambrian basement and its sedimentary cover to the southeast. As in Newfoundland, the Caledonides are cut by great strike-slip faults (notably the Great Glen fault), and internal divisions of the orogen are marked by

* This comparison is now the subject of a number of papers. See Kay (1969) and other papers in Am. Assoc. Petroleum Geologists Mem. 12, Dewey and Kay (1968), Anderson and Owen (1968) for the British Caledonides, and Schwab (1974).

major faults (the Highland Boundary fault and the Southern Upland fault which broke up and separated parts of the belt after the main orogenic phases.)

Large-scale thrust sheets (the Moine thrust) carry units northwest from the central part of the orogen to rest on the Precambrian basement, reminiscent of the allochthonous masses in western Newfoundland; however, the age of the basement in these two is substantially different. The Newfoundland basement was affected by a thermal event about 1 billion years ago, but the Lewisian basement of northwestern Scotland is nearly twice that age.

The Caledonide orogen can be readily subdivided into a northwestern belt which is metamorphosed (shows polyphase deformation) and was strongly affected by a late Cambrian to early Ordovician orogenic event (a little earlier than the Taconian) and a southern belt in which the rocks are less metamorphosed and were affected by an orogeny after the Silurian (somewhat before the Acadian). Thus the subdivision of the central zone in the two orogens is similar, but the timing of the orogenic events is not exactly the same. The Avalon platform is very similar to its British counterpart, which likewise has a deformed sedimentary basement overlain by clastic

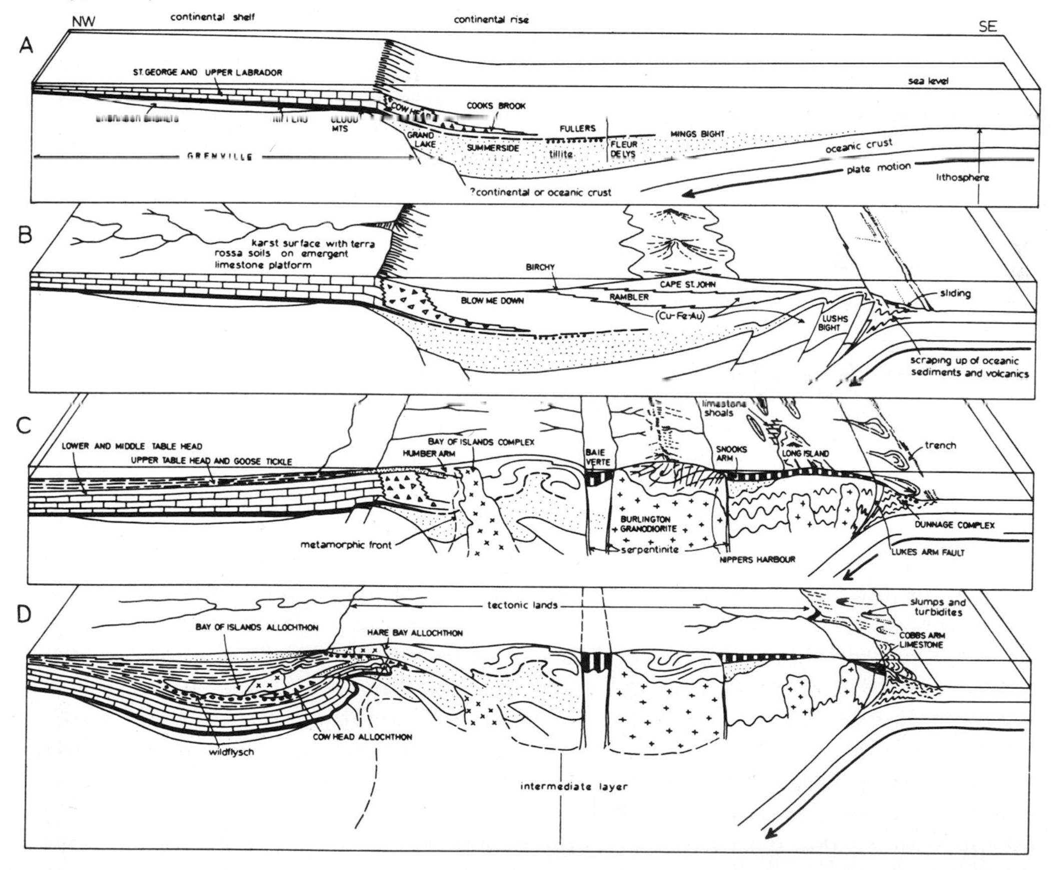

FIGURE 22-5 Schematic block diagrams illustrating the pre-Humberian and Humberian Taconic evolution of the continental margin of North America in Newfoundland: (*a*) pre-Humberian; (*b*) initiation of trench and island arc; (*c*) early Humberian; (*d*) late Humberian. (*From Bird and Dewey, 1970.*)

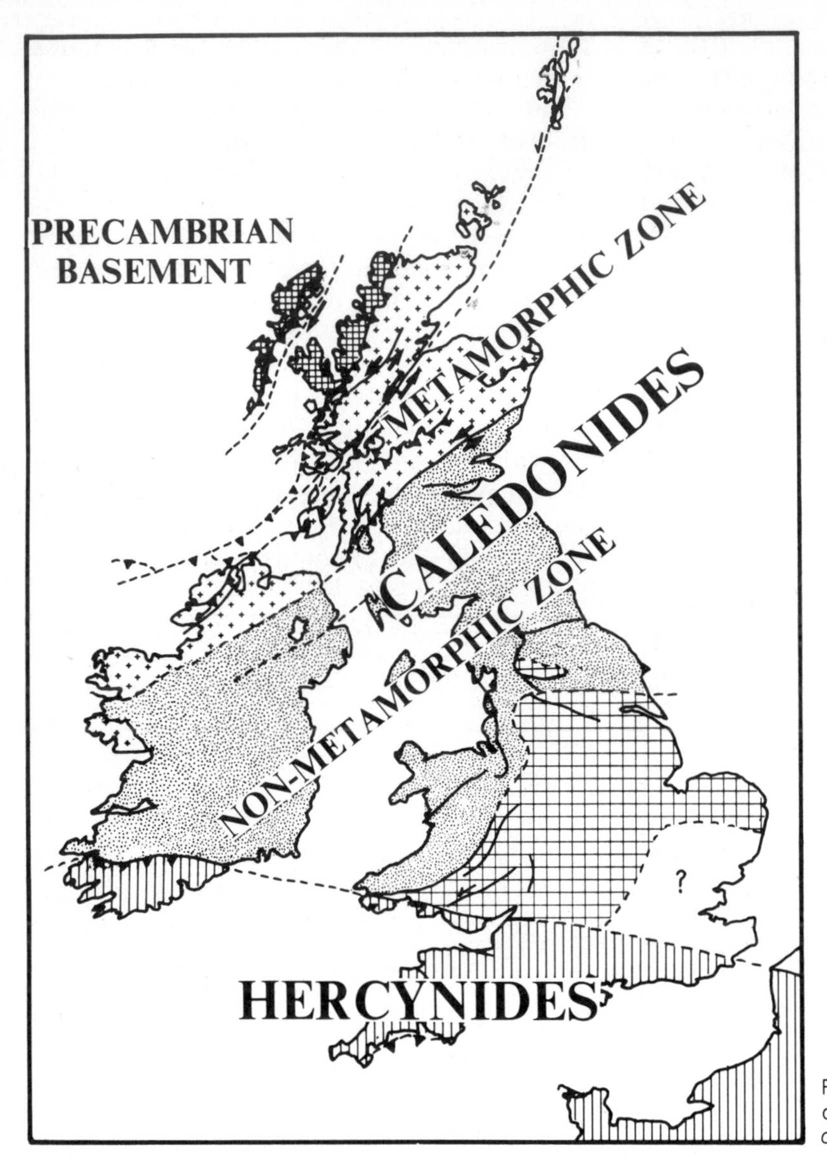

FIGURE 22-6 Tectonic sketch map of the British Isles. (*After the Geological Map of Great Britain.*)

sediments. Thus in at least gross aspects the northern Appalachians and the southern Caledonides are very similar.

Elements of the Caledonide orogen occur in northeastern Greenland (Haller, 1971) and in western Scandinavia (see Strand and Kulling, 1972). Externally directed tectonic transport of internal elements is apparent in both places. The Scandinavian Caledonides are a direct northern continuation of the British Caledonides, and the Greenland belt would lie northwest of Scandinavia in a closed Atlantic. Thus these northern elements would have been located between the Canadian Shield and the Scandinavian Shield during the orogenic history, and the main body of the Appalachians would have been located between the North American craton and the African shields, Fig. 22-1. Most attempts to interpret the Paleozoic history of these orogens employ plate-theory interpretation of continental collisions or uplift over areas of subduction and thus require the existence of an early Paleozoic Atlantic Ocean that closed and reopened in the Mesozoic.

THE SOUTHERN APPALACHIANS

Major Tectonic Elements of the Southern Appalachians*

The longitudinal division of the southern Appalachians into a western miogeosynclinal foreland fold belt, a central basement massif (the Blue Ridge anticlinorium), and an eastern eugeosynclinal belt (the Piedmont) provides a clear large-scale tectonic framework which can be used to describe the tectonic evolution and characteristic structural features of this region. The Blue Ridge basement massif does not separate the foreland from the Piedmont at either the northern or southern ends of this section. The metasedimentary rocks of the southern Piedmont are thrust directly onto the Paleozoic sedimentary rocks of the miogeosyncline in Alabama and western Georgia, and at its northern end the Blue Ridge plunges to the north. The northern end of the southern Appalachians is complicated by the presence of a series of Triassic basins, the Newark, Gettysburg, and Culpeper basins. The Newark basin covers the connection between the northern Piedmont and the crystalline complexes to the north which are exposed continuously from New York City to Newfoundland, Fig. 22-7. This configuration leaves a narrow belt of slightly metamorphosed lower Paleozoic rocks east of the Gettysburg basin and east of the basement ridge in contact with Piedmont rocks (much of this contact lies along a fault known as the Martic line). High-angle faults associated with the Triassic deformation complicate the configuration of the earlier structures in New Jersey and Pennsylvania.

* The Appalachians are sometimes divided into northern, central, and southern parts. The division between central and southern is placed at about the James River gap, where frontal thrusting of the Blue Ridge is evident. The central and southern sections are combined in this section.

Structural Features of the Foreland Fold Belt*

The foreland of the Appalachians is one of the best known fold and thrust belts in the world. The stratigraphic level exposed by erosion is deeper toward the southeast and exposures are good in most places. In addition, exploration for oil and gas have provided subsurface data, especially in the Appalachian basin.

The folds and thrusts in this belt involve rocks from late Precambrian to Pennsylvanian age in the south, although only the early Paleozoic units are so deformed in the narrow belt along the Hudson and Champlain valleys. In general, the folding is more asymmetric, more intense, and is characterized by more overturning and shorter wavelengths along the southeastern side of the belt. Folds become simpler, more nearly upright, more symmetrical, longer, and open toward the northwest, away from the Precambrian uplifts. The vast majority of asymmetric folds are inclined or overturned toward the northwest, and most thrust faults in the region dip southeast; so, it is clear that the direction of tectonic transport throughout the belt is toward the northwest. Folding of the middle and upper Paleozoic units occurred from Alabama, where the belt passes under the Coastal Plain sediments, to Pennsylvania, where the folding of these younger rocks dies out in the Catskill Mountains. Farther north the Cambrian and Ordovician units which lie between the Adirondack and Green Mountains are strongly deformed by both folding and thrusting of the Taconic orogeny. Thrust faults are a common feature of foreland belts. They are most numerous in their surface exposure in two sections of the Appalachians—at the southern end where

* Several examples of Appalachian folds and thrust faults are described in Chaps. 8, 11, and 14.

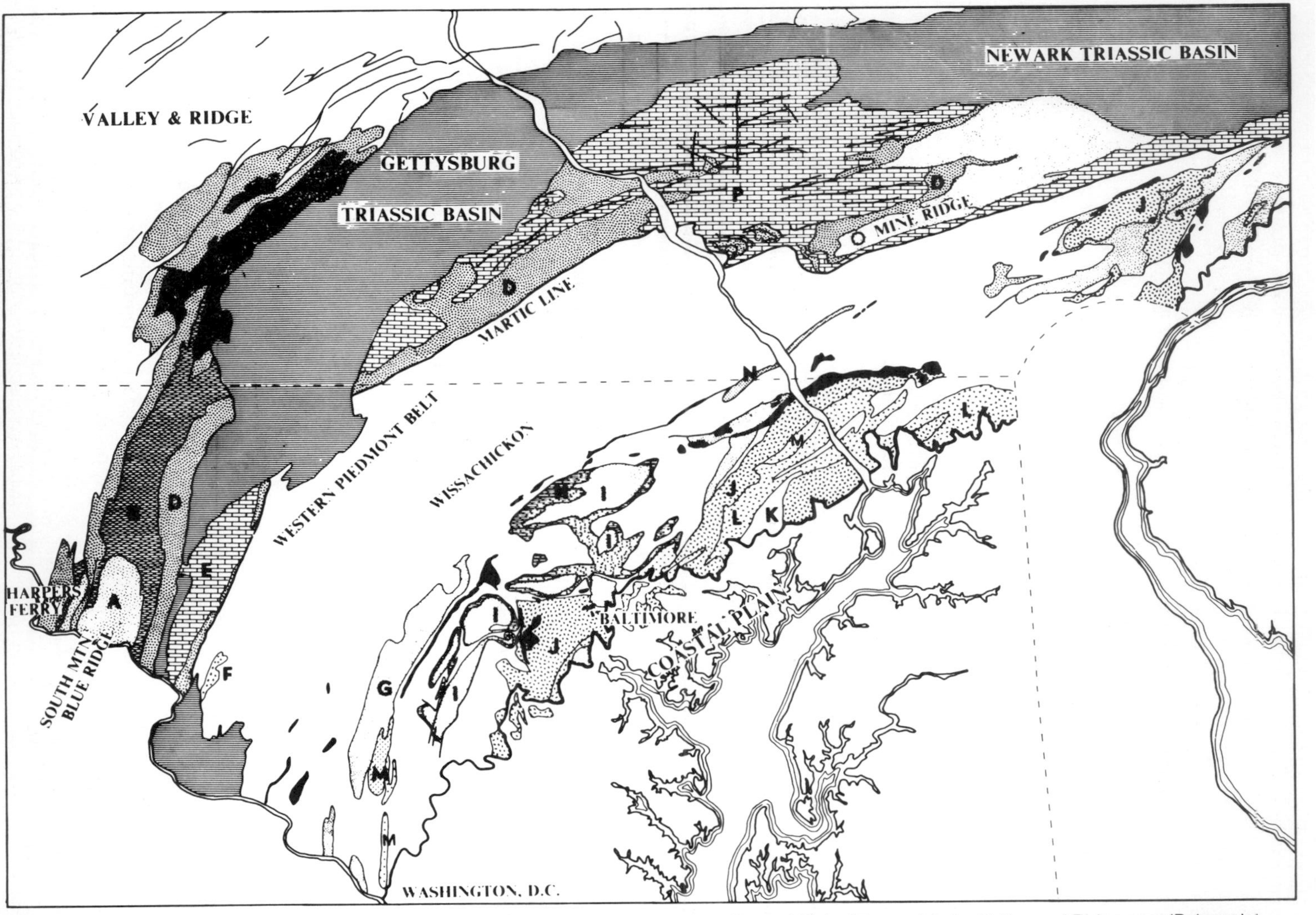

FIGURE 22-7 Sketch map of the Maryland-Pennsylvania-Piedmont area. Rocks of the Coastal Plain (Mesozoic), the Valley and Ridge area (Paleozoic), and the metasedimentary rocks of the northern Piedmont are all left blank. The letters are keyed as follows: A—Blue Ridge basement; B—Late Precambrian Catactin formation; C— Late Precambrian rhyolite; D—Chilhowee; E—Frederick Valley carbonates; F—Sugarloaf Mountain quartzite G—Boulder conglomerate in the Wissahickon; H—Setters and Cockyesville formations; I—Precambrian gneiss domes; J—Gabbro complexes; K—James Run volcanics; L & M—Paleozoic intrusions (granite and diorite); N—Peach Bottom syncline; O—Mine Ridge anticline (Precambrian basement); P—Paleozoic carbonates. (*After the state geologic maps of Maryland, 1968, and Pennsylvania, 1960.*)

they are exceptionally developed in Tennessee and southwestern Virginia and in western New England and southern Quebec. In both places imbricated thrust zones appear. That in the south is about 80 km wide and occurs immediately west of the Great Smoky Mountain region and southern sections of the Blue Ridge, where the Precambrian massif is itself clearly thrust to the west. A lower Cambrian shale, the Rome formation, is usually exposed above the fault, indicating that movement of the thrust sheets commonly took place within this unit, probably as a bedding fault, before cutting up in the section across more competent rocks. These steep rises are evident from the steep dips now seen on the faults in so many places. The belt of closely spaced thrusts dies out in central Virginia, although a few thrusts continue north.

The belt of thrusts in the south abruptly terminates, with thrusts along the side of the Sequatchee valley and the Pine Mountain thrust having a sharp boundary against unfaulted, flat-lying rocks to the west. In contrast the area west of the basement massifs from central Virginia to eastern Pennsylvania is folded not only close to the massif, but far west into the Appalachian Basin, although the folding does become simpler and more gentle, at least in surface exposure to the west, Fig. 22-2.

Folds and Related Faults in the Valley and Ridge and Allegheny Synclinorium

A relatively sharp topographic division can be recognized between the Valley and Ridge province and the plateaus to the west. Rocks of the Cumberland plateau in the south are flat lying and contrast strongly with the folded units in the Valley and Ridge but many of the strata in the Allegheny synclinorium are gently folded.

Upper Paleozoic rocks crop out over most of the Appalachian basin, and their structure, as evidenced by mapping, is that of a broad asymmetrical synclinorium. This synclinorium passes into the nearly flat-lying units of the Catskill Mountains to the north and on the flanks of the Cincinnati arch to the west. The southeast side of this region, sometimes referred to as the *Appalachian structural front,* is defined by the exposure of steep west limbs of anticlines or thrust faults. Surface maps reveal a system of doubly plunging folds with trends which parallel the folds and faults of the Valley and Ridge, sweeping in a broad arc through Pennsylvania. The folds at the surface die out toward the northwest. No prominent thrust faults are exposed in this region, and the strata are essentially horizontal, particularly in the western portion.

The amplitude of the anticlines decreases to the northwest, and most of the anticlines are asymmetrical toward the northwest, although a few are slightly asymmetric toward the southeast. Many of the major folds are over 100 km long, but eventually they do plunge. Lines drawn through the points at which these plunges occur define northwest-trending lines, suggesting transverse dislocations similar to those in the Jura Mountains, Figs. 22-8 and 23-5.

Wells drilled through units which are only broadly arched at the surface encounter fault breccias, repeated sections, and other unexpected complications at depth, Fig. 8-5. These complications are now thought to form as a result of *décollement* at depth.

The Burning Springs anticline located along the West Virginia–Ohio border is an important key to the *décollement* explanation of the structure of the synclinorium. The anticline is asymmetrical toward the west at the surface, but in one of the wells drilled on the eastern flank of the fold, the lower Devonian is repeated four times. Subsurface control indicates a zone of imbricate southeast-dipping thrusts parallel to the fold axis. No thrusts are known west of the fold in either the surface or subsurface, nor do those thrusts in the fold

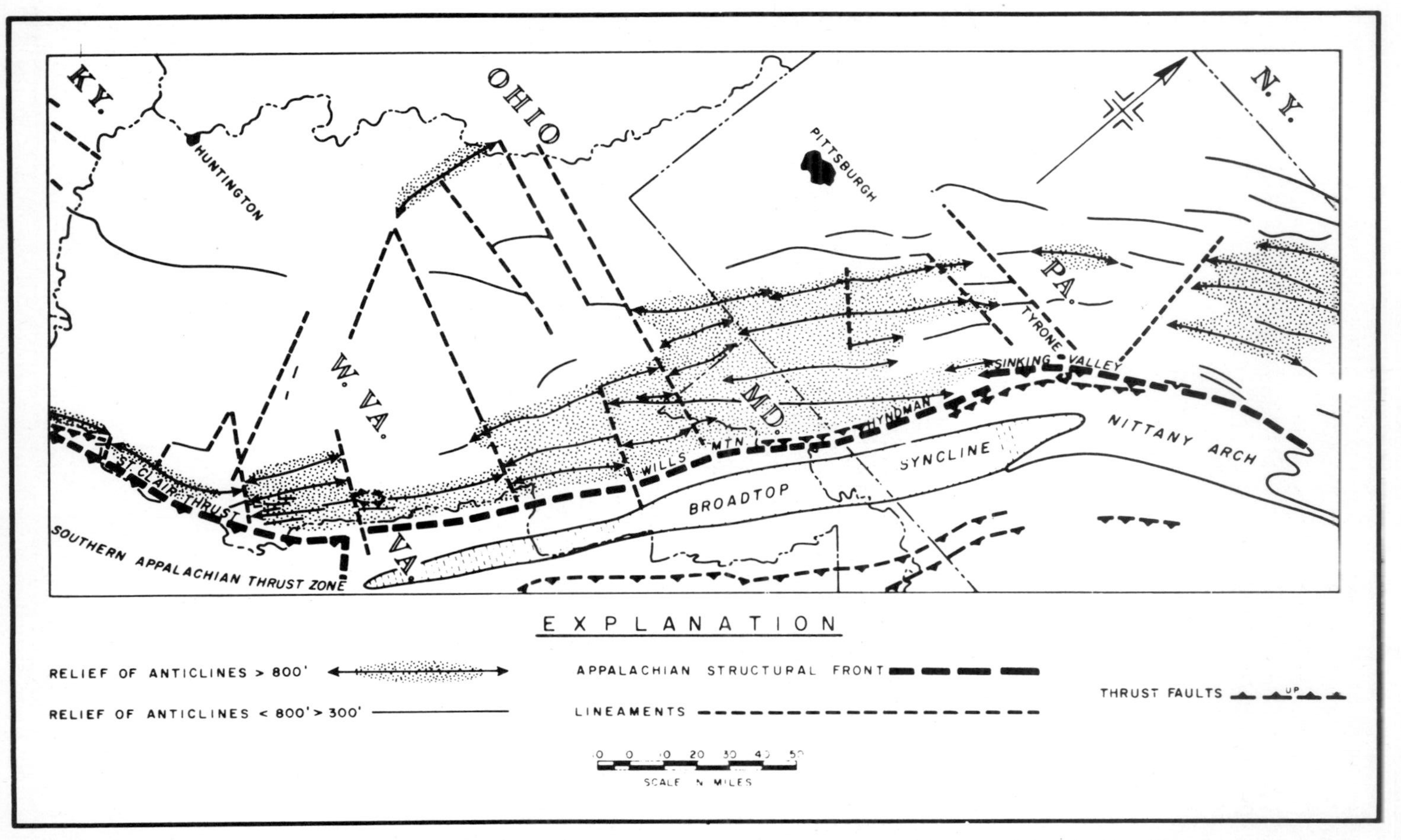

FIGURE 22-8 Map of surface anticlinal axes and structural relief, lineaments, and important faults of the central Appalachian Plateau and the northwestern Valley and Ridge provinces. (*From Gwinn, 1967.*)

appear to extend below Silurian units. The western edge of the Salinas salt beds of Silurian age occurs just east of the Burning Springs anticline. Rodgers (1964) attributes the development of the faulting and folding in this particular place to its location at the edge of the salt beds, which could well have provided a lubricating medium for *décollement* here as it did in the Jura Mountains.

Interpretation of the mechanics of the origin of the structural pattern in the plateau region revolves around what happens at depth. Little data have been published concerning the basement in the region, and not too surprisingly, arguments center around the role of basement in the deformation. Basement gneisses are exposed in the Blue Ridge, but nowhere farther west across the foreland region. Few deep wells are drilled in the Valley and Ridge, and very few reach the basement farther west in the Allegheny synclinorium. Gravity and magnetic work indicate basement depths greater than 10 km just west of the Blue Ridge, and new basement maps show that the basement rises gently to the west across the synclinorium. Early students of the Appalachians reasoned that the stresses that uplifted and strongly deformed the basement in the Blue Ridge and the sedimentary veneer in the Valley and Ridge region could not have been transmitted far enough through relatively flat-lying Paleozoic sedimentary sequences to cause folding in the synclinorium. They reasoned that folds seen at the surface continued downward, ultimately involving the basement, in the manner described in Chap. 14, Fig. 14-1. Many of the faults found on the flanks of folds in the Valley and Ridge were also projected down into the basement, where displacements were presumed to originate.

The alternative to basement involvement is popularly known as the theory of *thin-skinned deformation,* in which the folds and thrusts seen in the Paleozoic sedimentary units are confined to the cover and do not involve the underlying basement west of the Blue Ridge uplift where basement is obviously involved. The concept of thin-skinned or cover deformation is similar in some respects to the distinction made between infrastructure and suprastructure, but it is derived most directly from the *décollement* tectonics as represented in the Jura Mountains, where the Mesozoic sedimentary cover is folded and thrusts as a result of movement which caused lateral shortening of the cover which became detached from the underlying Hercynian crystalline basement and slipped over it in a basal section of Triassic salt deposits. The thin-skinned hypothesis had been applied to the Pine Mountain thrust sheet before its more general application to the Appalachian Valley and Ridge and Allegheny synclinorium in general as proposed by Rodgers.

Décollement models for the Appalachians involve several levels of bedding-plane movements, including the lower Cambrian Rome shale from which the imbricate thrusts of the southern Appalachians rise and the Silurian Salinas salt formation which lies east of the Burning Springs anticline. Other potential levels of *décollement* marked by incompetent beds are found in the lower Cambrian Chilhowee group and in the upper Ordovician shales. Gwinn (1967) outlined areas of probable dominance of specific sole-thrust or *décollement* zones throughout the foreland region. He postulated that the level of the *décollement* rises from lower Cambrian to Ordovician glide zones approximately along the Appalachian structural front, and then farther west it rises again to the Salina glide zone. It would appear in this interpretation that the sole fault gradually rises in the section toward the west, as described earlier in the Pine Mountain thrust sheet, Fig. 8-9.

Gwinn proposed an explanation for many of the complex structures encountered at depth beneath what appears to be simple anticlinal structures at the ground surface. This model is described in Chap. 8, Fig. 8-5, and involves uplift and folding of the sedimentary layers above ramps where bedding faults rise from one *décollement* zone to another, thickening the cover over the ramp and sometimes leading to thrusts on the flanks of the resulting fold. Additional well data and seismic sections have become available in recent years. Several of these sections are located across the western side of the Valley and Ridge in western Virginia. Cross sections based on these data have been devised by Jacobeen and Kane (1975), and their inferred development of these structures is illustrated, Fig. 22-9. This model, like Gwinn's, involves *décollement,* ramp rises, and back-limb thrusting on the southeastern limbs of the anticlines. It is also interesting to note that this model shows that the lowest sole thrusts are localized over high-angle faults in the underlying basement.

Changes in Structural Style Across the Appalachians

Gradual but strongly developed changes in structural style of deformation occur across the Appalachians at the northern end of the Blue Ridge as described by Cloos, Fig. 22-10. The Paleozoic units in the Valley and Ridge are folded into open, upright, symmetrical folds

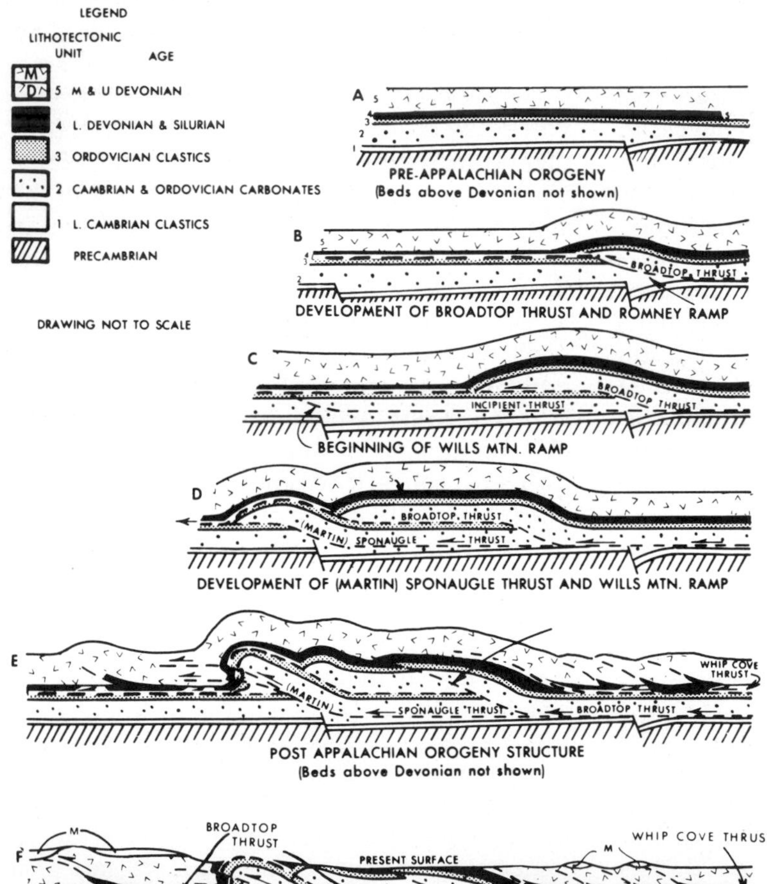

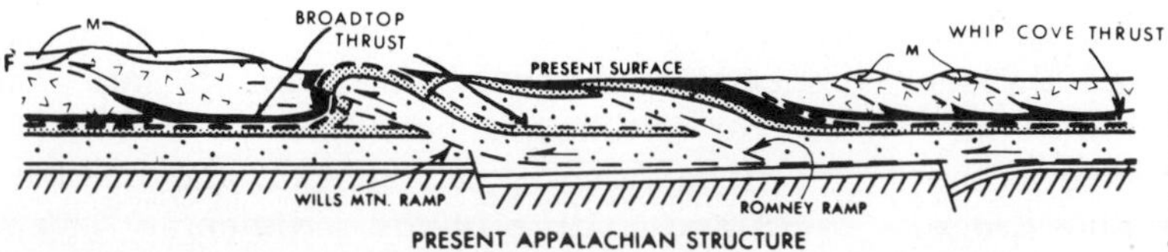

FIGURE 22-9 Stages of development in formation of Broadtop and Martin thrusts. Sequence illustrates that the most westerly thrust was last to form. (*From Jacobeen and Kanes, 1976.*)

which show only fracture cleavage. As the Blue Ridge is approached a flow cleavage becomes increasingly developed, and oolites show the strain effects of what becomes a penetrative deformation close to the Blue Ridge. This penetrative southeast-dipping slaty cleavage is prominent in the finer sedimentary rocks some of which are altered to phyllites (Harpers formation), and other strain effects show up in various lithologies. Lenticular blebs of chlorite appear in the late Precambrian greenstones, elongated sericite blebs are found in Precambrian metarhyolites; even the basement gneisses show strong lineations of micaceous minerals. Higher in the section, elongated pebbles, lumps, quartz rods, streaks of other minerals, and oolites all bear evidence of the orientation of the strain. The direction of tectonic transport is invariably toward the northwest. The lineations have that bearing and plunge southeast; cleavages strike northeast and dip southeast. The Blue Ridge anticlinorium is strongly asymmetric in this region, but evidence of large-scale thrusting along the western margin such as that described for the Smoky Mountain region is missing.

FIGURE 22-10 Cross section across the Appalachians showing variation in structural style from the Piedmont across the Blue Ridge to the plateau. (*From Ernst Cloos.*)

The Blue Ridge Anticlinorium

The Blue Ridge uplift, the largest of the Precambrian exposures in the Appalachians, is located between the foreland fold belt to the northwest and the Piedmont to the southeast. Northern continuations of this structural element appear in the Reading Hills, the New Jersey highlands, the Hudson highlands, the Berkshire highlands, the Green Mountains, and in the Long Range in Newfoundland. The Precambrian basement of these elements is composed of gneissic complexes metamorphosed to high grade during a thermal event that took place about 1 billion years ago in the Grenville province and produced granulite-facies metamorphism in many parts of these basement elements. A great variety of rocks lie unconformably on this old basement complex, including late Precambrian rocks in some places and early Cambrian units in others, and these rocks are an integral part of the Blue Ridge topographically and structurally.

The anticlinal structure of the Blue Ridge is most apparent at its northern end, where

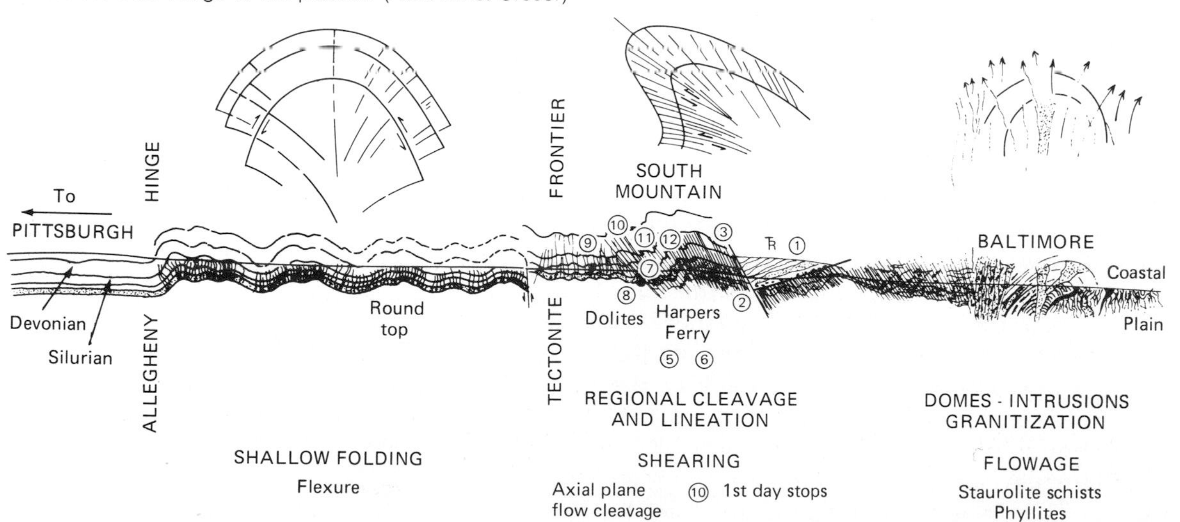

lower Cambrian clastic rock of the Chilhowee series wrap around the plunging Grenville rocks of the Blue Ridge basement complex, Fig. 22-7. The Chilhowee lies, unconformably, on the basement all along the western edge of the Blue Ridge, but older sedimentary and volcanic sequences occur in the interval between the basement and the Chilhowee on the eastern flank of the anticlinorium, and rocks of the Chilhowee group are not present everywhere on the east. The stratigraphy of the eastern flank is made more complex by strong changes in the facies along the basement and finally by both complicated structure and polyphase metamorphic alteration of the rocks. Toward the northern end a prominent volcanic sequence—the Catoctin—lies above the basement and below the Chilhowee. The Catoctin includes tuffaceous and clastic sediments (dated at approximately 800 million years) and in places thick plateau-type lava flows now altered to greenstone which can be traced as far south as the James River in central Virginia. Farther south a thick section of volcanics and interlayered sediments, the Mount Roger's sequence, occupies much the same stratigraphic position. These are in turn underlain by schists or clastic rocks (Ijamsville phyllite in the north, Lynchburg schist in central Virginia, the Ashe formation, Grandfather Mountain formation and Ocoee group farther south). The Ocoee is especially noteworthy because of its clastic character and great thickness, ±15 km, which requires a prominent source area of quartzose rocks. Thus the zone across the Blue Ridge uplift was the site of very significant facies changes during the late Precambrian. Shelf-type sedimentation took place in the west during Chilhowee time, but the region to the west was probably emergent earlier while the thick, extensive, and varied stratigraphic section on the east flank of the Blue Ridge formed.

The boundary between the east flank of the anticlinorium and the Piedmont is highly varied. The Triassic of the Culpepper basin forms this boundary in the north, and the Brevard fault zone defines it in the south. The intervening boundary in central Virginia is formed by the James River synclinorium—a belt of younger, Paleozoic, metasedimentary rocks now broken by many northeast-trending high-angle faults, Fig. 22-11.

The role of the Blue Ridge in the orogenic history of the southern Appalachians is not completely established, but it obviously is a very important element and its structure provides some important insights to the deformation. The time of the initial uplift and movement of this massif is uncertain, but it can be shown that it moved west during the late Paleozoic. A second important question is the importance of the massif in formation of the folds and faults to the northwest. It has been suggested by some geologists that the foreland folds originated as a result of gravity-driven sliding or spreading off the uplifts to the southeast. Others consider it more likely that the foreland folds result from lateral pressure which arose as the crystalline rocks of the Blue Ridge and Piedmont were forced northwest. The Blue Ridge massif would have acted as a relatively strong brittle element throughout most of its length in this model. According to this second model the Blue Ridge must have been uplifted before or during the foreland folding, and it must be allochthonous if its movement is to account for any substantial part of the shortening recorded in the foreland.

FIGURE 22-11 Interpretive geologic map of part of the crystalline Appalachians south of lat. 40° N. Most plutonic rocks are omitted as well as Paleozoic rocks younger than Middle Ordovician. All Precambrian and Paleozoic rocks have been metamorphosed. Numbered localities are: 1. Conowingo dam; 2. Area of the James Run formation; 3. Frederick, Maryland; 4. Contact between the Ijamville phyllite and the Wissahickon formation; 5. Area of the Chopawamsic formation; 6. Mt. Rogers; 7. Grandfather Mountain; 8. Locality of ? *Paradoxides carolinaensis;* 9. Murphy belt; 10. Martinsville, Va. S-W belt = Sugarloaf-Westminster belt; KM belt = Kings Mountain belt. (*From Rankin, 1975.*)

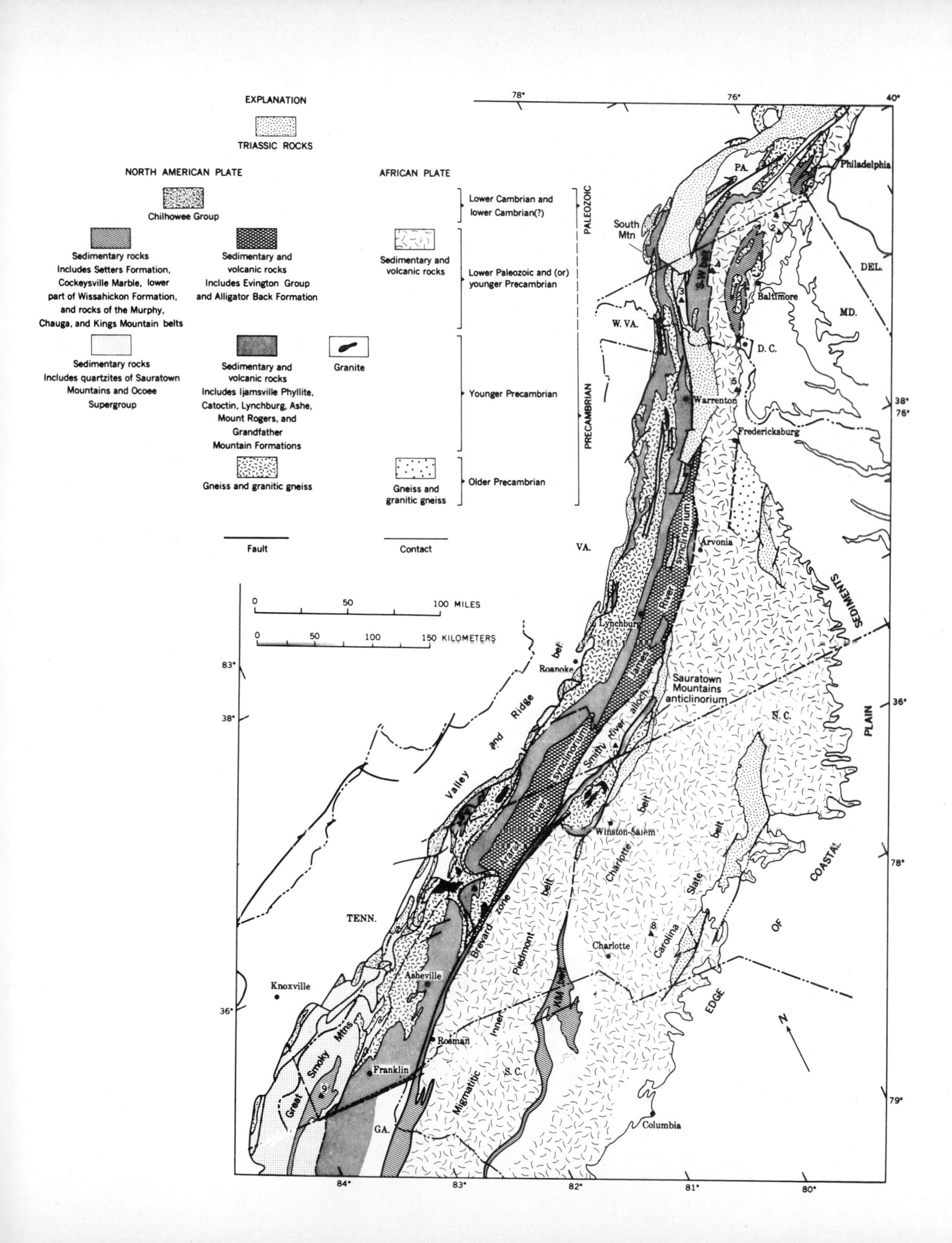
EXPLANATION
TRIASSIC ROCKS
NORTH AMERICAN PLATE
AFRICAN PLATE
Chilhowee Group
Lower Cambrian and lower Cambrian(?)
PALEOZOIC
Sedimentary rocks
Includes Setters Formation, Cockeysville Marble, lower part of Wissahickon Formation, and rocks of the Murphy, Chauga, and Kings Mountain belts
Sedimentary and volcanic rocks
Includes Evington Group and Alligator Back Formation
Sedimentary and volcanic rocks
Lower Paleozoic and (or) younger Precambrian
Sedimentary rocks
Includes quartzites of Sauratown Mountains and Ocoee Supergroup
Sedimentary and volcanic rocks
Includes Ijamsville Phyllite, Catoctin, Lynchburg, Ashe, Mount Rogers, and Grandfather Mountain Formations
Granite
Younger Precambrian
PRECAMBRIAN
Gneiss and granitic gneiss
Gneiss and granitic gneiss
Older Precambrian
Fault
Contact
0 50 100 MILES
0 50 100 150 KILOMETERS
78°
76°
40°
38°
36°
83°
84°
82°
81°
80°
79°
PA.
Philadelphia
South Mtn
DEL.
Baltimore
MD.
W. VA.
D. C.
Warrenton
Fredericksburg
VA.
Arvonia
James River synclinorium
Lynchburg
Roanoke
Valley and Ridge belt
Sauratown Mountains anticlinorium
SEDIMENTS
N. C.
PLAIN
Smith River alloch.
Ararat River synclinorium
Winston-Salem
Charlotte belt
Carolina Slate belt
COASTAL
OF
EDGE
TENN.
Brevard zone
Inner Piedmont belt
Charlotte
Knoxville
Asheville
Rosman
Franklin
Great Smoky Mtns
Migmatitic
S. C.
GA.
Columbia
N

Evidence that the Blue Ridge is thrust to the northwest is now well documented at its northern end as shown in Fig. 22-12 (Root, 1970); it is clearly thrust to the northwest on nearly horizontal thrusts south of the James River gap, Fig. 22-13 (Spencer, 1968, 1972). The amount of horizontal transport appears to increase even more to the southwest, and it is possible that the section between central Virginia and Pennsylvania is allochthonous as well, although the thrusts do not lie along the edge of the Blue Ridge in that sector.

The southern section of the Blue Ridge uplift, Fig. 22-2, provides interesting insights to the character of the Appalachians and especially to the relationship between the Paleozoic foreland fold belt and the Precambrian Blue Ridge. Everywhere the contact between these two provinces is marked by southeast-dipping thrust faults, and in a section west of the Smoky Mountains rocks as young as Mis sissippian are affected by the thrusts; so, a post-Mississippian age is established for at least some of the movement.

Proof of the northwestward movement of the Blue Ridge uplift over rocks of the miogeosyncline is found in a series of windows along the Great Smoky fault where Paleozoic rocks are overlain by late Precambrian rocks. Even greater westward transport is indicated by the appearance of unmetamorphosed Paleozoic rocks overlain by basement rocks at the Grandfather Mountain window, Fig. 22-14. The Blue Ridge would have to be moved over 30 km southeast to bring its northwestern edge back to the east of the Grandfather Mountain window.

The approximate age of this thrusting can be estimated from studies of the degree and age of its metamorphism. Rocks within the Grandfather Mountain window are essentially unmetamorphosed. Rocks immediately outside the window are metamorphosed to various grades by a thermal event that is dated at about 350 million years (Devonian). Therefore, at least some of the thrusting was clearly post-Devonian. Because Mississippian rocks are folded along the western edge of the Blue Ridge, it appears that thrusting was post-Mississippian as well.

The Grandfather Mountain window is also

FIGURE 22-12 Generalized geologic cross section from T. E. Nesbitt #1 well in the Valley and Ridge to Pennsylvania Survey core-hole in Blue Ridge. This section is located in southern Pennsylvania. *(From Root, 1970.)*

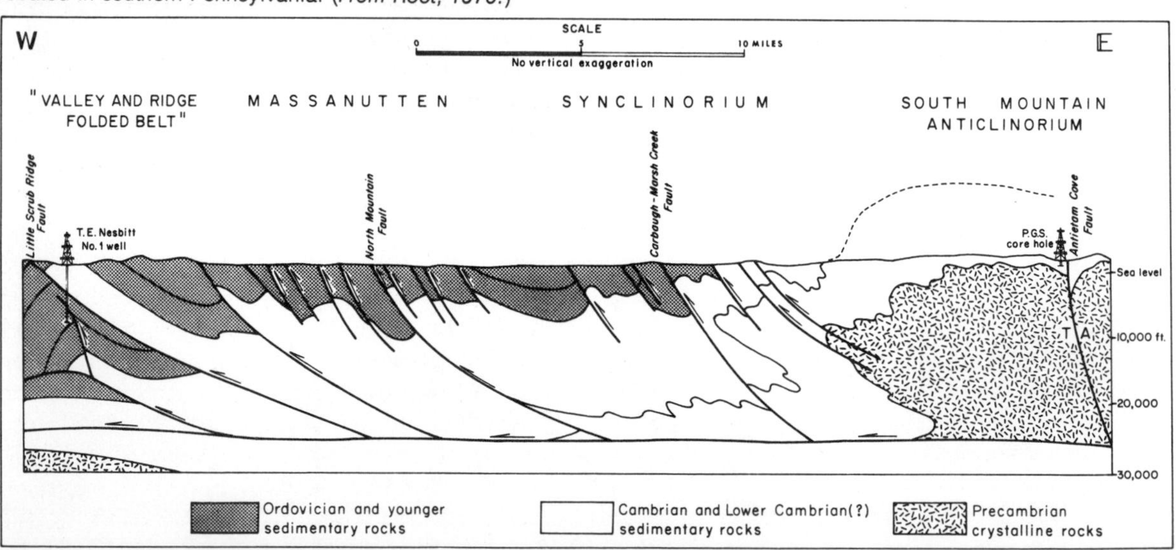

important as the outcrop area for another major fault system, the *Fries fault* which lies along the western edge of the *Ararat River synclinorium,* Fig. 22-14 (Rankin, Espenshade, and Shaw, 1973).

This fault carries both Grenville basement and the overlying cover of late Precambrian and possibly even lower Paleozoic metasediments west. The Fries can be traced from the Brevard zone around the Grandfather Mountain window northward to the west flank of the Blue Ridge anticlinorium; it may continue to the south as well, and it is certainly one of the major structural features of the Blue Ridge province.

The metamorphism in the southern Blue Ridge section also provides evidence of an early Paleozoic tectonic event. The isograd lines are undeflected as they cross some of the thrusts in the Smoky Mountain area (e.g., Greenbrier fault), proving that an episode of thrusting took place before the metamorphism. This metamorphism altered large parts of the Ocoee and other rock units, especially on the eastern side of the Blue Ridge.

Bouguer gravity anomalies also suggest westward displacement of the southern section of the Blue Ridge. Bouguer gravity anomalies are nearly parallel to the surface structural trends for long distances. The strongest positive anomalies tend to lie over the Blue Ridge and Piedmont, while negative values lie over the foreland fold belt, but this pattern does not hold toward the southern end of the Blue Ridge where the negative anomalies lie over the Smoky Mountain region and even extend into the Piedmont farther south. This suggests that these older rocks overlie lower density, presumably younger, materials. The gradient between the positive and negative anomalies forms a narrow belt which runs the length of the Appalachians. The trough of this belt follows the western edge of the Precambrain uplifts from central Virginia northward. The divergence of the base of this gradient coincides approximately with the development of low-angle thrust faults along the northwestern edge of the Blue Ridge, and the number of thrusts increases markedly to the south (King, 1959; Watkins, 1964).

The Piedmont

The Piedmont province of the southern Appalachians is covered everywhere along its eastern side by Cretaceous or Tertiary sediments which obscure the eastern extension of the complex of metasedimentary and igneous rocks. The rocks in the Piedmont are highly varied in age and composition, and it is obvious that much of the region has a complex tectonic history. A few domes of Grenville-age basement crop out in the northern section of the Piedmont, especially around Baltimore, a number of mid-Paleozoic plutonic bodies can be readily recognized in the south, and it is easy to identify the Triassic basins, but the age of much of the metasedimentary section in the Piedmont has long been a subject of debate. The problem arises in part because the stratigraphic sequence within the deformed metasedimentary rocks is unclear—fossils are absent, top and bottom criteria are difficult to identify, and sequences are inverted and thrust-faulted in many places.

The question of the age of the Piedmont rocks is crucial to our understanding of Appalachian history. If the Piedmont rocks are largely Precambrian (and they would have to be late Precambrian since they rest on Grenville basement), then the Piedmont could have been a source area for the rocks deposited west of the Blue Ridge in the foreland during the Paleozoic. In addition the Piedmont rocks would represent an earlier cycle of deposition and possibly orogeny as discussed by Hopson (1964). If, on the other hand, the Piedmont metasedimentary rocks are not Precambrian, then they are time equivalents to some parts of the stratigraphic sequences exposed in the

FIGURE 22-13 Cross section across the northwestern flank of the Blue Ridge, right, and its contact with the miogeosyncline in central Virginia (*After Spencer, 1971.*)

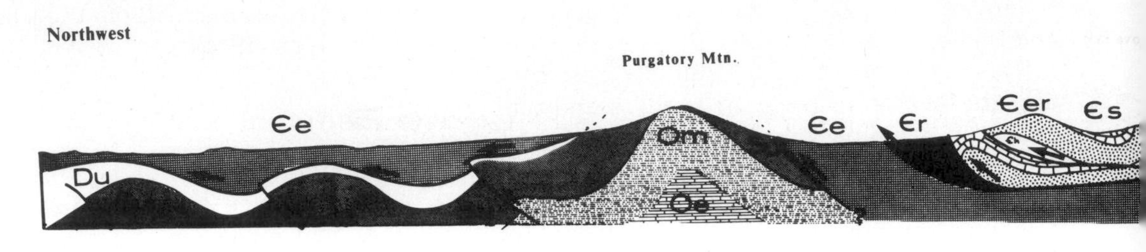

foreland, and we can thus hope to correlate events in the foreland with those in the Piedmont.

The Piedmont of Maryland and Pennsylvania

The best chance to solve this problem of the age relationship between the foreland and the Piedmont by stratigraphic correlation lies in the northern section of the Piedmont where the Grenville basement is exposed and where some lower Palezoic sediments can be identified east of the Blue Ridge, Fig. 22-7. This part of the Piedmont has been studied longer and in greater detail than any comparable area further south. The Grenville basement is exposed in a group of domes at Baltimore, and these are unconformably overlain by a sequence, known as the Glenarm series, that consist of a thin but discontinuous orthoquartzite, the Setters quartzite, a marble, the Cockeysville marble (±300 m thick), and a very thick sequence of mica schist, metagraywacke, and metamorphosed diamictite, the Wissahickon formation, for which thickness estimates of 5 to 10 km have been made. Fisher (1976) reports the results of recent mapping and stratigraphic studies of the Wissahickon. These studies indicate that the lower part of the Wissahickon can probably be traced into early Cambrian clastics in the Blue Ridge areas (parts of the Chilhowee group) and that facies in the upper Wissahickon suggest that it was deposited from a southeastern source and it is composed of diamictite near the source and turbidites farther northwest. Some of the clast in the diamictite contain contorted fragments of the lower Wissahickon formation, boulders of gneiss, mafic rocks apparently derived from a volcanic unit dated at 500 million years old, and gabbro probably derived from the Baltimore Gabbro complex* (see Crowley, 1976). If this inference is correct, then the upper Wissahickon is younger than 550 million years. Wissahickon rocks are now traced south into Virginia where they are overlain by fossil-bearing slates (Arvonia formation) of Late Ordovician age. Thus the Glenarm appears to record Cambrian and Ordovician events east of the Blue Ridge, and it reveals a transition from nonorogenic shelf sedimentation (Setters-Cockeysville) to synorogenic clastic sedimentation in later Cambrian and probably Ordovician in a eugeosyncline.

Progress is also being made in analysis of the complex internal structure of the northern Piedmont. Higgins (1973) recognizes five gen-

*The gabbro complexes are located along the eastern side of the Piedmont and may be in thrust contact with rocks to the west.

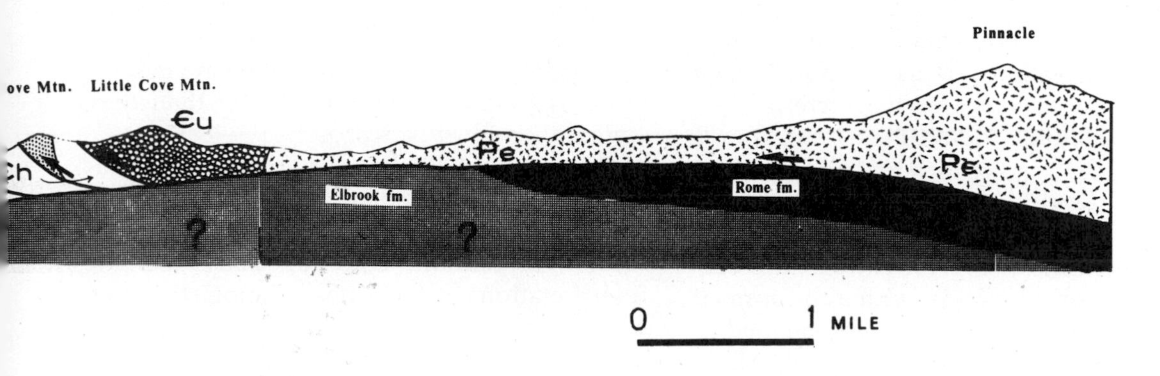

FIGURE 22-14 Zones of Paleozoic regional metamorphism in the Blue Ridge and adjacent Piedmont between lat. 35 and 37° N. Interpreted from Hadley and Nelson (1971) and Rankin, Espenshade, and Shaw (1973). Pervasive retrogression is present in and near the Brevard zone. (*From Rankin, 1976.*)

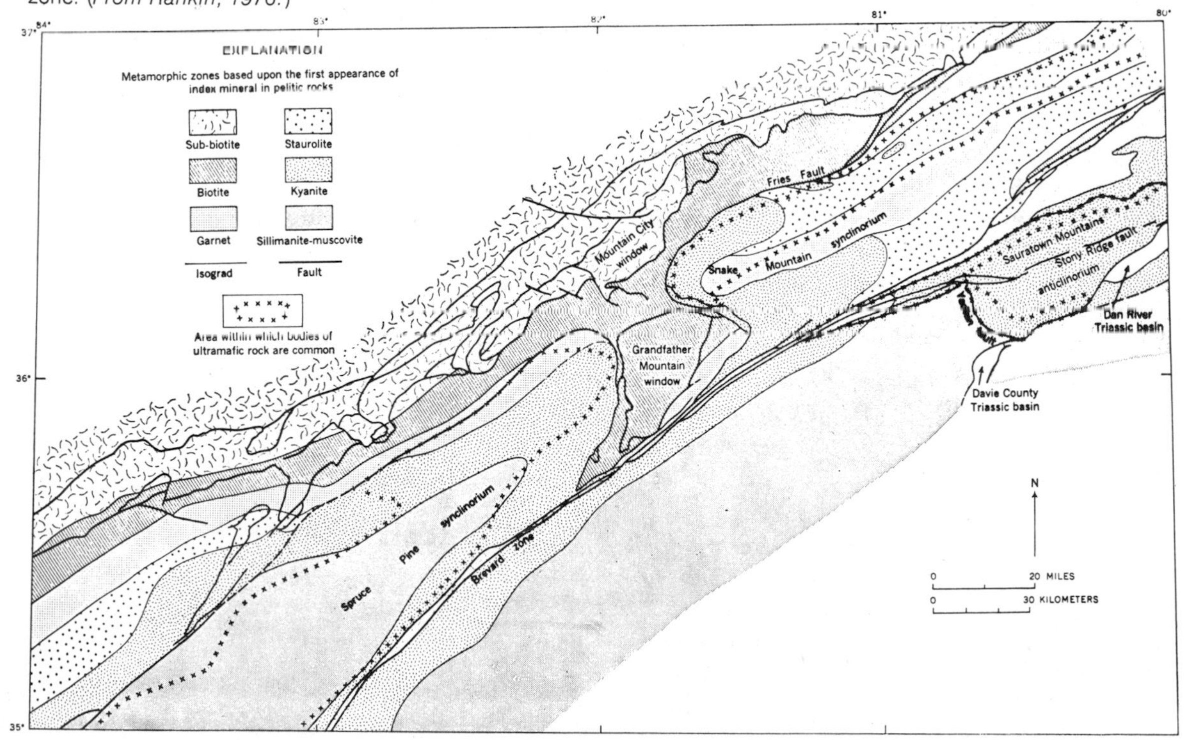

erations of folds as described below; and he considers it most likely that some of the later folds are products of Acadian tectonic activity.

The first generation of folds probably formed in autochthonous rocks of the eastern Piedmont during uplift, detachment, and movement of large masses of rock toward the northwest. These events may have been contemporaneous with deposition of the Upper Ordovician Martinsburg formation. The second generation of folds probably formed within these nappes and became flattened as deformation in the nappes continued. Metamorphic conditions rose into the amphibolite facies within the Piedmont during this flattening stage. Plutons intruded during this folding and metamorphism have radiometric ages of about 425 million years. Thus these events were part of the Acadian orogeny, as were the third and fourth generation of folds. The last generation of folds is probably an expression of the Alleghanian orogeny.

The Southern Piedmont

The Piedmont is more difficult to decipher than its northern counterpart in New England for several reasons—much of the Piedmont has a thick residual soil cover and relatively few good outcrops, and it has been affected by late Paleozoic orogeny in addition to the older Paleozoic events so well documented in New England. In addition, until recently, relatively few geologists were working on the southern part of the Piedmont, large areas are still not well known, and many correlations between the widely separated belts are uncertain. But significant progress is being made, and models of the region are now much better than they were only a few years ago.

The southern Piedmont consists of a number of elongate belts that trend northeast-southwest as depicted in Figs. 22-2 and 22-11. While correlations and especially tectonic interpretations differ, most geologists agree on a basic structural framework for the region which consists of the following belts:

1. *The Inner Piedmont belt anticlinorium:* This is a large lensoid-shaped belt in Georgia, Alabama, and the Carolinas composed of highly mobilized and migmatic rocks, including granitic gneisses, amphibolite gneisses, metagraywacke, and schists, all of uncertain age, but probably including at least some Precambrian basement and late Precambrian metasedimentary rocks. These rocks lie just east of the Brevard fault zone, and consist of a pile of nappes according to the work of Griffin (1974) and Hatcher (1972), Figs. 22-15 and 22-16. The high-temperature deformation of the Inner Piedmont took place in the early Paleozoic orogenic deformation; the last deformation in the Alleghanian orogeny consisted of nonpenetrative deformation which is expressed by microbreccias and block-like uplift and lateral movements (Griffin, 1974).

2. *The Charlotte belt anticlinorium:* This is exposed from Georgia through Virginia and is a large anticlinorium in which high-grade (some even sillimanite) gneisses, schists, and amphibolites are now

Taconian	1st generation	Isoclinal folds on the limbs of the second-generation folds
(Early Acadian?)	2d generation	Tight to isoclinal long-limbed folds with axial surfaces parallel to schistosity. These are greatly flattened folds
Acadian	3d generation	Folds of the schistosity with a moderately developed axial-surface cleavage
Acadian	4th generation	Gentle to tight folds with upright axial surfaces
	5th generation	Folds associated with an axial-planar fracture cleavage that transects all earlier features

exposed at the surface. These gneissic complexes are intruded by both pre- and postmetamorphic plutonic bodies of granitic and basic rock, Fig. 22-2. The age of the gneissic countryrock is uncertain but may well be as old as the Grenville basement of the Blue Ridge and the inferred basement in the Inner Piedmont; in addition, probable late Precambrian metasedimentary rocks occur in the belt along with a great many younger (mid-Paleozoic) plutonic bodies.

3. *The Sauratown Mountains anticlinorium and the Smith River allochthon:* These features are located along the Virginia–North Carolina border at the northern end of the Brevard zone within or along the southeastern side of the Ararat River–James River synclinoria. The Sauratown Mountains are a domal uplift in which Grenville basement is exposed overlain by a basal clastic sequence (Butler and Dunn, 1968). The Smith River allochthon is interpreted by Conley and Henika (1973) as a mass of granitic gneiss with plutonic inclusion dated at about 1 billion years and metasedimentary rocks which have been carried west over the Sauratown Mountains and have come to rest on the metasedimentary rocks of the James River synclinorium in a manner analogous to the nappes of the Inner Piedmont.

4. *Belts of younger rocks in synclinoria:* A number of narrow belts separate the older rocks in the anticlinoria of the Blue Ridge, Inner Piedmont, and Charlotte belts, and others lie within some of the above uplifted areas. These synclinoria contain metasedimentary rocks that may be either late Precambrian or early Paleozoic in age. In any case they are now composed typically of schists, phyllites, metasiltstones, quartzites, and impure marbles along with some amphibolites. The belts included in this category are the Talladega belt of Alabama and Georgia, the Brevard (Chauga belt), the Murphy Marble belt, the James River synclinorium, and we might include the rocks of the Ararat synclinorium now displaced westward into the Blue Ridge and the rocks of the Slate belt along the eastern side of the Piedmont.

The Carolina slate belt is one of the largest and least altered of these synclinoria. It is composed of low-grade metasedimentary and metavolcanic rocks, including metagraywacke, tuffaceous argillite, quartzite, and is intruded by younger plutonic bodies, including one dated at approximately 600 million years (Glover and others, 1971). Thus the slate Belt rocks may also be made up mainly of late Precambrian and possibly earliest Cambrian rocks. Thickness estimates in the slate belt indicate a section of between 4 and 7 km thickness.

5. *The Arvonia slate:* This outcrops in a small basin in the eastern Virginia Piedmont and deserves attention because it is the location of slates that contain Late Ordovician fossils, the youngest metasedimentary rocks known in the Piedmont.

The Enigmatic Brevard Zone

The Brevard zone is a long, narrow (1.5- to 6-km-wide) band of graphic and cataclastic schist which can be traced nearly 600 km from the southern edge of the Appalachians in Alabama to the Sauratown Mountain anticlinorium in North Carolina. Its northern continuation is uncertain. At both its southern and northern ends it is located within the Piedmont, but the central section forms the eastern edge of the Blue Ridge anticlinorium; so, it separates the Blue Ridge from the Inner Piedmont.

The Brevard has been the subject of considerable detailed analysis and much debate since it was first described as an isoclinal syncline (Keith, 1905). Some of these have arisen from mesoscopic structural analysis which suggested great strike-slip movement on the zone (Reed and Bryant, 1964), and for a time the Brevard was likened to an eastern San Andreas fault. The zone has been tentatively connected with the Martic line of Pennsylvania by a number of students, and it has been suggested both as a root zone for the southern part of the Blue Ridge allochthon and a suture line in plate theory separating rocks of the North American plate to the west from rocks of the African plate to the southeast. (Rankin, 1975). The role of this feature in the larger tectonic sense is still uncertain, but several de-

FIGURE 22-15 Tectonic map of northwest South Carolina showing the relationship of the Inner Piedmont to the Brevard zone and the nappe structure within the Inner Piedmont. (*From Griffin, 1974.*)

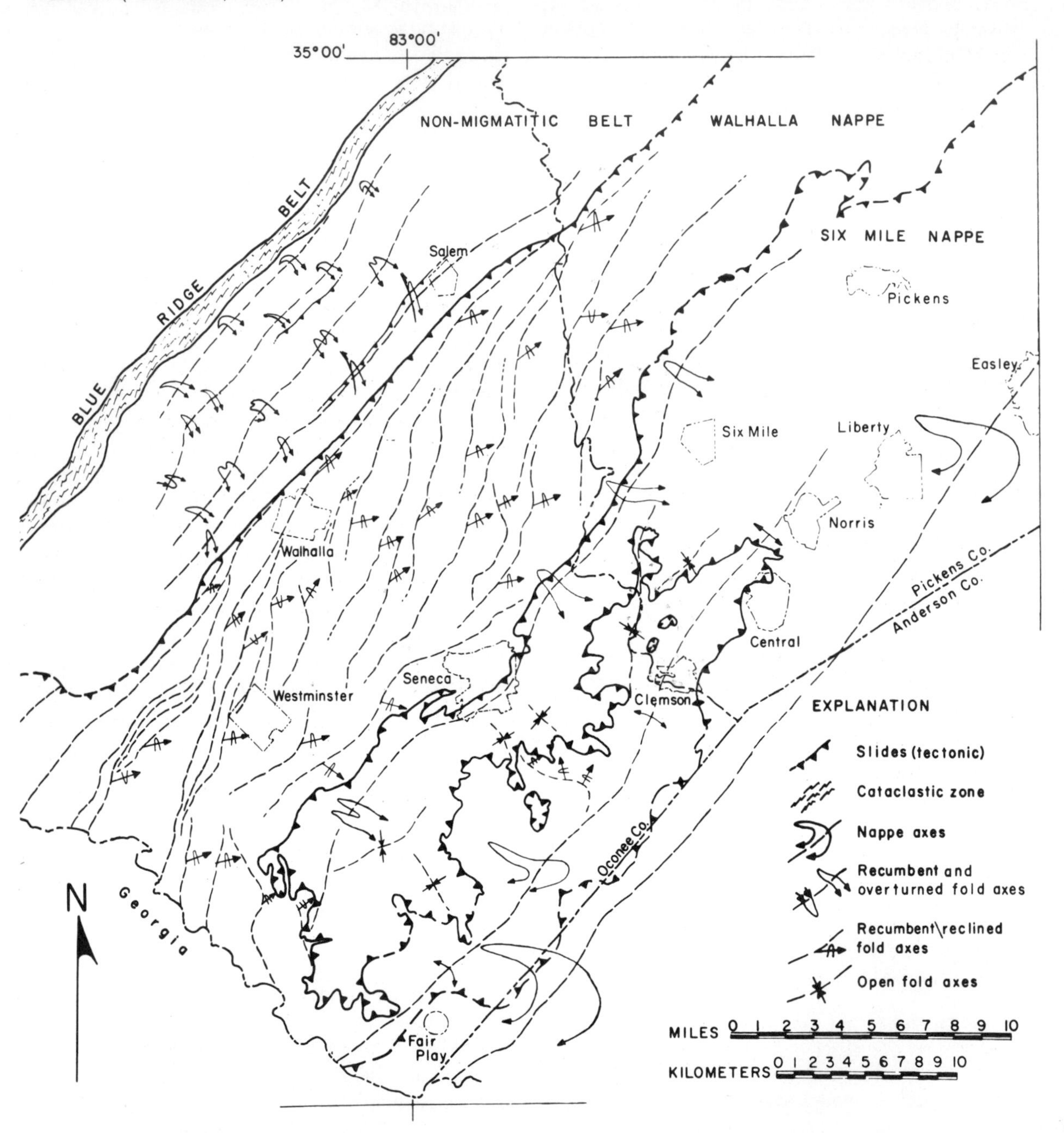

tailed studies now give a clearer picture of its structure (Roper and Dunn, 1973; Roper and Justus, 1973; Stirewalt and Dunn, 1973). Roper characterizes the rocks in the Brevard by (1) intense shearing except for lenses of less deformed rock, (2) representation of at least two phases of folding, (3) a strong horizontal lineation parallel to the fold axis is present, (4) one or both sides of the zone are fault contacts, (5) foliation dipping 20 to 60° southeast, and (6) polymetamorphic textures are present in the rocks, reaching garnet grade followed by retrograde metamorphism to greenschist facies.

Studies of the microscopic and mesoscopic features in the Brevard strongly suggest that all features of the zone can be interpreted as resulting from compressive straining perpendicular to the length of the zone (Stirewalt and Dunn, 1973). They recognize two phases of folding—an early isoclinal phase producing folds overturned to the northwest and a later coaxial open folding which is best developed at the northern end of the zone at the Sauratown Mountain anticlinorium, Figs. 22-17 and 22-18. Both of these phases are exhibited in the fabric of the rocks along and across the zone, suggesting that the Brevard zone does not mark a major structural discontinuity such as a suture in plate theory. It is also notable in this regard that the ultrabasic rocks are not present in the zone. A polytectonic evaluation for the Brevard is illustrated, Fig. 22-19, as suggested by Roper and Justus (1973).

FIGURE 22-16 A geologic cross section across the Inner Piedmont of South Carolina. (*From Griffin, 1974.*)

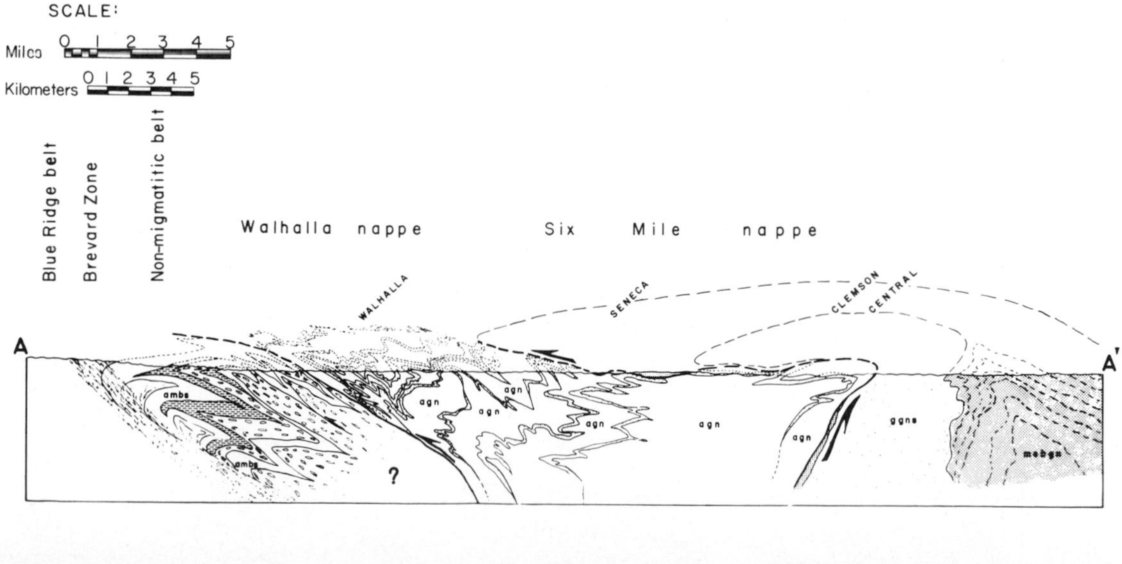

Evolution of the Southern Appalachian Crystalline Complexes

As is often the case, it is easier to recognize and date the younger events in the southern Appalachians than it is the older ones. Following this rational we can say that orogenic activity had ceased by Cretaceous time because flat-lying undeformed Cretaceous strata overlie both the foreland and the Piedmont. Some dikes of probable Jurassic age occur in the Piedmont but no Jurassic sedimentary rocks crop out. Triassic fault-bounded, graben-like, basins and high-angle faults cut the rocks within the Piedmont and across the northern end of the Blue Ridge where earlier folds and thrusts are abruptly cut.

Alleghanian folds and thrusts are extensively developed in the foreland region where *décollement* tectonics prevailed. Most of the foreland folds and faults are of this age, and it is apparent that movements took place at this time on the thrusts along the western flank of the Great Smoky region of the Blue Ridge and this is probably the age of most folding and thrusting west of the Blue Ridge massif, Fig. 22-20. Movements on the Brevard at this time appear to take place under relatively brittle conditions. The Piedmont was already a crystalline complex, but some radiometric dates probably as young as 320 to 400 million years

FIGURE 22-17 Suggested relations between megascopic polyphase folding on the Sauratown Mountains anticlinorium and adjacent James River synclinorium. Positions of phase I closures are schematic, and shearing as phase I folds overturned northwestward is likely. Megascopic phase II folding is indicated by warping of phase I axial plane schistosity (S_2) as shown. (*From Stirewalt and Dunn, 1973.*)

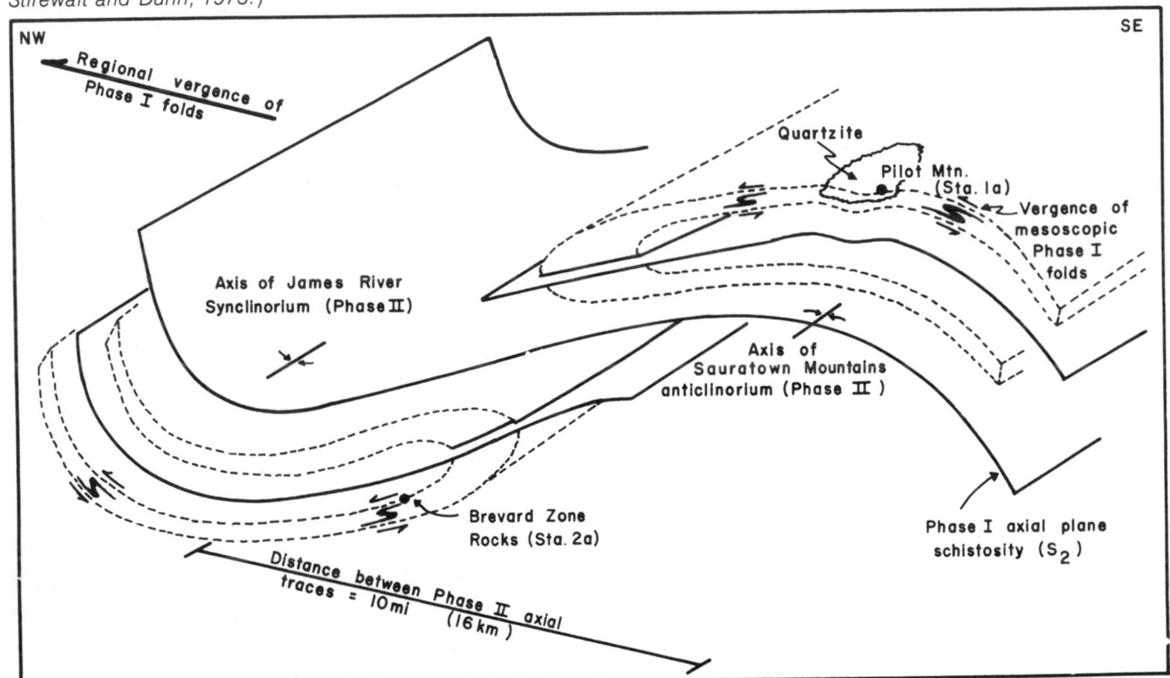

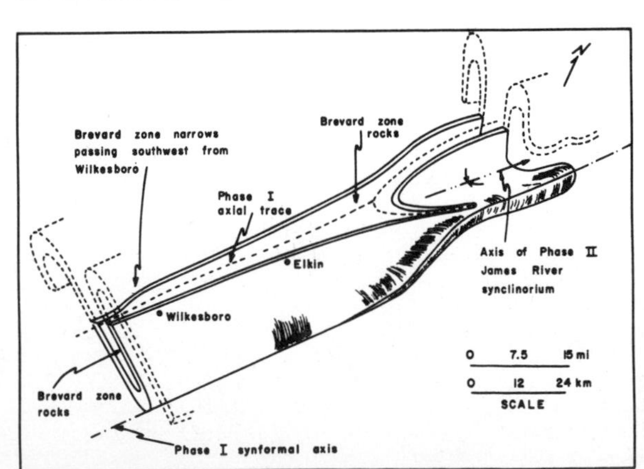

FIGURE 22-18 Interpretation of the structural geometry of the Brevard zone in North Carolina. The diverging outcrop pattern of the northeastern part of the zone is caused by erosion of superposed phase I and phase II folds as shown. Heavily mylonitized rocks along the Brevard zone largely occur southwest of Wilkesboro. Thrust-type displacements readily fit into the interpretation. (*From Stirewalt and Dunn, 1973.*)

are obtained from rocks in the James River synclinorium, the Maryland Piedmont, and other parts of the Piedmont, indicating thermal activity in the Devonian and Carboniferous.

Most of the metamorphic complex in the Piedmont is currently thought to be part of either the Precambrian basement (possibly the same as that in the Blue Ridge) or its late Precambrian and early Paleozoic cover. The older basement rocks occur in the anticlinal structures—the Inner Piedmont, Charlotte belt, Sauratown Mountains, Smith River allochthon, and the Baltimore domes. One of the most difficult problems now is establishment of a stratigraphy in the cover rocks, all of which appear to be older than late Ordovician, and most of which may be earliest Cambrian or late Precambrian as shown by the age of plutons which intrude some of the apparently young metasediments (e.g., Quantico slate is cut by a pluton 550 million years old). Although the country rock is older, most of the plutons in the Piedmont are of mid- to late Paleozoic age.

The early Paleozoic Taconic orogeny was probably the time of formation of many of the metamorphic and structural features we see in the Piedmont and Blue Ridge. This is the most likely age for the isoclinal folding, mobilization of basement rocks, and regional metamorphism and it was accompanied by intrusive activity. This would have been the time of uplift and nappe formation in the Inner Piedmont, probably the time of thrusting on the Fries fault in the Blue Ridge.

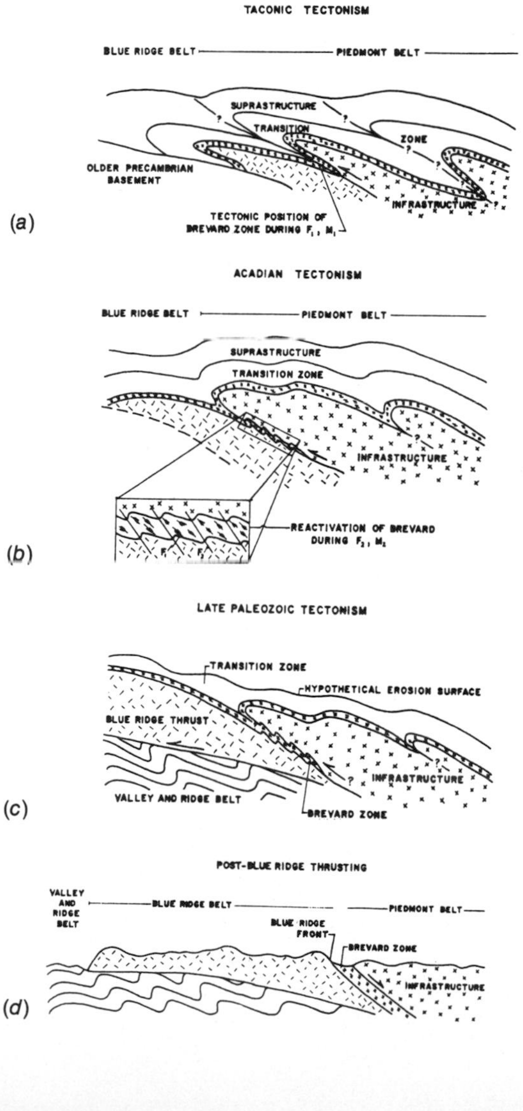

FIGURE 22-19 Evolution of the Brevard zone as interpreted by Roper and Justus. (*a*) Schematic diagram illustrating environment of first folding and metamorphism tectonism of Brevard lithologies within the detachment zone during the Taconic orogeny. This orogenesis may be the product of an island-arc–island-arc collision resulting in the eastern part of the Blue Ridge, Piedmont, and Slate belt island arc (not included in diagram) being accreted onto the North American continent. Plutonism not illustrated in diagram. (*b*) Illustrates reactivation of Brevard detachment zone producing F_2 folding and M_2 metamorphism during Devonian tectonism. (*c*) The Blue Ridge and western Inner Piedmont being thrusted over the Valley and Ridge province together as a continuous sheet during the late Paleozoic. Decoupling of the Blue Ridge from the Inner Piedmont occurred along parts of the Brevard zone during this tectonism, commencing the first stage of development of the Brevard fault zone. (*d*) The last major tectonic movements along the Brevard zone are postulated as normal faulting near the eastern edge of the underlying Valley and Ridge province where late Paleozoic overthrusting originated. This faulting may have begun in the late Permian or early Triassic and locally may enhance the relief of the Blue Ridge front. The fault zone is superimposed upon the fold zone in this illustration. (*From Roper and Justus, 1973.*)

A considerable effort is being made to interpret southern Appalachian history in terms of plate theory as is done by Hatcher (1972) and Rankin (1975). The crust from the Blue Ridge west is clearly part of the North American plate, and the rocks of the Piedmont exhibit characteristics associated with island arcs or eugeosynclinal belts, but one of the important questions that still remains is whether the eastern Piedmont is a part of the North African plate welded to the North American plate following an early Paleozoic plate collision as suggested by Rankin (1975), or was the entire Piedmont a part of the eastern North American continental margin in early Paleozoic time?

FIGURE 22-20 Section showing the present structural configuration across a portion of the southern Appalachians, as interpreted by Hatcher. TFN—Tallulah Falls nappe. BZ—Brevard zone. TF—Towaliga fault. P-Ꞓb—earlier Precambrian basement rocks. P-Ꞓbms—earlier to late Precambrian basement and metasedimentary and metavolcanic rocks. P-Ꞓo—Ocoee series. P-Ꞓmsv—late Precambrian metasedimentary and metavolcanic rocks. P-Ꞓ-Ꞓms and P-Ꞓ and Ꞓmsv—late Precambrian and Cambrian metasedimentary and metavolcanic rocks. PZi—Paleozoic intrusive rocks. Ꞓch—Chilhowee group. Ꞓs—Shady dolomite. Ꞓr—Rome formation. OꞒk-Ꞓc—Knox and Conasauga groups. M.O.—middle and Ordovician Chickamauga group rocks. KT—Cretaceous and Tertiary sediments. (*From Hatcher, 1975.*)

THE OUACHITA MOUNTAINS—A WESTWARD CONTINUATION OF THE APPALACHIANS*

The Ouachita Mountains in Arkansas and Oklahoma and the Marathon Mountains of

*King, *in* Nairn and Stehli (1975), gives an excellent account of the geology of this belt.

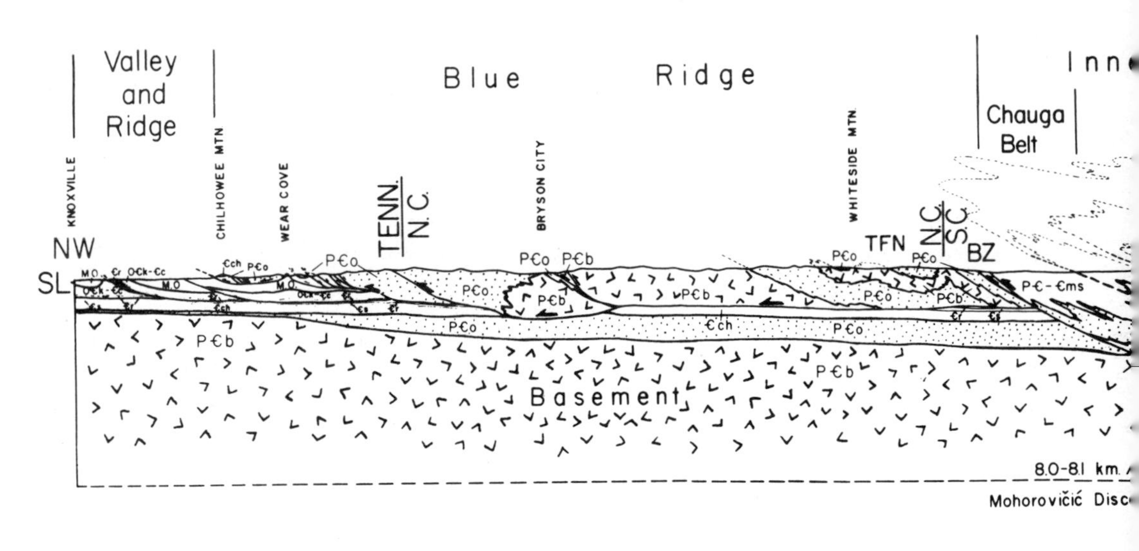

western Texas are almost certainly a westward continuation of the Appalachian orogen, Fig. 22-21. The rocks involved in all three systems are Paleozoic in age, the direction of thrusting and tectonic transport is toward the continental interior in all three belts, the style of deformation is similar, and the timing of orogenic deformation is similar though not identical, Fig. 22-22. The intervening areas are covered by Coastal Plain sediments which obscure the connection, and the southern side of the Ouachitas and Marathons, but rocks like those in the Appalachian foreland have been penetrated in drilling along a northwest-trending belt in Mississippi and Alabama. Metamorphic rocks are penetrated by wells located south of this belt. The exposed portions of the Ouachita and Marathon Mountains are very much alike but consist of a thinner lower Paleozoic section than that in the Valley and Ridge, and it has little limestone in it; the Carboniferous section is a very thick (10- to 12-km maximum) mass of flysch and turbidite deposits which are now telescoped by imbricated thrusts, the age of which is established by the involvement of Pennsylvanian units.

The Marathon trends terminate against thrusts of the Cordilleran orogen which trend northwest in eastern Mexico, and continuation of the Paleozoic orogen farther south is uncertain, although this has been postulated by Lopez-Ramos (1969) and de Cserna (1960).

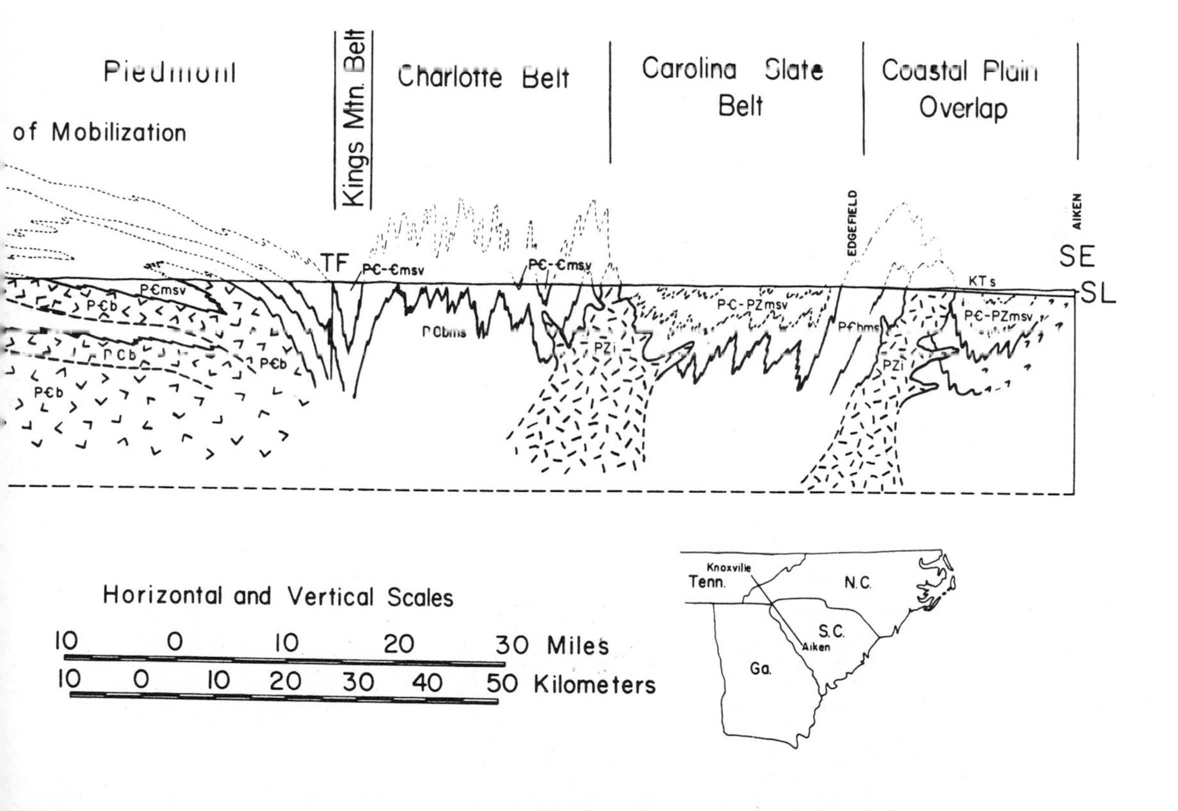

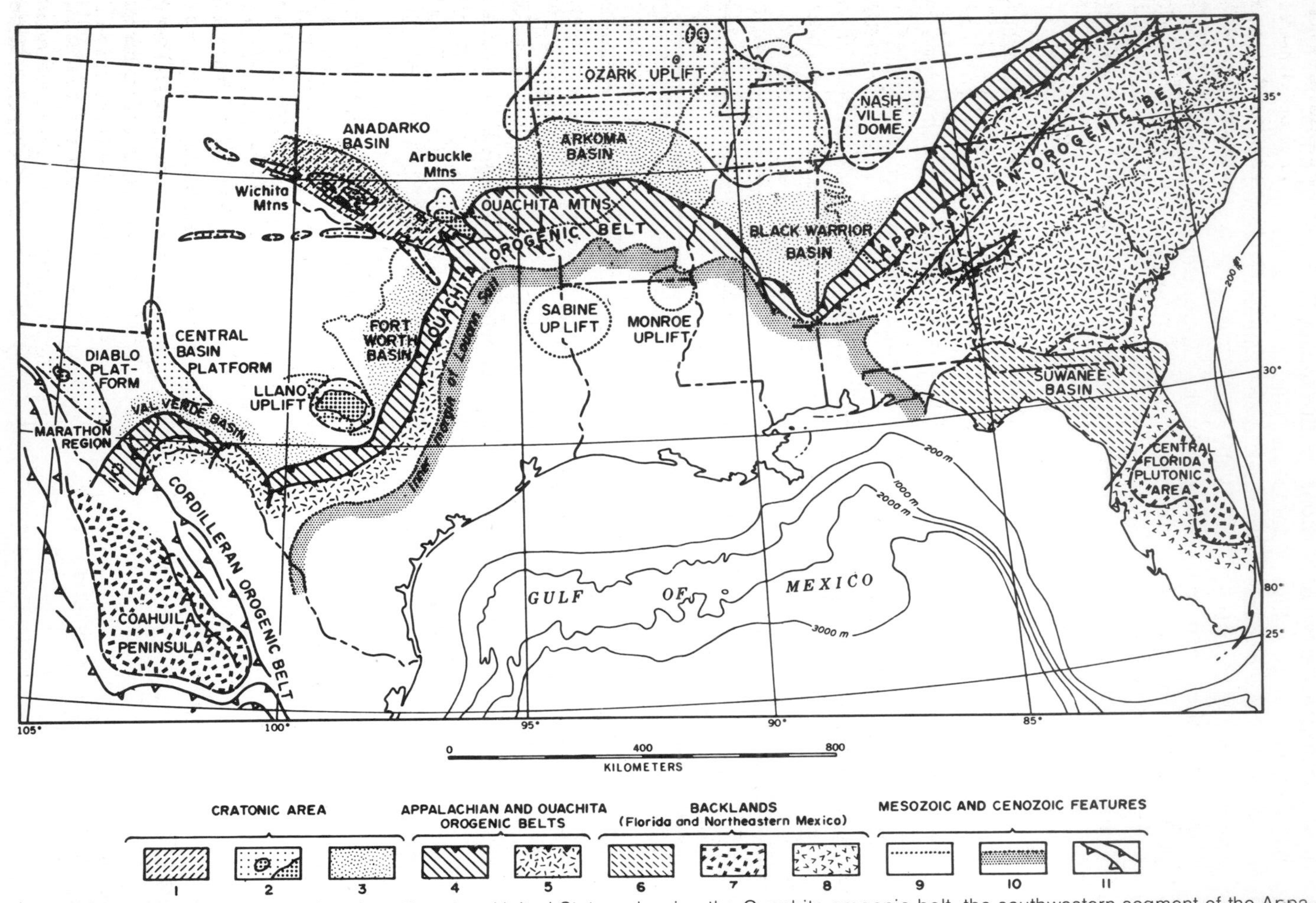

FIGURE 22-21 Map of south-central and southeastern United States, showing the Ouachita orogenic belt, the southwestern segment of the Appalachian orogenic belt, and adjoining tectonic units to the north, south, and west. Explanation of patterns: Cratonic area: (1) Late Proterozoic-early Cambrian aulocogene, in Wichita Mountains belt. (2) Positive areas, with small outcrops of Precambrian basement in darker shading. (3) Basins or foredeeps bordering the orogenic belts. Appalachian and Ouachita orogenic belts: (4) Deformed sedimentary rocks. (5) Metamorphic and plutonic rocks. Backlands in Florida and northeastern Mexico: (6) Flat-lying Paleozoic strata. (7) Plutonic rocks. (8) Volcanic rocks, part or all of Mesozoic age. Mesozoic and Cenozoic features: (9) Inner margin of Cretaceous and Tertiary Coastal Plain deposits. (10) Inner margin of Louann salt, of early Jurassic age. (11) Frontal structures of Cordilleran orogenic belt. (*From King, 1975.*)

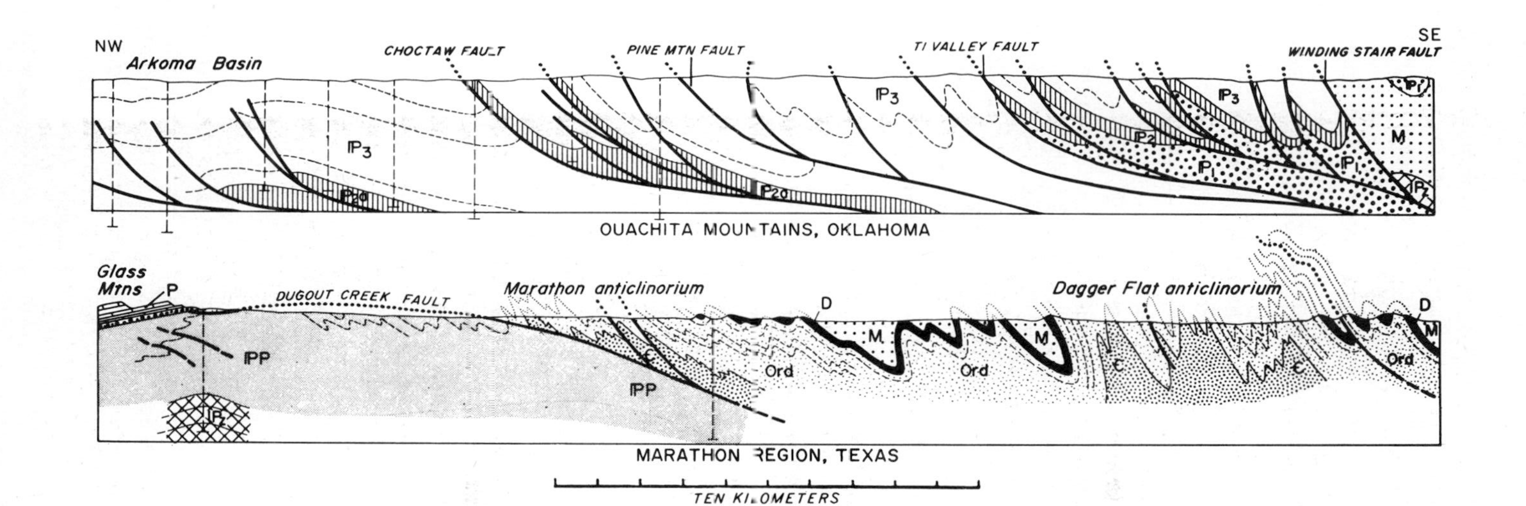

FIGURE 22-22 Structure sections comparing the frontal thrusts of the Ouachita Mountains and Marathon segments of the Ouachita orogenic belt. Above: Frontal thrusts between Ouachita Mountains and Arkoma basin, eastern Oklahoma. Based on Berry and Trumbly (1968, section AA′). Below: Dugout Creek thrust of Marathon region. Based on King (1937, section EE′, pl. 21). Well control in both sections is indicated by vertical dashed lines. Letter symbols as follows: Above: lPz, lower Paleozoic of interior Ouachitas; M, Stanley shale; IP1, Jackfork sandstone; IP2, Johns Valley shale; IP2, equivalent strata of frontal Ouachitas, including Wapanucka limestone; IP3, Atoka formation. Below: Є, Upper Cambrian; Ord, Ordovician; D, Caballos novaculite; 1Pz, lower Paleozoic of cratonic sequence: M, Tesnus formation; IPP, upper Pennsylvanian and lower Wolfcampian foredeep deposits; P, upper Wolfcampian (Lenox Hills formation). (*From King, 1975.*)

REFERENCES

Abdel-Khalek, M. Lofti, and **Khoury, S. G.,** 1971, Structural and metamorphic evolution of the Otter River area, west-central Piedmont, Virginia: Geol. Soc. America Bull., v. 82, no. 3, p. 707ff.

Amenta, R. V., 1974, Multiple deformation and metamorphism from structural analysis in the eastern Pennsylvania Piedmont: Geol. Soc. America Bull., v. 85, no. 10, p. 1647.

Anderson, J. G. C., and **Owen, T. R.,** 1968, The structure of the British Isles: Pergamon, Oxford.

Bearce, D. N., 1969, Geology of the southwestern Bald Mountains in the Blue Ridge province of Tennessee: Southeastern Geology, v. 11, no. 1, p. 21–36.

Billings, M. P., 1945, Mechanics of igneous intrusion in New Hampshire: Am. Jour. Sci., v. 243A, Daly vol., p. 41–68.

Bird, J. M., and **Dewey, J. F.,** 1970, Lithosphere plate–continental margin tectonics and the evolution of the Appalachian orogen: Geol. Soc. America Bull., v. 81, no. 4, p. 1031–1060.

Bowes, D. R., 1969, An orogenic interpretation of the Lewisan of Scotland: Int. Geol. Cong., 23rd, Czech. Rep., Sect. 4, Proc., p. 225–236.

Brown, G. C., and **Hughes, D. J.,** 1973, Great Glen fault and timing of granite intrusion on the proto-Atlantic continental margin: Nature, Phys. Sci., v. 244, no. 139, p. 129–132.

Bryant, Bruce, and **Reed, J. C., Jr.,** 1970, The significance of lineation and minor folds near major thrust faults in the southern Appalachians, and the British and Norwegian Caledonides: Geol. Mag., v. 106, no. 5, p. 412–429.

Bucher, W. H., 1957, Taconic klippe—A stratigraphic-structural problem: Geol. Soc. America Bull., v. 68, no. 6, p. 657–674.

Bullard, E. C., Everett, J. E., and **Smith, A. G.,** 1965, The fit of continents around the Atlantic: Phil. Trans. Roy. Soc. London Ser. A., v. 258, p. 41.

Butler, J. R., 1972, Age of Paleozoic regional metamorphism in the Carolinas, Georgia, and Tennessee southern Appalachians: Am. Jour. Sci., v. 272, no. 4, p. 319–333.

——— 1973, Paleozoic deformation and metamorphism in part of the Blue Ridge thrust sheet, North Carolina: Am. Jour. Sci., Cooper vol., p. 72–88.

——— and **Dunn, D. E.,** 1968, Geology of the Sauratown Mountains anticlinorium and vicinity, *in* Guidebook for field excursions, Geol. Soc. America, Southeastern Sec., Durham, N. C., April 1968, Spec. Pub. 1.

Cady, Wallace, 1960, Stratigraphy and geotectonic relationships in northern Vermont and southern Quebec: Geol. Soc. America Bull., v. 71, p. 531–576.

Carpenter, R. H., 1970, Metamorphic history of the Blue Ridge province of Tennessee and North Carolina: Geol. Soc. America Bull., v. 81, no. 3, p. 749–762.

Cebull, S. E., and others, 1976, Possible role of transform faults in the development of apparent offsets in the Ouachita–southern Appalachian tectonic belt: Jour. Geology, v. 84, p. 107–114.

Cloos, Ernst, 1947, Oolite deformation in the South Mountain fold, Maryland: Geol. Soc. America Bull., v. 58, p. 843–918.

——— and **Hopson, C. A.,** 1964, *in* Cloos, Ernst, ed., The geology of Howard and Montgomery counties: Maryland Geol. Survey.

Conley, J. F., and **Henika, W. S.,** 1973, Field trip across the Blue Ridge anticlinorium, Smith River allochthon, and Sauratown Mountains anticlinorium near Martinsville, Virginia, *in* Virginia Minerals, Division of Mineral Resources, v. 19, no. 4.

Cooper, B. N., 1964, Relation of stratigraphy to structure in the southern Appalachians, *in* Lowry, W. D., ed., Tectonics of the southern Appalachians, V. P. I. Dept. Geol. Sci. Mem. 1.

Craddock, J. C., 1957, Stratigraphy and structure of the Kinderhook Quadrangle, New York and the "Taconic klippe": Geol. Soc. America Bull., v. 68, no. 6, p. 675–724.

Crowley, W. P., 1976, The geology of the crystalline rocks near Baltimore and its bearing on the evolution of the eastern Maryland Piedmont: Rept. of Inv. 27, Maryland Geol. Surv.

Dallmeyer, R. D., 1974, Tectonic setting of the northeastern Reading prong: Geol. Soc. America Bull., v. 85, no. 1, p. 131–134.

De Cserna, Zoltan, 1960, Orogenesis in time and space in Mexico: Geol. Rundschau, Band 50.

Dewey, J. F., and **Kay, G. M.,** 1968, Appalachian and Caledonian evidence for drift in the North Atlantic, *in* Phinney, R. A., ed., The history of the earth's crust: a symposium: Goddard Inst. Space Stud., Conf. (1966), Contr.

Dietrich, R. V., 1964, Igneous activity in the southern

Appalachians, *in* Lowry, W. D., ed., Tectonics of the southern Appalachians: V. P. I. Dept. Geol. Sci. Mem. 1.

Drake, C. L., and **Woodward, H. P.**, 1963, Appalachian curvature, wrench faulting, and offshore structures: New York Acad. Sci. Trans., ser. II, v. 26, p. 48–63.

Eardley, A. J., 1962, Structural geology of North America: New York, Harper & Row.

Eskola, P., 1949, The problem of mantled gneiss domes: Quart. Jour. Geol. Soc. London, v. 104.

Faill, R. T., 1973, Tectonic development of the Triassic Newark-Gettysburg basin in Pennsylvania: Geol. Soc. America Bull., v. 84, no. 3, p. 725–740.

Faul, Henry, and others, 1963, Ages of intrusions and metamorphism in the northern Appalachians: Am. Jour. Sci., v. 261.

Fisher, G. W., 1976, The geologic evolution of the northeastern Piedmont of the Appalachians: Geol. Soc. America, Abs. with programs, v. 8, no. 2.

Flawn, P. T., and others, 1961, The Ouachita system: Texas Univ. (Bur. Econ. Geol.) Publ. 6120.

Freedman, J., **Wise, D. U.**, and **Bentley, R. D.**, 1964, Pattern of folded folds in the Appalachian Piedmont along Susquehanna River: Geol. Soc. America Bull., v. 75, p. 621–638.

Frey, M. G., 1973, Influence of Salina salt on structure in New York, Pennsylvania part of Appalachian Plateau: Am. Assoc., Petroleum Geologists Bull., v. 57/6.

Fullagar, P. D., 1971, Age and origin of the plutonic intrusions in the southern Appalachians: Geol. Soc. America Bull., v. 82, no. 11, p. 2845–2862.

Fyson, W. K., 1964, Folds in the Carboniferous rocks near Walton, Nova Scotia: Am. Jour. Sci., v. 262, p. 513–522.

Glover, Lynn III, and **Sinha, Akhaury Krishna**, 1973, The Virginia deformation, a late Precambrian to Early Cambrian orogenic event in the Central Piedmont of Virginia and North Carolina: Am. Jour. Sci., Cooper vol., p. 234–251.

———, ———, and **Higgins, M. W.**, 1971, Virginia phase [Precambrian and early Cambrian(?)] of the Avalonian orogeny in the central piedmont of Virginia (abs.): Geol. Soc. America, Abs., v. 3, no. 7.

Graham, S. A., **Dickinson, W. R.**, and **Ingersoll, R. V.**, 1975, Himalayan-Bengal model for flysch dispersal in the Appalachian-Ouachita systems: Geol. Soc. America Bull., v. 86, p. 273–286.

Griffin, V. S., 1969, Inner Piedmont tectonics in the vicinity of Walhalla, South Carolina: Geologic Notes, v. XIV, no. 1, p. 15–28.

Griffin, V. S., Jr., 1971a, Fabric relationships across the Catoctin Mtn.–Blue Ridge Anticlinorium in Central Virginia: Geol. Soc. America Bull., v. 82, no. 2, p. 417ff.

——— 1971b, The Inner Piedmont belt of the southern crystalline Appalachians: Geol. Soc. America Bull., v. 82, no. 7, p. 1885ff.

——— 1974, Analysis of the Piedmont in northwest South Carolina: Geol. Soc. America Bull., v. 85, no. 7, p. 1123–1138.

Gwinn, V. E., 1964, Thin-skinned tectonics in the plateau and northwestern valley and ridge provinces of the central Appalachians: Geol. Soc. America Bull., v. 75, p. 863–900.

——— 1967, Lateral shortening of layered rock sequences in the foothills regions of major mountain systems: Mineral Industries, v. 36.

Hadley, J. B., 1964, Correlation of isotopic ages, crustal heating and sedimentation in the Appalachian Region, *in* Lowry, W. D., ed., Tectonics of the southern Appalachians: V. P.I. Dept. Geol. Sci. Mem. 1.

——— and **Goldsmith, R.**, 1963, Geology of the eastern Great Smoky Mountains, Tennessee and North Carolina: U.S. Geol. Survey Prof. Paper 349–B.

Hall, J., 1859, Geological survey of New York: Paleontology, v. iii, introd.

Haller, John, 1971, Geology of the east Greenland Caledonides: New York, Wiley.

Harris, L. D., and **Zieta, Isidore**, 1962, Development of Cumberland overthrust block in vicinity of Chestnut Ridge fenster in southwest Virginia: Am. Assoc. Petroleum Geologists Bull., v. 46, p. 2148–2160.

Hatcher, R. D., Jr., 1972, Developmental model of the southern Appalachians: Geol. Soc. America Bull., v. 83, no. 9, p. 2735ff.

Hess, H. H., 1946, Appalachian peridotite belt: Its significance in sequence of events in mountain building (abs.): Geol. Soc. America Bull., v. 51.

Higgins, M. W., 1973, Superimposition of folding in the NE Maryland Piedmont and its bearing on the history and tectonics of the central Appalachians: Am. Jour. Sci., Cooper vol., p. 150–195.

Hopson, C. A., 1964, The crystalline rocks of Howard and Montgomery Counties, *in* Cloos, Ernst, ed., Geology of Howard and Montgomery Counties: Maryland Geol. Survey.

Jacobeen, Frank, Jr., and **Kanes, Wm. H.**, 1974,

Structure of broad-top synclinorium and its implications for Appalachian structural style: Am. Assoc. Petroleum Geologists Bull., v. 58, no. 3, p. 362–375.
——— 1975, Structure of Broadtop synclinorium, Wills Mountain anticlinorium, and Allegheny frontal zone: Am. Assoc. Petroleum Geologists Bull., v. 59, no. 7, p. 1136–1150.

James, D. E., Smith, T. J., and **Steinhart, J. S.,** 1968, Crustal structure of the Middle-Atlantic states: Jour. Geophys. Res., v. 73, no. 6, p. 1983–2007.

Kay, Marshall, 1951, North American geosynclines: Geol. Soc. America Mem. 48.
——— 1967, Stratigraphy and structure of northeastern Newfoundland bearing on drift in North Atlantic: Am. Assoc. Petroleum Geologists Bull., v. 51, no. 4, p. 579–600.
———, 1969, North Atlantic; geology and continental drift: Amer. Assoc. Petrol. Geol., Mem. no. 12.

Keith, Arthur, 1905, Description of the Mount Mitchell quadrangle (N.C.–Tenn.): U.S.G.S. Geol. Atlas Mount Mitchell folio, no. 124.
——— 1923, Outlines of Appalachian structure: Geol. Soc. America Bull., v. 23.

Kennedy, M. J., 1975, Repetitive orogeny in the northeastern Appalachians—New plate models based upon Newfoundland examples: Tectonophysics, v. 28, p. 39–87.

King, P. B., 1951, The tectonics of middle North America: Princeton, N. J., Princeton Univ.
——— 1959, The evolution of North America: Princeton, N. J., Princeton Univ.
——— 1964, Geology of the central Great Smoky Mountains, Tennessee: U.S. Geol. Survey Prof. Paper 349-C.
——— 1975, The Ouachita and Appalachian orogenic belts, *in* Nairn, A. E., and Stehli, F. G., eds., The ocean basins and margins, v. 3, The Gulf of Mexico and the Caribbean: New York, Plenum Press.

Lopez-Ramos, Ernesto, 1969, Marine Paleozoic rocks of Mexico: Am. Assoc. Petr. Geol. Bull., v. 53, no. 12.

Lowry, W. D., 1974, North American geosynclines—Test of continental-drift theory: Am. Assoc. Petroleum Geologists Bull., v. 58, no. 4, p. 575–620.

Milici, R. C., 1975, Structural patterns in the southern Appalachians—Evidence for a gravity slide mechanism for Alleghanian deformation: Geol. Soc. America Bull., v. 86, p. 1316–1320.

Miller, R. L., 1973, Where & why of Pine Mtn. and other major fault planes, Va., Ky., and Tenn.: Am. Jour. Sci., Cooper vol., p 353–371.
——— and **Fuller, J. O.,** 1954, Geology and oil resources of the Rose Hill district, Lee County Virginia: Virginia Div. Mineral Research Bull., v. 71.

Morgan, B. A., 1972, Metamorphic map of the Appalachians, scale 1:2,500,000: U.S. Geol. Survey Misc. Geol. Invest., Map. I-724.

Mohr, D. W., 1973, Stratigraphy and structure of part of the Great Smoky and Murphy belt groups, western N. C.: Am. Jour. Sci., Cooper vol., p. 41–71.

Nicholas, R. L. and **Rozendal, R. A.,** 1975, Subsurface positive elements with Ouachita foldbelt in Texas and their relation to Paleozoic cratonic margin: Am. Assoc. Petroleum Geologists Bull., v. 59, no. 2, p. 193–216.

Odom, A. L., and **Fullagar, P. D.,** 1973, Geochronologic and tectonic relationships between the inner Piedmont, Brevard zone, and Blue Ridge belts, N.C.: Am. Jour. Sci., Cooper vol., p. 133–149.

Osberg, P. H., 1965, Structural geology of the Knowlton-Richmond area, Quebec: Geol. Soc. America Bull., v. 76, p. 223–250.

Pettijohn, F. J., 1957, Sedimentary rocks, 2d ed.: Harper & Row, New York.

Poole, W. H., 1969, Tectonic evolution of Appalachian region of Canada, *in* Collected papers on geology of the Atlantic region, Hugh Lilly mem. vol.: Geol. Assn. Can. Spec. Paper no. 4, p. 9–51.

Rankin, D. W., 1975, The continental margin of eastern North America in the southern Appalachians: The opening and closing of the proto-Atlantic Ocean: Am. Jour. Sci., v. 275-A, p. 298–336.
———, **Espenshade, G. H.,** and **Shaw, K. W.,** 1973, Stratigraphy and structure of the metamorphic belt in NW N.C. and SW Virginia: A study from the Blue Ridge across the Brevard zone to the Sauratown Mtns. anticlinorium: Am. Jour. Sci., Cooper vol., p. 1–40.

Reed, J. C., Jr., 1955, Catoctin formation near Luray, Virginia: Geol. Soc. America Bull., v. 66, p. 871–896.
——— and **Bryant, Bruce,** 1964, Evidence for strike-slip faulting along the Brevard zone in North Carolina: Geol. Soc. America Bull., v. 75, p. 1177–1196.

Rodgers, John, 1953a, Geologic map of east Tennessee with explanatory text: Tennessee Div. Geology Bull., v. 58, pt. 2.
——— 1953b, The folds and faults of the Appalachian Valley and Ridge province: Kentucky Geol. Survey Spec. Pub. 1, p. 150–166.

——— 1963, Mechanics of Appalachian foreland folding in Pennsylvania and West Virginia: Am. Assoc. Petroleum Geologists Bull., v. 47, p. 1527–1536.

——— 1964, Basement and no-basement hypotheses in the Jura and the Appalachian Valley and Ridge, *in* Lowry, W. D., ed., Tectonics of the south Appalachians: V.P.I. Dept. Geol. Sci. Mem. 1.

——— 1970, The tectonics of the Appalachians: New York, Wiley-Interscience.

——— 1971, The Taconic orogeny: Geol. Soc. America Bull., v. 82, no. 5, p. 1141ff.

Rogers, G. S., and **Rogers, W. B.**, 1843, On the physical structure of the Appalachian chain, as exemplifying the laws which have regulated the elevation of great mountain chains generally: Am. Jour. Sci., v. 44.

Root, Samuel I., 1970, Structure of the northern terminus of the Blue Ridge in Pennsylvania: Geol. Soc. America Bull., v. 81, no. 3, p. 815–830.

Roper, P. J., and **Dunn, D. E.**, 1973, Superposed deformation and polymetamorphism, Brevard zone, South Carolina: Geol. Soc. America Bull., v. 84, no. 10, p. 3373ff.

——— and **Justus, P. S.**, 1973, Polytectonic evolution of the Brevard zone: Am. Jour. Sci., Copper vol., p. 105–132.

St. Julien, Pierre, and **Hubert, Claude**, 1975, Evolution of the Taconian orogen in the Quebec Appalachians: Am. Jour. Sci., v. 275-A, p. 337–362.

Schenik, P. E., 1970, Regional variation of the flysch-like Meguma group (Lower Paleozoic) of Nova Scotia, compared to recent sedimentation off the Scotian shelf: Geol. Assoc. Canada Spec. Paper no. 7, p. 127–153.

Schwab, F. L., 1974, Ancient geosynclinal sedimentation, paleogeography, and provinciality: A plate tectonics perspective for British Caledonides and Newfoundland Appalachians, *in* Ross, Charles A., ed., Paleogeographic provinces and provinciality: Soc. Economic Paleontologists and Mineralogists Spec. Pub. No. 21.

Spencer, E. W., 1968, Geology of the Natural Bridge, Sugarloaf Mountain, Buchanan, and Arnold Valley quadrangles, Virginia: Rept. of Inv. 13, Va. Div. Mineral Resources.

——— 1972, Structure of the Blue Ridge front in central Virginia, *in* Lessing, P., and others, eds., Appalachian structures, origin, evolution, and possible potential for new exploration frontiers: W. Va. Geol. and Eco. Survey.

Stille, Hans, 1941, Einführung in den Bau Amerikas: Berlin, Borntraeger.

Stirewalt, G. L., and **Dunn, D. E.**, 1973, Mesoscopic fabric and structural history of Brevard zone and adjacent rocks, N.C.: Geol. Soc. America Bull., v. 84, no. 5, p. 1629–1650.

Stose, G. W., and **Stose, A. J.**, 1946, Ocoee series of the southern Appalachians (United States) (abs.): Geol. Soc. America Bull., v. 57, p. 1233.

Strand, T., and **Kulling, O.**, 1972, Scandinavian Caledonides: New York, Wiley-Interscience.

Thomas, W. A., 1973, Southwestern Appalachian structural system beneath the Gulf Coastal Plain: Am. Jour. Sci., v. 273, Byron N. Cooper Memorial Number, p. 372–390.

Thompson, J. B. Jr., **Robinson, Peter**, **Clifford, T. N.**, and **Trask, N. J., Jr.**, 1968, Nappes and gneiss domes in west-central New England, *in* Zen, E-An, and others, eds., Studies of Appalachian geology: New York, Wiley-Interscience.

Tobisch, Othmar T., and **Glover, Lynn III**, 1971, Nappe formation in part of the southern Appalachian Piedmont: Geol. Soc. America Bull., v. 82, no. 8, p. 2209ff.

Tomlinson, C. W., 1959, Ouachita problems, *in* Cline, L. M., and others, eds., The geology of the Ouachita Mountains; a symposium: Dallas Geol. Soc. and Ardmore Geol. Soc., p. 1–19.

Watkins, J. S., 1964, Regional geologic implications of the gravity and magnetic fields of a part of eastern Tennessee and southern Kentucky: U. S. Geol. Survey Prof. Paper 516-A.

Wickham, J. S., 1972, Structural history of a portion of the Blue Ridge, northern Virginia: Geol. Soc. America Bull., v. 83, no. 3, p. 723ff.

Williams, Harold, 1964, The Appalachians in northeastern Newfoundland—A two-sided symmetrical system: Am. Jour. Sci., v. 262, p. 1137–1158.

Wilson, C. W., Jr., 1958, Structure of the Cumberland plateau: Geol. Soc. America Bull., v. 69.

Wilson, J. T., 1966, Did the Atlantic close and then reopen?: Nature, v. 211, no. 5050, p. 676–681.

Zen, E-An, 1961, Stratigraphy and structure at the north end of the Taconic Range in west central Vermont: Geol. Soc. America Bull., v. 72, p. 293–338.

——— 1967, Time and space relationships of the Taconic allochthon and autochthon: Geol. Soc. America Spec. Paper 97, p. 107.

23

The Alpine-Himalayan Orogen

THE ALPINE-HIMALAYAN OROGEN

The Alpine-Himalayan orogen includes the complex of deformed belts which extends from Spain and northwestern Africa to the Indonesian island arcs—nearly one-third the circumference of the earth. The European Alps and the Himalayas are the best known and most distinctive parts of it, Figs. 23-1 and 23-2. This mobile belt involves the rocks formed in large part in a Mesozoic seaway known as the *Tethys geosyncline*, and various parts of it have been involved repeatedly in Mesozoic and Cenozoic deformation continuing up to the present. This belt can be considered as a single crustal element despite the internal diversity of its features owing to its physical continuity and the approximate synchronous development of regional crustal movements initially as a zone of subsidence and deposition and later as the site of intermittent orogenic deformation in the late Mesozoic and Cenozoic.

This belt is particularly significant from the point of view of global tectonics—it is one of the two longest mountain belts in the world, it is the youngest of the topographically high orogenic belts, and it is the site of Cenozoic orogeny and Pleistocene uplift. Nappe theory originated in the Alpine region, and it is here that we must look for those features commonly referred to as *Alpine-type deformation* or *Alpine-type orogeny*. Finally the present relationship between the Tethys orogenic belt and the adjacent crustal elements is strikingly different from that of the older Appalachian-Caledonide belt or the Cordilleran–Rocky Mountain–Andean belt. The Appalachian orogen is now located along a continental margin, but as we saw in the last chapter its position during the Paleozoic orogenies is uncertain, although the favored opinion now is that the orogenies took place during a time when an earlier proto-Atlantic ocean was closing as subduction took place along the continental margin, eventually leading to one or more continental collisions. In contrast, the Cordilleran orogen, including the Andean sector, has and is experiencing

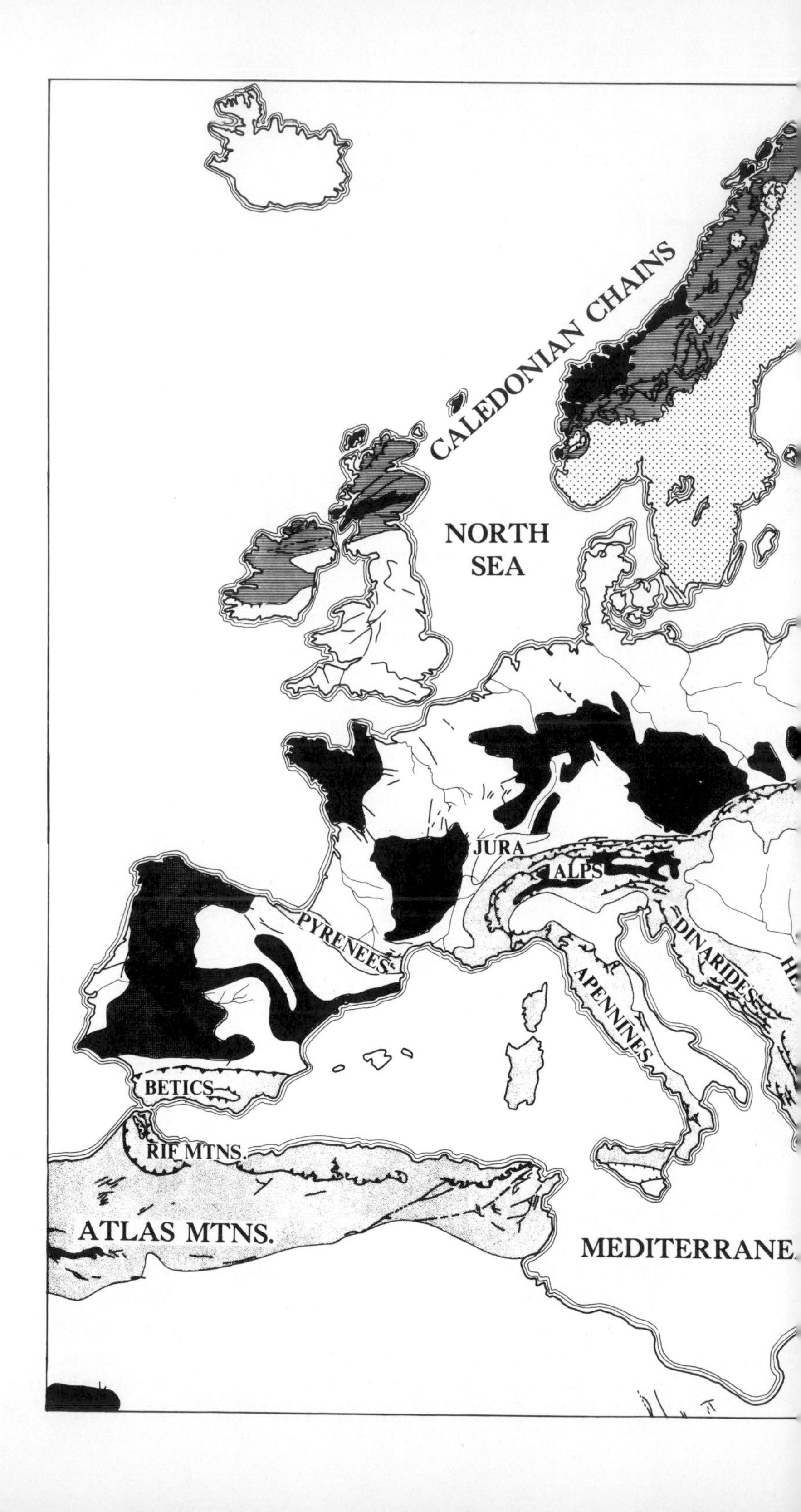
CALEDONIAN CHAINS
NORTH
SEA
JURA
ALPS
PYRENEES
DINARIDES
APENNINES
BETICS
RIF MTNS.
ATLAS MTNS.
MEDITERRANE

FIGURE 23-1 Tectonic sketch map of Europe showing major elements of the Tethy (Alpine-Himalayan) orogen. Areas of crystalline, largely Precambrian, basement are shown black, and dark shading covers areas deformed by Hercynian deformation and which were unaffected by Alpine deformation. The Caledonian orogen is indicated by horizontal lines and the Alpine belt by a fine dot pattern. Some of the major faults within the Alpine belt are shown. (*Simplified after the International Map of Europe and the Mediterranean Region, 1: 5,000,000, 1971, UNESCO.*)

orogeny in the Mesozoic and Cenozoic, and though no continental collision is taking place there, subduction is prominent along the coast of South America and is thought to have taken place along the west coast of the United States during the Mesozoic. The Alpine-Himalayan orogen is far more complex in its tectonic setting.

Geography of the Orogen

The Alpine-Himalayan system is composed of many major mountain chains—the Betics of Spain; the Rif and Atlas Mountains of Africa; the Pyrenees, Alps, Apennines, Dinarides, and Carpathians, of Europe; the Caucasus, Taurides, Anatolides, and Zagros Mountains of the Middle East; and the Quetta, Hindu Kush, Karakorum, Himalayas, and Arakan Yoma Mountains of Asia are all part of this system.

Relationship of the Orogen to Its Surroundings

The Tethys belt appears unrelated to previously developed structures of the basement on which it formed. Hercynian basement is exposed in the massifs in central Europe, in the Spanish Meseta, in the Anti-Atlas mountains of Africa, and across southern Russia. This basement is composed primarily of crystalline rocks that owe their structure and metamorphism to mid- to late Paleozoic orogenesis, the Hercynian orogeny. Although some phases of the Hercynian continued until late Paleozoic time, close to the time of the inception of Tethynian sedimentation, a clear break in time occurs between the two. A long interval of erosion with ultimate development of an erosion surface of low relief followed the Hercynian orogeny, and rocks once recrystallized at depth were exposed at the surface. The first indication of the new extensive basins of sedimentation or geosynclines out of which the Alpine chains were to emerge appeared in the early Mesozoic.

The setting of the orogen varies along its great length. Hercynian and Precambrian crystalline basement variously covered with thin platform deposits and some deeper basins lie north of the orogen along much of its length. In the west the orogen lies between Europe and Africa, but the Mediterranean, Black, and Caspian Seas lie astride the orogen and obscure continuity and relationships between many elements of the belt there, Fig. 23-2. The eastern half of the orogen has complex boundaries. The central part of this end, the Himalayas, lies between the Precambrian basement of the Indian Shield and the Hercynian and Precambrian fold belts and intervening basins of central Asia. Northern parts of the Indian Ocean, the Arabian Sea and the Bay of Bengal, lie on the south side of the orogen to the east and west of India.

At the western end of the orogen many of the structural elements pass into and appear to be terminated at the edge of the Atlantic Ocean, but at the eastern end the structural trends pass into the island-arc systems of Indonesia.

The orogen as a whole may best be described as a complex of discontinuous orogenic belts. The continuity these elements may have originally had is now obscured by rotation of some elements relative to others, by large-scale displacements on strike-slip and thrust faults, by development of large basins within the orogen, and by submergence of large areas. Inferences of subduction and oceanization processes are also called on to explain why some elements are missing.

The principal methods used to establish continuity of structural elements are similarity of deformational style, associations of unusual rock types (e.g., ultramafics, metamorphic facies), paleogeologic and paleogeographic reconstructions. The last can often be done with a relatively great degree of precision because the units are young, and narrow age ranges can be established. In terms of structural style a great range of variation is found from the foreland folding—a thin-skinned folding and thrusting of the sedimentary veneer over a relatively undeformed basement—to deformation in which the underlying crystalline basement becomes mobilized as a brittle mass in some cases and as ductile material that is involved in the folding and formation of nappes in others.

FIGURE 23-2 Regional setting of the Himalayan orogen. Major thrust zones are indicated by barbed lines, modern volcanoes are represented by stars; check patterns show outcrop of Precambrian basement in shield areas, and prominent structural lineaments are shown as dashed lines. (*After Gansser, 1966.*)

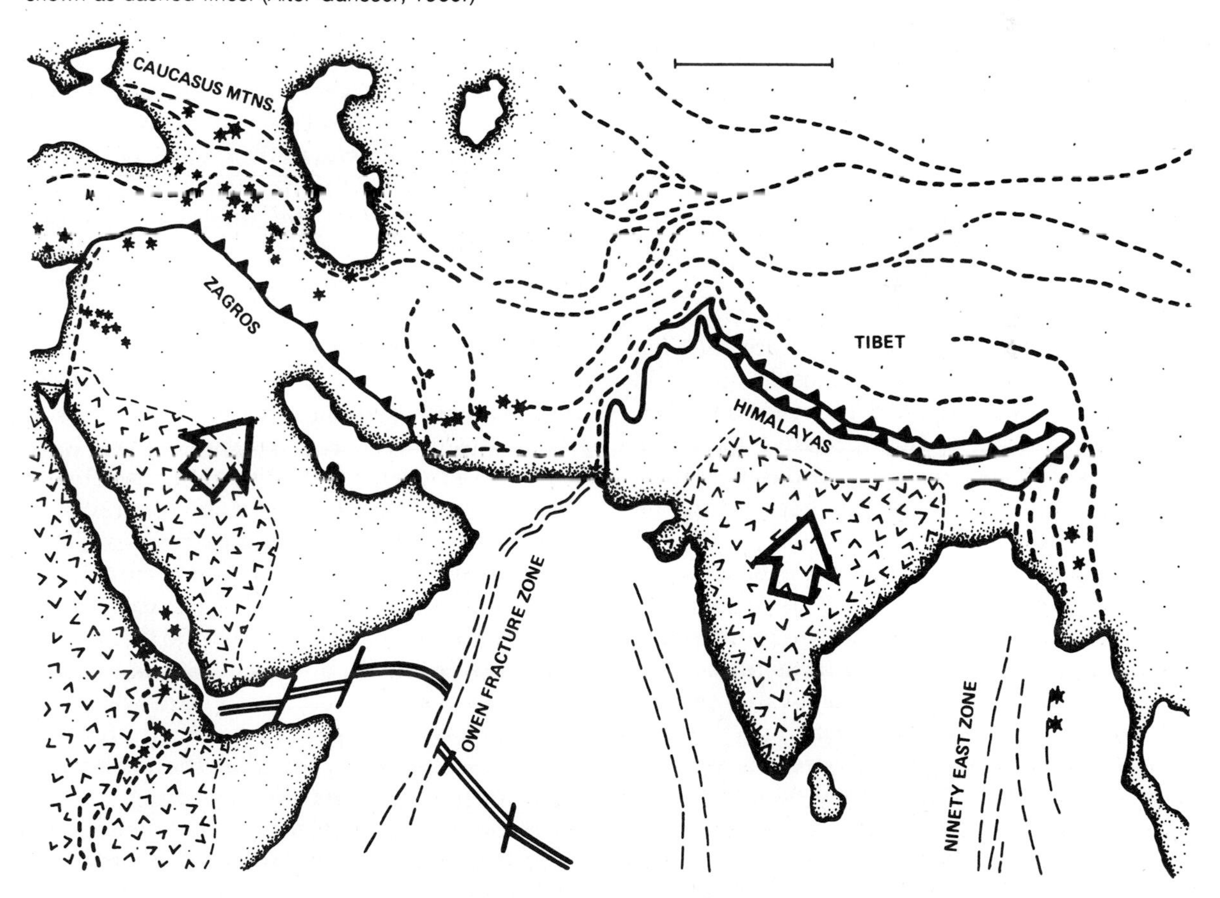

The Pattern of the Orogen

The Alpine-Himalayan orogen is the most highly contorted fold belt in the world. The pattern is not only sinuous, but "meandering" in places, Fig. 23-1. An important example of this is found in the Betic Cordillera of Spain and the Rif Mountains of Morocco. These are essentially a continuous structural unit which swings in trend through 180° at the western end of the Mediterranean Sea. Throughout this great curve the direction of tectonic transport is externally directed almost everywhere. So, thrusting is toward the Spanish Hercynian basement in central Spain on the north and toward the African Shield in the Rif and Atlas Mountains. Are such meanderings primary, or is the orogenic belt itself folded? Is this a large drag feature as a result of movement on a megashear (the Guadalquivir fault in Spain), or is the externally directed movement away from a now lost central uplift? Such are the questions these bends raise.

The Pyrenees occur as an east-west-trending fold belt located between the Hercynian basements of Spain and France. Folding, high uplift, major thrusting, and gravity gliding are all features of this belt, but it lacks any distinct asymmetry. It has been suggested that this belt was deformed as a result of the counterclockwise rotation of Spain relative to France which led to the opening of the Bay of Biscay, a view supported by matching geology and by paleomagnetic studies.

How to connect the Pyrenees, Betics, and Atlas Mountains with the Alps across the Mediterranean has long been one of the problems of Alpine geology. The Atlas Mountains appear to connect to the folds and thrusts across Sicily, but tectonic transport through most of Italy is directed from west to east, Fig. 23-3. This trend disappears under coarse sediments, a molasse, on the south side of the Po basin in northern Italy. The folds and faults that occur west and north of the Po basin are directed westward and northward and are part of the long continuous belt through central and eastern Europe. Thus it is very difficult indeed to trace the Betics, the Atlas, the Pyrenees, or the Apennines into the European Alps. The European Alps swing in a meandering arc from southern France into Austria. The externally directed tectonic transport of this region continues in the Carpathians as the fold and thrust belts continue around the Pannonian basin through the Balkan Chain and under the Black Sea to the Caucasus where the northern transport dies out.

Emerging from the region immediately north and east of the Po basin is a zone of southerly directed tectonic transport. This zone is separated from the northerly directed movement of the European Alps by a major fault with at least some transcurrent component of movement. Major folds and thrusts in the Dinarides and across Greece exhibit southwesterly movement. The orogen then passes into an island-arc system in the eastern Mediterranean and is prominent again in southern Turkey, in the Zagros Mountains of Iran, and along the southern border of the Himalayan chains, all of which show southerly directed tectonic transport.

The southerly directed thrust systems along the edge of the Himalayas terminate at both ends in complex zones of folds and strike-slip faults, the Quetta and Arakan Yoma zones, Fig. 23-2, which trend north-south. The Quetta zone lines up with the Owen and Murray fracture zones of the western Indian Ocean, and the Arakan Yoma zone is directly on trend with the Ninety East ridge of the eastern Indian Ocean. The large-scale picture this evokes and one supported by much data is that the Indian Shield has been forced northward into the Himalayan orogen along major fracture zones at either side and that the shield has bent the orogen, stretched the margins, and underridden the highly deformed contents of the orogen which make up the Himalayas.

The sinuousity of the orogen, the sharp changes in trend, the meandering of the fold belts, and the juxtaposition of oppositely directed movement plans make the Alpine-Himalayas orogen very difficult to interpret. The much older Appalachian orogen seems simple by comparison.

Orogenic History

Evidence of the timing of orogenic events in the orogen is found in the age of volcanism, metamorphic events, intrusions, unconformities, and drastic changes in sedimentation. Such evidence indicates that folding and faulting started as early as the Triassic in places and that volcanic activity and metamorphism occurred from the Jurassic to the present. Many modern volcanos stand today, and the orogen is still active seismically throughout its length.

An important phase of deformation affected much of the orogen during the Cretaceous. The area involved extended from the European Alps, through the Dinarides, across Turkey and Iran, into the Himalayas, and into the Indonesian arc. Most of this part of the belt was again affected by orogeny in the Miocene and/or the Pliocene and Pleistocene. In addition, the Betics, Atlas, Apennines, and Dinarides were involved in the Miocene. A few parts of the system have been deformed in a single phase of deformation, but the more general condition is that of repeated phases of different character as we will see in the case of the European Alps.

EUROPEAN ALPS

The Alps of France, Switzerland, and Austria form one of the most intensively studied structural units of the earth's crust, Fig. 23-3. Here the mountains rise to 4,000 m and have been deeply dissected by Pleistocene glaciation, producing superb exposures. Vast amounts of field work done here make this area necessarily of critical importance to our understanding of the Tethys orogenic belt and a logical starting point for any discussion of Alpine-type orogenesis.

The Limits and Major Structural Elements

The Hercynian basement crops out in the central Vosges, Schwarzwald, and Bohemian massifs north and west of the Alps, and the basement of the intervening areas on this side of the Alps is overlain by a little-deformed sedimentary cover. The Rhine graben extends northward between the Vosges and Schwarzwald massifs, and the Bresse graben is parallel to the Alpine (Jura Mountain) front in France. Both of these grabens contain undeformed Oligocene and younger sediments. Along the northern border, coarse continental sediments (molasse) derived from the uplifted Alps lie in foredeeps north of the frontal structures. Another molasse-filled basin, the Po basin of northern Italy, lies to the south, inside the great curved arc of the western Alps. Toward the east, sediments of the Vienna basin cover the Alpine structures which emerge and continue to the east in the Carpathians.

It is convenient to discuss the internal structure of the Alpine orogen in terms of the characteristics of distinctive structural elements. These include the basins which contain mainly molasse, the Jura foreland belt, the fold and thrust belts in the high mountains—the Helvetic Alps, the Prealps, the Hercynian massifs, the basement nappes of the Pennine and Austroalpine Alps, and the root zone from which these nappes originated.

The Molasse Basins

The term *molasse* was originally used in the Alps, and it is applied there to a highly varied assemblage of detrital sediments of both marine and freshwater origin. Sandstone and conglomerates are abundant, and coal and shelly limestone are sometimes present. Molasse is formed on land or in shallow brackish

FIGURE 23-3 Tectonic sketch map of the European Alps showing the basement massifs outside the Alpine belt, major structural elements and divisions in the orogen, and several important localities within the orogen identified as follows:

1. Zone of east-west folds in the subalpine chains
2. Argentera-Mercantour massifs
3. Pelvoux massif
4. Belledonne massif
5. Aiguilles Rouge and Mont Blanc massifs
6. Aar and Gotthard massifs
7. Penninic nappes (grid)
8. Austroalpine nappes (diagonal line)
9. Southern Alps
10. Bergell granite pluton
11. Bresse graben
12. Jura Mountains
13. Rhine graben
14. Northern limestone Alps

Fine stippled pattern is Tertiary rocks, the fine grid defines metamorphic rocks within the orogen, including basement massifs and metamorphosed cover. (*Simplified after the International Geological Map of Europe: 1:1,500,000, 1972, UNESCO.*)

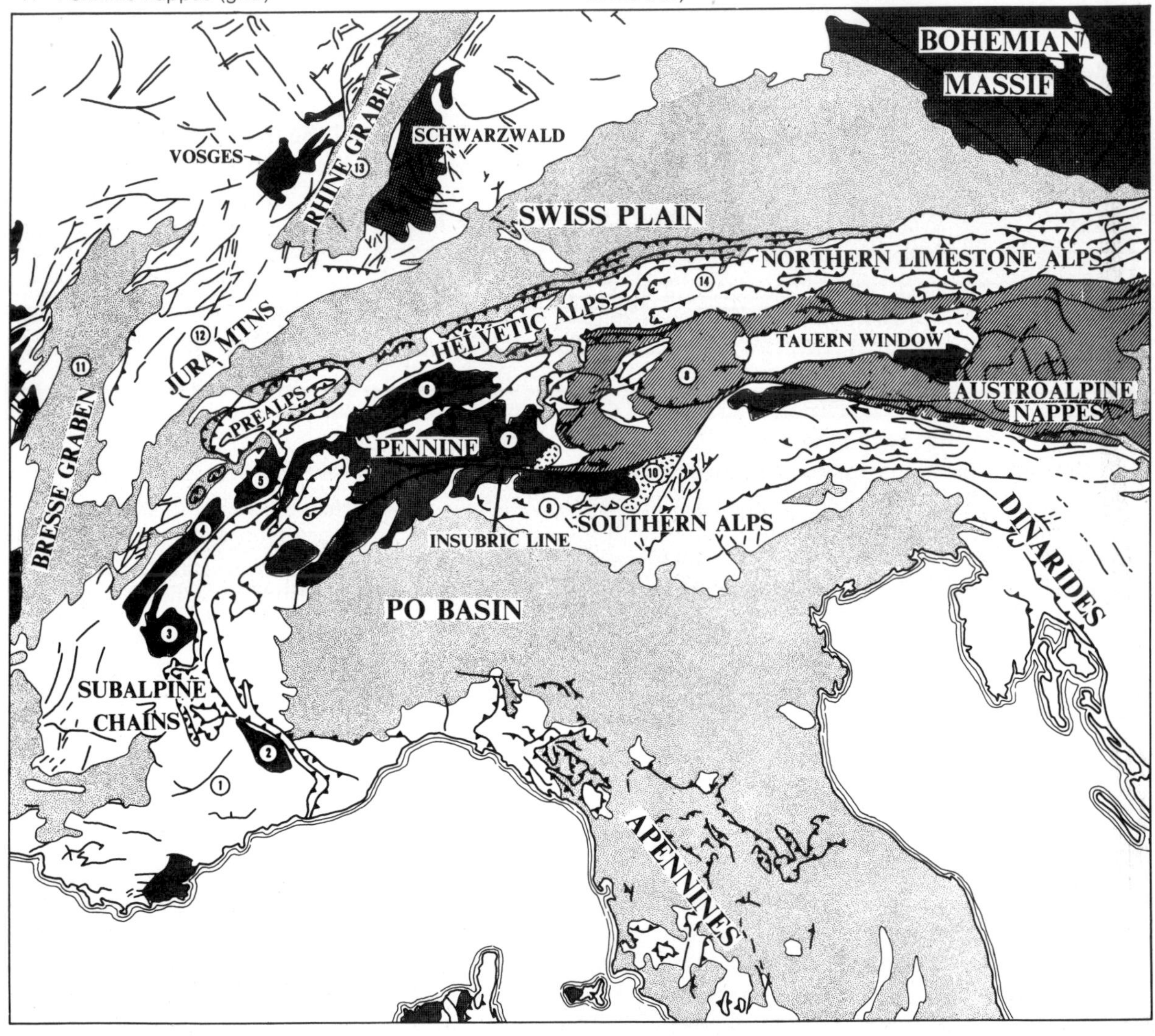

water and contains deltaic and other shoreline deposits. Many of the molasse beds are thick conglomeratic layers containing cobbles and occasionally boulders up to 50 m in diameter. These deposits clearly formed on the flanks of high mountains in postorogenic basins that are unrelated to the preorogenic sedimentary troughs.

Oligocene-Pliocene molasse deposits are extensively exposed primarily in basins that lie in positions marginal to the present high ranges. Molasse occurs in a broad open syncline in northern Switzerland, the Swiss Plain, which narrows toward the southwest in France. It separates the foreland fold belt of the Jura Mountains from the main alpine frontal structure. The molasse overlaps the Hercynian crystalline basement to the north and east in Germany. These molasse deposits of the Swiss Plain show little evidence of deformation except along the inner (southeastern) side of the syncline where folding, thrusting, and tilting of the coarse conglomerates are evident, Fig. 23-4. This clearly shows the effects of post-molasse movements along the alpine front toward the north. Control on the subsurface configuration of the molasse is poor, but depths of 4,500 m are indicated in some places. This, in a plain now 900 m above sea level, suggests that marked subsidence accompanied deposition in the molasse basins. The Po basin, which is located on the inner side of the Alpine arc, contains thicknesses of molasse estimated to be in the range 10 to 20 km. Subsidence here is thus over twice the amount of the elevation in the higher alps today.

The Vienna basin is remarkable for its position astride the alpine nappes which otherwise appear to continue from the Austrian Alps into the Carpathians. Subsidence controlled by movements on high-angle faults has allowed the deposition of more than 5,000 m of Miocene marine and Pliocene to Recent lake and stream deposits which remain undeformed. Failure of this portion of the alpine belt to rise along with that east and west of it remains an unresolved problem.

The Jura Mountains*

The Jura Mountains are situated in the foreland of the Alps but are separated from the more highly deformed Alps by the Swiss Plain, which is underlain by little deformed Oligocene-Pliocene molasse.

The region of the Jura Mountains is crescent-shaped, following the broad curve of the Alps and being wider (about 70 km) in its central part than at either end, Fig. 23-5. The Juras are readily divided into two parts: (1) the outer (northwestern) zone, a plateau reflecting the underlying flatness of the strata, which are little deformed but which have been tilted and cut by high-angle faults which cause displace-

* Primary sources and suggested readings are Pierce (1966), Laubscher (1972), and Rutten (1969).

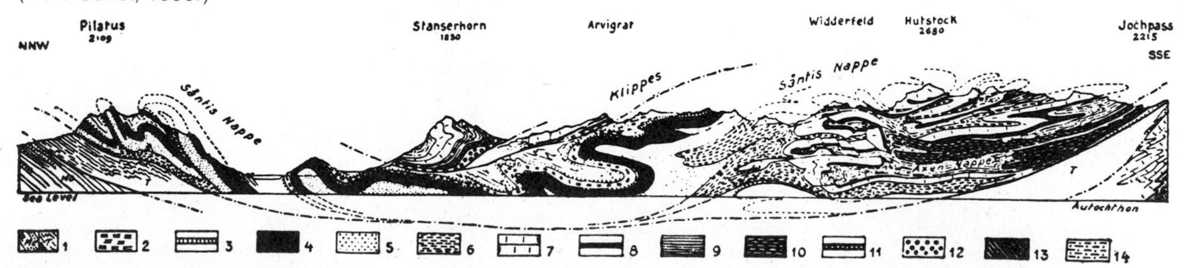

FIGURE 23-4 Geologic section across the High Calcareous Alps of central Switzerland. The symbols are identified by numbers as follows: 1 = molasse, flysch, and nummulitic limestone; 2 = Upper Cretaceous; 3 = Gault; 4 = Barremian; 5 = Hauterivian; 6 = Valanginian; 7 = Upper Jurassic; 8 = Argozian-Oxfordian; 9 = Dogger; 10 = Lias; 11 = Trias; 12 = Trias in the Prealps; 13 = crystalline schists; 14 = Cretaceous; T = Tertiary. *(From Collet, 1935.)*

ments in the Hercynian basement; and (2) an inner zone of folds broken in places by thrusting and offset by transverse faults. Folds in this belt are numerous, approximately 160 anticlines have been counted, and are reflected in the topography as anticlinal ridges and synclinal valleys. The folds die out toward either end of the belt.

The basement rock in central Europe is composed of deformed Paleozoic (Hercynian) rocks which are exposed north of the Jura Mountains. This basement outcrop is disconnected and separated by basins and major fault valleys such as the Rhine graben. The covering sedimentary rocks on this basement are Mesozoic and Tertiary rocks which show a general thickening from north to south and from east to west. The aggregate thickness of the sediments, all of which represent shelf deposits in the northwest, is approximately 3,000 m compared with 4,600 m in the southeast. The Triassic sediment in these sequences is composed mainly of conglomerates, sands, shales, and marls, a large part of which was deposited in shallow water or in continental environments. The Jurassic and Cretaceous beds are predominantly limestones of shallow-water origin, and the overlying Tertiary is molasse, partly continental and partly marine in origin. The sedimentary pile thus consists of rocks of varying competencies interstratified in a sequence of variable but great thickness. One part of the Triassic, the Muschelkalk, deserves special mention because it is composed in part

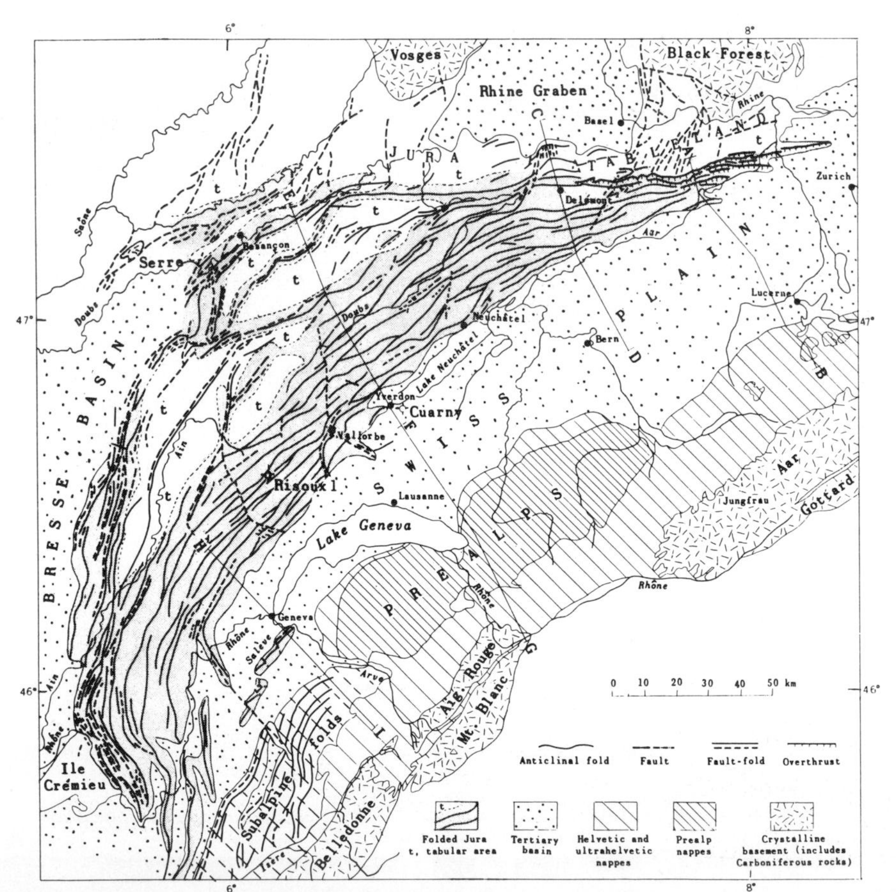

FIGURE 23-5 Tectonic map of the Jura Mountains and adjacent area. (*From Pierce, 1966; compiled from Bersier, 1934; Christ, 1934; and Lugeon, 1941; tabular areas from Dreyfuss, 1960.*)

of beds of salt, anhydrite, and gypsum which are the oldest units identified along the axes of the anticlines and along thrust faults. This has been interpreted to mean that older rocks are not involved in the faulting and folding. Buxtorf suggested that this zone of salts is a plane of *décollement* (literally, ungluing) over which the sedimentary veneer folded and faulted in response to a lateral shove (from the Alps) and beneath which the basement was unaffected by the deformation, Fig. 23-6. The level of the salt is well below the ground surface over most of the region, so this basement-veneer relationship cannot be directly observed. The forces necessary to cause the Jura folding and thrusting in the *décollement* hypothesis are laterally directed from the High Alps toward the foreland and must be transmitted through the Molasse basin of the Swiss Plain. This has caused concern among some geologists because of the generally undeformed nature of the Swiss Plain. Age of the Jura deformation appears compatible with the idea that the forces were transmitted through the rocks of the Swiss Plain. This age is established as post middle Pliocene immediately following the uplift of the basement massifs in the High Alps.

The *décollement* theory has been greatly strengthened as a result of drilling in the region of the western edge of the Jura near Arc Ledonien, France. As reported by Lienhardt (1960), wells drilled along the border of the Saone graben show that the Mesozoic epidermis of the Jura is thrust over Pliocene deposits in the graben. The movement history revealed is as follows: Permian and Mesozoic sediments were deposited over a block-faulted basement. By Eocene time the region was elevated and Eocene deposits were tilted and eroded. This was followed by strong renewal of block faulting in the Oligocene when the Rhine graben was also active. This terrain was then worn low by erosion and epidermis-type (*décollement*) thrusting and folding of the Jura proceeded over this surface. The thrusting occurred on a subhorizontal plane, and the overthrust plate moved at least 3 to 6 km over molasse in the graben. Exposures are not adequate to reveal the length of the thrust, but thrusts are now known at least locally along the northern and western edge of the Jura, and thrusts also occur to the west in the subalpine chains of southern France.

The large-scale plan of the deformed belt is illustrated by a tectonic map in Fig. 23-5, on which the following characteristics of the tectonic plan are shown:

1. Anticlines are far more numerous in the central part than at the ends of the belt. Folds are much more prominent on the southeastern side of the Jura than on the northwest.

2. Many of the anticlines split, forming two or more anticlinal ridges which follow subparallel trends.

3. The longer folds have arcuate axial traces.

4. Many of the folds have strong changes in trend where they meet transverse faults.

5. Transverse faults cut across the fold belt, usually

FIGURE 23-6 Section showing *décollement* of the folded Jura. (*After Buxtorf, 1916; from Bailey, 1935.*)

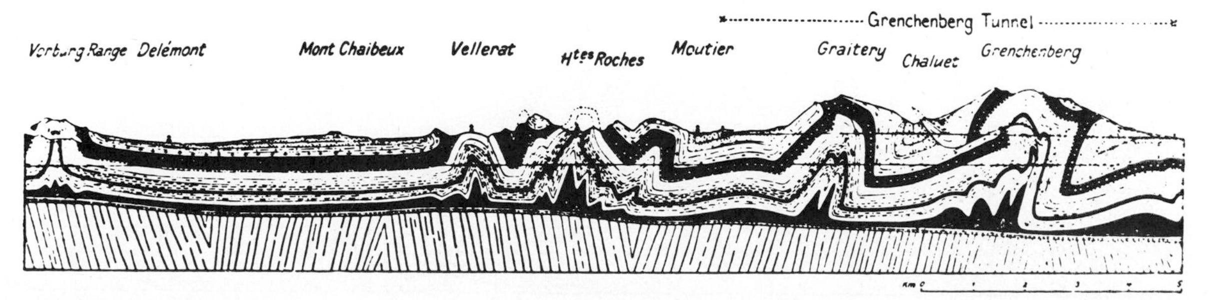

intersecting the folds at angles of about 45°. The most prominent of these have left-lateral displacement. Some of these faults are not simple strike-slip faults in that the structures on one side cannot be brought back into alignment with those on the other by a simple horizontal translational movement. Instead the faults mark boundaries across which the folding developed differently and often without great offset.

6. A second system of high-angle faults and fractures trending north-northeast occurs in the western portion of the Jura and is subparallel to the fault system of the Rhine graben. Many of the faults and fractures underwent movements contemporaneous with those in the Rhine graben.

7. Thrust faults emerge at the narrow northeast end of the fold belt, where movement has been consistently from south to north. Thrust faults have also been discovered by drilling on the west side of the Jura in the "Arc Ledonien."

Apart from the origin of the fold system, the geometry of the Jura folds is of great interest. Box- and fan-shaped folds as well as the more common symmetrical and asymmetrical shapes are beautifully exposed in the Jura Mountains. The character of the Jura folds is best seen in a few deep valleys cut transverse to their strike, Fig. 23-7. Many of the folds are demonstrably disharmonic, and in extreme cases synclines are situated above anticlines. Often the folds are broken on one limb by faults, and movements on some faults have brought older units up along bedding thrusts. These faults are often folded showing at least a continuation of folding after thrusting and suggesting the possibility that thrusting preceded the folding.

A number of students of the Juras have proposed hypotheses to explain the origin of the folds in terms of offsets in the basement, notably Aubert, Wegmann, Pavoni, and Laubscher. Different types of movement plans have been suggested, one of which proposes lateral movements in the basement along transverse fault zones, resulting in drag and faulting in the overlying sedimentary veneer. Certainly the basement is faulted where it is exposed, and movements in the basement have occurred within the right time interval to be the cause of the Jura folding. Even if the main movements responsible for the folding did originate by basement faulting, the salt beds still must have played an important role in the deformation through development of local *décollement* and disharmonic folding.

Helvetic Nappes of the High Calcareous Alps

It was in a portion of the High Calcareous Alps at Glarus that the nappe theory was first established as the significant tool for interpretation of Alpine structure. The High Calcareous Alps are now viewed as a massive pile of folded thrust sheets. These structures are composed primarily of the Mesozoic sedimentary deposits laid down on a Hercynian basement in a shelf environment. Some of the nappes are apparently far removed from their point of

FIGURE 23-7 Cross section across the Wannenfluh from Oensinger to Balsthal (first ridge in the Jura Mountains west of the Swiss Plain). Nos. 1–3 are Jurassic Dogger, Nos. 4–5 are Jurassic Malm, No. 6 is Tertiary molasse. (*After C. Wiedenmayer, 1921.*)

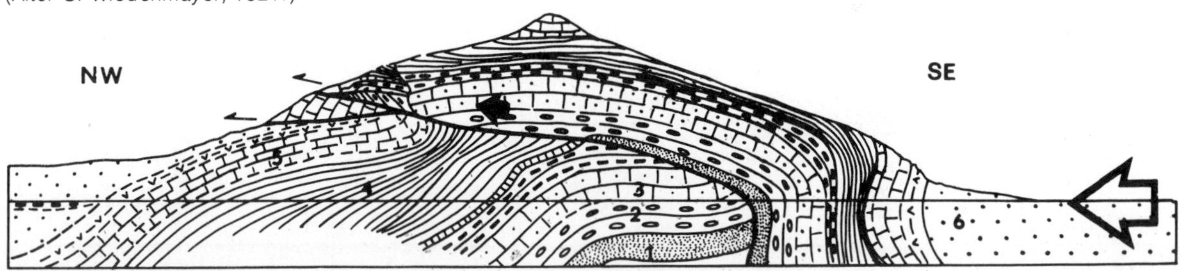

origin or roots, but others can be traced back to their original position. The Helvetic nappes are exposed in a long arcuate belt which includes large parts of the French subalpine chains as well as the High Calcareous Alps. The nappes are located mainly north and west of the basement massifs (e.g., Aar, Aiguilles Rouges, Belledonne, Argentera), and the direction of tectonic transport has been toward the north and west.

Characteristic features of the structure of some of the Helvetic nappes are brought out in Figs. 23-4 and 23-8. The superposition of one nappe on another is clear, but the complexity of the relationships is not immediately apparent. Several very different mechanisms of deformation are involved in the structure of these nappes. Some are due to effects of shearing and thrusting which arises out of the underlying basement, but far more movement of the cover is involved than is required for these basement faults.

At the Jungfrau, Paleozoic crystalline rocks are on top of Jurassic sediments which can be followed upside down under the crystalline complex. Toward the northwest the Jurassic sequence which mantles the crystalline rocks is folded into a complex recumbent syncline, and down the mountainside the Jurassic sequence is repeated beneath this recumbent syncline four times. The repetitions show through the preservation of thin, drawn-out remnants of Cretaceous and lower Tertiary units on the upper layers of the Jurassic. These lower faults are subparallel to bedding, but some of them can be traced back into zones of shearing that involve the basement complex. This shearing out of the basement and the development of inverted recumbent folds is shown in Fig. 23-8. Problems in interpretation arise in trying to decide where each nappe originated and in integrating the individual nappes into a general movement picture. Solutions to such problems have been found largely in stratigraphic studies. As an example of this approach, consider Fig. 23-9, which shows a structural cross section at top. Much of this can be seen in the valley walls with other areas filled in from aerial mapping.

FIGURE 23-8 Cross section across the Jungfrau (capped by granite) which is thrust over Mesozoic rocks (Cretaceous is black, Jurassic is brick pattern) which are repeated several times as a result of thrusting and tight folding. Authochthonous crystalline rocks are left blank at the bottom of the section. (*After Collet and Paréjas, 1920.*)

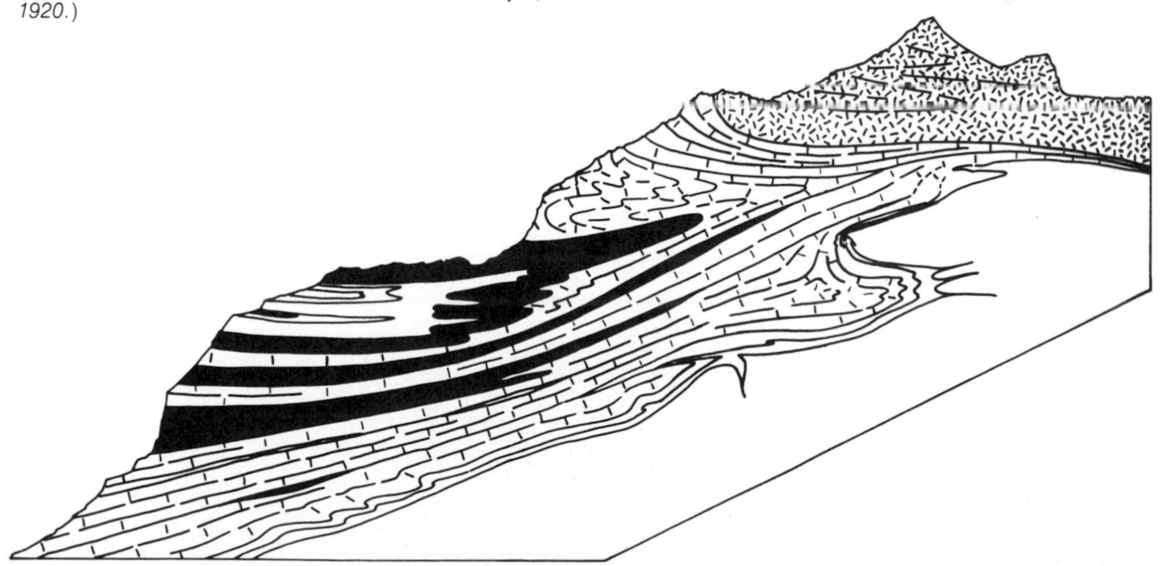

Trümpy (1971) has made a palinspastic reconstruction of the area, predeformation, by matching facies and thicknesses of units in the various fault slices in the nappes as shown in the bottom section, Fig. 23-9.

Source areas for the Helvetic nappes pose interesting problems with important tectonic implications. The rocks show little metamorphism; so, they must have been deformed at shallow depth, and it seems clear that some of the nappes developed from shearing of the sedimentary veneer off the basement blocks along faults which can be traced back into the basement, Fig. 23-10. It is clear that the basement massifs are not nearly wide enough to provide the necessary space for all of the reconstructed cover. They could not have come from much farther south because thin stratigraphic equivalents there are of a different facies. Thus the most likely source area for the remainder is found in zones between the basement blocks. One such zone is the Chamonix zone, a narrow belt of steeply dipping Triassic and Jurassic strata located between the Mont Blanc and Aiquilles Rouges massifs. The sedimentary veneer here has been sheared out and crushed, and a similar zone is found between the Aar and Gotthard massifs. This presents an important space problem—one of having more sedimentary rock in the nappes than there is basement from which they could have been derived. This has prompted the suggestion that the basement has disappeared, and since it is not at the surface it is proposed that it moved down into the mantle—by subduction.

FIGURE 23-9 Structural section and corresponding palinspastic section through the Glarus Alps. Scale 1:100,000; vertical scale not exaggerated in structural section, exaggerated 2½ times in palinspastic section. Am, Aar massif; TM, Tavetsch massif. Au, Autochthonous cover; SH, Subhelvetic slices; Wa, Wageten slice; G, Glarus nappe; M, Mürtschen nappe; A, Axen nappe; B, Bächistock slice; Wi, Wiggis; and Rä, Räderten subunits of Drusberg-Säntis nappe. Sm, Subalpine molasse; NFL, lower and NFU, upper sheets of north-Helvetic flysch; E, Einsiedeln slipsheets. (*From Trümpy, 1973.*)

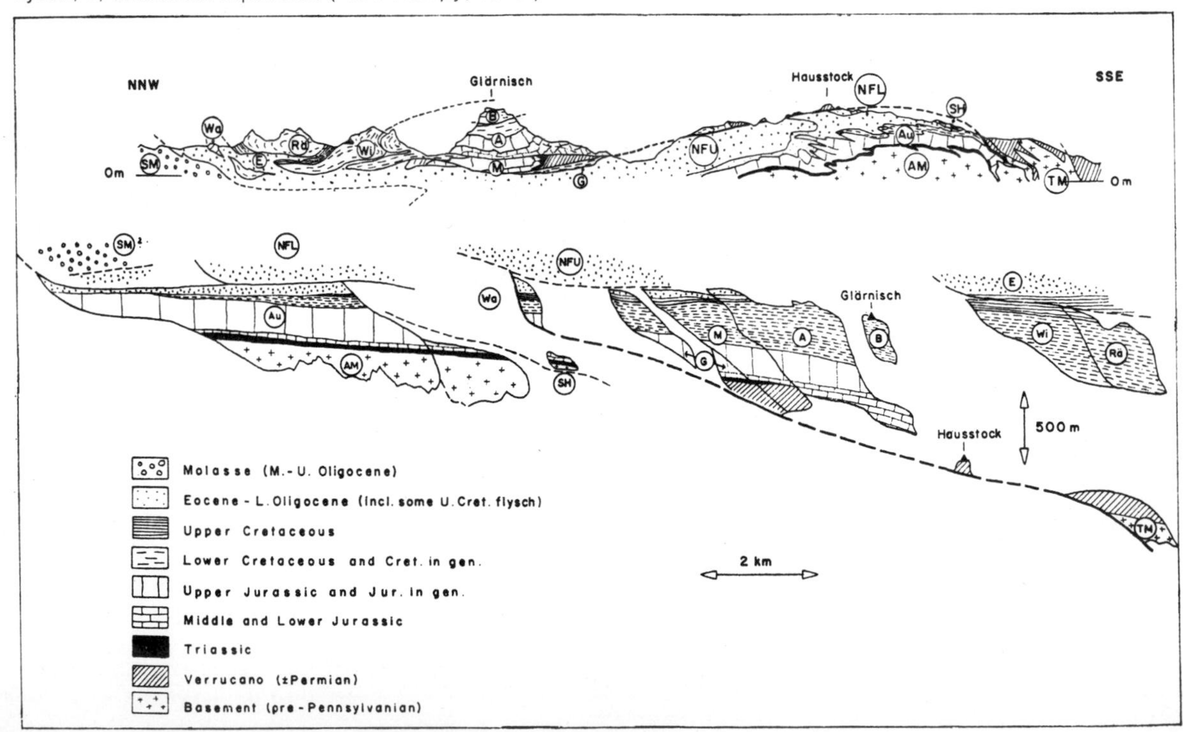

The Prealps

The Prealps south and east of Lake Geneva stand as a frontal projection of the High Alps into the Swiss Plain (Fig. 23-11). The intensely folded and faulted rocks which compose these mountains are of Mesozoic and early Tertiary age and younger than most of the Tertiary molasse deposits of the Swiss Plain. Furthermore, these Mesozoic units are not structurally or stratigraphically continuous with Mesozoic rocks of the High Calcareous Alps, which are immediately adjacent both to the southeast and along the northwestern edge of the Prealps. Detailed comparison shows that the Triassic and Jurassic units of the Prealps are of different facies and thickness from those of the same age in the nearby High Calcareous Alps.

Hans Schardt (1898) was the first to conceive that the Prealps, 120 km long and 40 km wide, is a great klippe of rocks derived from some place far to the south and now resting on a base of Helvetic nappes. Five main nappes, the Simme, Breccia, Klippe (also Median), Niesen, and Col, in descending order, make up the Prealps.

The highest nappe, the Breccia, crops out mainly at the southwest end of the Prealps, where it lies in fault contact partially on rocks of the High Calcareous Alps and partially on the Prealps Klippe nappe. The position of the Breccia nappe on a nappe of the High Calcareous Alps is important because it demonstrates clearly that the Breccia nappe was emplaced after the formation of the High Calcareous Alps nappes and must therefore have

FIGURE 23-10 Section across the Zone of Chamonix. (*After E. Paréjas.*) 1 = Mont Blanc massif; 2 = crystalline wedges; 3 = Aiguilles Rouges massif; 4 = Trias; 5 = lower Lias; 6 = middle Lias; 7 = upper Lias; 8 = Dogger (Bajocian-Bathonian); 9 = Argovian; 10 = Upper Jurassic; 11 = Lower Cretaceous; 12 = Tertiary. (*From Collet, 1935.*)

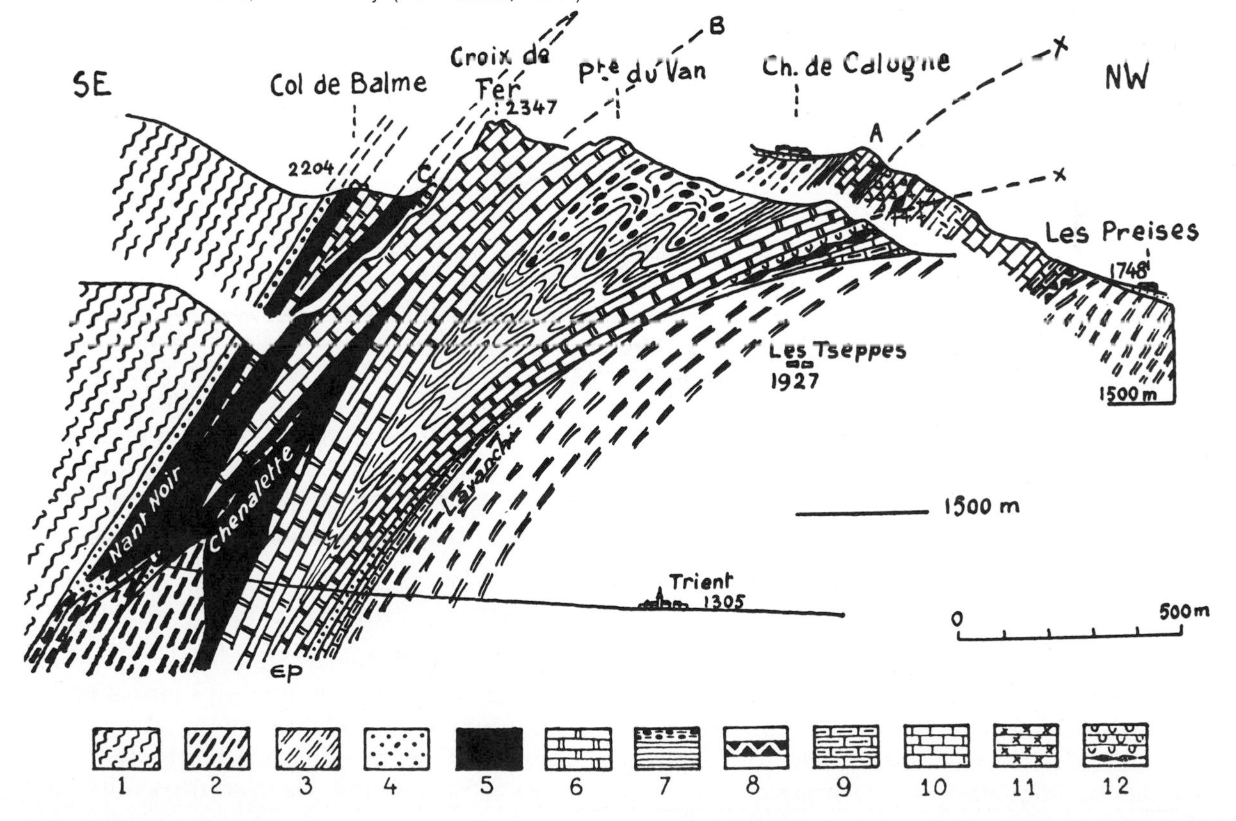

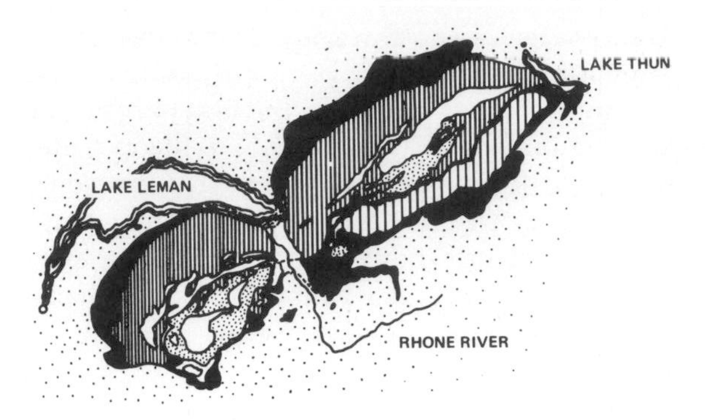

FIGURE 23-11 Map of the Prealps showing superimposed nappes, Simme nappe (white), Breccia nappe (dot pattern near center), Prealps Medianes nappes (closely spread vertical line), and Niesen nappe (open vertical pattern); ultrahelvetic units (black). (*After De Jong, 1973.*)

come from a great distance south. This interpretation is supported by the significant differences between facies of the rocks in the Breccia nappe and those of comparable ages nearby. Along the northwestern edge of the Col nappes, flysch is faulted onto the younger molasse, and that flysch is in turn overridden by Jurassic strata of the Klippe nappe. Thus, northwestward movement clearly postdates deposition of at least some of the molasse.

Such questions as the direction of movement of the nappes, how far they have moved, and what mechanism accomplished this movement have been long debated, but it is interesting that modern interpretations resemble those originally set forth by Schardt, who interpreted the nappes as having come from the south, traveling great distances by gravity sliding. He envisioned the Klippe nappe as having originated either as a fold which became faulted as movement proceeded or as a clean-cut thrust which was elevated, broke away from its root, and began gliding downhill, down the northwestern slope of the Pennine Alps which were moving northward at the same time. Support for the concept of gravity sliding is found in the character of the flysch deposits associated with the nappes of the Prealps. The Niesen flysch, for example, contains fragments of pre-Triassic schist of a type found in the Pennine Alps far to the south. The flysch in the Prealps contains conglomerates which develop into breccias containing huge angular blocks of granite and gneiss interbedded with fossiliferous shale. Schardt came to interpret these as having developed as landslides and screes along the advancing edge of the Klippe nappe. As the mass moved, it presumably sheared off the crystalline schist from the tops of exposed ridges in the Pennines. It is interesting to note the similarities between the origin proposed for the Prealps and that described earlier for the Taconic klippe.

Deformation of the Hercynian Basement

It is often impossible to determine the true age of the rocks which were recrystallized in the Hercynian orogeny because radioactive "clocks" were largely "reset" at this time. All this basement was tectonized during the Hercynian orogeny; so the crystalline rocks contained old folds, shear zones, and other metamorphic structures before Alpine events affected them.

Basement crops out at present in massifs north and west of the Alps, where structures related in age to Alpine deformation are grabens and other high-angle faults; in the central massifs, where a number of basement block-like domes were uplifted in Plio-Pleistocene time; as overthrust sheets and folds-basement nappes known as the Pennine and Austroalpine nappe complexes including their root zones; and as tilted blocks in the southern Alps—south of the Insubric-Tonale fault.

The Hercynian Massifs

The High Calcareous Alps, Hercynian basement massifs, and Pennine Alps form a pattern of concentric zones which swing in a great arc through more than 90° from the Mediterranean coast at Nice to the Austrian Alps. The basement is exposed as the Argentera, Pelvoux, Belledonne, Aiquilles Rouges, Mont Blanc, Aar, and Gotthard massifs, Fig. 23-3. These blocks are prominent today and form many of the highest glacier-covered peaks near the Alpine front. These massive blocks are composed of pre-Carboniferous metamorphic and igneous rocks containing small areas of infolded and highly sheared, late Paleozoic sedimentary rocks. These massifs are surrounded and separated from one another by a highly deformed cover of post-Paleozoic sediments which form a much wider outcrop belt on the north and west sides of the zone of massifs. The Helvetic nappes consist of the sedimentary cover stripped off the massifs.

The larger massifs are separated from one another by depressions, as between the Aar-Gotthard and Aiquilles Rouge–Mont Blanc massifs. The plunge of these massifs and the overlying cover are most important because they provide the exposures which allow inferences about the deeper structure within the deformed cover nappes. The down-structure method of viewing maps and cross sections has been used in the Alps to construct cross sections which depict the complexity of the nappes. Of course, the validity of this method rests on the assumption of lateral continuity of exposed structural elements, and this has been called into question with regard to many Alpine structural elements in recent years.

A number of the massifs are divided longitudinally by narrow zones of post-Paleozoic sediments. A good example is the zone between the Aar and Gotthard massifs. Sediments in these zones are highly deformed—faulted and sheared out—indicating a considerable amount of movement between the massifs. The massifs have generally been considered to be in place (autochthonous), but it appears that at least the more internal elements (e.g., Gotthard) may have moved some distance laterally and may not be firmly rooted where they stand today. The zone between the massifs looks like the most probable source area for much of the cover now found in the Helvetic nappes (Trümpy, 1963). If this is the source area, the basement on which this cover was deposited has disappeared, presumably at depth, going into the mantle (i.e., as in a subduction zone). Thus this was probably the area in which some of the crustal shortening in the Alps took place. Despite their current topographic prominence, the basement massifs are young features formed after the thrusting of the Helvetic nappes. The thrust planes of these nappes are arched over the massifs. Gravity-induced sliding of the nappes occurred concurrently with and following this uplift. Effects are most pronounced to the north and west of the massifs.

The Pennine Alps

The Pennine Alps are distinctive from the previously discussed divisions in several respects. They were derived from a deeper portion of the orogenic belt, which was overlain by a sedimentary cover that formed contemporaneously with the shelf and miogeosynclinal sediments now found in the Helvetic nappes. The cover of the Pennines was originally deposited in a eugeosyncline, and it contains basic and ultramafic intrusive and extrusive rocks. Penninic rocks were strongly affected by Alpine metamorphism in the Tertiary.

Rocks of the Pennine Alps are exposed in an arc extending almost from the Mediterranean Sea into eastern Switzerland, where rocks of the Austroalpine nappes have been thrust over the Pennine nappes. Proof that Pennine rocks lie under these Austroalpine nappes is found in the Tauern window of central Austria, where erosion has exposed Penninic nappes.

The large nappes of the Pennine Alps are composed of crystalline (Hercynian) cores mantled by schists or sedimentary rocks, *Schistés lustrés* (also *Bündner Schïefer*)—schistose rocks exhibiting varying degrees of metamorphism. Fossils found in these metamorphic rocks demonstrate that they range in age from Triassic to Eocene.

Large masses of granite and gneiss surrounded by covers of *Schistés lustrés* and Mesozoic sedimentary rocks make up the major structural elements of the Pennine Alps, Fig. 23-12. The relationships of the basement rocks and this cover can be seen at many places in the highly dissected ranges, as at the Matterhorn where *Schistés lustrés* underlie highly deformed, older basement gneisses. Here in the Pennines the basement floor of the geosyncline was actively deformed; it was folded and sheared out along faults with movements amounting to at least several miles. This much can be clearly demonstrated. Arguments concerning the Pennines center largely on the degree and amount of basement mobilization and the mechanisms of movement.

It was the driving of the Simplon tunnel (1895 to 1905) that set the stage for the view of large-scale basement folding. The tunnel was driven through a nappe, demonstrating that the basement rocks were actually involved in large-scale recumbent folds.*

What has come to be regarded as the classic synthesis of Pennine nappe structure was conceived by Argand in 1916. He envisioned each of the major pre-Triassic basement elements in the Pennines as large-scale nappes in which the basement gneisses had become mobilized, moved into the overlying cover of Mesozoic or Tertiary rocks, and flowed with some cover many kilometers northwest, Fig. 23-13. He reconstructed the sequence of events as the pile of nappes developed, explaining in a most ingenious manner the great variety of structural detail.

Recently, students of the Pennines have not been so enthusiastic about this grand synthesis which was based on the down-structure method of viewing geologic maps. They tend to think instead that movements have been over smaller distances, that much of the movement has consisted of gravity gliding, that basement movements have been accomplished more frequently by shearing than by actual folding, and that the individual nappe continuity is not of such a grand regional scale.

* See Milnes (1974) for a new analysis of this section.

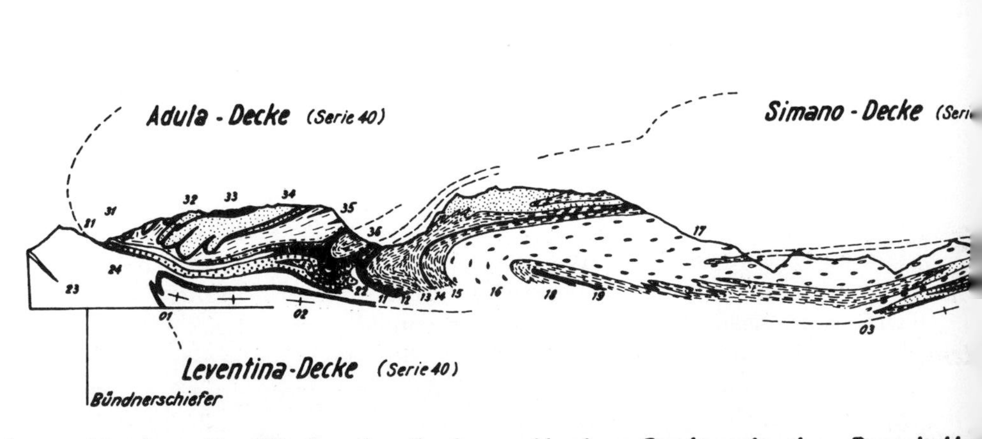

FIGURE 23-12 Schematic profile through part of the Pennine nappes in Switzerland and the Tonale line (root zone). (*From Milnes, 1974.*)

Salt and gypsum deposits of the Triassic, some of which outcrop for tens of kilometers along their strike, and the *Schistés lustrés* have played an important role as lubricants in the movement of nappes in the outer part of the Pennines. As the folding and faulting of the basement occurred, the salt beds were rolled up and lubricated the movement of the basement masses as they pushed northward. Masses of other rock became caught up in the salt which now completely encases them.

Alpine Metamorphism, Plutonic Activity, and Basement Deformation

Alpine-age metamorphism is most marked in the Pennine zone, although metamorphism has slightly affected the Mesozoic cover outside the Pennines. Even within the Pennines, metamorphism is highly variable. Mesozoic rocks in southern France are only slightly altered, but to the northwest the *Schistés lustrés* are more affected. In Switzerland the metamorphic grade is higher, and biotite, staurolite, kyanite, and garnet are present in the Jurassic rocks of the Pennine nappes. Bearth (1962) documents two main phases of metamorphism—an earlier one due to depth of burial which produced lawsonite and pumpellyite in the outer Pennine zones and glaucophane and choloritoid in the inner zone, and a later regional metamorphism which produced kyanite and sillimanite in the inner zone. The isograds cut across the nappe structure; so, the high-temperature recrystallization must be later, and the emplacement of the Pennine basement nappes must be associated with the earlier metamorphic phase. Widespread syntectonic granites, granitic gneisses, and migmatites of Alpine age are not present. Thus deep-seated folding and extensive flowage of the basement during development of the nappes is ruled out (Milnes, 1969). The nappes deformed primarily by shearing along relatively narrow zones. This is seen in the Simplon region, where several thin basement nappes each with its cover are piled one on another. The movement is confined to a thin shear zone and the cover is relatively undeformed. A foliation developed parallel to the nappe margins, and porphyroblasts grew

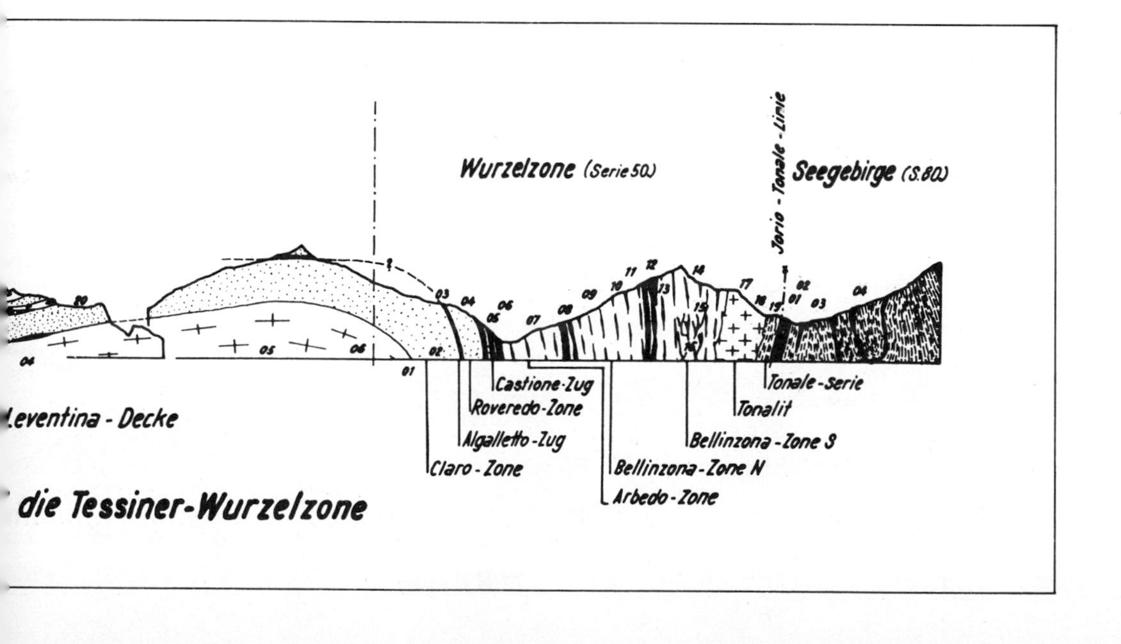

postectonically with respect to this main Alpine fabric (Milnes, 1969).

Igneous activity is not a prominent aspect of Alpine orogenesis, nor was deformation in the Alps accompanied by high temperatures and production of melt, at least not at the levels now exposed. These characteristics remind us of the plate-theory model of continental collision which is marked by mechanical deformation but with relatively little subduction or igneous activity. A few granitic intrusions of Tertiary age are known, but these are locally developed, Fig. 23-3. They appear to have been derived from remobilized granite basement during the height of Alpine metamorphism (Wenk, 1962). Boulders of this granite appear in Middle Oligocene molasse. Thus it must have been emplaced before this time. It cuts discordantly through the Pennine nappes on its northern side, showing that it is clearly postnappe movement.

Austroalpine Nappes

The Austroalpine nappes consist of a thick and complex system of basement nappes which were derived from south of the Pennine nappes

FIGURE 23-13 Cross section through the Alps. (*After Argand, 1916; from Collet, 1935.*)

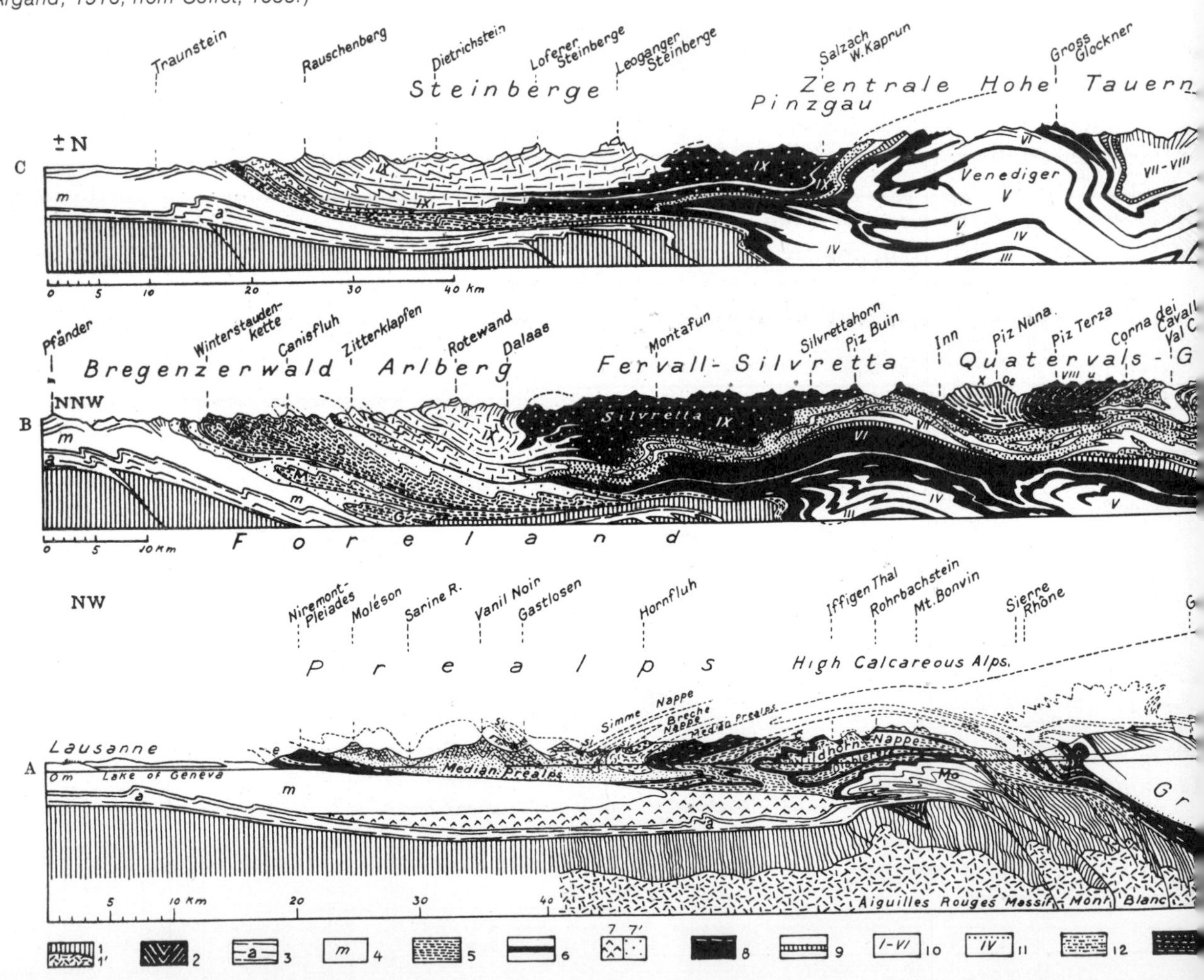

and are now found structurally over the Pennine rocks. The Penninic nappes plunge under the Austroalpine nappes along the Swiss-Austrian border, but reappear in a window, the Tauern window, in central Austria. The lower part of the Austroalpine rocks contain much more granite, granitic gneiss, diorite, and gabbro than the Pennine rocks.

The Austroalpine nappes are thought to have their roots in a narrow zone of vertical crystalline rocks in southern Austria, similar to the root zone of the Pennine nappes in southern Switzerland. The root zones are located north of a system of faults, the Insubric-Tonale-Pusteria faults which demark the southern edge of the Austroalpine nappes. The western end of the root zone is covered by younger sediments in the Po basin, and the eastern end is covered in the Vienna basin. The number of major divisions within the Austroalpine system and details of its paleogeographic restoration are still unresolved, as are the questions of the extent of nappe structure and how much the Pennine rocks in the Tauern window have moved. (Refer to Zentrale Hohe Tauern in Fig. 23-13*c*.)

As in the Pennine nappes, large-scale deformation of the crystalline basement and its

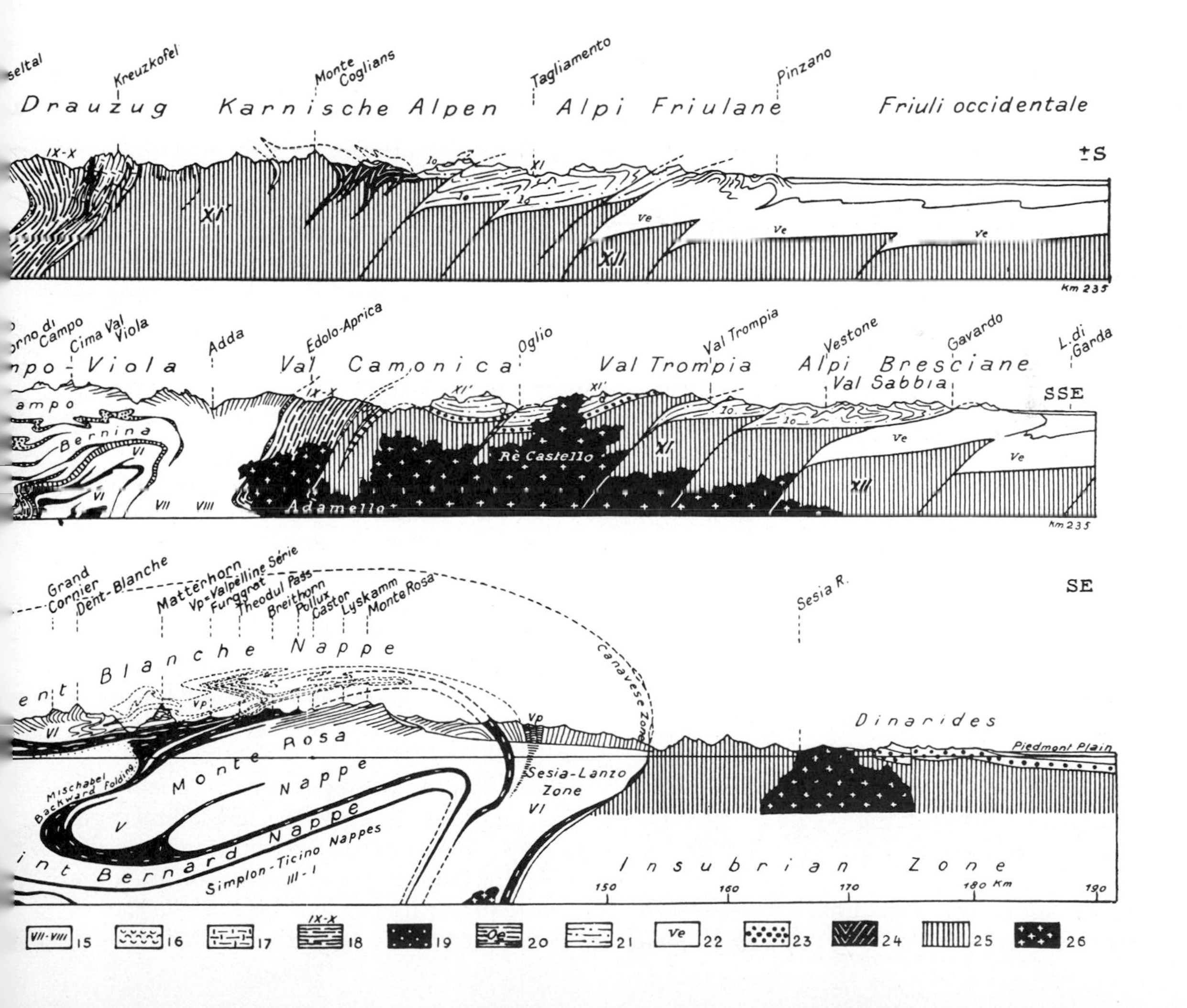

cover is involved in the Austroalpine nappes. The lowest of the main Austroalpine structural elements, the Arose nappe, consists of a series of thin superimposed nappes consisting of sheared crystalline rocks and cover, exposed along their contact with Pennine nappes near the Swiss-Austrian border, but it is not found to the east. Three distinct longitudinal divisions may be recognized in the Austroalpine region—the Northern Limestone Alps, composed mainly of Triassic and Jurassic limestone; the Graywacke zone—an outcrop belt of Paleozoic and Mesozoic rocks which are gray in color (not the rock graywacke); and a predominantly crystalline rock complex on the south.

Some students of this area contend that these three divisions are a single major nappe system in which the Mesozoic of the Northern Limestone Alps and the Paleozoic of the Graywacke zone are the sheared off cover of the southern crystalline complex. Others, notably Tollman (1965), say that the remnants of Paleozoic and Mesozoic facies found on the southern crystalline complex are too different to be a part of the units found to the north. Thus he concludes that the rocks of the Northern Limestone Alps and Graywacke zone must have been derived from a position south of those now on the crystalline nappes. An additional 100 km of thrusting and presumably crustal shortening is necessary in this hypothesis.

The Root Zone—The Insubric Line—The Ivrea Zone

The classical concept of the root zone is that of a zone from which the nappes were squeezed. The origin of the nappes, particularly the basement nappes, is a difficult problem. The source area for the nappes in southern France is obscured by the molasse fill of the Po basin, and in the area where the best exposures are found in southern Switzerland a sharp fault-line break, the Insubric-Tonale line, occurs on the southern side of the Pennine nappes. Where the Insubric line is exposed it is a narrow zone of intensely contorted and sheared rock with layers of banded mylonite. Significantly, the fault plane contains horizontal slickensides which indicate at least late strike-slip movement. Several students of the fault (Laubscher, 1971; Gansser, 1968) interpret it as a major strike-slip fault. Basement rock is exposed across the Insubric line in the southern Alps, but it has been unaffected by the Alpine metamorphism; so the rocks to the south of the fault are over 300 million years old, while only a few hundred meters across the line, high-grade metamorphic rocks with ages of 15 to 25 million years occur. Clearly the root zone here coincides with a major fault along which movement has taken place after the formation of the basement nappes and after emplacement of the Bergeller granite, which is sheared and drawn out along its southern margin where it parallels the Insubric line.

Several large and deeply glaciated valleys oriented north-south cut across the root zone. The large Pennine nappes can be seen in these valley walls where the layering is flat lying in the north and gradually folds over into subvertical layers. The zone along which these subvertical layers occur is often several kilometers wide and it is known as the *root zone*. Actually, it is a complex of zones of different lithologies (all crystalline, although some are marbles and calc-schists), some of which can be traced into particular Pennine and Austroalpine nappes; others have either no connection or at best tenuous connections with rocks exposed to the north in the nappes. Rocks within the root zone show pronounced effects of compression, with mylonites, strong lineations, drawn-out lenticular structures, and a few large folds, Figs. 23-12 and 23-13.

The root zone and the Insubric-Tonale fault are cut and offset 80 km with a left-lateral separation by a major northeast-trending fault, the Giudicaria fault; but an eastern continuation of the Insubric line called the *Pusteria line* or *Gailtal line* is found in Austria, and the Austroalpine nappes appear to emerge from root zones just north of this fault, Fig. 23-13.

The western end of the Insubric line is difficult to follow. Several major faults with a westerly trend branch off it, and the prominent differences in metamorphic effects observed across the line farther east begin to die out. Gansser (1968) and many other students of this area place the main fault just north of a belt of basic rocks known as the *Ivrea zone*.* This zone is particularly significant because it is composed in large part of ultrabasic and basic rocks which have been projected downward to a depth of 50± km on the basis of gravity and seismic work. Thus the rocks exposed in the Ivrea zone may very well be a part of the upper mantle thrust up onto crustal rocks and exposed at the surface. Gravity anomalies associated with these dense rocks can be traced under the Po basin almost to the Mediterranean coast, but the anomaly dies out there. The zone appears cut off by the Insubric line, and the anomaly is absent toward the east, Fig. 23-14.

* See special series of articles on this zone in Schweiz. Miner. Petrogr. Mitt., v. 48, 1968.

Evolution of the Western Alps*

To this point we have been concerned primarily with identification of the major geographic divisions and description of the most important structural elements. The temporal

* See Trümpy (1973).

FIGURE 23-14 The main tectonic elements of the western Alps and the northern Apennines (*Simplified after Laubscher, 1971.*)

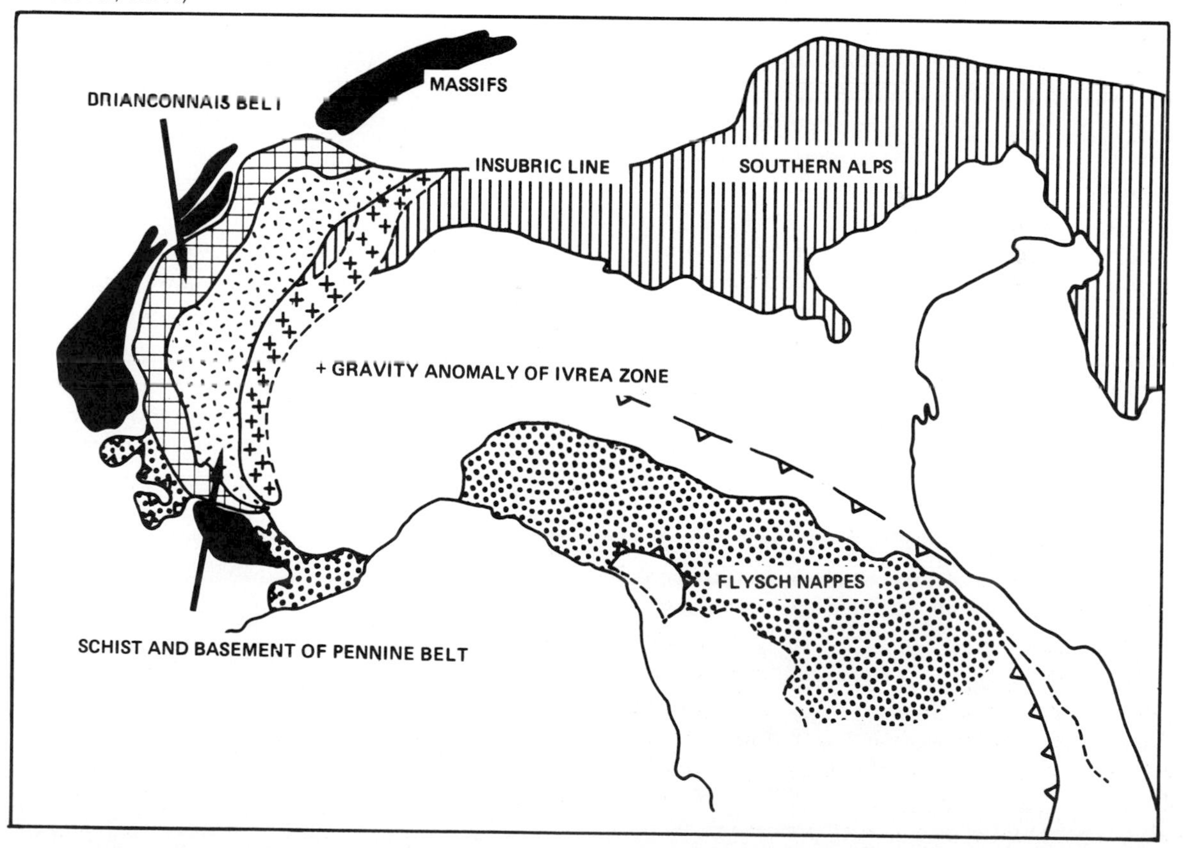

evolution of the orogen is another important aspect of tectonics. This involves removal of deformational effects, restoration of structural elements to their original positions, and reconstruction of the paleogeography. The tools used to accomplish this are largely those of stratigraphy and involve the use of thicknesses and sedimentary facies to identify former sedimentary basins and environments of deposition. Paleontology and geochronology are in turn used to sort out ages. Knowledge of structural geology and petrology is employed to decipher various mechanisms of deformation and importance of metamorphic effects in the deformation.

Paleogeography of the alpine geosyncline at various stages in its development has been reconstructed despite the structural complexity of the rocks involved. A shelf region and miogeosyncline (Helvetic realm) persisted in the external parts of the Alps throughout the depositional history. The internal zones consisted of two deep geosynclines—the Valais and Piemont from which the Penninic nappes would eventually evolve—separated in the west by a platform, the Brianconnaise. Farther south lies the source area of the Austroalpine nappes. The evolution of this sector of the Alpine system has been synthesized by Trümpy (1973) as outlined below.

Orogenic Evolution of the Alps

Mid-Paleozoic: Main Hercynian orogeny (ca. 310 my.).

Permian: Late Hercynian disturbances, emplacement of granite plutons, and volcanic activity.

Triassic: Initiation of geosynclinal subsidence, but no deep-water sediments of this age are known.

Lower Jurassic: Acceleration of subsidence—development of deep-sea conditions in geosyncline. Tensional faulting along the margins of Valais and Piemont troughs and Brianconnaise platforms.

Late middle Jurassic to Lower Cretaceous: Deep oceans exist in the internal zones of the Alpine geosyncline. Ophiolites (ultrabasic magmas) emplaced in Valais and Piemont troughs.

Mid- to Late Cretaceous: Paleo-alpine deformation in several episodes (30–50 km crustal shortening). Folding, thrusting, metamorphism (in the southeast). Flysch sedimentation accompanied deformation which started on the southern side of each geosynclinal trough and migrated northward. No more ophiolites emplaced.

Paleocene to Late Eocene: General vertical uplift probably followed by some gravity sliding but little or no lateral compression.

Late Eocene and Lower Oligocene: Meso-Alpine orogeny. Major lateral compression—300 km crustal shortening across geosyncline; main Alpine metamorphism; end of flysch sedimentation; intrusion of Periadriatic granitic plutons; initial movements on Insubric fault. Prealpine, Pennine, and Austroalpine nappes piled up.

Mid–Upper Oligocene: Postorogenic uplift of the Meso-Alps accompanied and followed by gravity gliding—notably of Prealpine nappes and Pennine flysch nappes. Movements continue on Insubric and related faults. Start of molasse sedimentation. Cooling of metamorphic rocks in Pennine and Austroalpine nappes.

Miocene: Neo-Alpine orogeny—formation of the Helvetic nappe complex with thrusting over molasse. Crustal shortening in area now occupied by the basement massifs (e.g., Aar-Gotthard). Continued uplift in Swiss Alps.

Late Miocene–Pliocene: Uplift of external basement massifs and accompanying gravity gliding. End of molasse sedimentation north of Alps.

Pliocene: Jura folding. Erosion in Alps; sedimentation in the Po basin.

Late Pliocene–Pleistocene: Renewed uplift.

The history of alpine deformation as summarized above leads to a number of interesting observations about the course of orogeny in the Alps:

1. The deformation is polyphase, consisting of three quite distinct and short-lived orogenic epi-

sodes. The oldest episode affected the most internal part of the belt, with younger deformation affecting successively more external parts.

2. Each of the three main phases is accompanied by apparent crustal shortening. That is, the width of preserved basement is not great enough to accommodate the amount of deformed cover. It thus appears necessary to remove the basement; the only possible place for it to go being down into the mantle.

3. Uplift followed and was not contemporaneous with each of the main phases. Thus the mountain formation followed crustal shortening, which apparently took place without notable development of high relief. When uplift did occur, gravity gliding on incompetent materials (shales, salt, etc.) took place. Thus gravity tectonics seems separated in part at least from crustal shortening.

4. Igneous and volcanic activity played a minor role in development of the structures now exposed in the Alps. High-grade metamorphism of a regional type is confined to the internal parts of the geosyncline. The granitic intrusions are not of great regional extent and they were emplace following nappe deformation. Great crustal compression is evident but without exceptionally high temperatures.

PLATE-TECTONIC INTERPRETATIONS OF THE ALPINE-HIMALAYAN SYSTEM

The size, variety, and structural complexity of this system all suggest that tectonic models may be complex as well; and indeed they are, and only a bare outline can be considered in the space available here. Plate-tectonic interpretations of the Alpine orogen involve subdivision of this orogen into a large number of small plates located between the African and Eurasian plates in the west and between the Indian and Eurasian plates in the east. These plates are bounded by subduction zones, divergent junctions, and major strike-slip faults which are interpreted as transform faults. Movements and adjustments along these plate

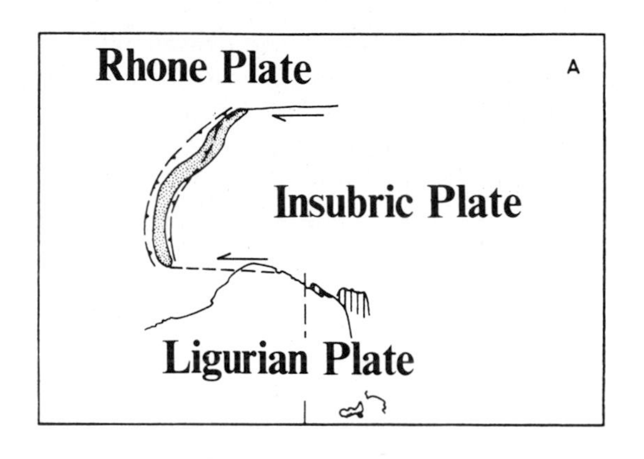

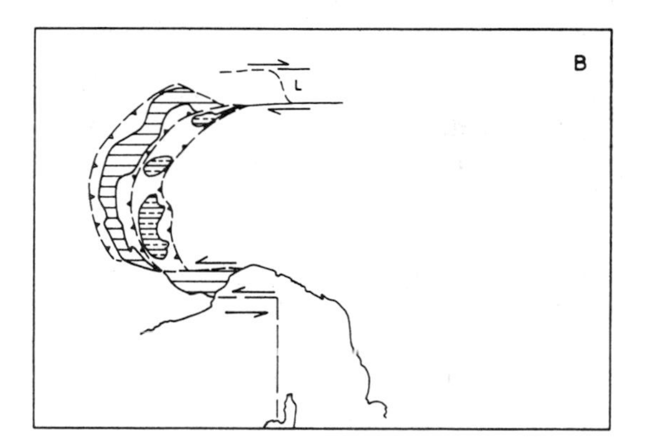

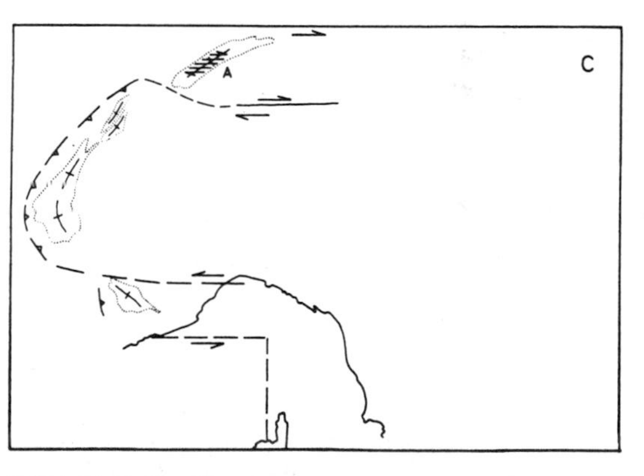

FIGURE 23-15 The major post-Oligocene kinematic elements of the western Alps (Insubric plate). (*a*) The Austroalpine element; (*b*) the Pennine element; L = transverse zone of the Lepontine Alps; (*c*) the Dauphiné (massifs externes) element; A = dextral shearing in the Aar massif. These elements, taken together, constitute the "Insubric plate," whose boundary zone apparently becomes more diffuse in the domain of (*b*) and (*c*) (*From Laubscher, 1971.*)

boundaries are related to the overall relative movement of the African, Eurasian, and Indian plates.

Plate Models for the Western Alps and Northern Apennines

Laubscher's (1971, 1975) analysis of the Alpine orogen in southern Europe is a good example of the plate-theory approach to interpretation of the complex regional structure of a part of the Alps. The major tectonic elements of this region are postulated to be parts of three large subplates—the Insubric plate, the Rhone plate, and the Ligurian plate, Fig. 23-14. The region north of the Insubric line and west of the basement massifs makes up the Rhone plate. Thus the Jura Mountains, the Swiss Plain, the Prealps, the Subalpine chains, and even the High Calcareous Alps and parts of the Pennine Alps are located on the Rhone plate. The Rhone plate is separated from the Insubric plate by the Insubric line, interpreted as a strike-slip fault on the north, and by thrusts located west of some of the basement massifs, Fig. 23-15. The southern boundary is formed by a strike-slip fault zone postulated to exist at depth.

The Ligurian plate consists of the Cretaceous flysch nappes located in the Apennines of northern Italy. This plate is necessary because the direction of tectonic transport in the Apennines is toward the north and east. The kinematics of the plate movements needed to obtain the present geometry, Fig. 23-16, satisfy the requirements of the presently observed directions of thrusting. Because the basement is not involved in the overthrusting, it is necessary to strip the sedimentary veneer off its basement, which is thus postulated to sink or be subducted.

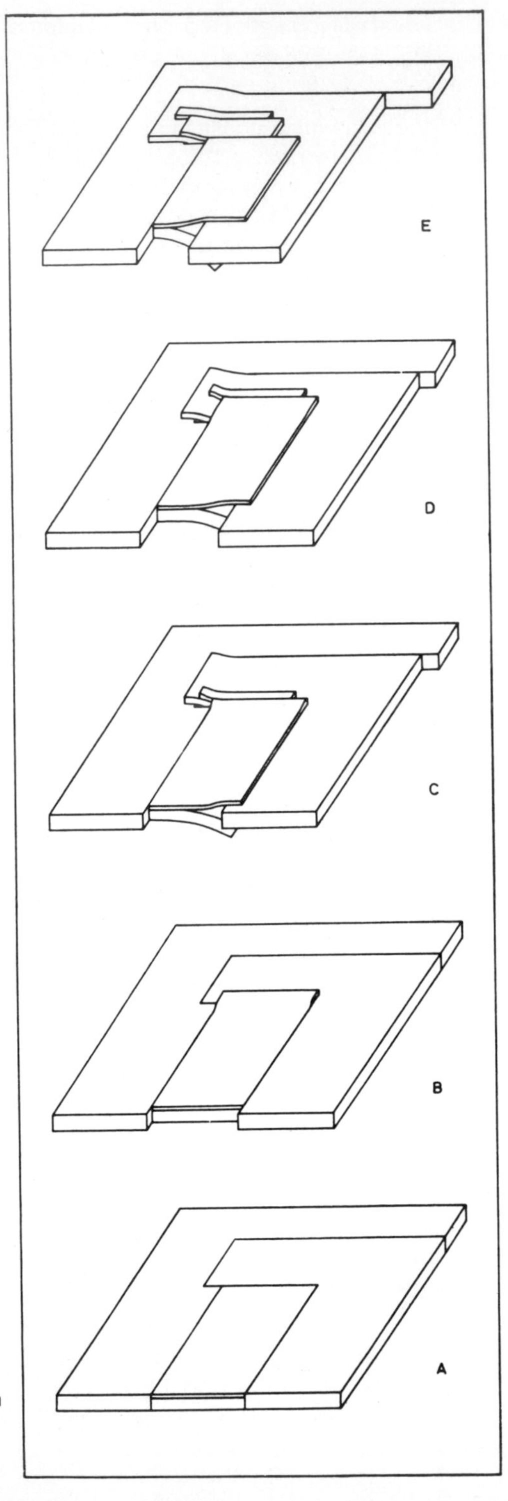

FIGURE 23-16 Qualitative sketch of the three-plate kinematics, involving disappearance to the west of the Ligurian lithosphere. Post-Oligocene thrusting of Ligurian masses (Corsica), which is possible but certainly of minor importance has been neglected. (*From Laubscher, 1971.*)

Dewey, Pitman, Ryan, and Bonin (1973) have attempted an even more ambitious plate-tectonic synthesis of the Alpine System. They obtain major restraints on plate positions by use of magnetic anomalies in the Atlantic Ocean from which they obtain restrictions of the relative positions of the Eurasian, North American, and African plates at various times over the last 180 million years. Motions of the smaller plates between Africa and Eurasia are inferred from the structural styles observed along their margins. The evolution of the Tethys traced in this study is exceedingly complex and involves "a constantly evolving mosaic of subsiding continental margins, migrating mid-oceanic ridges, transform faults, trenches, island arcs, and marginal seas (black-arc basins)" (Dewey and others, 1973, p. 3137). Although the evolution of the system as a whole is complex, some elements are capable of relatively straight forward interpretations in terms of plate theory, and we will examine one of these next.

A Plate Model for the Arabian-Zagros Area

The structure of the region made up largely of the Arabian Peninsula affords one of the best examples of the way plate theory can be used to synthesize tectonics. This plate consists of a continental lithosphere block which is bounded by the Gulf of Aden, the Red Sea, the Levant fault system, and the Zagros Mountains, Fig. 23-17. Both the Gulf of Aden and the Red Sea are well known and are often cited as examples of young spreading centers. Their margins demark the lines along which Arabia and Africa rifted as Arabia moved relatively toward the northeast away from Africa. Such a movement is compatible with the structural features observed from the northern end of the

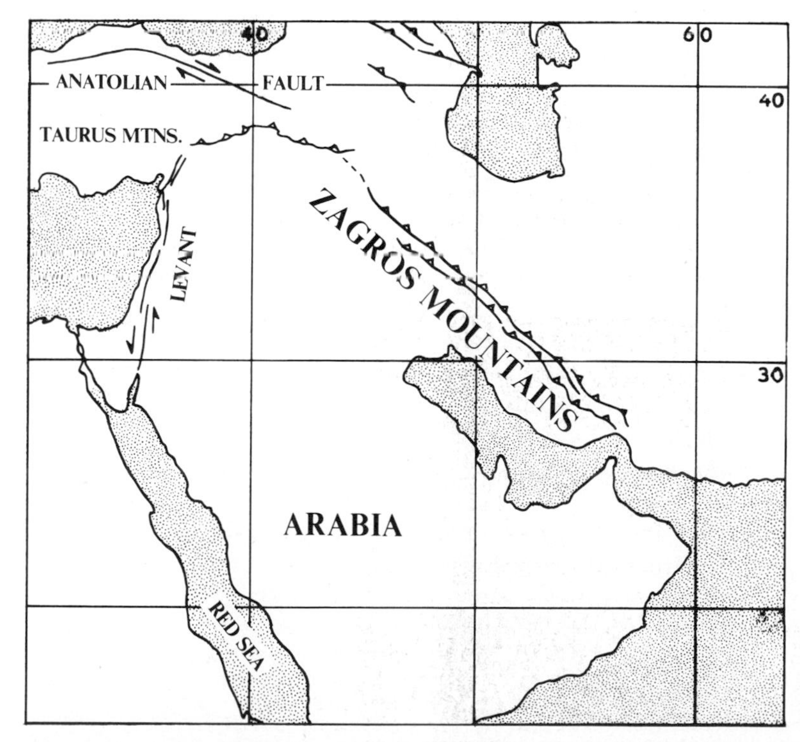

FIGURE 23-17 Major tectonic elements of the region near the Zagros Mountains, including the Red Sea, Levant, Zagros thrust and fold belt, and the Anatolian fault.

Red Sea into southern Turkey, where the Levant fault system consists of a complex of faults and folds which formed in the sedimentary cover over a deep-seated left-lateral strike-slip fault zone as described earlier in Chap. 10. The Zagros Mountains form the remaining boundary of the Arabian plate.

The Arabian plate has a crystalline basement overlain by a relatively thin, slightly folded, sedimentary cover of Mesozoic and Cenozoic age. Folding of this cover is most pronounced near the Zagros Mountains, which may be subdivided into several more or less continuous belts. The slightly folded belt passes into a broad zone of imbricated northeast-dipping thrust faults. The northern side of the thrust zone is called the *trench zone* because the strongly deformed rocks here include deep-water sediments with some radiolarian cherts. Blocks of metamorphic rocks and limestones found in this zone are interpreted as parts of mélange deposits formed by submarine slump into the trench from its northern side. Lenticular pods of ultramafic rocks aligned parallel to the orogen also occur in the trench zone, and these appear to have been emplaced as thrust slices. Farther to the north the trench zone rocks are overlain by a wide belt of Cretaceous limestones which are severely tectonized along their southwestern margin.

The Zagros Mountains represent a suture zone marking the former position of a subduction zone according to plate-tectonic interpretations. A model of the evolution of this zone is beautifully illustrated, Fig. 23-18, by Haynes and McQuillan (1974). They infer that a segment of oceanic crust separated the Arabian plate from a stable block in central Persia early in the Mesozoic. Later the plate begins to move northeast, a trench forms, and this trench becomes the site first of subduction and deep-water sedimentation and later the site of intense shearing, emplacement of slices of ultramafic rock, and imbricate overthrusting.

Thus the plate model affords a synthesis not only of the structure of the Zagros Mountains but of the independently inferred structure of the Red Sea and Levant fault zone.

PLATE MODEL FOR THE HIMALAYAS

The idea that the Himalayas were formed by the collision of India with Asia was suggested long before the advent of modern plate theory,* so, it is not surprising that when plate theory did evolve the Himalayas became the type example for the continental-collision model of orogeny. Indeed the overall configuration of the region strongly suggests that an angular block, the Indian plate, was forcefully shoved into the weaker rocks of the Tethys geosyncline which yielded folding and breaking as the Indian plate was forced northward, Fig. 23-2. The angular edges of the Indian plate are seen as deflecting and bending the folds of the orogen and ultimately displacing the whole trend of the fold system and thrusting under the uplifted geosynclinal rocks.

Although field work in the Himalayan region is still in its early stages, enough work has been completed to provide the details needed to evaluate this hypothesis.† India is composed of continental crust with Precambrian cratonic rocks exposed in the south and covered by relatively undeformed sedimentary cover in the north, the upper part of which is composed of thick molasse deposits derived from the high mountains.

The simplicity of the structural features in the cover gives way abruptly along the sides of the plate to zones of complex folding and fault-

* Carey (1958), Argand (1924).

† Fuchs (1969), Gansser (1964), Powell and Conaghan (1973), and Lefort (1975).

FIGURE 23-18 Schematic model for development of the Zagros suture zone. Relative movements between plates assume a fixed Persian plate. (*From Haynes and McQuillan, 1974.*)

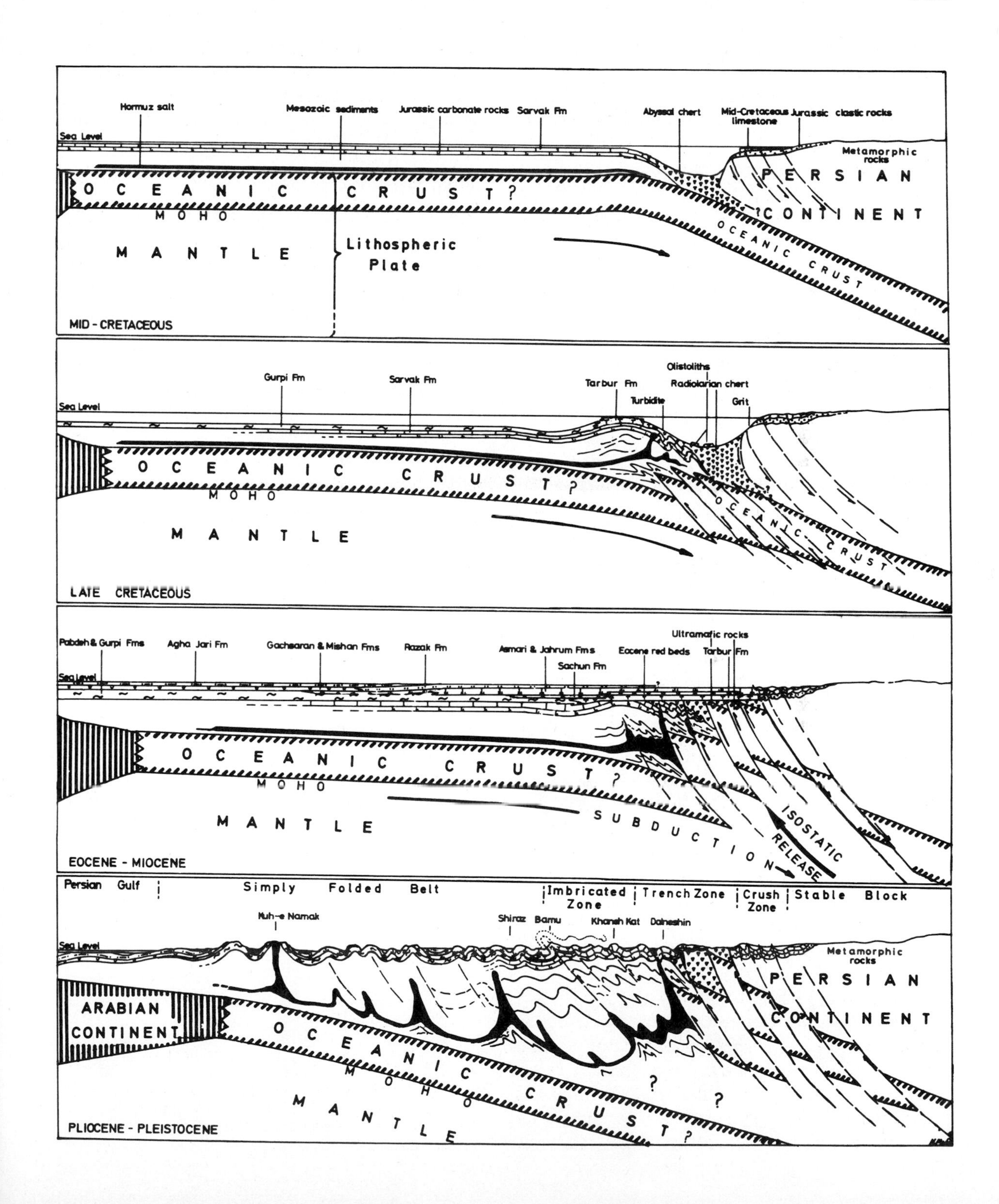
Hormuz salt
Mesozoic sediments
Jurassic carbonate rocks
Sarvak Fm
Abyssal chert
Mid-Cretaceous limestone
Jurassic clastic rocks
Sea Level
Metamorphic rocks
OCEANIC CRUST?
PERSIAN CONTINENT
MOHO
MANTLE
Lithospheric Plate
OCEANIC CRUST
MID-CRETACEOUS
Gurpi Fm
Sarvak Fm
Olistoliths
Tarbur Fm
Radiolarian chert
Turbidite
Grit
Sea Level
OCEANIC CRUST?
MOHO
MANTLE
OCEANIC CRUST
LATE CRETACEOUS
Pabdeh & Gurpi Fms
Agha Jari Fm
Gachsaran & Mishan Fms
Razak Fm
Asmari & Jahrum Fms
Eocene red beds
Ultramafic rocks
Tarbur Fm
Sachun Fm
Sea Level
OCEANIC CRUST?
MOHO
MANTLE
SUBDUCTION
ISOSTATIC RELEASE
EOCENE - MIOCENE
Persian Gulf
Simply Folded Belt
Imbricated Zone
Trench Zone
Crush Zone
Stable Block
Kuh-e Namak
Shiraz
Bamu
Khaneh Kat
Dalneshin
Sea Level
Metamorphic rocks
PERSIAN CONTINENT
ARABIAN CONTINENT
OCEANIC CRUST?
MOHO
MANTLE
PLIOCENE - PLEISTOCENE

ing—the Arakan Yoma and the Quetta on the sides and the thrust zones of the Himalayas to the north. The two lateral zones are aligned with strong lineaments in the floor of the Indian Ocean—the Owen-Murray fracture zones which are aligned with the Quetta and the Ninety East ridge which lines up with the Arakan Yoma. Both of these lateral zones are relatively narrow, and the internal structures are strongly aligned with the zones, including prominent strike-slip faults which have displacements, which indicate that India has moved north relative to both flanks.

The northern end of the Quetta is especially interesting because it is here that the Quetta passes into the folds and thrust of the Himalayas, Fig. 23-19, around the sharp protruding corner of the Indian plate. Structure of the Quetta is dominated here by a system of parallel left-lateral strike-slip faults everywhere except in the corner itself, where a conjugate right-lateral fault appears. Folds in the Quetta form a belt which bends sharply and opens out in the Himalayas. Many of the folds are *décollement*-type folds formed where the sedimentary veneer has been sheared off its under-

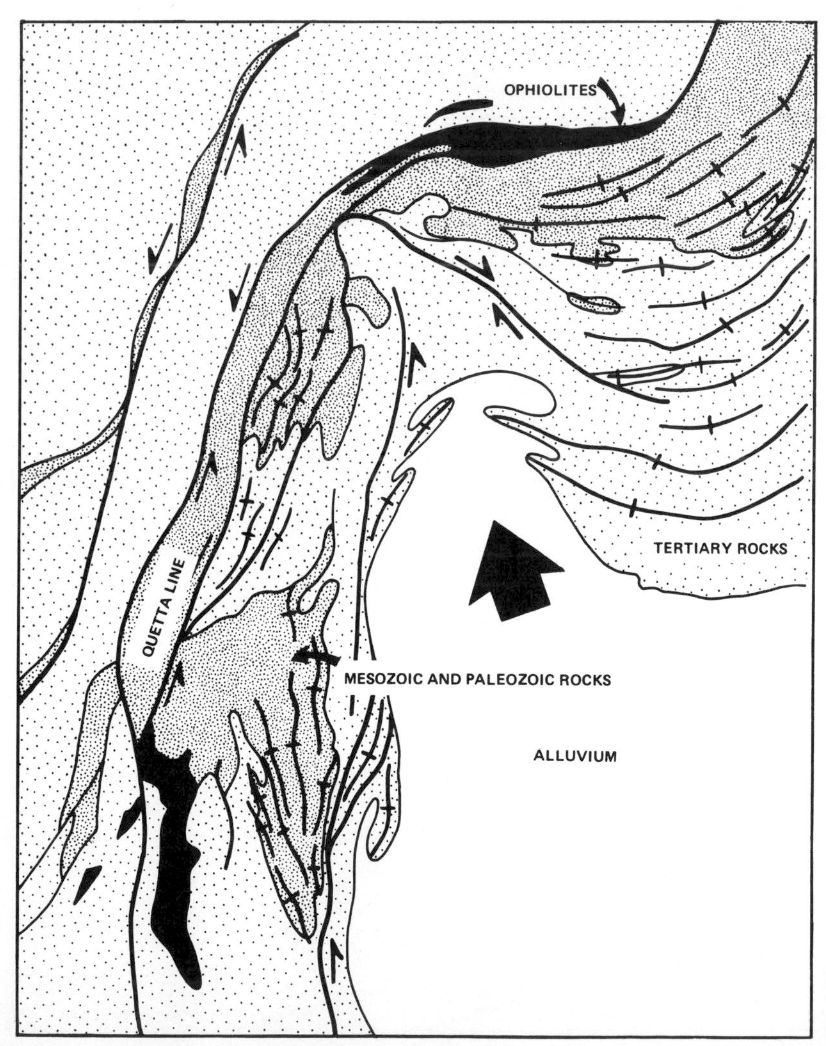

FIGURE 23-19 Regional structure along the Quetta Line. This figure illustrates the present structural configuration at the northwestern corner of the Indian plate where it is in contact with the Alpine-Himalayan orogen. (*After Gansser, 1964.*)

lying basement. It is important to note that both the Quetta and the Arakan Yoma are sites of ophiolite bodies, which support the idea that these zones cut very deep into the mantle.

The Himalayas* sweep in a great arc for more than 2,000 km along the northern edge of the Indian plate and can be subdivided into a succession of concentric zones from south to north as follows, Fig. 23-20—the molasse foredeep, the Lower Himalayas, the High Himalayas, the Tethyan Himalayas, the Indus Suture zone, and the Tibetan Plateau. The molasse thickens and becomes increasingly deformed as one goes toward the Himalayas from the south. The contact between the molasse foredeep and the Lower Himalayas is formed by a major, long, north-dipping thrust, the Main Boundary thrust, which carries Paleozoic and older sedimentary and metasedimentary rocks onto the Tertiary-age molasse. The rocks in the Lower Himalayas are largely Paleozoic and Precambrian in age, but large sections of them have been affected by a Tertiary metamorphism which accompanied the Himalayan orogeny. This metamorphism produced extensive migmatites and ultimately granites which invade Mesozoic rocks of the region.

The Lower Himalayas are essentially a pile of nappes or overthrust sheets, some of which are large recumbent folds. In general, progressively older rocks are involved in the thrusting toward the more internal part of the belt. Despite the strong young deformation some older patterns are still evident. Notably the young thrusting cuts across northeast-trending facies boundaries and structural features which reflect the same structural grain as the shield rock of India. Thus, the features produced by the Himalayan deformation are not coincident with the earlier patterns.

One of the interesting features of the Lower Himalayan region is the presence of regionally metamorphosed rocks which are of low grade or even unmetamorphosed at lower elevations but increase in grade with elevation—a reversed regional metamorphism. In one notable example slates at low elevations pass upward into schists and finally gneiss. Such sections are usually interpreted as being structurally inverted and part of the lower limb of a large recumbant fold, but Gansser (1964) points out a similar reversed metamorphism in some normal stratigraphic sections.

The Main Central thrust separates the Lower from the Higher Himalayas in most sections. This thrust brings a crystalline sheet sometimes exceeding 10 km in thickness onto the rocks of the Lower Himalayas. The metamorphism in this great crystalline thrust sheet is normal but its age is uncertain. The crystalline basement is overlain by Lower Paleozoic sedimentary rocks in conformable sequence in some sections (Gansser, 1964), but Jurassic fossils are reported in the gneiss in other places. In either case it is very difficult to connect the gneisses of the High Himalayas with the reversed metamorphic sequences of the Lower Himalayas. The highest peaks in the Himalayas are held up by the crystalline sheet and its cover, which has a general northward inclination so that younger unmetamorphosed

* See Gansser (1964).

FIGURE 23-20 Structural cross section across the Himalayas from south (left) to north (right). No. 1, Precambrian Shield; No. 2, granites; No. 3, ophiolitic rocks; No. 4, gneiss and migmatite; No. 5, sedimentary rocks of Paleozoic and older ages; No. 6, Tethys late Precambrian and lower Paleozoic; No. 7, Upper Paleozoic and Mesozoic rocks; No. 8 Tertiary rocks. (*After Gansser, 1964.*)

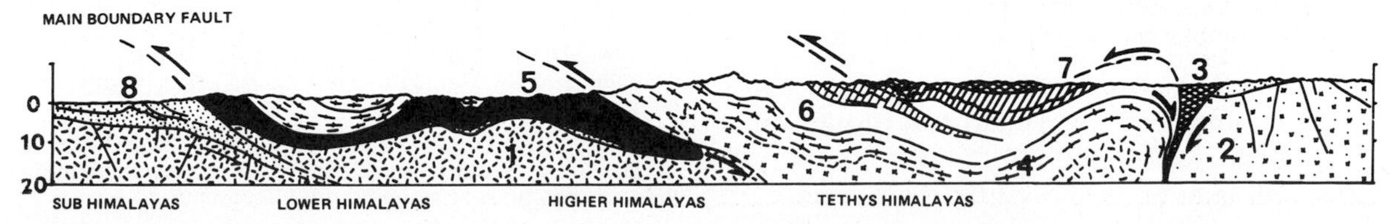

and less resistent rocks are exposed north of the high peaks. It is in this belt far north of the southern margin of the orogen that the Mesozoic sedimentary rocks laid down in the Tethys geosyncline first form a prominent part of the orogen. Clearly the orogen's development was not confined to the geosyncline alone but involved the northern margin of the Indian plate. The portion of the Himalayas in which the Tethys sedimentary sequences are prominent is often called the Tethys or Tibetian Himalayas (Gansser, 1964, p. 253):

Only in the Tibetian Himalayas do we find a geosynclinal influence in the sediments formed in a shallow, gradually sinking Tethys sea. We thus realize that the main Himalayan range has not developed from a geosyncline. . . . With the beginning of the Himalayan orogeny in middle and upper Cretaceous a deep seated tectonic disturbance occurred in the Tibetian Himalayas, and was responsible for the outflow of a large amount of ultrabasic rocks.

These ultrabasic rocks are confined to a long, but narrow zone known as the Indus Suture line which is often compared with the Insubric line in the Alps. The rocks in the Indus Suture include Cretaceous shallow-water sediments which were succeeded by a mélange of shales, radiolarites, marine flysch-type orogenic sediments, ultrabasics, and exotic blocks of Permian- and Paleogene-age sedimentary rocks thought to have formed in a deeper geosycline belt which is no longer found (Gansser, 1964, p. 254):

The northern exotic flysch belt must have been compressed by the underthrusting of the Indian Shield against the Tibetian mass, which at the same time was elevated. Large areas of geosynclinal deposits must have buckled down and disappeared, and this would explain the fact that the depositional areas of the exotic blocks remain so far undiscovered.

Thus deformation in the Indus Suture zone marked collision of the Asian and Indian continental blocks, but the uplift of the mountains came still later as is shown by the presence of shallow water Nummulitic limestones in the Himalayan region. The seas regressed in the Eocene, and the terrigenous molasse-like clastics began to flood southward in the Miocene. Uplift clearly continued into the Pleistocene and may well be going on today. A model for the collision interpretation has been described by Powell and Conaghan (1973).

THE EASTERN MEDITERRANEAN*

Geophysical studies, detailed bathymetric mapping and coring in the Mediterranean, in combination with continuing land-based studies, have begun to illuminate the fascinating but complex structure of this great basin. Gilbert Smith (1971) has created a predrift reconstruction of the continental crustal elements in the Mediterranean by use of a computer fit designed to join the separate blocks at the 1,000-m contour, Fig. 23-21. The relative positions of Europe and Africa are based on earlier Atlantic reassemblies. This requires a clockwise rotation† of most of the elements to effectively close the sea, but this can be done only by opening a great sea to the west, the Tethys. Carey (1958) resolved the problem by postulating global expansion, but few of the more recent studies of the region have followed this suggestion.

Gilbert Smith concludes from Atlantic-spreading data that the opening of the Mediterranean has been accomplished in three phases—an eastward movement of Africa in the Lower Jurassic, a westward movement of Africa in the Upper Cretaceous, and finally a northward movement from Eocene to the present. This northward movement is most compatible with the plate-tectonics model of the

* Refer to Rabinowitz and Ryan (1970), Comninakis and Papazachos (1972), Papazachos and Comninakis (1971), Ninkovich and Hays (1972), Lort (1971), Alvarez (1973), Klemme (1958), Woodside and Bowin (1970), Wong and others (1971).

† Lowrie and Alvarez (1974) report paleomagnetic data which indicate at least a 43° counterclockwise rotation of the Italian Peninsula since Upper Cretaceous.

eastern Mediterranean, where an island arc and accompanying subduction zone are postulated.

The arcuate form of the submarine topography is clear on the most recent bathymetric maps, a simplified version of which is shown in Fig. 23-22, but the form is more complicated than that associated with the typical Pacific arcs. Several major submarine morphologic features sweep in the great arc from the southern end of Italy to the eastern end of the Mediterranean—the Herodotus abyssal plain connects along the African coast to the Ionian Sea in the central Mediterranean. This deep is separated from an arcuate zone of great deeps, called the Hellenic trench, just south of Peloponnesus, Crete, and Rhodes by a broad ridge called the Eastern Mediterranean ridge. Detailed topographic and subsurface soundings reveal a system of folds that are parallel to the arc.

A large number of the features we associate with island arcs are clearly present here—an arcuate chain of active volcanos, a trench-like series of deeps, external with respect to the volcanos and coincident with a large negative free-air gravity anomaly belt, and a concentration of earthquake epicenters. Even low regional heat-flow valves are found here as in other subduction zones.

This arc bears an especially interesting rela-

FIGURE 23-21 Provisional reassembly along the 1,000-m submarine contour of the continental areas around the Mediterranean, traced from a computer-drawn Mercator map with the latitude and longitude of Eurasia in their present-day positions. The fit assumes that the 1,000-m contour marks the boundary between oceans and continents, and that the wedge-shaped oceans in the western Mediterranean have been formed by turning adjacent fragments about a point near the apex of each wedge. Thus, the Balearic Islands have been fitted to Spain; Corsica-Sardinia to Italy; Corsica-Sardinia-Italy to France-Balearic Islands-Africa; Greece-Turkey to Italy-Sicily-Africa. The positions of Africa, North America, Eurasia, and Spain have been taken from the Atlantic reassembly of Bullard and others (1965). The position of Arabia has been taken from Smith and Hallam (1970). The northern and eastern boundaries of all the fragments around the Mediterranean are uncertain but must lie within some part of the Alpine orogenic belts. In particular, it is not known to where the fragments that are believed to have once been attached to the Crimea and the southern parts of the Black Sea have moved. Coastlines are the thicker contours, lines within the coasts are present-day latitude-longitude lines at intervals of 10 degrees; the 500-fathom or 1,000-m contours are continuous, dashed, or dotted thin lines. Present land areas deformed during the Alpine orogeny are stippled. (*From Gilbert Smith, 1971.*)

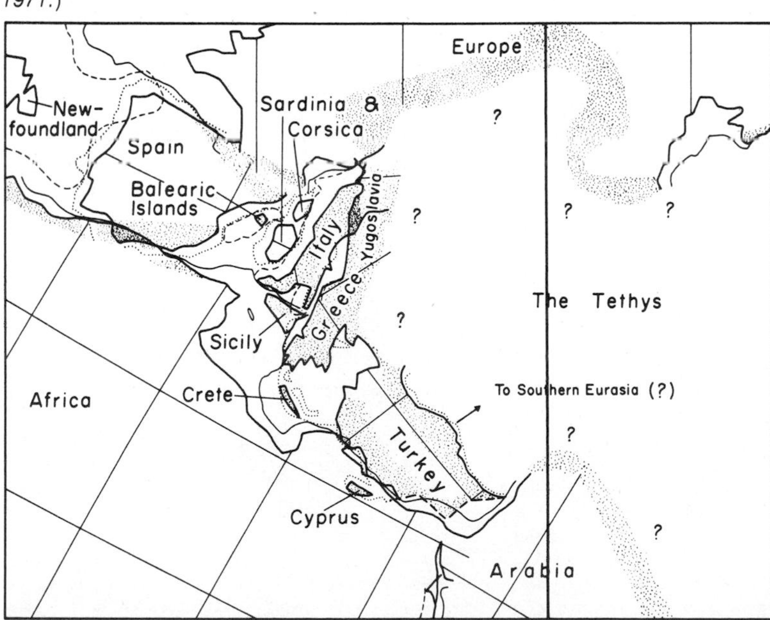

tionship to the belt of Alpine folds and thrusts which are parallel to the Hellenic trench, Fig. 23-22. The folds and thrusts in the Hellenids of Greece exhibit movement toward the convex side of the arc, again suggesting an overriding of the arc over the Mediterranean floor, or underthrusting, subduction, of the sea floor. Boccaletti, Manetti, and Peccerillo (1974) give a good overview of the more internal parts of the system. They point out the existence of a high-angle strike-slip fault system that exhibits the geometry of a conjugate shear set with the principal stress oriented horizontally in a north-south direction, just what one might predict on the basis of other structures in the region. It is interesting to note the northwardly directed thrusting just south of Bucharest which they interpret as back-arc thrusting, caused by the collision of the African plate with the island arc. Notably these back-arc thrusts do not involve great lateral transport; nor are they associated with ophiolites, or radiolarite-type rocks, all of which suggest that they are of shallow origin. Bocaletti and others further interpret the internal structure as resulting from a southward migration of the arc-trench system during the Tertiary. This migration would be accomplished by shifting of the position of the subduction zone to more

FIGURE 23-22 Map of the eastern Mediterranean Sea showing the major physiographic features. Note especially the deeps located along the southern edge of Peloponnesus and Crete, the Hellenic trough. Historically active volcanoes are represented by large black dots (left). Free-air gravity anomalies contoured at 20-mgal intervals are shown (right). The heavy line indicates the trend of alpine structural features through the islands. (*Modified after Rabinowitz and Ryan, 1970.*)

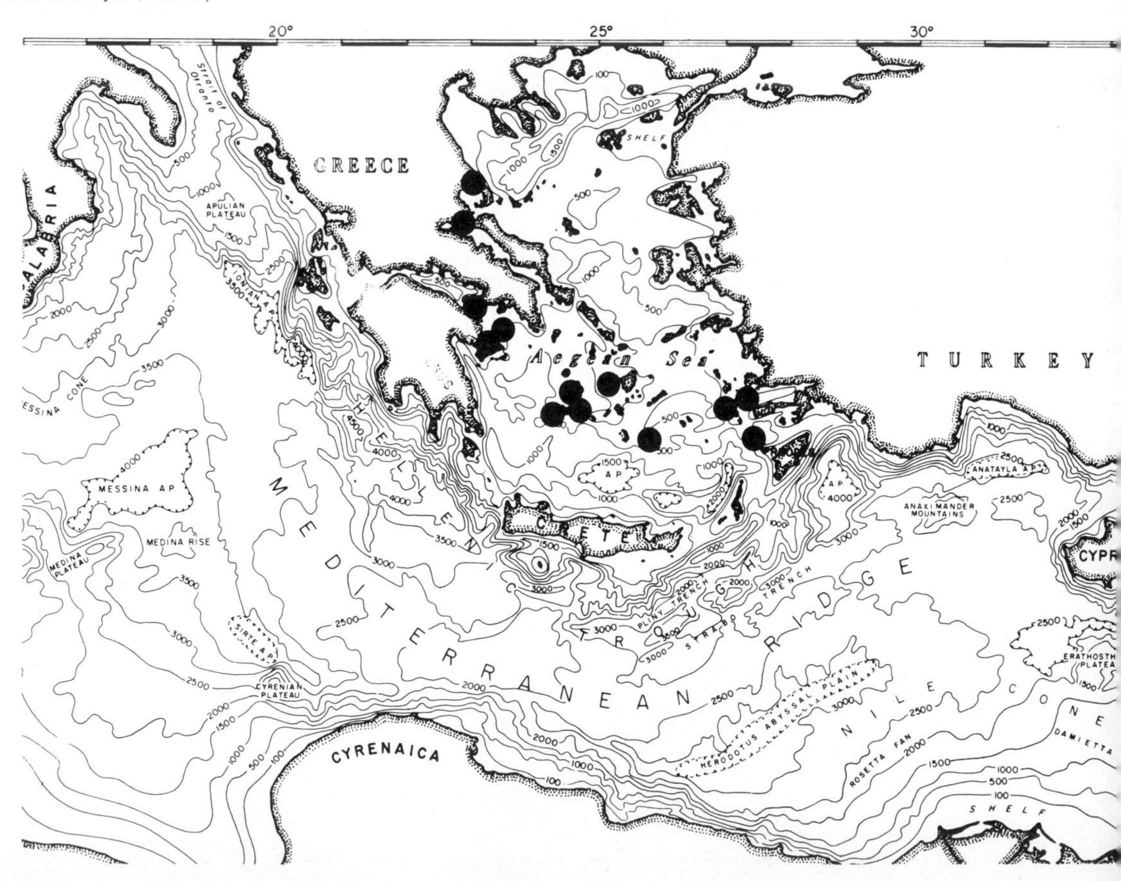

southerly positions and has been accompanied by extension in the Aegean Sea which has increased the curvature of the present Hellenic arc.

According to the present plate-theory model of the eastern Mediterranean we may envision that the sea is now being closed as Africa moves north relative to Asia. The floor of the Mediterranean is sinking into the mantle along a north-dipping subduction zone. Woodside and Bowin (1970) have concluded on the basis of gravity analysis that the subduction zone dips about 7° to the north and that the crust has been thickened by nearly 50 percent as a result of the subduction. These two great continental blocks are close together, and the Mediterranean ridge may well be a large fold formed as the two plates approach collision. Seismicity is concentrated along the East Mediterranean ridge, and fault-plane solutions here show that the axis of compression is horizontal and perpendicular to the ridge, again supporting the plate model.

THE WESTERN MEDITERRANEAN*

The tectonics of the western end of the Mediterranean is as interesting as that in the east, and here too we confront an arcuate feature, the Gibraltar arc; but this one certainly appears very different from the Hellenic arc in the east. The Gibraltar arc is an orogenic element of Alpine age situated between the Her-

* See Fontboté (1970), Mattauer (1963), Durand-Delga and others (1962).

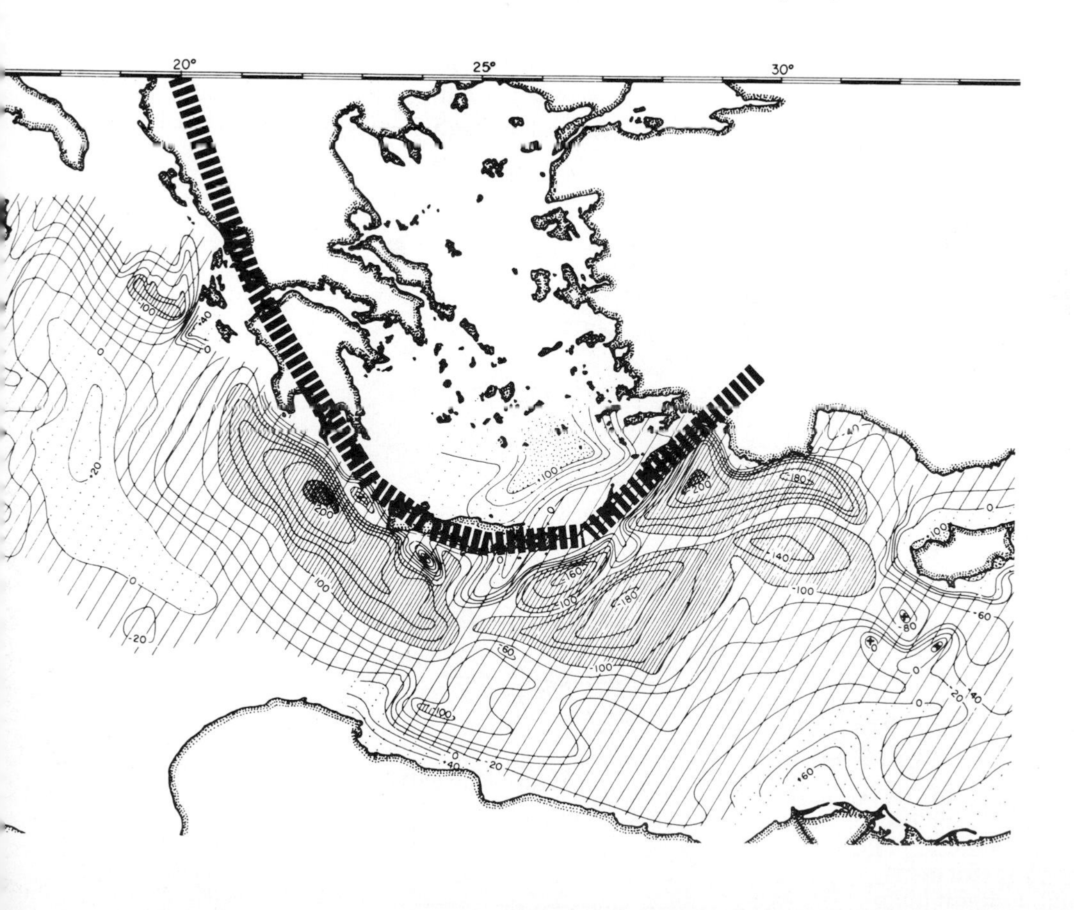

cynian basement complexes of the Iberian Peninsula and the African plate, Fig. 23-23. It is separated from the stable platform of central Spain by a strong lineament, the Guadalquivir fault, along which a right lateral movement is postulated. In the south the frontal structural features of the Gibraltar arc are bounded by and superimposed on the Moroccan Meseta (platform) and the Atlas Mountains.

The Gibraltar arc is in many ways a remarkably symmetrical orogen, with the plane of symmetry passing through the strait of Gibraltar. This symmetry is expressed both in the surficial geology and in the pattern of the gravity anomalies both in the Mediterranean and in the Atlantic, Fig. 23-24. One of the most remarkable features of the Gibraltar arc is that the direction of tectonic transport is directed externally (to the north in Spain, west at Gibraltar, and south in Africa) all around the arc; so the appearance is that of all elements moving outward and away from the Alboran Sea, Fig. 23-25. Folds in the most external belts, the Prebetics and the Atlas, exhibit trends that are generally parallel to the fronts of the external nappes of the Subbetics and the Intrarif zone where movement of the cover was facilitated by the presence of Triassic salt. Large *décollement*-type structural features as well as thrusts characterize the structure in the Subbetics. Allochthonous Triassic elements were carried onto Middle Miocene sediments in the Prebetic zone, establishing the age of at least one major episode of the deformation.

Rocks of the Subbetics are predominantly made up of Mesozoic and early Tertiary carbonates to the north and flysch toward the south, including the great thickness of Cretaceous-Eocene flysch at Gibraltar. Some of these rocks in the Intrarif zone exhibit schistosity.

The internal zone of the Gibraltar arc consists of a highly complex system of nappes

FIGURE 23-23 Schematic tectonic map of the Gibraltar arc. Nappes of the internal zones are outlined by solid barbed lines; external massifs with schistosity (dashed lines); external front of the thrusts (barbed line). (*Modified after Andrieux, Fontboté, and Mattauer, 1971.*)

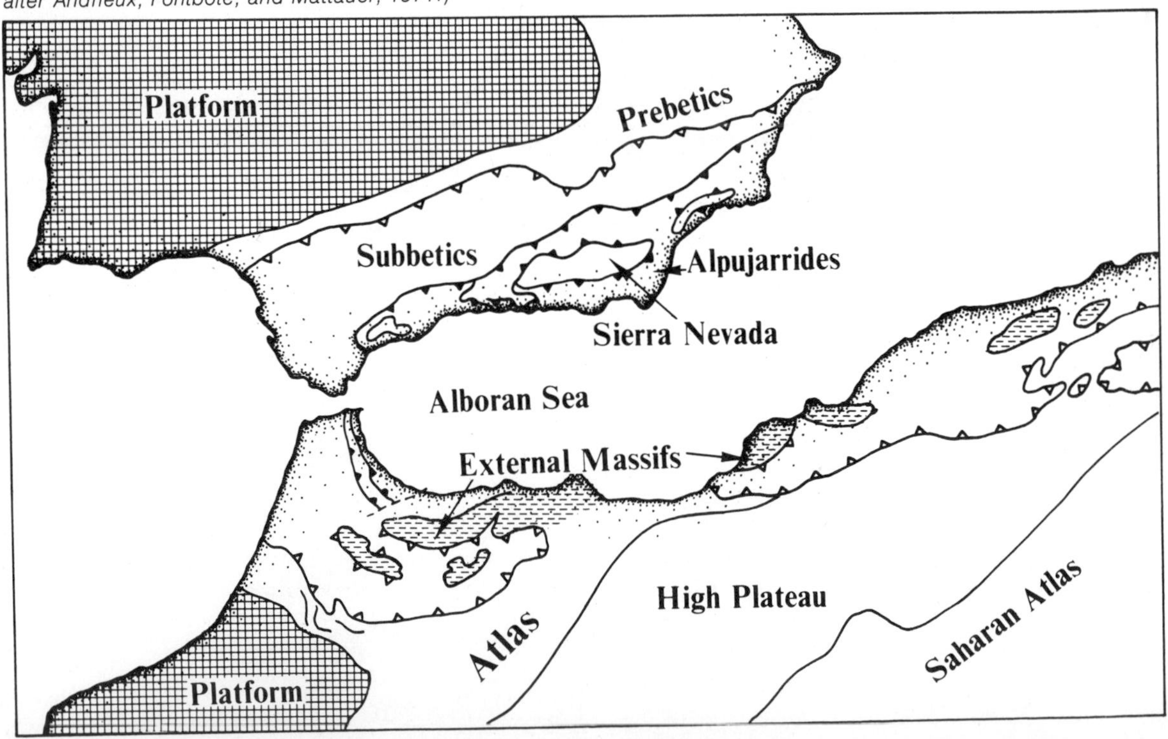

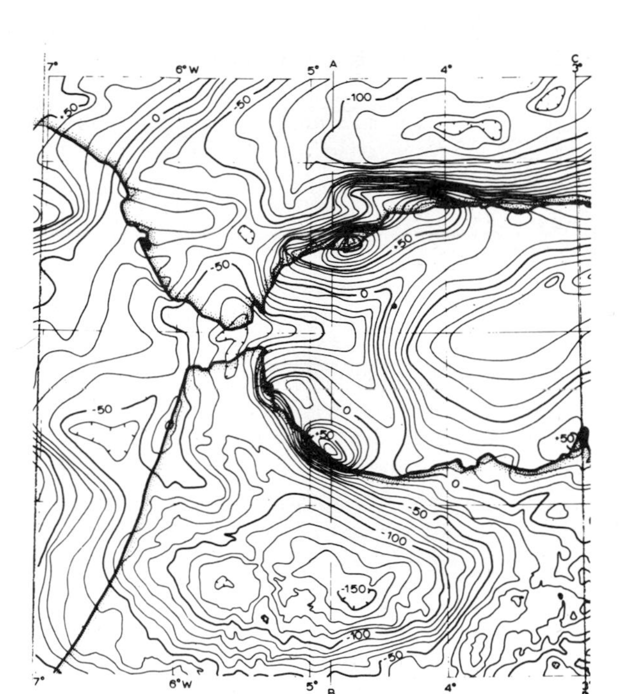

FIGURE 23-24 Bouguer gravity anomaly map of the region around the Strait of Gibraltar. (*From Bonini, Loomis, and Robertson, 1973.*)

which are described as being similar in both the Betic and Rif zones. These nappes are composed largely of metasedimentary rocks and include some peridotites and other ultrabasics (notably the Ronda complex). The nappe systems are well exposed on the flanks of the Sierra Nevada Mountains and along the coast of Spain where the lower nappes (Lujar) are thrust northward over the possibly autochthonous schists and amphibolites of uncertain age in the Sierra Nevada Mountains. The Lujar rocks are overridden in turn by the Alpujarride and the Malaga nappes, all directed externally.

Several authors have advanced interpretations of the Gibraltar arc in terms of plate tectonics. Andrieux, Fontboté, and Mattauer (1971) suggest that the curved externally directed tectonic movement could be due to the presence of a subplate in the Alboran Sea area which remained relatively stable while the African and European plates moved relatively eastward. Araña and Vegas (1974) suggest that the Alboran Sea has been the site of two phases of opening and closing. The initial opening is postulated for the early Mesozoic followed by a closing of the sea by subduction of the sea floor toward the north which would account for the distribution of calcalkaline and potassic volcanism in southern Spain. This could be accompanied by the type of movement indicated by Andrieux and others and followed by a final opening of the Alboran Sea.

ROLE OF MANTLE DIAPIRISM IN ALPINE OROGENY

Attempts to explain the geology of the Alpine system by plate-tectonic models alone have not been universally accepted by students of the region, although some corollaries of plate theory like movement of rigid continental fragments and opening and closing of parts of the orogen appear in almost all hypotheses. The modification and in some hypotheses the alternative to plate theory most frequently proposed in recent years is that of mantle diapirism.* As an example of how mantle diapirs may be involved in Alpine orogeny it is instructive to consider the Gibraltar arc area described in the preceding section. The large ultrabasic massifs exposed on the inside of the arc are interpreted as high-temperature mantle diapirs that penetrated the continental crust around the edge of an even larger diapir which was located in the area of the Alboran Sea, Fig. 23-26 (Van Bemmelen, 1973; Loomis, 1975). The high positive gravity anomalies that occur over these ultrabasic bodies also extend over a much larger area inside the arc. They delineate the near-surface extent of the mantle and demonstrate that the crust has been thinned over the mantle presumably as a result of crustal extension. The age of the intrusions is Miocene according to Loomis and thus coincides with the main thrusting and tectonic deformation in the more external parts of the Gibraltar

* Van Bemmelen (1973), Loomis (1975), Alvarez (1973), Maxwell (1970).

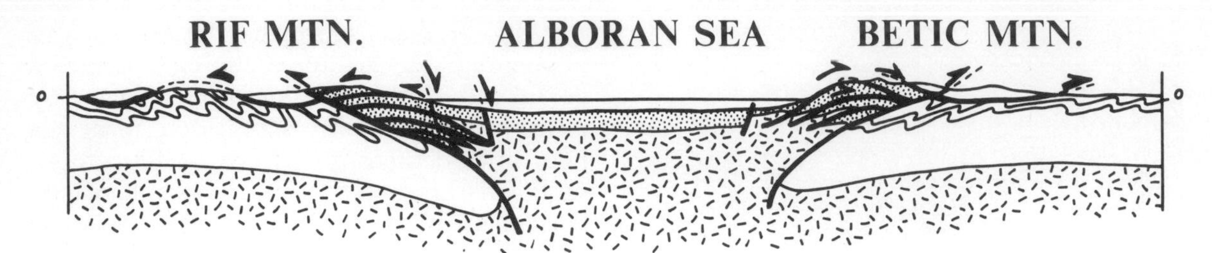

FIGURE 23-25 Schematic cross section across the Betic-Rif Mountain system showing inferred deep structure (*From Andrieux, Fontboté, and Mattauer, 1971.*)

arc and with the movement of the allochthonous flysch units.

The diapiric model proposes that a large mantle diapir, possibly originating in the low-velocity zone, rose in the Alboran Sea, displacing the rock masses around its margin by lateral, outwardly directed pressure initially and finally by outwardly directed gravity sliding down the sides of the diapir. Ductile shale units and evaporite beds facilitated these movements. The crust over the diapir would be thermally metamorphosed, plastically thinned at depth, and extended by brittle failure at higher levels. Finally cooling of the upwelled mantle could cause a volume decrease in the top of the diapir, with coincident subsidence and formation of the Alboran Sea.

The Gibraltar arc is perhaps the best example of a possible mantle diapir in the Alpine system but by no means the only one. Van Bemmelen (1973) postulates other sites of mantle diapirs in the Adriatic Sea, Tyrrhenian Sea, Ionian Sea, the Po basin in northern Italy, and the Pannonian basin inside the great arc of the Carpathian Mountains, where the crust in the basin is nearly 10 km thinner than normal, possibly as a result of thinning over a diapir.

FIGURE 23-26 Schematic cross section of the Betic-Rif orogenic system at the end of the Oligocene. Fine arrows at the surface indicate the direction of sliding and rapid sediment transport, and dashed lines represent flattening (foliation). (*From Loomis, 1975.*)

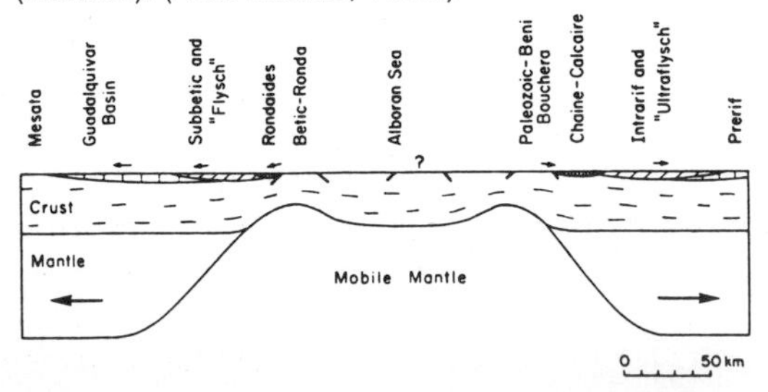

The concept of mantle diapirism offers a number of advantages over plate theory in explaining at least some features of the Alpine orogen—(1) it explains why spreading centers flanked by magnetic anomaly patterns have not been discovered in the Mediterranean (they do not form over a diapir); (2) a diapiric uplift with gravity sliding provides a mechanically simple explanation for the radial, outwardly directed thrusting seen in the strongly curved orogenic belts of the western Alpine orogen (plate models require complex mechanical solutions in these areas); (3) diapirs provide satisfactory explanations for ultrabasic bodies in the nappes, for forceful lateral pressure, for a source of uplift and subsidence, and for the origin of extensional features. Like the more popular plate-tectonic models, mantle diapirism is a powerful tectonic concept worthy of careful consideration in the evaluation of Alpine orogeny.

REFERENCES

Agar, D. V., and **Evamy, B. D.**, 1963, Geology of the southern French Jura: Proc. Geol. Assoc., v. 74.

Alvarez, W., 1973, The application of plate tectonics to the Mediterranean region, *in* Implications of continental drift to the earth sciences: London, Academic, v. 2.

Andrieux, J., **Fontboté, J. M.**, and **Mattauer, M.**, 1971, Sur un modele explicatif de L'Arc de Gibraltar: E. Planet. Sci. Letters, v. 12, p. 191–198.

Araña, Vicente, and **Vegas, Ramon,** 1974, Plate tectonics and volcanism in the Gibraltar arc: Tectonophysics, v. 24, no. 3, pp. 197–212.

Arbentz, P., 1912, Der Gebirgsbau der Zentralschweiz. Verh. der Schweizer: Naturforsch. Ges. 95 Jahresvers., Altdorf, Teil 2.

——— 1934, Die helvetische Region, *in* Gagnebin, E., and Christ, P., eds., Geologischer Führer der Schweiz II. Wepf, Basel, pp. 96–120.

Argand, Emile, 1912, Les nappes de Recouvrement des Alps Occidentales Carte Structurale, 1/500,000. Matériaux Carte Geol. Suisse, N. S. Livraison 31, Planche 1 (Carte Spéciale no. 64), Berne.

——— 1916, Sur Láre des Alpes Occidentales: Eclogae geol. Helv., v. 14, p. 145–191.

——— 1924, La tectonique de l'Asie, C.R. XIIIème Cong. Geol. Intern., Bruxelles 1922.

Bailey, E. B., 1935, Tectonic Essays: Oxford, Clarendon.

Bearth, P., 1962, Versuch Eines Gliederung Alpin Metamorpher Serien der Westalpen: Min. Petrogr. Mitt., Schweiz, v. 42, p. 127–137.

Berry, M. J., and **Knopoff, L.,** 1967, Structure of the upper mantle under the western Mediterranean basin: Jour. Geophys. Res. v. 72, no. 14, p. 3613–3626.

Bersier, Arnold, 1934, Carte tectonique du Jura, Fasc. I, pl. 3, *in* Guide Géologique de la Suisse: Soc. Géol. Suisse.

Bertrand, M., 1897, Structure des Alpes Françaises et Récurrence de Certain Faciès Sédimentaires: Internat. Geol. Cong., 6th Sess., 1894, Comptes rendus, p. 163–177.

Boccaletti, M., Manetti, P., and **Peccerillo, A.,** 1974, The Balkanids as an instance of back-arc thrust belt: Possible relation with the Hellenids: Geol. Soc. America Bull., v. 85, no. 7, p. 1077–1084.

Bonini, W. E., Loomis, T. P., and **Robertson, J. D.,** 1973, Gravity anomalies, ultramafic intrusions, and the tectonics of the region around the Strait of Gibraltar: Jour. Geophys. Res., v. 78, no. 8, p. 1372–1382.

Bullard, E. C., Everett, J. E., and **Smith, A. G.,** 1965, Fit of continents around Atlantic, *in* Blackett, P. M. S., Bullard, E. C., and Runcorn, S. K., eds., A symposium on continental drift: Roy. Soc. London, Phil. Trans., ser. A., v. 258.

Butler, J. R., 1971, Comparative tectonics of some Alpine and southern Appalachian structures: Southeast. Geology, v. 12, no. 4, p. 203–221.

Buxtorf, A., 1908, Geologische Beschreibung, des Weissensteintunnels und Seiner Umgebung, Beitr. Geol. Karte Schweiz, N. F. XXI, Bern.

——— 1916, Prognosen und Befunde Beim Hausenstein-basis und Grechenberg Tunnel, und die Bedeutung der letzeren für die Geologie des Juragebirges: Basel, Naturf. Gesell, Verh., v. 27.

Caputo, M., Panza, G. F., and **Postipischl, D.,** 1970, Deep structure of the Mediterranean basin: Jour. Geophys. Res., v. 75, no. 26, p. 4919ff.

Carey, S. W., 1958, The tectonic approach to continental drift, *in* Continental drift—A symposium: Geol. Dept. Univ. Tasmania, p. 177–355.

Collet, L. W., 1935, The structure of the Alps, 2d ed.: London, Arnold.

——— and **Paréjas, E.,** eds., 1920, Le Chapeau de Sédimentaire des Aiguilles Rouges de Chamonix et le Trias du Massif Aiguilles Rouges-Gastern: C. R. Soc. Phys. Hist. Nat., v. 37, no. 2, Genève, 1920.

Comninakis, P. E., and **Papazachos, B. C.,** 1972, Seismicity of the eastern Mediterranean and some tectonic features of the Mediterranean ridge: Geol. Soc. America Bull., v. 83, no. 4, p. 1093ff.

Contescu, L. R., 1974, Geologic history and paleogeography of eastern Carpathians: Example of Alpine geosynclinal evolution: Am. Assoc. Petroleum Geologists Bull., v. 58, no. 12, p. 2436–2476.

Crawford, A. R., 1972, Iran, continental drift plate tectonics:, Internat. Geol. Cong., 24th, India, Sec. 3, p. 106–112.

Curray, J. R., and **Moore, D. G.,** 1971, Growth of the Bengal deep-sea fan and denudation of the Himalayas: Geol. Soc. America Bull., v. 82, no. 3, p. 563ff.

Debelmas, J., and **Lemoine, M.,** 1971, The western Alps, paleogeography and structure: Earth-Sci. Rev., v. 6, no. 4, p. 221–256.

De Jong, K. A., 1973, Mountain building in the Mediterranean region, *in* De Jong, K. A., and Scholten, R., eds., Gravity and tectonics: New York, Wiley, p. 125–139.

De Sitter, L. U., 1953, Essai de Géologie Structurale Comparative de Trois Chaînes Tertiaries, Alpes Pyrénés, et Hout-Atlas: Soc. Belge. Geologie, 13.t., 62, f. 1, p. 38–58.

Dewey, J. F., Pitman, W. C. III, Ryan, W. B. F., and **Bonnin, Jean,** 1973, Plate tectonics and the evolution of the Alpine system: Geol. Soc. America Bull., v. 84, p. 3137ff.

Durand-Delga, M., and others, 1962, Données

actuelles sur la structure du Rif, Livre Mem. Paul Fallot, v. 1, p. 399–422.

Ernst, W. G., 1973, Interpretative synthesis of metamorphism in the Alps: Geol. Soc. America Bull., v. 84, no. 6, p. 2053.

Fitch, T. J., 1970, Earthquake mechanisms in the Himalayan, Burmeses, and Andaman regions and continental tectonics in Central Asia: Jour. Geophys. Res., v. 75, no. 14, p. 2699ff.

Fontboté, J. M., 1970, Sobre la historia preorogenica de las Cordilleras Béticas. Cuad.: Geol. Univ. de Granada, 1.

Fuchs, G., 1969, The geological history of the Himalayas: Int. Geol. Cong., 23rd, Czech., Rep., Sect. 3, Proc., p. 161–174.

Gansser, Augusto, 1964, Geology of the Himalayas: New York, Interscience.

——— 1968, The Insubric line, a major geotectonic problem: Schweiz. Mineral. Petrog. Mitt., v. 48.

Gilbert Smith, A., 1971, Alpine deformation and the oceanic areas of the Tethys, Mediterranean, and Atlantic: Geol. Soc. America Bull., v. 82, no. 8.

Haynes, S. J., and **McQuillan, Henry**, 1974, Evolution of the Zagros suture zone, southern Iran: Geol. Soc. America Bull., v. 85, p. 739–744.

Heim, Albert, 1878, Untersuchungen über den Mechanismus der Gerirgbildung: Basel.

——— and **Schmidt, C.**, 1911, Geologische Karte der Schweiz, 1/500,000, 2e Auflage, Neudruck.

Hsü, K. J., 1971, Origin of the Alps and western Mediterranean: Nature, v. 233, no. 5314, p. 44–48.

——— and **Schlanger, S. O.**, 1971, Ultrahelvetic flysch sedimentation and deformation related to plate tectonics: Geol. Soc. America Bull., v. 82, no. 5, p. 1207ff.

Johnson, M. R. W., 1973, Displacement of the Insubric line: Nature: Phys. Sci., v. 24, no. 110, p. 116–117.

Julivert, Manuel, 1971, Décollement tectonics in the Hercynian Cordillera of northwest Spain: Am. Jour. Sci., v. 270, no. 1, p. 1–29.

Klemme, H. D., 1958, Regional geology of circu-Mediterranean region: Am. Assoc. Petroleum Geologists Bull., v. 42.

Kohli, G., and others, 1970, Himalayan and Alpine orogeny: Intl. geol. Congress, 22nd, India, 1964, Rep., pt. 11.

Krasser, L. M., 1939, Der Bau der Alpen, Ein Hilfsbuch zur Einfuhrung, viii: Berlin, Gebruder Bertraeger.

Laubscher, H. P., 1971, The large-scale kinematics of the western Alps and the northern Apennines and its palinspastic implications: Am. Jour. Sci., v. 271, no. 3, p. 193–226.

——— 1972, Some overall aspects of Jura dynamics: Am. Jour. Sci., v. 272, no. 4, p. 293–304.

——— 1975, Plate boundaries and microplates in Alpine history: Am. Jour. Sci., v. 275, no. 8, p. 865–876.

——— 1975, Viscous components in Jura folding: Tectonophysics, v. 27, p. 239–254.

LeFort, Patrick, 1975, Himalayas: The collided range. Present knowledge of the continental arc: Am. Jour. Sci., v. 275-A, p. 1–44.

Lienhardt, Georges, 1960, La faille-pli de Bornay Jura; preuve de la passivité locale du socle vis-a-vis de la couverture: Soc. Géol. Nord, Ann. t. 80, lv. 2.

Liniger, Hans, 1958, Vem bau der Alpen: Muchen, Ott Verlag Thun., p. 236.

Lombard, A. E., 1948, Appalachian and Alpine structures—A comparative study: Am. Assoc. Petroleum Geologists Bull., v. 32, p. 709–744.

Loomis, T. P., 1975, Tertiary mantle diapirism, orogeny, and plate tectonics east of the Strait of Gibraltar: Am. Jour. Sci., v. 275, p. 1–30.

Lort, J. M., 1971, The tectonics of the eastern Mediterranean: A geophysical review: Rev. Geophys. Space Physics, v. 9, no. 2.

———, **Limond, W. Q.**, and **Gray, F.**, 1974, Preliminary seismic studies in the eastern Mediterranean: Earth Planet. Sci. Letters, v. 21, p. 355–366.

Lowrie, W., and **Alvarez, W.**, 1974, Rotation of the Italian peninsula: Nature, v. 251, no. 5473, p. 285–288.

——— 1975, Paleomagnetic evidence for rotation of the Italian peninsula: Jour. Geophys. Res., v. 80, p. 1579–1592.

Lugeon, M., and **Gagnebin, E.**, 1941, Observations et Vues Nouvelles sur la Géologie des Prealpes Romades: Univ. Lausanne Lab. Geol. Bull. 72.

McGinnis, L. D., 1971, Gravity fields and tectonics in the Hindu Kush: Jour. Geophys. Res., v. 76, no. 4, p. 1894ff.

McKenzie, D. P., 1970, Plate tectonics of the Mediterranean region, Nature, v. 226, p. 239–243.

Mattauer, M., 1963, Le style tectonique des chaînes tellienne et rifaine: Geol. Rundsch., v. 53, p. 296–313.

Maxwell, J. C., 1970, The Mediterranean, ophiolites, and continental drift, *in* Johnson, H., ed., The mega-

tectonics of continents and oceans: Rutgers Univ., New Brunswick, N. J.
Miljush, Petar, 1973, Geologic-tectonic structure and evolution of outer Dinarides and Adriatic area: Am. Assoc. Petroleum Geologists Bull., v. 57, no. 5, p. 913ff.
Milnes, A. G., 1969, On the orogenic history of the central Alps: Jour. Geology, v. 77, no. 1, p. 108–112.
——— 1973, Structural reinterpretation of the classic Simplon tunnel section of the central Alps: Geol. Soc. America Bull., v. 84, no. 1, p. 269–274.
——— 1974, Structure of the Pennine zone (central Alps): A new working hypothesis: Geol. Soc. America Bull., v. 85, p. 1727–1732.
Mitchell, A. H. G., and **McKerrow, W. S.,** 1975, Analogous evolution of the Burma orogen and the Scottish Caledonides: Geol. Soc. America Bull., v. 86, p. 305–315.
Ninkovich, Dragoslav, and **Hays, J. D.,** 1972, Mediterranean island arcs and origin of high potash volcanoes: Earth Planet. Sci. Letters, v. 16.
Nowroozi, A. A., 1971, Seismo-tectonics of the Persian plateau, eastern Turkey, Caucasus, and Hindu-Kush regions: Seis. Soc. America Bull., v. 61, no. 2, p. 317–341.
Oberholtzer, J., 1933, Geologie der Glarneralpen: Bern, Switzerland (privately published).
Oxburgh, E. R., 1969, An outline of the geology of the central eastern Alps: Geo. Assn. London Proc., v. 79, pt. 1, p. 1–46.
Papazachos, B. C., and **Comninakis, P. E.,** 1971, Geophysical and tectonic features of the Aegean arc: Jour. Geophys. Res., v. 76, no. 35.
Pierce, W. G., 1966, Jura tectonics as a décollement: Geol. Soc. America Bull., v. 77, p. 1265–1276.
Ponikorov, V. P., and others, 1970, Some aspects of the development of the Alpine-Himalayan folded region in the late Precambrian and Paleozoic: Geotectonics, no. 1, p. 46–49.
Powell, C. McA., and **Conaghan, P. J.,** 1973, Plate tectonics and the Himalayas: Earth Planet. Sci. Letters, v. 20, no. 1, p. 1–12.
Qureshy, M. N., Venkatachalam, S., and **Subrahmanyam, C.,** 1974, Vertical tectonics in the Middle Himalayas—An appraisal from recent gravity data: Geol. Soc. America Bull., v. 85, p. 921–926.
Rabinowitz, P. D., and **Ryan, W. B. F.,** 1970, Gravity anomalies and crustal shortening in the eastern Mediterranean: Tectonophysics, v. 10.
Ramsay, J. G., 1963, Stratigraphy, structure and metamorphism in the Western Alps: Geol. Assoc. Proc., v. 74.
Rutten, M. G., 1969, The geology of Western Europe: Elsevier Pub. Co., Amsterdam.
Roeder, D. H., 1973, Subduction and orogeny: Jour. Geophys. Res., v. 78, no. 23, p. 5005–5024.
Saxena, M. N., 1972, The crystalline axis of the Himalaya; the Indian shield and continental drift: Tectonophysics, v. 12, no. 6, p. 433–447.
Schardt, Hans, 1898, Les Régions Exotiques du Versant Nord des Alpes Suisses: Soc. Vaudoise Sci. Natl. Bull., v. 34, p. 113–219.
Smith, A. G., 1971, Alpine deformation and the oceanic areas of the Tethys, Mediterranean, and Atlantic: Geol. Soc. America Bull., v. 82, p. 2039–2070.
——— and **Hallam, A.,** 1970, The fit of the southern continents: Nature, v. 225.
Sougy, J., 1962, West African fold belt: Geol. Soc. America Bull., v. 73.
Stegena, L., Geczy, B., and **Horvath, F.,** 1974, Late Cenozoic evolution of the Pannonian basin: Tectonophysics, v. 26, p. 71–90.
Stocklin, Jovan, 1968, Structural history and tectonics of Iran; A review; Am. Assoc. Petroleum Geologists Bull., v. 52, no. 7, p. 1229–1258.
Strand, Trygve, 1961, The Scandinavian Caledonides—A review: Am. Jour. Sci., v. 259, p. 161–172.
Tchalenko, J. S., and **Berberian, M.,** Dasht-e Bayaz fault, Iran: Earthquake and earlier related structures in bed rock: Geol. Soc. America Bull., v. 86, p. 703–709.
Tollmann, A., 1963, Ostalpensynthese: Franz Deuticke Press, Vienna.
——— 1965, Faziesanalyse der alpidischen serien der Ostalpen: Verh. geol. Bundesanst., Sonderheft, Wien.
Trümpy, Rudolf, 1960, Paleotectonic evolution of the central and western Alps: Geol. Soc. America Bull., v. 71, p. 843–908.
——— 1963, Sur les racines ders nappes helvétiques: Fallot Memorial Vol., Soc. Géol. France, t. 2.
——— 1971, Stratigraphy in mountain belts: Geol. Soc. London, Quart. Jour., v. 126.
——— 1973, The timing of orogenic events in the central Alps, *in* De Jong, K. A., and Scholten, R., eds., Gravity and tectonics, New York, Wiley, p. 229–251.
Udias, A., and **Lopez Arroyo, A.,** 1972, Plate tec-

tonics and the Azores-Gibraltar region: Nat. Phys. Sci., v. 237, no. 74, p. 67–69.

Van Bemmelen, R. W., 1973, Geodynamic models for the Alpine type of orogeny: Tectonophysics, v. 18, no. 1–2.

——— 1974, Geodynamic models for the Alpine type of orogeny: Tectonophysics, v. 18, no. 1–2, p. 33–79.

Van der Voo, R., and **Boessenkool, A.**, 1973, Permian paleomagnetic result from the western Pyrenees delineating the plate boundary between the Iberian peninsula and stable Europe: Jour. Geophys. Res., v. 78, no. 23, p. 5118ff.

Weber, F., 1904, Über den Kali-Syenit des Piz Giuf und Umgebung (Östl. Aarmassiv): Beitr. geol. Karte Schweiz, N. F. 14 e Lief 44, Bern.

Wegmann, Eugene, 1961, Anatomie comparée des Hypothéses sur les plissements de couverture (le Jura plissé): Uppsala Univ. Geol. Inst. Bull., v. 40, p. 169–182.

Wenk, E., 1962, Das reaktivierte Grundgebirge der Zentralalpen: Geol. Rundschau, v. 52.

Wiedenmayer, C., 1931, Geologic map of Oensingen-Balsthal 1:10,000: Switzerland.

Wilson, H. H., 1969, Late Cretaceous eugeosynclinal sedimentation, gravity tectonics, and ophiolite emplacement in Oman Mtns., SE Arabia: Am. Assoc. Petroleum Geologists Bull., v. 53, no. 3, p. 626–671.

Wong, H. K., **Zarddzki, E. F. K.**, **Phillips, J. D.**, and **Giermann, G. K. F.**, 1971, Some geophysical profiles in the eastern Mediterranean: Geol. Soc. America Bull., v. 82, no. 1, p. 91–100.

Woodside, John, and **Bowin, Carl,** 1970, Gravity anomalies and inferred crustal structure in the eastern Mediterranean Sea: Geol. Soc. America Bull., v. 81, no. 4, p. 1107–1122.

24

The Cordilleran Orogen

THE CORDILLERAN OROGENS OF WESTERN NORTH AND SOUTH AMERICA

The world's longest orogen is that diverse, though essentially continuous, complex of structural elements which can be traced from the Aleutian arc and western Alaska southward across the western third of North America into Mexico and Central America. The system is caught up in the Caribbean arc where the structural trends are strongly deflected to the east only to reappear along the northern coast of South America and continue a southerly trend along the western margin of South America until at the southern tip they are deflected once again eastward in the Scotia arc before passing into the Palmer Peninsula and becoming lost beneath the ice cover of Antarctica.

Consideration of this complex of structural elements as a single orogen is based primarily on lateral continuity, lost only where the deformed belts pass into the Caribbean and Scotia arcs, and on the broad similarity in the timing (Mesozoic and Cenozoic) of deformation within the orogen. In at least one respect this orogen is strikingly different from both the Appalachian and the Alpine systems previously considered. The Cordilleran orogen is now and has been located along the marginal zone between cratonic and oceanic crust during the entire orogenic history of the belt. The Alpine orogen is now largely located between cratons, and if reconstructions of the Atlantic to the time of the Appalachian orogenies are correct, the Appalachian orogen was located between cratons at that time.

THE TECTONIC FRAMEWORK OF THE CORDILLERAN SYSTEM

The cratons of North America, South America, and Antarctica are tied together along the western edge by an orogenic belt. The orogenic belt is most continuous and best exposed along the margins of the cratons, and the features are least continuous and most dif-

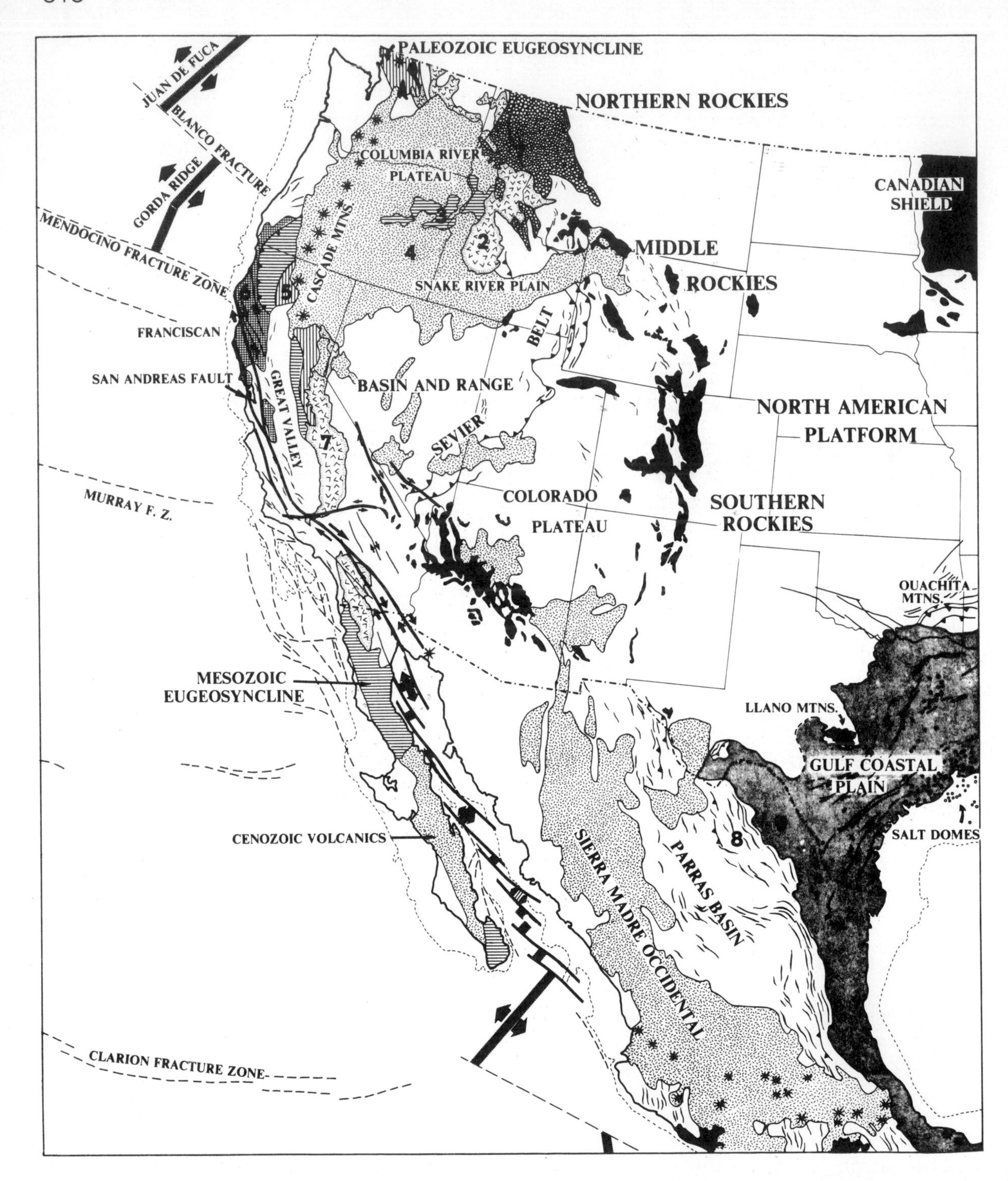

PALEOZOIC EUGEOSYNCLINE
NORTHERN ROCKIES
JUAN DE FUCA
BLANCO FRACTURE
GORDA RIDGE
MENDOCINO FRACTURE ZONE
COLUMBIA RIVER
PLATEAU
CASCADE MTNS
SNAKE RIVER PLAIN
MIDDLE
ROCKIES
CANADIAN
SHIELD
FRANCISCAN
SAN ANDREAS FAULT
GREAT VALLEY
BASIN AND RANGE
BELT
SEVIER
NORTH AMERICAN
PLATFORM
MURRAY F. Z.
COLORADO
PLATEAU
SOUTHERN
ROCKIES
OUACHITA
MTNS.
MESOZOIC
EUGEOSYNCLINE
LLANO MTNS.
GULF COASTAL
PLAIN
SALT DOMES
CENOZOIC VOLCANICS
SIERRA MADRE OCCIDENTAL
PARRAS BASIN
CLARION FRACTURE ZONE
1
2
3
4
5
6
7
8

ficult to understand in the intervening sections where modern island arcs are located. The greatest complexity is found in the Central American–Caribbean region, where the East Pacific rise approaches the continents. It is in this area that North and South America are offset with an apparent left-lateral displacement, and it was along this zone that rifting and separation of cratonic plates took place according to continental-drift theory.

The western margin of the orogen, especially that part between the oceanic and continental plates, is varied and complex. The Aleutian arc occurs at the northern end, but it terminates in the Gulf of Alaska, and the orogen bends sharply to the south, its western edge defined by the Fairweather fault, Fig. 24-2. Elements of the East Pacific rise as identified by magnetic anomaly patterns are found along the western coast of North America as far south as Cape Mendocino, but this part of the oceanic ridge system is broken, offset, and apparently concealed in part by the craton to the east. From Cape Mendocino south to the Gulf of California the ridge crest is not apparent. The ridge crest is found again in the Gulf of California and can be traced south and west across the Pacific. The western margin of the orogen changes character dramatically where the East Pacific rise crest is offshore. A trench largely filled with sediment forms the marginal zone between the Gulf of California and the Cocos ridge. A second deep is located off the coast between the Cocos and the Carnegie ridge. Finally a long trench parallels the coastal margin for much of the distance to the Chile rise. Thus the western margin south of the Gulf of California is essentially a long trench broken and interrupted where a series of submarine ridges come into the land.

What happens at both the northern and southern ends of the orogen is obscure. The orogen is wider in Alaska than it is farther south, and it consists of several major structural elements separated from one another by important strike-slip zones. The southernmost element passes into the Aleutian arc at the Alaska Peninsula, Fig. 21-18; the more northerly elements, including the Brooks Range, passes into the shelf areas of the Bering Sea and the Arctic Ocean but are not readily correlated with similar elements in northern Asia. The southern end of the orogen in Antarctica is largely covered by ice.

FIGURE 24-1 Sketch map of the North American sector of the Cordilleran system. The numbered localities are:
1. Belt (late Precambrian) rocks of the Northern Rockies
2. Idaho batholith
3. Mesozoic eugeosynclinal deposits (horizontal line)
4. Tertiary basalts (fine slash pattern)
5. Paleozoic eugeosynclinal deposits (vertical line)
6. Franciscan (grid pattern)
7. Sierra Nevada batholith
8. Late Mesozoic miogeosynclinal deposits affected by Laramide orogeny
9. Cenozoic Gulf Coastal Plain deposits

(After the North American Tectonic Map compiled by King, 1969.)

THE NORTH AMERICAN CORDILLERA

The North American Cordillera deserves our special attention because its structure and history are known in relatively great detail, and it is the type locality for one of the plate-tectonic models of orogeny, Figs. 24-1 and 24-2. The orogenic system is both vast in size and complex in history and structural configuration. Many sectors have been subjected to several orogenic and postorogenic tectonic events, and this superposition of deformations has made it difficult to trace the lateral continuity of some of the earlier features. Younger events in the Cordilleran region have obscured older geology in much the same way the Coastal Plains deposits of the Atlantic and Gulf Coasts blanket the older features there. Large areas are covered by Cenozoic volcanics, especially from the Columbia River Plateau into Mexico. In addition, substantial areas are covered by Quaternary alluvium and glacial deposits, notably in the numerous structural depressions located throughout the orogen where young block faulting has occurred.

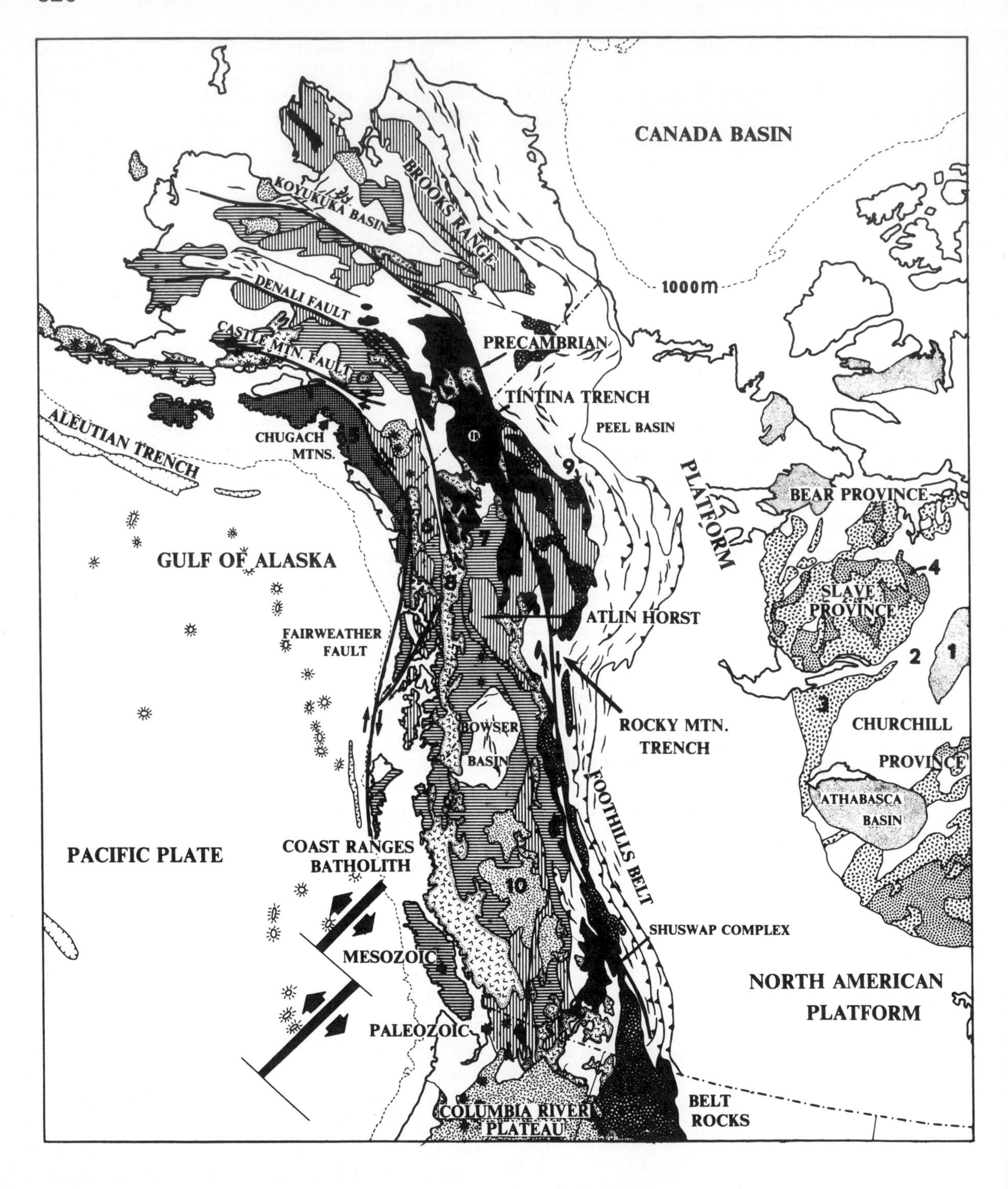
CANADA BASIN
BROOKS RANGE
KOYUKUKA BASIN
DENALI FAULT
CASTLE MTN. FAULT
1000m
PRECAMBRIAN
TINTINA TRENCH
PEEL BASIN
ALEUTIAN TRENCH
CHUGACH MTNS.
PLATFORM
BEAR PROVINCE
GULF OF ALASKA
SLAVE PROVINCE
ATLIN HORST
FAIRWEATHER FAULT
ROCKY MTN. TRENCH
CHURCHILL PROVINCE
BOWSER BASIN
ATHABASCA BASIN
FOOTHILLS BELT
PACIFIC PLATE
COAST RANGES BATHOLITH
SHUSWAP COMPLEX
MESOZOIC
NORTH AMERICAN PLATFORM
PALEOZOIC
COLUMBIA RIVER PLATEAU
BELT ROCKS

Divisions of the North American Cordillera

For purposes of this discussion of regional tectonics, the Cordilleran region, in Canada and the western states, is subdivided into the following structural divisions:

1. North American cratonic platform
   a. Undeformed sectors
   b. Deformed platform of the Middle and Southern Rocky Mountains
2. Eastern Cordilleran fold belt
   a. Northern Rocky Mountains
   b. Core area—the Omineca belt
   c. Sevier orogen of western United States
3. Intermontane belt of Canada
4. Basin and Range province
   a. Antler orogen
   b. Sonoma orogen
   c. Nevadan features
5. Western Cordilleran fold belt
   a. Paleozoic orogens (western Antler and Sonoma)
   b. Nevadan orogen
   c. Franciscan belt (western Sevier)

The North American Platform

The North American Cordillera is bounded by a continental platform all the way from the Yukon to Texas. The Great Plains is the most typical physiographic expression of the platform area, and the Great Plains has a sharp western boundary against the Foothills belt of the Northern Rocky Mountains from the Arctic Ocean into Montana, where the Northern Rockies end and the western edge of the platform is deformed. This deformed section of the platform includes the Middle and Southern Rocky Mountains and the Colorado Plateau, Fig. 24-3.

The surface of the platform has relatively flat or gently rolling topography which rises from elevations of over 100 m in the east to elevations of near 1,500 m (almost 5,000 ft) along the mountain front. This vast area is underlain by thin but widespread sedimentary rocks which contrast sharply in thickness with contemporaneous units in the miogeosyncline to the west as depicted for the Cambrian in Fig. 24-4. Most strata on the platform layers thin gradually toward the interior of the continent as a result of sedimentary pinch outs, and others have been beveled by erosion. The Paleozoic, Triassic, and Jurassic rocks are mostly marine, but the Cretaceous units are mixed continental and marine, and all Cenozoic units are continental strata that thin rapidly eastward and are made up of the erosion products from the high mountains to the west.

Structural features in the undeformed platform area consist of broad basins and arches, several of which are parallel to the mountain front and deepen toward the mountains as do the Peel and Albert basins in Canada and the Powder River and Denver basins in the south.

FIGURE 24-2 Sketch map of the northern part of the North American Cordilleran system. The numbered localities are:
1. Middle Proterozoic sedimentary basins in shield area
2. Hudsonian fold belt gneisses and plutons
3. Lower Proterozoic granites (double slash marks)
4. Archean metasediments (heavy dot)
5. Late Mesozoic eugeosynclinal deposits (grid)
6. Paleozoic eugeosynclinal deposits (vertical lines)
7. Triassic and Jurassic eugeosynclinal deposits (horizontal lines)
8. Jurassic and Cretaceous plutons (granitic) (v pattern)
9. Belt rocks (dark pebbly pattern)
10. Tertiary volcanics (fine slash)
11. Precambrian basement rocks in the orogen = black

(After the North American Tectonic Map compiled by King, 1969.)

The Deformed North American Platform—The Rocky Mountains

Most of the deformed region of the Middle and Southern Rockies has been part of the western continental shelf during most of the Paleozoic and Mesozoic, although arches and elongate basins formed in some areas during Paleozoic time. This region includes some of the highest of the Rocky Mountains. The present front of the Rocky Mountains in the United States is along the eastern edge of these high moun-

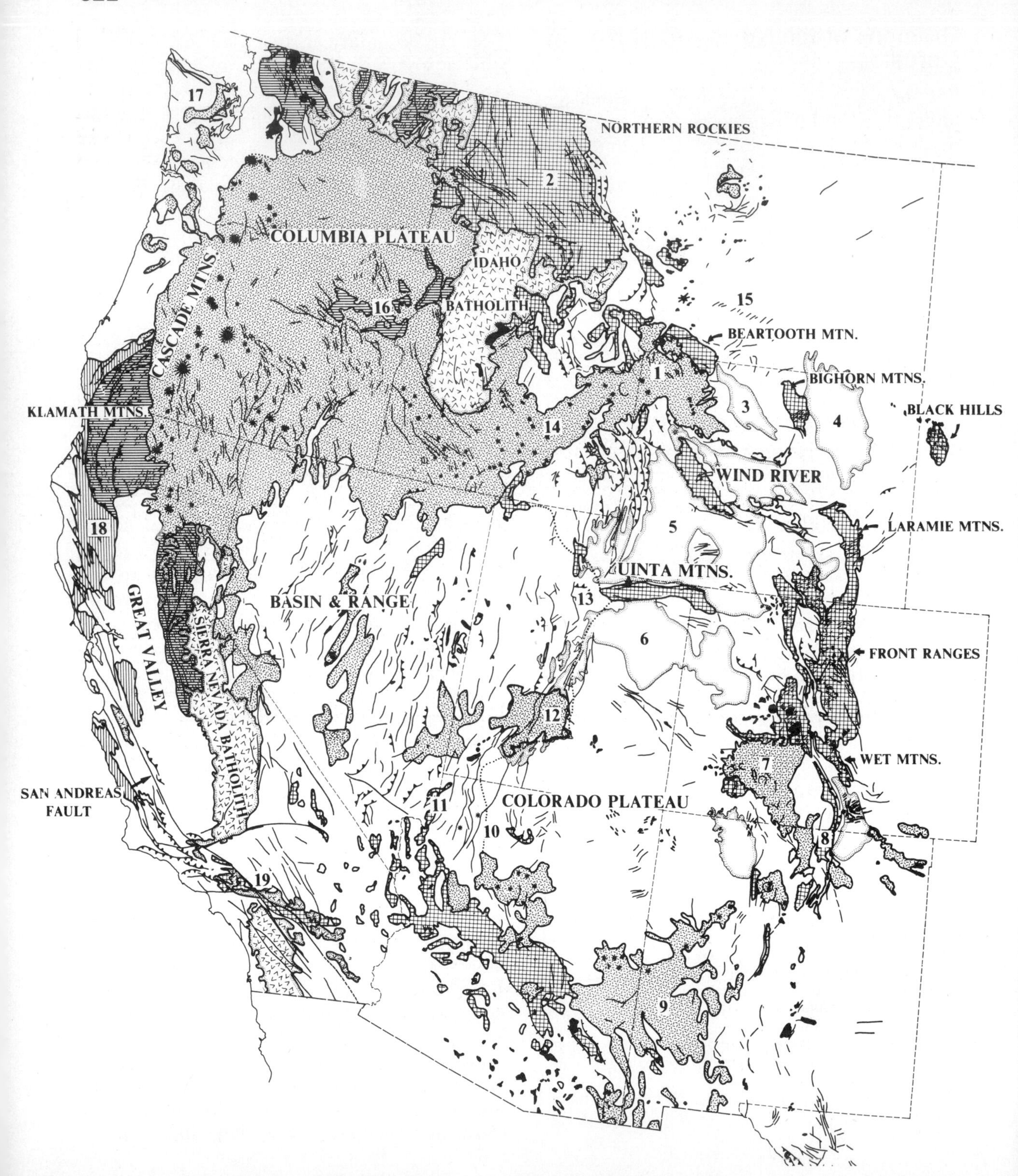

NORTHERN ROCKIES
COLUMBIA PLATEAU
IDAHO
BATHOLITH
CASCADE MTNS
BEARTOOTH MTN.
BIGHORN MTNS.
KLAMATH MTNS.
BLACK HILLS
WIND RIVER
LARAMIE MTNS.
UINTA MTNS.
BASIN & RANGE
GREAT VALLEY
SIERRA NEVADA BATHOLITH
FRONT RANGES
WET MTNS.
SAN ANDREAS FAULT
COLORADO PLATEAU
1
2
3
4
5
6
7
8
9
10
11
12
13
14
15
16
17
18
19

tains, but there is no structural counterpart of this deformed shelf region in Canada or in the Appalachians. On paleogeographic and structural grounds the southern continuation of the Canadian Rocky Mountain fold and thrust belt lies west of all these ranges. We may expect for this reason that the structures in the Middle and Southern Rockies are the product of unique circumstances in the pattern of development or the mechanism of deformation as it affected this part of the Cordilleran geosyncline.

This shelf region occupied part of the ancient continental craton and has crustal thicknesses greater than those to the west, Fig. 16-14. Seismic refraction profiles indicate crustal thicknesses of 40 to 50 km in the shelf area and very much thinner crust to the west, especially under the Basin and Range.

The deformed platform begins in south-central Montana, where the northern edge appears to coincide with a major deep-seated zone of transcurrent faulting, oriented almost east-west. Left-lateral movement is indicated by the *en échelon* faults in the Lake Basin fault zone, Fig. 24-3, but the fault zone passes into the Boulder and Idaho batholiths toward the west and it is covered toward the east.

The western edge of the deformed shelf is formed primarily by the former hinge between the shelf and the Cordilleran miogeosyncline, but later events have made that obscure. The border is marked by an irregular curved zone of thrusts just east of the Idaho batholith. These thrusts are covered by Tertiary volcanics west of Yellowstone Park but reemerge along the Wyoming-Idaho border, where the thrusts in the miogeosyncline are juxtaposed against undeformed sediments in the Green River basin. This simple pattern is broken up at the western end of the Uinta Mountains, where the thrust belt curves to the southwest and a

FIGURE 24-3 Tectonic map of the western United States. Precambrian complexes are shown by a grid pattern; Tertiary intrusives are solid black; Mesozoic and Cenozoic volcanics are represented by a slash pattern; Mesozoic plutons are shown by a v pattern; Mesozoic and Paleozoic metamorphic complexes are covered by a horizontal-line pattern; and Franciscan rocks are covered by a vertical line. Keyed numbers locate: 1. Yellowstone; 2. Belt rocks of the Northern Rockies; 3. Bighorn basin; 4. Powder River basin; 5. Green River basin; 6. Uinta basin; 7. San Juan Mountains; 8. Sangre de Cristo Mountains; 9. Datil volcanic area; 10. Toroweap fault; 11. Grand Wash fault; 12. High Plateau volcanic area; 13. Wasatch fault block; 14. Snake River plain; 15. Lake Basin fault zone; 16. Ochoco-Blue Mountains; 17. Olympic uplift; 18. Franciscan rocks; 19. Transverse ranges. (*After the Tectonic Map of the United States, 1961.*)

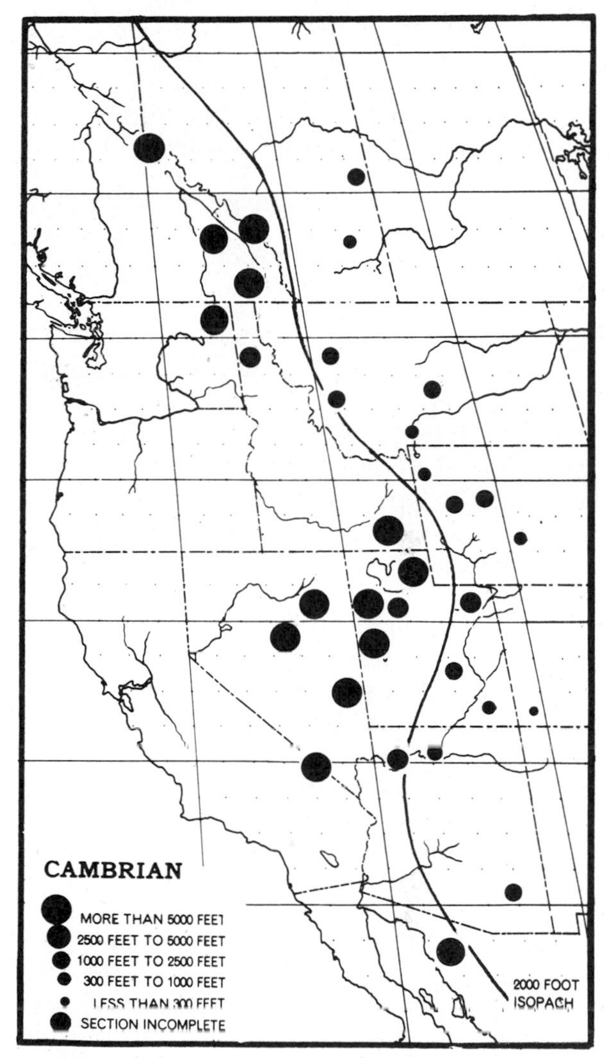

FIGURE 24-4 Distribution of Cambrian thicknesses in western North America. (*From Kay, 1951.*)

prominent zone of younger, high-angle, normal faults which trend north northeast is superimposed on the former shelf edge. These west-dipping, high-angle faults (Wasatch, Toroweap, Hurrican faults) form a set of step-like blocks which separate the higher Colorado Plateau from the Basin and Range to the west.

The Basin and Range also borders the deformed platform along its southern edge in Arizona and New Mexico where the Precambrian basement of the platform is cut and displaced to form grabens and horsts. This zone of fragmented basement blocks trends northwest and its border becomes increasingly obscure to the south where it is covered by Tertiary volcanic fields and Quaternary alluvial fills.

The eastern border of the deformed platform is the most prominent of all because it is defined by high spectacular basement uplifts which begin with the Bighorn Mountains in Wyoming and reemerge in an almost continuous belt consisting of the Laramie, Front Ranges, Wet, and Sangre de Cristo Mountains. Most of these basement blocks are bounded on at least one side by high-angle faults, and toward the southern end the basement blocks are long, narrow horsts oriented north-south and located between the southern Great Plains and the Colorado Plateau. These big basement blocks lie in a great arc starting at the southern end of the Northern Rocky Mountains and extending to Mexico. With the exception of the enigmatic east-west trending Uinta Mountains, they form a zone along the eastern side of the deformed platform. The basement blocks are separated from one another by large basins which contain thick Tertiary deposits, material eroded from the adjacent mountains. The sediments in these basins are deformed by folding and faulting that is primarily a result of stresses that originated in connection with the uplift of adjacent basement.

Basement Tectonics

Movements of essentially rigid basement blocks have controlled much of the deformation within the platform area. These blocks are defined by shatter zones of Precambrian age that became reactivated as faults in the Laramide orogeny. Most basement exposures are bounded by high-angle faults on one or more sides; however, some of these curve and become thrusts. The structural relief measured from the surface of the Precambrian in the mountains to the top of the Precambrian in adjacent basins is frequently 10,000 m (10 km) or more. Thus the faults of whatever shape have a large vertical component. Tight folding of the type described in the folded and thrust-faulted orogenic belts is absent; we find instead structural terraces, monoclines, and homoclines, and all greater fold complexity is confined to local structural features.

Among the lines of evidence which point to vertical tectonics in this region, the margins of the Bighorn and Beartooth Mountains* are particularly significant. Local thrusts dipping under the mountains are located on both the eastern and western sides of the Bighorn Mountains. Similar structures occur in the Beartooth Range, where the northeastern front is thrust to the north, the eastern front is thrust to the east, and the southwestern corner of the range is thrust to the southwest. These opposite-opposed thrusts almost preclude the idea that these are regional thrust faults like those commonly found in foreland belts. These thrusts consistently dip under the basement uplifts irrespective of their orientation, showing nothing of the strong alignment found in the Canadian Rockies, the Appalachians, the Alps, and other orogenic belts. These faults must extend to depths of at least 10 to 15 km in order to account for the vertical separation of the Precambrian basement, but the way the

* See Foose (1973).

fault surfaces are shaped at depth is not obvious. Hypotheses have been based on stress theory and on the presence of near-vertical faults in the same region along the border of other ranges, the nature of subsidiary structures along the mountain fronts, seismic evidence, and a few wells which have been drilled through the overthrust plate, Fig. 9-10. The most generally accepted hypotheses is that these are block-fault uplifts along a fault which changes from a thrust near the surface to a high-angle reverse fault and eventually a vertical fault at depth; but it has also been suggested that the basement blocks have been uplifted along low-angle thrusts with inclinations of 30° or less and that they are uplifts in which the structure developed by a combination of folding and thrusting, Fig. 9-15. Each can be used to explain such geologic observation as: (1) the low-angle thrust at the northeast corner of the Beartooth Range which is displaced along tears, Fig. 8-3, (2) the curvature of the cover on the overthrust plate such as that along the flanks of the Bighorn Mountains, (3) the normal faulting resulting from extension in the sedimentary cover such as that on the southern margin of the Wind River Range, Figs. 9-9, and (4) the presence of sedimentary cover beneath the overthrust Precambrian crystallines.

An analysis of the gravity anomalies along a traverse across the margin of the Wind River Range supports the interpretation that the thrust along the margin of this range extends as much as 30 km under the range at relatively low angles, Fig. 9-9, but the preferred model does show the fault steepening at a depth of about 3,050 m (about 10,000 ft) below sea level.

Vertical and near-vertical faults such as those along the Sangre de Cristo Mountains also clearly exist. A particularly good example of vertical faults is found in the Pryor Mountains of Wyoming, where the uplift is composed of four block-shaped units. Each is bounded on two or three sides by high-angle faults. The blocks are tilted, suggesting the opening of four trap doors. The faults intersect at the corners at nearly right angles, a configuration that is hard to obtain by movements along any but very steep faults. At the corner of one of these blocks, Precambrian rocks are brought up into fault contact with Cretaceous units. The throw decreases laterally along the fault until the covering Paleozoics can be seen to form a passive *drape-like fold* over the fault. This fold then passes along its axis into a monocline and eventually disappears.

The Eastern Cordilleran Fold Belt

The eastern border of this division is clearly defined in Canada, where the folds and thrusts of the Northern Rocky Mountains are sharply juxtaposed against relatively undeformed sedimentary units on the platform, Figs. 24-1 and 24-2. This boundary is poorly defined in the United States, where the continuity of the folds and thrusts from the north are truncated by a transverse fault zone, the Lake Basin zone, and covered by the younger volcanic rocks of the Snake River plain.

Most of the Eastern Cordilleran fold belt occupies the site of a Paleozoic miogeosyncline which was situated along the continental margin and in which great thicknesses of sediment accumulated during the Paleozoic and Mesozoic Eras. The contents of the geosyncline were strongly uplifted, folded, and thrust toward the continental platform in an orogenic phase which began in the Mesozoic and lasted into the early Tertiary. The name *Sevier orogeny* (Harris, 1959) was initially applied to deformation represented by the folds and thrusts on the east side of the Cordilleran fold belt in Utah and Nevada, but the name has been extended both to western Nevada and laterally to the north and south. The age of this orogeny is not sharply defined, but intrusions which cut some of these folds are as old as the Jurassic. Thus, the Sevier orogeny was a protracted crustal deformation beginning in the Jurassic and continuing to the end of the Mesozoic. The *Laramide orogenic phase* is the name applied to the Late Cretaceous–Eocene deformation which affected the easternmost part of the orogenic belt and the adjacent platform.

The Canadian Sector of the Eastern Cordilleran Fold Belt—The Northern Rocky Mountains and Their Core Zone*

The well layered, anisotropic miogeosynclinal-shelf and clastic wedge sequences, characteristic of the Rocky Mountains, are deformed principally by a vast array of generally southwest-dipping, concave-upward, locally folded thrust faults. The sedimentary rocks have been stripped along the layering from their underlying basement. Relative to the basement, they have been thrust northeastward . . . as much as 200 km. to where they are now stacked up in a series of thrust sheets on the edge of the North American Craton. . . . The stacked thrust sheets have produced surficial tectonic thickening of 8 km. above the passive basement.

The structural style and density of thrusting in the Southern Rocky Mountains are related to the competence and anisotropy of the rocks which, in turn, are governed by the stratigraphic level exposed. Consequently, thrust faults are most numerous in the weak Cretaceous rocks of the clastic wedges underlying most of the Rocky Mountain foothills [see Fig. 24-5]. They are fewer westward in the Front Ranges, composed of the Devonian to Jurassic miogeosynclinal sequence, and are comparatively rare in the thick, competent lower Paleozoic beds that comprise the eastern Main Ranges. The style of folding in the Western Main Ranges changes to the penetrative slip type in the shaly, less anisotropic facies of Cambrian and older rocks from the concentric type characteristic of the highly anisotropic layering in the ranges to the east. . . . The first thrust faults apparently formed in the west, probably in late Jurassic time, and progressively younger ones developed farther east.*

The Rocky Mountain marginal zone is separated from the core zone of the Eastern Cordilleran fold belt by a prominent morphotectonic feature—the Rocky Mountain trench.

Rocky Mountain Trench

The Rocky Mountain trench can be traced from Flathead Lake in Montana as a nearly straight lineament over 1,600 km (about 1,000 miles) long through British Columbia. The Tintina trench is almost on the same trend in the Yukon Territory, and it is traced 725 km (450 miles) into Alaska. Both of these trenches have characteristics often associated with strike-slip

* A plate-tectonic model for the Canadian Cordillera has been developed by Monger, Souther, and Gabrielse (1972).

* As described by Wheeler (1970, p. 155).

FIGURE 24-5 Cross section across the Northern Rocky Mountains from the Selkirk Mountains to the Great Plains, including the Foothills, Front Ranges, Main Ranges, Rocky Mountain trench, and the Selkirk Mountains. (*From R. A. Price* in *Wheeler, 1970.*)

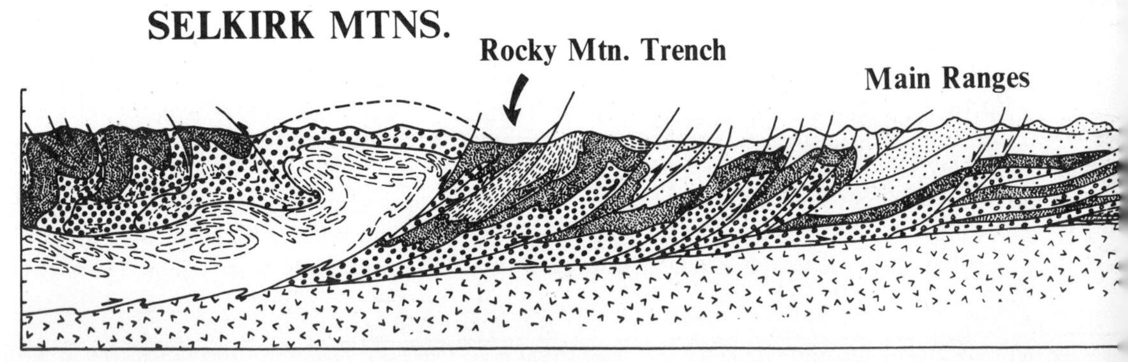

faults: (1) trench-like topographic form, (2) linearity, (3) separation across the trench of similar lithologies, and (4) termination of structural zones which intersect the trench. However, it has not yet been conclusively demonstrated that specific structural or lithologic zones are actually offset as a result of strike-slip movements along the trench, and other evidence is difficult to explain in terms of strike-slip movement. The trench is a sinuous zone near Cranbrook, British Columbia; it is asymmetrical in cross section; the eastern flank appears to be a youthful fault scarp, and it cannot be traced south of Flathead Lake, Montana. The trench has been attributed to erosion of a major thrust and strike-slip fault zone, to normal faulting, and to erosion of zones of weak strata.

Sheared and highly deformed rocks outcrop in the trench, including many that are incompetent. What role these weak, easily eroded strata play as compared with erosion along a fault is still uncertain.

A seismic reflection profile taken across the trench near the southern end shows that both sides of the trench are flanked by mountains underlain by eastward-dipping Beltian strata and that the trench is characterized by block faulting, with blocks tilted toward the east. Bally, Gordy, and Stewart (1966, p. 356) make the following interpretation of the seismic line and regional geological data:

> Consideration of these geophysical and geological points suggests the propositions (1) that the Trench was formed in Tertiary time and after the main thrusting phase, (2) that the Trench is underlain by an undisturbed gently westward dipping basement, and (3) that location and strike of the Trench is dictated by a complex system of curved low-angle normal faults.

The trenches may be of fundamental importance in the tectonics of the Cordillera. They do mark a boundary between two provinces that differ in lithologic character and structure. It is essentially a boundary between the unmetamorphosed strata now deformed into great thrust sheets in the Laramide orogenic belt to the east, the long-term miogeosyncline, and the somewhat older orogenic belt to the west with its metamorphic and plutonic rocks. The trench may be located approximately along the original edge of the Precambrian craton. If it is a thrust it is one of the longest in the world. If it is a zone of strike-slip movement it may be related to the bend in the Nevadan belt in Washington and Oregon and the transcurrent movements along the West Coast.

The Core of the Northern Rocky Mountains Eastern Fold Belt

The core zone of the eastern Cordilleran fold belt in Canada is made up of a number of complex structural elements* often referred to collectively as the Omineca geanticline. Most of the core zone is located west of the Rocky Mountain trench, and it can be traced more or

* Including the Cassiar platform, Shuswap complex, Cariboo Mountains, Kootenay arc, and Purcell anticlinorium.

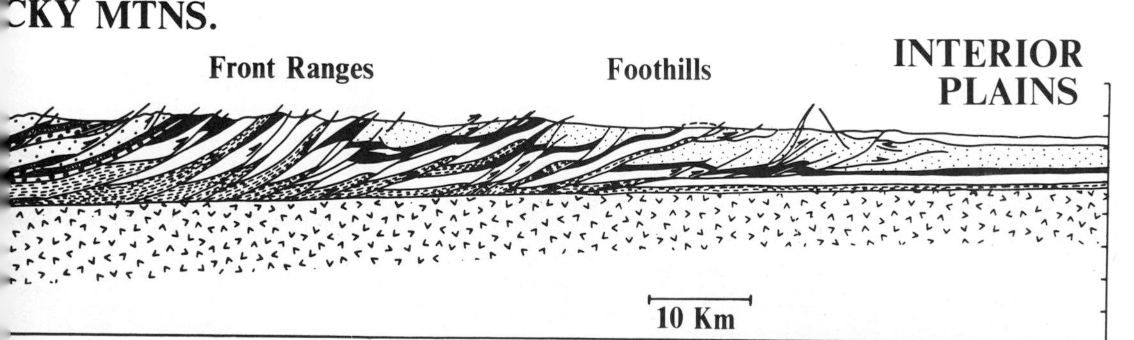

less continuously from Washington State, where it disappears under the volcanics of the Columbia River Plateau, to Alaska, where it splits. The rocks of the core zone include large amounts of Precambrian basement and its late Precambrian clastic cover as well as Paleozoic and Mesozoic units, all of which have been affected by Mesozoic metamorphism. The core zone in Canada occupies a position comparable to the more internal parts of the Sevier belt, south of the Columbia River Plateau.

The Omineca belt was located on the edge of the North American craton during the late Precambrian and throughout most of the Paleozoic, and it was the site of accumulation of slope deposits laid down on the border of a continental type basement. Toward the end of the Paleozoic, volcanic products became a prominent part of the sediment in the Omineca belt, which was probably the site of a volcanic island arc until the Mesozoic, when large granitic plutonic bodies were emplaced, and the older rocks were metamorphosed, uplifted, and thrust toward the east. One of the largest of these thrusts, the Purcell, is found exposed in the western wall of the Rocky Mountain trench. This mid- to late Mesozoic deformation was not the first, but certainly it has dominated the present configuration. Superimposed folds, the earliest of which are probably of mid-Paleozoic age, are common in the Omineca belt. Among the younger deformations, easterly directed thrust faults and asymmetrical recumbent folds are predominant. One part of the Omineca belt, the Shuswap complex, consists of a series of mantled gneiss domes with cores of migmatitic granitoid gneisses enveloped in mantles of metasedimentary gneisses that are located at about 80-km (50-mile) intervals.

Tracing the Eastern Cordilleran Fold Belt South—Sevier Orogen

Equivalents of the Canadian Rockies and its core, the Omineca zone, are found in the United States, where the name Sevier belt is commonly applied. The eastern limits of the belt generally defined by a system of west-dipping thrusts and asymmetric folds with eastern vergence are relatively clear, but the position and character of the western edge is still debated. The belt is severely deformed by Cenozoic block faulting in the Basin and Range physiographic province which makes it difficult to piece together the original continuity of the thrusts, but it now seems clear that the thrusts of the Northern Rockies pass just east of the Idaho batholith, along the western border of Wyoming and then along the western edge of the Colorado Plateau. Their continuity across Arizona into Mexico is not clear. A broad system of northwest-trending folds of late Mesozoic age is found in central Mexico, but these are Laramide features probably contemporaneous with deformation of the platform area of the Central and Southern Rockies.

Roberts and Crittenden (1975, p. 416–418) characterize the eastern fold belt (Sevier) as being composed of a deep infrastructure, the site of early stage deformation, and a more surficial suprastructure which was deformed somewhat later. These are described as follows:

> The early-stage deformation of the Sevier Belt was characterized by polyphase penetrative deformation, low-angle faulting, plutonism, and metamorphism . . . where the infrastructure has been brought to the surface, the patterns of deformation are extremely complex. Imbricate thrusts, recumbent folds, and penetrative deformation characteristic of the infrastructure have been described by Compton (1971). . . . Deformation in three sub-stages is recognized: [two sets of overturned folds]. . . . Both sets of folds and related thrusts are truncated by the decollement which separates them from the little-deformed rocks of the Oquirrh formation (Pennsylvanian), which form the suprastructure. . . .
>
> Late-stage deformation in the Sevier belt was characterized by large-scale uplift and eastward movement of gravity plates, resulting in attenuation and partial denudation of the hinterland. . . . Concomitant with maximum uplift and transport of de-

bris, plates moved eastward under the influence of gravity, ultimately reaching the craton.

Burchfiel and Davis (1975) have extended the scope of the Sevier orogenic belt to include a large part of the Cordilleran system, Fig. 24-6. They propose that the eastern belt of west-dipping thrusts, the great granitic plutons of the Cordilleran, and the east-dipping thrust belts found in the Insular zone of Canada and in the Klamath Mountains of California and Oregon are all parts of the same orogenic system. In this case the eastern (continental) part of the orogen is separated from the portion along the west coast, and on oceanic crust, by a broad zone of relatively undeformed rocks both in the Intermontane belt of Canada and in the western United States, Figs. 24-6.

The Intermontane Belt of Canada

The eastern and western Cordilleran fold belts are separated in Canada by a morphotectonic division called the *Intermontane belt,* a division not clearly developed south of the Columbia River Plateau. Much of this division in Canada is a plateau between the higher mountainous regions on either side. Stratigraphic studies of the Intermontane belt indicate that it was west of the continental platform and underlain by oceanic crust until the Triassic, when it became the site of volcanic islands, and later large basins formed in which eugeosynclinal deposits accumulated. Rocks of the Intermontane belt escaped the strong orogenic deformation that affected the areas on both sides and show only moderate deformation, primarily gentle folds that formed early in the Jurassic. The region has been uplifted, but very much less than adjacent mountainous areas. The final shaping of the present tectonic pattern consisted primarily of movement on high-angle faults some of which are normal and others transcurrent.

The Basin and Range Province

The location of the Basin and Range area is somewhat analogous to the Intermontane belt of Canada, but the histories of the two differ in many respects and especially in the degree to which the Basin and Range has been subjected to block faulting and Tertiary volcanic activity. Despite all its complexities it is here that many clues to the early history of the Cordilleran system have been found. Evidence for the Paleozoic history of the Cordilleran system is widely scattered, and therefore, it is difficult to interpret, but it is clear that parts of the Cordilleran region experienced orogenic deformation about the middle of the Paleozoic in an orogeny called the *Antler orogeny,* Fig. 24-6. Evidence for this event is found mainly in central Nevada where a large thrust sheet, the *Roberts Mountain allochthon,* composed of siliceous materials formed in deep water, was thrust possibly 100 km to the east over contemporaneous platform carbonates late in the Devonian, Fig. 24-7. Evidence for the Antler orogeny is also found in the Klamath Mountains, where an ophiolite sequence was thrust west above probable oceanic crust (Davis, 1968) in an area which had apparently been part of an island-arc system since the Ordovician. Sediments derived from the uplifted region in the Antler orogeny were spread eastward into eastern Nevada and Utah and eventually covered the Roberts Mountain allochthon.

A second orogenic episode, *the Sonoma orogeny,* affected the Cordilleran region again at the end of the Paleozoic, Fig. 24-6. Like the earlier Antlers event, oceanic sediments were thrust eastward across the continental shelf in Nevada, and again volcanic activity of this age is evident in the Klamath Mountains, which suggest that the mechanism for these two orogenic episodes was similar. Burchfiel and Davis (1975) propose that the mechanism consisted of initial subduction of oceanic crust along the west coast followed by obduction resulting from plate convergence and closure

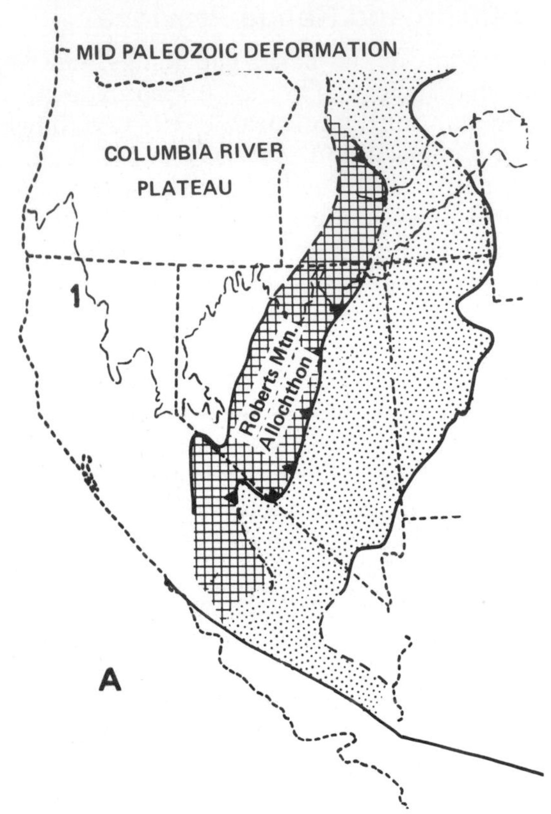

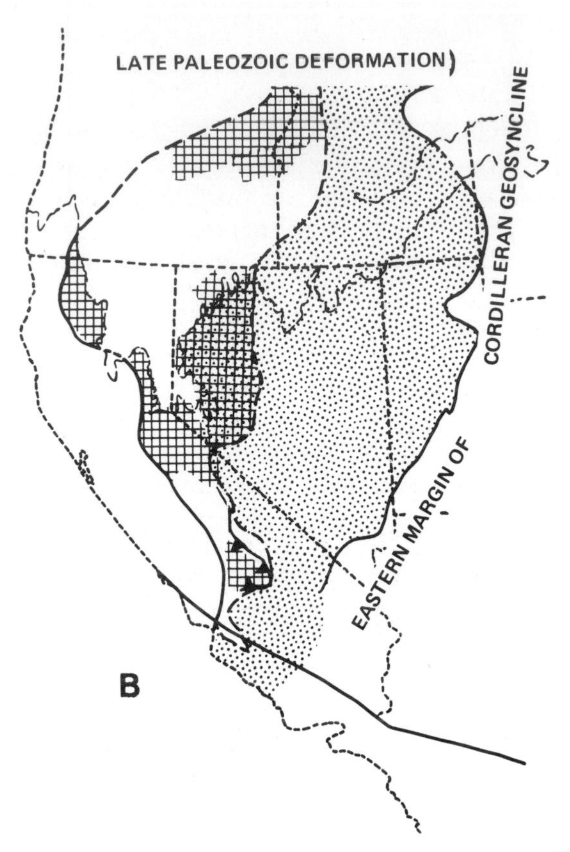

FIGURE 24-6 Paleogeographic sketch maps showing areas of continental and oceanic crust and areas affected by deformation for four major time intervals. The eastern margin of the geosyncline is shown by a solid line along the eastern edge of the map areas. The dot patterns show approximately the areas within the geosyncline which were underlain by continental crust that was unaffected by deformation during the time interval indicated. A grid pattern over the dot pattern indicates areas of continental crust that were strongly affected by deformation. Areas of oceanic crust strongly affected by deformation are covered by the grid pattern only. The large blank areas within the geosyncline are areas where post-Mesozoic sediments and volcanics obscure the relationship. The area keyed 1 is an area of Ordovician-Devonian volcanic rocks. (*Simplified after Burchfiel and Davis, 1975.*)

of the marginal basin in Nevada with resulting eastward thrusting in the basin.

The final phase of thrusting to affect the Basin and Range area came in the Mesozoic (largely Jurassic), when all the earlier formed structures were uplifted and again thrust to the east as part of the *Nevadan orogeny,* Fig. 24-6*c*.

THE WESTERN CORDILLERAN BELT

The Canadian section of this belt is made up largely of huge granitic plutonic complexes in which large sections of the roof rock are preserved. Most of these plutons were emplaced late in the Cretaceous and constitute part of the Mesozoic plutonic complexes that occur in the western fold belts all the way from the Yukon into Mexico and include the Coast Range, Idaho, and Sierra Nevada batholiths, Fig. 24-1.

McTaggart (1970) suggests that the coast plutonic complexes and the northern Cascade Ranges represent two different levels of exposure in the same belt. The upper level is exposed in the Cascades, which are composed of upper Paleozoic sediments that were metamorphosed early and again late in the Mesozoic and granite plutons emplaced late during the Mesozoic and again late in the Tertiary.

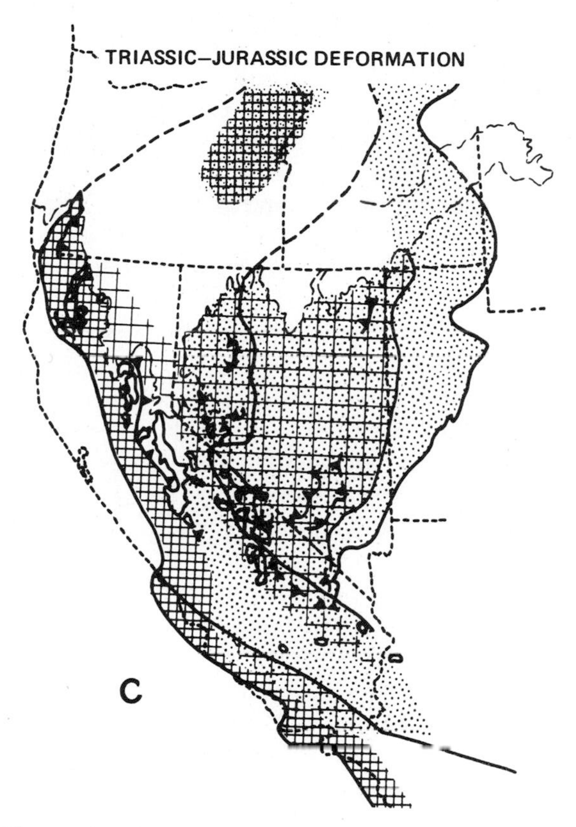

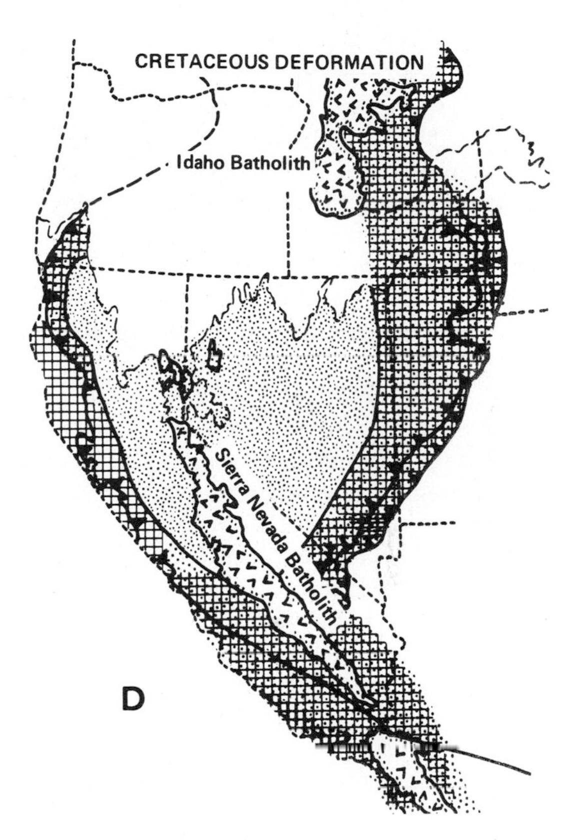

Thrust faults occur on both flanks of the northern Cascades and dip toward the axial zone of metamorphism and intrusion on both sides. The larger thrusts on both sides contain ultramafic rocks and some blueschist metamorphics, the earmarks of subduction zones in plate theory.

Eugeosynclinal conditions have persisted in the western part of the Cordillera since early in the Paleozoic. Three distinct groupings of eugeosynclinal rocks are shown on the North American tectonic map. The oldest of these were involved in orogenic episodes during the Paleozoic. Notable among these episodes were the *Cariboo orogeny* in Canada and the Antler and Sonoma orogenies in Nevada. Only relatively small portions of this oldest belt are still preserved, but it is interesting to note that all the remaining segments are located closer to the craton than the younger eugeosynclinal belts.

The Nevadan Eugeosynclinal Belt

The second eugeosynclinal belt contains rocks of late Paleozoic to late Jurassic age, and large areas of these rocks are metamorphosed and intruded by granitic batholiths, some of which were emplaced in the Jurassic. These rocks were affected by an orogenic episode, known as the *Nevadan orogeny,* that took place in the Jurassic. Parts of this belt occur as fragments as far south as the tip of Baja California, and larger areas are exposed in northern Baja, along the western edge of the Sierra Nevada batholith and in the Klamath Mountains. Exposures also occur far to the east in Oregon and Idaho, and they are broadly exposed especially in the Omineca and Intermontane belts of Canada. The trends of folds and faults within the metamorphic complex as well as the distribution of outcrops suggest that the orogen swings in an arc (called the Columbian arc), Figs. 24-2 and

24-3. The direction of tectonic transport in this orogen is not uniform. We have already discussed the eastward movements in the Omineca belt, but considerable evidence suggests that the movement in the Klamath Mountains is toward the west (Irwin, 1965; Dott, 1965).

The Klamath Mountains are selected to illustrate the nature of this eugeosynclinal belt because they have been examined in some detail and knowledge of them has been recently synthesized (Dott, 1965; Irwin, 1960, 1965; Burchfiel and Davis, 1975). Rocks of eugeosynclinal facies have formed in the area now exposed in the Klamath Mountains since the early Paleozoic. The region can be divided longitudinally into several arcuate belts—a western Jurassic belt and eastern and western Paleozoic to Triassic belts separated by a central metamorphic belt (Fig. 24-8). Granitic plutons and ultramafic rocks occur in all these belts. The plutonic and metamorphic activity culminated in the Late Jurassic between 135 and 145 million years ago in the Nevadan orogeny. Each of these outcrop belts appears to be separated from the next by a fault or a zone of ultramafic or granitic rocks. The most prominent rock types in the Paleozoic sedimentary belts are greenstones, graywacke sandstones, mudstones, and bedded chert; a little limestone and rhyolite occur in several parts of the section. The central metamorphic belt contains hornblende and mica schists; the western Jurassic belt is mainly composed of slaty mudstones and graywacke.

Irwin (1965) postulates that the boundaries between these belts are thrust faults and that the arcuate character of the Klamath structural trends is in part a reflection of this westward-directed thrusting in which each belt is moved over the adjacent western belt (Fig. 24-8), with the resultant development of outliers and windows.

The lithologies, paleogeography, and structural features of the Klamath area are now interpreted as being those associated with an island-arc system that developed on oceanic crust in the early Paleozoic and became the site of repeated episodes of subduction tectonics that now appear to be contemporaneous with deformation in the Antlers, Sonoma, and Nevadan orogenies. The site of subduction shifted progressively to the west after each orogeny, and after the Nevadan it jumped even farther west to the present coast.

FIGURE 24-7 Restored cross sections across the Basin and Range. Top, Early Permian section showing sediments deposited across the Roberts Mountain thrust. Bottom, Late Permian section showing Golconda thrust which carried rocks east over the Antlers orogenic structures during the Sonoma orogeny. (*After Roberts and Thomasson, 1964.*)

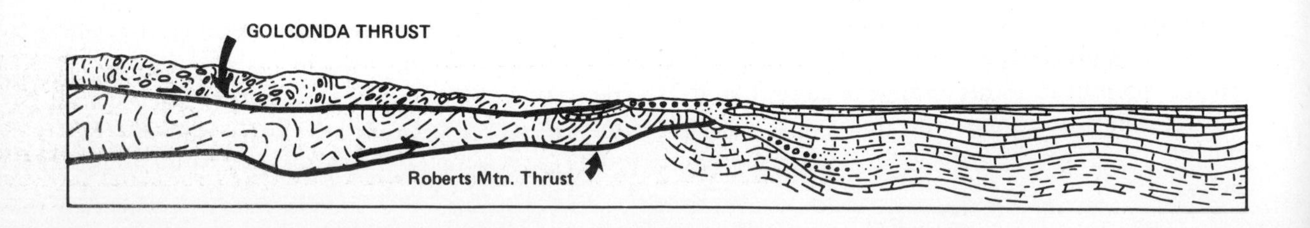

The Franciscan Eugeosynclinal Belt*

The youngest of the eugeosynclinal belts includes largely unmetamorphosed graywacke of Late Jurassic and Cretaceous age, known as the *Franciscan group*. Most of these rocks formed after the Nevadan orogeny, and generally the two groups are separated by faults. The Franciscan is exposed in the Coast Ranges along the California coast, and similar rocks occur along the Alaskan coast. The Franciscan is predominantly graywacke, but shale, altered mafic volcanic rock, chert, and minor limestone are a part of the assemblage. Its total thickness cannot be determined by normal stratigraphic methods, but is probably more than 20 km. Sedimentary structures indicate rapid deposition of unsorted material, presumably by turbidity currents. Altered mafic rocks consist of pillows, tuffs, or breccias resulting from submarine eruptions, and some of the more massive units may be intrusive. These volcanics range from a few meters to kilometers in thickness. Chert and a distinctive shale occurring with it are present in minor quantities and are thought to be chemical precipitates formed by the reaction of magma and sea water under considerable hydrostatic pressure.

* See Bailey, Irwin, and Jones (1964).

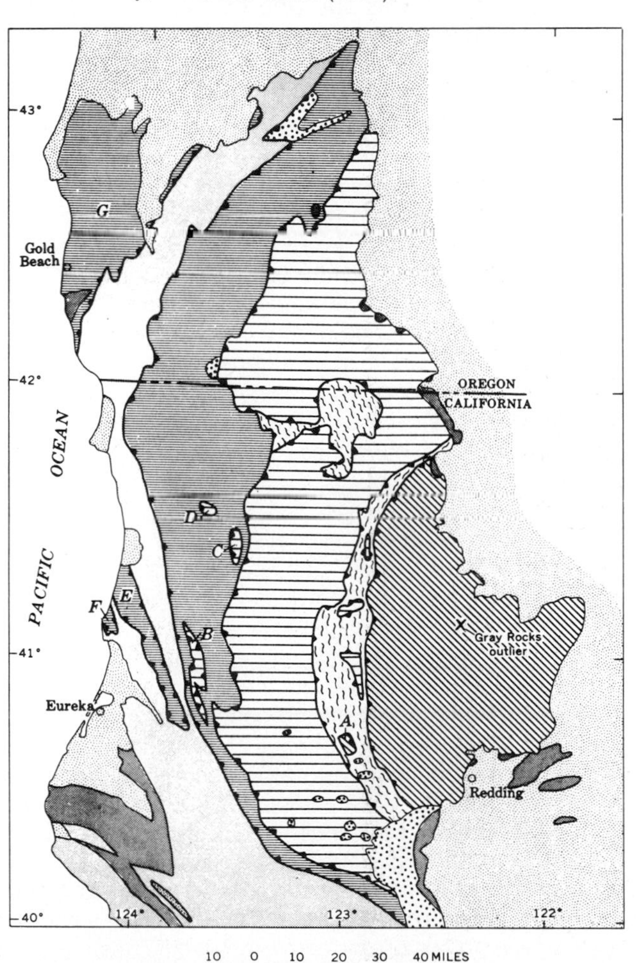

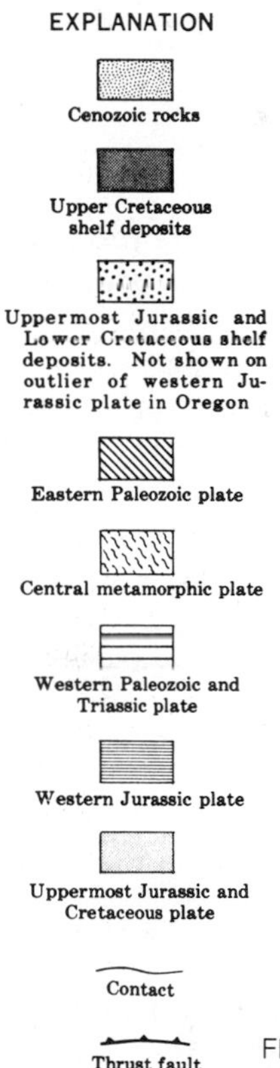

FIGURE 24-8 Principal postulated thrust plates of the Klamath Mountains and adjacent Coast Ranges. Thrust outliers are indicated by letter symbol: E = Redwood Mountain. (*From Irwin, 1975.*)

The cherts are associated with serpentinite and other ultramafic rocks that occur widely through the Franciscan rocks. Many of the masses are tabular in form, the largest is 100 km long and several kilometers wide, and most of them are concordant with the country rock. These ultramafics are highly sheared, show no contact metamorphic effects, and thus were probably emplaced as solidified slices.

Metamorphic rocks, including zeolite, blueschist, and eclogite facies, are present though not in great quantity. The blueschist-facies rocks occur as small isolated patches within and gradational into unmetamorphosed graywacke, as areas several kilometers wide and tens of kilometers long, suggesting regional metamorphism, and as isolated rounded masses of schists up to 100 m in diameter surrounded by nonmetamorphic rocks. The eclogite facies occurs only in this latter form, and these rounded masses are found in shear zones, in serpentine, and in unaltered Franciscan rocks, and they probably represent tectonic inclusions. Experimental studies suggest that the blueschists develop under high pressure (5 kbars) but relatively low temperature (under 300°C).

The basement of the Franciscan is not exposed, but since the inclusions brought up in the ultramafic masses are all Franciscan rock types, it is probable that the rocks of the group were deposited on a basaltic or peridotite crust (Fig. 24-9). The Franciscan is highly deformed, but structures within it cannot generally be ascertained because of its persistent heterogeneity and lack of key beds. Most folds trend northwest. The major faults are shear zones that contain large blocks of Franciscan rocks in a sheared matrix and include tectonic inclusions of schist and sheared masses of serpentine. A fault separates the Franciscan rocks of the northern Coast Ranges from the crystalline rocks of the Klamath Mountains. South of the Klamath Mountains the fault is covered by Tertiary sediments of the Great Valley of California.

FIGURE 24-9 Tectonic model for the relation between the Great Valley sequence and Franciscan assemblage. No vertical exaggeration. Note the depression of the 150°C isotherm shown as a dashed line. (*After Ernst, 1970.*)

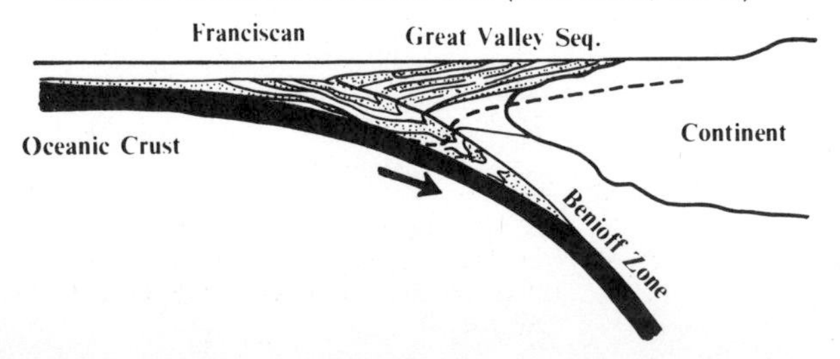

Modern interpretations of these western belts are generally based on plate-theory models, such as the one described below by Hamilton (1969, p. 2409):

The Mesozoic evolution of California is interpreted as dominated by the underflow of oceanic mantle beneath the continental margin. Underflow during part of Late Cretaceous time of more than *200 km of the eastern Pacific plate* seems required by the marine magnetic data. Correspondingly, varied oceanic environments—abyssal hill, island arc, trench, oceanic crust, and upper mantle, perhaps also continental rise and abyssal plain—appear to be represented in the eugeosynclinal terranes of California. *The rock juxtapositions accord with the concept that these materials were scraped off against the continent as the oceanic plate slid beneath it along Mesozoic Benioff seismic zones, which are now seen as serpentine belts separating profoundly different rock assemblages.*

The chaotic Franciscan Formation of coastal California consists of deep-ocean Late Jurassic to Late Cretaceous sedimentary, volcanic, crustal, and mantle materials. As open-ocean abyssal oozes and the oceanic crust beneath them were swept into the Benioff-zone trench at the continental margin, they were covered by terrigenous clastic sediments, and the entire complex was carried beneath the correlative continental-shelf and continental-slope deposits (Great Valley sequence) and the older Mesozoic complexes. [Fig. 24-9.]

The other eugeosynclinal terranes of California can be interpreted, albeit with less confidence, in similar terms of underflow of Pacific mantle. [As is illustrated by Hsü (1971), Fig. 24-10.]

Reversal of Cenozoic extension, strike-slip faulting, and volcanic crustal growth in the western United States reveals a Cretaceous tectonic pattern strikingly like the modern pattern of the Andes, so the paleotectonic setting of North America can be inferred from the South American present. The Mesozoic batholiths of North America, like the late Cenozoic volcanic belt of the central Andes, are products of the same rapid motion of oceanic plates that carried oceanic sediments against the continent to form eugeosynclinal terranes. Magmas generated in the Benioff zones formed the batholiths and the volcanic fields which initially capped them.

Block Faulting in the Basin and Range

The present configuration of the Basin and Range physiographic province is due largely to Cenozoic block faulting which has continued into Pleistocene and Recent times and suggests the possibility of a connection between this block faulting and Cenozoic transcurrent movements along the West Coast, Fig. 24-11.

Cenozoic block faulting bears no apparent relation to the boundary of the older geosynclinal margins. Extensive faulting along predominantly northwest-southeast lines occurs in the basalts of the Columbia River Plateau in Oregon, southern Idaho, northern California, and Nevada where faults cut across the trend of the metamorphic eugeosynclinal belt of the Nevadan orogeny, Fig. 24-12. Block faulting is largely responsible for the north-south elongation of the ranges and valleys of Nevada, the largest of which, the Sierra Nevada Range, is

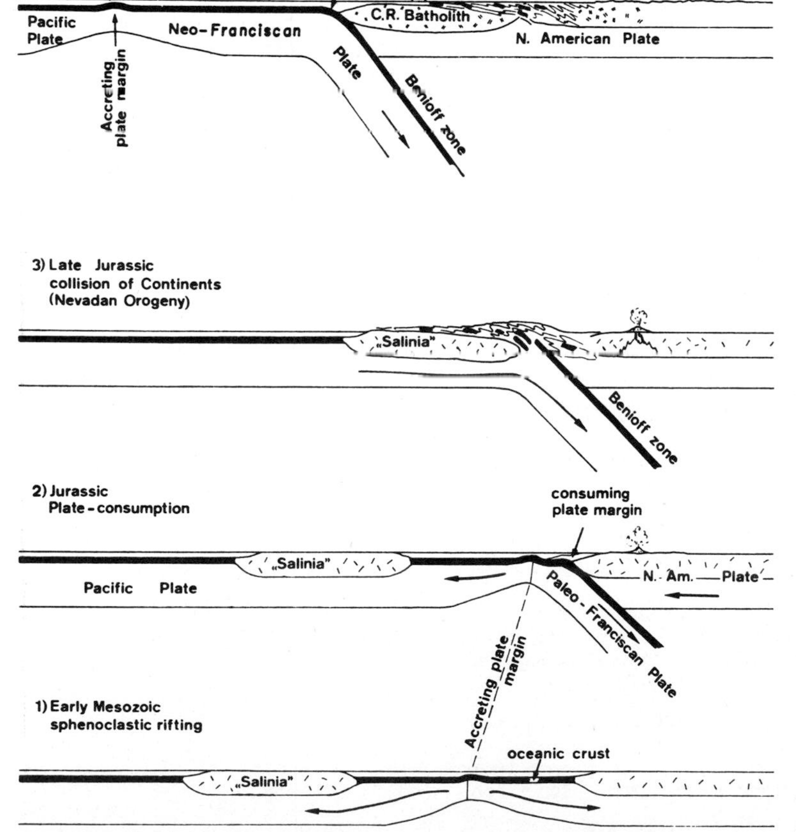

FIGURE 24-10 Evolution of the Franciscan geology interpreted in the framework of plate tectonics. (*From Hsü, 1971.*)

being uplifted along its eastern edge by Cenozoic fault movements. Christensen (1966) describes the geometry of this movement by means of a structure contour map constructed by use of uplifted Tertiary river channels and volcanic rocks. The latest movements, he finds, began during the Pliocene and have amounted to over 2,000 m along the eastern (faulted) margin since that time. The Sierra Nevada block has moved as a large, rigid unit and differs in this respect from the more closely spaced faulted blocks to the east. Cenozoic uplift on high-angle faults is by no means confined to the Basin and Range; many faults in the shelf region, as in Wyoming, have been active in the Pleistocene also, but the effects have not been nearly as drastic in modifying the regional tectonic pattern as they have been in the Basin and Range. Block faulting also occurred in the Appalachians after the last major orogeny (the Triassic basins), but the orientation of the basins, the spacing, and the details of the fault patterns are hardly comparable to those in the west. Most of the faults in the Basin and Range region are normal; the structure is essentially that of graben and horst block faulting. Thick tectonic breccias such as those associated with thrust and reverse faults are absent. Drags in the sedimentary and volcanic cover are generally absent; however, folds are sometimes traced into faults or found over faults as described in the Middle Rocky Mountains. The faults typically dip toward the basins at an angle of 45 to 80°, often as a series of steps. Thrust faults of Cenozoic age are not found associated with the high-angle faults,

FIGURE 24-11 Schematic cross section of the Basin and Range province of North America showing thin crust under the Basin and Range province and high heat flow in this area. (*From Scholtz, Barazangi, and Sbar, 1971.*)

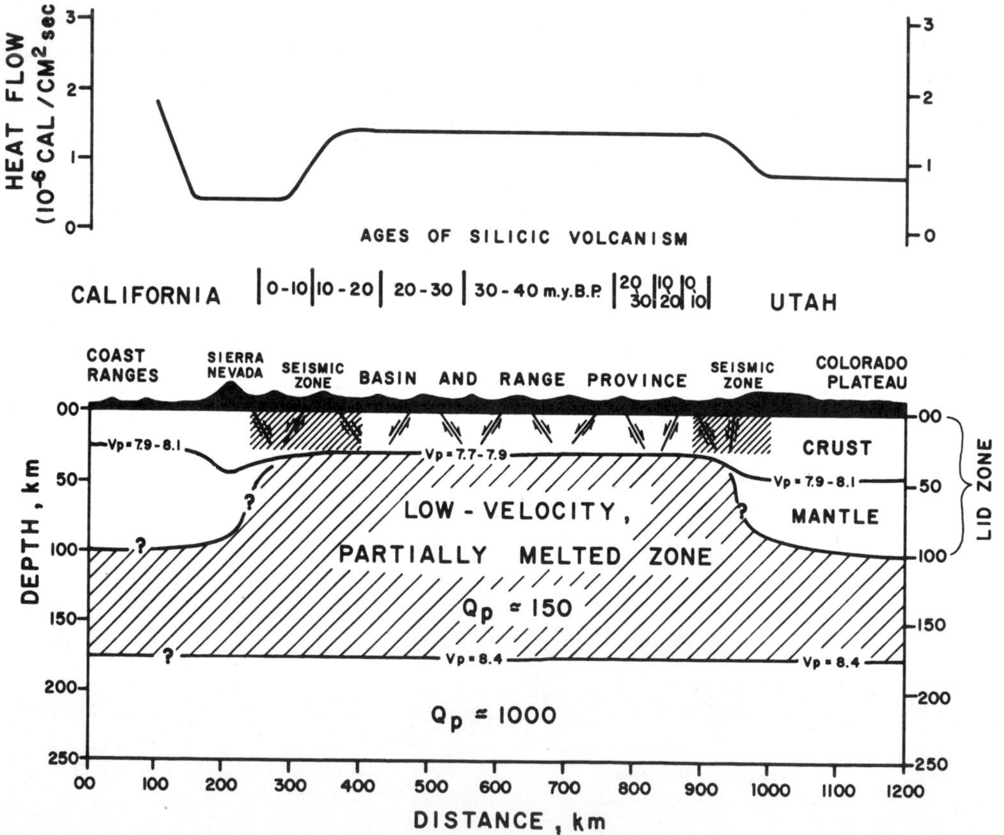

and in southern Oregon, small volcanic vents are found aligned parallel to the fault direction (Fuller and Waters, 1929), as might be expected if the movement is essentially extensional.

Geophysical studies made across the basins reveal steep faults similar to those on the surface and show considerable variation in throw (Donath and Kuo, 1962). Gravity surveys run over Dixie and Fairview valleys, Nevada, indicate that these valleys contain over 1,000 m of low-density (Cenozoic) sediments (Thompson, 1959); so, the structural relief caused by the faulting here is on the order of 4 km. Detailed analysis of recent movements made by use of a carefully surveyed triangulation network allow us to estimate the total horizontal separation by knowing the amount of vertical separation and the number of faults. On this basis it is estimated that the Basin and Range has extended about 50 km in the last 15 million years (Thompson, 1959.)

Most students of the Basin and Range province accept the interpretation of the structural pattern as being due to extension. If that extension is taking place approximately at right angles to the normal faults, the crust is being extended along northeast-southwest lines in Oregon and along east-west lines in Nevada. The pattern is not as clear in southern Arizona, and to the west in California and the Baja Peninsula, transcurrent movements dominate. The necessary extension could arise in several ways—regional uplift followed by partial collapse, regional extension across the geosyncline, or extension produced as a secondary effect of major transcurrent movements parallel to the elongation of the geosyncline as explained in the next section.

Hypothesis of Strike-Slip Control of Block Faulting

A number of zones of strike-slip faulting have been identified in the Basin and Range, and many more lineaments are probable zones of strike-slip movement (Fig. 21 13). Among the largest of these are Walker Lane, which trends northwest-southeast in western Nevada and probably continues to the south as the Las Vegas shear zone, and northwest as the Death Valley fault zone (Fig. 24-14). All these are right-lateral strike-slip zones, as are the San

FIGURE 24-12 Cross sections across parts of the Basin and Range showing present structural configuration. Note that the older (Paleozoic) thrusts are offset by Cenozoic normal faults. Precambrian = slash pattern, lower Paleozoic = dots, late Paleozoic = diagonal lines, and Cenozoic = light dots. (*From Roberts and Crittenden, 1975.*)

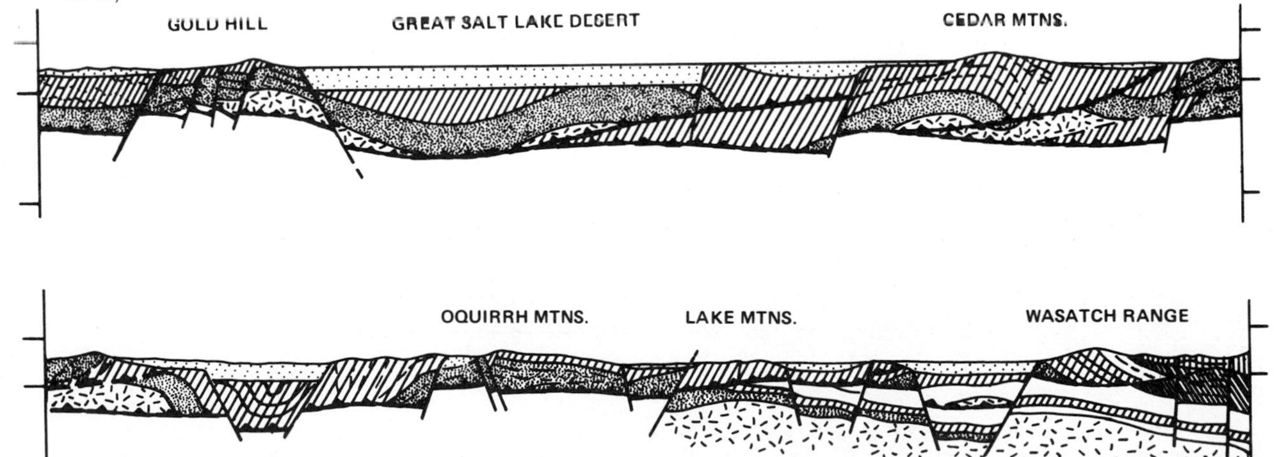

Andreas and other major faults along the West Coast.

Shawe (1965), while analyzing historical faulting in Nevada, discovered that the recent faults occur in a well-defined arc which transgresses several of the fault-block mountains, and he suggests that these are not formed independently. Faults in the southern portion of this arc are arranged *en échelon,* indicating a possible deep-seated shear of right-lateral movement. Some of the faults bounding valleys are dip-slip, others are mainly strike-slip, and a systematic change from dip-slip to strike-slip occurs from north to south. Shawe suggests that the Basin and Range structures have been created as near-surface adjustments in the crust to movements along deep-seated transcurrent faults, perhaps extending into the upper mantle. Stratigraphic evidence also supports a right-lateral movement in the Basin

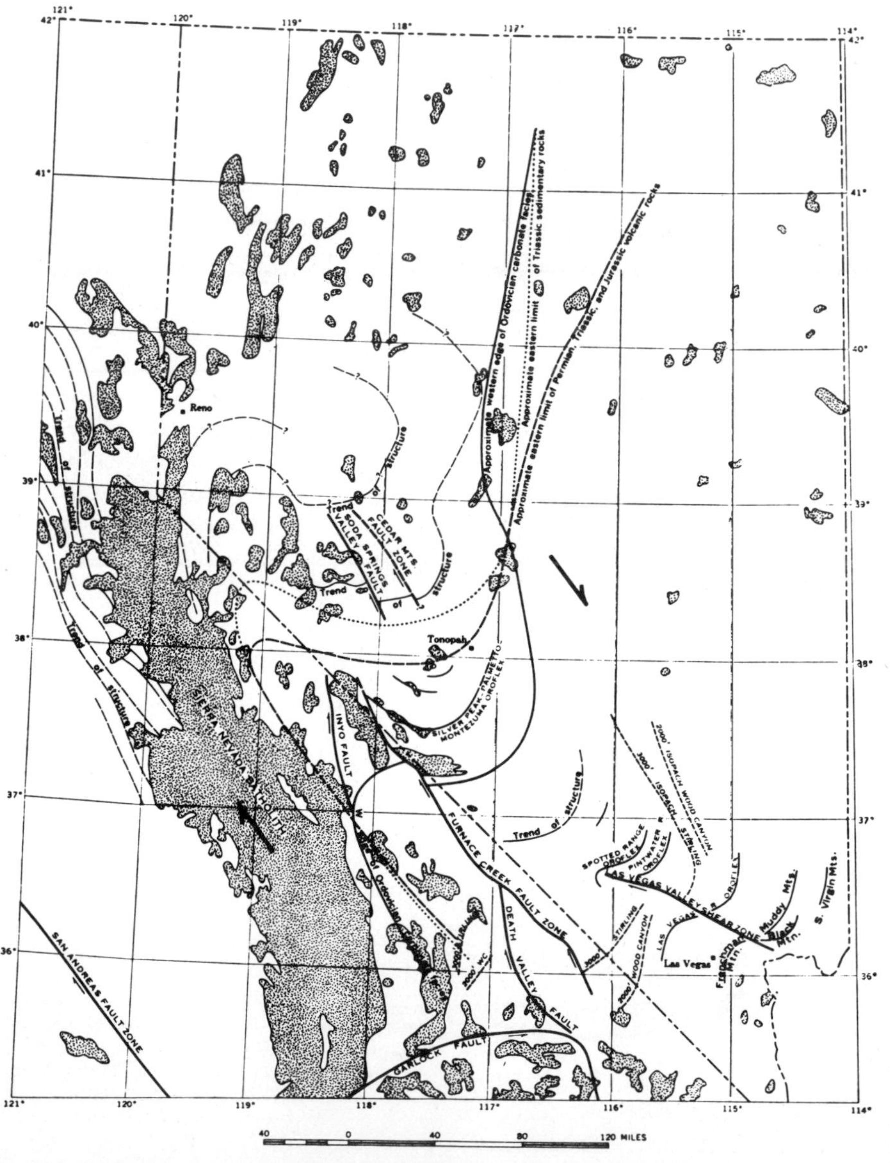

FIGURE 24-13 Summary of structural features along the mobile belt east of the Sierra Nevada. Stippled areas are granitic plutonic rocks. (*From Albers, 1967.*)

and Range as Albers (1967) has shown by documenting the pronounced sigmoidal bends in the limits of Permian, Triassic, and Jurassic volcanics and Ordovician carbonates. These bends coincide with the fault zones along the southwestern border of Nevada, Fig. 24-13.

Basin and Range Faults—Effects of a Rotational Couple?

The block faulting in the Basin and Range may be directly related to contemporaneous strike-slip faulting as suggested by Carey (1958) and

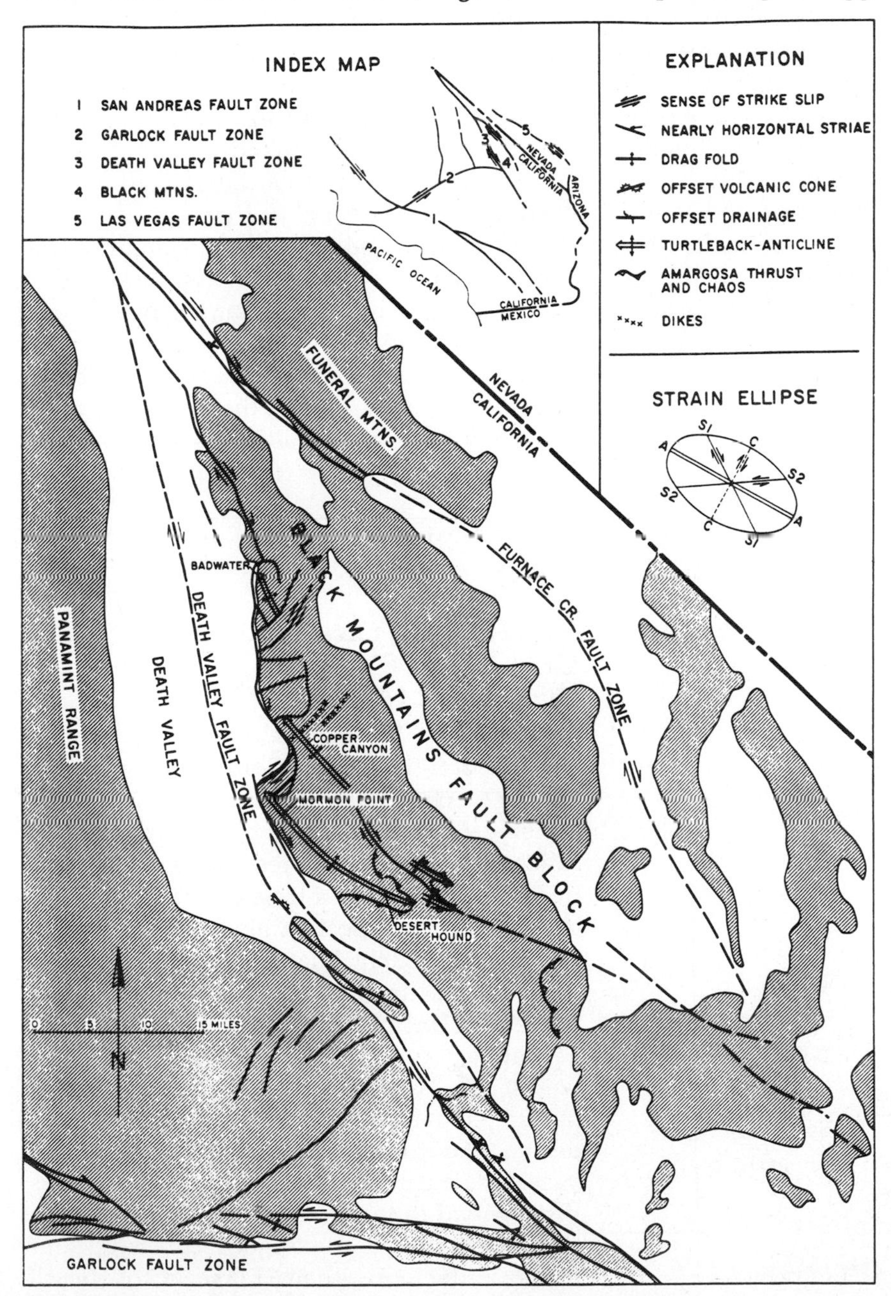

FIGURE 24-14 Tectonic sketch map, Death Valley region, California, showing major strike-slip faults and related folds. (*From Hill and Troxel, 1966.*)

Wise (1963b), who point out the possible connection between the Columbian arc and the block faulting. If the Rocky Mountain trench is a strike-slip fault, then movement along that fault, the west coast faults, and bending of the Columbian arc may be due to a great rotational (clockwise) couple. In this interpretation the Columbia arc is a folded orogen (an orocline) and the bend is essentially a large-scale drag feature. This would account for the presence of the block faulting just south of the bend in the orogen as indicated by the strain ellipse, Fig. 24-15, and its absence farther north and south. It would seem plausible to explain the Pleistocene uplift in the Rocky Mountains as a response to these lateral shifts due to the couple as well.

Basin and Range as an Ensialic Interarc Basin

Plate-tectonic theory has been applied (Scholz and others, 1971) to explain both the near-surface structure and a rapidly growing body of geophysical data about the lithosphere under the Basin and Range. Geophysical studies indicate that the region has an unusually high heat flow, that the crust is thinner than that either east or west of it, and that the upper mantle has unusually low velocity characteristics. All this suggests the possibility that the region is being extended over rising hot mantle material. Scholz concludes that this may be similar to the spreading found behind some island arcs and that the spreading is occurring over a slab of crust that was being subducted in the subduction zone along the west coast early in the Tertiary. When this subduction stopped, the slab became bouyant and began to rise. A mantle diapir over it moved upward to the base of the crust where it began to spread laterally, rifting the overlying brittle crust and allowing basaltic volcanic materials from this mantle diapir to move up along some of the faults and pour out on the surface to form the extensive late Tertiary lava flows in the Basin and Range, Fig. 24-11.

FIGURE 24-15 Hypothetical stress distribution in the Cordilleran region. California and Mexico batholith zone on the south. Canadian batholith zone on the north. Strain-ellipsoid orientation for Basin and Range province and for Pacific Northwest are indicated. Rotational sense of marine magnetic pattern indicated at northwest. Small arrows in Wyoming indicate combined compression and shear to produce parallelogram pattern. (*From Wise, 1963b.*)

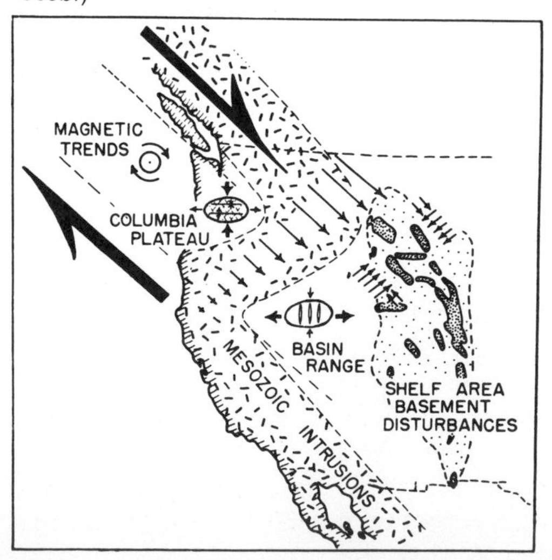

PLATE TECTONICS AND THE CENOZOIC EVOLUTION OF WESTERN NORTH AMERICA

The Cenozoic has been a time of widespread volcanic activity in the Cordilleran system, a time of block faulting, fragmentation, and displacement of the orogenic belts that were consolidated and incorporated into the continent in earlier times, and a time of extensive right-lateral shearing especially along the West Coast but to some extent inland as well.

Volcanic activity has been most pronounced in western United States and Mexico, but some Tertiary volcanic fields occur in Canada. The largest area of volcanics is found in the Columbia River Plateau, where flows of basalt, andesite, and dacite erupted and spread over vast areas during the Miocene and Pliocene especially and in the Eocene in the Coast Ranges of Washington and Oregon, covering some areas with more than 1,000 m of basalt.

Volcanic activity has continued through the Quaternary in the Snake River downwarp area south of the Idaho batholith where these extrusives can be traced into the Yellowstone Park region.* The high and young volcanos of the Cascades which erupt olivine basalt stand out as a gently curved arc from northern Washington into California, forming the most prominent features of Quaternary volcanism.

Other large areas covered by Tertiary volcanics are found in the Sierra Madre Occidental mountains of Mexico, along the southern and western sides of the Colorado Plateau, in Baja California, and a large Quaternary volcanic field cuts across Mexico passing through Guadalajara and Veracruz, Fig. 24-1.

Plate-tectonic theory has provided a new approach to the problems of Cenozoic deformation especially along the coast, as exemplified by a unifying model proposed by Atwater in 1970. The key to the use of this model lay in relating the magnetic anomaly patterns found in the sea floor of the eastern and northern Pacific to deformation patterns and history in the western part of the Cordilleran system.

The magnetic anomaly patterns in the eastern Pacific were described by Raff and Mason (1961) even before the sea-floor-spreading hypothesis was well established. But unlike the symmetrical patterns subsequently defined along the mid-Atlantic ridge, those off the West Coast are not symmetrically disposed on either side of a clearly defined spreading ridge, Fig. 20-4. Later as the magnetic anomaly patterns became better known and particular anomalies identified and dated, it was established that the anomalies in the eastern Pacific lie on one side (the west or Pacific side) of a spreading ridge. Symmetrically disposed anomalies could be identified on both sides of the East Pacific rise, and these could be traced into the mouth of the Gulf of California, but no ridge crest could be found from the Gulf northward until the Gorda rise is encountered just north of the Mendocino fracture zone. The Gorda rise trends north a short distance until it is cut by the Blanco fracture zone north of which another ridge, the Juan de Fuca ridge, is offset to the west. The Juan de Fuca is in turn offset to the west at its northern end, and finally the ridge and associated magnetic anomalies end abruptly as they pass into the continental margin just north of Vancouver Island. Farther west the magnetic anomaly belts which trend generally north-south in the eastern Pacific bend sharply to the west and pass into the Aleutian trench.

The gross pattern of ridge crests and magnetic anomalies along the West Coast suggest that the North American continent is situated over much of the eastern side of an originally symmetrical spreading ridge and actually lies over parts of the ridge crest itself—in other words we appear to see a collision of a continent with a spreading ridge here. The magnetic anomaly patterns allow us to at least postulate precollision reconstructions of the ridge crest, and that has been done as shown in Fig. 20-12 for a time about 100 million years ago. At that time the spreading center was much closer to the central Pacific and the ocean was divided into four large plates, the Pacific, Kula, Farallon, and Phoenix, which joined in two triple junctions. The Kula plate was being subducted in trenches along the Aleutian and western Pacific arcs; the Farallon plate was being subducted along the western coast of North America, and the Phoenix plate was moving into the trenches along the South American coast. During the Tertiary the spreading ridge crests moved north and east relative to the North American plate so that the area of the Kula and Farallon plates grew progressively smaller, and eventually the Kula plate and the northern triple junction disappeared into subduction zones in the northeast Pacific. The reconstructions as postulated by Atwater (1970) on the basis of restoring anomaly patterns are shown in Fig. 24-16. The north-

* See Suppe, Powell, and Berry (1975) for a recent summary of Quaternary volcanism and tectonics in the western United States.

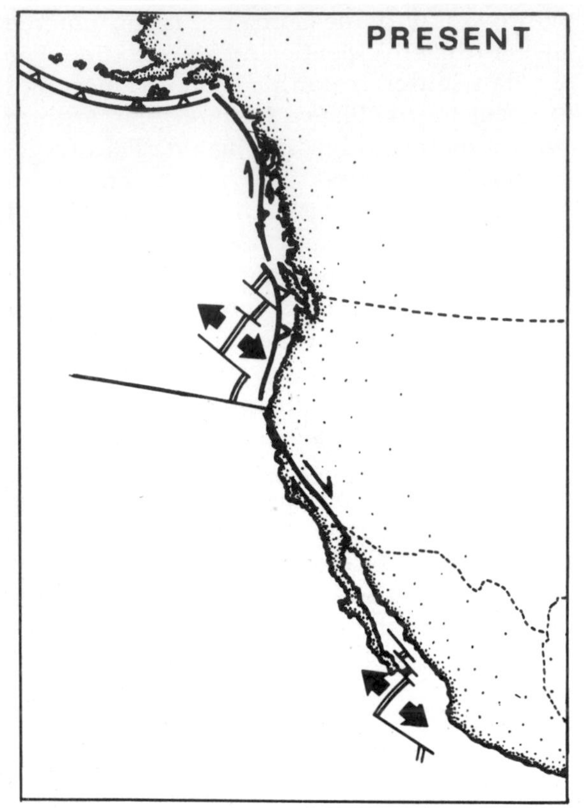

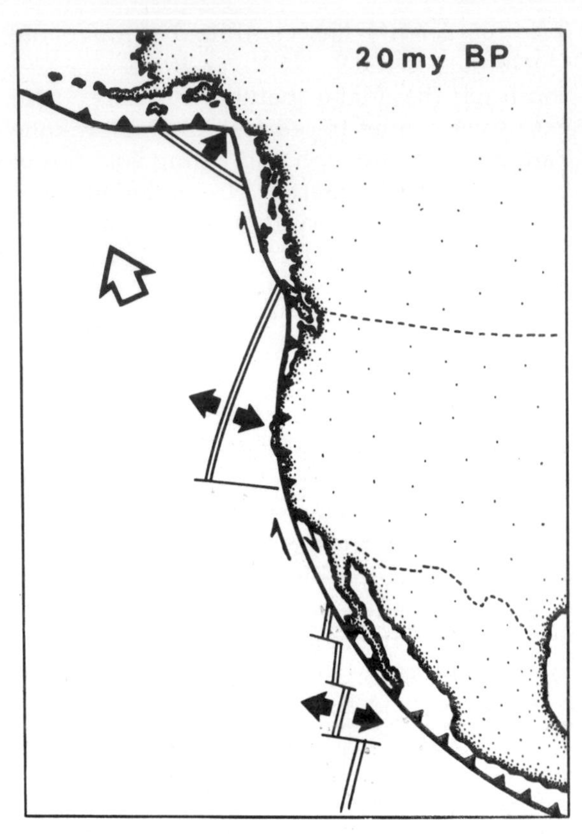

FIGURE 24-16 Plate-tectonic interpretation of the eastern Pacific and western coast of North America during the last 60 million years. Subduction zones (barbs) are located approximately relative to present coast. Spreading ridges are double lines. See text for discussion. (*Redrawn and simplified after T. Atwater, 1970.*)

ern triple point migrates northward as the Kula and Farallon plates gradually disappear. Movement between the Kula and North America plates occurs along a right-lateral shear at the continental marginal. This shear extends southward to the intersection of the spreading center between the Kula and Farallon plates. The Farallon plate at the same time is being subducted along the coast. This system gradually moves up along the coast until finally the northern triple junction is subducted along with the Kula plate in the Aleutian trench, but just before this happens, a section of the East Pacific spreading ridge between the Pacific and Farallon plates, which is offset along a number of major transform faults such as the Mendocino and Murray fracture zones, reaches the continental margin somewhere south of the present location of Los Angeles.

Atwater postulates that once the ridge crest impinges on the continental margin, spreading along that section stops, although it continues both to the north and south. Continued northward movement of the Pacific plate and relative overriding of North America over the Farallon plate gradually increases the separation of the spreading ridge crest along the continental margin. Because the Pacific plate is moving north relative to North America, a strike-slip fault system must form along this part of the coast. In this fashion, plate theory can explain the San Andreas fault system. The fault is developed within the continent just inland from a former subduction zone. The Mendocino fracture zone is presumably a much older transform which connects to the

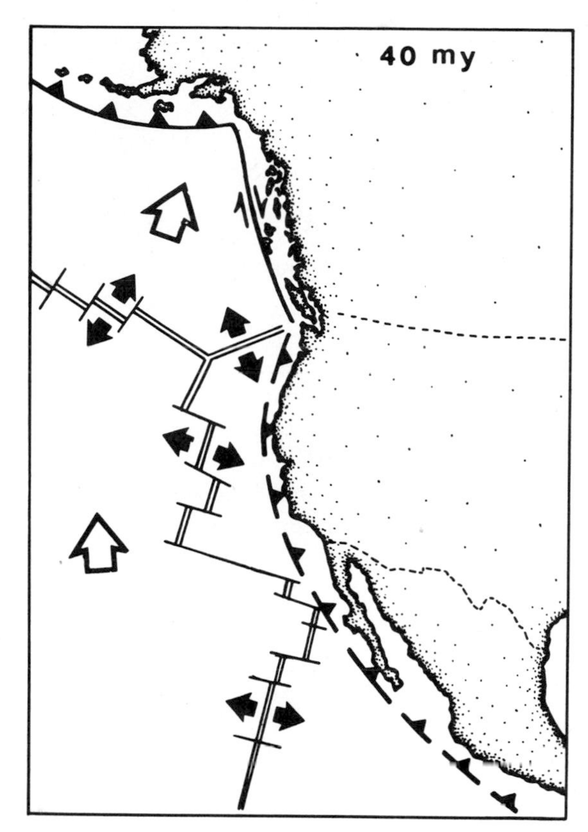

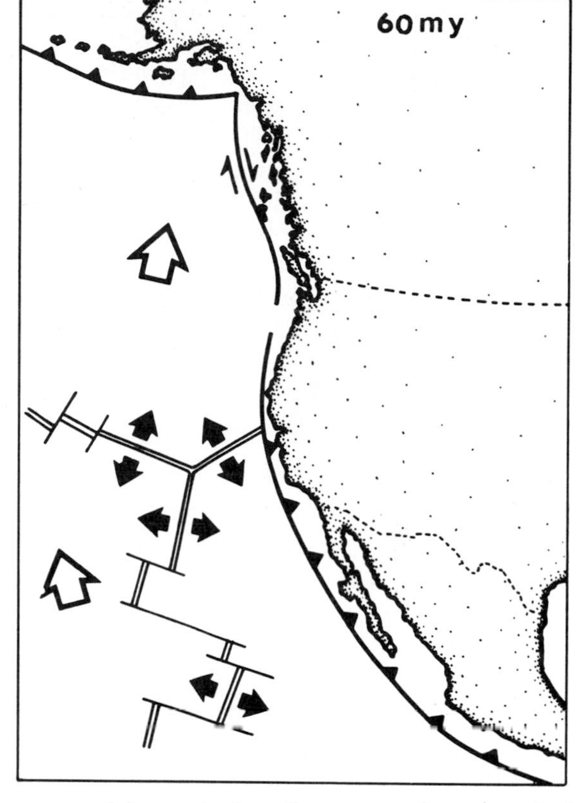

San Andreas fault in the north. The San Andreas, which separates a coastal strip that is moving with the Pacific plate from the continental interior, passes back into a ridge crest which is now located at the northern end of the Gulf of California. Anomalies in the Gulf indicate that it is only 4 or 5 million years old; so the spreading center must have moved eastward into that area at about that time.

The Juan de Fuca plate appears to be a remanent of the Farallon plate, and its boundaries are well defined by earthquakes. The coastal section from Cape Mendocino to Vancouver Island must be a subduction zone in this plate model. The volcanism in the Cascades can thus be explained as that associated with a subduction zone, as is the Cenozoic deformation in the northern Coast Ranges. The postulated subduction zone here is not well defined by earthquake foci nor is it marked by a trench, although a few deep-focus earthquakes have occurred here. As in other cases the appeal of the plate model here is that it allows a coherent synthesis of many divergent observations, and it provides a basis for predicting the evolution of the Cenozoic diastrophism along the west coast even where data are not available. Quaternary tectonics farther east in the Cordilleran region may also be related to plate movements.

Mason L. Hill (1974), a leading opponent of interpreting the San Andreas as a transform fault, cites the following arguments against that interpretation. (1) The San Andreas fault is but one of many northwest-trending right-lateral strike-slip faults in the continental crust of the western United States (e.g., Las Vegas fault), yet it is the only one interpreted as a transform fault. (2) Faults of the San Andreas zone seem to be active in Sonora, Mexico, an area beyond the hypothetical end of the East Pacific rise. (3) Structural features associated with the San Andreas are neither parallel nor perpendicular to the fault zone as might be expected if they were related to spreading centers. (4) The east-west orientation of the Mendocino and other faults in the Pacific crust are hard to reconcile with the northwest trend of the San Andreas. (5) Evidence from matching offset features along the San Andreas indicates that movements date back into the Mesozoic, but some plate-theory interpretations place an age of only 29 million years on the fault (Atwater, 1970). (6) Transform faults recognized in the Gulf of California do not extend into adjacent continental crust.

Mantle Hot Spots

Suppe and others (1975) have advanced the idea that Quaternary volcanism and much of the Cenozoic uplift east of the west coast are related to two mantle hot spots fixed in position relative to the Hawaiian hot spot and their traces. The hot spots are presently located at Yellowstone Park and near Raton, New Mexico, and are moving northeast at a rate of 2.5 cm/year relative to North America. The trace of the northern hot spot extends from Yellowstone across the Snake River downwarp of southern Idaho and into the Cascades of Oregon. The southern hot-spot trace runs from the northeastern corner of New Mexico to southwestern Arizona; thus they bound much of the area of exceptionally high topography in the Rocky Mountain region. Suppe suggests the possibility that Quaternary uplift along the line between these hot spots may be related to the migration of the hot spots and mantle processes giving rise to the hot spots.

Subplates in the Western United States

Smith and Sbar (1974) conclude that modern seismic activity in the Basin and Range and Rocky Mountain region may be related to movements along subplate boundaries in the North American plate. The main boundary delimited by modern seismicity, which is predominantly shallow, forms an arc extending from the southern end of the Northern Rocky Mountains south, closely paralleling the structural boundary along the eastern edge of the Sevier fold belt to southwestern Utah where it trends west. A second seismic zone is parallel to the east-west trend of the Lake Basin fault zone in Montana and separates the Northern Rockies from the Great Basin. These zones define the proposed Great Basin and North Rocky Mountains subplates. Fault-plane solutions indicate that these subplates are moving west relative to the North American plate and that north-south extension is taking place between the subplates. It is suggested that these motions are related to the mantle plume rising under Yellowstone.

THE CENTRAL AMERICAN SECTOR

The Laramide fold belt passes through Mexico, but the eastern side is covered by Cenozoic sediments of the Gulf Coastal Plain and the western side by Tertiary volcanics, Figs. 24-1 and 24-17. An old metamorphic (schist, gneisses, migmatite) basement complex emerges from beneath this cover along the southwestern coast of Mexico. Rocks of Precambrian and Paleozoic age make up this complex which is exposed almost continuously into Guatemala, where the basement complex is cut, sheared, and displaced (left-lateral) along a major northeast-trending strike-slip fault which parallels the Bartlett fault and the Cayman trench. This fault coincides with the sharp bend of the Laramide fold belt in Guatemala, which swings into alignment with the Cayman ridge, suggesting a drag-like bend in the orogen. The old metamorphic basement is extensively exposed in Honduras but is covered by Tertiary volcanics farther south, if indeed it continues south at all.

A chain of modern volcanos line the west coast of Central America from southern Mex-

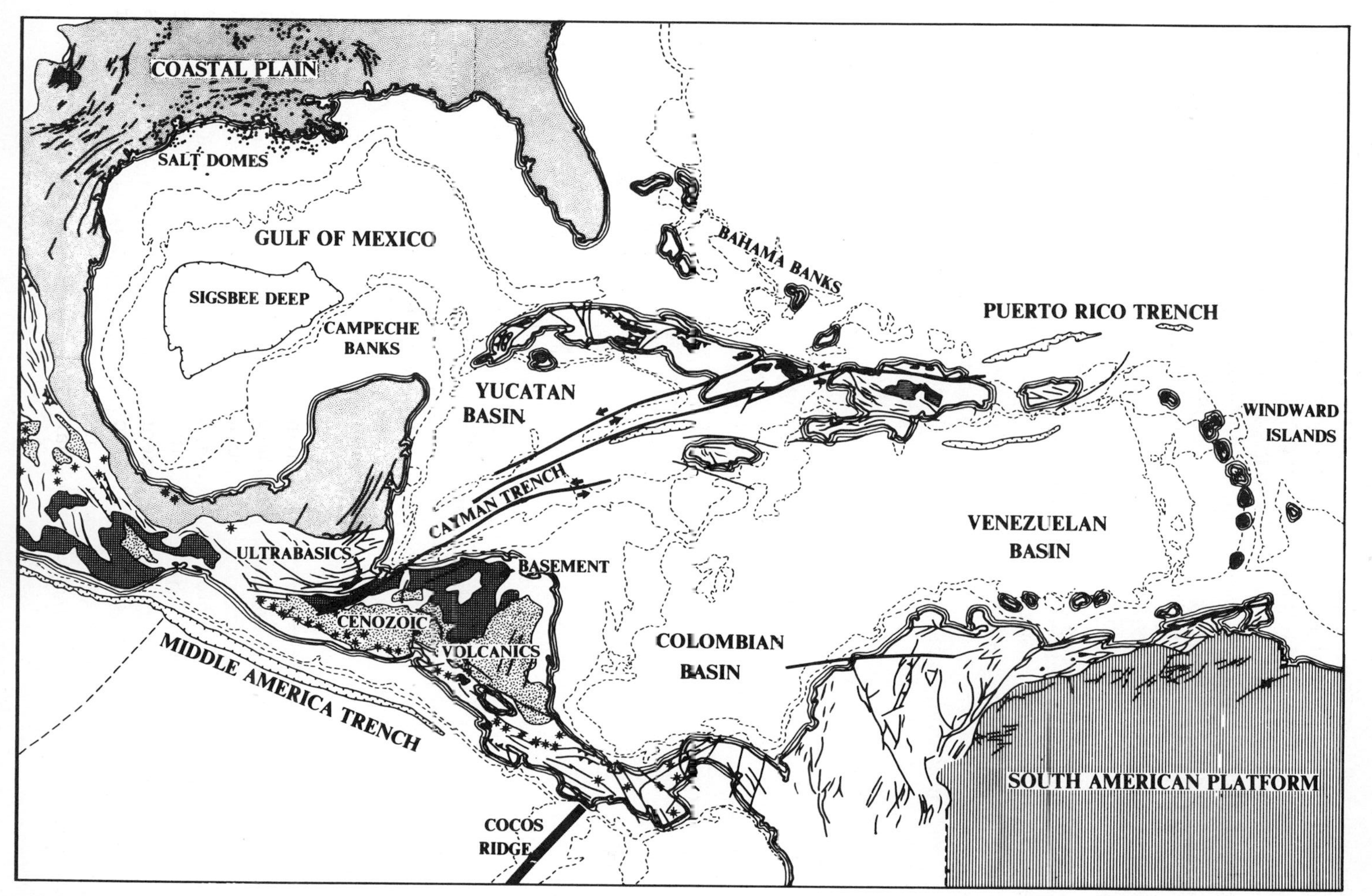

FIGURE 24-17 Tectonic map of Central America–Caribbean. (*After the Tectonic map of North America compiled by King, 1969.*)

ico to Panama. These are aligned in a sinuous curve which is nearly parallel with the southern portion of the Middle America trench. The structural pattern here is similar to that of many Pacific island arcs. A line of normal faults forming step patterns and grabens is parallel with the volcanos; the chain is also cut by a number of high-angle normal faults, and most of the folds in this sector are broad open arch-like features.

Volcanic activity is less pronounced in Panama, and here the underlying rocks are better exposed. These are mainly composed of late Mesozoic eugeosynclinal deposits overlain by early Tertiary miogeosynclinal units that have been folded or arched into broad open structures and cut by a complex system of high-angle (probably strike-slip) faults. These folds are aligned essentially parallel with the present coast and continue toward the south as part of the Andes.

THE CARIBBEAN REGION

Central America and the Caribbean arc provide the connecting links between the North and South American orogens. However, the Caribbean region is different in several important respects. No craton is situated east of this segment of the orogen, and perhaps most importantly, this area is largely occupied by North and South America on most predrift reconstructions of the Atlantic region, Fig. 22-1. If these are right then much of the area of the Gulf of Mexico and the Caribbean Sea has formed since the end of the Paleozoic, the larger islands are likely fragments of the continents, and we must expect that most of the major structural elements are expressions of the Mesozoic and Cenozoic movements which accompanied opening of the area between the two large cratons. Alternatives to this picture are that the seas represent areas where continental crust has been transformed into oceanic crust, or that the region has been a long-term seaway between the cratons and that drift has not taken place.

The Caribbean has strongly developed morphological forms and trends to which structural significance is attributed, although Cenozoic sedimentary deposits blanket and obscure some of these features. The Florida shelf and the Great Bahama Banks separate the arc from the North American craton. The large, high islands of Cuba, Hispaniola, and Puerto Rico form the northern flank of the arc, which continues as the Lesser Antilles and forms a strongly curved lineament ultimately leading to Tobago and Trinidad and into Venezuela.

Several deep submarine trenches in the region are thought to mark the position of important faults, Fig. 24-17. These include the Cayman trench (the Bartlett trough is within the Cayman trench) south of Cuba; the Anegado passage, which separates Puerto Rico from the Lesser Antilles; the Puerto Rico trench, located on the Atlantic side of Puerto Rico and the northernmost Lesser Antilles; and the Leeward trench, north of the Venezuelan shelf.

Gravity Anomalies*

A weak negative-gravity anomaly over southern Mexico may be interpreted as part of an east-west belt extending across the southern Gulf (Fig. 24-18) and the northern part of Yucatan and into Cuba. A strong negative anomaly emerges from the eastern end of Cuba and lies over the southern part of the Bahama Banks, the Puerto Rico trench, and farther south over the Barbados ridge, Tobago, Trinidad, and northeastern Venezuela, where it dies out. The striking characteristic here is that the anomaly lies over both trench and elevated ridge and mountains—indicating a deep source. A second weaker negative anomaly of east-west trend lies along the southern side of

* See Hess (1938) and Bush and Bush (1969).

the Cayman trench, Jamaica, and the southern part of Hispaniola. A third negative anomaly lies over the Leeward trench north of Venezuela. Positive anomalies are present over the major basins, over the Windward Island, and over the Cayman trench.

Seismic Profiles*

The basins in the Caribbean are floored by crustal sections that are of intermediate character between continental-type and oceanic-type sections, Fig. 24-19. The thickest crust is located under the Antilles Islands and the Nicaragua rise (22 km±). The crust in general is closer to oceanic type, but sediment thickness is greater and the velocities obtained at depth are mixed. Many mantle velocities are lower than the 8.0 to 8.2 km/sec associated with mantle materials, and the observed velocities (7.4 to 8.0 km/sec) are commonly interpreted as altered mantle. Still lower velocities (6.6 to 7.4 km/sec) are found at the same level under islands of the island arcs and under trenches. These velocities could represent a number of types of material: gabbro or other igneous rocks of intermediate to basic composition, hydrated peridotite, or serpentine.

Seismic sections across the Cayman trench (Ewing and others, 1960) reveal a relatively thin crust. The Moho is 10 km deep in the bottom of the trench and drops off to 20 km or so both to the north under the Yucatan basin and toward the south under the Nicaraguan rise.

* See Fox and others (1969), Officer and others (1957), and Ewing and others (1960).

FIGURE 24-18 Gravity anomalies in the Caribbean. The black and diagonally lined areas are negative anomalies where there is a mass deficiency. (*After Hess, 1938.*)

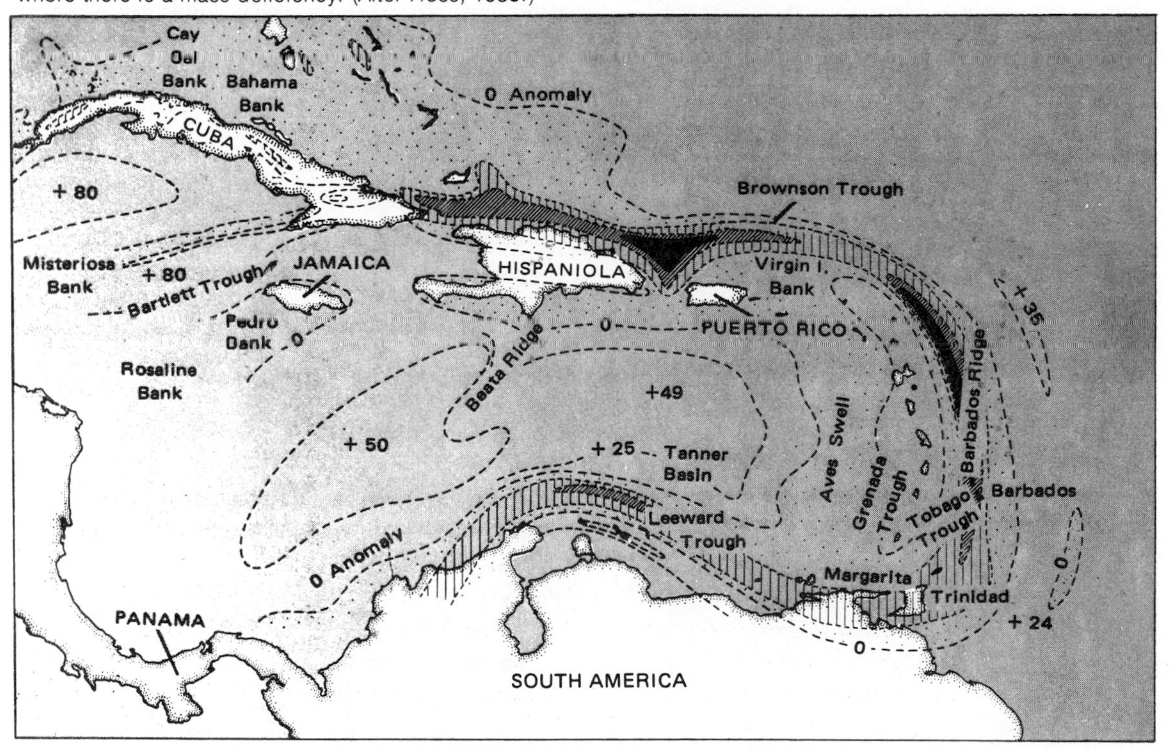

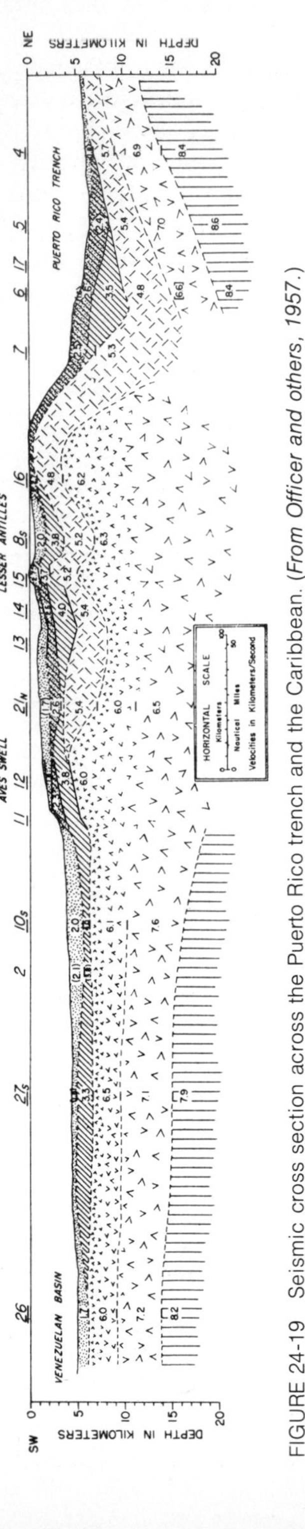

FIGURE 24-19 Seismic cross section across the Puerto Rico trench and the Caribbean. (*From Officer and others, 1957.*)

The Large Islands of the Caribbean—Cuba, Hispaniola, Puerto Rico, Jamaica*

Despite their present distance from other parts of the orogen, the large islands of the Caribbean bear strong resemblances in structure and orogenic history to central and northern South America.

The basement rocks on all four of these islands are composed of Mesozoic eugeosynclinal sedimentary materials, Fig. 24-20. Small portions of these are strongly metamorphosed, but neither the sediments nor their metamorphic equivalents are widely exposed, nor do they form a distinct belt. Miogeosynclinal rocks of Jurassic and Cretaceous age are exposed, especially along the northwestern coast of Cuba. The basement rocks commonly show effects of three distinct phases of deformation, the earliest of which occurred in the late Mesozoic and corresponds approximately in timing to the Laramide orogeny. This was followed in the Tertiary (usually Eocene) by a second deformation, which affected early Tertiary eugeosynclinal and miogeosynclinal deposits as well as the older basement. The third and most recent deformational phase is characterized by block and transcurrent faults, some of which remain active at present.

The deformational history of these large islands is interesting in that the eugeosynclinal character of parts of the region have persisted since the Mesozoic. Large volumes of ultrabasic materials, much of which is now transformed into serpentine, occur on these islands, especially on Cuba. A number of these bodies are closely associated with the Mesozoic basement which is metamorphosed, and most of the serpentine is thought to be of Mesozoic age, although the age of some of it is uncertain. Large quantities of ultramafic mate-

* References include: *General*—Eardley (1962), Maxwell (1948), Butterlin (1956), Donnelly (1964, 1970), Mattson (1972), Dengo (1969), McBirney and Bass (1967). *Cuba*—Hill (1959), Wassall (1957). *Hispaniola*—Hess and Maxwell (1953). *Puerto Rico*—Kaye (1957), Berryhill and others (1960), Mattson (1966, 1973).

rials are associated with thrust sheets in Cuba. Kozary (1968) suggests that some of these thrust sheets are best described as flow features. He envisions the lower thrust sheets as being thrown into isoclinal folds that are invaded by pseudoviscous serpentine, which pierced the cores of the folds and eventually became separated into thin slivers.

Mattson (1973) also describes Cretaceous nappes, consisting of radiolarian chert, serpentinized peridotite, pillow lava, and metamorphosed tholeiitic rocks, in Puerto Rico which he interprets as having formed on oceanic crust. Subsequently, the masses moved northward, probably as a gravitational sliding mass directed toward an ancient trench, and volcanic activity broke out between the slides and the trench axis.

Large-scale thrusting is best developed along the northern coast of Cuba, but folding is evident on all four of the larger Antilles. Large relatively open folds are oriented approximately parallel with the alignment of the islands. Folding and thrusting in the late Mesozoic were directed toward the north and carried volcanic and clastic sedimentary eugeosynclinal facies northward over carbonate facies in Cuba. This was later broken by high-angle faults, displaced, and partially eroded. Tectonic transport was again directed toward the north during the Eocene orogeny in the islands; however, much of the movement may be gravity gliding down paleoslopes.

Pliocene and Pleistocene deformation has been dominated by faulting. Systems of high-

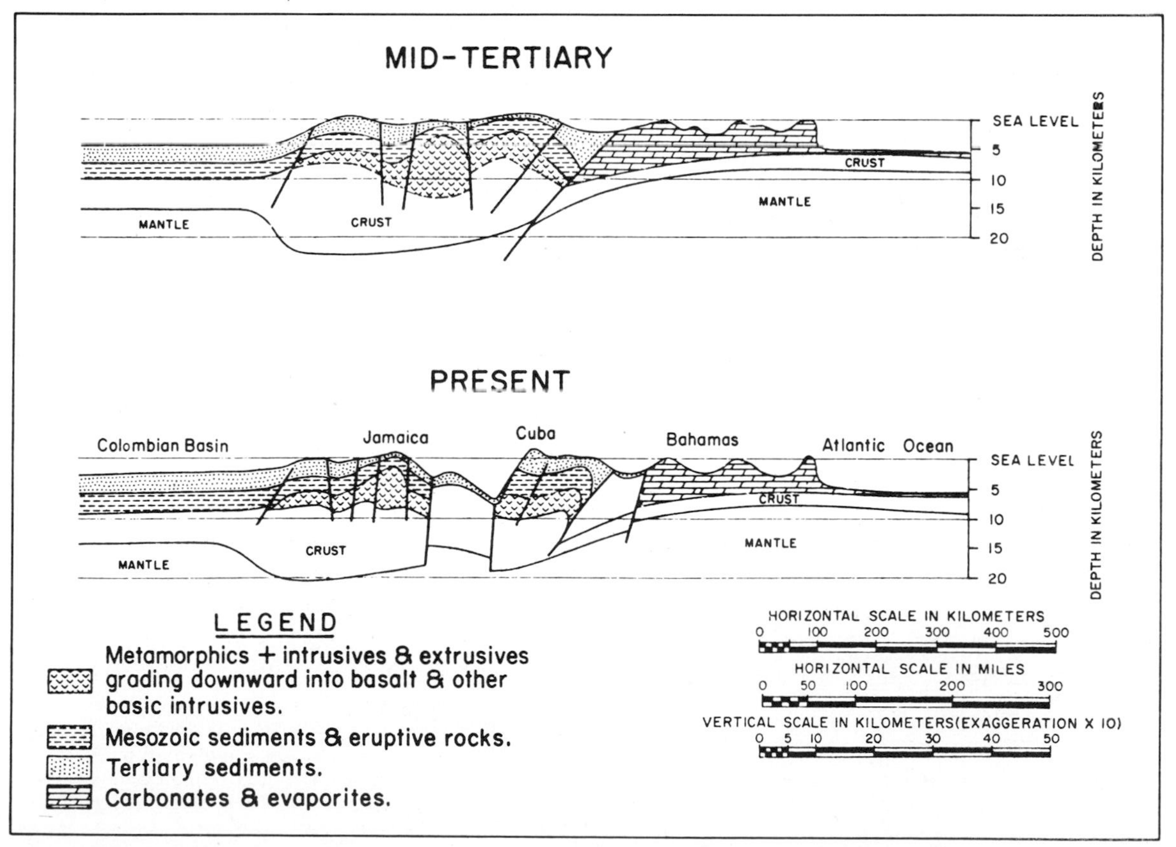

FIGURE 24-20 Interpretive cross sections illustrating the origin and history of the Nicaraguan rise and Greater Antilles. (*From Arden, 1969.*)

angle faults cut across the older folds and thrust masses. Many of these may be directly related to strike-slip movement on major faults such as the Bartlett fault and its presumed landward extensions. The prominent fault pattern in Puerto Rico, Fig. 24-21, may be explained as an *en échelon* pattern of tears developed in the vicinity of a right-lateral regional shear.

The Lesser Antilles

The Greater Antilles are separated from the Lesser Antilles by a deep narrow submarine trough, which is probably a transcurrent fault as suggested by Hess and Maxwell (1953). The islands south of this break are much smaller in size and are made up almost entirely of volcanic materials. Both inner volcanic and outer nonvolcanic arcs are present, as in many other modern island arcs. The islands of the inner arc are sites of Recent andesitic volcanism, although some mid-Tertiary volcanics are also present. The volcanoes are associated with a submarine swell, but their basement is not exposed, and they are not on the crest of the swell as might be expected if the volcanism were the result of tension over the crest of the ridge.

The eastern nonvolcanic islands are largely composed of Tertiary-Quaternary limestone. Where a basement is exposed on St. Martin, it is strongly folded, slightly metamorphosed tuff with intrusions of diorite. The basement of most of these islands is composed of volcanic debris, tuffs, with lava flows intruded by diorite. This old volcanic activity ended in the Oligocene, and the islands have not been much deformed since (Christman, 1953).

Barbados lies east of the volcanic arc and is separated from the volcanic islands by a shallow submarine trough; it lies on a ridge that is directly in line with the Puerto Rico trench to the north, and the negative gravity anomalies over the trench can be traced directly over the Barbados ridge. The oldest rocks on Barbados are strongly folded and faulted Eocene detrital sedimentary rocks that are overlain by a thick sequence of deep-water sediments, including radiolarian oozes, suggesting that the region was submerged rather rapidly (Senn, 1940). Subsequently, uplift and folding have taken place and continue into the Pleistocene.

Tobago appears to lie on the same ridge as Barbados, but rocks exposed in Tobago are much older and may indicate the character of the basement of this outer ridge. Tobago and Trinidad both contain outcrops of Mesozoic basement, folded schists of metavolcanic origin, and intermediate intrusives with some ultramafics. A Late Cretaceous age is probable for the deformation, but the intrusions may be a little younger than the folding. Miocene and younger units lie unconformably and undeformed on the basement in Tobago.

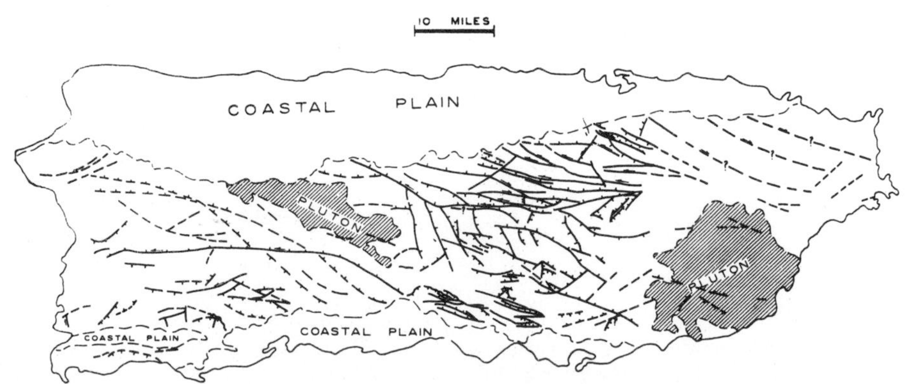

FIGURE 24-21 Simplified structural map of Puerto Rico. (*From Donnelly, 1964, after Briggs, 1961.*)

FIGURE 24-22 Plate-tectonic interpretation of the evolution of the Caribbean region. Subduction zones (barbed lines), strike-slip faults, and general direction of movement of plates (arrows) are depicted. (*After Malfait and Dinkelman, 1972.*)

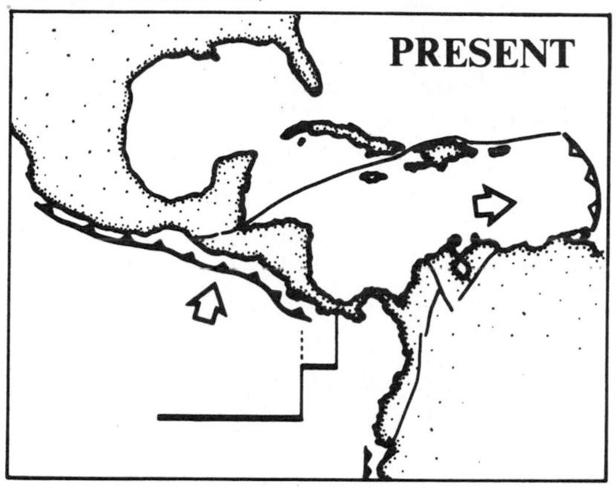

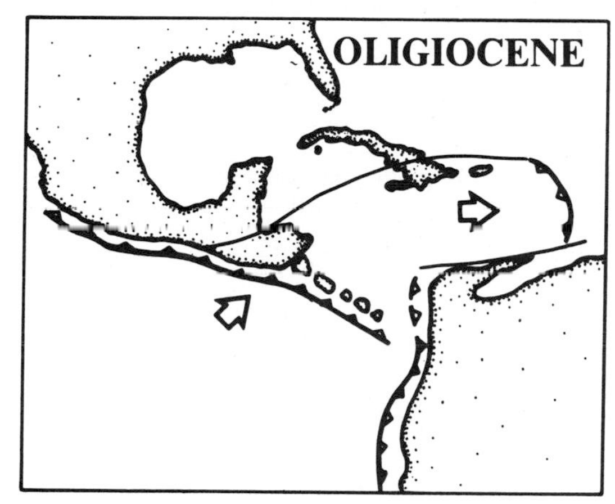

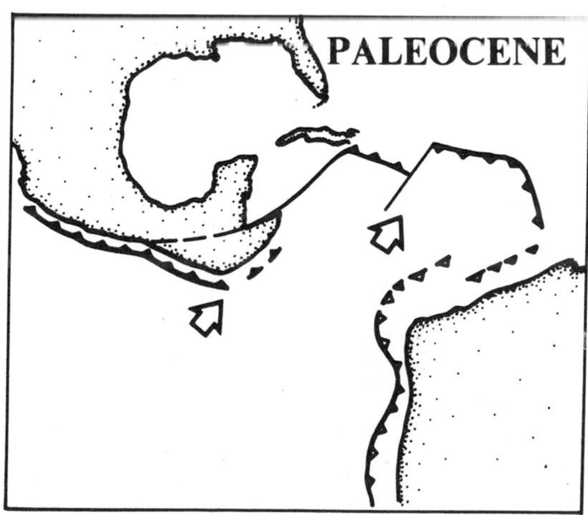

PLATE-TECTONIC INTERPRETATION OF CARIBBEAN STRUCTURE

A considerable body of plate-tectonic interpretation of the Caribbean has now been developed.* One of the most comprehensive syntheses has been proposed by Malfait and Dinkelman (1972), whose model of the evolution of the region serves well to illustrate how salient features of Caribbean geology are being interpreted in terms of plate theory, Fig. 24-22. The most prominent feature of this model is the northeastward and later eastward movement of a section of the East Pacific plate into the gap between the North and South American cratons. This was accompanied by convergence of these continents near the end of the Mesozoic in a movement pattern which accounts for Laramide orogeny in the region. The Cayman trough marks the position of a major transform fault on the north, and the Dolores, Oca, and El Pilar faults represent the main movement zone on the south of the Caribbean plate. The Puerto Rico trench forms the eastern boundary, and the eastern edge of it represents a modern subduction zone. The subduction zone of the Middle Americas trench had extended south to separate the Caribbean and East Pacific plates by the middle of the Tertiary, and a volcanic arc—Middle America—formed behind this subduction zone.

The Caribbean plate is bordered by subduction zones on both the east and west sides, by a transform on the north, and by a complex zone of deformation in the south. If subduction zones have the effect of pinning down plate movement as has been suggested, then the net movement of the Caribbean plate should be small—a conclusion independently reached by Jordan (1975) on the basis of an

* See Chase and Bunce (1969), Dengo (1969), Bell (1972), Bracey and Vogt (1970), Ball and Harrison (1969), Maley and others (1974), Maresch (1974), Mattson (1973), Krause (1971), Silver (1975).

analysis of hot-spot data and several absolute-motion models using data from adjacent plates.

NORTH COAST OF SOUTH AMERICA*

This sector presents an especially complex regional structural pattern. The volcanically active arc in the Lesser Antilles ends at the edge of the South American continental shelf north of Trinidad. The orogenic zone, which has an east-west trend in Trinidad and along the coast of Venezuela, becomes wider to the west, and the major structural elements become reoriented parallel with, and pass into, the Colombian Andes. This orogen wraps around the northern and western margins of the South American craton made up of a Precambrian core and surrounding platform. The platform is covered by a relatively thin sedimentary veneer with a few deeper basins near the edge of the orogen, and it has been little deformed, except for some block faulting. Structures within the orogen include some thrusting, generally directed toward the craton. These thrusts carry Precambrian and Paleozoic metamorphic rocks, metamorphosed and unmetamorphosed Mesozoic eugeosynclinal and miogeosynclinal units, and early Tertiary miogeosynclinal deposits toward and onto the craton platform deposits in various places. In addition, many of the uplifts within the orogen are bounded by high-angle reverse faults (upthrusts) on both sides.

The evolution of the Venezuelan coast ranges, Fig. 24-23, exemplifies the history of this region. Jurassic and Cretaceous sediments were deposited along the edge of the craton, where thicknesses of 20 to 30 km were reached in places along the coast, and offshore, where maximum thicknesses likely coincide with a free-air gravity minimum. Metamorphism and uplift began in the Cretaceous along this coastal zone, with sediments flooding off toward the craton. Uplift along high-angle faults and southward-directed thrusting and formation of allochthonous masses began in the early Tertiary. Block faulting and movements on major strike-slip faults continued into the Pleistocene, Fig. 24-24.

The role of strike-slip faults (megashears) in this region has long been debated, but large-scale movements on demonstrated transcurrent faults are increasingly called on to explain the overall tectonic pattern. One such fault separates the metamorphosed Mesozoic rocks from unaltered rocks in Trinidad, Fig. 24-17. This fault (a right-lateral fault) lines up with another right-lateral strike-slip zone near Caracas which bends toward the southwest, east of the Maracaibo basin. A second megashear continues west across the northern part of the Maracaibo basin. A third prominent zone of high-angle faults, possibly over another shear zone, Fig. 24-24, lies west of the

FIGURE 24-23 Schematic representation of the overall evolution of the Venezuelan coast ranges. T = Tertiary, J = Jurassic, Ku = Cretaceous undivided, Kv = Cretaceous volcanics, and Kg = Cretaceous granites. (*After Bell, 1974.*)

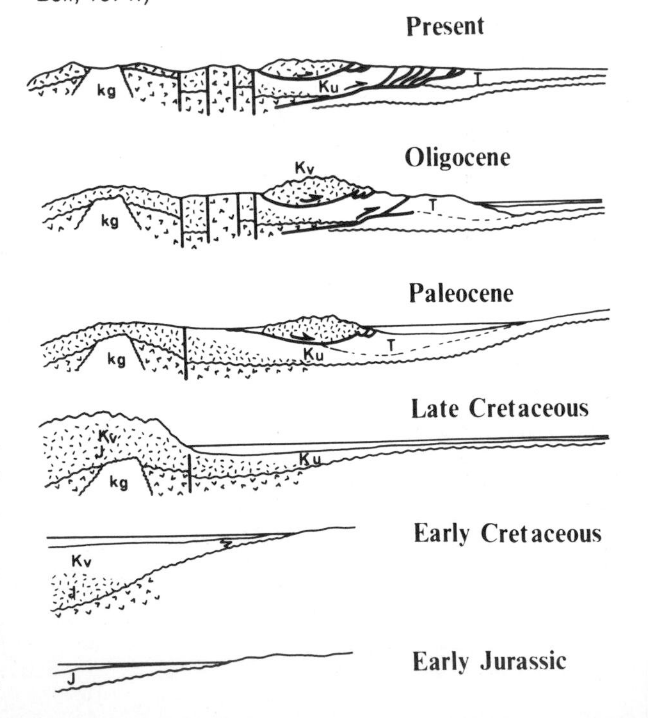

* See Saunders, Bell, and Campbell, *in* Spencer (1974).

basin. These three zones form a triangular frame around the Maracaibo basin, one of several large fault-bounded basins in Venezuela and Colombia which contain great thicknesses of late Tertiary and Quaternary deposits. Strike-slip movements clearly are a major component of the tectonics of this region, but proof of the total displacement is rare. Some interpretations call for lateral displacements of hundreds of kilometers.

THE ANDES

The Andean mountain system emerges from the Caribbean at Trinidad and extends as a narrow but continuous deformed belt to the southern tip of South America, where the trends bend east into the Scotia arc. This system is almost as long as its northern counterpart; however it is a much narrower belt, and it is confined throughout its length to the western edge of the South American shields. In many respects the Andes lack some of the complexities of the North American Cordillera. In part this may be due to the fact that it is less accessible and less intensively studied.

The Andes constitutes one of the distinctive types of continental margins. Its characteristic features are (1) the presence of a trench along the Pacific coast, (2) the narrow, young, volcanically, seismically, and tectonically active margin, and (3) the stable interior Precambrian craton. This margin is interpreted in plate-tectonic theory as a modern subduction zone where oceanic crust is moving under the South American plate with consequent seismic activity, volcanism, and tectonism. A Benioff zone, Fig. 21-6, dips under the Andes, and the marginal zone is one characterized by andesite volcanism.

FIGURE 24-24 Simplified tectonic sketch map of northern South America showing the platform of the Brazilian Shield area, the major uplifts that bring pre-Tertiary rocks up (diagonal line) and Tertiary basins (dot pattern). (*Modified after Case, in Burk and Drake, 1974.*)

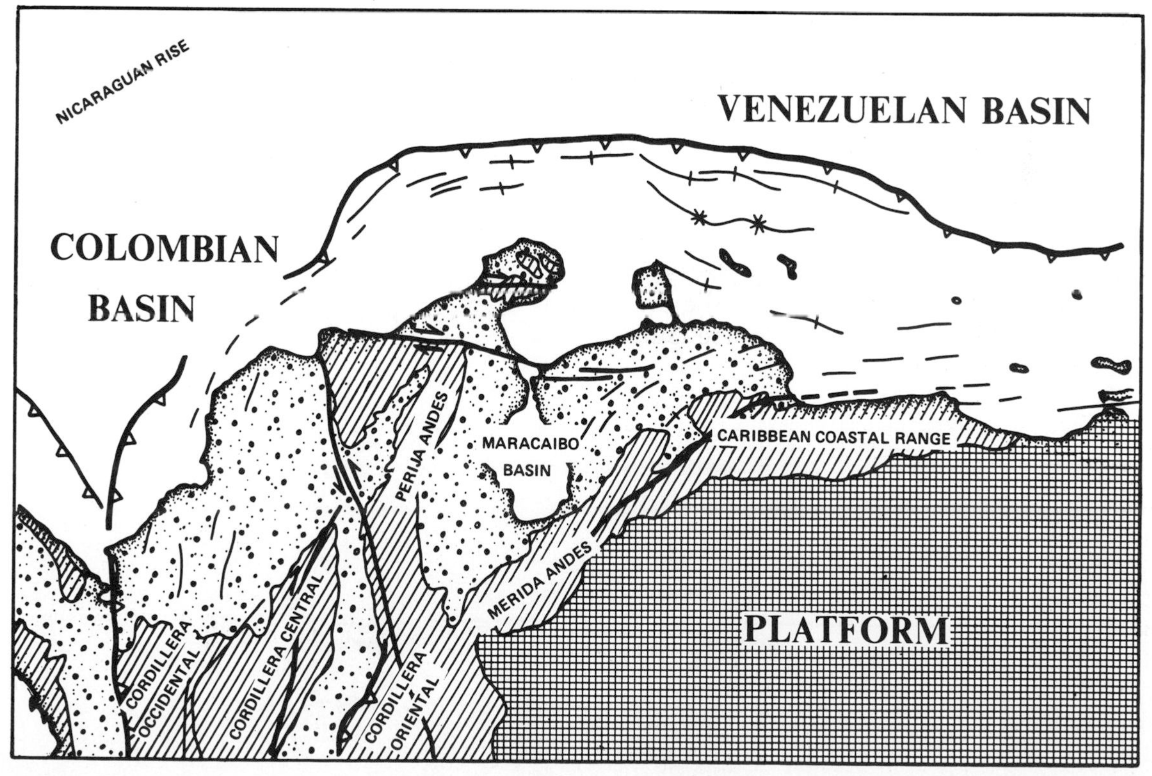

The Peru-Chile Trench—A Problem for Plate Tectonics?

This trench is known primarily through sonic profiles* and soundings. It is one of the longest deep-sea trenches, extending as a continuous feature from the Carnegie ridge to the Chile rise. The trench follows the curved coast of South America and is consistently separated from the Andes by a narrow continental shelf.

Profiler records and interpretative sections for two sections depict some characteristic features of the trench and its relationship to the Andes, Fig. 24-25. Notably, the trench floor contains a thick pile of undeformed sedimentary rock, primarily turbidites. These sediments have been sampled in places and are composed of late Cenozoic deposits derived from the land areas to the east. The age, thickness, and undeformed character of these deposits pose one of the serious problems of plate tectonics. If the trench marks the surface trace of a subduction zone in which oceanic crust is being consumed, how can we account for the fact that the sediments in the trench have escaped deformation during the scraping off process, and how can sediments 20 million years old be left in the trench? Estimates of spreading rates and subduction rates for this trench based on magnetic studies range from 5 to 10 cm/year for the last 10 to 15 million years.

Scholl and others (1970) evaluate the subduction hypothesis by comparing the quantity of sediment within and adjacent to the trench with the quantity of sediment estimated to have been deposited along the margin as a result of denudation of the Andes. We may assume that any surplus of sediment supplied over the amount present today has been subducted. Scholl finds that the quantity of Cretaceous and early Tertiary sediment estimated to have resulted from denudation is, in fact, less than the amount present, but for the later portion of the Cenozoic the amount of sedi-

* Scholl and others, 1970.

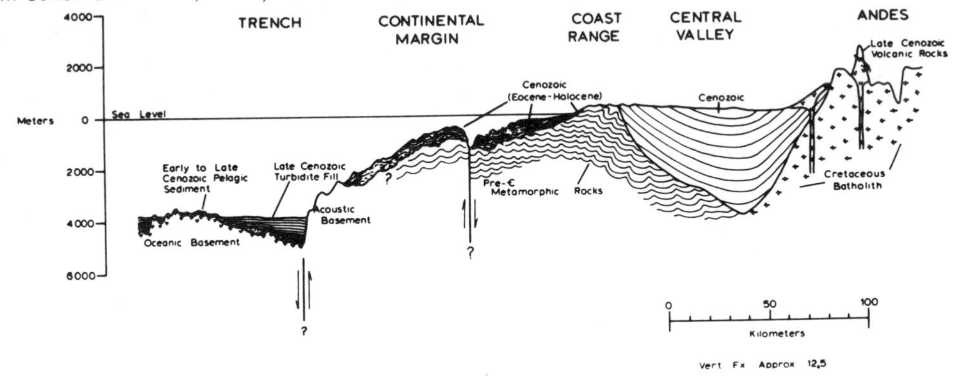

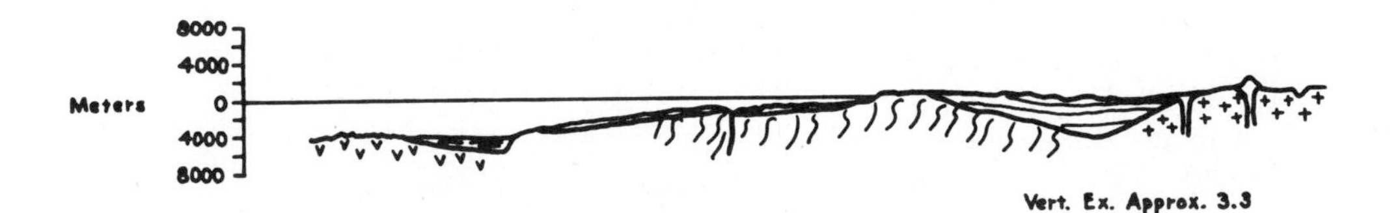

FIGURE 24-25 Idealized structural sections from Pacific sea floor to crest of Andes. Structure of offshore area is based on our interpretation of acoustic reflection profiles. (*From Scholl and others, 1970.*)

ment present is very close to the amount thought to have been removed by denudation. This despite the rapid rate of subduction calculated from magnetic anomaly studies.

Thus it seems that we must conclude that either subduction here has not involved removal of sediment or subduction has not taken place.

Structural Framework of the Andes

The Andes differ from the North American Cordillera in many respects; they are similar in others. They are similar in their marginal position relative to the eastern craton, in their great length relative to their width, in their long orogenic history which dates back to the Paleozoic in both orogens, in the general parallelism of structural elements in the belt with the trend of the belt as a whole, in their modern volcanic and seismic activity (both regions have extensive covers of Cenozoic volcanic rocks), and in their termination in island arcs. However, their differences are also impressive. Structural provinces appear to be less continuous in the Andes than are those in the north. The Andes cannot be subdivided into a succession of elements such as craton–miogeosynclinal foreland fold belt–eugeosyncline–ocean. There is no Andean counterpart of the Rocky Mountain region, where the craton and its platform are involved in the deformation. Wrench faulting does not appear to be nearly as prominent a feature of Andean tectonics (except in the northern sections near the Caribbean) as it is in North America. Block faulting is prominent, and vertical movements appear to be far more prevalent than the thrusting and wrenching found in the north. Large areas of the west coast of South America are composed of Precambrian and lower Paleozoic rocks. Few such rocks are found along the western coast of North America. While the western part of North America is probably made up of materials formed on oceanic crust, the Andes are situated on a Precambrian basement.

Evolution of the Colombian Andes

A general impression of the Andes may be obtained through a comparison of the descriptions of several segments of the orogen.

The Colombian Andes* can be subdivided into three major longitudinal cordillera separated by basins, Fig. 24-26. The complex pattern in section reflects the long and complex deformation history described by Campbell in Spencer (1974, p. 711) as follows:

> Present evidence suggests that the evolution of the Colombian Andes was largely accomplished by the progressive westerly migration of sedimentary provinces and tectonic events through Phanerozoic time. . . .
>
> Geosynclinal subsidence occurred in the East Andean region during the early Palaeozoic and was brought to a close by . . . [Caledonian-Devonian] orogenic movements. The geosynclinal rocks were isoclinally folded and faulted by compressive E-W forces, mildly metamorphosed, uplifted and eroded.
>
> During the late Palaeozoic a new geosyncline is thought to have formed in the Central Andes. . . . This cycle was brought to a close by early Triassic movements that were responsible for the metamorphism of the geosynclinal prism and for the emplacement of batholiths in both the Central and Eastern Andes. . . .
>
> During the early Mesozoic, geosynclinal conditions prevailed in the [Central and Western Andes]. . . . The cycle was brought to a close by Nevadan orogenic movements that metamorphosed the geosyncline. . . .
>
> During the late Mesozoic, miogeosynclinal conditions set in to the east . . . while eugeosynclinal conditions, characterized by the extrusion of submarine lavas, persisted to the west. . . . The central

* See Campbell, *in* Spencer (1974).

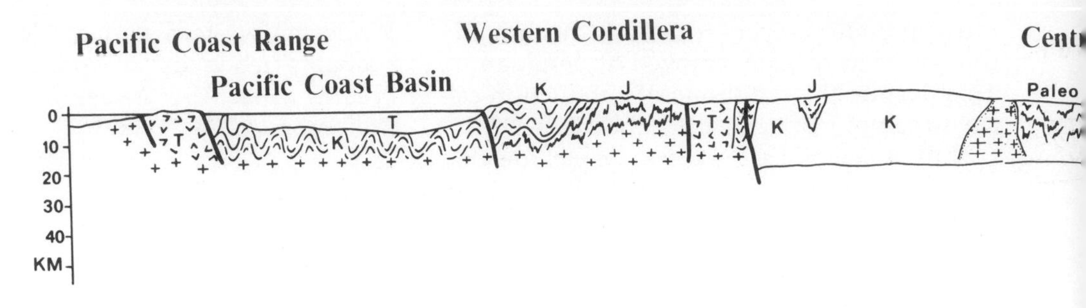

FIGURE 24-26 A cross section across the Colombian Andes. (*From C. J. Campbell*, in *A. M. Spencer, 1974.*)

Andes formed a partly submerged welt between the two geosynclines and was characterized by the emplacement of late Cretaceous batholiths. . . . The late Mesozoic tectonic cycle persisted until the outbreak of Laramide orogenic disturbances in the . . . Eocene. These movements were most intense in the West Andean region where they were accompanied by igneous activity. . . .

[Localized thrusting, uplift, and prominent volcanism in the Central Andes occurred following the Laramide deformation.]

The main Andean orogenic phase of late Miocene to Pliocene age is responsible for the greater part of the present structure. In contrast to earlier orogenic movements, the Andean phase was more intense in the east than the west. The Eastern Andes were moderately folded and cut by marginal thrusts. The chain is bilaterally symmetrical suggesting that deformation was due mainly to vertical movements of the crust.

FIGURE 24-27 Section through the Eastern Cordillera, Colombia, South America. T = Tertiary, K = Cretaceous, J = Jurassic, p-Є = Precambrian, + signs indicate highly altered rocks; Jurassic and Paleozoic highly deformed rocks are shown by intricate fold pattern. (*After Campbell and Bürge, 1965.*)

Andean Orogenic Belts of Peru and Chile

The orogenic history of the Andean orogen farther south bears a close resemblance to that of the Colombian Andes. Taken as a whole the orogen contains an early Paleozoic orogenic belt, a late Paleozoic orogenic belt, a Mesozoic orogenic belt, belts of Tertiary to recent volcanic activity, and much of the orogen has been affected by late Tertiary to Recent uplift, Fig. 24-27. However, these belts are not uniformly present, nor do they retain consistent relative positions within the orogen.

The Precambrian basement is exposed in uplifts along the coast of Peru, and this basement is broadly exposed in the southern part of South America, where the orogen becomes much narrower. The Paleozoic fold belts are largely confined to the western edge of the craton. The earlier Paleozoic fold belt is best developed in the northern parts of the Andes, while the later Paleozoic fold belt is best exposed in the south. Mesozoic miogeosynclinal and eugeosynclinal fold belts form the most

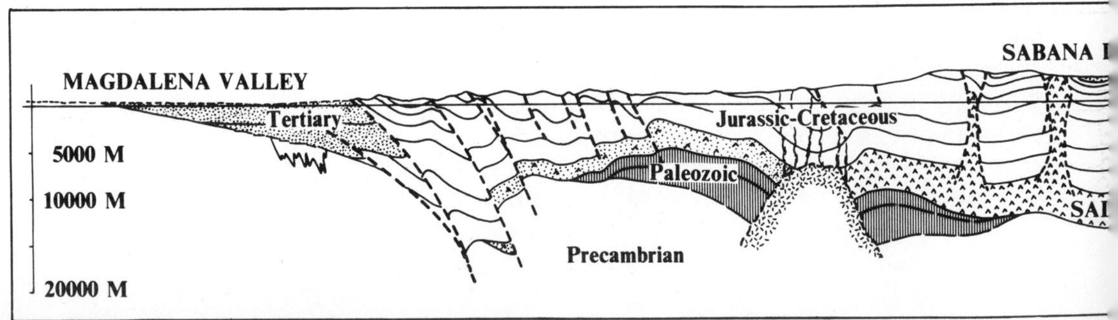

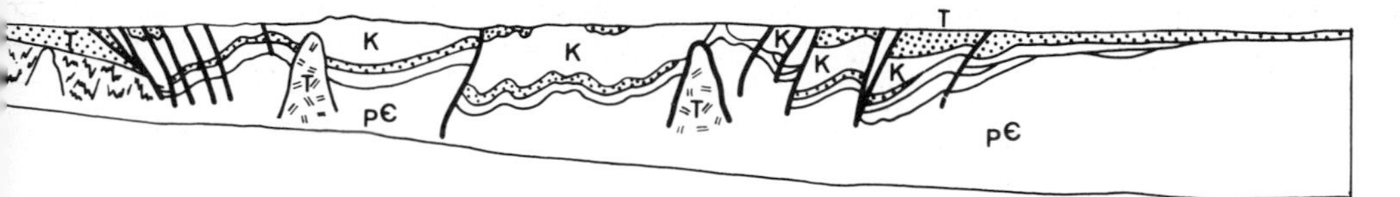

continuous elements of the orogen and extend the entire length of the continent. The great pile of volcanic materials is most extensively developed in the central part of the orogen. Thus, the Andean orogen shares the property of many other orogenic belts of having a long and complex history of deformation. Like others, the depositional phases preceding orogeny are best characterized by a miogeosynclinal section (located internally) and a more external eugeosynclinal belt.

Thrusting is locally important in the Andes, and folding and metamorphism are evident, but many students of this region characterize the deformation as being predominantly controlled by vertical movements rather than the large-scale lateral transport seen in some other orogens (Cobbing, 1972). The presence of Precambrian sialic basement along the coast suggests that the Andes developed on continental crust rather than on oceanic crust. The linear character of so many of the cordillera and other structural elements points to the importance of control by deep-seated and probably vertical faults. The uplift which accounts for the present elevation of the Andes is apparently largely governed by vertical movements. Cobbing indicates that sedimentation has repeatedly been controlled by block-faulting parallel to the continental margin, and that these faults (probably of Precambrian age) also determined the position of the major batholithic intrusions, at least in Peru.

The contrast between the Andes and the Alpine, North American Cordilleran, and Appalachian orogens should now be apparent. These differences constitute the basis for considering the Andean margin as a distinctive type of continental margin and orogenic belt.

A Plate Model for the Central Andes

The plate-tectonic model proposed for the central Andes (James, 1971) is considerably different from those advanced for either the island arcs or western North America. Evidence of lateral compression is relatively minor in the Andes; so, the model is designed primarily to explain the trench offshore and inland vertical

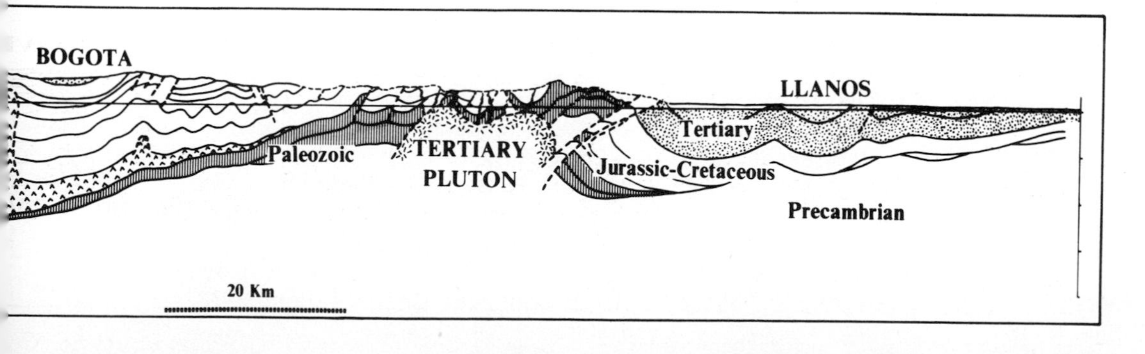

movements and igneous activity. According to this model the trench marks the contact between a subducting oceanic plate and an old cratonic margin. James interprets the seismicity of the region to mean that a 50-km-thick oceanic plate (Pacific) is moving down and under a 200-to-300-km thick South American plate. This underthrusting began in Triassic time and has been accompanied by igneous activity which gradually migrated eastward, culminating in Pliocene-Pleistocene volcanism. Partial melting of rock within or over the subduction zone and consequent volume increases are called on to account for the observed vertical uplift and continentward-directed compression seen in the fold and thrust belt on the eastern side of the Andes. It is notable that ophiolites and eugeosynclinal rocks are not observed in the Central Andes.

REFERENCES

Albers, J. P., 1967, Belt of sigmoidal bending and right-lateral faulting in the western Great Basin: Geol. Soc. America Bull., v. 78, p. 143–156.

Almy, Charles C., Jr., 1969, Sedimentation and tectonism in the upper Cretaceous Puerto Rican portion of the Caribbean island arc: Trans., Gulf Coast Assoc. Geol. Soc., v. XIX, p. 269–279.

Anderson, R. E., 1971, Thin skin distension in tertiary rocks of southeastern Nevada: Geol. Soc. America Bull., v. 82, no. 1, p. 43–58.

Antoine, J., and **Ewing, J.**, 1963, Seismic refraction measurements on the margins of the Gulf of Mexico: Jour. Geophys. Res., v. 68, p. 1975–1996.

Arden, Daniel D., Jr., 1969, Geologic history of the Nicaraguan rise: Trans., Gulf Coast Assoc. Geol. Soc., v. XIX, p. 295–309.

Atwater, T., 1970, Implications of plate tectonics for the Cenozoic tectonics of western North America: Geol. Soc. America Bull., v. 81, no. 12, p. 3513–3536.

Bailey, E. H., 1970, Late Mesozoic tectonic development of western California: Geotectonics, no. 3, p. 148–154.

———, **Irwin, W. P.**, and **Jones, D. L.**, 1964, Franciscan and related rocks and their significance in the geology of western California: California Div. Mines and Geol. Bull., v. 183.

Baird, A. K., **Morton, D. M.**, **Woodford, A. O.**, and **Baird, K. W.**, 1974, Transverse ranges province: A unique structural-petrochemical belt across the San Andreas fault system: Geol. Soc. America Bull., v. 85, no. 2, p. 163–174.

Ball, M. M., and **Harrison, C. G. A.**, 1969, Origin of the Gulf and Caribbean and implications regarding ocean ridge extension, migration, and shear: Trans., Gulf Coast Assoc. Geological Societies, v. XIX, p. 287–294.

Bally, A. W., **Gordy, P. L.**, and **Stewart, G. A.**, 1966, Structure, seismic data, and orogenic evolution of southern Canadian Rocky Mtns.: Canadian Petroleum Geol. Bull., v. 14, p. 337–381.

Barbat, W. F., 1971, Megatectonics in the Coast Ranges, California: Geol. Soc. America Bull., v. 82, no. 6, p. 1541ff.

Bell, J. S., 1971, Tectonic evolution of the central part of the Venezuelan coast ranges, *in* Caribbean geophysical, tectonic, and petrologic studies: Geol. Soc. America Mem. no. 130.

——— 1972, Geotectonic evolution of the southern Caribbean area, *in* Studies in earth and space sciences, Geol. Soc. America Mem. no. 132.

——— 1974, Venezuelan coast ranges, *in* Spencer, A. M., ed., Mesozoic-Cenozoic orogenic belts: Edinburgh, Scottish Acad. Press.

Berg, R. R., 1962, Mountain flank thrusting in Rocky Mtn. foreland, Wyoming and Colorado: Am. Assoc. Petroleum Geologists Bull., v. 46, p. 2019–2032.

——— and **Romberg, R. E.**, 1966, Gravity profile across the Wind River Mtns., Wyoming: Geol. Soc. America Bull., v. 77, p. 647–656.

Berry, F. A. F., 1973, High fluid potentials in California Coast Ranges and their tectonic significance: Am. Assoc. Petroleum Geologists Bull., v. 57/7, p. 1219ff.

Berryhill, H. L., Jr., and others, 1960, Stratigraphy, sedimentation, and structure of late Cretaceous rocks in eastern Puerto Rico: Prelim. Rept., Am. Assoc. Petroleum Geologists Bull., v. 44.

Boos, C. M., and **Boos, M. F.**, 1957, Tectonics of eastern flank and foothills of Front Range, Colorado: Am. Assoc. Petroleum Geologists Bull., v. 41, no. 12, p. 2603–2676.

Bostrom, R. C., 1967, Ocean-ridge system in northwest America: Am. Assoc. Petroleum Geologists Bull., v. 51, no. 9, p. 1816–1832.

Bowin, C. O., 1968, Geophysical study of the Cayman trough: Jour. Geophys. Res., v. 73, no. 16, p. 5159ff.

———, **Nalwalk, A. J.**, and **Hersey, J. B.**, 1966, Serpentinized peridotite from the north wall of the Puerto Rico trench: Geol. Soc. America Bull., v. 77, p. 257–270.

Brace, W. F., **Ernst, W. G.**, and **Kallberg, R. W.**, 1970, An experimental study of tectonic overpressure in Franciscan rocks: Geol. Soc. America Bull., v. 81, no. 5, p. 1325–1338.

Bracey, D. R., and **Vogt, P. R.**, 1970, Plate tectonics in the Hispaniola area: Geol. Soc. America Bull., v. 81.

Bucher, W. H., 1952, Geologic structures and orogenic history of Venezuela: Geol. Soc. America Mem. 49.

Bunce, E. T., and **Hersey, J. B.**, 1966, Continuous seismic profiles of the outer ridge and Nares basin north of Puerto Rico: Geol. Soc. America Bull., v. 77, no. 8, p. 803–812.

Burchfiel, B. C., and **Davis, G. A.**, 1975, Nature and controls of Cordilleran orogenesis, western United States; extensions of an earlier synthesis: Am. Jour. Sci., v. 275-A, p. 363–396.

——— and **Stewart, J. H.**, 1966, "Pull-apart" origin of the central segment of Death Valley, California: Geol. Soc. America Bull., v. 77, no. 4, p. 439–442.

Burk, C. A., 1965, Geology of the Alaska peninsula —Island arc and continental margin, pt. 1: Geol. Soc. America Mem. 99.

Bush, S. A., and **Bush, P. A.**, 1969, Isostatic gravity map of the eastern Caribbean region: Trans., Gulf Coast Assoc. Geological Societies, v. XIX, p. 281–285.

Butterlin, J., 1956, La Constitution Géologique et la Structure des Antilles. Paris, Centre National de la Recherche Scientifique.

Campbell, C. J., 1974, Colombian Andes, *in* Spencer, A. M., ed.: Mesozoic-Cenozoic Orogenic Belts: The Geol. Soc. London, London.

——— and **Bürge, H.**, 1965, Section through the eastern Cordillera of Colombia, South America: Geol. Soc. America Bull., v. 76, p. 567–590.

Caner, B., **Cannon, W. H.**, and **Livingstone, C. E.**, 1967, Geomagnetic depth sounding and upper mantle structure in the Cordilleran region of western North America: Jour. Geophys. Res., v. 72, no. 24, p. 6335–6351.

Carey, S. W., 1958, A tectonic approach to continental drift, *in* Continental drift, a symposium: Geol. Dept. Univ. Tasmania.

Case, J. E., 1974, Major basins along the continental margin of northern South America, *in* Burk, C. A., and Drake, C. L., eds., The geology of continental margins: New York, Springer-Verlag.

Castillo, Lus Del, and **Moore, G. W.**, 1974, Tectonics evolution of the southern Gulf of Mexico: Geol. Soc. America Bull., v. 85, p. 607–618.

Chase, R. L., and **Bunce, E. T.**, 1969, Underthrusting of the eastern margin of the Antilles by the floor of the western North Atlantic Ocean, and the origin of the Barbados ridge: Jour. Geophys. Res., v. 74, no. 6, p. 1413–1420.

Christensen, M. N., 1966, Late Cenozoic crustal movements in the Sierra Nevada of California: Geol. Soc. America Bull., v. 77, no. 2, p. 163–182.

Christman, R. A., 1953, Geology of St. Bartholomew, St. Martin and Anguilla, Lesser Antilles: Geol. Soc. America Bull., v. 94, p. 65–96.

Churkin, M., Jr., 1962, Facies across Paleozoic miogeosynclinal margin of central Idaho: Am. Assoc. Petroleum Geologists Bull., v. 46, no. 5, p. 569–591.

Clark, L. D., 1960, Foothills fault system western Sierra Nevada, California: Geol. Soc. America Bull., p. 483–496.

Coats, R. R., 1962, Magma type and crustal structure in the Aleutian arc, *in* The crust of the Pacific basins: Am. Geophys. Union Mon. 6, p. 92–109.

Cobbing, N. J., 1972, Tectonic elements of Peru and the evolution of the Andes, *in* Tectonics—Tectonique, Section 3, Int. Geol. Congr., Proc. Congr. Geol. Int., Programme, no. 24.

Compton, R. R., 1966, Analyses of Pliocene-Pleistocene deformation and stresses in northern Santa Lucia Range, California: Geol. Soc. America Bull., v. 77, p. 1361–1380.

Coney, P. J., 1971, Structural evolution of the Cordillera Huayhuash, Andes of Peru: Geol. Soc. America Bull., v. 82, no. 7, p. 1863ff.

——— 1972, Cordilleran tectonics and North American plate motion: Am. Jour. Sci., v. 272, no. 7, p. 603–628.

——— 1973, Non-collision tectogenesis in western North America: Implications of continental drift to the earth sciences, v. 2, p. 713–727.

Cowan, D. S., 1974, Deformation and metamorphism of the Franciscan subduction zone complex northwest of Pacheco pass, California: Geol. Soc. America Bull., v. 85, no. 10, p. 1623.

Crowell, J. C., 1963, Displacements along the San Andreas fault, California: Geol. Soc. America Spec. Paper 71.

Curray, J. R., and **Nason, R. D.**, 1967, San Andreas

fault north of Point Arena, California: Geol. Soc. America Bull., v. 78, p. 413–418.
Dahlstrom, C. D. A., 1970, Structural geology in the eastern margin of the Canadian Rocky Mountains: Can. Petrol. Geol. Bull., v. 18, p. 332–406.
Dalziel, I. W. D., Caminos, R., Palmer, K. F., Nullo, F., and **Casanova, R.,** 1974, South extremity of Andes: Geology of Isla de los Estados, Argentine Tierra del Fuego: Am. Assoc. Petroleum Geologists Bull., v. 58, no. 12, p. 2502–2512.
Davis, G. A., 1968, Westward thrusting in the south-central Klamath Mtns., Calif.: Geol. Soc. America Bull., v. 79.
De Cserna, Zoltan, 1960, Orogenesis in time and space in Mexico: Geol. Rundsch., Band 50, p. 595–605.
——— 1971, Taconian (early Caledonian) deformation in the Huastecan structural belt of eastern Mexico: Am. Jour. Sci., v. 271, no. 5, p. 544–550.
Dengo, Gabriel, 1969, Problems of tectonic relations between Central America and the Caribbean: Trans., Gulf Coast Assoc. Geol. Soc., v. XIX, p. 311–320.
——— 1975, Paleozoic and Mesozoic tectonic belts in Mexico and Central America, *in* Nairn, A. E., and Stehli, F. G., eds., The ocean basins and margins, v. 3, The Gulf of Mexico and the Caribbean: New York, Plenum Press.
Dickinson, W. R., 1966, Structural relationships of San Andreas fault system, Cholame valley and Castle Mtn. Range, California: Geol. Soc. America Bull., v. 77, p. 707–726.
Donath, F., 1962, Analysis of basin-range structure, south-central Oregon: Geol. Soc. America Bull., v. 73.
——— and **Kuo, J.,** 1962, Seismic-refraction study of block faulting south-central Oregon: Geol. Soc. America Bull., v. 73.
Donnelly, T. W., 1964, Evolution of eastern Greater Antillean island arc: Am. Assoc. Petroleum Geologists Bull., v. 48, no. 5, p. 680–696.
——— 1970, International Field Institute guidebook to the Caribbean island-arc system: Am. Geol. Institute.
Dott, R. H., 1965, Mesozoic-Cenozoic tectonic history of the SW Oregon coast in relation to Cordilleran orogenesis: Jour. Geophys. Res., v. 70, no. 18, p. 4687–4707.
Durán, S. L. G., and **Lopéz, R. A.,** 1971, Interpretatión tectonofísica tentativa del Caribe Colombiano: Primer Seminario Nacional de Ciencias del Mar, Cartegena, p. 79–91.
Eardley, A. J., 1960, Igneous and tectonic provinces of the western United States: Internat. Geol. Cong. Rept., 21st, Norden, Pt. XIII, p. 18–28.
——— 1962, Structural geology of North America, 2d ed.: New York, Harper & Row.
Ernst, W. G., 1970, Tectonic contact between the Franciscan mélange and the Great Valley sequence-crustal expression of a late Mesozoic Benioff zone: Jour. Geophys. Res., v. 75, no. 5, p. 886–901.
Ewing, W. M., 1960, Earth's crust below the oceans and in continents: Am. Geophys. Union Trans., v. 41.
Fisher, R. V., 1967, Early Tertiary deformation in north-central Oregon: Am. Assoc. Petroleum Geologists Bull., v. 51, no. 1, p. 111–123.
Fitzgerald, E. L., 1968, Structure of British Columbia foothills, Canada: Am. Assoc. Petroleum Geologists Bull., v. 52, no. 4, p. 641–664.
Fleck, R. J., 1970, Tectonic style, magnitude, and age of deformation in the Sevier orogenic belt in southern Nevada and eastern California: Geol. Soc. America Bull., v. 81, no. 6, p. 1705–1720.
Foose, R. M., 1973, Vertical tectonism and gravity in the Big Horn basin and surrounding ranges of the Middle Rocky Mountains, *in* De Jong, K. A., and Scholten, R., eds., Gravity and tectonics: New York, Wiley-Interscience.
——— and others, 1961, Structural geology of the Beartooth Mountains, Montana and Wyoming: Geol. Soc. America Bull., v. 72, p. 1143–1172.
Fox, F. G., 1959, Structure and accumulation of hydrocarbons in southern foothills, Alberta, Canada: Am. Assoc. Petroleum Geologists Bull., v. 43, no. 5, p. 992–1025.
——— 1969, Some principles governing interpretation of structure in the Rocky Mountains orogenic belt, *in* Kent, P. E., and others, eds., Time and place in orogeny: Geol. Soc. London, Spec. Publ., no. 3.
Fuller, R. E., and **Waters, A. C.,** 1929, The nature and origin of the horst and graben structure of southern Oregon: Jour. Geology, v. 37.
Gabrielse, H., and **Wheeler, J. Q.,** 1961, Tectonic framework of So. Yukon and NW British Columbia: Geol. Surv. Canada Paper 60–24.
Garfunkel, Zvi, 1974, Model for the late Cenozoic tectonic history of the Mojave Desert, California, and for its relation to adjacent regions: Geol. Soc. America Bull., v. 85, p. 1931–1944.
Gilluly, James, 1963, The tectonic evolution of the

western United States: Geol. Soc. London Quart. Jour., v. 119, p. 133.

——— 1965, Volcanism, tectonism and plutonism in the western United States: Geol. Soc. America Spec. Paper 80.

Hamilton, Warren, 1962, L. Cenozoic structure of west central Idaho: Geol. Soc. America Bull., v. 73.

——— 1969, Mesozoic California and the underflow of Pacific mantle: Geol. Soc. America Bull., v. 80, no. 12, p. 2409–2429.

——— and **Myers, W. B.**, 1965, Cenozoic tectonics of the western United States, *in* The world rift system: Geol. Surv. Canada Paper 66–14, p. 291.

Harris, H. D., 1959, A late Mesozoic positive area in western Utah: Am. Assoc. Petroleum Geologists Bull., v. 43, no. 11.

Harrison, J. E., Griggs, A. B., and **Wells, J. D.**, 1974, Tectonic features of the Precambrian belt basin and their influence on post-belt structures: Geol. Survey Prof. Paper 866.

Hatherton, T., 1969, Geophysical anomalies over the eu- and mio-geosynclinal systems of California and New Zealand: Geol. Soc. America Bull., v. 80, no. 2, p. 213–230.

Hayes, D. E., 1966, A geophysical investigation of the Peru-Chile trench: Marine Geology, v. 4, p. 309–351.

Heezen, B. C., 1975, Geology of the Caribbean crust, *in* Nairn, A. E., and Stehli, F. G., eds., The ocean basins and margins, v. 3, The Gulf of Mexico and the Caribbean: New York, Plenum Press.

Henderson, G. G. L., and **Dahlstrom, C. D. A., 1959,** First-order nappe in Canadian Rockies: Am. Assoc. Petroleum Geologists Bull., v. 43, p. 641–654.

Henyey, T. L., and **Bischoff, J. L.**, 1973, Tectonic elements of the northern part of the Gulf of California: Geol. Soc. America Bull., v. 84, no. 1, p. 315–330.

Hess, H. H., 1938, Gravity anomalies and island arc structure with particular reference to the West Indies: Am. Philos. Soc. Proc., v. 70, p. 71–96.

——— and **Maxwell, J. C.**, 1953, Caribbean research project: Geol. Soc. America Bull., v. 64, p. 1–6.

Hill, M. L., 1974, Is the San Andreas a transform fault?: Geology, v. 2, no. 11.

——— and **Dibblee, T. W., Jr.**, 1953, San Andreas, Garlock, and Big Pine faults, California: Geol. Soc. America Bull., v. 64, p. 443–458.

——— and **Troxel, B. W.**, 1966, Tectonics of Death Valley region, California: Geol. Soc. America Bull., v. 77, p. 435–438.

Hill, P. A., 1959, Geology and structure of the northwest Trinidad Mountains, Las Villas Province, Cuba: Geol. Soc. America Bull., v. 70, no. 11.

Holcombe, T. L., Vogt, P. R., Matthews, J. E., and **Murchison, R. R.**, 1973, Evidence for sea-floor spreading in the Cayman trough: Earth Planet. Sci. Letters, v. 20, p. 357–371.

Hoppin, Richard, 1961, Precambrian rocks and their relationship to Laramide structure along the east flank of Bighorn Mtns., near Buffalo, Wyoming: Geol. Soc. America Bull., v. 72, p. 351–368.

——— and **Palmquist, J. C.**, 1965, Basement influence on later deformation: The problem, techniques of investigation, and examples from Bighorn Mtns., Wyoming: Am. Assoc. Petroleum Geologists Bull., v. 49, no. 7.

Horsfield, W. T., 1975, Quaternary vertical movements in the Greater Antilles: Geol. Soc. America Bull., v. 86, p. 933–938.

Howard, J. H., 1966, Structural development of the Williams Range thrust, Colorado: Geol. Soc. America Bull., v. 77, p. 1247–1264.

Hose, R. K., and **Zdenko, F. D.**, 1973, Development of the late Mesozoic to early Cenozoic structures of the eastern Great Basin, *in* De Jong, K. A., and Scholten, R., eds., Gravity and tectonics: New York, Wiley, p. 429–441.

Hsü, K. J., 1971, Franciscan mélanges as a model for eugeosynclinal sedimentation and underthrusting tectonics: Jour. Geophys. Res., v. 76, no. 5.

——— 1973a, Franciscan mélanges as a model for eugeosynclinal sedimentation and underthrusting tectonics, *in* Plate tectonics: Washington, D. C., Amer. Geophys. Union, p. 418–426.

——— 1973b, Mesozoic evolution of the California Coast Ranges: A second look, *in* De Jong, K. A., and Scholten, R., eds., Gravity and tectonics: New York, Wiley, p. 379–396.

Hoffman, O. F., 1972, Lateral displacement of upper Miocene rocks and the Neogene history of offset along the San Andreas fault in central California: Geol. Soc. America Bull., v. 83, no. 10, p. 2913ff.

Irving, E. M., 1975, Structural evolution of the northernmost Andes, Colombia, Geol. Survey Prof. Paper 846.

Irwin, W. P., 1960, Geologic reconnaissance of the northern Coast Ranges and Klamath Mtns., Califor-

nia, with a summary of the mineral resources: California Div. of Mines (San Francisco) Bull. 179.
——— 1965, Late Mesozoic orogenies in the ultramafic belts of northwestern California and southwestern Oregon: U. S. Geol. Survey Prof. Paper 501-C, p. C1–C9.
James, D. E., 1971, Plate tectonic model for the evolution of the Central Andes: Geol. Soc. America Bull., v. 82, no. 12, p. 3325ff.
Johnson, J. G., 1971, Timing and coordination of orogenic, epeirogenic and eustatic events: Geol. Soc. America Bull., v. 82, p. 3263–3298.
Jones, P. B., 1971, Folded faults and sequence of thrusting in Alberta foothills: Am. Assoc. Petroleum Geologists Bull., v. 55, no. 2, p. 292–306.
Jordan, T. H., 1975, The present-day motions of the Caribbean plate: Jour. Geophys. Res., v. 80, no. 32, p. 4433–4439.
Kay, Marshall, 1951, North American geosynclines: Geol. Soc. America Mem. 48.
——— and **Crawford, J. P.**, 1964, Paleozoic facies from the miogeosynclinal to the eugeosynclinal belt in thrust slices, Central Nevada: Geol. Soc. America Bull., v. 75.
Kaye, C. A., 1957, Notes of the structural geology of Puerto Rico: Geol. Soc. America Bull., v. 68, no. 1.
Keller, F., Jr., and others, 1954, Aeromagnetic surveys in the Aleutian, Marshall and Bermuda Islands: Am. Geophys. Union Trans., v. 35, p. 558–572.
Ken, J. W., 1962, Paleo. sequences and thrust of the Seetoya Mtns. Independence Range, Nevada: Geol. Soc. America Bull., v. 73.
Koch, J. G., 1963, Late Mesozoic orogenesis and sedimentation, Klamath province, southwest Oregon coast: Unpubl. Ph.D. dissertation, Univ. Wisconsin.
Kozary, M. T., 1968, Ultramafic rocks in thrust zones of northwestern Oriente province, Cuba: Am. Assoc. Petroleum Geologists Bull., v. 52, no. 12, p. 2298–2317.
Krause, D. C., 1971, Bathymetry, geomagnetism and tectonics of the Caribbean Sea north of Colombia: Geol. Soc. America Mem., v. 130, p. 35–54.
Kurie, A. E., 1966, Recurrent structural disturbance of Colorado Plateau margin near Zion National Park, Utah: Geol. Soc. America Bull., v. 77, no. 8, p. 867–872.
LaFountain, L. J., 1975, Unusual polyphase folding in a portion of the northeastern Front Range, Colorado: Geol. Soc. America Bull., v. 86, p. 1725–1732.
Lipman, P. W., 1964, Structure and origin of an ultramafic pluton in the Klamath Mountains, California: Am. Jour. Sci., v. 262, p. 199–222.
Loczy, Louis de, 1971, Role of transcurrent faulting in South American tectonic framework: Am. Assoc. Petroleum Geologists Bull., v. 54, no. 11, p. 2111–2119.
Lofgen, B. E., 1960, Crustal structure in the California-Nevada region: Jour. Geophys. Res., v. 65, no. 3.
Lohmann, H. H., 1970, Outline of tectonic history of Bolivian Andes: Am. Assoc. Petroleum Geologists Bull., v. 54, no. 5, p. 735–757.
Lopez-Ramos, Ernesto, 1969, Marine Paleozoic rocks of Mexico: Am. Assoc. Petroleum Geologists Bull., v. 53, no. 12, p. 2399–2417.
Lowry, W. D., and **Baldwin, E. M.**, 1952, Late Cenozoic geology of the Lower Columbia River Valley, Oregon and Washington: Geol. Soc. America Bull., v. 63.
McBirney, A. R., and **Bass, I. G.**, 1967, Relations of oceanic volcanic rocks to mid-oceanic rises and heat flow: Earth Planet. Sci. Lett., v. 2.
McCrossan, R. G., and **Glaister, R. P.**, eds., 1964, Geological history of western Canada: Alberta Soc. Petroleum Geologists Calgary, Alberta.
MacDonald, G. J. F., 1961, Gravity measurements over the southern Rocky Mountain trench area of British Columbia: Jour. Geophys. Res., v. 66, no. 8.
McTaggart, K. C., 1970, Tectonic history of the northern Cascade Mountains: Geol. Assoc. Canada, Special Paper no. 6, p. 137–148.
Maley, T. S., and others, 1974, Topography and structure of the western Puerto Rico trench: Geol. Soc. America Bull., v. 85, no. 4, p. 513–518.
Malfait, B. T., and **Dinkelman, M. G.**, 1972, Circum-Caribbean tectonic and igneous activity and the evolution of the Caribbean plate: Geol. Soc. America Bull., v. 83, no. 2, p. 254–272.
Malin, P. E., and **Dillon, Wm. P.**, 1973, Geophysical reconnaissance of the western Cayman ridge: Jour. Geophys. Res., v. 78, no. 32, p. 7769ff.
Maresch, W. V., 1974, Plate tectonics origin of the Caribbean Mountain system of northern South America: Discussion and proposal: Geol. Soc. America Bull., v. 85, no. 5, p. 669–682.
Martin, L. J., 1963, Tectonics of northern Cordillera in Canada: Am. Assoc. Petroleum Geologists Mem. 2, p. 243–251.
Mattson, P. H., 1966, Geological characteristics of

Puerto Rico, *in* Continental margins and island arcs —Internat. Upper Mantle Comm., Symposium, Ottawa, 1965: Canada Geol. Survey Paper 66-15.

——— 1972, Plate tectonics in the Caribbean: Nature, v. 235.

——— 1973, Middle Cretaceous nappe structures in Puerto Rican ophiolites and their relation to the tectonic history of the Greater Antilles: Geol. Soc. America Bull., v. 84, no. 1, p. 21–38.

Maxwell, J. C., 1948, Geology of Tobago, B.W.I.: Geol. Soc. America Bull., v. 59.

——— 1974, Early western margin of the United States: Preprint—chap. *in:* Geology of continental margins.

Meyerhoff, A. A., and **Meyerhoff, H. A.**, 1972, Continental drift, 4, The Caribbean "plate": Jour. Geology, v. 80, p. 34–60.

Moench, R. H., and others, 1962, Precambrian folding in the Idaho Springs—Central city area, Front Range, Colorado: Geol. Soc. America Bull., v. 73.

Molnar, P., and **Sykes, L. R.**, 1969, Tectonics of the Caribbean and Middle America regions from focal mechanisms and seismicity: Geol. Soc. America Bull., v. 80, p. 1639–1684.

Monger, J. W. H., Souther, J. G., and **Gabrielse, H.**, 1972, Evolution of the Canadian Cordillera: A plate tectonic model: Am. Jour. Sci., v. 272, no. 7, p. 577–602.

Moore, D. G., 1973, Plate-edge deformation and crustal growth, Gulf of California province: Geol. Soc. America Bull., v. 84, no. 6, p. 1883.

Moore, G. W., and **Castillo, Luis Del**, 1974, Tectonic evolution of the southern Gulf of Mexico, Geol. Soc. America Bull., v. 85, no. 4, p. 607–618.

Murphy, A. J., Sykes, L. R., and **Donnelly, T. W.**, 1970, Preliminary survey of the microseismicity of the northeastern Caribbean: Geol. Soc. America Bull., v. 81, no. 8, p. 2459–2464.

Nolan, T. B., 1943, The basin and range province in Utah, Nevada, California: U. S. Geol. Survey Prof. Paper 197D, p. 141–196.

Officer, C. B., 1957, Geophysical investigations in the eastern Caribbean: Venezuelan Basin, Antilles island arc, and Puerto Rico trench: Geol. Soc. America Bull., v. 68, p. 359–378.

Packer, D. R., Brogan, G. E., Stone, D. B., 1975, New data on plate tectonics of Alaska: Tectonophysics, v. 29, no. 1, p. 87–102.

Page, B. M., 1970, Sur-Nacimiento fault zone of California: continental margin tectonics: Geol. Soc. America Bull., v. 81, no. 3, p. 667–690.

Peter, G., 1972, Geologic structure offshore north-central Venezuela: Trans. Caribbean Geol. Conf., 6th, p. 283–294.

Peterman, Z. E., and **Hedge, C. E.**, 1968, Age of Precambrian events in the NE Front Range, Colorado: Jour. Geophys. Res., v. 73, no. 6, p. 2277–2296.

Peterson, J. A., 1965, Rocky Mountain sedimentary basins—Introduction: Am. Assoc. Petroleum Geologists Bull., v. 49, no. 11, p. 1779–1780.

Porterm S. C., 1966, Stratigraphy and deformation of Paleozoic section at Anaktuvuk Pass, Central Brooks Range, Alaska: Am. Assoc. Petroleum Geologists Bull., v. 50, no. 5, p. 952–980.

Price, R. A., 1972, Tectonics and structural geology in Canada: Earth Sci. Rev., v. 8, no. 1, p. 137–138.

Prucha, J. J., Graham, J. A., and **Nickelsen, R. P.**, 1965, Basement-controlled deformation in Wyoming, province of Rocky Mountains foreland: Am. Assoc. Petroleum Geologists Bull., v. 49, no. 7, p. 966–992.

Rabinowitz, P. D., and **Ryan, W. B. F.**, 1970, Gravity anomalies and crustal shortening in the eastern Mediterranean: Tectonophysics, v. 10, p. 585–608.

Raff, A. D., and **Mason, R. G.**, 1961, Magnetic survey off the west coast of North America 40° N latitude to 50° N latitude: Geol. Soc. America Bull., v. 72.

Richards, H. G., 1974, Tectonic evolution of Alaska: Am. Assoc. Petroleum Geologists Bull., v. 58, no. 1.

Roberts, R. J., 1972, Evolution of the Cordilleran fold belt: Geol. Soc. America Bull., v. 83, no. 7, p. 1989ff.

——— and **Crittenden, M. D., Jr.**, 1975, Orogenic mechanisms, Sevier orogenic belt, Nevada and Utah, *in* De Jong, D. A., and Scholten, R., eds., Gravity and tectonics: New York, Wiley.

——— and **Thomasson, M. R.**, 1964, Comparison of late Paleozoic depositional history of northern Nevada and central Idaho: Geol. Survey Prof. Paper 475-D, article 122.

Roddick, J. A., 1967, Tintina trench: Jour. Geology, v. 75, no. 1, p. 23–33.

Rodgers, John, 1970, The tectonics of the Appalachians: New York, Interscience.

Rubey, W. W., and **Hubbert, M. K.**, 1959, Role of fluid pressure in mechanics of overthrust faulting, II: Geol. Soc. America Bull., v. 70.

Rutland, R. W. R., 1971, Andean orogeny and ocean

floor spreading: Nature, v. 233, no. 5317, p. 252–255.

Savage, J. C., and **Burford, R. O.,** 1973, Geodetic determination of relative plate motion in Central California: Jour. Geophys. Res. v. 78, no. 5, p. 832ff.

Scholl, D. W., and others, 1970, Peru-Chile trench sediments and sea-floor spreading: Geol. Soc. America Bull., v. 81, p. 1339–1360.

Scholten, Robert, 1958, Paleozoic evolution of the geosynclinal margin north of the Snake River plain: Geol. Soc. America Bull., v. 68.

Scholz, C. H., Barazangi, Muawia, and **Sbar, M. L.,** 1971, Late Cenozoic evolution of the great basin, western U.S., as an ensialic interarc basin: Geol. Soc. America Bull., v. 82, no. 11, p. 2979ff.

———, **Wyss, M.,** and **Smith, S. W.,** 1969, Seismic and aseismic slip on the San Andreas fault: Jour. Geophys. Res., v. 74, no. 8, p. 2049–2069.

Senn, A., 1940, Paleogens of Barbados and its bearing on history and structure of Antillean-Caribbean region: Am. Assoc. Petroleum Geologists Bull., v. 24, no. 9, p. 1548–1610.

Shagam, R., 1973, Andean research project, Venezuela, principal data and tectonic implications, *in* Studies in earth and space sciences: Geol. Soc. America Mem. 132, p. 449–463.

——— 1975, The northern termination of the Andes, *in* Nairn, A. E., and Stehli, F. G., eds., The ocean basin and margins, v. 3, The Gulf of Mexico and the Caribbean: New York, Plenum Press.

Shaw, E. W., 1963, Canadian Rockies—Orientation in time and space: Am. Assoc. Petroleum Geologists Mem. 2, Backbone of the Americas, p. 231–242.

Shawe, D. R., 1965, Strike-slip control of basin-range structure indicated by historical faults in western Nevada: Geol. Soc. America Bull., v. 76, no. 12, p. 119–162.

Silver, E. A., 1971, Transitional tectonics and late Cenozoic structure of the continental margin off northernmost California: Geol. Soc. America Bull., v. 82, no. 1, p. 1–22.

——— 1975, Speculations of plate boundaries and movements in the Caribbean region, submitted *to* Geology.

———, **Case, J. E.,** and **MacGillavry, H. J.,** 1975, Geophysical study of the Venezuelan borderland: Geol. Soc. America Bull., v. 86, p. 213–226.

Skvor, Vladimir, 1969, The Caribbean area: A case of destruction and regeneration of continent: Geol. Soc. America Bull., v. 80, no. 6, p. 961–968.

Smith, R. B., and **Sbar, M. L.,** 1974, Contemporary tectonics and seismicity of the western United States with emphasis on the intermountain seismic belt: Geol. Soc. America Bull., v. 85, p. 1205–1218.

Solomon, S. C., and **Sleep, N. H.,** 1974, Some simple physical models for absolute plate motions: Jour. Geophys. Res., v. 79, p. 2557–2567.

Spencer, A. M., 1974, Mesozoic-Cenozoic orogenic belts: The Geological Soc. London Spec. Publ., no. 4.

Stewart, J. H., 1971, Basin and range structure: A system of horsts and grabens produced by deep seated extension: Geol. Soc. America Bull., v. 82, no. 4, p. 1019ff.

——— and **Poole, F. G.,** 1975, Extension of the Cordilleran miogeosynclinal belt to the San Andreas fault, southern California: Geol. Soc. America Bull., v. 86, p. 205–212.

———, **Evernden, J. F., Snelling, N. J.,** 1974, Age determinations from Andean Peru: A reconnaissance survey: Geol. Soc. America Bull., v. 85, no. 7, p. 1107–1116.

Suppe, J., Powell, C., and **Berry, R.,** 1975, Regional topography, seismicity, Quaternary volcanism and the present-day tectonics of the western United States: Am. Jour. Sci., v. 275-A, p. 397–436.

Sykes, L. R., and **Ewing, M.,** 1965, The seismicity of the Caribbean region: Jour. Geophys. Res., v. 70, no. 20, p. 5065–5074.

Tanner, W. F., 1962, Surface structural patterns obtained from strike-slip models: Jour. Geology, v. 70, p. 101–107.

Thompson, G. A., 1959, A study of basin and range structures: Jour. Geophys. Res., v. 64, no. 2.

——— and **Talwani, M.,** 1964, Geology of the crust and mantle, western United States: Science, v. 146, p. 1539–1549.

Tobisch, O. T., 1968, Gneissic amphibolite at Las Palmas, Puerto Rico, and its significance in the early history of the Greater Antilles island arc: Geol. Soc. America Bull., v. 79, no. 5, p. 557–574.

Travers, W. B., 1973, A trench off central California in late Eocene-Early Oligocene time: Geol. Soc. America Bull., Mem. no. 132, p. 173–182.

Von Herzen, R. P., Simmins, G., and **Folinsbee, A.,** 1970, Heat flow between the Caribbean Sea and the mid-Atlantic ridge: Jour. Geophys. Res., v. 75, no. 11, p. 1973–1984.

Wallace, R. E., 1949, Structure of a portion of the San Andreas rift in southern California: Geol. Soc. America Bull., v. 60, p. 781–806.

Wassall, H., 1957, The relation of oil and serpentine in Cuba: Geologia del Petroleo, Sec. III: Mexico, Cong. Geol. Internac.

Wheeler, J. O., 1970, Introduction, *in* Structure of the southern Canadian Cordillera, Geol. Assoc. Can., Spec. Paper no. 6.

White, W. H., 1959, Cordilleran tectonics in British Columbia: Am. Assoc. Petroleum Geologists Bull., v. 43, p. 60–101.

Whitten, C. A., 1956, Crustal movement in California and Nevada: Am. Geophys. Union Trans., v. 37, no. 4.

Willis, B., 1938a, The San Andreas rift in California: Jour. geology, v. 46, p. 793–827.

——— 1938b, The San Andreas rift in southwestern California: Jour. Geology, v. 46, p. 1017–1057.

Wise, D. U., 1963a, Keystone faulting and gravity sliding driven by basement uplift of Owl Creek Mountains, Wyo.: Am. Assoc. Petroleum Geologists Bull., v. 47, p. 586–598.

——— 1963b, An outrageous hypothesis for the tectonic pattern of North American Cordillera: Geol. Soc. America Bull., v. 74.

Yeats, R. S., 1968, Southern California structure, sea-floor spreading, and history of the Pacific basin: Geol. Soc. America Bull., v. 79, no. 12, p. 1693–1702.

A

Stress Theory

PLANE STRESS—ANALYSIS OF STRESS IN TWO DIMENSIONS

The analysis of stress conditions in two dimensions is used here because it is much simpler to treat mathematically, and the form of the stress relationships in two and three dimensions is similar. But we must remember that all stress conditions in the earth are three dimensional. The cube of the three-dimensional case is reduced to a square with unit depth in the z direction, Fig. A-1, and stress components acting perpendicular to the plane are zero. Only eight stress components are involved, and for homogeneous and equilibrium stress conditions, the shearing stresses on opposite surfaces are equal as are those on adjacent sides:

τ_{yx} (top) = τ_{xy} (bottom)
$\tau_{yx} = \tau_{xy}$

Note also that adjacent shear stress components act either away from or toward each other—never in the same direction, since that would denote rotation. As a result of the above relations, the stresses on the square are fully defined if we know σ_x, σ_y, and any shear stress component. (In three dimensions σ_x, σ_y, σ_z, τ_{xy}, τ_{yz}, and τ_{xz} define the stress conditions.) Thus the force on the right side of this square is

F (right side)
$= \text{area of side} \times \sigma_x + \text{area of side} \times \tau_{xy}$

F (left side)
$= \text{area of side} \times \sigma_x - \text{area of side} \times \tau_{xy}$

Finding Equations of Stress for a Surface of any Orientation in Terms of Known Stress Components

The coordinate axes are selected so that they are oriented parallel to the sides of the square just described, but the values of the stress components vary with orientation of the plane considered. In the following analysis, general equations are derived which make it possible to find the normal and shear components on any plane in terms of those on the reference square.

The orientation of this arbitrary plane is expressed in terms of the angle ϕ which the plane makes with the y coordinate axis, Fig. A-2. The state of stress for a square with sides oriented parallel to a given coordinate system is defined as previously described and illustrated, Fig. A-1. The state of stress can also be found when the square is not oriented parallel to the coordinate system. (The square is shown in Fig. A-2, top.) That state of stress is defined when we know the normal and shearing stresses on each side of the square. A state of homogeneous and equilibrium stress condition is assumed. The normal and shear stresses across any arbitrarily selected plane AB (one side of the square) may be calculated as follows:

1. The plane AB intersects the coordinate axes, forming a triangle OAB, but unit thickness in the z direction is assumed as one of the conditions of plane stress; so, the sides of the triangle may be thought of as having area. The areas are

$OA = AB \sin \phi$

$OB = AB \cos \phi$

2. It is possible to determine the sum of the stress components acting parallel to the coordinate axes,

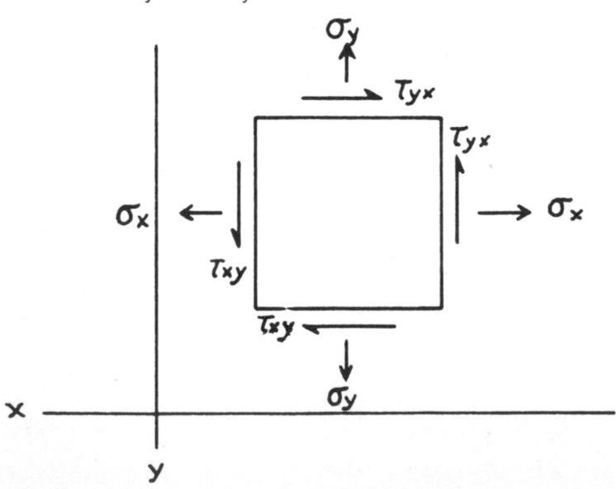

FIGURE A-1 Normal and shearing stress components on the sides of a square (for plane stress condition) oriented with reference on arbitrarily selected coordinate system indicated by x and y axes.

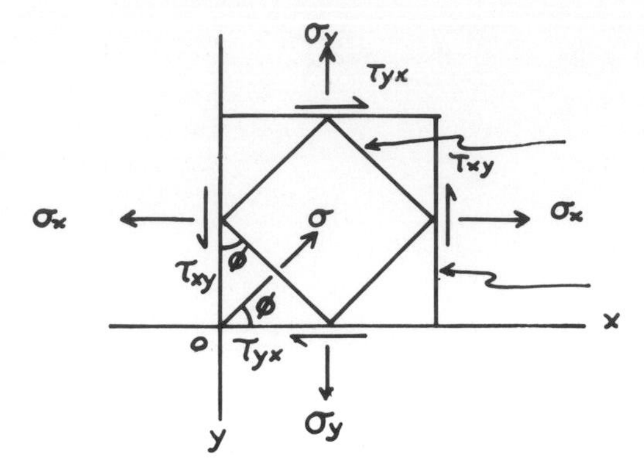

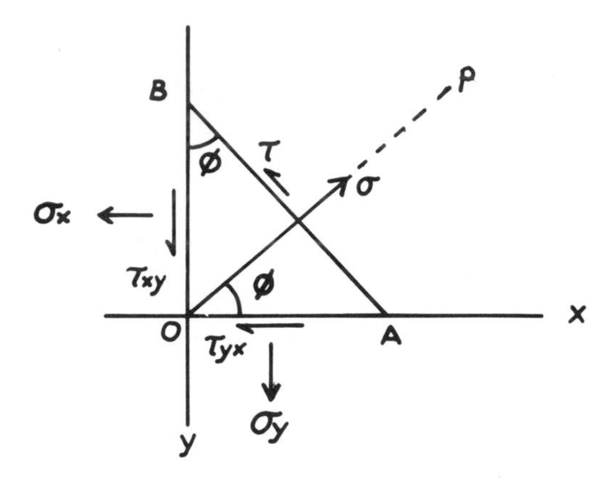

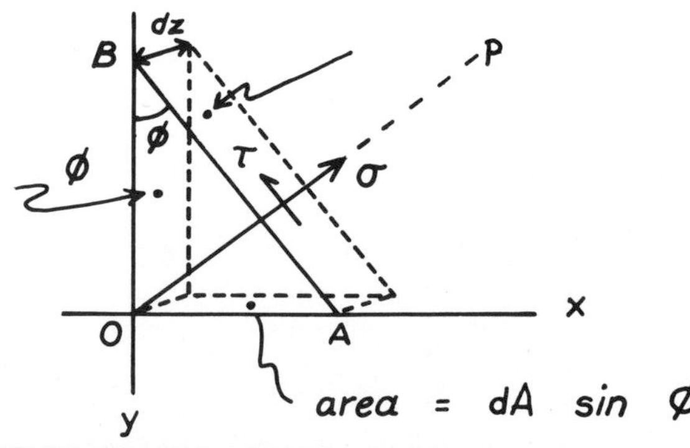

FIGURE A-2 Three alternate representations of normal and shearing stress components on a plane oriented at an angle ϕ to the coordinate axes. (See text for discussion.)

and these must be balanced against the components of the force vector that are parallel to the coordinate axes and labeled S_x and S_y. Those acting in the *x* direction consist of σ_x acting across area *OB* and τ_{yx} acting on *OA*. These are balanced by S_x:

$$S_x = \sigma_x AB \cos \phi + \tau_{yx} AB \sin \phi$$

Those acting in the *y* direction consist of σ_y acting across area *OA* and τ_{xy} acting on *OB*. Thus

$$S_y = \sigma_y AB \sin \phi + \tau_{xy} AB \cos \phi$$

3. Resolve S_x and S_y onto *OP*, the normal to *AB*, to find the component of force, hence the stress acting normal to the plane *AB*:

$$AB\ \sigma = AB \sin \phi\ (\tau_{xy} \cos \phi + \sigma_y \sin \phi) + AB \cos \phi\ (\sigma_x \cos \phi + \tau_{yx} \sin \phi)$$

$$\sigma = \sin \phi\ (\tau_{xy} \cos \phi + \sigma_y \sin \phi) + \cos \phi\ (\sigma_x \cos \phi + \tau_{yx} \sin \phi)$$

4. Similarily, resolve S_x and S_y in the direction of *AB* to determine the component of shearing stress in that direction:

$$\tau = (\sigma_y - \sigma_x) \sin \phi \cos \phi + \tau_{yx} (\cos^2 \phi - \sin^2 \sigma)$$

5. Since $\tau_{xy} = \tau_{yx}$, the above equations may be rewritten as:

$$\sigma = \sigma_x \cos^2 \phi + 2\tau_{xy} \sin \phi \cos \phi + \sigma_y \sin^2 \phi \quad \text{(A-1)}$$

$$\tau = (\sigma_y - \sigma_x) \sin \phi \cos \phi + \tau_{xy} (\cos^2 \phi - \sin^2 \phi) \quad \text{(A-2)}$$

These equations give a description of the way in which the normal and shearing stresses vary on a plane of any orientation specified by the angle ϕ. The values of the normal and shearing stresses are specified in terms of stress components (e.g., σ_x) relative to some known coordinate axes.

Variation of Shearing and Normal Stresses on an Inclined Plane

Normal stresses on planes of various orientations through a point vary from the maximum principal stress value to the least principal stress value. These occur at orientations of 0 and 90° to the maximum principal stress direction (normal stresses to a plane are maximum when that plane is perpendicular to the maximum principal stress). Shearing stresses disappear at these two orientations.

The value of σ and τ for any orientation of a plane can be determined in terms of the principal stresses, Fig. A-3. Consider the stresses on various planes through a bar under compression. The plane is inclined at an angle α to the long axis of the bar, and the normal to the plane makes an angle ϕ with the axis of the bar above. If the inclined plane is of unit length then $AO = 1 \cos \phi$ and $OB = 1 \sin \phi$.

Components of the stresses σ_1 and σ_2 acting normal to *AB* can be found as follows:

1. σ_1 acts on a surface with area equal to that of *OA*.
2. If *AB* has an area of unity, then $OA = 1 \cos \phi$.
3. The force on *OA* due to $\sigma_1 = \sigma_1$ (area of *OA*) $= \sigma_1 \cos \phi$.
4. If $\sigma_1 \cos \phi$ is the force acting in the direction of *OB*, then the component of this acting normal to *AB* is $(\sigma_1 \cos \phi) \cos \phi$ because $\cos \phi = \sigma$ (component)/*OB*.
5. Similarly the normal component on *AB* due to σ_2 can be shown to equal $\sin \phi\ (\sigma_2 \sin \phi)$.
6. Thus the total normal force acting on *AB* is

$$\sigma = (\sigma_1 \cos \phi) \cos \phi + (\sigma_2 \sin \phi) \sin \phi = \sigma_1 \cos^2 \phi + \sigma_2 \sin^2 \phi$$

or from trigonometry,

$$\sigma = \frac{\sigma_1 + \sigma_2}{2} + \frac{\sigma_1 - \sigma_2}{2} \cos 2\phi$$

FIGURE A-3 Shearing and normal components are shown on a plane *AB* oriented at an angle α to the maximum principal stress direction. (See text for discussion.)

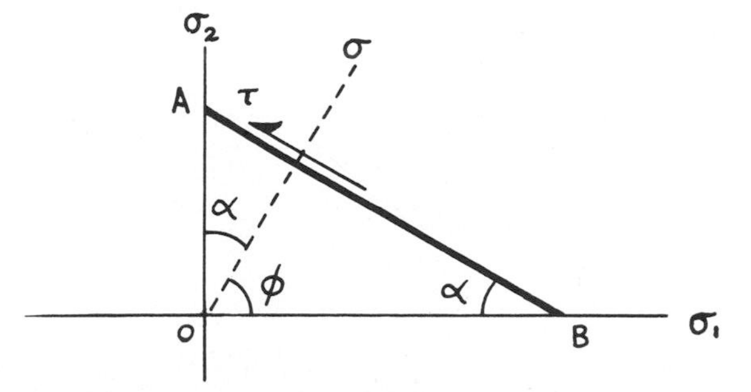

Similarly, the force can be resolved into shearing components along *AB:*

$$\tau = \cos\phi\,(\sigma_1 \sin\phi) - \sin\phi\,(\sigma_2 \cos\phi)$$
$$= (\sigma_1 - \sigma_2) \sin\phi \cos\phi$$
$$= \frac{\sigma_1 - \sigma_2}{2} \sin 2\phi$$

These equations are plotted, Fig. A-4, which shows that shearing stress is a maximum at 45 and 135° orientations of the inclined plane, and normal stress is a maximum of 0 and 90° orientations.

The relationships developed above have important implications for the structural geologists. On the basis of stress theory we can calculate at least to a rough approximation the stress conditions in a mass of rock subjected to given forces. It is sometimes possible to estimate the orientation, if not the magnitude, of the principal stress directions (e.g., the maximum principal stress is generally at right angles to the long axis of a fold). Also the orientation of the maximum shearing stresses is an important consideration in analysis of rock failure (see Chap. 5).

FIGURE A-4 Variations of normal stress σ and shear stress τ with changes in ϕ. *(From Ramsay, 1967.)*

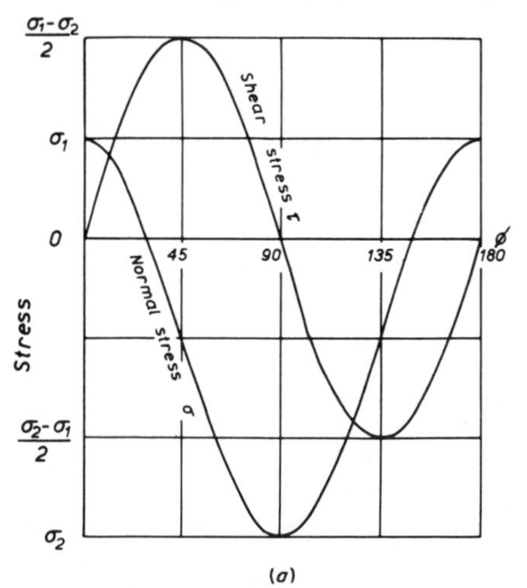

THE CONCEPT OF PRINCIPAL STRESSES AND PRINCIPAL STRESS AXES

Equations (A-1) and (A-2) show how normal and shearing stresses vary with orientation. Orientation for maximum and minimum normal stresses and the relationship between normal and shearing stress values can be determined from these equations.

If σ_x, σ_y, and τ_{xy} are known and are constant at a particular location at which stress conditions are being examined, then from Equation (A-2) σ is a function of ϕ. A necessary condition that a function of one variable have a maximum and/or minimum value is that its derivative with respect to the independent variable (ϕ in this case) be equal to zero. Thus $d\sigma/d\phi = 0$ (condition for maximum or minimum σ). Differentiation of the equation for the normal stress to a plane (σ) with respect to the angle of inclination of the plane (ϕ), Equation (A-2), yields

$$\frac{d\sigma}{d\phi} = 2\,(\sigma_y - \sigma_x) \sin\phi \cos\phi + 2\,\tau_{xy}\,(\cos^2\phi - \sin^2\phi) = 0 \quad \text{(A-3)}$$

Compare this with the equation for τ, Equation (A-1), and we see that

$$\frac{d\sigma}{d\phi} = 2\,\tau \quad \text{(and this must be equal to zero for maximum or minimum normal stress)}$$

Hence the conclusion that *shearing stresses are zero on those planes for which normal stresses are maximum or minimum.* The orientation of the planes of maximum and minimum normal stress can be solved from Equation (A-3). The normal stress is a maximum or a minimum when ϕ has the following value (Jaeger, 1969):

$$\tan 2\phi = \frac{2\tau_{xy}}{\sigma_x - \sigma_y}$$

This equation yields two angles (or directions) at right angles to one another for which the normal stress is a maximum and a minimum when the shear stresses across them are zero. These two directions, one a maximum and one a minimum normal stress direction for which shear stresses are zero, are called the *principal axes of stress,* and the stresses in these directions are called the *principal stresses.*

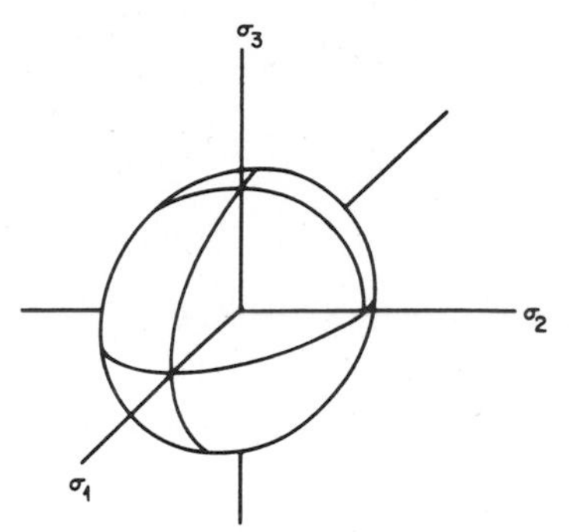

FIGURE A-6 Stress ellipsoid showing principal stress axes.

THREE-DIMENSIONAL STRESS

One principal stress axis is a maximum, the other a minimum in plane stress, but in the three-dimensional case, it can be shown that three mutually perpendicular stress axes exist which are designated as follows:

σ_1 = maximum principal stress
σ_2 = intermediate principal stress
σ_3 = least principal stress

FIGURE A-5 Stress components acting on a tetrahedron oriented obliquely with respect to an arbitrarily oriented coordinate system *x, y, z.* (*From Nádai, 1950.*)

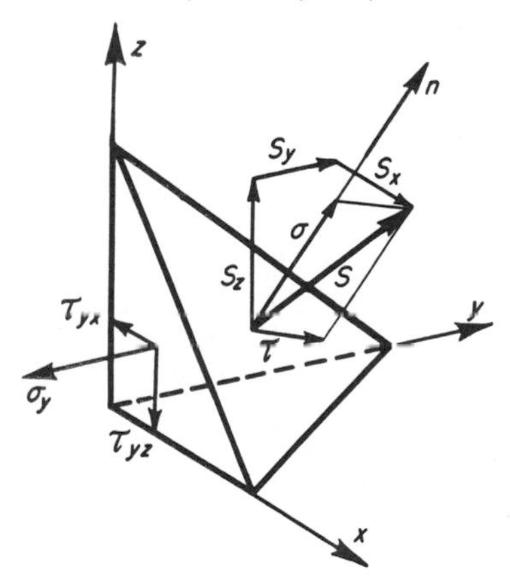

The simplest way to define the state of stress at a point is to describe the principal stresses acting at that point. This has been shown by Nádai (1950). Instead of a triangle, as in plane stress, a tetrahedron is used, and it is shown that three stress axes can be used to define the stress conditions on the tetrahedron, Fig. A-5. The equation relating these stresses is in the form of the equation of an ellipsoid, and it is called the *stress ellipsoid.* Thus, the three-dimensional state of stress at a point can be defined in terms of three mutually perpendicular normal stresses which are related to one another as the axes of an ellipsoid, Fig. A-6.

REFERENCES

Jaeger, J. C., 1969, Elasticity, fracture, and flow: London, Methuen.

Nádai, A., 1950, Theory of flow and fracture of solids: New York, McGraw-Hill.

Ramsay, J. G., 1967, Folding and fracturing of rocks: New York, McGraw-Hill.

B

Structural Maps and Cross Sections

STRUCTURE CONTOUR MAPS

A structure contour map depicts the shape and elevations on a given surface by means of contour lines. These maps often depict a stratigraphic marker horizon such as the top of a bed, but they may be drawn on fault surfaces, the margins of a salt plug, or an igneous intrusion. Structure contour maps are used extensively as a means of depicting structures where subsurface control is available. *Form-line contours* can be drawn to approximate the shape of the structure even when subsurface information is lacking or sparse. These maps are used to portray the geometry of structural features in much the same way that topographic maps portray the shape of the ground surface. Structure contour and form-line maps provide a two-dimensional picture of the three-dimensional shape of the surface on which they are drawn. Profiles of this surface may be prepared from them along any line. If structure contour maps are prepared for several surfaces in the same area, cross sections can be prepared from them, and changes in structure with depth can be detected.

Properties of Structure Contours

The properties of structure contours are similar to those of topographic contours, and the shapes of the contoured surfaces are sometimes similar. As in making and interpreting topographic maps, the difference between good and mediocre analysis is found in attention given to details, knowledge of the forms to be expected, ability to recognize those forms, and understanding of the way they appear on a contoured map of the structure. Experience in this type of work is indispensable, and the beginning student will learn a great deal by carefully studying published structure contour maps.* Structure contours and structure contour maps exhibit the following properties:

* The United States Tectonic Map and the North American Tectonic Map are recommended for study.

1. A structure contour connects points of equal elevation on some surface.

2. The elevation of the contour is specified. The datum may be mean sea level, but many structure contour maps are based on other datum levels. Negative values may be used on contours. The negative sign indicates that the contour is drawn below the datum level. A note may indicate that all values are depths below the datum, in which case higher numerical designations signify lower elevations.

3. The contour interval is the vertical distance between contours. Most maps are drawn with a single contour interval, but some may have variable intervals reflecting variation in the amount of control available. When variation in contour interval is used, additional contours are inserted between the larger interval used over the entire map, and the additional contours may end abruptly. Structure contours on the Tectonic Map of the United States are drawn on different contacts in different parts of the country. The regions contoured on a certain contact are separated by dotted lines along which the stratigraphic horizons used are indicated. In a few areas structure contours are drawn on two different stratigraphic horizons.

4. Structure contours can cross themselves and other contours. Contours merge or come close together on the steep limb of an *asymmetrical fold*. If the fold is *overturned* the contours merge and cross on the overturned limb, and they cross for any *recumbent fold*. Contours on the surface of salt structures may merge and cross at the overhanging edge of a salt dome. Contours on a bed cut by a reverse fault also cross if contours are shown on both blocks.

5. Spacing between contours is a function of the inclination of the surface on which they are drawn, the contour interval selected, and the scale of the map. True and apparent dips can be measured from structure contour maps. At any given point a tangent to a structure contour is the strike of the contact. True dip must be measured at right angles to that tangent. Apparent dips can be determined in any direction. The dip may be found graphically by using the vertical and horizontal distance between two contours in the selected direction, reading the difference in elevation, and solving the right triangle.

6. Special notations are placed along some contours to aid in interpretation. Every fifth contour is sometimes made heavier to make reading more rapid. Closed depressions are indicated by hachures on the contours.

7. When the contoured surface is faulted, the fault should be indicated. A gap is produced in the contoured pattern for normal faults. Reverse faults produce overlapping contour patterns. The fault may also be shown by contours (made heavier or lighter to allow distinction from the others).

Structure Contours on a Plane Dipping Surface

An inclined plane surface is defined by its *strike* and *dip*. The intersection of the inclined plane in any horizontal surface is the strike of the plane and is shown by a straight line with a specific compass direction, Fig. B-1. The dip of the inclined plane can be viewed in a cross section drawn at right angles to the strike. If the vertical and horizontal scales are the same, the true dip will appear undistorted. One way

FIGURE B-1 Structure contour map for a plane dipping formation top based on elevation data obtained in five wells (circled points).

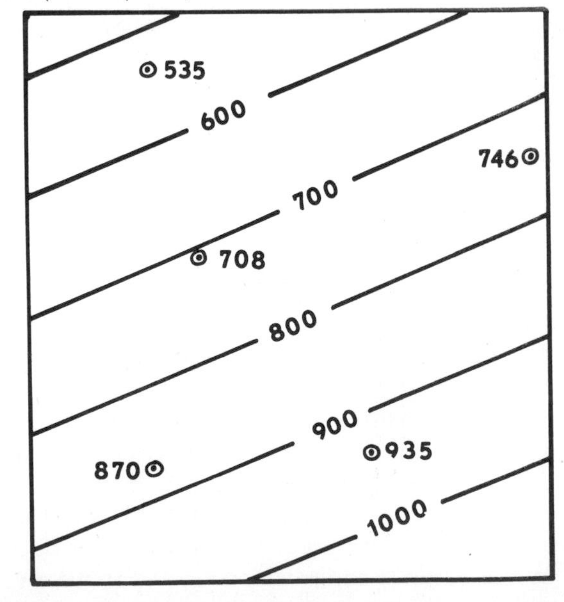

of representing this is illustrated in Fig. B-2. The cross section is drawn along a line which is perpendicular to the strike and in the direction of the dip. The horizon on which the strike is shown appears as a line in the plane of the cross section. Any number of other horizontal planes at different elevations could be represented in this section. A set of horizons can be selected at a fixed interval of vertical separation, the *contour interval.* The point where each of these planes crosses the inclined contact in the section marks the position of that structure contour. The projection of these points of intersection to the original horizon indicates where structure contours will appear. All structure contours are parallel to the contact line since they are all strike lines.

Finding the Outcrop Pattern of a Dipping Surface

The outcrop pattern of a plane-dipping contact in an irregular topography can be determined if a topographic map is available, the strike and dip are known, and at least one outcrop of the contact is known. Draw a strike line through the known outcrop to the edge of the map. Prepare a cross section using the same vertical and horizontal scale. Align the cross section along a line perpendicular to the strike and so that the strike line intersects the cross section at the point in that section which has the same elevation as the outcrop. Now any point along the original strike line which has the same elevation as the outcrop will also

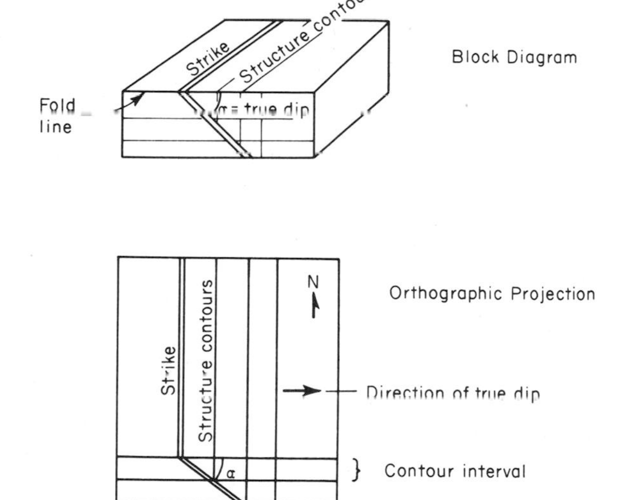

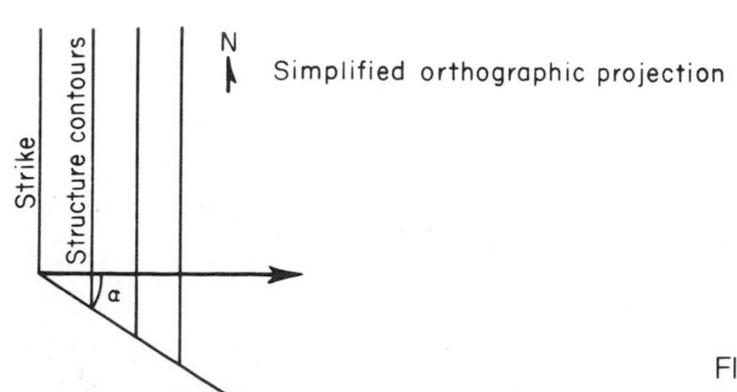

FIGURE B-2 Block diagram of a dipping stratum and an orthographic projection of the same.

contain the contact. Additional structure contours can be constructed for as many elevations as are needed, and points falling along those contours, which have the same topographic elevation as the structure contour, will contain contacts.

Preparation of a Structure Contour Map

The method used to prepare a structure contour map depends on the amount and type of information available. The contact to be mapped may or may not outcrop at the ground surface. Topographic and surface geologic maps may be available. Strike and dip information may be obtained from surface outcrops or from wells which have penetrated the contact. Good subsurface control may be available from wells which penetrate the contact. Seismic lines showing profiles of a contact or adjacent contacts may give control. For reconnaissance purposes, form-line maps based entirely on photogeologic studies may be constructed. A good mapping horizon is selected and traced, strikes and dip on it are estimated, and form lines are constructed. Badgley (1959) outlines methods used in preparing maps when various types of data are available:

1. When both topographic and geologic maps are available, the elevation of the contact may be read directly by placing a transparent overlay of the contact on the topographic map. Each point where the elevation is determined becomes a control point.

2. Strike and dip information on the contact may be projected for some distance on either side. The assumption is made that the dip remains relatively constant. This assumption is usually relatively safe if dips are known in overlying and underlying formations and they remain constant. On large *monoclines*, *homoclines*, and other broad flexures, this assumption may be safe for great distances. If the elevation of the contact, the dip at the contact, the scale of the map, and the contour interval are known, a set of multiple dividers or a ruler may be used to lay off a series of points above and below the contact through which structure contours will pass if the dip is constant.

3. Many structure contour maps are drawn on the basis of well logs. Lithologic, electric, and radioactivity logs all may be used to pick formational contacts. These logs show the depth of the contact below the ground surface. In plotting the data, all readings must first be reduced to the elevation above or below the selected datum. Once the figures are reduced and plotted, contours may be drawn.

4. Seismic data may be extremely valuable. Older seismic surveys provided spot determinations of depth and dip, but some of the modern methods allow construction of continuous profiles. In either case the data are reduced to the datum level in use and are plotted on the map as a series of control points.

Drawing Structure Contours

Structure contours are drawn on the basis of elevation data on a given stratigraphic boundary. The data points are spread, usually unevenly, over an area, and from this array of points at which the elevation of a chosen marker is known, geologists try to prepare a map which depicts the shape of the marker. Any set of numerical values can be contoured in a great many different ways, and even the most experienced geologist may change initial contours many times as he or she begins to envision the structure represented by the data points. Just as streams help guide topographic contouring, the use of knowledge of the shapes of various known structural features help guide structure contouring. For example, it may be possible to pick and use as a guide the crest or trough of a fold, a fault, reefs, or salt structures. Unless control suggests otherwise, contours should be drawn as smooth, easy flowing lines, not as intricate patterns. Similarly, contour spacing should be even, and changes in spacing should appear as gradual changes rather than sudden breaks, unless control dictates abrupt change (this may be indicative of a fault).

A structure contour map drawn to represent a plane, uniformly dipping surface consists of a set of parallel evenly spaced lines, Fig. B-1. The closer the lines are together, the steeper the plane dips. The strike of the plane is everywhere tangent to the contour lines. An example of a plunging fold is illustrated, Fig. B-3.

CROSS SECTIONS

The construction of a cross section is only partially a straightforward mechanical process. The initial steps are easily set forth, but projection of contacts at depth requires decisions concerning the geometry of the fold form, continuity of beds, uniformity of thickness, and other factors which may or may not be well known. The initial step is preparation of a profile of the ground surface along the line of the cross section. If the cross section is not to be exaggerated, the same vertical scale is used as the horizontal scale of the map. Vertical exaggeration should be used only when dips are low or when the section is long. The profile is drawn by laying a piece of graph paper with a labeled vertical scale along the line. Where contours cross the edge of the graph paper, the elevations are read and plotted. Enough points must be taken to make it possible to draw a smooth line.

Second, place the contacts of formations in their proper position along the profile. If the cross section is drawn at right angles to the strike of the formations, the dip of each unit can be laid off at the point of the contact and drawn as a short line. If the cross section is not drawn perpendicular to the strike, the true dips given must be converted to apparent dips (this can be done by use of orthographic or sterographic projections and by means of prepared nomographs, i.e., U.S. Geological Survey Professional Paper 120-G by H. S. Palmer). The apparent dip is then drawn below the profile from the point of contact.

From this point, cross sections lacking subsurface control must be drawn on the basis of some assumption regarding the style of deformation and its extension at depth. There is little difficulty when it is reasonable to assume that formational contacts or existing faults may be represented by a straight line, but when folding is involved, the geometry of the folding must be judged and drawn accordingly. If the section includes igneous or metamorphic rocks, they must be drawn according to some assumption regarding their shape. In the absence of control, hypothetical boundaries of such bodies are used. The shape of these is based on previous experiences or known similar features. One of the most complete descriptions of cross-section preparation is found in Busk (1929). Other valuable sources are Badgley (1959), Ragan (1973), Dennison (1968), and Donn and Shimer (1958).

PARALLEL FOLDING

A geometrical method is available for use when it can be assumed that the formations within the region of the cross section were originally uniform in thickness and remained uniform in thickness during folding. This method is reasonably accurate in the case of flexural (concentric) folding. The procedure following construction of the profile, location of contacts, and insertion of true or apparent dip line along the profile is as follows:

1. Perpendiculars are erected to each dip symbol plotted in cross section along the profile. These are drawn long enough to intersect the lines drawn from adjacent dips.

2. Using the points of intersection of adjacent lines, arcs are drawn with a compass. These arcs are drawn through the contacts and between the lines connecting the two dips with the point of intersection.

3. The compass is then moved to the next point of intersection and the contacts are continued using the new radius for the arc. In this fashion the contacts are drawn across the cross section (Fig. B-4).

FIGURE B-3 A block diagram (top) of a plunging anticline and syncline. Block diagram (middle) showing the contoured horizon in perspective. Structure contour map (bottom). Note thickness and depth measurements are shown at top. (*Redrawn after Durán, 1951.*)

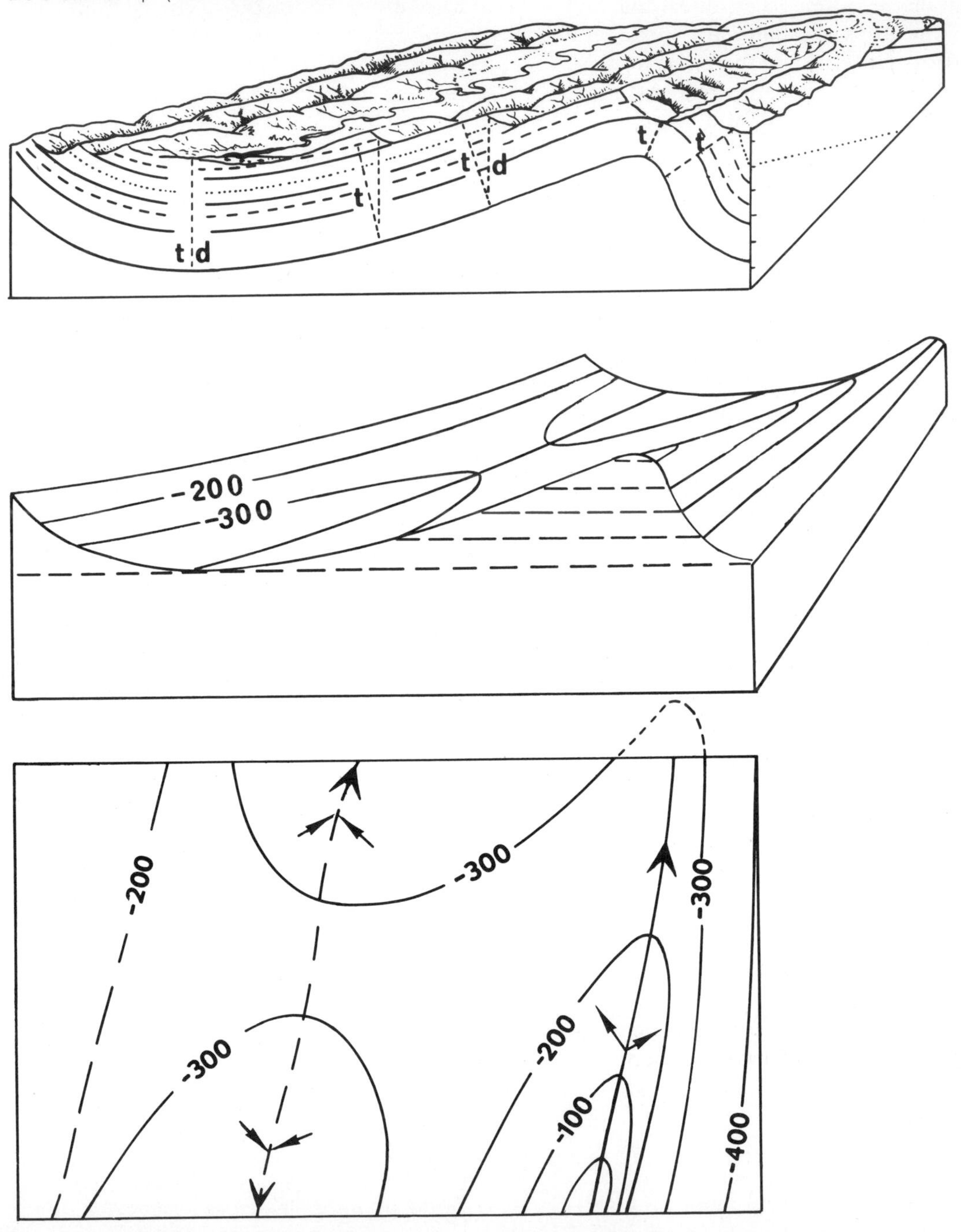

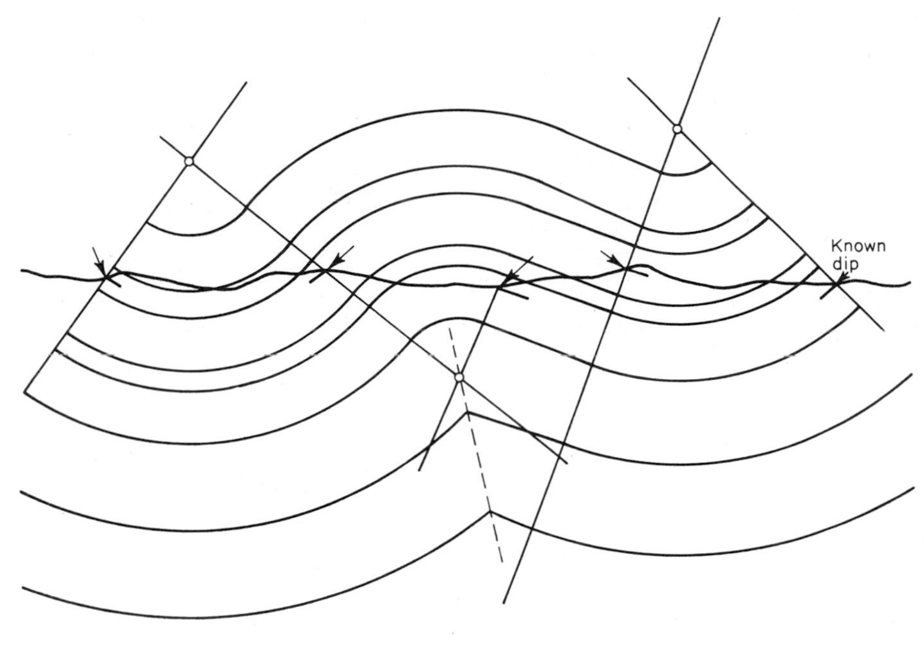

FIGURE B-4 Construction of parallel folds using dip data shown along the ground surface. Normals are drawn to adjacent dips, and the point where these cross is used as the center of circular arcs representing contacts between the normals.

Folding dies out with depth, and the shape of the folds changes both upward and downward in parallel folding.

COMPACTION FOLDING

When compaction is the principal process responsible for formation of folds, the *boundary-ray method* may be used. This method is based on the observation that the amount of compaction within a formation at a given point is a function of the dip at that point. A method for drawing cross sections through compacted structures is given in Coates (1945), Gill (1953), and Badgley (1959).

COMBINATIONS OF FOLD TYPES

Various combinations of flexure and flow are found. The arc method of cross-section construction described above will not produce a true representation of the structure when beds are not uniform in thickness. How far off it will be depends on the amount of thickening and thinning. Few geometrical construction techniques have been devised that can be applied in the same mechanical way the arc method is used.* Instead the geologist must look for field evidence that suggests the nature of the fold and must then try to interpret that data in a cross section that is consistent with the data. This is done freehand or by a combination of the arc method and freehand drawing. In stratified sequences consisting of massive units interbedded with weaker materials such as shale, the shale may be expected to behave differently from the more massive formations.

* A numerical technique is described by Phillips and Byrne (1969).

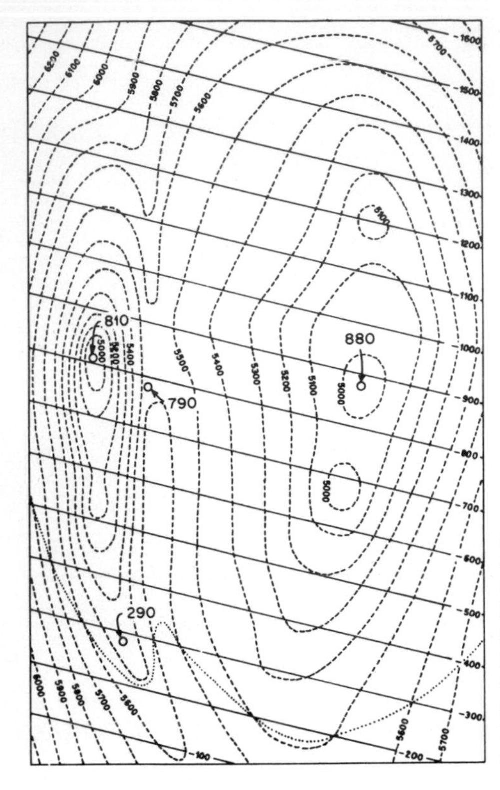

PROBLEM 1 Given a structure contour map, dotted lines, on a marker horizon. Four holes have been drilled (at circles) and the depth to the top of the next marker horizon is shown. Assume a regional uniform change in the thickness and draw a structure contour map on the second marker. (Procedure: Determine contours showing change of thickness at 100-ft intervals; superpose these on the structure contour map; determine the elevation on the second marker at each point where the two sets of contours cross; contour the second marker.)

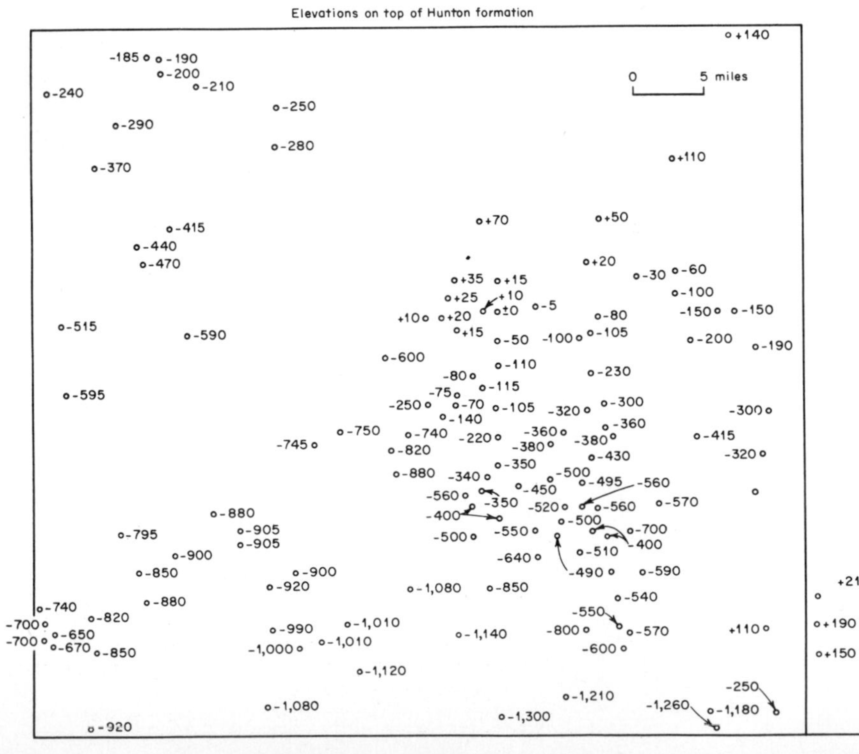

PROBLEM 2 Given a map of parts of De Witt and McLean Counties, Ill., showing subsurface elevations on the Hunton formation. Draw a structure contour map of the Hunton and describe the structure. (*After Ill. Geol. Survey Circ. 377.*)

R10E R11E A′

◎ Producing well
○ Faulted out
∘ Top of *P*

R10E R11E A′ A

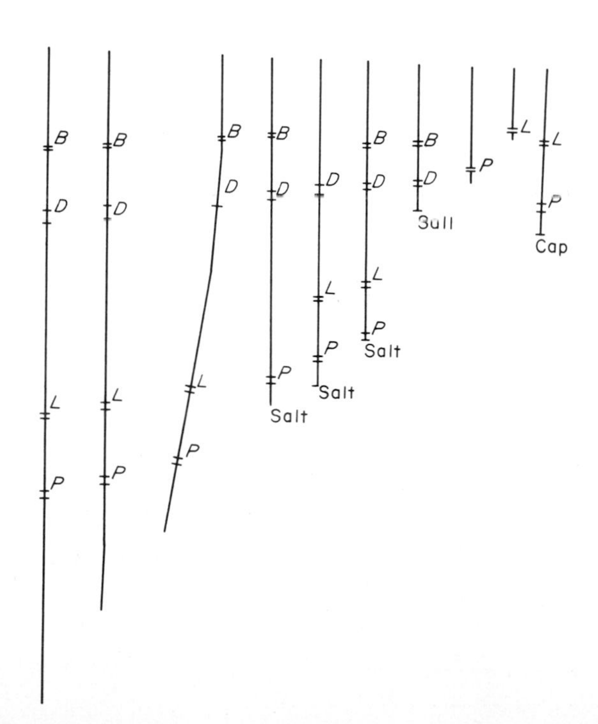

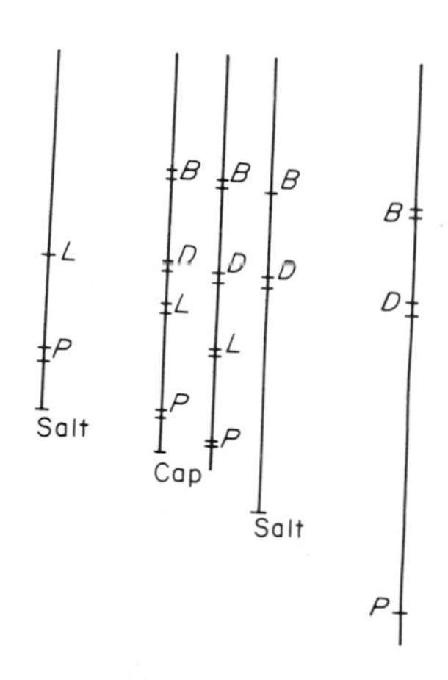

PROBLEM 3 Given elevations on top of the salt in the Bayou Blue salt dome (right), elevations and location of producing wells on top of the P horizon (left), and a cross section showing wells and the levels in each at which marker horizons have been picked. Draw structure contours on the salt and on the P horizon. Then complete the cross section, indicating the positions of faults. Describe the mechanics of intrusion of the salt and the origin of the structure over the dome. (*From Mais, 1957.*)

PROBLEM 4 The wells from the Cement-Chickasha area of Oklahoma are shown on a base map. Elevations in each well are given for the top of the Hoxbar, Deese, Atoka, and Morrow formations, all of Pennsylvanian age. All elevations are negative and taken with reference to sea level. (*a*) Draw a structure contour map for each horizon. (*b*) Determine isochore maps for the Hoxbar, Deese, and Atoka formations. (*c*) Draw cross sections across the structure along the lines A–A′ and B–B′.

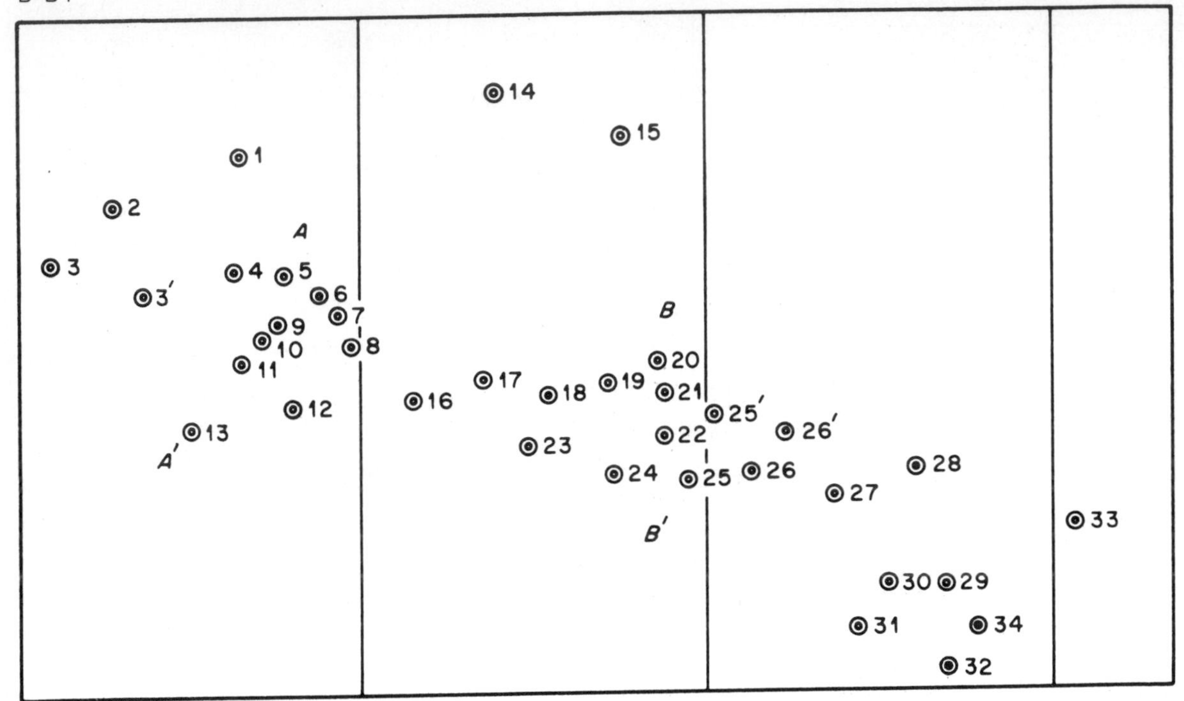

No.*	Hoxbar	Deese	Atoka	Morrow
1	6,794	10,000		
2	5,588	8,983		
3	4,444	6,964		
4	3,910	Breccia	Breccia	
5	5,119	Breccia	Breccia	
6		4,573	Breccia	12,950
7	Breccia	Breccia	6,500	
8	Breccia	4,292		
9	Breccia	Breccia		8,600
10	Breccia	Breccia	9,272	11,100
11	3,615	6,560	10,400	12,350
12		7,415		
13	5,596	8,876	11,746	13,316
14	6,848	10,358		
15	6,803	10,390		
16	3,492	5,025		
17	3,286	5,023		
18	2,887	5,117	8,209	9,299
19	3,409	Breccia		
20	6,740	11,103	12,173	12,583
21	2,260	Breccia	9,500	9,900
22	2,100	3,282		4,600
23	3,545	6,088		
24	2,960	6,276		
25	2,920	5,410	7,400	8,300
26	2,769	4,555		
27	2,480		6,200, est.	4,452
28	6,945	Breccia		Breccia
29		4,200 est.	4,740	5,400
30	2,156	4,304	6,800	7,800
31	3,856	7,151	10,400	11,600
32		3,639	6,000	7,000
33	6,959	10,594		
34	2,900		10,000	12,000
26′	5,993	Breccia		Breccia
3′	3,402	5,900		
25′	4,224	Breccia		Breccia

*All values negative.
Source: Herrmann (1961).

REFERENCES

Badgley, P. C., 1959, Structural methods for the exploration geologist: New York, Harper & Row.

Busk, H. G., 1929, Earth flexures: Cambridge, Cambridge Univ.

Coates, J., 1945, The construction of geological sections: Geol. Mining Metall. Soc. India Quart. Jour., v. 17.

Compton, R. R., 1962, Manual of field geology: New York, Wiley.

Crowell, J. C., 1948, Template for spacing structure contours: Am. Assoc. Petroleum Geologists Bull., v. 32, p. 2290–2294.

Dennison, J. M., 1968, Analysis of geologic structures: New York, Norton.

Donn, W. L., and **Shimer, J. A.**, 1958, Graphic methods in structural geology: New York, Appleton-Century-Crofts.

Durán S. L. G., 1948, Structural contour maps; "Planificando estratos discordantes": Petroleo del Mundo, New York. (Unconformable or converging beds)

——— 1951, Structural contour maps; "Trigonometric and graphic solution of problems in structural mapping": World Oil. (General case; conformable beds)

Gabriel, V. G., and **Dotson, J. C.**, 1953, The use of V-concept in structural geology: Am. Geophys. Union Trans., v. 34, no. 6.

Gill, W. D., 1953, Construction of geological sections of folds with steep limb attenuation: Am. Assoc. Petroleum Geologists Bull., v. 37, p. 2389–2406.

Harrington, J. W., 1951, The elementary theory of subsurface structural contouring: Am. Geophys. Union Trans., v. 32, p. 77–80.

Herrmann, L. A., 1961, Structural geology of Cement-Chickasha area, Caddo and Grady Counties, Oklahoma: Am. Assoc. Petroleum Geologists Bull., v. 45, no. 12.

Howard, Richard, 1965, Niagaran reef dolomites DeWitt-McLean Co., Illinois: Illinois Geol. Survey Circ. 377, App. 1.

Iglehart, Charles F., 1970, Descriptive classification of subsurface correlative tops: Amer. Assoc. Petroleum Geologists Bull., v. 54, no. 9, p. 1697–1705.

Lahee, F. H., 1961, Field geology, 6th ed.: New York, McGraw-Hill.

LeRoy, L. W., 1950, Subsurface geologic methods: Golden, Colorado, Colorado School of Mines.

Low, J. W., 1957, Geologic field methods: New York, Harper & Row.

Mackin, J. W., 1950, The down-structure method of viewing geologic maps: Jour. Geology, v. 58, p. 55–72.

Mais, W. R., 1957, Peripheral faulting at Bayou Blue salt dome, Iberville Parish, Louisiana: Am. Assoc. Petroleum Geologists Bull., v. 41, p. 1915–1951.

Miller, V. C., 1950, Rapid dip estimation in photogeological reconnaissance: Am. Assoc. Petroleum Geologists Bull., v. 34, p. 1739–1743.

Palmer, H. S., 1918, New graphic method for determining the depth and thickness of strata and the projection of dip: U.S. Geol. Survey Prof. Paper 120.

Phillips, W. E. A., and **Byrne, J. G.**, 1969. The construction of sections in areas of highly deformed rocks, *in* Kent, P. E., and others, eds., Time and place in orogeny: Geol. Soc. London, Spec. Publ., no. 3.

Pierce, W. G., and others, 1947, Structure contour map of the Big Horn basin, Wyoming and Montana: U.S. Geol. Survey Oil and Gas Inv. Map 74.

Ragan, D. M., 1973, Structural geology, an introduction to geometrical techniques, 2d ed.: New York, Wiley.

Reeves, R. G., 1969, Structural geologic interpretations from radar imagery: Geol. Soc. America Bull., v. 80, no. 11, p. 2159–2164.

Satin, L. R., 1960, Apparent-dip computer: Geol. Soc. America Bull., v. 71, no. 2, p. 231–234.

Shearer, E. M., 1957, Stereo-structural contouring: Am. Assoc. Petroleum Geologists Bull., v. 41, p. 1694–1703.

Simpson, Brian, 1960, Geological map exercises: London, Philip.

Stockwell, C. H., 1950, The use of plunge in the construction of cross-sections of folds: Geol. Assoc. Canada Proc., v. 3, p. 97–121.

Sugden, W., 1962, Structural analysis, and geometrical prediction for change of form with depth, of some Arabian Plains-type folds: Am. Assoc. Petroleum Geologists Bull., v. 46, p. 2213–2228.

C

Orthographic Projection

The *orthographic projection* is a means of representing regular geometrical shapes (i.e., a dipping surface, an inclined line) on a plane (a piece of paper) in such a way that their dimensions and angular relations are not distorted. This is done by folding vertical planes containing the features into the plane of the projection which is horizontal. The perspective of block diagrams is lost, but the projection is drawn to a consistent scale.

Many structural problems consist of determining relationships among lines, points, and planes. Many formational contacts, fractures, faults, unconformities, dikes, and sills can be approximated by planes. The intersection of any two of these features forms a line. Movements within fault zones are often translational and along well-defined lines of movement. When structures can be approximated by lines and planes, problems involving angles or distances among various elements can readily be solved by use of the orthographic projection.

Vertical or inclined features are projected in much the same way a cardboard box may be flattened, by cutting all edges and folding the sides and any interior partitions down. The lines along which the folding is done are called *fold lines* (Fig. C-1). In the following examples some common problems encountered in geology are solved by orthographic projections. Orthographic projections are useful when:

1. The elements of the structure can be approximated by lines and planes.
2. Quantitative distance measurements are required. (The stereographic projection offers an easier way to solve problems involving angular relations only.)
3. A true scale model is required.

The plane of the projection is almost always a horizontal surface simulating the flat ground surface, a level in a mine, or a given level through a structure. Because most geologic problems involve features with strike or bearing, the projection surface approximates a map and should be oriented with respect to a compass. North should always be labeled, and all lines actually lying within the plane of the projection should be laid off with their strike direction or bearing. A scale must be selected which will allow a convenient working-size model.

Orthographic Projection of a Single Plane

Any horizontal plane or line lies within the plane of the projection. It is possible to project a line from within some other plane into the plane of the projection. Going back to the model of the cardboard box and assuming that the bottom of the box is the plane of the projection, the top edges of the box all lie within a plane parallel to the bottom but at a distance above the bottom which is equal to the height of the box. These edges can be placed within the bottom plane by folding the sides of the box down (Fig. C-1). Fold lines should be shown on the plane of the projection, and vertical and horizontal scales are the same.

A vertical plane shows up within the plane of the projection as a straight line. Any portion of the vertical plane can be rotated into the plane of the projection by using this straight line as a fold line.

The general case is that of an inclined plane —a dipping contact, for example. This plane or contact appears in the projection as a straight line (the strike line of the contact). The direction of dip may be shown by a line drawn perpendicular to the strike line in the direction of dip. The amount of dip is shown by folding the vertical plane which contains a right-angle cross section of the dipping plane into the plane of the projection. The fold line in this case is the line showing the direction of dip. Use of any other line would have shown an *apparent dip* which would have been less than the true dip. The inclined surface is completely defined by the strike line, the line showing direction of dip, and the sectional view of the dipping surface folding into the plane of the projection. It is often desirable to show, in addition, structure contours or some particular structure contour. A contour at some given depth below the level of the projection can be found by reference to the cross section. Be-

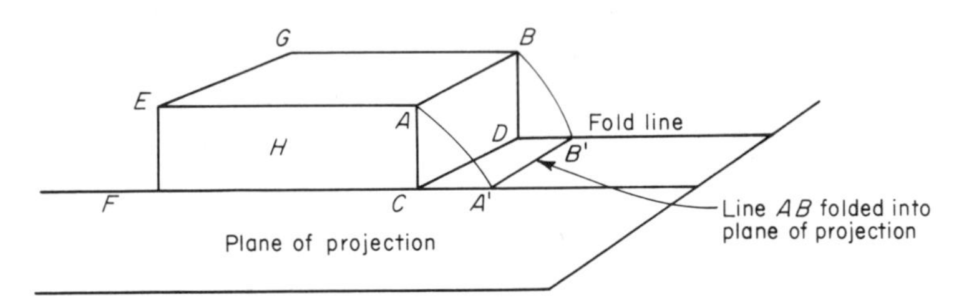

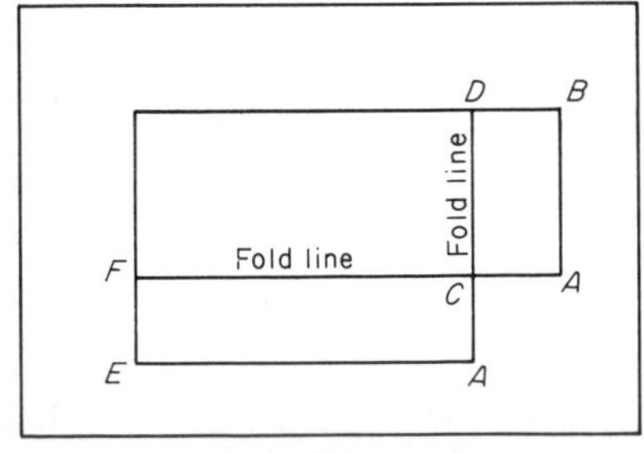

FIGURE C-1 Representation of a block by orthographic projection.

cause this is a vertical distance, it is measured perpendicularly from the fold line down to the projected dipping line. The structure contour for any depth is parallel to the strike line (Fig. B-2).

Finding an Apparent Dip

When the true strike and dip are given, the plane can be shown on the projection. To find the angle of apparent dip along some given line, the bearing of this line is drawn on the projection (if this line is not specified as to location it may be drawn anywhere so that it crosses the strike line of the inclined surface). Since the apparent dip of the inclined plane could now be seen if a section were cut along this new line, the line in the direction of the apparent dip is used as a fold line. One or more structure contours are drawn. The depth to the inclined surface below any point where one of these structure contours crosses the fold line of the apparent dip is known. This depth, drawn to scale, is laid off perpendicular to the line along which the apparent dip is to be found. Then a line representing the cross section of the inclined plane is drawn from this depth back to the point of intersection of the strike line and the line along which the apparent dip is desired. The apparent angle of dip can then be measured directly, Fig. C-2.

Finding True Strike and Dip from Apparent Dips

Situations may arise in which apparent dips can be measured but true strike and dip are not known. The lines along which apparent

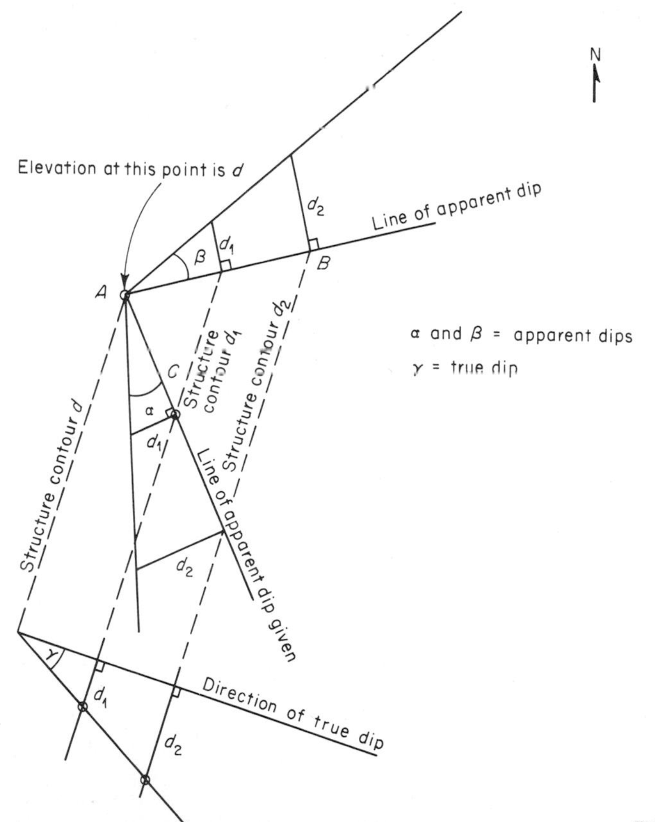

FIGURE C-2 Apparent dip determination.

dips are known are drawn from a point, Fig. C-2, and the apparent dips are drawn along these lines, using them as fold lines. Structure contours are then constructed for any depths, and these indicate the strike direction. The true dip is then found by constructing a cross section perpendicular to the strike direction.

It is also possible to find the true dip when the strike and one apparent dip are given. The method involved is a slight modification of that outlined above. The main difference is that it is unnecessary to find the strike.

Thickness Determination with Orthographic Projections

If a unit is plate-like, dips vertically, and outcrops on a flat surface, its thickness can be measured directly on the ground surface along a line perpendicular to the upper and lower contacts. If such a unit is horizontal, the thickness can be found by determining the difference between the elevation of the upper and lower contacts. More often, however, the ground surface is sloping, and the unit is dipping.

CASE I. GROUND HORIZONTAL, UNIT DIPPING

The thickness and width of outcrop form two sides of a right triangle in which one angle is the angle of dip (Fig. C-3). The orthographic projection may also be used in this case. A fold line is drawn perpendicular to the two contacts in the plane of the projection. A cross-sectional view is then drawn by laying off the dips of the two contacts and extending them. If the dips of the two contacts are not identical, the wedge shape of the unit is reflected in differences in thickness. (*Note:* In the case of a wedge, the orthographic projections of struc-

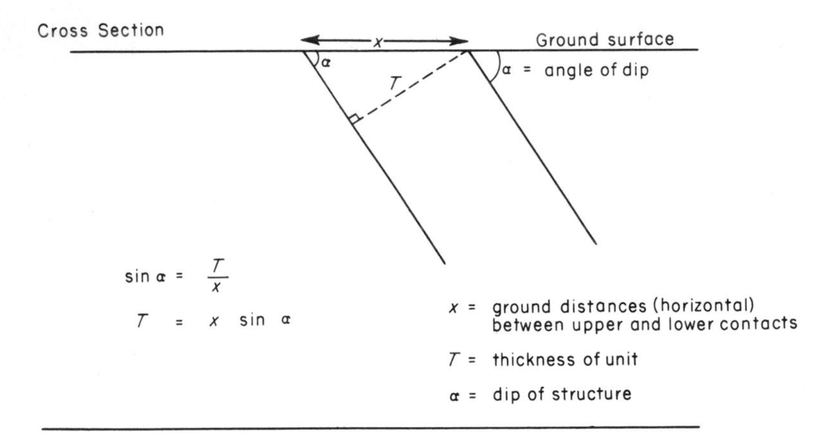

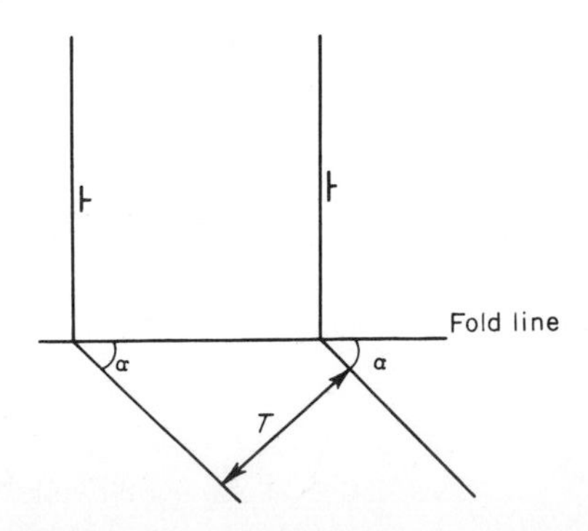

FIGURE C-3 Thickness measurement, case I, in cross section and by orthographic projection.

ture contours on the two contacts are not parallel unless thickness increases directly down dip.)

CASE II. GROUND SURFACE INCLINED

A cross-sectional view perpendicular to the strike of the unit is prepared. This view must show, to scale, the direction and amount of slope of the ground surface as it appears along the fold line (perpendicular to the strike), the location of the contacts along this slope, and the angle of dip of the contacts, Fig. C-4. Once this construction is completed, the thickness can be measured. Apparent thicknesses along any line (e.g., a vertical well hole) can also be measured directly. These thicknesses can also be calculated by using trigonometric relations.

Intersection of Planes; Pitch*

The field geologist is often confronted with problems involving the intersection of two planes. Ore or oil may be concentrated at such intersections, as when mineralizing solutions moving up a dike or fault hit a limestone formation. The position of this line of intersection then has practical importance. The solution of fault problems and the analysis of fracture systems also employ the methods outlined here.

The most general case involves two inclined

Plunge: The plunge of a line is the angle between the line and its horizontal projection measured in a vertical plane.

Pitch: The pitch of a line in a plane is the angle between the line and a horizontal line in the plane measured in the plane.

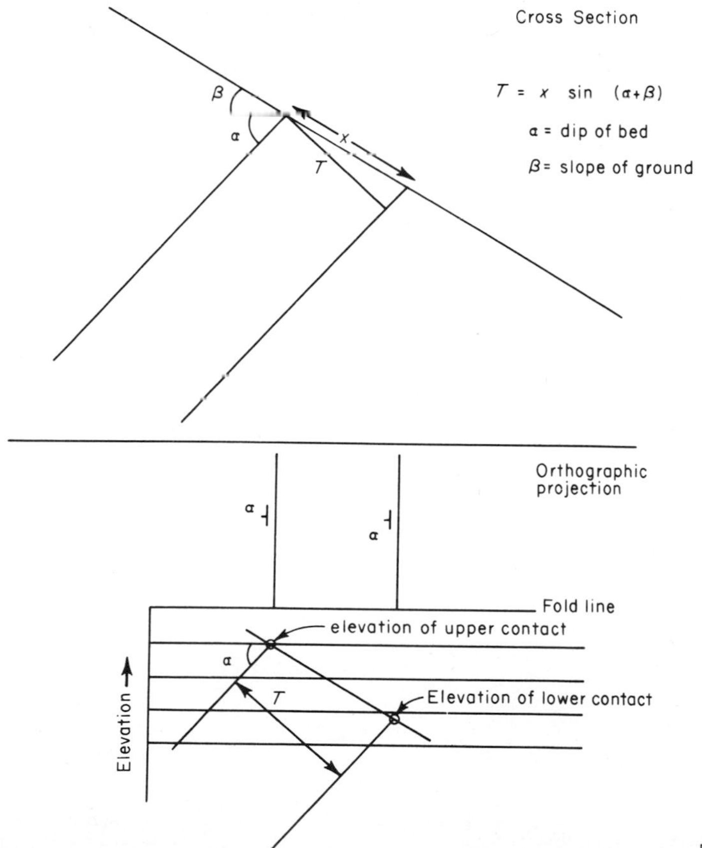

FIGURE C-4 Thickness determination, case II.

planes of different strike and dip. Enough field data must be obtained to determine the strike, dip, and location of the two planes. Once the two have been plotted to scale with orthographic projections of the dips shown, the line of intersection is found by use of constructed structure contours. One point on the line of intersection occurs within the plane of the projection where the two strike lines cross. A second is located by constructing structure contours for the same elevation on the two planes. The horizontal projection of this second point is located where the two structure contours cross. The line drawn through these two points defines the *horizontal projection* of the line of intersection, and its bearing can be measured directly. The *plunge* of this line can then be determined by using it as a fold line, rotating the line of intersection to the plane of the projection. This is accomplished, as before, by measuring or knowing the depth of the second point (where the structure contours crossed) on the line below the horizon and plotting that distance to scale on a line drawn perpendicular to the fold line and through the point on the fold line which lies vertically above the point of known depth (Fig. C-5). A line drawn from this point through the point of intesection of the contacts is the orthographic projection of the line of plunge. The angle of plunge is then measured with a protractor (Fig. C-5).

It is sometimes desirable to measure the angle of *pitch* of a line within a plane. The best way to measure this directly with the orthographic projection is to rotate the plane containing the lineation into a horizontal position so that a protractor can be used. To find out what happens to a line within a plane that is rotated back to the horizon around its strike, project one or two points of known depth on that line into the section which shows the true dip of the plane (use structure contours). When the plane is rotated, these points will be rotated to a position perpendicular to the original strike line at a distance from the strike line

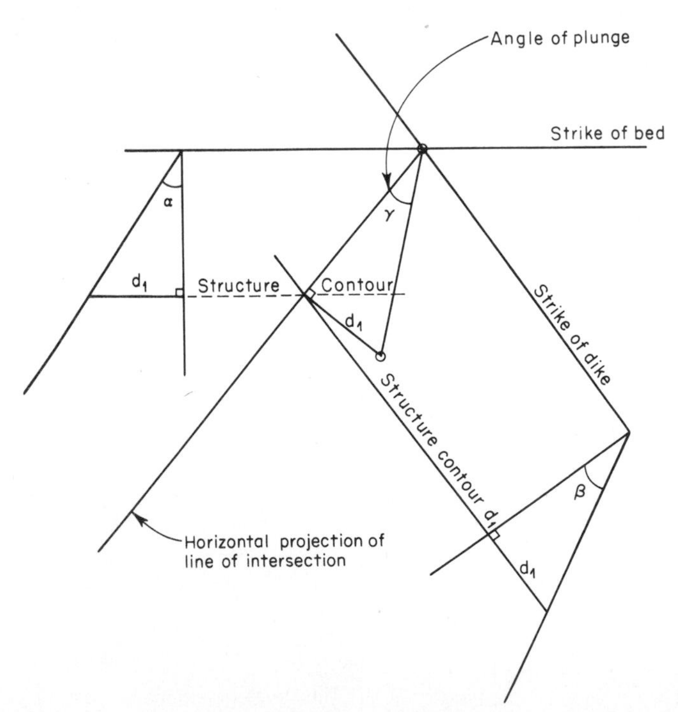

FIGURE C-5 Finding the line of intersection of two planes.

equal to their former distance from the strike line measured in the true dip section (Fig. C-6).

GRAPHICAL SOLUTION OF FAULT PROBLEMS

The orthographic projection is uniquely adapted to the solution of fault problems. Fault problems involving rotations of the blocks on either side of the fault relative to one another are most easily solved by use of orthographic and stereographic projections together. The stereographic projection is used to determine changes in strike and dip after rotation, and the orthographic projection is used to depict the problem to scale and to make distance measurements which cannot be made with the stereographic projection.

Faults with translational movements involve four variable elements:

1. Attitude of the fault
2. Attitude of the beds, assumed to remain constant on both sides of the fault
3. Amount of displacement, which is uniform in all parts of the fault.
4. Direction of movement, which is uniform in all parts of the fault

Problems are most likely to involve finding:

1. The dip of the fault
2. The position of a displaced vein, dike, or contact
3. The direction of displacement
4. The amount of displacement

Care must be taken at the start of any fault problem to consider what information has been given and to be sure that all pertinent information is indicated on the initial drawing of the problem. If the attitude of the fault, a contact, a dike, or a vein is given, it should be represented by orthographic projection on the drawing. It will usually be helpful at the start to draw the trace or traces of any contact in the plane of the fault (e.g., the trace of the contact of a dike, formation, or vein in the fault).

If movements within the fault plane are translational, all points which were originally adjacent on opposite sides of the fault are displaced by the same amount and in the same direction, Figs. C-7 and C-8. If slickensides are present, they indicate the last direction of that movement (show the bearing of the slickensides in the fault plane). Movements in translational faults are strike-slip, dip-slip, or oblique-slip. The amount and direction of

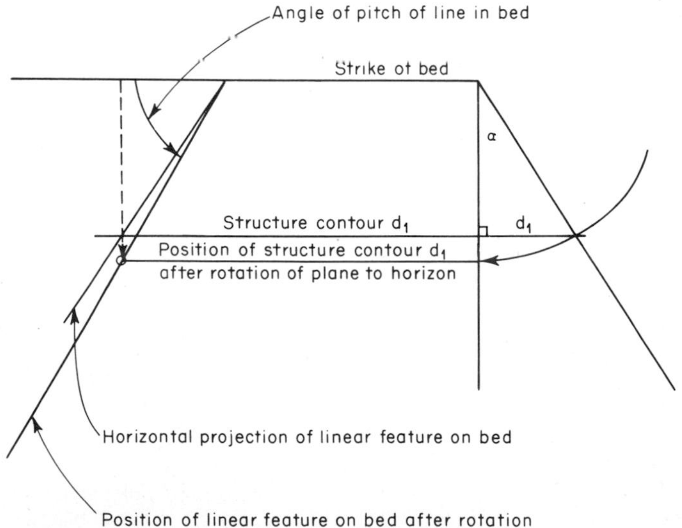

FIGURE C-6 Determining the pitch of a line in a plane.

movement in the fault can be found if any two points which were adjacent before faulting can be identified and located. For example, the intersection of two dikes or a dike and a bed form a line. This line intersects the fault plane at a point. If this line is cut by the fault, the points which were originally together on opposite sides of the fault may be located and their displacement measured.

Tips on the Solution of Translational Fault Problems

1. *Dip-slip movement.* Movement takes place directly down the fault, perpendicular to its strike. The amount of movement can be measured in any section showing the true dip of the fault, Fig. C-9*c*.

Any inclined contacts in the plane of projection which are on the upthrown side of the fault and not parallel to the fault will have migrated in the direc-

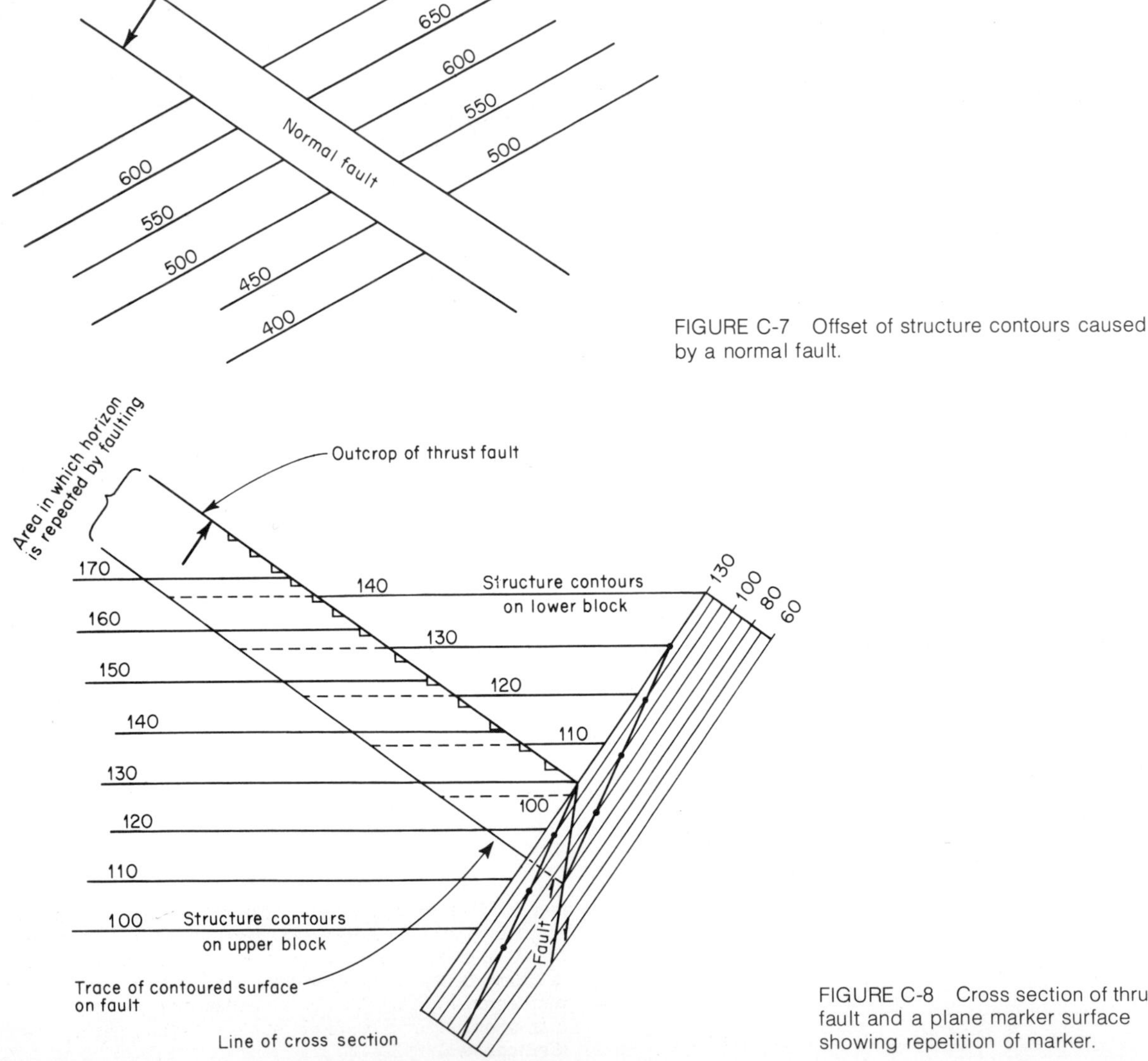

FIGURE C-7 Offset of structure contours caused by a normal fault.

FIGURE C-8 Cross section of thrust fault and a plane marker surface showing repetition of marker.

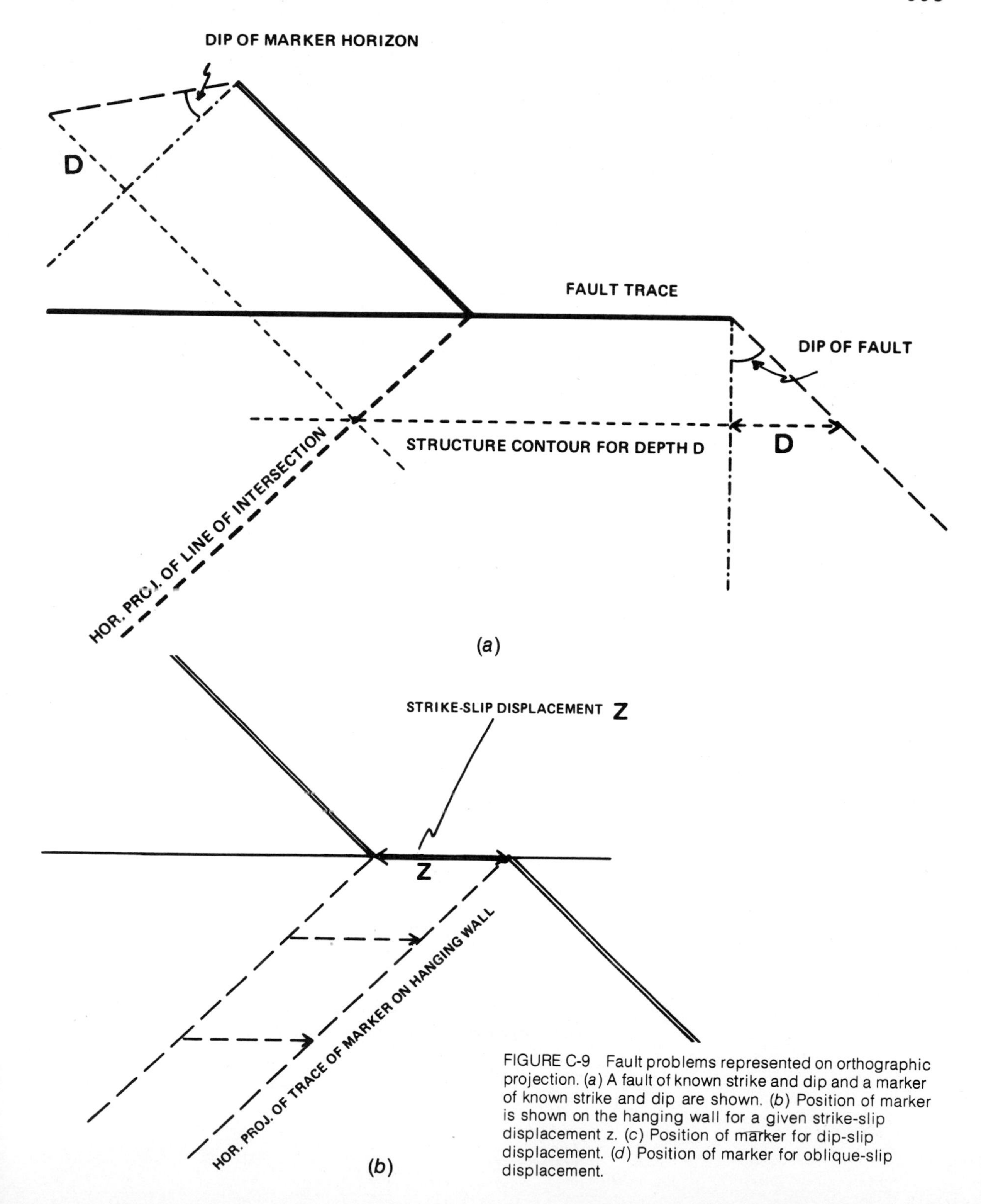

FIGURE C-9 Fault problems represented on orthographic projection. (*a*) A fault of known strike and dip and a marker of known strike and dip are shown. (*b*) Position of marker is shown on the hanging wall for a given strike-slip displacement z. (*c*) Position of marker for dip-slip displacement. (*d*) Position of marker for oblique-slip displacement.

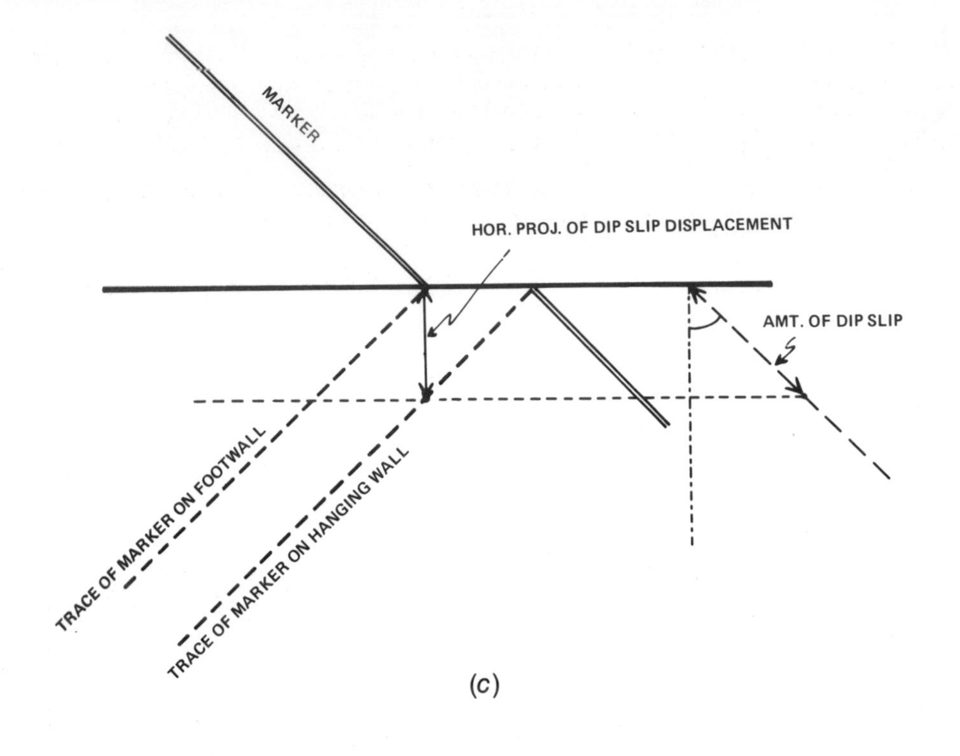

(*c*)

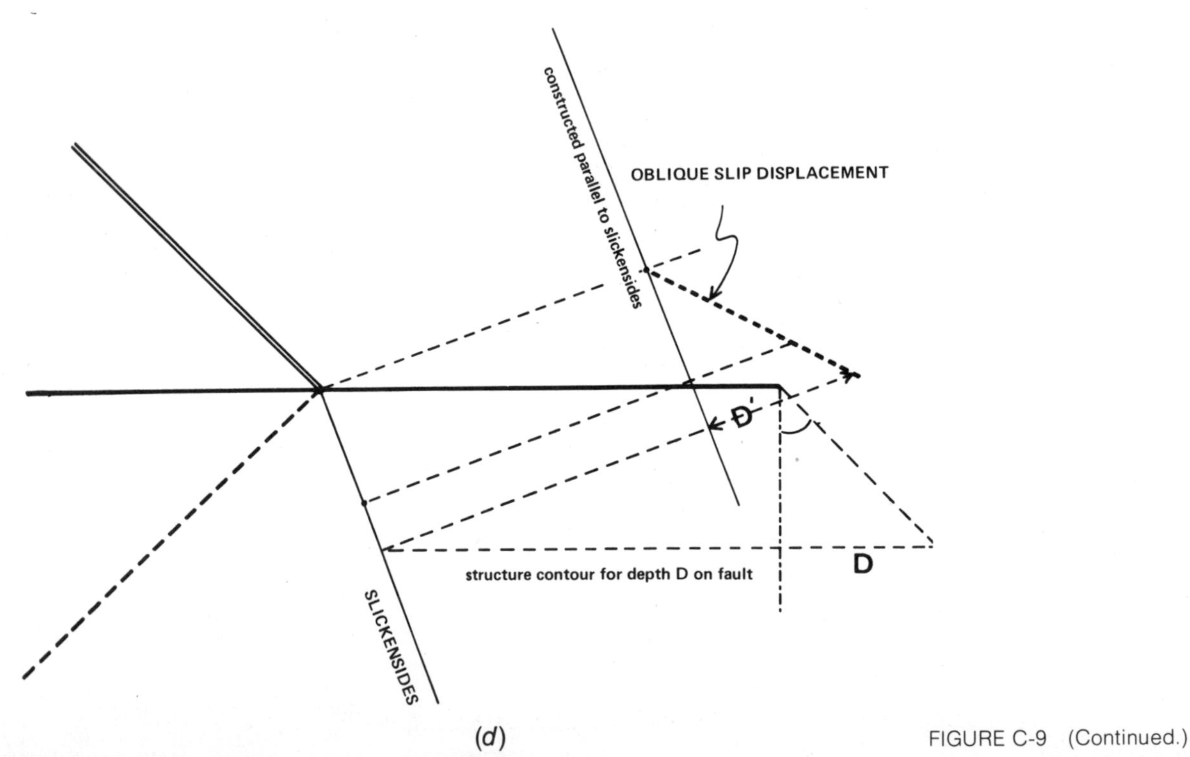

(*d*)

FIGURE C-9 (Continued.)

tion of their dip relative to those on the downthrown side. The lower the dip, the greater the amount of migration. If an intersection of a bed with the fault is known at one point on the surface, and if the trace of the displaced bed on the fault is known, the amount of movement is found by measuring the distance (within the fault plane and along a line down the dip of the fault) between the two.

2. *Strike-slip movement.* Originally adjacent points are displaced laterally, Fig. C-9*b*. Displacement of outcrops in the plane of the projection and along the traces of contacts on the fault (strike separation) is due solely to fault movement.

FIGURE C-10 Set up for a fault problem involving separation of two intersecting markers. Point Y is the point of intersection of the two dikes on the footwall block. Point Z is the intersection of the two dikes on the hanging-wall block. The net slip is the distance from Y to Z measured in the fault plane as shown in the cross section of the net slip.

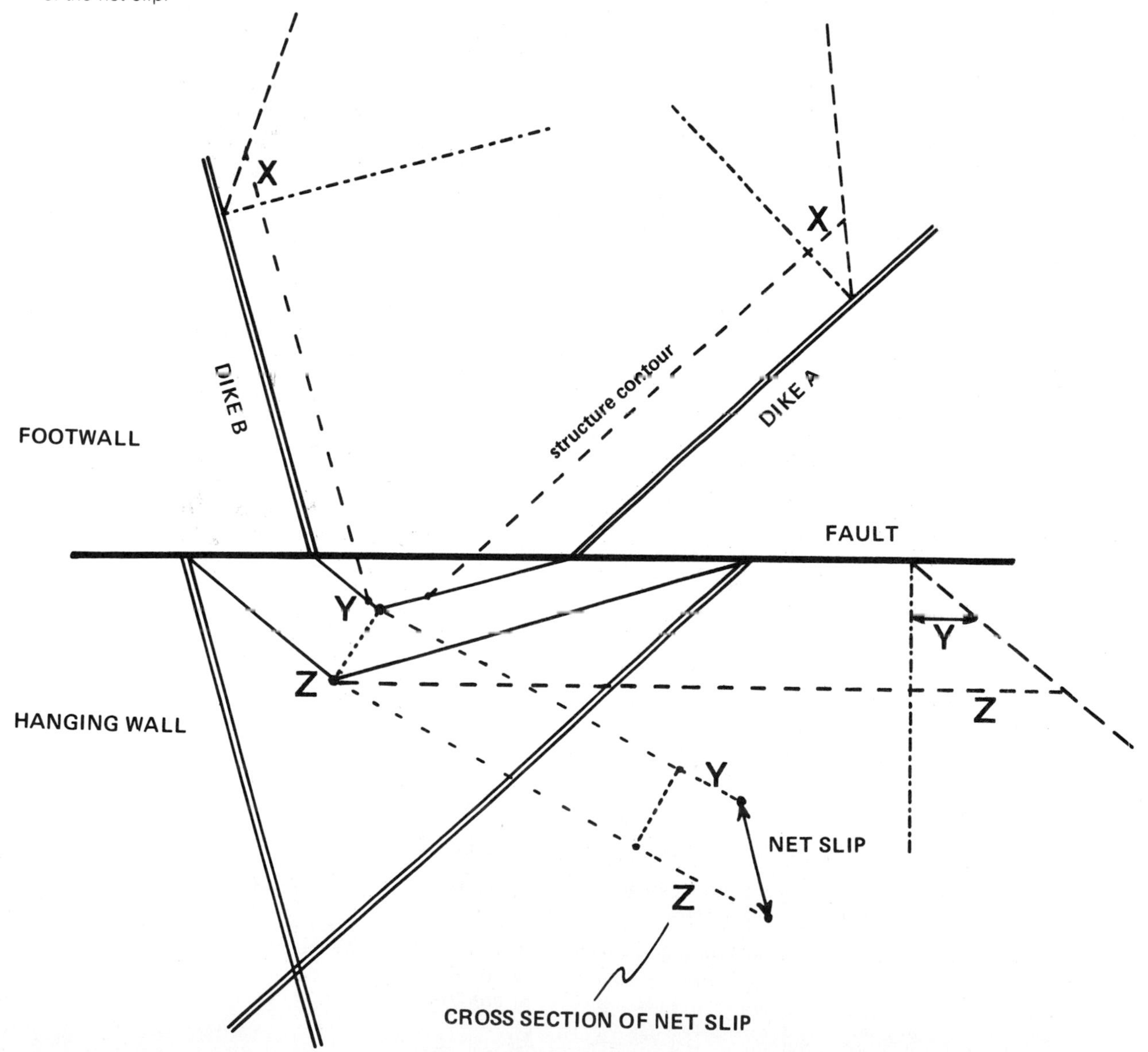

3. *Oblique-slip movement.* The amount of displacement must be measured within the fault plane but not in the section at right angles to the fault, Fig. C-9*d*. It must be measured in a section that includes the horizontal projection of the line of movement (e.g., the horizontal projection of slickensides). If the bearing of this section is known, its angle of plunge can be drawn by using structure contours. The amount of movement may be found from separation of contacts or displacement of points that were originally adjacent. If the problem calls for location of a displaced contact and gives the amount and direction of movement, then it is well to remember that the traces of the contact on either side of the fault are parallel, and the directions of strike and dip of the displaced contact are parallel on either side of the fault.

4. *Faulting of two intersecting planes* (Fig. C-10). This condition arises when two dikes or veins, or a dike or vein and a formation, are faulted. This type of problem can be solved with less than the usual amount of information. The intersection of any three different planes occurs only in a point. Plot all contacts and the traces of these contacts on the fault. A unique point is located where the traces of two contacts on the same side of the fault intersect. If the same thing can be done on the other side of the fault, two originally adjacent points are known and the distance between them can then be measured in a section constructed along the horizontal projection of the line connecting them.

Three-Point Problems

It is possible to determine the strike and dip of any plane if the elevation and location of any three points are known on that plane. The orthographic projection, Fig. C-11, can be used to solve this type of problem as follows:

1. First locate or plot the points on a map.

2. Determine the elevation of the three points (the difference in elevation may be used also).

3. Note the elevation of intermediate value (point *A*). A structure contour (strike line) for this elevation will cross the line between the higher and lower elevations somewhere.

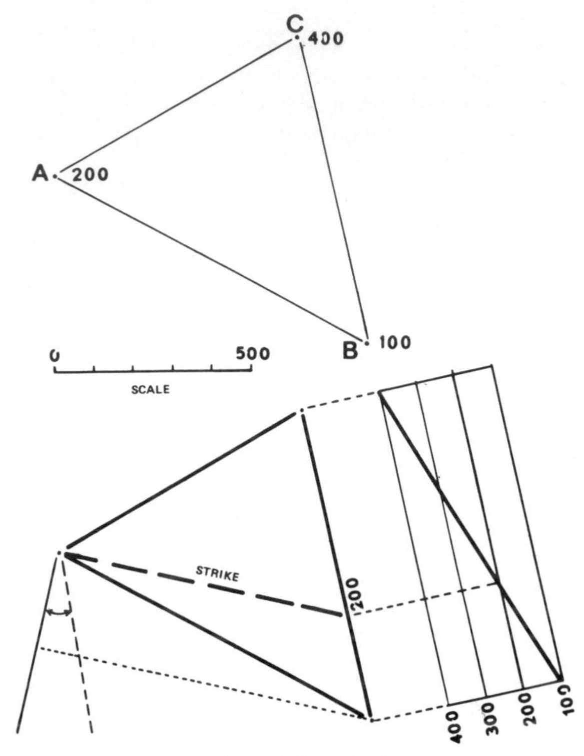

FIGURE C-11 Set up for solution of a three-point problem as described in the text.

4. Draw a cross section along the line between the high and low elevation showing the marker bed.

5. Now find the elevation of the intermediate point (*A*) and its projection along the line *BC*. The strike is now known.

6. The dip is measured normal to this strike line and is found by projecting one other structure contour onto it. (See Fig. B-2.)

EXERCISES

Problem 1. A contact between two rock units strikes N 50° E, dips 40° SE. Find the apparent dip along lines bearing
N 10° E
N 20° W
EW

Problem 2. A contact has a strike of N 20° W. The apparent dip along a line bearing N 10° E is 25°. Find the true dip.

Problem 3. Apparent dips on a contact are determined to be 30° NE along a line bearing N 60° E and 50° NW along a line bearing N 10° W. Find the true strike and dip.

Problem 4. A dike striking N 15° E dips 40° SE and cuts across a sedimentary contact, striking N 40° W, dipping 30° NE. Find:

(*a*) The bearing and plunge of the line of intersection
(*b*) The pitch of the dike in the sedimentary contact
(*c*) The pitch of the sedimentary contact in the dike

Problem 5. An angular unconformity has been tilted so that the erosion surface strikes N 20° E, dips 18° SE. Beds below the unconformity now strike N 30° W, dip 42° NE. Assume that the erosion surface was nearly horizontal when it formed, and determine the strike and dip of beds below the unconformity at that time.

Problem 6. Outcrops of the basal contact of the Tuscarora formation are located at the three places shown on the topographic map, but no strike or dip information was obtained. Assume that the Tuscarora is homoclinal in this region, and construct an orthographic projection of the unit. Determine the strike and dip of the basal contact.

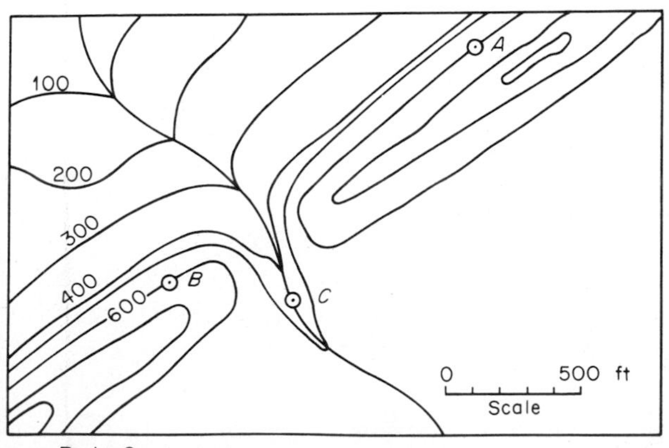

Prob. 6

Problem 7. Three outcrops of the basal contact of the Beekmantown dolomite are shown. Determine the strike and dip of the contact, assuming that the contact is homoclinal in this area. Use the orthographic projection. Check your results using an alternate method.

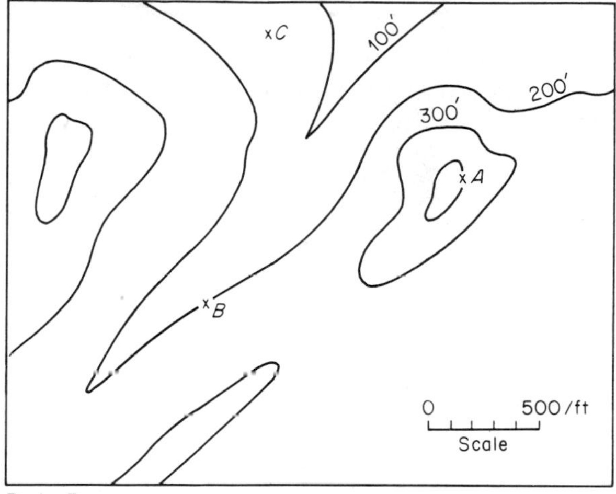

Prob. 7

Problem 8. A dike with strike of N 30° W, dip 25° SW can be traced until it is cut by an east-west-trending fault. South of the fault the dike is found offset 300 m to the east. The fault is exposed in one place, where it dips 60° S, and slickensides on the fault indicate dip-slip movement. There is no indication of rotation during faulting. What is the amount of displacement and the type of fault?

Problem 9. A marker horizon is found to strike N 50° E and dip 65° NW on the north side of an east-west-trending fault which dips 40° S. A well is drilled in the position shown and the top of the marker horizon is found at a depth of 150 m. Locate the outcrop position of the marker on the southern side of the fault trace and determine the amount of strike and

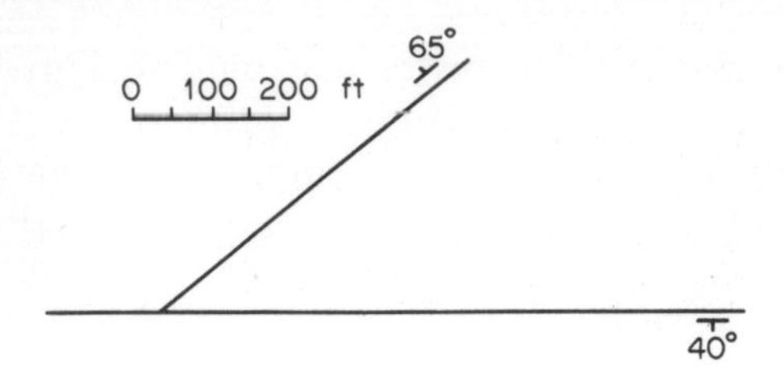

Prob. 9

dip separation. How would you classify this fault?

Problem 10. The strike and dip of two dikes and a fault which offsets them are given on the map. Determine the displacement, strike separation of each dike, dip separation, vertical component of the dip separation of each dike, horizontal component of the dip separation of each dike, and normal separation.

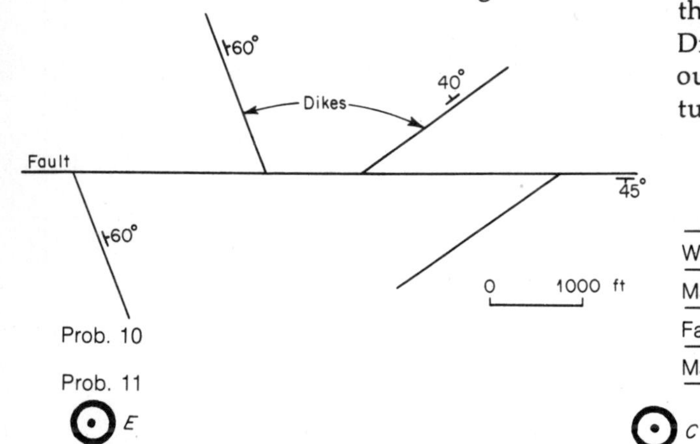

Prob. 10

Problem 11. Given: six drill holes located in the pattern shown. Assume that the ground is flat. The depth to a specific marker horizon is given in each well. Faults are encountered in three of the wells at the depths shown. Make an interpretation of the structure, assuming that the marker horizon has remained planar. Draw a geologic map in this area showing the outcrop of the marker horizon; draw a structure contour map on the marker and a struc-

Well	A, m	B, m	C, m	D ,m	E, m	F, m
Marker	135	60	115	254		104
Fault	295	105			30	
Marker					104	

Prob. 11

E C A

F

ture contour map on the fault. What type of fault is this? Determine the dip separation at the point on the ground where the fault intersects the outcrop of the marker horizon and determine its horizontal and vertical components.

REFERENCES

Badgley, P. C., 1959, Structural methods for the exploration geologist: New York, Harper & Row.

Donn, W. L., and **Shimer, J. A.**, 1958, Graphic methods in structural geology: New York, Appleton-Century-Crofts.

D

Stereographic Projection

The stereographic projection,* Fig. D-1, is a means of showing three-dimensional linear and planar relations on a single, horizontal plane. This representation is achieved essentially by projecting meridians and parallels from a hemisphere (one produced by cutting a sphere through its poles) onto a flat surface. These lines are drawn on the plane of the projection as they would appear if viewed from a position on the sphere vertically above the center, Fig. D-2. This projection is used to determine angles between lines and planes, but it cannot be used to measure distances or show spacial relationships other than angular ones. In order to measure angles between these elements, they must have a common point of intersection, the center of the projection (this point is represented by a dot in the center).

Lines in the stereographic projection are laid off as lines of longitude and latitude would be. The circle representing the outer edge of the hemisphere is divided into four quadrants, and these are divided into nine 10° intervals, as is the distance from the center of the projection to the edge. This choice of divisions allows strike, dip, bearing, and plunge measurements to be plotted easily. One of the poles of the hemisphere is labeled north, the other south; east is 90° to the right of north, and west is on the opposite side. The outer circle is a representation of the horizontal plane. Any line which lies in the horizontal plane appears as a straight line crossing the circle through the center with a given bearing. A vertical line passes through the center and appears as a dot.

An inclined plane, Fig. D-3, appears as a straight line passing through the center of the projection to its edges and an arcuate line connecting the ends of the straight line. That arcuate line represents the intersection of the plane with the outer edge of the hemisphere. Since the straight line passing through the center of the projection is horizontal, it represents the

* Bucher (1944), Donn and Shimer (1958), Badgley (1959), Ragan (1973), and Dennison (1968) summarize the characteristics of the projection and cite a number of examples which demonstrate its application.

strike of the plane. The dip is measured at right angles to the strike. Angles of dip, inclination, or plunge are read along either the north-south or the east-west axis of the projection.

It is convenient to use a piece of tracing paper attached to the projection by a pin placed through the center. North is recorded on the tracing paper so that the paper can be reoriented each time a bearing or strike direction is to be read. Drawing a plane of given strike and dip is accomplished by drawing a straight line across the projection with the given strike direction, while north on the tracing is oriented over north on the projection. The tracing is then rotated until the line of strike lies along the north-south axis. In this position the dip direction is along the east-west line. If the plane dips east of north, the amount of dip is plotted along the east-west line on the right side of the center by counting the number of degrees of dip from the outside of the projection toward the center, Fig. D-3. A meridian line passing through this point is used to trace the arcuate outline of the line of intersection of the plane with the hemisphere.

FIGURE D-1 Stereographic projection.

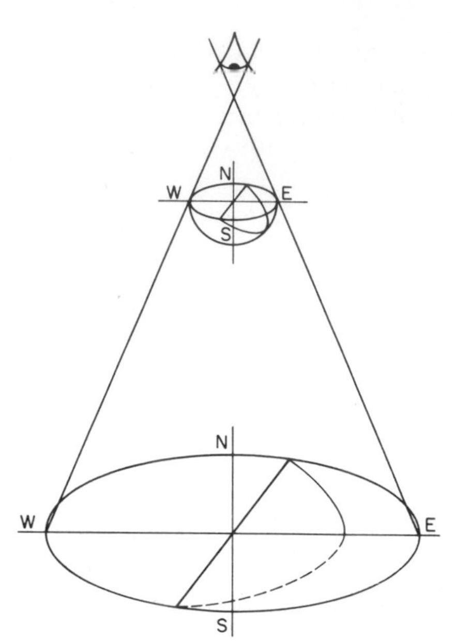

FIGURE D-2 Stereographic projection of a dipping plane. The plane appears the way it would if it passed through a hemisphere viewed from above.

FIGURE D-3 Representation of a plane by stereographic projection.

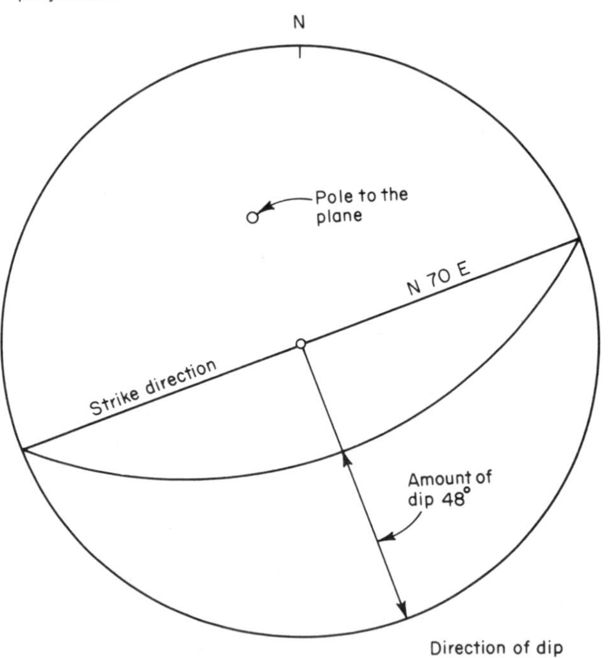

With the plane in this position, it is easy to plot a single point which represents that particular plane. This point is the point of intersection of a line perpendicular to the plane with the hemisphere, called the *pole* of the plane. This lies on the east-west axis 90° from the point initially located to determine the dip, Fig. D-3.

Summary

1. The outer circle functions as a compass. With north properly oriented, all bearings and strike directions are read from it.

2. Dip is measured from the outside of the projection toward the center. This is done along the north-south or east-west axis because these are the only two radial lines drawn on the projection. (*Note:* A properly scaled ruler can be used to plot dip amount and direction.)

3. Intersections of planes with the outside of the hemisphere are conveniently drawn by use of the meridian lines which represent such intersections for 2° intervals of dip. (Lines of parallel, except for the center parallel, do not pass through the center of the projection and thus cannot be used.)

4. Poles to planes can be constructed by allowing a point representation of a plane. This is useful in some of the more complicated problems.

5. The movements of points on the edge of a plane can be followed along lines in the projection if the plane is rotated around a vertical axis or around the north-south horizontal axis. For example, take a vertical plane with a strike of due north. This plane appears as a straight line on the projection. If this plane is rotated about the north-south axis so that it starts to dip east, movement of the points along its intersection with the hemisphere can be followed along the parallels, Fig. D-4. Rotation of a plane about its strike must be accomplished by first rotating that plane about a vertical axis until its strike line lies along the north-south axis.)

APPLICATION OF THE STEREOGRAPHIC PROJECTION

1. *To find apparent dip.* (*a*) Draw the plane in the projection; (*b*) draw the line along which the apparent dip is desired (often given as a bearing). The amount of the apparent dip is the angle between the horizon (the outside of the circle) and the point of intersection of this line with the line of intersection of the original plane with the hemisphere Fig. D-5. This angle is measured along one of the two axes of the projection.

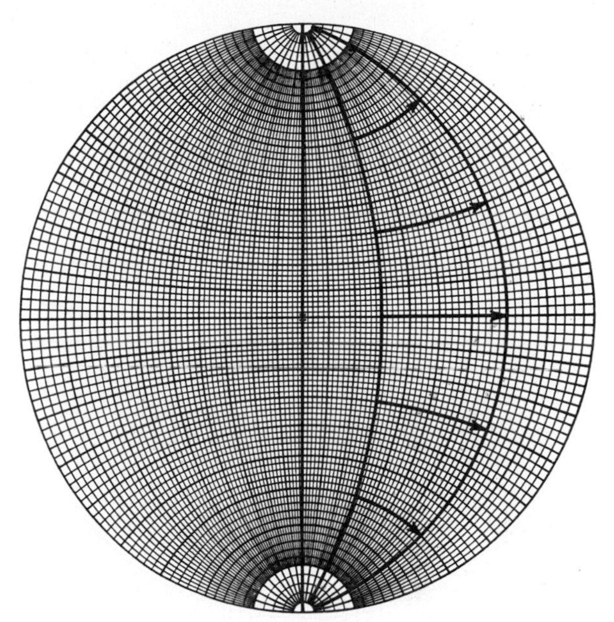

FIGURE D-4 Rotation of a plane by 40° around its strike line.

2. *To find true strike and dip from apparent dips.* When two apparent dips are given, the bearings of the lines are drawn, and points are located along them at distances which correspond to the apparent dips. These two points lie on the intersection of the plane with the hemisphere; thus a meridian line which passes through them defines the true plane. That meridian is found by rotating the points on the tracing overlay until they lie along the same meridian. The two ends of that meridian define the strike of the plane, and its true dip can then be measured (Fig. D-6).

3. *To find the bearing and plunge of the line of intersection of two planes.* Sketch the two planes in the projection. Unless they have identical strikes, the lines of intersection of the two planes with the hemisphere will cross at some point. A straight line between the center of the projection and this point of intersection defines the line of intersection. The

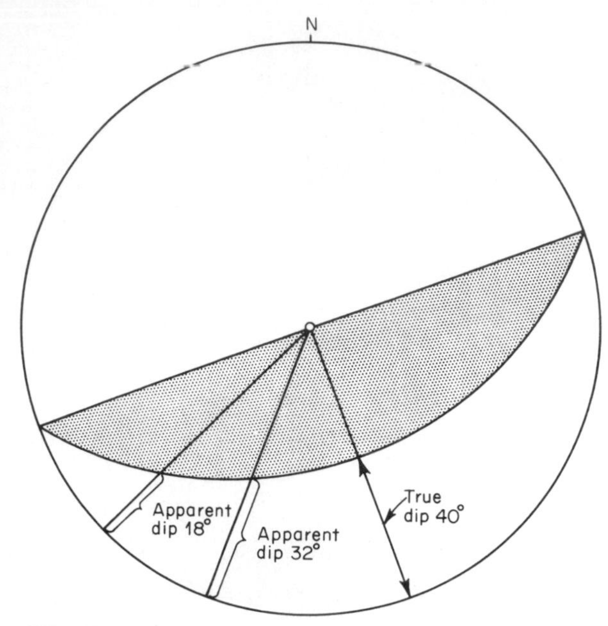

FIGURE D-5 Apparent dip determination.

straight line is the horizontal projection of an inclined line. Its bearing can be read by extending the line out to the edge of the projection. Its plunge can be read by rotating the line to an axis and counting the degrees between the horizon and the point of intersection of the plunging line with the hemisphere, Fig. D-7.

4. *To find the pitch of a line in a plane.* The angle lies within the plane and cannot be measured in the horizon unless that plane is horizontal. However, angles within dipping planes can be measured along the line of intersection of the plane with the hemisphere. Since the pitch of a line in a plane is the angle between the horizon and that line measured in the plane, it must be measured along a meridian line (the one which contains the arcuate intersection of the plane with the hemisphere) (Fig. D-7).

5. *To find the acute angle of intersection of two planes.* This angle lies in a plane which is mutually perpendicular to the other two. The plane which is mutually perpendicular to two vertical planes is a horizontal plane, and the angle between two such planes may be measured directly on the projection. Generally the two planes involved are both inclined,

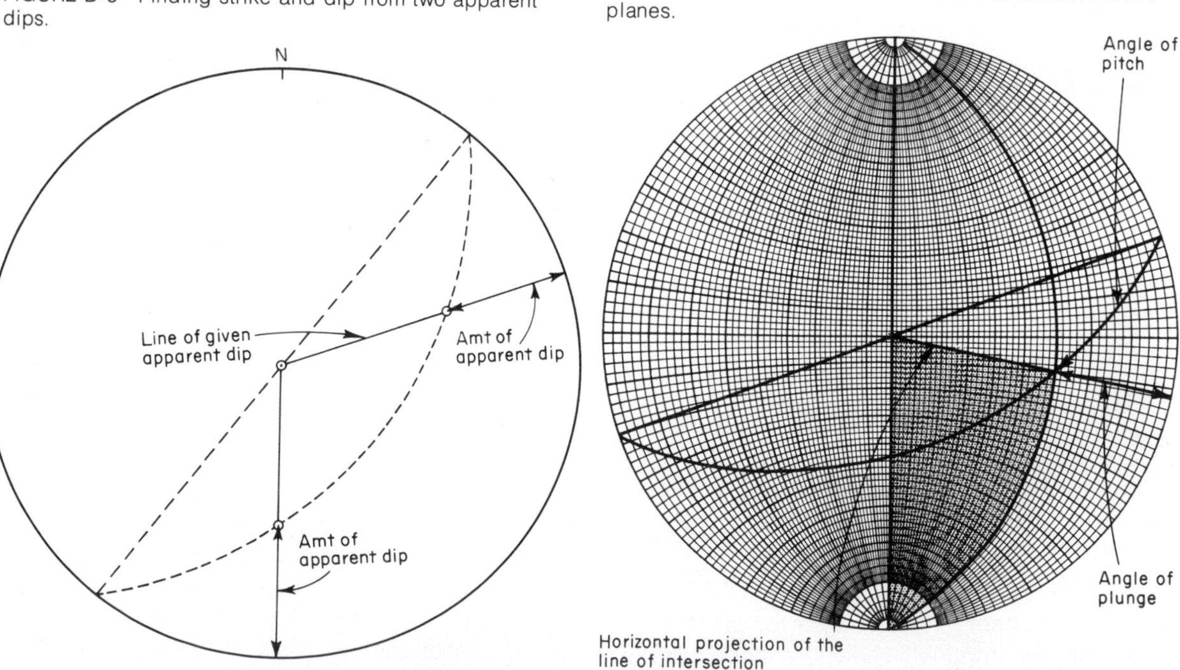

FIGURE D-6 Finding strike and dip from two apparent dips.

FIGURE D-7 Determination of line of intersection of two planes.

and the plane that is at right angles to them is also inclined. The mutually perpendicular plane must also be perpendicular to the line of intersection of the two planes involved. Start by drawing the line of intersection and a horizontal line at right angles to it. Measure 90° along the meridian for each plane starting from the point of intersection. These two points lie on the mutually perpendicular plane. The acute angle between the two planes can be measured along the meridian line between the points of intersection of the two original planes with the mutually perpendicular plane (Fig. D-8).

The above problem can be rapidly solved by finding the pole to each of the two planes involved and then rotating the two poles until they lie along the same meridian. The angle between the two planes can then be read directly along that meridian line.

6. *Rotation of a plane about its strike line.* An inclined plane with a line of known pitch lying in it is rotated about the strike line of the plane. Draw the projection of the plane and the line in it. Rotate the plane until its strike line lies on the north-south axis. (The direction and amount of rotation must be given in the problem.) Each point on the line of intersection of the plane with the hemisphere may be traced as rotation proceeds by moving the point the correct number of degrees in the right direction along the parallels of projection. The point of intersection of a line of known pitch in a given plane with the hemisphere will be among the points moved in this way. Its new position may be found and its new bearing and plunge measured; its pitch, of course, remains unchanged by this operation.

FIGURE D-8 Finding a plane normal to two other planes (angle between two planes).

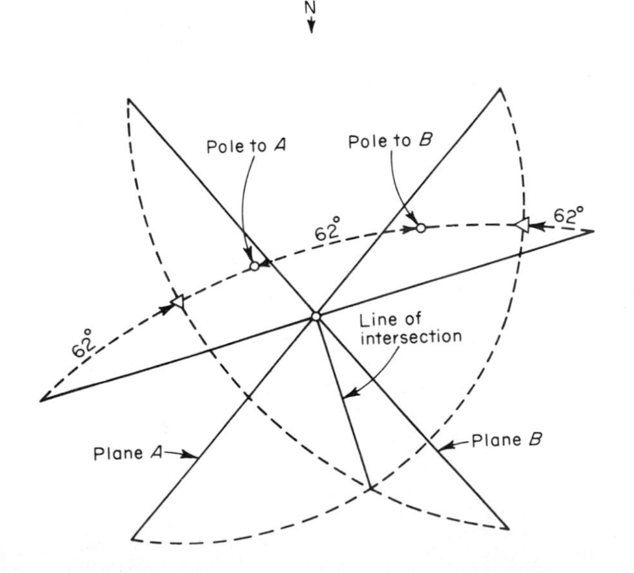

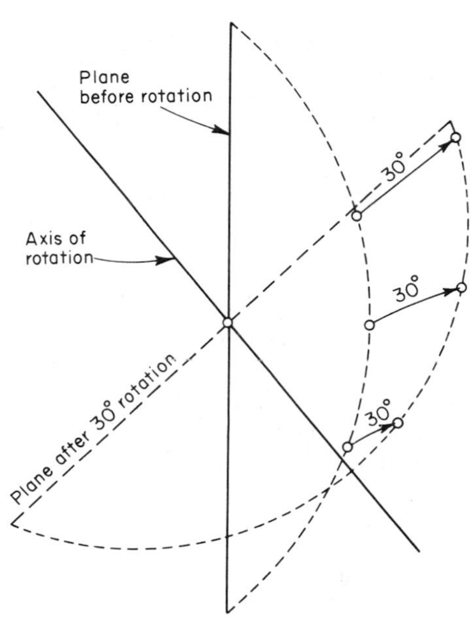

FIGURE D-9 Rotation of a plane about a line.

7. *Rotation of a plane about any line.* The principle is the same as above except that some line other than the strike line is the line of rotation. Place the line around which rotation is to occur along the north-south axis. When rotation of a given direction and amount occurs around this line, all points on the given plane rotate by the same amount and direction. Locate two points on the line of intersection of the plane with the hemisphere, and rotate them by moving them along parallels passing through them in the given direction and by the given number of degrees. These two points can now be used to redraw the line of intersection of the entire rotated plane and the hemisphere by moving them until they lie on a single meridian (Fig. D-9).

8. *Differential rotation of intersecting planes.* A number of problems (e.g., rotational faults and restoration of beds below an unconformity) involve finding the new strike and dip of a plane which has been rotated around an axis perpendicular to another inclined plane. Consider, for example, a rotational fault problem in which both the fault and the

formations that were cut by the fault are inclined. What will be the new strike and dip of beds rotated by a given amount and direction within the fault plane (around an axis perpendicular to the fault)? This can be solved as follows (Fig. D-10):

a. Draw the fault and a formation contact as it appears before rotation.

b. Rotate the fault to a horizontal position around its strike line.

c. At the same time rotate the beds to their new position resulting from the rotation of the fault.

d. Redraw the formational contact in a position so that its strike is changed by the number of degrees indicated by the amount of rotational faulting. The dip remains the same.

e. Return the fault to its original position, tracing out the new position of the formational contact as it is rotated with the fault. You are now ready to measure the new strike and dip of the beds on the rotated block of the fault.

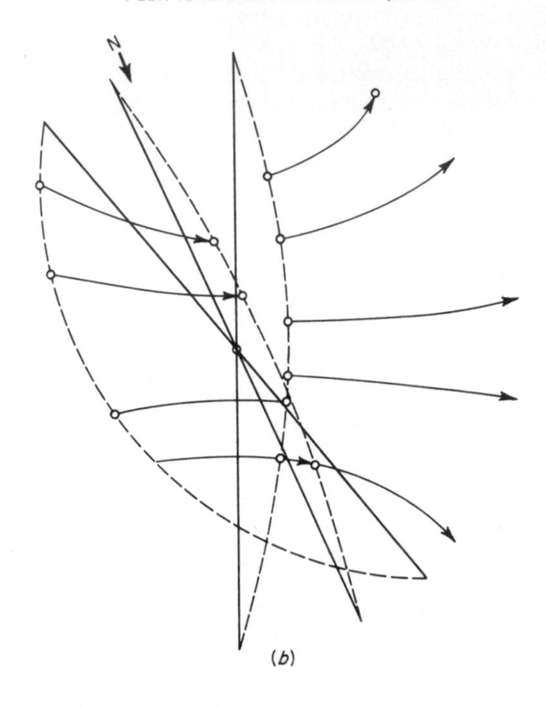

FIGURE D-10 Rotational fault problem. The steps illustrated are described in the text.

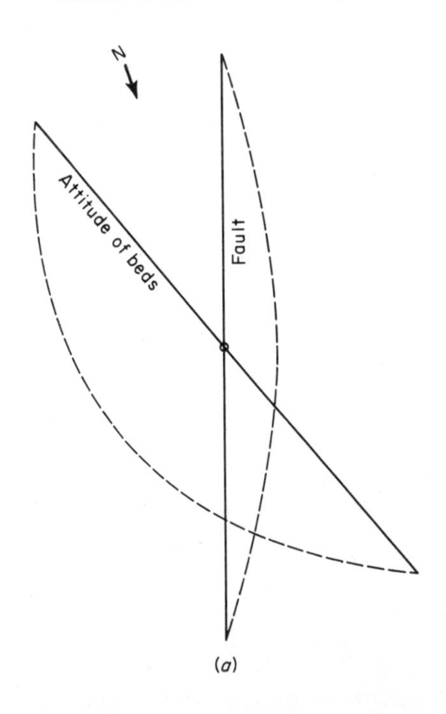

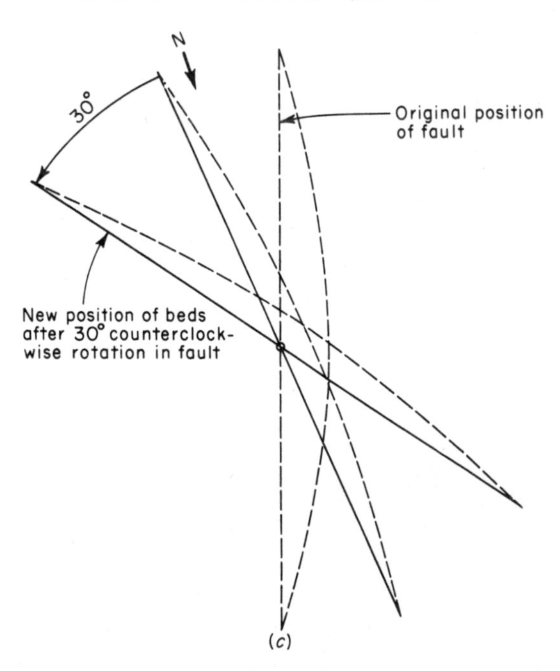

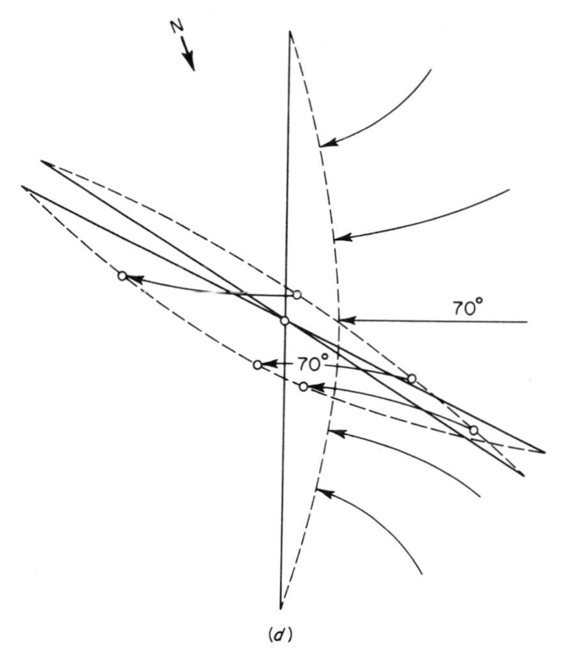

(d)

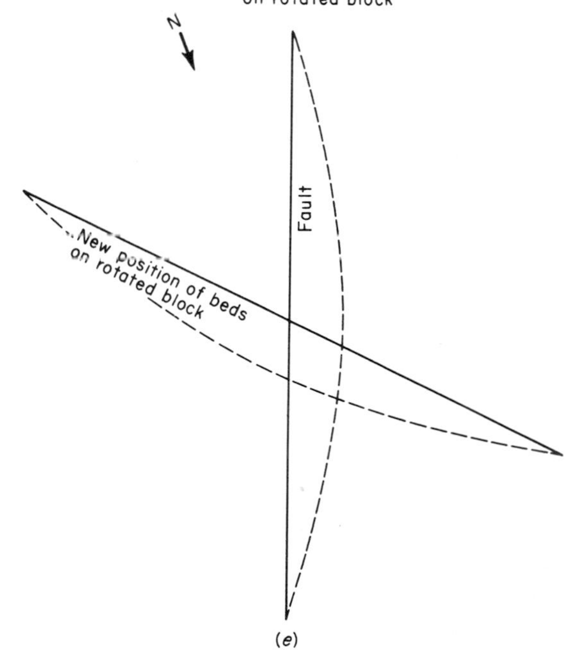

(e)

EQUAL-AREA PROJECTION

Since planes and lines can be represented on stereographic projections, the projection is uniquely suitable for statistical analysis of a population of measurements of planar or linear structures such as cleavage, bedding, foliation, fractures, or lineations. A common problem arising in areas where there has been multiple deformation is distinguishing the structural elements that belong to each of the deformations. It is sometimes possible to make this distinction by measuring a large number of the elements, plotting them on a stereonet as poles, and analyzing them, first to see whether they fall into distinct groups and, if they do, to determine the geometric relations among the various groups to see which might logically belong together. This type of analysis makes it possible to eliminate or at least reduce the chance of erroneous subjective judgment. Only rarely do all structures belonging to a single set have exactly the same strike and dip. If two sets appear in a single outcrop, selection of the attitude of each set may be difficult unless they are usually consistent and at large angles to one another. If three or more sets are present, it may be virtually impossible to separate them into related groups unless a large number are measured and analyzed. The stereonet and equal-area projection provide simple ways of making this analysis.

The stereonet has one particular shortcoming for use in this type of study. The area of a degree of latitude and longitude in the center of the hemisphere is quite different in size from that of a degree near the edge of the hemisphere. If the concentration of points on a projection is to be used as a means of determining to which group a given point belongs, projections of areas of identical size should be equal. The Lambert equal-area projection, also called a Schmidt net (Fig. D-11), is designed to meet this need.

Analysis of the population of measurments is done through three mechanical steps:

1. Plotting the points (often poles to planes)
2. Counting the points
3. Contouring the counts

Points may be plotted in either the upper or the lower hemisphere. The lower hemisphere is usually used in structural studies. A method for plotting points in the lower hemisphere has already been described, but this method, involving drawing the plane first, is too laborious for plotting of a large sample, and the final product would be a maze of lines. A number of methods are available for rapid plotting. The following technique, which requires only tracing paper and a Schmidt net, can be used. The 10° indicators in the northern half of the net are labeled as shown in Fig. D-11. North is marked in its correct position on the tracing paper. To plot the pole to a plane of given strike and dip, rotate the north point on the tracing to the strike direction indicated (on the projection). The dip reading is then measured along the newly labeled east-west axis of the projection (the usual north-south axis). Whether the dip is measured along the top or the bottom half of that axis depends on two things—the direction of dip and the hemisphere in which the point is being plotted. The pole to a plane with attitude N 45° E, 40° SE will appear in the northwestern quadrant of a

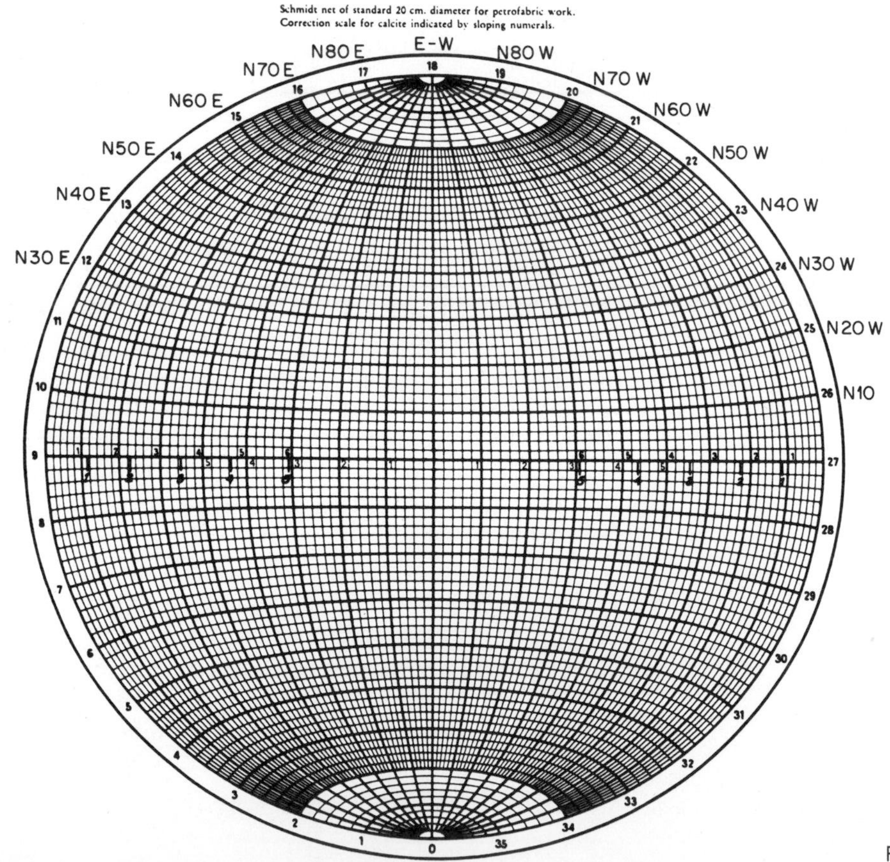

FIGURE D-11 Equal-area net.

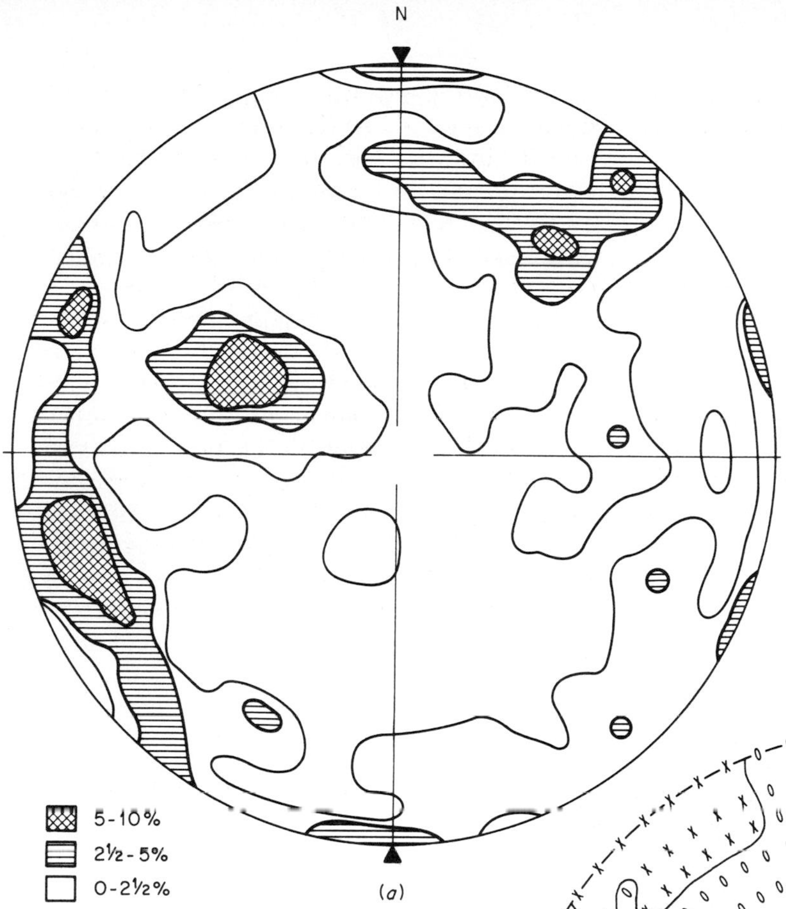

N

5-10%

2½-5%

0-2½%

(a)

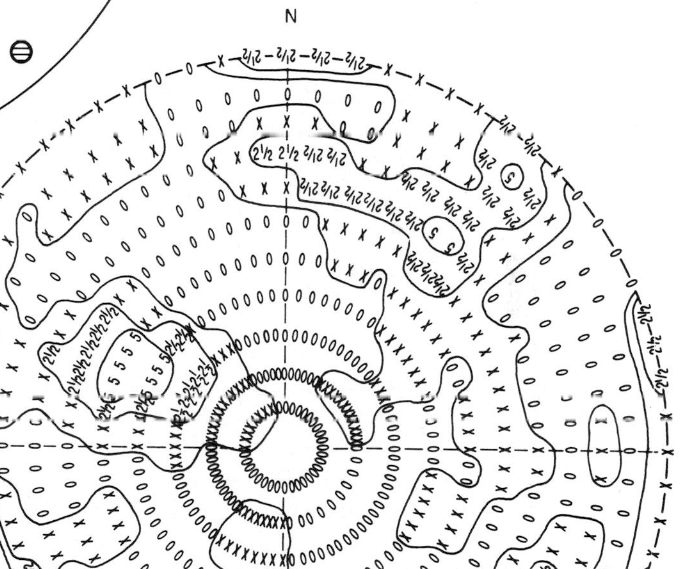

N

(b)

FIGURE D-12 (*a*) Count of plotted points in percent of sample representing joints in a basalt dike; (*b*) contoured point diagram of joints in a basalt dike. (*Data and plot by Arie Poldervaart.*)

lower hemisphere projection but in the southeastern quadrant of an upper hemisphere projection. The upper hemisphere is preferred by some because the points appear in the quadrant toward which the planes actually dip. In either case measurment of the dip is done in the reverse manner from that used to find the dip of a plane. The pole to a plane that is vertical will appear at the outer edge of the projection circle (the normal zero dip). Since poles to planes are perpendicular to the plane, the complement of the dip must be used. This is simplified by locating the pole on a scale which reads the complement of the dip (i.e., one that is 0° at the center of the projection and 90° at the outer edge).

Counting points may also be done by several techniques. Counting is necessary in order to contour the points so that exact concentrations can be located. It is not always possible to pick the centers of a cluster of points, and as the clustering becomes less pronounced the difficulty of doing this successfully increases. Most Schmidt nets are printed at a scale such that the diameter of the net is exactly 20 cm. At this scale a circle in the projection that has a radius of 1 cm represents 1 percent of the total area of the projection. This simplifies counting because the tracing paper with all the points plotted on it can be placed over a piece of graph paper laid out in 1-cm squares. The grid intersection can be used for locating the center of a counting circle of 1-cm radius. This counting circle can be moved along the grid, and the number of points in each circle can be counted and recorded on the tracing either as the actual number of points or as the percentage of the total number. As the counting circle is moved along, it will overlap previously measured points. This system must be slightly modified along the edge of the projection. For counting these points a peripheral counter is used. The number of points to be counted for a marginal position on the projection is the total of those within the half circles on opposite edges. This total is recorded at both edges. Once all grid corners are tabulated, contouring may begin.

An alternative to the above methods of counting has been described by Duschatko (1955). This method requires a plastic template with a series of overlapping spherical circles, and when many point diagrams are to be analyzed it is much faster than the method above. Each of the spherical circles is a 1 percent circle. Counting is done by rotating the template (a 5° rotation each time is generally sufficient).

Contouring of point diagrams is facilitated by drawing in the highest concentration contours first. Selection of the contour interval may be determined in part by the magnitude of the maxima, but as a general rule it should be 2 percent. The one way in which contouring of point diagrams differs from that of other types of numerical distributions is the unusual shape of the projection. At the edge of the projection any contour which runs off the edge must appear 180° away on the other side, as in Fig. D-12.

EXERCISES

Problem 1. A bed strikes N 40° E and dips 60° SE. What is its apparent dip in a vertical plane trending EW?

Problem 2. The faces of a quarry trend EW and NS. A coal seam has apparent dips of 20° N in the NS wall and 40° W in the EW wall. What is the true strike and dip of the coal seam?

Problem 3. Two dikes with orientations (1) N 60° E, 30° SE and (2) N 10° W, 60° SW intersect.

(*a*) What is the bearing and plunge of the line of intersection?

(*b*) What is the pitch of dike 1 in dike 2?

Problem 4. Locate the poles to each of the dikes in Problem 3.

Problem 5. What is the acute angle of intersection between the two dikes in Problem 3?

Problem 6. What would be the strike and dip of dike 1 (Problem 3) if dike 2 were rotated until it is horizontal?

Problem 7. Solve angular relations for the problems in Appendix C.

REFERENCES

Bucher, W. H., 1944, The stereographic projection, a handy tool for the practical geologist: Jour. Geology, v. 52, p. 191–212.

Clark, R. H., and **McIntyre, D. B.**, 1951, The use of the terms pitch and plunge: Am. Jour. Sci., v. 249, p. 591–599.

Dennison, J. M., 1968, Analysis of geologic structures: New York, Norton.

Donn, W. L., and **Shimer, J. A.**, 1958, Graphic methods in structural geology: New York, Appleton-Century-Crofts.

Duschatko, Robert, 1955, Mechanical aid for plotting and counting out pole diagrams: Geol. Soc. America Bull., v. 66, p. 1521–1524.

Prager, G. D., 1975, Stereographic technique for determination of minor fold sense: Geol. Soc. America Bull., v. 86, p. 316–318.

Ragan, D. M., 1973, Structural geology, An introduction to geometrical techniques, 2d ed.: New York, Wiley.

E

Structural Notation

The description of many features of deformed rocks is simplified by the use of notations which indicate the orientation of the feature and its relation to other parts of the structure. Notations are used especially in analyses of rock fabric and when several generations of structures are superimposed. In general the notations can be applied to any linear or planar structures, but unfortunately, some of the notations have been used in more than one way. Ideally they should be applied in a purely descriptive manner, but at times certain lines and planes have been widely thought to bear definite relations to the movements which took place during deformation, and the notations have therefore been used to indicate the movement picture. Cloos (1946) outlines the history of the usage of various notations. Those proposed by Sander (1926, 1930) are among the most commonly used ones.

Sander's Notations

1. *s surface.* A planar structure (any plane of mechanical inhomogeneity). The term is used in a purely descriptive, nongenetic way. It may be applied to bedding, the individual layers in a cross-bedded deposit, foliation, schistosity, banding, cleavages, joints, etc. When more than one planar structure exists in the field of study, the various *s* surfaces may be designed $s_1, s_2, s_3, \ldots, s_n$, in the sequence of their formation.

2. *Coordinate axes b, a, c.* Three mutually perpendicular axis defined as follows:

b = the direction of orientation of the most prominent fold axes.

a = the direction perpendicular to b and lying in the plane of movement, the direction of tectonic transport. In cases of schistosity it lies in the plane of the schistosity. The direction may be visualized in terms of the direction of maximum expansion in a block of clay compressed in a vise.

c = the direction perpendicular to the ab plane.

3. *B.* Used to designate a fold axis actually measured on the fold where it can be positively identified: $B = b$.

4. *β.* Used to designate a line constructed by the intersection of two planes. For example, in the case of cylindrical folding, the lines formed

by the intersection of planes tangent to bedding are parallel to the fold axis. The tangent planes correspond to strike and dip measurements on the fold limbs.

5. π. Used to designate the pole to a π circle as plotted on a stereographic projection.

6. π circle. A circle or arc defined on the stereographic projection by the plot of a number of poles. The poles to a large number of bedding-plane measurements on a folded bed define great circles in cylindrical-type folding or small circles in conical folding.

7. *Lineations.* Sander originally used *b* to designate lineations. Many lineations do form parallel to fold axes, but because other lineations are not parallel to *b*, it seems advisable to follow Cloos's suggestion and designate lineations by *L*. When more than one lineation is present, the lineations become L_1, L_2, L_3, etc. Linear features arise in a variety of ways. Flow in igneous rocks commonly gives rise to lineations parallel or perpendicular to flow direction. Fold axes are lineations, and where many small folds with parallel axes are formed the lineation is quite prominent. The intersection of bedding with cleavage, foliation, or fractures, or more generally the intersection of any two *s* surfaces, produces a lineation. Slickensides, oriented elongate minerals, and smeared-out crystals also form lineations.

Turner and Weiss Usage

The rules set forth by Turner and Weiss (1963) for the application of the notations *a, b,* and *c* are widely used. The notations *a, b,* and *c* are three orthogonal fabric axes which can be selected only in a fabric that is at least in part homogeneous. As a purely descriptive notation they are used according to the following rules:

1. In a fabric dominated by a prominent planar structure *s*, *s* = the *ab* plane, and any regular lineation in *s* is called *b*.

2. In fabrics with two or more intersecting planar structures such that they intersect in a common axis, the most prominent planar structure is called *ab*, and the axis of intersection is *b*.

3. In fabrics with two or more intersecting planar structures that do not intersect in a common axis, the most prominent planar structure is called *ab*, and the intersection with the second most prominent planar structure is called *b*.

4. In fabrics dominated by a strong lineation, that lineation is called *b*.

Unhappily even this use of notation involves a subjective element, as when the observer must decide which of two planar elements is more prominent.

Turner and Weiss restrict the use of axes *a, b,* and *c* to describe the kinematic (or movement-picture) axes to fabrics having monoclinic symmetry and expressible in terms of gliding upon a prominent structural discontinuity such as a set of foliations or cleavages. In such cases,

a = glide direction
b = normal to *a* and lies in the glide plane
c = normal to the *ab* plane
ab = glide plane
ac = deformation plane

REFERENCES

Cloos, Ernst, 1946, Lineation, a critical review and annotated bibliography: Geol. Soc. America Mem. 18.

Sander, B., 1926, Zur Petrographisch-Tektonischen Analyse III. Jahrb. Geol. Bundesanstalt (Austria), 76.

——— 1930, Gefügekunde der Gesteine, Vienna.

Turner, F. J., and **Weiss, L. E.,** 1963, Structural analysis of metamorphic tectonites: New York, McGraw-Hill.

Indexes

Name Index

Abdel-Khalek, M. Lofti, 470
Abendanon, E. C., 121, 390, 392
Achenbach, Abo-Zona, 173
Adams, Frank, 47, 70
Agar, D. V., 512
Aki, K., 300
Ala, M. A., 283
Albers, J. P., 538, 539, 558
Alexander, S. S., 296, 300, 329
Alldredge, L. R., 393
Allen, J. R. L., 12, 13, 18
Allen, T. D., 392
Almy, Charles C., Jr., 558
Alvarez, W., 506, 512, 514
Amenta, R. V., 470
Amery, G. B., 368
Anderson, C. N., 395
Anderson, D. E., 264
Anderson, D. L., 289, 300
Anderson, E. M., 110, 111, 121, 159, 165, 173, 225
Anderson, J. G. C., 44, 470
Anderson, R. E., 558
Anderson, Roger N., 379, 380, 392, 406, 422
Andrade, E. N. C., 38, 42, 70
Andrews, Donald I., 283
Andrieux, 510–512
Anhaeusser, C. R., 335–338, 353
Antoine, J. W., 268, 283, 368, 558
Applin, P. L., 283, 364, 368
Araña, V., 511, 513
Arbentz, P., 513
Arbolelya, M. L., 192, 204, 226
Arden, D. D., Jr., 549, 558
Argand, Emile, 492, 494, 502, 513
Arthaud, F. et Matte, Ph., 173
Ashraf, Asaf, 369
Asmus, H., 284
Atwater, G. I., 283
Atwater, Tanya, 329, 541, 542, 544, 558
Aubert, D., 486
Aubouin, Jean, 305, 329
Avé Lallement, H. G., 58, 70

Babcock, E. A., 94, 99
Bader, H., 225
Badgley, P. C., 5, 576, 577, 579, 583, 599, 601
Bail, T. K., 203
Bailey, E. B., 5, 225, 256, 265, 485, 513
Bailey, E. H., 533, 558
Bailey, S. W., 70
Bain, J. H. C., 421, 422
Baird, A. K., 558
Baird, K. W., 558
Balacic, J. D., 62, 71
Baldwin, E. M., 562
Balk, Robert, 92, 97, 225, 272, 283
Ball, Mahlon M., 393, 551, 558
Bally, A. W., 527, 558
Barazangi, Muawia, 410, 422, 424, 536, 564
Barbat, W. F., 558
Barber, A. M., 141
Baron, J. G., 330
Barton, D. C., 283
Bartram, J. G., 151, 156
Bass, I. G., 548, 562
Bates, F. W., 273, 274, 283
Bayaz, Dasht E., 163
Bayly, M. B., 203, 229, 244, 265
Bearce, D. N., 470
Bearth, P., 493, 513
Behre, C. H., Jr., 213, 225
Behrendt, J. C., 368
Bell, J. A., 71
Bell, J. F., 53, 70

Bell, J. S., 551, 552, 558
Beloussov, V. V., 5, 322, 323, 329, 420, 422
Belyakov, L. V., 141
Belyayevskiy, N. A., 409
Ben-Avraham, Zvi, 422
Benioff, Hugo, 300, 313, 329, 401, 403, 422
Benjamins, J. Y., 98
Bennison, G. M., 265
Bentley, C. R., 422
Bentley, R. D., 93, 99, 471
Berberian, M., 165, 175, 515
Berg, R. R., 153, 156, 558
Bergantino, R. N., 369
Berry, F. A. F., 558
Berry, G. F., Jr., 283
Berry, M. J., 353, 513
Berry, R., 541, 564
Berryhill, H. L., Jr., 548, 558
Bersier, A., 484, 513
Bertrand, Marcel, 254, 513
Bick, K. F., 126, 141
Big, Chingchang, 422
Billings, M. P., 6, 119–121, 199, 202, 203, 264, 470
Biot, Maurice A., 229, 234, 237–239, 244
Birch, A. F., 300
Birch, Francis, 141
Bird, J. M., 310, 316–319, 321, 329, 418, 423, 443–445, 470
Bird, Peter, 329
Bischoff, J. L., 386, 393, 561
Bishop, D. G., 162, 173, 174
Bjork, R., 101, 102, 104
Blake, M. C., Jr., 329
Boccaletti, M., 508, 513
Boessenkool, A., 516
Bombolakis, E. G., 27, 97, 104
Bonatti, Enrico, 392
Bonham, L. C., 97
Bonini, W. E., 511, 513
Bonnin, Jean, 501, 513
Boos, C. M., 558
Boos, M. F., 558
Borg, Iris Y., 51, 52, 54, 57, 60, 64, 70–72
Bornhauser, Max, 283
Bostrom, R. C., 558
Bowes, D. R., 470
Bowin, Carl O., 506, 509, 516, 558
Boyer, R. E., 121
Brace, W. F., 18, 26, 42, 78, 80, 97, 121, 122, 559
Bracey, D. R., 395, 404, 442, 551, 559
Bradley, W. H., 9, 18
Braun, L. T., 191, 203, 265
Braunstein, Jules, 283
Brewster, David, 232
Briggs, L. I., 550
Brogan, E., 563
Brooks, J. A., 422
Brothers, R. N., 424
Brouwer, H. A., 422
Brown, G. C., 470
Bruce, Clemont H., 149, 151, 156
Brune, J. N., 298–300
Brush, L. M., Jr., 18
Bryan, G., 284
Bryant, Bruce, 461, 470, 472
Bryant, W. R., 268, 283
Bucher, W. H., 6, 62, 70, 97, 137, 138, 141, 188, 201, 203, 231, 232, 245, 265, 297, 300, 329, 440, 441, 470, 559, 601, 611
Buddington, A. F., 193, 203, 347, 348, 353
Buessem, W. R., 215
Buffington, T. R. A., 424
Bullard, E. C., 351, 428, 470, 507, 513
Bullen, K. E., 288, 300
Bunce, E. T., 368, 375, 408, 422, 559
Burchfiel, B. C., 419, 422, 529, 530, 532, 559
Burckle, L. H., 382, 392
Burford, R. O., 564
Bürge, H., 556, 559
Burk, C. A., 415, 422, 439, 553, 559
Burnett, C. M., 157
Burrell, H. C., 184
Bush, P. A., 546, 559
Bush, S. A., 546, 559
Busk, H. G., 203, 577, 583
Butler, J. R., 461, 470, 513
Butler, L. W., 368
Butterlin, J., 548, 559
Buxtorf, A., 495, 513
Byerlee, J. D., 70, 122
Byrne, J. G., 579, 583

Cady, W. M., 310, 329, 470
Cambell, J. D., 203
Caminos, R., 560
Campbell, C. J., 552, 555, 556, 559
Campbell, J. W., 203
Caner, B., 559
Cannon, W. H., 559
Caputo, M., 513
Carey, S. W., 38, 39, 42, 68–70, 119, 122, 159, 173, 174, 188, 198, 199, 202, 203, 251, 261, 265, 304, 327–329, 390, 392, 420, 422, 502, 506, 513, 539, 559
Carl, J. D., 156
Carlisle, Donald, 141
Carpenter, R. H., 470
Carr, J. J., 401, 402, 422, 425
Carter, F. W., 281, 283
Carter, N. L., 52–54, 56–58, 65, 70, 72, 229, 243, 245
Carver, R. E., 156
Casanova, R., 560
Case, J. E., 271, 284, 553, 559, 564
Casella, C. J., 97, 125, 141, 156
Castillo, Lus Del, 559, 563
Cate, A. S., 258, 260, 265
Cebull, S. E., 470
Chalmers, R. M., 265
Chamberlin, T. C., 201, 320, 329
Chao, E. C. T., 101, 105
Chapman, C. A., 97
Chapman, R. E., 141
Chapple, W. M., 112, 122, 229, 240, 242, 245, 246
Chase, C. G., 353
Chase, R. L., 384, 408, 422, 551, 559
Chaudhuri, A. K., 203
Chidester, A. H., 310, 329
Chiliggar, G., 97
Chinnery, M. A., 122, 171, 172, 174
Christ, C. L., 484
Christensen, M. N., 424, 536, 559
Christie, J. M., 56–58, 65, 70

Christman, R. A., 550, 559
Churkin, M., Jr., 559
Clark, L. D., 559
Clark, R. H., 199, 203, 611
Clark, S. K., 122
Clark, S. P., Jr., 35, 42, 51, 70, 290, 300, 308, 309, 329
Clayton, Lee, 164, 166, 174
Cleary, J. M., 97
Clifford, P., 203
Clifford, T. N., 442, 473
Clinton, N. J., 89, 99
Cloos, Ernst, 18, 28, 33, 43, 76, 92, 97, 147, 157, 188, 190, 203, 221–223, 225, 452, 453, 470, 613, 614
Cloos, Hans, 6, 117, 119, 203
Clough, C. T., 218
Coates, J., 579, 583
Coats, R. P., 283
Coats, R. R., 559
Cobbing, N. J., 557, 559
Cohen, R., 394
Coleman, P. J., 422
Coleman, R. G., 310, 329, 397, 419, 422
Collet, L. W., 483, 487, 489, 494, 513
Comninakis, P. E., 506, 513, 515
Compton, R. R., 528, 559, 583
Conaghan, P. J., 502, 506, 515
Condie, K. C., 353
Coney, P. J., 329, 559
Conley, J. F., 461, 470
Contescu, L. R., 513
Cook, P. J., 18
Cooke, H. C., 353
Coons, R. L., 353
Cooper, B. N., 18, 141, 225, 248, 249, 265, 470
Couch, R. W., 376, 392
Cowan, D. S., 559
Coward, M. P., 337, 339, 353
Cox, Allan, 310, 311, 329
Craddock, Campbell, 350, 353
Craddock, J. C., 470
Crampton, C. B., 70
Crawford, A. R., 513
Crawford, J. P., 562
Cray, E. J., Jr., 354
Crittenden, M. P., Jr., 537, 563
Crook, K. A. W., 225
Crosby, Gary W., 33, 225, 368
Crowell, J. C., 11, 18, 108, 122, 160, 174, 559, 583
Crowley, W. P., 458, 470
Cummings, David, 174
Curray, J. R., 513, 559
Currie, J. B., 203, 233–237, 245, 265, 271, 275, 283

Dahlstrom, C. D. A., 182, 200, 203, 257, 265, 560, 561
Dale, T. N., 225
Dallemyer, R. D., 470
Dalrymple, G. B., 329
Daly, R. A., 6, 201, 309, 329
Dalziel, I. W. D., 225, 560
Dana, J. D., 202, 263, 329
Danes, Z. F., 350, 354
Davies, R., 18
Davis, G. A., 419, 422, 529, 530, 532, 559, 560
Davis, G. H., 142
Dearnely, R., 329
De Beaumont, Elie, 304, 329
Debelmas, J., 513
DeBoer, Jelle, 369
De Cserna, Zoltan, 467, 470, 560
De Golyer, E., 267, 283
Dehlinger, P., 376, 392
DeJong, Kees A., 142, 329, 490, 513
Den, N., 424
Dence, M. R., 101, 102, 105
Dengo, Gabriel, 548, 551, 560
Denison, R. E., 354
Dennison, J. M., 255, 265, 577, 583, 601, 611
DeSaussure, Horace, 6
De Sitter, L. U., 6, 131, 142, 188, 203, 209, 217, 224, 225, 513
Dewey, J. F., 218, 219, 225, 315–319, 321, 329, 418, 423, 443–445, 470, 501, 513
Dibblee, T. W., Jr., 561
Dickinson, W. R., 330, 382, 393, 423, 471, 560
Dickson, G. O., 330, 393
Dieterich, J. H., 71, 229, 243, 245
Dietrich, R. V., 470
Dietz, F. T., 368
Dietz, R. S., 101, 105, 320, 330, 381, 392
Dillon, Wm. D., 562
Dinkelman, M. G., 551, 562
Dixon, J. M., 43, 283
Dixon, K. P., 273, 274, 283
Dobrin, M. B., 278–283, 288, 300
Doell, R. R., 329
Dolginov, Ye. A., 353
Doll, C. G., 194, 203
Donath, F. A., 42, 52, 55, 59, 62, 71, 73, 98, 157, 194–196, 202, 203, 245, 537, 560
Donn, W. L., 577, 583, 599, 601, 611
Donnelly, T. W., 548, 550, 560, 563
Dotson, J. C., 583
Dott, R. H., 532, 560
Drake, C. L., 146, 157, 361, 363–365, 368, 422, 439, 471, 553, 559
Drever, J. I., 344, 354
Dreyfuss, M., 484
Dunn, D. E., 98, 461–464, 470, 473
Dupouy-Camet, J., 283, 284
Durán, S. L. G., 583, 560
Durand-Delga, M., 509, 513
Duschatko, Robert, 610, 611
Duska, Leslie, 203
Dwerryhouse, A. R., 265

Eardley, A. J., 6, 142, 471, 548, 560
Egyed, L., 327, 328, 330
Elders, W. A., 386, 387, 393
Ellen, S. D., 183, 204, 265
Elles, G. L., 265
Elliott, D., 32, 43, 138, 139, 142, 203, 419, 423
Elsasser, W. M., 290, 300, 321, 330
Elston, D. P., 281, 283
Elter, P., 134, 142
Elver, D., 424
Emery, K. O., 141, 142, 358–362, 365–369, 422
Emmons, E., 201
Engel, A. E. J., 330, 354

Engel, C. G., 336, 338, 339, 340, 352–354
Engelder, J. L., 122
Engels, G. G., 390, 391, 393
Epstein, J. B., 213, 215, 225
Erickson, B. H., 423
Ericson, D. B., 18
Ernst, W. G., 330, 514, 534, 559, 560
Escher, B. G., 274, 284
Escher van der Linth, A., 253, 254, 265
Eskola, P., 442, 471
Espenshade, G. H., 457, 459, 472
Etheridge, M. A., 225
Evamy, B. D., 512
Everett, J. E., 470, 513
Evernden, J. F., 564
Ewing, J. I., 365, 368
Ewing, John, 368, 393, 558
Ewing, Maurice, 18, 157, 288, 294, 295, 300, 301, 361, 368, 369, 392, 402, 409, 425, 564
Ewing, W. M., 547, 560
Eyring, H., 71

Faill, R. T., 55, 71, 255, 265, 471
Fairbairn, H. W., 54, 71
Fairburn, H. W., 6
Fairhurst, C., 98
Faul, Henry, 471
Feden, R. H., 369
Fedynskiy, V. V., 409
Firman, R. J., 98
Fisher, G. W., 458, 471
Fisher, R. L., 308, 330, 375, 382, 393, 394, 424
Fisher, R. V., 560
Fitch, T. J., 423, 514
Fitzgerald, E. L., 125, 142, 191, 203, 265, 560
Flawn, P. T., 471
Fleck, R. J., 560
Fleming, H. S., 369
Fleuty, M. J., 199–201, 203
Folinsbee, A., 564
Fontbote, J. M., 509–514
Foose, R. M., 152, 157, 275, 279, 284, 524, 560
Forman, M. J., 283
Forristall, G. Z., 142
Forsyth, D., 331
Fox, F. G., 547, 560
Fox, P. J., 300, 377, 393
Fox, P. P., 126, 142
Francheteau, Jean, 332, 368
Frazier, D. E., 18
Freedman, J., 471
French, B. M., 105
Freund, Raphael, 174
Frey, M. G., 471
Friedman, Melvin, 43, 50–53, 58, 65, 71, 94, 98, 245, 265
Frocht, M. M., 232, 245
Fruth, L. S., Jr., 71
Fuchs, G., 502, 514
Fuchs, Karl, 353
Fullagar, P. D., 471, 472
Fuller, J. D., 128, 129, 142
Fuller, J. O., 472
Fuller, R. E., 157, 537, 560
Fyfe, W. S., 330
Fyson, W. K., 203, 471

Gabriel, V. G., 583
Gabrielse, H., 526, 560, 563
Gagnebin, E., 514
Gansser, Augusto, 479, 496, 497, 502, 504–506, 514
Garbarini, G. S., 152, 157
Garfunkel, Z., 174, 321, 330
Garretty, M. D., 184
Garrison, L. E., 368
Garson, M. S., 98
Gast, P. W., 393
Gastil, G., 340, 341, 354
Gastil, R. G., 386, 387, 393
Gault, D. E., 101, 105
Gay, N. C., 98, 245
Gay, S. P., 95, 98
Geczy, B., 515
Geigle, J. E., 215, 226
Geikie, James, 6
Gemperle, M., 376, 392
Gere, J. M., 43, 246
Ghosh, S. K., 43, 203, 241, 245
Gibson, I. L., 373, 393
Giermann, G. K. F., 516
Gilbert Smith, A., 506, 507, 514
Gill, J. E., 120, 122, 354
Gill, W. D., 579, 583
Gilliland, W. N., 394
Gilluly, James, 330, 560
Gilmer, T. H., 353
Girdler, R. W., 146, 157
Glaister, R. P., 562
Glikson, A. Y., 354
Glover, Lynn, III, 263, 265, 461, 471, 473
Godfrey, J. D., 204
Goguel, Jean, 6
Goldberg, M., 18
Goldrich, S. S., 354
Goldsmith, R., 471
Gonzaley-Bonorino, F., 225
Goranoon, R. W., 71
Gordy, P. L., 527, 558
Goryatchev, A. V., 423
Gough, D. I., 368
Graham, J. A., 563
Graham, S. A., 154, 471
Gramberg, J., 82, 84, 85, 96, 98
Gray, F., 514
Green, Harry W., II, 320
Gregory, J. W., 157, 390, 393
Griffin, V. S., 471
Griffin, V. S., Jr., 254, 265, 460, 462, 471
Griffith, A. A., 79, 81, 98
Griggs, A. B., 561
Griggs, D. T., 6, 48, 53–59, 61–63, 65, 67, 70, 71, 73, 75–77, 98, 117, 122, 321, 330
Grim, P. J., 423
Groshong, R. H., Jr., 71
Gross, Barton, 353
Grow, J. A., 401, 404, 406, 407, 423
Gümbel, C. W. Von, 263
Gunn, B. M., 174
Gustafson, J. K., 184
Gutenberg, Beno, 289, 295, 300, 372, 401
Gwinn, V. E., 126–128, 142, 450, 451, 471
Gzovsky, M. V., 71, 393

Hadley, J. B., 459, 471
Hafner, W., 112–115, 122, 155, 156
Hager, D. S., 157
Hager, R. V., Jr., 42, 43, 62, 71
Hahn, S. J., 71
Hall, James, 305, 306, 330, 431, 433, 471
Hallam, A., 507, 515
Haller, J., 254, 265, 446, 471
Hamblin, W. K., 147, 157
Hamilton, Warren, 534, 561
Hancock, P. L., 225
Handin, J. W., 6, 38, 42, 43, 51, 52, 54, 55, 59–67, 71, 72, 75–77, 98
Hanna, M. A., 270, 284
Hansen, Edward, 71
Hara, Ikuo, 204
Harbison, R. N., 425
Hardin, F. R., 157
Hardin, G. C., Jr., 157
Harding, J. L., 368
Harding, T. P., 159, 168
Harker, Alfred, 285
Harper, P. A., 283
Harrington, J. W., 583
Harris, H. D., 525, 561
Harris, J. F., 98
Harris, L. D., 129, 142, 471
Harrison, C. G. A., 393, 551, 558
Harrison, J. E., 561
Hastie, L. M., 157
Hatcher, R. D., Jr., 460, 466, 471
Hatherton, Trevor, 330, 423, 561
Haughton, S., 225
Hayes, C. W., 142
Hayes, D. E., 330, 381, 393, 404, 423, 561
Haynes, S. J., 502, 514
Hays, J. D., 506, 515
Head, J. W., 101, 105
Healy, J. H., 301
Heard, H. C., 48, 53–55, 60–62, 65, 67, 71–73, 78, 98
Hedberg, H. D., 14, 18, 368
Hedge, C. E., 563
Heezen, B. C., 18, 157, 361, 374, 375, 393, 561
Heim, A., 232, 514
Heirtzler, J. R., 330, 364, 368, 381, 393, 394, 404, 423
Heiskanen, W. A., 309, 330
Henderson, G. G. L., 257, 265
Henika, W. S., 461, 470
Henyey, T. L., 386, 393, 561
Heritsch, F., 203, 253, 263–265
Hermes, J. J., 419, 425
Herrmann, L. A., 583
Herron, E. M., 330
Hersey, J. B., 358, 559
Hershey, Garland, 353
Hess, H. H., 308, 310, 311, 320, 330, 381–383, 393, 402, 417, 423, 471, 546–548, 550, 561
Hewitt, D. F., 345, 346, 354
Higgins, M. W., 64, 72, 458, 471
Higgs, D. V., 72
Hilenberg, O. C., 327
Hill, M. J., 165, 166, 174
Hill, M. L., 109, 110, 122, 330, 539, 543, 561
Hill, P. A., 93, 98, 548, 561
Hills, E. S., 6, 184, 199, 202, 204
Hills, J. M., 118, 121
Hinze, W. J., 350, 351, 354
Hirshman, J., 364, 368
Hobbs, B. E., 72, 229, 239–241, 245
Hodgson, J. H., 300, 376, 393
Hodgson, R. A., 84, 95, 98
Hoeppener, R., 213
Hoffman, O. F., 561
Holcombe, T. L., 561
Holden, J. C., 330
Holinquist, P. J., 225
Holmes, A., 202, 263, 327, 330
Hooke, Robert. 34, 38
Hoppin, R. A., 95, 98, 561
Hopson, C. A., 457, 470, 471
Horai, Ki-iti, 300
Horsfield, W. T., 561
Horvath, F., 515
Horz, Friedrich, 72
Hose, R. K., 561
Hoskins, H., 19
Hough, V. N. D., 94
Howard, J. H., 116, 122, 142, 156, 157, 561
Howard, Richard, 583
Howell, B. F., Jr., 288, 300
Hoy, R. B., 275, 279, 284
Hsü, K. Kinghwa, 142, 331, 514, 534, 535, 561
Hubbard, N. J., 393
Hubbert, M. K., 15, 19, 72–83, 98, 112, 122, 136–138, 142, 143, 230, 245, 563
Hubert, Claude, 473
Hudleston, P. J., 204, 245, 300
Hughes, Dudley J., 157, 271, 275–277, 284, 470
Hurley, P. M., 335, 351, 352, 354
Hutchinson, R. M., 91, 98, 390, 391, 393

Iglehart, Charles F., 583
Illies, J. H., 354, 390, 393
Ingersoll, R. V., 471
Ingerson, E., 72
Irvine, T. N., 6, 157
Irwin, E. M., 533
Irwin, W. P., 329, 532, 533, 558, 561
Isacks, Bryan, 301, 310, 313, 329, 331, 372, 393, 405, 410, 413, 422–424
Itson, S. P., 354

Jackson, E. D., 425
Jackson, R. E., 98
Jacobeen, Frank, Jr., 131, 142, 452, 471
Jaeger, J. C., 570, 571
Jaeger, J. E., 27, 43, 99, 112, 122
Jahns, R. H., 92, 99
James, D. E., 472, 557, 562
James, P. R., 353
Jeffreys, H., 6, 304, 331
Jenkins, David A. L., 142, 423
Jensky, Wallace, 386, 387, 393
Johnson, A. M., 183, 204, 254, 265
Johnson, G. L., 383, 395, 425
Johnson, J. G., 562
Johnson, M. R. W., 514
Jones, Bill, 368
Jones, D. L., 533, 558
Jones, P. B., 562

Jordan, Tom H., 289, 300, 551, 562
Jordon, Pascual, 328, 331
Julian, B. R., 332
Julivert, Manuel, 261, 262, 265, 514
Juo, J. T., 157
Justus, P. S., 462, 463, 465, 473

Kallberg, R. W., 559
Kamb, W. B., 72
Kanamori, Hiroo, 300
Kanes, W. H., 131, 142, 452, 471
Karig, D. E., 398, 401, 407, 418, 419, 423, 425
Katili, J. A., 423
Katsumata, Mamoru, 423
Kaula, W. M., 300, 331
Kay, G. Marshall, 6, 202, 264, 305, 306, 331, 433, 438, 439, 443, 444, 470, 472, 523, 562
Kay, R., 393
Kaye, C. A., 548, 562
Keep, C. E., 137, 142
Kehle, R. O., 132, 142
Keith, Arthur, 461, 472
Kelleher, John, 423
Keller, F., Jr., 562
Kelley, V. C., 89, 99, 122
Kelm, D. L., 330, 354
Ken, J. W., 562
Kennedy, M. J., 472
Kennedy, W. Q., 160, 174
Khattri, Kailash, 408, 423
Khoury, S. G., 470
King, Chi-Yu, 122
King, E. R., 368, 393
King, P. B., 130, 142, 331, 348, 432, 457, 466, 468, 469, 472, 519, 521, 545
Kirk, H. K., 425
Klemme, H. D., 506, 514
Knoff, E. B., 72
Knopff, L., 39, 513
Koch, J. G., 562
Kogan, M. G., 424
Kohli, G., 514
Korn, H., 151
Kovach, R. L., 162, 174
Kozak, S. J., 95, 99
Kozary, M. T., 133, 135, 142, 549, 562
Krasser, L. M., 514
Krause, D. C., 368, 424, 551, 562
Krebs, Wolfgang, 424
Krenkle, E., 390, 393
Krutikhoskaya, Z. A., 354
Kuenen, P. H., 9, 16–18, 193, 204, 245, 274, 284
Kulling, O., 446, 473
Kulm, L. D., 400, 424
Kuo, J., 537, 560
Kupfer, D. H., 274, 278, 284
Kurbatova, N. S., 246
Kurie, A. E., 562
Kutina, Jan, 99

Lachenbush, A. H., 99
La Forge, 119, 201
LaFountain, L. J., 98, 562
Lahee, F. H., 201, 583
Lajtai, E. Z., 174
Lanbert, R. St. J., 301
Landes, K. K., 283, 284
Langseth, M. G., Jr., 297, 301, 383, 393, 394
Lanphere, M. A., 424
Larson, R. L., 331, 382, 384, 386, 393
Laubscher, H. P., 142, 483, 486, 496, 497, 499, 500, 514
Laudon, T. S., 422
Laughton, A. S., 157
Lawver, L. A., 425
Lebedeva, N. B., 264
Lee, M. F., 225
Lee, W. H. K., 301
Lees, G. M., 172–174
LeFort, 502, 514
Lehmann, I., 289, 301
Lehner, Peter, 284
Leith, C. K., 199, 208, 225
Lemoine, M., 513
Lensen, G. J., 157, 174
Le Pichon, X., 312, 330, 331, 368, 372, 378, 393
Le Roy, L. W., 583
Leyden, R., 284, 392
Lidiak, E. G., 354
Lienhardt, G., 485, 514
Lillie, A. R., 162, 163, 174, 265, 424
Limond, W. Q., 514
Lindgren, W., 118–121
Liniger, Hans, 514
Lipman, P. W., 562
Livingstone, C. E., 559
Loczy, Louis de, 174, 562
Lofgen, B. E., 562
Logan, J. M., 245, 265
Lohest, M., 221, 225
Lohmann, H. H., 562
Lombard, A. E., 514
Lomnitz, C., 72
Longwell, C. R., 123, 142
Loomis, T. P., 511–514
Lopez, Arroyo A., 515
Lopez-Ramos, Ernesto, 467, 472, 562
Lort, J. M., 506, 514
Lotze, Franz, 6
Louden, Keith, 425
Love, A. E. H., 40, 43
Lovering, T. S., 99, 142
Low, J. E., 583
Lowdon, J. A., 354
Lowell, James D., 155, 157, 164, 170, 174, 394
Lowrie, W., 506, 514
Lowry, W. D., 18, 331, 472, 562
Ludwig, W. J., 368. 404, 423, 424
Lugeon, M., 484, 514
Luyendyk, B. P., 424

McBirney, A. R., 245, 330, 548, 562
McCallien, W., 200, 201
McConnell, R. B., 388, 390, 394
McCrossan, R. G., 562
MacDonald, D. C., 424
MacDonald, G. J. F., 298, 301, 562
McDowell, A. N., 278, 284
MacGillavry, H. J., 564
McGinnis, L. D., 514
MacGregor, A. M., 354
McIntyre, D. B., 197, 199, 200, 203, 204, 226, 611
McKee, E. D., 18
McKenzie, D. P., 321, 331, 410, 424, 514

McKerrow, W. S., 301, 515
Mackin, J. H., 258, 265
Mackin, J. W., 583
McKinstry, H. E., 99, 174
McMaster, Robert L., 369
McQuillan, Henry, 502, 514
McTaggart, K. C., 530, 562
Magnusson, N. H., 354
Mais, W. R., 581, 583
Malahoff, Alexander, 142
Maley, T. S., 551, 562
Malfait, B. F., 551, 562
Malin, P. E., 562
Manetti, P., 508, 513
Mantura, A. J., 331
Marauchi, S., 394
Marcos, Alberto, 17, 18, 192, 204, 208, 212, 225, 261, 265
Maresch, W. V., 551, 562
Marlow, M. S., 424
Marshall, C. E., 19
Marshall, P., 424
Martin, L. J., 562
Martin, R. G., 271, 284
Martinez, J. D., 284
Mason, R. C., 331, 353, 376, 381, 394
Mason, R. G., 394, 424, 541, 563
Massingill, J. V., 369
Matsuda, Tykihiko, 331
Mattauer, M., 509, 510–512, 514
Matthews, D. H., 311, 333, 381, 395
Matthews, J. E., 561
Mattox, R. B., 284
Mattson, P. H., 548, 549, 551, 562
Maxwell, J. C., 14, 19, 70, 142, 174, 213–215, 226, 310, 322, 323, 331, 514, 548, 550, 561, 563
Mead, W. J., 226
Melton, F. A., 94, 99
Menard, H. W., 331, 374, 378, 393, 394, 424, 425
Mertie, J. B., 204
Meyer, R. P., 351, 354
Meyerhoff, A. A., 563
Meyerhoff, H. A., 563
Middleton, G. V., 10, 19
Migliorini, C. I., 9, 18
Mikhaylov, A. Y., 331
Milanovsky, E. E., 157, 394
Milici, R. C., 472
Miljush, Petar, 515
Miller, R. L., 128–130, 142, 472
Miller, V. C., 583
Miller, W. B., 71
Milnes, A. G., 492–494, 515
Milson, John, 424
Milton, D. J., 101, 103, 105
Minear, J. W., 331, 332, 407, 424
Mitchell, A. H. G., 331, 515
Mitronovas, Walter, 424
Miyashiro, Akiho, 320, 331, 425
Moench, R. H., 563
Mogi, Kiyoo, 72
Mohanty, B. B., 99
Mohr, D. W., 472
Mohr, Otto, 24–26, 43, 77, 79–82, 85, 97, 136
Mollard, J. D., 88, 95, 99
Molnar, Peter, 292, 301, 331, 423, 563
Monger, J. W. H., 526, 563
Moody, J. D., 165, 166, 174
Moore, D. G., 19, 386–388, 394, 513, 563
Moore, G. W., 559, 563
Moore, J. C., 215, 226, 424
Moore, Walter, 143
Morgan, B. A., 472
Morgan, W. J., 315, 321, 331, 394
Morton, D. M., 558
Mountjoy, E. W., 138, 139, 143
Mualchin, Lalliana, 425
Muehlberger, W. R., 99, 121, 272, 284, 354
Mueller, P. A., 204
Mueller, Stephan, 301
Mügge, O., 72
Murauchi, S., 404, 424
Murchison, R. R., 561
Murphy, A. J., 563
Murray, G. E., 268, 284
Myers, W. B., 561

Nádai, A., 38, 40, 41, 43, 72, 511
Nafe, J. E., 368
Nagy, B., 215
Nairn, A. E., 466
Nalwalk, A. J., 559
Nason, R. D., 559
Naugler, F. P., 424
Nelson, A., 265, 459
Nettleton, L. L., 231, 245, 275, 278, 280, 284, 288, 301
Nevin, C. M., 6, 16, 19, 201, 202
Newton, Sir Isaac, 35
Nicholas, R. L., 472
Nicholson, J. T., 70
Nickelsen, R. P., 94, 99, 154, 563
Ninkovich, Dragoslav, 506, 515
Nolan, T. B., 563
Nowroozi, A. A., 515
Nullo, F., 560
Nur, A., 162, 174, 300

Oberbeck, R., 105
Oberbeck, V. R., 101, 105
Oberholtzer, J., 515
O'Brien, G. D., 283
O'Bryan, J. W., 394
Ocamb, R. D., 150, 157
Ocola, L. C., 351, 354
Odé, Helmer, 79, 81, 99, 122
Odom, A. L., 472
O'Driscoll, E. S., 34, 43, 204
Oertel, Gerhard, 204, 226
Off, Theodore, 160, 174
Officer, C. B., 547, 563
Ogden, T. A., 404, 422
Oliver, Jack, 292, 301, 313, 331, 372, 393, 405, 410, 413, 423, 424
Omara, S., 284
O'Neill, B. J., Jr., 275, 279, 284
Oppel, G., 43
Orowan, E., 72, 99, 116, 117, 122, 331
Osanik, A., 18
Osberg, P. H., 472
Ostenso, N. A., 395
Owen, T. R., 444, 470
Oxburgh, E. R., 301, 321, 332, 383, 394, 395, 407, 411, 424, 515

Packer, D. R., 563
Page, B. M., 563

Page, David, 120
Pakiser, L. C., 293, 294, 298, 301
Palmer, H. S., 577, 583
Palmer, K. F., 560
Palmquist, J. C., 95, 98, 561
Panza, G. F., 513
Papazachos, B. C., 506, 513, 515
Parejas, E., 487, 513
Parker, J. M., 94, 99
Parker, R. B., 194–196, 202, 203, 245
Parker, R. L., 331
Parker, T. J., 278, 284
Parmentier, E. M., 301
Paschall, R. H., 160, 174
Pashkevich, I. K., 354
Paterson, M. S., 52, 55, 62, 65, 71, 72, 226
Patnode, H. W., 203, 223, 234–237, 245, 265
Pavoni, N., 486
Peccerillo, A., 508, 513
Pekeris, C. L., 307, 331
Peter, G., 404, 409, 424, 425, 563
Peterman, Z. E., 563
Peterson, J. A., 563
Peterson, J. J., 300, 393
Pettijohn, F. L., 10, 19, 307, 344, 345, 354, 437, 472
Phillips, J. D., 331, 394, 516
Phillips, W. E. A., 579, 583
Picard, L., 173, 174, 390, 394
Pierce, W. G., 132, 133, 143, 483, 484, 515, 583
Pincus, H. J., 87, 99
Pitman, W. C., III, 330–332, 382, 384, 385, 392–394, 501, 513
Platt, L. B., 143
Poldervaart, Arie, 6, 301
Ponikarov, V. P., 353, 515
Poole, F. G., 564
Poole, W. H., 332, 424, 472
Porterm, S. C., 563
Postipischl, D., 513
Potter, P. E., 10, 19
Powell, C. McA., 226, 502, 506, 515, 541, 564
Power, W. Robert, 119
Powers, Sidney, 19
Prager, G. D., 611
Pratt, D. M., 361
Press, Frank, 171, 174, 288, 290, 294, 300, 301, 332
Preston, F. W., 94, 99
Price, N. J., 81, 99, 229, 235, 241, 244, 245
Price, R. A., 136, 138, 139, 143, 563
Prince, R. A., 400, 424, 526
Printz, Martin, 85, 93, 99
Proctor, McKeown, 122
Prucha, J. J., 154, 157, 284, 563

Quaide, W. L., 105
Quennell, A. M., 172, 174
Qureshy, M. N., 515

Rabinowitz, P. D., 506, 508, 515, 563
Raff, A. D., 331, 376, 381, 394, 395, 541, 563
Ragan, D. M., 34, 43, 265, 577, 583, 601, 611
Raistrick, Arthur, 19
Raitt, R. W., 424
Raleigh, C. B., 50, 52–54, 62, 70, 72, 143, 243
Ramberg, Hans, 43, 184, 185, 193, 204, 221, 222, 226, 229, 233, 237, 239–241, 245, 326, 327, 332
Ramsay, J. G., 6, 27, 32, 37, 38, 43, 183, 184, 186–190, 197, 200, 204, 224, 226, 229, 245, 515, 570, 571
Rand, J. R., 335, 352, 354
Rankin, D. W., 454, 457, 459, 461, 466, 472
Ravich, M. G., 354
Read, H. H., 192, 193, 204
Read, J. L., 284
Reading, H. G., 331
Ree, Taikyue, 71
Reed, B. L., 424
Reed, J. C., Jr., 461, 470, 472
Reeves, R. G., 583
Reid, H. F., 118–122
Reina, H. L., 560
Reiner, M., 43
Reynolds, Osborne, 36
Rhodes, R. C., 101, 105
Richards, H. G., 563
Richey, J. E., 85, 99
Richter, C. F., 288, 295, 300, 372, 401
Richter-Bernberg, G., 284
Rickard, M. J., 122, 204
Ridlon, J. B., 424
Riecker, R. E., 27, 43, 78, 80
Ringwood, A. E., 290, 301, 308, 309, 320, 329, 417
Roberts, A., 43, 265
Roberts, David, 217, 226, 233, 245
Roberts, J. C., 99
Roberts, R. K., 528, 532, 537, 563
Robertson, E. C., 66, 72
Robertson, J. D., 511, 513
Robinson, L. H., Jr., 67, 72
Robinson, Peter, 442, 473
Roddick, J. A., 563
Rodgers, John, 332, 437, 438, 451, 472, 563
Rodolfo, K. S., 404, 424
Roeder, D. H., 332, 515
Rogers, G. S., 473
Rogers, W. B., 473
Romberg, F. E., 153, 156, 558
Rona, Peter A., 369
Root, Samuel I., 456, 473
Roper, Paul J., 332, 462, 463, 465, 473
Rose, R. L., 350
Ross, D. A., 390, 394
Rowan, L. C., 204
Rozendal, R. A., 472
Rubey, W. W., 15, 19, 62, 72, 136–138, 142, 143, 563
Rudman, A. J., 350, 351, 354
Runcorn, S. K., 332
Rutland, R. W. R., 563
Rutten, M. G., 483, 515
Rutter, E. H., 72
Ryan, W. B. F., 331, 501, 506, 508, 513, 515, 563

Sacks, S., 368
Saemundsson, Kristjan, 394

St. John, V. P., 424
St. Julien, Pierre, 473
Saito, T., 392
Salisbury, R. D., 201
Sammis, Charles, 289, 296, 300, 329
Sander, B., 6, 204, 207, 226, 613, 614
Sanders, J. E., 19
Sanderson, D. J., 226
Sanford, A. R., 115, 117, 118, 122, 154–157, 246
Satin, L. R., 583
Saul, R. B., 164, 174
Saunders, J. B., 552
Savage, J. C., 99, 157, 564
Saxena, M. N., 515
Sbar, M. L., 536, 544, 564
Schardt, Hans, 489, 490, 515
Scheidegger, A. E., 39
Schenik, P. E., 473
Schlanger, S. O., 514
Schlee, John, 390, 394
Schmid, S. M., 72
Schmidt, C., 514
Schmidt, Walter, 6
Schneider, E. D., 382, 395
Schock, R. N., 72
Scholl, D. W., 424, 554, 564
Scholten, Robert, 142, 329, 564
Scholz, C. H., 67, 72, 536, 540, 564
Schott, Wolfgang, 284
Schreiber, E., 300, 393
Schubert, G., 332
Schuchert, Charles, 328, 332
Schwab, F. L., 444, 473
Sclater, J. G., 332, 394, 410, 424
Scott, W. B., 119
Scruton, P. C., 19
Seager, W. R., 143
Secor, D. T., Jr., 99
Sederholm, J. J., 192, 204
Seely, D. R., 159, 168
Seifert, K. E., 72
Senn, A., 550, 564
Shagam, R., 564
Shand, S. J., 157
Sharman, G. F., 418, 419, 423
Sharpe, D., 226
Shaw, E. W., 394, 564
Shaw, H. R., 321, 332, 383, 425
Shaw, K. W., 457, 459, 472
Shawe, D. R., 174, 538, 564
Shearer, E. M., 583
Shelton, J. W., 157
Sheridan, R. E., 361, 369
Sherrill, R. E., 16, 19
Sherwin, Jo-Ann, 229, 246
Shimer, J. A., 577, 583, 601, 611
Shoemaker, E., 101, 102, 105, 301
Shor, C. G., 409, 425
Shor, G. G., 375, 393
Short, N. M., 101–103, 105
Shouldice, J. R., 143
Shrock, R. R., 10, 19
Siddens, A. W. B., 207, 226
Silver, E. A., 425, 551, 564
Simmins, G., 564
Simonenko, T. N., 354
Simpson, Brian M., 255, 265, 583
Simpson, E. S. W., 368
Sinha, Akhaury Krishna, 471
Sjostrom, Hakan, 332
Skvor, Vladimir, 564
Sleep, N. H., 329, 564
Sloan, R. E., 350, 354
Sloss, L. L., 332
Small, W. M., 143, 281, 284
Smith, A. G., 470, 507, 513, 515
Smith, Derrell A., 284
Smith, G. I., 174
Smith, J. G., 140, 143, 425
Smith, R. B., 226, 544, 564
Smith, S. M., 393, 394, 564
Smith, T. J., 472
Smoluchowski, M., 135, 136, 143
Smythe, D. K., 229, 241, 244, 246
Snelling, N. J., 564
Snider, H. I., 304
Soloman, S. C., 564
Sonder, R. A., 6
Sorby, H. C., 211, 226
Sougy, J., 515
Souther, J. G., 526, 563
Spang, J. H., 72, 245
Spencer, A. M., 397, 552, 555, 556, 564
Spencer, E. W., 87, 95, 99, 456, 458, 473
Spry, Allen, 29, 99, 226
Stauder, William, 314, 332, 425
Stauffer, Mel R., 204
Stearns, D. W., 71, 88, 89, 90, 99, 256, 265
Stearns, R. G., 105
Stegena, L., 515
Stehli, Albert C., 466
Steinhart, J. S., 472
Stephansson, O., 245
Stephens, D. R., 72
Stevenson, J. S., 354
Stewart, G. A., 527, 558
Stewart, J. H., 559, 564
Stickney, D. M., 354
Stille, Hans, 306, 332, 433, 473
Stillwell, F. L., 191, 204
Stirewalt, G. L., 225, 462–464, 473
Stočes, Bohuslave, 6, 200
Stockard, Henry, 365, 368
Stocklin, Jovan, 515
Stockwell, C. H., 200, 343, 354, 583
Stoiber, R. E., 401, 422, 425
Stone, D. B., 404, 425, 563
Stose, A. J., 473
Stose, G. W., 473
Stover, C. W., 394
Strand, T., 446, 473, 515
Strömgård, Karl-Erik, 226, 233, 245, 246
Subrahmanyam, C., 515
Suess, Edward, 332, 397, 425
Sugden, W., 583
Suggate, R. P., 174
Sugimura, A., 397, 402, 425
Summerson, C. H., 350, 351, 354
Suppe, J., 541, 544, 564
Sutton, G. H., 361, 368
Sutton, J., 354
Sychev, P. M., 328, 332
Sykes, L. R., 313, 331, 332, 372, 376, 393, 394, 401, 423, 425, 563, 564

Talbot, C. J., 332
Talbot, J. L., 54, 72
Talwani, Manik, 332, 379, 385, 394, 402, 409, 425, 564
Tan, B. K., 43

Tanner, W. F., 174, 328, 332, 564
Tarling, D. H., 332
Taylor, F. B., 304
Tchalenko, J. S., 162, 165, 174, 175, 515
Thiel, E. C., 353
Thom, W. T., Jr., 175
Thomas, W. A., 473
Thomasson, M. R., 532, 563
Thompson, G. A., 537, 564
Thompson, G. H., 157
Thompson, J. B., Jr., 442, 473
Thompson, J. E., 350, 354
Thorarisson, Sigurdur, 373, 395
Threet, R. L., 122
Timoshenko, S. P., 43, 246
Tobin, D. G., 55, 71, 73, 425
Tobisch, Othmar T., 204, 226, 263, 265, 473
Toksöz, M. N., 329, 331, 332, 407, 424
Tollman, A., 496, 515
Tomlinson, C. W., 473
Torrence, K. E., 301
Trask, N. J., Jr., 442, 473
Travers, W. B., 564
Treagus, S. H., 229, 241, 246
Trevisan, L., 134, 142
Troxel, B. W., 539, 561
Trump, R. P., 203, 233–237, 245, 265
Trümpy, R., 488, 491, 497, 498, 515
Trusheim, F., 272, 278, 279, 281, 284, 290
Tucker, D. R., 157
Tullis, Terry E., 73, 226
Turcotte, D. L., 301, 321, 332, 383, 394, 395, 407, 411, 424
Turner, F. J., 6, 38, 39, 48, 52, 53, 61, 65, 70, 71, 73, 99, 179, 180, 184, 193, 197, 202, 204, 213, 224, 226, 263, 614

Uchupi, Elazar, 358, 369
Udias, A., 515
Uhlig, Johannes, 390, 395
Ulrich, E. D., 441
Umbgrove, J. H. F., 6, 332, 425
Uyeda, Seiya, 331, 395, 397, 402, 425

Vacquier, V., 395
Vagas, R., 511, 513
Valencio, D. A., 333
Van Andel, T. H., 19, 395
Van Bemmelen, R. W., 310, 322–326, 332, 416, 422, 425, 511, 512, 515
Van Der Lee, Joyceanne, 72
Van der Voo, R., 516
Van Hilten, D., 333
Van Vlack, L. H., 73
Varnes, D. J., 23, 43, 79, 99
Vening, Meinesz F. A., 307–309, 313, 330, 333, 402, 404, 419, 425
Venkatachalam, S., 515
Verhoogen, Jean, 38, 73
Ver Steeg, K., 99
Vikhert, A. V., 246
Vilas, J. F., 333
Viljoen, M. J., 353
Viljoen, R. P., 353
Vine, F. J., 311, 333, 395
Vinogradov, S. D., 99
Visser, W. A., 419, 425
Vogt, P. R., 100, 382, 395, 551, 559, 561
Voight, Barry, 43, 143
Von Herzen, R. P., 19, 395, 424, 564
Von Huene, R., 424, 425
Vyeda, S., 425

Wageman, J. M., 424, 425
Walcott, R. I., 301
Walker, G. P. L., 373, 393
Wallace, R. E., 100, 162, 565
Walls, R., 226
Walsh, J. B., 122, 175
Ward, H. L., 137, 142
Warren, D. H., 301
Warren, R. E., 395
Wassall, H., 548, 565
Watanabe, Torijolp, 402, 425
Waters, A. C., 157, 537, 560
Watkins, J. S., 457, 473
Watts, A. B., 425
Weber, F., 516
Weeks, L. A., 404, 425
Weertman, J., 301
Wegener, Alfred, 297, 301, 304, 305, 333, 420
Wegmann, C. E., 221, 226, 355
Wegmann, Eugene, 486, 516
Weiss, L. E., 6, 39, 52, 72, 98, 99, 179, 180, 184, 193, 197, 202, 204, 213, 224, 226, 245, 263, 614
Wellman, H. W., 157, 163, 175, 333
Wells, J. D., 561
Wendlandt, E. A., 149, 157
Wenk, Hans-Rudolf, 494, 516
Wesson, P. S., 321, 333, 383, 395
Wheeler, J. Q., 526, 560, 565
White, C. H., 6, 200
White, W. H., 565
White, W. S., 209, 226, 435
Whiten, E. H. T., 6, 43, 197, 204
Whitten, C. A., 565
Wickham, J. S., 473
Wiedenmayer, C., 486, 516
Wilcox, R. E., 159, 165, 168, 175
Wilhelm, Oscar, 369
Williams, Emyr, 189, 205
Williams, Harold, 443, 473
Williams, P. F., 226
Willis, Bailey, 6, 38, 55, 73, 157, 205, 231, 232, 246, 265, 390, 395, 565
Willis, Robin, 6, 265
Wilson, C. W., Jr., 100, 175, 473
Wilson, Gilbert, 205, 219, 220, 226
Wilson, H. H., 143, 516
Wilson, J. T., 159, 175, 312, 315, 316, 321, 333, 395, 444, 473
Wilson, M. E., 227, 355
Windisch, C. C., 394
Windley, B. F., 354
Wise, D. U., 95, 100, 152, 156, 157, 174, 328, 333, 471, 540, 565
Wong, H. K., 506, 516

Wood, D. S., 73
Woodford, A. O., 558
Woods, R. D., 284
Woodside, John, 506, 509, 516
Woodward, H. P., 363, 368, 471
Woodward, L. A., 143
Woodworth, J. B., 84, 100
Woollard, G. P., 291, 292, 301, 333, 350, 353
Worzel, J. L., 268, 368, 402, 409, 425
Wright, L., 353
Wyllie, P. J., 309, 333, 339, 355
Wynne-Edwards, H. R., 198, 205, 227, 265
Wyrick, R. F., 368
Wyss, M., 564

Yeats, R. S., 565
Yellin, M., 424
Yoshii, Toshikatsu, 404, 425

Zarudzki, E. F. K., 368, 516
Zdenko, F. D., 561
Zembruscki, S., 284
Zen, E-An, 333, 439, 441, 442, 473
Zieta, Isidore, 471
Zietz, Isidor, 129, 142, 293, 294, 298, 301, 355, 393

Subject Index

"ab" plane, 244
Absaroka Mountains, 132
Acadian orogeny, 437–438
Accelerating creep, 66
Accordian folds, 200
Adirondack dome in New York, 150–151
Affine deformation, 38
Africa, 152
Age of fractures, determination of, 84
Alleghanian orogeny, 429, 436
Allegheny synclinorium, folds and related faults in, 449
Allochthon, 253, 263
Allochthonous, 202
Alpine basement deformation, 493
Alpine fault, 171
Alpine-Himalayan Mountains:
  belt region, 140
  of Europe and Asia, 250
Alpine-Himalayan orogen, 475
  geography of, 478
  history of, 481
  pattern of, 480
  relationship to its surroundings, 478
Alpine-Himalayan system, plate and tectonic interpretations of, 499
Alpine metamorphism, 493
Alpine orogeny, role of the mantle diapirism, 511
Alpine plutonic activity, 493
Alpine-type deformation, 475
Alpine-type orogeny, 475
American Gulf Coast, 148
Andaman-Nicobar Islands, negative anomalies, 402
Andean orogenic belts of Peru and Chile, 556
Andes orogen, 553
Angle of internal friction, 80
Angular cross-bedding, 13
Annealing recrystallization, 63
Antarctica, Atlantic-type margins, 357
Anticline, 200
Anticlinorium, 202, 251, 263
Antidune ripples, 10
Antiform, 200
Antithetic faults, 118, 146, 154, 271
Antler orogeny, 529
Apennine Mountains, 134
Appalachian-Caledonian system:
  of eastern North America, 250
  of Scandinavia, 250
  of western Great Britain, 250
Appalachian orogen:
  geosynclinal model, 431
  major divisions of, 429
  orogenic belts, 94
  tectonic history of, 433
  tectonic setting of, 427
  thrust faults, 128
Apparent dip, 586
  finding of, 587
  finding true strike and dip, 587
Arabian Peninsula, 173
Arabian-Zagros area, plate model for, 501
Ararat River synclinorium, 457
Arc-trench gap, 400
Arvonia slate, 461
Asia, 152
Asthenosphere, 289, 324
Astroblemes, 101
Asymmetrical fold, 182, 200
Asymmetrical (current) ripples, 10

Atlantic margin:
  plate-tectonic interpretation of, 364
  shallow structure of, 358
Atlantic Ocean, opening and growth of, 384
Atlantic-type margin, 357
Australia, Atlantic-type margins, 357
Austroalpine nappes, 494
Autochthon, 202, 253, 268
Avalon platform, 443
Avery Island salt dome, 270
Axial fractures, 81, 96

Back thrusts, 126
Banda arc, 399
Barbados, 550
Basement, 429
Basement fracture zones, 95
Basement tectonics, 524
Basin, 201
Basin and range, 299
  as an ensialic interarc basin, 540
  faults, 539
Bear Province of the Canadian Shield, 341
Beartooth Mountains, 85, 93, 125, 152
  Precambrian dikes, 93
Bed tops, 10
Bedding fault, 110, 121
Bend gliding, 49
Bendigo, Australia, 191
Bending fold, 184
Benioff zones, 401
Bering Sea, plate motions, 314
Biextensional strain, 40, 96
Bighorn Mountains, Wyoming:
  basement tectonics, 524
  deformed North American platform, 524
Bingham body, 40, 41
  viscoplastic material, 36, 41
Blake Plateau, Cenozoic sediments, 362
Block faulting:
  in basin and range, 535
  in continental cratonic areas, 152
  hypothesis of strike-slip control, 537
  in orogenic belts, 152
Blue Ridge anticlinorium, 453
Body forces, 21, 28
Boudinage, 221
Boundary-ray method, 579
Box fold, 200
Brent, Colorado, 101
Brevard zone, 461
British Caledonides, comparison with Newfoundland, 444
Brittle behavior, 38, 96
Brittle fracture, Griffith's theory of, 81
Brittle fracturing in experiments, 81
Broadtop anticline, 131
Bronson Hill anticlinorium, 440
  nappes and mantled gneiss domes, 442
Buckle fold, 184
Bulk modulus, 41
Bündner Schïefer, 491
Burma, 137
Bushveld, South Africa, 101

Calaveras fault zone, California, 162, 164, 165
Calcite, deformation of, 53
California, 137, 152
Canadian Rockies, 139
Canadian Shield, 84, 340
  large-scale refolding, 263
  Pleistocene deposits, 95
  structural provinces of, 341
Cape Cod, 141
Cape Mendocino, 378
Caribbean region, 546
  large islands of, 548
Caribbean structure, plate-tectonic interpretation of, 551
Cariboo orogeny, 531
Carnegie ridge, deep sea trenches, 554
Caryn Seamount, 141
Cataclasis, 46, 76, 96
Cataclastic flow, 96
Catskill Mountains, 94
Cayman trench, Bartlett trough, 546
Central America, 544
Central Andes, plate model for, 557
Central mobile belt of Newfoundland, 443
Channels:
  cut and fill structure, 11
  grooves, 11
  lobate rill marks, 11
  rill marks, 11
  tidal channels, 11
Charlotte belt anticlinorium in Appalachian orogen, 460, 465
Chattanooga shale, 129
Chestnut ridge anticline, 127
Chevron fold, 184, 196, 200
Churchill Province, Canada, 341
Clearwater Lake, Quebec, 102
Cleavage:
  definition of, 224
  fracture, 207–209
  metamorphic origin in Pennsylvania, 215
  occurrence of, 217
  penetrative, 208
  relationship to folds, 215, 217
  schistosity, 208–212
  slaty, 208, 213–216
  strain-slip, 208–209
  superposed, 209
  unconsolidated sediment, 214
Cleavage mullions, 219
Clipperton zone, 374
Closure, 201
Cohesion, 42
Cold working, 40
Colorado Plateau, 86, 146
Columbian Andes, evolution of, 555
Columnar fractures, 92, 96
Compaction, 13
Compaction folding, 579
Competent rocks, 38, 56

Compressibility, 41
Compressive strength, 59, 96
Concentric folds, 182, 196, 200
Concepts of strain, 27–34
Confining pressure, 38, 46
Congruous folds, 201
Conical folds, 178, 200
Conjugate folds, 200
Conjugate sets, 218
Conjugate shears, 159
Connecticut-Gaspé synclinorium, 440
Connecticut Valley, 184
Contemporaneous faults, 148
Continent-continent collision, 318
Continent–island-arc collision, 318
Continental cratons, 335
Continental crust, divergent junctions in, 386
Continental drift hypothesis, 304
Continental margins:
  Atlantic-type, 357
  deeper structure of, 361
Continental shields, 298
Contraction hypothesis, 304
Convection current hypothesis, 307, 320–321, 383
Convolute bedding, 16, 18
Cordilleran orogen:
  in Alaska, 261
  of Western, North, and South America, 517
    tectonic framework of, 517
  in the Yukon, 261
Cordilleran-type orogen, 317
County Cork, Ireland, Carboniferous slates, 218
Cratons, 335
  definition of, 336
  Precambrian evolution, 338
Creep, 38, 67
  accelerating, 66
  steady-state, 66
  transient, 66
Creep test, 47, 66
Crenulations, 209
Cross-bedding:
  angular, 13
  torrential, 13
Cross folds, 201
Cross fracture, 96
Cross laminations, 13
Cross sections, 359, 577
Crust, 289
  defining by gravity anomalies, 291
  structural divisions, 295
  upper mantle, 291
Crustal elements, classification of, 298
Crustal rebound, 47
Cryptoexplosive features, 101
Cryptovolcanic features, 101
Crystal defects, 49
Crystalline aggregates, deformation of, 54
Cuba, 546
Culmination, 202, 263
Cumberland Mountains:
  monocline, 129
  overthrust, 128, 130
Cumberland Plateau, 130
Curvilinear folds, 182, 200
Curviplanar fractures, 96
Cut and fill structure, 11, 13
Cylindrical folds, 178–182, 200

Dasht E Bayaz, Iran, 165
Dead Sea, 172, 173
Deborah mine, 191
Décollement, 127, 128, 189, 252, 262
  of Burning Springs anticline, 449
  in Cambrian Chilhowee group, 451
  of Cambrian and Ordovician sediments, 442
  of foreland belts, 130–131
  in Himalayas, 504
  in Jura Mountains, 449–451, 485
  model for Appalachians, 451
  tectonic characteristics of Appalachians, 433
Deformation:
  affine, 38
Deformation:
  of igneous rocks, 58
  of metamorphic rocks, 58
  rotational, 28
Deformation band, 49
Deformation lamella, 49
Deformation mechanisms:
  of calcite, 53
  of crystalline aggregates, 54
  deformation band, 49
  deformation lamella, 49
  dolomite, 54
  fracture, 49
  granulation, 49
  kinking, 49
  mica, 54
  olivine, 54
  orthopyroxene, 54
  plagioclase, 54
  pyroxene, 54
  of quartz, 53
  in single crystals, 48
  slip, 49
  slip band, 49, 57
  translation gliding, 49
  in twin gliding, 49
Deformed North American platform, 521
Detachment surfaces, 189
Deviatoric stresses, 26, 38
Devil's Tower, 93
Dextral strike slip, 119, 120
Diagonal fracture, 96
Diapiric fold, 201
Diapirs, 267–282
Diastrophism, 5, 8
Differential compaction, 13
Differential stress, 46
Dilation, 27, 30, 40
Dip fault, 110, 121
Dip separation, 120
Dip slip, 108, 119
Dip-slip fault, 121
Dipping surface, finding outcrop pattern of, 575
Direct stratification, 9
Disharmonic fold, 200
Dispersion, 293
Dispersion curves, 293

Displacement, 119
  net slip, 108
Distortional strain, 40
Divergent junctions:
  in continental crust, 386
  divergent plate boundaries, 151, 371
Dome, 201
Donath-Parker fold, classification of, 195
Down-structure method of viewing structures, 258
Drag, 126, 190, 201
Ductile behavior, 96
  brittle transition, 56
Ductile materials, 38
Ductility, 195
Ductility contrast, 195, 196
Dunnage mélange, 444
Dynamic structural geology, 521
Dynamics, 40, 46

Earth craters, 101
East Africa, 173
  Rhine graben, 146, 151
  rift systems of, 388
  rift valleys, 304
East Pacific rise, 377
East Pakistan, 137
Easter Island, oceanic ridge system, 372
Eastern coastal belt, 440
Eastern Cordilleran:
  Canadian sector of, 526
  Sevier orogen, 528
    fold belt, 525
    tracing fold belt south, 528
Eastern Mediterranean, 506
Elastic aftereffect, 41
Elastic afterworking, 41
Elastic behavior, 34
Elastic limit, 41
Elastic moduli, 41
Elastic plastic, 37
Elastic strain, 40, 77
Elastic substance, 40, 41
Elasticoviscous solid, 40, 41
Electrical conductivity, 296
Elk basin, oil field of Wyoming, 151
En echelon features, 84, 96, 117, 155, 162, 164, 172
  deformed North American platform, 523
  faults, 165, 167
  folds, 165, 178, 202, 252, 263
  fractures, 165
  in Puerto Rico, 546
Equal-area projection, 607
Eugeosyncline, 306, 433
European Alps, 481
  limits and major structural elements, 481
  molasse basins, 481
  thrust faults, 128
Exotic blocks, 9
  of crystalline aggregates, 54–69
  environmental conditions, 47
  of igneous and metamorphic rocks, 58
  of marble and limestone, 55–57
  of sandstone, 57–58
  of single crystals, 48–54
Experimental rock deformation, summary, 69
Extension fracture, 76, 96
External force, 39

Fabric, 7, 224
Failure, 96
  Coulomb's criterion, 79, 96
  Griffith's criterion, 79
  Tresca's and Von Mise's maximum shear-stress criterion, 79
  Varnes criterion, 79
False cleavage, 207
Fan fold, 200
Fault:
  definition of, 76, 107, 117
  relation to bedding orientation, 110
Fault breccia, 118
Fault classifications:
  general terms, 121
  normal fault, 109
  reverse fault, 109
  strike-slip fault, 109, 121
  thrust fault, 109
Fault fold, 118
Fault movement, displacement, slip, and separation, 107
Fault orientation from displacement fields, 115
Fault plane, 118
Fault problems, graphical solutions of, 591
Fault scarp, 118
Fault trace, 118
Fault zone, 118
  character of, 152
Faulting, maximum depth of, 116
Faults:
  on folds, 151
  and stress orientations, 110
Fayette County, Pennsylvania, 127
Fenster, 19
Firmoviscous substance, 41
Fishtailing, 15
Flexure fold, 187, 202
Flow, 42
  gliding, 42
  pseudoviscous, 42
Flow casts, 17
Flow cleavage, 224
Flow fold, 202
Flute cast, 17, 18
Flysch, 253
Fold dimensions:
  height, 200
  interlimb angle, 200
  wavelength, 200
Fold generation, 202, 263
Fold mullions, 219
Fold systems:
  definition of, 263
  features of, 202
  occurrence and tectonic associations, 247
Fold types, combination of, 579
Folded rocks:
  cleavage and foliation, 207
  mesoscopic features, 207
Folding:
  analysis of strain, 239
  classic model experiments, 231
  by elastic buckling, 233
  large-scale superimposed folds, 261

Folding:
- parallel, 194
- passive, 195
- in principal and experiment, 229
- Rheid, 39, 202
- scale-model experiments in, 230
- syndepositional, 248
- variables in, 230
- variation in stress and strain, 240
- of viscoelastic materials, 236

Folds:
- analysis of, 254
- angular, 191
- asymmetric, 182, 200, 241
- axis of, 179
- bending, 184
- buckle, 184
- chevron, 184, 186, 200
- concentric, 182, 194, 200
- curvilinear, 182, 200
- cylindrical, 178–182, 200
- divergence of limbs, 200
- Donath-Parker classification, 195
- *en échelon,* 165, 178, 202
- fault-related, 201
- flattening of, 187
- flexures, 187, 201
- genetic classification, 201
- inclined, 200
- irregular flow, 184
- isoclinal, 182
- nappe structures, 202, 253
- nontectonic, 247
- occurrence and tectonic associations, 247
- overfold, 200
- overturned, 183, 200
- parallel, 194
- parts of, 199
- profile shape, 201
- ptygmatic, 192
- real, 185
- recognition of, 178
- recrystallization in, 191
- recumbent, 182, 200
- relation among, 201
- shear, 194

Folds:
- similar, 183, 184, 201
- slippage in, 188
- steeply plunging, 163
- superposed, 195
- systems of, 202
- of tectonic origin, 250
- true thickness, 182
- wedging of beds, 190

Foliation, 207
- definition of, 224

Footwall, 119

Force, 38

Forces:
- body, 21
- surface, 21

Foreland fold belts, 251, 439

Fracture cleavage, 207–209
- definition of, 224

Fracture density, 88

Fracture patterns:
- determination of, 86
- regional, 94
- related to folds, 89

Fracture spacing, 88

Fracture surfaces, features on, 84

Fractures, 49, 75–76, 97
- age of, 84
- in basement rock, 95
- in dikes, sills, and flows, 92
- as a guide to fault orientation, 111
- occurrence of, 82
- in plutons, 91
- primary system of, 91

Franciscan eugeosynclinal belt, Franciscan group, 533

French River:
- Greenville belt, 345
- plunging folds, 345

Fries fault, 457

Fundamental strength, 35, 42

Gailtal line, 496

Gap, 120

Geanticline, 202, 251, 263

Geostatic pressure, 38

Geosynclinal theory, 305

Geosyncline, 202, 251, 263, 306

Geotectonic hypothesis, roles of gravity in, 309

Ghost River, Alberta, 125

Giants Causeway, Ireland, 93

Glide surfaces, 133

Gliding flow, 42

Global expansion hypothesis, 327

Global tectonics:
- concepts of, 303
- origin of stress in lithosphere, 303

Goose Greek, Virginia, 124
- thrust faults, 124

Gosses Bluff, 103

Gouge fault, 107

Graben, 119, 155

Graded bedding, 9

Grampian Highlands, 162

Grand Canyon, 84, 146
- basement rock, 84, 95
- fracture sets, 95

Grand Saline, salt dome, 274

Granulation, 49

Gravity, 121, 402

Gravity anomalies, 548

Gravity-glide fault, 121, 310

Gravity gliding, 131, 133, 150
- characteristics of, 131

Gravity measurements over oceanic ridges, 379

Gravity spreading, 133

Great fault zones of Pacific basin, 374

Great Glen fault, Scotland, displacement, evidence of, 160

Great Plains:
- North American platform, 521
- regional fracture patterns, 94

Green River:
- basins, 154
- shales, 9

Greenstone belts, 336

Grenville belt, Canada:
- character of structures in, 344
- French River, 345
- Harvey-Cardiff arch, 345
- Stark complex, 345

Grenville event, 340

Grenville province, 341

Grooves, 11
Gross Bluff, Australia, 101
Ground horizontal, unit dipping, 588
Ground surface inclined, 589
Group velocity, 293
Growth faults, 148, 271
Guatemala, Laramide fold belt, 544
Gulf of Aden, 173, 390
  East African rift system, 388
  mid-Atlantic ridge, 390
  oceanic ridge system, 372
Gulf of Akaba, 173
Gulf of California, 152, 160, 386
Gulf Coast, 147
Gulf Coastal Plain, 128, 146, 148
Gulf of Eilat, 172
Gutenberg low-velocity zone, 295

Hade, 119
Hanging wall, 108, 119
Harvey-Cardiff arch, metamorphic and igneous structures, 345
Hawkins field, salt dome, 149
Heart Mountain thrust, Wyoming, 132
Heat flow, 296
Heat-flow measurements:
  of island arcs, 405
  over oceanic ridges, 379
Heave, 119
Helvetic nappes of High Calcareous Alps, 486
Hercynian basement, deformation of, 490
Hercynian massifs, 490
High-angle reverse faults, 152
High Calcareous Alps, Helvetic nappes of, 486
Himalayas, plate model for, 502
Homoclines, 201
Homogeneous strain, 27, 38
Hooke's law, 41
Hookian elastic body, 36, 41
Horizontal dip slip, 120
Horizontal projection, 590
Horizontal separation, 120
Horst, 119
Hydroplastic, 16
Hydroplasticity, definition of, 224
Hydrostatic condition, 26
Hydrostatic stress, 23

Iceland:
  grabens, 151
  mid-Atlantic ridge, 373
Ideal materials, 34, 38
Idu earthquake in Japan, 171
Igneous rocks, deformation of, 58
Illinois, Rough Creek zone, 152
Imbricate faults, 126
Impact craters:
  characteristic structural features of, 103
  and structures, 101
Impact metamorphism, 104
Impressed force, 39
Inclined fold, 200
Incompetent rocks, 56
Incompressibility modulus, 41
Incongruous folds, 201
India, Atlantic-type margins, 357
Indian Ocean, Carlsberg Ridge, 311
Infinitesimal strains, 27
Infrastructure, 251
Inner Piedmont belt anticlinorium, 460, 465
Insubric line, 496
Interference ripples, 10
Intermontane belt of Canada, 592
Internal friction, angle of, 80
Iran, Dasht E Bayaz fault zone, 165
Irregular flow folds, 184
Irregular mullions, 219
Island arcs, 299
  anomalous zones in mantle under, 405
  distribution of, 397
  emplacement of ultrabasics in, 417
  heat flow, 405
  geological features of, 410
  geophysical characteristics of, 401
Island arcs:
  magnetic studies of, 403
  physiography-bathymetry of, 400
  seismic cross sections, 404
  southeastern Pacific, complexities of, 419
  subduction-zone model of, 413
Isochromatics, 233
Isoclinal fold, 182, 200
Isotropic substance, 39
Ivrea zone, 496

Jefferson County, Colorado, 153
Joint-drag kink bands, 218
Joints (*see* Fractures)
Jordan Valley, 172
Jura Mountains, 131, 252, 483
  décollement, 485
  salt structures, 282

Kelvin body, 37, 41
Kelvin firmoviscous, 36
  viscoelastic, 37
Kentucky, 152
Khaur field, Pakistan, 137
Kinematic structural geology, 5
Kinematic viscosity, 35
Kinematics, 39
Kink bands, 49–53, 252
  definition of, 224
  strain in, 51
  types of, 218
Kink folds, 196
Kinking, 49
Klamath Mountains, Nevadan eugeosynclinal belt, 531
Klippe, 119, 124

Lake Basin fault zone, Montana, 165, 172
Lake Huron, collapse structures in salt basins, 283
Lake Michigan, collapse structures in salt basins, 283
Lake Superior, negative gravity anomalies, 348

Late Precambrian orogeny, evidence of, 437
Lateral spreading, mechanics of, 137
Left lateral slip, 120
Lesser Antilles, 550
Lineations, 222–224, 614
Linguoid ripples, 11
Lithosphere:
- definition of, 289
- major structural elements of, 287

Lithostatic pressure, 26, 39
Load casts, 16–18
Lobate rill marks, 11
Logan's line, 427
Longitudinal fault, 110, 121
Longitudinal fracture, 97
Longitudinal strains, 30
Louisiana, 137
Low-angle thrusts, mode of occurrence of, 130

Magnetic anomalies, 380
Mantle convection and oceanic ridges, 383
- in plate-tectonic theory, 320

Mantle diapirism, 322
- orogenic belts, 326

Mantle hot spots, 544
- Yellowstone, 544

Mantle plumes, 321
Maracaibo basin, 552
Marathon Mountains, 128
- craton platform, 348

Material behavior:
- elastic, 35
- ideal models, 34
- viscous, 35

Maxwell body, 37
Maxwell elasticoviscous, 36
Maxwell liquid, 41
Megashear, 119, 159
Mendocino fault, 160
Merrimack synclinorium, 440
Metaripples, 11
Meteor Crater, Arizona, 101, 102
Meteorite impact, 46
- features, 101

Mica, 54
Microlithons, 188, 209
- definition of, 224

Mid-Atlantic ridge, 373
Mid-continent gravity high, 348
Mid-continent or shield extensions, 298
Middle Rocky Mountains, 154
Middlesboro syncline, 130
Milford Sound, New Zealand, 162
Miogeosyncline, 433
Mobile belts, 336
Modern tectonic hypothesis, 308, 309
Mohr circle, 39, 78
Mohr envelope, 78, 80, 81, 97
Mohr's representation of stress, 24, 77
Molasse basins, 481
Monocline, 201
Mount Airy quarry, North Carolina, 92
Mount Cook, New Zealand, 162
Mount Hosmer, Canada, 254
Mount Maurice fault, Montana, 125
Movements:
- rotational, 109
- translational, 109
- types of, 109

Mud cracks, 11
Muddy Mountains, Nevada, 123
Mullion:
- cleavage, 219
- definition of, 225
- fold, 219
- irregular, 219

Mylonite, 107, 119

Nain province, Canada, 341
Nappe, 202, 203, 253
- definition of, 264

Nappes and mantled gneiss domes of Bronson Hill anticlinorium, 442
Net slip, 108, 120
Neutral surface in folds, 239
Nevadan eugeosynclinal belt, 531
Nevadan orogeny, 530–531

New England granites, 92
Newfoundland, comparison with the British Caledonides, 444
Newfoundland structure, plate-theory interpretation of, 444
Newtonian fluid, 36, 41
New York, Adirondack dome, 150–151
New Zealand:
- Alpine fault, 160, 162
- Milford Sound, 162

New Zealand Alps, 162, 163, 170
Nondiastrophic strain, 8
Nonhomogeneous strain, 27
Nonnewtonian fluids, 35
Nonrotational strain, 97
Nontectonic folds, 247
Normal fault zones, character of, 145
Normal faults:
- mechanics of, 147
- related to thrust faults, 151
- over salt domes, 148, 150
- structural patterns, 145
- tectonic settings for, 148, 150

Normal horizontal separation, 120
Normal stress, 22, 39
- variations of, on an inclined plane, 569

Normal traction, 39
North American continental margin, surface configuration, 357
North American Cordillera, 519
- divisions of, 521

North American platform, 346, 521
Northern Alps, plate models for, 500
Northern Appalachians, 439
- Bronson Hill anticlinorium, 440
- eastern coastal belt, 440
- Merrimack synclinorium, 440
- foreland belt, 439
- western fold belt, 440
- zone of basement uplifts, 440

Northern Germany, salt structures, 278–281
Northern Rocky Mountains, core zone, 526

Nuclear explosion, 101
Nye-Bowler zone, Montana, 167

Obduction, emplacement of ultrabasics in island arcs, 417
Oblique fault, 110, 121
Oblique folds, 201
Oblique slip, 108, 120
Oblique-slip fault, 121
Oblique-slip movement, 596
Ocean basins, 299
  methods used to study, 371
  seismic evidence for crustal structure, 376
Ocean ridge system, 300, 372
Oceanic ridges:
  heat-flow measurements, 379
  mantle convection, 383
  petrology of, 382
  seismic activity along, 376
Ocean trenches, 300
Oceanization, 323–326
Offset, 120
Oklahoma, 52, 165, 167, 172
  thrust belts, 94
Olivine, deformation mechanisms, 54
Oolites, 76, 190
Ordovician Martinsburg formation in the folded Appalachians, 213
Organic markings, 12
  scolithus, 12
Oriskany formation in Pennsylvania, 258
Orocline, 202, 251, 264
Orogen:
  continent-continent collision, 318
  continent–island-arc collision, 318
  Cordilleran-type, 317
  island-arc type, 316
Orogenic belts:
  Alpine-Himalayan orogen, 475–512
  Appalachian orogen, 427–469
  Cordilleran orogen, 517–558
Orthographic projection, 585
  of a single plane, 586
  thickness determination, 588
Orthopyroxene, deformation mechanisms, 54
Ottawa, grabens, 152
Ouachita Mountain fold belt, 94
  craton platform, 348
  regional fracture patterns, 94
Ouachita Mountains, westward continuation of the Appalachians, 466
Overfold, 200
Overlap, 120
Overthrow fold, 200
Overthrust, 121, 123
Overturned fold, 183
Ozark Mountains, fault zone, 348

Pakistan, 133
Palisades sill, New Jersey, 93
Para-autochthonous, 203, 253, 264
Parallel faults, 119
Parallel folding, 194, 577
Passive folding, 195
Pelitic strain bands, 218
Penetrative cleavage, 208
Pennine Alps, 491
Peripheral faults, 119
Perpendicular slip, 120
Peru-Chile trench, problem for plate tectonics, 554
Petrofabrics, 5
Phase velocity, 294
Photoelastic strain analysis, 233
  techniques, 233
Photoelasticity, 39
Piedmont, the, 457
  of Maryland and Pennsylvania, 458
Piedmont domes, 269
Pine Mountain fault, 128–130
Pitch, 589, 590
Plagioclase, deformation mechanisms, 54
Plane stress, analysis of stress in two dimensions, 567
Plane stress condition, 39
Planes, intersection of, 589
Plastic behavior, 37
  elastic plastic, 37
Plastic flow, 97
Plastic strain, 40
Plastic substance, 41
Plasticoviscous, 37, 40, 41
  Bingham body, 37, 40
Plate motion, 314
Plate movements, geometry of, 313
Plate-tectonic theory, mantle convection in, 320
Plate tectonics, 310–314
  causes of orogeny, 315–322
  interpretation: of Alpine orogen, 500–501
    of Atlantic-type margins, 364
    of Caribbean structure, 551
    of island arcs, 406–419
    of northern Appalachians, 444–447
    of Zagros and Himalayan orogens, 501–506
Plunge, 589, 590
Plutons, fractures in, 91
Poisson's ratio, 41
Pole of relative motion, 314
Pore pressure, 16, 46–47, 59–61, 137
  effects in deformation of sand, 61
  neutral stress, 225
Porosity, 14
Powell Valley, anticline, 129
Practical strength, 42
Prealps, the, 489
Precambrian evolution of cratons, 338
Predrift configurations, restoration of cratons to, 352
Pressure fringe, 29
Pressure gradient, 39
Pressure-release jointing, 97
Primary flat joint, 97
Primary fracture systems, 91
Primary structural features, 7
Principal axes of strain, 32, 40
Principal stress axes, concept of, 570
Principal stress directions, 39

Principal stresses, 22, 78, 570
hydrostatic, 23
Pseudoviscous flow, 42
Ptygmatic fold, 192
Puerto Rico, 546
Pusteria line, 496
Pyrenees Mountains, tectonic activity, 283
Pyroxene, deformation mechanisms, 54

Quartz:
deformation of, 53
syntectonic recrystallization of, 65
Quartz rods, 221
Quasi-flexural folds, 195, 202
Quincy Massachusetts, sheeting, 92

Radial faults, 119
Raindrop impressions, 11
Ramp rises along thrusts, 129, 131
Rattlesnake Mountain, Wyoming, drape fold, 257
Real folds, 185
Recess, 202, 252, 264
Recrystallization, 62, 191
annealing, 63
syntectonic, 63
Recumbent fold, 182, 202–203, 253
Red Sea, 146, 151, 172, 173, 390
East African rift system, 388
mid-Atlantic ridge, 390
rift valleys, 304
Refolded fold, 201
Regional fracture patterns, 94
Regional fractures in foreland regions, 94
Relaxation time, 41, 42
Residual salt features, turtle structure in northern Germany, 269
Resultant force, 39
Retardation time, 37
Reverse drag, 146, 190
Reynolds' number, 36, 42
Rheid, 39, 69
Rheid concept, 68
Rheid folding, 39, 202
Rheidity, 39, 69
Rheology, 39
Riecke's principle, 212
definition of, 225
Ries depression, Germany, 102
Rift systems:
of East Africa, 388
hypothesis on formation of, 390
Rifts, 159
Right-lateral slip, 120
Rill marks, 11
Ripples:
antidune, 10
asymmetrical (current), 10
interference, 10
linguoid, 11
metaripples, 11
symmetrical, 10
Roberts Mountain allochthon, 529
Rock behavior, composite models, 38
Rock failure, theory and experiment, 75
Rock strength, 35
fundamental, 35
ultimate, 35
Rocky Mountain front, 125, 156
Rocky Mountain trench, 526
Rocky Mountains, deformed North American platform, 521
Rodding, 219
definition of, 225
Romanche zone, 374
Rome formation, 129
Root, 203
definition of, 264
Root zone, alpine, 496
Rotational deformation, 28
Rotational movement, 109, 119
Rough Creek fault zone, 152

s surface, 613
Saddle reef, 189
Sag, 201
St. Venant plastic, 36, 41
Salient, 202
definition of, 264
Salt basins, collapse structures in, 283
Salt domes, 267
deformational pattern within, 272
formation of, 275
Salt glaciers, 279
Salt pillows, 269, 279
Salt structures, 267
basement involvement in, 270
fault patterns, 271
of Gulf Coast, 267–281
of northern Germany, 278
of tectonic origin, 281
Salt swells, 269
San Andreas fault, 160–162, 166, 172
evidence of motion, 160
transform faults, 386
Sander's notation, 613
Sandstone, experimental deformation of, 57
San Fernando earthquake, 171
Sangre de Cristo Mountains, deformed North American platform, 524
Sauratown Mountains anticlinorium, 46, 465
Sawtooth Range, Montana, 89
Schisosity, 208, 212
definition of, 225
hypothesis of the formation of, 211
Riecke's principle, 212
Schistés lustrés, 491–493
Scolithus, 12
Scottish Highlands, 188
Sea-floor spreading, plate tectonics, 310
Secondary structural features, 7
Sediment distribution and sea-floor spreading, 380
Sedimentary orogenic cycle:
concept of, 306
stages of, 307
Segregation kink bands, 218
Seismic activity along ocean ridges, 376

Seismic discontinuities, 289
  crust, 289
Seismic profiles, 547
Seismic shear zones, 401
Seismic studies, internal structure of earth, 288
Separation, 20
  dip separation, 108, 120
  horizontal separation, 108, 120
  stratigraphic separation, 108, 121
  strike separation, 108, 121
  vertical separation, 108
Separations and slips, classification based on, 110
Shatter zones, 95, 97
Shear, 97
Shear fold, 194, 202
Shear fracture, 76, 97
Shear kink bands, 219
Shear modulus, 42
Shear stress, 22, 39
  variation of, on an inclined plane, 569
Shear zone, 119
Sheeting structure, 92, 97
Shield, 336
Shift, 120
Sierra Nevada, batholith, 161
Similar fold, 183, 184, 196
Simple shear, 30, 34, 35, 40
Sinai, 172
Sinistral, 121
Sinistral strike slip, 120
Slaty cleavage, 208, 213–216
  definition of, 225
  of eastern Pennsylvania, 213
  hypothesis of the formation of, 211
  Riecke's principle, 212
Slave province, Canada, 341
Slickensides, 119, 189
Slickensides grooving, 219
Sliding, 131
Slip, 42, 49, 120
  classification based on, 110
  dip slip, 108, 119
  net slip, 108, 120
  oblique slip, 108
  strike slip, 108, 120
Slip band, 49, 51
Slip fold, 202
Slump structure, 16, 18
Snowball garnet, 29
Soft sediment, 13, 16
Soft-sediment structural features, 16
Solenhofen limestone, 55
Sonoma orogeny, 529
South America:
  north coast of, 552
  subduction zones, 313
Southern Appalachians:
  evolution of crystalline complexes, 463
  major tectonic elements of, 447
Southern Piedmont, 460
Spherulites, 33
Stable craton, 439
State of stress, 22
Staunton-Pulaski fault, Virginia, 124
Steady-state creep, 66
Step faults, 119, 155, 271
Stereographic projection, 601
  applications of, 603
Strain, 27, 40
  biextensional, 40
  cold working, 40
  concepts of, 27–34
  dilation, 27, 40
  distortional strain, 40
  elastic strain, 40, 77
  homogeneous strain, 27, 40
  infinitesimal strain, 27, 40
  longitudinal, 30
  nondiastrophic, 8
  nonhomogeneous, 27
  plastic strain, 40
  principal axes, 32
  principal directions of, 40
  pure compression, 40
  pure extension, 40
  pure rotation, 40
  pure shear, 40
  pure strain, 40
  pure translation, 27, 40
  simple shear, 40
  strain hardening, 40
  uniextensional, 39
  work hardening, 40
Strain bands, 218
  definition of, 225
Strain ellipse, 31
Strain ellipsoid, 33, 40
Strain energy, 76
Strain hardening, 37, 40
  cold working, 40
  work hardening, 40
Strain-rate effects, 67
Strain-rate test, 47, 66
Strain-slip cleavage, 207–209
  definition of, 225
Strain softening, 61
Stratification, 8
  direct, 9
Stratigraphic separation, 121
Stratigraphic throw, 271
Strength, 42, 78
  cohesion, 42
  fundamental, 42
  of a material, 42
  practical, 42
  ultimate, 42
Stress, 39
  analysis of, 567
  deviatoric, 26
  distribution and faulting, 112
  within the earth, 24
  ellipsoid, 39, 571
  field, 39
  finding equations of, 567
  Mohr's representation of, 24
  normal stress, 22
  shearing stress, 22
  state of, 22
  superimposed, 25
  tensile stress, 39
  theory, 567
  three-dimensional, 571
  trajectories, 39
Strewn islands, 399
Strike fault, 121
Strike separation, 121
Strike slip, 108, 120
Strike-slip fault, 159
  mechanical considerations, 165
  San Andreas fault, 160–162, 166, 172
  structural pattern, 162
  tectonic settings of, 172
  vertical displacement, 170

Strike-slip movement, 595
Structural divisions of crust, 295
Structural feature(s), 5
  nontectonic, 7–19
  primary, 7
  secondary, 7
  soft sediment, 13, 16
Structural geology, 3, 5
  dynamic, 5, 21
  kinematic, 5
Structural notation, 207, 224, 613, 614
Structural terrace, 201
Structure contour maps, 573
  preparation of, 576
Structure contours:
  drawing of, 576
  on a plane dipping surface, 574
  properties of, 513
Subduction zones, 313–320
  of Appalachian orogen, 444, 496, 500–503
  of island arcs, 397–421
  of South America, 313
Subplates in western United States, 544
Sudbury, Ontario, 101
Summit field Pennsylvania:
  North Summit pool, 127
  South Summit pool, 127
Superimposed folds, 262
Superimposed stress, 25
Superior province, Canada, 341
Superposed cleavages, 209
Superposed deformations, determination of, 197
Superposed folds, 195, 196
Suprastructure, 251
Supratenuous fold, 15, 202
Surface forces, 21, 39
Surface traction, 39
Surface wave:
  dispersion, 293
  dispersion curves, 293
  group velocity, 293
  interpretation of, 292
  phase velocity, 294
Swell, 201, 251
Symmetrical fold, 200
Symmetrical ripples, 10
Syncline, 201
Synclinorium, 202, 251, 264
Syndepositional folding, 248
Synform, 201
Syntectonic recrystallization, 63
  of quartz, 65
Synthetic faults, 107, 119, 154

Taconic klippe, 440
Taconic orogeny, 437
Tanganyika plateau, rift valleys, 388
Tango earthquake in Japan, 171
Tasmania, 93
Tear fault, 121, 125
Tectogene hypothesis, 307, 308
Tectonic setting, 156
  of strike-slip faults, 172
Tectonic theories:
  classical geosynclinal theory, 305
  continental drift hypothesis, 304
  contraction hypothesis, 304
  convection current hypothesis, 307
  plate theory, 320
  sea-floor spreading, 310
  sedimentary orogenic cycles, 306
  tectogene hypothesis, 307
Tectonics, 5, 300
Tectonite, 39
Tehuantepec ridge, 374
Ten Yard coal seam, 15
Tensile strength, 97
Tensile stress, 39
Tension, 97
Tension fracture, 76
Tensors, 21
Tethys geosyncline, 475
Texture, 7
Thixotropy, 36
Three-point problems, 596
Throw, 120
Thrust, 121
Thrust fault zones, characteristics of, 123
Thrust faults, 109, 123
  orientation of fault surface, 124
  structural patterns, 124
Thrusting:
  mechanical considerations, 134
  stratigraphic controls, 128
  tectonic setting, 140
Tidal channels, 11
Tilts and strains, 296
Torrential cross-bedding, 13
Torsion modulus, 42
Trace slip, 120
Trace-slip fault, 120
Transcurrent, 121
Transform fault, 159, 162, 314
Transient creep, 66
Translation gliding, 49
  bend gliding, 49
Translational fault problems, tips on solution of, 592
Translational movement, 109, 119
Transverse or tear fault, 110, 121, 159
Transverse fold, 201
Trenches (*see* Island arcs)
Triassic basins in eastern United States, 152
Triaxial tests, 47
Trinidad, 137, 552
Triple junctions, 312
  and Pacific plates, 384
Trough of folds, 199
Trough line, 199
Trough surface, 200
Turner Valley fault, 126
Turner and Weiss structural notation, 614
Twin gliding, 49–52

Ultimate strength, 35, 42
Ultrabasic rocks, proof of mantle involvement in orogeny, 310
Undation hypothesis of Van Bemmelen, 323
Underthrust, 123, 126
Uniaxial compression, 97
Uniextensional strain, 40, 97
Uniform flow, 76, 97
Upper mantle, 291
Upright fold, 200
Upthrust, 121, 152
  dip, 153
  dip of, 152

Upthrust:
  gravity analysis, 153
  mechanical considerations of, 155
  structural patterns along, 152
Utah, block faulting, 152

Varve, 9
  in the Green River shales, 9
Vector, 40
Venezuela, fault-bounded basins, 553
Vertical movements due to differentiation in the mantle, 323
Vertical slip, 120
Vertical tectonic models, 322
Viscoelastic material, folding of, 236
Viscoplastic material, 41
Viscosity, 42
  nonnewtonian, 42
Viscous behavior, 35
Viscous mediums:
  buckling of elastic plates in, 237
  folding of viscous plates, 237
Volume elasticity, 42
Vredefort feature in South Africa, 102

Weeks Island salt dome, structural features, 275
Wells Creek, Tennessee, 102, 104
West Hawk Lake, Canada, 101
Western Alps:
  evolution of, 497–498
  plate models for, 500
Western Cordilleran belt, 530
Western Mediterranean, 509
Western North America, plate tectonics and Cenozoic evolution of, 540
Whip Cove, 131
Williams Range, Colorado, 116, 156
Willow Creek fault, Montana, 125
Wind River Mountains, 153–155
Winnfield salt dome, structural features, 275
Work hardening, 40
Wrench fault, 121, 159
Wyoming, 152
  Elk basin, 151

Yellowstone Park, mantle hot spot, 544
Yield point, 34, 35, 40
Young's modulus, 42
Yule marble, 55

Zambesi River, South Africa, East African fault system, 388
Zigzag fold, 201